Inquire Interact Inspire Invent

This iScience Interactive Student Textbook Belongs to:

Name

Teacher/Class

Where am I located?

The dot on the map shows where my school is.

McGraw Hill Education

Glenc

Snow Leopard, *Uncia uncia*

The snow leopard lives in central Asia at altitudes of 3,000m–5,500 m. Its thick fur and broad, furry feet are two of its adaptations that make it well suited to a snowy environment. Snow leopards cannot roar but can hiss, growl, and make other sounds.

The McGraw·Hill Companies

Education

Send all inquiries to:
McGraw-Hill Education
8787 Orion Place
Columbus, OH 43240-4027

ISBN: 978-0-07-660223-0
MHID: 0-07-660223-0

Printed in the United States of America.

2 3 4 5 6 7 8 9 10 QDB 15 14 13 12 11

Flori

ida Teacher Advisory Board

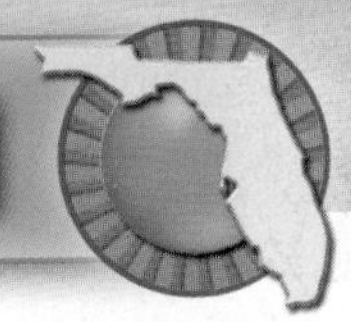

The Florida Teacher Advisory Board provided valuable input in the development of the © 2012 Florida student textbooks.

Ray Amil
Union Park Middle School
Orlando, FL

Maria Swain Kearns
Venice Middle School
Venice, FL

Ivette M. Acevedo Santiago, MEd
Resource Teacher
Lake Nona High School
Orlando, FL

Christy Bowman
Montford Middle School
Tallahassee, FL

Susan Leeds
Department Chair
Howard Middle School
Orlando, FL

Rachel Cassandra Scott
Bair Middle School
Sunrise, FL

Authors and Contributors

Authors

American Museum of Natural History
New York, NY

Michelle Anderson, MS
Lecturer
The Ohio State University
Columbus, OH

Juli Berwald, PhD
Science Writer
Austin, TX

John F. Bolzan, PhD
Science Writer
Columbus, OH

Rachel Clark, MS
Science Writer
Moscow, ID

Patricia Craig, MS
Science Writer
Bozeman, MT

Randall Frost, PhD
Science Writer
Pleasanton, CA

Lisa S. Gardiner, PhD
Science Writer
Denver, CO

Jennifer Gonya, PhD
The Ohio State University
Columbus, OH

Mary Ann Grobbel, MD
Science Writer
Grand Rapids, MI

Whitney Crispen Hagins, MA, MAT
Biology Teacher
Lexington High School
Lexington, MA

Carole Holmberg, BS
Planetarium Director
Calusa Nature Center and Planetarium, Inc.
Fort Myers, FL

Tina C. Hopper
Science Writer
Rockwall, TX

Jonathan D. W. Kahl, PhD
Professor of Atmospheric Science
University of Wisconsin-Milwaukee
Milwaukee, WI

Nanette Kalis
Science Writer
Athens, OH

S. Page Keeley, MEd
Maine Mathematics and Science Alliance
Augusta, ME

Cindy Klevickis, PhD
Professor of Integrated Science and Technology
James Madison University
Harrisonburg, VA

Kimberly Fekany Lee, PhD
Science Writer
La Grange, IL

Michael Manga, PhD
Professor
University of California, Berkeley
Berkeley, CA

Devi Ried Mathieu
Science Writer
Sebastopol, CA

Elizabeth A. Nagy-Shadman, PhD
Geology Professor
Pasadena City College
Pasadena, CA

William D. Rogers, DA
Professor of Biology
Ball State University
Muncie, IN

Donna L. Ross, PhD
Associate Professor
San Diego State University
San Diego, CA

Marion B. Sewer, PhD
Assistant Professor
School of Biology
Georgia Institute of Technology
Atlanta, GA

Julia Meyer Sheets, PhD
Lecturer
School of Earth Sciences
The Ohio State University
Columbus, OH

Michael J. Singer, PhD
Professor of Soil Science
Department of Land, Air and Water Resources
University of California
Davis, CA

Karen S. Sottosanti, MA
Science Writer
Pickerington, Ohio

Paul K. Strode, PhD
I.B. Biology Teacher
Fairview High School
Boulder, CO

Jan M. Vermilye, PhD
Research Geologist
Seismo-Tectonic Reservoir Monitoring (STRM)
Boulder, CO

Judith A. Yero, MA
Director
Teacher's Mind Resources
Hamilton, MT

Dinah Zike, MEd
Author, Consultant, Inventor of Foldables
Dinah Zike Academy; Dinah-Might Adventures, LP
San Antonio, TX

Margaret Zorn, MS
Science Writer
Yorktown, VA

Authors and Contributors

Consulting Authors

Alton L. Biggs
Biggs Educational Consulting
Commerce, TX

Ralph M. Feather, Jr., PhD
Assistant Professor
Department of Educational Studies and Secondary Education
Bloomsburg University
Bloomsburg, PA

Douglas Fisher, PhD
Professor of Teacher Education
San Diego State University
San Diego, CA

Edward P. Ortleb
Science/Safety Consultant
St. Louis, MO

Series Consultants

Science

Solomon Bililign, PhD
Professor
Department of Physics
North Carolina Agricultural and Technical State University
Greensboro, NC

John Choinski
Professor
Department of Biology
University of Central Arkansas
Conway, AR

Anastasia Chopelas, PhD
Research Professor
Department of Earth and Space Sciences
UCLA
Los Angeles, CA

David T. Crowther, PhD
Professor of Science Education
University of Nevada, Reno
Reno, NV

A. John Gatz
Professor of Zoology
Ohio Wesleyan University
Delaware, OH

h Gille, PhD
ssor
ity of California San
La

Dav
Profe se, PhD
North sics
Raleigh tate University

Janet S. Herman, PhD
Professor
Department of Environmental Sciences
University of Virginia
Charlottesville, VA

David T. Ho, PhD
Associate Professor
Department of Oceanography
University of Hawaii
Honolulu, HI

Ruth Howes, PhD
Professor of Physics
Marquette University
Milwaukee, WI

Jose Miguel Hurtado, Jr., PhD
Associate Professor
Department of Geological Sciences
University of Texas at El Paso
El Paso, TX

Monika Kress, PhD
Assistant Professor
San Jose State University
San Jose, CA

Mark E. Lee, PhD
Associate Chair & Assistant Professor
Department of Biology
Spelman College
Atlanta, GA

Linda Lundgren
Science writer
Lakewood, CO

Keith O. Mann, PhD
Ohio Wesleyan University
Delaware, OH

Charles W. McLaughlin, PhD
Adjunct Professor of Chemistry
Montana State University
Bozeman, MT

Katharina Pahnke, PhD
Research Professor
Department of Geology and Geophysics
University of Hawaii
Honolulu, HI

Jesús Pando, PhD
Associate Professor
DePaul University
Chicago, IL

Hay-Oak Park, PhD
Associate Professor
Department of Molecular Genetics
Ohio State University
Columbus, OH

David A. Rubin, PhD
Associate Professor of Physiology
School of Biological Sciences
Illinois State University
Normal, IL

Toni D. Sauncy
Assistant Professor of Physics
Department of Physics
Angelo State University
San Angelo, TX

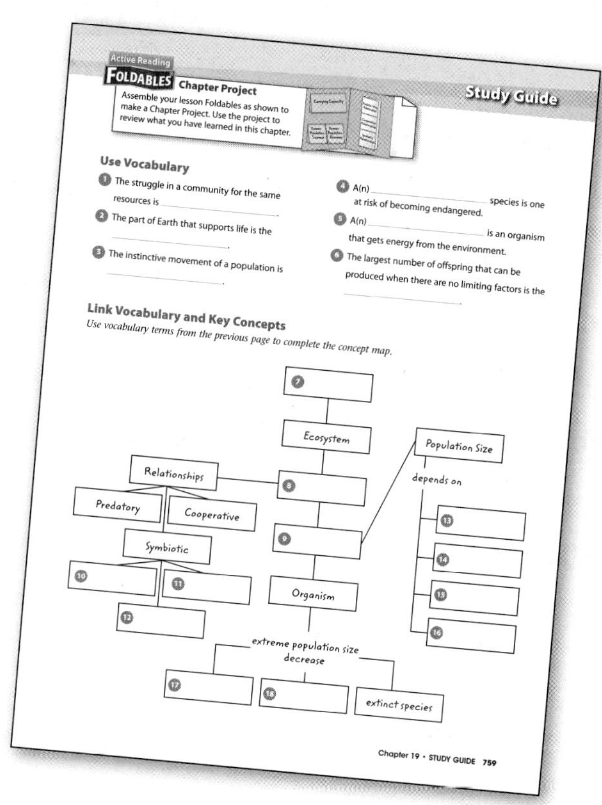

Active Reading

FOLDABLES Chapter Project

Assemble your lesson Foldables as shown to make a Chapter Project. Use the project to review what you have learned in this chapter.

Study Guide

Use Vocabulary

1. The struggle in a community for the same resources is ____________.
2. The part of Earth that supports life is the ____________.
3. The instinctive movement of a population is ____________.
4. A(n) ____________ species is one at risk of becoming endangered.
5. A(n) ____________ is an organism that gets energy from the environment.
6. The largest number of offspring that can be produced when there are no limiting factors is the ____________.

Link Vocabulary and Key Concepts

Use vocabulary terms from the previous page to complete the concept map.

Concept Map

Each chapter's **Concept Map** gives you a place to show all the connections you've learned.

Notes Pages

Notes Pages in each chapter are the place to write down other important information you want to remember.

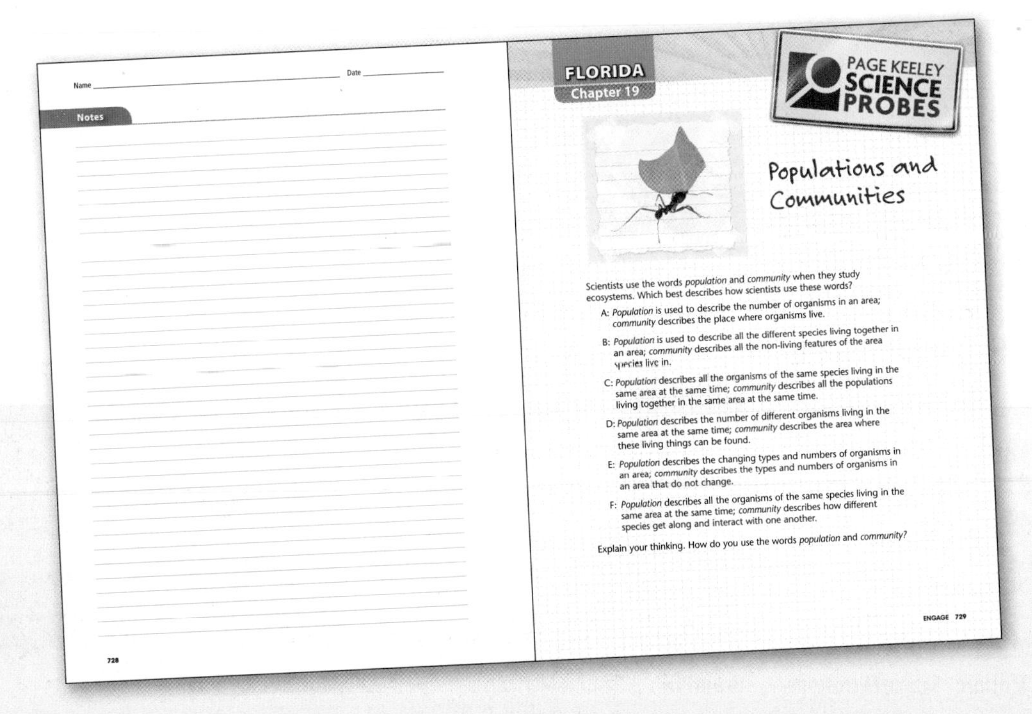

Name ____________ Date ____________

Notes

FLORIDA Chapter 19

PAGE KEELEY SCIENCE PROBES

Populations and Communities

Scientists use the words *population* and *community* when they study ecosystems. Which best describes how scientists use these words?

A: *Population* is used to describe the number of organisms in an area; *community* describes the place where organisms live.

B: *Population* is used to describe all the different species living together in an area; *community* describes all the non-living features of the area species live in.

C: *Population* describes all the organisms of the same species living in the same area at the same time; *community* describes all the populations living together in the same area at the same time.

D: *Population* describes the number of different organisms living in the same area at the same time; *community* describes the area where these living things can be found.

E: *Population* describes the changing types and numbers of organisms in an area; *community* describes the types and numbers of organisms in an area that do not change.

F: *Population* describes all the organisms of the same species living in the same area at the same time; *community* describes how different species get along and interact with one another.

Explain your thinking. How do you use the words *population* and *community*?

TABLE OF CONTENTS

Check It! ☐ Lesson 1 ☐ Lesson 2 ☐ Lesson 3

Inquiry LAB STATION

Skill Practice: How can you build your own scientific instrument?

Inquiry Lab: How can you help develop a bioreactor?

Your online portal to everything you need!
Video • Audio • Review • Lab Station • WebQuests • Assessment • Concepts in Motion • Personal Tutors • Virtual Labs

Here are some of the exciting digital activities you will find in this chapter!

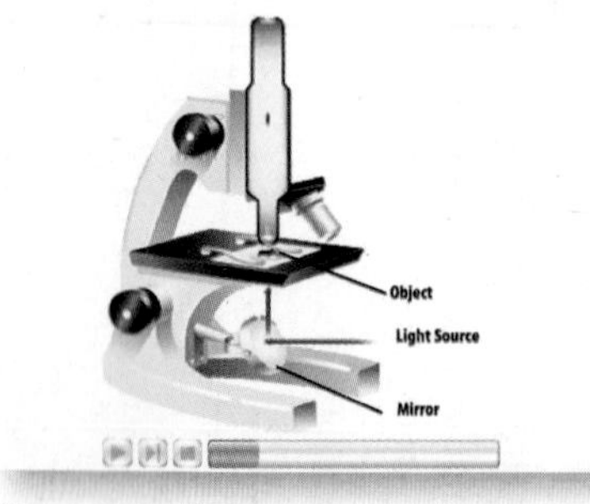

Concepts in Motion: Simple Microscope

BrainPOP: Scientific Methods

Page Keeley Science Probe

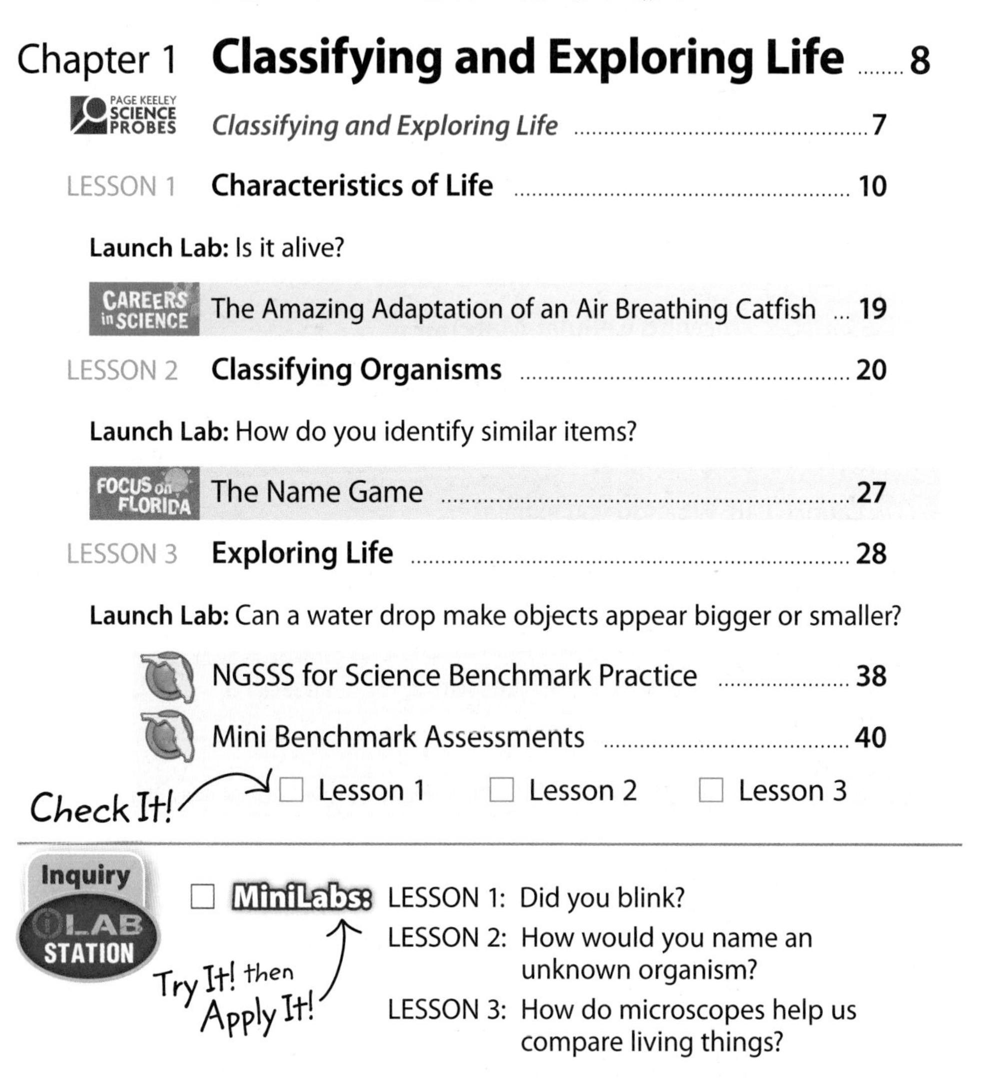

Unit 1 LIFE: STRUCTURE AND FUNCTION

Check It! ☐ Lesson 1 ☐ Lesson 2 ☐ Lesson 3

Inquiry iLAB STATION

☐ **MiniLabs:** LESSON 1: Did you blink?
LESSON 2: How would you name an unknown organism?
LESSON 3: How do microscopes help us compare living things?

Try It! then Apply It!

Skill Practice: How can you identify a beetle?
Inquiry Lab: Constructing a Dichotomous Key

Your online portal to everything you need!
Video • Audio • Review • iLab Station • WebQuests • Assessment • Concepts in Motion • Personal Tutors • Virtual Labs

Here are some of the exciting digital activities you will find in this chapter!

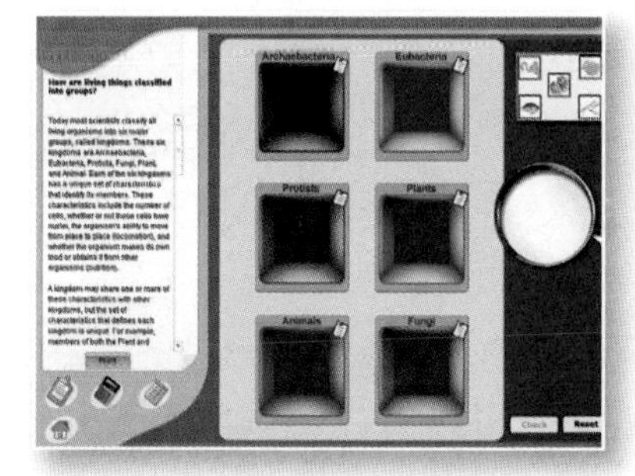

Virtual Lab: How are living things classified into groups?

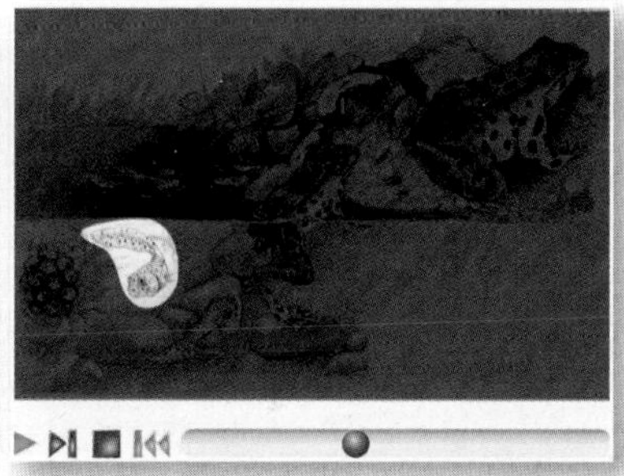

Concepts in Motion: Frog's Life Cycle

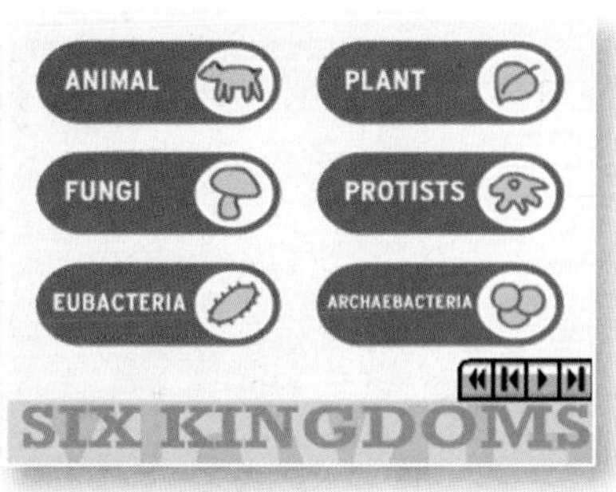

BrainPOP: Kingdoms

FLORIDA SCIENCE LIFE

TABLE OF CONTENTS

Inquiry LAB STATION

☐ **MiniLabs:** LESSON 1: How can you observe DNA?
LESSON 2: How do eukaryotic and prokaryotic cells compare?
LESSON 3: How is a balloon like a cell membrane?

Try It! then Apply It!

Skill Practice: How are plant and animal cells similar and how are they different?

Skill Practice: How does an object's size affect the transport of materials?

Inquiry Lab: Photosynthesis and Light

Your online portal to everything you need!
Video • Audio • Review • Lab Station • WebQuests • Assessment • Concepts in Motion • Personal Tutors • Virtual Labs

Here are some of the exciting digital activities you will find in this chapter!

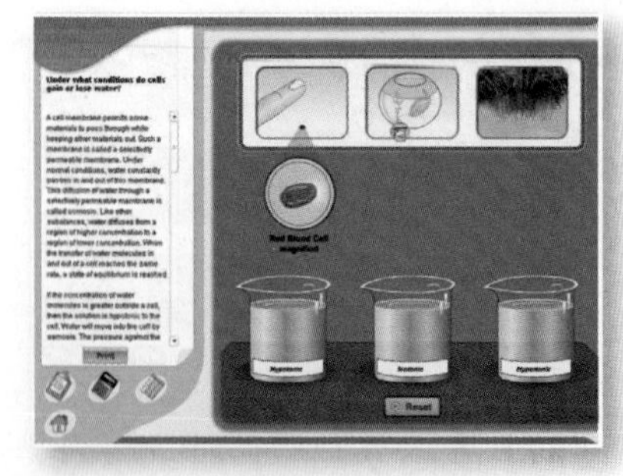

Virtual Lab: Under what conditions do cells lose and gain water?

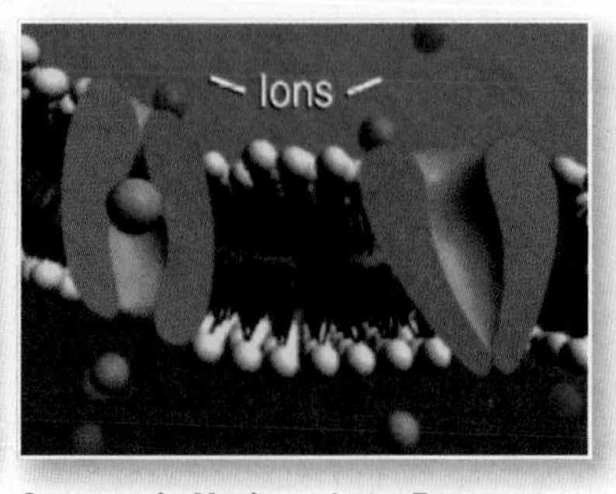

Concepts in Motion: Active Transport

BrainPOP: Cell Structures

FLORIDA SCIENCE LIFE

Chapter 3 From a Cell to an Organism 92

Check It! ☐ Lesson 1 ☐ Lesson 2

Inquiry LAB STATION

☐ **MiniLabs:** LESSON 1: How does mitosis work?
LESSON 2: How do cells work together to make an organism?

Try It! then Apply It!

Inquiry Lab: Cell Differentiation

Your online portal to everything you need!
Video • Audio • Review • ⓘLab Station • WebQuests • Assessment • Concepts in Motion • Personal Tutors • Virtual Labs

Here are some of the exciting digital activities you will find in this chapter!

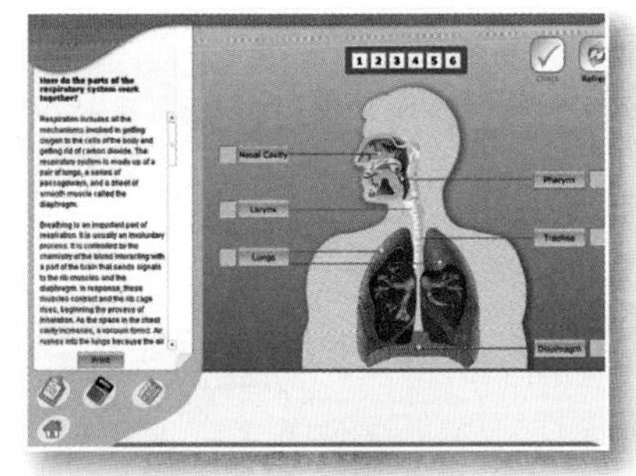

Virtual Lab: How do the parts of the respiratory system work together?

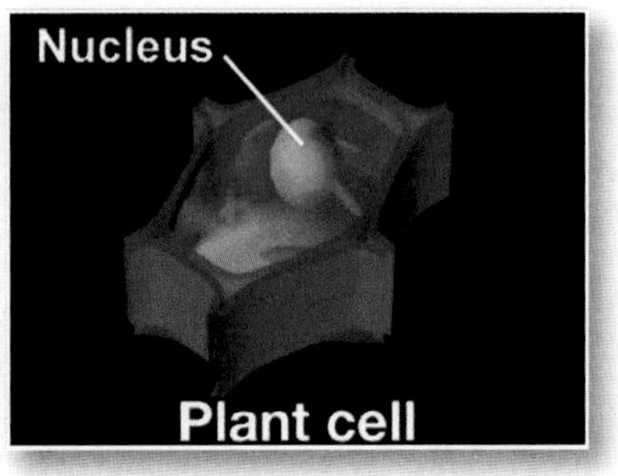

Concepts in Motion: Cell Organization

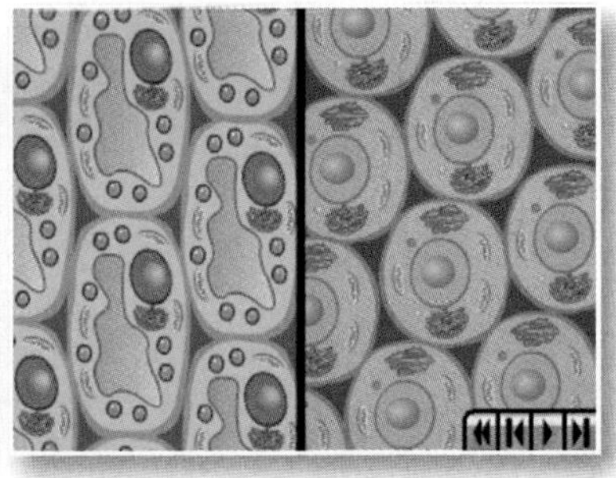

BrainPOP: Cell Specialization

TABLE OF CONTENTS

Your online portal to everything you need!

Video • Audio • Review • ⓘLab Station • WebQuests • Assessment • Concepts in Motion • Personal Tutors • Virtual Labs

Here are some of the exciting digital activities you will find in this chapter!

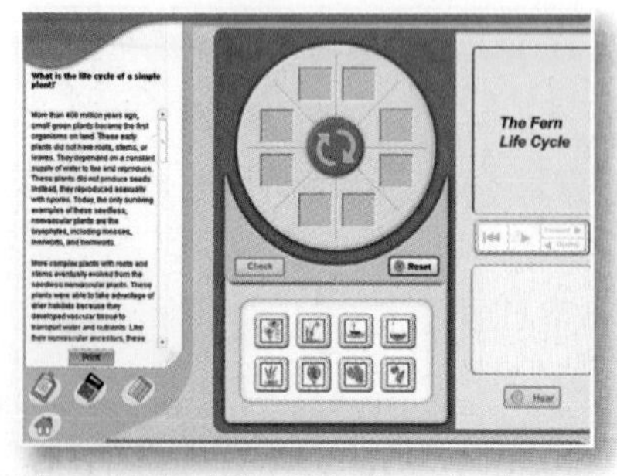

Virtual Lab: What is the life cycle of a simple plant?

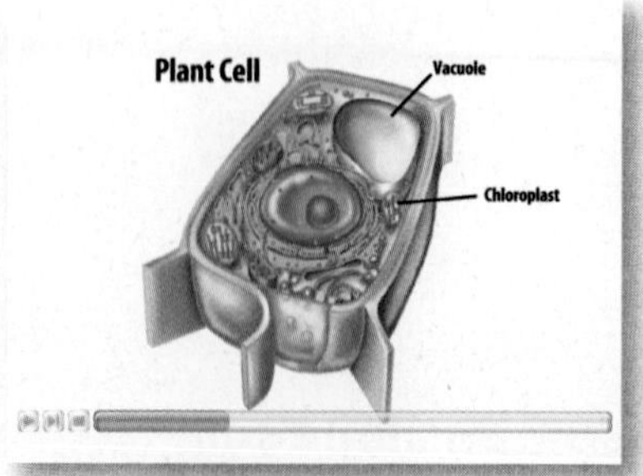

Concepts in Motion Meiosis

BrainPOP: Fertilization and Birth

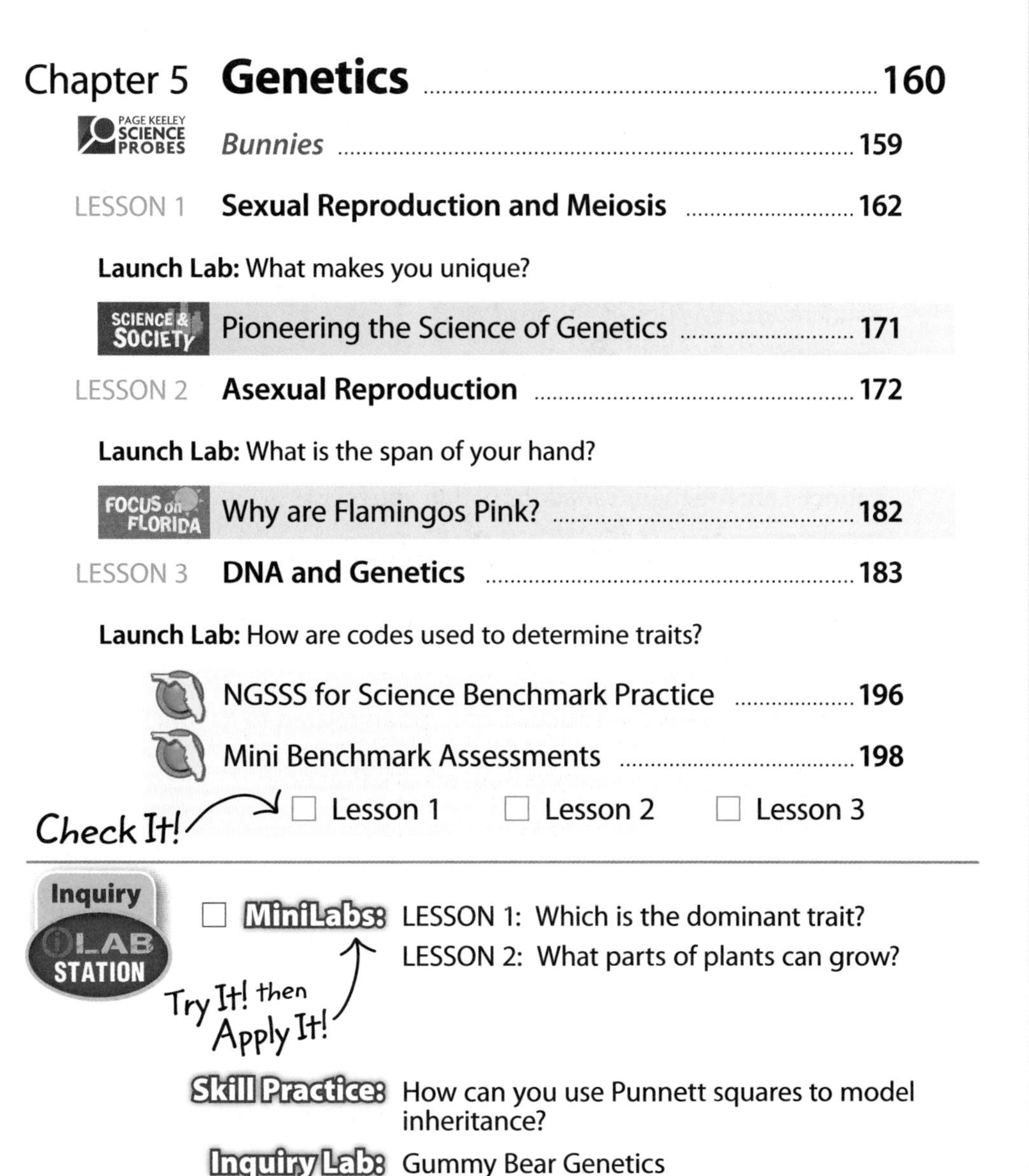

Check It! ☐ Lesson 1 ☐ Lesson 2 ☐ Lesson 3

Inquiry LAB STATION

☐ **MiniLabs:** LESSON 1: Which is the dominant trait?
LESSON 2: What parts of plants can grow?

Try It! then Apply It!

Skill Practice: How can you use Punnett squares to model inheritance?

Inquiry Lab: Gummy Bear Genetics

ConnectED **Your online portal to everything you need!**
Video • Audio • Review • iLab Station • WebQuests • Assessment • Concepts in Motion • Personal Tutors • Virtual Labs

Here are some of the exciting digital activities you will find in this chapter!

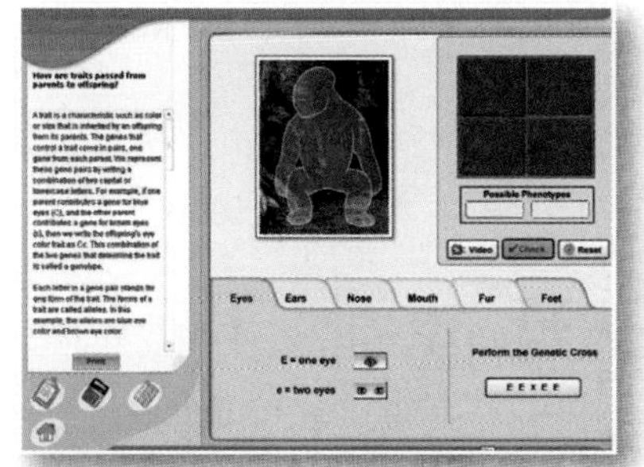

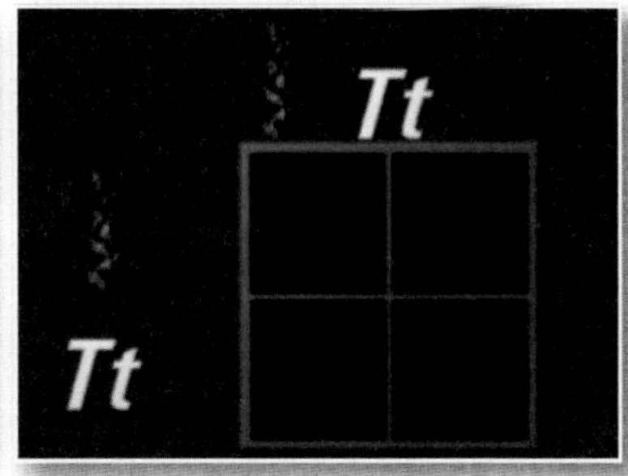

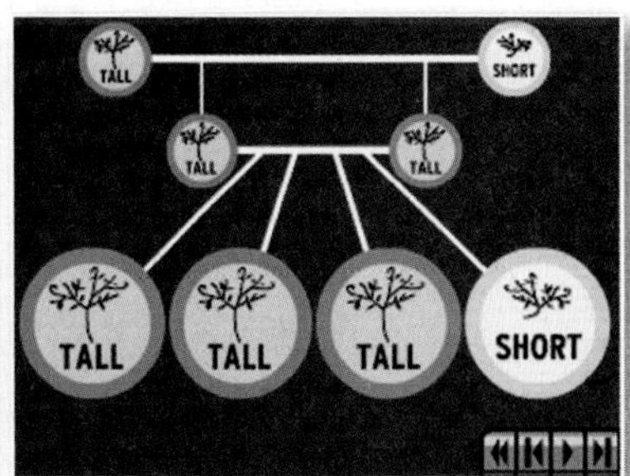

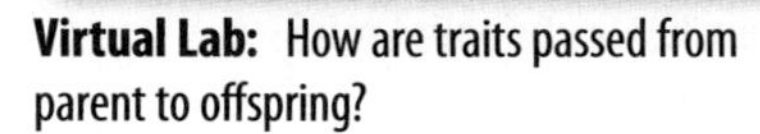

Virtual Lab: How are traits passed from parent to offspring?

Concepts in Motion: Punnett Square

BrainPOP: Heredity

TABLE OF CONTENTS

Your online portal to everything you need!
Video • Audio • Review • ⓘLab Station • WebQuests • Assessment • Concepts in Motion • Personal Tutors • Virtual Labs

Here are some of the exciting digital activities you will find in this chapter!

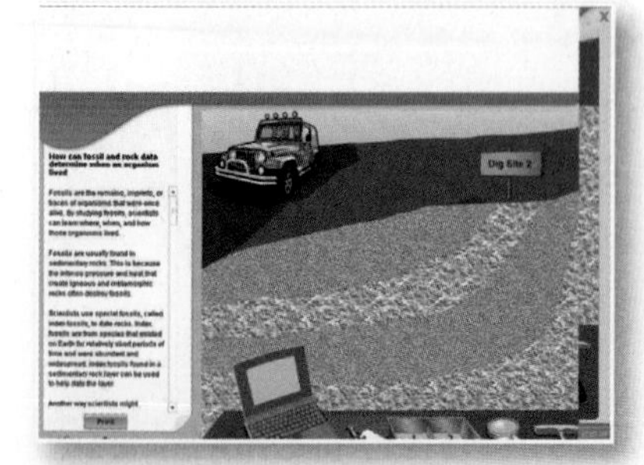

Virtual Lab: How can fossil and rock data determine when an organism lived?

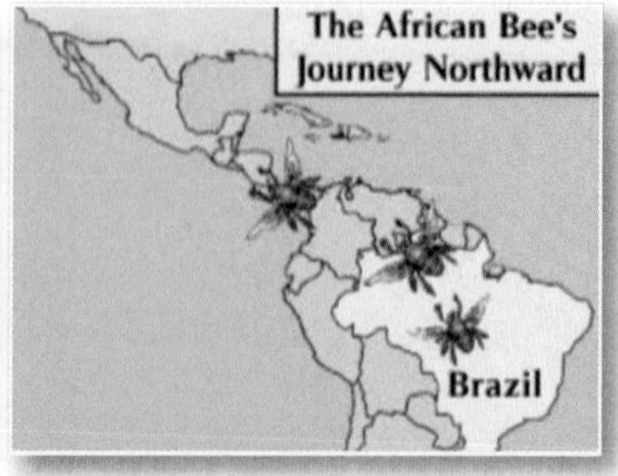

What's Science Got to Do With It? Killer Bees

Page Keeley Science Probe

FLORIDA
SCIENCE
LIFE

Unit 2 FROM BACTERIA TO ANIMALS

Check It! ☐ Lesson 1 ☐ Lesson 2 ☐ Lesson 3

Inquiry LAB STATION

☐ **MiniLabs:** LESSON 1: How does a slime layer work?
LESSON 2: Can decomposition happen without oxygen?
LESSON 3: How do antibodies work?

Try It! then Apply It!

Skill Practice: How do lab techniques affect an investigation?
Inquiry Lab: Bacterial Growth and Disinfectants

ConnectED **Your online portal to everything you need!**
Video • Audio • Review • Lab Station • WebQuests • Assessment • Concepts in Motion • Personal Tutors • Virtual Labs

Here are some of the exciting digital activities you will find in this chapter!

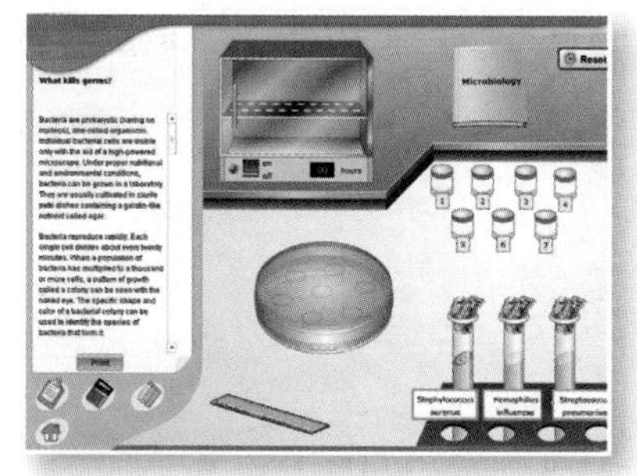

Virtual Lab: What kills germs?

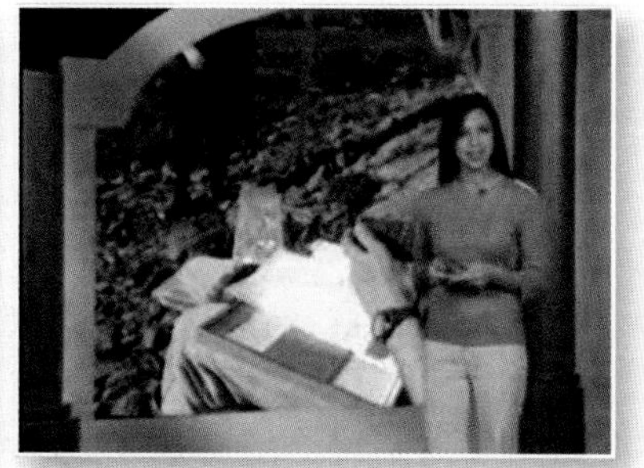

What's Science Got to Do With It? Cleaning Crew

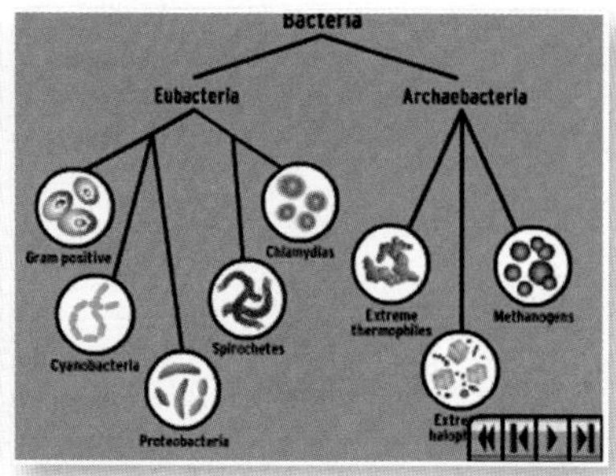

BrainPOP: Bacteria

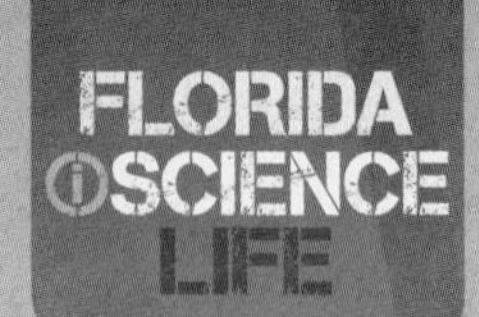

TABLE OF CONTENTS

Check It! ☐ Lesson 1 ☐ Lesson 2

Inquiry LAB STATION

☐ **MiniLabs:** LESSON 1: How can you model the movement of an amoeba?
LESSON 2: What do fungal spores look like?

Try It! then Apply It!

Inquiry Lab: What does a lichen look like?

Your online portal to everything you need!
Video • Audio • Review • Lab Station • WebQuests • Assessment • Concepts in Motion • Personal Tutors • Virtual Labs

Here are some of the exciting digital activities you will find in this chapter!

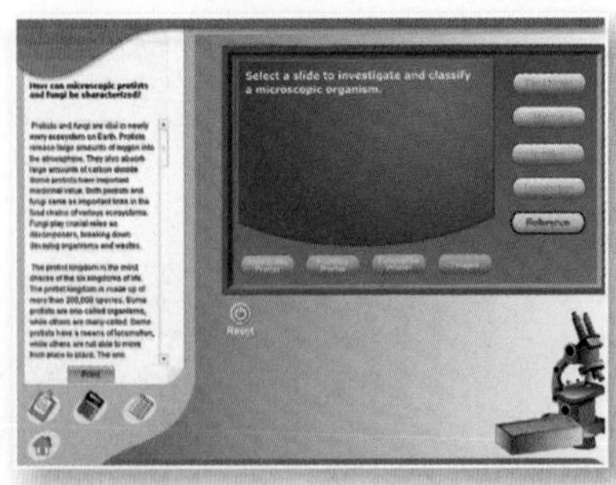

Virtual Lab: How can microscopic protists and fungi be characterized?

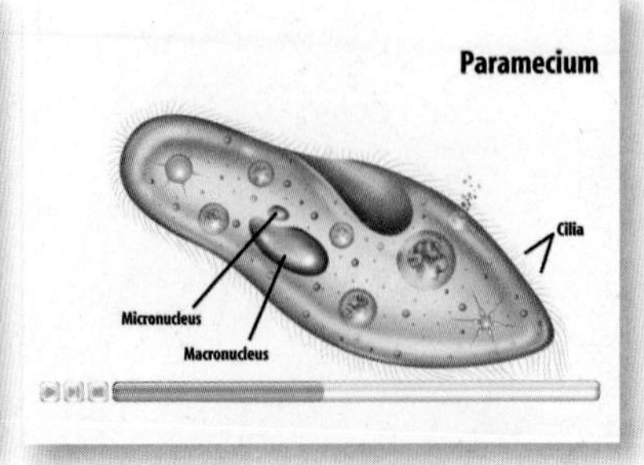

Concepts in Motion: Paramecium

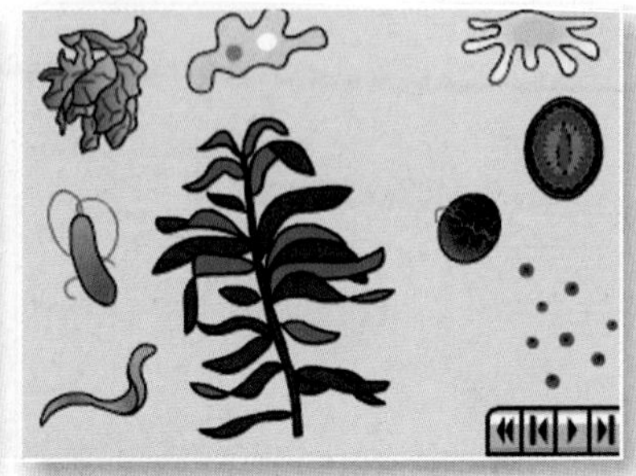

BrainPOP: Protists

Your online portal to everything you need!
Video • Audio • Review • ⓘLab Station • WebQuests • Assessment • Concepts in Motion • Personal Tutors • Virtual Labs

Here are some of the exciting digital activities you will find in this chapter!

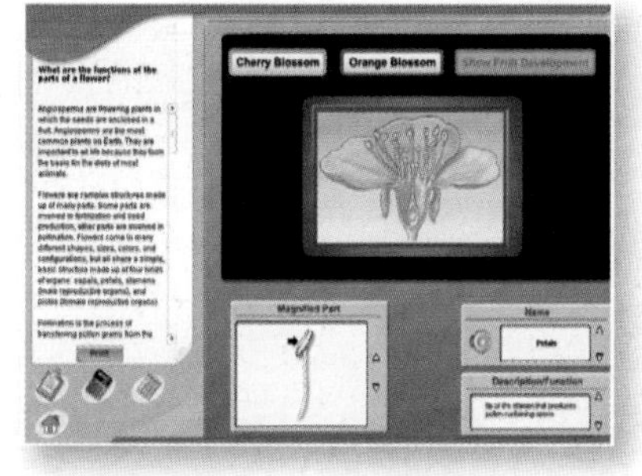

Virtual Lab: What are the functions of the parts of a flower?

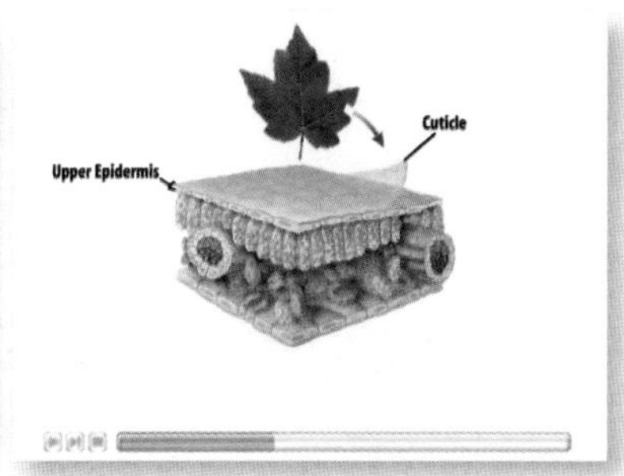

Concepts in Motion: Leaf Structure

BrainPOP: Plant Growth

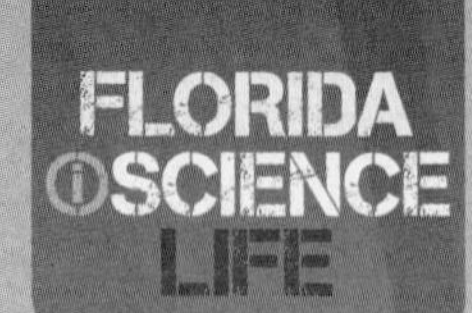

TABLE OF CONTENTS

Inquiry ⓘLAB STATION

☐ **MiniLabs:** LESSON 1: Can you observe plant processes?
LESSON 2: When will plants flower?
LESSON 3: Can you model a flower?

Try It! then Apply It!

Skill Practice: What happens to seeds if you change the intensity of light?

Inquiry Lab: Design a Stimulating Environment for Plants

Your online portal to everything you need!
Video • Audio • Review • ⓘLab Station • WebQuests • Assessment • Concepts in Motion • Personal Tutors • Virtual Labs

Here are some of the exciting digital activities you will find in this chapter!

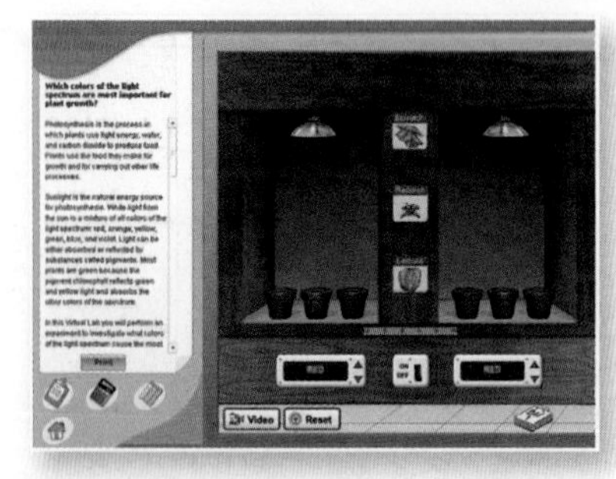

Virtual Lab: Which colors of the light spectrum are most important for plant growth?

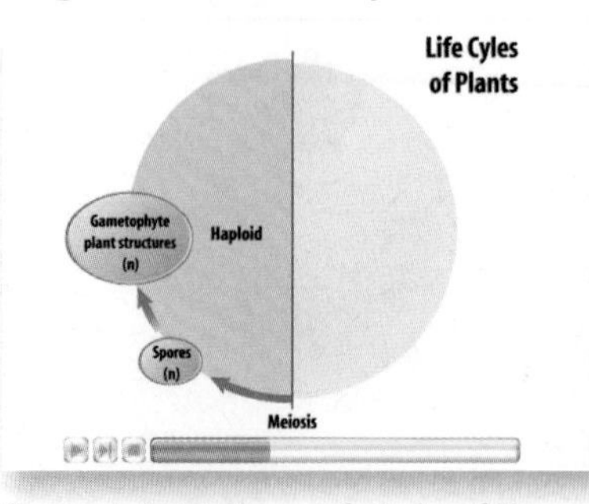

Concepts in Motion: Life Cycles of Plants

Page Keeley Science Probe

Your online portal to everything you need!
Video • Audio • Review • iLab Station • WebQuests • Assessment • Concepts in Motion • Personal Tutors • Virtual Labs

Here are some of the exciting digital activities you will find in this chapter!

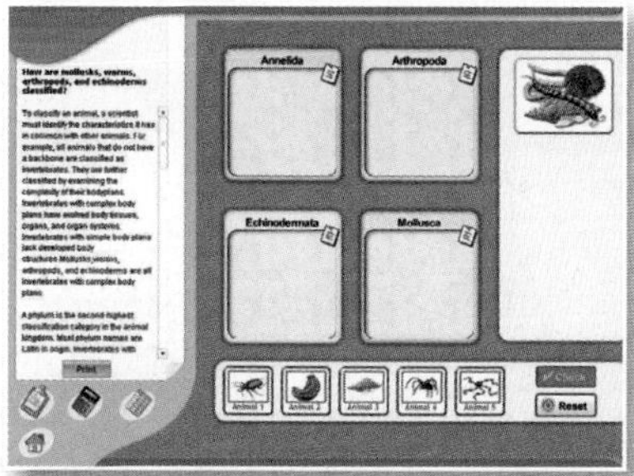

Virtual Lab: How are mollusks, worms, arthropods, and echinoderms classified?

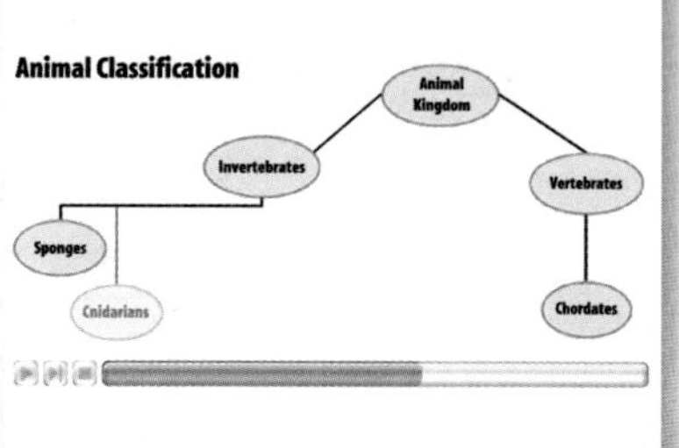

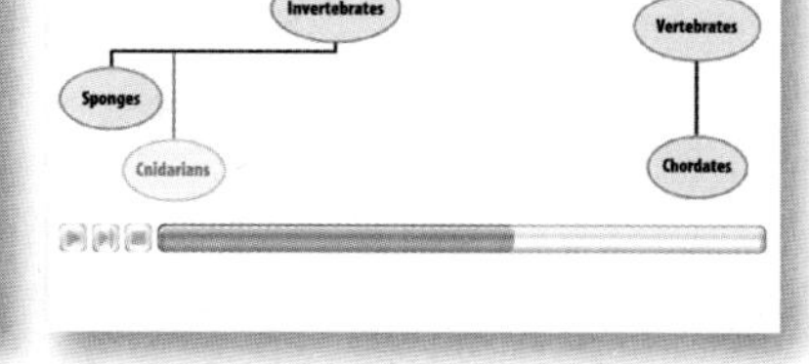

Concepts in Motion: Animal Classification

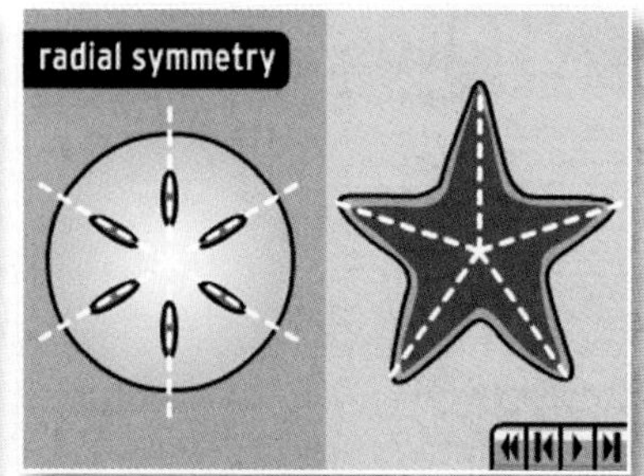

BrainPOP: Invertebrates

TABLE OF CONTENTS

Unit 3 HUMAN BODY SYSTEMS

Check It! ☐ Lesson 1 ☐ Lesson 2 ☐ Lesson 3

Inquiry LAB STATION

☐ **MiniLabs:** LESSON 1: How does the skeleton protect organs?
LESSON 2: How strong are your hand muscles?
LESSON 3: Why are you sweating?

Try It! then Apply It!

How do the three types of muscle cells compare?

Inquiry Lab: Dissect a Chicken Wing

Your online portal to everything you need!
Video • Audio • Review • Lab Station • WebQuests • Assessment • Concepts in Motion • Personal Tutors • Virtual Labs

Here are some of the exciting digital activities you will find in this chapter!

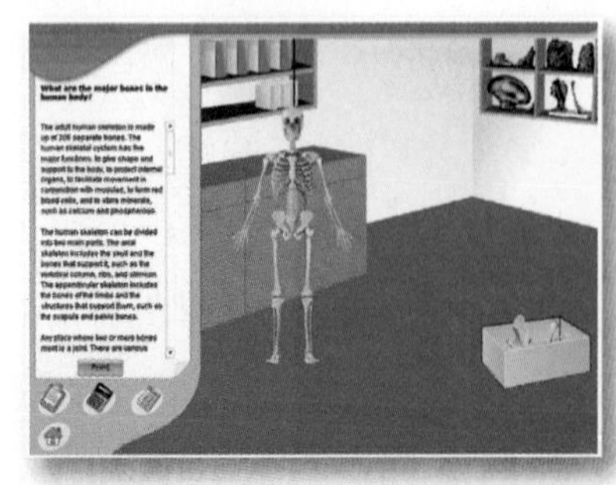

Virtual Lab: What are the major bones in the human body?

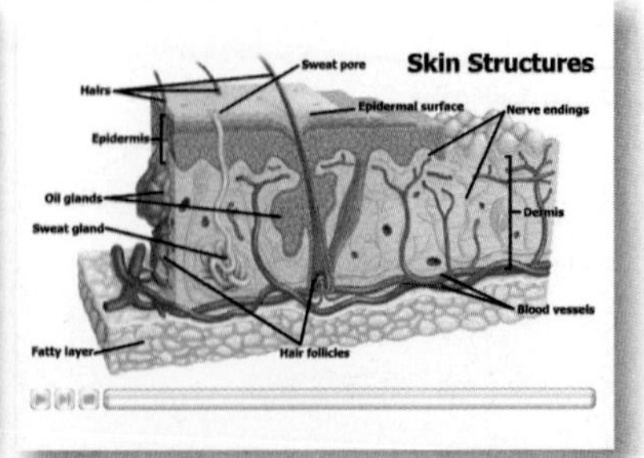

Concepts in Motion: Skin Structures

What's Science Got to Do With It? Modern Mummies

Chapter 13 Digestion and Excretion

Check It! ☐ Lesson 1 ☐ Lesson 2 ☐ Lesson 3

☐ **MiniLabs:** LESSON 1: What nutrients are in foods?
LESSON 2: How can you model digestion?
LESSON 3: How can you model the function of a kidney?

Try It! then Apply It!

Skill Practice: How do foods compare?
Inquiry Lab: Model Digestion from Start to Finish

ConnectED **Your online portal to everything you need!**
Video • Audio • Review • ⓘLab Station • WebQuests • Assessment • Concepts in Motion • Personal Tutors • Virtual Labs

Here are some of the exciting digital activities you will find in this chapter!

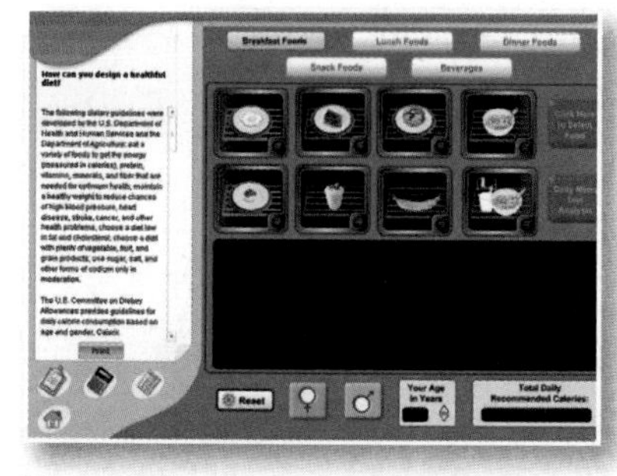

Virtual Lab: How can you design a healthful diet?

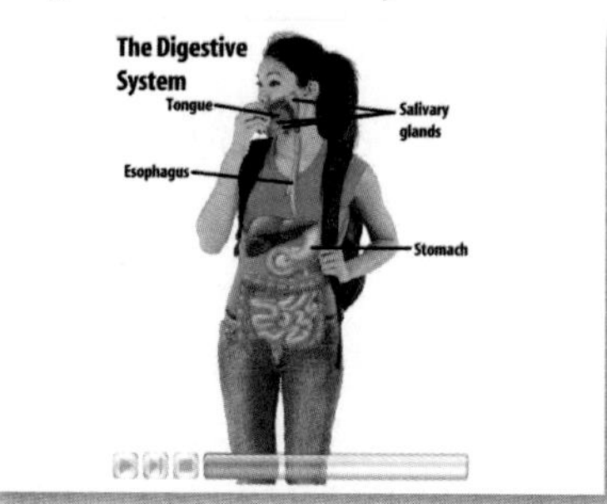

Concepts in Motion: The Digestive System

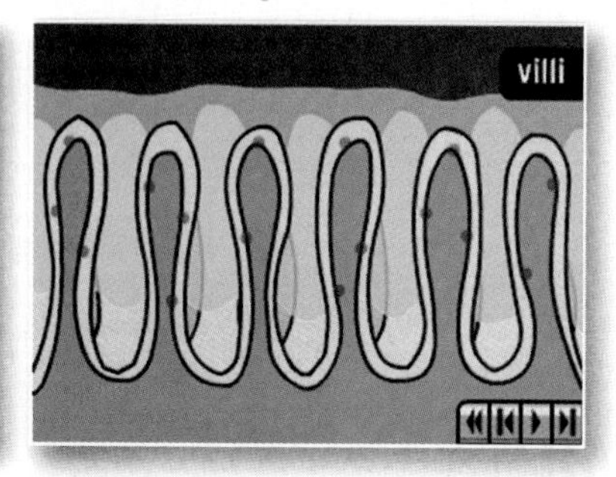

BrainPOP: Digestion

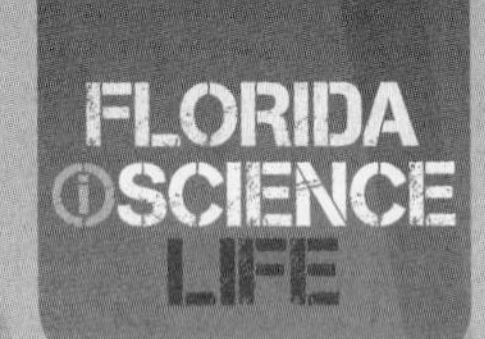

TABLE OF CONTENTS

Your online portal to everything you need!
Video • Audio • Review • Lab Station • WebQuests • Assessment • Concepts in Motion • Personal Tutors • Virtual Labs

Here are some of the exciting digital activities you will find in this chapter!

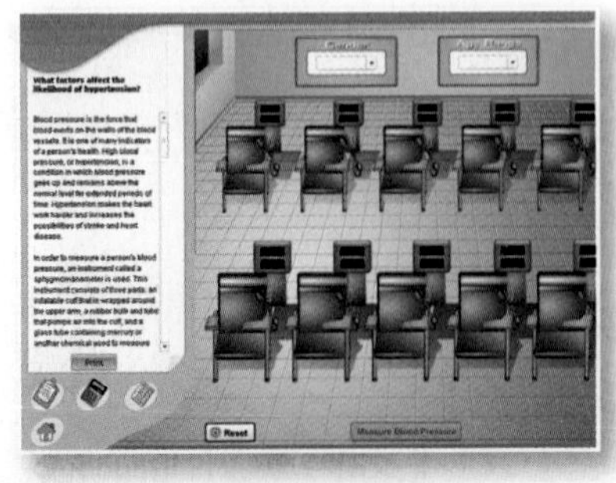

Virtual Lab: What factors affect the likelihood of hypertension?

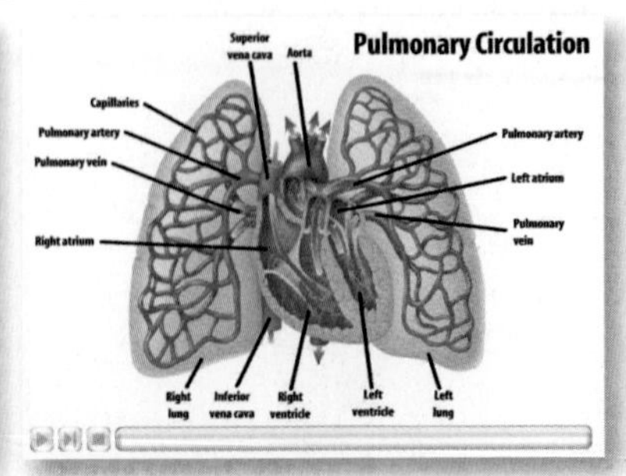

Concepts in Motion: Pulmonary Circulation

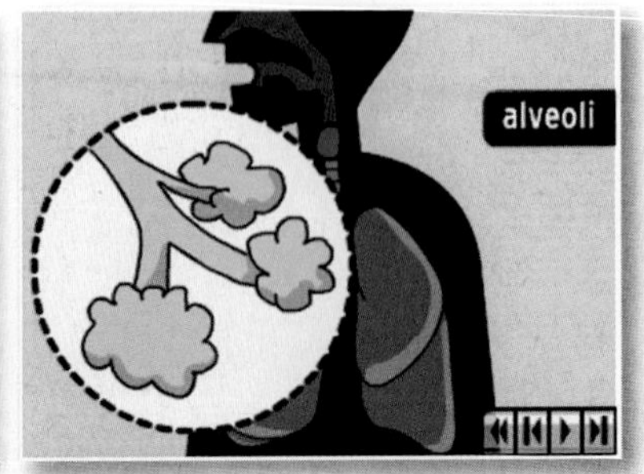

BrainPOP: Respiratory System

Chapter 15 Immunity and Disease

Check It! ☐ Lesson 1 ☐ Lesson 2 ☐ Lesson 3

Inquiry LAB STATION

☐ **MiniLabs:**
- LESSON 1: How does an infectious disease spread throughout a population?
- LESSON 2: How do different layers of your skin protect your body?
- LESSON 3: How clean are your hands?

Try It! then Apply It!

Skill Practice: How would you prepare a work area for procedures that require aseptic techniques?

Inquiry Lab: Can one bad apple spoil the bunch?

ConnectED **Your online portal to everything you need!**
Video • Audio • Review • ⓘLab Station • WebQuests • Assessment • Concepts in Motion • Personal Tutors • Virtual Labs

Here are some of the exciting digital activities you will find in this chapter!

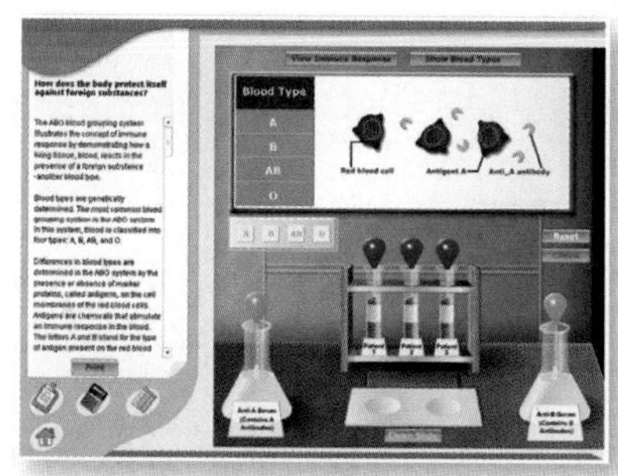

Virtual Lab: How does the body protect itself against foreign substances?

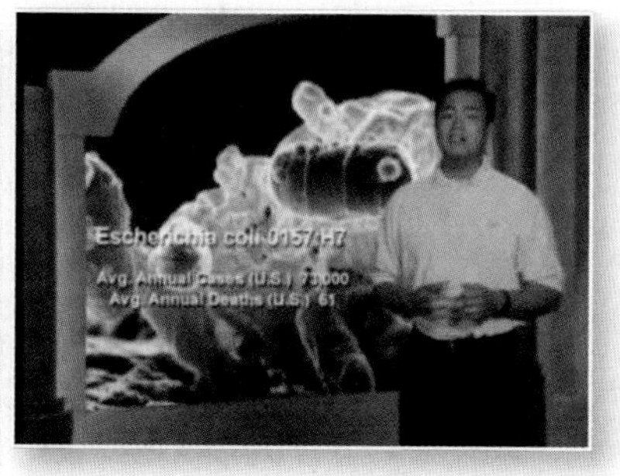

What's Science Got to do With It? Epidemic

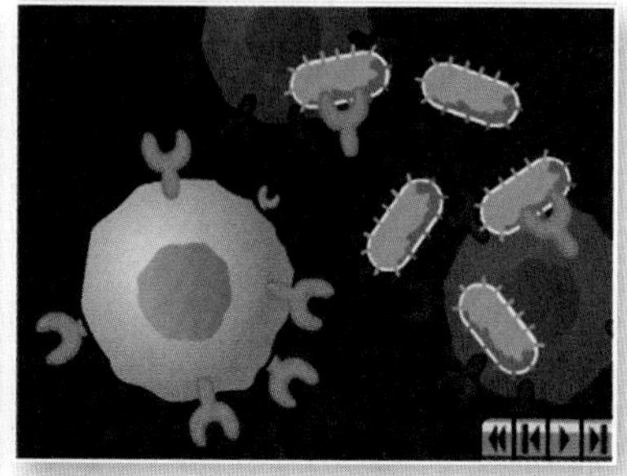

BrainPOP: Immune System

TABLE OF CONTENTS

ConnectED **Your online portal to everything you need!**
Video • Audio • Review • Lab Station • WebQuests • Assessment • Concepts in Motion • Personal Tutors • Virtual Labs

Here are some of the exciting digital activities you will find in this chapter!

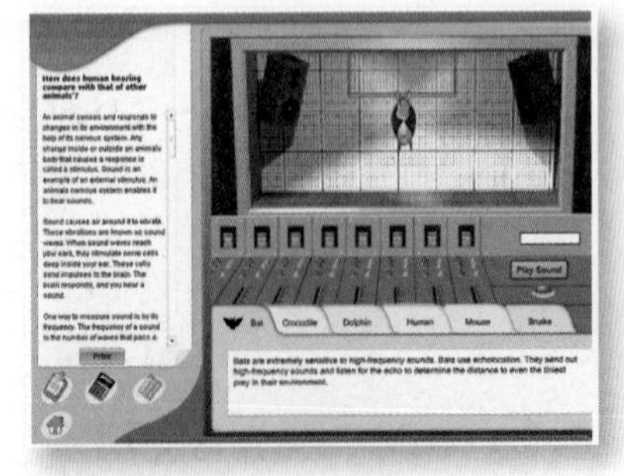

Virtual Lab: How does human hearing compare to that of other animals?

What's Science Got to do With It? Access Denied

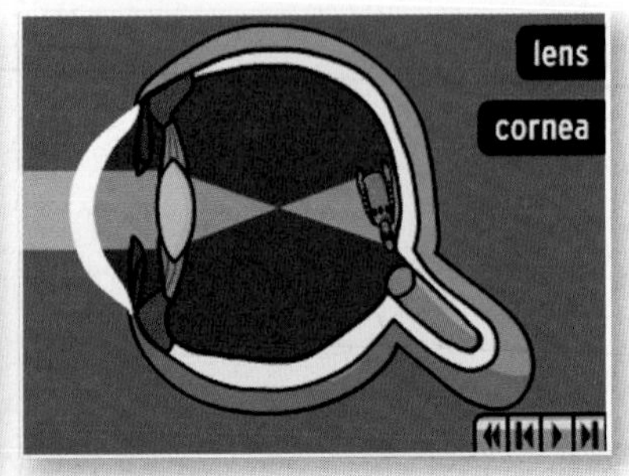

BrainPOP: Eye

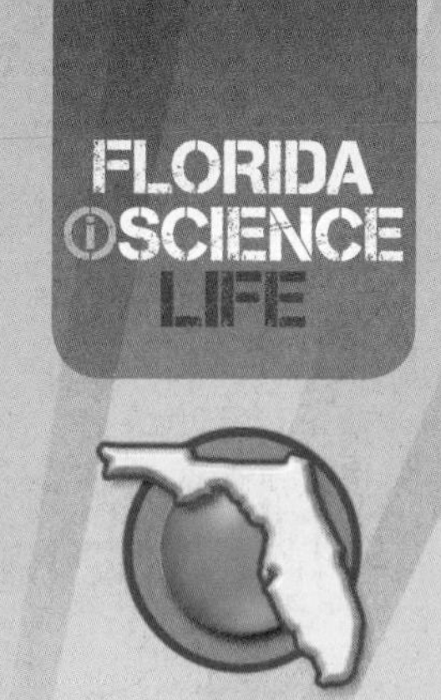

Chapter 17 **Reproduction and Development**

Inquiry ⓘLAB STATION

☐ **MiniLabs:** LESSON 1: What occurs in menstrual cycle phases?
LESSON 2: How do life stages after birth differ?

Try It! then Apply It!

Inquiry Lab: Educating Extraterrestrials About Human Development

Your online portal to everything you need!
Video • Audio • Review • ⓘLab Station • WebQuests • Assessment • Concepts in Motion • Personal Tutors • Virtual Labs

Here are some of the exciting digital activities you will find in this chapter!

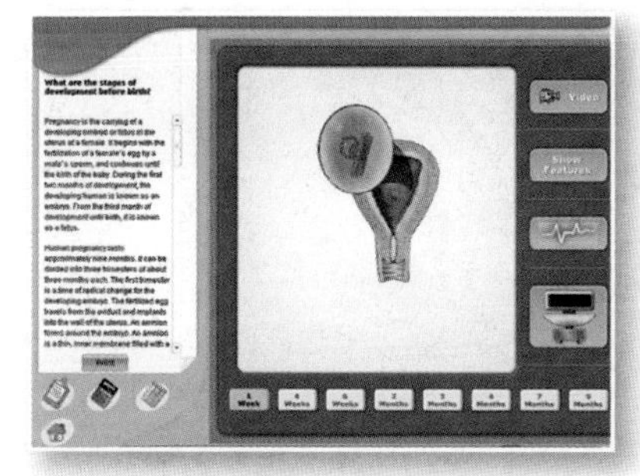

Virtual Lab: What are the stages of development before birth?

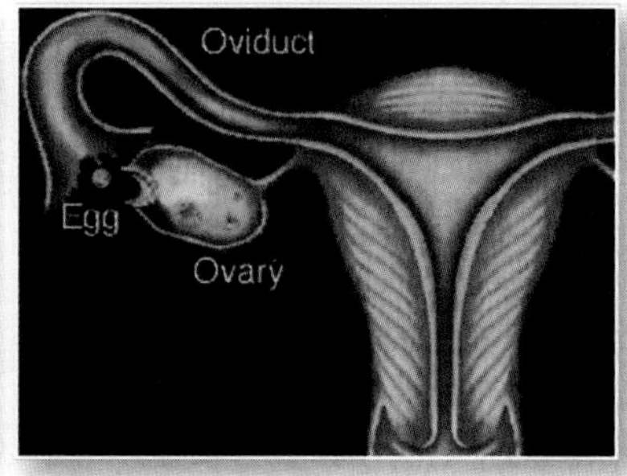

Concepts in Motion: Female Reproductive System

BrainPOP: Fertilization and Birth

TABLE OF CONTENTS

Unit 4 INTERACTIONS OF LIFE

Your online portal to everything you need!
Video • Audio • Review • ⓘLab Station • WebQuests • Assessment • Concepts in Motion • Personal Tutors • Virtual Labs

Here are some of the exciting digital activities you will find in this chapter!

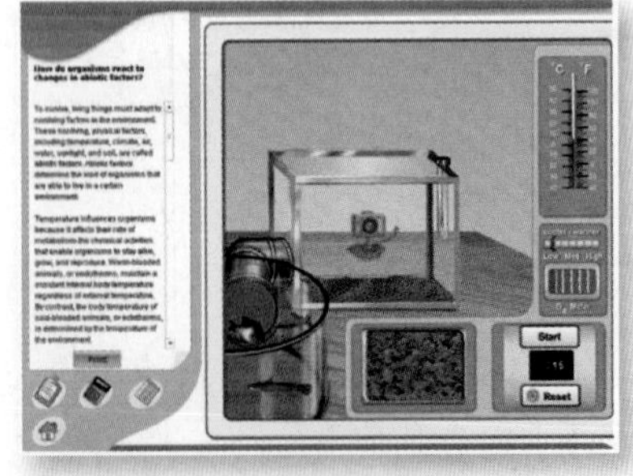

Virtual Lab: How do organisms react to changes in abiotic factors?

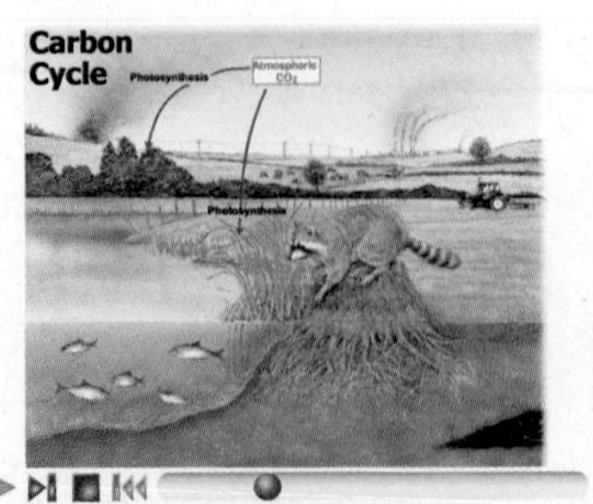

Concepts in Motion: Carbon Cycle

BrainPOP: Water Cycle

Chapter 19 Populations and Communities

Check It! ☐ Lesson 1 ☐ Lesson 2 ☐ Lesson 3

Inquiry LAB STATION

☐ **MiniLabs:** LESSON 1: What are limiting factors?
LESSON 2: How does migration affect population size?
LESSON 3: How can you model a food web?

Try It! then Apply It!

Skill Practice: How do popululations change in size?

Inquiry Lab: How can you model a symbiotic relationship?

Your online portal to everything you need!
Video • Audio • Review • Lab Station • WebQuests • Assessment • Concepts in Motion • Personal Tutors • Virtual Labs

Here are some of the exciting digital activities you will find in this chapter!

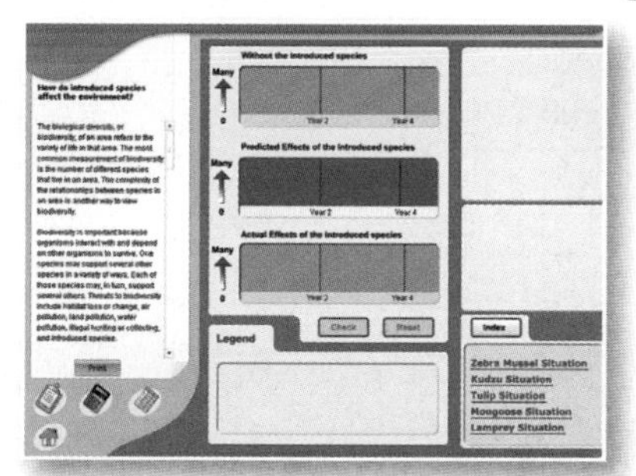

Virtual Lab: How do introduced species affect the environment?

Concepts in Motion: A Food Web

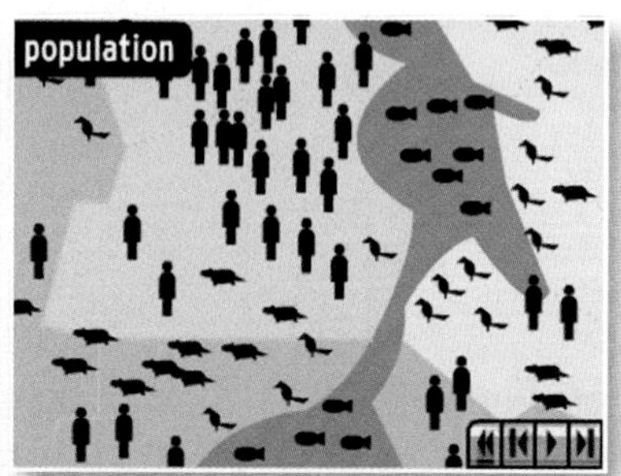

BrainPOP: Population Growth

TABLE OF CONTENTS

Your online portal to everything you need!
Video • Audio • Review • iLab Station • WebQuests • Assessment • Concepts in Motion • Personal Tutors • Virtual Labs

Here are some of the exciting digital activities you will find in this chapter!

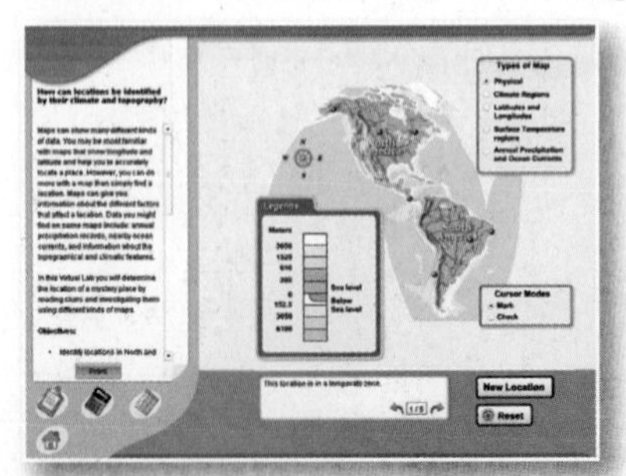

Virtual Lab: How can locations be identified by their climate and topography?

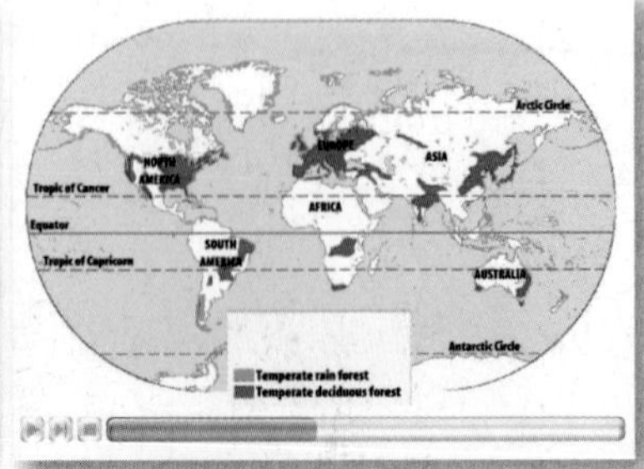

Concepts in Motion: Land Biomes

What's Science Got to do With It?
Fighting Fire with Fire

Scientific Explanations

An explanation helps provide answers to a question a scientist might be wondering about. Which of the following do you think involves providing a scientific explanation? Select the best response.

A. hypothesis

B. scientific theory

C. scientific law

D. hypothesis and scientific theory

E. scientific theory and scientific law

F. hypothesis, scientific theory, and scientific law

G. None of the above. An explanation is something else.

Explain your thinking. Describe how explanations are used in science.

Scientific EXPLANATIONS

Nature of Science

This chapter begins your study of the nature of science, but there is even more information about the nature of science in this book. Each unit begins by exploring an important topic that is fundamental to scientific study. As you read these topics, you will learn even more about the nature of science.

FLORIDA BIG IDEAS

1 The Practice of Science

2 The Characteristics of Scientific Knowledge

3 The Role of Theories, Laws, Hypotheses, and Models

4 Science and Society

The Big Idea

Think About It! How can science provide answers to your questions about the world around you?

These two divers are collecting data about corals. They are marine biologists, scientists who study living things in oceans and other saltwater environments.

1. What information about corals do you think these scientists are collecting?

____

____

2. How do you think science can provide answers to their questions and your questions?

____

____

Florida NGSSS

SC.6.N.2.1 Distinguish science from other activities involving thought.
SC.6.N.2.2 Explain that scientific knowledge is durable because it is open to change as new evidence or interpretations are encountered.
SC.6.N.2.3 Recognize that scientists who make contributions to scientific knowledge come from all kinds of backgrounds and possess varied talents, interests, and goals.
SC.6.N.3.1 Recognize and explain that a scientific theory is a well-supported and widely accepted explanation of nature and is not simply a claim posed by an individual. Thus, the use of the term theory in science is very different than how it is used in everyday life.
SC.6.N.3.2 Recognize and explain that a scientific law is a description of a specific relationship under given conditions in the natural world. Thus, scientific laws are different from societal laws.
SC.6.N.3.3 Give several examples of scientific laws.
SC.7.N.1.2 Differentiate replication (by others) from repetition (multiple trials).
SC.7.N.1.3 Distinguish between an experiment (which must involve the identification and control of variables) and other forms of scientific investigation and explain that not all scientific knowledge is derived from experimentation.
SC.7.N.1.4 Identify test variables (independent variables) and outcome variables (dependent variables) in an experiment.
SC.7.N.1.5 Describe the methods used in the pursuit of a scientific explanation as seen in different fields of science such as biology, geology, and physics.
SC.7.N.1.6 Explain that empirical evidence is the cumulative body of observations of a natural phenomenon on which scientific explanations are based.
SC.7.N.1.7 Explain that scientific knowledge is the result of a great deal of debate and confirmation within the science community.
SC.7.N.2.1 Identify an instance from the history of science in which scientific knowledge has changed when new evidence or new interpretations are encountered.
SC.7.N.3.1 Recognize and explain the difference between theories and laws and give several examples of scientific theories and the evidence that supports them.
SC.8.N.1.3 Use phrases such as "results support" or "fail to support" in science, understanding that science does not offer conclusive 'proof' of a knowledge claim.
SC.8.N.1.4 Explain how hypotheses are valuable if they lead to further investigations, even if they turn out not to be supported by the data.
SC.8.N.1.6 Understand that scientific investigations involve the collection of relevant empirical evidence, the use of logical reasoning, and the application of imagination in devising hypotheses, predictions, explanations and models to make sense of the collected evidence.
SC.8.N.2.2 Discuss what characterizes science and its methods.
SC.8.N.3.2 Explain why theories may be modified but are rarely discarded.
SC.8.N.4.1 Explain that science is one of the processes that can be used to inform decision making at the community, state, national, and international levels.
SC.8.N.4.2 Explain how political, social, and economic concerns can affect science, and vice versa.
Also covers: LA.6.2.2.3, MA.6.A.3.6, SC.6.N.1.2, SC.6.N.1.3, SC.6.N.1.5, SC.7.N.1.1

There's More Online!
Video • Audio • Review • ⓘLab Station • WebQuest • Assessment • Concepts in Motion • Multilingual eGlossary

Lesson 1

Understanding SCIENCE

ESSENTIAL QUESTIONS

- What is scientific inquiry?
- What are the results of scientific investigations?
- How can a scientist prevent bias in a scientific investigation?

Vocabulary

science p. NOS 4
observation p. NOS 6
inference p. NOS 6
hypothesis p. NOS 6
prediction p. NOS 7
technology p. NOS 8
scientific theory p. NOS 9
scientific law p. NOS 9
critical thinking p. NOS 10

What is science?

The last time that you watched squirrels play in a park or in your yard, did you realize that you were practicing science? Every time you observe the natural world, you are practicing science. **Science** *is the investigation and exploration of natural events and of the new information that results from those investigations.*

When you observe the natural world, you might form questions about what you see. While you are exploring those questions, you probably use reasoning, creativity, and skepticism to help you find answers to your questions. People use these behaviors in their daily lives to solve problems, such as keeping a squirrel from eating birdseed, as shown in **Figure 1.** Similarly, scientists use these behaviors in their work.

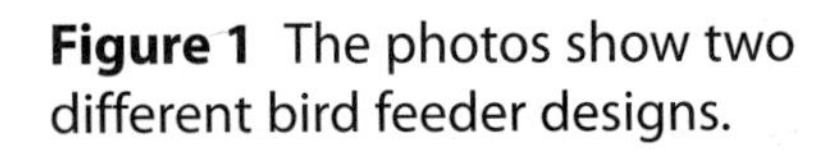

Figure 1 The photos show two different bird feeder designs.

Active Reading **1. Express** Someone used

____

and ______________________ to design each of these squirrel-proof bird feeders. These skills can be used to solve other problems. However, some solutions don't work.

Active Reading **2. Recall** Describe a time when you used science to gain new information.

____

____

Branches of Science

No one person can study the entire natural world. Therefore, people tend to focus their efforts on one of the three fields or branches of science—life science, Earth science, or physical science, as described below. Then people or scientists can seek answers to specific problems within one field of science.

Word Origin

biology
from Greek *bios*, means "life"; and *logia*, means "study of"

Active Reading **3. Analyze** In the boxes below, write examples of the types of questions scientists might ask.

Life Science

Biology, or life science, is the study of all living things. This forest ecologist studies interactions in forest ecosystems, investigating lichens growing on Douglas firs.

Life Science Questions:

Earth Science

The study of Earth and space, including landforms, rocks, soil, and forces that shape Earth's surface, are Earth science. These Earth scientists are studying soil samples in Africa.

Earth Science Questions:

Physical Science

The study of chemistry and physics is physical science. Physical scientists study the interactions of matter and energy. This chemist is preparing antibiotic solutions.

Physical Science Questions:

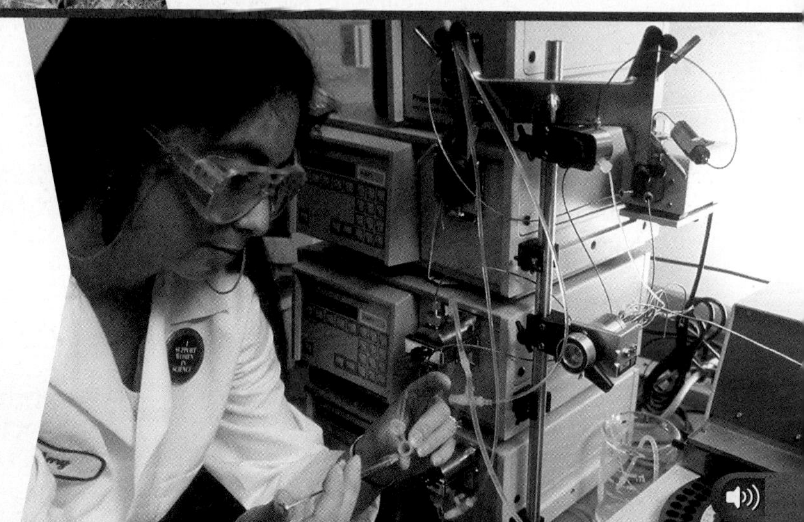

Scientific Inquiry

The chart in **Figure 2** shows a sequence of the skills or methods that a scientist might use in an investigation. However, sometimes, not all of these skills are performed in an investigation, or performed in this order. Scientists practice scientific inquiry—a process that uses a variety of skills and tools to answer questions or to test ideas about the natural world.

Ask Questions

Like a scientist, you use scientific inquiry in your life. Suppose you plant a garden. As you plant the vegetable seeds, you water the seeds. Then, you mix fertilizer into the soil. After a few weeks, you observe that some plants are growing better than others. An **observation** *is using one or more of your senses to gather information and take note of what occurs.*

Observations are the beginning of the process of inquiry and can lead to questions such as "Why are some plants growing better than others?" As you make observations and ask questions, you recall from science class that plants need water and sunlight to grow. You infer that some plants are receiving more water than others and are growing better. An **inference** *is a logical explanation of an observation that is drawn from prior knowledge or experience.*

Active Reading **4. Define** Complete the information about a hypothesis.

A hypothesis is a

____

about an

____

that can be tested by

____

______________________.

Hypothesize

After making observations and inferences, you develop a hypothesis and investigate why some plants are growing better than others. *A possible explanation about an observation that can be tested by scientific investigations is a* **hypothesis.** Your hypothesis might be: Some plants are growing taller and more quickly than others because they are receiving more water. Or, your hypothesis might be: The plants that are growing quickly received fertilizer because fertilizer helps plants grow.

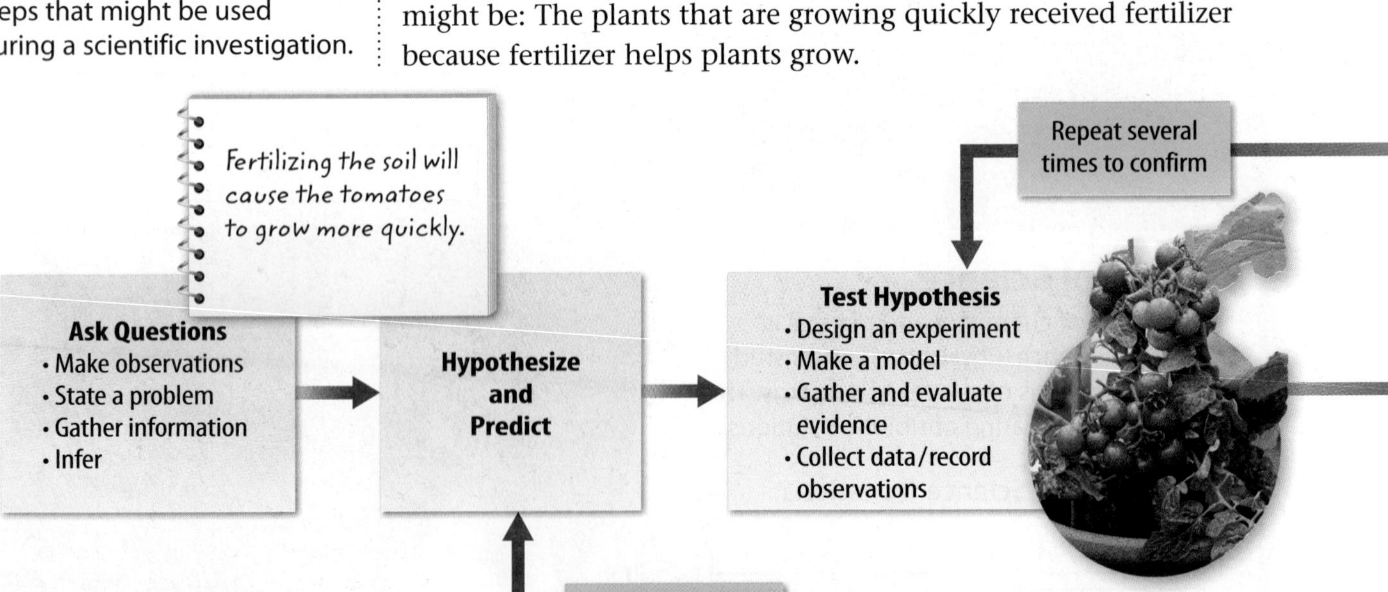

Figure 2 This flowchart shows steps that might be used during a scientific investigation.

Predict

After you state a hypothesis, make a prediction to help you test your hypothesis. *A* **prediction** *is a statement of what will happen next in a sequence of events.* Based on your hypothesis, you might predict that if plants receive more water, or fertilizer, then they will grow taller and more quickly.

Test Your Hypothesis

When you test a hypothesis, you test your predictions. For example, you might design an experiment to test your hypothesis on the fertilizer. Set up an experiment in which you plant seeds and add fertilizer to only some of them. Your prediction is that the plants that get fertilizer will grow more quickly. If your prediction is confirmed, it supports your hypothesis. If your prediction is not confirmed, your hypothesis might need revision.

Analyze Results

As you test your hypothesis, collect data about the plants' rates of growth and how much fertilizer each plant receives. It might be difficult to recognize patterns and relationships in data. Your next step is to organize and analyze your data. You can create graphs, classify information, or make models.

Once data are organized, you more easily can study the data and draw conclusions. Some methods of testing a hypothesis and analyzing results are shown in **Figure 2.**

Draw Conclusions

Now you must decide whether your data support your hypothesis and then draw conclusions. A conclusion is a summary of the information gained from testing a hypothesis. If your hypothesis is supported, you can repeat your experiment several times to confirm your results. If your hypothesis is not supported, you can modify it and repeat the scientific inquiry process.

Communicate Results

An important step in scientific inquiry is communicating results to others. Scientists write scientific articles, speak at conferences, or exchange information on the Internet. This part of scientific inquiry is important because scientists use new information in their research or perform other scientists' investigations to verify results.

Active Reading **5. Define** Underline the term scientific inquiry and its meaning.

Active Reading **6. Label** Fill in the blanks in the flowchart below.

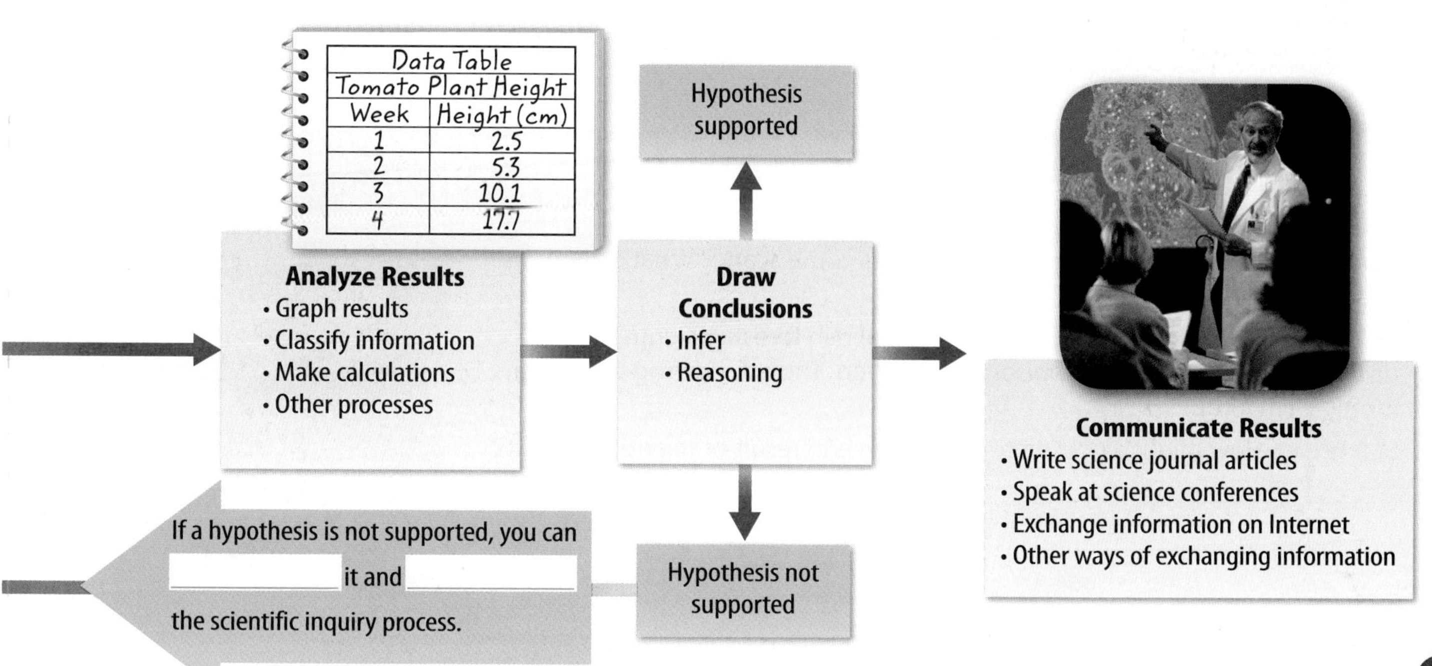

Results of Scientific Inquiry

In science, both you and scientists perform scientific inquiry to find answers to questions. There are many outcomes of scientific inquiry, such as technology, materials, and explanations, as shown below.

Active Reading **7. Express** Fill in the blanks below with some results of the scientific investigations.

Technology

The practical use of scientific knowledge, especially for industrial or commercial use, is **technology.** Televisions, MP3 players, and computers are examples of technology. The C-Leg is one design of computer-aided limbs. The prosthetic leg has sensors that anticipate the user's next move, which prevents him or her from stumbling or tripping. In addition, the C-Leg can enable the user to walk, stand, or even ride a bike.

Infer How is the C-Leg a form of technology that resulted from scientific inquiry?

____

____

____

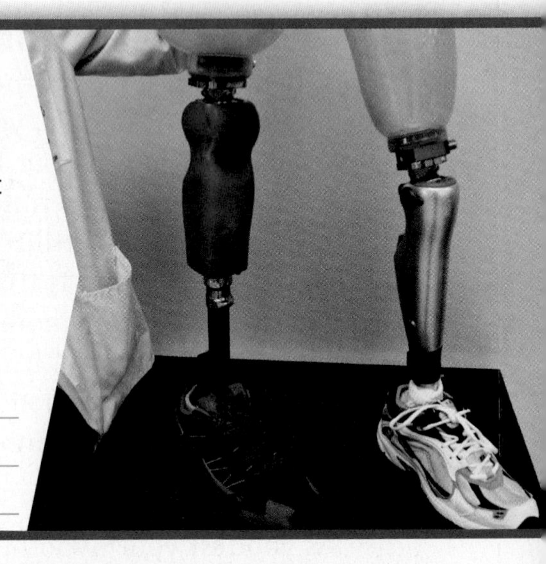

New Materials

Using technology, scientists have developed a bone bioceramic. A bioceramic is a natural calcium-phosphate mineral complex that is part of bones and teeth. This synthetic bone mimics natural bone's porous structure. It can be shaped into implants treated with certain cells from the patient's bone marrow and then implanted into the patient's body to replace missing bone.

Summarize How is bone bioceramics a form of new material?

____

____

____

Possible Explanations

Scientific investigations often answer the questions *who, what, when, where,* or *how.* For example, who left fingerprints at a crime scene? What organisms live in rain forests?

In 2007, while exploring in Colombia's tropical rain forests, scientists discovered a new species of poisonous tree frog. The golden frog of Supatá is only 2 cm long.

Describe How might a new discovery be the end result of scientific inquiry?

____

____

____

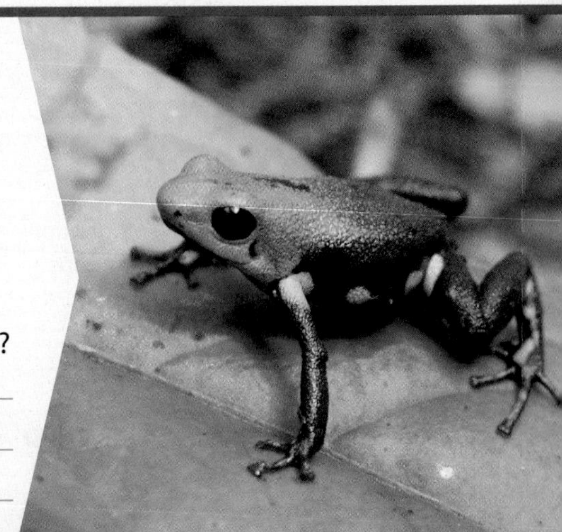

Scientific Theory and Scientific Laws

Scientists often repeat scientific investigations to verify that the results for a hypothesis or a group of hypotheses are correct. This can lead to a scientific theory.

Scientific Theory The everyday meaning of the word *theory* is an untested idea or an opinion. However, a **scientific theory** *is an explanation of observations or events based on knowledge gained from many observations and investigations.* About 300 years ago, scientists began looking at samples through the first microscopes. They noticed tinier units, or cells, as shown in **Figure 3.** As more scientists observed cells in other organisms, their observations became known as the cell theory. This theory explains that all living things are made of cells. A scientific theory is assumed to be the best explanation of observations unless it is disproved. The cell theory will continue to explain the makeup of all organisms until an organism is discovered that is not made of cells.

Scientific Laws Scientific laws are different from societal laws, which are agreements on a set of behavior. A **scientific law** *describes a pattern or an event in nature that is always true.* A scientific theory might explain how and why an event occurs. But a scientific law states only that an event in nature will occur under specific conditions. For example, the law of conservation of mass states that the mass of materials will be the same before and after a chemical reaction. This scientific law does not explain why this occurs—only that it will occur. **Table 1** compares a scientific theory and a scientific law.

Active Reading **9. Recall** Fill in the table below to compare a scientific theory and a scientific law.

Table 1 Comparing Scientific Theory and Scientific Law

Scientific Theory	Scientific Law
A scientific theory is based on repeated ______ and ______.	Scientific laws are observations of similar events that have been ______.
If new information does not support a ______, it will be modified or rejected.	If many new observations do not follow ______, it is rejected.
A scientific theory attempts to explain ______.	A scientific law states that ______.
A ______ usually is more complex than a ______ and contains many well-supported hypotheses.	A ______ usually contains one well-supported hypothesis that states that something will happen.

Active Reading **8. Summarize** Why does a scientist repeat scientific investigations? Provide an example.

______

______

______

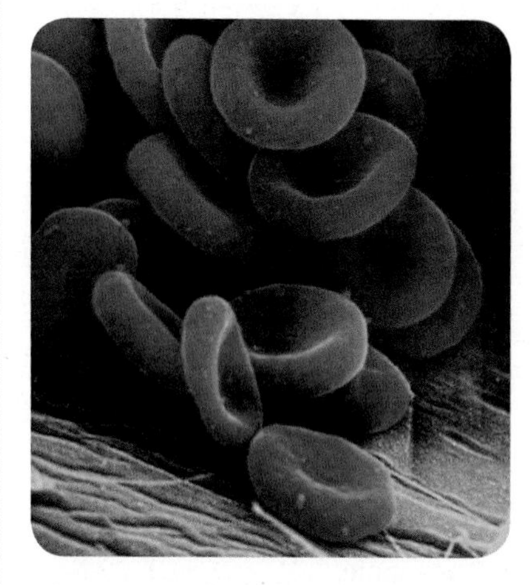

Figure 3 When you view blood using a microscope, you will see that it contains red blood cells.

Active Reading FOLDABLES® LA.6.4.2.2

Make a vertical two-column chart book. Label it as shown. Use it to organize your notes on scientific investigations.

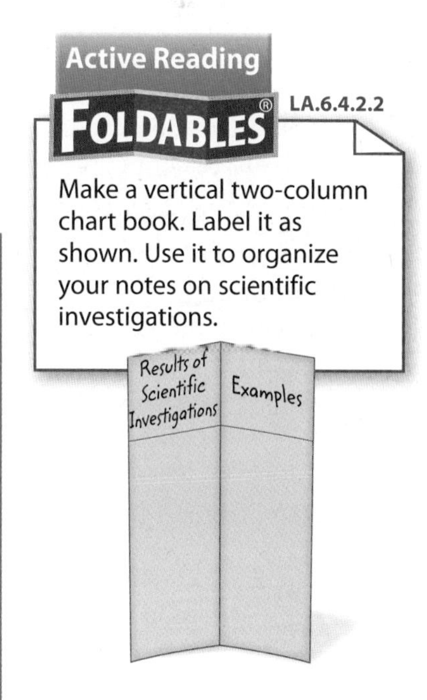

Active Reading **10. Explain** Why is it important to be skeptical about new information?

Skepticism in Media

When you see scientific issues in newspapers, radio, television, and magazines, it is important to be skeptical and question information. It is essential to question statements made by people outside their area of expertise and to question claims that are based on vague statements. To determine whether information is scientific or pseudoscience (information incorrectly represented as scientific), be skeptical and identify facts and opinions.

Evaluating Scientific Evidence

An important skill in scientific inquiry is critical thinking. **Critical thinking** *is comparing what you already know with the information you are given in order to decide whether you agree with it.* To identify and prevent bias in an investigation, sampling, repetition, and blind studies can be helpful.

1 Sampling

Studying small amounts of something in order to learn about the larger whole is sampling. A sample should be a random representation of the whole.

3 Blind Study

In a blind study, the investigator, subject, or both do not know which item they are testing. Then, personal biases cannot affect an investigation.

2 Bias

Bias is intentional or unintentional prejudice toward a specific outcome. Sources of bias in an investigation can include equipment choices, hypothesis formation, and prior knowledge.

Active Reading **11. Express** Describe how the prior knowledge of the cost of cereal might bias a participant in a taste test.

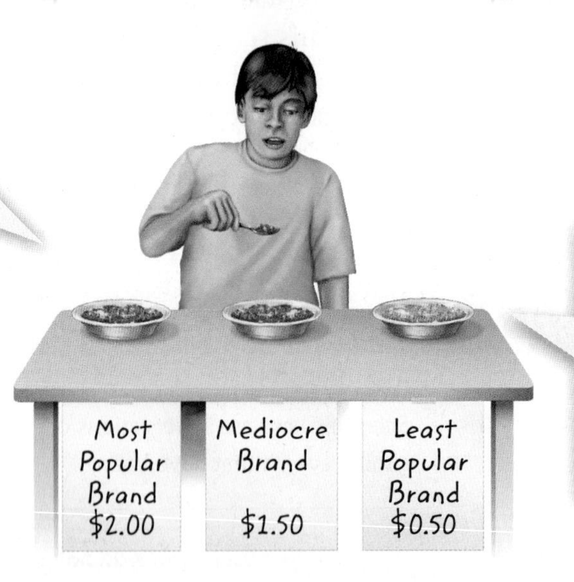

4 Repetition

If your results differ greatly when you repeat an investigation, then one of the investigations probably was flawed. Repetition of experiments helps reduce bias.

Active Reading **12. Analyze** Complete the table below with information on how a scientist can help prevent bias in a scientific investigation.

Sampling	Bias	Blind Study	Repetition

Science cannot answer all questions.

Many questions can be answered through a scientific investigation. But there are some questions that science cannot answer, such as the one posed in **Figure 4.** Questions about personal opinions, values, beliefs, and feelings cannot be answered scientifically.

Safety in Science

Scientists follow safety procedures when they conduct investigations. You too should follow safety procedures during any experiment. You should wear appropriate safety equipment, listen to your teacher's instructions, learn to recognize potential hazards, and know the meaning of safety symbols.

Ethics are important when using living things during investigations. Animals should be treated properly. Scientists should tell research participants about the potential risks and benefits of the research.

Active Reading 13. **Analyze** Infer three reasons why ethics are important when using animals or humans in research.

Figure 4 Science cannot answer questions based on opinions or feelings, such as which paint color is the prettiest.

Lesson Review 1

Use Vocabulary

1. **Use the terms** *scientific law* and *scientific theory* in a complete sentence. SC.7.N.3.1

2. **Compare** critical thinking and inference. SC.8.N.2.2

Understand Key Concepts

3. **Discuss** three ways a scientist can minimize bias in scientific investigations.

4. **Which** should NOT be part of scientific inquiry? SC.8.N.2.2
   (A) bias
   (B) analysis
   (C) hypothesis
   (D) testing

Interpret Graphics

5. **List** an example of how to test a hypothesis using scientific inquiry. SC.6.2.2.3

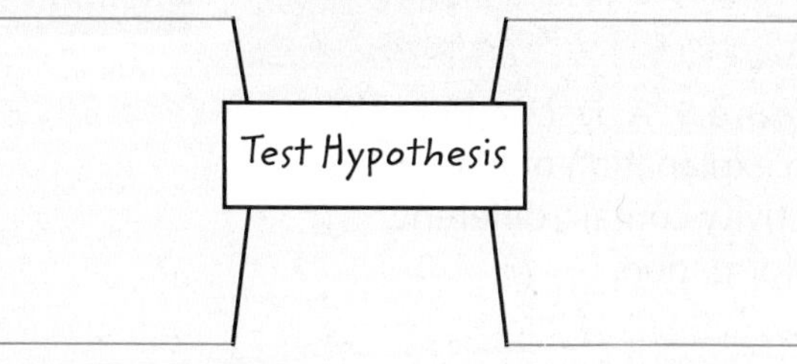

Critical Thinking

6. **Evaluate** Two teams of scientists attempted to answer the same question and came to opposite conclusions. How do you decide which investigation was valid?

Lesson 2

Measurement and SCIENTIFIC TOOLS

ESSENTIAL QUESTIONS

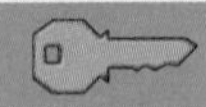

- What is the difference between accuracy and precision?
- Why should you use significant digits?
- What are some tools used by life scientists?

Vocabulary

description p. NOS 12
explanation p. NOS 12
International System of Units (SI) p. NOS 12
accuracy p. NOS 14
precision p. NOS 14
significant digits p. NOS 15

Description and Explanation

How would you describe the squirrel's activity in **Figure 5?** A **description** *is a spoken or written summary of observations.* You might describe the squirrel as burying five acorns near a large tree. A qualitative description uses your senses (sight, sound, smell, touch, taste) to describe an observation. *A large tree* is a qualitative description. A quantitative observation uses tools, such as a ruler, balance, or thermometer and numbers to describe the observation. *Five acorns* is a quantitative description.

How would you explain the squirrel's activity? An **explanation** *is an interpretation of observations.* You might explain that the squirrel is storing acorns for food. When you describe something, you report what you observe. But when you explain something, you try to interpret your observations.

Figure 5 A description and an explanation of a squirrel's activity contain different information.

Active Reading **1. Apply** Write a brief description of a spider. Include both qualitative and quantitative descriptors.

______

______

______

______

______

Active Reading **2. Analyze** Why do scientists around the world agree to use the same system of measurements?

______

______

______

______

______

The International System of Units

Suppose you observed a squirrel and recorded that it traveled about 200 ft from its nest. Someone who measures distances in meters might not understand how far the squirrel traveled. The scientific community solved this problem in 1960. It adopted *an international system for measurement called the* **International System of Units (SI).**

SI Base Units and Prefixes

Like many others around the world, you probably use the SI system in your classroom. All SI units are derived from seven base units, as listed in **Table 2.** The base unit most commonly used to measure length is the meter. However, you have probably made measurements in kilometers or millimeters before.

A prefix can be added to a base unit's name to indicate either a fraction or a multiple of that base unit. The prefixes are based on powers of ten, such as 0.01 and 100, as shown in **Table 3.** For example, one centimeter (1 cm) is one-hundredth of a meter, and a kilometer (1 km) is 1,000 meters.

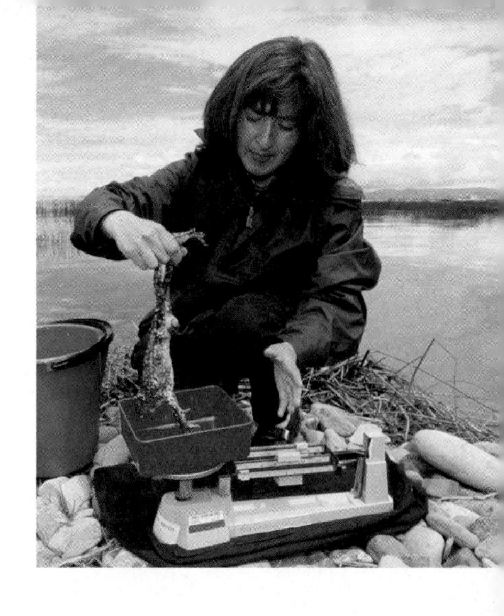

Table 2 SI Base Units

Quantity Measured	Unit (symbol)
Length	meter (m)
Mass	kilogram (kg)
Time	second (s)
Electric current	ampere (A)
Temperature	Kelvin (K)
Substance amount	mole (mol)
Light intensity	candela (cd)

Table 3 Prefixes

Prefix	Meaning
Mega– (M)	1,000,000 (10^6)
Kilo– (k)	1,000 (10^3)
Hecto– (h)	100 (10^2)
Deka– (da)	10 (10^1)
Deci– (d)	0.1 (10^{-1})
Centi– (c)	0.01 (10^{-2})
Milli– (m)	0.001 (10^{-3})
Micro– (μ)	0.000 001 (10^{-6})

Conversion

It is easy to convert from one SI unit to another. You either multiply or divide by a power of ten. You also can use proportion calculations to make conversions. For example, a biologist measures an emperor goose in the field. Her triple-beam balance shows the goose has a mass of 2.8 kg. She could perform the calculation below to find its mass in grams, X.

$$\frac{X}{2.8\ \text{kg}} = \frac{1{,}000\ \text{g}}{1\ \text{kg}}$$

$$(1\ \text{kg})X = (1{,}000\ \text{g})(2.8\ \text{kg})$$

$$X = \frac{(1{,}000\ \text{g})(2.8\ \cancel{\text{kg}})}{1\ \cancel{\text{kg}}}$$

$$X = 2{,}800\ \text{g}$$

Notice that the answer has the correct units.

Active Reading **3. Complete** Apply the correct prefix to the base units below.

Measurement Prefixes

________ meter	0.01 of a meter
________ gram	1,000 grams
________ gram	0.01 of a gram
________ meter	1,000,000 meters

Active Reading **4. Illustrate** Label each target with the term from the green box that accurately describes the image.

Accurate and precise
Accurate
Not accurate or precise
Precise but not accurate

An arrow in the center indicates high accuracy.

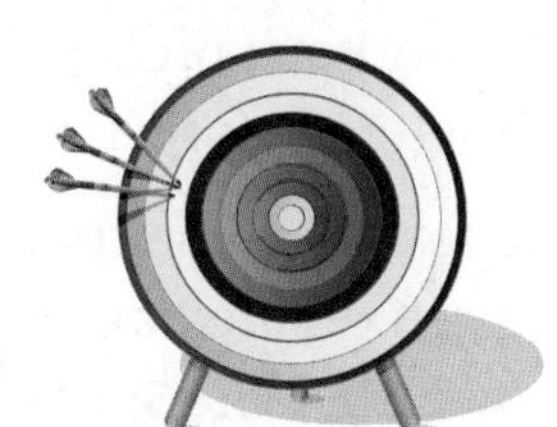
Arrows far from the center indicate low accuracy. Arrows close together indicate high precision.

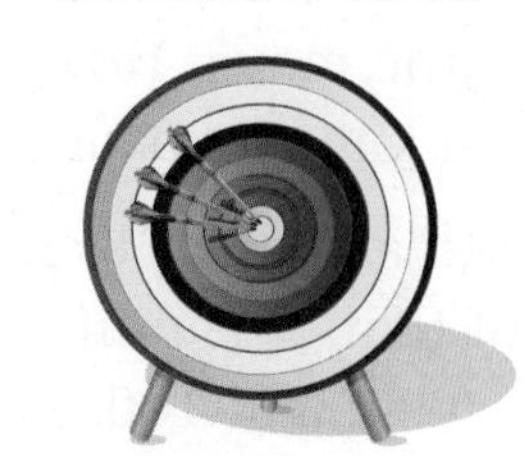
Arrows in the center indicate high accuracy. Arrows close together indicate high precision.

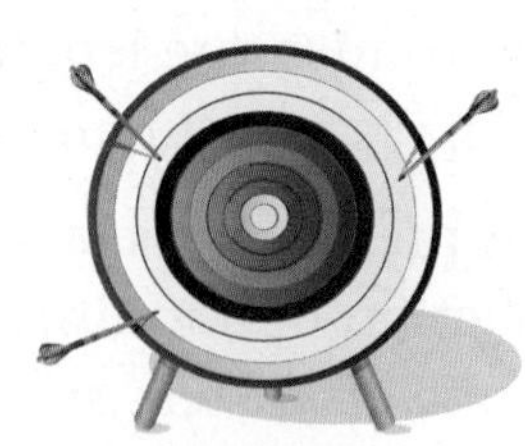
Arrows far from the center indicate low accuracy. Arrows far apart indicate low precision.

Figure 6 The archery target illustrates accuracy and precision.

Precision and Accuracy

Suppose your friend Simon tells you that he will call you in one minute, but he calls you a minute and a half later. Sarah tells you that she will call you in one minute, and she calls exactly 60 seconds later. What is the difference? Sarah is accurate, and Simon is not. **Accuracy** *is a description of how close a measurement is to an accepted or true value.* However, if Simon always calls about 30 seconds later than he says he will, then Simon is precise. **Precision** *is a description of how similar or close measurements are to each other,* as shown in **Figure 6.**

Table 4 illustrates the difference between precise and accurate measurements. Students were asked to find the melting point of sucrose, or table sugar. Each student took three temperature readings and calculated the mean, or average, of his or her data. As the recorded data in the table shows, student A had more accurate data. The melting point mean, 184.7°C, is closer to the scientifically accepted melting point, 185°C. Although not accurate, Student C's measurements are the most precise because they are similar in value.

Active Reading LA.6.4.2.2

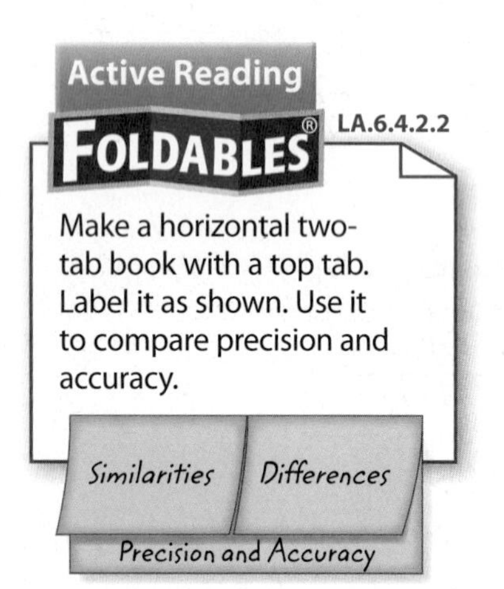

Active Reading **5. Identify** Fill in the correct letter as you determine which student's data is more accurate and which is more precise.

Table 4 The data taken by student ______ are more accurate because each value is close to the accepted value. The data taken by student ______ are more precise because the data are similar.

Table 4 Student Melting Point Data

	Student A	Student B	Student C
Trial 1	183.5°C	190.0°C	181.2°C
Trial 2	185.9°C	183.3°C	182.0°C
Trial 3	184.6°C	187.1°C	181.7°C
Mean	184.7°C	186.8°C	181.6°C
Sucrose Melting Point (accepted value) 185°C			

Measurement and Accuracy

The tools used to take measurements can limit the accuracy of the measurements. Suppose you are measuring the temperature at which sugar melts, and the thermometer's measurements are divided into whole numbers. If your sugar sample melts between 183°C and 184°C, you can estimate the temperature between these two numbers. But, if the thermometer's measurements are divided into tenths, and your sample melts between 183.2°C and 183.3°C, your estimate between these numbers would be more accurate.

Significant Digits

In the second example above, you know that the temperature is between 183.2°C and 183.3°C. You could estimate that the temperature is 183.25°C. When you take any measurement, some digits you know for certain and some digits you estimate. **Significant digits** *are the number of digits in a measurement that are known with a certain degree of reliability.* The significant digits in a measurement include all digits you know for certain plus one estimated digit. Therefore, your measurement of 183.25°C would contain five significant digits, as explained in **Table 5.** Using significant digits lets others know how certain your measurements are. **Figure 7** shows an example of rounding to 3 significant digits.

6. Summarize Why would you use significant digits?

____

____

Figure 7 Since the ruler is divided into tenths, you know the rod is between 5.2 cm and 5.3 cm. You can estimate that the rod is 5.25 cm.

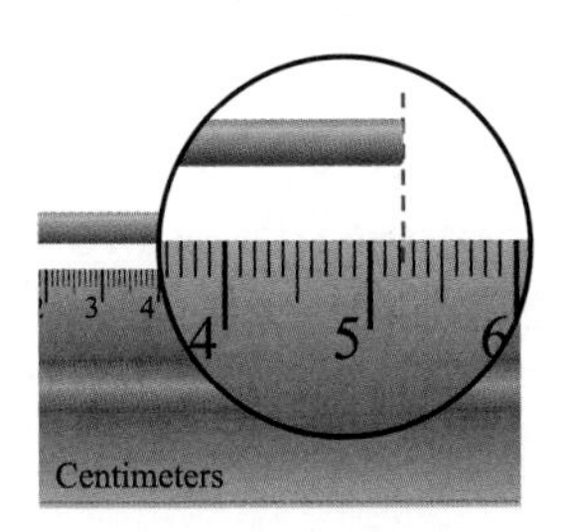

Math Skills MA.6.A.3.6

Significant Digits

The number 5,281 has 4 significant digits. Rule 1 in **Table 5** below states that all nonzero numbers are significant.

Practice

7. Use the rules in **Table 5** to Underline the significant digits in each of the following numbers:

2.02 0.0057 1,500 0.500

SCIENCE USE V. COMMON USE

digital

Science Use of, pertaining to, or using numbers (numerical digits)

Common Use of or pertaining to a finger

Table 5 Significant Digits

Rules

1. All nonzero numbers are significant.
2. Zeros between significant digits are significant.
3. All final zeros to the right of the decimal point are significant.
4. Zeros used solely for spacing the decimal point are not significant. The zeros indicate only the position of the decimal point.

* The blue numbers in the examples are the significant digits.

Example	Significant Digits	Applied Rules
1.234	4	1
0.200	3	1, 3
0.012	2	1, 4
50,600	3	1, 2, 4

Scientific Tools

Scientific inquiry often requires the use of specific tools. Scientists, including life scientists, might use the following tools. You might use one or more of them during a scientific inquiry, too.

Science Journal ▶

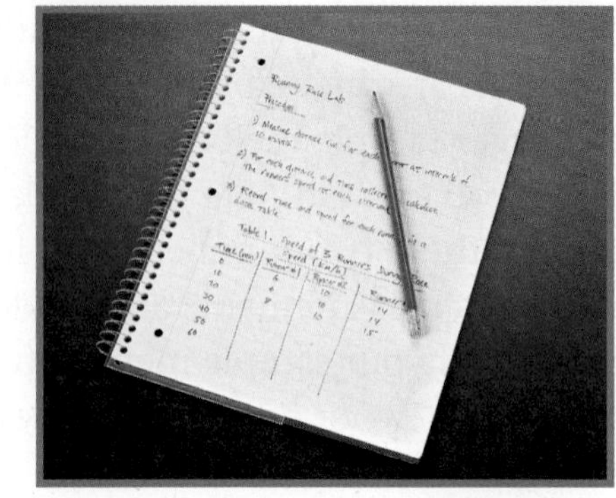

In a science journal, you can record descriptions, explanations, plans, and steps used in a scientific inquiry. A science journal can be any notebook. Keep your science journal organized so that you can find information when you need it. Make sure your notes are legible, thorough, and accurate records.

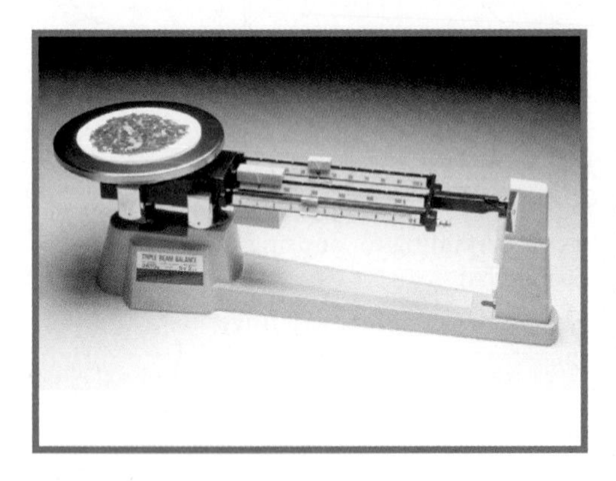

◀ Balances

You can use a triple-beam balance or an electric balance to measure mass. Mass usually is measured in kilograms (kg) or grams (g). When using a balance, do not let objects drop heavily onto the balance. Gently remove an object after you record its mass.

Thermometer ▶

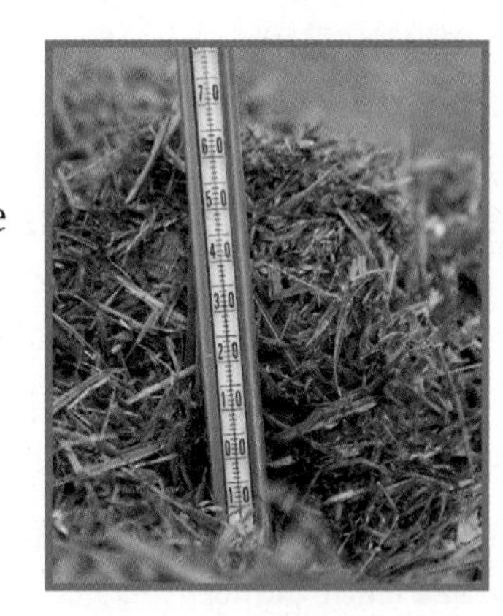

A thermometer measures the temperature of substances. Although the Kelvin (K) is the SI unit for temperature, in the science classroom, you measure temperature in degrees Celsius (°C). Never use a thermometer to stir anything. Handle glass thermometers gently so that they do not break. If a thermometer does break, tell your teacher immediately. Do not touch the broken glass or the thermometer's liquid.

◀ Glassware

Laboratory glassware is used to hold, pour, heat, and measure liquids. Most labs have many types of glassware. For example, flasks, beakers, petri dishes, test tubes, and specimen jars are used as containers. To measure the volume of a liquid, you use a graduated cylinder. The unit of measure for liquid volume is the liter (L) or milliliter (mL), a formula of length times width times height.

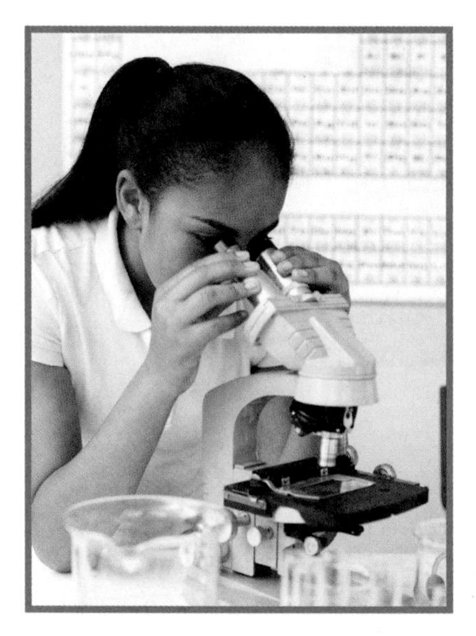

◀ Compound Microscope

Microscopes enable you to observe small objects that you cannot observe with just your eyes. Usually, two types of microscopes are available in science classrooms—dissecting microscopes and compound light microscopes, such as the one shown to the left. The girl is looking through two eyepieces to observe a magnified image of a small object or organism. However, some microscopes have only one eyepiece.

Microscopes can be damaged easily. Follow your teacher's instructions when carrying and using a microscope.

Active Reading **8. Point out** In the chart below, suggest a safety rule for each scientific tool discussed.

Scientific Tool	Safety Rule
Balance	
Thermometer	
Glassware	
Microscope	

Computers—Hardware and Software ▶

Computers process information. In science, you can use computers to compile, retrieve, and analyze data for reports. You also can use them to create reports and other documents, to send information to others, and to research information.

The physical components of computers, such as monitors and keyboards, are called hardware. The programs that you run on computers are called software. These programs include word processing, spreadsheets, and presentation programs. When scientists write reports, they use word processing programs. They use spreadsheet programs for organizing and analyzing data. Presentation programs can be used to explain information to others.

Active Reading **9. Consider** Underline why a scientist might use a data spreadsheet. Circle what type of program a scientist might use to write a report.

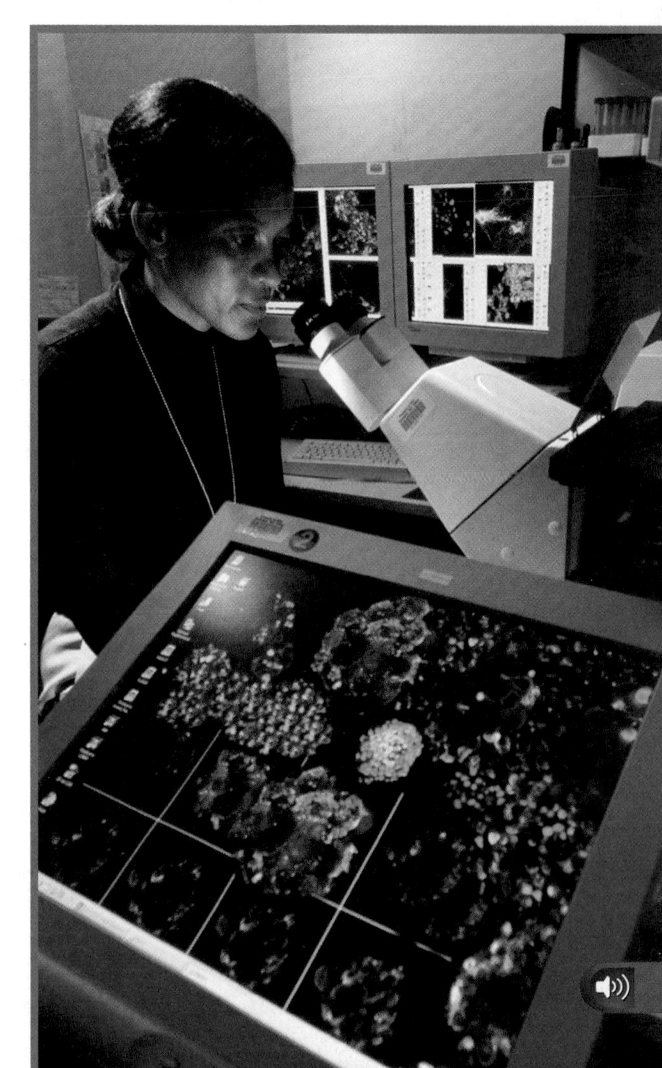

Active Reading **10. Apply** State an example of a situation in which you might use each scientific tool.

Pipette:

Glass slide:

Magnifying lens:

Dissecting tools:

Tools Used by Life Scientists

Magnifying Lens

A magnifying lens is a handheld lens that enlarges an image of an object. It is useful when great magnification is not needed. Magnifying lenses also can be used outside the lab where microscopes might not be available.

Slide

To observe an item using a compound light microscope, you must place the item on a thin, rectangular piece of glass called a slide.

Dissecting Tools

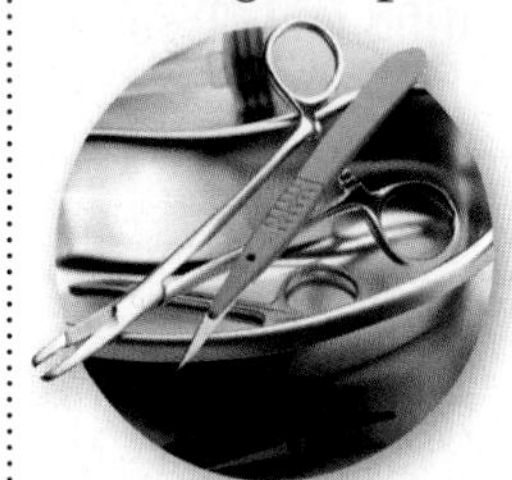

Scientists use dissecting tools, such as scalpels and scissors, to examine tissues, organs, or prepared organisms. The tools can be sharp, so use caution when handling them.

Pipette

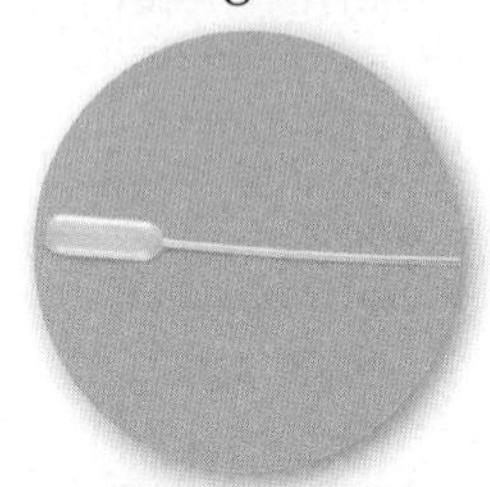

A pipette is similar to a dropper. It is a small tube used to draw up and transfer liquids.

Lesson Review 2

Use Vocabulary

1. **Define** description and explanation in your own words.

2. **Use the term** *International System of Units (SI)* in a sentence to show its scientific meaning.

Understand Key Concepts

3. **Describe** the difference between accuracy and precision.

4. **Explain** why scientists use significant digits.

Interpret Graphics

5. **Synthesize** Complete the graphic organizer with the missing information about each SI base unit, substance measured, and symbol. LA.6.2.2.3

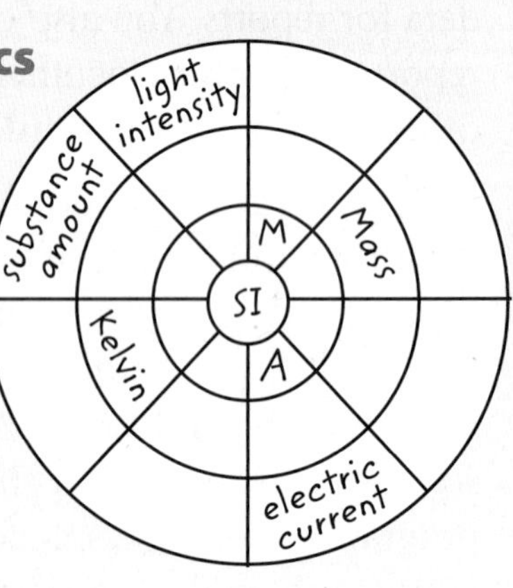

Critical Thinking

6. **Recommend** ways that computers can assist life scientists in their work.

Math Skills MA.6.A.3.6

7. **Suppose** the mass of a book is 420.0890 g. Underline the significant digits in this measurement.

Science & Engineering

The Design Process

Create a Solution

Scientists investigate and explore natural events and then interpret data and share information learned from those investigations. How do engineers differ from scientists? Engineers design, construct, and maintain things that do not occur naturally. Roads, submarines, toys and games, microscopes, medical equipment, and amusement park rides all

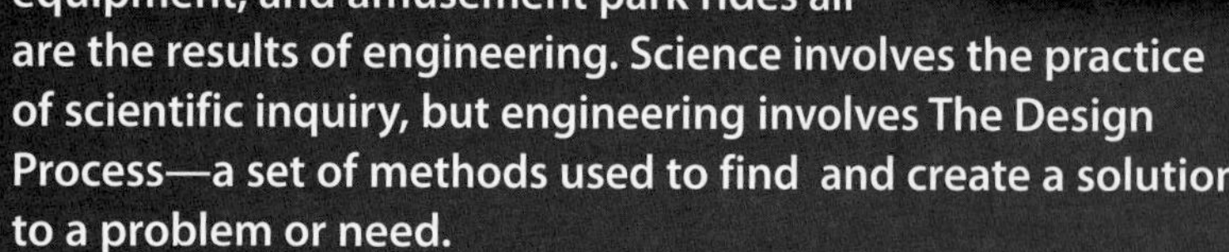

are the results of engineering. Science involves the practice of scientific inquiry, but engineering involves The Design Process—a set of methods used to find and create a solution to a problem or need.

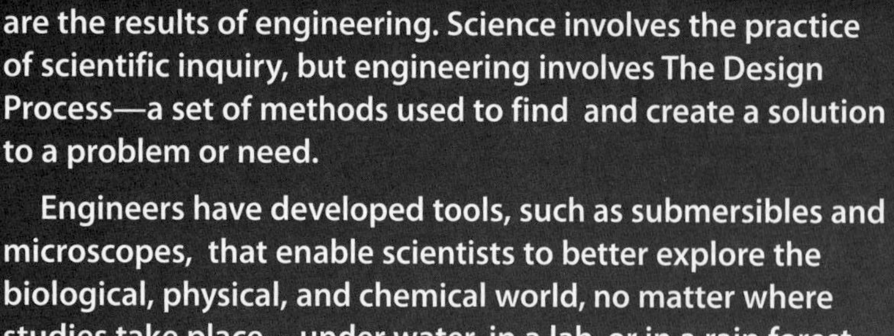

Engineers have developed tools, such as submersibles and microscopes, that enable scientists to better explore the biological, physical, and chemical world, no matter where studies take place—under water, in a lab, or in a rain forest.

Alvin, a deep-sea submersible, has been in operation since 1964. It makes about 200 deep-sea dives each year and has helped scientists discover human artifacts, deep-sea organisms, and seafloor processes. Microscopes enable scientists to examine closely things that are not visible to the naked eye.

The Design Process

1. Identify a Problem or Need
- Determine a problem or need
- Document all questions, research, and procedures throughout the process

2. Research and Development
- Brainstorm solutions
- Research any existing solutions that address the problem or need
- Suggest limitations

3. Construct a Prototype
- Develop possible solutions
- Estimate materials, costs, resources, and time to develop the solutions
- Select the best possible solution
- Construct a prototype

4. Test and Evaluate Solutions
- Use models to test the solution
- Use graphs, charts, and tables to evaluate results
- Analyze the solution's strengths and weaknesses

5. Communicate Results and Redesign
- Communicate the design process and results to others
- Redesign and modify the solution
- Construct the final solution

DESIGN PROCESS LAB Design a Magnifying Tool.

Lesson 3

Case STUDY

ESSENTIAL QUESTIONS

How do independent and dependent variables differ?

How is scientific inquiry used in a real-life scientific investigation?

Vocabulary

variable p. NOS 20

dependent variable p. NOS 20

independent variable p. NOS 20

constants p. NOS 20

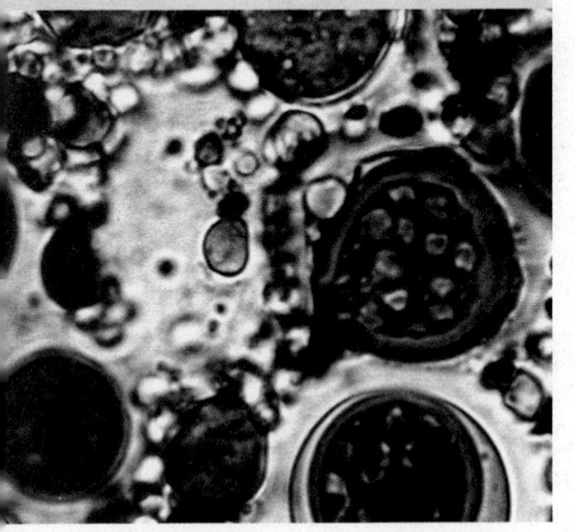

Figure 8 Florida companies experiment to develop larger, more cost-effective methods to produce biodiesel fuel using a variety of microalgae. Microalgae are plantlike organisms that can make oils.

Active Reading **1. Analyze** Summarize the relationship between independent and dependent variables.

____

____

____

____

____

Biodiesel from Microalgae

For the last few centuries, fossil fuels have been the main sources of energy for industry and transportation. But, studies have shown that burning fossil fuels negatively affects the environment. In addition, there are concerns about using up the world's reserves of fossil fuels.

Florida scientists have explored using protists to produce biodiesel. Biodiesel is fuel made primarily from living organisms. Protists, shown in **Figure 8,** are a group of microscopic organisms that usually live in water or moist environments. Microalgae are plantlike protists that make their own food using sunlight, water, and carbon dioxide in a process called photosynthesis.

Designing a Controlled Experiment

Scientists use scientific inquiry to investigate the use of protists to make biodiesel. They design controlled experiments to test their hypotheses. Scientists conducting the studies practice scientific inquiry. The notebook pages contain information that a scientist might write in a science journal.

A controlled experiment is a scientific investigation that tests how one variable affects another. A **variable** *is any factor in an experiment that can have more than one value.* In controlled experiments, there are two types of variables. The **dependent variable** *is the factor measured or observed during an experiment.* The **independent variable** *is the factor that you want to test. It is changed by the investigator to observe how it affects a dependent variable.* **Constants** *are the factors in an experiment that remain the same.*

A controlled experiment has two groups—an experimental group and a control group. The experimental group is used to study how a change in the independent variable changes the dependent variable. The control group contains the same factors as the experimental group, but the independent variable is not changed. Without a control, it is difficult to know whether your experimental observations result from the variable you are testing or from another factor.

Biodiesel

The idea of engines running on fuel made from plant or plantlike sources is not entirely new. Rudolph Diesel, shown in **Figure 9,** invented the diesel engine. He used peanut oil to demonstrate how his engine worked. However, when petroleum was introduced as a diesel fuel source, it was preferred over peanut oil because it was cheaper.

Figure 9 Rudolph Diesel invented the first diesel engine in the early 1900s.

Oil-rich food crops, such as soybeans, can be used as a source of biodiesel. However, some people are concerned that crops grown for fuel sources will replace crops grown for food. If farmers grow more crops for fuel, then the amount of food available worldwide will be reduced. Because of food shortages in many parts of the world, replacing food crops with fuel crops is not a good solution.

2. Identify (Circle) the product Rudolph Diesel used as fuel to demonstrate how his new engine worked.

Aquatic Species Program

In the late 1970s, the U.S. Department of Energy began funding its Aquatic Species Program (ASP) to investigate ways to remove air pollutants. Coal-fueled power plants produce carbon dioxide (CO_2), a pollutant, as a by-product. In the beginning, the study examined all aquatic organisms that use CO_2 during photosynthesis—their food-making process. During the studies, the project leaders noticed that some microalgae produced large amounts of oil. The program's focus soon shifted to using microalgae to produce oils that could be processed into biodiesel.

3. Analyze Highlight the original purpose for which scientists studied microalgae.

Scientific investigations often begin when someone observes an event in nature and wonders why or how it occurs.

A hypothesis is a tentative explanation that can be tested by scientific investigations. A prediction is a statement of what someone expects to happen next in a sequence of events.

Observation A:
While testing microalgae to discover if they would absorb carbon pollutants, ASP project leaders noticed that some species of microalgae had high oil content.

Hypothesis A:
Some microalgae species can be used as a source of biodiesel fuel because the microalgae produce oil.

Prediction A:
If the correct species is found and the growing conditions are correct, then large amounts of oil will be produced.

One way to test a hypothesis is to design an experiment, collect data, and test predictions.

Design an Experiment and Collect Data:
Develop a rapid screening test to discover which species produce the most oil.
Independent Variable: amount of nitrogen available
Dependent Variable: amount of oil produced
Constants: the growing conditions of algae (temperature, water quality, exposure to the Sun, etc.)

During an investigation, observations, hypotheses, and predictions are often made when new information is discovered.

Observation B:
Based on previous studies, starving microalgae of nutrients could produce more oil.
Hypothesis B:
Microalgae grown with inadequate amounts of nitrogen alter their growth processes and produce more oil.
Prediction B:
If microalgae receive inadequate amounts of nitrogen, then they will produce more oil.

Figure 10 Green microalgae and diatoms showed the most promise during testing for biodiesel production.

Which Microalgae?

Microalgae are microscopic organisms that live in marine (salty) or freshwater environments. Like many plants and other plantlike organisms, they use photosynthesis and make sugar. The process requires light energy. Microalgae make more sugar than they can use as food. They convert excess sugar to oil. Scientists focused on these microalgae because their oil then could be processed into biodiesel.

Research began by collecting and identifying microalgae species. Focusing on microalgae in shallow, inland, saltwater ponds, scientists predicted that these microalgae were more resistant to changes in temperature and salt content in the water.

A test was developed to identify microalgae with high oil content. Soon, 3,000 microalgae species had been collected. Scientists checked these samples for tolerance to acidity, salt levels, and temperature and selected 300 species. Green microalgae and diatoms, as shown in **Figure 10,** showed the most promise. However, it was obvious that no one species was going to be perfect for all climates and water types.

Active Reading **4. Confirm** Highlight three factors for microalgae production that scientists considered in their research.

Oil Production in Microalgae

Scientists also began researching how microalgae produce oil. Some studies suggested that starving microalgae of nutrients, such as nitrogen, could increase the amount of oil they produced. However, starving microalgae also caused them to be smaller, resulting in no overall increase in oil production.

Outdoor Testing v. Bioreactors

Scientists first grew microalgae in outdoor ponds in New Mexico. However, outdoor conditions were very different from those in the laboratory. Cooler temperatures in the outdoor ponds resulted in smaller microalgae. Native algae species invaded the ponds, forcing out the experimental, high-oil-producing microalgae species.

Florida scientists continue to focus on growing microalgae in open ponds, similar to the one shown in **Figure 11.** Many scientists believe these open ponds are better for producing large quantities of biodiesel from microalgae. Some researchers are now growing microalgae in closed glass containers called bioreactors, also shown in **Figure 11.** Inside these bioreactors, organisms live and grow under controlled conditions. However, bioreactors are more expensive than open ponds.

Biofuel companies have been experimenting with low-cost bioreactors. Scientists hypothesize they could use long plastic bags, shown in **Figure 11,** instead of closed glass containers.

Active Reading **5. Identify** In the boxes provided, state an advantage and a disadvantage for each microalgae environment shown.

Open Ponds	
Advantage:	
Disadvantage:	

Open ponds are less expensive than bioreactors for growing microalgae.

Figure 11
These three methods of growing microalgae are examples of three different hypotheses that are being tested in controlled experiments.

Microalgae grown in plastic bags are very expensive to harvest.

Plastic Bags	
Advantage:	
Disadvantage:	

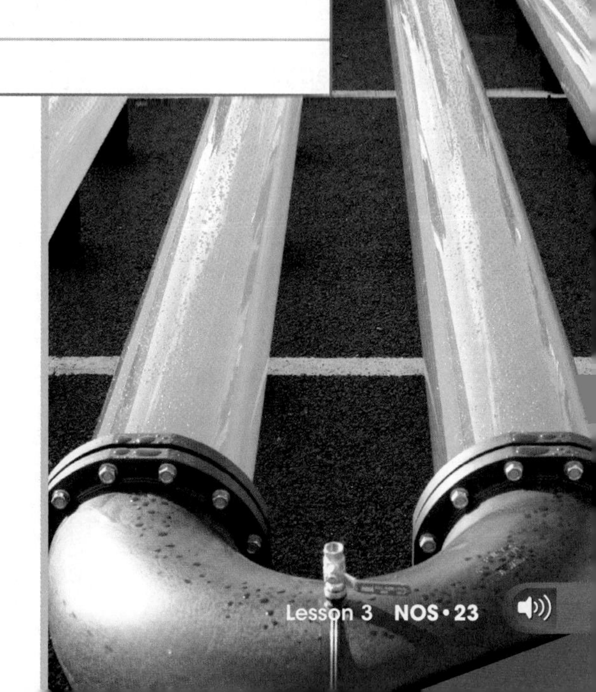

Microalgae grow under controlled conditions in glass bioreactors.

Glass Bioreactor Tubes	
Advantage:	
Disadvantage:	

Why So Many Hypotheses?

According to Dr. Richard Sayre, a biofuel researcher, research is based on forming hypotheses. Dr. Sayre says, "It is hypothesis-driven. You just don't go in and say 'Well, I have a feeling this is the right way to do it.' You propose a hypothesis. Then you test it."

Dr. Sayre added, "Biologists have been trained to develop research strategies based on hypotheses. You don't get research support by saying, 'I'm going to put together a system, and it's going to be wonderful.' You have to come up with a question. You propose some strategies for answering the question. What are your objectives? What outcomes do you expect for each objective?"

Observation C:
Microalgae use light energy, water, and carbon dioxide to make sugar, which they can convert to oil.

Hypothesis C:
Microalgae will produce more oil if light is distributed evenly throughout the pond because they need light energy to grow and produce more oil.

Prediction C:
If light is distributed more evenly, then more microalgae will grow, and more oil will be produced.

Active Reading 6. **Analyze** Explain why it is important for a scientific researcher to develop a good hypothesis.

____

____

Increasing Oil Yield

Scientists from a biofuel company in Washington State thought of another way to increase oil production. Researchers knew microalgae use light energy, water, and carbon dioxide and make sugar. The microalgae eventually convert sugar into oil. The scientists wondered if they could increase microalgae oil production by distributing light to all microalgae. The experimental lab setup to test this idea is shown in **Figure 12.**

Active Reading 7. **Fill in** the blanks to complete the caption.

Figure 12 Acrylic rods distribute ____________ to microalgae below the ____________. If microalgae receive ____________, they can photosynthesize and produce more ____________. Without light, microalgae are not productive.

Bringing Light to Microalgae

Normally microalgae grow near the surface of a pond. Any microalgae about 5 cm below the pond's surface do not grow well. Why is this? First, water blocks light from reaching deep into a pond. Second, microalgae at the top of a pond block light from reaching microalgae below them. Only the top part of a pond is productive.

Experimental Group

Researchers assembled a team of engineers to design a light-distribution system. Light rods distribute artificial light to microalgae in a bioreactor. The bioreactor controls the environmental conditions that affect how the microalgae grow. These conditions include temperature, nutrient levels, carbon dioxide level, airflow, and light.

Active Reading **8. Identify** Discuss what conditions are held constant in a bioreactor.

a. ____________________

____

____

____

Why is this essential for maximum microalgae growth and oil production?

b. ____________________

____

____

____

Data from their experiments showed scientists how microalgae in well-lit environments grow compared to how microalgae grow in dimmer environments. Using solar data for various parts of the country, the scientists concluded that the light rod would significantly increase microalgae growth and oil production in outdoor ponds. These scientists next plan to use the light-rod growing method in outdoor ponds.

Scientists tested their hypothesis, collected data, analyzed the data, and drew conclusions.

Analyze Results:
Results showed that microalgae produce more oil using a light-rod system than by using just sunlight.
Draw a Conclusion:
The light-rod system greatly increased microalgae oil production.

Research scientists and scientists in the field rely on scientific methods and scientific inquiry to solve real-life problems. When a scientific investigation lasts for several years and involves many scientists, such as this study, many hypotheses can be tested. Some hypotheses are supported, and other hypotheses are not. However information is gathered, hypotheses are refined and tested many times. This process of scientific inquiry results in a better understanding of the problem and possible solutions.

Field Testing

Scientists plan to take light to microalgae instead of moving microalgae to light. Dr. Jay Burns is chief microalgae scientist at a biofuel company. He said, "What we are proposing to do is to take the light from the surface of a pond and distribute it throughout the depth of the pond. Instead of only the top 5 cm being productive, the whole pond becomes productive."

Active Reading **9. Point Out** Highlight the benefit of the light-distribution system.

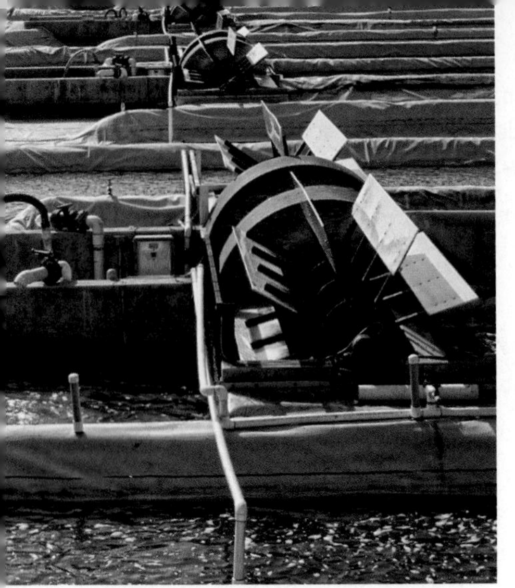

Figure 13 During cultivation, paddlewheels bring microalgae to the surface and expose them to sunlight.

Another Way to Bring Light to Microalgae

Light rods are not the only way to bring light to microalgae. Paddlewheels, as shown in **Figure 13,** can be used to keep the microalgae's locations changing. Paddlewheels continuously rotate microalgae to the surface. This exposes the organisms to more light.

Active Reading **10. Analyze** Underline the purpose of using paddlewheels in an algae pond.

Active Reading **11. Revisit** Identify several ways in which scientific inquiry was used in this case study.

____

Why Grow Microalgae?

While the focus of this case study is microalgae growth for biodiesel production, there are other benefits of growing microalgae, as shown in **Figure 14.** Power plants that burn fossil fuels release carbon dioxide into the atmosphere, contributing to global warming. During photosynthesis, microalgae use this carbon dioxide and water, release oxygen, and produce sugar, which they convert to oil. Not only do microalgae produce a valuable fuel, they also remove pollutants from and add oxygen to the atmosphere.

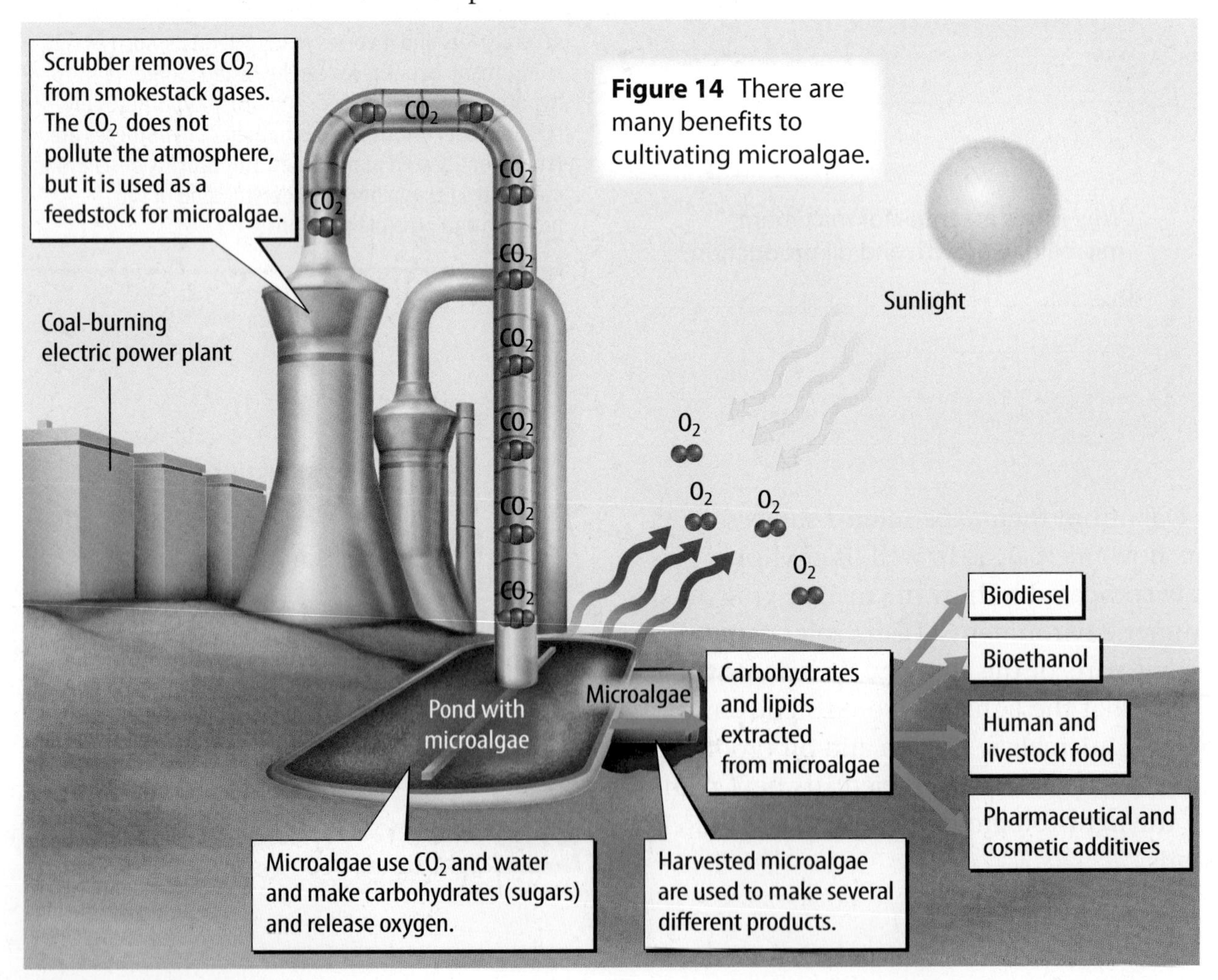

Figure 14 There are many benefits to cultivating microalgae.

Microalgae Research in Florida

A company located in Florida has invested in growing microalgae for biodiesel oil production using environmental-simulation chambers which were first designed by NASA for a mission to Mars. The chambers controlled the environment for growing vegetables and fruit in space. Now they are being used to control the environment for growing algae for biodiesel.

These Florida researchers focus on specific microalgae species that are very greasy. Approximately half of the algae's weight is in lipids, or oily substances, with higher than average oil production. Scientists are exploring ways to produce biodiesel in larger quantities, at a faster rate, and at lower costs to help meet energy needs.

Similar companies across the country have different focuses. Some companies are developing strategies to capture carbon dioxide from smokestacks and feed it to algae in bioreactors. The algae can then be used for animal feed or liquid fuel. In warm areas of the country, open ponds are being used to cultivate a wide variety of algal species. Each species can grow using a different spectrum of light. The algae are then separated from the water and harvested.

Scientists face many challenges in their quest to produce biodiesel from microalgae. Scientists from Florida to California are involved in the search for readily available, low-cost, clean, biodiesel to help meet the world's energy needs without using food crops. Through the combined efforts of government-funded programs and commercial biodiesel companies, microalgae-based biodiesel is becoming more realistic.

Figure 15 This microalgae-cultivating facility is reducing the amount of carbon dioxide pollution in the atmosphere while at the same time producing a valuable, affordable, renewable energy source.

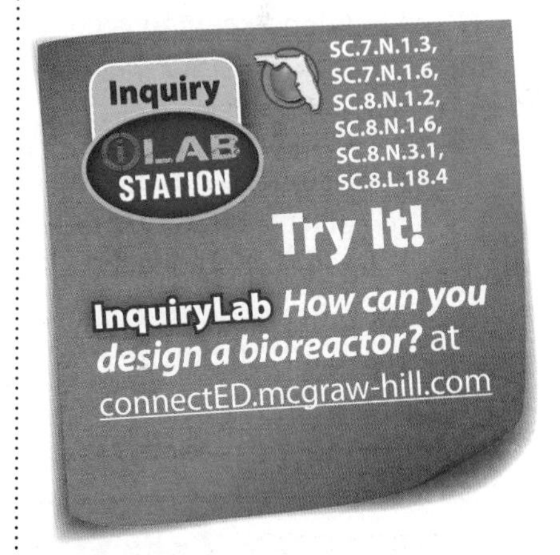

Lesson Review 3

Use Vocabulary

1. **Define** *variable* in your own words. SC.7.N.1.4

Understand Key Concepts

2. Which factor does the investigator change during an investigation? SC.7.N.1.4
   - (A) constant
   - (B) dependent variable
   - (C) independent variable
   - (D) variable

Interpret Graphics

3. **Organize Information** Fill in the graphic organizer with the three methods of oil production discussed in the study. LA.6.4.2.2

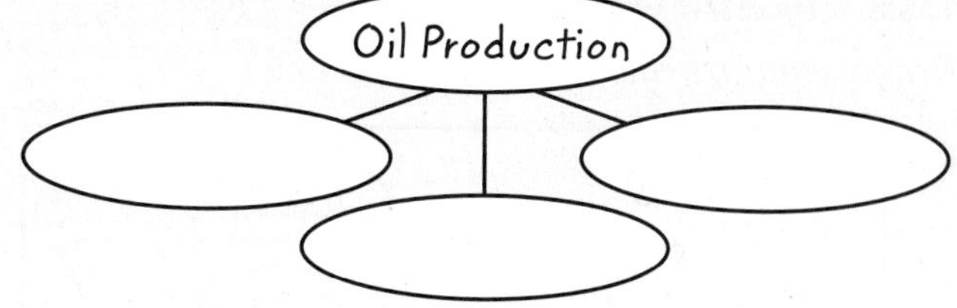

Critical Thinking

4. **Hypothesize** a method to increase the oil from microalgae for biodiesel production.

Scientific Explanations Study Guide

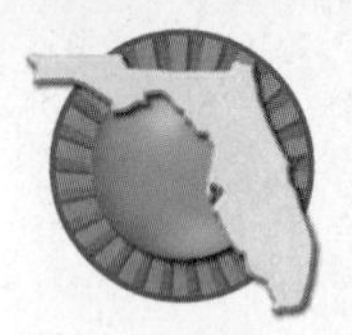

Think About It! **You can answer questions about your world by using scientific inquiry to creatively construct investigations and collect and evaluate appropriate data through observations and inferences.**

Key Concepts Summary

LESSON 1 Understanding Science

- Scientific inquiry, also known as scientific methods, is a collection of skills that scientists use in different combinations to perform scientific investigations.
- Scientific investigations often result in new **technology,** new materials, newly discovered objects or events, or answers to questions.
- A scientist can help prevent bias in a scientific investigation by taking random samples, doing blind studies, repeating an experiment several times, and keeping accurate and honest records.

Vocabulary

science p. NOS 4
observation p. NOS 6
inference p. NOS 6
hypothesis p. NOS 6
prediction p. NOS 7
technology p. NOS 8
scientific theory p. NOS 9
scientific law p. NOS 9
critical thinking p. NOS 10

LESSON 2 Measurement and Scientific Tools

- **Precision** is a description of how similar or close measurements are to each other. **Accuracy** is a description of how close a measurement is to an accepted value.
- **Significant digits** communicate the precision of the tool used to make measurements.
- Life scientists use many tools, such as science journals, microscopes, computers, magnifying lenses, slides, and dissecting tools.

description p. NOS 12
explanation p. NOS 12
International System of Units (SI) p. NOS 12
accuracy p. NOS 14
precision p. NOS 14
significant digits p. NOS 15

LESSON 3 Case Study: Biodiesel from Microalgae

- The **independent variable** is a factor in an experiment that is manipulated or changed by the investigator to observe how it affects a dependent variable. The **dependent variable** is the factor measured or observed during an experiment.
- Scientific inquiry is used to gain information and find solutions to real-life problems and questions.

variable p. NOS 20
dependent variable p. NOS 20
independent variable p. NOS 20
constants p. NOS 20

Use Vocabulary

Briefly contrast each set of terms below.

1

Scientific law	
Scientific theory	

2

Description	
Explanation	

3

International System of Units (SI)	
Significant digits	

4

Variable	
Constant	

Scientific Explanations Review

Fill in the correct answer choice.

Understand Key Concepts

1. Which is a quantitative observation? **SC.7.N.1.1**
   Ⓐ 15 m long
   Ⓑ red color
   Ⓒ rough texture
   Ⓓ strong odor

2. Which is one way scientists indicate how precise and accurate their experimental measurements are? **SC.7.N.1.5**
   Ⓐ They keep accurate, honest records.
   Ⓑ They make sure their experiments can be repeated.
   Ⓒ They use significant figures in their measurements.
   Ⓓ They record small samples of data.

Critical Thinking

3. **Identify** the next step in the scientific inquiry process below. **SC.7.N.1.5, SC.8.N.2.2**

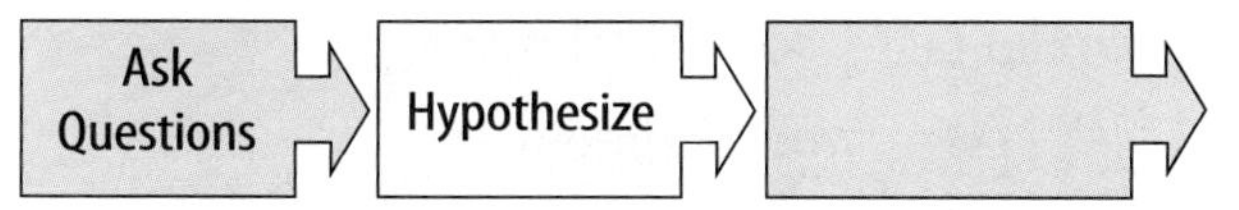

4. In the paragraph below, circle the experimental group, underline the control group, and highlight the constants. Explain your reasoning. **SC.7.N.1.3**

   A scientist tests a liquid cough medicine by giving it to participants who have colds. The scientist gives another group with colds a liquid and tells them it is cough medicine. The people in both groups are women between the ages of 20 and 30 who normally are in good health.

Writing in Science

5. On a separate sheet of paper, write a paragraph that includes examples of how bias can be intentional or unintentional and how scientists can help avoid bias. Be sure to include topic and concluding sentences in your paragraph. **LA.7.2.2.3**

Big Idea Review

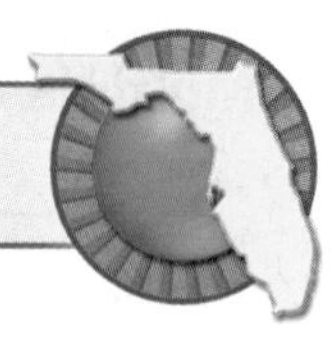

6. Use the space below to design a flowchart of possible steps scientists might perform during scientific inquiries. **SC.8.N.2.2, LA.7.2.2.3**

Math Skills **SC.7.N.1.5, MA.6.A.3.6**

7. **Determine** the significant digits in the following numerals:

   0.00840:

   15.7:

   13.040:

Multiple Choice *Bubble the correct answer.*

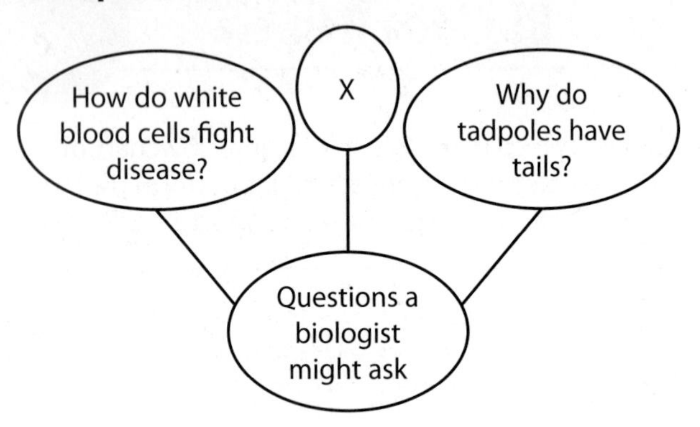

1. In the graphic organizer above, which goes in X? **SC.7.L.17.3**

- (A) Does atomic structure affect an atom's electrical charge?
- (B) How much rain falls annually in the Mojave Desert?
- (C) Which minerals are most commonly found in Maine?
- (D) Why does moss grow on some rocks but not others?

2. Organizing results into a table, graph, or chart is done during which step of a scientific inquiry? **SC.7.N.1.5**

- (F) analyze results
- (G) ask questions
- (H) communicate results
- (I) draw conclusions

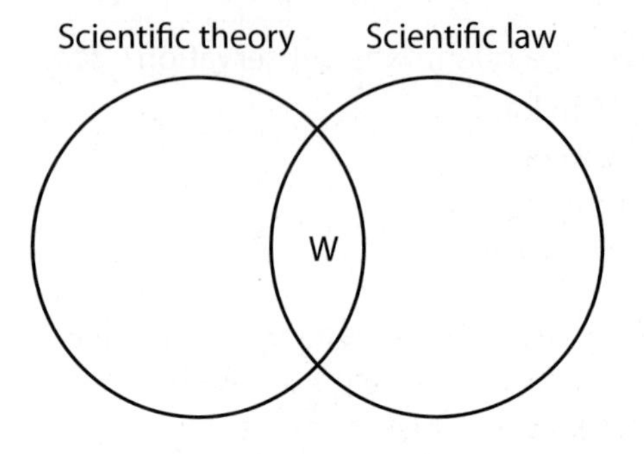

3. In the Venn diagram above, which can go in W? **SC.7.N.3.1**

- (A) based on repeated observations
- (B) contains one well-supported hypothesis
- (C) explains why something happens
- (D) states that something will happen

4. After a hypothesis has been confirmed, which is the next step a good scientist would take? **SC.6.N.1.2, SC.7.N.1.5**

- (F) design an experiment
- (G) modify or revise the hypothesis
- (H) repeat the test several times
- (I) speak at a science conference

Name ______________________ Date ______________ Class __________

Multiple Choice *Bubble the correct answer.*

R	06.309 cm
S	102.00 cm
T	105.07 cm
U	00.054 cm

1. Which measurement above contains the greatest number of significant digits? **MA.6.S.6.2**

(A) R

(B) S

(C) T

(D) U

2. Water boils at 100°C. Kayla measures the temperature of boiling water three times and receives the following results: 96.6°C, 96.8°C, and 96.5°C. Which best describes her measurements? **SC.7.N.1.1**

(F) They are more accurate than precise.

(G) They are more precise than accurate.

(H) They are both precise and accurate.

(I) They are neither precise nor accurate.

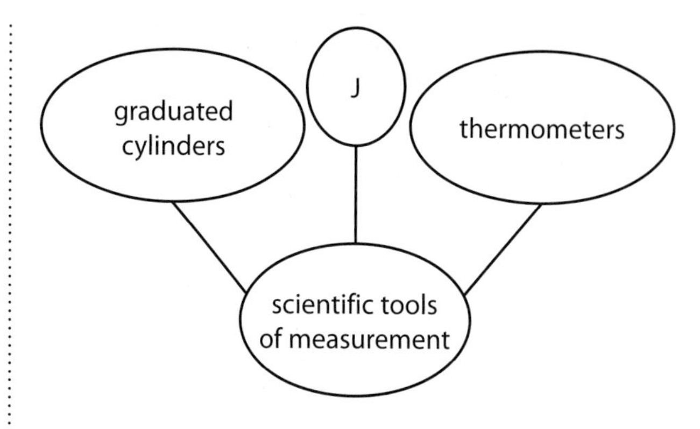

3. In the graphic organizer above, which can go in J? **LA.6.4.2.2**

(A) balance

(B) computer

(C) microscope

(D) pipette

4. Which is the SI base unit used when measuring mass? **SC.7.N.1.5**

(F) kilogram

(G) meter

(H) mole

(I) ounce

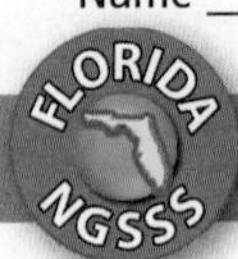

Effect of Depth on Algae Growth	
Depth Below the Water's Surface	Change in Size
3 cm	10%
7 cm	2%
12 cm	0%

1. Using data from the table above, which is a conclusion that cannot be drawn? **SC.7.N.1.1**

Ⓐ Algae grown at a depth of 7 cm changed in size a small amount.

Ⓑ Algae grown at the greatest depth will eventually change in size.

Ⓒ Algae grown farthest from the water's surface had no change.

Ⓓ Algae grown nearest the water's surface changed the most.

2. When a scientific investigation runs for several years, which is most likely true? **SC.7.N.1.7**

Ⓕ The data are difficult to analyze, so no conclusion drawn.

Ⓖ The end result will be a new scientific law.

Ⓗ Many different hypotheses will be tested.

Ⓘ New technologies will be developed in the process.

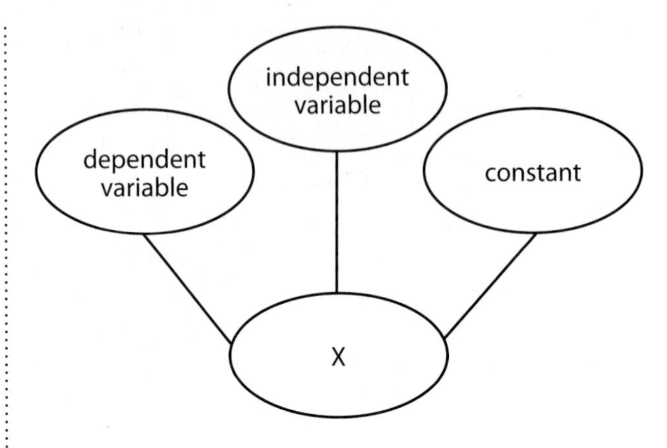

3. The graphic organizer above shows three aspects of which? **SC.7.N.1.4**

Ⓐ controlled experiment

Ⓑ critical thinking

Ⓒ modified hypothesis

Ⓓ scientific inquiry

4. Algae can be grown in glass bioreactors, plastic bags, or open ponds. Which is a good hypothesis to begin testing to find the best method? **SC.7.N.1.6**

Ⓕ Algae will grow best in open ponds.

Ⓖ Where do algae grow best?

Ⓗ Algae will grow better in a plastic bag than in a glass bioreactor.

Ⓘ Will algae grow in glass bioreactors, plastic bags, or open ponds?

Name ______________________ Date ______________ Class ________

Notes

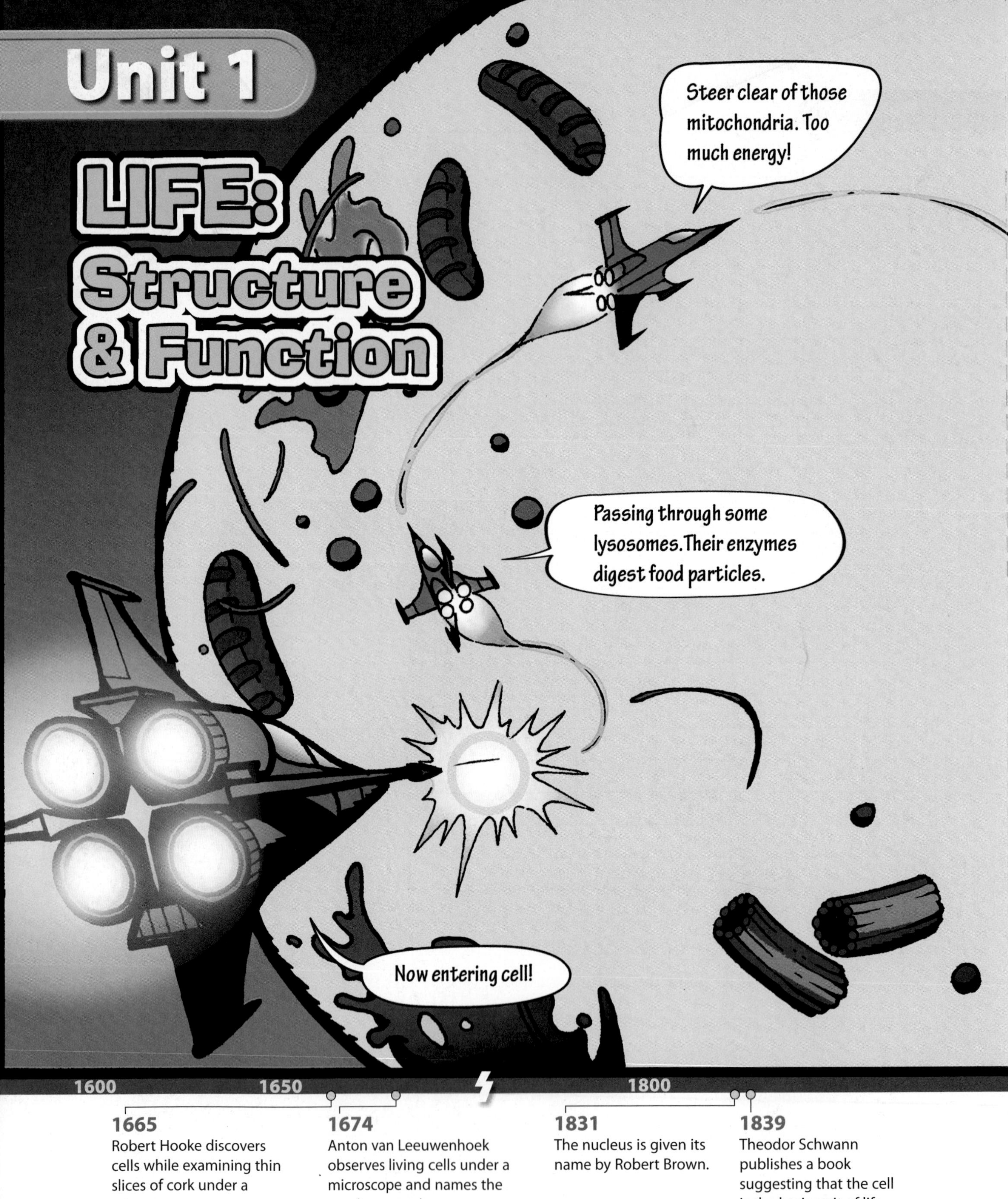

1665
Robert Hooke discovers cells while examining thin slices of cork under a microscope.

1674
Anton van Leeuwenhoek observes living cells under a microscope and names the moving organisms *animalcules.*

1831
The nucleus is given its name by Robert Brown.

1839
Theodor Schwann publishes a book suggesting that the cell is the basic unit of life.

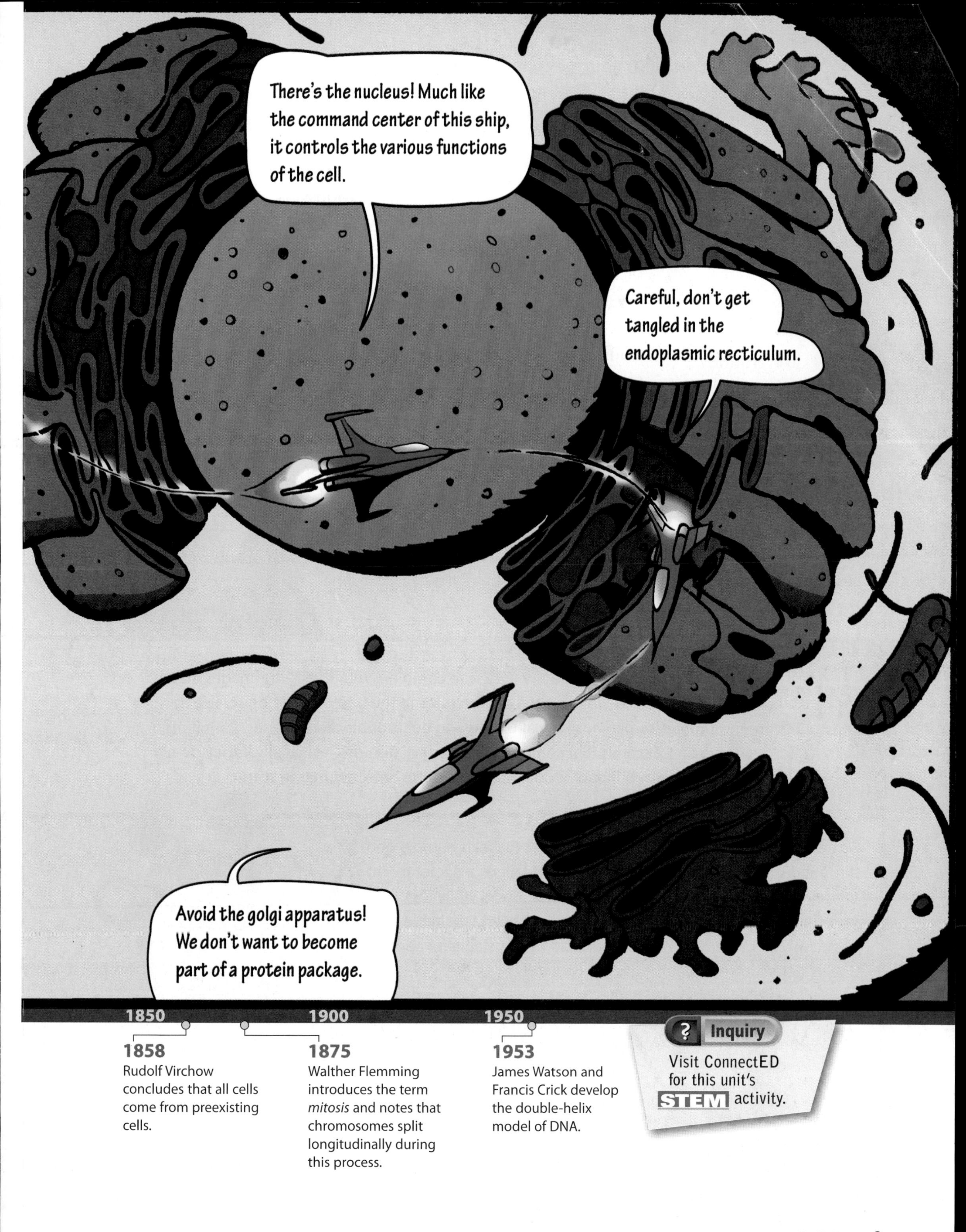
There's the nucleus! Much like the command center of this ship, it controls the various functions of the cell.
Careful, don't get tangled in the endoplasmic recticulum.
Avoid the golgi apparatus! We don't want to become part of a protein package.
1850
1900
1950
1858
Rudolf Virchow concludes that all cells come from preexisting cells.
1875
Walther Flemming introduces the term mitosis and notes that chromosomes split longitudinally during this process.
1953
James Watson and Francis Crick develop the double-helix model of DNA.
Inquiry
Visit ConnectED for this unit's STEM activity.

Models SC.7.N.1.5, SC.7.N.3.2

What would you do without your heart—one of the most important muscles in your body? Worldwide, people on donor lists wait for heart transplants because their hearts are not working properly. Today, doctors can diagnose and treat heart problems with the help of models.

A **model** is a representation of an object, a process, an event, or a system that is similar to the physical object or idea being studied. Models are used to study things that are too big or too small, happen too quickly or too slowly, or are too dangerous or expensive to study directly.

Active Reading **1. Determine** Underline the term *model* and its definition.

A magnetic resonance image (MRI) is a type of model created by using a strong magnetic field and radio waves. MRI machines produce high-resolution images from a series of images of layers. An MRI model of the heart helps cardiologists diagnose heart disease or damage. To obtain a clear MRI, the patient must be still. Even the beating of the heart can limit the ability of an MRI to capture clear images.

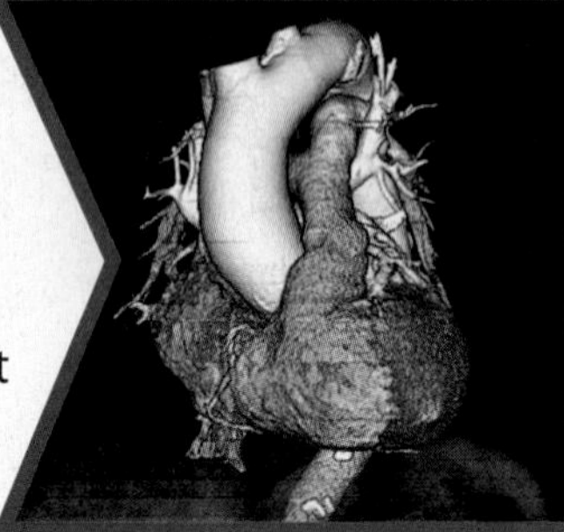

A computer tomography (CT) scan combines multiple X-ray images into a detailed 3-D visual model of structures in the body. Cardiologists use this model to diagnose a malfunctioning heart or blocked arteries. A limitation of a CT scan is that some coronary artery diseases, especially if they do not involve a buildup of calcium, may not be detected by the scan.

An artificial heart is a physical model of a heart that can pump blood throughout the body. For a patient with heart failure, a doctor might temporarily replace the heart with an artificial model while they wait for a transplant. Because of its size, the artificial heart is suitable for about only 50 percent of the male population. It is stable for about 2 years before it wears out.

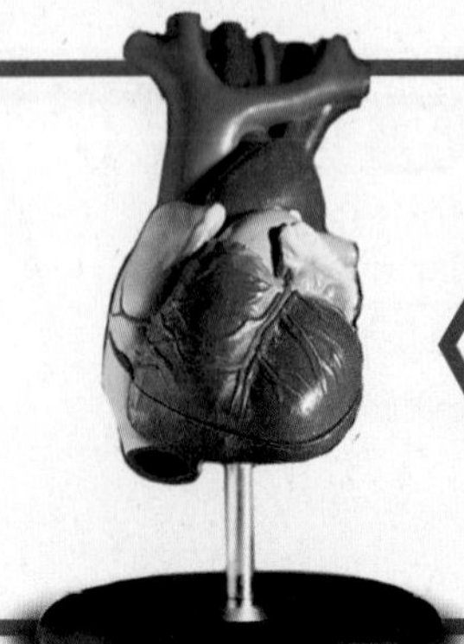

A cardiologist might use a physical model of a heart to explain a diagnosis to a patient. The parts of the heart can be touched and manipulated to explain how a heart works and the location of any complications. However, this model does not function like a real heart, and it cannot be used to diagnose disease.

Maps as Models

One way to think of a computer model, such as an MRI or a CT scan, is as a map. A map is a model that shows how locations are arranged in space. A map can be a model of a small area, such as your street. Or, maps can be models of very large areas, such as a state, a country, or the world.

Biologists study maps to understand where different animal species live, how they interact, and how they migrate. Most animals travel in search of food, water, specific weather, or a place to mate. By placing small electronic tracking devices on migrating animals biologists can create maps of their movements, such as the map of elephant movement in **Figure 1.** These maps are models that help determine how animals survive, repeat the patterns of their life cycle, and respond to environmental changes.

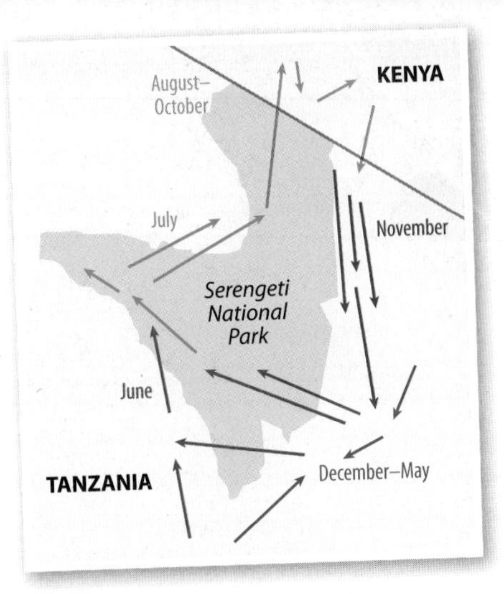

Figure 1 This map is a model of elephants' movements. The colored lines show the paths of three elephants that were equipped with tracking devices for a year.

Limitations of Models

It is impossible to include all the details about an object or an idea in one model. A map of elephant migration does not tell you whether the elephant is eating, sleeping, or playing with other elephants. Scientists must consider the limitations of the models they use when drawing conclusions about animal behavior.

All models have limitations. When making decisions about a patient's diagnosis and treatment, a cardiologist must be aware of the information each type of model does and does not provide. Scientists and doctors consider the purpose and limitations of the models they use to ensure that they draw the most accurate conclusions possible.

Active Reading **2. Revisit** Review all the information about models and their uses. (Circle) limitations and restrictions to the use of models in research, diagnosis, and general use.

Apply It!
After you complete the lab, answer these questions.

1. **Propose** How might you use a computer program to help model the design of a wildlife enclosure?

2. **Recall** List several types of models you have constructed and how you used them.

Name ______________________ Date ______________

Notes

Classification Systems

Mrs. Kenner's life science class discussed their ideas about classification systems. The class has different ideas about how classification systems are developed. Here are their ideas:

Group A thinks classification systems are developed naturally by the way organisms are grouped in nature.

Group B thinks classification systems are developed according to scientists' purposes for grouping organisms.

Which group best matches your thinking about classification? Explain why you agree with that group.

Classifying and Exploring LIFE

FLORIDA BIG IDEAS

1 **The Practice of Science**
2 **The Characteristics of Scientific Knowledge**
14 **Organization and Development of Living Organisms**
15 **Diversity and Evolution of Living Organisms**

Think About It!

What are living things, and how can they be classified?

At first glance, you might think someone dropped dinner rolls on a pile of rocks. These objects might look like dinner rolls, but they're not.

1. What do you think the objects are? Do you think they are alive?

2. Why do you think they look like this?

3. What are living things, and how can they be classified?

Get Ready to Read

What do you think about exploring life?

Before you read, decide if you agree or disagree with each of these statements? As you read this chapter, see if you change your mind about any of the statements.

	AGREE	DISAGREE
1. All living things move.	☐	☐
2. The Sun provides energy for almost all organisms on Earth.	☐	☐
3. A dichotomous key can be used to identify an unknown organism.	☐	☐
4. Physical similarities are the only traits used to classify organisms.	☐	☐
5. Most cells are too small to be seen with the unaided eye.	☐	☐
6. Only scientists use microscopes.	☐	☐

There's More Online!
Video • Audio • Review • ⓘLab Station • WebQuest • Assessment • Concepts in Motion • Multilingual eGlossary

Lesson 1

Characteristics of LIFE

ESSENTIAL QUESTION

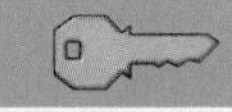

What characteristics do all living things share?

Vocabulary

organism p. 11
cell p. 12
unicellular p. 12
multicellular p. 12
homeostasis p. 15

Florida NGSSS

LA.6.2.2.3 The student will organize information to show understanding (e.g., representing main ideas within text through charting, mapping, paraphrasing, summarizing, or comparing/contrasting);

LA.6.4.2.2 The student will record information (e.g., observations, notes, lists, charts, legends) related to a topic, including visual aids to organize and record information and include a list of sources used;

SC.6.N.1.5 Recognize that science involves creativity, not just in designing experiments, but also in creating explanations that fit evidence.

SC.6.N.2.1 Distinguish science from other activities involving thought.

SC.6.N.2.3 Recognize that scientists who make contributions to scientific knowledge come from all kinds of backgrounds and possess varied talents, interests, and goals.

SC.6.L.14.3 Recognize and explore how cells of all organisms undergo similar processes to maintain homeostasis, including extracting energy from food, getting rid of waste, and reproducing.

SC.7.L.17.1 Explain and illustrate the roles of and relationships among producers, consumers, and decomposers in the process of energy transfer in a food web.

Inquiry Launch Lab

SC.6.N.1.5
SC.6.L.14.3

15 minutes

Is it alive?

Living organisms have specific characteristics. Is a rock a living organism? Is a dog? What characteristics describe something that is living?

Procedure

1. Read and complete a lab safety form.
2. Place three pieces of **pasta** in the bottom of a **clear plastic cup.**
3. Add **carbonated water** to the cup until it is 2/3 full.
4. Observe the contents of the cup for 5 minutes. Record your observations below.

Data and Observations

Think About This

1. Think about living things. How do you know they are alive?

2. Which characteristics of life do you think you are observing in the cup?

3. Key Concept Is the pasta alive? How do you know?

Inquiry **What do you think is missing?**

1. This toy looks like a dog and can move, but it is a robot. What characteristics are missing that would make it alive?

Characteristics of Life

Look around your classroom and then at **Figure 1.** You might see many nonliving things, such as lights and books. Look again, and you might see many living things, such as your teacher, your classmates, and plants. What makes people and plants different from lights and books?

People and plants, like all living things, have all the characteristics of life. All living things are organized, grow and develop, reproduce, respond, maintain certain internal conditions, and use energy. Nonliving things might have some of these characteristics, but they do not have all of them. Books might be organized into chapters, and lights use energy. However, only those things that have all the characteristics of life are living. *Things that have all the characteristics of life are called* **organisms.**

Active Reading **2. Recall** Underline how living things differ from nonliving things.

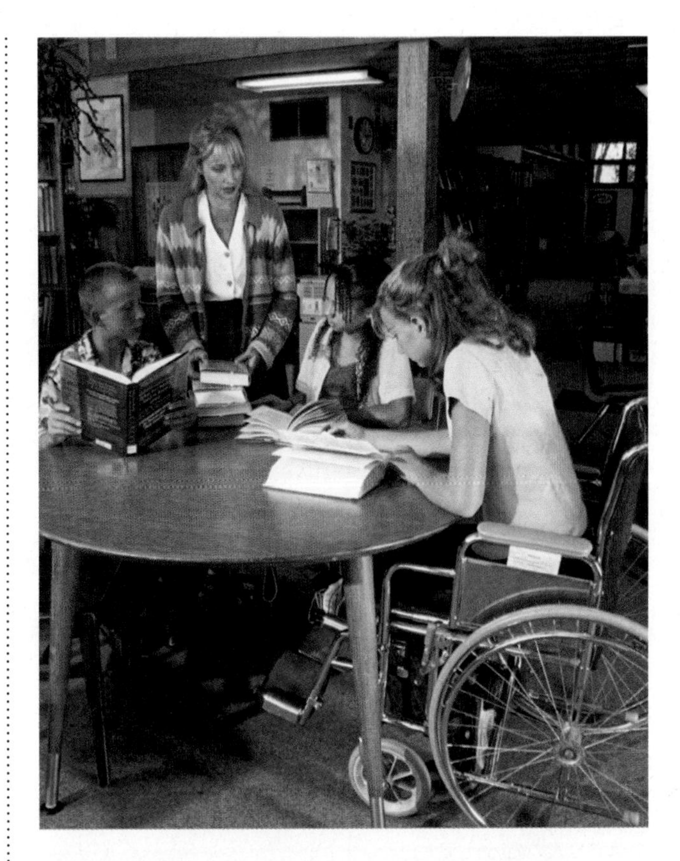

Figure 1 A classroom might contain living and nonliving things.

Active Reading

FOLDABLES® LA.6.2.2.3

Fold a sheet of paper into a half book. Label it as shown. Use it to organize your notes on the characteristics of living things.

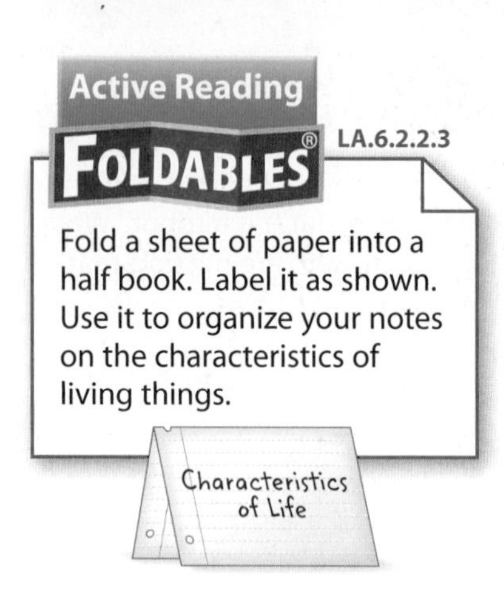

Organization

Your home is probably organized in some way. For example, the kitchen is for cooking, and the bedrooms are for sleeping. Living things are also organized. Whether an organism is made of one **cell**—*the smallest unit of life*—or many cells, all living things have structures that have specific functions.

Living things that are made of only one cell are called **unicellular** *organisms.* Within a unicellular organism are structures with specialized functions just like a house has rooms for different activities. Some structures take in nutrients or control cell activities. Other structures enable the organism to move.

Living things that are made of two or more cells are called **multicellular** *organisms.* Some multicellular organisms only have a few cells, but others have trillions of cells. The different cells of a multicellular organism usually do not perform the same function. Instead, the cells are organized into groups that have specialized functions, such as digestion or movement.

Growth and Development

The tadpole in **Figure 2** is not a frog, but it will soon lose its tail, grow legs, and become an adult frog. This happens because the tadpole, like all organisms, will grow and develop. When organisms grow, they increase in size. A unicellular organism grows as the cell increases in size. Multicellular organisms grow as the number of their cells increases.

Figure 2 A tadpole grows in size while developing into an adult frog.

Growth and Development

3. Visual Check State What characteristics of life can you identify in this figure?

1. **A frog egg develops into a tadpole.**
2. **As the tadpole grows, it develops legs.**

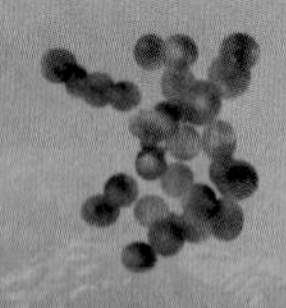

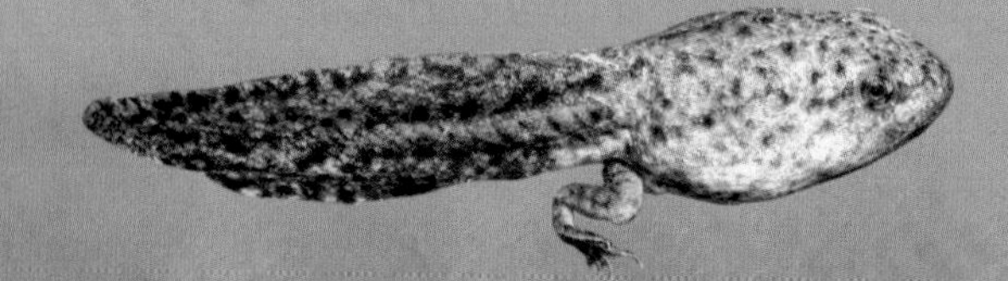

Changes that occur in an organism during its lifetime are called development. In multicellular organisms, development happens as cells become specialized into different cell types, such as skin cells or muscle cells. Some organisms undergo dramatic developmental changes over their lifetime, such as a tadpole developing into a frog.

Active Reading 4. **Find** (Circle) what happens in development.

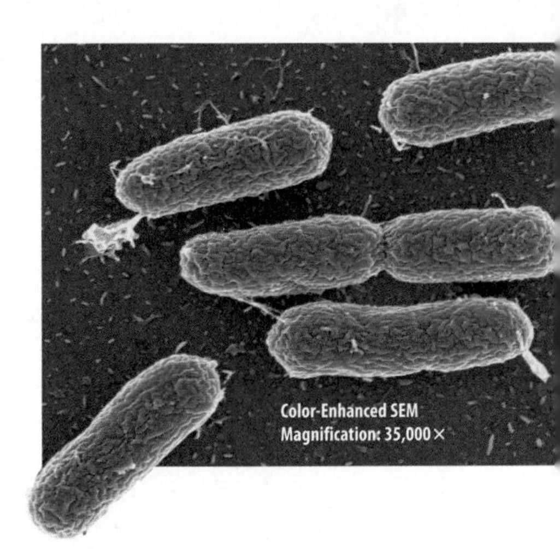

Figure 3 Some unicellular organisms, like the bacteria shown here, reproduce by dividing. The two new organisms are identical to the original organism.

Reproduction

As organisms grow and develop, they usually are able to reproduce. Reproduction is the process by which one organism makes one or more new organisms. In order for living things to continue to exist, organisms must reproduce. Some organisms within a population might not reproduce, but others must reproduce if the species is to survive.

Organisms do not all reproduce in the same way. Some organisms, like the ones in **Figure 3,** can reproduce by dividing and become two new organisms. Other organisms have specialized cells for reproduction. Some organisms must have a mate to reproduce, but others can reproduce without a mate. The number of offspring produced varies. Humans usually produce only one or two offspring at a time. Other organisms, such as the frog in **Figure 2,** can produce hundreds of offspring at one time.

Responses to Stimuli

If someone throws a ball toward you, you might react by trying to catch it. This is because you, like all living things, respond to changes in the environment. These changes can be internal or external and are called stimuli (STIHM yuh li).

Figure 4 The leaves and stems of plants like this one will grow toward a light source.

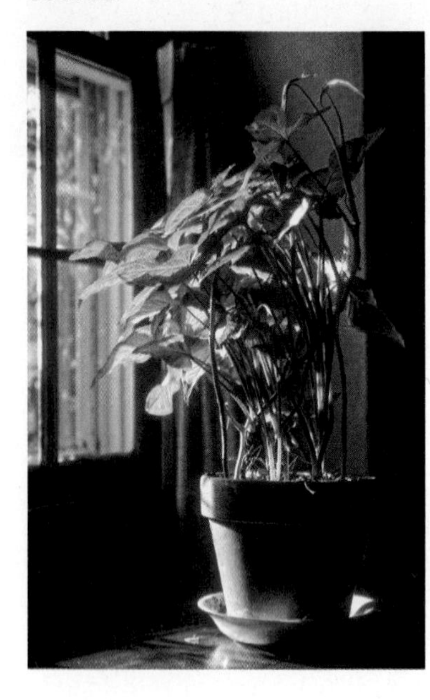

Internal Stimuli

You respond to internal stimuli (singular, stimulus) every day. If you feel hungry and then look for food, you are responding to an internal stimulus—the feeling of hunger. The feeling of thirst that causes you to find and drink water is another example of an internal stimulus.

External Stimuli

Changes in an organism's environment that affect the organism are external stimuli. Some examples of external stimuli are light and temperature.

Many plants, like the one in **Figure 4,** will grow toward light. You respond to light, too. Your skin's response to sunlight might be to darken, turn red, or freckle.

Some animals respond to changes in temperature. The response can be more or less blood flowing to the skin. For example, if the temperature increases, the diameter of an animal's blood vessels increases. This allows more blood to flow to the skin, cooling an animal.

Apply It! After you complete the lab, answer these questions.

1. Think of another time you had a reflex. Write it below.

2. How was your reaction similar to the ball toss?

3. How was it different?

Homeostasis

Have you ever noticed that if you drink more water than usual, you have to go to the bathroom more often? That is because your body is working to keep your internal environment under normal conditions. *An organism's ability to maintain steady internal conditions when outside conditions change is called* **homeostasis** (hoh mee oh STAY sus).

WORD ORIGIN

homeostasis
from Greek *homoios,* means "like, similar"; and *stasis,* means "standing still"

The Importance of Homeostasis

Are there certain conditions you need to do your homework? Maybe you need a quiet room with a lot of light. Cells also need certain conditions to function properly. Maintaining certain conditions—homeostasis—ensures that cells can function. If cells cannot function normally, then an organism might become sick or even die.

Methods of Regulation

A person might not survive if his or her body temperature changes more than a few degrees from 37°C. When your outside environment becomes too hot or too cold, your body responds. It sweats, shivers, or changes the flow of blood to maintain a body temperature of 37°C.

Unicellular organisms, such as the paramecium in **Figure 5,** also have ways of regulating homeostasis. A structure called a contractile vacuole (kun TRAK tul • VA kyuh wohl) collects and pumps excess water out of the cell.

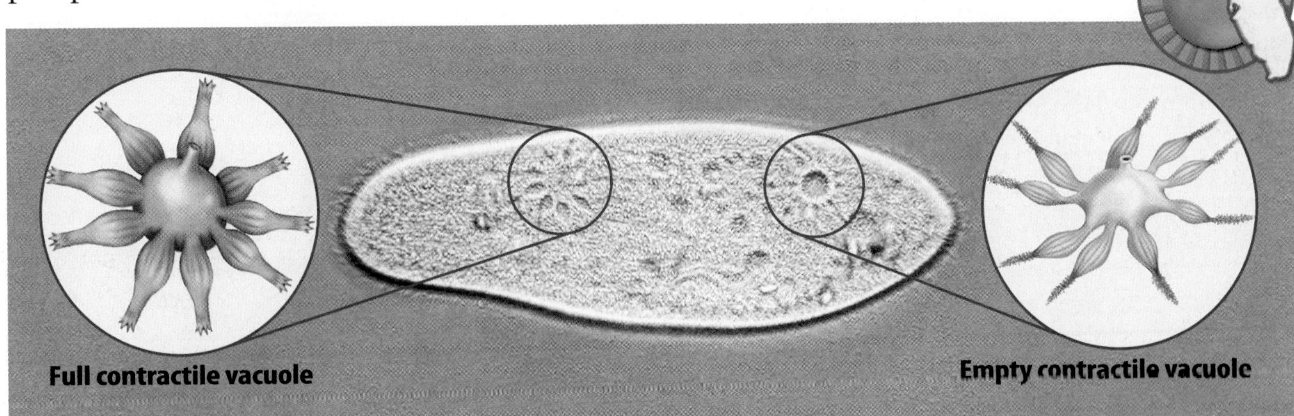

Figure 5 This paramecium lives in freshwater in Florida and many other states. Water continuously enters its cell and collects in contractile vacuoles. The vacuoles contract and expel excess water from the cell. This maintains normal water levels in the cell.

There is a limit to the amount of change that can occur within an organism. For example, you are able to survive only a few hours in water that is below 10°C. No matter what your body does, it cannot maintain steady internal conditions, or homeostasis, under these circumstances. As a result, your cells lose their ability to function.

5. **NGSSS Check** **Infer** Underline why maintaining homeostasis is important to organisms. SC.6.L.14.3

Energy

Everything you do requires energy. Digesting your food, sleeping, thinking, reading and all of the characteristics of life shown in **Table 1** on the next page require energy. Cells continuously use energy to transport substances, make new cells, and perform chemical reactions. Where does this energy come from?

For most organisms, this energy originally came to Earth from the Sun, as shown in **Figure 6.** For example, energy in the cactus came from the Sun. The squirrel gets energy by eating the cactus, and the coyote gets energy by eating the squirrel.

6. List What characteristics do all living things share?

Energy Use

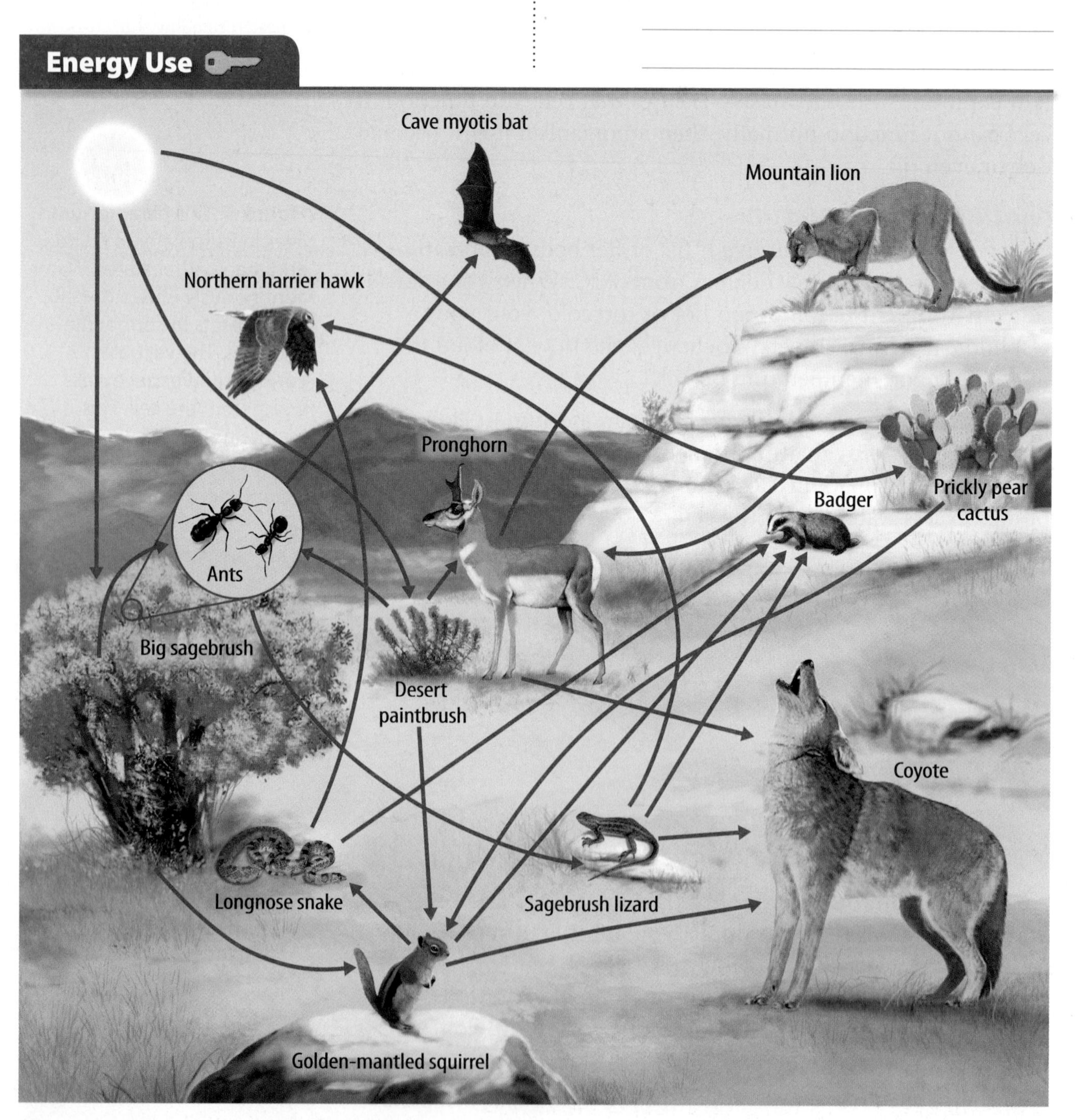

Figure 6 All organisms require energy to survive. In this food web, energy passes from one organism to another and to the environment.

7. Visual Check Identify (Circle) From which food sources does the badger get energy?

Table 1 Characteristics of Life

Characteristic	Definition	Example
Organization	Living things have specialized structures with specialized functions. Living things with more than one cell have a greater level of organization because groups of cells function together.	
Growth and development	Living things grow by increasing cell size and/or increasing cell number. Multicellular organisms develop as cells develop specialized functions.	
Reproduction	Living things make more living things through the process of reproduction.	
Response to stimuli	Living things adjust and respond to changes in their internal and external environments.	
Homeostasis	Living things maintain a stable internal environment.	
Use of energy	Living things use energy for all the processes they perform. Living things get energy by making their own food, eating food, or absorbing food.	

Lesson Review 1

Visual Summary

An organism has all the characteristics of life.

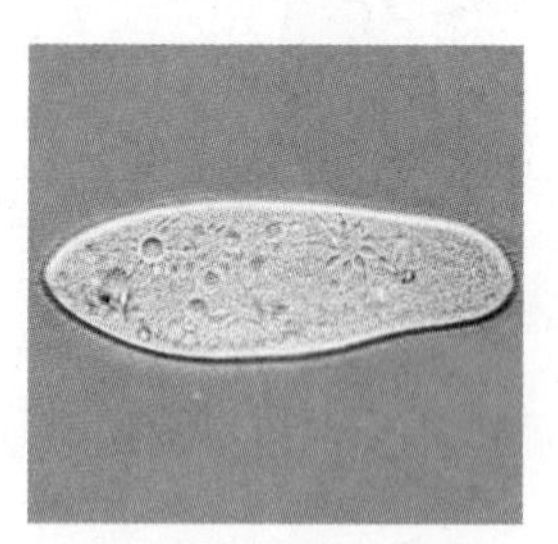

Unicellular organisms have specialized structures, much like a house has rooms for different activities.

Homeostasis enables living things to maintain a steady internal environment.

Use Vocabulary

1. A(n) ___________ is the smallest unit of life. SC.6.L.14.3

2. **Distinguish** between unicellular and multicellular. SC.6.L.14.3

3. **Define** the term *homeostasis* in your own words. SC.6.L.14.3

Understand Key Concepts

4. Which is NOT a characteristic of all living things?
   - (A) breathing
   - (B) growing
   - (C) reproducing
   - (D) using energy

5. **Compare** the processes of reproduction and growth.

6. **Choose** the characteristic of living things that you think is most important. Explain why you chose that characteristic.

7. **Critique** the following statement: A candle flame is a living thing.

Interpret Graphics

8. **Summarize** Fill in the graphic organizer below to summarize the characteristics of living things. LA.6.2.2.3

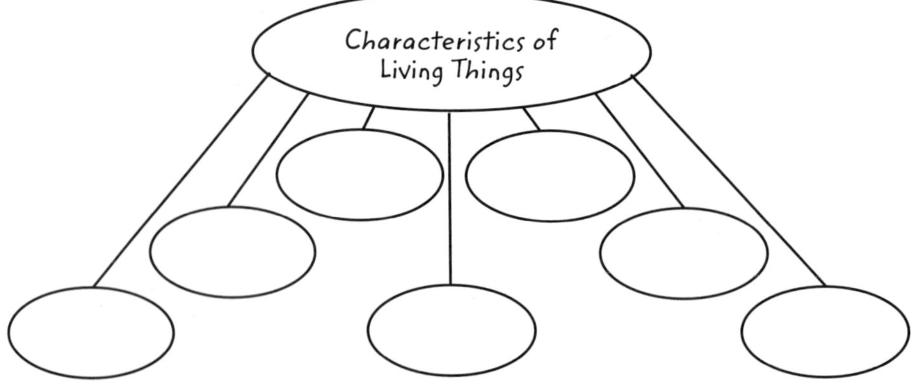

Critical Thinking

9. **Hypothesize** what would happen if living things could not reproduce.

AMERICAN MUSEUM OF NATURAL HISTORY

CAREERS in SCIENCE

The Amazing Adaptation of an Air-Breathing Catfish

Discover how some species of armored catfish breathe air.

Have you ever thought about why animals need oxygen? All animals, including you, get their energy from food. When you breathe, the oxygen you take in is used in your cells. Chemical reactions in your cells use oxygen and change the energy in food molecules into energy that your cells can use. Mammals and many other animals get oxygen from air. Most fish get oxygen from water. Either way, after an animal takes in oxygen, red blood cells carry oxygen to cells throughout its body.

Adriana Aquino is an ichthyologist (IHK thee AH luh jihst) at the American Museum of Natural History in New York City. She discovers and classifies species of fish, such as the armored catfish in family Loricariidae from South America. It lives in freshwater rivers and pools in the Amazon. Its name comes from the bony plates covering its body. Some armored catfish can take in oxygen from water and from air!

Some armored catfish live in fast-flowing rivers. The constant movement of the water evenly distributes oxygen throughout it. The catfish can easily remove oxygen from this oxygen-rich water.

But other armored catfish live in pools of still water, where most oxygen is only at the water's surface. This makes the pools low in oxygen. To maintain a steady level of oxygen in their cells, these fish have adaptations that enable them to take in oxygen directly from air. These catfish can switch from removing oxygen from water through their gills to removing oxygen from air through the walls of their stomachs. They can only do this when they do not have much food in their stomachs. Some species can survive up to 30 hours out of water!

Meet an Ichthyologist

Aquino examines hundreds of catfish specimens. Some she collects in the field, and others come from museum collections. She compares the color, the size, and the shape of the various species. She also examines their internal and external features, such as muscles, gills, and bony plates.

Some armored catfish remove oxygen from air.

It's Your Turn

BRAINSTORM Work with a group. Choose an animal and list five physical characteristics. Brainstorm how these adaptations help the animal be successful in its habitat. Present your findings to the class. LA.6.4.2.2

Lesson 2

Classifying ORGANISMS

ESSENTIAL QUESTIONS

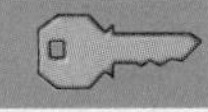
What methods are used to classify living things into groups?

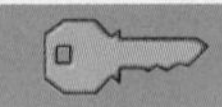
Why does every species have a scientific name?

Vocabulary

binomial nomenclature p. 23
species p. 23
genus p. 23
dichotomous key p. 24
cladogram p. 25

Florida NGSSS

LA.6.2.2.3 The student will organize information to show understanding (e.g., representing main ideas within text through charting, mapping, paraphrasing, summarizing, or comparing/contrasting);

SC.6.L.15.1 Analyze and describe how and why organisms are classified according to shared characteristics with emphasis on the Linnaean system combined with the concept of Domains.

SC.6.N.1.1 Define a problem from the sixth grade curriculum, use appropriate reference materials to support scientific understanding, plan and carry out scientific investigation of various types, such as systematic observations or experiments, identify variables, collect and organize data, interpret data in charts, tables, and graphics, analyze information, make predictions, and defend conclusions.

SC.6.N.1.4 Discuss, compare, and negotiate methods used, results obtained, and explanations among groups of students conducting the same investigation.

SC.6.N.2.1 Distinguish science from other activities involving thought.

SC.6.N.2.2 Explain that scientific knowledge is durable because it is open to change as new evidence or interpretations are encountered.

SC.6.N.2.3 Recognize that scientists who make contributions to scientific knowledge come from all kinds of backgrounds and possess varied talents, interests, and goals.

SC.7.N.2.1 Identify an instance from the history of science in which scientific knowledge has changed when new evidence or new interpretations are encountered.

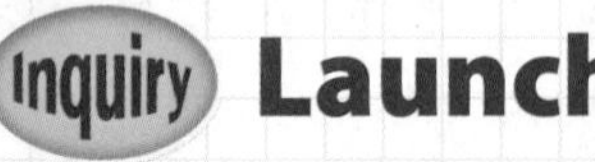

Inquiry Launch Lab

SC.6.L.15.1, SC.6.N.1.1

15 minutes

How do you identify similar items?

Do you separate your candies by color before you eat them? When your family does laundry, do you sort the clothes by color first? Identifying characteristics of items can enable you to place them into groups.

Procedure

1. Read and complete a lab safety form
2. Examine twelve **leaves.** Choose a characteristic that you could use to separate the leaves into two groups. Record the characteristic below.
3. Place the leaves into two groups, A and B, using the characteristic you chose in step 2.
4. Choose another characteristic that you could use to further divide group A. Record the characteristic, and divide the leaves.
5. Repeat step 4 with group B.

Data and Observations

Think About This

1. What types of characteristics did other groups in class choose to separate the leaves?

2. Key Concept Why would scientists need rules for separating and identifying items?

Inquiry **Do you think they are alike?**

1. In a band, instruments are organized into groups, such as brass and woodwinds. The instruments in a group are alike in many ways. In a similar way, living things are classified into groups. Why are living things classified?

Classifying Living Things

How would you find your favorite fresh fruit or vegetable in the grocery store? You might look in the produce section, such as the one shown in **Figure 7.** Different kinds of peppers are displayed in one area. Citrus fruits such as oranges, lemons, and grapefruits are stocked in another area. There are many different ways to organize produce in a grocery store. In a similar way, there have been many different ideas about how to organize, or classify, living things.

A Greek philosopher named Aristotle (384 B.C.–322 B.C.) was one of the first people to classify organisms. Aristotle placed all organisms into two large groups, plants and animals. He classified animals based on the presence of "red blood," the animal's environment, and the shape and size of the animal. He classified plants according to the structure and size of the plant and whether the plant was a tree, a shrub, or an herb.

Figure 7 The produce in this store is classified into groups.

2. **Visual Check** **State** What other ways can you think of to classify and organize produce?

Determining Kingdoms

In the 1700s, Carolus Linnaeus, a Swedish physician and botanist, classified organisms based on similar structures. Linnaeus placed all organisms into two main groups, called **kingdoms.** Over the next 200 years, people learned more about organisms and discovered new organisms. In 1969 American biologist Robert H. Whittaker proposed a five-kingdom system for classifying organisms. His system included kingdoms Monera, Protista, Plantae, Fungi, and Animalia.

Science Use v. Common Use

kingdom

Science Use a classification category that ranks above phylum and below domain

Common Use a territory ruled by a king or a queen

Determining Domains

The classification system of living things is still changing. The current classification method is called systematics. Systematics uses all the evidence that is known about organisms to classify them. This evidence includes an organism's cell type, its habitat, the way an organism obtains food and energy, structure and function of its features, and the common ancestry of organisms. Systematics also includes molecular analysis—the study of molecules such as DNA within organisms.

Using systematics, scientists identified two distinct groups in Kingdom Monera—Bacteria and Archaea (ar KEE uh). This led to the development of another level of classification called domains. All organisms are now classified into one of three domains—Bacteria, Archaea, or Eukarya (yew KER ee uh)—and then into one of six kingdoms, as shown in **Table 2.**

3. NGSSS Check Select Underline evidence used to classify living things into groups. SC.6.L.15.1

Table 2 Domains and Kingdoms

Domain	Bacteria	Archaea	Eukarya			
Kingdom	**Bacteria**	**Archaea**	**Protista**	**Fungi**	**Plantae**	**Animalia**
Example						
Characteristics	Bacteria are simple unicellular organisms.	Archaea are simple unicellular organisms that often live in extreme environments.	Protists are unicellular and are more complex than bacteria or archaea.	Fungi are unicellular or multicellular and absorb food.	Plants are multicellular and make their own food.	Animals are multicellular and take in their food.

Scientific Names

Suppose you did not have a name. What would people call you? All organisms, just like people, have names. When Linnaeus grouped organisms into kingdoms, he also developed a system for naming organisms. This naming system, called binomial nomenclature (bi NOH mee ul · NOH mun klay chur), is the system we still use today.

Binomial Nomenclature

Linneaus's naming system, **binomial nomenclature,** *gives each organism a two-word scientific name,* such as *Ursus arctos* for a brown bear. This two-word scientific name is the name of an organism's species (SPEE sheez). *A* **species** *is a group of organisms that have similar traits and are able to produce fertile offspring.* In binomial nomenclature, the first word is the organism's genus (JEE nus) name, such as *Ursus. A* **genus** *is a group of similar species.* The second word might describe the organism's appearance or its behavior.

How do species and genus relate to kingdoms and domains? Similar species are grouped into one genus (plural, genera). Similar genera are grouped into families, then orders, classes, phyla, kingdoms, and finally domains, as shown for the grizzly bear in **Table 3.**

Word Origin

genus
from Greek *genos,* means "race, kind"

Table 3 The classification of the brown bear or grizzly bear shows that it belongs to the order Carnivora.

Table 3 Classification of the Brown Bear

Taxonomic Group	Number of Species	Examples
Domain Eukarya	About 4–10 million	
Kingdom Animalia	About 2 million	
Phylum Chordata	About 50,000	
Class Mammalia	About 5,000	
Order Carnivora	About 270	
Family Ursidae	8	
Genus *Ursus*	4	
Species *Ursus arctos*	1	

4. **Visual Check** Identify What domain does the brown bear belong to?

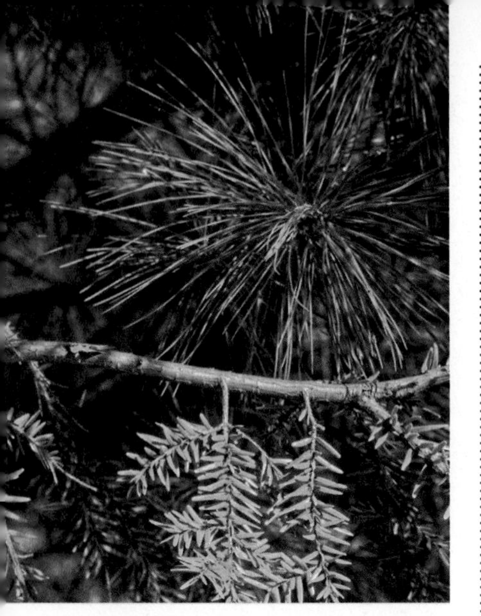

Figure 8 These trees are two different species. *Pinus alba* has long needles, and *Tsuga canadensis* has short needles.

Uses of Scientific Names

When you talk about organisms, you might use names such as bird, tree, or mushroom. However, these are common names for a number of different species. Sometimes there are several common names for one organism. The animal in **Table 3** on the previous page might be called a brown bear or a grizzly bear, but it has only one scientific name, *Ursus arctos*.

Other times, a common name might refer to several different types of organisms. For example, you might call both of the trees in **Figure 8** pine trees. But these trees are two different species. How can you tell? Scientific names are important for many reasons. Each species has its own scientific name. Scientific names are the same worldwide. This makes communication about organisms more effective because everyone uses the same name for the same species.

5. **NGSSS Check** **Explain** Why does every species have a scientific name? SC.6.L.15.1

____

____

Classification Tools

Suppose you go fishing and catch a fish you don't recognize. How could you figure out what type of fish you have caught? There are several tools you can use to identify organisms.

Dichotomous Keys

A **dichotomous key** *is a series of descriptions arranged in pairs that lead the user to the identification of an unknown organism.* The chosen description leads to either another pair of statements or the identification of the organism. Choices continue until the organism is identified. The dichotomous key shown in **Figure 9** identifies several species of fish.

Dichotomous Key

Dichotomous Key
1. a. This fish has a mouth that extends past its eye. It is an arrow goby. **1. b.** This fish does not have a mouth that extends past its eye. Go to step 2.
2. a. This fish has a dark body with stripes. It is a chameleon goby. **2. b.** This fish has a light body with no stripes. Go to step 3.
3. a. This fish has a black-tipped dorsal fin. It is a bay goby. **3. b.** This fish has a speckled dorsal fin. It is a yellowfin goby.

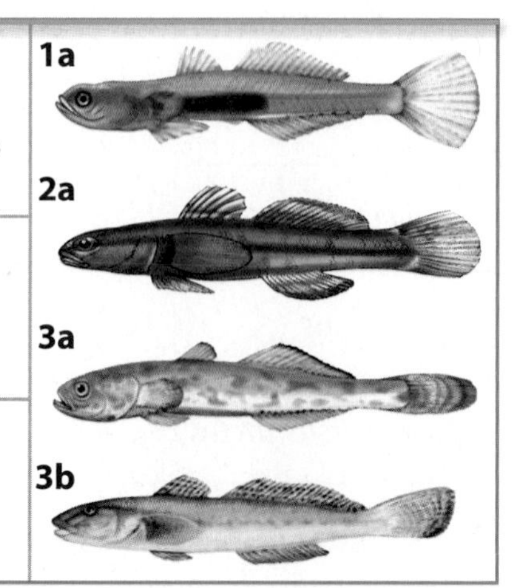

Figure 9 Dichotomous keys include a series of questions to identify organisms.

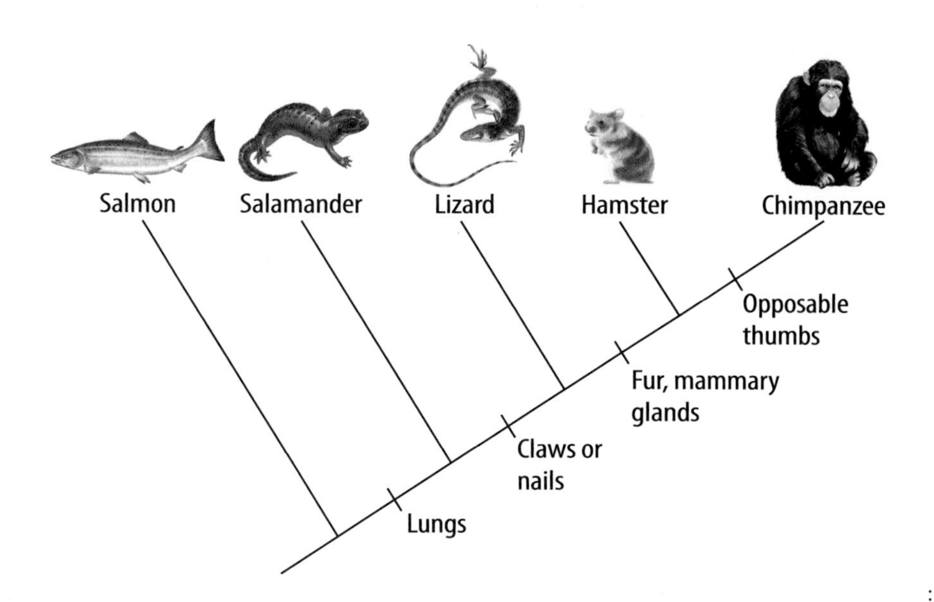

Figure 10 A cladogram shows relationships among species. In this cladogram, salamanders and lizards are more closely related to each other than they are to salmon.

Cladograms

A family tree shows the relationships among family members, including common ancestors. Biologists use a similar diagram, called a cladogram. *A* **cladogram** *is a branched diagram that shows the relationships among organisms, including common ancestors.* A cladogram, as shown in **Figure 10,** has a series of branches. Notice that each branch follows a new characteristic. Each characteristic is observed in all the species to its right. For example, the salamander, lizard, hamster, and chimpanzee have lungs, but the salmon does not. Therefore, they are more closely related to each other than they are to the salmon.

Active Reading **FOLDABLES**® LA.6.2.2.3

Make a horizontal two-tab book to compare two of the tools scientists use to identify organisms—dichotomous keys and cladograms.

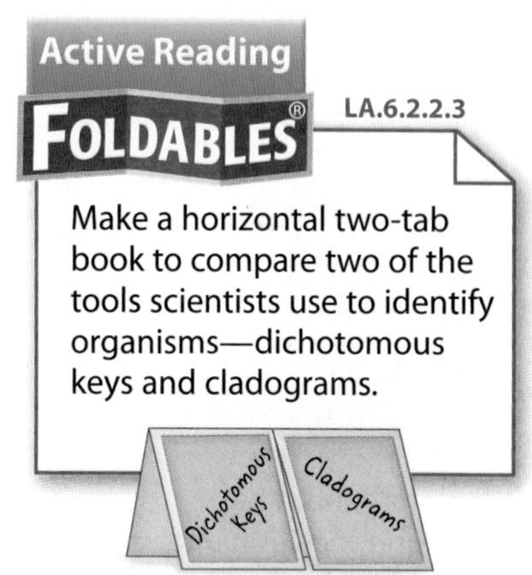

MiniLab *How would you name an unknown organism?* at connectED.mcgraw-hill.com

Apply It! After you complete the lab, answer this question.

1. Explain how you would place the organisms from the MiniLab into a cladogram. Use the example in **Figure 10** to help you.

Lesson Review 2

Visual Summary

All organisms are classified into one of three domains: Bacteria, Archaea, or Eukarya.

Every organism has a unique species name.

A dichotomous key helps to identify an unknown organism through a series of paired descriptions.

Use Vocabulary

1. A naming system that gives every organism a two-word name is ______________________________. SC.6.L.15.1

2. **Use the term** *dichotomous key* in a sentence. SC.6.L.15.1

____

3. **Organisms** of the same __________ are able to produce fertile offspring.

Understand Key Concepts

4. **Describe** how you write a scientific name. SC.6.L.15.1

____

5. Which is NOT used to classify organisms? SC.6.L.15.1
   (A) ancestry
   (B) habitat
   (C) age of the organism
   (D) molecular evidence

Interpret Graphics

6. **Organize Information** Fill in the graphic organizer below to show how organisms are classified. SC.6.L.15.1

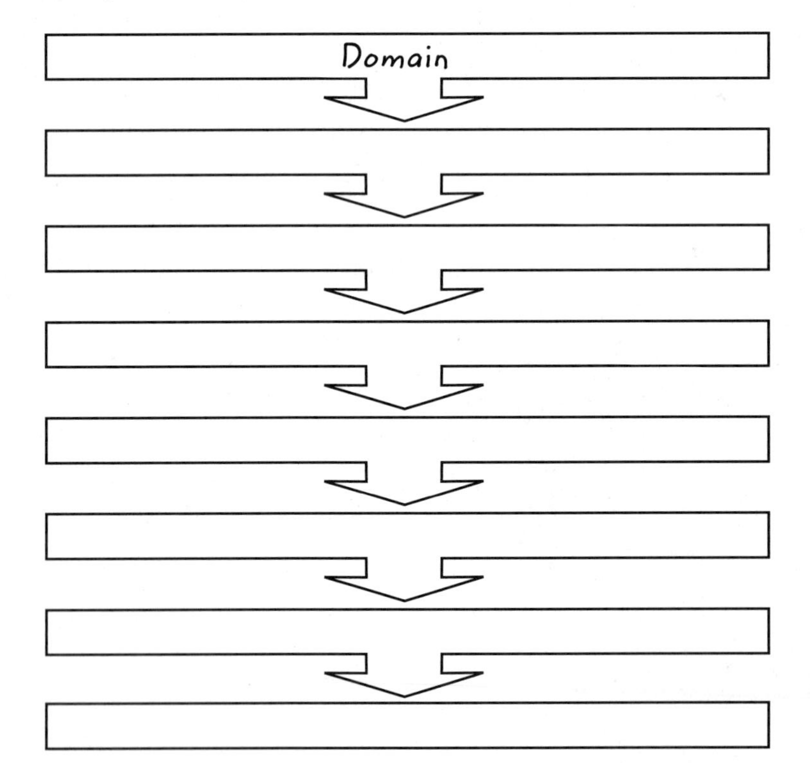

Critical Thinking

7. **Evaluate** the importance of scientific names. SC.6.L.15.1

____

____

The Name Game

Taxonomy, the science of organism identification and classification, is a discipline in which conclusions change as advances in technology result in new information.

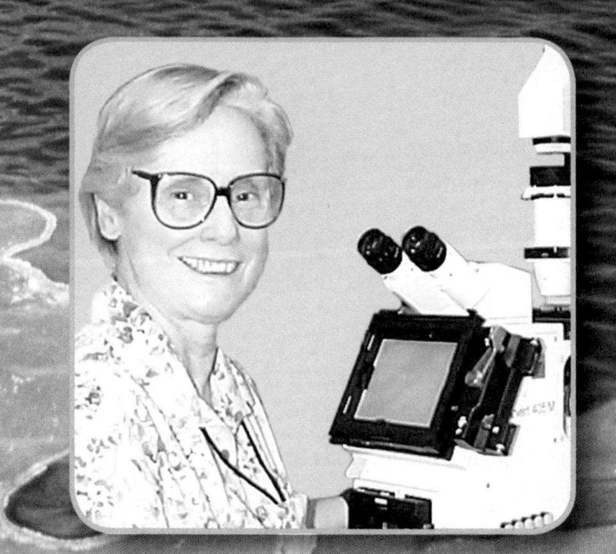

Dr. Karen Steidinger is a taxonomist at the Fish and Wildlife Research Institute in St. Petersburg, Florida. Using improved technology, she revises species descriptions and classifications as new organism characteristics are discovered. Dr. Steidinger works with the toxic *Karenia brevis* shown below. This organism can produce algal blooms with harmful effects on public health and marine resources. Due to these effects, assigning this organism to the accurate taxonomic category is especially important. The genus *Karenia* was established in November 2000 to honor Dr. Steidinger. The species in this genus is the Florida red tide organism, *Karenia brevis*.

K. brevis "Tide Line"

1948: *Gymnodinium brevis* was named by Charles C. Davis from cells collected during a red tide event in Florida in 1946–1947.

1979: Dr. Steidinger transferred the species to the genus *Ptychodiscus* because she discovered it has a resistant cell covering.

1989: Scientists agreed that the original name, *Gymnodinium breve*, would be used until the type species of *Gymnodinium*, *Gyrodinium*, *Balechina*, and *Ptychodiscus* were investigated further to adequately characterize them.

2000: Scientists came to a conclusion about the need for a new genus based on algal cell pigments. The genus *Karenia* was established the following November.

2010: *K. brevis* continues to be classified under the genus *Karenia*. Dr. Steidinger currently investigates Florida Red Tide levels all around the coast each week.

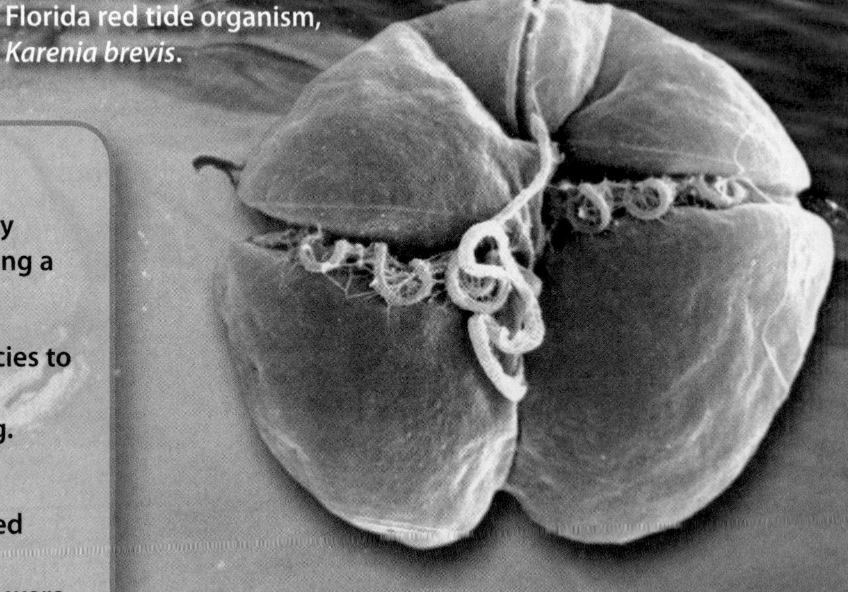

Karenia brevis

It's Your Turn

RESEARCH an organism that has been reclassified based on new scientific evidence. What was the reason for the reclassification? Report your results to the class.

Lesson 3

EXPLORING LIFE

ESSENTIAL QUESTIONS

 How did microscopes change our ideas about living things?

 What are the types of microscopes, and how do they compare?

Vocabulary

light microscope p. 30
compound microscope p. 30
electron microscope p. 31

Florida NGSSS

LA.6.2.2.3 The student will organize information to show understanding (e.g., representing main ideas within text through charting, mapping, paraphrasing, summarizing, or comparing/contrasting).

MA.6.A.3.6 Construct and analyze tables, graphs, and equations to describe linear functions and other simple relations using both common language and algebraic notation.

SC.6.N.1.5 Recognize that science involves creativity, not just in designing experiments, but also in creating explanations that fit evidence.

SC.6.N.2.1 Distinguish science from other activities involving thought.

SC.6.N.2.2 Explain that scientific knowledge is durable because it is open to change as new evidence or interpretations are encountered.

SC.6.N.2.3 Recognize that scientists who make contributions to scientific knowledge come from all kinds of backgrounds and possess varied talents, interests, and goals.

SC.6.L.15.1 Analyze and describe how and why organisms are classified according to shared characteristics with emphasis on the Linnaean system combined with the concept of Domains.

SC.7.N.1.3 Distinguish between an experiment (which must involve the identification and control of variables) and other forms of scientific investigation and explain that not all scientific knowledge is derived from experimentation.

SC.7.N.2.1 Identify an instance from the history of science in which scientific knowledge has changed when new evidence or new interpretations are encountered.

SC.6.N.1.5

Inquiry Launch Lab

15 minutes

Can a water drop make objects appear bigger or smaller?

For centuries, people have been looking for ways to see objects in greater detail. How can something as simple as a drop of water make this possible?

Procedure

1. Read and complete a lab safety form.
2. Lay a sheet of **newspaper** on your desk. Examine a line of text, noting the size and shape of each letter. Record your observations.
3. Add a large drop of **water** to the center of a piece of **clear plastic.** Hold the plastic about 2 cm above the same line of text.
4. Look through the water at the line of text you viewed in step 2. Record your observations.

Data and Observations

Think About This

1. Describe how the newsprint appeared through the drop of water.

2. Key Concept How might microscopes change your ideas about living things?

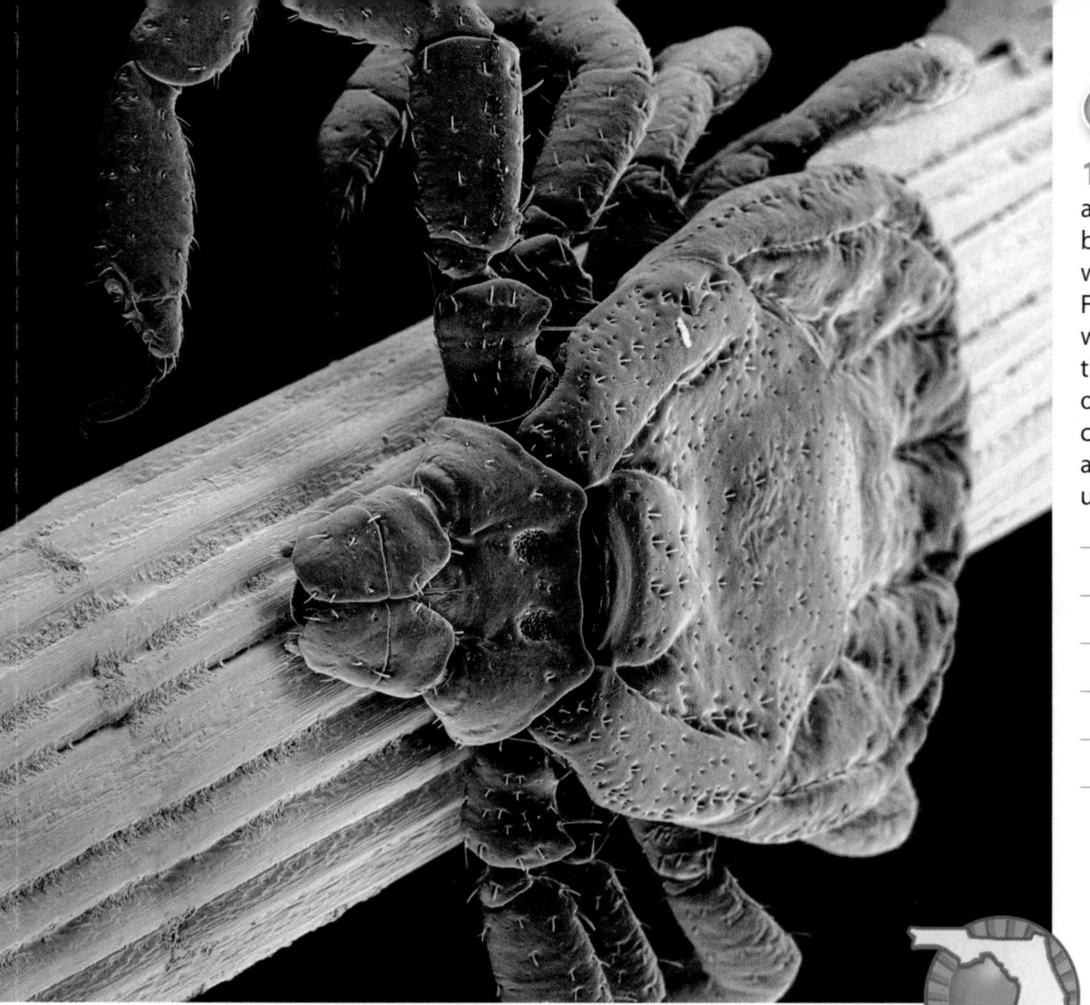

Inquiry Giant Insect?

1. Although this might look like a giant insect, it is a photo of the brown dog tick, common in warmer climate states such as Florida. This photo was taken with a high-powered microscope that can enlarge an image of an object up to 200,000 times. How can seeing an enlarged image of a living thing help you understand life?

____
____
____
____
____
____

The Development of Microscopes

Have you ever used a magnifying lens to see details of an object? If so, then you have used a tool similar to the first microscope. The invention of microscopes enabled people to see details of living things that they could not see with the unaided eye. The microscope also enabled people to make many discoveries about living things.

In the late 1600s the Dutch merchant Anton van Leeuwenhoek (LAY vun hook) made one of the first microscopes. His microscope, similar to the one shown in **Figure 11,** had one lens and could magnify an image about 270 times its original size. Another inventor of microscopes was Robert Hooke. In the early 1700s Hooke made one of the most significant discoveries using a microscope. He observed and named cells. Before microscopes, people did not know that living things are made of cells.

Active Reading **2. Suggest** Underline how microscopes changed our ideas about living things.

Figure 11 Anton van Leeuwenhoek observed pond water and insects using a microscope like the one shown above.

Math Skills MA.6.A.3.6

Use Multiplication

The magnifying power of a lens is expressed by a number and a multiplication symbol (×). For example, a lens that makes an object look ten times larger has a power of 10×. To determine a microscope's magnification, multiply the power of the ocular lens by the power of the objective lens. A microscope with a 10× ocular lens and a 10× objective lens magnifies an object 10 × 10, or 100 times.

Practice

3. What is the magnification of a compound microscope with a 10× ocular lens and a 4× objective lens?

Types of Microscopes

One characteristic of all microscopes is that they magnify objects. Magnification makes an object appear larger than it really is. Another characteristic of microscopes is resolution—how clearly the magnified object can be seen. The two main types of microscopes—light microscopes and electron microscopes—differ in magnification and resolution.

Light Microscopes

If you have used a microscope in school, then you have probably used a light microscope. **Light microscopes** *use light and lenses to enlarge an image of an object.* A simple light microscope has only one lens. *A light microscope that uses more than one lens to magnify an object is called a* **compound microscope.** A compound microscope magnifies an image first by one lens, called the ocular lens. The image is then further magnified by another lens, called the objective lens. The total magnification of the image is equal to the magnifications of the ocular lens and the objective lens multiplied together.

Light microscopes can enlarge images up to 1,500 times their original size. The resolution of a light microscope is about 0.2 micrometers (µm), or two-millionths of a meter. A resolution of 0.2 µm means you can clearly see points on an object that are at least 0.2 µm apart.

Light microscopes can be used to view living or nonliving objects. In some light microscopes, an object is placed directly under the microscope. For other light microscopes, an object must be mounted on a slide. In some cases, the object, such as the white blood cells in **Figure 12,** must be stained with a dye in order to see any details.

Active Reading 4. **Find** Highlight some ways an object can be examined under a light microscope.

Compound Light Microscope

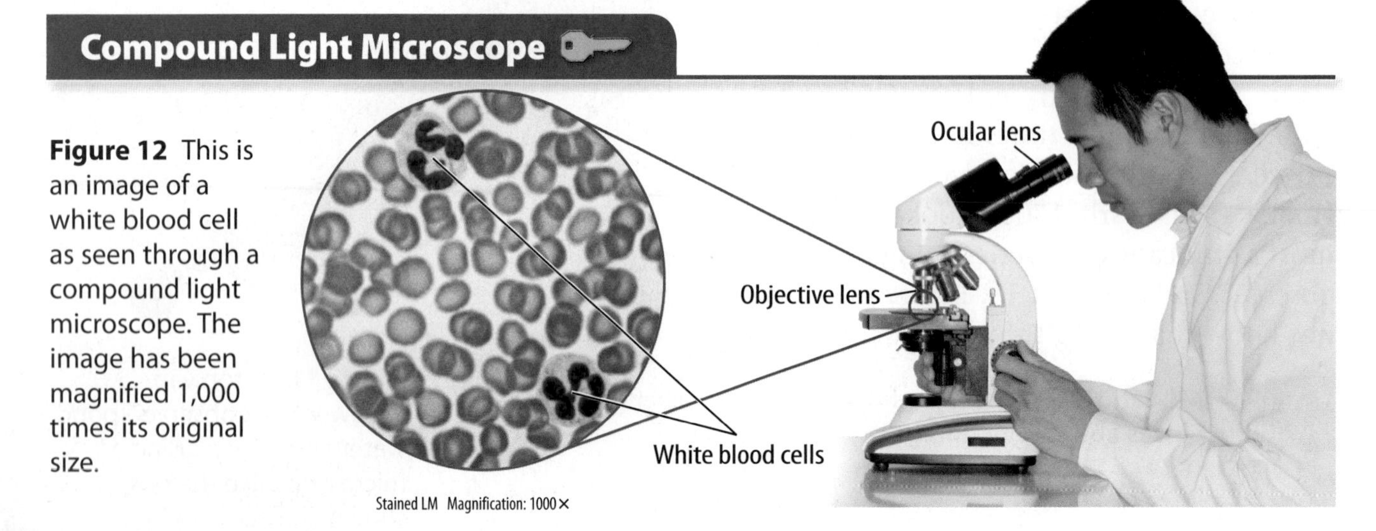

Figure 12 This is an image of a white blood cell as seen through a compound light microscope. The image has been magnified 1,000 times its original size.

Electron Microscopes

You might know that electrons are tiny particles inside **atoms.** **Electron microscopes** *use a magnetic field to focus a beam of electrons through an object or onto an object's surface.* An electron microscope can magnify an image up to 100,000 times or more. The resolution of an electron microscope can be as small as 0.2 nanometers (nm), or two-billionths of a meter. This resolution is up to 1,000 times greater than a light microscope. The two main types of electron microscopes are transmission electron microscopes (TEMs) and scanning electron microscopes (SEMs).

TEMs are usually used to study extremely small things such as cell structures. Because objects must be mounted in plastic and then very thinly sliced, only dead organisms can be viewed with a TEM. In a TEM, electrons pass through the object and a computer produces an image of the object. A TEM image of a white blood cell is shown in **Figure 13.**

SEMs are usually used to study an object's surface. In an SEM, electrons bounce off the object and a computer produces a three-dimensional image of the object. An image of a white blood cell from an SEM is shown in **Figure 13.** Note the difference in detail in this image compared to the image in **Figure 12** of a white blood cell from a light microscope.

Active Reading **5. Group** Underline the types of microscopes and how they compare.

Review Vocabulary

atom
the building block of matter that is composed of protons, neutrons, and electrons

Active Reading

Electron Microscopes

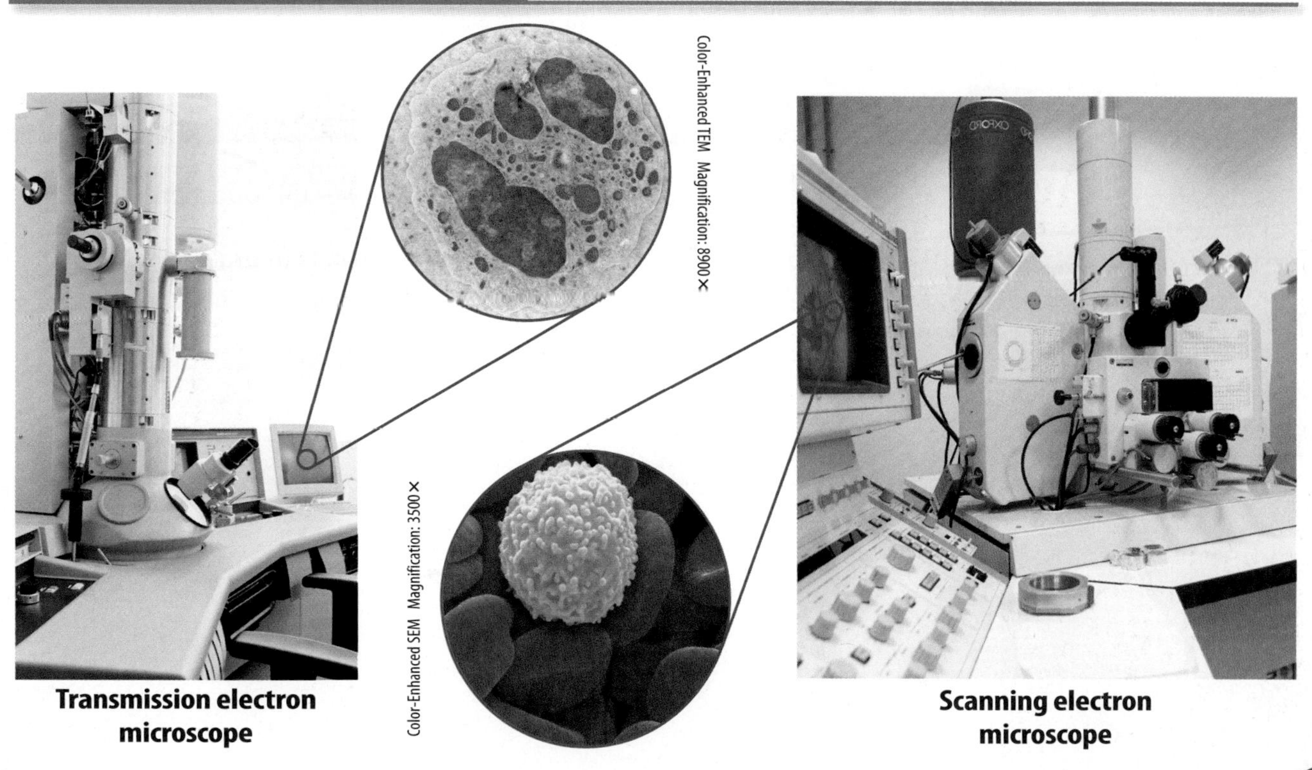

Figure 13 A TEM greatly magnifies thin slices of an object. An SEM is used to view a three-dimensional image of an object.

Using Microscopes

The **microscopes** used today are more advanced than the microscopes used by Leeuwenhoek and Hooke. The quality of today's light microscopes and the invention of electron microscopes have made the microscope a useful tool in many fields.

Health Care

People in health-care fields, such as doctors and laboratory technicians, often use microscopes. Microscopes are used in surgeries, such as cataract surgery and brain surgery. They enable doctors to view the surgical area in greater detail. The area being viewed under the microscope can also be displayed on a TV monitor so that other people can watch the procedure. Laboratory technicians use microscopes to analyze body fluids, such as blood and urine. They also use microscopes to determine whether tissue samples are healthy or diseased.

ACADEMIC VOCABULARY

identify
(verb) to determine the characteristics of a person or a thing

WORD ORIGIN

microscope
from Latin *microscopium,* means "an instrument for viewing what is small"

Other Uses

Health care is not the only field that uses microscopes. Have you ever wondered how police determine how and where a crime happened? Forensic scientists use microscopes to study evidence from crime scenes. The presence of different insects can help identify when and where a homicide happened. Microscopes might be used to **identify** the type and age of the insects.

People who study fossils might use microscopes. They might examine a fossil and other materials from where the fossil was found.

Some industries also use microscopes. The steel industry uses microscopes to examine steel for impurities. Microscopes are used to study jewels and identify stones. Stones have some markings and impurities that can be seen only by using a microscope.

6. List What are some uses of microscopes?

____

____

Apply It! After you complete the lab, answer this question.

1. Have you used a microscope to view other objects or organisms? Record them below.

Lesson Review 3

Visual Summary

Living organisms can be viewed with light microscopes.

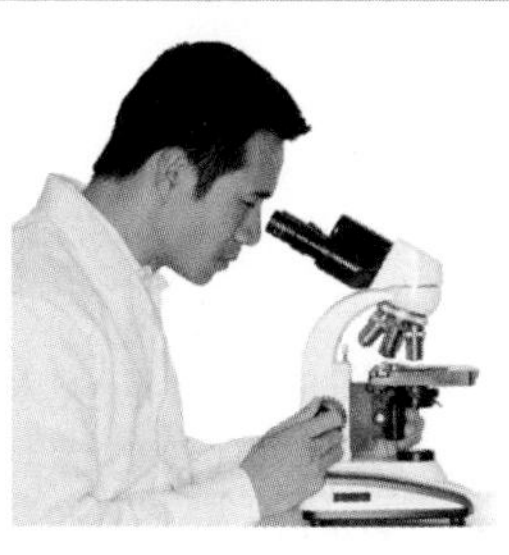

A compound microscope is a type of light microscope that has more than one lens.

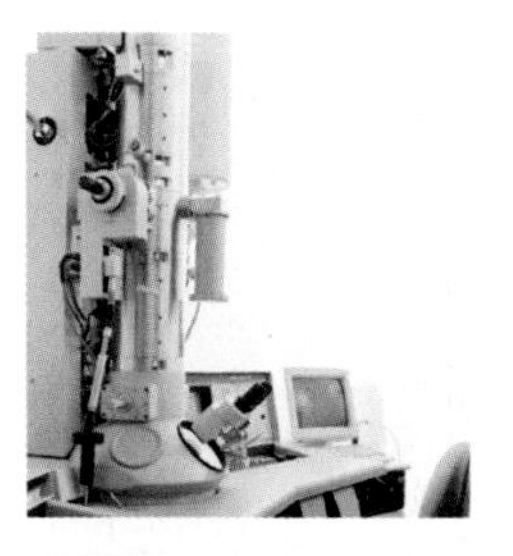

Living organisms cannot be viewed with a transmission electron microscope.

Inquiry Lab
Constructing a Dichotomous key at connectED.mcgraw-hill.com

Use Vocabulary

1. **Define** the term *light microscope* in your own words.

2. A(n) ________ focuses a beam of electrons through an object or onto an object's surface.

Understand Key Concepts

3. **Explain** how the discovery of microscopes has changed what we know about living things. SC.7.N.2.1

4. Which microscope would you use if you wanted to study the surface of an object?
   (A) compound microscope
   (B) light microscope
   (C) scanning electron microscope
   (D) transmission electron microscope

Interpret Graphics

5. **Identify** Fill in the graphic organizer below to identify four uses of microscopes. LA.6.2.2.3

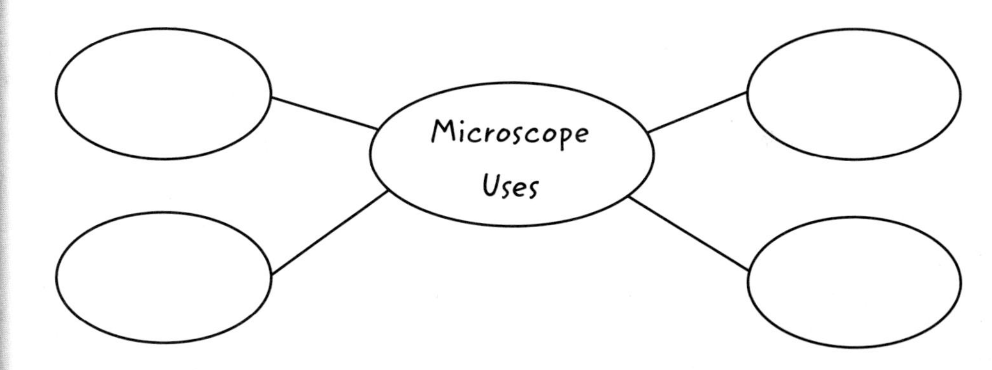

Critical Thinking

6. **Develop** a list of guidelines for choosing a microscope to use.

Math Skills MA.6.A.3.6

7. A student observes a blood sample with a compound microscope that has a 10× ocular lens and a 40× objective lens. How much larger do the blood cells appear under the microscope?

Chapter 1 Study Guide

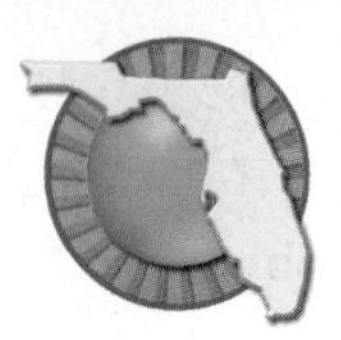

Think About It! All living things share certain characteristics and can be organized in a functional and structural hierarchy. The invention of the microscope has enabled us to explore life further, which has led to changes in classification.

Key Concepts Summary

LESSON 1 Characteristics of Life

- An **organism** is classified as a living thing because it has all the characteristics of life.
- All living things are organized, grow and develop, reproduce, respond to stimuli, maintain **homeostasis,** and use energy.

Vocabulary

organism p. 11
cell p. 12
unicellular p. 12
multicellular p. 12
homeostasis p. 15

LESSON 2 Classifying Organisms

- Living things are classified into different groups based on physical or molecular similarities.
- Some **species** are known by many different common names. To avoid confusion, every species has a scientific name based on a system called **binomial nomenclature.**

binomial nomenclature p. 23
species p. 23
genus p. 23
dichotomous key p. 24
cladogram p. 25

LESSON 3 Exploring Life

- The invention of microscopes allowed scientists to view cells, which enabled them to further explore and classify life.
- A **light microscope** uses light and has one or more lenses to enlarge an image up to about 1,500 times its original size. An **electron microscope** uses a magnetic field to direct beams of electrons, and it enlarges an image 100,000 times or more.

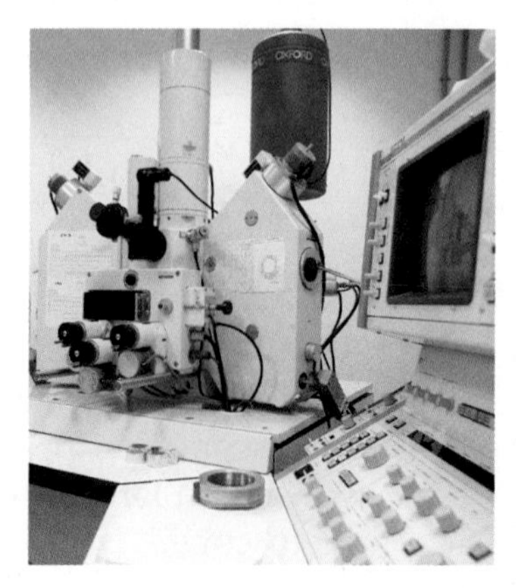

light microscope p. 30
compound microscope p. 30
electron microscope p. 31

Active Reading

FOLDABLES® Chapter Project

Assemble your lesson Foldables as shown to make a Chapter Project. Use the project to review what you have learned in this chapter.

Use Vocabulary

1. A(n) ______________ organism is made of only one cell.

2. Something with all the characteristics of life is a(n) ______________.

3. A(n) ______________ shows the relationships among species.

4. A group of similar species is a(n) ______________.

5. A(n) ______________ has a resolution up to 1,000 times greater than a light microscope.

6. A(n) ______________ is a light microscope that uses more than one lens to magnify an image.

Link Vocabulary and Key Concepts

Use vocabulary terms from the previous page to complete the concept map.

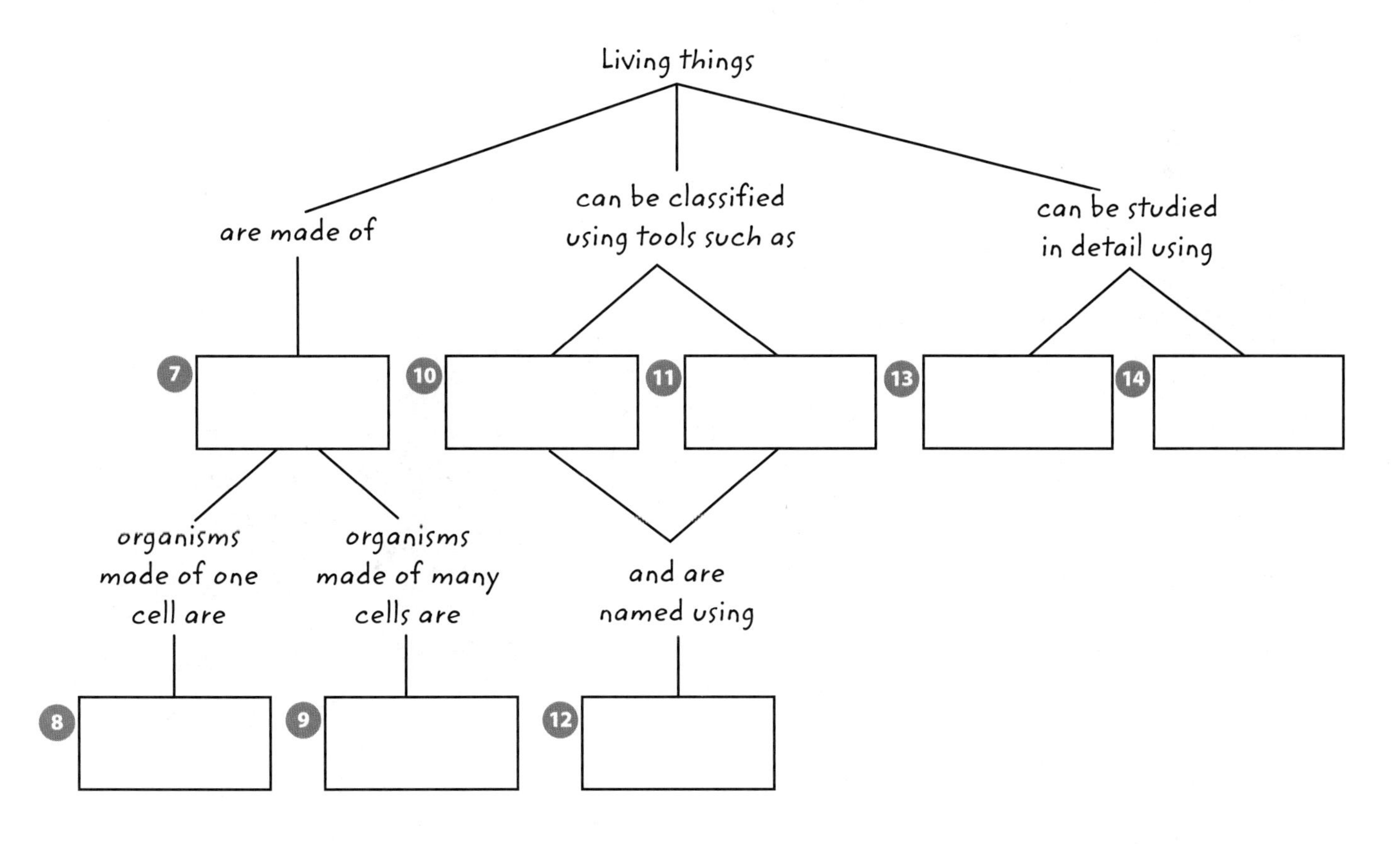

Chapter 1 Review

Fill in the correct answer choice.

Understand Key Concepts

1. Which is an internal stimulus? **SC.6.L.14.3**
   Ⓐ an increase in moisture
   Ⓑ feelings of hunger
   Ⓒ number of hours of daylight
   Ⓓ the temperature at night

2. Which is an example of growth and development? **SC.6.L.14.3**
   Ⓐ a caterpillar becoming a butterfly
   Ⓑ a chicken laying eggs
   Ⓒ a dog panting
   Ⓓ a rabbit eating carrots

3. Based on the food web below, what is an energy source for the mouse? **SC.6.L.14.3**

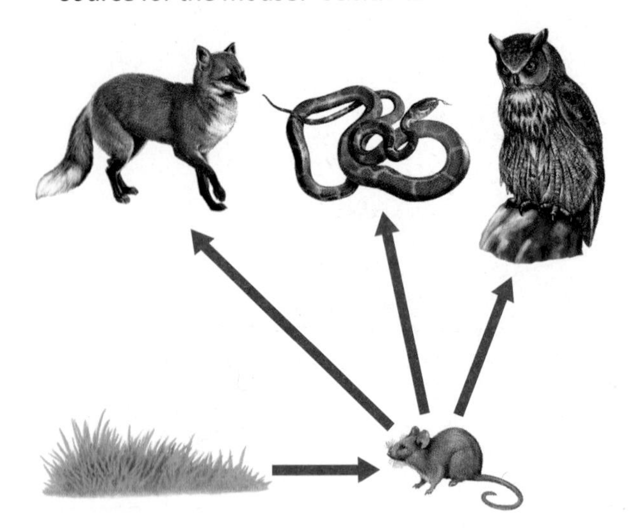

   Ⓐ fox
   Ⓑ grass
   Ⓒ owl
   Ⓓ snake

4. Which shows the correct order for the classification of species? **SC.6.L.15.1**
   Ⓐ domain, kingdom, class, order, phylum, family, genus, species
   Ⓑ domain, kingdom, phylum, class, order, family, genus, species
   Ⓒ domain, kingdom, phylum, class, order, family, species, genus
   Ⓓ domain, kingdom, phylum, order, class, family, genus, species

Critical Thinking

5. **Distinguish** between a unicellular organism and a multicellular organism. **SC.6.L.14.3**

6. **Critique** the following statement: An organism that is made of only one cell does not need organization. **SC.6.L.14.3**

7. **Infer** In the figure below, which plant is responding to a lack of water in its environment? Explain your answer. **SC.6.L.14.3**

8. **Explain** how using a dichotomous key can help you identify an organism. **SC.6.L.15.1**

9. **Describe** how the branches on a cladogram show the relationships among organisms. **SC.6.L.15.1**

10 **Assess** the effect of molecular evidence on the classification of organisms. SC.6.L.15.1

11 **Compare** light microscopes and electron microscopes.

12 **State** how microscopes have changed the way living things are classified.

13 **Compare** magnification and resolution.

14 **Evaluate** the impact microscopes have on our daily lives.

Writing in Science

15 **Write** a five-sentence paragraph on a separate sheet of paper explaining the importance of scientific names. Be sure to include a topic sentence and a concluding sentence in your paragraph. LA.6.2.2.3

Big Idea Review

16 **Define** the characteristics that all living things share. SC.6.L.14.3

Math Skills MA.6.A.3.6

Use Multiplication

17 A microscope has an ocular lens with a power of 5× and an objective lens with a power of 50×. What is the total magnification of the microscope?

18 A student observes a unicellular organism with a microscope that has a 10× ocular lens and a 100× objective lens. How much larger does the organism look through this microscope?

Florida NGSSS Benchmark Practice

Fill in the correct answer choice.

Multiple Choice

1 What feature of living things do the terms *unicellular* and *multicellular* describe?

Ⓐ how they are organized
Ⓑ how they reproduce
Ⓒ how they maintain temperature
Ⓓ how they produce macromolecules

Use the diagram below to answer question 2.

2 Which characteristic of life does the diagram show? **SC.6.L.14.3**

Ⓕ homeostasis
Ⓖ organization
Ⓗ growth and development
Ⓘ response to stimuli

3 A newly discovered organism is 1 m tall, multicellular, green, and it grows on land and performs photosynthesis. To which kingdom does it most likely belong? **SC.6.L.15.1**

Ⓐ Animalia
Ⓑ Fungi
Ⓒ Plantae
Ⓓ Protista

4 Unicellular organisms are members of which kingdoms? **SC.6.L.15.1**

Ⓕ Animalia, Archaea, Plantae
Ⓖ Archaea, Bacteria, Protista
Ⓗ Bacteria, Fungi, Plantae
Ⓘ Fungi, Plantae, Protista

5 Which are living things made of only one cell?

Ⓐ multicellular organisms
Ⓑ unicellular organisms
Ⓒ tricellular organisms
Ⓓ offspring

Use the diagram below to answer questions 6 and 7.

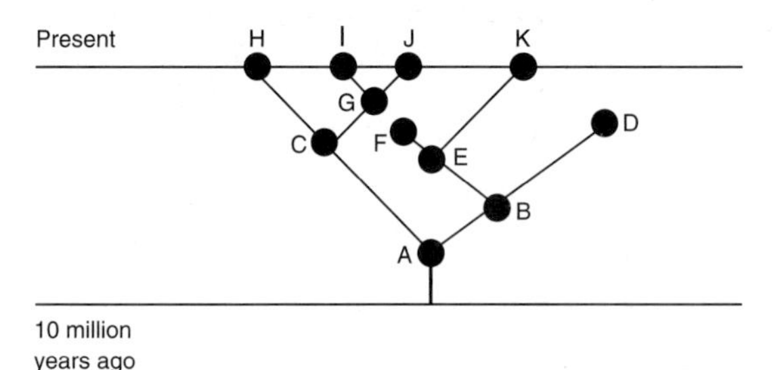

6 What type of diagram is shown in the illustration above?

Ⓕ dichotomous key
Ⓖ cladogram
Ⓗ DNA profile
Ⓘ pedigree

7 In the illustration, I and J are close together. What does this indicate? **SC.6.L.15.1**

Ⓐ They share few characteristics.
Ⓑ They are closest to the common ancestor.
Ⓒ They share many characteristics.
Ⓓ They share the same characteristics as H and K.

Use the diagram below to answer question 8.

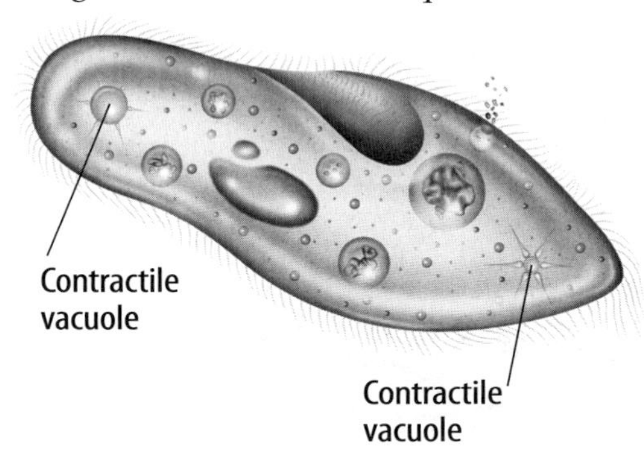

8 Which is the function of the structures in this paramecium? **SC.6.L.14.3**

Ⓕ growth

Ⓖ homeostasis

Ⓗ locomotion

Ⓘ reproduction

9 Which sequence is from the smallest group of organisms to the largest group of organisms? **SC.6.L.15.1**

Ⓐ genus → family → species

Ⓑ genus → species → family

Ⓒ species → family → genus

Ⓓ species → genus → family

10 Which information about organisms is excluded in the study of systematics? **SC.6.L.15.1**

Ⓕ calendar age

Ⓖ molecular analysis

Ⓗ energy source

Ⓘ normal habitat

11 Which of the following characteristics of life is shared by a tree growing a new branch and a duck having babies? **SC.6.L.14.3**

Ⓐ response to stimuli

Ⓑ homeostasis

Ⓒ organization

Ⓓ reproduction

12 Which statement is NOT true? **SC.6.L.15.1**

Ⓕ Binomial names are given to all unknown organisms.

Ⓖ Binomial names are less precise than common names.

Ⓗ Binomial names are different from common names.

Ⓘ Binomial names enable scientists to communicate accurately.

Use the diagram below to answer question 13.

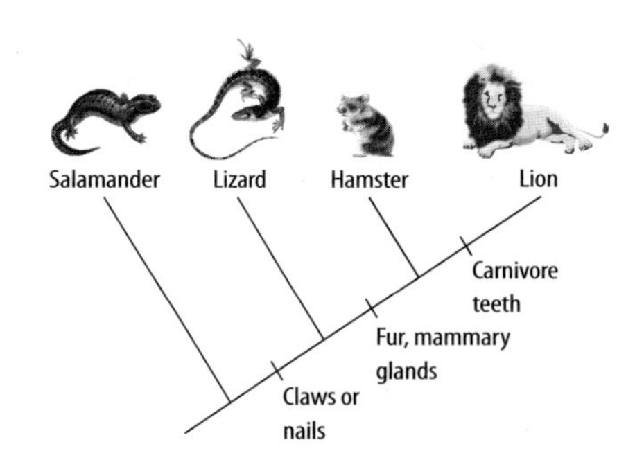

13 According to the diagram above, which of the following all have claws? **SC.6.L.15.1**

Ⓐ lizard, lion, salamander

Ⓑ hamster, salamander, lion

Ⓒ lizard, hamster, salamander

Ⓓ lizard, hamster, lion

NEED EXTRA HELP?

If You Missed Question...	1	2	3	4	5	6	7	8	9	10	11	12	13
Go to Lesson...	1	1	2	2	1	2	2	1	2	2	1	2	2

Name ______________________________ Date ______________

Benchmark Mini-Assessment Chapter 1 • Lesson 1

FLORIDA NGSSS

mini BAT

Multiple Choice *Bubble the correct answer.*

Use the image to answer questions 1–3.

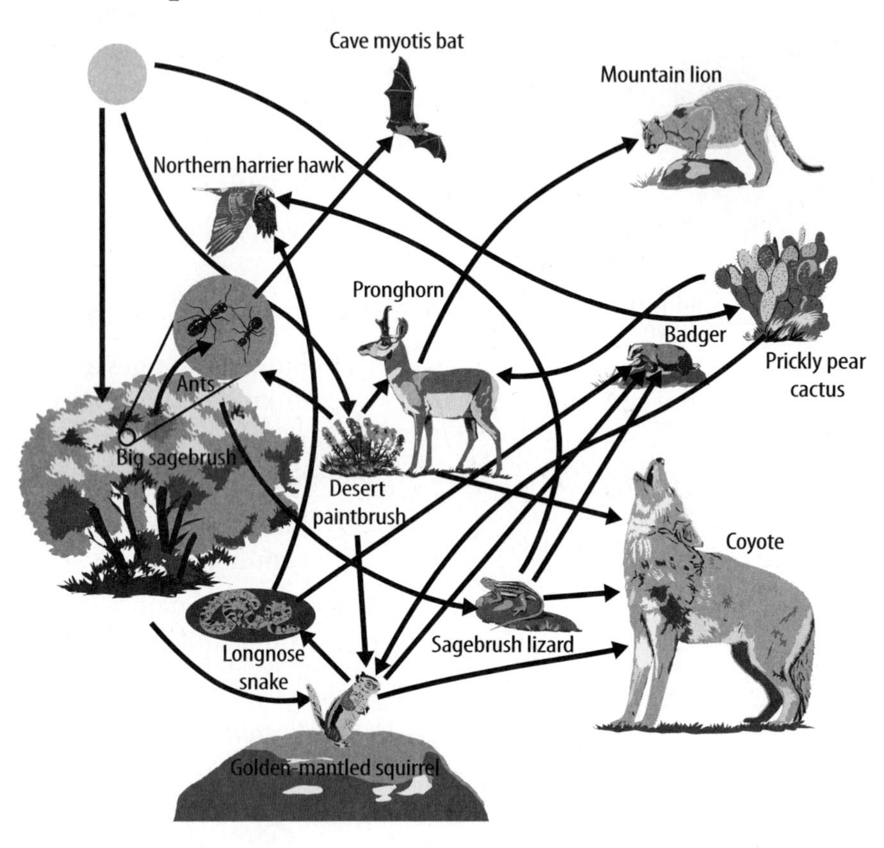

1. Which part of the food web is NOT a living thing? **SC.6.L.14.3**

Ⓐ an ant

Ⓑ a bat

Ⓒ soil bacteria

Ⓓ the Sun

2. Which characteristic of all living things is most obvious in the food web? **SC.6.L.14.3**

Ⓕ All living things reproduce.

Ⓖ All living things use energy.

Ⓗ All living things grow and develop.

Ⓘ All living things have homeostasis.

3. Which is a source of energy for a coyote? **SC.6.L.14.3**

Ⓐ an ant

Ⓑ a cactus

Ⓒ a pronghorn

Ⓓ the Sun

Name ____________________ Date ____________

Benchmark Mini-Assessment Chapter 1 • Lesson 2

Multiple Choice *Bubble the correct answer.*

1. The first question on a dichotomous key is: "1A. Are wings covered by an exoskeleton? vs. 1B. Are wings not covered by an exoskeleton?" Which organism is separated from the others by this question? **SC.6.L.15.1**

(A) 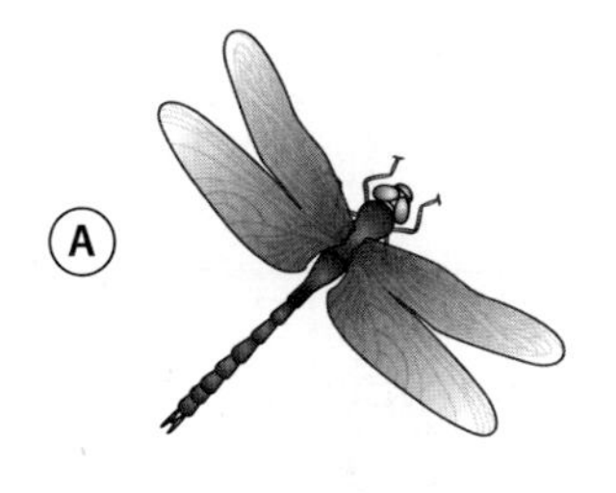

(B) 

(C) 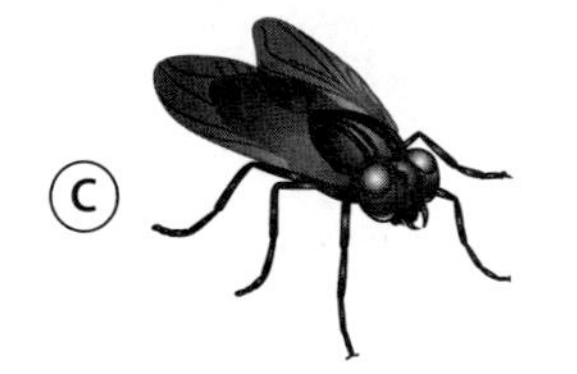

(D) 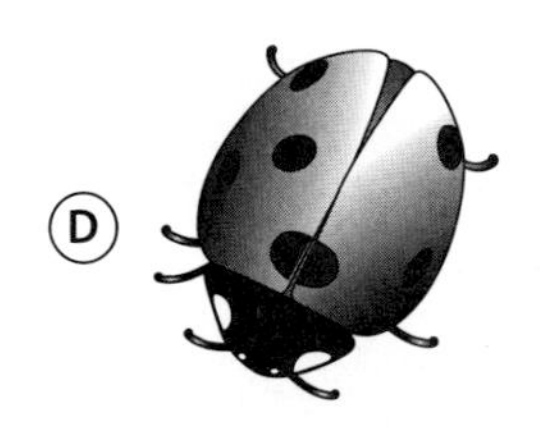

2. In which series is the order of classification correct? **SC.6.L.15.1**

(F) domain, class, kingdom, species

(G) domain, kingdom, class, species

(H) domain, kingdom, species, class

(I) domain, species, kingdom, class

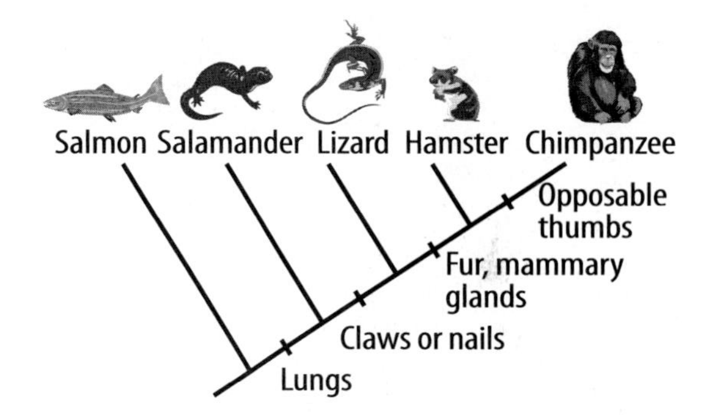

3. In the cladogram above, which pair of species is most closely related to each other? **SC.6.L.15.1**

(A) hamster and salmon

(B) lizard and salamander

(C) salamander and chimpanzee

(D) salmon and lizard

Name ______________________ Date ______________

Multiple Choice *Bubble the correct answer.*

1. Which tool would be used to examine the workings of a living unicellular organism? **SC.6.L.15.1**

Ⓐ 

Ⓑ 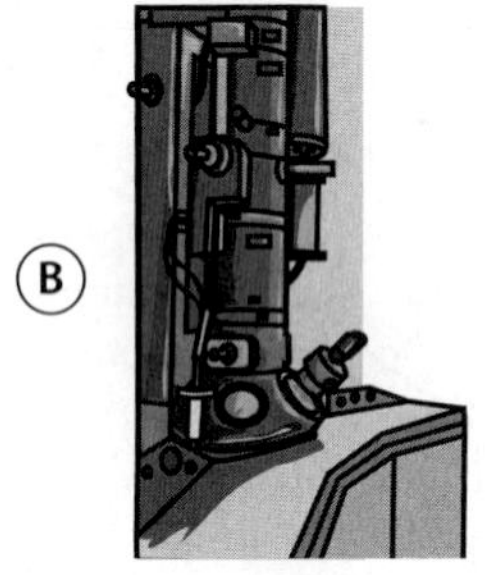

Ⓒ 

Ⓓ 

2. Which magnifying device provides the most detail in its images? **SC.7.N.1.1**

Ⓕ compound magnifying device

Ⓖ electron magnifying device

Ⓗ hand-lens magnifying device

Ⓘ single-lens magnifying device

Microscope	Magnification	Resolution
Leeuwenhoek's invention	270X	2 micrometers
Modern light microscope	1500X	0.2 micrometers
Electron microscope	100,000+X	0.2 nanometers

3. The table above shows the different levels of magnification provided by different microscopes. About how much larger is the image from an electron microscope than an image from the first microscope invented by Anton van Leeuwenhoek? **MA.6.A.3.6**

Ⓐ about 50 times

Ⓑ about 100 times

Ⓒ about 200 times

Ⓓ about 500 times

Name ______________________ Date ______________

Notes

Name ____________ Date ____________

Notes

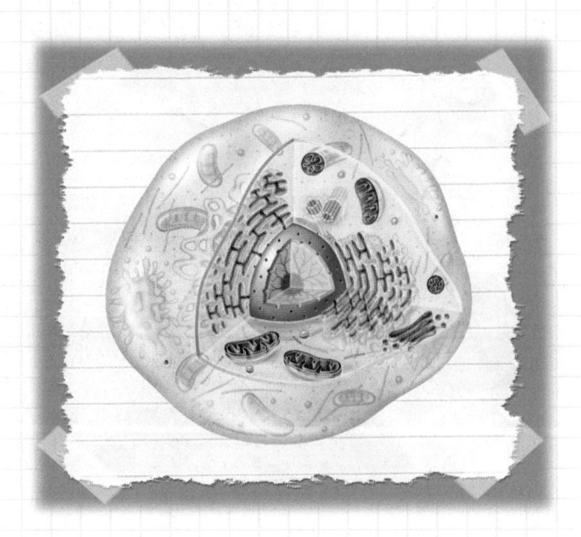

The Basic Unit of Life

The cell is called the basic unit of life. What do you think that means? Circle the answer that best matches your thinking.

A. I think it means the cell is the smallest part of matter.

B. I think it means the cell is the smallest part of mass.

C. I think it means the cell is the smallest part of volume.

D. I think it means the cell is the smallest part of mass and volume.

E. I think it means the cell is the smallest part of energy.

F. I think it means the cell is the smallest part of structure.

G. I think it means the cell is the smallest part of structure and function.

H. I think it means the cell is the smallest part of matter, structure, and function

I. I think it means the cell is the smallest part of matter, energy, and structure.

Explain your answer. Describe your thinking about the cell as a basic unit of life.

Cell Structure and FUNCTION

FLORIDA BIG IDEAS

1 **The Practice of Science**
2 **The Characteristics of Scientific Knowledge**
3 **The Role of Theories, Laws, Hypotheses, and Models**
14 **Organization and Development of Living Organisms**
18 **Matter and Energy Transformations**

The Big Idea

Think About It!

How do the structures and processes of a cell enable it to survive?

You might think this unicellular organism looks like something out of a science-fiction movie. Although it looks scary, the hairlike structures in its mouth enable the organism to survive.

1. What do you think the hairlike structures do?

2. How might the shape of the hairlike structures relate to their function?

3. How do you think the structures and processes of a cell enable it to survive?

Get Ready to Read

What do you think about cells?

Before you read, decide if you agree or disagree with each of these statements. As you read this chapter, see if you change your mind about any of the statements.

Statement	AGREE	DISAGREE
1 Nonliving things have cells.	☐	☐
2 Cells are made mostly of water.	☐	☐
3 Different organisms have cells with different structures.	☐	☐
4 All cells store genetic information in their nuclei.	☐	☐
5 Diffusion and osmosis are the same process.	☐	☐
6 Cells with large surface areas can transport more than cells with smaller surface areas.	☐	☐
7 ATP is the only form of energy found in cells.	☐	☐
8 Cellular respiration occurs only in lung cells.	☐	☐

There's More Online!
Video • Audio • Review • ⓘLab Station • WebQuest • Assessment • Concepts in Motion • Multilingual eGlossary

Lesson 1

CELLS AND LIFE

ESSENTIAL QUESTIONS

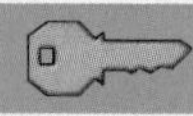 How did scientists' understanding of cells develop?

 What basic substances make up a cell?

Vocabulary

cell theory p. 50
macromolecule p. 51
nucleic acid p. 52
protein p. 53
lipid p. 53
carbohydrate p. 53

Florida NGSSS

LA.6.2.2.3 The student will organize information to show understanding (e.g., representing main ideas within text through charting, mapping, paraphrasing, summarizing, or comparing/contrasting);

SC.6.L.14.2 Investigate and explain the components of the scientific theory of cells (cell theory): all organisms are composed of cells (single-celled or multi-cellular), all cells come from pre-existing cells, and cells are the basic unit of life.

SC.6.N.1.4 Discuss, compare, and negotiate methods used, results obtained, and explanations among groups of students conducting the same investigation.

SC.6.N.2.2 Explain that scientific knowledge is durable because it is open to change as new evidence or interpretations are encountered.

SC.6.N.2.3 Recognize that scientists who make contributions to scientific knowledge come from all kinds of backgrounds and possess varied talents, interests, and goals.

SC.6.N.3.1 Recognize and explain that a scientific theory is a well-supported and widely accepted explanation of nature and is not simply a claim posed by an individual. Thus, the use of the term theory in science is very different than how it is used in everyday life.

SC.7.N.2.1 Identify an instance from the history of science in which scientific knowledge has changed when new evidence or new interpretations are encountered.

SC.7.N.3.1 Recognize and explain the difference between theories and laws and give several examples of scientific theories and the evidence that supports them.

Inquiry Launch Lab

SC.6.L.14.2
SC.7.N.1.1

10 minutes

What's in a cell?

Most plants grow from seeds. A seed began as one cell, but a mature plant can be made up of millions of cells. How does a seed change and grow into a mature plant?

Procedure

1. Read and complete a lab safety form.
2. Use a **toothpick** to gently remove the thin outer covering of a **bean seed** that has soaked overnight.
3. Open the seed with a **plastic knife,** and observe its inside with a **magnifying lens.** Draw the inside of the seed in the space below.
4. Gently remove the small, plantlike embryo, and weigh it on a **balance.** Record its mass below.
5. Gently pull a **bean seedling** from the soil. Rinse the soil from the roots. Weigh the seedling, and record the mass.

Data and Observations

Think About This

1. How did the mass of the embryo and the bean seedling differ?

2. Key Concept If a plant begins as one cell, where do all the cells come from?

Inquiry Two of a Kind?

1. At first glance, the plant and animal in the photo might seem like they have nothing in common. The plant is rooted in the ground, and the iguana can move quickly. Are they more alike than they appear? How can you find out?

____

____

____

____

____

Understanding Cells

Have you ever looked up at the night sky and tried to find other planets in our solar system? It is hard to see them without using a telescope. This is because the other planets are millions of kilometers away. Just like we can use telescopes to see other planets, we can use microscopes to see the basic units of all living things—cells. But people didn't always know about cells. Because cells are so small, early scientists had no tools to study them. It took hundreds of years for scientists to learn about cells.

More than 300 years ago, an English scientist named Robert Hooke built a microscope. He used the microscope to look at cork, which is part of a cork oak tree's bark. What he saw looked like the openings in a honeycomb, as shown in **Figure 1.** The openings reminded him of the small rooms, called cells, where monks lived. He called the structures cells, from the Latin word *cellula* (SEL yuh luh), which means "small rooms."

Figure 1 To Robert Hooke, the cells of cork looked like the openings in a honeycomb.

The Cell Theory

After Hooke's discovery, other scientists began making better microscopes and looking for cells in many other places, such as pond water and blood. The newer microscopes enabled scientists to see different structures inside cells. Matthias Schleiden (SHLI dun), a German scientist, used one of the new microscopes to look at plant cells. Around the same time, another German scientist, Theodor Schwann, used a microscope to study animal cells. Schleiden and Schwann realized that plant and animal cells have similar features. You'll read about many of these features in Lesson 2.

Almost two decades later, Rudolf Virchow (VUR koh), a German doctor, proposed that all cells come from preexisting cells, or cells that already exist. The observations made by Schleiden, Schwann, and Virchow were combined into one **theory.** As illustrated in **Table 1,** *the* **cell theory** *states that all living things are made of one or more cells, the cell is the smallest unit of life, and all new cells come from preexisting cells.* After the development of the cell theory, scientists raised more questions about cells. If all living things are made of cells, what are cells made of?

REVIEW VOCABULARY

theory
explanation of things or events based on scientific knowledge resulting from many observations and experiments

2. NGSSS Check Explain How did scientists' understanding of cells develop? SC.6.L.14.2

____

____

____

____

Table 1 Scientists developed the cell theory after studying cells with microscopes.

Active Reading 3. Summarize Review the three principles of the cell theory by completing the table below.

Table 1 The Cell Theory

Principle	Example
All living things are made of ______________ cells.	Leaf cells
The cell is the ______________ unit of life.	This unicellular amoeba is surrounding an algal cell to get food and energy. Algal cell; Amoeba
All new cells come from ______________ cells.	Existing cell → Cell dividing → New cells

Basic Cell Substances

Have you ever watched a train travel down a railroad track? The locomotive pulls train cars that are hooked together. Like a train, many of the substances in cells are made of smaller parts that are joined together. *These substances, called* **macromolecules,** *form by joining many small molecules together.* As you will read later in this lesson, macromolecules have many important roles in cells. But macromolecules cannot function without one of the most important substances in cells—water.

Word Origin
macromolecule
from Greek *makro–,* means "long"; and Latin *molecula,* means "mass"

The Main Ingredient—Water

The main ingredient in any cell is water. It makes up more than 70 percent of a cell's volume and is essential for life. Why is water such an important molecule? In addition to making up a large part of the inside of cells, water also surrounds cells. The water surrounding your cells helps to insulate your body, which maintains homeostasis, or a stable internal environment.

The structure of a water molecule makes it ideal for dissolving many other substances. Substances must be in a liquid to move into and out of cells. A water molecule has two areas:

- An area that is more negative (–), called the negative end; this end can attract the positive part of another substance.
- An area that is more positive (+), called the positive end; this end can attract the negative part of another substance.

Examine **Figure 2** to see how the positive and negative ends of water molecules dissolve salt crystals.

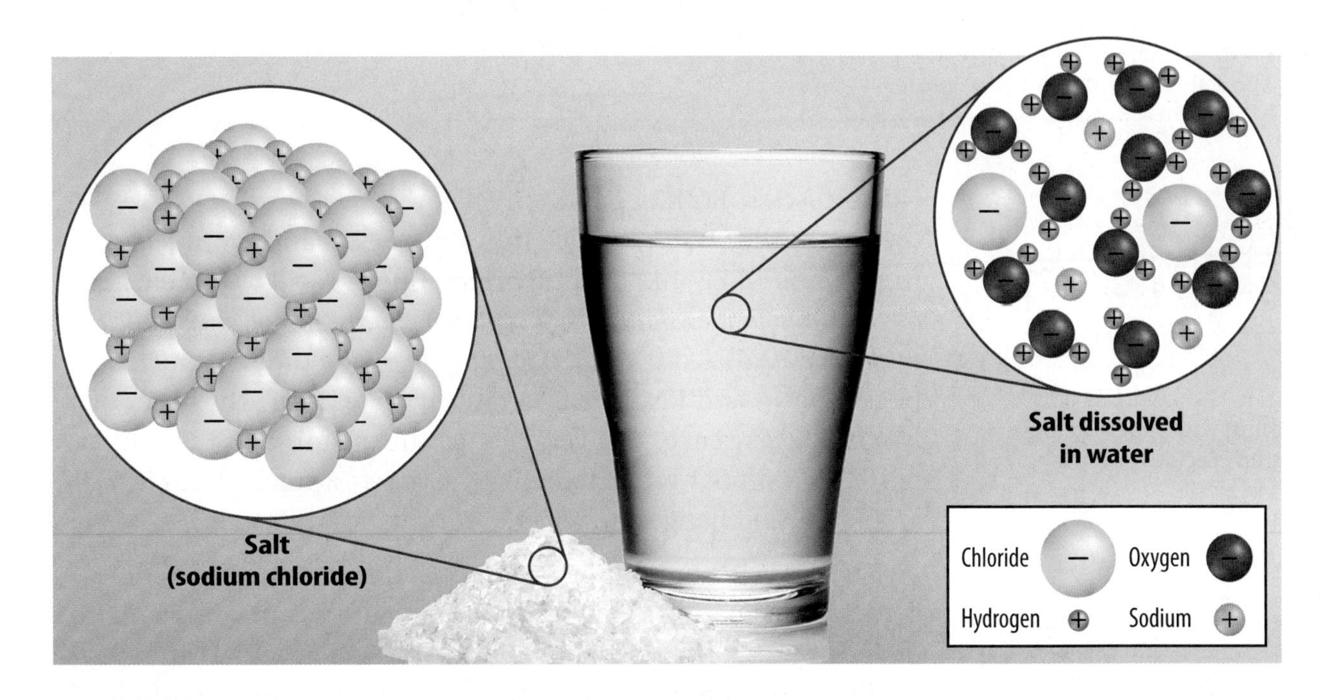

Figure 2 The positive and negative ends of a water molecule attract the positive and negative parts of another substance, similar to the way magnets are attracted to each other.

Active Reading **4. Identify** Circle the part of the salt crystal that is attracted to the oxygen in the water molecule.

Macromolecules

Although water is essential for life, all cells contain other substances that enable them to function. Recall that macromolecules are large molecules that form when smaller molecules join together. As shown in **Figure 3,** there are four types of macromolecules in cells: nucleic acids, proteins, lipids, and carbohydrates. Each type of macromolecule has unique functions in a cell. These functions range from growth and communication to movement and storage.

Active Reading **5. Identify** In **Figure 3,** (circle) the names of the macromolecules that provide support to a cell.

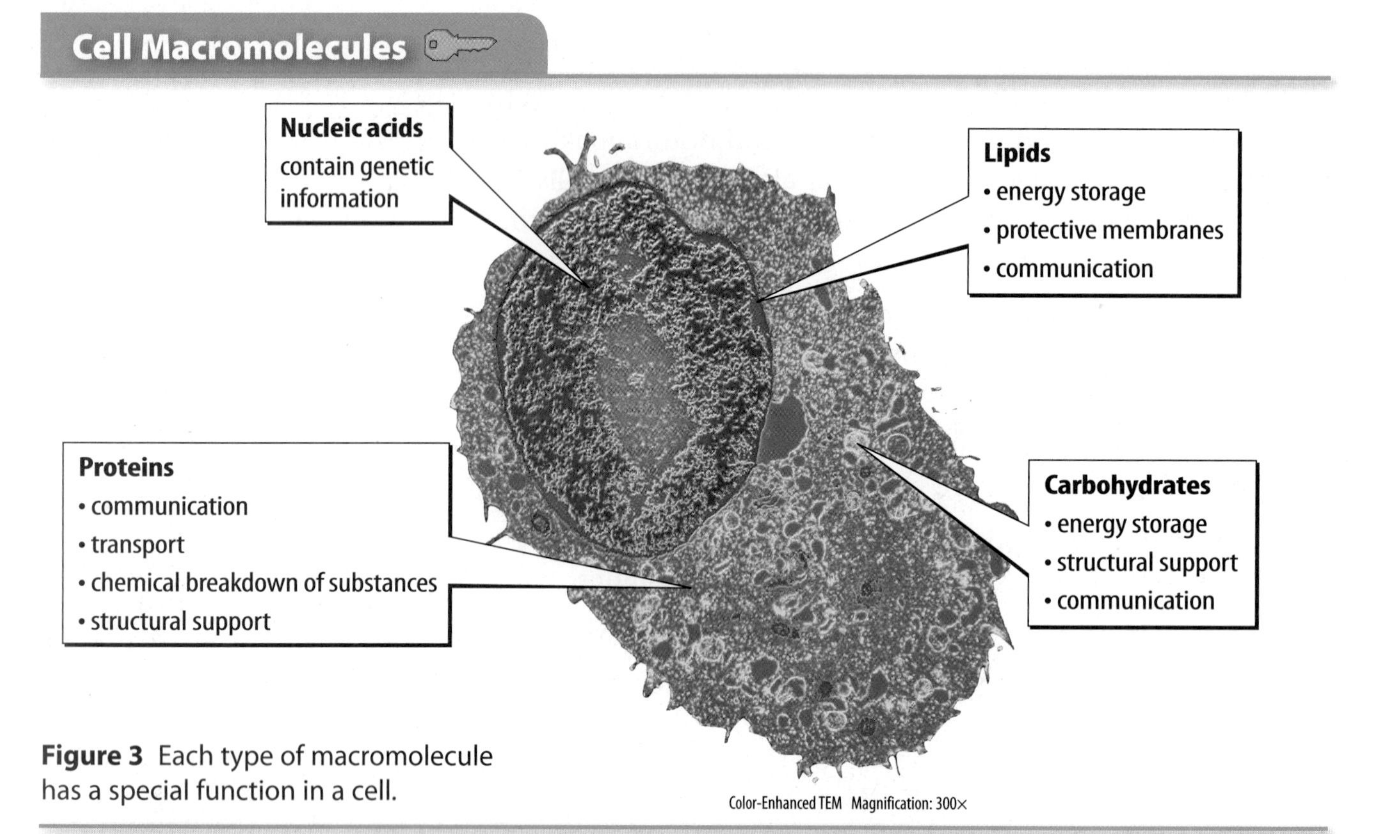

Figure 3 Each type of macromolecule has a special function in a cell.

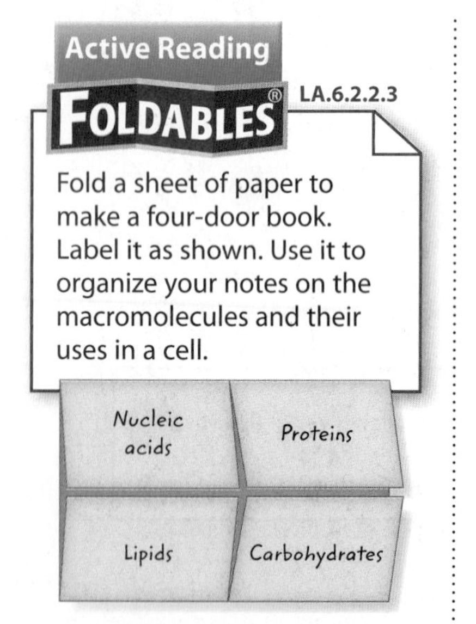

Nucleic Acids Both deoxyribonucleic (dee AHK sih ri boh noo klee ihk) acid (DNA) and ribonucleic (ri boh noo KLEE ihk) acid (RNA) are nucleic acids. **Nucleic acids** *are macromolecules that form when long chains of molecules called nucleotides (NEW klee uh tidz) join together.* The order of nucleotides in DNA and RNA is important. If you change the order of words in a sentence, you can change the meaning of the sentence. In a similar way, changing the order of nucleotides in DNA and RNA can change the genetic information in a cell.

Nucleic acids are important in cells because they contain genetic information. This information can pass from parents to offspring. DNA includes instructions for cell growth, cell reproduction, and cell processes that enable a cell to respond to its environment. DNA is used to make RNA. RNA is used to make proteins.

Proteins The macromolecules necessary for nearly everything cells do are proteins. **Proteins** *are long chains of amino acid molecules.* You just read that RNA is used to make proteins. RNA contains instructions for joining amino acids together.

Cells contain hundreds of proteins. Each protein has a unique function. Some proteins help cells communicate with each other. Other proteins transport substances around inside cells. Some proteins, such as amylase (AM uh lays) in saliva, help break down nutrients in food. Other proteins, such as keratin (KER uh tun)—a protein found in hair, horns, and feathers—provide structural support.

Active Reading **7. Identify** Highlight four functions of proteins.

Lipids Another group of macromolecules found in cells is lipids. *A* **lipid** *is a large macromolecule that does not dissolve in water.* Because lipids do not mix with water, they play an important role as protective barriers in cells. They are also the major part of cell membranes. Lipids play roles in energy storage and in cell communication. Examples of lipids are cholesterol (kuh LES tuh rawl), phospholipids (fahs foh LIH pids), and vitamin A.

Active Reading **6. Determine** Underline why lipids are important to cells.

Carbohydrates *One sugar molecule, two sugar molecules, or a long chain of sugar molecules make up* **carbohydrates** (kar boh HI drayts). Carbohydrates store energy, provide structural support, and are needed for communication between cells. Sugars and starches are carbohydrates that store energy. Fruits contain sugars. Breads and pastas are mostly starch. The energy in sugars and starches can be released quickly through chemical reactions in cells. Cellulose is a carbohydrate in the cell walls in plants that provides structural support.

Active Reading **8. Recall** What basic substances make up a cell?

____

____

____

____

Apply It! After you complete the lab, answer the questions below.

1. Why do you think DNA was not visible in all of the cells you observed?

2. What other macromolecules do you think you could observe in the onion root-tip cells? Explain.

Lesson Review 1

Visual Summary

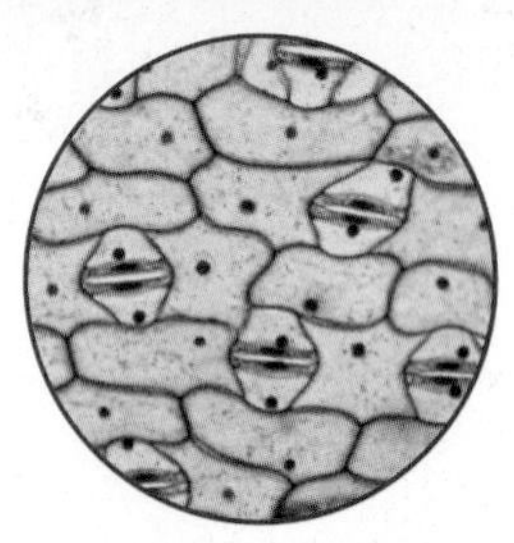

The cell theory summarizes the main principles for understanding that the cell is the basic unit of life.

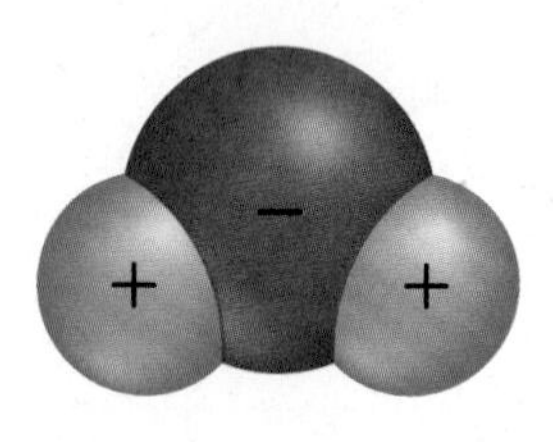

Water is the main ingredient in every cell.

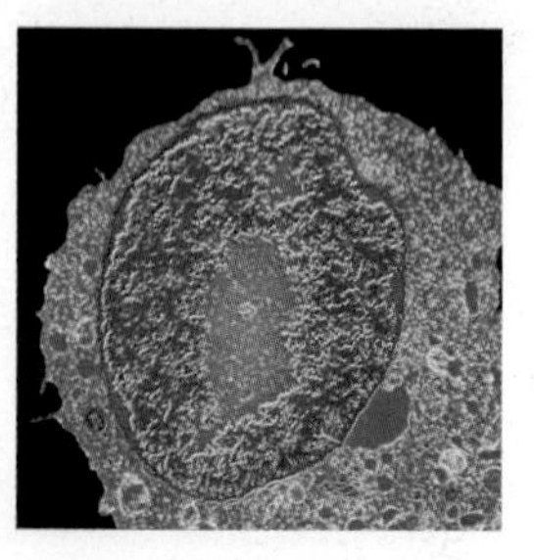

A nucleic acid, such as DNA, contains the genetic information for a cell.

Use Vocabulary

1. The ______________ states that the cell is the basic unit of all living things. SC.6.L.14.2

2. **Use the term** *nucleic acid* in a sentence.

Understand Key Concepts

3. Which macromolecule is made from amino acids?
   - (A) lipid
   - (B) protein
   - (C) carbohydrate
   - (D) nucleic acid

4. **Describe** how the invention of the microscope helped scientists understand cells.

5. **Compare** the functions of DNA and proteins in a cell.

Interpret Graphics

6. **Summarize** Fill in the graphic organizer below to summarize the main principles of the cell theory. SC.6.L.14.2

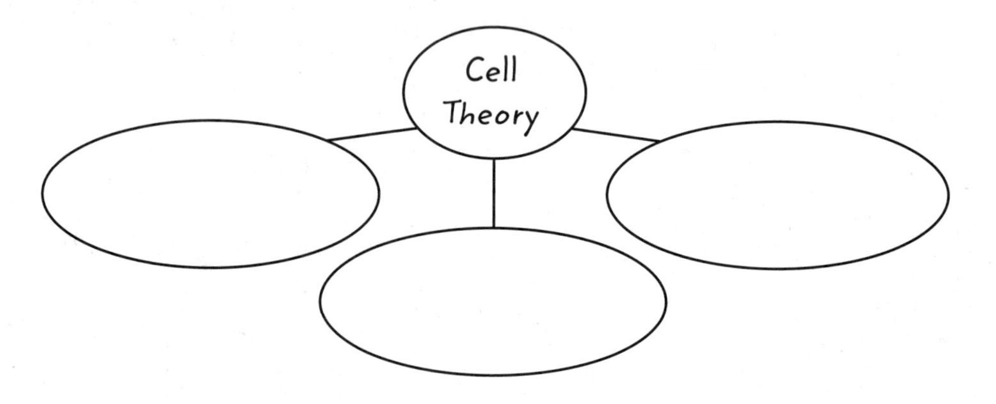

Critical Thinking

7. **Summarize** the functions of lipids in cells.

8. **Hypothesize** why carbohydrates are found in plant cell walls.

HOW IT WORKS

A Very Powerful Microscope

Using technology to look inside cells

If Robert Hooke had used an atomic force microscope (AFM), he would have observed more than just cells. He would have seen the macromolecules inside them! An AFM can scan objects that are only nanometers in size. A nanometer is one one-billionth of a meter. That's 100,000 times smaller than the width of a human hair. AFM technology has enabled scientists to better understand how cells function. It also has given them a three-dimensional look at the macromolecules that make life possible. This is how it works.

Photodiode

1 A probe moves across a sample's surface to identify the sample's features. The probe consists of a cantilever with a tiny, sharp tip. The tip is about 20 nm in diameter at its base.

2 The cantilever can bend up and down, similar to the way a diving board can bend, in response to pushing and pulling forces between the atoms in the tip and the atoms in the sample.

3 A laser beam senses the cantilever's up and down movements. A computer converts these movements into an image of the sample's surface.

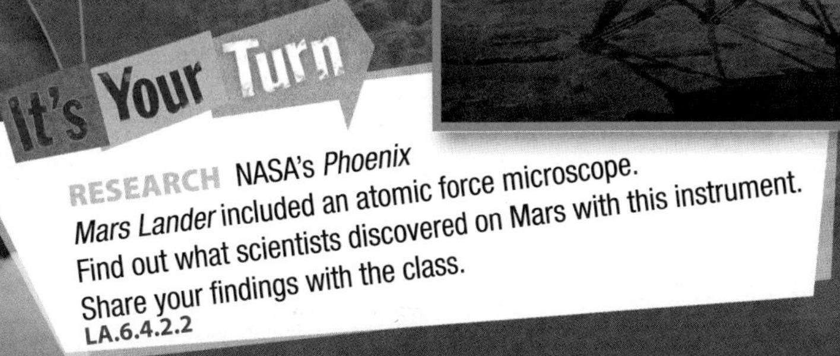

It's Your Turn

RESEARCH NASA's *Phoenix Mars Lander* included an atomic force microscope. Find out what scientists discovered on Mars with this instrument. Share your findings with the class.
LA.6.4.2.2

Lesson 2

THE CELL

ESSENTIAL QUESTIONS

How are prokaryotic cells and eukaryotic cells similar, and how are they different?

What do the structures in a cell do?

Vocabulary

cell membrane p. 58

cell wall p. 58

cytoplasm p. 59

cytoskeleton p. 59

organelle p. 60

nucleus p. 61

chloroplast p. 63

Florida NGSSS

SC.6.L.14.4 Compare and contrast the structure and function of major organelles of plant and animal cells, including cell wall, cell membrane, nucleus, cytoplasm, chloroplasts, mitochondria, and vacuoles.

SC.6.N.1.1 Define a problem from the sixth grade curriculum, use appropriate reference materials to support scientific understanding, plan and carry out scientific investigation of various types, such as systematic observations or experiments, identify variables, collect and organize data, interpret data in charts, tables, and graphics, analyze information, make predictions, and defend conclusions.

SC.6.N.1.5 Recognize that science involves creativity, not just in designing experiments, but also in creating explanations that fit evidence.

SC.7.N.1.3 Distinguish between an experiment (which must involve the identification and control of variables) and other forms of scientific investigation and explain that not all scientific knowledge is derived from experimentation.

SC.7.N.2.1 Identify an instance from the history of science in which scientific knowledge has changed when new evidence or new interpretations are encountered.

Also covers: LA.6.2.2.3, LA.6.4.2.2

SC.6.N.1.5

Inquiry Launch Lab

10 minutes

Why do eggs have shells?

Bird eggs have different structures, such as a shell, a membrane, and a yolk. Each structure has a different function that helps keep the egg safe and assists in development of the baby bird inside of it.

Procedure

1. Read and complete a lab safety form.
2. Place an **uncooked egg** in a bowl.
3. Feel the shell, and record your observations below.
4. Crack open the egg. Pour the contents into the bowl.
5. Observe the inside of the shell and the contents of the bowl. Record your observations.

Data and Observations

Think About This

1. What do you think is the role of the eggshell?

2. Are there any structures in the bowl that have the same function as the eggshell? Explain.

3. Key Concept What does the structure of the eggshell tell you about its function?

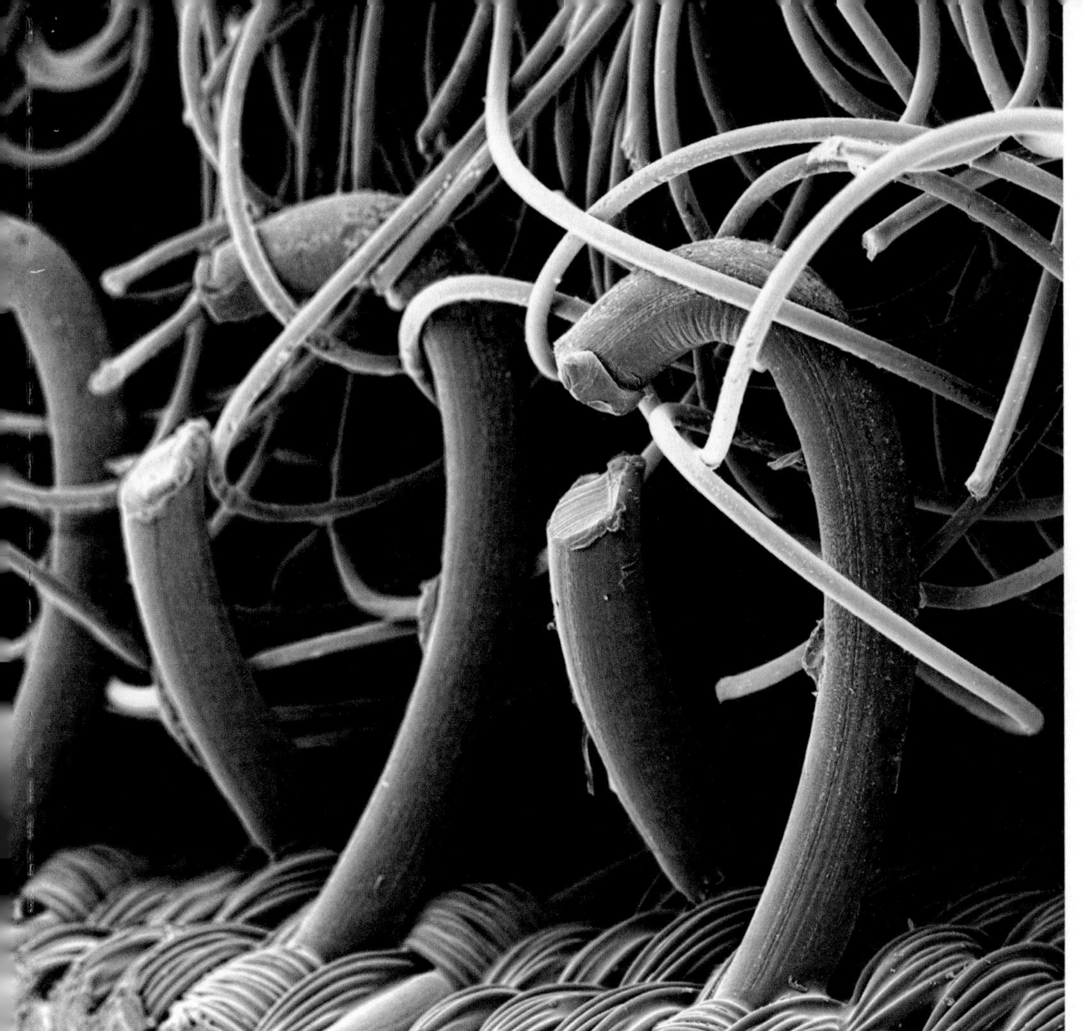

Hooked Together?

1. What happens when one of the hooks in the photo at left goes through one of the loops? The two sides fasten together. The shapes of the hooks and loops in the hook-and-loop tape are suited to their function—to hold the two pieces together. How do you think the shape of a cell also might be suited to its function?

Cell Shape and Movement

You might recall from Lesson 1 that all living things are made up of one or more cells. As illustrated in **Figure 4,** cells come in many shapes and sizes. The size and shape of a cell relates to its job or function. For example, a human red blood cell cannot be seen without a microscope. Its small size and disk shape enable it to pass easily through the smallest blood vessels. The shape of a nerve cell enables it to send signals over long distances. Some plant cells are hollow and make up tubelike structures that carry materials throughout a plant.

The structures that make up a cell also have unique functions. Think about how the players on a football team perform different tasks to move the ball down the field. In a similar way, a cell is made of different structures that perform different functions that keep a cell alive. You will read about some of these structures in this lesson.

Figure 4 The shape of a cell relates to the function it performs.

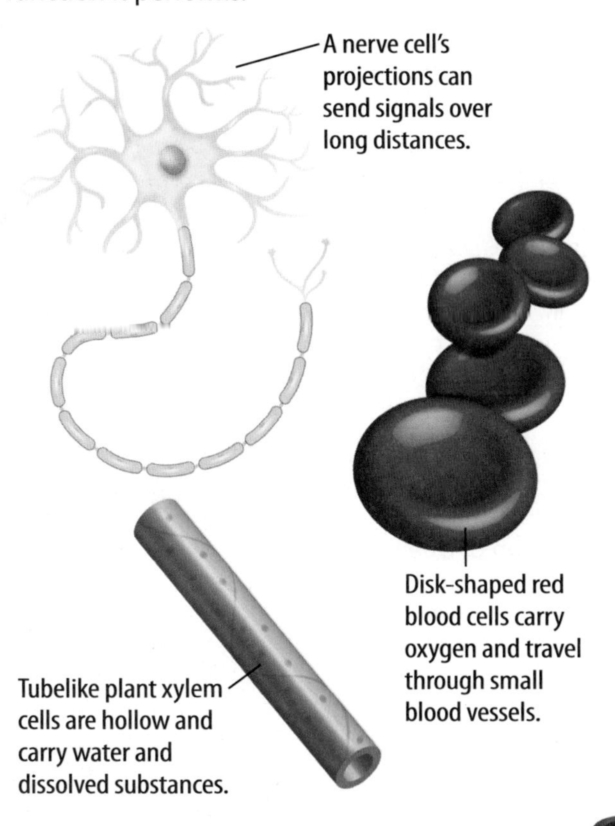

Plant Cell

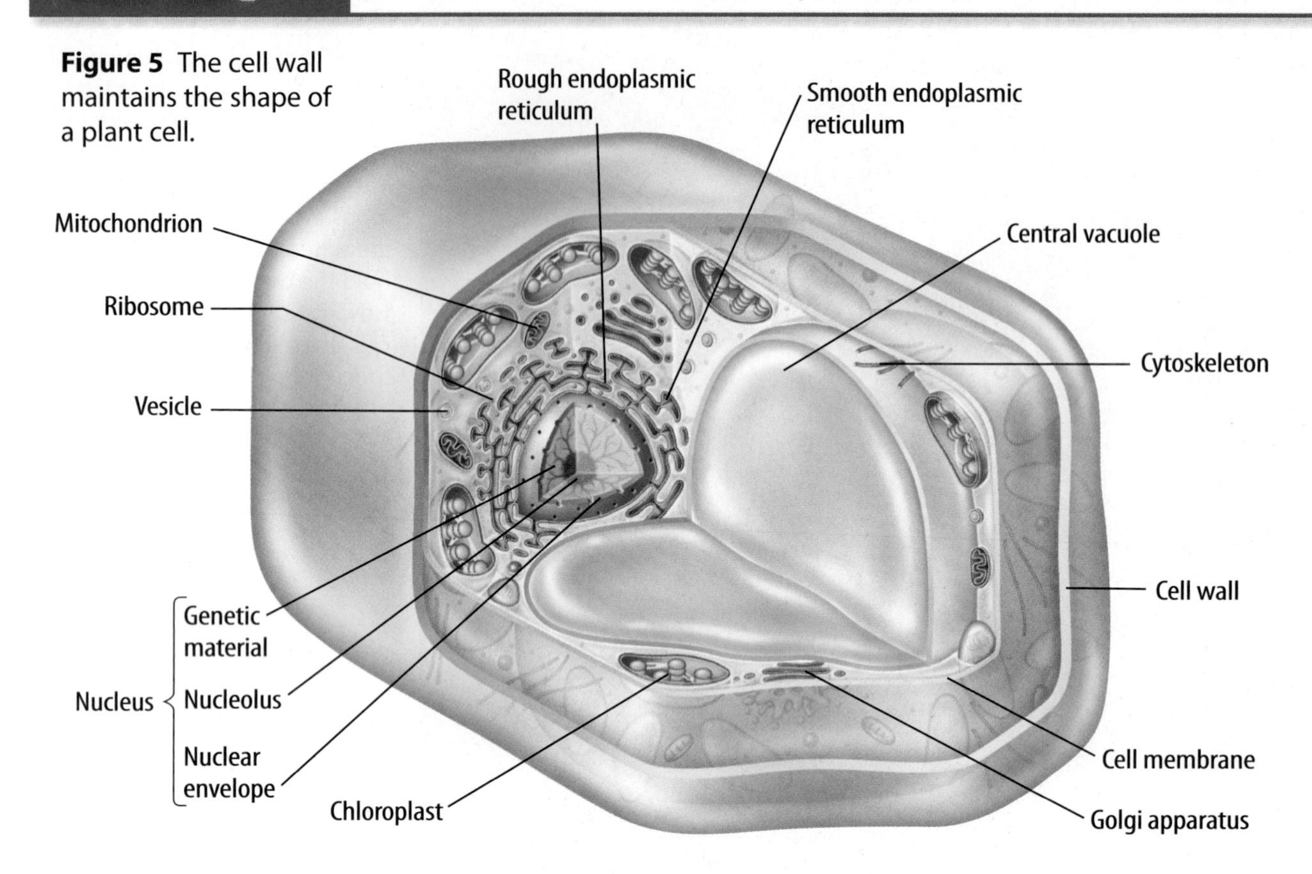

Figure 5 The cell wall maintains the shape of a plant cell.

Cell Membrane

Although different types of cells perform different **functions,** all cells have some structures in common. As shown in **Figure 5** and **Figure 6,** every cell is surrounded by a protective covering called a membrane. *The* **cell membrane** *is a flexible covering that protects the inside of a cell from the environment outside a cell.* Cell membranes are mostly made of two different macromolecules—proteins and a type of lipid called phospholipids. Think again about a football team. The defensive line tries to stop the other team from moving forward with the football. In a similar way, a cell membrane protects the cell from the outside environment.

ACADEMIC VOCABULARY

function

(noun) the purpose for which something is used

Active Reading **2. Identify** Underline the two main macromolecules that make up cell membranes.

Cell Wall

Every cell has a cell membrane, but some cells are also surrounded by a structure called the cell wall. Plant cells such as the one in **Figure 5,** fungal cells, bacteria, and some types of protists have cell walls. *A* **cell wall** *is a stiff structure outside the cell membrane.* A cell wall protects a cell from attack by viruses and other harmful organisms. In some plant cells and fungal cells, a cell wall helps maintain the cell's shape and gives structural support.

Active Reading **3. Define** Describe the structure and function of a cell wall.

Animal Cell

Figure 6 The cytoskeleton maintains the shape of an animal cell.

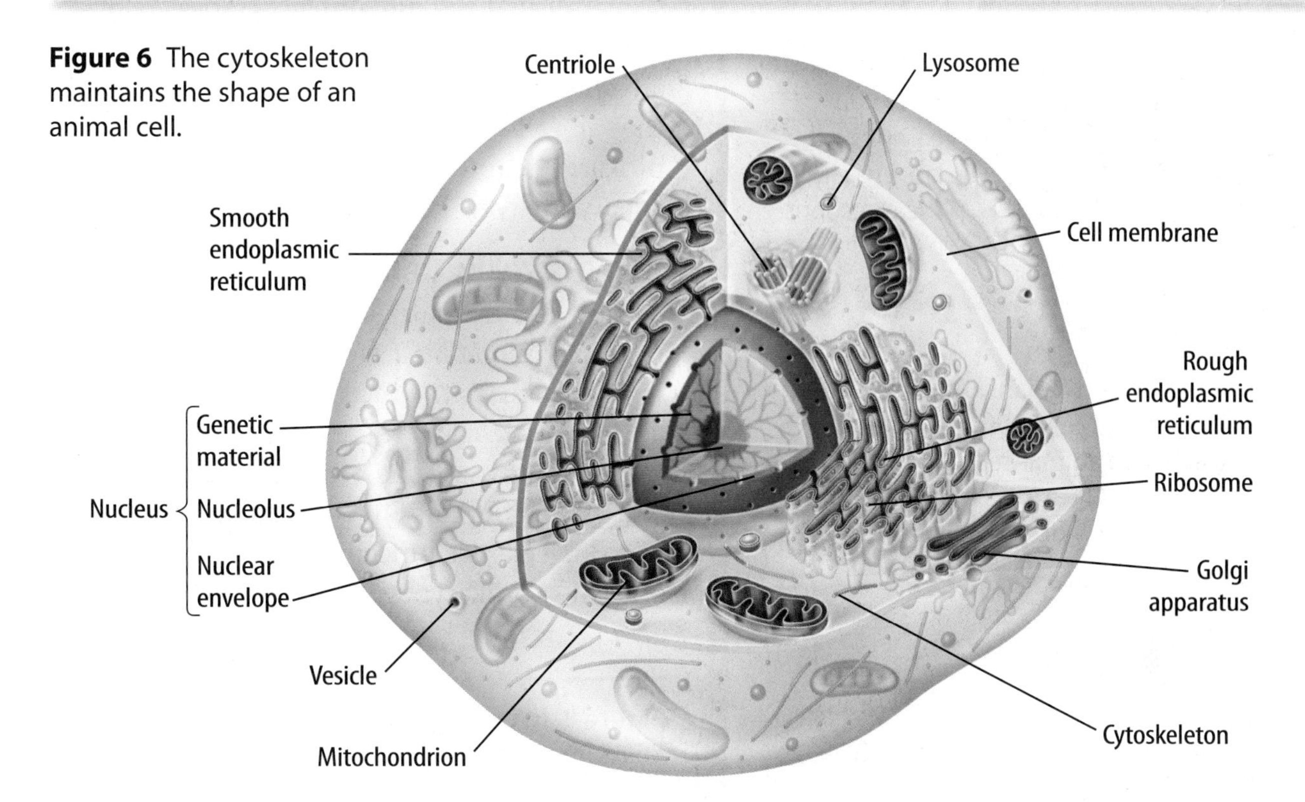

Active Reading **4. Identify** (Circle) the names of two parts in the plant cell in **Figure 5** that are not in the animal cell in **Figure 6.**

Cell Appendages

Arms, legs, claws, and antennae are all types of appendages. Cells can have appendages too. Cell appendages are often used for movement. Flagella (fluh JEH luh; singular, flagellum) are long, tail-like appendages that whip back and forth and move a cell. A cell can also have cilia (SIH lee uh; singular, cilium) like the ones shown in **Figure 7.** Cilia are short, hairlike structures. They can move a cell or move molecules away from a cell. A microscopic organism called a paramecium (pa ruh MEE shee um) moves around its watery environment using its cilia. The cilia in your windpipe move harmful substances away from your lungs.

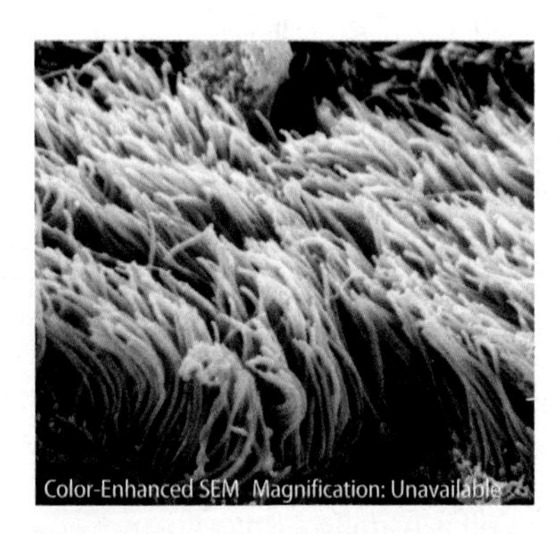

Figure 7 Lung cells have cilia that help move fluids and foreign materials.

Cytoplasm and the Cytoskeleton

In Lesson 1, you read that water is the main ingredient in a cell. Most of this water is in the **cytoplasm,** *a fluid inside a cell that contains salts and other molecules.* The cytoplasm also contains a cell's cytoskeleton. *The* **cytoskeleton** *is a network of threadlike proteins that are joined together.* The proteins form a framework inside a cell. This framework gives a cell its shape and helps it move. Cilia and flagella are made from the same proteins that make up the cytoskeleton.

Word Origin
cytoplasm
from Greek *kytos,* means "hollow vessel"; and *plasma,* means "something molded"

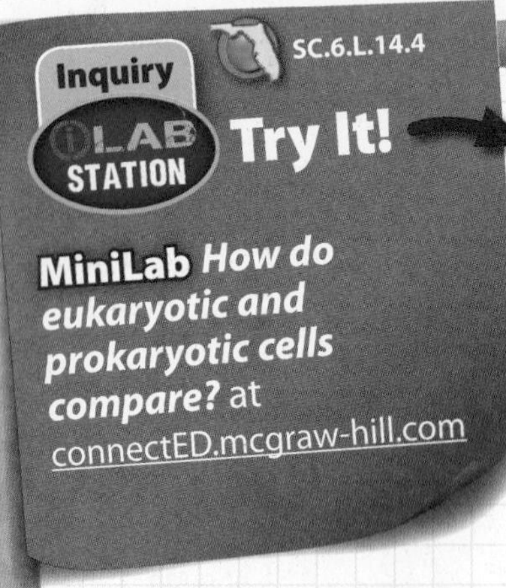

Apply It!

After you complete the lab, answer the questions below.

1. If you modeled a plant cell, compare it to the animal cell in **Figure 6.** If you modeled an animal cell, compare it to the plant cell in **Figure 5.** How do the cells compare? How do they differ?

2. Why do you think prokaryotic cells have fewer cell parts than eukaryotic cells?

5. Organize Classify cells as prokaryotic or eukaryotic by writing *P* or *E* in the right-hand column of the table below.

Characteristic	Cell Type
Cell's genetic material is surrounded by a membrane.	
Cell is usually a unicellular organism.	
Cell contains organelles.	

Active Reading **6. Compare and Contrast** How are prokaryotic cells and eukaryotic cells similar, and how are they different?

Cell Types

Recall that the use of microscopes enabled scientists to discover cells. With more advanced microscopes, scientists discovered that all cells can be grouped into two types—prokaryotic (proh ka ree AH tihk) cells and eukaryotic (yew ker ee AH tihk) cells.

Prokaryotic Cells

The genetic material in a prokaryotic cell is not surrounded by a membrane, as shown in **Figure 8.** This is the most important feature of a prokaryotic cell. Prokaryotic cells also do not have many of the other cell parts that you will read about later in this lesson. Most prokaryotic cells are unicellular organisms and are called prokaryotes.

Figure 8 In prokaryotic cells, the genetic material floats freely in the cytoplasm.

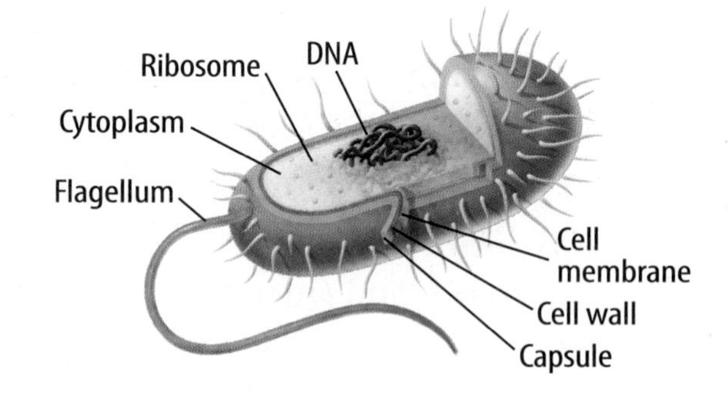

Eukaryotic Cells

Plants, animals, fungi, and protists are all made of eukaryotic cells, such as the ones shown in **Figure 5** and **Figure 6,** and are called eukaryotes. With few exceptions, each eukaryotic cell has genetic material that is surrounded by a membrane. Every eukaryotic cell has *other structures, called* **organelles,** *which have specialized functions. Most organelles are surrounded by membranes.* Eukaryotic cells are usually larger than prokaryotic cells. About ten prokaryotic cells would fit inside one eukaryotic cell.

Cell Organelles

As you have just read, organelles are eukaryotic cell structures with specific functions. Organelles enable cells to carry out different functions at the same time. For example, cells can obtain energy from food, store information, make macromolecules, and get rid of waste materials all at the same time because different organelles perform the different tasks.

Active Reading

FOLDABLES® LA.6.2.2.3

Fold a sheet of paper into a vertical half book. Use it to record information about cell organelles and their functions.

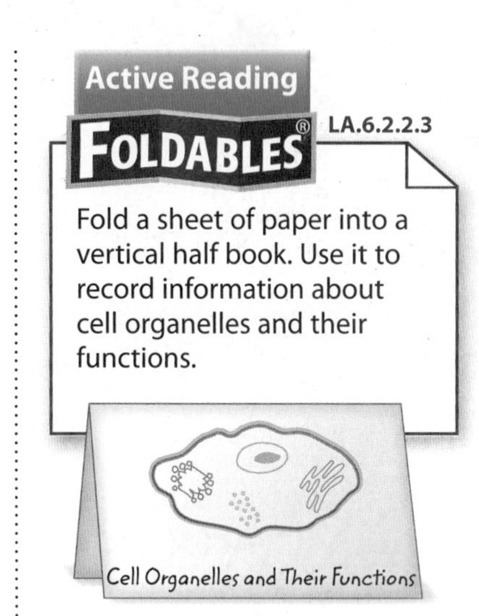

The Nucleus

The largest organelle inside most eukaryotic cells is the nucleus, shown in **Figure 9.** *The* **nucleus** *is the part of a eukaryotic cell that directs cell activities and contains genetic information stored in DNA.* DNA is organized into structures called chromosomes. The number of chromosomes in a nucleus is different for different species of organisms. For example, kangaroo cells contain six pairs of chromosomes. Most human cells contain 23 pairs of chromosomes.

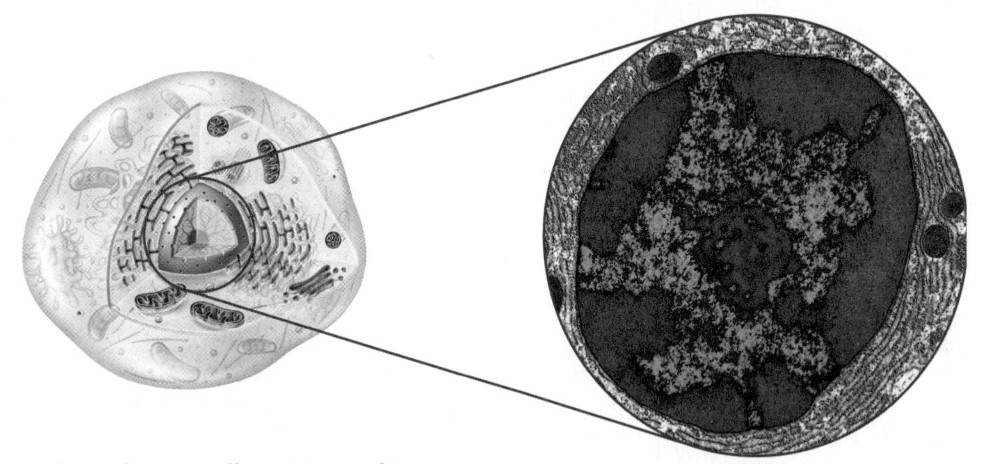

Figure 9 The nucleus directs cell activity and is surrounded by a membrane.

In addition to chromosomes, the nucleus contains proteins and an organelle called the nucleolus (new KLEE uh lus). The nucleolus is often seen as a large dark spot in the nucleus of a cell. The nucleolus makes ribosomes, organelles that are involved in the production of proteins. You will read about ribosomes later in this lesson.

Surrounding the nucleus are two membranes that form a structure called the nuclear **envelope.** The nuclear envelope contains many pores. Certain molecules, such as ribosomes and RNA, move into and out of the nucleus through these pores.

Active Reading 7. **Describe** What is the nuclear envelope?

____

____

SCIENCE USE V. COMMON USE

envelope

Science Use an outer covering

Common Use a flat paper container for a letter

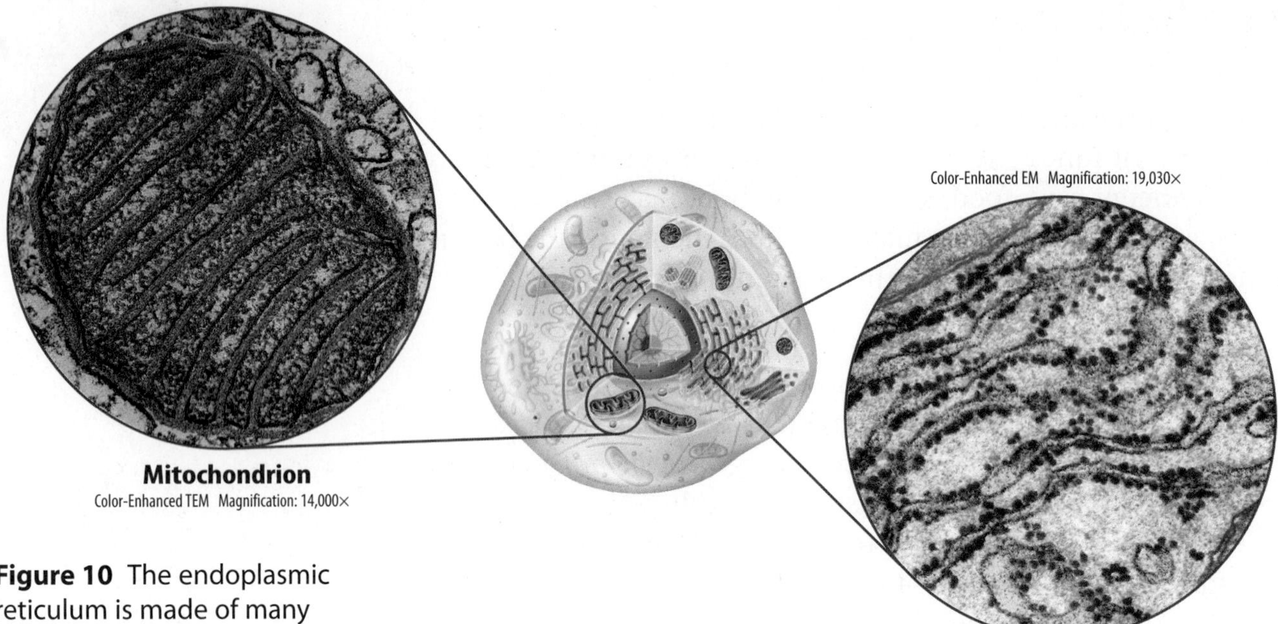

Figure 10 The endoplasmic reticulum is made of many folded membranes. Mitochondria provide a cell with usable energy.

Manufacturing Molecules

You might recall from Lesson 1 that proteins are important molecules in cells. Proteins are made on small structures called ribosomes. Unlike other cell organelles, a ribosome is not surrounded by a membrane. Ribosomes are in a cell's cytoplasm. They also can be attached to a weblike organelle called the endoplasmic reticulum (en duh PLAZ mihk • rih TIHK yuh lum), or ER. As shown in **Figure 10,** the ER spreads from the nucleus throughout most of the cytoplasm. ER with ribosomes on its surface is called rough ER. Rough ER is the site of protein production. ER without ribosomes is called smooth ER. It makes lipids such as cholesterol. Smooth ER is important because it helps remove harmful substances from a cell.

Active Reading **8. Contrast** Underline the description and function of smooth ER. Highlight the description and function of rough ER.

Processing Energy

All living things require energy in order to survive. Cells process some energy in specialized organelles. Most eukaryotic cells contain hundreds of organelles called mitochondria (mi tuh KAHN dree uh; singular, mitochondrion), shown in **Figure 10.** Some cells in a human heart can contain a thousand mitochondria.

Like the nucleus, a mitochondrion is surrounded by two membranes. Energy is released during chemical reactions that occur in the mitochondria. This energy is stored in high-energy molecules called ATP—adenosine triphosphate (uh DEH nuh seen • tri FAHS fayt). ATP is the fuel for cellular processes such as growth, cell division, and material transport.

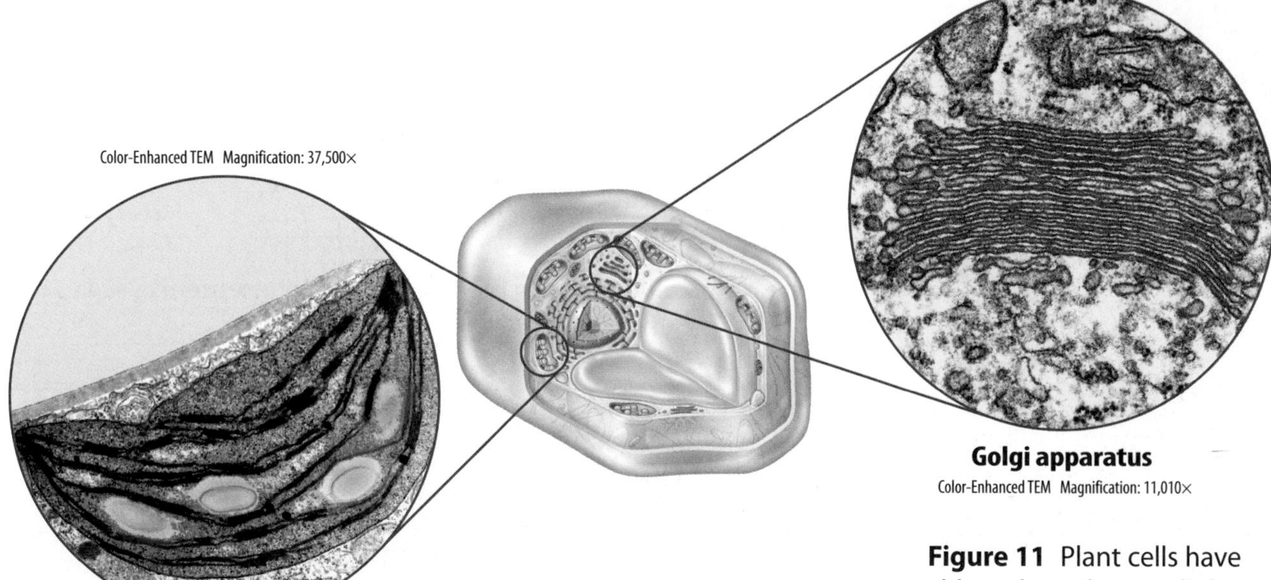

Figure 11 Plant cells have chloroplasts that use light energy and make food. The Golgi apparatus packages materials into vesicles.

Plant cells and some protists, such as algae, also contain organelles called chloroplasts (KLOR uh plasts), shown in **Figure 11. Chloroplasts** *are membrane-bound organelles that use light energy and make food—a sugar called glucose—from water and carbon dioxide in a process known as photosynthesis* (foh toh SIHN thuh sus). The sugar contains stored chemical energy that can be released when a cell needs it. You will read more about photosynthesis in Lesson 4.

Active Reading **9. Identify** Underline the types of cells that contain chloroplasts.

Processing, Transporting, and Storing Molecules

Near the ER is an organelle that looks like a stack of pancakes. This is the Golgi (GAWL jee) apparatus, shown in **Figure 11.** It prepares proteins for their specific jobs or functions. Then it packages the proteins into tiny, membrane-bound, ball-like structures called vesicles. Vesicles are organelles that transport substances from one area of a cell to another area of a cell. Some vesicles in an animal cell are called lysosomes. Lysosomes contain substances that help break down and recycle cellular components.

Some cells also have saclike structures called vacuoles (VA kyuh wohlz). Vacuoles are organelles that store food, water, and waste material. A typical plant cell usually has one large vacuole that stores water and other substances. Some animal cells have many small vacuoles.

Active Reading **10. Determine** Highlight the function of the Golgi apparatus.

Lesson Review 2

Visual Summary

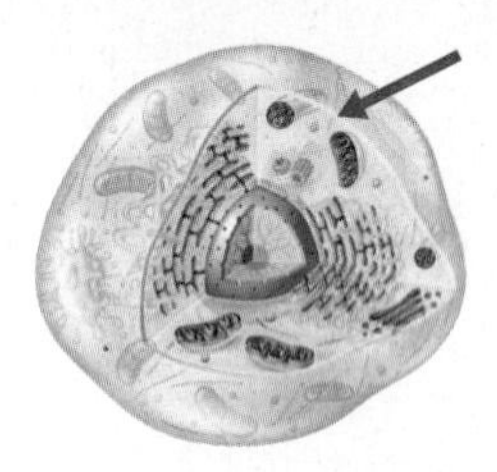

A cell is protected by a flexible covering called the cell membrane.

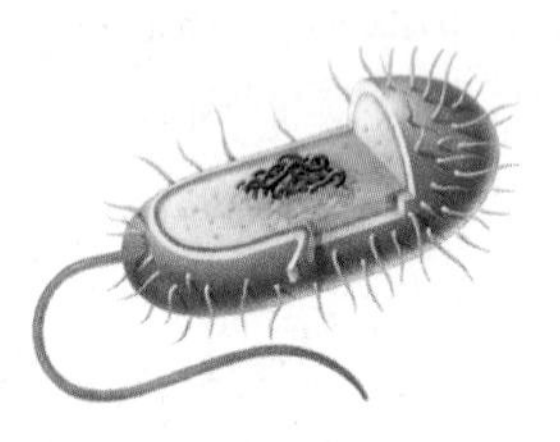

Cells can be grouped into two types—prokaryotic cells and eukaryotic cells.

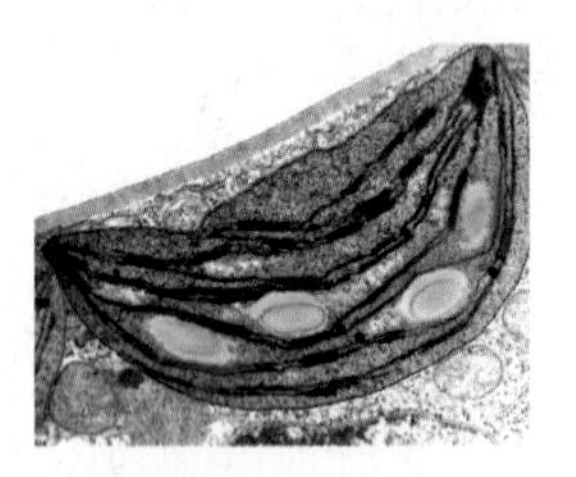

In a chloroplast, light energy is used for making sugars in a process called photosynthesis.

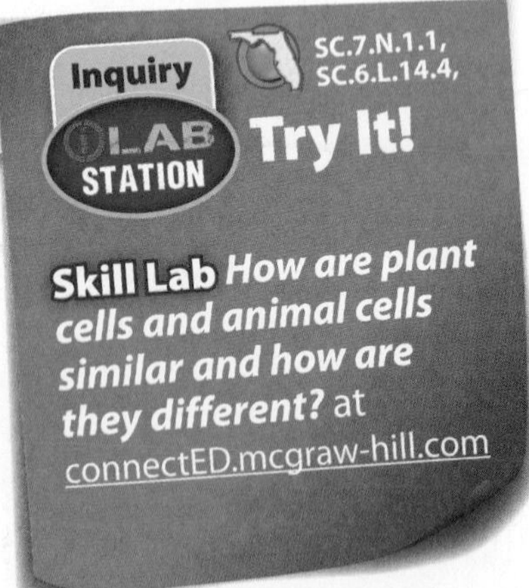

Use Vocabulary

1. **Distinguish** between the cell wall and the cell membrane. SC.6.L.14.4

2. **Use the terms** *mitochondria* and *chloroplasts* in a sentence. SC.6.L.14.4

3. **Define** *organelle* in your own words.

Understand Key Concepts

4. Which organelle is used to store water? SC.6.L.14.4
   (A) chloroplast
   (B) lysosome
   (C) nucleus
   (D) vacuole

5. **Draw** a prokaryotic cell and label its parts.

Interpret Graphics

6. **Explain** how the structure of the cells below relates to their function. SC.6.N.1.5

Critical Thinking

7. **Analyze** Why are most organelles surrounded by membranes?

8. **Compare** the features of eukaryotic and prokaryotic cells.

Active Reading **Classify** information about organelles. In the right-hand column, indicate whether the organelle is in a plant cell, an animal cell, or both.

Organelle	Function	Plant, Animal, or Both?
Nucleus	**1.**	
Ribosome	**2.**	
Endoplasmic reticulum	**3.**	
Mitochondria	**4.**	
Chloroplast	**5.**	
Golgi apparatus	**6.**	
Vesicle	**7.**	
Central vacuole	**8.**	
Lysosome	**9.**	

Lesson 3

Moving Cellular MATERIAL

ESSENTIAL QUESTIONS

 How do materials enter and leave cells?

 How does cell size affect the transport of materials?

Vocabulary

passive transport p. 67
diffusion p. 68
osmosis p. 68
facilitated diffusion p. 69
active transport p. 70
endocytosis p. 70
exocytosis p. 70

Florida NGSSS

LA.6.2.2.3 The student will organize information to show understanding (e.g., representing main ideas within text through charting, mapping, paraphrasing, summarizing, or comparing/contrasting);

SC.6.N.1.5 Recognize that science involves creativity, not just in designing experiments, but also in creating explanations that fit evidence.

SC.6.N.2.3 Recognize that scientists who make contributions to scientific knowledge come from all kinds of backgrounds and possess varied talents, interests, and goals.

SC.6.L.14.4 Compare and contrast the structure and function of major organelles of plant and animal cells, including cell wall, cell membrane, nucleus, cytoplasm, chloroplasts, mitochondria, and vacuoles.

MA.6.A.3.6 Construct and analyze tables, graphs, and equations to describe linear functions and other simple relations using both common language and algebraic notation.

Inquiry Launch Lab

SC.6.L.14.4, SC.6.N.1.1, SC.6.N.1.5

5 minutes

What does the cell membrane do?

All cells have a membrane around the outside of the cell. The cell membrane separates the inside of a cell from the environment outside a cell. What else might a cell membrane do?

Procedure

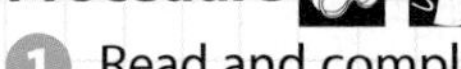

1. Read and complete a lab safety form.
2. Place a square of **wire mesh** on top of a **beaker.**
3. Pour a small amount of **birdseed** on top of the wire mesh. Record your observations below.

Data and Observations

Think About This

1. What part of a cell does the wire mesh represent?
2. What happened when you poured birdseed on the wire mesh?
3. Key Concept How do you think the cell membrane affects materials that enter and leave a cell?

Inquiry Why the Veil?

1. A beekeeper often wears a helmet with a face-covering veil made of mesh. The openings in the mesh are large enough to let air through, yet small enough to keep bees out. In a similar way, some things must be allowed in or out of a cell, while other things must be kept in or out. How do you think the right things enter or leave a cell?

Passive Transport

Recall from Lesson 2 that membranes are the boundaries between cells and between organelles. Another important role of membranes is to control the movement of substances into and out of cells. A cell membrane is semipermeable. This means it allows only certain substances to enter or leave a cell. Substances can pass through a cell membrane by one of several different processes. The type of process depends on the physical and chemical properties of the substance passing through the membrane.

Small molecules, such as oxygen and carbon dioxide, pass through membranes by a process called passive transport. **Passive transport** *is the movement of substances through a cell membrane without using the cell's energy.* Passive transport depends on the amount of a substance on each side of a membrane. For example, suppose there are more molecules of oxygen outside a cell than inside it. Oxygen will move into that cell until the amount of oxygen is equal on both sides of the cell's membrane. Since oxygen is a small molecule, it passes through a cell membrane without using the cell's energy. The different types of passive transport are explained on the following pages.

Active Reading **2. Define** What is a semipermeable membrane?

Active Reading FOLDABLES® LA.6.2.2.3

Fold a sheet of paper into a two-tab book. Label the tabs as shown. Use it to organize information about the different types of passive and active transport.

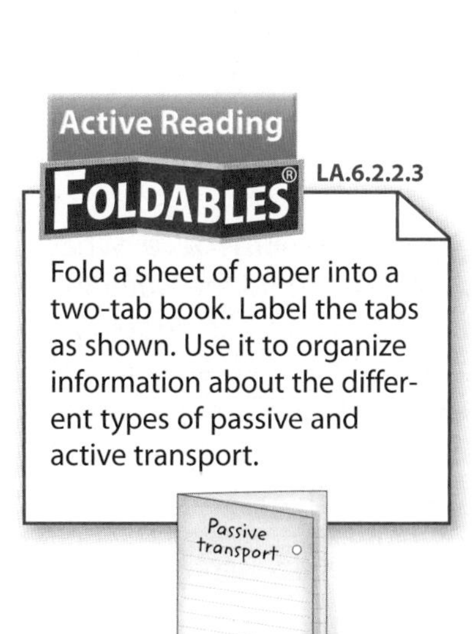

Diffusion

What happens when the concentration, or amount per unit of volume, of a substance is unequal on each side of a membrane? The molecules will move from the side with a higher concentration of that substance to the side with a lower concentration. **Diffusion** *is the movement of substances from an area of higher concentration to an area of lower concentration.*

Word Origin
diffusion
from Latin *diffusionem,* means "scatter, pour out"

Usually, diffusion continues through a membrane until the concentration of a substance is the same on both sides of the membrane. When this happens, a substance is in equilibrium. Compare the two diagrams in **Figure 12.** What happened to the red dye that was added to the water on one side of the membrane? Water and dye passed through the membrane in both directions until there were equal concentrations of water and dye on both sides of the membrane.

Figure 12 Over time, the concentration of dye on either side of the membrane becomes the same.

Diffusion

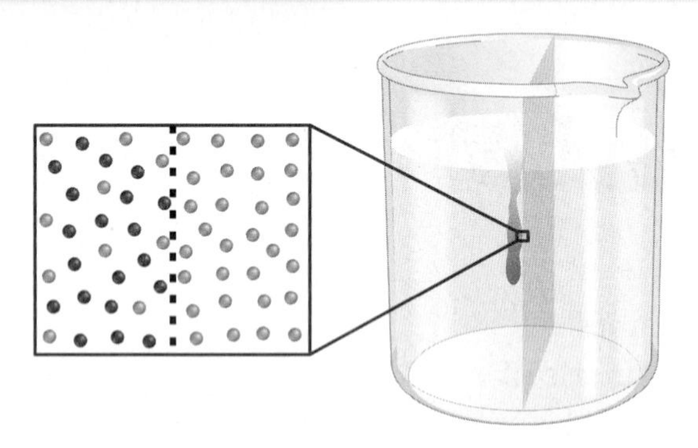
Dye added to water

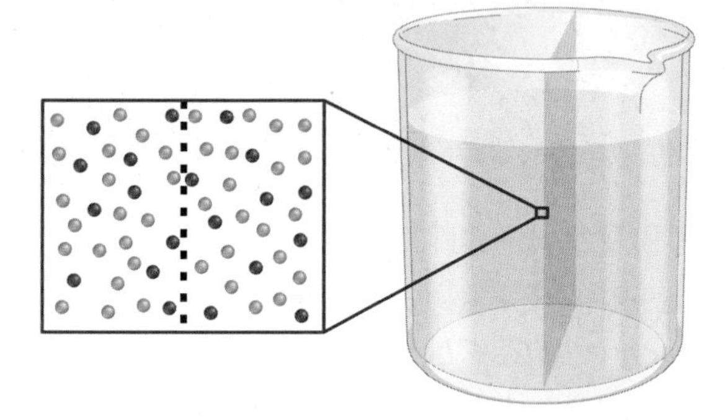
After 30 minutes

Active Reading **3. Illustrate** What would the water in the beaker on the right look like if the membrane did not let anything through? Draw your answer below.

Osmosis—The Diffusion of Water

Diffusion refers to the movement of any small molecules from higher to lower concentrations. However, **osmosis** *is the diffusion of water molecules only through a membrane.* Semipermeable cell membranes also allow water to pass through them until equilibrium occurs. For example, the amount of water stored in the vacuoles of plant cells can decrease because of osmosis. That is because the concentration of water in the air surrounding the plant is less than the concentration of water inside the vacuoles of plant cells. Water will continue to diffuse into the air until the concentrations of water inside the plant's cells and in the air are equal. If the plant is not watered to replace the lost water, it will wilt and eventually die.

Facilitated Diffusion

Some molecules are too large or are chemically unable to travel through a membrane by diffusion. *When molecules pass through a cell membrane using special proteins called transport proteins, this is* **facilitated diffusion.** Like diffusion and osmosis, facilitated diffusion does not require a cell to use energy. As shown in **Figure 13,** a cell membrane has transport proteins. The two types of transport proteins are carrier proteins and channel proteins. Carrier proteins carry large molecules, such as the sugar molecule glucose, through the cell membrane. Channel proteins form pores through the membrane. Atomic particles, such as sodium ions and potassium ions, pass through the cell membrane by channel proteins.

4. Describe How do materials move through the cell membrane in facilitated diffusion?

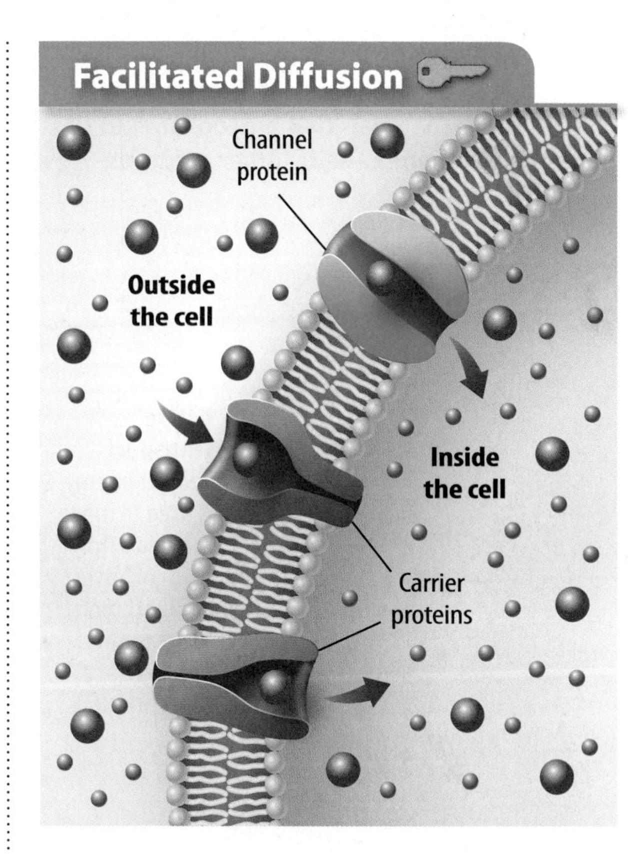

Figure 13 Transport proteins are used to move large molecules into and out of a cell.

MiniLab ***How is a balloon like a cell membrane?*** at connectED.mcgraw-hill.com

Apply It! After you complete the lab, answer these questions.

1. What type of passive transport occurred in this lab—diffusion, osmosis, or facilitated diffusion? Explain your answer.

2. What do you think the advantages of passive transport are?

Figure 14 Active transport is most often used to bring needed nutrients into a cell. Endocytosis and exocytosis move materials that are too large to pass through the cell membrane by other methods.

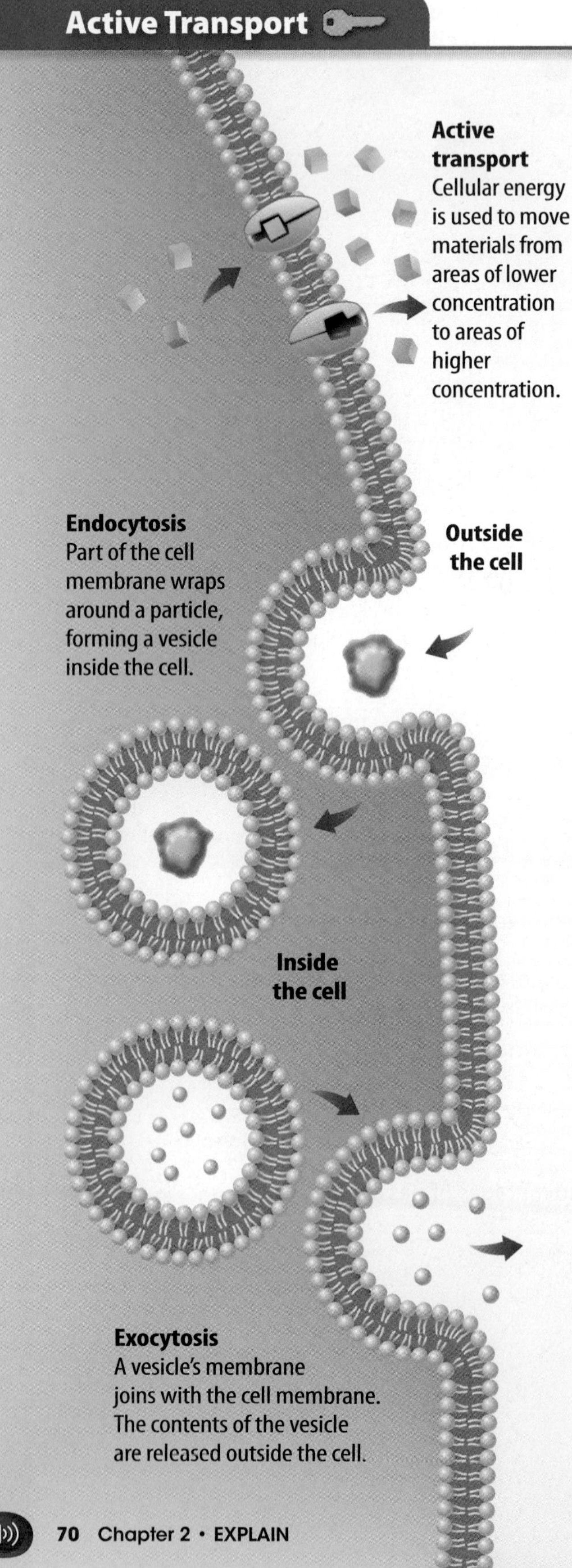

Active Transport

Sometimes when cellular materials pass through membranes it requires a cell to use energy. **Active transport** *is the movement of substances through a cell membrane only by using the cell's energy.*

Recall that passive transport is the movement of substances from areas of higher concentration to areas of lower concentration. However, substances moving by active transport move from areas of lower concentration to areas of higher concentration, as shown in **Figure 14.**

Cells can take in needed nutrients from the environment through carrier proteins by using active transport. This occurs even when concentrations of these nutrients are lower in the environment than inside the cell. Some other molecules and waste materials also leave cells by active transport.

Endocytosis and Exocytosis

Some substances are too large to enter a cell membrane by diffusion or by using a transport protein. These substances can enter a cell by another process. **Endocytosis** (en duh si TOH sus), shown in **Figure 14,** *is the process during which a cell takes in a substance by surrounding it with the cell membrane.* Some cells take in bacteria and viruses using endocytosis.

Some substances are too large to leave a cell by diffusion or by using a transport protein. These substances can leave a cell another way. **Exocytosis** (ek soh si TOH sus), shown in **Figure 14,** *is the process during which a cell's vesicles release their contents outside the cell.* Proteins and other substances are removed from a cell through this process.

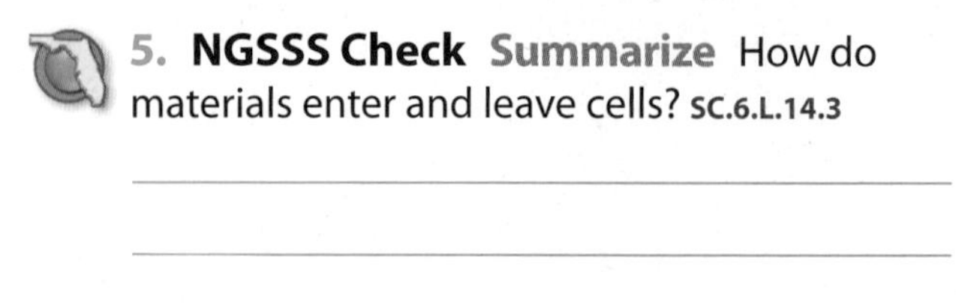

5. NGSSS Check **Summarize** How do materials enter and leave cells? **SC.6.L.14.3**

Cell Size and Transport

Recall that the movement of nutrients, waste material, and other substances into and out of a cell is important for survival. For this movement to happen, the area of the cell membrane must be large compared to its volume. The area of the cell membrane is the cell's surface area. The volume is the amount of space inside the cell. As a cell grows, both its volume and its surface area increase. The volume of a cell increases faster than its surface area. If a cell were to keep growing, it would need large amounts of nutrients and would produce large amounts of waste material. However, the surface area of the cell's membrane would be too small to move enough nutrients and wastes through it for the cell to survive.

Active Reading **6. Analyze** How does cell size affect the transport of materials?

Math Skills MA.6.A.3.6

Use Ratios

A ratio is a comparison of two numbers, such as surface area and volume. If a cell were cube-shaped, you would calculate surface area by multiplying its length (ℓ) by its width (w) by the number of sides (6).

You would calculate the volume of the cell by multiplying its length (ℓ) by its width (w) by its height (h).

To find the surface-area-to-volume ratio of the cell, divide its surface area by its volume.

In the table below, surface-area-to-volume ratios are calculated for cells that are 1 mm, 2 mm, and 4 mm per side. Notice how the ratios change as the cell's size increases.

Surface area $= \ell \times w \times 6$

Volume $= \ell \times w \times h$

$\frac{\text{Surface area}}{\text{Volume}}$

	1 mm 1 mm 1 mm	2 mm 2 mm 2 mm	4 mm 4 mm 4 mm
Length	1 mm	2 mm	4 mm
Width	1 mm	2 mm	4 mm
Height	1 mm	2 mm	4 mm
Number of sides	6	6	6
Surface area ($\ell \times w \times$ no. of sides)	1 mm × 1 mm × 6 = 6 mm^2	2 mm × 2 mm × 6 = 24 mm^2	4 mm × 4 mm × 6 = 96 mm^2
Volume ($\ell \times w \times h$)	1 mm × 1 mm × 1 mm = 1 mm^3	2 mm × 2 mm × 2 mm = 8 mm^3	4 mm × 4 mm × 4 mm = 64 mm^3
Surface-area-to-volume ratio	$\frac{6\ mm^2}{1\ mm^3} = \frac{6}{1}$ or 6:1	$\frac{24\ mm^2}{8\ mm^3} = \frac{3}{1}$ or 3:1	$\frac{96\ mm^2}{64\ mm^3} = \frac{1.5}{1}$ or 1.5:1

Practice

7. What is the surface-area-to-volume ratio of a cell whose six sides are 3 mm long?

Lesson Review 3

Visual Summary

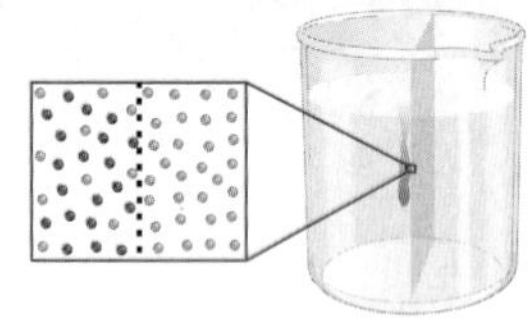

Small molecules can move from an area of higher concentration to an area of lower concentration.

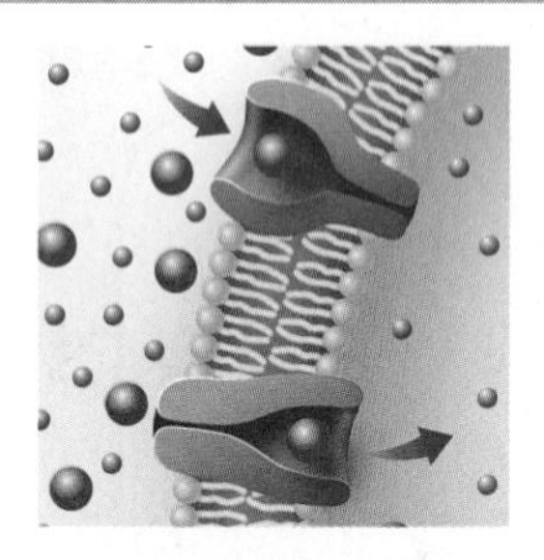

Proteins transport larger molecules through a cell membrane.

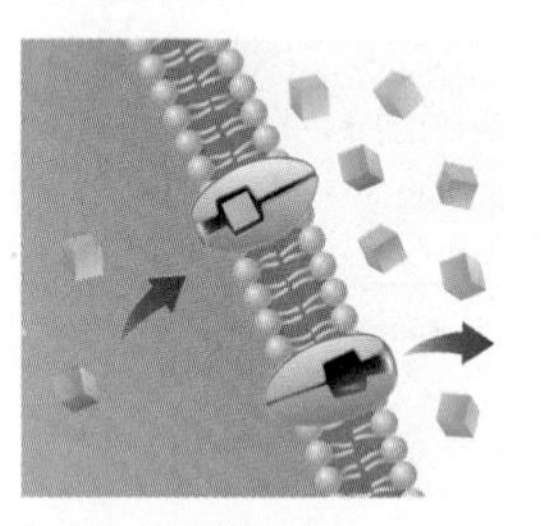

Some molecules move from areas of lower concentration to areas of higher concentration.

Use Vocabulary

1. **Distinguish** between active transport and passive transport. SC.6.L.14.4

2. The process by which vesicles move substances out of a cell is ____________.

Understand Key Concepts

3. **Summarize** the function of endocytosis.

4. What is limited by a cell's surface-area-to-volume ratio?
   - (A) cell shape
   - (B) cell size
   - (C) cell surface area
   - (D) cell volume

Interpret Graphics

5. Fill in the graphic organizer below to describe ways that cells transport substances. SC.6.L.14.3

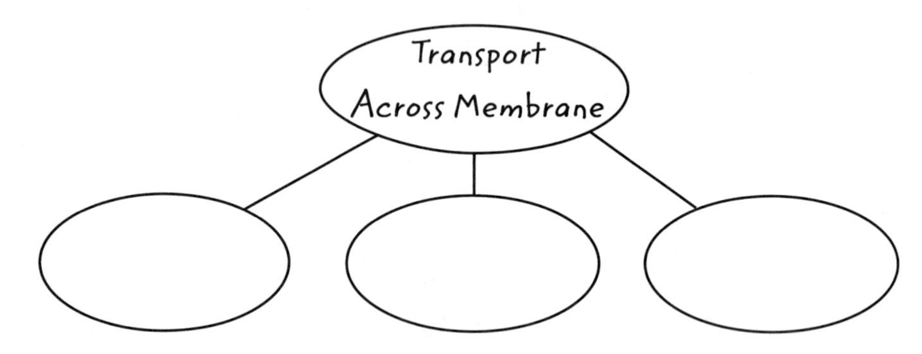

Critical Thinking

6. **Relate** the surface area of a cell to the transport of materials.

Math Skills MA.6.A.3.6

7. **Calculate** the surface-area-to-volume ratio of a cube whose sides are 6 cm long.

Explain the process of facilitated diffusion.

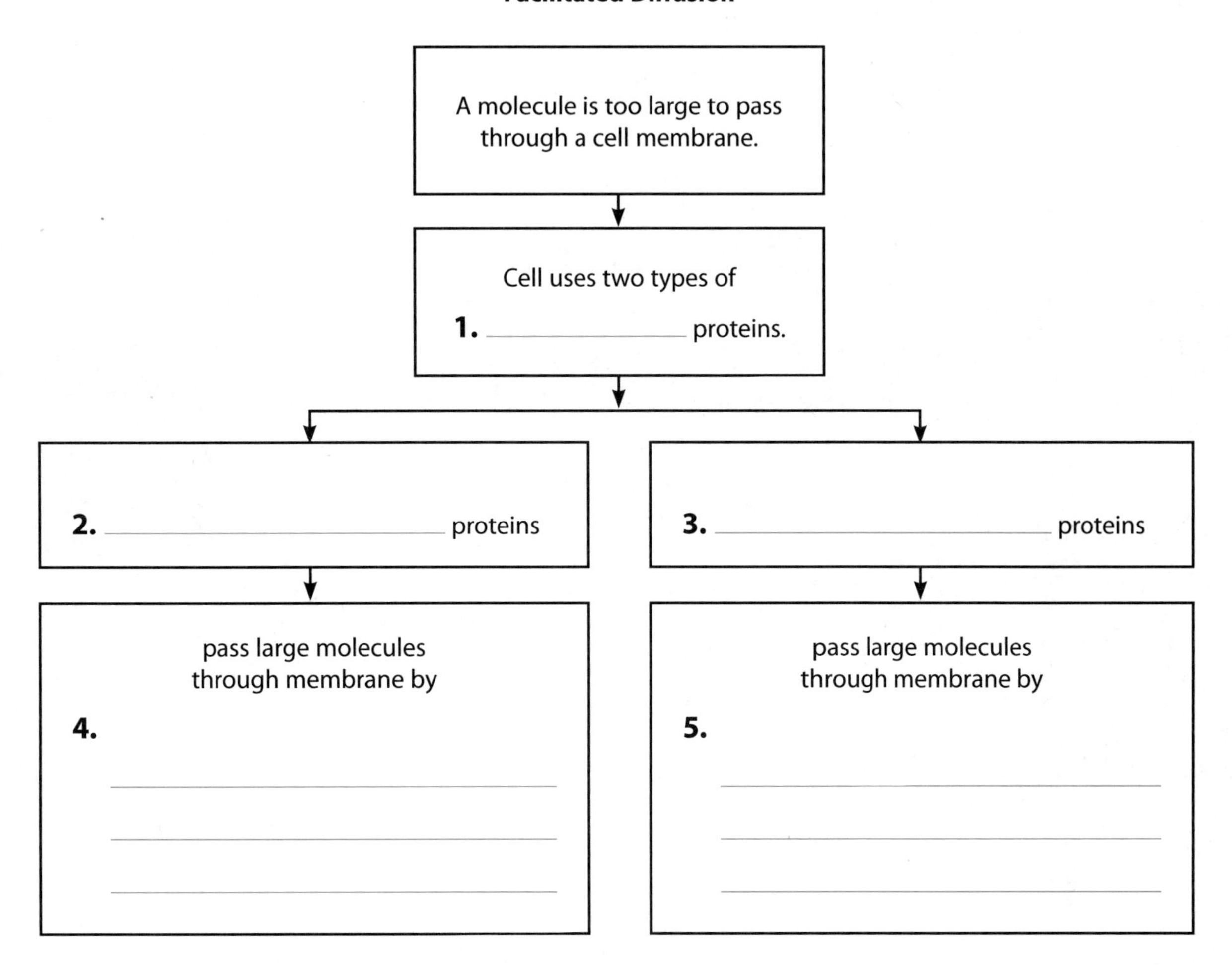

Organize information about active transport.

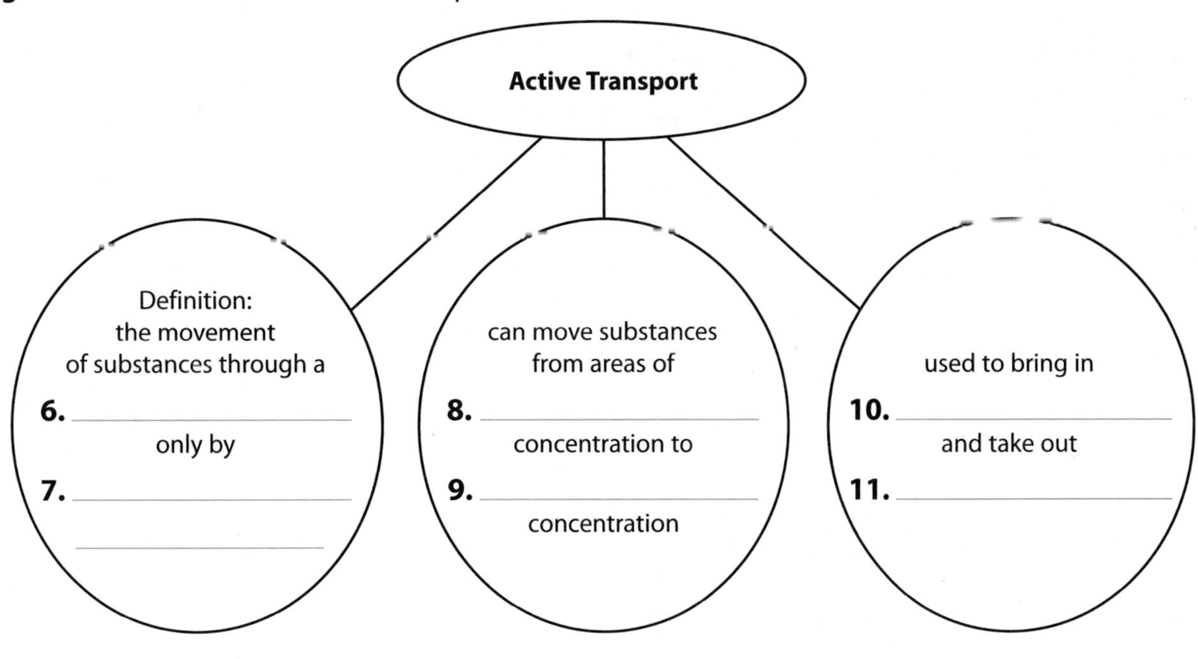

Lesson 4

Cells and ENERGY

ESSENTIAL QUESTIONS

 How does a cell obtain energy?

 How do some cells make food molecules?

Vocabulary

cellular respiration p. 75
glycolysis p. 75
fermentation p. 76
photosynthesis p. 77

Florida NGSSS

SC.6.N.1.5 Recognize that science involves creativity, not just in designing experiments, but also in creating explanations that fit evidence.

SC.6.N.2.1 Distinguish science from other activities involving thought.

SC.6.L.14.3 Recognize and explore how cells of all organisms undergo similar processes to maintain homeostasis, including extracting energy from food, getting rid of waste, and reproducing.

SC.8.N.1.1 Define a problem from the eighth grade curriculum using appropriate reference materials to support scientific understanding, plan and carry out scientific investigations of various types, such as systematic observations or experiments, identify variables, collect and organize data, interpret data in charts, tables, and graphics, analyze information, make predictions, and defend conclusions.

SC.8.N.1.6 Understand that scientific investigations involve the collection of relevant empirical evidence, the use of logical reasoning, and the application of imagination in devising hypotheses, predictions, explanations and models to make sense of the collected evidence.

SC.8.L.18.1 Describe and investigate the process of photosynthesis, such as the roles of light, carbon dioxide, water and chlorophyll; production of food; release of oxygen.

SC.8.L.18.2 Describe and investigate how cellular respiration breaks down food to provide energy and releases carbon dioxide.

Also covers: LA.6.2.2.3, LA.6.4.2.2, MA.6.S.6.2

SC.7.N.1.1

Launch Lab

5 minutes

What do you exhale?

Does the air you breathe in differ from the air you breathe out?

Procedure

1. Read and complete a lab safety form.
2. Unwrap a **straw.** Use the straw to slowly blow into a small **cup** of **bromthymol blue.** Do not splash the liquid out of the cup.
3. Record any changes in the solution.

Data and Observations

Think About This

1. What changes did you observe in the solution?

2. What do you think caused the changes in the solution?

3. Key Concept Why do you think the air you inhale differs from the air you exhale?

Inquiry Why are there bubbles?

1. Have you ever seen bubbles on a green plant in an aquarium? Where do you think the bubbles come from?

Cellular Respiration

When you are tired, you might eat something to give you energy. All living things, from one-celled organisms to humans, need energy to survive. Recall that cells process energy from food into the energy-storage compound ATP. **Cellular respiration** *is a series of chemical reactions that convert the energy in food molecules into a usable form of energy called ATP.* Cellular respiration is a complex process that occurs in two parts of a cell—the cytoplasm and the mitochondria.

Reactions in the Cytoplasm

The first step of cellular respiration, called glycolysis, occurs in the cytoplasm of all cells. **Glycolysis** *is a process by which glucose, a sugar, is broken down into smaller molecules.* As shown in **Figure 15,** glycolysis produces some ATP molecules. It also uses energy from other ATP molecules. You will read on the following page that more ATP is made during the second step of cellular respiration than during glycolysis.

Active Reading 2. **Determine** What is produced during glycolysis?

Glycolysis

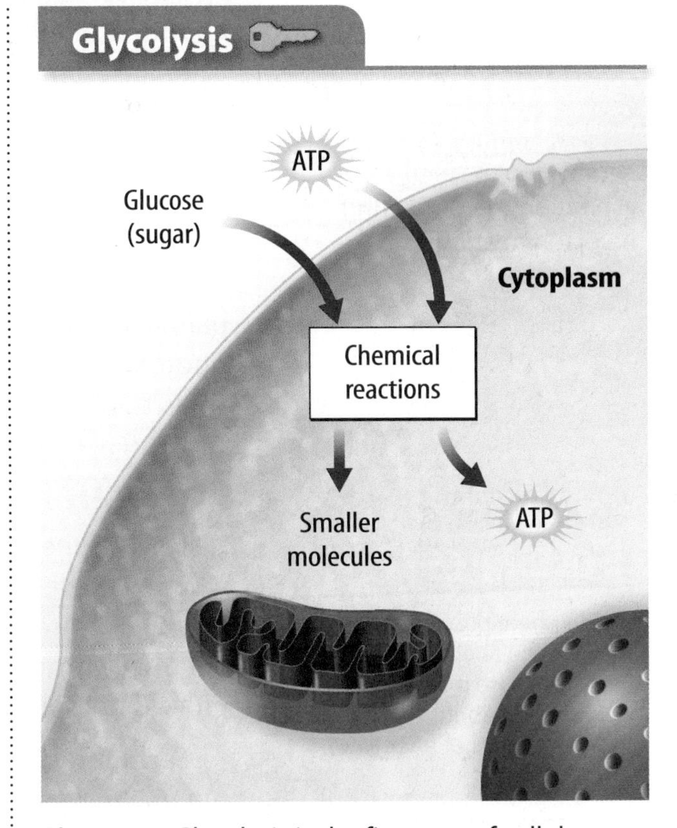

Figure 15 Glycolysis is the first step of cellular respiration.

3. **Visual Check Explain** Why is ATP shown twice in the figure?

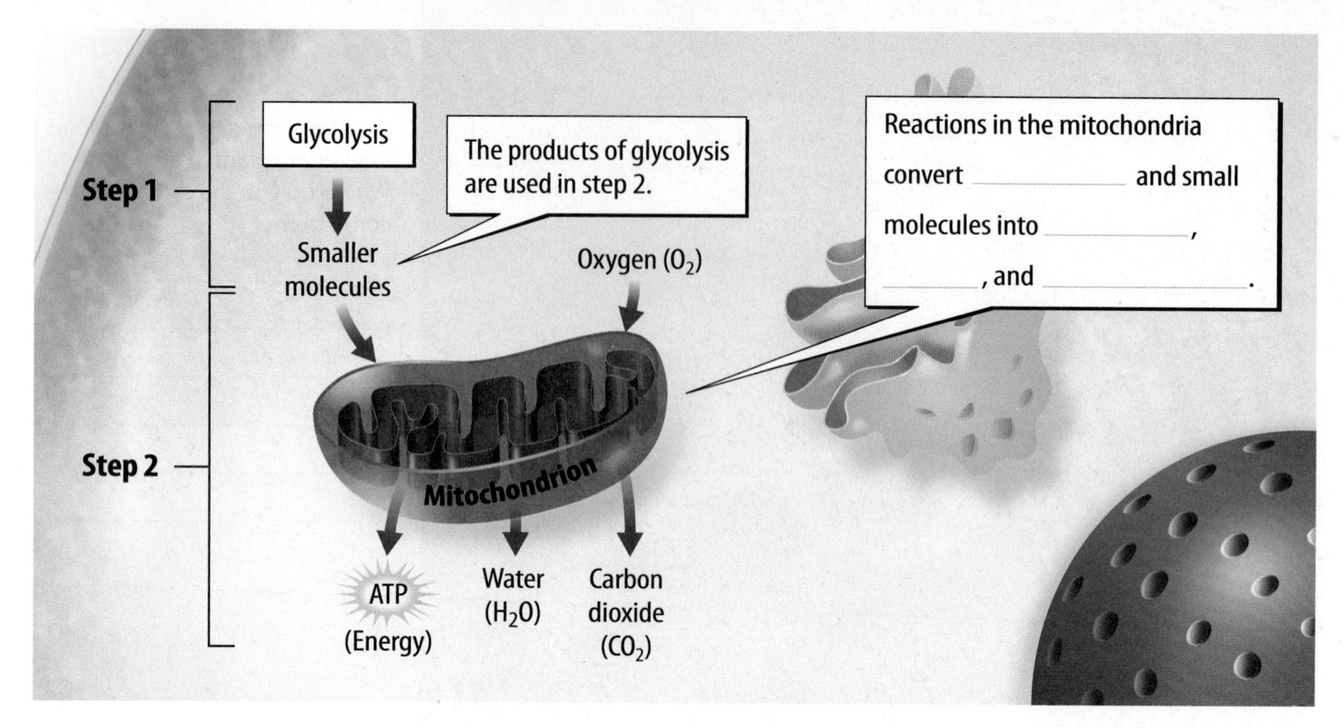

Figure 16 After glycolysis, cellular respiration continues in the mitochondria.

Active Reading **4. Apply** Summarize the second step of cellular respiration by completing the diagram above.

Reactions in the Mitochondria

The second step of cellular respiration occurs in the mitochondria of eukaryotic cells, as shown in **Figure 16.** This step of cellular respiration requires oxygen. The smaller molecules made from glucose during glycolysis are broken down. Large amounts of ATP—usable energy—are produced. Cells use ATP to power all cellular processes. Two waste products—water and carbon dioxide (CO_2)—are given off during this step.

The CO_2 released by cells as a waste product is used by plants and some unicellular organisms in another process called photosynthesis. You will read more about the chemical reactions that take place during photosynthesis in this lesson.

Fermentation

Have you ever felt out of breath after exercising? Sometimes when you exercise, your cells don't have enough oxygen to make ATP through cellular respiration. Then, chemical energy is obtained through a different process called fermentation. This process does not use oxygen.

Fermentation *is a reaction that eukaryotic and prokaryotic cells can use to obtain energy from food when oxygen levels are low.* Because no oxygen is used, fermentation makes less ATP than cellular respiration does. Fermentation occurs in a cell's cytoplasm, not in mitochondria.

5. NGSSS Check **Explain** How does a cell obtain energy? SC.6.L.14.3

______

______

______

Types of Fermentation

One type of fermentation occurs when glucose is converted into ATP and a waste product called lactic acid, as illustrated in **Figure 17.** Some bacteria and fungi help produce cheese, yogurt, and sour cream using lactic-acid fermentation. Muscle cells in humans and other animals can use lactic-acid fermentation and obtain energy during exercise.

Some types of bacteria and yeast make ATP through a process called alcohol fermentation. However, instead of producing lactic acid, alcohol fermentation produces an alcohol called ethanol and CO_2, also illustrated in **Figure 17.** Some types of breads are made using yeast. The CO_2 produced by yeast during alcohol fermentation makes the dough rise.

Figure 17 Your muscle cells and yeast cells can convert glucose to energy through fermentation.

Active Reading **6. Recall** Identify the waste products of the two types of fermentation by completing the figure below.

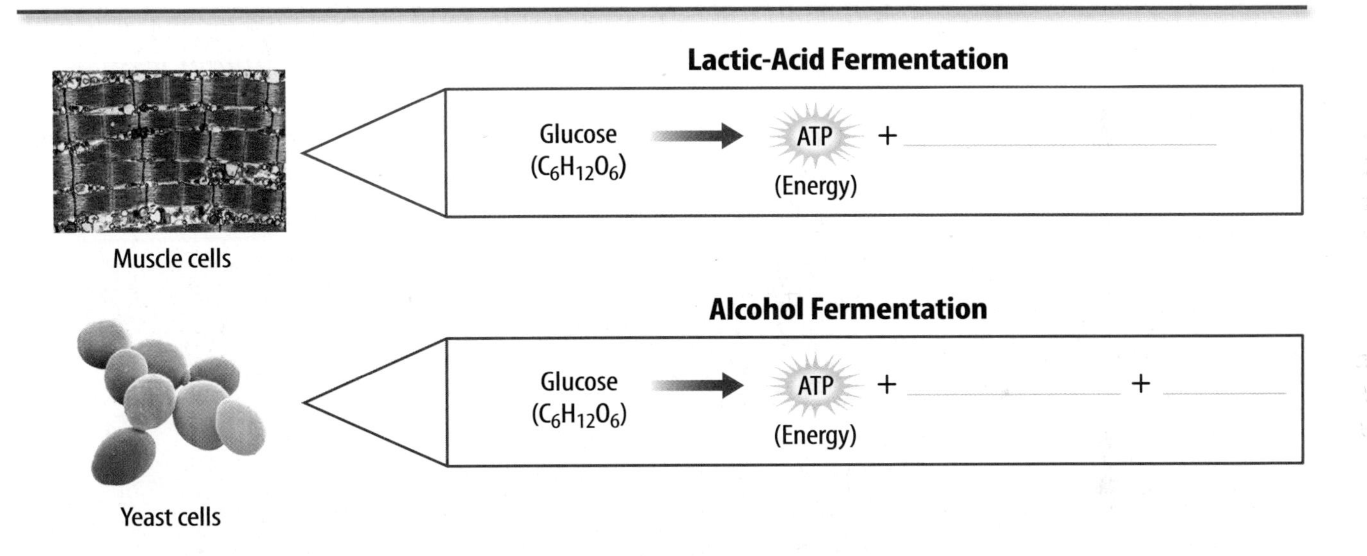

Photosynthesis

Humans and other animals convert food energy into ATP through cellular respiration. However, plants and some unicellular organisms obtain energy from light. **Photosynthesis** *is a series of chemical reactions that convert light energy, water, and* CO_2 *into the food-energy molecule glucose and give off oxygen.*

Active Reading **7. Recall** Underline the name of the food-energy molecule produced through photosynthesis.

WORD ORIGIN

photosynthesis
from Greek photo, means "light"; and synthesis, means "composition"

Lights and Pigments

Photosynthesis requires light energy. In plants, pigments such as chlorophyll absorb light energy. When chlorophyll absorbs light, it absorbs all colors except green. Green light is reflected as the green color seen in leaves. However, plants contain many pigments that reflect other colors, such as yellow and red.

Reactions in Chloroplasts

The light energy absorbed by chlorophyll and other pigments powers the chemical reactions of photosynthesis. These reactions occur in chloroplasts, the organelles in plant cells that convert light energy to chemical energy in food. During photosynthesis, light energy, water, and carbon dioxide combine and make sugars. Photosynthesis also produces oxygen that is released into the atmosphere, as shown in **Figure 18.**

Active Reading **8. Determine** Highlight the substances that some cells use to make food.

Importance of Photosynthesis

Recall that photosynthesis uses light energy and CO_2 and makes food energy and releases oxygen. This food energy is stored in the form of glucose. When an organism, such as the bird in **Figure 18,** eats plant material, such as fruit, it takes in food energy. An organism's cells use the oxygen released during photosynthesis and convert the food energy into usable energy through cellular respiration. **Figure 18** illustrates the important relationship between cellular respiration and photosynthesis.

Figure 18 The relationship between cellular respiration and photosynthesis is important for life.

Active Reading **9. Classify** Label the chemical reactions below as either *cellular respiration* or *photosynthesis.*

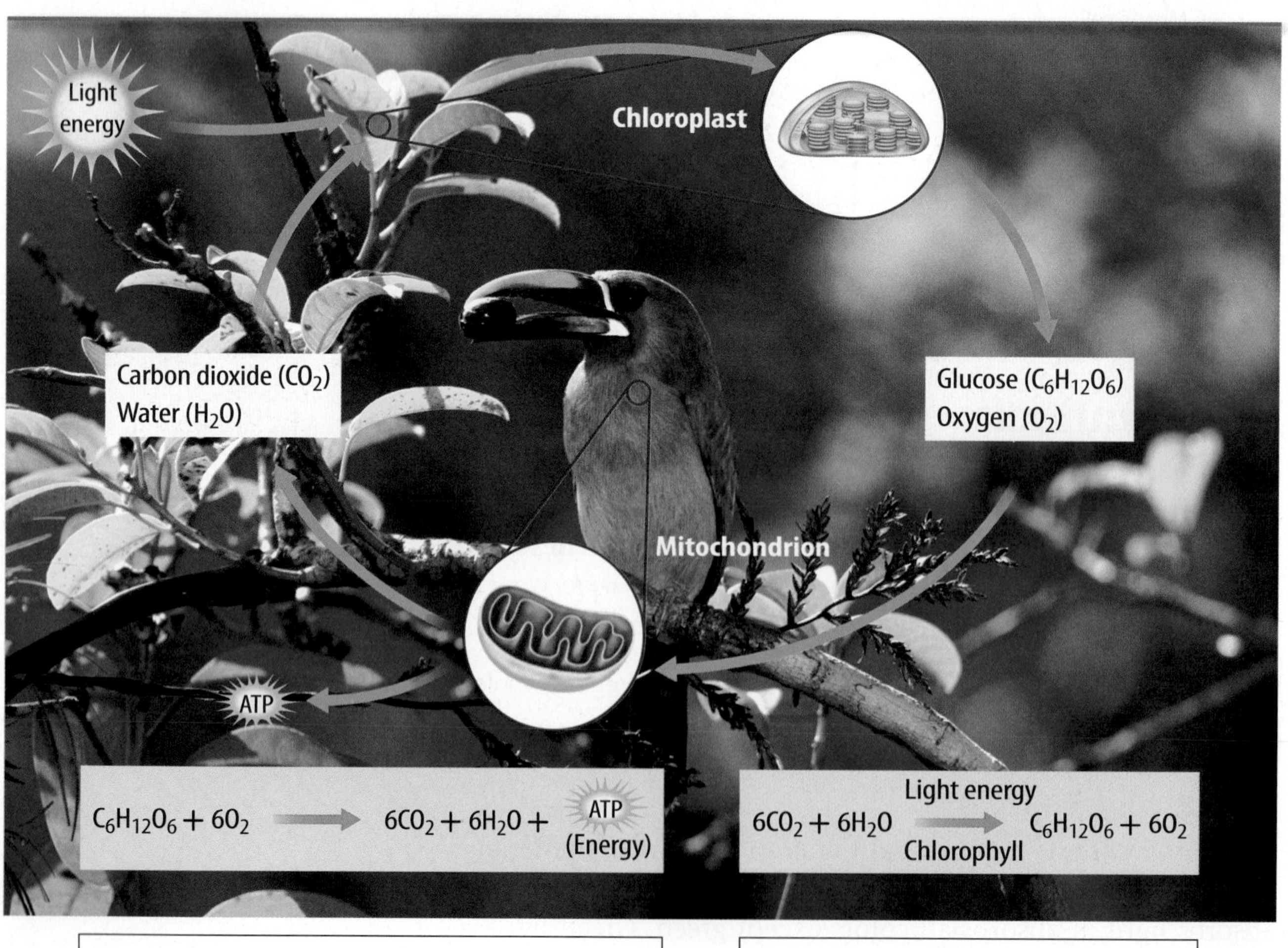

Lesson Review 4

Visual Summary

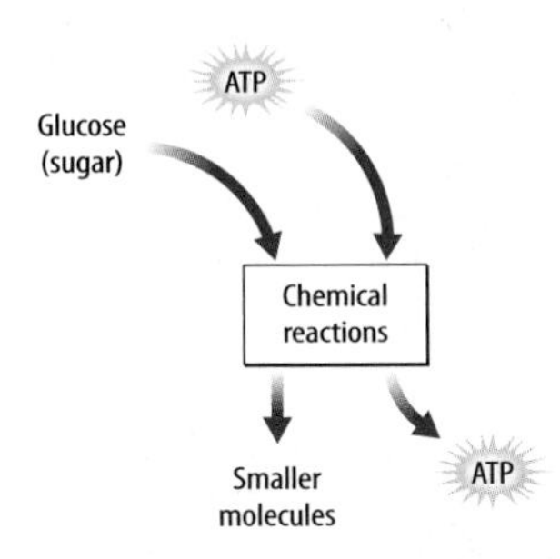

Glycolysis is the first step in cellular respiration.

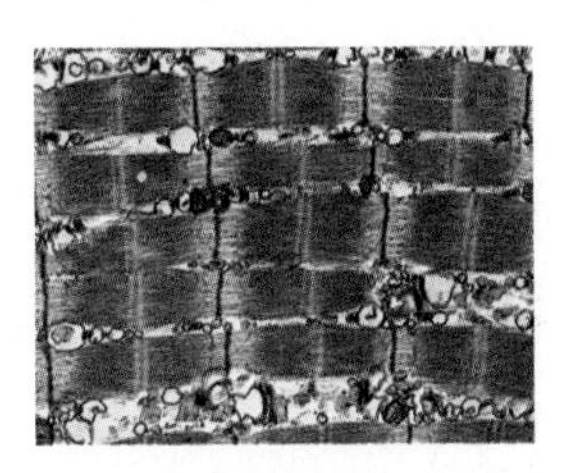

Fermentation provides cells, such as muscle cells, with energy when oxygen levels are low.

Light energy powers the chemical reactions of photosynthesis.

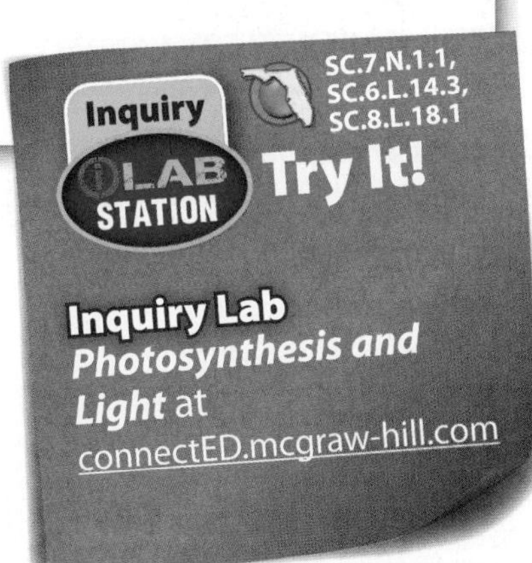

Use Vocabulary

1. **Define** *glycolysis* using your own words.

2. **Distinguish** between cellular respiration and fermentation. SC.6.L.14.3

3. A process used by plants to convert light energy into food energy is ______. SC.6.L.14.3

Understand Key Concepts

4. Which contains pigments that absorb light energy?
   - (A) chloroplast
   - (B) mitochondrion
   - (C) nucleus
   - (D) vacuole

5. **Relate** mitochondria to cellular respiration.

6. **Describe** the role of chlorophyll in photosynthesis.

Interpret Graphics

7. Fill in the boxes with the substances used and produced during photosynthesis. SC.8.L.18.1

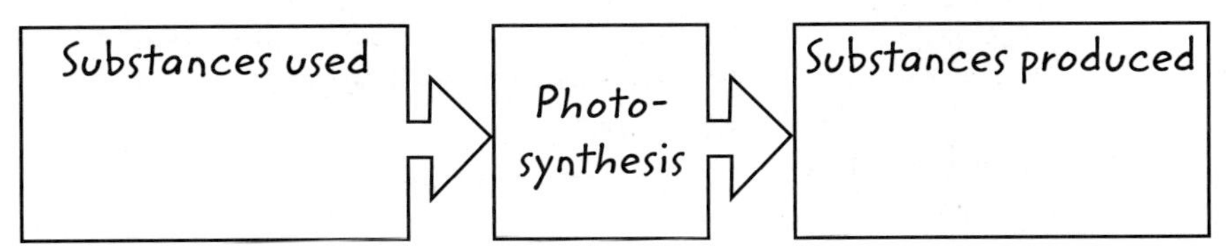

Critical Thinking

8. **Design** a concept map on a separate sheet of paper to show the relationship between cellular respiration in animals and photosynthesis in plants. SC.6.L.14.3

9. **Summarize** the roles of glucose and ATP in energy processing. SC.6.L.14.3

Chapter 2 Study Guide

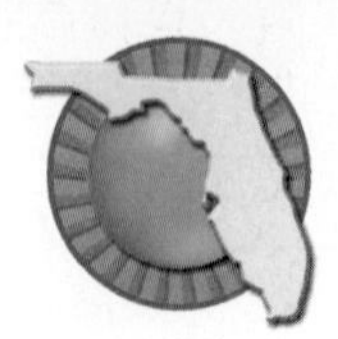

Think About It! **A cell is made up of various structures that are essential for growth, reproduction, and homeostasis. They provide support and movement, process energy, and transport materials.**

Key Concepts Summary

LESSON 1 Cells and Life

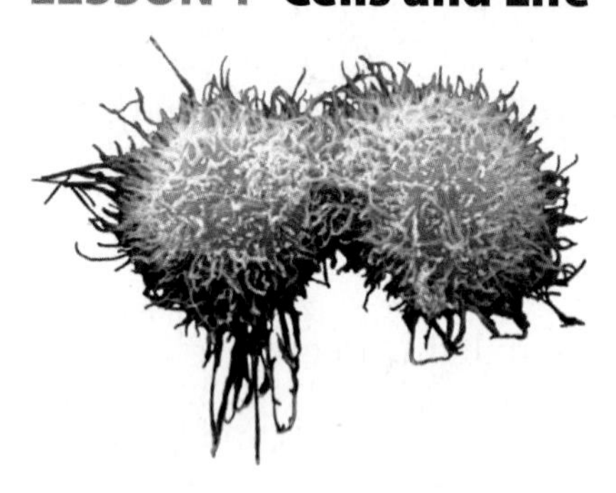

- The invention of the microscope led to discoveries about cells. In time, scientists used these discoveries to develop the **cell theory,** which explains how cells and living things are related.
- Cells are composed mainly of water, **proteins, nucleic acids, lipids,** and **carbohydrates.**

Vocabulary

cell theory p. 50
macromolecule p. 51
nucleic acid p. 52
protein p. 53
lipid p. 53
carbohydrate p. 53

LESSON 2 The Cell

- Cell structures have specific functions, such as supporting a cell, moving a cell, controlling cell activities, processing energy, and transporting molecules.
- A prokaryotic cell lacks a nucleus and other **organelles,** while a eukaryotic cell has a nucleus and other organelles.

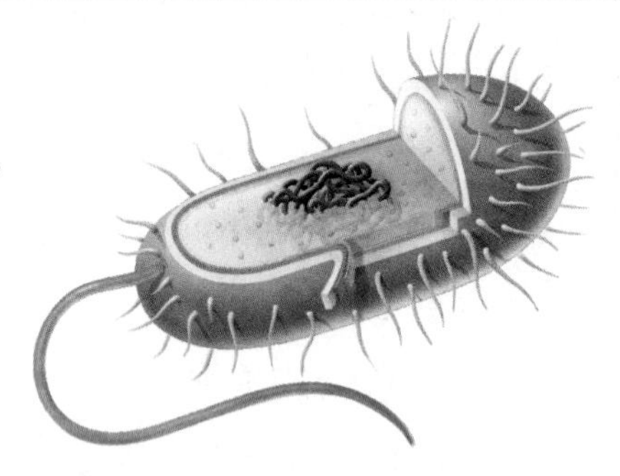

cell membrane p. 58
cell wall p. 58
cytoplasm p. 59
cytoskeleton p. 59
organelle p. 60
nucleus p. 61
chloroplast p. 63

LESSON 3 Moving Cellular Material

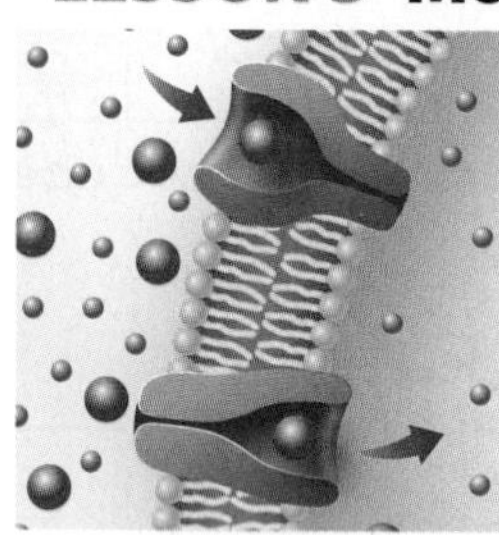

- Materials enter and leave a cell through the cell membrane using **passive transport** or **active transport, endocytosis,** and **exocytosis.**
- The ratio of surface area to volume limits the size of a cell. In a smaller cell, the high surface-area-to-volume ratio allows materials to move easily to all parts of a cell.

passive transport p. 67
diffusion p. 68
osmosis p. 68
facilitated diffusion p. 69
active transport p. 70
endocytosis p. 70
exocytosis p. 70

LESSON 4 Cells and Energy

- All living cells release energy from food molecules through **cellular respiration** and/or **fermentation.**
- Some cells make food molecules using light energy through the process of **photosynthesis.**

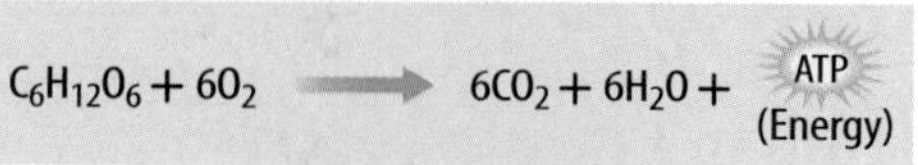

$$C_6H_{12}O_6 + 6O_2 \longrightarrow 6CO_2 + 6H_2O + \text{ATP (Energy)}$$

Cellular respiration

$$6CO_2 + 6H_2O \xrightarrow[\text{Chlorophyll}]{\text{Light energy}} C_6H_{12}O_6 + 6O_2$$

Photosynthesis

cellular respiration p. 75
glycolysis p. 75
fermentation p. 76
photosynthesis p. 77

Active Reading

FOLDABLES® Chapter Project

Assemble your lesson Foldables as shown to make a Chapter Project. Use the project to review what you have learned in this chapter.

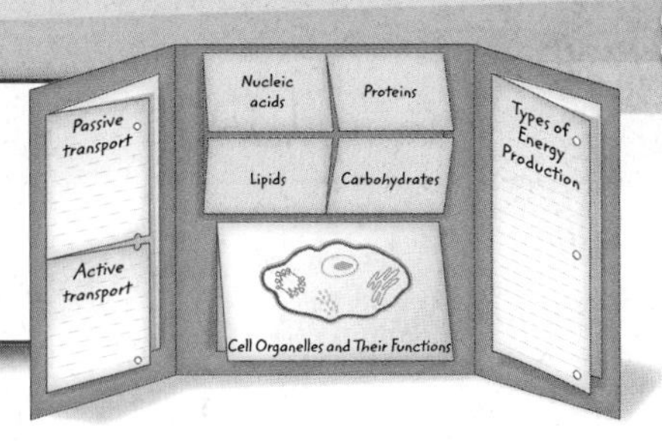

Use Vocabulary

1. Substances formed by joining smaller molecules together are called ____________.

2. The ____________ consists of proteins joined together to create fiberlike structures inside cells.

3. The movement of substances from an area of high concentration to an area of low concentration is called ____________.

4. A process that uses oxygen to convert energy from food into ATP is ____________ ____________.

Link Vocabulary and Key Concepts

Use vocabulary terms from the previous page to complete the concept map.

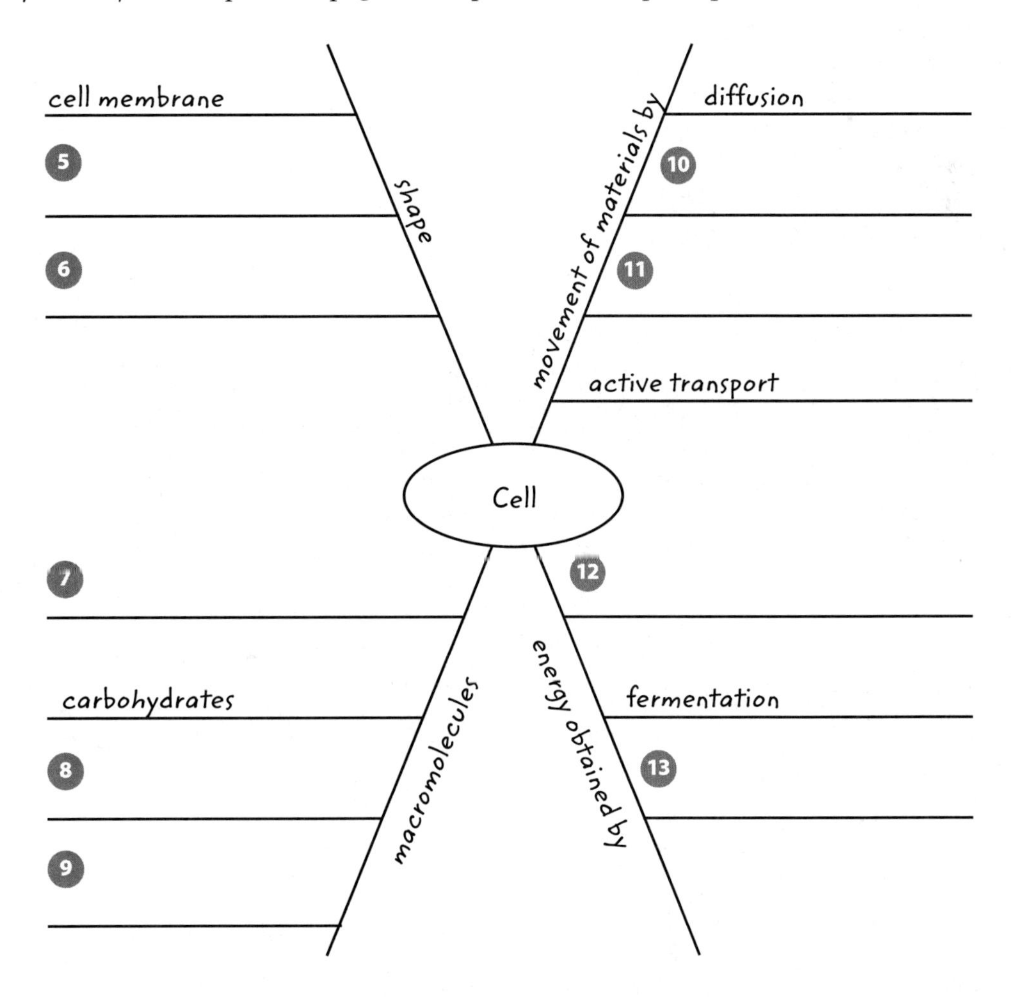

Chapter 2 Review

Fill in the correct answer choice.

Understand Key Concepts

1. Cholesterol is which type of macromolecule? **SC.6.L.14.2**
   Ⓐ carbohydrate
   Ⓑ lipid
   Ⓒ nucleic acid
   Ⓓ within the crust.

2. Genetic information is stored in which macromolecule? **SC.6.L.14.2**
   Ⓐ DNA
   Ⓑ glucose
   Ⓒ lipid
   Ⓓ starch

3. The arrow below is pointing to which cell part? **SC.6.L.14.4**

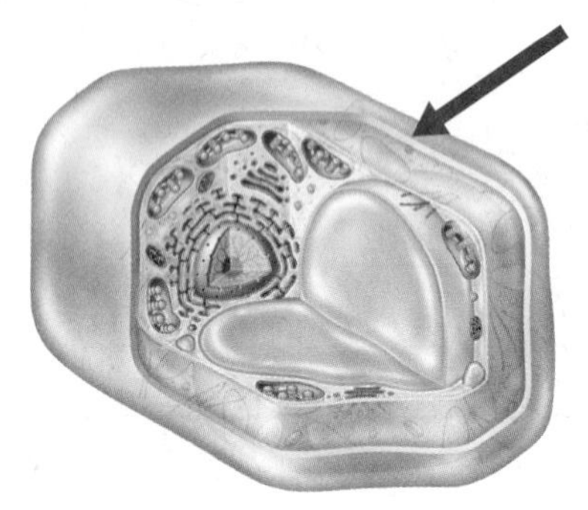

   Ⓐ chloroplast
   Ⓑ mitochondrion
   Ⓒ cell membrane
   Ⓓ cell wall

4. Which best describes vacuoles? **SC.6.L.14.4**
   Ⓐ lipids
   Ⓑ proteins
   Ⓒ contained in mitochondria
   Ⓓ storage compartments

5. Which is true of fermentation? **SC.6.L.14.3**
   Ⓐ does not generate energy
   Ⓑ does not require oxygen
   Ⓒ occurs in mitochondria
   Ⓓ produces lots of ATP

Critical Thinking

6. **Evaluate** the importance of the microscope to biology. **SC.6.L.14.2**

7. **Summarize** the role of water in cells. **LA.6.2.2.3**

8. **Infer** Why do cells need carrier proteins that transport glucose? **SC.6.L.14.3**

9. **Compare** the amounts of ATP generated in cellular respiration and fermentation. **LA.6.2.2.3**

10. **Hypothesize** how air pollution like smog affects photosynthesis. **SC.6.L.14.3**

11 **Compare** prokaryotes and eukaryotes by filling in the table below. SC.6.L.14.3

Structure	Prokaryote (yes or no)	Eukaryote (yes or no)
Cell membrane		
DNA		
Nucleus		
Endoplasmic reticulum		
Golgi apparatus		
Cell wall		

Writing in Science

12 **Write** On a separate sheet of paper, write a five-sentence paragraph relating the cytoskeleton to the walls of a building. Be sure to include a topic sentence and a concluding sentence in your paragraph. SC.6.L.14.4

Big Idea Review

13 How do the structures and processes of a cell enable it to survive? As an example, explain how chloroplasts help plant cells. SC.6.L.14.4

14 The photo at the beginning of the chapter shows a protozoan. What structures enable it to get food into its mouth? SC.6.L.14.4

Math Skills MA.6.A.3.6

Use Ratios

15 A rectangular solid measures 4 cm long by 2 cm wide by 2 cm high. What is the surface-area-to-volume ratio of the solid?

16 At different times during its growth, a cell has the following surface areas and volumes:

Time	Surface area (μm)	Volume (μm)
1	6	1
2	24	8
3	54	27

What happens to the surface-area-to-volume ratio as the cell grows?

Florida NGSSS Benchmark Practice

Fill in the correct answer choice.

Multiple Choice

1 Which is NOT a principle of cell theory? **SC.6.L.14.2**

Ⓐ All living things are made of one or more cells.

Ⓑ The cell is the smallest unit of life.

Ⓒ All cells are exactly the same.

Ⓓ All new cells come from preexisting cells.

Use the diagram below to answer question 2.

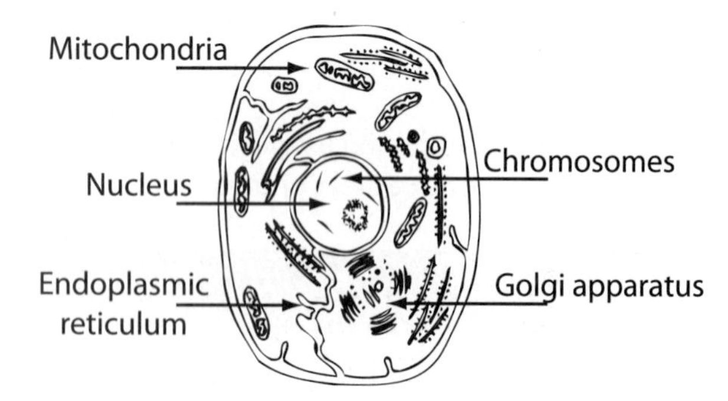

2 Where does respiration occur in the cell? **SC.6.L.14.4**

Ⓕ endoplasmic reticulum

Ⓖ nucleus

Ⓗ Golgi apparatus

Ⓘ mitochondria

3 Which macromolecule in a cell is responsible for storing energy, providing structural support, and is needed for communication among cells? **SC.6.L.14.3, SC.6.L.14.4**

Ⓐ proteins

Ⓑ lipids

Ⓒ nucleic acids

Ⓓ carbohydrates

Use the diagram below to answer questions 4 and 5.

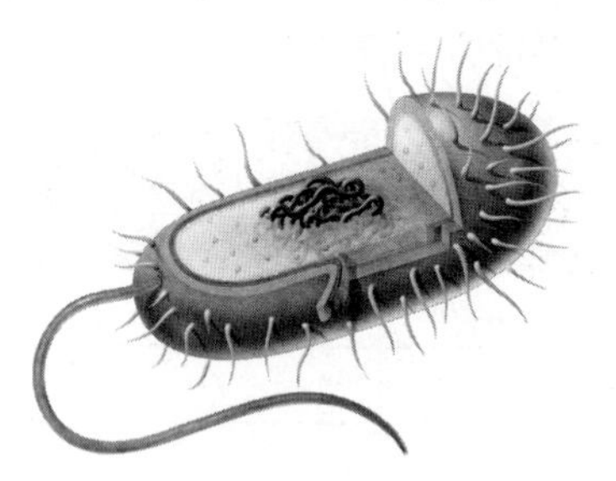

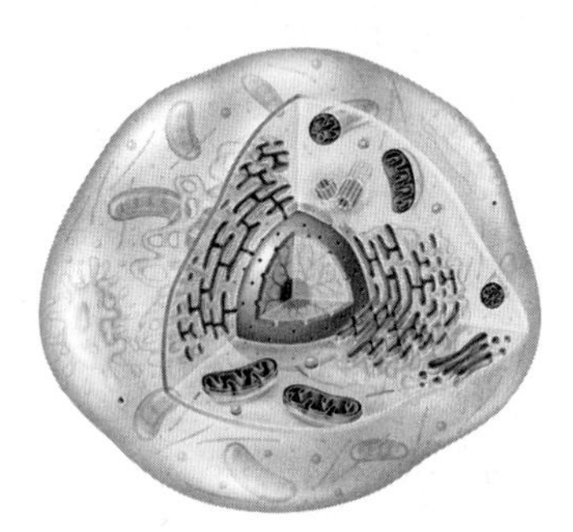

4 Which structure does the arrow point to in the eukaryotic cell? **SC.6.L.14.4**

Ⓕ cytoplasm

Ⓖ lysosome

Ⓗ nucleus

Ⓘ ribosome

5 Which feature does a typical prokaryotic cell have that is missing from some eukaryotic cells, like the one above? **SC.6.L.14.4**

Ⓐ cytoplasm

Ⓑ DNA

Ⓒ cell membrane

Ⓓ cell wall

6 Which cell organelle uses light energy to make food during photosynthesis? SC.6.L.14.4

Ⓕ nucleus

Ⓖ mitochondria

Ⓗ vacuole

Ⓘ chloroplast

Use the diagram below to answer question 7.

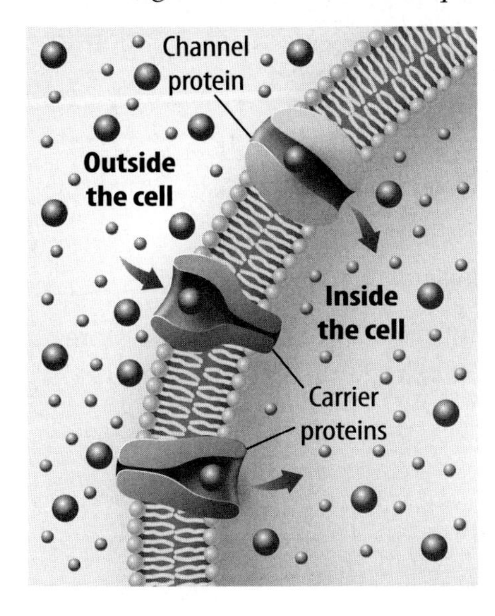

7 What process uses channel proteins and carrier proteins to control the movement of substances into and out of cells? SC.6.L.14.3

Ⓐ osmosis

Ⓑ facilitated diffusion

Ⓒ passive transport

Ⓓ diffusion

8 Which structure in an animal cell is a network of threadlike proteins that are joined together to provide a framework inside the cell? SC.6.L.14.4

Ⓕ cytoskeleton

Ⓖ cell wall

Ⓗ cytoplasm

Ⓘ cell membrane

9 What process do cells use when they need to take in nutrients from a lower concentration to a higher concentration? SC.6.L.14.3

Ⓐ diffusion

Ⓑ active transport

Ⓒ passive transport

Ⓓ exocytosis

10 Which organism has a cell wall? SC.6.L.14.4

Ⓕ mouse

Ⓖ pine tree

Ⓗ lion

Ⓘ snake

11 What is the main ingredient in any cell? SC.6.L.14.4

Ⓐ protein

Ⓑ nucleic acids

Ⓒ lipids

Ⓓ water

NEED EXTRA HELP?

If You Missed Question...	1	2	3	4	5	6	7	8	9	10	11
Go to Lesson...	1	2	1	2	2	2	3	2	3	2	1

Name ______________________ Date ______________

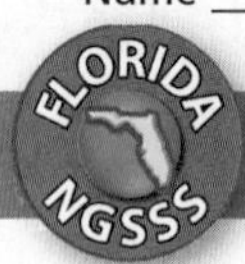

Multiple Choice *Bubble the correct answer.*

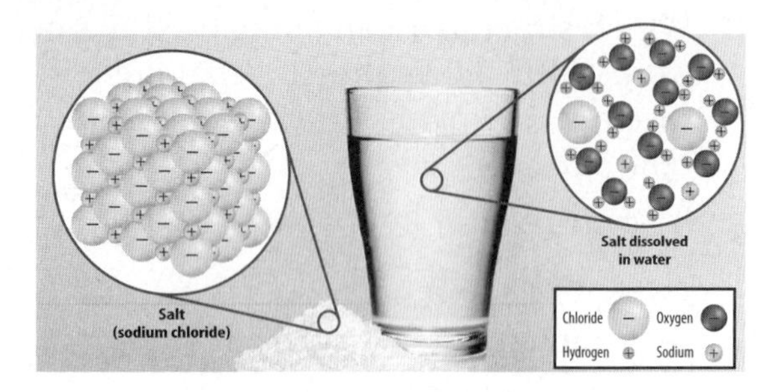

1. The picture above shows salt (NaCl) dissolved in water (H_2O). Which statement is true? **LA.6.2.2.3**

Ⓐ Chloride is attracted to the hydrogen atoms in a water molecule.

Ⓑ Sodium is attracted to the hydrogen atoms in a water molecule.

Ⓒ Neither sodium nor chloride is attracted to the atoms in a water molecule.

Ⓓ Sodium and chloride are attracted to the oxygen atoms in a water molecule.

2. Which type of macromolecule helps a cell break down food? **SC.6.L.14.2**

Ⓕ lipids

Ⓖ proteins

Ⓗ carbohydrates

Ⓘ nucleic acids

3. An amoeba can divide and form two new identical amoebas. Which macromolecules are copied and pass genetic information to the new cells? **SC.6.L.14.2**

Ⓐ lipids

Ⓑ proteins

Ⓒ carbohydrates

Ⓓ nucleic acids

4. One example of a carbohydrate that stores energy is **SC.6.L.14.2**

Ⓕ fat.

Ⓖ hair.

Ⓗ keratin.

Ⓘ sugar.

Name ______________________ Date __________

mini BAT

Multiple Choice *Bubble the correct answer.*

1. Which cell belongs to a prokaryote? **SC.6.L.14.3**

Ⓐ
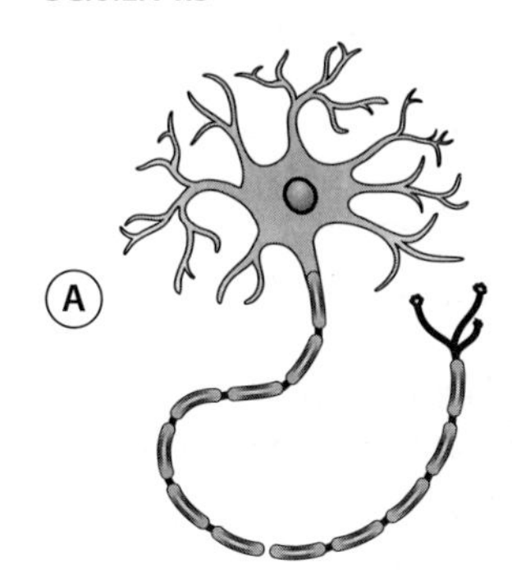

Ⓑ
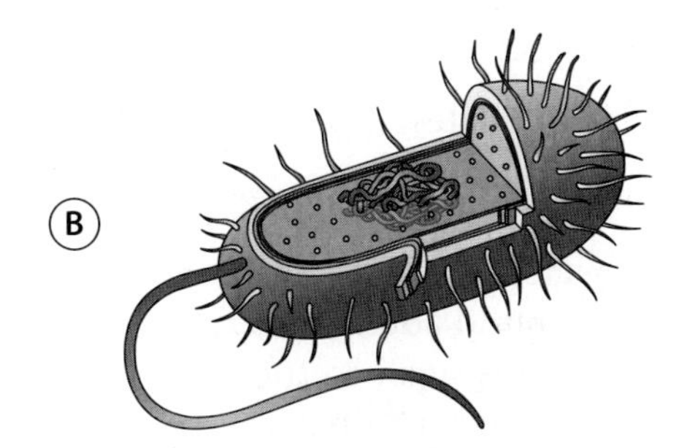

Ⓒ
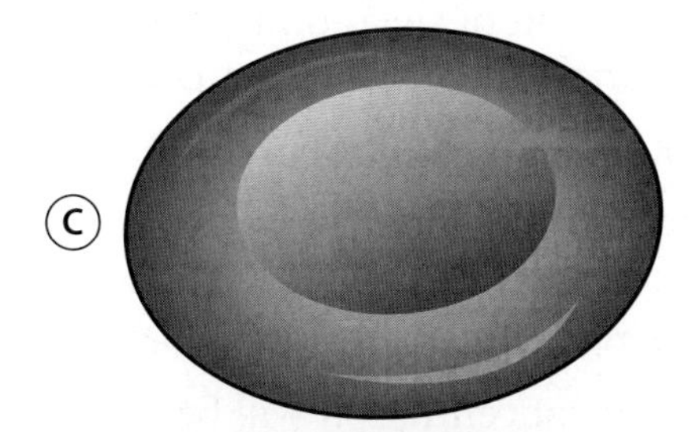

Ⓓ
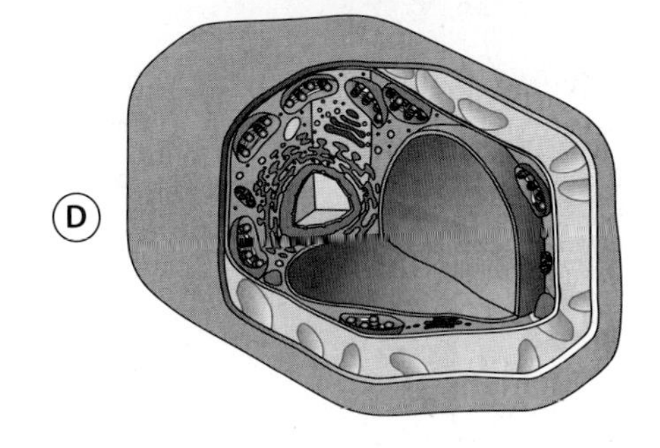

2. Your body is protected by multiple layers of skin cells. What shape would you expect a cell on the outer layer of skin to have? **SC.6.L.14.4**
   - Ⓕ a flat shape
   - Ⓖ a round shape
   - Ⓗ a long, branching shape
   - Ⓘ a short, hollow shape

3. Which cell structure could be called the packaging center of the cell? **SC.6.L.14.4**
   - Ⓐ lysosome
   - Ⓑ mitochondrion
   - Ⓒ endoplasmic reticulum
   - Ⓓ Golgi apparatus

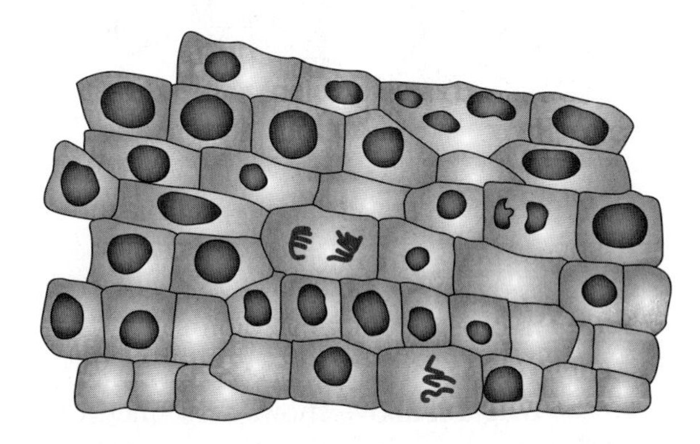

4. Rosa uses a microscope to look at a group of cells, as shown above. She sees that the cells are joined together, so she knows that they are from one organism. She also sees that all of them have cell walls. Rosa could be looking at **SC.6.L.14.3**
   - Ⓕ bacterial cells.
   - Ⓖ human cells.
   - Ⓗ mouse cells.
   - Ⓘ mushroom cells.

Name ______________________ Date ____________

Multiple Choice *Bubble the correct answer.*

Use the image below to answer questions 1 and 2.

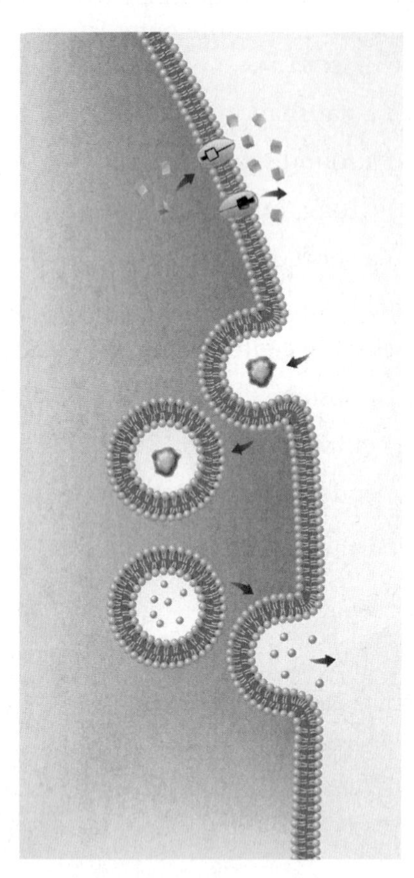

1. The diagram above shows three methods that require energy to move materials across the cell membrane. Which process do cells use to take in bacteria and viruses? **SC.6.L.14.3**

Ⓐ endocytosis

Ⓑ exocytosis

Ⓒ active transport

Ⓓ passive transport

2. Which process removes proteins and hormones from a cell? **SC.6.L.14.3**

Ⓕ endocytosis

Ⓖ exocytosis

Ⓗ active transport

Ⓘ passive transport

3. Which type of macromolecule helps move molecules into a cell through the process of facilitated diffusion? **SC.6.L.14.3**

Ⓐ lipids

Ⓑ proteins

Ⓒ carbohydrates

Ⓓ nucleic acids

4. Sasha places a semipermeable membrane in a beaker and adds water to the beaker. She then adds sugar to the water on the left side of the beaker. What will happen to the concentration of sugar on both sides of the membrane after 30 minutes? **LA.6.2.2.3**

Ⓕ All the sugar will be concentrated on the left side of the membrane.

Ⓖ All the sugar will be concentrated on the right side of the membrane.

Ⓗ Sugar concentration will be equal on both sides.

Ⓘ Sugar concentration will be high on the left side and low on the right side.

Name ______________________________ Date ______________

Multiple Choice *Bubble the correct answer.*

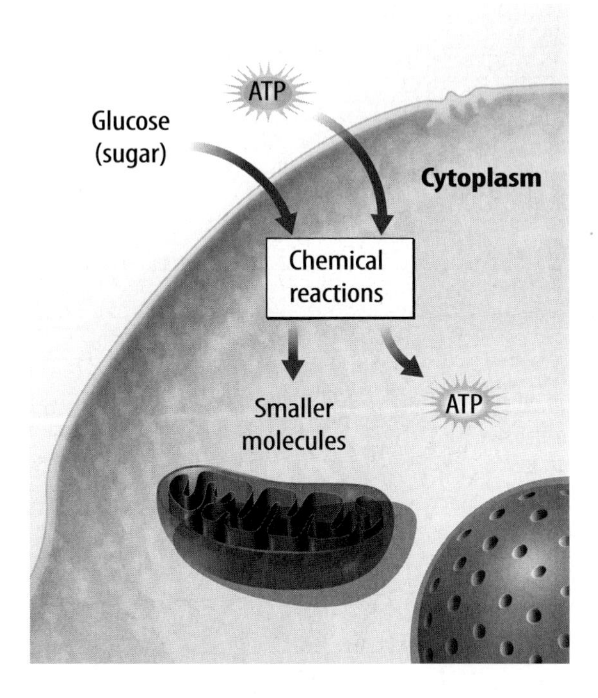

1. Which process is illustrated in the image above? **SC.8.L.18.2**

(A) glycolysis

(B) photosynthesis

(C) alcohol fermentation

(D) lactic-acid fermentation

2. Which process produces the most ATP? **SC.6.L.14.3**

(F) Plant cells convert sunlight to sugar through photosynthesis.

(G) Yeast cells produce ethanol through alcohol fermentation.

(H) Human stomach cells convert sugar to energy through cellular respiration.

(I) Human muscle cells generate energy at the end of a race through lactic-acid fermentation.

3. Which substance is a product of cellular respiration? **SC.6.L.14.4**

(A) glucose

(B) lactic acid

(C) oxygen

(D) water

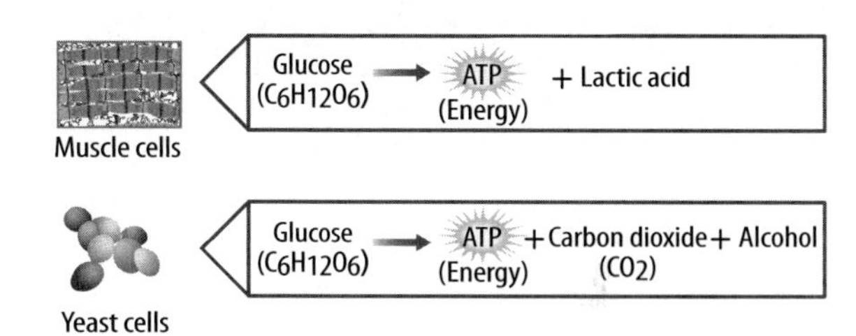

4. Which processes are shown in the image above? **SC.6.L.14.3**

(F) two types of cellular respiration

(G) two types of fermentation

(H) two types of glycolysis

(I) two types of photosynthesis

Name ______ Date ______

Notes

Getting Bigger

A small baby snake grows into a larger adult snake. Choose the explanation that best describes why a small baby snake grows into a larger adult snake.

A. The baby snake's cells divide.

B. The baby snake's cells grow into longer cells.

C. The baby snake's cells grow into much larger cells.

D. The baby snake's body parts stretch out and get longer.

E. The food the baby snake eats makes it grow big and strong.

F. The baby snake's cells differentiate into different types of cells.

Explain your thinking. Describe your ideas about growth.

From a Cell to an ORGANISM

The Big Idea

FLORIDA BIG IDEAS

1 The Practice of Science

3 The Role of Theories, Laws, Hypotheses, and Models

14 Organization and Development of Living Organisms

Think About It! How can one cell become a multicellular organism?

From the outside, a chicken egg looks like a simple oval object. But big changes are taking place inside the egg. Over several weeks, the one cell in the egg will grow and divide and become a chick.

1. How do you think the original cell changed over time?

2. What might have happened to the chick's cells as the chick grew?

3. How do you think one cell can become a multicellular chick?

Get Ready to Read What do you think about organism development?

Before you read, decide if you agree or disagree with each of these statements. As you read this chapter, see if you change your mind about any of the statements.

	AGREE	DISAGREE
1 Cell division produces two identical cells.	☐	☐
2 Cell division is important for growth.	☐	☐
3 At the end of the cell cycle, the original cell no longer exists.	☐	☐
4 Unicellular organisms do not have all the characteristics of life.	☐	☐
5 All the cells in a multicellular organism are the same.	☐	☐
6 Some organs work together as part of an organ system.	☐	☐

There's More Online!
Video • Audio • Review • ⓘLab Station • WebQuest • Assessment • Concepts in Motion • Multilingual eGlossary

Lesson 1

The Cell Cycle and Cell DIVISION

ESSENTIAL QUESTIONS

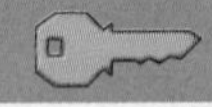 What are the phases of the cell cycle?

 Why is the result of the cell cycle important?

Vocabulary

cell cycle p. 95
interphase p. 96
sister chromatid p. 98
centromere p. 98
mitosis p. 99
cytokinesis p. 99
daughter cell p. 99

Launch Lab

SC.6.N.1.5

15 minutes

Why isn't your cell like mine?

All living things are made of cells. Some are made of only one cell, while others are made of trillions of cells. Where do all those cells come from?

Procedure

1. Read and complete a lab safety form.
2. Ask your team members to face away from you. Draw an animal cell on a sheet of **paper.** Include as many organelles as you can.
3. Use **scissors** to cut the cell drawing into equal halves. Fold each sheet of paper in half so the drawing cannot be seen.
4. Ask your team members to face you. Give each team member half of the cell drawing.
5. Have team members sit facing away from each other. Each person should use a **glue stick** to attach the cell half to one side of a sheet of paper. Then, each person should draw the missing cell half.
6. Compare the two new cells to your original cell.

Think About This

1. How did the new cells compare to the original cell?

2. **Key Concept** What are some things that might be done in the early steps to produce two new cells that are more like the original cell?

Florida NGSSS

LA.6.2.2.3 The student will organize information to show understanding (e.g., representing main ideas within text through charting, mapping, paraphrasing, summarizing, or comparing/contrasting);

MA.6.A.3.6 Construct and analyze tables, graphs, and equations to describe linear functions and other simple relations using both common language and algebraic notation.

SC.6.L.14.3 Recognize and explore how cells of all organisms undergo similar processes to maintain homeostasis, including extracting energy from food, getting rid of waste, and reproducing.

SC.7.L.16.3 Compare and contrast the general processes of sexual reproduction requiring meiosis and asexual reproduction requiring mitosis.

SC.6.N.1.5 Recognize that science involves creativity, not just in designing experiments, but also in creating explanations that fit evidence.

SC.6.N.2.1 Distinguish science from other activities involving thought.

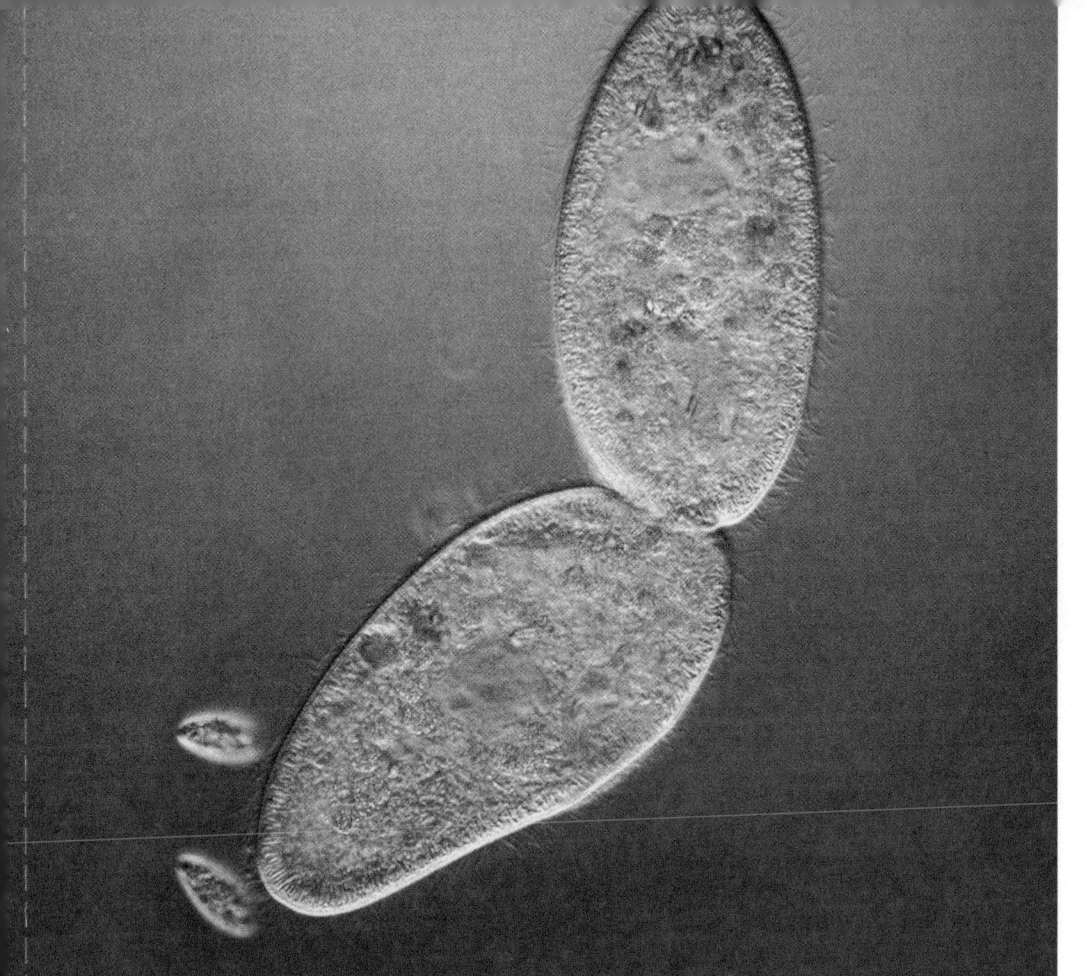

Inquiry Time to Split?

1. Unicellular organisms such as these reproduce when one cell divides into two new cells. The two cells are identical to each other. What do you think happened to the contents of the cell before it divided?

The Cell Cycle

No matter where you live, you have probably noticed that the weather changes in a regular pattern each year. Some areas experience four seasons—winter, spring, summer, and fall. In other parts of the world, there are only two seasons—rainy and dry. As seasons change, temperature, precipitation, and the number of hours of sunlight vary in a regular cycle.

These changes can affect the life cycles of organisms such as trees. Notice how the tree in **Figure 1** changes with the seasons. Like changing seasons or the growth of trees, cells go through cycles. *Most cells in an organism go through a cycle of growth, development, and division called the* **cell cycle.** Through the cell cycle, organisms grow, develop, replace old or damaged cells, and produce new cells.

Figure 1 This maple tree changes in response to a seasonal cycle.

2. **Visual Check** List What are the seasonal changes of this maple tree?

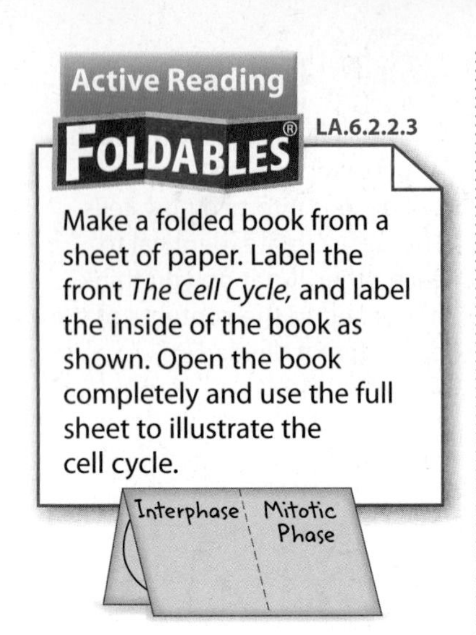

Phases of the Cell Cycle

There are two main phases in the cell cycle—interphase and the mitotic (mi TAH tihk) phase. **Interphase** *is the period during the cell cycle of a cell's growth and development.* A cell spends most of its life in interphase, as shown in **Figure 2.** During interphase, most cells go through three stages:

- rapid growth and replication, or copying, of the membrane-bound structures called organelles;
- copying of DNA, the genetic information in a cell; and
- preparation for cell division.

Interphase is followed by a shorter period of the cell cycle known as the mitotic phase. A cell reproduces during this phase. The mitotic phase has two stages, as illustrated in **Figure 2.** The nucleus divides in the first stage, and the cell's fluid, called the cytoplasm, divides in the second stage. The mitotic phase creates two new identical cells. At the end of this phase, the original cell no longer exists.

3. NGSSS Check Differentiate Underline the two main phases of the cell cycle. SC.6.L.14.3

The Cell Cycle

Figure 2 A cell spends most of its life growing and developing during interphase.

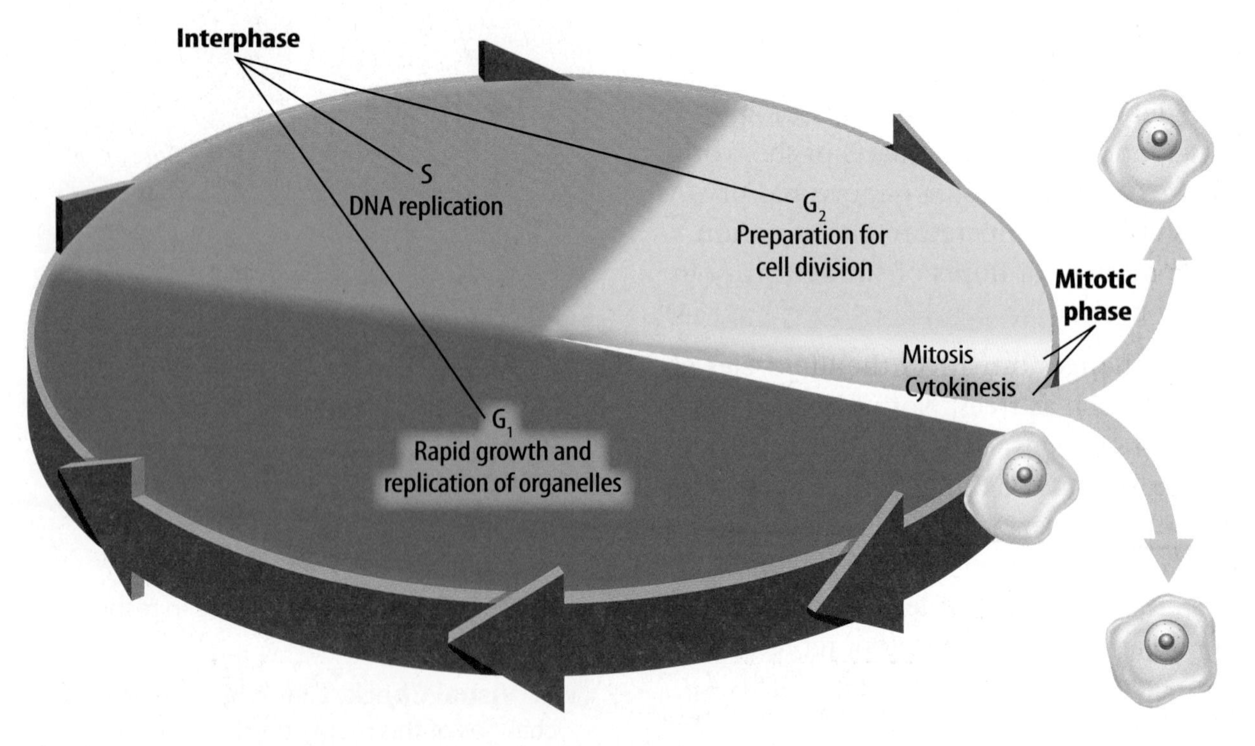

4. Visual Check Name Which stage of interphase is the longest? ______

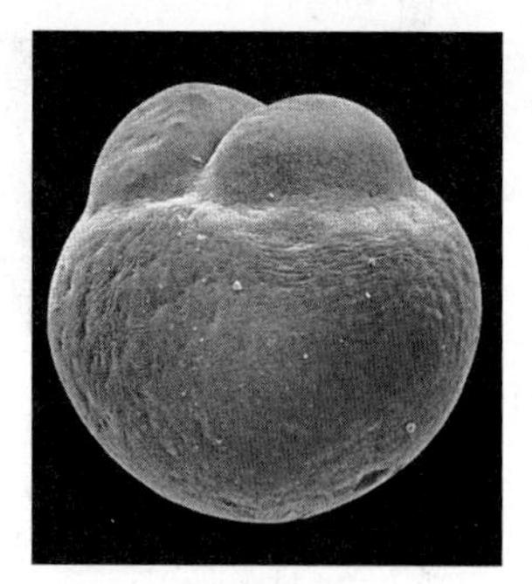
2-cell stage
SEM Magnification: 160×

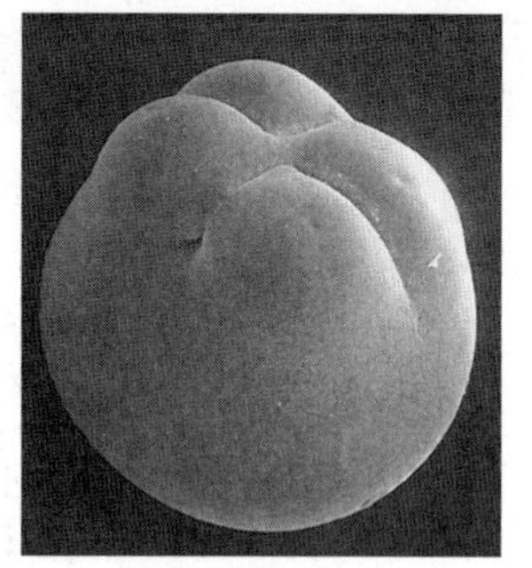
4-cell stage
SEM Magnification: 155×

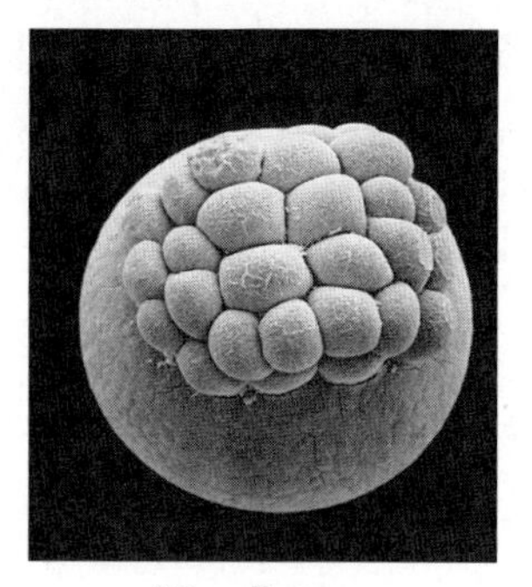
32-cell stage
SEM Magnification: 150×

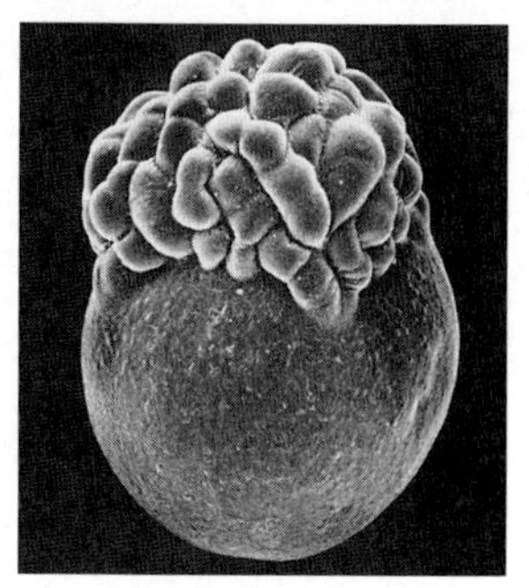
256-cell stage
SEM Magnification: 130×

Figure 3 The fertilized egg of a zebra fish divides into 256 cells in 2.5 hours.

Length of a Cell Cycle

The time it takes a cell to complete the cell cycle depends on the type of cell that is dividing. Recall that a **eukaryotic** cell has membrane-bound organelles, including a nucleus. For some eukaryotic cells, the cell cycle might last only eight minutes. For other cells, the cycle might take as long as one year. Most dividing human cells normally complete the cell cycle in about 24 hours. As illustrated in **Figure 3,** the cells of some organisms divide very quickly.

REVIEW VOCABULARY

eukaryotic
a cell with membrane-bound structures

Interphase

As you have read, interphase makes up most of the cell cycle. Newly produced cells begin interphase with a period of rapid growth—the cell gets bigger. This is followed by cellular activities such as making proteins. Next, actively dividing cells make copies of their DNA and prepare for cell division. During interphase, the DNA is called chromatin (KROH muh tun). Chromatin is long, thin strands of DNA, as shown in **Figure 4.** When scientists dye a cell in interphase, the nucleus looks like a plate of spaghetti. This is because the nucleus contains many strands of chromatin tangled together.

Figure 4 During interphase, the nuclei of an animal cell and a plant cell contain long, thin strands of DNA called chromatin.

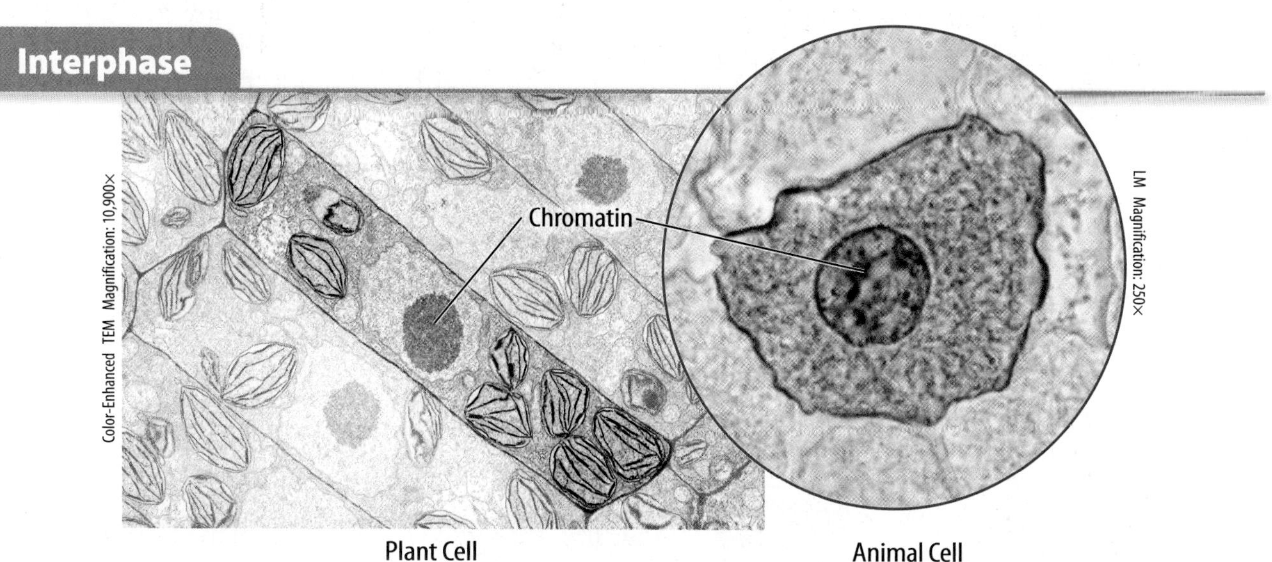

Table 1 Phases of the Cell Cycle

Phase	Stage	Description
Interphase		growth and cellular functions; organelle replication
		growth and chromosome replication; organelle replication
		growth and cellular functions; organelle replication
Mitotic phase		division of nucleus
		division of cytoplasm

Table 1 The two phases of the cell cycle can each be divided into different stages.

Active Reading **5. Discover** Fill in the missing information in the table.

Phases of Interphase

Scientists divide interphase into three stages, as shown in **Table 1.** Interphase begins with a period of rapid growth—the G_1 stage. This stage lasts longer than other stages of the cell cycle. During G_1, a cell grows and carries out its normal cell functions. For example, during G_1, cells that line your stomach make enzymes that help digest your food. Although most cells continue the cell cycle, some cells stop the cell cycle at this point. For example, mature nerve cells in your brain remain in the G_1 stage and do not divide again.

During the second stage of interphase—the S stage—a cell continues to grow and copies its DNA. There are now identical strands of DNA. These identical strands of DNA ensure that each new cell gets a copy of the original cell's genetic information. Each strand of DNA coils up and forms a chromosome. Identical chromosomes join together. The cell's DNA is now arranged as pairs of identical chromosomes. Each pair is called a duplicated chromosome. *Two identical chromosomes, called* **sister chromatids,** *make up a duplicated chromosome,* as shown in **Figure 5.** Notice that the *sister chromatids are held together by a structure called the* **centromere.**

The final stage of interphase—the G_2 stage—is another period of growth and the final preparation for mitosis. A cell uses energy copying DNA during the S stage. During G_2, the cell stores energy that will be used during the mitotic phase of the cell cycle.

Active Reading **6. Identify** Highlight what happens in the G_2 phase.

Figure 5 The coiled DNA forms a duplicated chromosome made of two sister chromatids connected at the centromere.

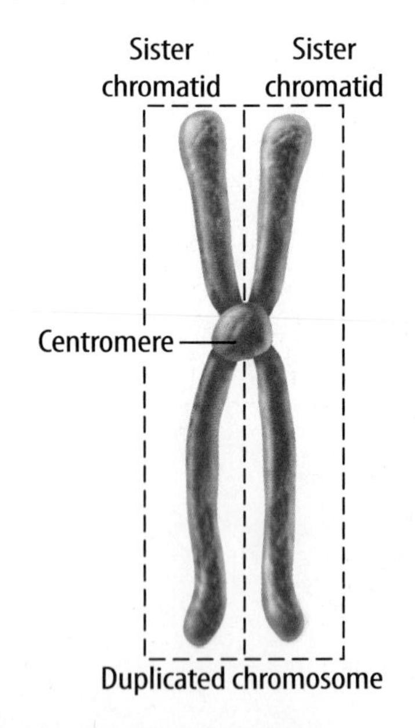

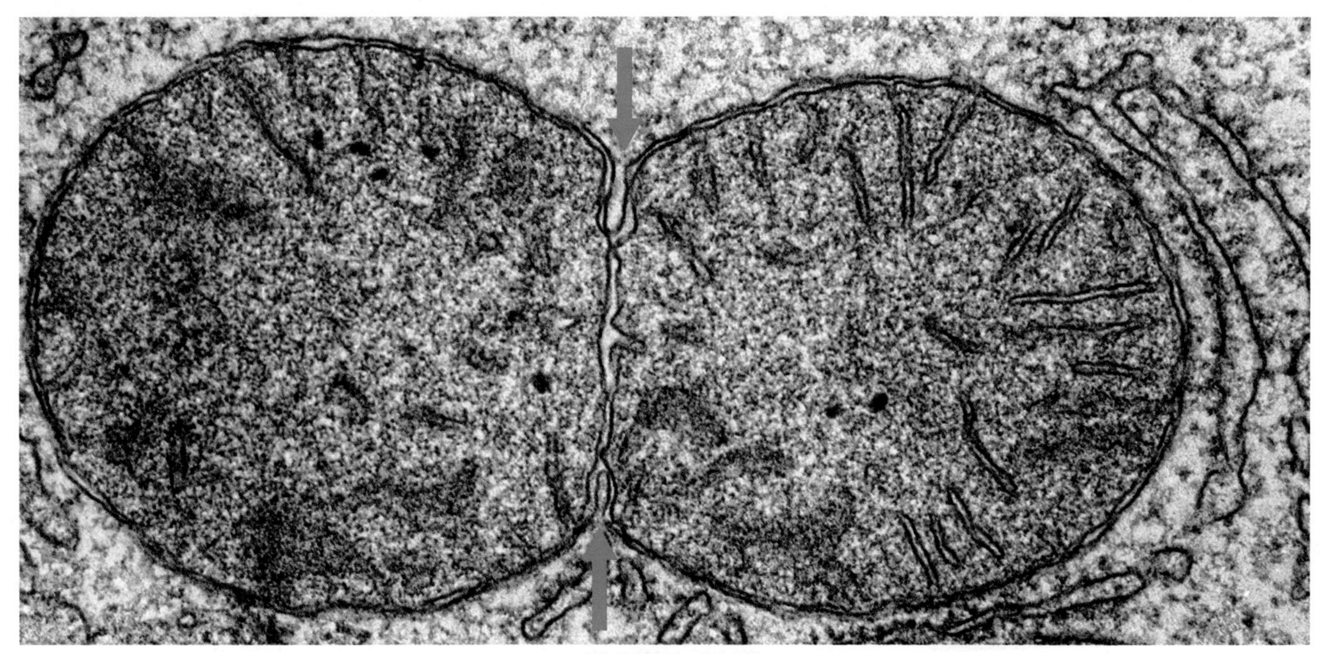
TEM Magnification: Unavailable

Figure 6 This mitochondrion is in the final stage of dividing.

Organelle Replication

During cell division, the organelles in a cell are distributed between the two new cells. Before a cell divides, it makes a copy of each organelle. This enables the two new cells to function properly. Some organelles, such as the energy-processing mitochondria and chloroplasts, have their own DNA. These organelles can make copies of themselves on their own, as shown in **Figure 6.** A cell produces other organelles from materials such as proteins and lipids. A cell makes these materials using the information contained in the DNA inside the nucleus. Organelles are copied during all stages of interphase.

The Mitotic Phase

The mitotic phase of the cell cycle follows interphase. It consists of two stages: mitosis (mi TOH sus) and cytokinesis (si toh kuh NEE sus). *In* **mitosis,** *the nucleus and its contents divide. In* **cytokinesis,** *the cytoplasm and its contents divide.* **Daughter cells** *are the two new cells that result from mitosis and cytokinesis.*

Word Origin

mitosis
from Greek *mitos,* means "warp thread"; and Latin *–osis,* means "process"

During mitosis, the contents of the nucleus divide, forming two identical nuclei. The sister chromatids of the duplicated chromosomes separate from each other. This gives each daughter cell the same genetic information. For example, a cell that has ten duplicated chromosomes actually has 20 chromatids. When the cell divides, each daughter cell will have ten different chromatids. Chromatids are now called chromosomes.

In cytokinesis, the cytoplasm divides and forms the two new daughter cells. Organelles that were made during interphase are divided between the daughter cells.

Phases of Mitosis

Like interphase, mitosis is a continuous process that scientists divide into different phases, as shown in **Figure 7.**

Prophase During the first phase of mitosis, called prophase, the copied chromatin coils together tightly. The coils form visible duplicated chromosomes. The nucleolus disappears, and the nuclear membrane breaks down. Structures called spindle fibers form in the cytoplasm.

Metaphase During metaphase, the spindle fibers pull and push the duplicated chromosomes to the middle of the cell. Notice in **Figure 7** that the chromosomes line up along the middle of the cell. This arrangement ensures that each new cell will receive one copy of each chromosome. It is important that new cells be identical. Incorrect genetic information in the new cell can lead to the uncontrolled division and growth of new cells, known as cancer.

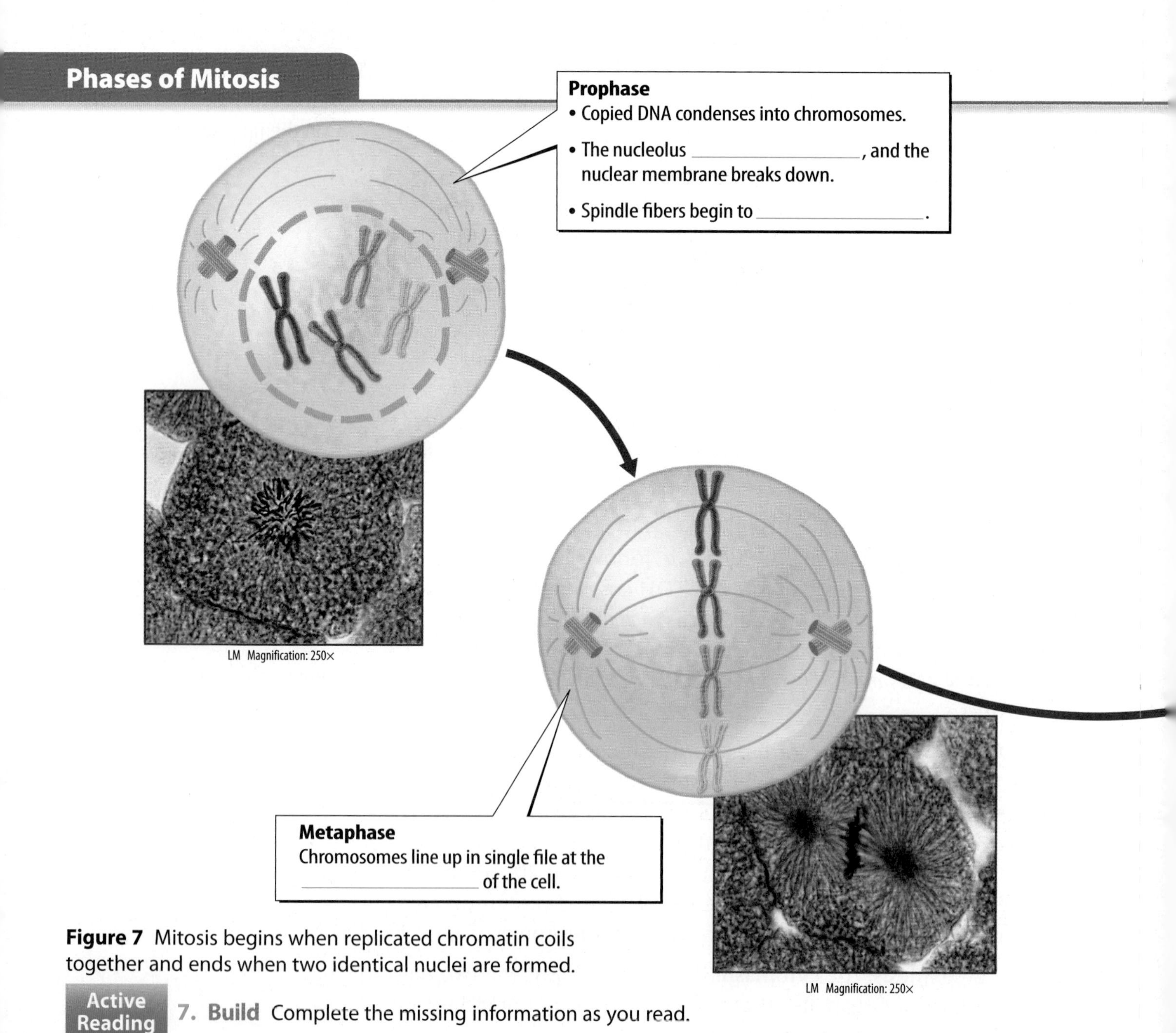

Figure 7 Mitosis begins when replicated chromatin coils together and ends when two identical nuclei are formed.

Active Reading **7. Build** Complete the missing information as you read.

Anaphase In anaphase, the third stage of mitosis, the two sister chromatids in each chromosome separate from each other. The spindle fibers pull them in opposite directions. Once separated, the chromatids are now two identical, single-stranded chromosomes. As they move to opposite sides of a cell, the cell begins to get longer. Anaphase is complete when the two identical sets of chromosomes are at opposite ends of a cell.

Telophase During telophase, the spindle fibers begin to disappear. Also, the chromosomes begin to uncoil. A nuclear membrane forms around each set of chromosomes at either end of the cell. This forms two new identical nuclei. Telophase is the final stage of mitosis. It is often described as the reverse of prophase because many of the processes that occur during prophase are reversed during telophase.

8. State What are the phases of mitosis?

____

____

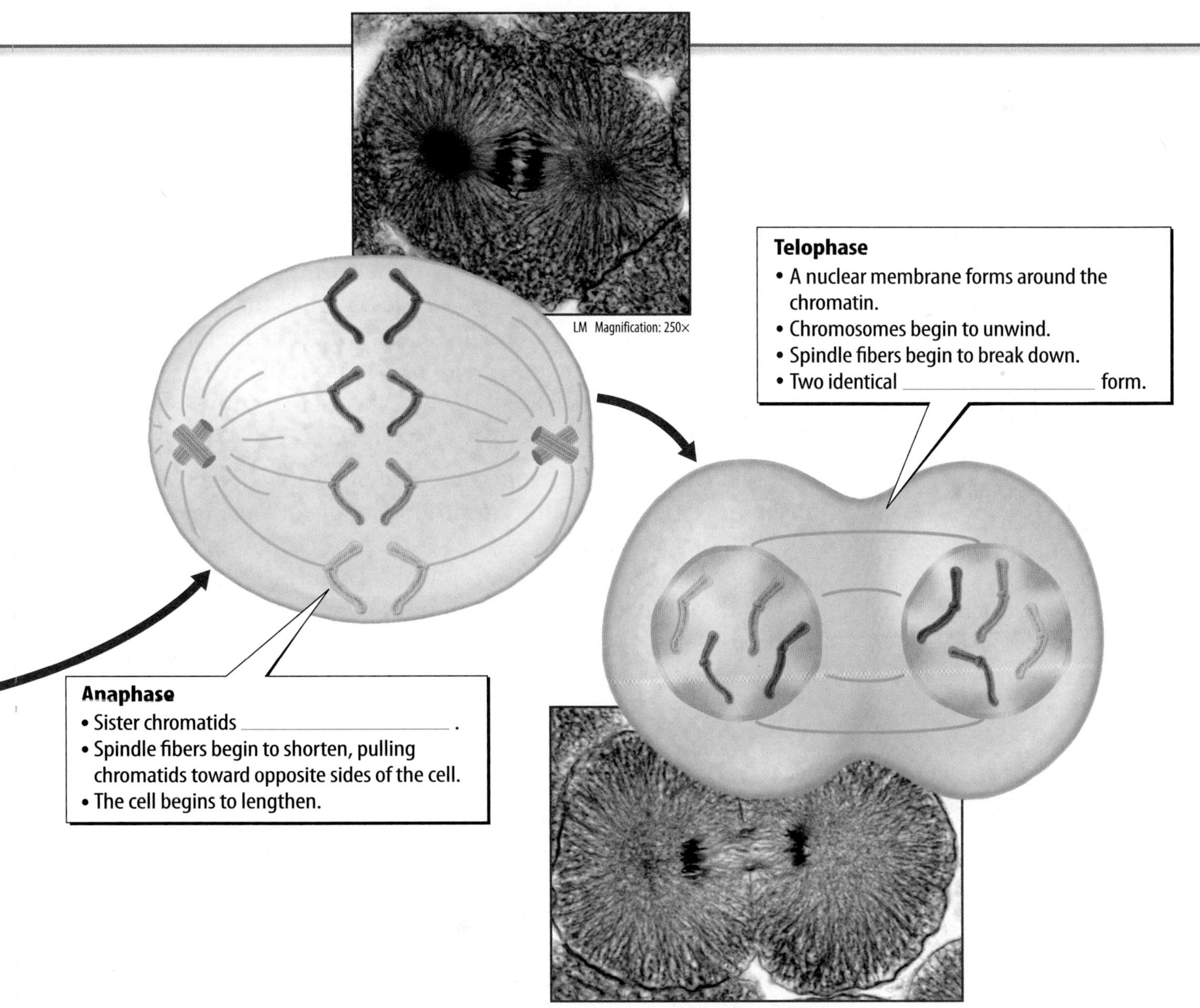

Cytokinesis

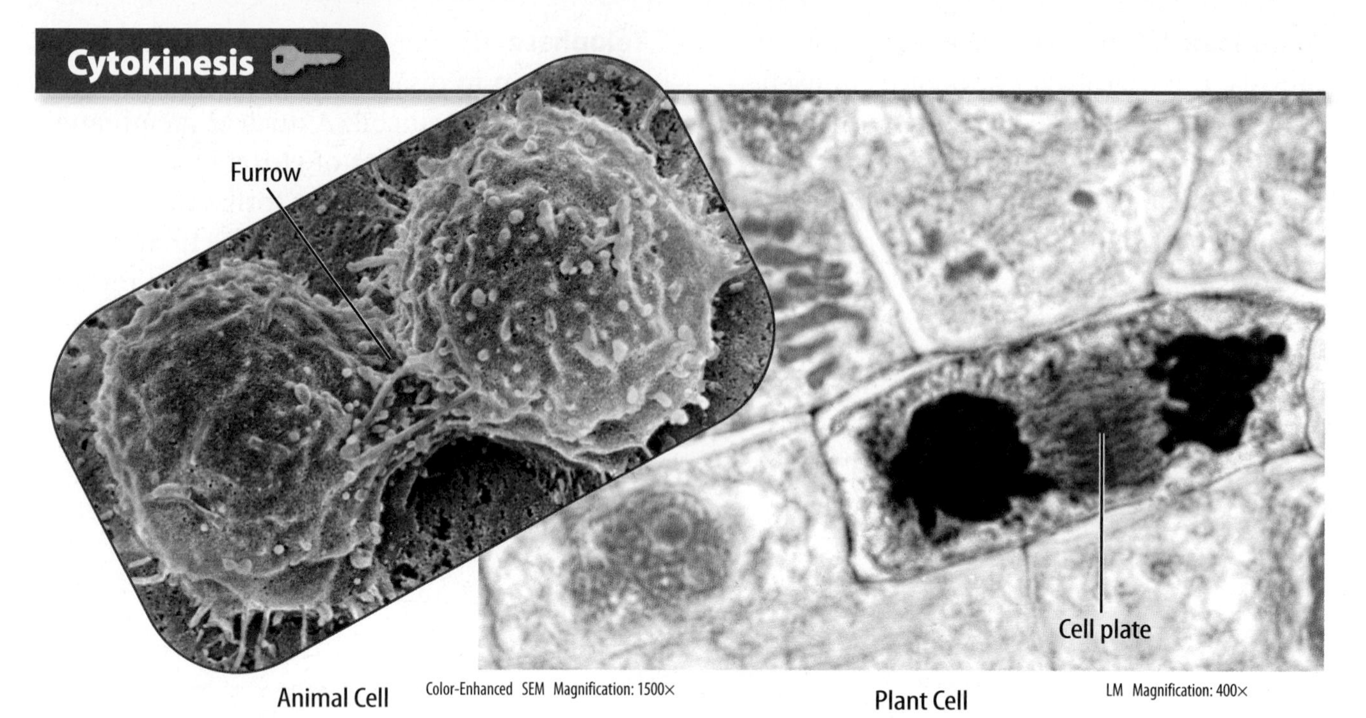

Figure 8 Cytokinesis differs in animal cells and plant cells.

Dividing the Cell's Components

Following the last phase of mitosis, a cell's cytoplasm divides in a process called cytokinesis. The specific steps of cytokinesis differ depending on the type of cell that is dividing. In animal cells, the cell membrane contracts, or squeezes together, around the middle of the cell. Fibers around the center of the cell pull together. This forms a crease, called a furrow, in the middle of the cell. The furrow gets deeper and deeper until the cell membrane comes together and divides the cell. An animal cell undergoing cytokinesis is shown in **Figure 8.**

Cytokinesis in plants happens in a different way. As shown in **Figure 8,** a new cell wall forms in the middle of a plant cell. First, organelles called vesicles join together to form a membrane-bound disk called a cell plate. Then the cell plate grows outward toward the cell wall until two new cells form.

Active Reading **9. Compare** Highlight the process of cytokinesis in plant and animal cells.

Math Skills MA.6.A.3.6

Use Percentages

A percentage is a ratio that compares a number to 100. If the length of the entire cell cycle is 24 hours, 24 hours equals 100%. If part of the cycle takes 6.0 hours, it can be expressed as 6.0 hours/24 hours. To calculate percentage, divide and multiply by 100. Add a percent sign.

$\frac{6.0}{24} = 0.25 \times 100 = 25\%$

Practice

10. Interphase in human cells takes about 23 hours. If the cell cycle is 24 hours, what percentage is interphase?

Results of Cell Division

Recall that the cell cycle results in two new cells. These daughter cells are genetically identical to each other and to the original cell that no longer exists. For example, a human cell has 46 chromosomes. When that cell divides, it will produce two new cells with 46 chromosomes each. The cell cycle is important for reproduction in some organisms, growth in multicellular organisms, replacement of worn out or damaged cells, and repair of damaged tissues.

Reproduction

In some unicellular organisms, cell division is a form of asexual reproduction. For example, an organism called a paramecium often reproduces by dividing into two new daughter cells or two new paramecia. Cell division is also important in other methods of reproduction in which the offspring are identical to the parent organism.

Growth

Cell division allows multicellular organisms, such as humans, to grow and develop from one cell (a fertilized egg). In humans, cell division begins about 24 hours after fertilization and continues rapidly during the first few years of life. It is likely that during the next few years you will go through another period of rapid growth and development. This happens because cells divide and increase in number as you grow and develop.

Replacement

Even after an organism is fully grown, cell division continues. This process replaces cells that wear out or are damaged. The outermost layer of your skin is always rubbing or flaking off. A layer of cells below the skin's surface is constantly dividing. This produces millions of new cells daily to replace the ones that are rubbed off.

Repair

Cell division is also critical for repairing damage. When a bone breaks, cell division produces new bone cells that patch the broken pieces back together.

Not all damage can be repaired, however, because not all cells continue to divide. Recall that mature nerve cells stop the cell cycle in interphase. For this reason, injuries to nerve cells often cause permanent damage.

11. **NGSSS Check** **Judge** Underline why the result of the cell cycle is important. **SC.6.L.14.3**

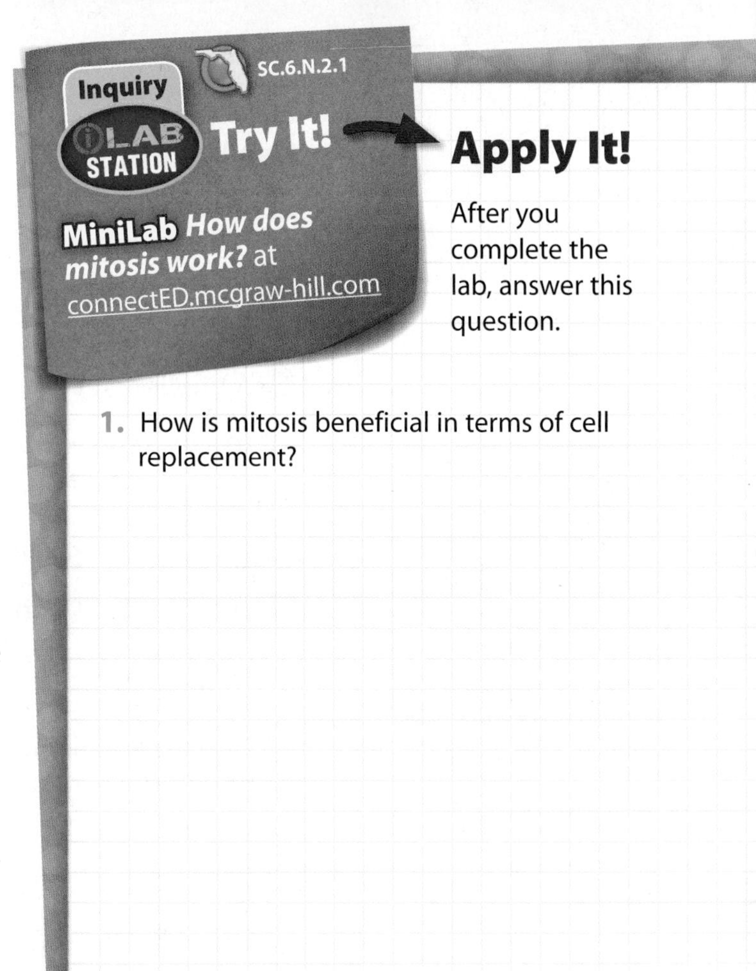

Active Reading

12. **Assess** Write a question about the main idea under each heading. Exchange quizzes with another student. Together, discuss the answers to the quizzes.

Lesson Review 1

Visual Summary

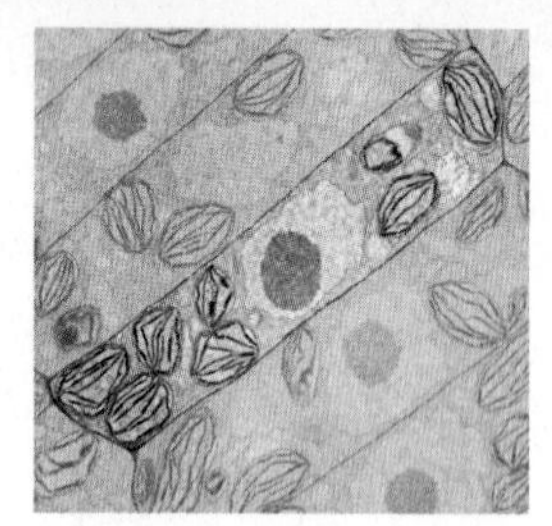

During interphase, most cells go through periods of rapid growth and replication of organelles, copying DNA, and preparation for cell division.

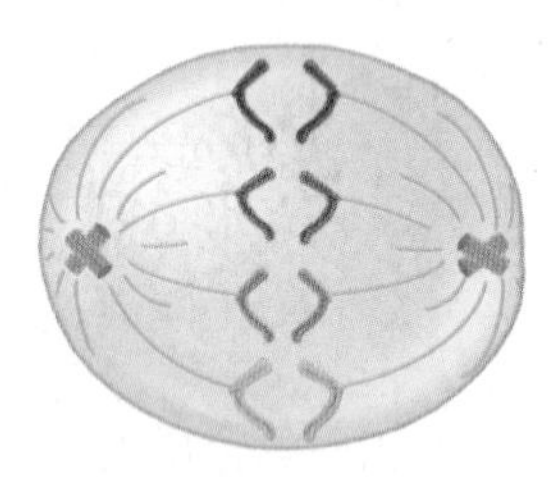

The nucleus and its contents divide during mitosis.

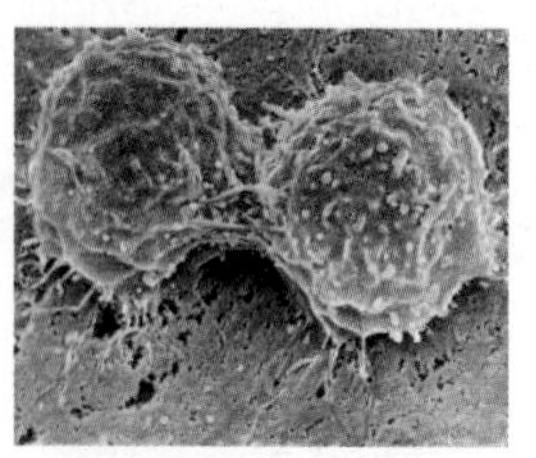

The cytoplasm and its contents divide during cytokinesis.

Use Vocabulary

1. **Distinguish** between mitosis and cytokinesis.

____

____

2. A duplicated chromosome is made of two _______________.

3. **Use the term** *interphase* in a sentence.

____

Understand Key Concepts

4. Which is NOT part of mitosis?
   - (A) anaphase
   - (B) interphase
   - (C) prophase
   - (D) telophase

5. **Give three examples** of why the result of the cell cycle is important. **SC.6.L.14.3**

____

____

Interpret Graphics

6. **Organize** Fill in the graphic organizer below to show the results of cell division. **LA.6.2.2.3**

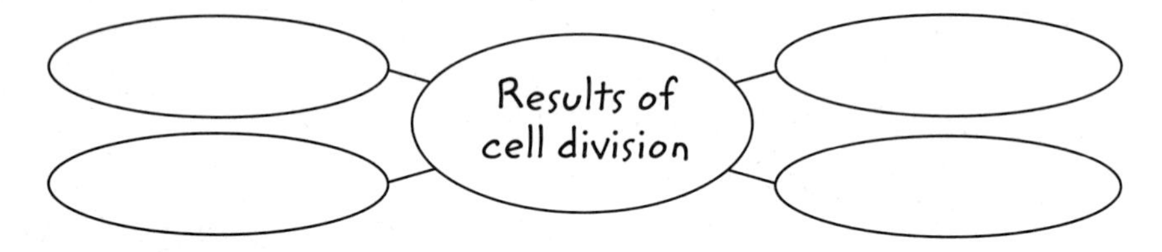

Critical Thinking

7. **Predict** what might happen to a cell if it were unable to divide by mitosis.

____

____

Math Skills

8. The mitotic phase of the human cell cycle takes approximately 1 hour. What percentage of the 24-hour cell cycle is the mitotic phase?

DNA Fingerprinting

SCIENCE & SOCIETY

DNA

Solving Crimes One Strand at a Time

Every cell in your body has the same DNA in its nucleus. Unless you are an identical twin, your DNA is entirely unique. Identical twins have identical DNA because they begin as one cell that divides and separates. When your cells begin mitosis, they copy their DNA. Every new cell has the same DNA as the original cells. That is why DNA can be used to identify people. Just as no two people have the same fingerprints, your DNA belongs to you alone.

Using scientific methods to solve crimes is called forensics. DNA fingerprinting is now a basic tool in forensics. Samples collected from a crime scene can be compared to millions of samples previously collected and indexed in a computer.

Every day, everywhere you go, you leave a trail of DNA. It might be in skin cells. It might be in hair or in the saliva you used to lick an envelope. If you commit a crime, you will most likely leave DNA behind. An expert crime scene investigator will know how to collect that DNA.

DNA evidence can prove innocence as well. Investigators have reexamined DNA found at old crime scenes. Imprisoned persons have been proven not guilty through DNA fingerprinting methods that were not yet available when the crime was committed.

DNA fingerprinting can also be used to identify bodies that were previously known only as a John or Jane Doe.

The Federal Bureau of Investigation (FBI) has a nationwide index of DNA samples called CODIS (Combined DNA Index System).

It's Your Turn

DISCOVER Your cells contain organelles called mitochondria. They have their own DNA, called mitochondrial DNA. Your mitochondrial DNA is identical to your mother's mitochondrial DNA. Find out how this information is used.

Lesson 2

Levels of ORGANIZATION

ESSENTIAL QUESTIONS

How do unicellular and multicellular organisms differ?

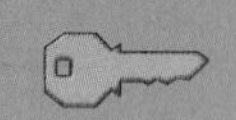
How does cell differentiation lead to the organization within a multicellular organism?

Vocabulary

cell differentiation p. 109
stem cell p. 110
tissue p. 111
organ p. 112
organ system p. 113

Florida NGSSS

LA.6.2.2.3 The student will organize information to show understanding (e.g., representing main ideas within text through charting, mapping, paraphrasing, summarizing, or comparing/contrasting).

SC.6.N.1.1 Define a problem from the sixth grade curriculum, use appropriate reference materials to support scientific understanding, plan and carry out scientific investigation of various types, such as systematic observations or experiments, identify variables, collect and organize data, interpret data in charts, tables, and graphics, analyze information, make predictions, and defend conclusions.

SC.6.N.1.5 Recognize that science involves creativity, not just in designing experiments, but also in creating explanations that fit evidence.

SC.6.N.2.1 Distinguish science from other activities involving thought.

SC.6.L.14.1 Describe and identify patterns in the hierarchical organization of organisms from atoms to molecules and cells to tissues to organs to organ systems to organisms.

SC.6.L.14.3 Recognize and explore how cells of all organisms undergo similar processes to maintain homeostasis, including extracting energy from food, getting rid of waste, and reproducing.

SC.6.N.2.1

Inquiry Launch Lab

15 minutes

How is a system organized?

The places people live are organized in a system. Do you live in or near a city? Cities contain things such as schools and stores that enable them to function on their own. Many cities together make up another level of organization.

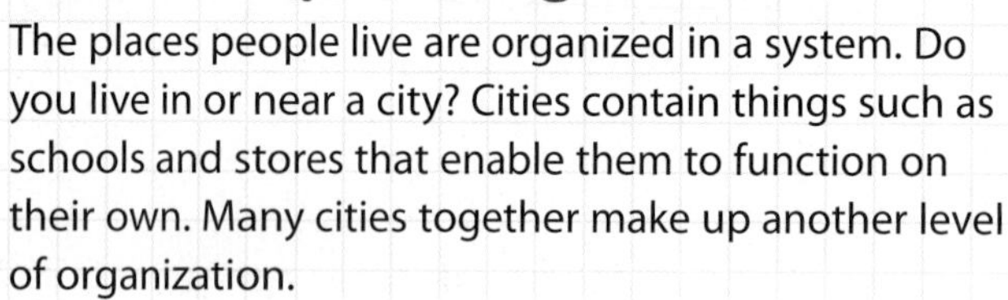

Procedure

1. Read and complete a lab safety form.
2. Using a **metric ruler** and **scissors,** measure and cut squares of **construction paper** that are 4 cm, 8 cm, 12 cm, 16 cm, and 20 cm on each side. Use a different color for each square.
3. Stack the squares from largest to smallest, and glue them together.
4. Cut apart the *City, Continent, Country, County,* and *State* labels your teacher gives you.
5. Use a **glue stick** to attach the *City* label to the smallest square. Sort the remaining labels from smallest to largest, and glue to the corresponding square.

Think About This

1. What is the largest level of organization a city belongs to?

2. Can any part of the system function without the others? Explain.

3. Key Concept How do you think the system used to organize where people live is similar to how your body is organized?

Scales on Wings?

1. This butterfly has a distinctive pattern of colors on its wings. The pattern is formed by a cluster of tiny scales. In a similar way, multicellular organisms are made of many small parts working together. How do these scales work together?

Life's Organization

You might recall that all matter is made of atoms and that atoms combine and form molecules. Molecules make up cells. A large animal, such as a Komodo dragon, is not made of one cell. Instead, it is composed of trillions of cells working together. Its skin, shown in **Figure 9,** is made of many cells that are specialized for protection. The Komodo dragon has other types of cells, such as blood cells and nerve cells, that perform other functions. Cells work together in the Komodo dragon and enable it to function. In the same way, cells work together in you and in other multicellular organisms.

Recall that some organisms are made of only one cell. These unicellular organisms carry out all the activities necessary to survive, such as absorbing nutrients and getting rid of wastes. But no matter their sizes, all organisms are made of cells.

Figure 9 Skin cells are only one of the many kinds of cells that make up a Komodo dragon.

Unicellular Organisms 

Figure 10 Unicellular organisms carry out life processes within one cell.

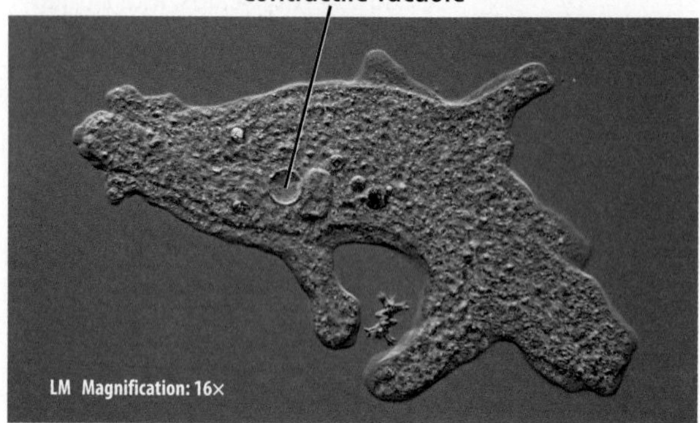

This unicellular amoeba captures a desmid for food.

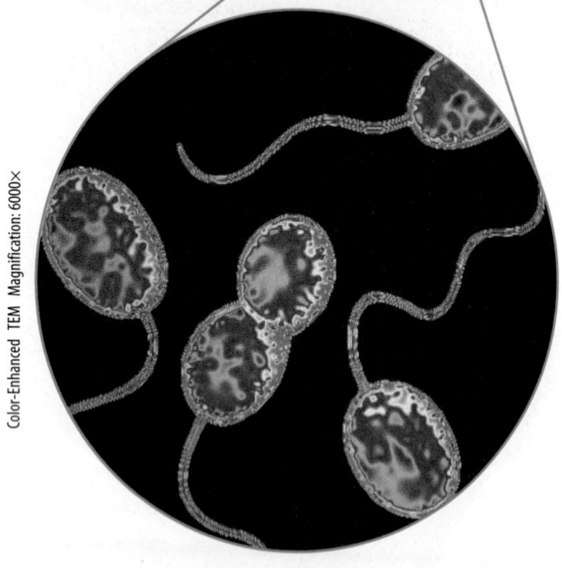

These heat-loving bacteria are often found in hot springs as shown here. They get their energy to produce food from sulfur instead of from light like plants.

Unicellular Organisms

As you read on the previous page, some organisms have only one cell. Unicellular organisms do all the things needed for their survival within that one cell. For example, the amoeba in **Figure 10** is ingesting another unicellular organism, a type of green algae called a desmid, for food. Unicellular organisms also respond to their environment, get rid of waste, grow, and even reproduce on their own. Unicellular organisms include both prokaryotes and some eukaryotes.

Prokaryotes

Recall that a cell without a membrane-bound nucleus is a prokaryotic cell. In general, prokaryotic cells are smaller than eukaryotic cells and have fewer cell structures. A unicellular organism made of one prokaryotic cell is called a prokaryote. Some prokaryotes live in groups called colonies. Some can also live in extreme environments, as shown in **Figure 10.**

Eukaryotes

You might recall that a eukaryotic cell has a nucleus surrounded by a membrane and many other specialized organelles. For example, the amoeba shown in **Figure 10** has an organelle called a contractile vacuole. It functions like a bucket that is used to bail water out of a boat. A contractile vacuole collects excess water from the amoeba's cytoplasm. Then it pumps the water out of the amoeba. This prevents the amoeba from swelling and bursting.

A unicellular organism that is made of one eukaryotic cell is called a eukaryote. There are thousands of different unicellular eukaryotes, such as algae that grow on the inside of an aquarium and the fungus that causes athlete's foot.

Active Reading **2. Identify** (Circle) an example of a unicellular eukaryotic organism.

Multicellular Organisms

Multicellular organisms are made of many eukaryotic cells working together, like the crew on an airplane. Each member of the crew, from the pilot to the mechanic, has a specific job that is important for the plane's operation. Similarly, each type of cell in a multicellular organism has a specific job that is important to the survival of the organism.

Active Reading **3. Determine** Highlight how unicellular and multicellular organisms differ.

Active Reading FOLDABLES® LA.6.2.2.3

Make a layered book from three sheets of notebook paper. Label it as shown. Use your book to describe the levels of organization that make up organisms.

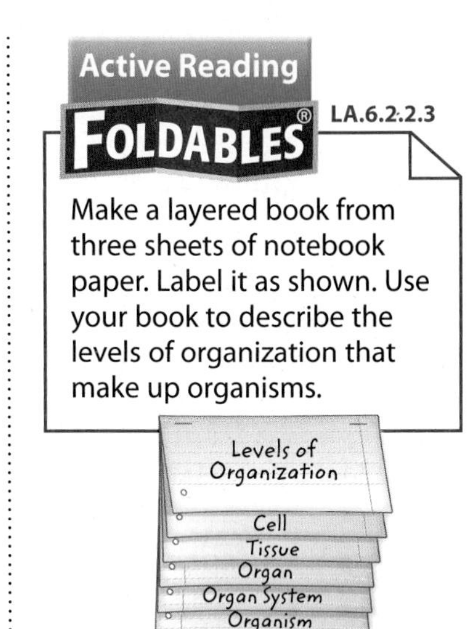

Cell Differentiation

As you read in the last lesson, all cells in a multicellular organism come from one cell—a fertilized egg. Cell division starts quickly after fertilization. The first cells made can become any type of cell, such as a muscle cell, a nerve cell, or a blood cell. *The process by which cells become different types of cells is called* **cell differentiation** (dihf uh ren shee AY shun).

You might recall that a cell's instructions are contained in its chromosomes. Also, nearly all the cells of an organism have identical sets of chromosomes. If an organism's cells have identical sets of instructions, how can cells be different? Different cell types use different parts of the instructions on the chromosomes. A few of the many different types of cells that can result from human cell differentiation are shown in **Figure 11.**

Figure 11 A fertilized egg produces cells that can differentiate into a variety of cell types.

Cell Differentiation in Eukaryotes

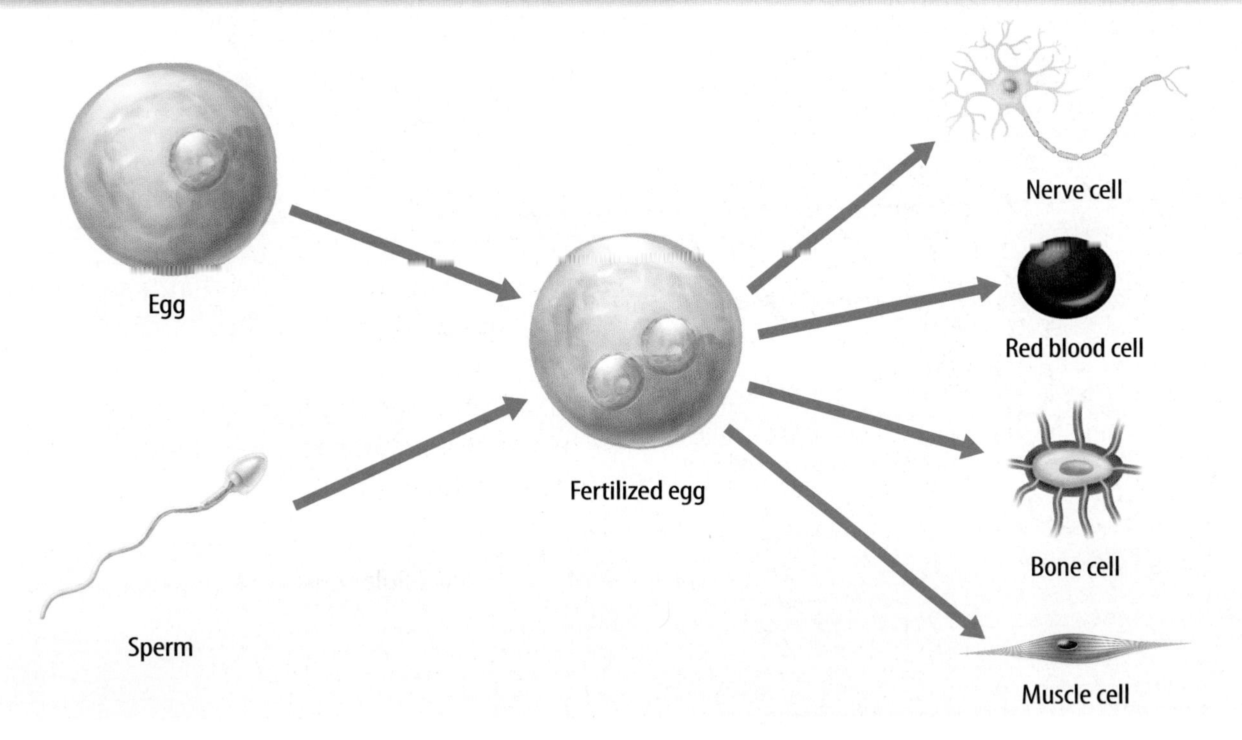

Animal Stem Cells Not all cells in a developing animal differentiate. **Stem cells** *are unspecialized cells that are able to develop into many different cell types.* There are many stem cells in embryos but fewer in adult organisms. Adult stem cells are important for the cell repair and replacement you read about in Lesson 1. For example, stem cells in your bone marrow can produce more than a dozen different types of blood cells. These replace ones that are damaged or worn out. Stem cells have also been discovered in skeletal muscles. These stem cells can produce new muscle cells when the **fibers** that make up the muscle are torn.

SCIENCE USE V. COMMON USE

fiber

Science Use a long muscle cell

Common Use a thread

Plant Cells Plants also have unspecialized cells similar to animal stem cells. These cells are grouped in areas of a plant called meristems (MER uh stemz). Meristems are in different areas of a plant, including the tips of roots and stems, as shown in **Figure 12.** Cell division in meristems produces different types of plant cells with specialized structures and functions, such as transporting materials, making food, storing food, or protecting the plant. These cells might become parts of stems, leaves, flowers, or roots.

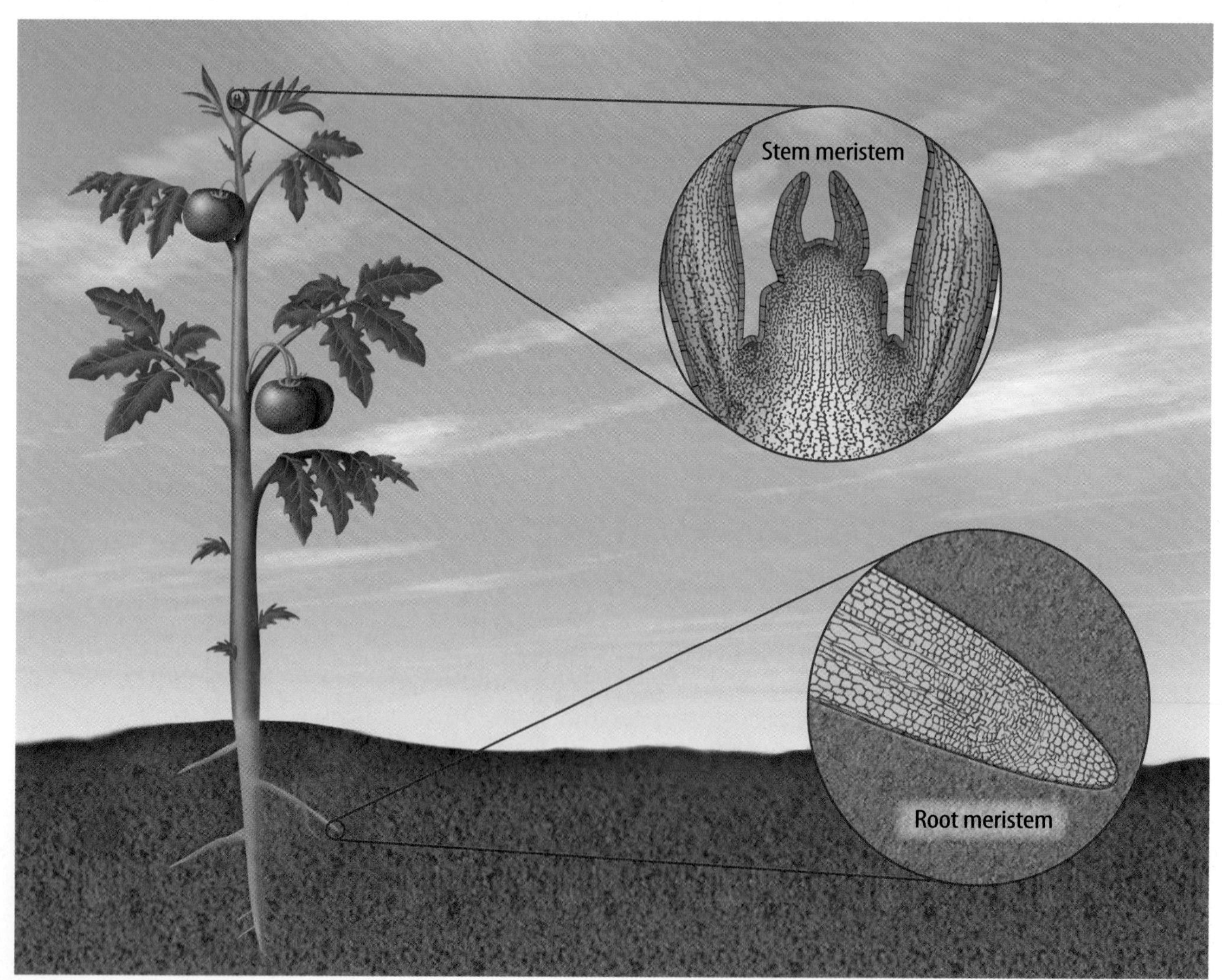

Figure 12 Plant meristems produce cells that can become part of stems, leaves, flowers, or roots.

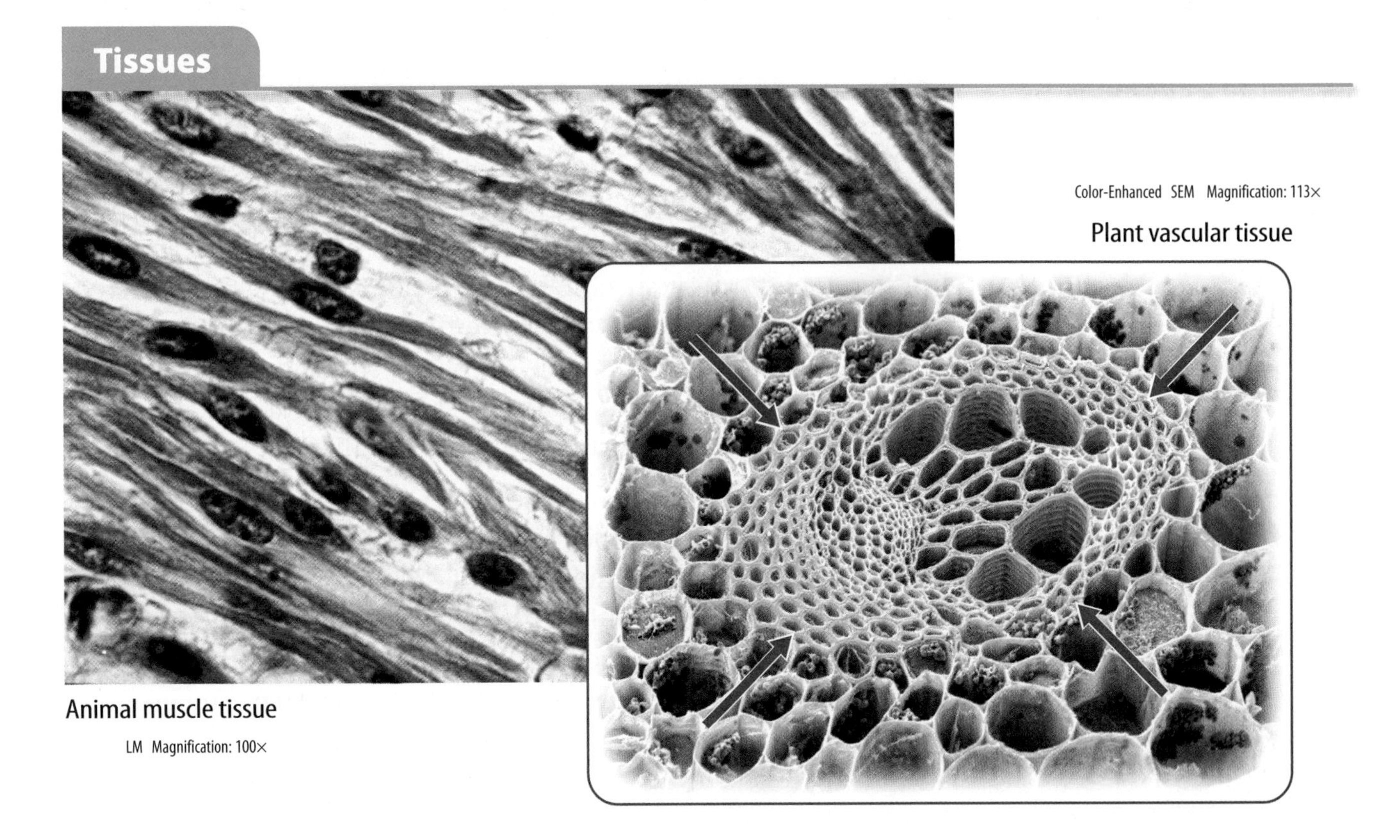

Figure 13 Similar cells work together and form tissues such as this animal muscle tissue that contracts the stomach to help digestion. Plant vascular tissue, indicated by red arrows, moves water and nutrients throughout a plant.

Tissues

In multicellular organisms, similar types of cells are organized into groups. **Tissues** *are groups of similar types of cells that work together to carry out specific tasks.* Humans, like most other animals, have four main types of tissue—muscle, connective, nervous, and epithelial (eh puh THEE lee ul). For example, the animal tissue shown in **Figure 13** is smooth muscle tissue that is part of the stomach. Muscle tissue causes movement. Connective tissue provides structure and support and often connects other types of tissue together. Nervous tissue carries messages to and from the brain. Epithelial tissue forms the protective outer layer of the skin and the lining of major organs and internal body cavities.

Plants also have different types of tissues. The three main types of plant tissue are dermal, vascular (VAS kyuh lur), and ground tissue. Dermal tissue provides protection and helps reduce water loss. Vascular tissue, shown in **Figure 13,** transports water and nutrients from one part of a plant to another. Ground tissue provides storage and support and is where photosynthesis takes place.

Word Origin

tissue
from Latin *texere,* means "weave"

Active Reading **4. Contrast** Highlight the different animal and plant tissues and their descriptions.

Organs

ACADEMIC VOCABULARY

complex
(adjective) made of two or more parts

Complex jobs in organisms require more than one type of tissue. **Organs** *are groups of different tissues working together to perform a particular job.* For example, your stomach is an organ specialized for breaking down food. It is made of all four types of tissue: muscle, epithelial, nervous, and connective. Each type of tissue performs a specific function necessary for the stomach to work properly. Layers of muscle tissue contract and break up pieces of food, epithelial tissue lines the stomach, nervous tissue sends signals to indicate the stomach is full, and connective tissue supports the stomach wall.

Plants also have organs. The leaves shown in **Figure 14** are organs specialized for photosynthesis. Each leaf is made of dermal, ground, and vascular tissues. Dermal tissue covers the outer surface of a leaf. The leaf is a vital organ because it contains ground tissue that produces food for the rest of the plant. Ground tissue is where photosynthesis takes place. The ground tissue is tightly packed on the top half of a leaf. The vascular tissue moves both the food produced by photosynthesis and water throughout the leaf and the rest of the plant.

Active Reading **6. List** What are the tissues in a leaf?

____

Figure 14 A plant leaf is an organ made of several different tissues.

5. Visual Check Assess Circle which plant tissue makes up the thinnest layer.

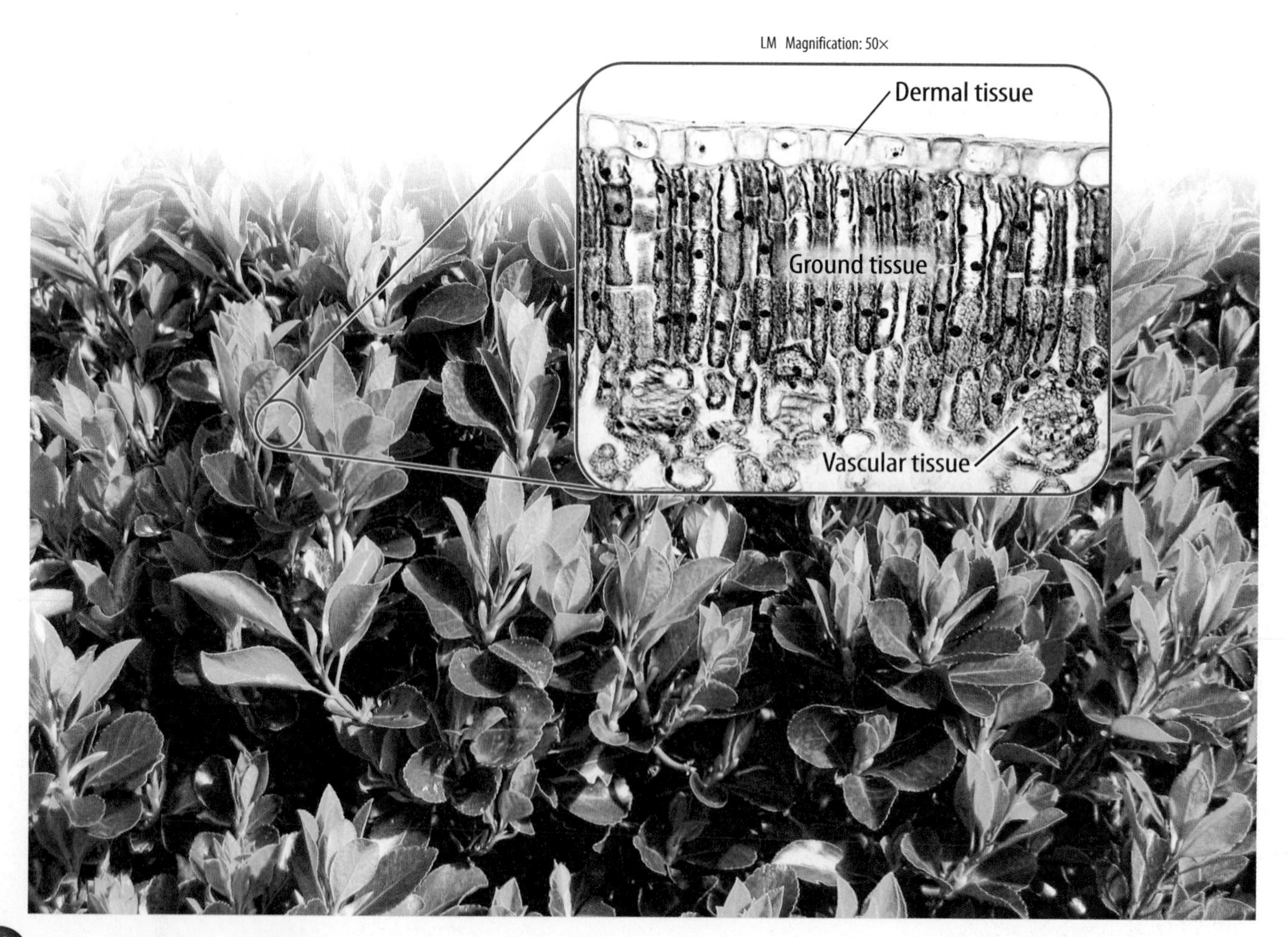

Organ Systems

Usually organs do not function alone. Instead, **organ systems** *are groups of different organs that work together to complete a series of tasks*. Human organ systems can be made of many different organs working together. For example, the human digestive system is made of many organs, including the stomach, the small intestine, the liver, and the large intestine. These organs and others all work together to break down food and take it into the body. Blood absorbs and transports nutrients from broken down food to cells throughout the body.

Plants have two major organ systems—the shoot system and the root system. The shoot system includes leaves, stems, and flowers. Food and water are transported throughout the plant by the shoot system. The root system anchors the plant and takes in water and nutrients.

Active Reading **7. Name** Write down the major organ systems in plants.

____

____

Sequence the organization of cells, tissues, organs, and organ systems in a multicellular organism.

8. Cells are organized in ______________.

9. Different ______________ working together to perform a particular job are called ______________.

10. Groups of ______________ that work together to complete a series of tasks are called ______________.

11. Many ______________ working together make up a(n) ______________.

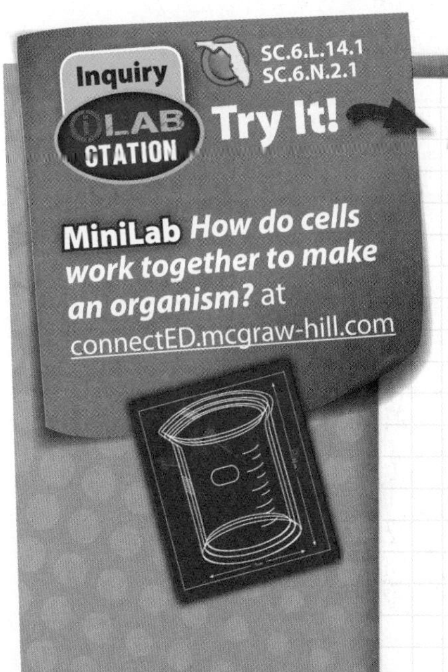

MiniLab *How do cells work together to make an organism?* at connectED.mcgraw-hill.com

Apply It! After you complete the lab, answer this question.

1. How do the organ systems of a plant work together to support the organism?

Organisms

Multicellular organisms usually have many organ systems. These systems work together to carry out all the jobs needed for the survival of the organisms. For example, the cells in the leaves and the stems of a plant need water to live. They cannot absorb water directly. Water diffuses into the roots and is transported through the stem to the leaves by the transport system.

In the human body, there are many major organ systems. Each organ system depends on the others and cannot work alone. For example, the cells in the muscle tissue of the stomach cannot survive without oxygen. The stomach cannot get oxygen without working together with the respiratory and circulatory systems. **Figure 15** will help you review how organisms are organized.

12. **NGSSS Check** **Reason** How does cell differentiation lead to the organization within a multicellular organism? SC.6.L.14.1

Bone cell

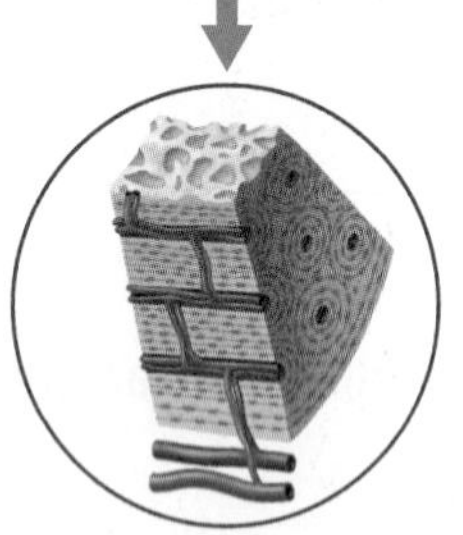

Bone tissue

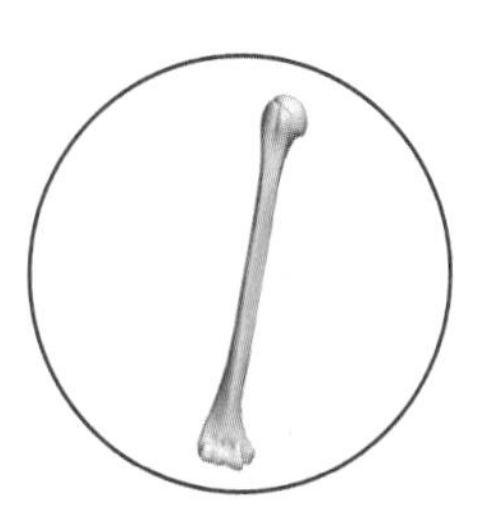

Bone (organ)

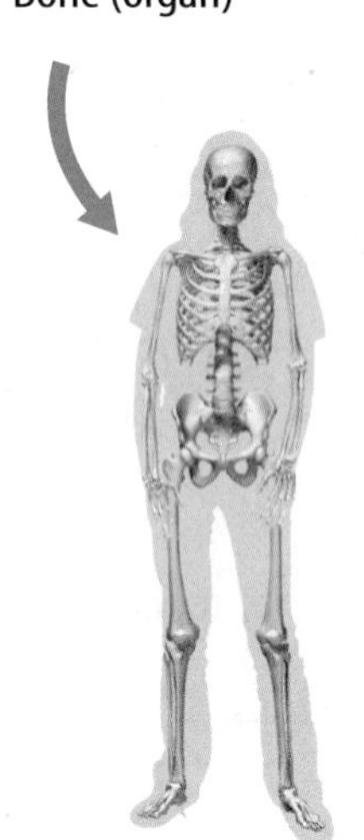

Skeletal system

Respiratory system

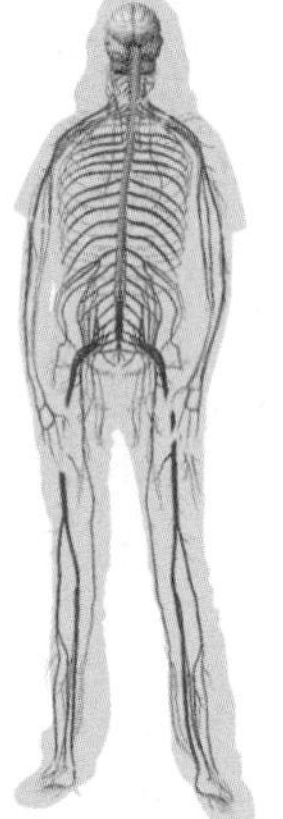

Nervous system

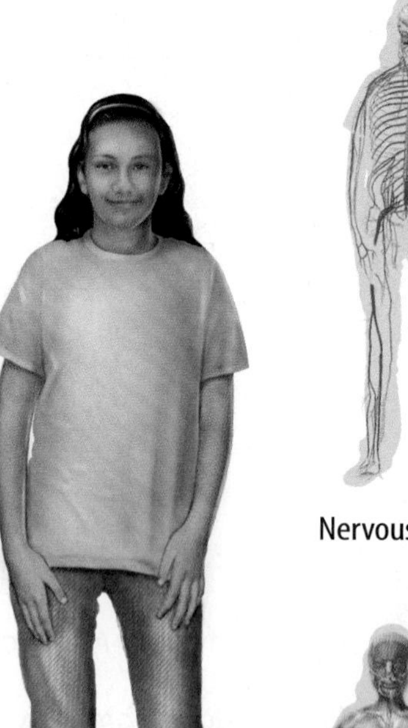

Person (organism)

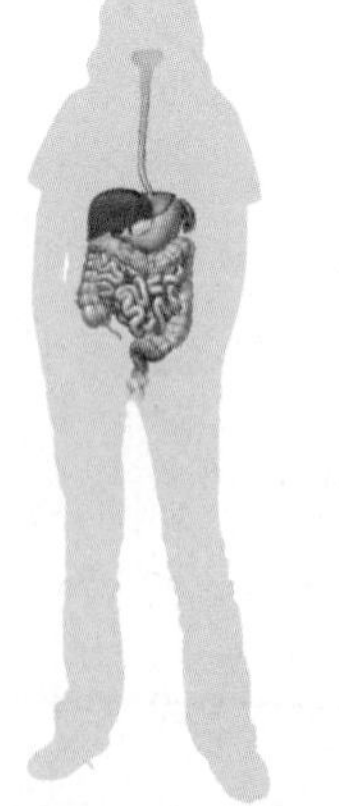

Digestive system

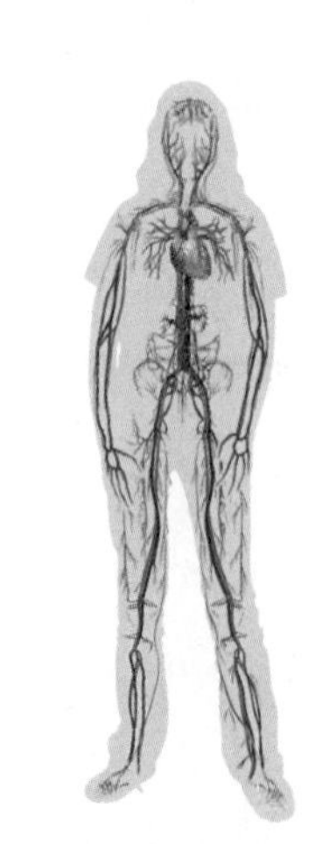

Circulatory system

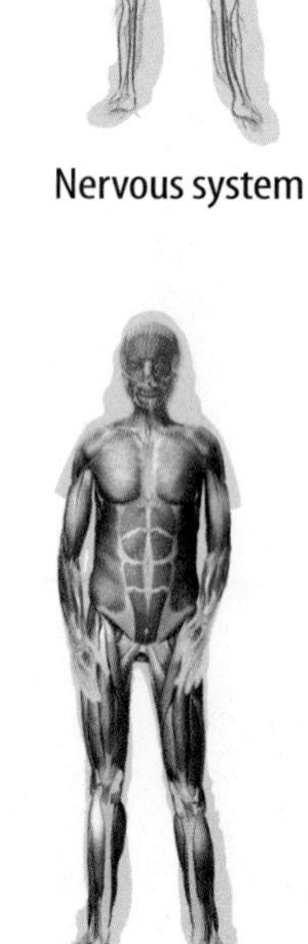

Muscular system

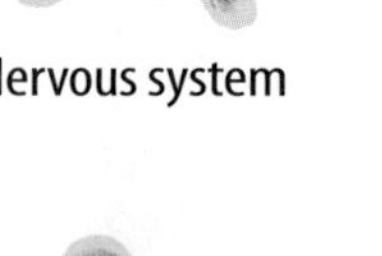

Figure 15 An organism is made of organ systems, organs, tissues, and cells that all function together and enable the organism to survive.

Lesson Review 2

Visual Summary

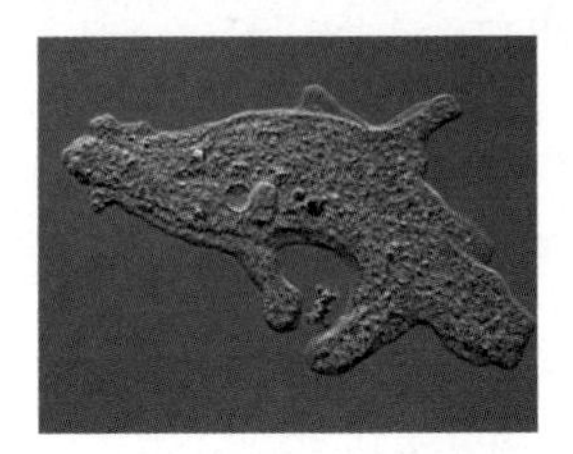

A unicellular organism carries out all the activities necessary for survival within one cell.

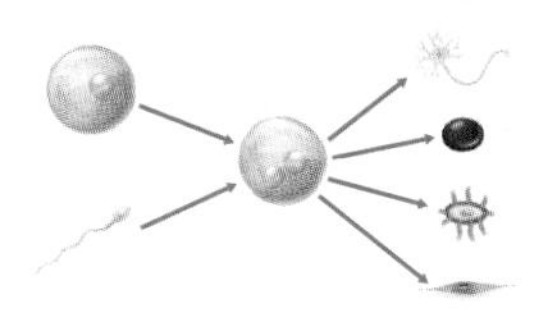

Cells become specialized in structure and function during cell differentiation.

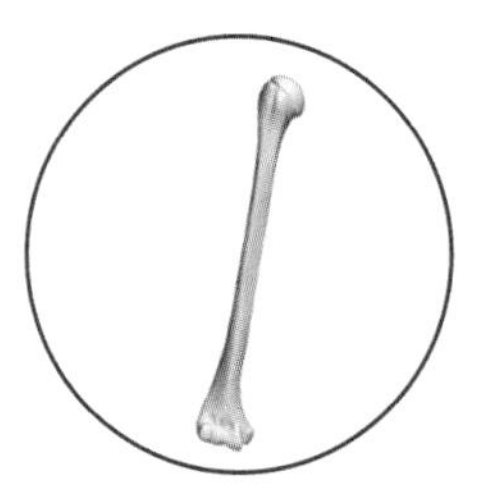

Organs are groups of different tissues that work together to perform a job.

Use Vocabulary

1. **Define** *cell differentiation* in your own words.

2. **Distinguish** between an organ and an organ system. SC.6.L.14.1

Understand Key Concepts 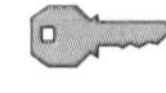

3. **Explain** the difference between a unicellular organism and a multicellular organism. SC.6.L.14.1

4. **Describe** how cell differentiation produces different types of cells in animals.

5. Which is the correct sequence of the levels of organization? SC.6.L.14.1
   - (A) cell, organ, tissue, organ system, organism
   - (B) organism, organ, organ system, tissue, cell
   - (C) cell, tissue, organ, organ system, organism
   - (D) tissue, organ, organism, organ system, cell

Interpret Graphics

6. **Organize** Fill in the table below to summarize the characteristics of unicellular and multicellular organisms. LA.6.2.2.3

Organism Characteristics	
Unicellular	Multicellular

Critical Thinking

7. **Predict** A mistake occurs during mitosis of a muscle stem cell. How might this affect muscle tissue?

8. **Compare** the functions of a cell to the functions of an organism, such as getting rid of wastes.

Chapter 3 Study Guide

Think About It! **Through various physiological functions essential for growth and reproduction, one cell can grow and develop into a multicellular organism.**

Key Concepts Summary

LESSON 1 The Cell Cycle and Cell Division

- The **cell cycle** consists of two phases. During **interphase,** a cell grows and its chromosomes and organelles replicate. During the mitotic phase of the cell cycle, the nucleus divides during **mitosis,** and the cytoplasm divides during **cytokinesis.**
- The cell cycle results in two genetically identical **daughter cells.** The original parent cell no longer exists.
- The cell cycle is important for growth in multicellular organisms, reproduction in some organisms, replacement of worn-out cells, and repair of damaged cells.

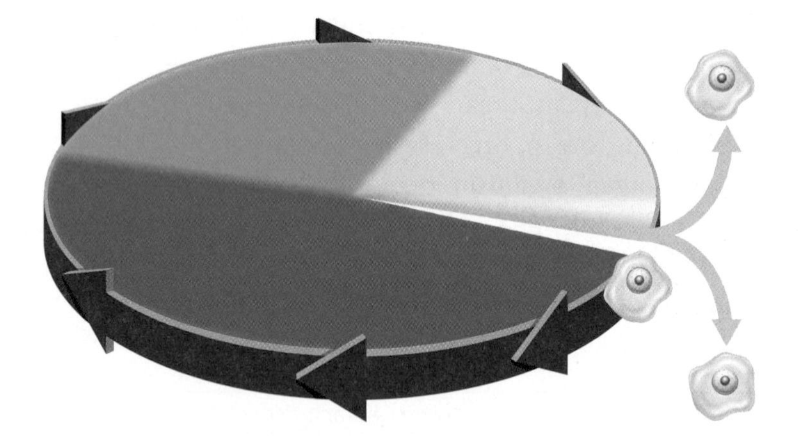

Vocabulary

cell cycle p. 95
interphase p. 96
sister chromatid p. 98
centromere p. 98
mitosis p. 99
cytokinesis p. 99
daughter cell p. 99

LESSON 2 Levels of Organization

- The one cell of a unicellular organism is able to obtain all the materials that it needs to survive.
- In a multicellular organism, cells cannot survive alone and must work together to handle the organism's needs.
- Through **cell differentiation,** cells become different types of cells with specific functions. Cell differentiation leads to the formation of **tissues, organs,** and **organ systems.**

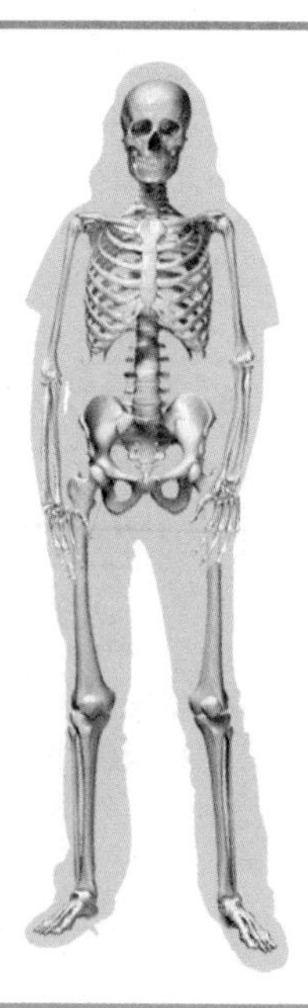

cell differentiation p. 109
stem cell p. 110
tissue p. 111
organ p. 112
organ system p. 113

Active Reading FOLDABLES® Chapter Project

Assemble your lesson Foldables as shown to make a Chapter Project. Use the project to review what you have learned in this chapter.

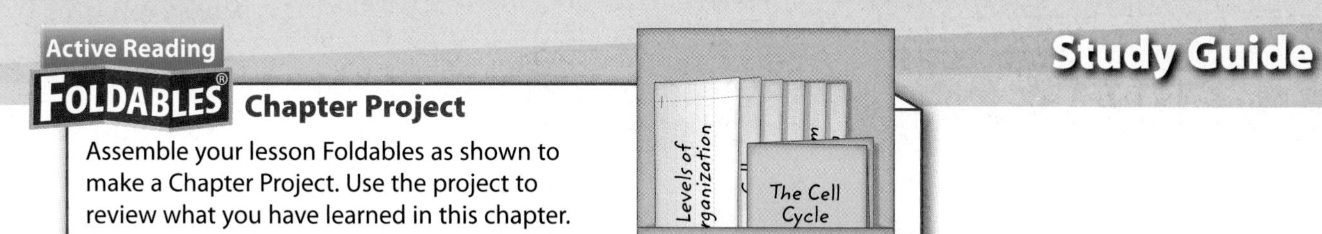

Use Vocabulary

1. Use the term *sister chromatids* in a sentence.

2. Define the term *centromere* in your own words.

3. The new cells formed by mitosis are called.

4. Use the term *cell differentiation* in a sentence.

5. Define the term *stem cell* in your own words.

6. Organs are groups of ______ working together to perform a specific task. SC.6.L.14.1

Link Vocabulary and Key Concepts

Use vocabulary terms from the previous page and from the chapter to complete the concept map. SC.6.L.14.3

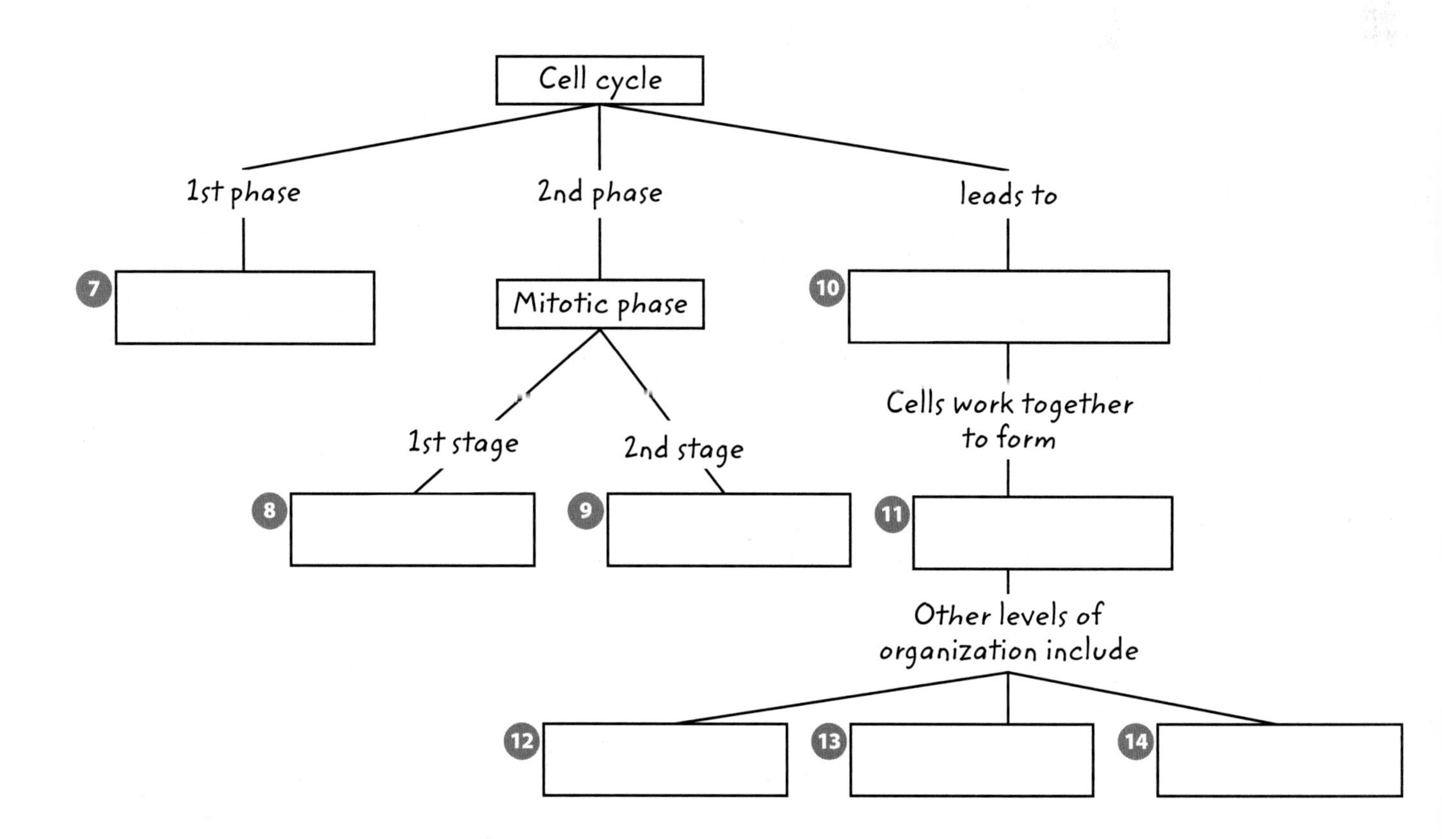

Chapter 3 Review

Fill in the correct answer choice.

Understand Key Concepts

1. Chromosomes line up in the center of the cell during which phase? **SC.7.L.16.3**
   Ⓐ anaphase
   Ⓑ metaphase
   Ⓒ prophase
   Ⓓ telophase

2. Which stage of the cell cycle precedes cytokinesis? **SC.7.L.16.3**
   Ⓐ G_1
   Ⓑ G_2
   Ⓒ interphase
   Ⓓ mitosis

Use the figure below to answer questions 3 and 4.

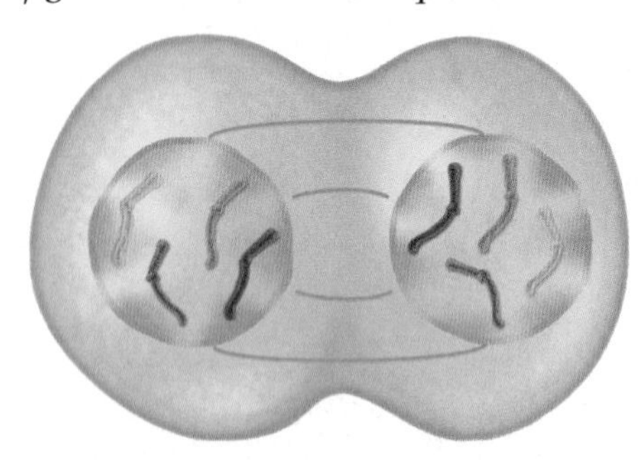

3. The figure represents which stage of mitosis? **SC.7.L.16.3**
   Ⓐ anaphase
   Ⓑ metaphase
   Ⓒ prophase
   Ⓓ telophase

4. What forms during this phase?
   Ⓐ centromere
   Ⓑ furrow
   Ⓒ sister chromatid
   Ⓓ two nuclei

5. What is the longest part of the cell cycle? **SC.6.L.14.3**
   Ⓐ anaphase
   Ⓑ cytokinesis
   Ⓒ interphase
   Ⓓ mitosis

Critical Thinking

6. **Sequence** the events that occur during the phases of mitosis. **SC.7.L.16.3**

7. **Infer** why the chromatin condenses into chromosomes before mitosis begins.

8. **Create** Use the figure below to create a cartoon that shows a duplicated chromosome separating into two sister chromatids.

9. **Classify** a leaf as a tissue or an organ. Explain your choice. **SC.6.L.14.1**

10. **Distinguish** between a tissue and an organ. **SC.6.L.14.1**

11. **Construct** a table that lists and defines the different levels of organization. SC.6.L.14.1

12. **Summarize** the differences between unicellular organisms and multicellular organisms. LA.6.2.2.3

Writing in Science

13. **Write** a five-sentence paragraph on a seperate piece of paper describing a human organ system. Include a main idea, supporting details, and a concluding statement. LA.6.2.2.3

Big Idea Review

14. Why is cell division important for multicellular organisms? SC.6.L.14.3

Math Skills

Use Percentages

15. During an interphase lasting 23 hours, the S stage takes an average of 8.0 hours. What percentage of interphase is taken up by the S stage?

Use the following information to answer questions 16 and 17.

During a 23-hour interphase, the G_1 stage takes 11 hours and the S stage takes 8.0 hours.

16. What percentage of interphase is taken up by the G_1 and S stages?

17. What percentage of interphase is taken up by the G_2 phase?

Fill in the correct answer choice.

Multiple Choice

1 Which statement describes the number of chromosomes in a newly formed cell after mitosis? **SC.6.L.14.3**

Ⓐ They are equal to the number of the parent cell.

Ⓑ They are half the number of the parent cell.

Ⓒ They are double the number of the parent cell.

Ⓓ They are double the number of the two parent cells.

Use the diagram below to answer question 2.

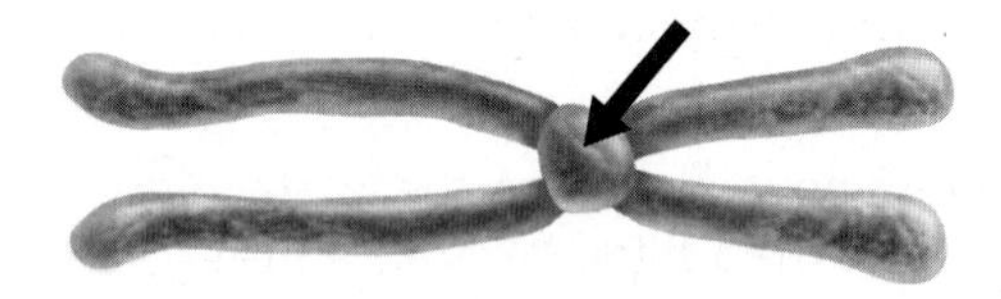

2 What is the function of the object indicated by the arrow? **SC.6.L.14.3**

Ⓕ provide energy to the cell

Ⓖ hold cytoplasm in the cell

Ⓗ hold the sister chromatids together

Ⓘ divide the organelles

3 Which is the correct sequence of organization in a multicellular organism? **SC.6.L.14.1**

Ⓐ cell → tissue → organ system → organ

Ⓑ organ system → cell → tissue → organ

Ⓒ tissue → cell → organ → organ system

Ⓓ cell → tissue → organ → organ system

4 What structures separate during anaphase? **SC.6.L.14.3**

Ⓕ centromeres

Ⓖ chromatids

Ⓗ nuclei

Ⓘ organelles

Use the diagram below to answer question 5.

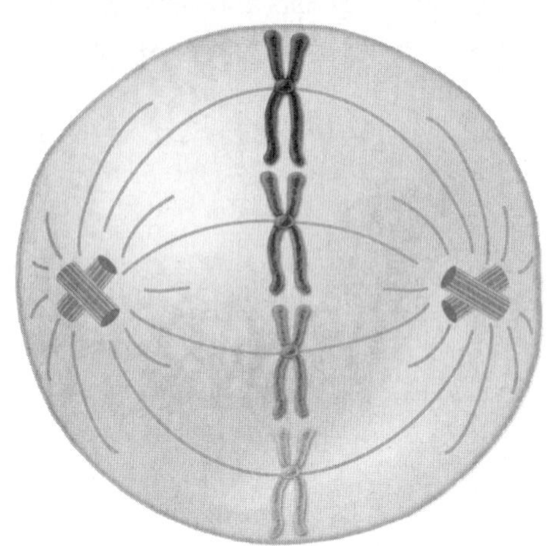

5 Which stage of mitosis does the image above represent? **SC.7.L.16.3**

Ⓐ anaphase

Ⓑ metaphase

Ⓒ prophase

Ⓓ telophase

6 Which mitotic phase occurs when a cell's cytoplasm and its contents are divided? **SC.6.L.14.3**

Ⓕ prophase

Ⓖ cytokinesis

Ⓗ interphase

Ⓘ telophase

7 Which is the most accurate description of a leaf or your stomach? **SC.6.L.14.1**

Ⓐ a cell

Ⓑ an organ

Ⓒ an organ system

Ⓓ a tissue

Use the figure below to answer question 8.

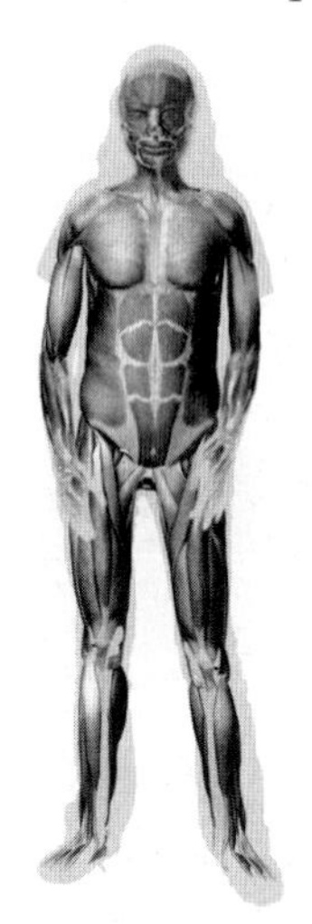

8 Which does this figure illustrate? SC.6.L.14.1

Ⓕ an organ

Ⓖ an organism

Ⓗ an organ system

Ⓘ a tissue

9 If a cell has 30 chromosomes at the start of mitosis, how many chromosomes will be in each new daughter cell?

Ⓐ 10

Ⓑ 15

Ⓒ 30

Ⓓ 60

10 Which are groups of similar types of cells that work together to carry out specific tasks? SC.6.L.14.1

Ⓕ stem cells

Ⓖ organs

Ⓗ organ systems

Ⓘ tissues

Use the figures below to answer questions 11 and 12

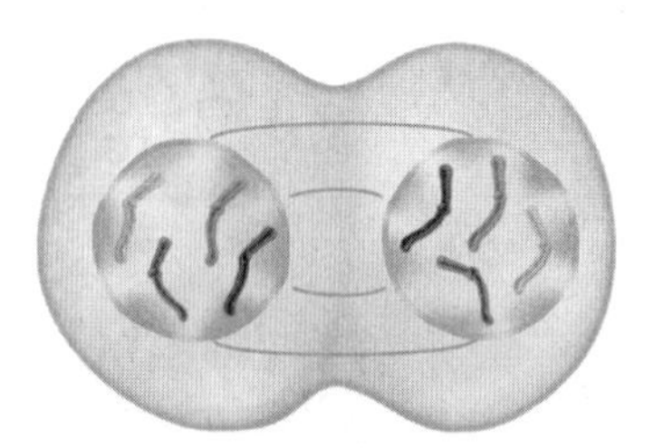

Figure A

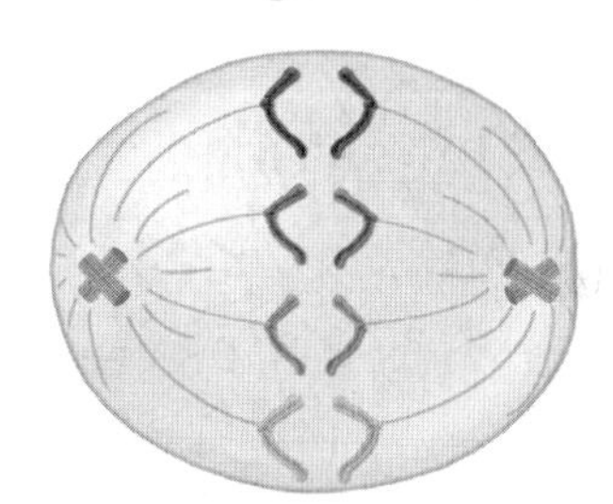

Figure B

11 In Figure A, during which stage of mitosis does a nuclear membrane form around each set of chromosomes? SC.6.L.14.3

Ⓐ prophase

Ⓑ metaphase

Ⓒ anaphase

Ⓓ telophase

12 What occurs in a cell undergoing anaphase, as shown in Figure B? SC.6.L.14.3

Ⓕ Duplicated chromosomes appear, and the nuclear membrane breaks down.

Ⓖ The sister chromatids separate.

Ⓗ The duplicated chromosomes line up in the middle of the cell.

Ⓘ A nuclear membrane develops around each set of chromosomes as they uncoil.

NEED EXTRA HELP?

If You Missed Question...	1	2	3	4	5	6	7	8	9	10	11	12
Go to Lesson...	1	1	2	1	1	1	2	2	1	2	1	1

Name ______________________ Date ______________

mini BAT

Multiple Choice *Bubble the correct answer.*

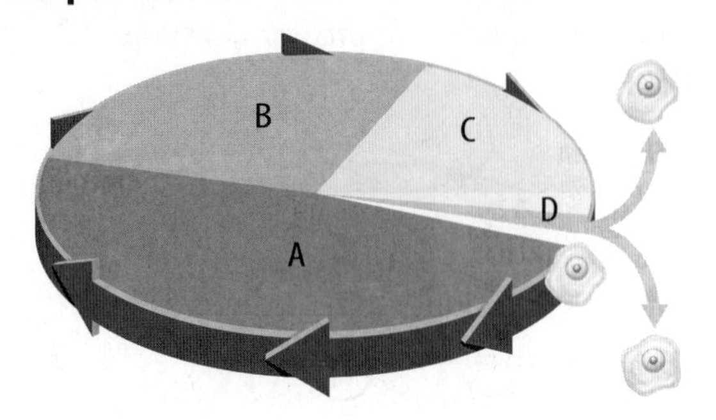

1. In the image above, which phase of a cell's life is represented by D? **SC.6.L.14.3**

Ⓐ DNA replication phase

Ⓑ mitotic phase

Ⓒ preparation for cell division phase

Ⓓ rapid growth and replication phase

2. The final phase of the cell cycle is cytokinesis. How does cytokinesis occur in plant cells? **SC.6.L.14.3**

Ⓕ A new cell wall forms in the middle of the cell.

Ⓖ The old cell wall grows inward, forming a new cell wall.

Ⓗ Fibers grow into the cell, dividing the cytoplasm and the nuclei.

Ⓘ Fibers tighten around the cell to divide the cytoplasm and nuclei.

3. Which image shows the first stage in the division of a fertilized egg? **SC.6.L.14.3**

Ⓐ
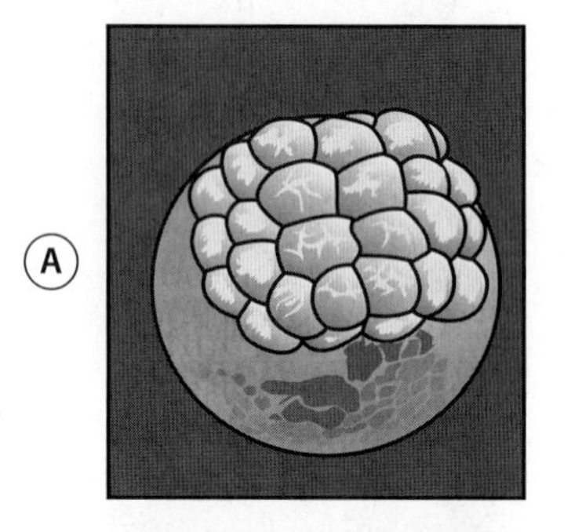

Ⓑ
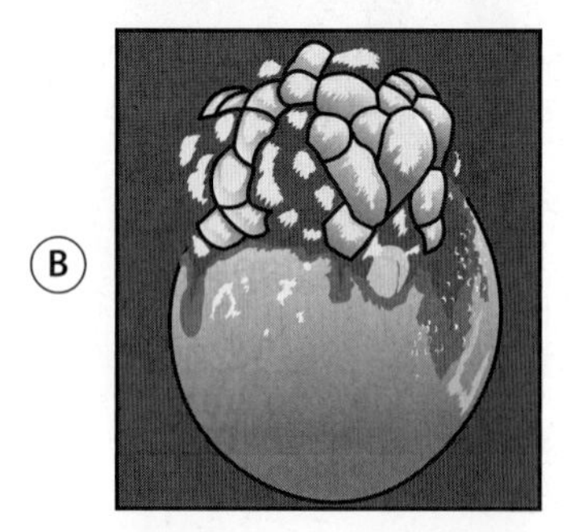

Ⓒ
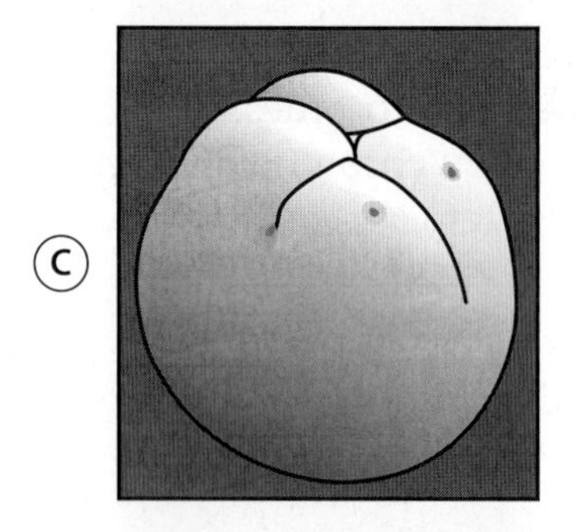

Ⓓ
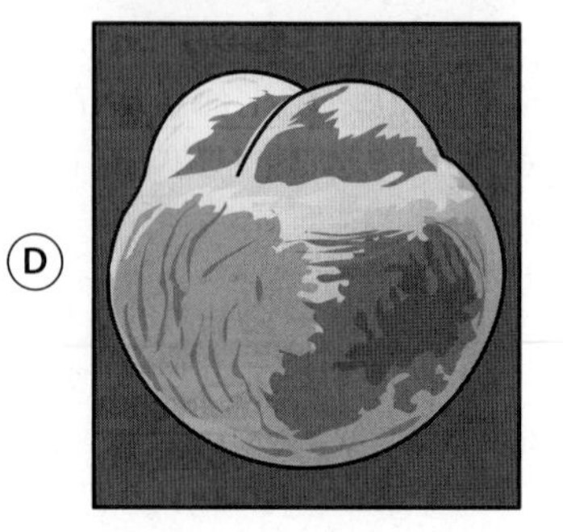

Name ____________________ Date ____________

Benchmark Mini-Assessment Chapter 3 • Lesson 2

FLORIDA NGSSS

mini BAT

Multiple Choice *Bubble the correct answer.*

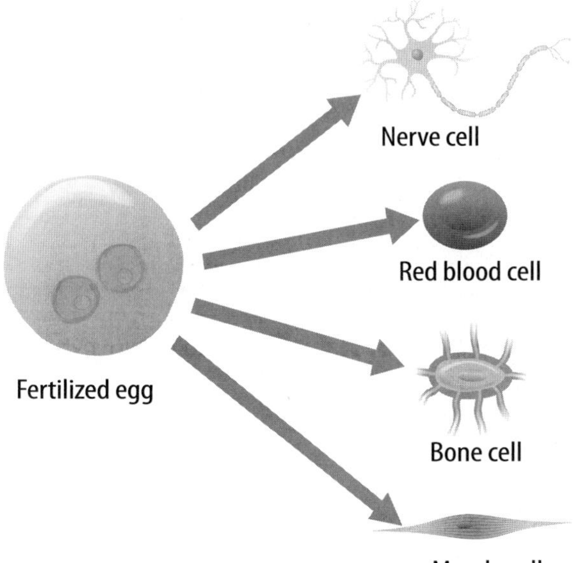

1. Which process is shown in the image above? **SC.6.L.14.3**

(A) metaphase

(B) photosynthesis

(C) cell differentiation

(D) organelle replication

2. Which of these is the most complex organization of cells in a multicellular organism? **SC.6.L.14.3**

(F) organ

(G) tissue

(H) differentiated cell

(I) organ system

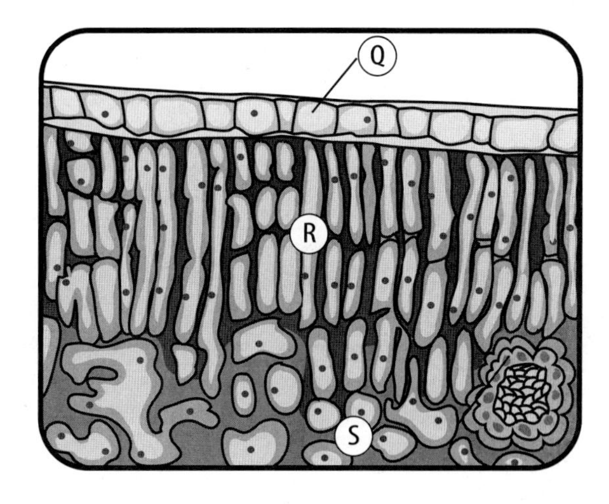

3. In the image of the plant leaf above, what do Q, R, and S represent? **SC.6.L.14.3**

(A) bacteria

(B) muscles

(C) roots

(D) tissues

Name ______________________ Date ____________

Notes

Reproduction

Some organisms reproduce by sexual reproduction. Put an *X* next to each organism you think uses sexual reproduction to produce offspring.

_____ frogs

_____ fish

_____ cats

_____ whales

_____ dogs

_____ chickens

_____ trees

_____ worms

_____ bean plants

_____ horses

_____ humans

_____ snakes

_____ carrots

_____ penguins

_____ weeds

Explain your thinking. Describe your ideas about the types of organisms that reproduce sexually.

Reproduction of ORGANISMS

FLORIDA BIG IDEAS

1 The Practice of Science

16 Heredity and Reproduction

The Big Idea

Think About It!

Why do living things reproduce?

Have you ever seen a family of animals, such as the one of manatees shown here? Notice the baby manatee beside its parents. Like all living things, manatees reproduce.

1. Do you think all living things have two parents?

2. What might happen if the manatees did not reproduce?

3. Why do living things reproduce?

Get Ready to Read

What do you think about reproduction?

Before you read, decide if you agree or disagree with each of these statements. As you read this chapter, see if you change your mind about any of the statements.

	AGREE	DISAGREE
1. Humans produce two types of cells: body cells and sex cells.	☐	☐
2. Environmental factors can cause variation among individuals.	☐	☐
3. Two parents always produce the best offspring.	☐	☐
4. Cloning produces identical individuals from one cell.	☐	☐
5. All organisms have two parents.	☐	☐
6. Asexual reproduction occurs only in microorganisms.	☐	☐

There's More Online!
Video • Audio • Review • ⓘLab Station • WebQuest • Assessment • Concepts in Motion • Multilingual eGlossary

Lesson 1

Sexual Reproduction and MEIOSIS

ESSENTIAL QUESTIONS

 What is sexual reproduction, and why is it beneficial?

 What is the order of the phases of meiosis, and what happens during each phase?

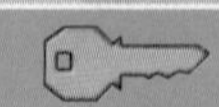 Why is meiosis important?

Vocabulary

sexual reproduction p. 129
egg p. 129
sperm p. 129
fertilization p. 129
zygote p. 129
diploid p. 130
homologous chromosomes p. 130
haploid p. 131
meiosis p. 131

Florida NGSSS

LA.6.2.2.3 The student will organize information to show understanding (e.g., representing main ideas within text through charting, mapping, paraphrasing, summarizing, or comparing/contrasting);

MA.6.A.3.6 Construct and analyze tables, graphs, and equations to describe linear functions and other simple relations using both common language and algebraic notation.

SC.7.N.1.3 Distinguish between an experiment (which must involve the identification and control of variables) and other forms of scientific investigation and explain that not all scientific knowledge is derived from experimentation.

SC.7.L.16.3 Compare and contrast the general processes of sexual reproduction requiring meiosis and asexual reproduction requiring mitosis.

SC.7.L.16.4 Recognize and explore the impact of biotechnology (cloning, genetic engineering, artificial selection) on the individual, society and the environment.

HE.6.C.1.4 Recognize how heredity can affect personal health.

SC.7.N.1.3

Launch Lab

15 minutes

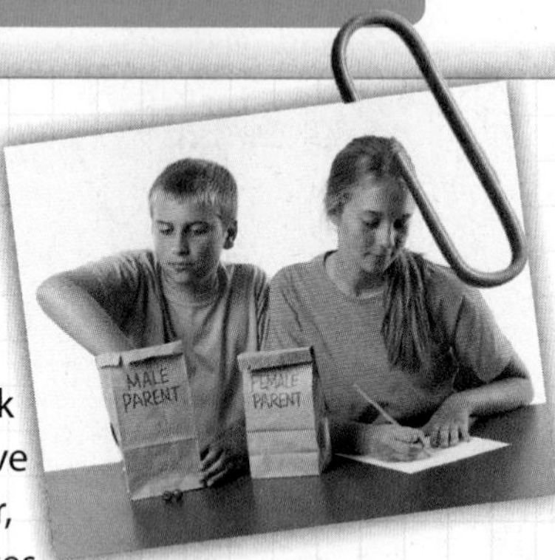

Why do offspring look different?

Unless you're an identical twin, you probably don't look exactly like any siblings you might have. You might have differences in physical characteristics, such as eye color, hair color, ear shape, or height. Why are there differences in the offspring from the same parents?

Procedure

1. Read and complete a lab safety form.
2. Open the **paper bag** labeled *Male Parent,* and, without looking, remove three **beads.** Record the bead colors and replace the beads.
3. Open the **paper bag** labeled *Female Parent,* and, without looking, remove three **beads.** Record the bead colors, and replace the beads.
4. Repeat steps 2 and 3 for each member of the group.
5. After each member has recorded his or her bead colors, study the results. Each combination of male and female beads represents an offspring.

Think About This

1. Compare your group's offspring to another group's offspring. What similarities or differences do you observe?

2. What caused any differences you observed? Explain.

3. Key Concept Why might this type of reproduction be beneficial to an organism?

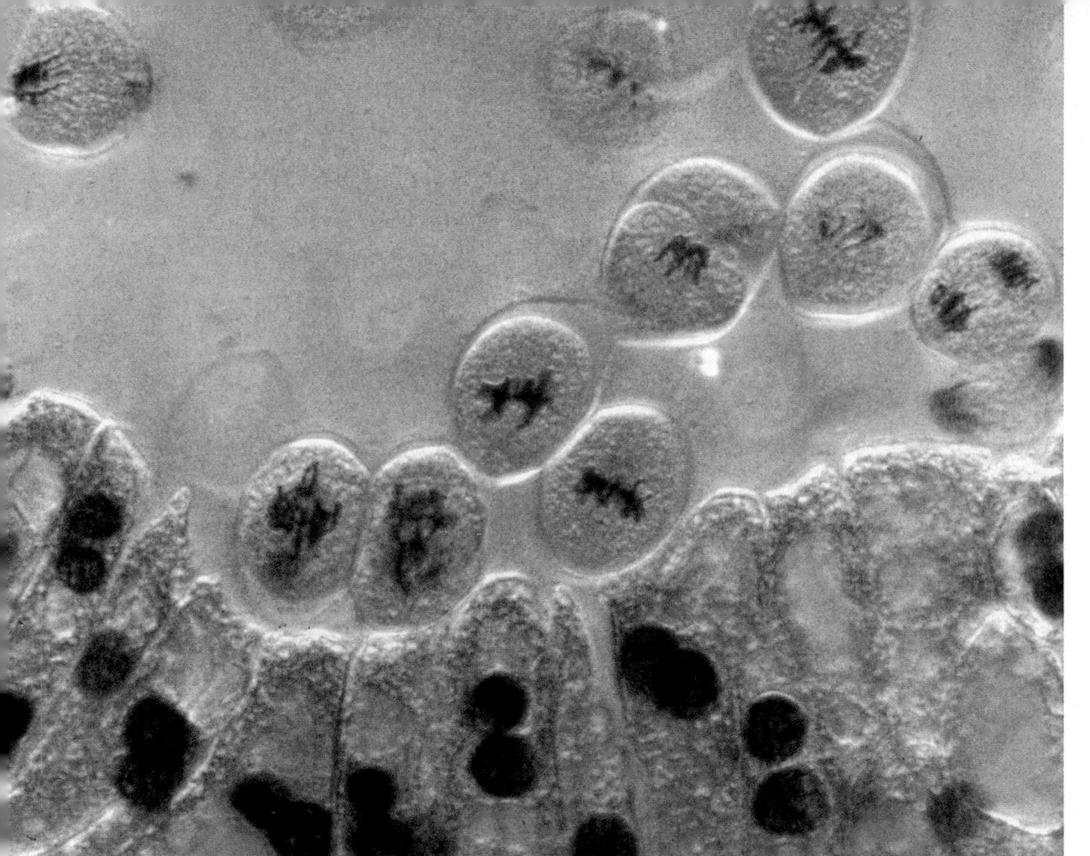

Inquiry Modern Art?

1. This photo looks like a piece of modern art. Look closely at the image. The cells are dividing by a process that occurs during the production of sex cells. From the photo, are you able to tell if these are animal cells or plant cells? Explain.

What is sexual reproduction?

Have you ever seen a litter of kittens? One kitten might have orange fur like its mother. A second kitten might have gray fur like its father. Still another kitten might look like a combination of both parents. How is this possible?

The kittens look different because of sexual reproduction. **Sexual reproduction** *is a type of reproduction in which the genetic materials from two different cells combine, producing an offspring.* The cells that combine are called sex cells. Sex cells form in reproductive organs. *The female sex cell, an* **egg**, *forms in an ovary. The male sex cell, a* **sperm**, *forms in a testis. During a process called* **fertilization** (fur tuh luh ZAY shun), *an egg cell and a sperm cell join together.* This produces a new cell. *The new cell that forms from fertilization is called a* **zygote**. As shown in **Figure 1,** the zygote develops into a new organism.

Active Reading 3. **Identify** Circle the name of the female sex cell in **Figure 1.** Place a box around the name of the male sex cell in **Figure 1.**

Active Reading 2. **Label** Fill in the blanks in **Figure 1** using these terms:
- egg
- sperm
- zygote

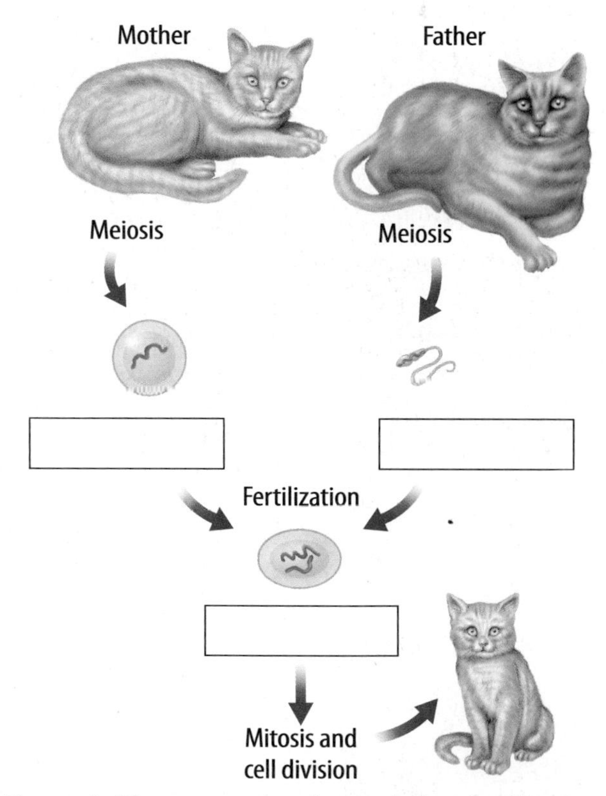

Figure 1 The zygote that forms during fertilization can become a multicellular organism.

Diploid Cells

Following fertilization, a zygote goes through mitosis and cell division. These processes produce nearly all the cells in a multicellular organism. Organisms that reproduce sexually form two kinds of cells—body cells and sex cells. In body cells of most organisms, similar chromosomes occur in pairs. **Diploid** *cells are cells that have pairs of chromosomes.*

Chromosomes

Pairs of chromosomes that have genes for the same traits arranged in the same order are called **homologous** (huh MAH luh gus) **chromosomes.** Because one chromosome is inherited from each parent, the chromosomes are not identical. For example, the kittens mentioned earlier in this lesson inherited a gene for orange fur color from their mother. They also inherited a gene for gray fur color from their father. So, some kittens might be orange, and some might be gray. Both genes for fur color are at the same place on homologous chromosomes, but they code for different colors.

Different organisms have different numbers of chromosomes. Recall that diploid cells have pairs of chromosomes. Notice in **Table 1** that human diploid cells have 23 pairs of chromosomes for a total of 46 chromosomes. A fruit fly diploid cell has 4 pairs of chromosomes, and a rice diploid cell has 12 pairs.

Table 1 An organism's chromosomes can be matched as pairs of chromosomes that have genes for the same traits.

Active Reading **4. Find** Complete the table with the correct information.

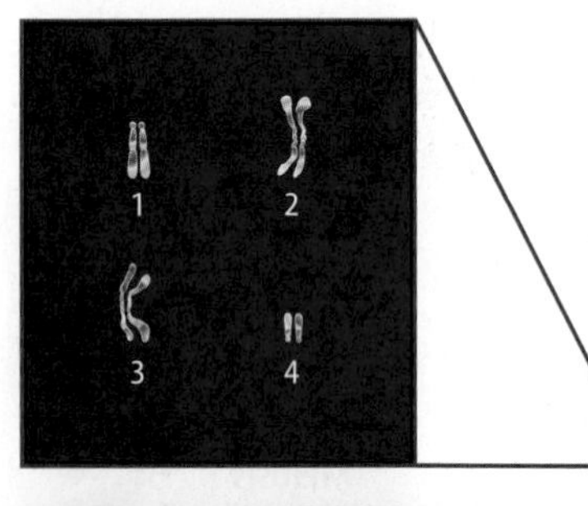

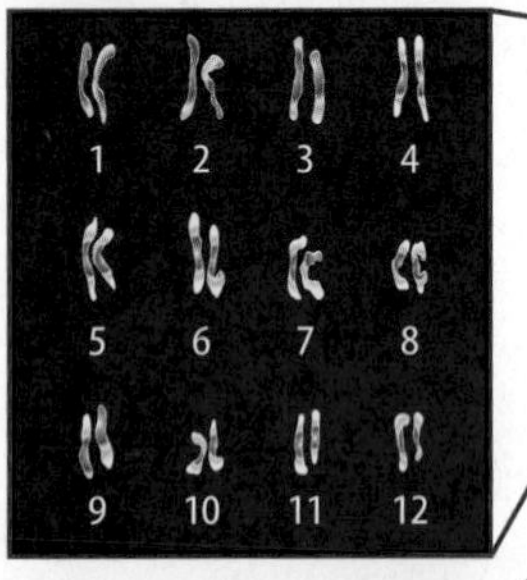

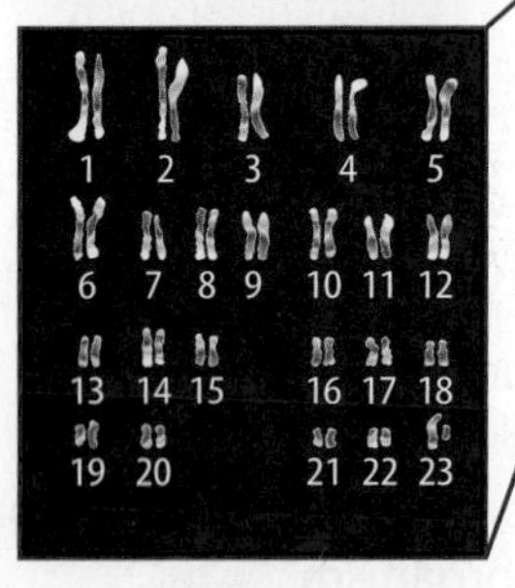

Table 1 Chromosomes of Selected Organisms

Organism	Number of Chromosomes	Number of Homologous Pairs
Fruit fly	8	
Rice	24	
Yeast	32	16
Cat	38	19
Human		23
Dog	78	39
Fern	1,260	630

Having the correct number of chromosomes is very important. If a zygote has too many or too few chromosomes, it will not develop properly. For example, a genetic condition called Down syndrome occurs when a person has an extra copy of chromosome 21. A person with Down syndrome can have short stature, heart defects, or mental disabilities.

Haploid Cells

Organisms that reproduce sexually also form egg and sperm cells, or sex cells. Sex cells have only one chromosome from each pair of chromosomes. **Haploid** *cells are cells that have only one chromosome from each pair.* Organisms produce sex cells using a special type of cell division called meiosis. *In* **meiosis,** *one diploid cell divides and makes four haploid sex cells.* Meiosis occurs only during the formation of sex cells.

WORD ORIGIN

haploid

from Greek *haploeides,* means "single"

Active Reading **5. Contrast** How do diploid cells differ from haploid cells?

Active Reading **6. Detail** Discuss the relationship between diploid cells and homologous chromosomes.

Active Reading **7. Explain** Define haploid cells, and explain how they are produced.

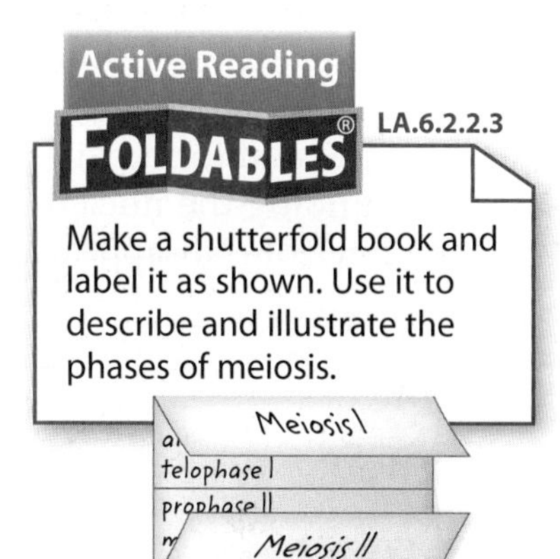

The Phases of Meiosis

Next, you will read about the phases of meiosis. Many of the phases might seem familiar to you because they also occur during mitosis. Recall that mitosis and cytokinesis involve one division of the nucleus and the cytoplasm. Meiosis involves two divisions of the nucleus and the cytoplasm called meiosis I and meiosis II. They result in four haploid cells with half the number of chromosomes as the original cell. When the number of chromosomes is reduced during cell division, it is called a reduction division.

Inquiry LAB STATION SC.7.N.1.3, SC.7.L.16.3

Try It! **MiniLab** *How does one cell produce four cells?* at connectED.mcgraw-hill.com

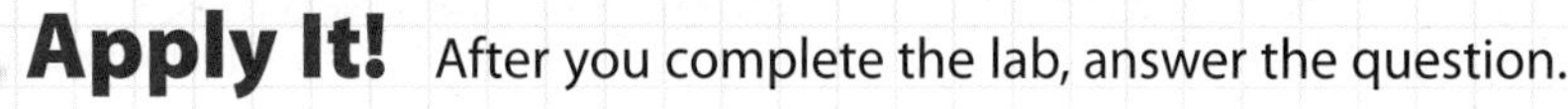

Apply It! After you complete the lab, answer the question.

1. What is the purpose of meiosis?

Phases of Meiosis I

A reproductive cell goes through interphase before beginning meiosis I, which is shown in **Figure 2**. During interphase, the reproductive cell grows and copies, or duplicates, its chromosomes. Each duplicated chromosome consists of two sister chromatids joined together by a centromere.

1. **Prophase I** In the first phase of meiosis I, duplicated chromosomes condense and thicken. Homologous chromosomes come together and form pairs. The membrane surrounding the nucleus breaks apart, and the nucleolus disappears.
2. **Metaphase I** Homologous chromosome pairs line up along the middle of the cell. A spindle fiber attaches to each chromosome.
3. **Anaphase I** Chromosome pairs separate and are pulled toward the opposite ends of the cell. Notice that the sister chromatids stay together.
4. **Telophase I** A nuclear membrane forms around each group of duplicated chromosomes. The cytoplasm divides through cytokinesis and two daughter cells form. Sister chromatids remain together.

Meiosis

Active Reading **8. Apply** Complete the graphic of meiosis I by labeling each phase.

Meiosis I

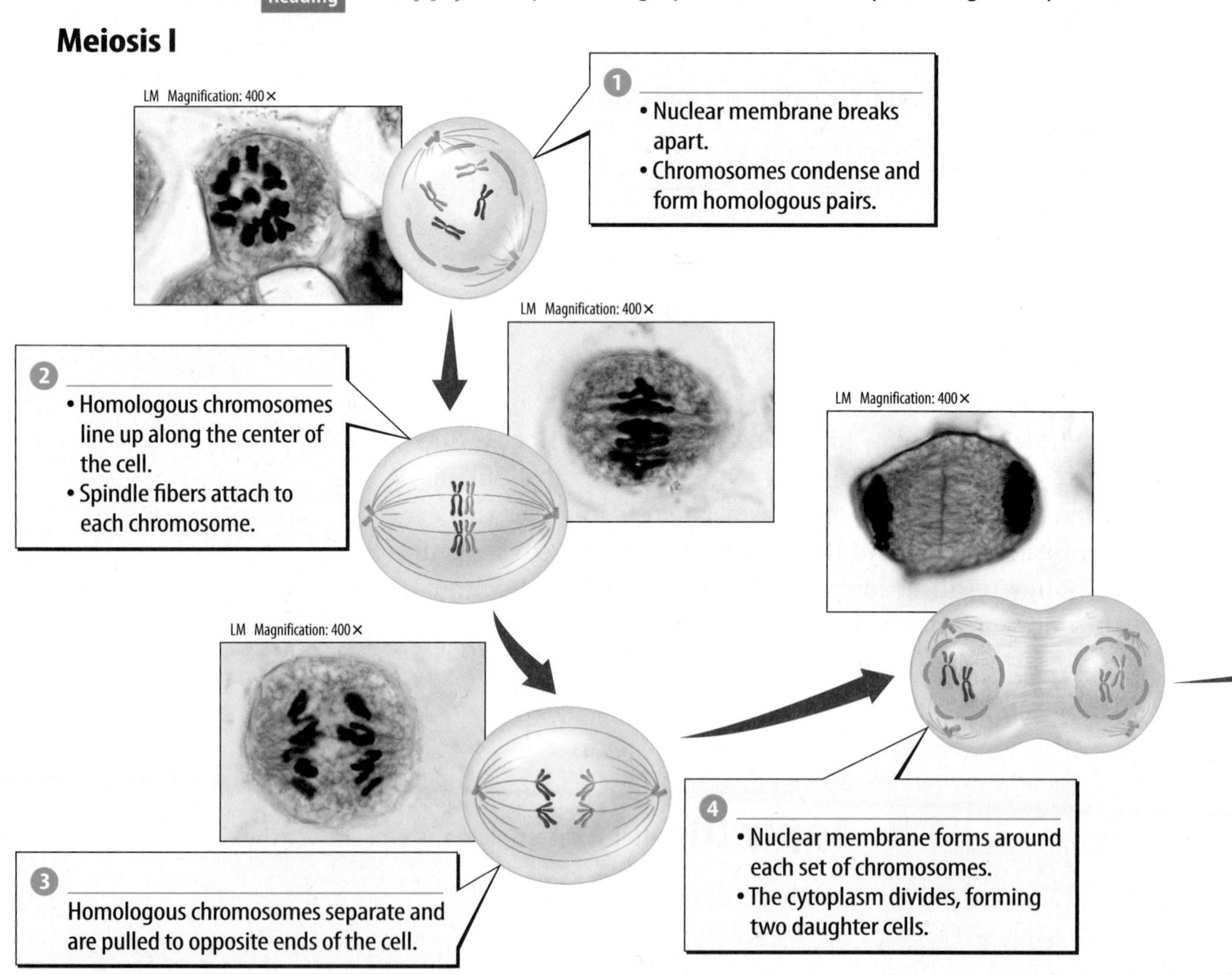

Figure 2 Unlike mitosis, meiosis involves two divisions of the nucleus and the cytoplasm.

Phases of Meiosis II

During meiosis II, the two cells formed previously go through a second division of the nucleus and the cytoplasm, as shown in **Figure 2**. This reduction division results in a haploid gamete or spore.

5 **Prophase II** Chromosomes are not copied again before prophase II. They remain as condensed, thickened sister chromatids. The nuclear membrane breaks apart, and the nucleolus disappears in each cell.

6 **Metaphase II** The pairs of sister chromatids line up along the middle of the cell in single file.

7 **Anaphase II** The sister chromatids of each duplicated chromosome are pulled away from each other and move toward opposite ends of the cells.

8 **Telophase II** During the final phase of meiosis—telophase II—a nuclear membrane forms around each set of chromatids, which are again called chromosomes. The cytoplasm divides through cytokinesis, and four haploid cells form.

Active Reading **9. Identify** List the phases of meiosis in order.

____

____

____

____

Active Reading **10. Apply** Complete the graphic of meiosis II by labeling each phase.

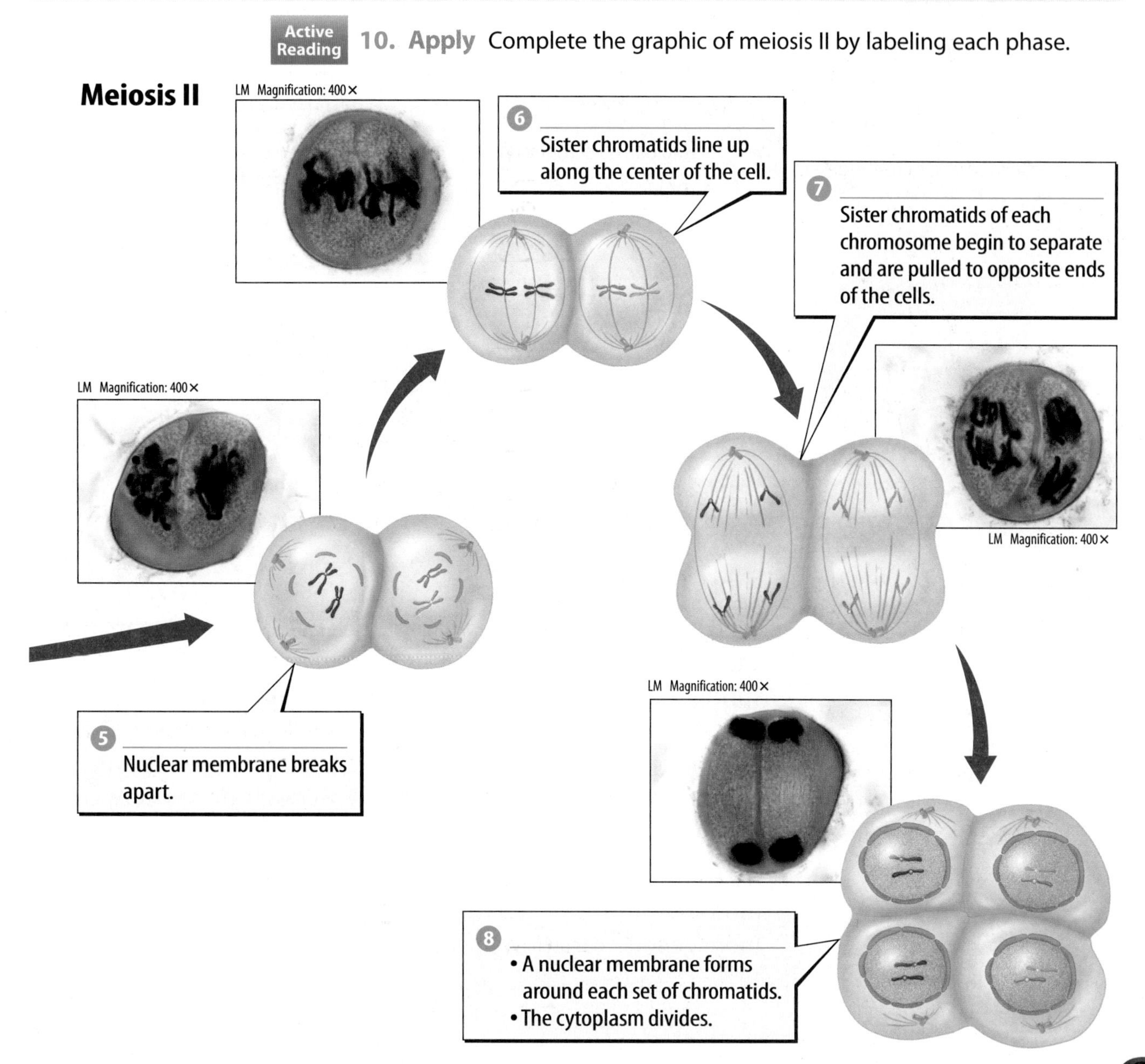

Why is meiosis important?

Meiosis forms sex cells with the correct haploid number of chromosomes. This maintains the correct diploid number of chromosomes in organisms when sex cells join. Meiosis also creates genetic variation by producing haploid cells.

Maintaining Diploid Cells

Recall that diploid cells have pairs of chromosomes. Meiosis helps to maintain diploid cells in offspring by making haploid sex cells. When haploid sex cells join together during fertilization, they make a diploid zygote, or fertilized egg. The zygote then divides by mitosis and cell division and creates a diploid organism. **Figure 3** illustrates how the diploid number is maintained in ducks.

Figure 3 Meiosis ensures that the chromosome number of a species stays the same from generation to generation.

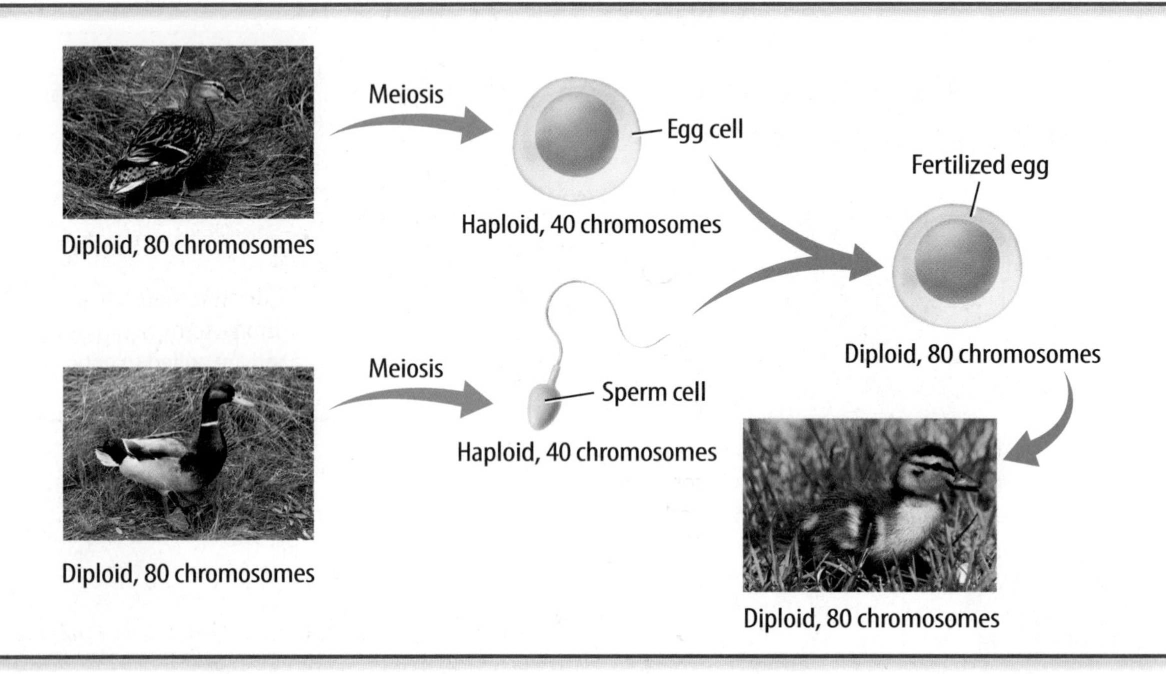

Active Reading **11. Explain** Why is meiosis important?

Creating Haploid Cells

The result of meiosis is haploid sex cells. This helps maintain the correct number of chromosomes in each generation of offspring. Haploid cell formation is important because it allows for genetic variation through a process called crossing over. During this process, chromosomal segments are exchanged between a pair of homologous chromosomes. The resulting sex cells contain new combinations of genes.

The genetic makeup of offspring is a combination of chromosomes from two sex cells. Variation in the sex cells results in more genetic variation in the next generation.

How do mitosis and meiosis differ?

Sometimes, it's hard to remember the differences between mitosis and meiosis. Use **Table 2** to review these processes.

During mitosis and cell division, a body cell and its nucleus divide once and produce two identical cells. These processes are important for growth and repair or replacement of damaged tissue. Some organisms reproduce by these processes. The two daughter cells produced by mitosis and cell division have the same genetic information.

During meiosis, a reproductive cell and its nucleus divide twice and produce four cells—two pairs of identical haploid cells. Each cell has half the number of chromosomes as the original cell. Meiosis happens in the reproductive organs of multicellular organisms. Meiosis forms sex cells used for sexual reproduction.

Active Reading **12. Compare and Contrast** Use **Table 2** to complete the Venn diagram below.

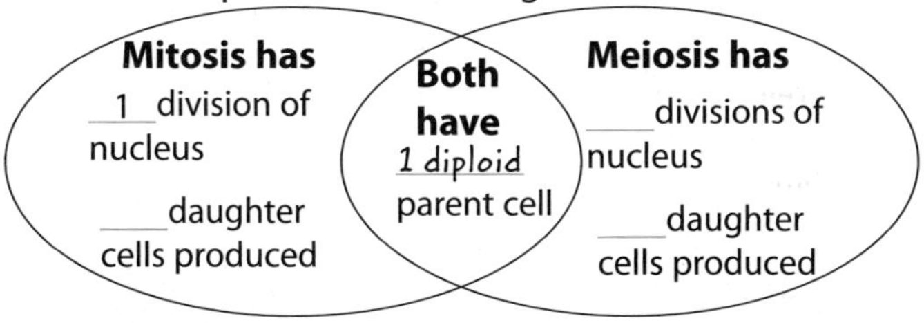

Mitosis has
1 division of nucleus
____ daughter cells produced

Both have
1 diploid parent cell

Meiosis has
____ divisions of nucleus
____ daughter cells produced

Table 2 Comparison of Types of Cell Division

Characteristic	Meiosis	Mitosis and Cell Division
Number of chromosomes in parent cell	diploid	diploid
Type of parent cell	reproductive	body
Number of divisions of nucleus	2	1
Number of daughter cells produced	4	2
Chromosome number in daughter cells	haploid	diploid
Function	forms sperm and egg cells	growth, cell repair, some types of reproduction

Active Reading **13. Compare** How many cells are produced during mitosis? During meiosis?

____

____

Math Skills MA.6.A.3.6

Use Proportions

An equation that shows that two ratios are equivalent is a proportion. The ratios $\frac{1}{2}$ and $\frac{3}{6}$ are equivalent, so they can be written as $\frac{1}{2} = \frac{3}{6}$.

You can use proportions to figure out how many daughter cells will be produced during mitosis. If you know that one cell produces two daughter cells at the end of mitosis, you can use proportions to calculate how many daughter cells will be produced by eight cells undergoing mitosis.

Set up an equation of the two ratios. $\frac{1}{2} = \frac{8}{y}$

Cross-multiply. $1 \times y = 8 \times 2$

$1y = 16$

Divide each side by 1. $y = 16$

Practice

14. You know that one cell produces four daughter cells at the end of meiosis. How many daughter cells would be produced if eight sex cells undergo meiosis?

Advantages of Sexual Reproduction

Did you ever wonder why a brother and a sister might not look alike? The answer is sexual reproduction. The main advantage of sexual reproduction is that offspring inherit half their DNA from each parent. Offspring are not likely to inherit the same DNA from the same parents. Different DNA means that each offspring has a different set of traits. This results in genetic variation among the offspring.

Review Vocabulary

DNA
the genetic information in a cell

Active Reading 15. **Identify** Why is sexual reproduction beneficial?

Genetic Variation

As you just read, genetic variation exists among humans. You can look at your friends to see genetic variation. Genetic variation occurs in all organisms that reproduce sexually. Consider the plants shown in **Figure 4.** The plants are members of the same species, but they have different traits, such as the ability to resist disease.

The inheritance of one trait does not influence the inheritance of another trait. This trend is called independent assortment. Independent assortment means genetic variation can vary widely. These differences might be an advantage if the environment changes. This helps some individuals survive unusually harsh conditions, such as drought or severe cold.

16. **Visual Check** **Describe** How does cassava mosaic disease affect cassava leaves?

Genetic Variation

Disease-resistant cassava leaves

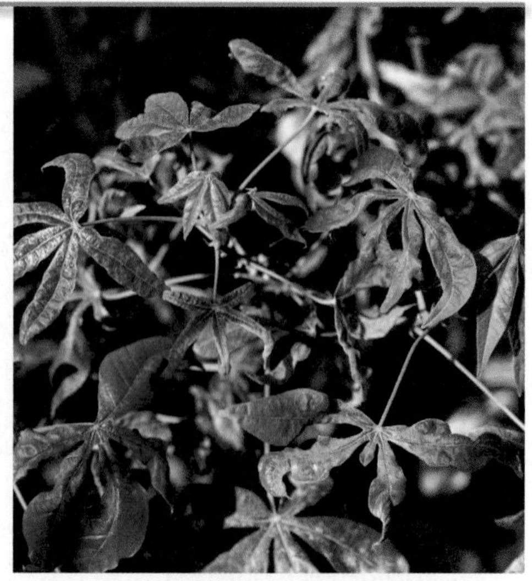
Cassava leaves with cassava mosaic disease

Figure 4 These plants belong to the same species. However, one is more disease-resistant than the other.

Selective Breeding

Did you know that broccoli, kohlrabi, kale, and cabbage all descended from one type of mustard plant? More than 2,000 years ago, farmers noticed that some mustard plants had different traits, such as larger leaves or bigger flower buds. The farmers started to choose which traits they wanted by selecting certain plants to reproduce and grow. For example, some farmers chose only the plants with the biggest flowers and stems and planted their seeds. Over time, the offspring of these plants became what we know today as broccoli, shown in **Figure 5.** This process is called selective breeding or artificial selection. Artificial selection is used to develop many types of plants and animals with desirable traits. It is another example of the benefits of sexual reproduction.

Active Reading **17. Explain** Write why genetic variation and selective breeding are advantages of sexual reproduction.

Advantage	Explanation
Genetic variation	
Selective breeding	

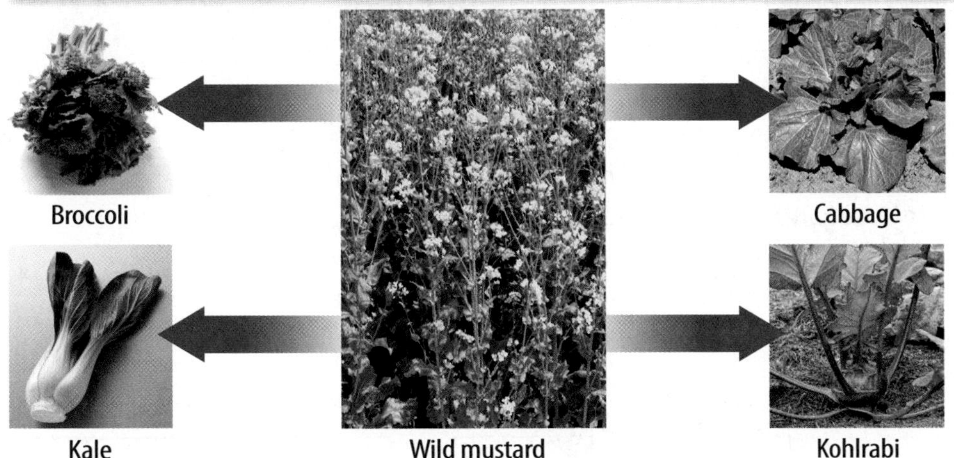

Figure 5 The wild mustard is the common ancestor to all these plants.

Disadvantages of Sexual Reproduction

Although sexual reproduction produces more genetic variation, it does have some disadvantages. Sexual reproduction takes time and energy. Organisms have to grow and develop until they are mature enough to produce sex cells. Then the organisms have to form sex cells—either eggs or sperm. Before they can reproduce, organisms usually have to find mates. Searching for a mate can take a long time and requires energy. The search for a mate might also expose individuals to predators, diseases, or harsh environmental conditions. In addition, sexual reproduction is limited by certain factors. For example, fertilization cannot take place during pregnancy, which can last as long as two years in some mammals.

Active Reading **18. Locate** Underline two disadvantages of sexual reproduction.

Lesson Review 1

Visual Summary

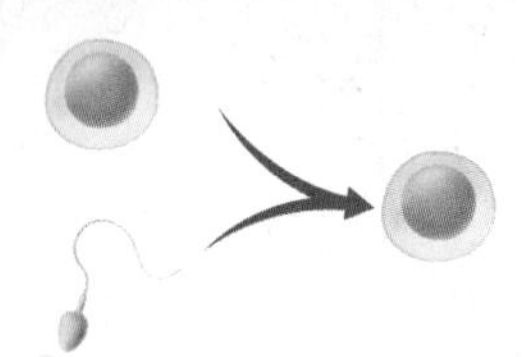

Fertilization occurs when an egg cell and a sperm cell join together.

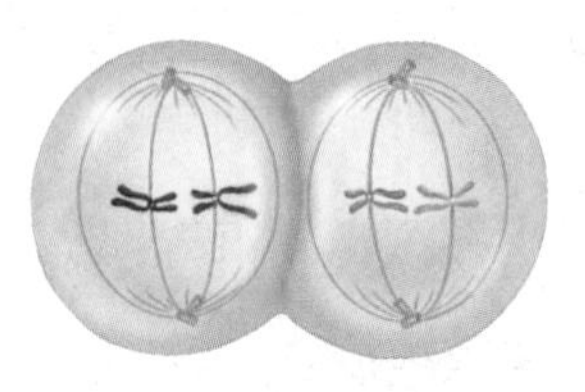

Organisms produce sex cells through meiosis.

Sexual reproduction results in genetic variation among individuals.

Use Vocabulary

1. **Extend** Use the terms *egg, sperm,* and *zygote* in a sentence.

2. **Distinguish** between haploid and diploid. SC.7.L.16.3

Understand Key Concepts

3. **Define** sexual reproduction. SC.7.L.16.3

4. Homologous chromosomes separate during which phase of meiosis?
   - (A) anaphase I
   - (B) anaphase II
   - (C) metaphase I
   - (D) metaphase II

Interpret Graphics

5. **Organize** Fill in the graphic organizer below to sequence the phases of meiosis I and meiosis II. LA.6.2.2.3

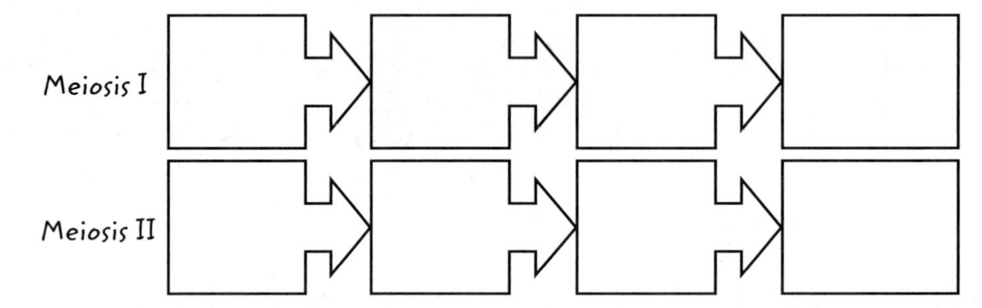

Critical Thinking

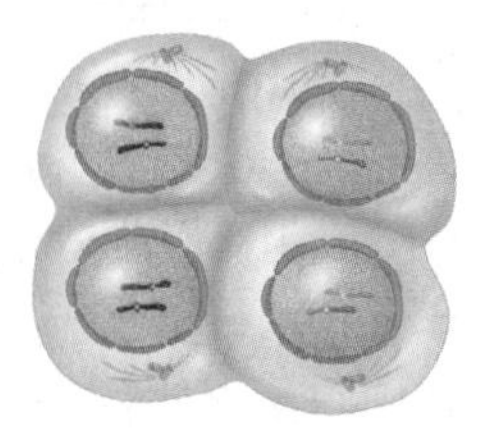

6. **Analyze** Why is the result of this stage of meiosis an advantage for organisms that reproduce sexually? SC.7.L.16.3

Math Skills MA.6.A.3.6

7. If 15 cells undergo meiosis, how many daughter cells will be produced?

CAREERS in SCIENCE

The Spider Mating Dance

Meet Norman Platnick, a scientist studying spiders.

Norman Platnick is fascinated by all spider species—from the dwarf tarantula-like spiders of Panama to the blind spiders of New Zealand. These are just two of the over 1,400 species he's discovered worldwide.

How does Platnick identify new species? One way is the pedipalps. Every spider has two pedipalps, but they vary in shape and size among the over 40,000 species. Pedipalps look like legs but function more like antennae and mouthparts. Male spiders use their pedipalps to aid in reproduction.

Getting Ready When a male spider is ready to mate, he places a drop of sperm onto a sheet of silk he constructs. Then he dips his pedipalps into the drop to draw up the sperm.

Finding a Mate The male finds a female of the same species by touch or by sensing certain chemicals she releases.

Courting and Mating Males of some species court a female with a special dance. For other species, a male might present a female with a gift, such as a fly wrapped in silk. During mating, the male uses his pedipalps to transfer sperm to the female.

What happens to the male after mating? That depends on the species. Some are eaten by the female, while others move on to find new mates.

▲ **Spiders reproduce sexually, so each offspring has a unique combination of genes from its parents. Over many generations, this genetic variation has led to the incredible diversity of spiders in the world today.**

◀ **Norman Platnick is an arachnologist (uh rak NAH luh just) at the American Museum of Natural History. Arachnologists are scientists who study spiders.**

It's Your Turn

RESEARCH Select a species of spider and research its mating rituals. What does a male do to court a female? What is the role of the female? What happens to the spiderlings after they hatch? Use images to illustrate a report on your research.

Lesson 2

Asexual REPRODUCTION

ESSENTIAL QUESTIONS

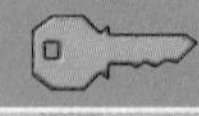 What is asexual reproduction, and why is it beneficial?

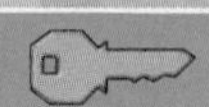 How do the types of asexual reproduction differ?

Vocabulary

asexual reproduction p. 141
fission p. 142
budding p. 143
regeneration p. 144
vegetative reproduction p. 145
cloning p. 146

Florida NGSSS

LA.6.2.2.3 The student will organize information to show understanding (e.g., representing main ideas within text through charting, mapping, paraphrasing, summarizing, or comparing/contrasting);

SC.7.N.1.1 Define a problem from the seventh grade curriculum, use appropriate reference materials to support scientific understanding, plan and carry out scientific investigation of various types, such as systematic observations or experiments, identify variables, collect and organize data, interpret data in charts, tables, and graphics, analyze information, make predictions, and defend conclusions.

SC.7.N.1.3 Distinguish between an experiment (which must involve the identification and control of variables) and other forms of scientific investigation and explain that not all scientific knowledge is derived from experimentation.

SC.7.L.16.3 Compare and contrast the general processes of sexual reproduction requiring meiosis and asexual reproduction requiring mitosis.

SC.7.L.16.4 Recognize and explore the impact of biotechnology (cloning, genetic engineering, artificial selection) on the individual, society and the environment.

SC.8.N.1.3 Use phrases such as "results support" or "fail to support" in science, understanding that science does not offer conclusive 'proof' of a knowledge claim.

SC.8.N.1.6 Understand that scientific investigations involve the collection of relevant empirical evidence, the use of logical reasoning, and the application of imagination in devising hypotheses, predictions, explanations and models to make sense of the collected evidence.

Launch Lab

SC.7.N.1.1, SC.7.N.1.3, SC.7.L.16.3

20 minutes

How do yeast reproduce?

Some organisms can produce offspring without meiosis or fertilization. You can observe this process when you add sugar and warm water to dried yeast.

Procedure

1. Read and complete a lab safety form.
2. Pour 125 mL of water into a **beaker.** The water should be at a temperature of 34°C.
3. Add 5 g of **sugar** and 5 g of **yeast** to the water. Stir slightly. Record your observations after 5 minutes.
4. Using a **dropper,** put a drop of the yeast solution on a **microscope slide.** Place a **coverslip** over the drop.
5. View the yeast solution under a **microscope.** On a separate sheet of paper draw what you see.

Data and Observations

Think About This

1. What evidence did you observe that yeast reproduce?

2. Key Concept How do you think this process differs from sexual reproduction?

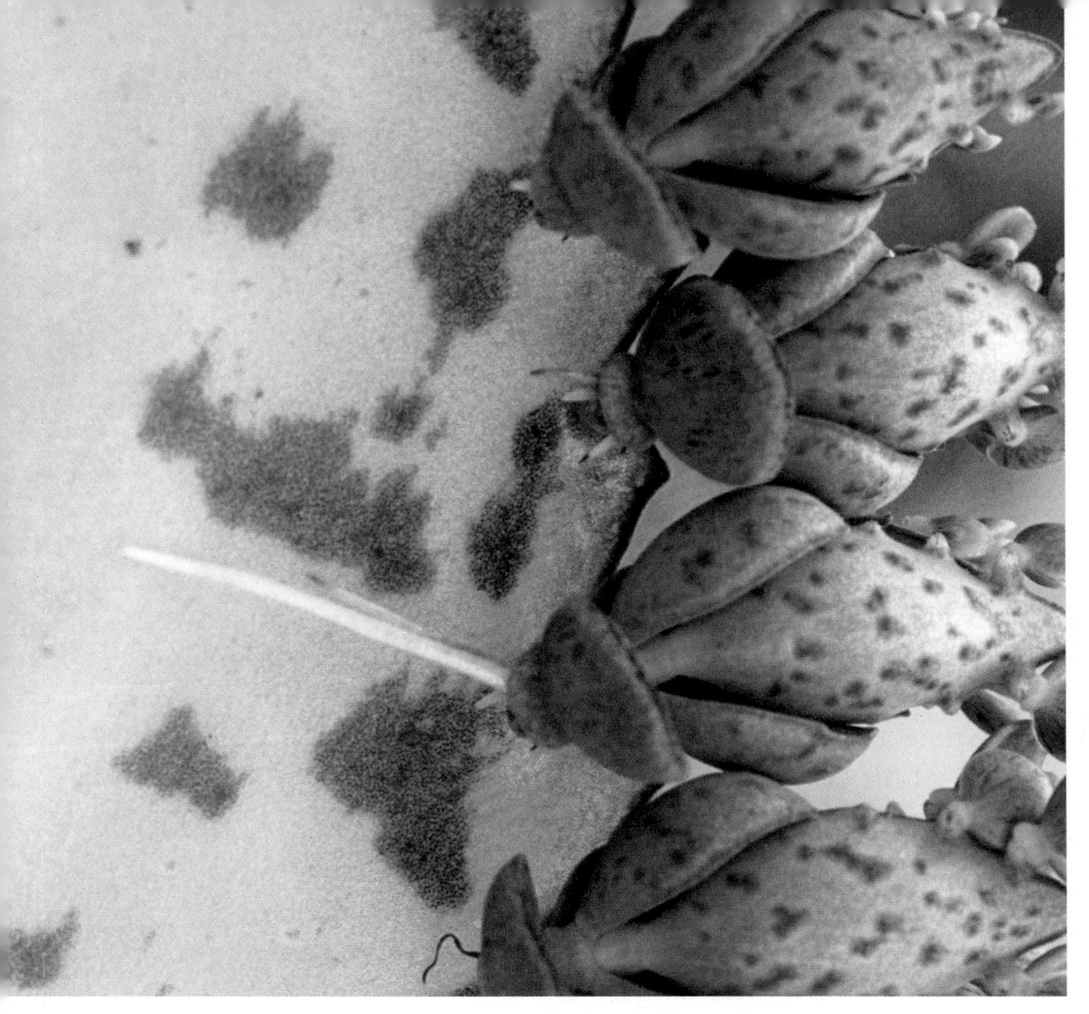

Inquiry Plants on plants?

1. Look closely at the edges of the plant's leaves. Tiny plants are growing there. How do you think this type of plant can reproduce without meiosis and fertilization?

______

______

______

What is asexual reproduction?

Lunch is over and you are in a rush to get to class. You wrap up your half-eaten sandwich and toss it into your locker. A week goes by before you spot the sandwich in the corner of your locker. The surface of the bread is now covered with fuzzy mold—not very appetizing. How did that happen?

The mold on the sandwich is a type of fungus (FUN gus). A fungus releases enzymes that break down organic matter, such as food. It has structures that penetrate and anchor to food, much like roots anchor plants to soil. A fungus can multiply quickly in part because generally a fungus can reproduce either sexually or asexually. Recall that sexual reproduction involves two parent organisms and the processes of meiosis and fertilization. Offspring inherit half their DNA from each parent, resulting in genetic variation among the offspring.

In **asexual reproduction,** *one parent organism produces offspring without meiosis and fertilization.* Because the offspring inherit all their DNA from one parent, they are genetically identical to each other and to their parent.

Active Reading FOLDABLES® LA.6.2.2.3, SC.7.L.16.3

Fold a sheet of paper into a six-celled chart. Label the front "Asexual Reproduction," and label the chart inside as shown. Use it to compare types of asexual reproduction.

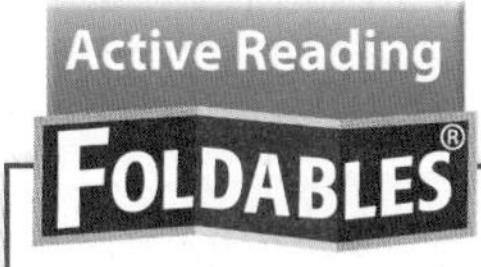

Active Reading 2. **Restate** Describe asexual reproduction in your own words.

______

______

______

______

______

Types of Asexual Reproduction

There are many different types of organisms that reproduce by asexual reproduction. In addition to fungi, bacteria, protists, plants, and animals can reproduce asexually. In this lesson, you will learn how organisms reproduce asexually.

Word Origin

fission
from Latin *fissionem,* means "a breaking up, cleaving"

Fission

Recall that prokaryotes have a simpler cell structure than eukaryotes. A prokaryote's DNA is not contained in a nucleus. For this reason, mitosis does not occur, and cell division in a prokaryote is a simpler process than in a eukaryote. *Cell division in prokaryotes that forms two genetically identical cells is known as* **fission.**

Active Reading **3. Explain** What advantage might asexual reproduction by fission have over sexual reproduction?

Fission begins when a prokaryote's DNA molecule is copied. Each copy attaches to the cell membrane. Then the cell begins to grow longer, pulling the two copies of DNA apart. At the same time, the cell membrane begins to pinch inward along the middle of the cell. Finally the cell splits and forms two new, identical offspring. The original cell no longer exists.

As shown in **Figure 6,** *E. coli,* a common bacterium, divides through fission. Some bacteria can divide every 20 minutes. At that rate, 512 bacteria can be produced from one original bacterium in about three hours.

Fission

Figure 6 Bacteria can divide very rapidly through fission.

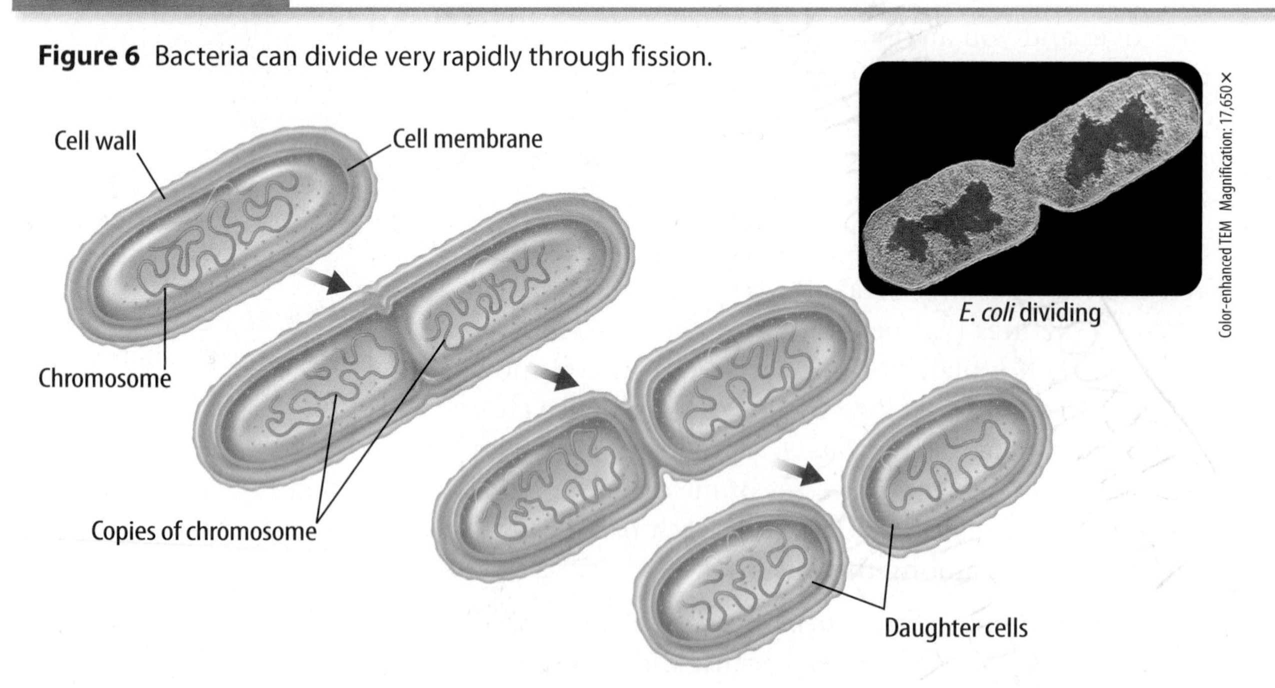

4. Visual Check **Analyze** What happens to the original cell's chromosones during fission?

Mitotic Cell Division

Many unicellular eukaryotes reproduce by mitotic cell division. In this type of asexual reproduction, an organism forms two offspring through mitosis and cell division. In **Figure 7,** an amoeba's nucleus has divided by mitosis. Next, the cytoplasm and its contents divide through cytokinesis, and two new amoebas form.

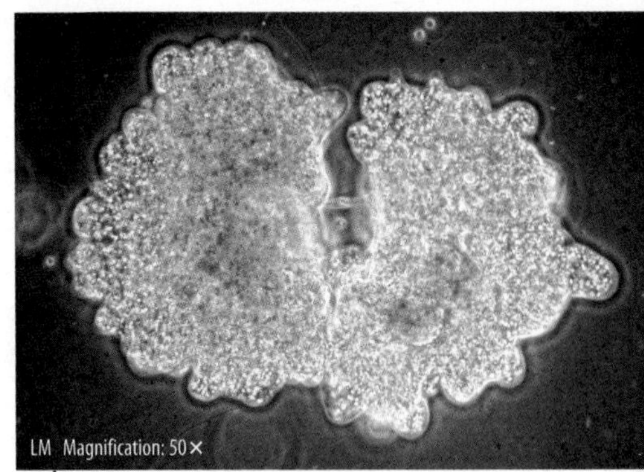

Figure 7 During mitotic cell division, an amoeba divides its chromosomes and cell contents evenly between the daughter cells.

Budding

In **budding,** *a new organism grows by mitosis and cell division on the body of its parent.* The bud, or offspring, is genetically identical to its parent. When the bud becomes large enough, it can break from the parent and live on its own. In some cases, an offspring remains attached to its parent and starts to form a colony. **Figure 8** shows a hydra in the process of budding. The hydra is an example of a multicellular organism that can reproduce asexually. Unicellular eukaryotes, such as yeast, can also reproduce through budding, as you saw in the Launch Lab.

Active Reading **5. Identify** What type of reproduction occurs by mitotic cell division?

Budding

Figure 8 The hydra bud has the same genetic makeup as its parent.

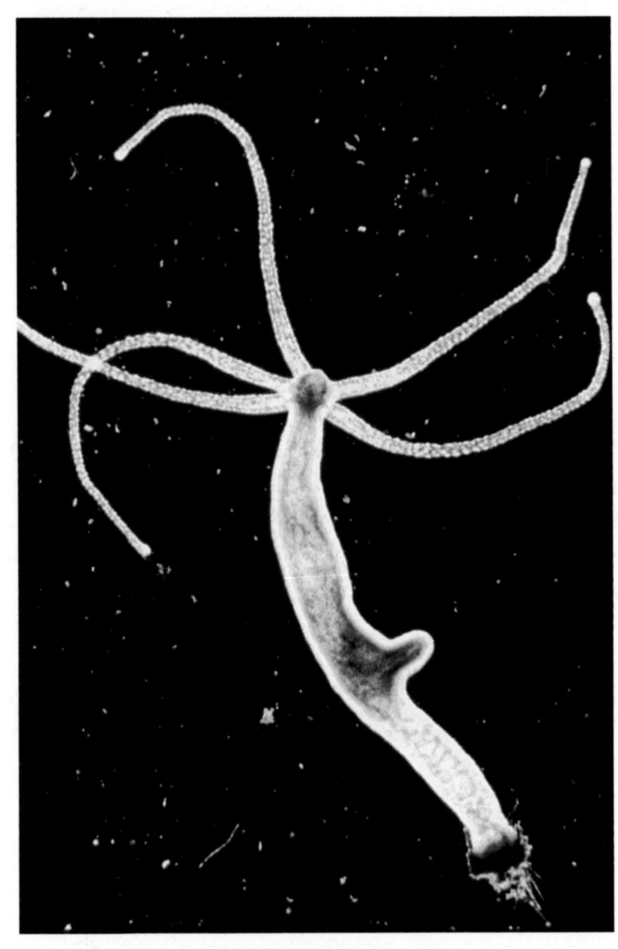

Bud forms.

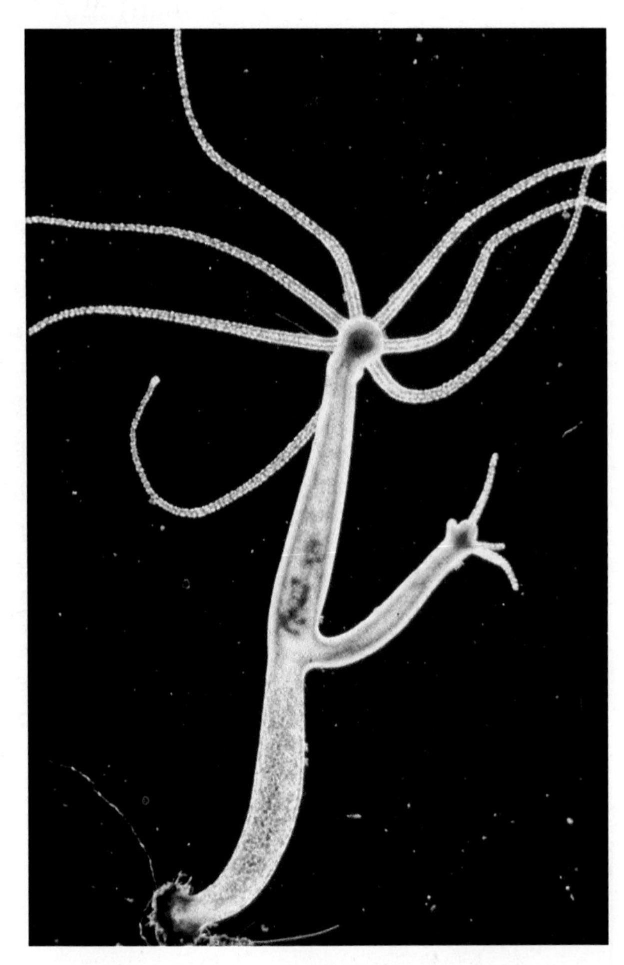

Bud develops a mouth and tentacles.

Figure 9 A planarian can reproduce through regeneration.

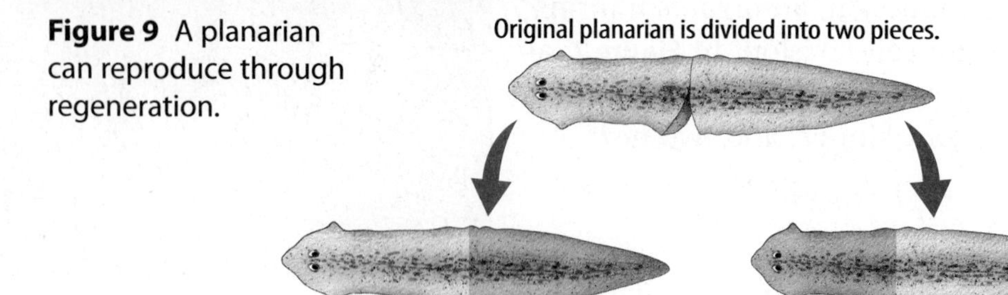

Animal Regeneration

Another type of asexual reproduction, **regeneration,** *occurs when an offspring grows from a piece of its parent.* The ability to regenerate a new organism varies greatly among animals.

Producing New Organisms Some sea stars have five arms. If separated from the parent sea star, each arm has the **potential** to grow into a new organism. To regenerate a new sea star, the arm must contain a part of the central disk of the parent. If conditions are right, one five-armed sea star can produce as many as five new organisms.

ACADEMIC VOCABULARY
potential
(noun) possibility

Sea urchins, sea cucumbers, sponges, and planarians, such as the one shown in **Figure 9,** can also reproduce through regeneration. Notice that each piece of the original planarian becomes a new organism. As with all types of asexual reproduction, the offspring is genetically identical to the parent.

Active Reading **6. Compare** What is true of all cases of asexual reproduction?

____

____

Producing New Parts When you hear the term *regeneration,* you might think about a salamander regrowing a lost tail or leg. Regeneration of damaged or lost body parts is common in many animals. Newts, tadpoles, crabs, hydra, and zebra fish are all able to regenerate body parts. Even humans are able to regenerate some damaged body parts, such as the skin and the liver. This type of regeneration, however, is not considered asexual reproduction. It does not produce a new organism.

Active Reading **7. Explain** Complete the graphic below to explain how animal regeneration can produce two results.

Animal regeneration
produces

new ______________________.	new ______________________ organisms.
A complete offspring ______________________ ______________________ ______________________.	An organism can grow a ______________________ ______________________ when ______________________.

Vegetative Reproduction

Plants can also reproduce asexually in a process similar to regeneration. **Vegetative reproduction** *is a form of asexual reproduction in which offspring grow from a part of a parent plant.* For example, the strawberry plants shown in **Figure 10** send out long, horizontal stems called stolons. Wherever a stolon touches the ground, it can produce roots. Once the stolons have grown roots, a new plant can grow—even if the stolons have broken off the parent plant. Each new plant grown from a stolon is genetically identical to the parent plant.

Vegetative reproduction usually involves structures such as the roots, the stems, and the leaves of plants. In addition to strawberries, many other plants can reproduce by this method, including raspberries, potatoes, and geraniums.

Active Reading **8. Identify** Write the correct terms to identify the structures of plants usually involved with vegetative reproduction.

______	______	______

Figure 10 The smaller plants were grown from stolons produced by the parent plant.

9. Visual Check Identify Which plants in **Figure 10** are the parent plants?

______

______

Apply It! After you complete the lab, answer the question.

1. **Infer** How is the process of vegetative reproduction similar to animal regeneration?

Cloning

Fission, budding, and regeneration are all types of asexual reproduction that can produce genetically identical offspring in nature. In the past, the term *cloning* described any process that produced genetically identical offspring. Today, however, the word usually refers to a technique developed by scientists and performed in laboratories. **Cloning** *is a type of asexual reproduction performed in a laboratory that produces identical individuals from a cell or from a cluster of cells taken from a multicellular organism.* Farmers and scientists often use cloning to make copies of organisms or cells that have desirable traits, such as large flowers.

Active Reading **10. Identify** Underline three advantages of using tissue cultures.

Plant Cloning Some plants can be cloned using a method called tissue **culture,** as shown in **Figure 11.** Tissue culture enables plant growers and scientists to make many copies of a plant with desirable traits, such as sweet fruit. Also, a greater number of plants can be produced more quickly than by vegetative reproduction.

SCIENCE USE V. COMMON USE

culture

Science Use the process of growing living tissue in a laboratory

Common Use the social customs of a group of people

Tissue culture also enables plant growers to reproduce plants that might have become infected with a disease. To clone such a plant, a scientist can use cells from a part of a plant where they are rapidly undergoing mitosis and cell division. This part of a plant is called a meristem. Cells in meristems are disease-free. Therefore, if a plant becomes infected with a disease, it can be cloned using meristem cells.

Figure 11 New carrot plants can be produced from cells of a carrot root using tissue culture techniques.

Plant Cloning

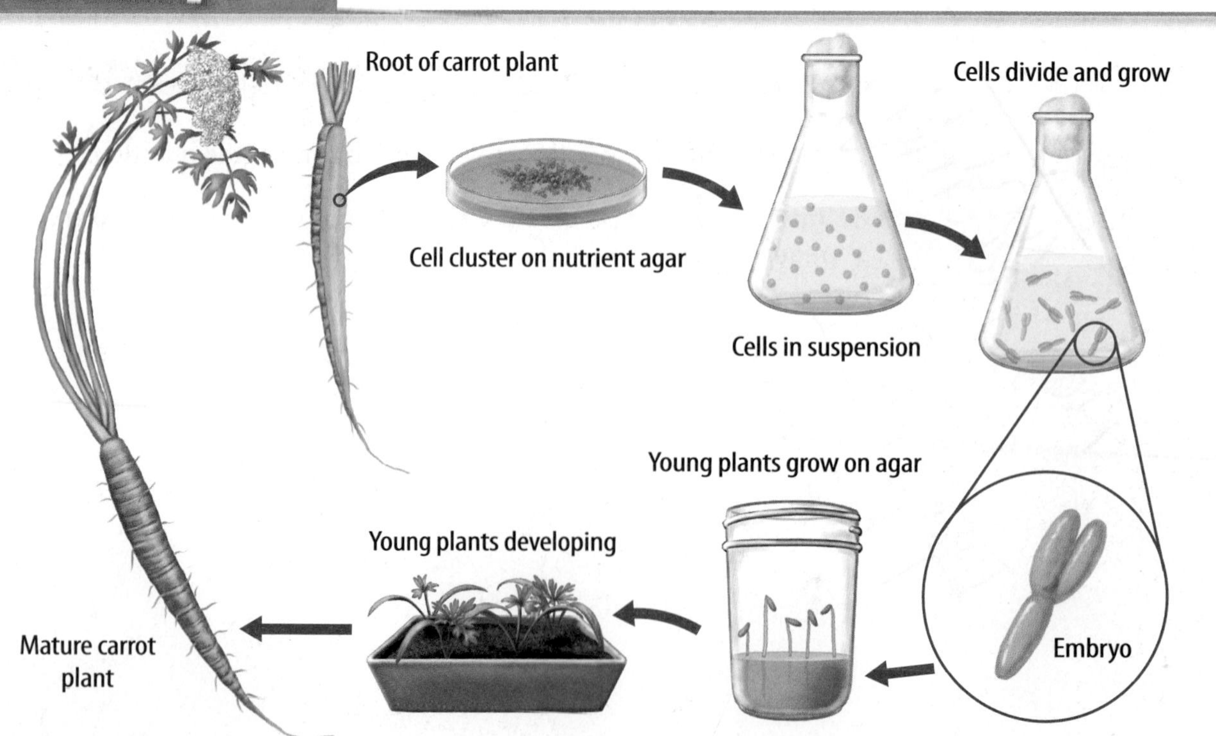

Animal Cloning In addition to cloning plants, scientists have been able to clone many animals. Because all of a clone's chromosomes come from one parent (the donor of the nucleus), the clone is a genetic copy of its parent. The first mammal cloned was a sheep named Dolly. **Figure 12** illustrates how this was done.

Scientists are currently working to save some endangered species from extinction by cloning. Although cloning is an exciting advancement in science, some people are concerned about the high cost and the ethics of this technique. Ethical issues include the possibility of human cloning. You might be asked to consider issues like this during your lifetime.

11. NGSSS Check Compare and Contrast Discuss the different types of asexual reproduction. **SC.7.L.16.3**

____

____

____

Animal Cloning

Figure 12 Scientists used two different sheep to produce the cloned sheep known as Dolly.

Active Reading **12. Identify** (Circle) the two sheep that are genetically identical.

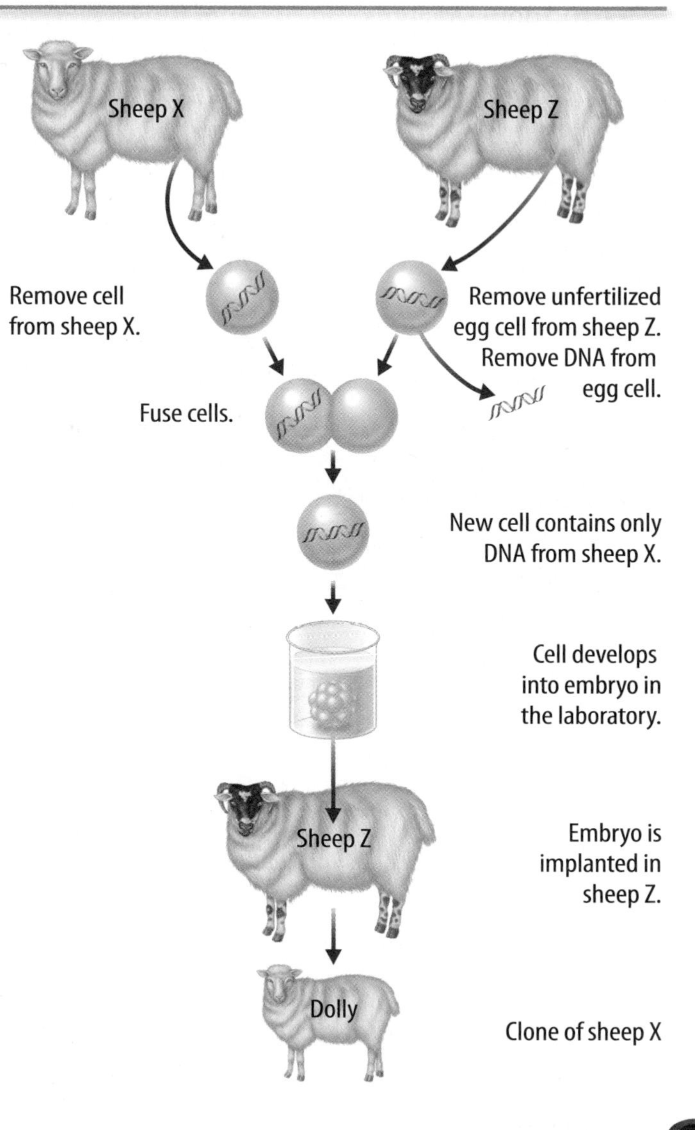

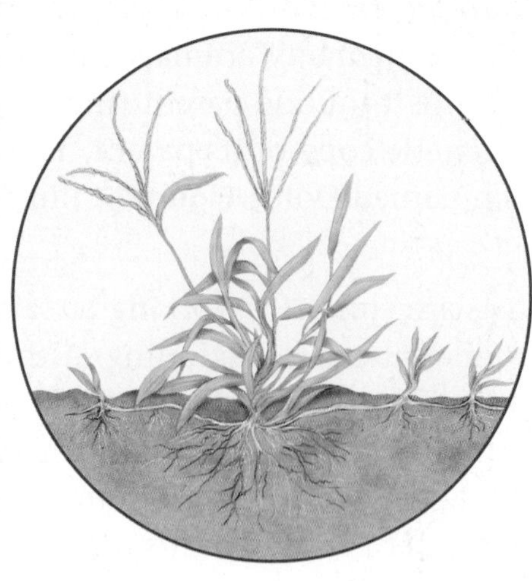

Figure 13 Crabgrass can spread quickly.

Active Reading **13. Explain** How can the crabgrass in **Figure 13** spread so quickly?

____

____

____

____

Active Reading **15. Define** Write a question about each vocabulary term in this lesson on a separate piece of paper. Exchange questions with another student. Together, discuss the answers to the questions.

Advantages of Asexual Reproduction

What are the advantages to organisms of reproducing asexually? Asexual reproduction enables organisms to reproduce without a mate. Recall that searching for a mate takes time and energy. Asexual reproduction also enables some organisms to rapidly produce a large number of offspring. For example, the crabgrass shown in **Figure 13** reproduces asexually by underground stems called stolons. This enables one plant to spread and colonize an area in a short period of time.

Active Reading **14. Locate** Underline one way in which asexual reproduction is beneficial.

Disadvantages of Asexual Reproduction

Although asexual reproduction usually enables organisms to reproduce quickly, it does have some disadvantages. Asexual reproduction produces offspring that are genetically identical to their parent. This results in little genetic variation within a population. Why is genetic variation important? Recall from Lesson 1 that genetic variation can give organisms a better chance of surviving if the environment changes. Think of the crabgrass. Imagine that all the crabgrass plants in a lawn are genetically identical to their parent plant. If a certain weed killer can kill the parent plant, then it can kill all the crabgrass plants in the lawn. This might be good for your lawn, but it is a disadvantage for the crabgrass.

Another disadvantage of asexual reproduction involves genetic changes, called mutations, that can occur. If an organism has a harmful mutation in its cells, the mutation will be passed to asexually reproduced offspring. This could affect the offspring's ability to survive.

Lesson Review 2

Visual Summary

In asexual reproduction, offspring are produced without meiosis and fertilization.

Cloning is one type of asexual reproduction.

Asexual reproduction enables organisms to reproduce quickly.

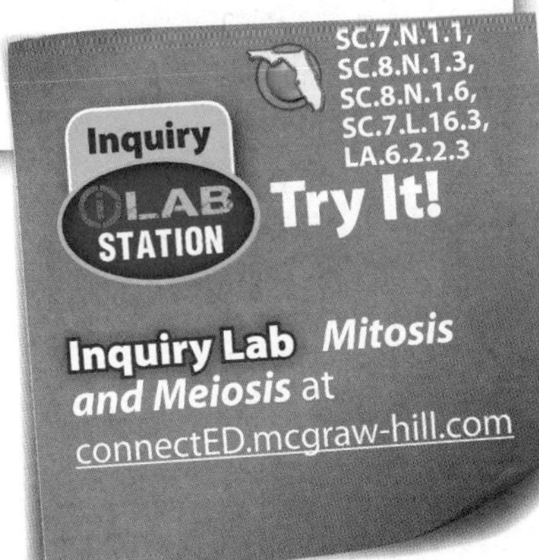

Inquiry Lab ***Mitosis and Meiosis*** at connectED.mcgraw-hill.com

Use Vocabulary

1. In ______________, only one parent organism produces offspring.

2. **Define** *cloning* in your own words. SC.7.L.16.4

3. **Use the term** *regeneration* in a sentence.

Understand Key Concepts 

4. **State** two reasons why asexual reproduction is beneficial.

5. Which is an example of asexual reproduction by regeneration?
   - (A) cloning sheep
   - (B) lizard regrowing a tail
   - (C) sea star arm producing a new organism
   - (D) strawberry plant producing stolons

Interpret Graphics

6. **Organize** Fill in the graphic organizer below to list the different types of asexual reproduction that occur in multicellular organisms. SC.7.L.16.3

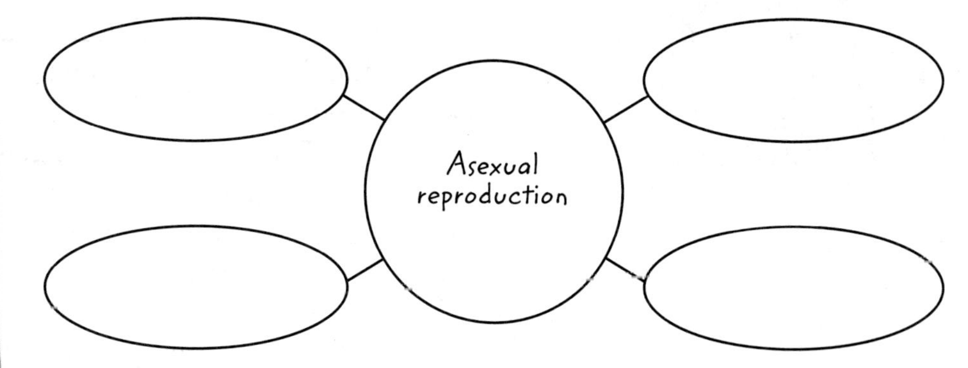

Critical Thinking

7. **Justify** the use of cloning to save endangered animals. SC.7.L.16.4

Chapter 4 Study Guide

Think About It! Reproduction is essential for the survival of species.

Key Concepts Summary

LESSON 1 Sexual Reproduction and Meiosis

- **Sexual reproduction** is the production of an offspring from the joining of a **sperm** and an **egg**.
- Division of the nucleus and cytokinesis happens twice in **meiosis**. Meiosis I separates homologous chromosomes. Meiosis II separates sister chromatids.
- Meiosis maintains the chromosome number of a species from one generation to the next.

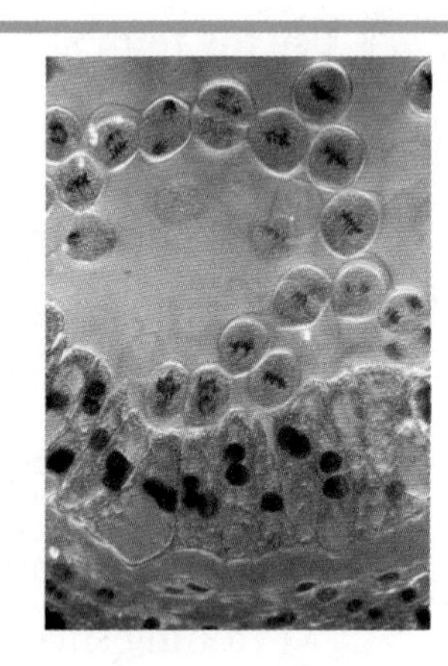

Vocabulary

sexual reproduction p. 129
egg p. 129
sperm p. 129
fertilization p. 129
zygote p. 129
diploid p. 130
homologous chromosomes p. 130
haploid p. 131
meiosis p. 131

LESSON 2 Asexual Reproduction

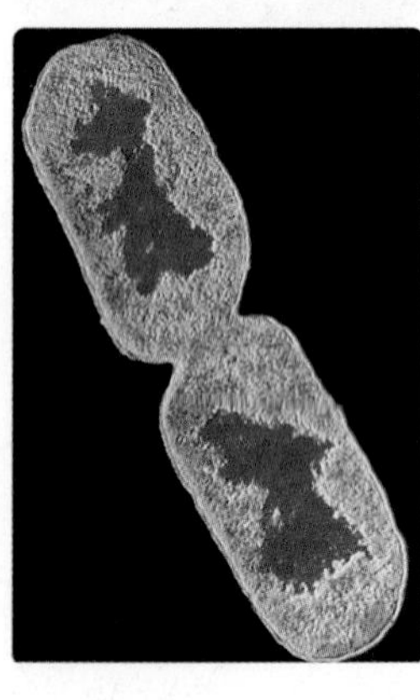

- **Asexual reproduction** is the production of offspring by one parent, which results in offspring that are genetically identical to the parent.
- Types of asexual reproduction include **fission**, mitotic cell division, **budding**, **regeneration**, **vegetative reproduction**, and **cloning**.
- Asexual reproduction can produce a large number of offspring in a short amount of time.

asexual reproduction p. 141
fission p. 142
budding p. 143
regeneration p. 144
vegetative reproduction p. 145
cloning p. 146

Active Reading

FOLDABLES® Chapter Project

Assemble your lesson Foldables as shown to make a Chapter Project. Use the project to review what you have learned in this chapter.

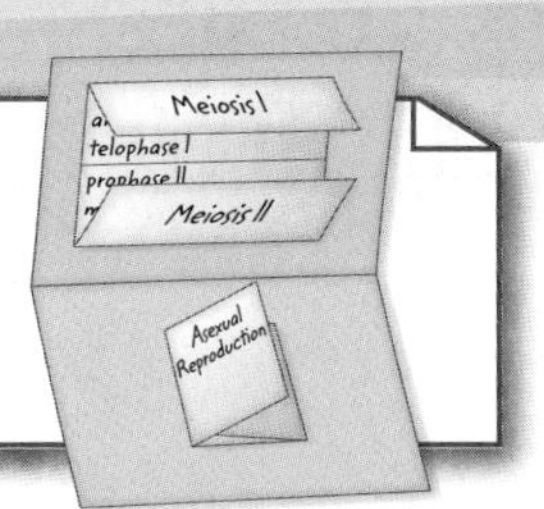

Use Vocabulary

1. Define meiosis in your own words.

2. Distinguish between an egg and a zygote.

3. Use the vocabulary words *haploid* and *diploid* in a sentence.

4. Cell division in prokaryotes is called ________________.

5. Define the term *vegetative reproduction* in your own words.

6. Distinguish between regeneration and budding.

7. A type of reproduction in which the genetic materials from two different cells combine, producing an offspring, is called ________________.

Link Vocabulary and Key Concepts

Use vocabulary terms from the previous page to complete the concept map.

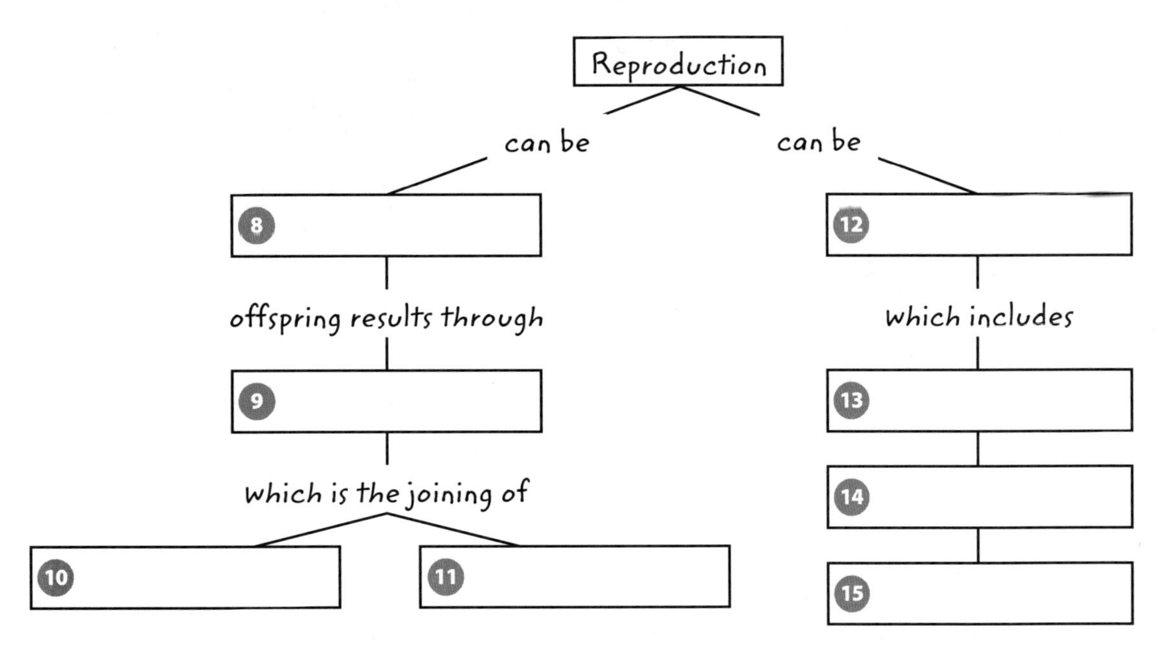

Chapter 4 Review

Fill in the correct answer choice.

Understand Key Concepts

1. Which is an advantage of sexual reproduction? **SC.7.L.16.3**
   Ⓐ Offspring are identical to the parents.
   Ⓑ Offspring with genetic variation are produced.
   Ⓒ Organisms don't have to search for a mate.
   Ⓓ Reproduction is rapid.

2. Which describes cells that have only one copy of each chromosome? **SC.7.L.16.3**
   Ⓐ diploid
   Ⓑ haploid
   Ⓒ homologous
   Ⓓ zygote

Use the figure below to answer questions 3 and 4.

3. Which phase of meiosis I is shown in the diagram? **SC.7.L.16.3**
   Ⓐ anaphase I
   Ⓑ metaphase I
   Ⓒ prophase I
   Ⓓ telophase I

4. Which phase of meiosis I comes after the phase in the diagram? **SC.7.L.16.3**
   Ⓐ anaphase I
   Ⓑ metaphase I
   Ⓒ prophase I
   Ⓓ telophase I

5. Tissue culture is an example of which type of reproduction? **SC.7.L.16.4**
   Ⓐ budding
   Ⓑ cloning
   Ⓒ fission
   Ⓓ regeneration

Critical Thinking

6. **Contrast** haploid cells and diploid cells. **SC.7.L.16.3**

7. **Model** Make a model of homologous chromosomes using materials of your choice. **SC.7.L.16.1**

8. **Form a hypothesis** about the effect of a mistake in separating homologous chromosomes during meiosis. **SC.7.L.16.3**

9. **Analyze** Crabgrass reproduces asexually by vegetative reproduction. Use the figure below to explain why this form of reproduction is an advantage for the crabgrass. **SC.7.L.16.3**

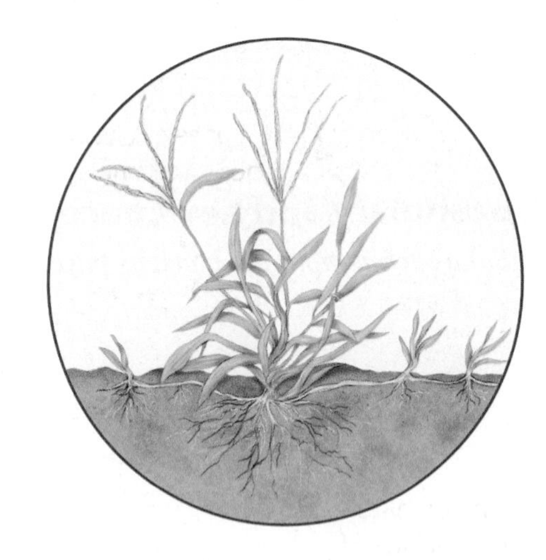

10 **Compare** budding and cloning. LA.7.2.2.3

11 **Compare and contrast** sexual reproduction and asexual reproduction. SC.7.L.16.3

Writing in Science

12 **Create** On a separate sheet of paper, create a plot for a short story that describes an environmental change and the importance of genetic variation in helping a species survive that change. Include characters, a setting, a climax, and an ending for your plot. SC.7.L.16.3

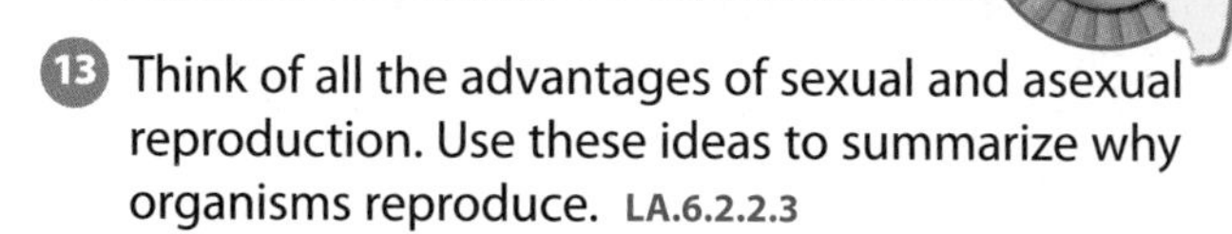

13 Think of all the advantages of sexual and asexual reproduction. Use these ideas to summarize why organisms reproduce. LA.6.2.2.3

14 The baby manatee below has a mother and a father. Do all living things have two parents? Explain. SC.7.L.16.3

Math Skills MA.6.A.3.6

Use Proportions

15 During mitosis, the original cell produces two daughter cells. How many daughter cells will be produced if 250 mouse cells undergo mitosis?

16 During meiosis, the original reproductive cell produces four daughter cells. How many daughter cells will be produced if 250 mouse reproductive cells undergo meiosis?

17 Two reproductive cells undergo meiosis. Each daughter cell also undergoes meiosis. How many cells are produced when the daughter cells divide?

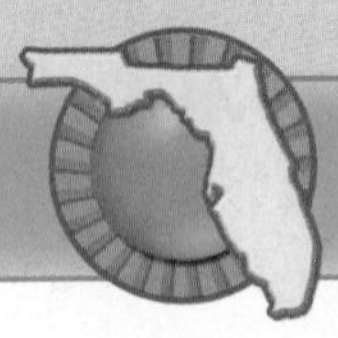

Florida NGSSS Benchmark Practice

Fill in the correct answer choice.

Multiple Choice

1 Which is a type of asexual reproduction performed in a laboratory that produces identical individuals from a cell, or from a cluster of cells, taken from a multicellular organism? **SC.7.L.16.4**

Ⓐ genetic engineering

Ⓑ cloning

Ⓒ gene therapy

Ⓓ artificial selection

Use the figure below to answer question 2.

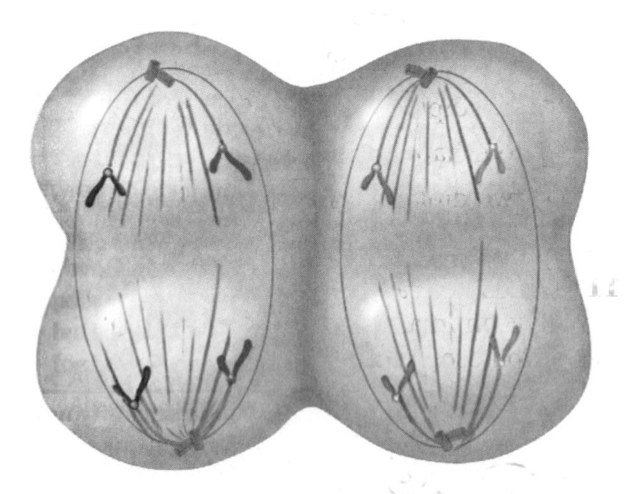

2 The image above represents a cell going through meiosis. How do you know the image does not represent mitosis? **SC.7.L.16.3**

Ⓕ Mitosis only has one cell division set. The image shows two cell division sets.

Ⓖ Mitosis only occurs in reproductive cells, such as the ones in the picture.

Ⓗ Mitosis is involved in the production of haploid sex cells.

Ⓘ Mitosis divides cells into eight cells, not four.

3 Why would scientists clone an animal? **SC.7.L.16.4**

Ⓐ to save the animal from dying

Ⓑ to study the sexual reproduction process

Ⓒ to save money

Ⓓ to save an endangered species

4 Hereditary information is coded in DNA on chromosomes. What happens when a zygote has an incorrect number of chromosomes? **SC.7.L.16.1**

Ⓕ It will not develop properly.

Ⓖ It will be unaffected.

Ⓗ It will be completely unable to develop.

Ⓘ Only human zygotes develop improperly.

Use the table below to answer question 5.

Comparison of Types of Cell Division		
Characteristic	**Meiosis**	**Mitosis**
Number of divisions of nucleus	2	A
Number of daughter cells produced	B	2

5 Which numbers should be inserted for A and B in the chart? **SC.7.L.16.3**

Ⓐ A=1 and B=2

Ⓑ A=1 and B=4

Ⓒ A=2 and B=2

Ⓓ A=2 and B=4

6 Which description best identifies characteristics of asexual reproduction? SC.7.L.16.3

Ⓕ two parents, offspring similar to parents, but not genetically identical

Ⓖ two parents, offspring genetically identical to parents

Ⓗ one parent, offspring similar to the parent, but not genetically identical

Ⓘ one parent, offspring genetically identical to the parent

7 How many homologous pairs of chromosomes will a diploid yeast cell have if the original cell has 32 chromosomes? SC.7.L.16.3

Ⓐ 8

Ⓑ 16

Ⓒ 32

Ⓓ 64

Use the figure below to answer question 8.

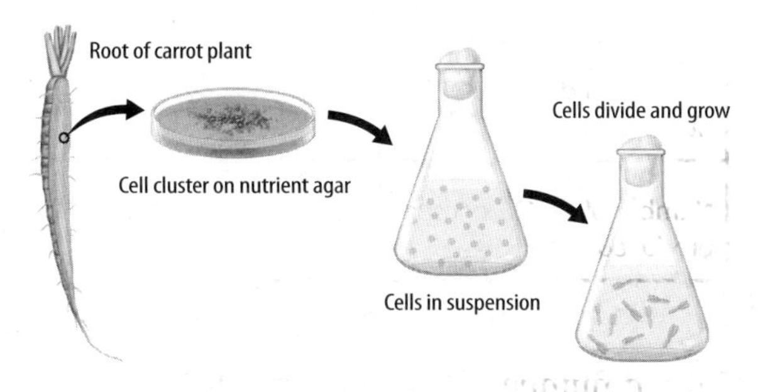

8 The figure illustrates the first four steps of which reproductive process? SC.7.L.16.4

Ⓕ animal cloning

Ⓖ regeneration

Ⓗ tissue culture

Ⓘ vegetative reproduction

9 Prokaryotes reproduce asexually through fission. Why does mitosis NOT occur in prokaryotes? SC.7.L.16.3

Ⓐ A prokaryote undergoes meiosis instead of mitosis.

Ⓑ A prokaryote's DNA is contained in the nucleus.

Ⓒ A prokaryote's DNA is not contained in the nucleus.

Ⓓ A prokaryote is a more complex organism than a eukaryote.

10 Which is the correct result of mitosis and meiosis? SC.7.L.16.3

Ⓕ four diploid cells and four haploid cells

Ⓖ two diploid cells and four haploid cells

Ⓗ four haploid cells and four haploid cells

Ⓘ two diploid cells and two haploid cells

11 Which diagram represents a reduction division in an organism that has a diploid chromosome number of eight? SC.7.L.16.3

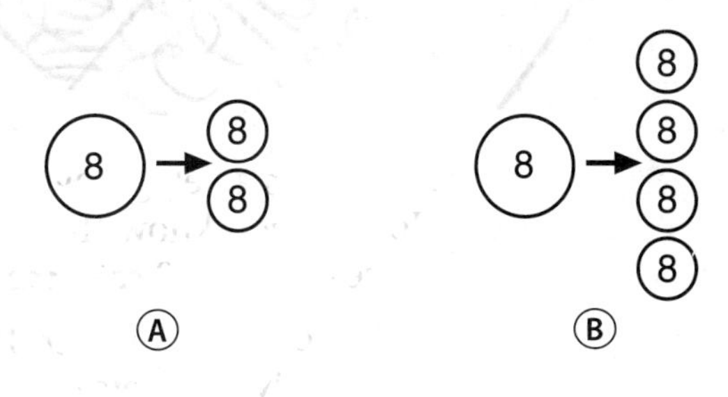

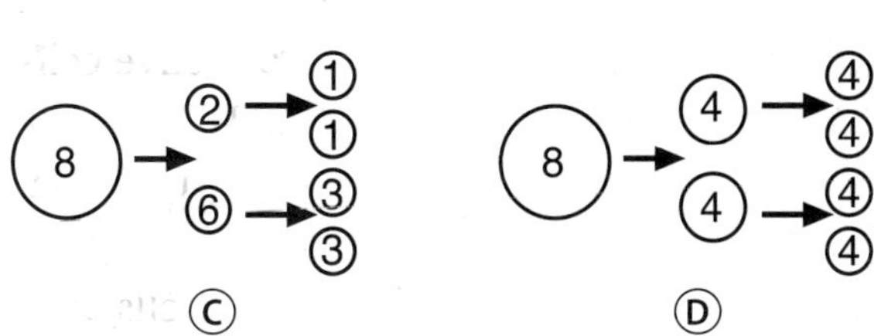

NEED EXTRA HELP?

If You Missed Question...	1	2	3	4	5	6	7	8	9	10	11
Go to Lesson...	2	1	2	1	1	2	1	2	2	1	1

Name ________________________ Date ____________

Bubble the correct answer.

Use the table below to answer questions 1 and 2.

Organism	Number of Diploid Chromosomes
Mosquito	6
Housefly	12
Tomato	24
Alligator	32
Apple	34
Human	46
Chicken	78
Crab	208

1. In which organism is the number of homologous pairs 12? **SC.7.L.16.3**

Ⓐ apple

Ⓑ housefly

Ⓒ mosquito

Ⓓ tomato

2. Which organism's haploid number is 16? **SC.7.L.16.3**

Ⓕ alligator

Ⓖ chicken

Ⓗ crab

Ⓘ human

3. Which image below shows a stage from meiosis II? **SC.7.L.16.3**

Ⓐ

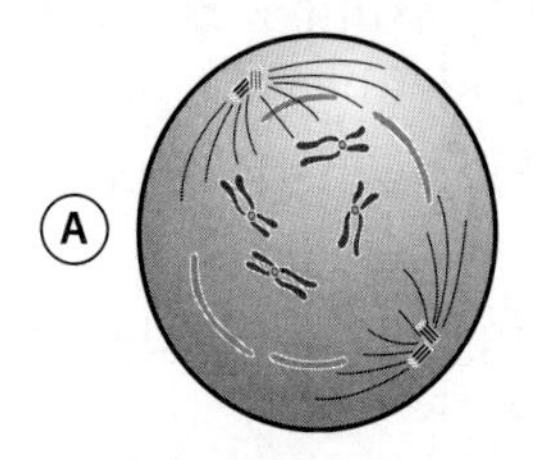

Ⓑ

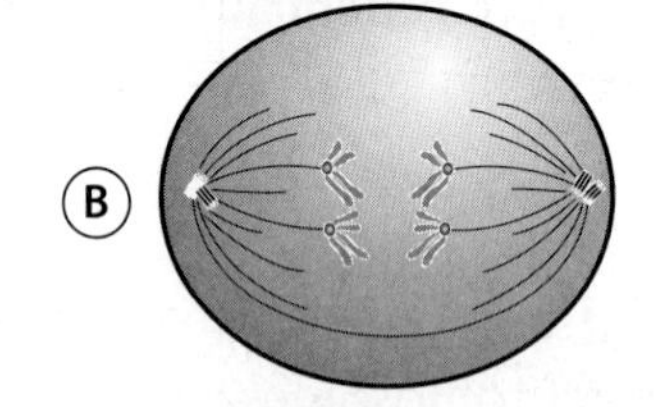

Ⓒ

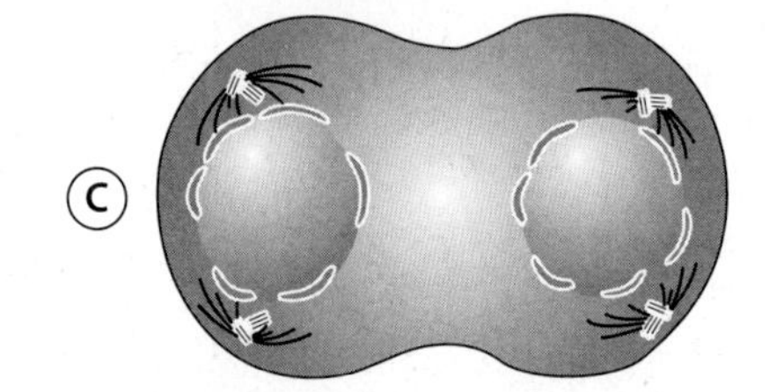

Ⓓ

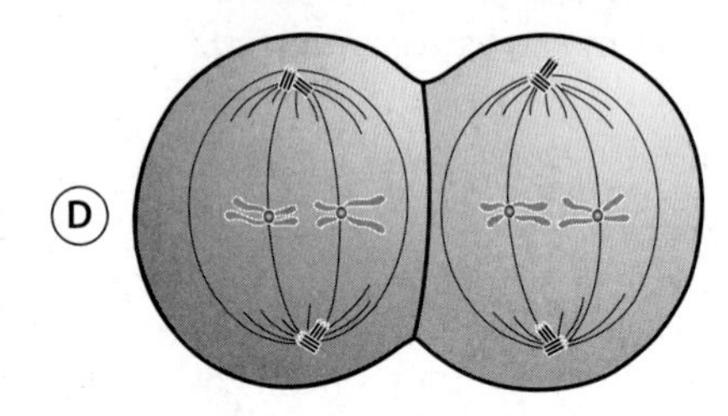

4. A zygote is formed during a process called **SC.7.L.16.3**

Ⓕ fertilization.

Ⓖ meiosis.

Ⓗ mitosis.

Ⓘ diploid.

Name ______________________ Date ____________

Multiple Choice *Bubble the correct answer.*

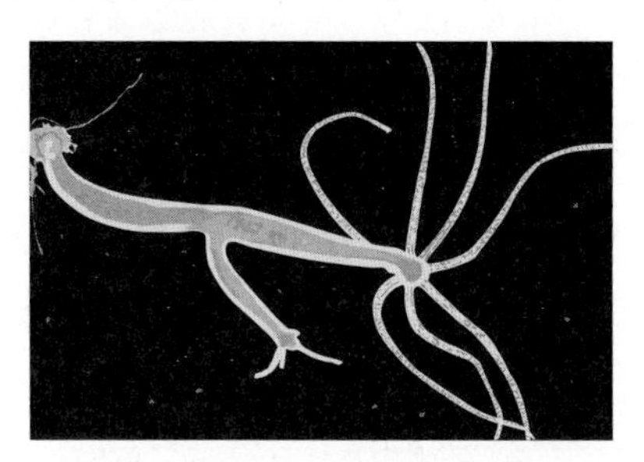

1. What type of asexual reproduction is shown in the image above? **SC.7.L.16.3**

(A) budding

(B) cloning

(C) fission

(D) regeneration

2. Which is an example of asexual reproduction by cloning? **SC.7.L.16.3**

(F) An amoeba divides into two new cells.

(G) A cut willow branch grows roots in moist soil.

(H) A harebell plant sends stolons underground.

(I) A severed starfish arm develops into a new starfish.

3. Which image below shows mitotic cell division? **SC.7.L.16.3**

(A)

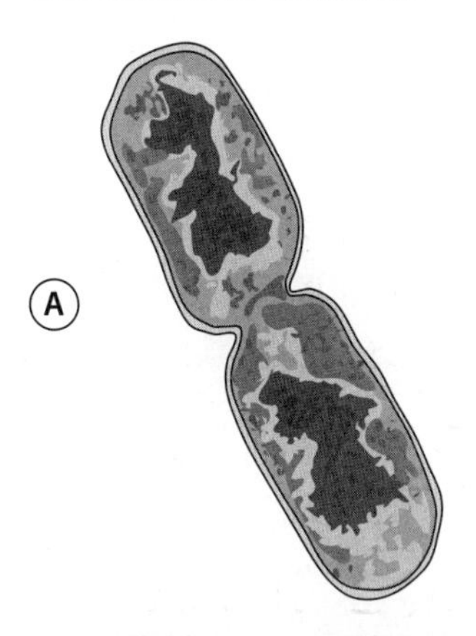

(B)

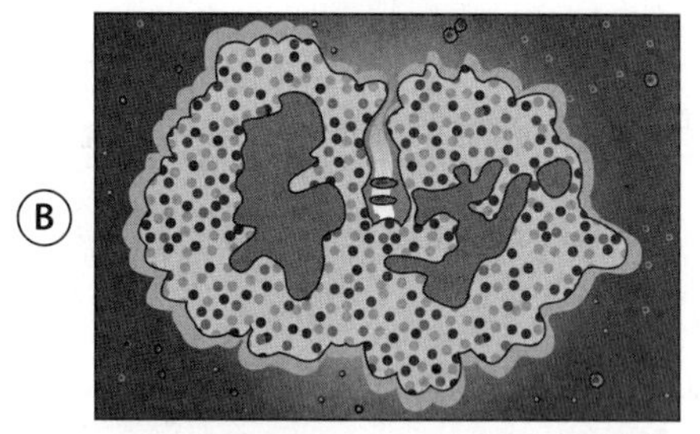

(C)

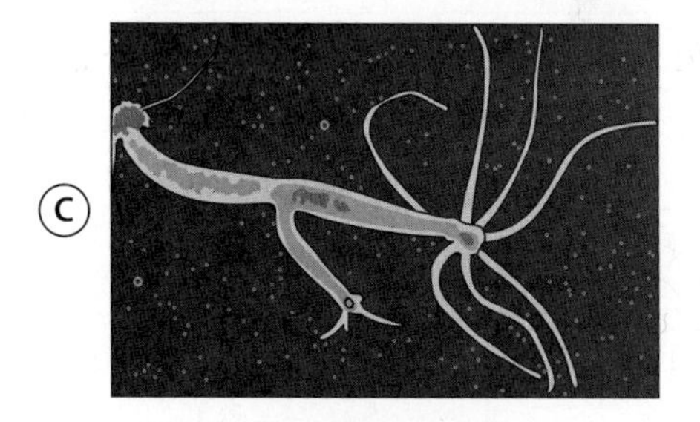

(D)

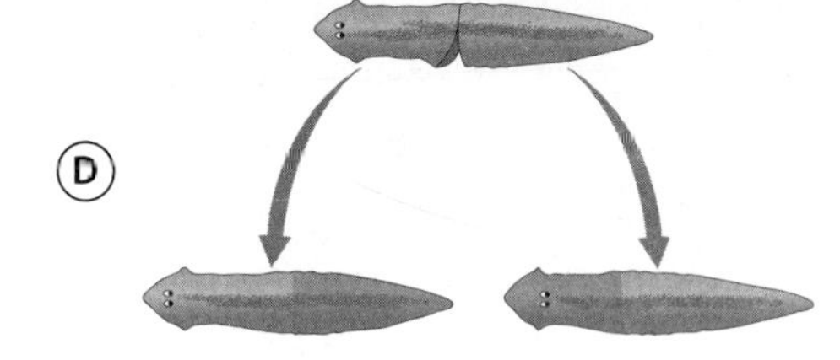

Name ______________________ Date ______________

Notes

Bunnies

Natalie's white rabbit had a litter of six bunnies. Four bunnies were black and two bunnies were white. The father of the bunnies is black. Natalie wondered why there were more black bunnies than white bunnies in the litter. Circle the response you think is true about the litter of bunnies.

A: All of the black bunnies are male and the white bunnies are female.

B: The fur color has nothing to do with the parents. It depends on the environment.

C: The bunnies got more traits for fur color from their father than from their mother.

D: There are fewer white bunnies because they do not survive as well as black bunnies.

E: There are more black bunnies because father rabbits have stronger traits for fur color than mother rabbits.

F: Each parent contributed the same amount of information about fur color during reproduction.

G: There must have been something wrong because black rabbits and white rabbits should produce grey rabbits.

Explain your thinking. Describe your ideas about how traits such as fur color are determined.

FLORIDA
Chapter 5
GENETICS

FLORIDA BIG IDEAS

1 **The Practice of Science**

2 **The Characteristics of Scientific Knowledge**

3 **The Role of Theories, Laws, Hypotheses, and Models**

16 **Heredity and Reproduction**

The Big Idea

Think About It!

How are traits passed from parents to offspring?

The color of this fawn is caused by a genetic trait called albinism. Albinism is the absence of body pigment. Notice that the fawn's mother has brown fur, the normal fur color of an adult whitetail deer.

1. Why do you think the fawn looks so different from its mother?
2. What do you think determines the color of the offspring?
3. How do you think traits are passed from generation to generation?

Get Ready to Read

What do you think about genetics?

Before you read, decide if you agree or disagree with each of these statements. As you read this chapter, see if you change your mind about any of the statements.

	AGREE	DISAGREE
1 Like mixing paints, parents' traits always blend in their offspring.	☐	☐
2 If you look more like your mother than your father, then you received more traits from your mother.	☐	☐
3 All inherited traits follow Mendel's patterns of inheritance.	☐	☐
4 Scientists have tools to predict the form of a trait an offspring might inherit.	☐	☐
5 Any condition present at birth is genetic.	☐	☐
6 A change in the sequence of an organism's DNA always changes the organism's traits.	☐	☐

There's More Online!
Video • Audio • Review • ⓘLab Station • WebQuest • Assessment • Concepts in Motion • Multilingual eGlossary

Lesson 1

Mendel and His PEAS

ESSENTIAL QUESTIONS

 Why did Mendel perform cross-pollination experiments?

What did Mendel conclude about inherited traits?

 How do dominant and recessive factors interact?

Vocabulary

heredity p. 163

genetics p. 163

dominant trait p. 169

recessive trait p. 169

Florida NGSSS

SC.6.N.1.5 Recognize that science involves creativity, not just in designing experiments, but also in creating explanations that fit evidence.

SC.6.N.2.2 Explain that scientific knowledge is durable because it is open to change as new evidence or interpretations are encountered.

SC.6.N.2.3 Recognize that scientists who make contributions to scientific knowledge come from all kinds of backgrounds and possess varied talents, interests, and goals.

SC.7.N.1.3 Distinguish between an experiment (which must involve the identification and control of variables) and other forms of scientific investigation and explain that not all scientific knowledge is derived from experimentation.

SC.7.N.1.5 Describe the methods used in the pursuit of a scientific explanation as seen in different fields of science such as biology, geology, and physics.

SC.7.N.1.6 Explain that empirical evidence is the cumulative body of observations of a natural phenomenon on which scientific explanations are based.

SC.7.N.1.7 Explain that scientific knowledge is the result of a great deal of debate and confirmation within the science community.

SC.7.N.2.1 Identify an instance from the history of science in which scientific knowledge has changed when new evidence or new interpretations are encountered.

Also covers: SC.7.L.16.1, LA.6.2.2.3, MA.6.A.3.6

SC.6.N.1.5

Inquiry Launch Lab

10 minutes

What makes you unique?

Traits such as eye color have many different types, but some traits have only two types. By a show of hands, determine how many students in your class have each type of trait below.

Student Traits		
Trait	Type 1	Type 2
Earlobes	Unattached	Attached
Thumbs	Curved	Straight
Interlacing fingers	Left thumb over right thumb	Right thumb over left thumb

Think About This

1. Why might some students have types of traits that others do not have?

2. If a person has dimples, do you think his or her offspring will have dimples? Explain.

3. **Key Concept** What do you think determines the types of traits you inherit?

Inquiry Same Species?

1. Have you ever seen a black ladybug? It is less common than the orange variety you might know, but both are the same species of beetle. So why do you think they look different?

Early Ideas About Heredity

Have you ever mixed two paint colors to make a new color? Long ago, people thought an organism's characteristics, or traits, mixed like colors of paint because offspring resembled both parents. This is known as blending inheritance.

Today, scientists know that **heredity** (huh REH duh tee)—*the passing of traits from parents to offspring*—is more complex. Every new generation of organisms requires a set of instructions. These instructions determine what an organism will look like. More than 150 years ago, Gregor Mendel, an Austrian monk, performed experiments that helped answer these questions and disprove the idea of blending inheritance. Because of his research, Mendel is known as the father of **genetics** (juh NEH tihks)—*the study of how traits are passed from parents to offspring.*

Active Reading **2. Summarize** Describe genetics, and explain why Gregor Mendel is known as the father of genetics.

WORD ORIGIN

genetics
from Greek *genesis*, means "origin"

Active Reading **3. Identify** Underline three characterisics that make pea plants ideal for genetic studies.

Mendel's Experimental Methods

During the 1850s, Mendel studied genetics by doing controlled breeding experiments with pea plants. Pea plants are ideal for genetic studies because

- they reproduce quickly. This enabled Mendel to grow many plants and collect a lot of data.
- they have easily observed traits, such as flower color and pea shape. This enabled Mendel to observe whether a trait was passed from one generation to the next.
- Mendel could control which pairs of plants reproduced. This enabled him to determine which traits came from which plant pairs.

Pollination in Pea Plants

To observe how a trait was inherited, Mendel controlled which plants pollinated other plants. Pollination occurs when pollen lands on the pistil of a flower. **Sperm** cells from the pollen then can fertilize **egg** cells in the pistil. Pollination in pea plants can occur in two ways. Self-pollination occurs when pollen from one plant lands on the pistil of a flower on the same plant, as shown in **Figure 1.** Cross-pollination occurs when pollen from one plant reaches the pistil of a flower on a different plant. Cross-pollination occurs naturally when wind, water, or animals such as bees carry pollen from one flower to another. Mendel allowed one group of flowers to self-pollinate. With another group, he cross-pollinated the plants himself.

REVIEW VOCABULARY

sperm
a haploid sex cell formed in the male reproductive organs

egg
a haploid sex cell formed in the female reproductive organs

Self-Pollination

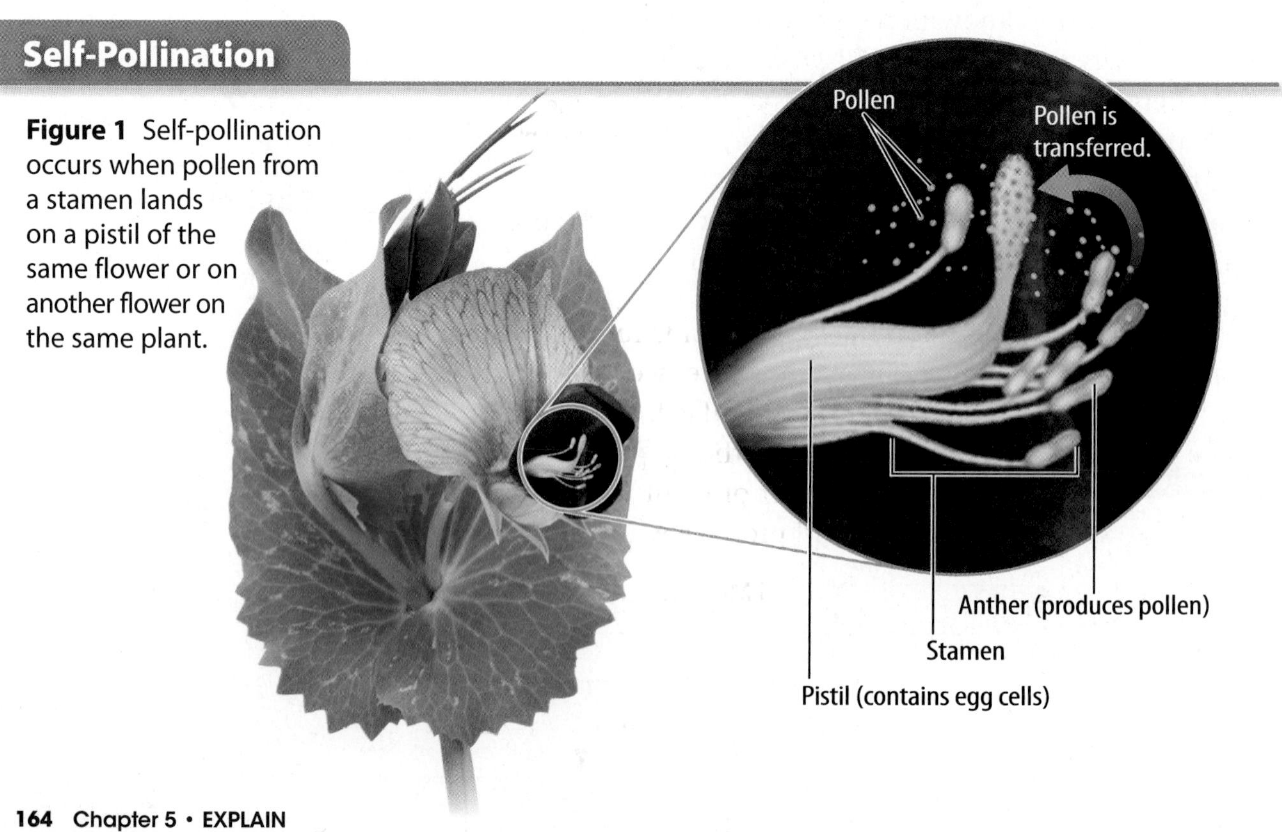

Figure 1 Self-pollination occurs when pollen from a stamen lands on a pistil of the same flower or on another flower on the same plant.

True-Breeding Plants

Mendel began his experiments with plants that were true-breeding for the trait he would test. When a true-breeding plant self-pollinates, it always produces offspring with traits that match the parent. For example, when a true-breeding pea plant with wrinkled seeds self-pollinates, it produces only plants with wrinkled seeds. In fact, plants with wrinkled seeds appear generation after generation.

Mendel's Cross-Pollination

By cross-pollinating plants himself, Mendel was able to select which plants pollinated other plants. **Figure 2** shows an example of a manual cross between a plant with white flowers and one with purple flowers.

Figure 2 To control pollination, Mendel removed the stamens of one flower and pollinated that flower with pollen from a flower of a different plant.

Active Reading **4. Identify** Circle the part of **Figure 2** that shows Mendel cross-pollinating two plants.

Cross-Pollination

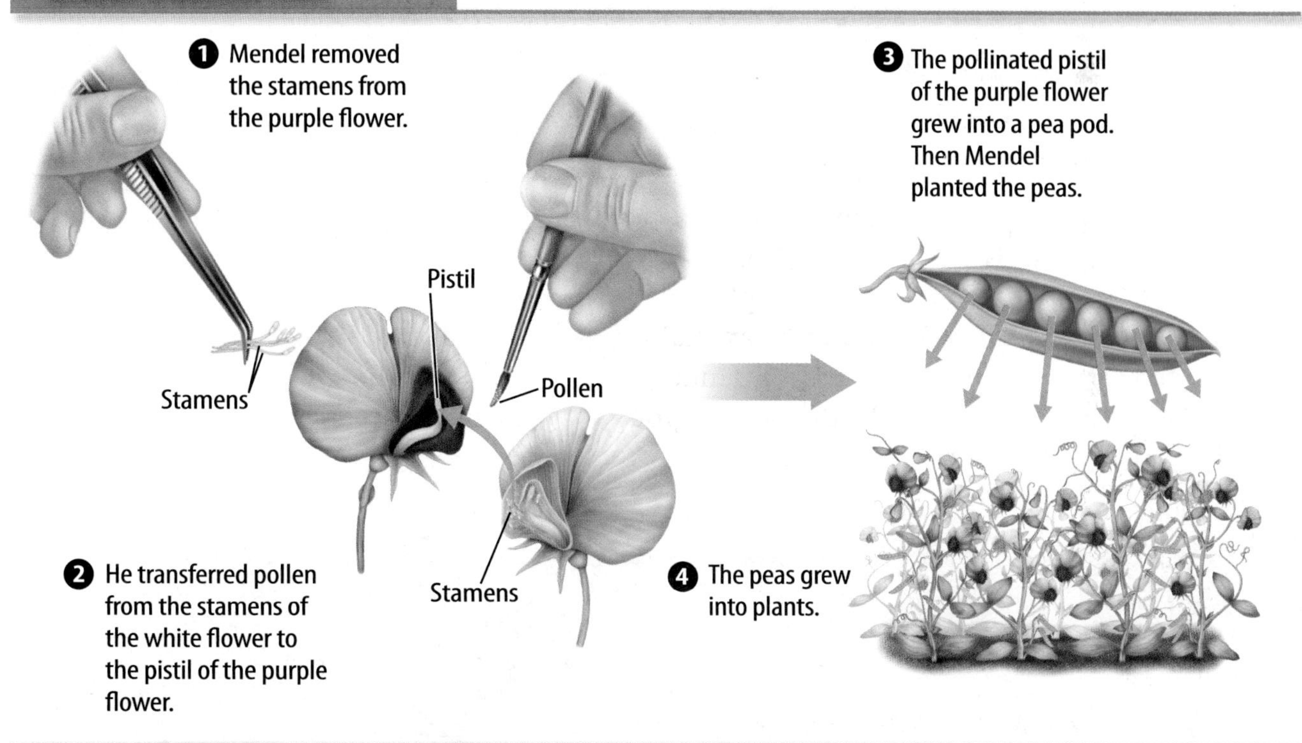

Mendel cross-pollinated hundreds of plants for each set of traits, such as flower color—purple or white; seed color—green or yellow; and seed shape—round or wrinkled. With each cross-pollination, Mendel recorded the traits that appeared in the offspring. By testing such a large number of plants, Mendel was able to predict which crosses would produce which traits.

5. NGSSS Check Explain Why did Mendel perform cross-pollination experiments? SC.7.L.16.1

Mendel's Results

Once Mendel had enough true-breeding plants for a trait that he wanted to test, he cross-pollinated selected plants. His results are shown in **Figure 3.**

First-Generation Crosses

A cross between true-breeding plants with purple flowers produced plants with only purple flowers. A cross between true-breeding plants with white flowers produced plants with only white flowers. But something unexpected happened when Mendel crossed true-breeding plants with purple flowers and true-breeding plants with white flowers—all the offspring had purple flowers.

New Questions Raised

The results of the crosses between true-breeding plants with purple flowers and true-breeding plants with white flowers led to more questions for Mendel. Why did all the offspring always have purple flowers? Why were there no white flowers? Why didn't the cross produce offspring with pink flowers—a combination of the white and purple flower colors? Mendel carried out more experiments with pea plants to answer these questions.

6. Analyze Use **Figure 3** to predict the offspring of a cross between two true-breeding pea plants with smooth seeds.

____

____

First-Generation Crosses

Figure 3 Mendel crossed three combinations of true-breeding plants and recorded the flower colors of the offspring.

Purple × Purple

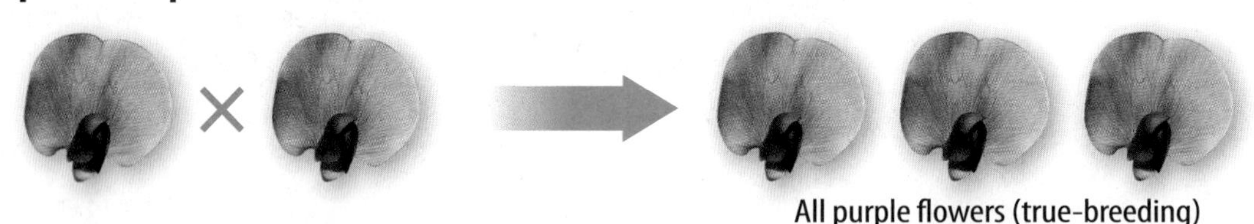

White × White

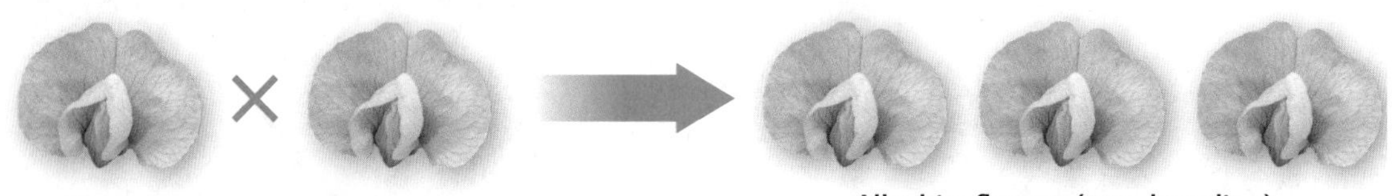

Purple (true-breeding) × White (true-breeding)

7. Visual Check **Apply** Suppose you cross hundreds of true-breeding plants with purple flowers with hundreds of true-breeding plants with white flowers. Based on the results of this cross in the figure above, would any offspring produce white flowers? Explain.

____

____

____

____

Second-Generation (Hybrid) Crosses

The first-generation purple-flowering plants are called **hybrid** plants. This means they came from true-breeding parent plants with different forms of the same trait. Mendel wondered what would happen if he cross-pollinated two purple-flowering hybrid plants.

As shown in **Figure 4,** some of the offspring had white flowers, even though both parents had purple flowers. The results were similar each time Mendel cross-pollinated two hybrid plants. The trait that had disappeared in the first generation always reappeared in the second generation.

The same result happened when Mendel cross-pollinated pea plants for other traits. For example, he found that cross-pollinating a true-breeding yellow-seeded pea plant with a true-breeding green-seeded pea plant always produced yellow-seeded hybrids. A second-generation cross of two yellow-seeded hybrids always yielded plants with yellow seeds and plants with green seeds.

Science Use v. Common Use

hybrid

Science Use the offspring of two animals or plants with different forms of the same trait

Common Use having two types of components that perform the same function, such as a vehicle powered by both a gas engine and an electric motor

Active Reading 8. **Explain** What is a hybrid plant?

____

____

____

Figure 4 When Mendel cross-pollinated first-generation hybrid offspring, the trait that had disappeared from the first generation reappeared in the second generation.

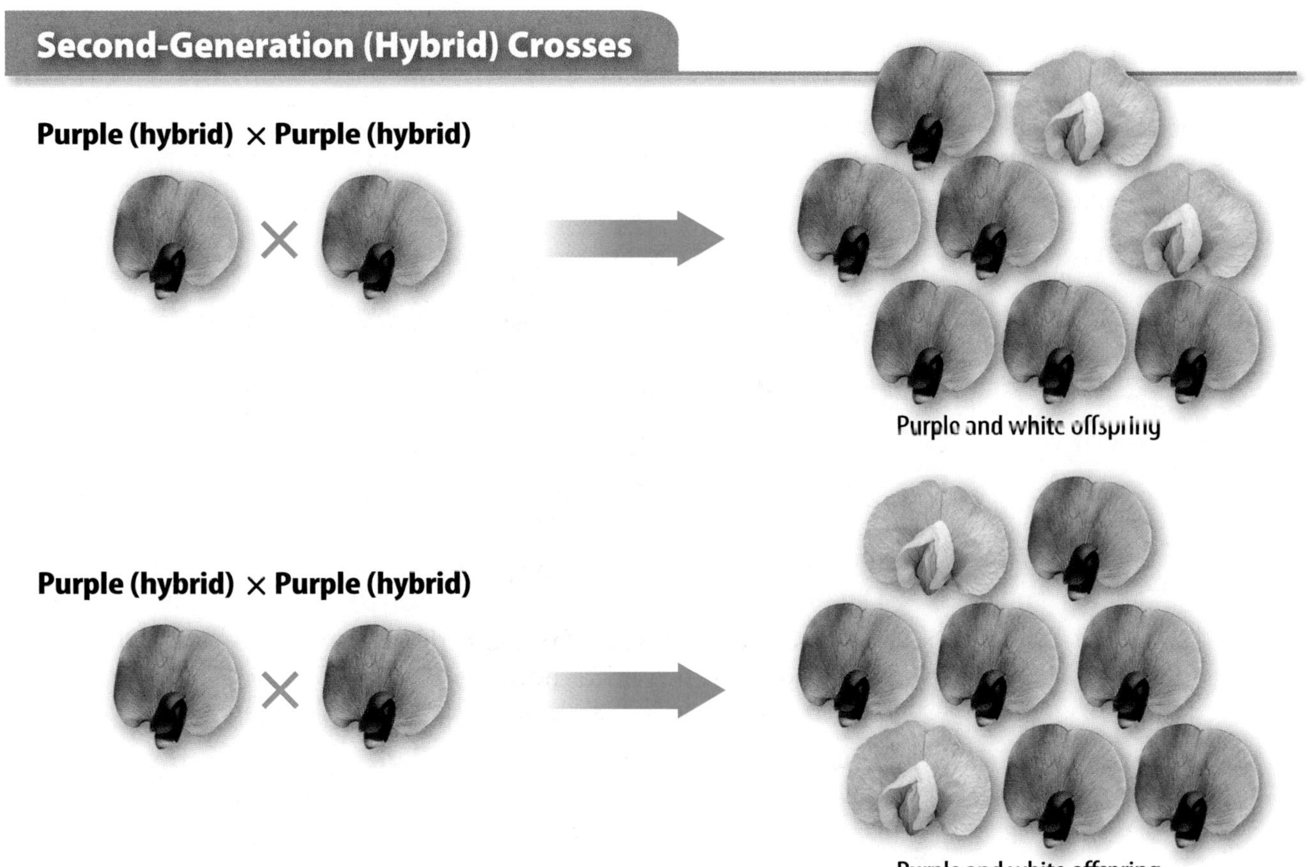

Table 1 When Mendel crossed two hybrids for a given trait, the trait that had disappeared then reappeared in a ratio of about 3:1.

Active Reading **9. Identify** In **Table 1,** (circle) the trait that appeared most often for each characteristic.

Table 1 Results of Hybrid Crosses

Characteristic	Trait and Number of Offspring	Trait and Number of Offspring	Ratio
Flower color	Purple 705	White 224	3.15:1
Flower position	Axial (Side of stem) 651	Terminal (End of stem) 207	3.14:1
Seed color	Yellow 6,022	Green 2,001	3.01:1
Seed shape	Round 5,474	Wrinkled 1,850	2.96:1
Pod shape	Inflated (Smooth) 882	Constricted (Bumpy) 299	2.95:1
Pod color	Green 428	Yellow 152	2.82:1
Stem length	Long 787	Short 277	2.84:1

Math Skills MA.6.A.3.6

Use Ratios

A ratio is a comparison of two numbers or quantities by division. For example, the ratio comparing **6,022** yellow seeds to **2,001** green seeds can be written as follows:

6,022 to **2,001** or

6,022 : 2,001 or

$$\frac{6{,}022}{2{,}001}$$

To simplify the ratio, divide the first number by the second number.

$$\frac{6{,}022}{2{,}001} = \frac{3}{1} \text{ or } 3:1$$

Practice

10. There are 14 girls and 7 boys in a science class. Simplify the ratio.

More Hybrid Crosses

Mendel counted and recorded the traits of offspring from many experiments in which he cross-pollinated hybrid plants. Data from these experiments are shown in **Table 1.** He analyzed these data and noticed patterns. For example, from the data of crosses between hybrid plants with purple flowers, he found that the ratio of purple flowers to white flowers was about 3:1. This means purple-flowering pea plants grew from this cross three times more often than white-flowering pea plants grew from the cross. He calculated similar ratios for all seven traits he tested.

Mendel's Conclusions

After analyzing the results of his experiments, Mendel concluded that two genetic factors control each inherited trait. He also proposed that when organisms reproduce, each reproductive cell—sperm or egg—contributes one factor for each trait.

Active Reading **11. Recall** Underline what Mendel concluded about inherited traits.

Dominant and Recessive Traits

Recall that when Mendel cross-pollinated a true-breeding plant with purple flowers and a true-breeding plant with white flowers, the hybrid offspring had only purple flowers. Mendel hypothesized that the hybrid offspring had one genetic factor for purple flowers and one genetic factor for white flowers.

Mendel also hypothesized that the purple factor is the only factor expressed because it blocks the white factor. *A genetic factor that blocks another genetic factor is called a* **dominant** (DAH muh nunt) **trait.** A dominant trait is observed when offspring have either one or two dominant factors. *A genetic factor that is blocked by the presence of a dominant factor is called a* **recessive** (rih SE sihv) **trait.** A recessive trait is observed only when two recessive genetic factors are present in offspring.

From Parents to Second Generation

For the second generation, Mendel cross-pollinated two hybrids with purple flowers. About 75 percent of the second-generation plants had purple flowers. These plants had at least one dominant factor. Twenty-five percent of the second-generation plants had white flowers. These plants had the same two recessive factors.

Active Reading **12. Describe** How do dominant and recessive factors interact?

Active Reading **FOLDABLES®** LA.6.2.2.3

Make a vertical two-tab book and label it as shown. Use it to organize your notes on dominant and recessive factors.

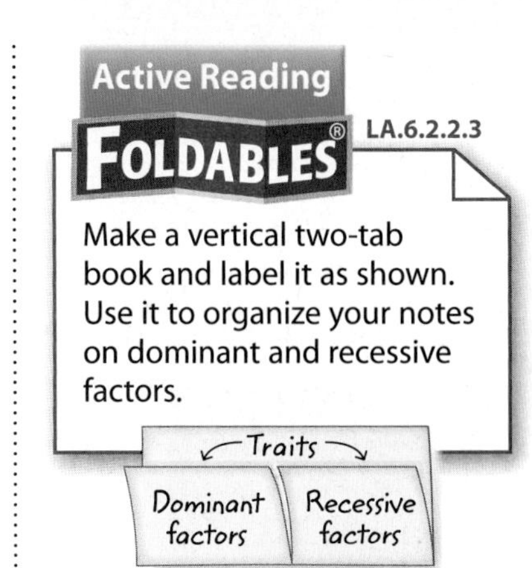

Inquiry **LAB STATION** SC.7.N.1.3, SC.7.L.16.1 **Try It!**

MiniLab ***Which is the dominant trait?*** at connectED.mcgraw-hill.com

Apply It! After you complete the lab, answer this question.

1. How do the data from the first and second generations help you determine that the terminal flower position is a recessive trait?

Lesson Review 1

Visual Summary

Genetics is the study of how traits are passed from parents to offspring.

Mendel studied genetics by doing cross-breeding experiments with pea plants.

Mendel's experiments with pea plants showed that some traits are dominant and others are recessive.

Use Vocabulary

1. Distinguish between heredity and genetics. SC.7.L.16.1

2. Define the terms *dominant* and *recessive*.

Understand Key Concepts

3. A recessive trait is observed when an organism has ______ recessive genetic factor(s).
   - (A) 0
   - (B) 1
   - (C) 2
   - (D) 3

4. **Summarize** Mendel's conclusions about how traits pass from parents to offspring. SC.7.L.16.1

Interpret Graphics

5. **Suppose** the two true-breeding plants shown at right were crossed. What color would the flowers of the offspring be? Explain.

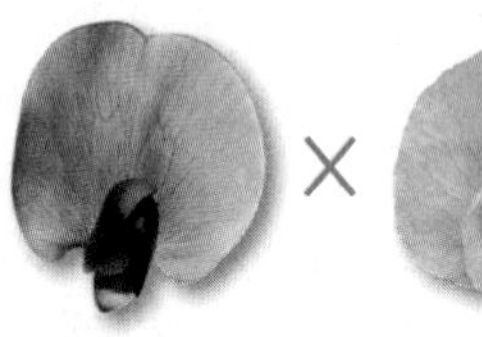

Critical Thinking

6. **Examine** how Mendel's conclusions disprove blending inheritance. SC.7.N.2.1

Math Skills MA.6.A.3.6

7. A cross between two pink camellia plants produced the following offspring: 7 plants with red flowers, 7 with white flowers, and 14 with pink flowers. What is the ratio of red to white to pink?

SCIENCE & SOCIETY

Pioneering the Science of Genetics

One man's curiosity leads to a branch of science.

Gregor Mendel—monk, scientist, gardener, and beekeeper—was a keen observer of the world around him. Curious about how traits pass from one generation to the next, he grew and tested almost 30,000 pea plants. Today, Mendel is called the father of genetics. After Mendel published his findings, however, his "laws of heredity" were overlooked for several decades.

In 1900, three European scientists, working independently of one another, rediscovered Mendel's work and replicated his results. Then, other biologists quickly began to recognize the importance of Mendel's work.

Gregor Mendel ▶

1902: American physician Walter Sutton demonstrates that Mendel's laws of inheritance can be applied to chromosomes. He concludes that chromosomes contain a cell's hereditary material on genes.

1906: William Bateson, a United Kingdom scientist, coins the term *genetics.* He uses it to describe the study of inheritance and the science of biological inheritance.

1952: American geneticists Martha Chase and Alfred Hershey prove that DNA transmits inherited traits from one generation to the next.

1953: Francis Crick and James Watson determine the structure of the DNA molecule. Their work begins the field of molecular biology and leads to important scientific and medical research in genetics.

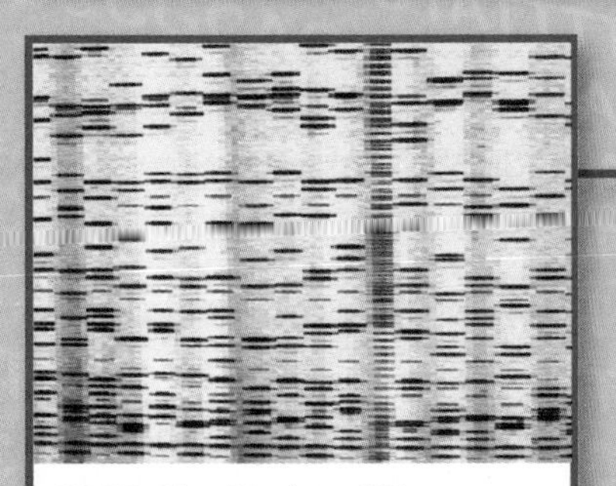

2003: The National Human Genome Research Institute (NHGRI) completes mapping and sequencing human DNA. Researchers and scientists are now trying to discover the genetic basis for human health and disease.

It's Your Turn

RESEARCH What are some genetic diseases? Report on how genome-based research might help cure these diseases in the future.

Lesson 2

Understanding INHERITANCE

ESSENTIAL QUESTIONS

What determines the expression of traits?

How can inheritance be modeled?

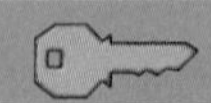
How do some patterns of inheritance differ from Mendel's model?

Vocabulary

gene p. 174
allele p. 174
phenotype p. 174
genotype p. 174
homozygous p. 175
heterozygous p. 175
Punnett square p. 176
incomplete dominance p. 178
codominance p. 178
polygenic inheritance p. 179

Florida NGSSS

SC.6.N.2.2 Explain that scientific knowledge is durable because it is open to change as new evidence or interpretations are encountered.

SC.7.N.1.3 Distinguish between an experiment (which must involve the identification and control of variables) and other forms of scientific investigation and explain that not all scientific knowledge is derived from experimentation.

SC.7.N.1.6 Explain that empirical evidence is the cumulative body of observations of a natural phenomenon on which scientific explanations are based.

SC.7.N.1.7 Explain that scientific knowledge is the result of a great deal of debate and confirmation within the science community.

SC.7.L.16.1 Understand and explain that every organism requires a set of instructions that specifies its traits, that this hereditary information (DNA) contains genes located in the chromosomes of each cell, and that heredity is the passage of these instructions from one generation to another.

SC.7.L.16.2 Determine the probabilities for genotype and phenotype combinations using Punnett Squares and pedigrees.

Also covers: HE.6.C.1.4, LA.6.2.2.3, MA.6.S.6.2

MA.6.S.6.2

Inquiry Launch Lab

15 minutes

What is the span of your hand?

Mendel discovered that some traits have a simple pattern of inheritance—dominant or recessive. However, some traits, such as eye color, have more variation. Is human hand span a Mendelian trait?

Procedure

1. Read and complete a lab safety form.
2. Use a **metric ruler** to measure the distance (in cm) between the tips of your thumb and little finger with your hand stretched out.
3. As a class, record everyone's name and hand span in a data table.

Data and Observations

Think About This

1. What range of hand span measurements did you observe?

2. Key Concept Do you think hand span is a simple Mendelian trait like pea plant flower color?

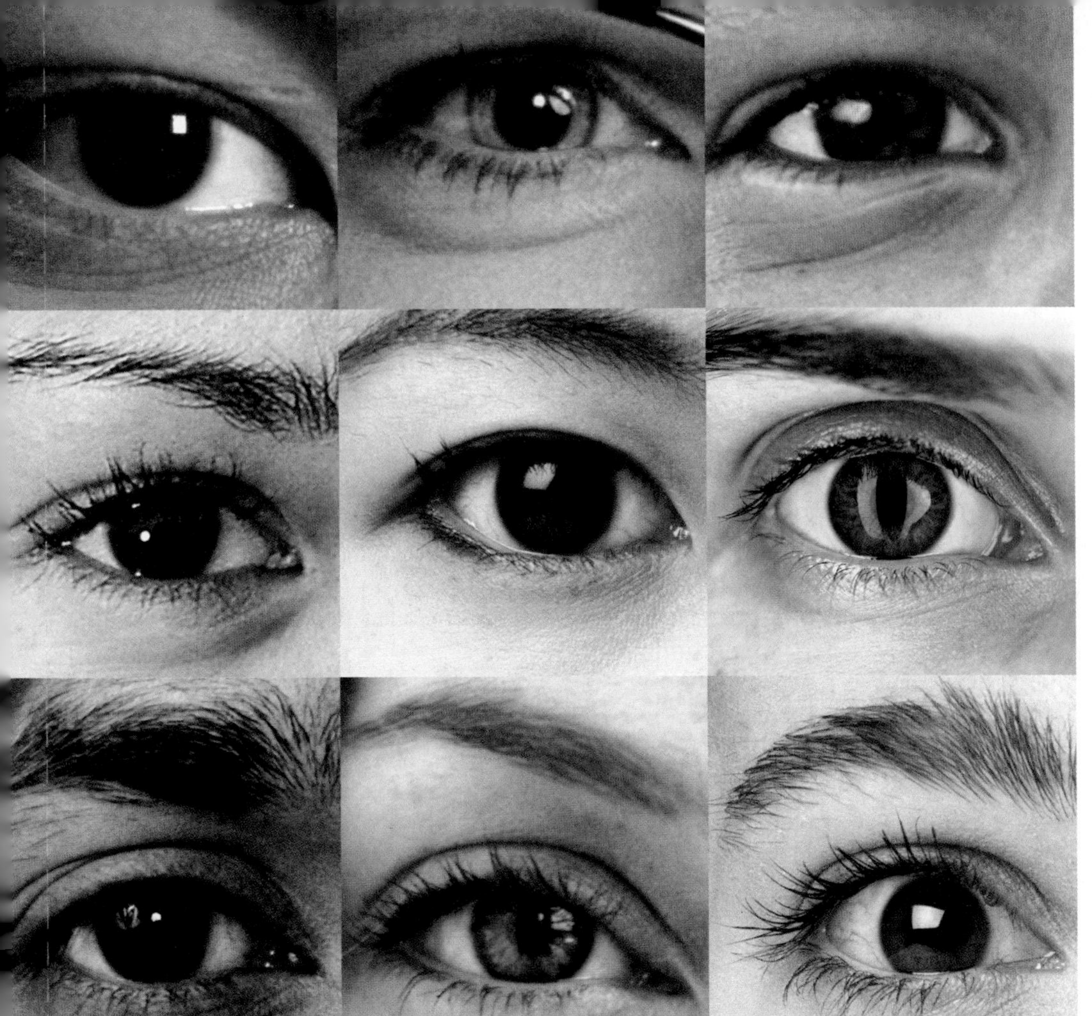

Make the Connection

1. Physical traits, such as those shown in these eyes, can vary widely from person to person. Take a closer look at the eyes on this page. What traits can you identify among them? How do they differ?

____

____

____

____

____

____

____

____

____

What controls traits?

Mendel concluded that two factors—one from each parent—control each trait. Mendel hypothesized that one factor came from the egg cell and one factor came from the sperm cell. What are these factors? How are they passed from parents to offspring?

Chromosomes

When other scientists studied the parts of a cell and combined Mendel's work with their work, these factors were more clearly understood. Scientists discovered that inside each cell is a nucleus that contains threadlike structures called chromosomes. Over time, scientists learned that chromosomes contain genetic information that controls traits. We now know that Mendel's "factors" are part of chromosomes and that each cell in offspring contains chromosomes from both parents. As shown in **Figure 5,** these chromosomes exist as pairs—one chromosome from each parent.

Figure 5 Humans have 23 pairs of chromosomes. Each pair has one chromosome from the father and one chromosome from the mother.

1 2 3 4 5
6 7 8 9 10 11 12
13 14 15 16 17 18
19 20 21 22 X Y

Genes and Alleles

Scientists have discovered that each chromosome can have information about hundreds or even thousands of traits. *A* **gene** (JEEN) *is a section on a chromosome that has genetic information for one trait.* For example, a gene of a pea plant might have information about flower color. Recall that an offspring inherits two genes (factors) for each trait—one from each parent. The genes can be the same or different, such as purple or white for pea flower color. *The different forms of a gene are called* **alleles** (uh LEELs). Pea plants can have two purple alleles, two white alleles, or one of each allele. In **Figure 6,** the chromosome pair has information about three traits—flower position, pod shape, and stem length.

Active Reading **2. Determine** How many alleles controlled flower color in Mendel's experiments?

Genotype and Phenotype

Look again at the photo at the beginning of this lesson. What human trait can you observe? You might observe that eye color can be shades of blue or brown. *Geneticists call how a trait appears, or is expressed, the trait's* **phenotype** (FEE nuh tipe). What other phenotypes can you observe in the photo?

WORD ORIGIN
phenotype
from Greek *phainein,* means "to show"

Mendel concluded that two alleles control the expression or phenotype of each trait. *The two alleles that control the phenotype of a trait are called the trait's* **genotype** (JEE nuh tipe). Although you cannot see an organism's genotype, you can make inferences about a genotype based on its phenotype. For example, you have already learned that a pea plant with white flowers has two recessive alleles for that trait. These two alleles are its genotype. The white flower is its phenotype.

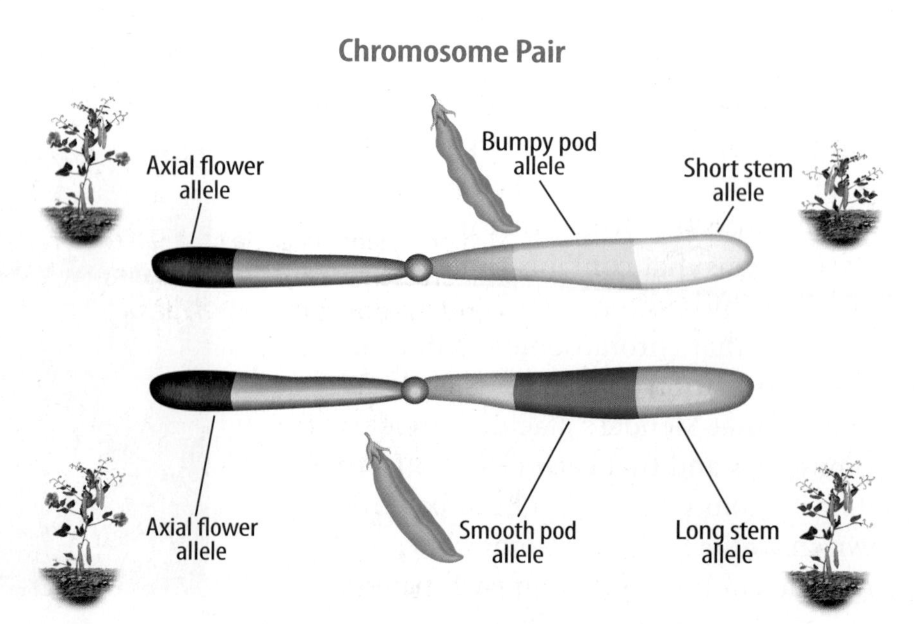

Figure 6 This chromosome pair has information about flower position, pod shape, and stem length.

3. Visual Check
Identify Which alleles are the same?

Which alleles are different?

Symbols for Genotypes Scientists use symbols to represent the alleles in a genotype. In genetics, uppercase letters represent dominant alleles and lowercase letters represent recessive alleles. **Table 2** shows the possible genotypes for both round and wrinkled seed phenotypes. Notice that the dominant allele, if present, is written first.

Table 2 Phenotype and Genotype

Phenotypes (observed traits)	Genotypes (alleles of a gene)
Round	Homozygous dominant *(RR)*
	Heterozygous *(Rr)*
Wrinkled	Homozygous recessive *(rr)*

A round seed can have two genotypes—*RR* and *Rr*. Both genotypes have a round phenotype. Why does *Rr* result in round seeds? This is because the round allele *(R)* is dominant to the wrinkled allele *(r)*.

A wrinkled seed has the recessive genotype, *rr*. The wrinkled-seed phenotype is possible only when the same two recessive alleles *(rr)* are present in the genotype.

Homozygous and Heterozygous *When the two alleles of a gene are the same, its genotype is* **homozygous** (hoh muh ZI gus). Both *RR* and *rr* are homozygous genotypes, as shown in **Table 2.**

If the two alleles of a gene are different, its genotype is **heterozygous** (he tuh roh ZI gus). *Rr* is a heterozygous genotype.

4. **NGSSS Check** Explain How do alleles determine the expression of traits? SC.7.L.16.1

Apply It! After you complete the lab, answer these questions.

1. What are the possible genotypes for a green dragon with four legs and short wings?

2. In a pea plant, an allele for a tall stem is dominant to an allele for a short stem. You see a pea plant with a short stem. What can you conclude about the genotype of this plant?

Punnett Square

Figure 7 A Punnett square can be used to predict the possible genotypes of the offspring. Offspring from a cross between two heterozygous parents can have one of three genotypes.

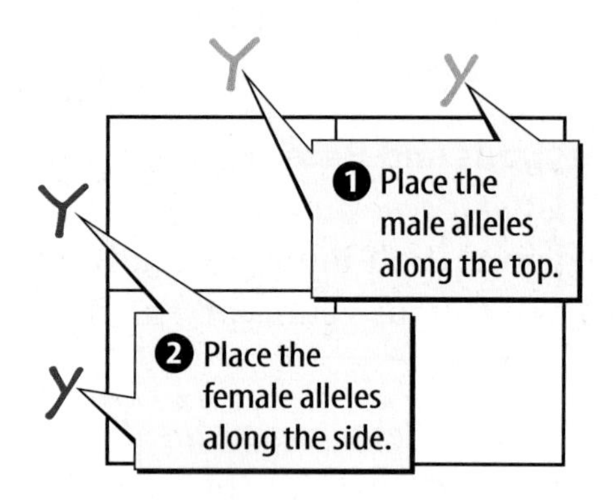

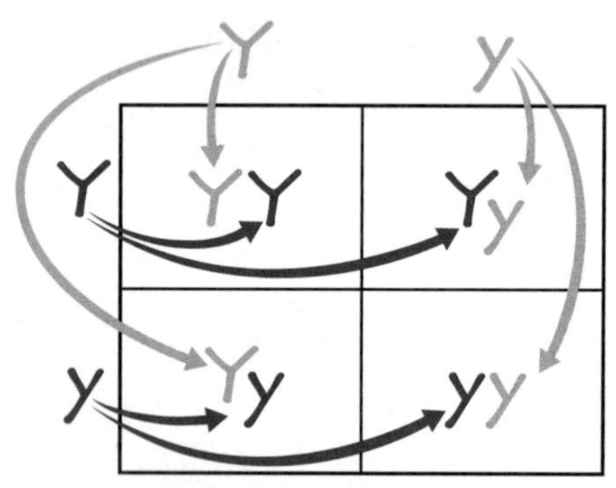

3 Copy female alleles across each row. Copy male alleles down each column. Always list the dominant trait first.

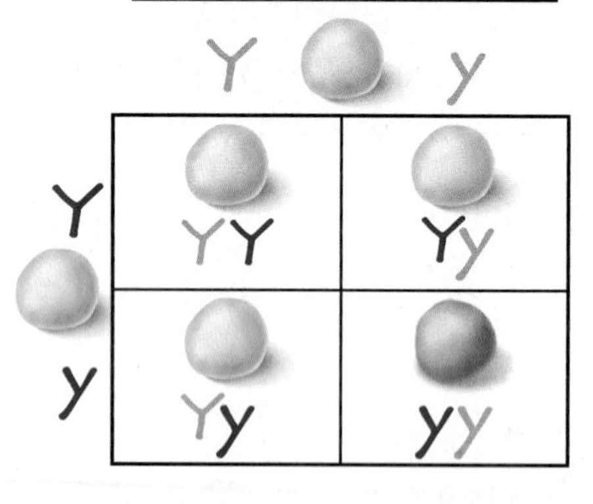

6. Visual Check Determine What phenotypes are possible for pea offspring of this cross?

Modeling Inheritance

Have you ever flipped a coin and guessed heads or tails? Because a coin has two sides, there are only two possible outcomes—heads or tails. You have a 50 percent chance of getting heads and a 50 percent chance of getting tails. The chance of getting an outcome can be represented by a ratio. The ratio of heads to tails is 50:50 or 1:1.

Active Reading **5. Interpret** What does a ratio of 2:1 mean?

Plant breeders and animal breeders use a method for predicting how often traits will appear in offspring that does not require performing the crosses thousands of times. Two tools—a Punnett square and a pedigree—can be used to predict and identify traits among genetically related individuals.

Punnett Squares

If the genotypes of the parents are known, then the different genotypes and phenotypes of the offspring can be predicted. *A* **Punnett square** *is a model used to predict possible genotypes and phenotypes of offspring.* Follow the steps in **Figure 7** to learn how to make a Punnett square.

Analyzing a Punnett Square

Figure 7 shows an example of a cross between two pea plants that are heterozygous for pea seed color—*Yy* and *Yy*. Yellow is the dominant allele—*Y*. Green is the recessive allele—*y*. The offspring can have one of three genotypes—*YY, Yy,* or *yy*. The ratio of genotypes is written as 1:2:1.

Because *YY* and *Yy* represent the same phenotype—yellow—the offspring can have one of only two phenotypes—yellow or green. The ratio of phenotypes is written 3:1. The probability that a cross between two heterozygous pea plants will produce yellow seeds is 75 percent, and the probability that it will produce green seeds is 25 percent.

Using Ratios to Predict

Given a 3:1 ratio, you can expect that an offspring from heterozygous parents has a 3:1 chance of having yellow seeds. But you cannot expect that a group of four seeds will have three yellow seeds and one green seed. This is because one offspring does not affect the phenotype of another offspring. In a similar way, the outcome of one coin toss does not affect the outcome of other coin tosses.

However, if you counted large numbers of offspring from a particular cross, the overall ratio would be close to the ratio predicted by a Punnett square. Mendel did not use Punnett squares. However, by studying nearly 30,000 pea plants, his ratios nearly matched those that would have been predicted by a Punnett square for each cross.

Pedigrees

Another tool that can show inherited traits is a pedigree. A pedigree shows phenotypes of genetically related family members. It can also help determine genotypes. In the pedigree in **Figure 8,** three offspring have a trait—attached earlobes—that the parents do not have. If these offspring received one allele for this trait from each parent, but neither parent displays the trait, the offspring must have received two recessive alleles.

7. NGSSS Check Summarize How can inheritance be modeled? SC.7.L.16.2

____

____

____

____

Pedigree

Figure 8 In this pedigree, the parents and two offspring have unattached earlobes—the dominant phenotype. Three offspring have attached earlobes—the recessive phenotype.

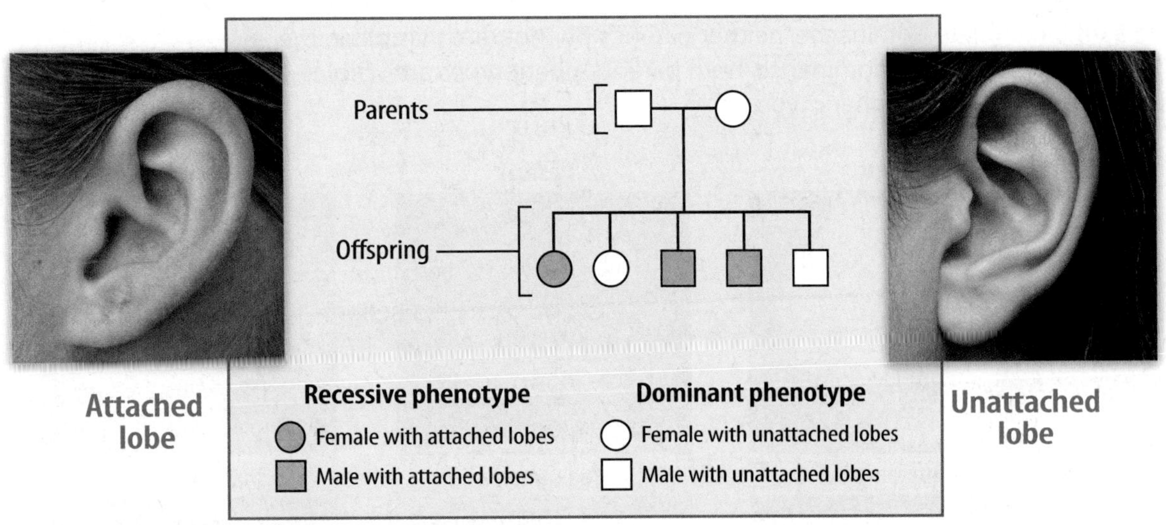

8. Visual Check Determine If the genotype of the offspring with attached lobes is *uu,* what is the genotype of the parents? How can you tell?

____

____

____

____

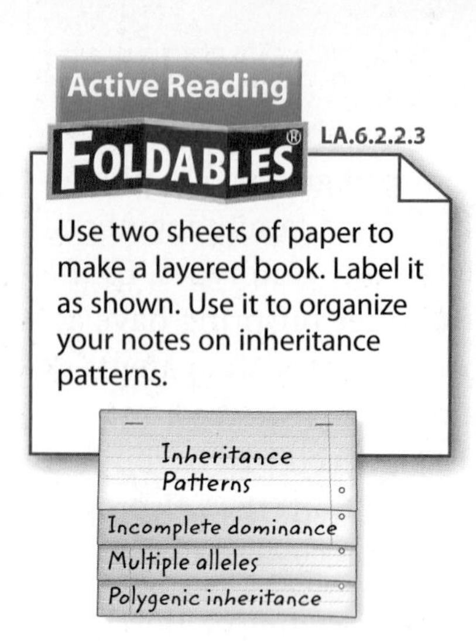

Complex Patterns of Inheritance

By chance, Mendel studied traits that are only influenced by one gene with two alleles. However, we know now that some inherited traits have complex patterns of inheritance.

Types of Dominance

Recall that for pea plants, the presence of one dominant allele produces a dominant phenotype. However, not all allele pairs have a dominant-recessive interaction.

Incomplete Dominance Sometimes traits appear to be combinations of alleles. *Alleles show* **incomplete dominance** *when the offspring's phenotype is a combination of the parents' phenotypes.* For example, a pink camellia, as shown in **Figure 9,** results from incomplete dominance. A cross between a camellia plant with white flowers and a camellia plant with red flowers produces only camellia plants with pink flowers.

Codominance The coat color of some cows is an example of another type of interaction between two alleles. *When both alleles can be observed in a phenotype, this type of interaction is called* **codominance.** If a cow inherits the allele for white coat color from one parent and the allele for red coat color from the other parent, the cow will have both red and white hairs.

Types of Dominance

Figure 9 In incomplete dominance, neither parent's phenotype is visible in the offspring's phenotype. In codominance, both parents' phenotypes are visible separately in the offspring's phenotype.

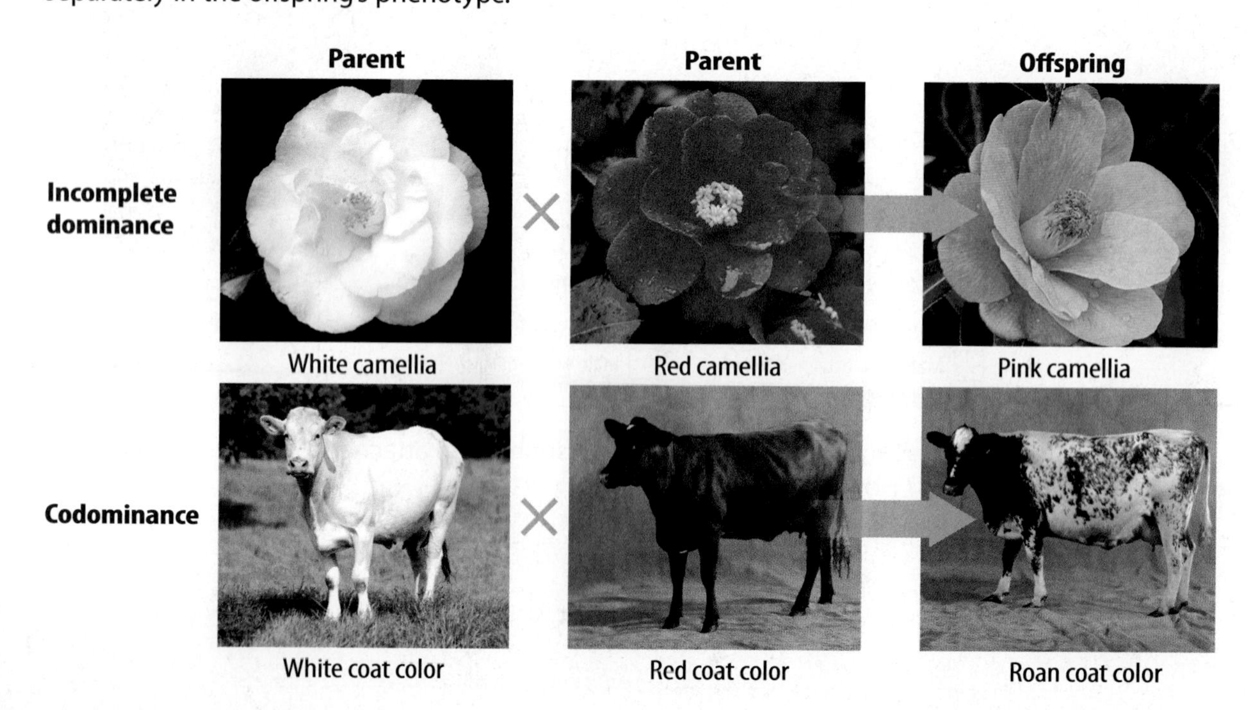

Multiple Alleles

Some genes have more than two alleles, or multiple alleles. Human ABO blood type is an example of a trait that is determined by multiple alleles. There are three different alleles for the ABO blood type—I^A, I^B, and i. The way the alleles combine results in one of four blood types—A, B, AB, or O. The I^A and I^B alleles are codominant to each other, but they both are dominant to the i allele. A person can inherit only two of these alleles—one from each parent, as shown in **Table 3.**

Table 3 Human ABO Blood Types

Phenotype	Possible Genotypes
Type A	I^AI^A or I^Ai
Type B	I^BI^B or I^Bi
Type O	ii
Type AB	I^AI^B

9. Visual Check Identify In **Table 3,** circle the blood type(s) that have the I^A allele.

Sex-Linked Traits

Sex chromosomes determine an organism's gender, or sex. Females have two X chromosomes, and males have an X and a Y chromosome. When the allele for a trait is on an X or a Y chromosome, it is called a sex-linked trait.

Polygenic Inheritance

Mendel **concluded** that each trait was determined by only one gene. However, we now know that a trait can be affected by more than one gene. **Polygenic inheritance** *occurs when multiple genes determine the phenotype of a trait.* Because several genes determine a trait, many alleles affect the phenotype even though each gene has only two alleles. Therefore, polygenic inheritance has many possible phenotypes. An example of polygenic inheritance is height, as shown in **Figure 10.**

Academic Vocabulary

conclude
(verb) to reach a logically necessary end by reasoning

Active Reading **10. Contrast** How does polygenic inheritance differ from Mendel's model?

Figure 10 The eighth graders in this class have different heights.

Genes and the Environment

You read earlier in this lesson that an organism's genotype determines its phenotype. Scientists have learned that genes are not the only factors that can affect phenotypes. An organism's environment can also affect its phenotype. For example, the flower color of one type of hydrangea is determined by the soil in which the hydrangea plant grows. **Figure 11** shows that acidic soil produces blue flowers and basic, or alkaline, soil produces pink flowers. Other examples of environmental effects on phenotype are also shown in **Figure 11.**

For humans, healthful choices can also affect phenotype. Many genes affect a person's chances of having heart disease. However, what a person eats and the amount of exercise he or she gets can influence whether heart disease will develop.

Active Reading **11. Identify** Underline the environmental factors that affect phenotype.

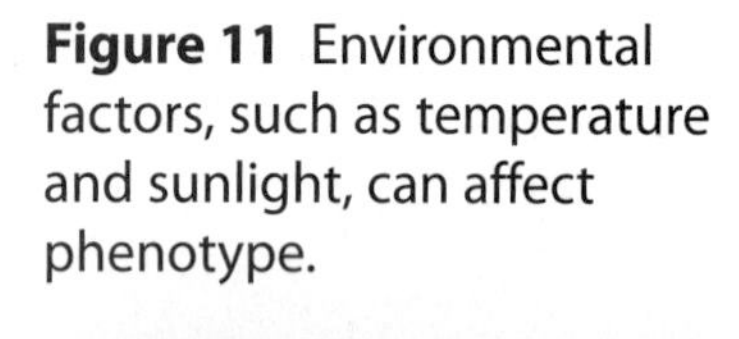

Figure 11 Environmental factors, such as temperature and sunlight, can affect phenotype.

◀ These hydrangea plants are genetically identical. The plant grown in acidic soil produced blue flowers. The plant grown in alkaline soil produced pink flowers.

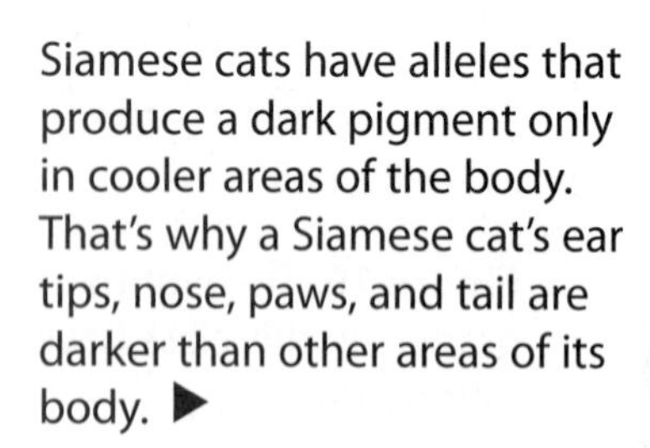

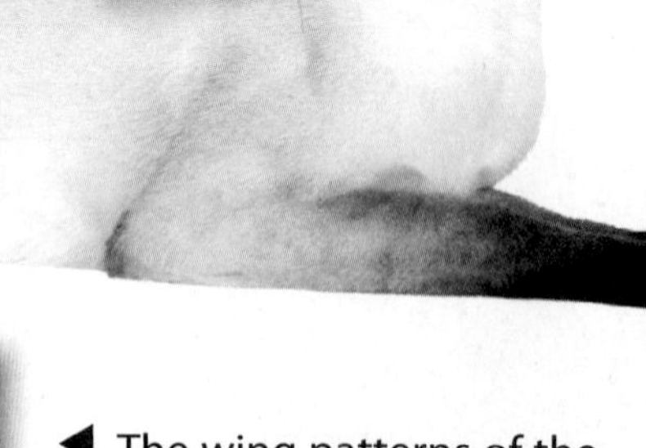

Siamese cats have alleles that produce a dark pigment only in cooler areas of the body. That's why a Siamese cat's ear tips, nose, paws, and tail are darker than other areas of its body. ▶

◀ The wing patterns of the map butterfly, *Araschnia levana,* depend on what time of year the adult develops. Adults that developed in the spring have more orange in their wings than those that developed in the summer.

Lesson Review 2

Visual Summary

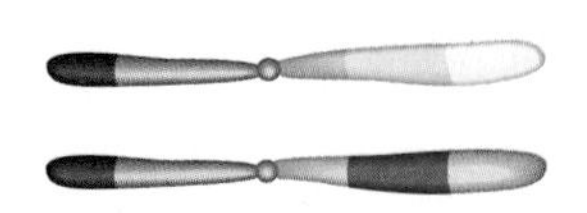

The genes for traits are located on chromosomes.

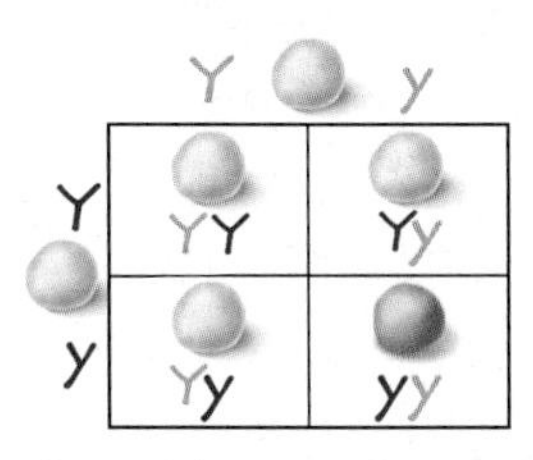

Geneticists use Punnett squares to predict the possible genotypes and phenotypes of offspring.

In polygenic inheritance, traits are determined by more than one gene and have many possible phenotypes.

Use Vocabulary

1. **Contrast** homozygous and heterozygous.

2. **Define** *incomplete dominance* in your own words.

Understand Key Concepts

3. How many alleles control a Mendelian trait, such as pea seed color?
   - (A) one
   - (B) two
   - (C) three
   - (D) four

4. **Explain** where the alleles for a given trait are inherited from. SC.7.L.16.1

5. **Describe** how the genotypes *RR* and *Rr* result in the same phenotype. SC.7.L.16.2

Interpret Graphics

6. **Analyze** this pedigree. If ■ represents a male with the homozygous recessive genotype *(aa)*, what is the mother's genotype? SC.7.L.16.2

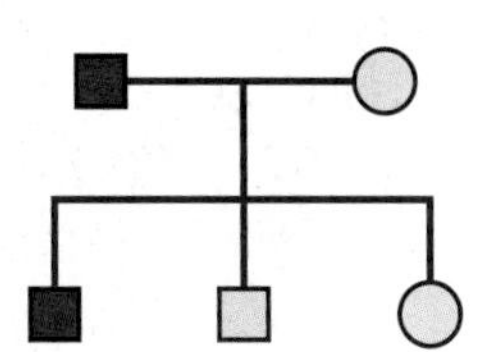

Critical Thinking

7. **Predict** the possible blood genotypes of a child, using the table at right, if one parent is type O and the other parent is type A. SC.7.L.16.2

Phenotype	Genotype
Blood Type O	*ii*
Blood Type A	I^AI^A or I^Ai

FOCUS on FLORIDA

Why are Flamingos Pink?

You are what you eat.

Have you ever seen a vibrant pink flamingo? At first it might seem like a flamingo's pink color is the result of genetics. However, flamingos are white or gray when they are born. They actually get their pink color from the food they eat.

In the wild, flamingos eat tiny, shrimplike animals called krill. These organisms contain pigments called beta carotene. These pigments are blue or green, but they turn the flamingo's feathers pink or red after they are digested. The brightness of the flamingo's feathers depends on how much beta carotene is in the bird's food. If a flamingo's diet does not contain much beta carotene, its feathers will be pale pink or even white.

Flamingos that live in captivity are often given special diets that help them keep their bright colors. These diets sometimes include extra beta carotene, as well as other nutrients. For example, the flamingo keeper at Sunken Gardens in St. Petersburg, Florida, feeds the flamingos a special pellet diet. These pellets contain all the nutrients flamingos need to stay bright and healthy.

It's Your Turn

RESEARCH the diet of an animal you might see in a zoo. Present your findings to the class.

Lesson 3

DNA and GENETICS

ESSENTIAL QUESTIONS

 What is DNA?

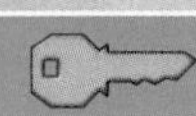 What is the role of RNA in protein production?

 How do changes in the sequence of DNA affect traits?

Vocabulary

DNA p. 184
nucleotide p. 185
replication p. 186
RNA p. 187
transcription p. 187
translation p. 188
mutation p. 189

Florida NGSSS

SC.7.L.16.1 Understand and explain that every organism requires a set of instructions that specifies its traits, that this hereditary information (DNA) contains genes located in the chromosomes of each cell, and that heredity is the passage of these instructions from one generation to another.

SC.7.L.16.2 Determine the probabilities for genotype and phenotype combinations using Punnett Squares and pedigrees.

SC.6.N.1.5 Recognize that science involves creativity, not just in designing experiments, but also in creating explanations that fit evidence.

SC.6.N.2.2 Explain that scientific knowledge is durable because it is open to change as new evidence or interpretations are encountered.

SC.7.N.1.3 Distinguish between an experiment (which must involve the identification and control of variables) and other forms of scientific investigation and explain that not all scientific knowledge is derived from experimentation.

SC.7.N.1.1 Define a problem from the seventh grade curriculum, use appropriate reference materials to support scientific understanding, plan and carry out scientific investigation of various types, such as systematic observations or experiments, identify variables, collect and organize data, interpret data in charts, tables, and graphics, analyze information, make predictions, and defend conclusions.

Also covers: HE.6.C.1.4, LA.6.2.2.3, SC.7.N.1.7, SC.7.N.2.1, SC.7.N.3.2, SC.8.N.1.3, SC.8.N.1.6, SC.8.N.3.1, MA.6.S.6.2

 SC.6.N.1.5

Inquiry Launch Lab

20 minutes

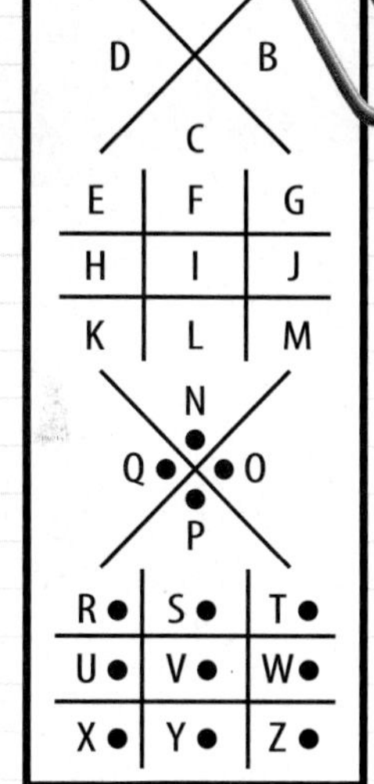

How are codes used to determine traits?

Interpret this code to learn more about how an organism's body cells use codes to determine genetic traits.

Procedure

1. Analyze the pattern of the simple code shown to the right. For example, ⟩|⟨ ∟ = DOG
2. Record the correct letters for the symbols in the code below.

Think About This

1. What do all codes, such as Morse code and Braille, have in common?

2. What do you think might happen if there is a mistake in the code?

3. Key Concept How do you think an organism's cells might use code to determine its traits?

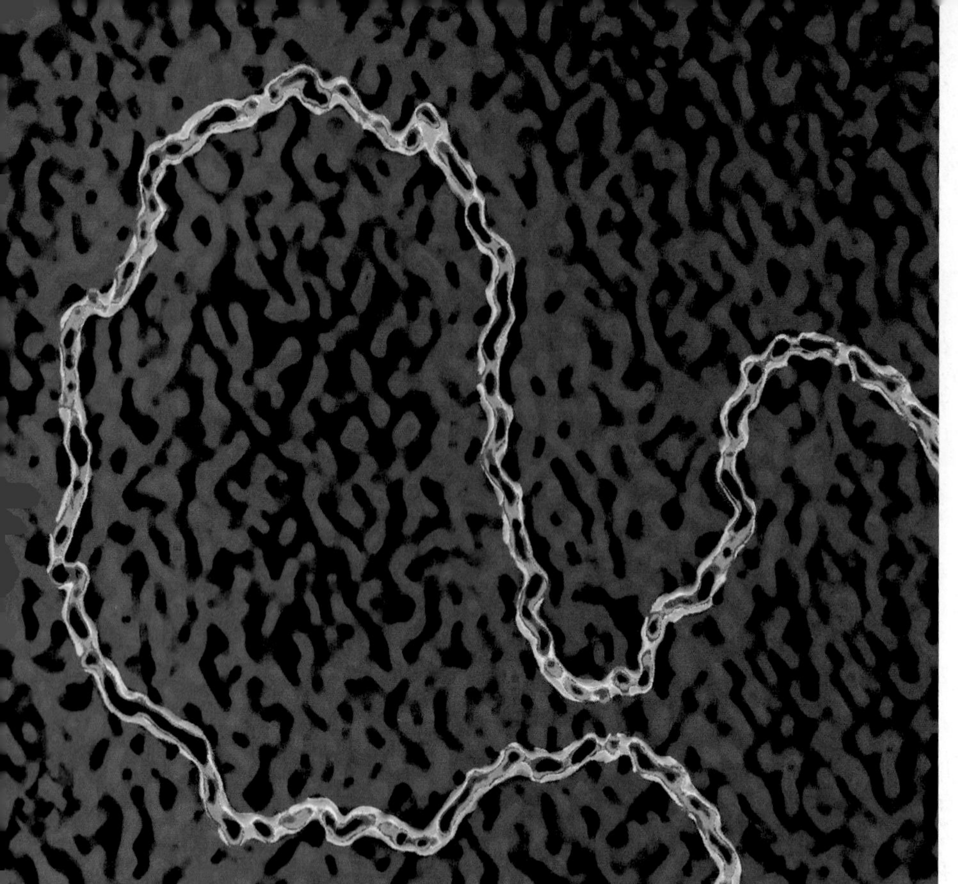

Inquiry What are these coils?

1. What color are your eyes? How tall are you? Traits are controlled by genes. But genes never leave the nucleus of the cell. How do you think a gene controls a trait?

The Structure of DNA

Have you ever put together a toy or a game for a child? If so, it probably came with directions. Cells put molecules together in much the same way you might assemble a toy. They follow a set of directions.

Genes provide directions for a cell to assemble molecules that express traits such as eye color or seed shape. Recall from Lesson 2 that a gene is a section of a chromosome. Chromosomes are made of proteins and deoxyribonucleic (dee AHK sih ri boh noo klee ihk) acid, or **DNA**—*an organism's genetic material.* A gene is a segment of DNA on a chromosome.

Cells and organisms contain millions of different molecules. Countless numbers of directions are needed to make all those molecules. How do all these directions fit on a few chromosomes? The information, or directions, needed for an organism to grow, maintain itself, and reproduce is contained in DNA. As shown in **Figure 12,** strands of DNA in a chromosome are tightly coiled, like a coiled spring. This coiling allows more genes to fit in a small space.

2. **NGSSS Check** Explain What is DNA?
SC.7.L.16.1

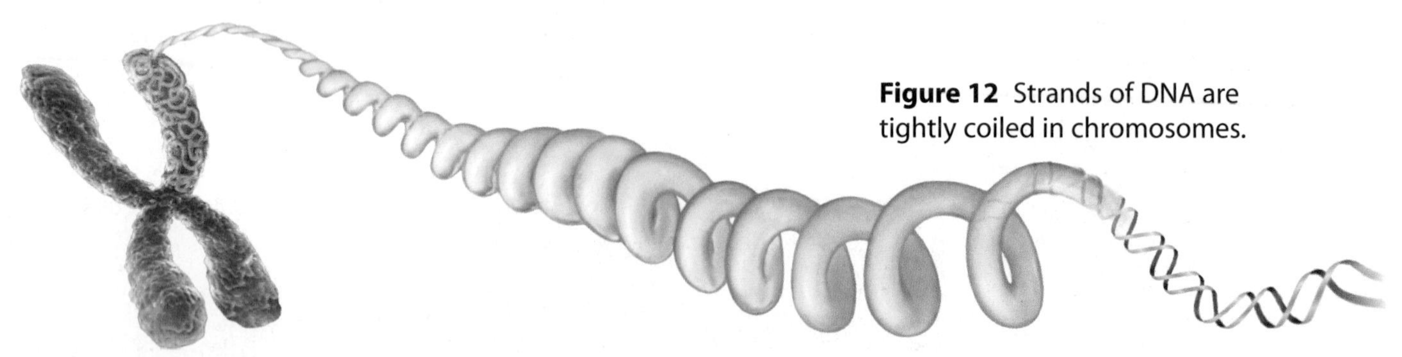

Figure 12 Strands of DNA are tightly coiled in chromosomes.

A Complex Molecule

What is the shape of the DNA molecule, and how does it fit into a chromosome? The work of several scientists revealed that DNA is like a twisted zipper. This twisted-zipper shape is called a double helix. A model of DNA's double helix structure is shown in **Figure 13.**

How did scientists make this discovery? Rosalind Franklin and Maurice Wilkins were two scientists in London who used X-rays to study DNA. Some of the X-ray data indicated that DNA has a helix shape.

American scientist James Watson visited Franklin and Wilkins and saw one of the DNA X-rays. Watson realized that the X-ray gave valuable clues about DNA's structure. Watson worked with an English scientist, Francis Crick, to build a model of DNA.

Watson and Crick based their work on information from Franklin's and Wilkins's X-rays. They also used chemical information about DNA discovered by another scientist, Erwin Chargaff. After several tries, Watson and Crick built a model that showed how the smaller molecules of DNA bond together and form a double helix.

Four Nucleotides Shape DNA

DNA's twisted-zipper shape is because of molecules called nucleotides. *A* **nucleotide** *is a molecule made of a nitrogen base, a sugar, and a phosphate group.* Sugar-phosphate groups form the sides of the DNA zipper. The nitrogen bases bond and form the teeth of the zipper. As shown in **Figure 13,** there are four nitrogen bases: adenine (A), cytosine (C), thymine (T), and guanine (G). A and T always bond together, and C and G always bond together.

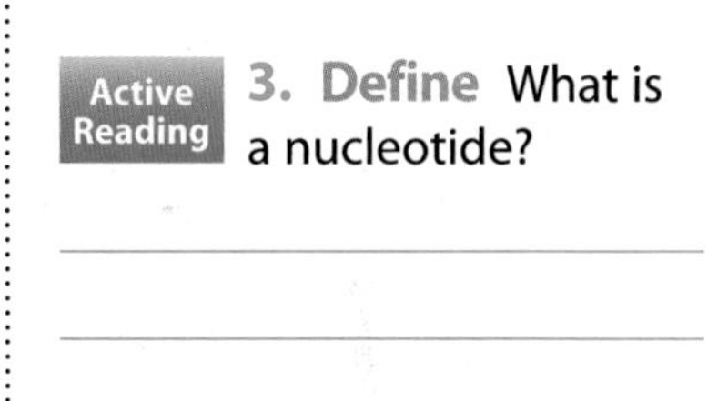

Figure 13 A DNA double helix is made of two strands of DNA. Each strand is a chain of nucleotides.

4. Visual Check Identify Write the names of the appropriate part of DNA in the spaces below.

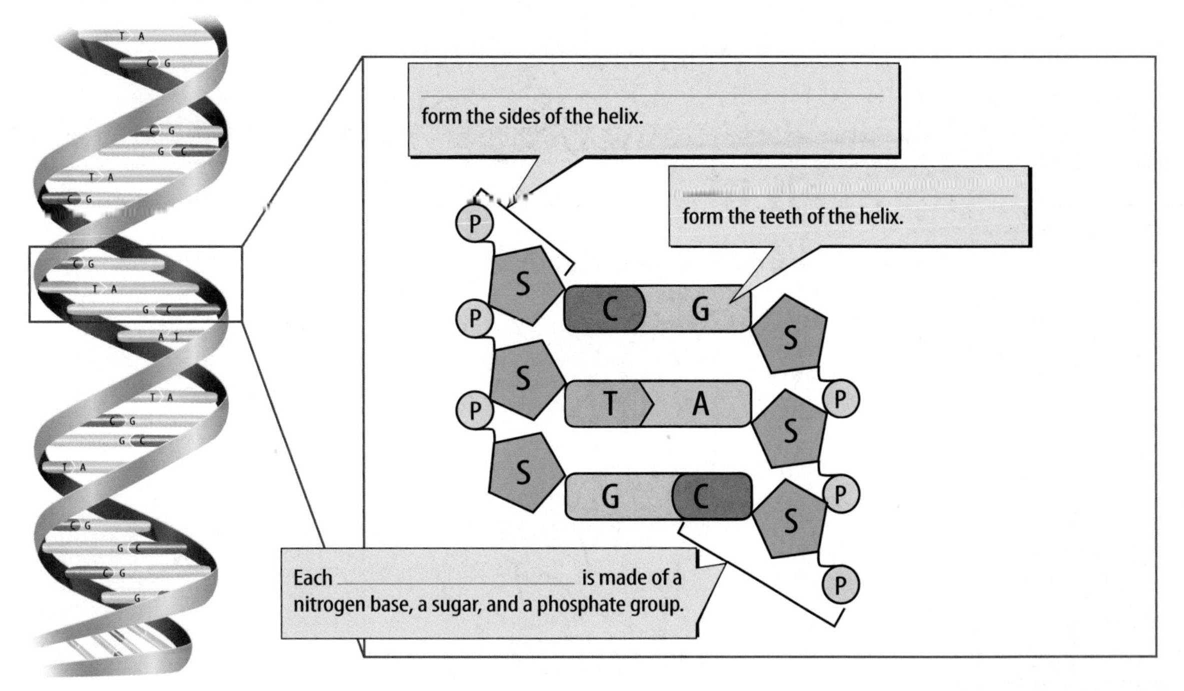

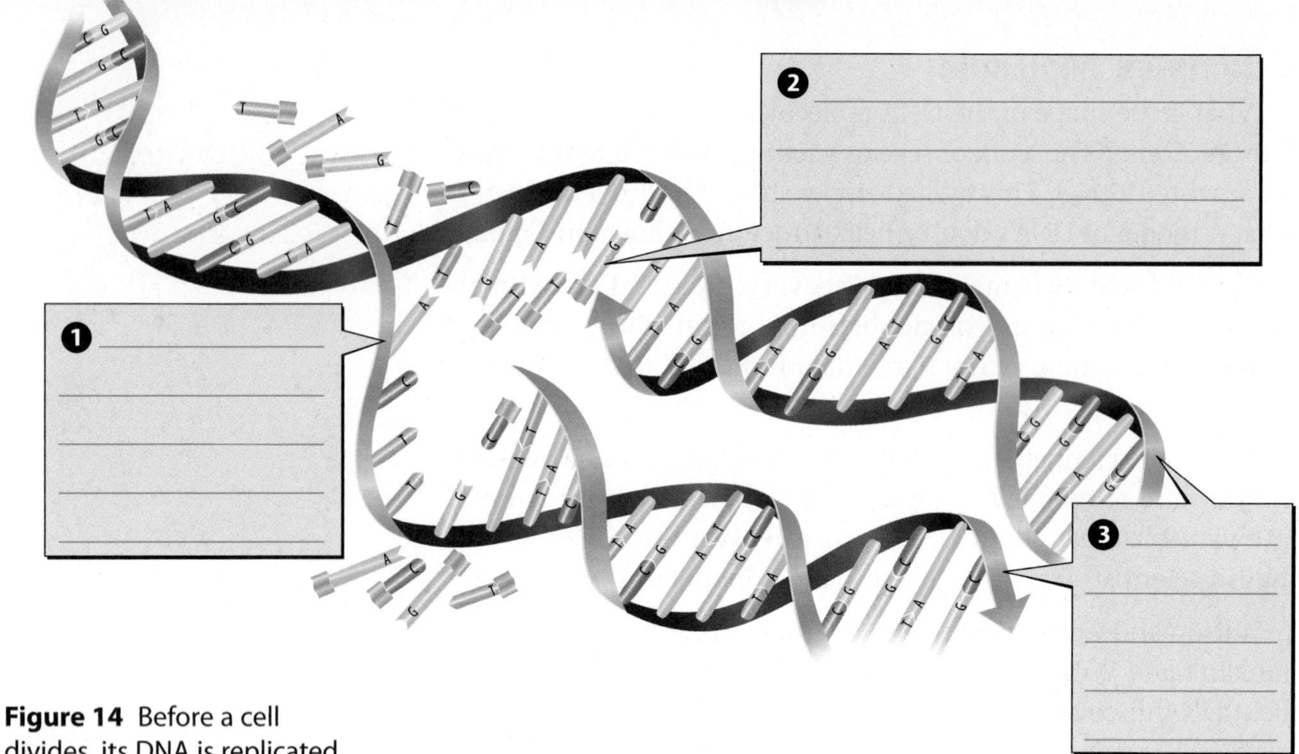

Figure 14 Before a cell divides, its DNA is replicated.

Active Reading **5. Describe** Fill in the steps of DNA replication above.

How DNA Replicates

Cells contain DNA in chromosomes. So, every time a cell divides, all chromosomes must be copied for the new cell. The new DNA is identical to existing DNA. *The process of copying a DNA molecule to make another DNA molecule is called* **replication**. You can follow the steps of DNA replication in **Figure 14.** First, the strands separate in many places, exposing individual bases. Then nucleotides are added to each exposed base. This produces two identical strands of DNA.

Active Reading **6. Explain** What is replication?

Inquiry SC.8.N.3.1

LAB STATION **Try It!**

MiniLab ***How can you model DNA?*** at connectED.mcgraw-hill.com

Apply It! After you complete the lab, answer this question.

1. How would you use your model to model DNA?

Making Proteins

Recall that proteins are important for every cellular process. The DNA of each cell carries a complete set of genes that provides instructions for making all the proteins a cell requires. Most genes contain instructions for making proteins. Some genes contain instructions for when and how quickly proteins are made.

Junk DNA

As you have learned, all genes are segments of DNA on a chromosome. However, you might be surprised to learn that most of your DNA is not part of any gene. For example, about 97 percent of the DNA on human chromosomes does not form genes. Segments of DNA that are not parts of genes are often called junk DNA. It is not yet known whether junk DNA segments have functions that are important to cells.

The Role of RNA in Making Proteins

How does a cell use the instructions in a gene to make proteins? Proteins are made with the help of ribonucleic acid **(RNA)**—*a type of nucleic acid that carries the code for making proteins from the nucleus to the cytoplasm.* RNA also carries amino acids around inside a cell and forms a part of ribosomes.

RNA, like DNA, is made of nucleotides. However, there are key differences between DNA and RNA. DNA is double-stranded, but RNA is single-stranded. RNA has the nitrogen base uracil (U) instead of thymine (T) and the sugar ribose instead of deoxyribose.

The first step in making a protein is to make mRNA from DNA. *The process of making mRNA from DNA is called* **transcription**. **Figure 15** shows how mRNA is transcribed from DNA.

Active Reading **7. Explain** What is the role of RNA in protein production?

____

____

____

____

Transcription

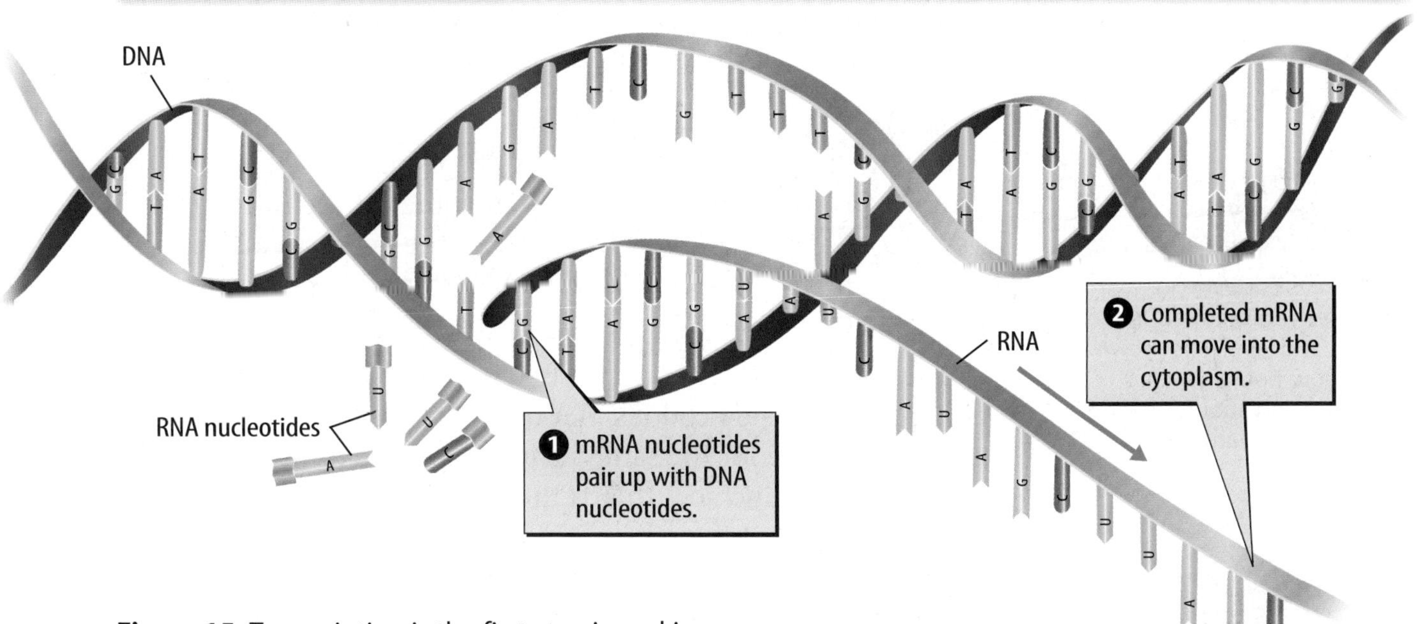

Figure 15 Transcription is the first step in making a protein. During transcription, the sequence of nitrogen bases on a gene determines the sequence of bases on mRNA.

Translation

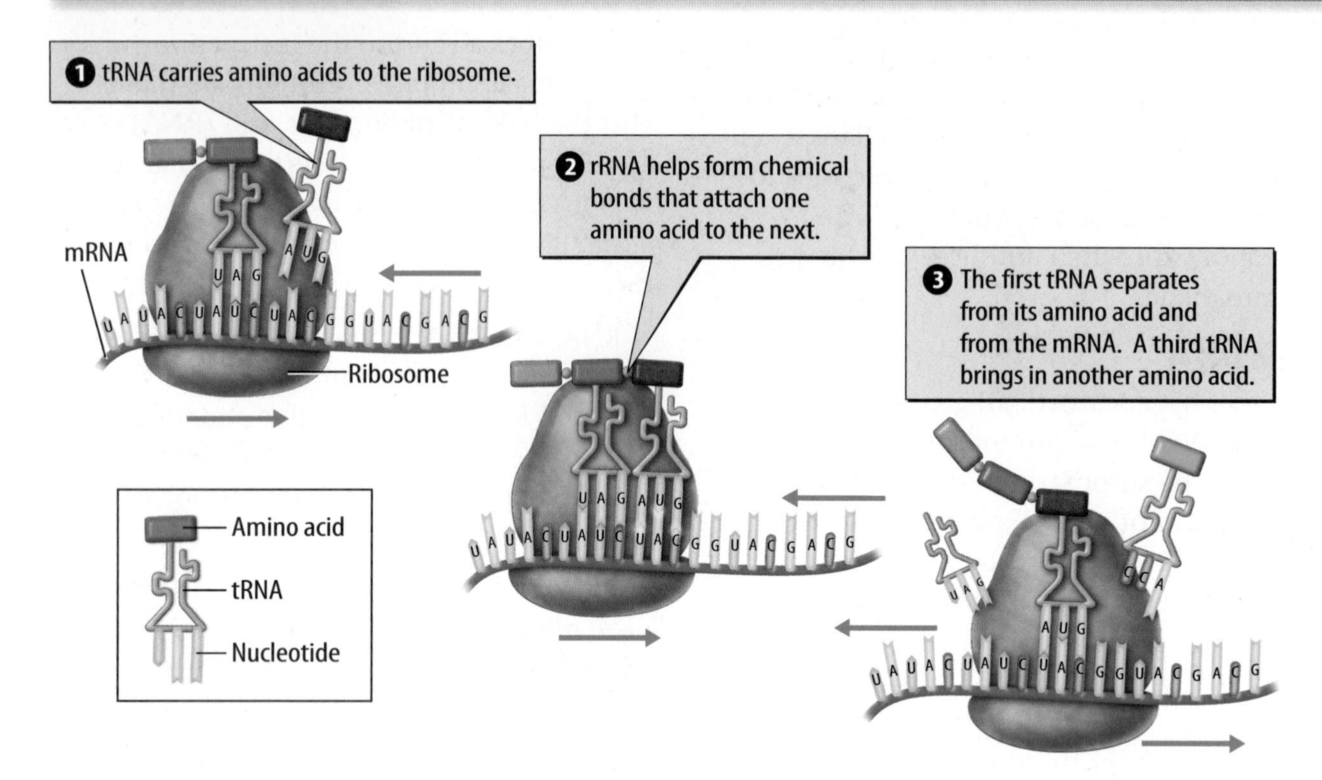

Figure 16 A protein forms as mRNA moves through a ribosome. Different amino acid sequences make different proteins. A complete protein is a folded chain of amino acids.

Three Types of RNA

On the previous page, you read about messenger RNA (mRNA). There are two other types of RNA—transfer RNA (tRNA) and ribosomal RNA (rRNA). **Figure 16** illustrates how the three work together to make proteins. *The process of making a protein from RNA is called* **translation.** Translation occurs in ribosomes. Recall that ribosomes are cell organelles that are attached to the rough endoplasmic reticulum (rough ER). Ribosomes are also in a cell's cytoplasm.

Translating the RNA Code

Making a protein from mRNA is like using a secret code. Proteins are made of amino acids. The order of the nitrogen bases in mRNA determines the order of the amino acids in a protein. Three nitrogen bases on mRNA form the code for one amino acid.

Each series of three nitrogen bases on mRNA is called a codon. There are 64 codons, but only 20 amino acids. Some of the codons code for the same amino acid. One of the codons codes for an amino acid that is the beginning of a protein. This codon signals that translation should start. Three of the codons do not code for any amino acid. Instead, they code for the end of the protein. They signal that translation should stop.

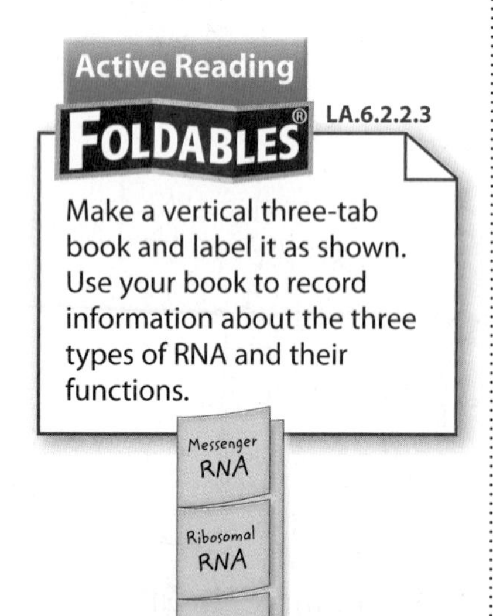

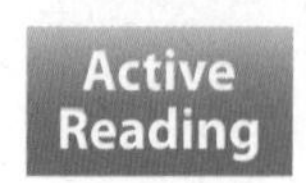

8. Identify Underline the definition of a codon.

Mutations

You have read that the sequence of nitrogen bases in DNA determines the sequence of nitrogen bases in mRNA, and that the mRNA sequence determines the sequence of amino acids in a protein. You might think these sequences always stay the same, but they can change. *A change in the nucleotide sequence of a gene is called a* **mutation.**

The 46 human chromosomes contain between 20,000 and 25,000 genes that are copied during DNA replication. Sometimes, mistakes can happen during replication. Most mistakes are corrected before replication is completed. A mistake that is not corrected can result in a mutation. Mutations can be triggered by exposure to X-rays, ultraviolet light, radioactive materials, and some kinds of chemicals.

Active Reading **9. Identify** Underline the causes of mutations.

WORD ORIGIN

mutation
from Latin *mutare,* means "to change"

Types of Mutations

There are several types of DNA mutations. Three types are shown in **Figure 17.** In a deletion mutation, one or more nitrogen bases are left out of the DNA sequence. In an insertion mutation, one or more nitrogen bases are added to the DNA. In a substitution mutation, one nitrogen base is replaced by a different nitrogen base.

Each type of mutation changes the sequence of nitrogen base pairs. This can cause a mutated gene to code for a different protein than a normal gene. Some mutated genes do not code for any protein. For example, a cell might lose the ability to make one of the proteins it needs.

Figure 17 Three types of mutations are substitution, insertion, and deletion.

10. Visual Check **Determine** Which base pairs were omitted during replication in the deletion mutation?

Mutations

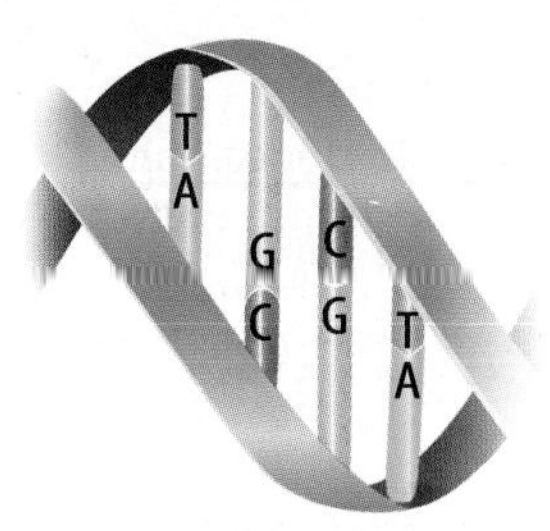

Original DNA sequence

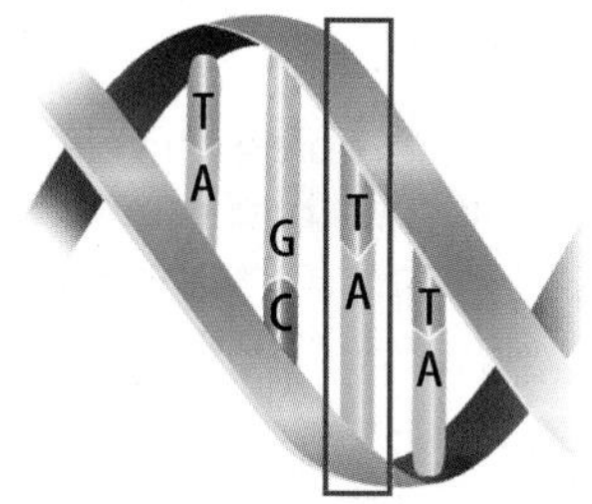

Substitution
The C-G base pair has been replaced with a T-A pair.

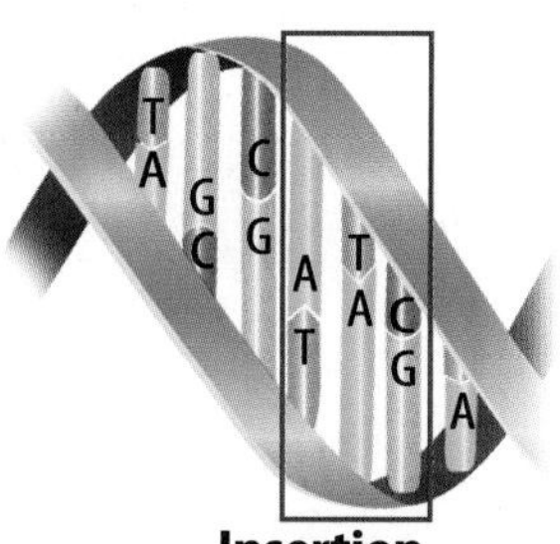

Insertion
Three base pairs have been added.

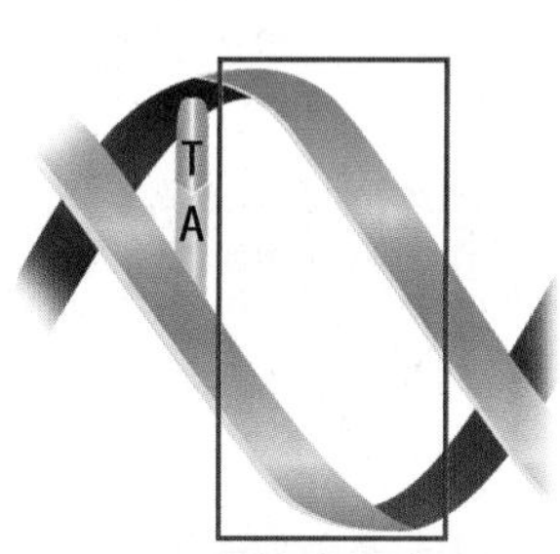

Deletion
Three base pairs have been removed. Other base pairs will move in to take their place.

Results of a Mutation

The effects of a mutation depend on where in the DNA sequence the mutation happens and the type of mutation. Proteins express traits. Because mutations can change proteins, they can cause traits to change. Some mutations in human DNA cause genetic disorders, such as those described in **Table 4.**

However, not all mutations have negative effects. Some mutations don't cause changes in proteins, so they don't affect traits. Other mutations might cause a trait to change in a way that benefits the organism.

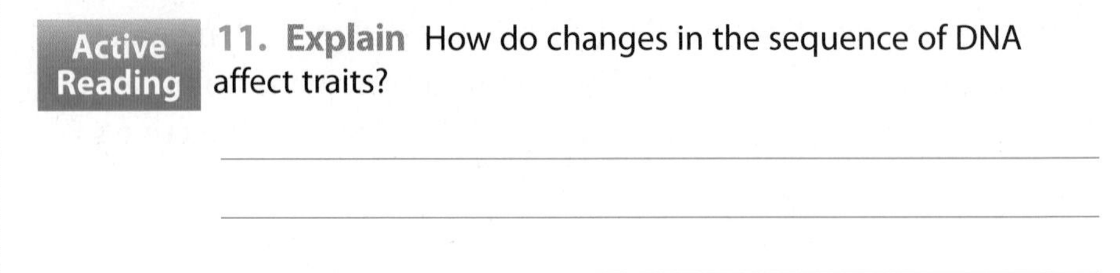

Active Reading **11. Explain** How do changes in the sequence of DNA affect traits?

Scientists still have much to learn about genes and how they determine an organism's traits. Scientists are researching and experimenting to identify all genes that cause specific traits. With this knowledge, we might be one step closer to finding cures and treatments for genetic disorders.

Table 4 Genetic Disorders

Defective Gene or Chromosome	Disorder	Description
Chromosome 12, PAH gene	Phenylketonuria (PKU)	People with defective PAH genes cannot break down the amino acid phenylalanine. If phenylalanine builds up in the blood, it poisons nerve cells.
Chromosome 7, CFTR gene	Cystic fibrosis	In people with defective CFTR genes, salt cannot move in and out of cells normally. Mucus builds up outside cells. The mucus can block airways in lungs and affect digestion.
Chromosome 7, elastin gene	Williams syndrome	People with Williams syndrome are missing part of chromosome 7, including the elastin gene. The protein made from the elastin gene makes blood vessels strong and stretchy.
Chromosome 17, BRCA 1; Chromosome 13, BRCA 2	Breast cancer and ovarian cancer	A defect in BRCA1 and/or BRCA2 does not mean the person will have breast cancer or ovarian cancer. People with defective BRCA1 or BRCA2 genes have an increased risk of developing breast cancer and ovarian cancer.

Lesson Review 3

Visual Summary

DNA is a complex molecule that contains the code for an organism's genetic information.

RNA carries the codes for making proteins.

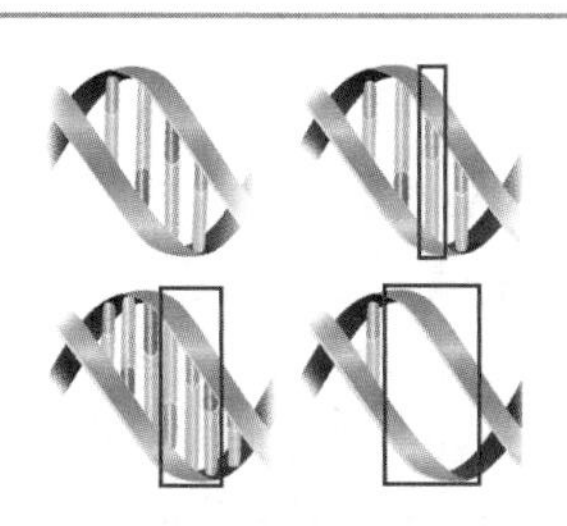

An organism's nucleotide sequence can change through the deletion, insertion, or substitution of nitrogen bases.

Use Vocabulary

1. **Use the terms** *DNA* and *nucleotide* in a sentence.

____

____

2. A change in the sequence of nitrogen bases in a gene is called a(n) ____.

Understand Key Concepts 

3. Where does the process of transcription occur?
   (A) cytoplasm (C) cell nucleus
   (B) ribosomes (D) outside the cell

4. **Illustrate** On a separate sheet of paper, make a drawing that illustrates the process of translation.

5. **Distinguish** between the sides of the DNA double helix and the teeth of the DNA double helix.

____

____

Interpret Graphics

6. **Sequence** Fill in the graphic organizer below about important steps in making a protein, beginning with DNA and ending with protein. LA.6.2.2.3

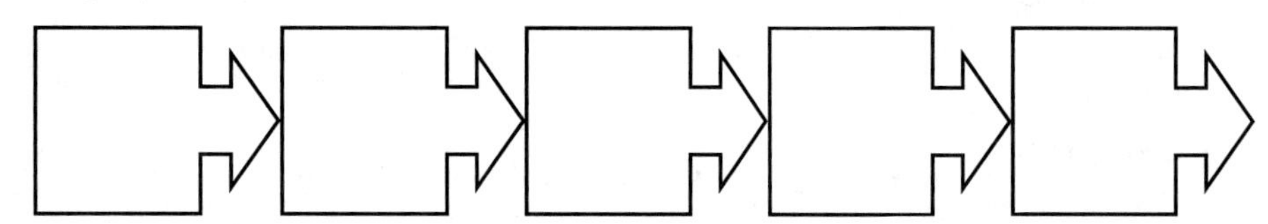

Critical Thinking

7. **Hypothesize** What would happen if a cell were unable to make mRNA?

____

____

8. **Assess** What is the importance of DNA replication occurring without any mistakes?

____

____

____

Chapter 5 Study Guide

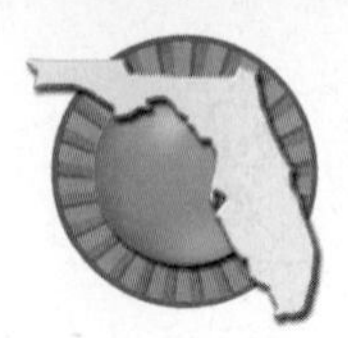

Think About It! Genetic information is passed from generation to generation by DNA; DNA controls the traits of an organism.

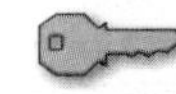

Key Concepts Summary

LESSON 1 Mendel and His Peas

- Mendel performed cross-pollination experiments to track which traits were produced by specific parental crosses.
- Mendel found that two genetic factors—one from a sperm cell and one from an egg cell—control each trait.
- **Dominant** traits block the expression of **recessive** traits. Recessive traits are expressed only when two recessive factors are present.

Vocabulary

heredity p. 163
genetics p. 163
dominant trait p. 169
recessive trait p. 169

LESSON 2 Understanding Inheritance

- **Phenotype** describes how a trait appears.
- **Genotype** describes alleles that control a trait.
- **Punnett squares** and pedigrees are tools to model patterns of inheritance.
- Many patterns of inheritance, such as **codominance** and **polygenic inheritance,** are more complex than Mendel described.

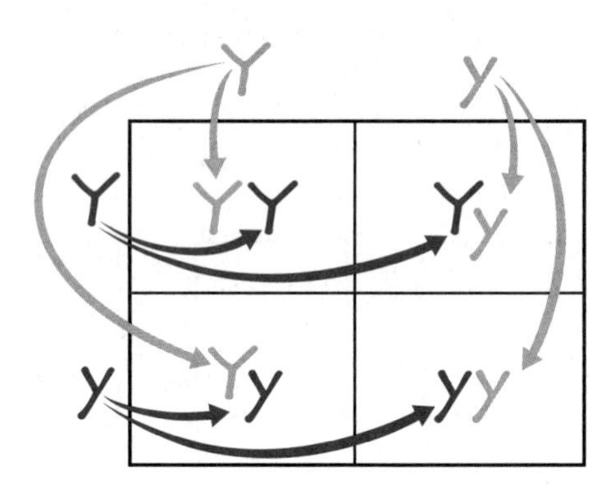

gene p. 174
allele p. 174
phenotype p. 174
genotype p. 174
homozygous p. 175
heterozygous p. 175
Punnett square p. 176
incomplete dominance p. 178
codominance p. 178
polygenic inheritance p. 179

LESSON 3 DNA and Genetics

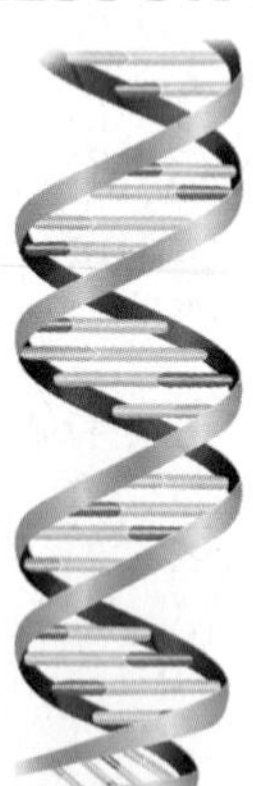

- **DNA** contains an organism's genetic information.
- **RNA** carries the codes for making proteins from the nucleus to the cytoplasm. RNA also forms part of ribosomes.
- A change in the sequence of DNA, called a **mutation,** can change the traits of an organism.

DNA p. 184
nucleotide p. 185
replication p. 186
RNA p. 187
transcription p. 187
translation p. 188
mutation p. 189

Active Reading

FOLDABLES® Chapter Project

Assemble your lesson Foldables as shown to make a Chapter Project. Use the project to review what you have learned in this chapter.

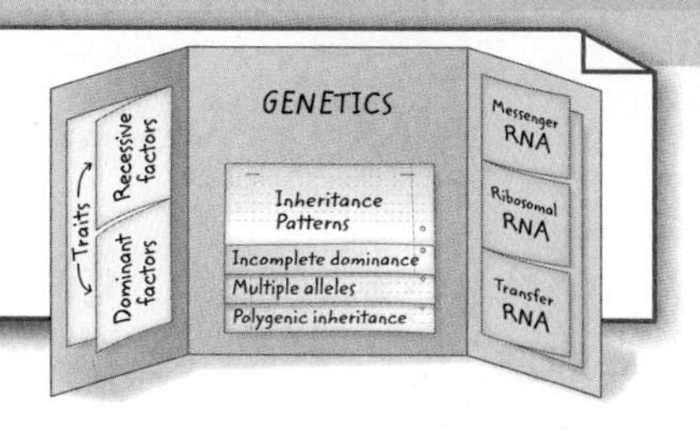

Use Vocabulary

1. The study of how traits are passed from parents to offspring is called ____________.

2. The passing of traits from parents to offspring is ____________.

3. Human height, weight, and skin color are examples of characteristics determined by ____________.

4. A helpful device for predicting the ratios of possible genotypes is a(n) ____________.

5. The code for a protein is called a(n) ____________.

6. An error made during the copying of DNA is called a(n) ____________.

Link Vocabulary and Key Concepts

Use vocabulary terms from the previous page to complete the concept map.

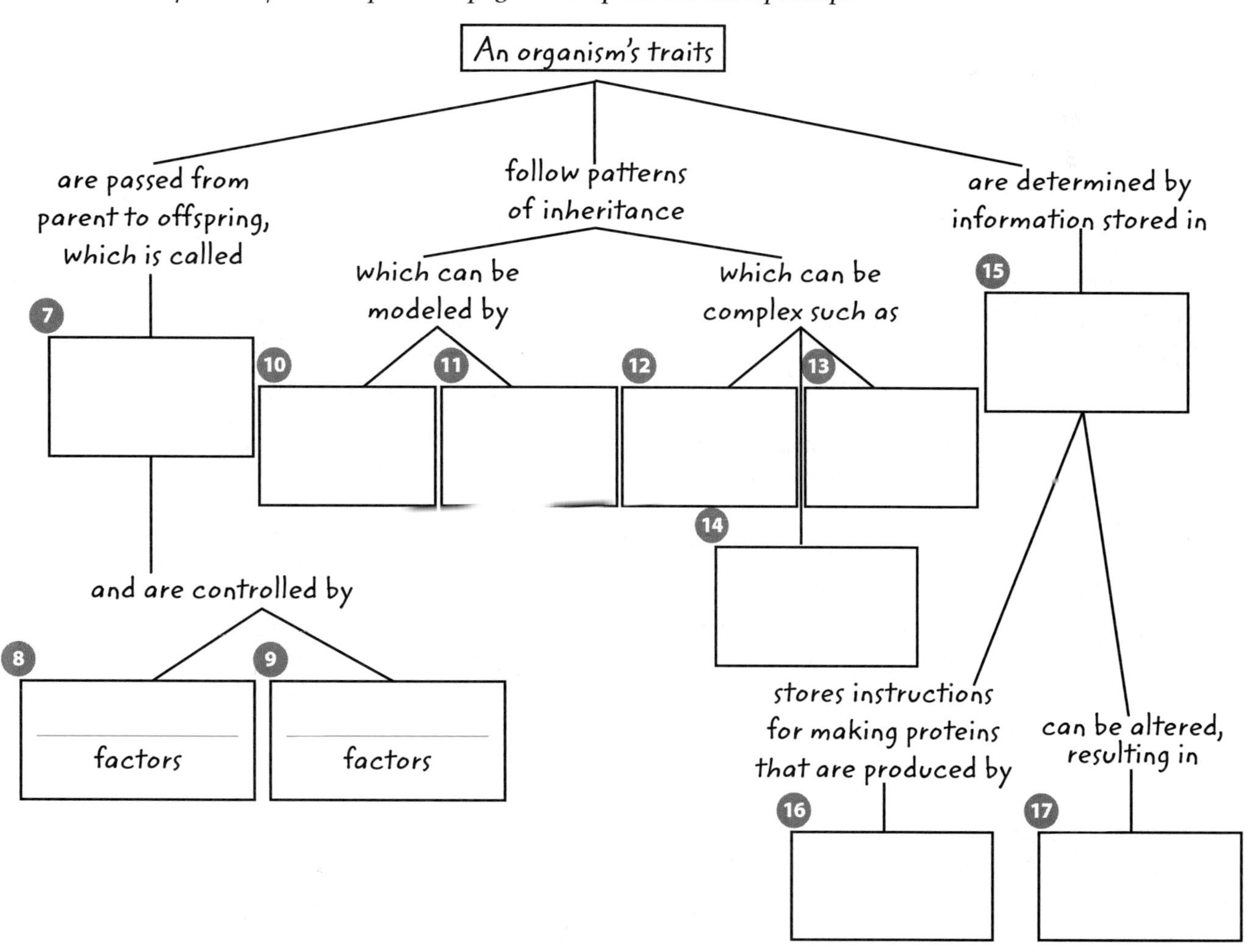

Chapter 5 Review

Fill in the correct answer choice.

Understand Key Concepts

1. The process shown below was used by Mendel during his experiments.

What is the process called? SC.7.N.2.1

Ⓐ cross-pollination
Ⓑ segregation
Ⓒ asexual reproduction
Ⓓ blending inheritance

2. Which statement best describes Mendel's experiments? SC.7.N.2.1

Ⓐ He began with hybrid plants.
Ⓑ He controlled pollination.
Ⓒ He observed only one generation.
Ⓓ He used plants that reproduce slowly.

3. Before Mendel's discoveries, which statement describes how people believed traits were inherited? SC.7.N.2.1

Ⓐ Parental traits blend like colors of paint to produce offspring.
Ⓑ Parental traits have no effect on their offspring.
Ⓒ Traits from only the female parent are inherited by offspring.
Ⓓ Traits from only the male parent are inherited by offspring.

4. Which term describes the offspring of a first-generation cross between parents with different forms of a trait? SC.7.L.16.1

Ⓐ genotype
Ⓑ hybrid
Ⓒ phenotype
Ⓓ true breeding

Critical Thinking

5. **Compare** heterozygous genotype and homozygous genotype. LA.6.2.2.3

6. **Distinguish** between multiple alleles and polygenic inheritance. LA.6.2.2.3

7. **Give an example** of how the environment can affect an organism's phenotype. LA.6.2.2.3

8. **Predict** In pea plants, the allele for smooth pods is dominant to the allele for bumpy pods. Predict the genotype of a plant with bumpy pods. Can you predict the genotype of a plant with smooth pods? Explain. SC.7.L.16.1

9. **Compare and contrast** characteristics of replication, transcription, translation, and mutation. Which of these processes takes place only in the nucleus of a cell? Which can take place in both the nucleus and the cytoplasm? How do you know? LA.6.2.2.3

10 **Interpret Graphics** In tomato plants, red fruit (R) is dominant to yellow fruit (r). Interpret the Punnett square below, which shows a cross between a heterozygous red plant and a yellow plant. Include the possible genotypes and corresponding phenotypes. SC.7.L.16.2

	R	r
r	Rr	rr
r	Rr	rr

Writing in Science

11 **Write** a paragraph contrasting the blending theory of inheritance with the current theory of inheritance. Include a main idea, supporting details, and a concluding sentence. LA.6.2.2.3

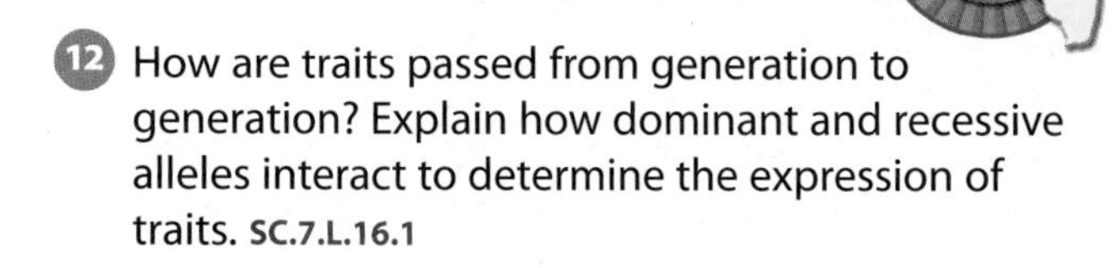

Big Idea Review

12 How are traits passed from generation to generation? Explain how dominant and recessive alleles interact to determine the expression of traits. SC.7.L.16.1

13 The photo at the beginning of the chapter shows an albino offspring from a non-albino mother. If albinism is a recessive trait, what are the possible genotypes of the mother, the father, and the offspring? SC.7.L.16.2

Math Skills MA.6.A.3.6

Use Ratios

14 A cross between two heterozygous pea plants with yellow seeds produced 1,719 yellow seeds and 573 green seeds. What is the ratio of yellow to green seeds?

15 A cross between two heterozygous pea plants with smooth green pea pods produced 87 bumpy yellow pea pods, 261 smooth yellow pea pods, 261 bumpy green pea pods, and 783 smooth green pea pods. What is the ratio of bumpy yellow to smooth yellow to bumpy green to smooth green pea pods?

16 A jar contains three red, five green, two blue, and six yellow marbles. What is the ratio of red to green to blue to yellow marbles?

Florida NGSSS Benchmark Practice

Fill in the correct answer choice.

Multiple Choice

Use the diagram below to answer questions 1 and 2.

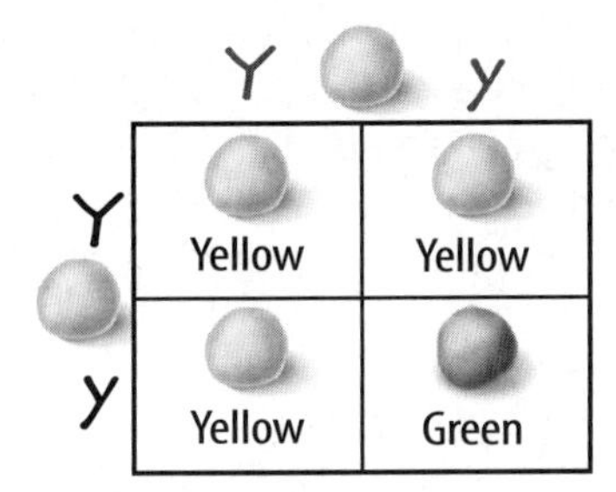

1 Which genotype belongs in the lower-right square? **SC.7.L.16.2**

Ⓐ YY

Ⓑ Yy

Ⓒ yY

Ⓓ yy

2 What percentage of plants from this cross will produce yellow seeds? **SC.7.L.16.2**

Ⓕ 25 percent

Ⓖ 50 percent

Ⓗ 75 percent

Ⓘ 100 percent

3 What is heredity? **SC.7.L.16.1**

Ⓐ the study of how traits are passed from parents to offspring

Ⓑ the study of how DNA replicates

Ⓒ the process of chromosomes mutating

Ⓓ the passing of traits from parents to offspring

4 Which can be determined by using a Punnett square or pedigree? **SC.7.L.16.2**

Ⓕ phenotypes of polygenic traits

Ⓖ phenotypes of dominant and recessive traits

Ⓗ phenotypes of codominant traits

Ⓘ genotype mutations

Use the chart below to answer questions 5 and 6.

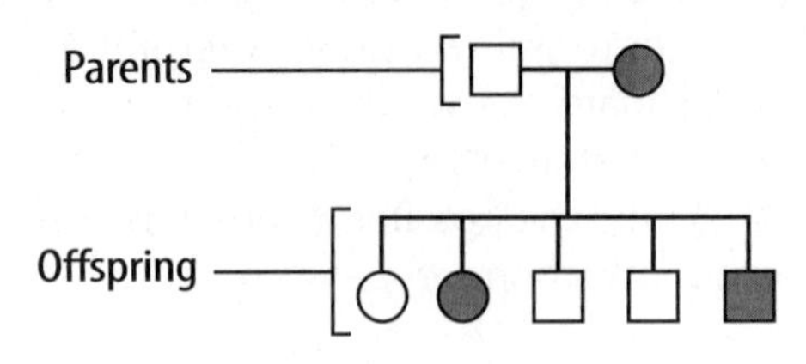

Phenotypes

○ Female, dominant ● Female, recessive

□ Male, dominant ■ Male, recessive

5 Based on the pedigree above, how many offspring from this cross had the recessive phenotype? **SC.7.L.16.2**

Ⓐ 1

Ⓑ 2

Ⓒ 3

Ⓓ 5

6 Based on the pedigree above, how many offspring from this cross are homozygous dominant? **SC.7.L.16.2**

Ⓕ 0

Ⓖ 1

Ⓗ 3

Ⓘ 5

7 Which is a section on a chromosome that has genetic information for one trait? **SC.7.L.16.1**

Ⓐ allele

Ⓑ genotype

Ⓒ phenotype

Ⓓ gene

8 Which is a model used to predict possible genotypes and phenotypes of offspring? **SC.7.L.16.2**

Ⓕ polygenic inheritance

Ⓖ ratio

Ⓗ Punnett square

Ⓘ incomplete dominance

Use the diagrams below to answer questions 9 and 10.

	R	r
R	RR	Rr
r	Rr	rr

Genotype	Phenotype
RR	Red
Rr	Pink
rr	White

9 According to the information in the diagrams above, what is the ratio of the offspring? **SC.7.L.16.2**

Ⓐ 0 red: 4 pink: 0 white

Ⓑ 1 red: 2 pink: 1 white

Ⓒ 3 red: 0 pink: 1 white

Ⓓ 4 red: 0 pink: 0 white

10 According to the information in the diagrams above, what is the phenotype of homozygous dominant offspring? **SC.7.L.16.2**

Ⓕ red

Ⓖ pink

Ⓗ white

Ⓘ yellow

11 Mendel crossed a true-breeding plant with round seeds and a true-breeding plant with wrinkled seeds. Which was true of every offspring of this cross? **SC.7.L.16.2**

Ⓐ They had the recessive phenotype.

Ⓑ They showed a combination of traits.

Ⓒ They were homozygous.

Ⓓ They were hybrid plants.

Use the diagram below to answer question 12.

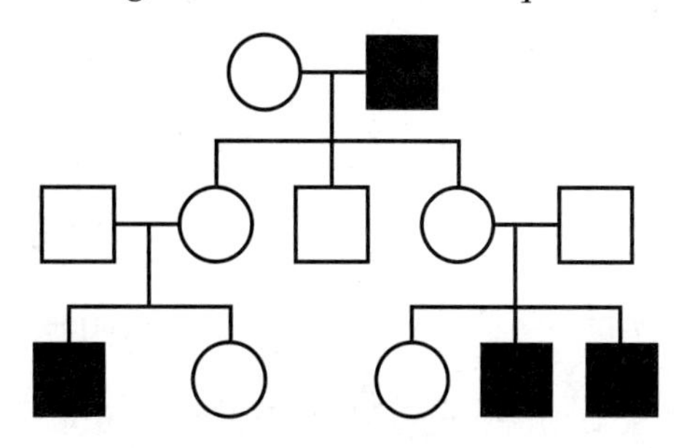

Phenotypes

○ female, dominant ● female, recessive
□ male, dominant ■ male, recessive

12 Using the diagram above, what can be determined about the genotypes of the first generation of offspring? **SC.7.L.16.2**

Ⓕ They are all homozygous recessive.

Ⓖ They are all homozygous dominant.

Ⓗ They are all heterozygous.

Ⓘ Two are heterozygous, and one is homozygous recessive.

13 If a pea plant with a homozygous dominant genotype for a trait is crossed with a pea plant with a heterozygous genotype for the same trait, what is the ratio of offspring? **SC.7.L.16.2**

Ⓐ 1 homozygous dominant: 2 heterozygous: 1 homozygous recessive

Ⓑ 2 homozygous dominant: 2 heterozygous: 0 homozygous recessive

Ⓒ 0 homozygous dominant: 2 heterozygous: 2 homozygous recessive

Ⓓ 1 homozygous dominant: 0 heterozygous: 3 homozygous recessive

NEED EXTRA HELP?

If You Missed Question...	1	2	3	4	5	6	7	8	9	10	11	12	13
Go to Lesson...	2	2	1	2	2	2	1	2	2	2	1	2	2

Name ______________________________ Date ______________

Benchmark Mini-Assessment Chapter 5 • Lesson 1

mini BAT

Multiple Choice *Bubble the correct answer.*

White

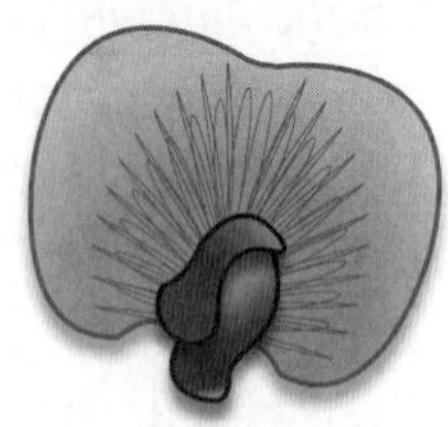

Purple

1. Look at the image above. What color flowers did Mendel discover were produced in the first generation? **SC.7.L.16.1**

Ⓐ The flowers were all blue.

Ⓑ The flowers were all pink.

Ⓒ The flowers were all purple.

Ⓓ The flowers were all white.

2. Which is NOT a reason that Mendel used pea plants for his experiments? **SC.7.L.16.1**

Ⓕ Pea plants do not self-pollinate.

Ⓖ Pea plants reproduce quickly.

Ⓗ Pea plants have many easily observed traits.

Ⓘ Pea plant reproduction could be controlled by Mendel.

Guinea Pig Fur Color		
Generation	White fur (number of offspring)	Black fur (number of offspring)
First	0	9
Second	6	19

3. Based on the table above, which statement is true? **SC.7.L.16.1**

Ⓐ Guinea pigs are not true-breeding for fur color.

Ⓑ The ratio of a hybrid cross in guinea pigs is about 2:1.

Ⓒ In the first generation, the trait for black fur is masked by the trait for white fur.

Ⓓ In guinea pigs, the trait for black fur is dominant, and the trait for white fur is recessive.

4. How did Mendel control pollination during his cross-pollination experiments involving the study of flower color in pea plants? **SC.7.L.16.1**

Ⓕ He allowed pollinators such as bees to pollinate the plant.

Ⓖ He removed the pistils from the plant being pollinated.

Ⓗ He removed the stamens from the plant being pollinated.

Ⓘ He transferred pollen from flower to flower on the same plant.

Name ______________________________ Date ______________

Multiple Choice *Bubble the correct answer.*

1. Which Punnett square shows a cross between two heterozygous parents? **SC.7.L.16.2**

Ⓐ

	M	m
M	MM	Mm
m	Mm	mm

Ⓑ

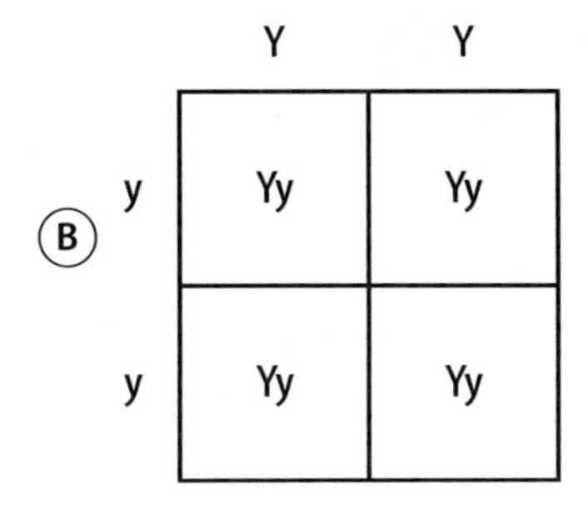

Ⓒ

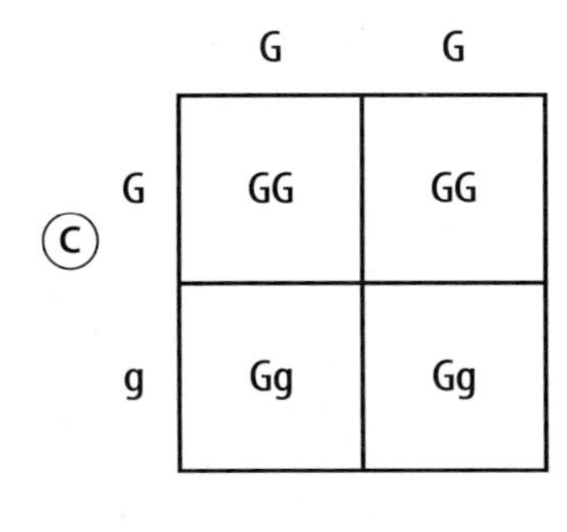

Ⓓ

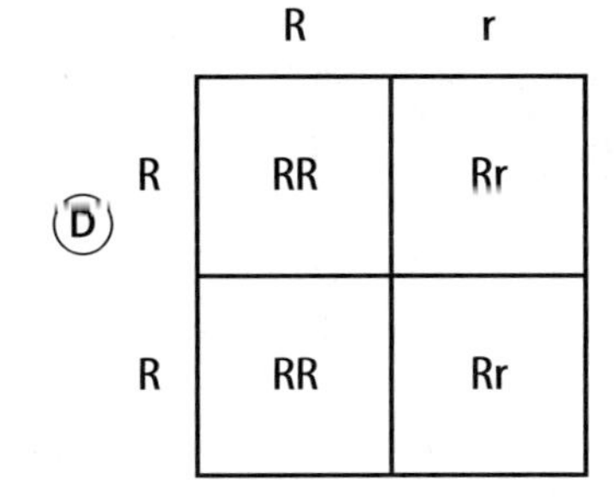

2. A cross between a plant with red flowers and a plant with white flowers produces offspring with pink flowers. What is this an example of? **SC.7.L.16.2**

Ⓕ codominance

Ⓖ dominance

Ⓗ incomplete dominance

Ⓘ polygenic inheritance

3. In one generation, two true-breeding plants produce offspring that look like one of the parents. In the second hybrid generation, the offspring have a ratio of 3:1 for the traits. Which statement is true of the parents of the first generation? **SC.7.L.16.2**

Ⓐ Both were heterozygous for the dominant trait.

Ⓑ Both were homozygous for the recessive trait.

Ⓒ One was homozygous for the dominant trait and the other was homozygous for the recessive trait.

Ⓓ One was homozygous for the dominant trait and the other was heterozygous for the dominant trait.

4. Which of these is a trait in humans that is determined by multiple alleles? **SC.7.L.16.1**

Ⓕ height

Ⓖ weight

Ⓗ blood type

Ⓘ skin color

Name ______________________________ Date ______________

Benchmark Mini-Assessment

mini BAT

Multiple Choice *Bubble the correct answer.*

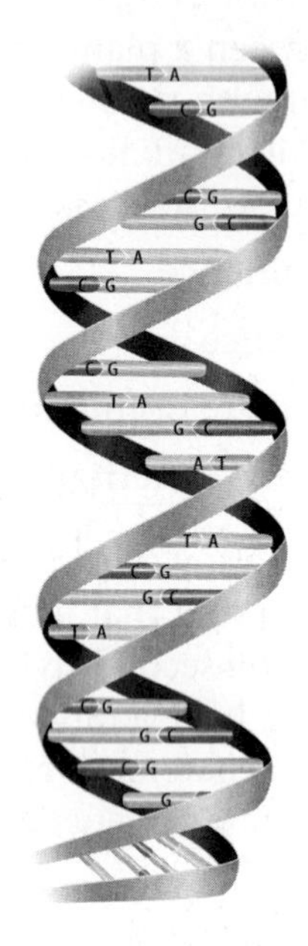

1. In the image above, what are *A, C, T,* and *G*? **SC.7.L.16.1**
   - (A) codons
   - (B) nucleotides
   - (C) amino acids
   - (D) nitrogen bases

2. How do mutations potentially cause harm? **SC.7.L.16.1**
   - (F) Mutations break down chromosomes.
   - (G) Mutations damage ribosomes.
   - (H) Mutations can cause a gene to code for a different protein than normal.
   - (I) Mutations prevent the formation of RNA, so translation does not take place.

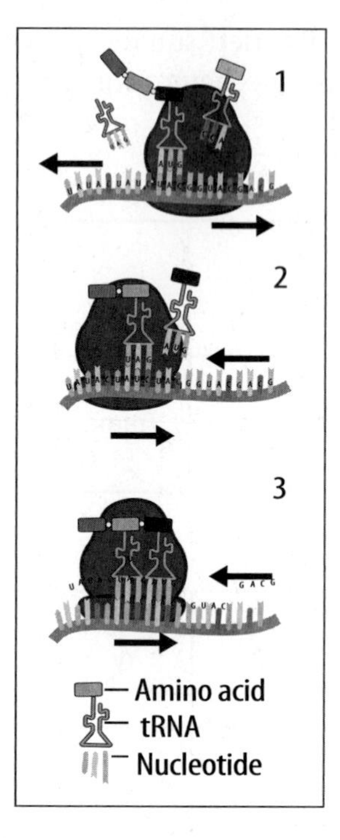

3. In the image above, what is the correct order of the steps shown? **SC.7.L.16.1**
   - (A) 1, 2, 3
   - (B) 1, 3, 2
   - (C) 2, 3, 1
   - (D) 3, 2, 1

4. RNA carries the code for making **SC.7.L.16.1**
   - (F) acids.
   - (G) bases.
   - (H) DNA.
   - (I) proteins.

Name ______________________ Date ____________

Notes

Name ______________________ Date ______________

Notes

Tree Snails

A population of a tree snail species has different patterns in their shells. These slight differences in appearance among the individual members of the species are called variations. How do you think these variations get passed on from one generation of tree snails to the next? Circle the answer that best matches your thinking.

A. from parents to offspring

B. through the environment

C. from both parents to offspring and through the environment

Explain your thinking. Describe how variations are passed on from one generation to the next.

The Environment and Change Over Time

The Big Idea

FLORIDA BIG IDEAS

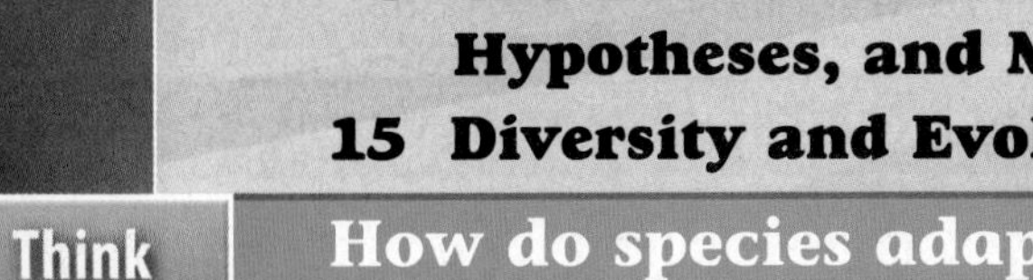

1 **The Practice of Science**
2 **The Characteristics of Scientific Knowledge**
3 **The Role of Theories, Laws, Hypotheses, and Models**
15 **Diversity and Evolution of Living Organisms**

Think About It! How do species adapt to changing environments over time?

A type of orchid plant, called a bee orchid, produces this flower. You might have noticed that the flower looks like a bee.

1. What do you think is the advantage to the plant to have flowers that look like bees?

2. How do you think the appearance of the flower developed over time?

3. How do you think species adapt to changing environments over time?

Get Ready to Read What do you think about changes in the environment?

Before you read, decide if you agree or disagree with each of these statements. As you read this chapter, see if you change your mind about any of the statements.

	AGREE	DISAGREE
1 Original tissues can be preserved as fossils.	☐	☐
2 Organisms become extinct only in mass extinction events.	☐	☐
3 Environmental change causes variations in populations.	☐	☐
4 Variations can lead to adaptations.	☐	☐
5 Living species contain no evidence that they are related to each other.	☐	☐
6 Plants and animals share similar genes.	☐	☐

There's More Online!
Video • Audio • Review • ⓘLab Station • WebQuest • Assessment • Concepts in Motion • Multilingual eGlossary

Lesson 1

Fossil Evidence of EVOLUTION

ESSENTIAL QUESTIONS

How do fossils form?

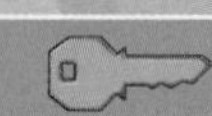
How do scientists date fossils?

How are fossils evidence of biological evolution?

Vocabulary

fossil record p. 207
mold p. 209
cast p. 209
trace fossil p. 209
geologic time scale p. 211
extinction p. 212
biological evolution p. 213

Florida NGSSS

LA.6.2.2.3 The student will organize information to show understanding (e.g., representing main ideas within text through charting, mapping, paraphrasing, summarizing, or comparing/contrasting);

MA.6.A.3.6 Construct and analyze tables, graphs, and equations to describe linear functions and other simple relations using both common language and algebraic notation.

SC.7.L.15.1 Recognize that fossil evidence is consistent with the scientific theory of evolution that living things evolved from earlier species.

SC.6.N.1.5 Recognize that science involves creativity, not just in designing experiments, but also in creating explanations that fit evidence.

SC.7.N.1.3 Distinguish between an experiment (which must involve the identification and control of variables) and other forms of scientific investigation and explain that not all scientific knowledge is derived from experimentation.

SC.6.N.2.1 Distinguish science from other activities involving thought.

SC.7.N.1.3

Launch Lab

20 minutes

How do fossils form?

Evidence from fossils helps scientists understand how organisms have changed over time. Some fossils form when impressions left by organisms in sand or mud are filled in by sediments that harden.

Procedure

1. Read and complete a lab safety form.
2. Place a **container of moist sand** on top of **newspaper.** Press a **shell** into the moist sand. Carefully remove the shell. Brush any sand on the shell onto the newspaper.
3. Observe the impression, and record your observations.
4. Pour **plaster of paris** into the impression. Wait for it to harden. ⚠ *The mix gets hot as it sets—do not touch it until it has hardened.*
5. Remove the shell fossil from the sand, and brush it off.
6. Observe the structure of the fossil, and record your observations.

Data and Observations

Think About This

1. What effect did the shell have on the sand?

2. Key Concept What information do you think someone could learn about the shell and the organism that lived inside it by examining the fossil?

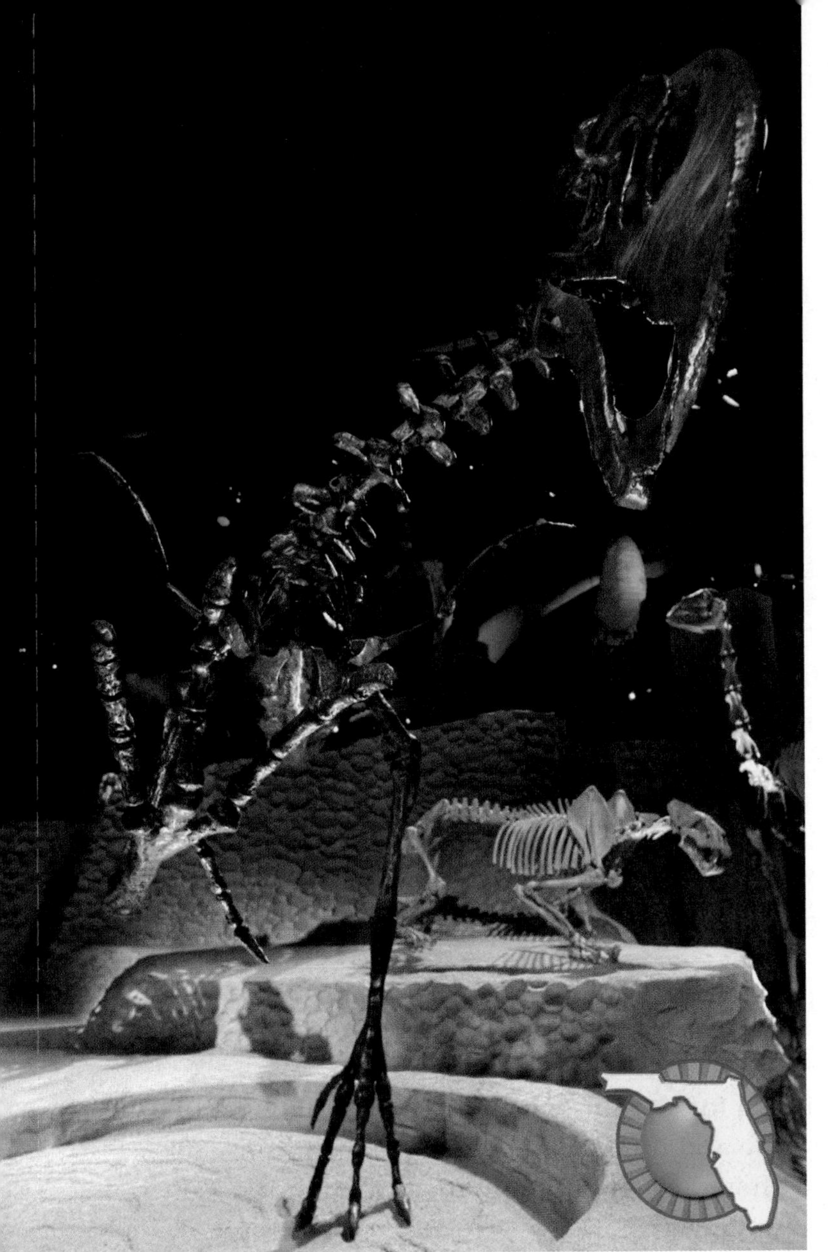

The Fossil Record

On your way to school, you might have seen an oak tree or heard a robin. Although these organisms shed leaves or feathers, their characteristics remain the same from day to day. It might seem as if they have been on Earth forever. However, if you were to travel a few million years back in time, you would not see oak trees or robins. You would see different species of trees and birds. That is because species change over time.

You might already know that fossils are the remains or evidence of once-living organisms. *The* **fossil record** *is made up of all the fossils ever discovered on Earth.* It contains millions of fossils that represent many thousands of species. Most of these species are no longer alive on Earth. The fossil record provides evidence that species have changed over time. Fossils help scientists picture what these species looked like. **Figure 1** shows how scientists think the giant bird *Titanus* might have looked when it was alive. The image is based on fossils that have been discovered and are represented in the photo on the left.

The fossil record is enormous, but it is still incomplete. Scientists think it represents only a small fraction of all the organisms that have ever lived on Earth.

Inquiry What do you think can be learned from fossils?

1. When scientists find fossils, like the one above at the Florida Museum of Natural History, they use them as evidence to try to answer questions about past life on Earth. When did this organism live? What did this organism eat? How did it move or grow? How did this organism die? To what other organisms is this one related?

Figure 1 Based on fossil evidence, scientists can re-create the physical appearance of species that are no longer alive on Earth.

Fossil Formation

SCIENCE USE V. COMMON USE

tissue

Science Use similar cells that work together and perform a function

Common Use a piece of soft, absorbent paper

If you have ever seen vultures or other animals eating a dead animal, you know they leave little behind. Any soft **tissues** that animals do not eat, bacteria break down. Only the dead animal's hard parts, such as bones, shells, and teeth, remain. In most instances, these hard parts also break down over time. However, under rare conditions, some become fossils. The soft tissues of animals and plants, such as skin, muscles, or leaves, can also become fossils, but these are even more rare. Some of the ways that fossils can form are shown in **Table 1.**

2. Infer (Circle) the correct answer: Fossils formed by mineralization are more likely to be found near a desert / stream.

Active Reading **3. Explain** Why is it rare for soft tissue to become a fossil?

____

____

Mineralization

After an organism dies, its body could be buried under mud, sand, or other sediments in a stream or river. If minerals in the water replace the organism's original material and harden into rock, a fossil forms. This process is called mineralization. Minerals in water also can filter into the small spaces of a dead organism's tissues and become rock. Most mineralized fossils are of shell or bone, but wood can also become a mineralized fossil, as shown in **Table 1.**

Table 1 Fossils form in several ways.

4. Identify What types of organisms or tissues are often preserved as carbon films?

____

____

____

Carbonization

In carbonization, a fossil forms when a dead organism is compressed over time and pressure drives off the organism's liquids and gases. As shown in **Table 1,** only the carbon outline, or film, of the organism remains.

Table 1 How Fossils Form

	Mineralization	Carbonization
Description	Rock-forming minerals in water filled in the small spaces in the tissue of these pieces of petrified wood. Water also replaced some of the wood's tissue. Mineralization can preserve the internal structures of an organism.	Fossil films made by carbonization are usually black or dark brown. Fish, insects, and plant leaves, such as this fern frond, are often preserved as carbon films.
Example		

Molds and Casts

Sometimes when an organism dies, its shell or bone might make an impression in mud or sand. When the sediment hardens, so does the impression. *The impression of an organism in a rock is called a* **mold.** Sediments can later fill in the mold and harden to form a cast. *A* **cast** *is a fossil copy of an organism in a rock.* A single organism can form both a mold and a cast, as shown in **Table 1.** Molds and casts show only external features of organisms.

Trace Fossils

Evidence of an organism's movement or behavior—not just its physical structure—also can be preserved in rock. *A* **trace fossil** *is the preserved evidence of the activity of an organism.* For example, an organism might walk across mud. The tracks, such as the ones shown in **Table 1,** can fossilize if they are filled with sediment that hardens.

Original Material

In rare cases, the original tissues of an organism can be preserved. Examples of original-material fossils include mammoths frozen in ice and saber-toothed cats preserved in tar pits. Fossilized remains of ancient humans have been found in bogs. Most of these fossils are younger than 10,000 years old. However, the insect encased in amber in **Table 1** is millions of years old. Scientists also have found original tissue preserved in the bone of a dinosaur that lived 70 million years ago (mya).

Active Reading **5. Identify** List the different ways fossils can form.

____

____

WORD ORIGIN

fossil
from Latin *fossilis,* means "to obtain by digging"

Active Reading **6. Explain** Why are trace fossils used as evidence of an organism's movement or behavior?

____

____

____

____

____

____

____

Molds and Casts	Trace Fossils	Original Material
When sediments hardened around this buried trilobite, a mold formed. Molds are usually of hard parts, such as shells or bone. If a mold is later filled with more sediments that harden, the mold can form a cast.	These footprints were made when a dinosaur walked across mud that later hardened. This trace fossil might provide evidence of the speed and weight of the dinosaur.	If original tissues of organisms are buried in the absence of oxygen for long periods of time, they can fossilize. The insect in this amber became stuck in tree sap that later hardened.

Dating Fossils

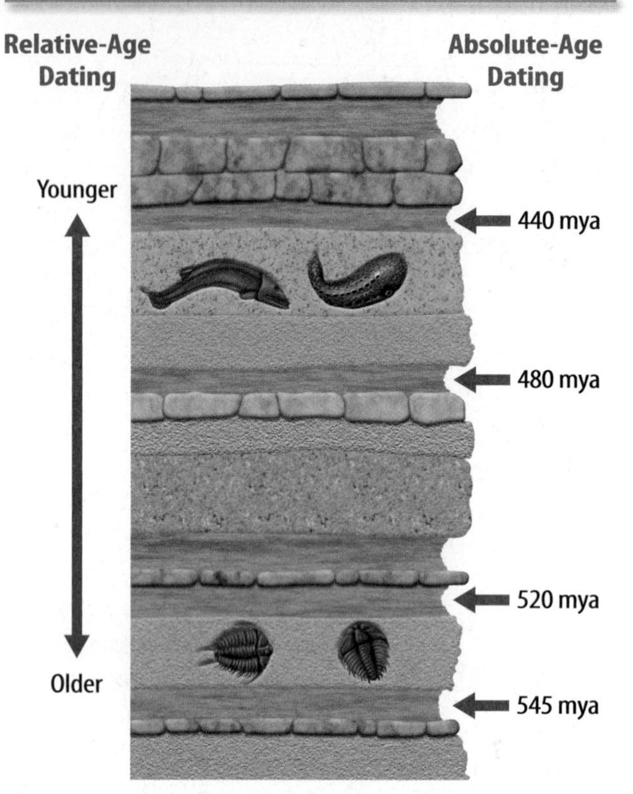

Figure 2 If the age of the igneous layers is known, as shown above, it is possible to estimate the age of the sedimentary layers—and the fossils they contain—between them.

7. Visual Check Infer What is the estimated age of the trilobite fossils (bottom layer of fossils)?

8. Visual Check Apply  Circle the layer in **Figure 2** that is approximately 450 million years old.

Active Reading FOLDABLES® LA.6.2.2.3

Make a small shutterfold book. Label it as shown. Under the left tab describe relative-age dating. Under the right tab describe absolute-age dating.

Determining a Fossil's Age

Scientists cannot date most fossils directly. Instead, they date the rocks the fossils are embedded inside. Rocks erode or are recycled over time. However, scientists can determine ages for most of Earth's rocks.

Relative-Age Dating

How does your age compare to the ages of those around you? You might be younger than a brother but older than a sister. This is your relative age. Similarly, a rock is either older or younger than rocks nearby. In relative-age dating, scientists determine the relative order in which rock layers were deposited. In an undisturbed rock formation, they know that the bottom layers are oldest and the top layers are youngest, as shown in **Figure 2.** Relative-age dating helps scientists determine the relative order in which species have appeared on Earth over time.

Active Reading 9. Explain How does relative-age dating help scientists learn about fossils?

Absolute-Age Dating

Absolute-age dating is more precise than relative-age dating. Scientists take advantage of radioactive decay, a natural clocklike process in rocks, to learn a rock's absolute age, or its age in years. In radioactive decay, unstable isotopes in rocks change into stable isotopes over time. Scientists measure the ratio of unstable isotopes to stable isotopes to find the age of a rock. This ratio is best measured in igneous rocks.

Igneous rocks form from volcanic magma. Magma is so hot that it is rare for parts of organisms in it to form fossils. Most fossils form in sediments, which become sedimentary rock. To measure the age of sedimentary rock layers, scientists calculate the ages of igneous layers above and below them. In this way, they can estimate the ages of the fossils embedded within the sedimentary layers, as shown in **Figure 2.**

Fossils over Time

How old do you think Earth's oldest fossils are? You might be surprised to learn that evidence of microscopic, unicellular organisms has been found in rocks 3.4 billion years old. The oldest fossils visible to the unaided eye are about 565 million years old.

The Geologic Time Scale

It is hard to keep track of time that is millions and billions of years long. Scientists organize Earth's history into a time line called the geologic time scale. *The* **geologic time scale** *is a chart that divides Earth's history into different time units.* The longest time units in the geologic time scale are eons. As shown in **Figure 3,** Earth's history is divided into four eons. Earth's most recent eon—the Phanerozoic (fa nuh ruh ZOH ihk) eon—is subdivided into three eras, also shown in **Figure 3.**

10. Locate Highlight the definition of the geologic time scale.

Dividing Time

You might have noticed in **Figure 3** that neither eons nor eras are equal in length. When scientists began developing the geologic time scale in the 1800s, they did not have absolute-age dating methods. To mark time boundaries, they used fossils. Fossils provided an easy way to mark time. Scientists knew that different rock layers contained different types of fossils. Some of the fossils scientists use to mark the time boundaries are shown in **Figure 3.**

Often, a type of fossil found in one rock layer did not appear in layers above it. Even more surprising, entire collections of fossils in one layer were sometimes absent from layers above them. It seemed as if whole communities of organisms had suddenly disappeared.

11. Recall What do scientists use to mark boundaries in the geologic time scale?

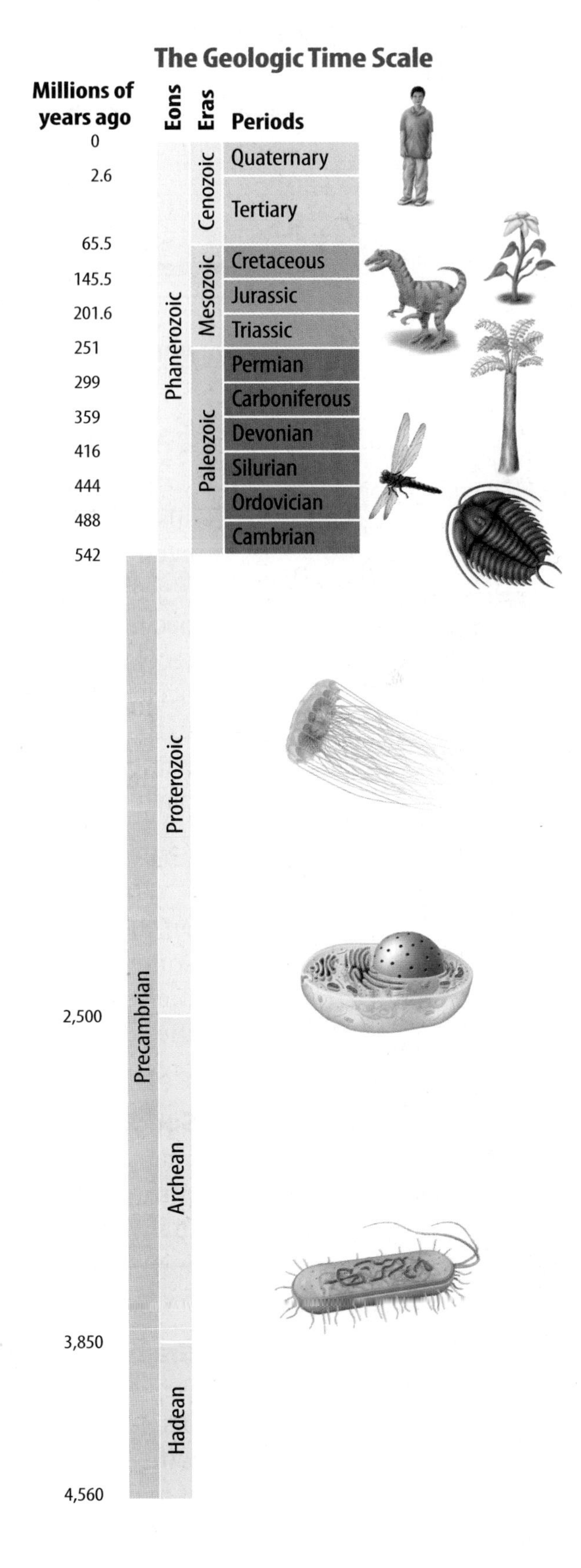

Figure 3 The Phanerozoic eon began about 540 million years ago and continues to the present day. It contains most of Earth's fossil record.

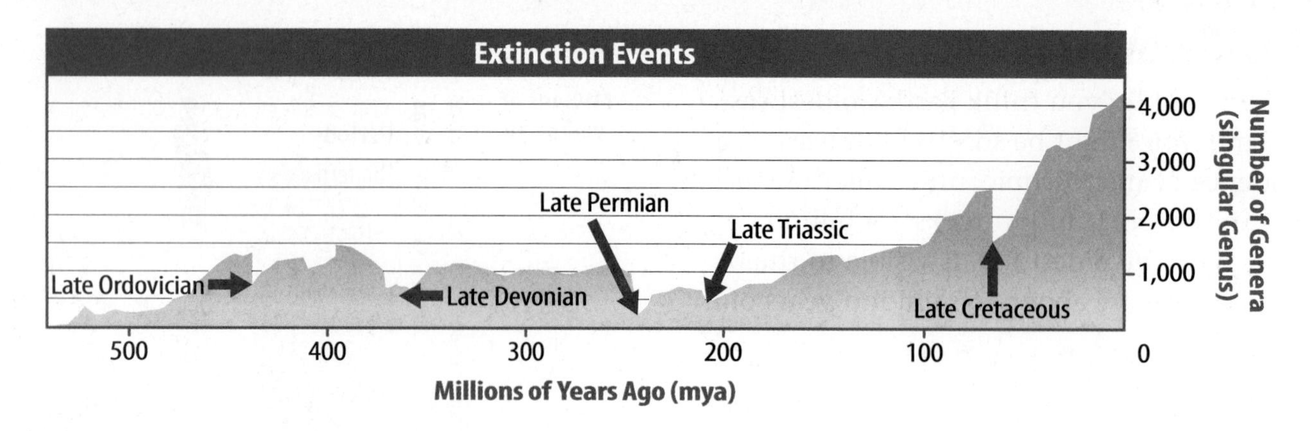

Figure 4 Arrows mark the five major extinction events of the Phanerozoic eon.

Extinctions

Scientists now understand that sudden disappearances of fossils in rock layers are evidence of extinction (ihk STINGK shun) events. **Extinction** *occurs when the last individual organism of a species dies.* A mass extinction occurs when many species become extinct within a few million years or less. The fossil record contains evidence that five mass extinction events have occurred during the Phanerozoic eon, as shown in **Figure 4.** Extinctions also occur at other times, on smaller scales. Evidence from the fossil record suggests that extinctions have been common throughout Earth's history.

Environmental Change

What causes extinctions? Populations of organisms depend on resources in their environment for food and shelter. Sometimes environments change. After a change happens, individual organisms of a species might not be able to find the resources they need to survive. When this happens, the organisms die, and the species becomes extinct.

Sudden Changes Extinctions can occur when environments change quickly. A volcanic eruption or a meteorite impact can throw ash and dust into the atmosphere, blocking sunlight for many years. This can affect global climate and food webs. Scientists hypothesize that the impact of a huge meteorite 65 million years ago contributed to the extinction of dinosaurs.

Gradual Changes Not all environmental change is sudden. Depending on the location, Earth's tectonic plates move between 1 and 15 cm each year. As plates move and collide with each other over time, mountains form and oceans develop. If a mountain range or an ocean isolates a species, the species might become extinct if it cannot find the resources it needs. Species also might become extinct if sea level changes.

Active Reading **13. Relate** What is the relationship between extinction and environmental change?

Math Skills MA.6.A.3.6

Use Scientific Notation

Numbers that refer to the ages of Earth's fossils are very large, so scientists use scientific notation to work with them. For example, mammals appeared on Earth about 200 mya or 200,000,000 years ago. Change this number to scientific notation using the following process.

Move the decimal point until only one nonzero digit remains on the left.

200,000,000 = 2.00000000

Count the number of places you moved the decimal point (8) and use that number as a power of ten.

200,000,000 = 2.0×10^8 years.

Practice

12. The first vertebrates appeared on Earth about 490,000,000 years ago. Express this time in scientific notation.

Extinctions and Evolution

The fossil record contains clear evidence of the extinction of species over time. But it also contains evidence of the appearance of many new species. How do new species arise?

Many early scientists thought that each species appeared on Earth independently of every other species. However, as more fossils were discovered, patterns in the fossil record began to emerge. Many fossil species in nearby rock layers had similar body plans and similar structures. It appeared as if they were related. For example, the series of horse fossils in **Figure 5** suggests that the modern horse is related to other extinct species. These species changed over time in what appeared to be a sequence. Change over time is evolution. **Biological evolution** *is the change over time in populations of related organisms.* Charles Darwin developed a theory about how species evolve from other species. You will read about Darwin's theory in the next lesson.

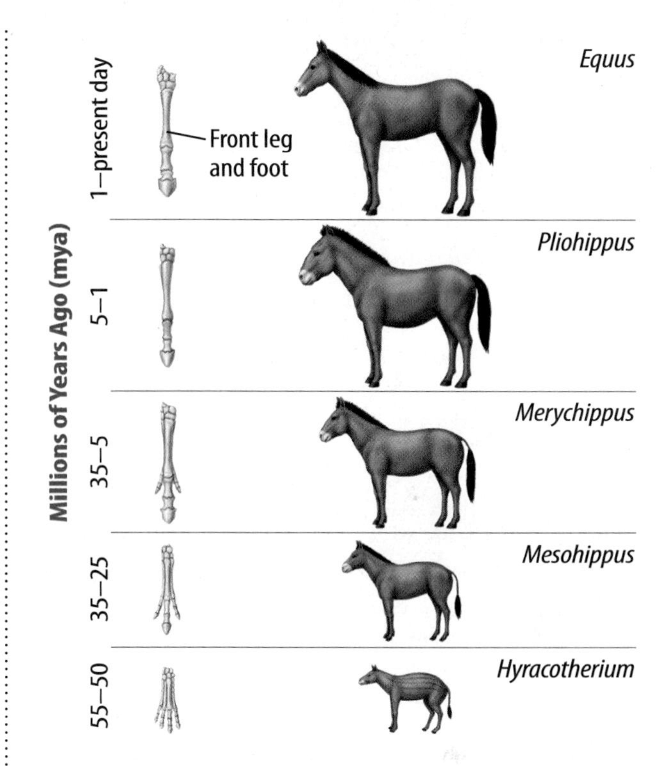

Figure 5 The fossil record is evidence that horses descended from organisms for which only fossils exist today.

14. NGSSS Check Summarize How are fossils evidence of biological evolution? SC.7.L.15.1

Inquiry SC.7.N.1.5, SC.7.L.15.1

LAB STATION **Try It!**

MiniLab ***How do species change over time?*** at connectED.mcgraw-hill.com

Apply It! After you complete the lab, answer these questions.

1. **Infer** Do you think the organism you drew in frame 6 of your comic strip is better suited for survival than its ancestor? Why or why not?

2. **Explain** How do scientists use fossils to determine if species are related to older species?

Lesson Review 1

Visual Summary

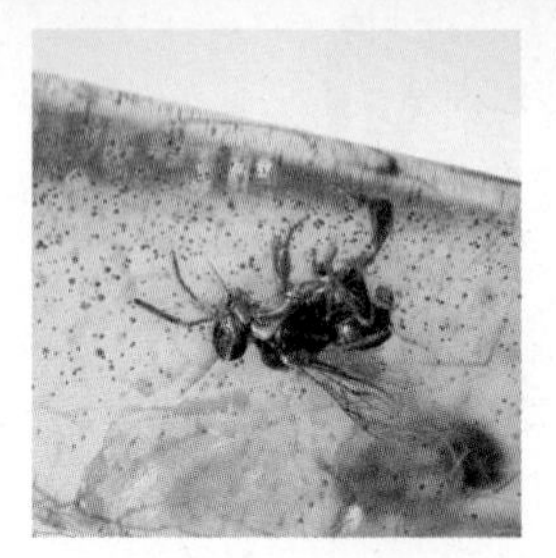

Fossils can consist of the hard parts or soft parts of organisms. Fossils can be an impression of an organism or consist of original tissues.

Scientists determine the age of a fossil through relative-age dating or absolute-age dating.

Scientists use fossils as evidence that species have changed over time.

Use Vocabulary

1. All of the fossils ever found on Earth make up the ____________.
2. If the last individual of a species cannot adapt to a changing environment and dies, ____________ occurs. SC.7.L.15.3
3. **Use the term** *biological evolution* in a sentence.

Understand Key Concepts

4. Which is the preserved evidence of the activity of an organism?
   - (A) cast
   - (B) mold
   - (C) fossil film
   - (D) trace fossil
5. **Explain** why fossil evidence is consistent with the scientific theory of evolution. SC.7.L.15.1

Interpret Graphics

6. **Identify** Fill in the table below to provide examples of changes that might lead to an extinction event. LA.6.2.2.3

Sudden changes	
Gradual changes	

Critical Thinking

7. **Infer** If the rock layers shown to the right have not been disturbed, what type of dating method would help you determine which layer is oldest? Explain. SC.6.N.1.5

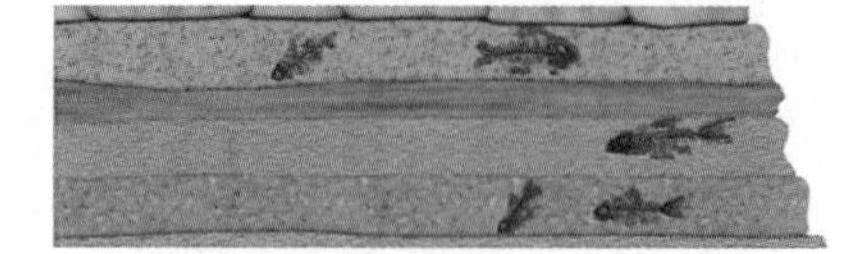

Math Skills MA.6.A.3.6

8. Dinosaurs became extinct about 65 mya. Express this number in scientific notation.

Megalodon

FOCUS on FLORIDA

The Oceans' Largest Predator

Imagine you have an ancient ancestor who was four times your size. Imagine your ancestor also had 276 razor-sharp teeth to devour its food, which included large whales, dolphins, porpoises, and giant sea turtles. Both of those things would be true if you were a great white shark.

This ancient-ancestor shark is called a Megalodon shark. Megalodon sharks lived approximately 25 to 1.5 million years ago during the Cenozoic era. Based on fossils discovered off the coast of Florida and other locations around the world, Megalodon was the oceans' largest predator. It could grow to a length of 18 to 20 meters! At that size, it is the largest shark ever known.

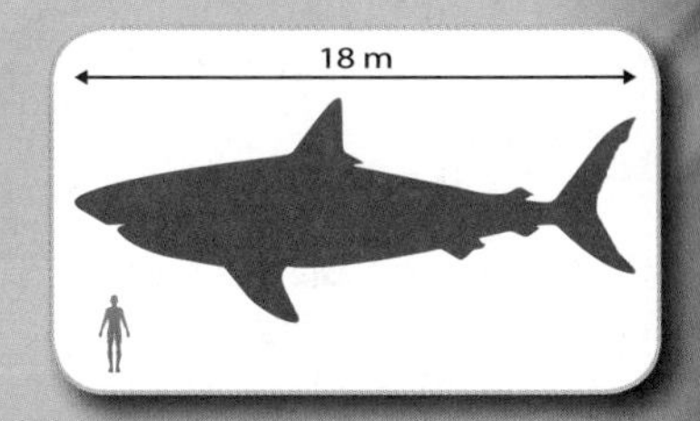

How can scientists determine the size of this ancient species? Fossils of Megalodon teeth and vertebrae are the only remaining evidence of this giant shark. Scientists use the fossilized teeth to estimate how big Megalodon was, and what it looked like. By comparing the size, shape, and angle of the teeth in the mouth, scientists can determine which current species of sharks are related to Megalodon sharks. Scientists currently hypothesize that the Megalodon shark is a distant, ancient relative of the great white shark.

Scientists can determine the biological evolution of sharks by comparing fossils for similarities and putting them in a sequence to trace how sharks have changed over time. By studying how populations of related organisms have changed over time, scientists can better understand how new species evolve from other species.

This Megalodon shark exhibit, created by the Florida Museum of Natural History in Gainesville, Florida, shows the huge size of Megalodon shark jaws. These jaws are big enough to easily swallow a human.

Growth series of Megalodon
30 to 60 feet long

It's Your Turn

RESEARCH The only remaining fossils of Megalodon sharks are their teeth. Determine which fossilization process created these amazing fossils. Conduct research about the biology of Megalodon sharks and determine why other parts of their bodies never turned into fossils. Summarize your information and present it to your class.

Lesson 2

Theory of Evolution by NATURAL SELECTION

ESSENTIAL QUESTIONS

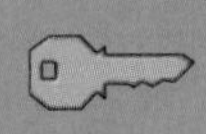
Who was Charles Darwin?

How does Darwin's theory of evolution by natural selection explain how species change over time?

How are adaptations evidence of natural selection?

Vocabulary

naturalist p. 217
variation p. 219
natural selection p. 220
adaptation p. 221
camouflage p. 222
mimicry p. 222
selective breeding p. 223

Florida NGSSS

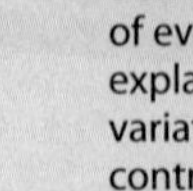

SC.7.L.15.2 Explore the scientific theory of evolution by recognizing and explaining ways in which genetic variation and environmental factors contribute to evolution by natural selection and diversity of organisms.

SC.7.L.15.3 Explore the scientific theory of evolution by relating how the inability of a species to adapt within a changing environment may contribute to the extinction of that species.

SC.6.N.2.1 Distinguish science from other activities involving thought.

SC.6.N.2.2 Explain that scientific knowledge is durable because it is open to change as new evidence or interpretations are encountered.

SC.6.N.2.3 Recognize that scientists who make contributions to scientific knowledge come from all kinds of backgrounds and possess varied talents, interests, and goals.

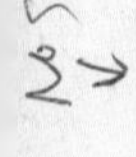

SC.6.N.3.1 Recognize and explain that a scientific theory is a well-supported and widely accepted explanation of nature and is not simply a claim posed by an individual. Thus, the use of the term theory in science is very different than how it is used in everyday life.

SC.7.N.1.7 Explain that scientific knowledge is the result of a great deal of debate and confirmation within the science community.

Also covers: SC.7.N.2.1, SC.7.N.3.1, SC.8.N.3.2, LA.6.2.2.3, MA.6.S.6.2, MA.6.A.3.6

Inquiry Launch Lab

SC.7.L.15.2, MA.6.S.6.2, MA.6.A.3.6

20 minutes

Are there variations within your class?

All populations contain variations in some characteristics of their members.

Procedure

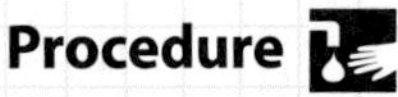

1. Read and complete a lab safety form.
2. Use a **meterstick** to measure the length from your elbow to the tip of your middle finger in centimeters. Record the measurement.
3. Add your measurement to the class list.
4. Organize all of the measurements from shortest to longest.
5. Break the data into regular increments, such as 31–35 cm, 36–40 cm, and 41–45 cm. Count the number of measurements within each increment.
6. Construct a bar graph using the data. Label each axis and give your graph a title.

Think About This

1. What are the shortest and longest measurements?

2. How much do the shortest and longest lengths vary from each other?

3. Key Concept Describe how your results provide evidence of variations within your classroom population.

1. Look closely at these zebras. Are they all exactly the same? How are they different? What accounts for these differences? How do the stripes help these organisms survive in their environments?

Active Reading **2. Explain** Who was Charles Darwin?

Charles Darwin

How many species of birds can you name? You might think of robins or chickens. Scientists estimate that about 10,000 species of birds live on Earth today. Each bird species has similar characteristics. Each has wings, feathers, and a beak. Scientists hypothesize that all birds evolved from an earlier, or ancestral, population of birdlike organisms. As this population evolved into different species, birds became different sizes and colors. They developed different songs and eating habits, but all retained similar bird characteristics.

How do birds and other species evolve? One scientist who worked to answer this question was Charles Darwin. Darwin was an English naturalist who, in the mid-1800s, developed a theory of how evolution works. *A* **naturalist** *is a person who studies plants and animals by observing them.* Darwin spent many years observing plants and animals in their natural habitats before developing his theory. Recall that a theory is an explanation of the natural world that is well supported by evidence. Darwin was not the first to develop a theory of evolution, but his theory is the one best supported by evidence today.

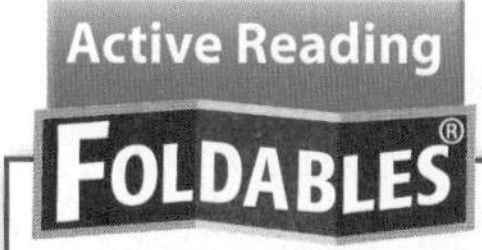

LA.6.2.2.3

Make a small, four-door shutterfold book. Use it to investigate the who, what, when, and where of Charles Darwin, the Galápagos Islands, and the theory of evolution by natural selection.

Intermediate Tortoise
- Shell shape is between dome and saddleback
- Can reach low and high vegetation

Saddleback Tortoise
- Large space between shell and neck
- Can reach high vegetation

3. **Visual Check** Infer Which characteristics does the Domed Tortoise have? Fill in the bullets below.

Domed Tortoise
- ______________________________
- ______________________________

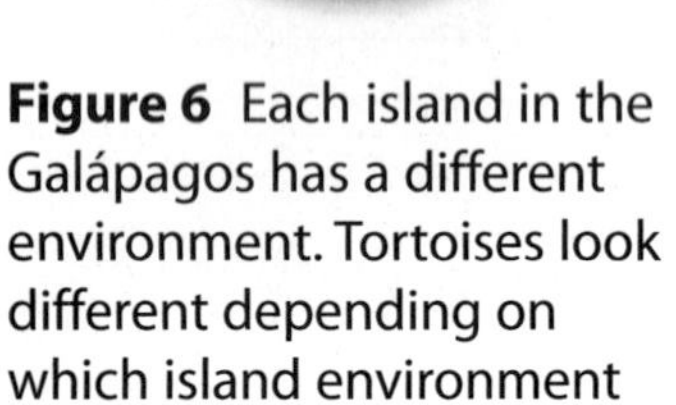

Figure 6 Each island in the Galápagos has a different environment. Tortoises look different depending on which island environment they inhabit.

Active Reading **4. Illustrate** Draw an example of the type of vegetation domed tortoises eat.

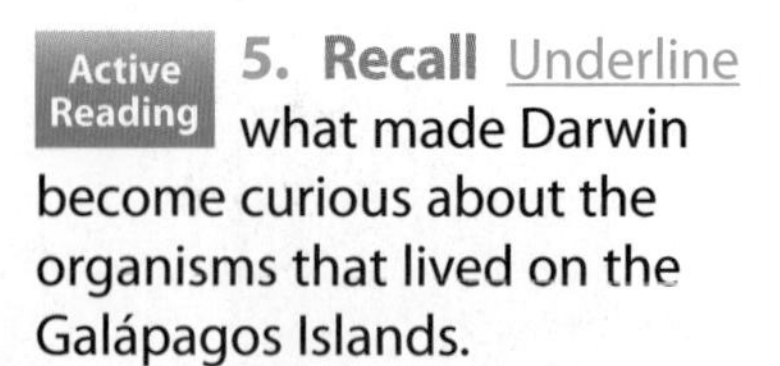

Active Reading **5. Recall** Underline what made Darwin become curious about the organisms that lived on the Galápagos Islands.

Voyage of the *Beagle*

Darwin served as a naturalist on the HMS *Beagle,* a survey ship of the British navy. During his voyage around the world, Darwin observed and collected many plants and animals.

The Galápagos Islands

Darwin was especially interested in the organisms he saw on the Galápagos (guh LAH puh gus) Islands. These islands are located 1,000 km off the South American coast in the Pacific Ocean. Darwin saw that each island had a slightly different environment. Some were dry. Some were more humid. Others had mixed environments.

Tortoises Giant tortoises lived on many of the islands. When a resident told him that the tortoises on each island looked different, as shown in **Figure 6,** Darwin became curious.

Mockingbirds and Finches Darwin also became curious about the variety of mockingbirds and finches he saw and collected on the islands. Like the tortoises, different types of mocking-birds and finches lived in different island environments. Later, he was surprised to learn that many of these varieties were different enough to be separate species.

Darwin's Theory

Darwin realized there was a relationship between each species and the food sources of the island it lived on. Look again at **Figure 6.** You can see that tortoises with long necks lived on islands that had tall cacti. Their long necks enabled them to reach high to eat the cacti. The tortoises with short necks lived on islands that had plenty of short grass.

Common Ancestors

Darwin became convinced that all the tortoise species were related. He thought they all shared a common ancestor. He suspected that a storm had carried a small ancestral tortoise population to one of the islands from South America millions of years before. Eventually, the tortoises spread to the other islands. Their neck lengths and shell shapes changed to match their islands' food sources. How did this happen?

Variations

Darwin knew that individual members of a species exhibit slight differences, or variations. *A* **variation** *is a slight difference in an inherited trait of individual members of a species.* Even though the snail shells in **Figure 7** are not all exactly the same, they are all from snails of the same species. You can also see variations in the zebras in the photo at the beginning of this lesson. Variations arise naturally in populations. They occur in the offspring as a result of sexual reproduction. You might recall that variations are caused by random mutations, or changes, in genes. Mutations can lead to changes in phenotype. Recall that an organism's phenotype is all of the observable traits and characteristics of the organism. Genetic changes to phenotype can be passed on to future generations.

Figure 7 The variations among the shells of a species of tree snail occur naturally within the population.

Active Reading **6. Describe** (Circle) three snail shells in the image above. Describe each variation below.

Natural Selection

Darwin realized variations were the key to how populations of tortoises and other organisms evolved. Darwin understood that food is a limiting resource. That means the food on each island could not support every tortoise that was born. Tortoises had to compete with each other for food. As the tortoises spread to other islands, some were born with random variations in neck length. If a variation benefited a tortoise, allowing it to compete for food better than other tortoises, the tortoise lived longer. Because it lived longer, it reproduced more. It passed on its variations to its offspring. This is Darwin's theory of evolution by natural selection, as illustrated in **Figure 8.**

Natural selection *is the process by which populations of organisms with variations that help them survive in their environments live longer, compete better, and reproduce more than those that do not have the variations.* Natural selection has four main ideas that explain how populations change as their environments change. First, individuals in a population show variations. Second, variations can be passed from parent to offspring. Third, organisms have more offspring that can survive on the available resources. Finally, the variations that help an organism survive and reproduce are more likely to be passed on to future generations.

7. NGSSS Check Summarize What role do variations have in the theory of evolution by natural selection? SC.7.L.15.2

____

____

____

____

Natural Selection

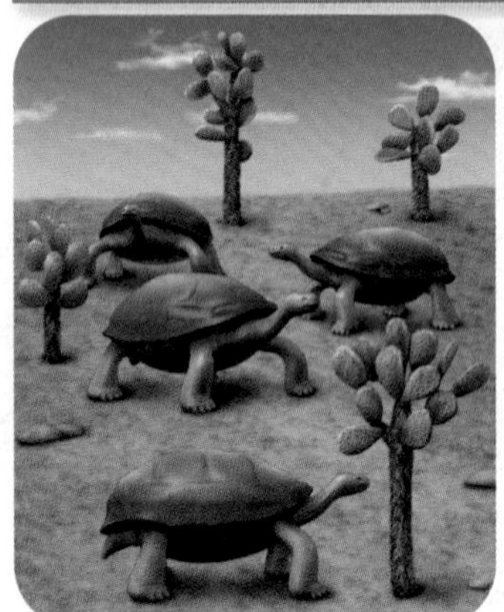

1 Reproduction A population of tortoises produces many offspring that inherit its characteristics.

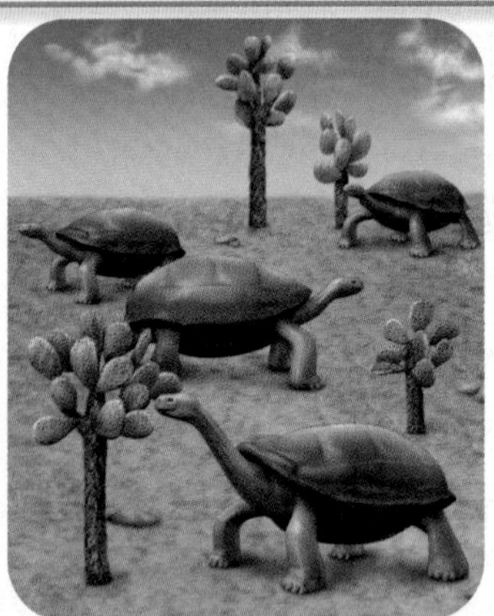

2 Variation A tortoise is born with a variation that makes its neck slightly longer.

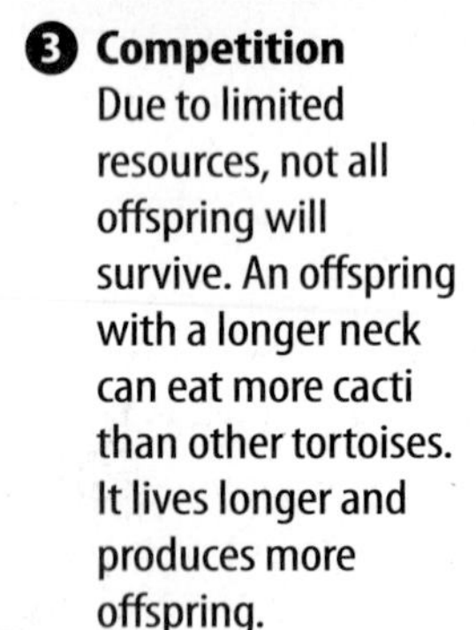

3 Competition Due to limited resources, not all offspring will survive. An offspring with a longer neck can eat more cacti than other tortoises. It lives longer and produces more offspring.

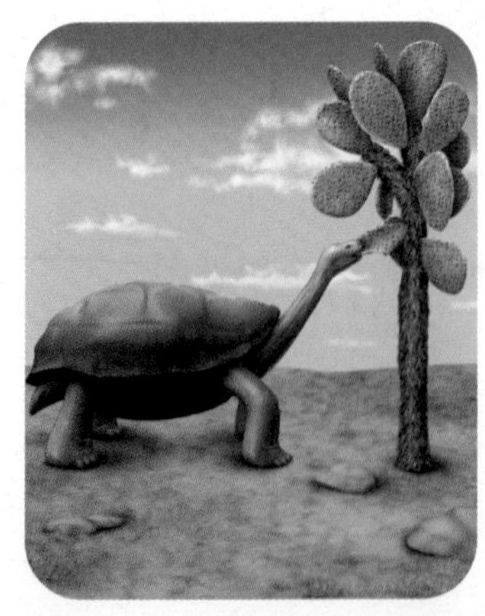

4 Selection Over time, the variation is inherited by more and more offspring. Eventually, all tortoises have longer necks.

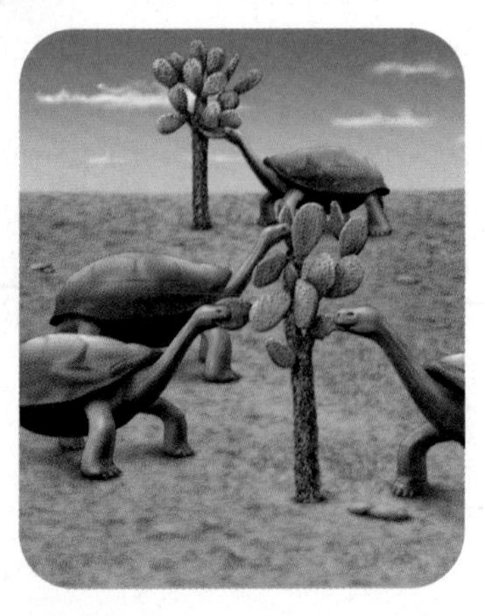

Figure 8 A beneficial variation in neck length spreads through a tortoise population by natural selection.

Adaptations

Natural selection explains how all species change over time as their environments change. Through natural selection, a helpful variation in one individual can be passed on to future members of a population. As time passes, more variations arise. The accumulation of many similar variations can lead to an adaptation (a dap TAY shun). *An* **adaptation** *is an inherited trait that increases an organism's chance of surviving and reproducing in its environment.* The long neck of certain species of tortoises is an adaptation to an environment with tall cacti.

WORD ORIGIN

adaptation
from Latin *adaptare,* means "to fit"

Types of Adaptations

Every species has many adaptations. Scientists classify adaptations into three categories: structural, behavioral, and functional. Structural adaptations involve color, shape, and other physical characteristics. The shape of a tortoise's neck is a structural adaptation. Behavioral adaptations involve the way an organism behaves or acts. Hunting at night and moving in herds are examples of behavioral adaptations. Functional adaptations involve internal body systems that affect biochemistry. A drop in body temperature during hibernation is an example of a functional adaptation. **Figure 9** illustrates examples of all three types of adaptations in the desert jackrabbit.

8. NGSSS Check
Explain How do variations lead to adaptions? SC.7.L.15.2

____

____

____

____

____

Figure 9 The desert jackrabbit has structural, behavioral, and functional adaptations. These adaptations enable it to survive in its desert environment.

Active Reading **9. Determine** Identify the type of adaptation for each jackrabbit in the spaces below.

The jackrabbit's powerful legs help it run fast to escape from predators. This is a ______________________ adaptation.

The jackrabbit stays still during the hottest part of the day, helping it conserve energy. This is a ______________________ adaptation.

The blood vessels in the jackrabbit's ears expand to enable the blood to cool before reentering the body. This is a ______________________ adaptation.

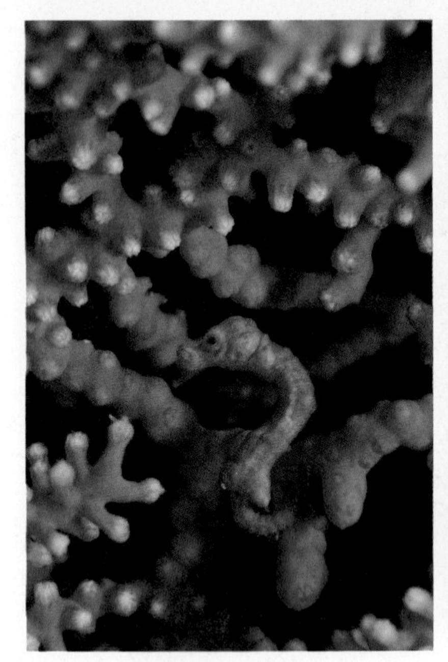

Sea Horse

Caterpillar

Pelican

Figure 10 Species evolve adaptations as they interact with their environments, which include other species.

Active Reading **10. Contrast** How do camouflage and mimicry differ?

____

____

____

Environmental Interactions

Have you ever wanted to be invisible? Many species have evolved adaptations that make them nearly invisible. The sea horse in **Figure 10** is the same color and has a texture similar to the coral it is resting on. This is a structural adaptation called camouflage (KAM uh flahj). **Camouflage** *is an adaptation that enables a species to blend in with its environment.*

Some species have adaptations that draw attention to them. The caterpillar in **Figure 10** resembles a snake. Predators see it and are scared away. *The resemblance of one species to another species is* **mimicry** (MIH mih kree). Camouflage and mimicry are adaptations that help species avoid being eaten. Many other adaptations help species eat. The pelican in **Figure 10** has a beak and mouth that are uniquely adapted to its food source—fish.

Environments are complex. Species must adapt to an environment's living parts as well as to an environment's nonliving parts. Nonliving things include temperature, water, nutrients in soil, and climate. Deciduous trees shed their leaves due to changes in climate. Camouflage, mimicry, and mouth shape are adaptations mostly to an environment's living parts. An extreme example of two species adapting to each other is shown in **Figure 11.**

Living and nonliving factors are always changing. Even slight environmental changes affect how species adapt. If a species is unable to adapt, it becomes extinct. The fossil record contains many fossils of species that were unable to adapt to change.

Figure 11 This orchid and its moth pollinator have evolved so closely together that one cannot exist without the other.

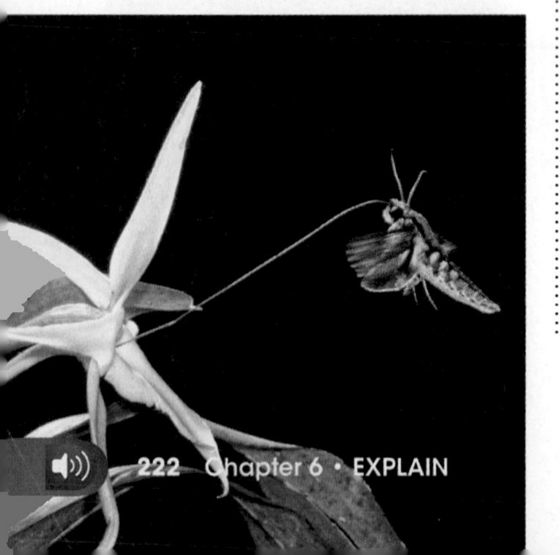

Artificial Selection

Adaptations provide evidence of how closely Earth's species match their environments. This is exactly what Darwin's theory of evolution by natural selection predicted. Darwin provided many examples of adaptation in *On the Origin of Species,* the book he wrote to explain his theory. Darwin did not write this book until 20 years after he developed his theory. He spent those years collecting more evidence for his theory by studying barnacles, orchids, corals, and earthworms.

Darwin also had a hobby of breeding domestic pigeons. He selectively bred pigeons of different colors and shapes to produce new, fancy varieties. *The breeding of organisms for desired characteristics is called* **selective breeding.** Like many domestic plants and animals produced from selective breeding, pigeons look different from their ancestors, as shown in **Figure 12.** Darwin realized that changes caused by selective breeding were much like changes caused by natural selection. Instead of nature selecting variations, humans selected them. Darwin called this process artificial selection.

Artificial selection explains and supports Darwin's theory. As you will read in Lesson 3, other evidence also supports the idea that species evolve from other species.

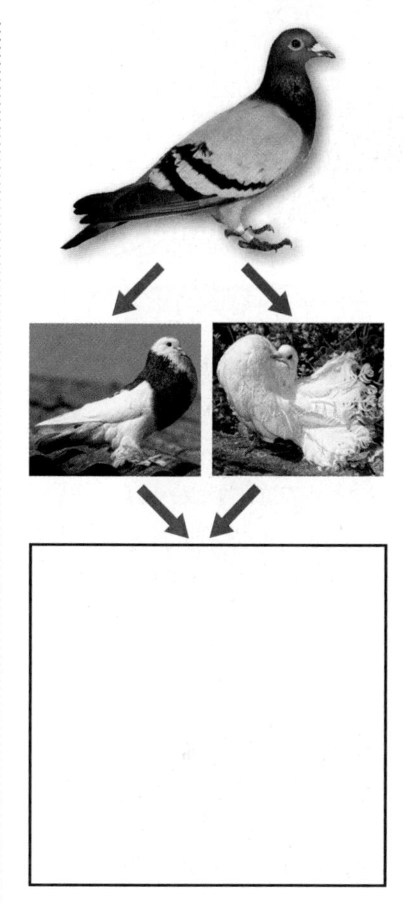

Figure 12 The pouter pigeon (above bottom left) and the fantail pigeon (above bottom right) were derived from the wild rock pigeon (above top).

Active Reading **11. Predict** What would happen if the pouter pigeon and the fantail pigeon were selectively bred? Draw the results in the space provided.

Inquiry SC.6.N.2.1 SC.7.L.15.2

LAB STATION **Try It!**

MiniLab ***Who survives?*** at connectED.mcgraw-hill.com

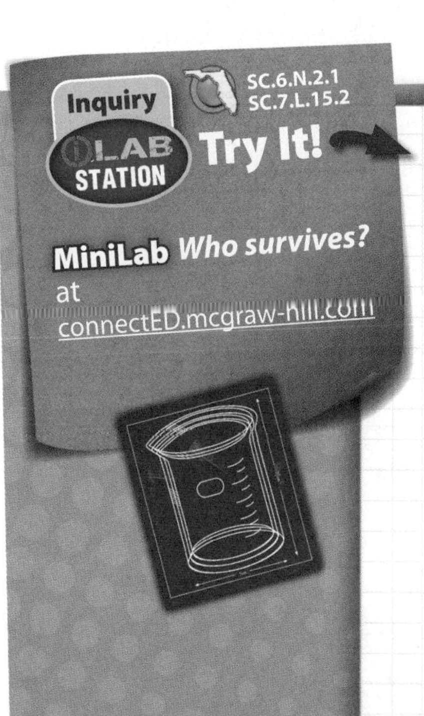

Apply It! After you complete the lab, answer these questions.

1. **Infer** For the moths that were not spotted, how did the color of their camouflage compare to the color of the environment where they were placed?

2. **Predict** What kind of camouflage would moths develop if they started living in the desert?

Lesson Review 2

Visual Summary

Charles Darwin developed his theory of evolution partly by observing organisms in their natural environments.

Natural selection occurs when organisms with certain variations live longer, compete better, and reproduce more often than organisms that do not have the variations.

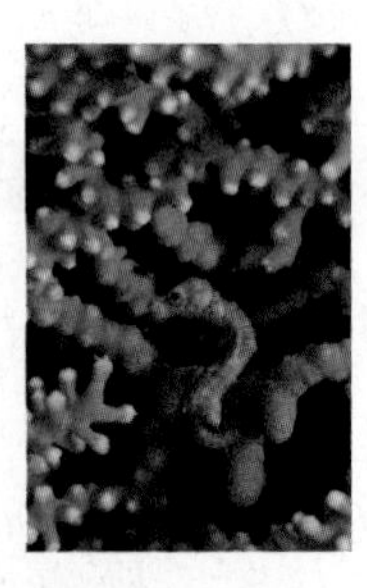

Adaptations occur when a beneficial variation is eventually inherited by all members of a population.

Use Vocabulary

1. A person who studies plants and animals by observing them is a(n) ________________.
2. Through __________, populations of organisms adapt to their environments. SC.7.L.15.2
3. Some species blend in to their environments through __________.

Understand Key Concepts

4. The observation that the Galápagos tortoises did not all live in the same environment helped Darwin SC.8.N.3.2
   - (A) develop his theory of adaptation.
   - (B) develop his theory of evolution.
   - (C) observe mimicry in nature.
   - (D) practice artificial selection.

5. **Assess** the importance of variations to natural selection. SC.7.L.15.2

6. **Compare and contrast** natural selection and artificial selection.

Interpret Graphics

7. **Sequence** Use the graphic organizer below to sequence the steps by which a population of organisms changes by natural selection. LA.6.2.2.3

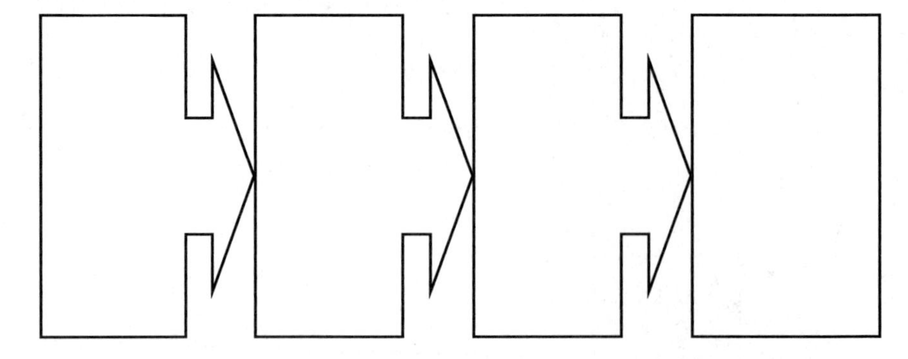

Critical Thinking

8. **Conclude** how Earth's birds developed their diversity through natural selection. SC.7.L.15.2

Peter and Rosemary Grant

CAREERS in SCIENCE

Observing Natural Selection

Charles Darwin was a naturalist during the mid-1800s. Based on his observations of nature, he developed the theory of evolution by natural selection. Do scientists still work this way—drawing conclusions from observations? Is there information still to be learned about natural selection? The answer to both questions is yes.

Peter and Rosemary Grant are naturalists who have observed finches in the Galápagos Islands for more than 30 years. They have found that variations in the finches' food supply determine which birds will survive and reproduce. They have observed natural selection in action.

The Grants live on Daphne Major, an island in the Galápagos, for part of each year. They observe and take measurements to compare the size and shape of finches' beaks from year to year. They also examine the kinds of seeds and nuts available for the birds to eat. They use this information to relate changes in the birds' food supply to changes in the finch species' beaks.

The island's ecosystem is fragile, so the Grants take great care not to change the environment of Daphne Major as they observe the finches. They carefully plan their diet to avoid introducing new plant species to the island. They bring all the freshwater they need to drink, and they wash in the ocean. For the Grants, it's just part of the job. As naturalists, they try to observe without interfering with the habitat in which they are living.

▲ Peter and Rosemary Grant make observations and collect data in the field.

▲ This large ground finch is one of the kinds of birds studied by the Grants.

It's Your Turn

RESEARCH AND REPORT Find out more about careers in evolution, ecology, or population biology. What kind of work is done in the laboratory? What kind of work is done in the field? Write a report to explain your findings.

Lesson 3

Biological Evidence of EVOLUTION

ESSENTIAL QUESTIONS

What evidence from living species supports the theory that species descended from other species over time?

How are Earth's organisms related?

Vocabulary

comparative anatomy p. 228

homologous structure p. 228

analogous structure p. 229

vestigial structure p. 229

embryology p. 230

Florida NGSSS

LA.6.2.2.3 The student will organize information to show understanding (e.g., representing main ideas within text through charting, mapping, paraphrasing, summarizing, or comparing/contrasting);

SC.6.N.1.5 Recognize that science involves creativity, not just in designing experiments, but also in creating explanations that fit evidence.

SC.6.N.3.1 Recognize and explain that a scientific theory is a well-supported and widely accepted explanation of nature and is not simply a claim posed by an individual. Thus, the use of the term theory in science is very different than how it is used in everyday life.

SC.7.L.15.2 Explore the scientific theory of evolution by recognizing and explaining ways in which genetic variation and environmental factors contribute to evolution by natural selection and diversity of organisms.

SC.8.N.1.6 Understand that scientific investigations involve the collection of relevant empirical evidence, the use of logical reasoning, and the application of imagination in devising hypotheses, predictions, explanations and models to make sense of the collected evidence.

SC.8.N.3.2 Explain why theories may be modified but are rarely discarded.

MA.6.A.3.6 Construct and analyze tables, graphs, and equations to describe linear functions and other simple relations using both common language and algebraic notation.

SC.6.N.1.5

Launch Lab

15 minutes

How is the structure of a spoon related to its function?

Would you eat your morning cereal with a spoon that had holes in it? Is using a teaspoon the most efficient way to serve mashed potatoes and gravy to a large group of people? How about using an extra large spoon, or ladle, to eat soup from a small bowl?

Procedure

1. Read and complete a lab safety form.
2. In a small group, examine your **set of spoons** and discuss your observations.
3. Sketch or describe the structure of each spoon below. Discuss the purpose that each spoon shape might serve.
4. Label the spoons with their purposes.

Think About This

1. Describe the similarities and differences among the spoons.

2. If spoons were organisms, what do you think the ancestral spoon would look like?

3. Key Concept Explain how three of the spoons have different structures and functions, even though they are related by their similarities.

Inquiry Do you think this bird can fly?

1. Some birds, such as the flightless cormorant above, have wings but cannot fly. Their wings are too small to support their bodies in flight. Why do you think they still have wings? What do you think scientists can learn about the ancestors of present-day birds that have wings but do not fly?

Active Reading **2. Confirm** Which horse is the common ancestor to all horse species in the graph in **Figure 13?** How do you know?

Evidence for Evolution

Recall the sequence of horse fossils from Lesson 1. The sequence might have suggested to you that horses evolved in a straight line—that one species replaced another in a series of orderly steps. Evolution does not occur this way. The diagram in **Figure 13** shows a more realistic version of horse evolution, which looks more like a bush than a straight line. Different horse species were sometimes alive at the same time. They are related to each other because each descended from a common ancestor.

Living species that are closely related share a close common ancestor. The degree to which species are related depends on how closely in time they diverged, or split, from their common ancestor. Although the fossil record is incomplete, it contains many examples of fossil sequences showing close ancestral relationships. Living species show evidence of common ancestry, too.

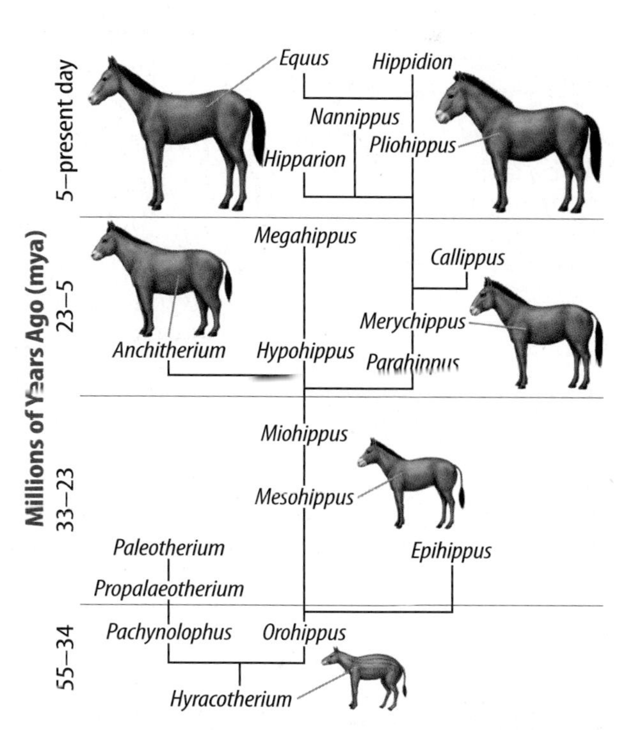

Figure 13 The fossil record indicates that different species of horses often overlapped with each other.

Homologous Structures

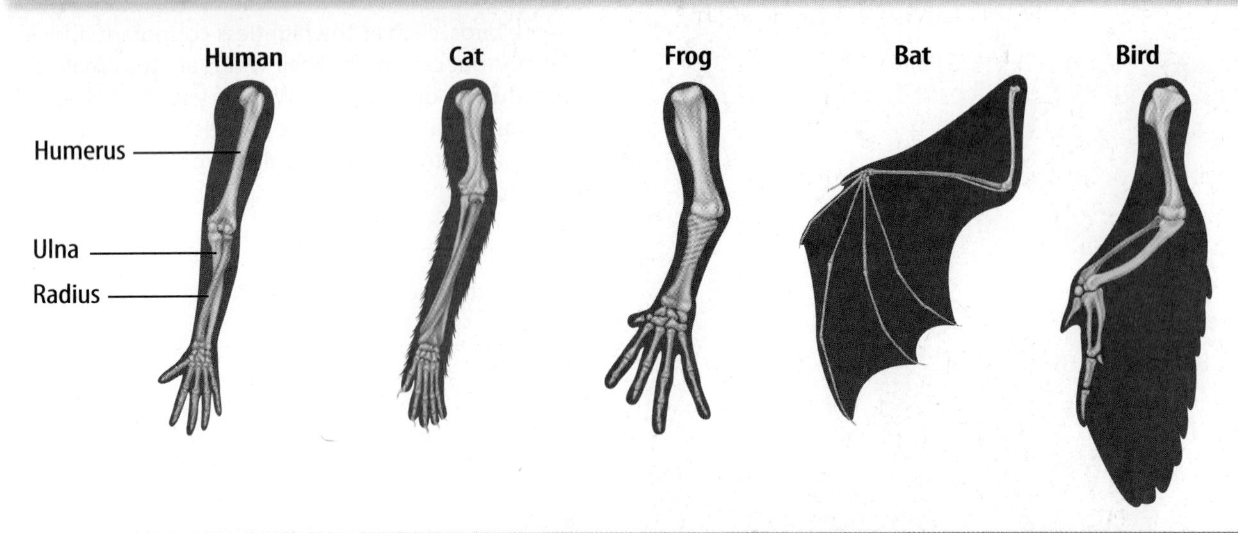

Figure 14 The forelimbs of these species are different sizes, but their placement and structure suggest common ancestry.

Comparative Anatomy

Common ancestry is not difficult to see in many species. For example, it might seem easy to tell that robins, finches, and hawks evolved from a common ancestor. They all have similar features, such as feathers, wings, and beaks. The same is true for tigers, leopards, and house cats. But how are hawks related to cats? How are both hawks and cats related to frogs and bats? Observations of structural and functional similarities and differences in species that do not look alike are possible through comparative anatomy. **Comparative anatomy** *is the study of similarities and differences among structures of living species.*

Homologous Structures Humans, cats, frogs, bats, and birds look different and move in different ways. Humans use their arms for balance and their hands to grasp objects. Cats use their forelimbs to walk, run, and jump. Frogs use their forelimbs to jump. Bats and birds use their forelimbs as wings for flying. However, the forelimb bones of these species exhibit similar patterns, as shown in **Figure 14. Homologous** (huh MAH luh gus) **structures** *are body parts of organisms that are similar in structure and position but different in function.*

Homologous structures, such as the forelimbs of humans, cats, frogs, bats, and birds, suggest that these species are related. The more similar two structures are to each other, the more likely it is that the species have evolved from a recent common ancestor.

Active Reading **3. Summarize** How do homologous structures provide evidence for evolution?

Active Reading
FOLDABLES® LA.6.2.2.3

Make a table with five rows and three columns. Label the rows and columns of the table as shown below. Give your table a title.

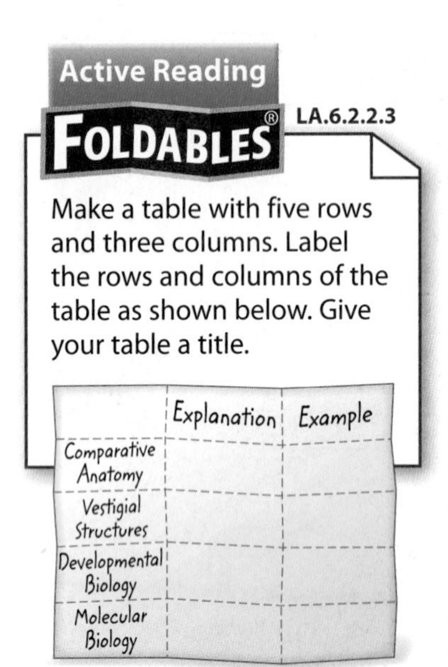

Analogous Structures Can you think of a body part in two species that serves the same purpose but differs in structure? How about the wings of birds and flies? Both wings in **Figure 15** are used for flight. But bird wings are covered with feathers. Fly wings are covered with tiny hairs. *Body parts that perform a similar function but differ in structure are* **analogous** (uh NAH luh gus) **structures.** Differences in the structure of bird and fly wings indicate that birds and flies are not closely related.

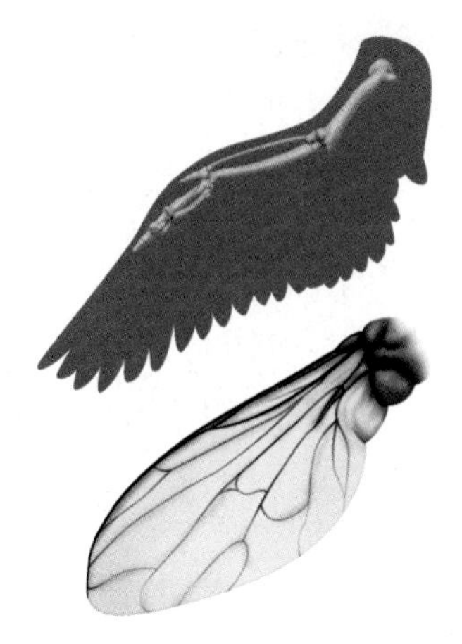

Figure 15 Though used for the same function—flight—the wings of birds (top) and insects (bottom) are too different in structure to suggest close common ancestry.

Vestigial Structures

The bird in the photo at the beginning of this lesson has short, stubby wings. Yet it cannot fly. The bird's wings are an example of vestigial structures. **Vestigial** (veh STIH jee ul) **structures** *are body parts that have lost their original function through evolution.* The best explanation for vestigial structures is that the species with a vestigial structure is related to an ancestral species that used the structure for a specific purpose.

The whale shown in **Figure 16** has tiny pelvic bones inside its body. The presence of pelvic bones in whales suggests that whales descended from ancestors that used legs for walking on land. The fossil evidence supports this conclusion. Many fossils of whale ancestors show a gradual loss of legs over millions of years. They also show, at the same time, that whale ancestors became better adapted to their watery environments.

Active Reading **4. Check** How are vestigial structures evidence of descent from ancestral species?

____

____

Figure 16 Present-day whales have vestigial structures in the form of small pelvic bones.

Between 50–40 million years ago, this mammal breathed air and walked clumsily on land. It spent a lot of time in water, but swimming was difficult because of its rear legs. Individuals born with variations that made their rear legs smaller lived longer and reproduced more. This mammal is an ancestor of modern whales.

Pelvis

Ambulocetus natans

Vestigial pelvis

Modern toothed whale

After 10–15 million more years of evolution, the ancestors of modern whales could not walk on land. They were adapted to an aquatic environment. Modern whales have two small vestigial pelvic bones that no longer support legs.

Pharyngeal Pouches

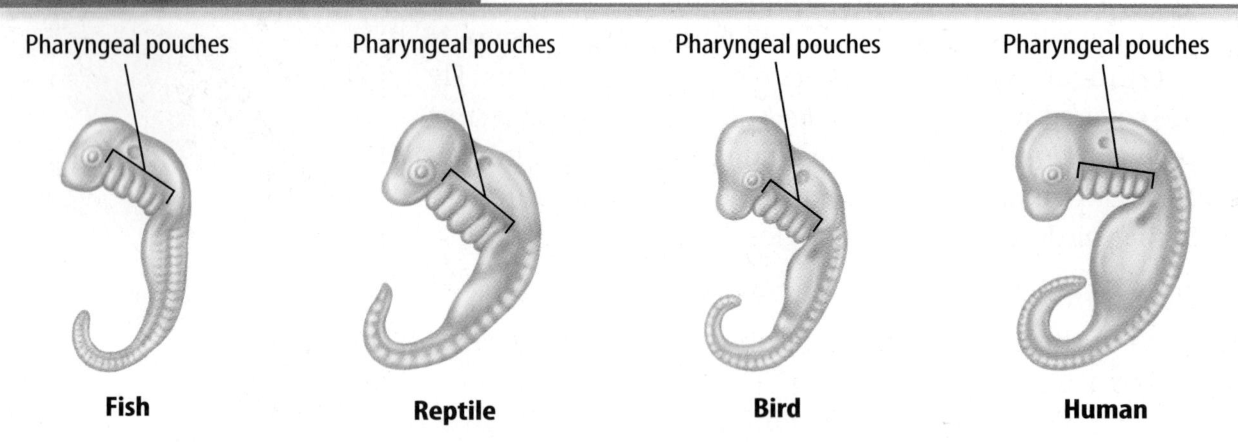

Figure 17 All vertebrate embryos exhibit pharyngeal pouches at a certain stage of their development. These features, which develop into neck and face parts, suggest relatedness.

WORD ORIGIN

embryology
from Greek *embryon,* means "to swell," and from Greek *logia,* means "study of"

Active Reading **5. Summarize** How do pharyngeal pouches provide evidence of relationships among species?

____
____
____
____
____
____

Developmental Biology

You have just read that studying the internal structures of organisms can help scientists learn more about how organisms are related. Studying the development of embryos can also provide scientists with evidence that certain species are related. *The science of the development of embryos from fertilization to birth is called* **embryology** (em bree AH luh jee).

Pharyngeal Pouches Embryos of different species often resemble each other at different stages of their development. For example, all vertebrate embryos have pharyngeal (fuh rihn JEE ul) pouches at one stage, as shown in **Figure 17.** This feature develops into different body parts in each vertebrate. Yet, in all vertebrates, each part is in the face or neck. For example, in reptiles, birds, and humans, part of the pharyngeal pouch develops into a gland in the neck that regulates calcium. In fish, the same part becomes the gills. One function of gills is to regulate calcium. The similarities in function and location of gills and glands suggest a strong evolutionary relationship between fish and other vertebrates.

Active Reading **6. Sequence** the development of the pharyngeal pouch in different species. Express the conclusion of scientists in embryology.

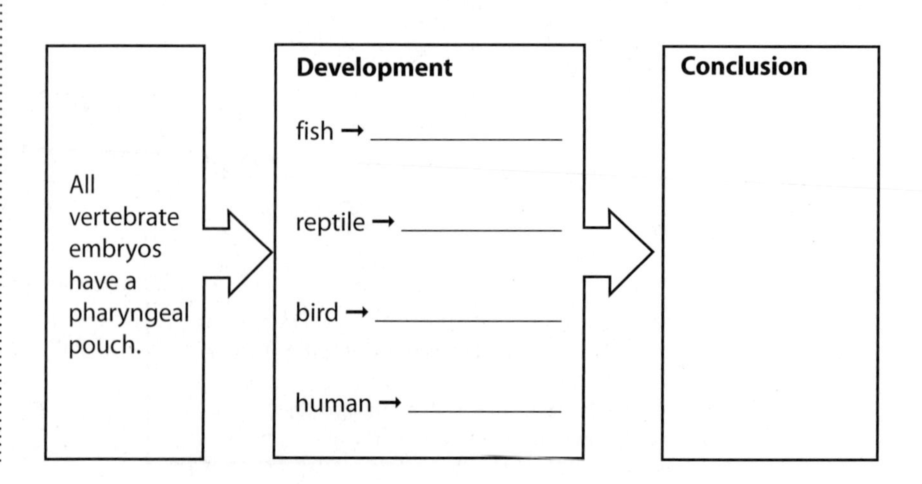

Molecular Biology

Studies of fossils, comparative anatomy, and embryology provide support for Darwin's theory of evolution by natural selection. Molecular biology is the study of gene structure and function. Discoveries in molecular biology have confirmed much of the data collected about the theory of evolution. Darwin did not know about genes, but scientists today know gene mutations are the source of variations upon which natural selection acts. Genes provide powerful support for evolution.

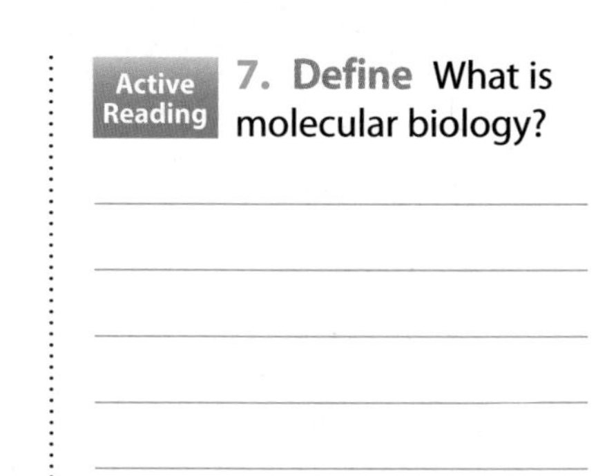

Active Reading **7. Define** What is molecular biology?

Comparing Sequences All organisms on Earth have genes. All genes are made of DNA, and all genes work in similar ways. This supports the idea that all organisms are related. Scientists can study relatedness of organisms by comparing genes and proteins among living species. For example, nearly all organisms contain a gene that codes for cytochrome *c*, a protein required for cellular respiration. Some species, such as humans and rhesus monkeys, have nearly identical cytochrome *c*. The more closely related two species are, the more similar their genes and proteins are.

Active Reading **8. Explain** How is molecular biology used to determine relationships among species?

Divergence Scientists have found that some stretches of shared DNA mutate at regular, predictable rates. Scientists use this "molecular clock" to estimate at what time in the past living species diverged from common ancestors. For example, as shown in **Figure 18,** molecular data indicate that whales and porpoises are more closely related to hippopotamuses than they are to any other living species.

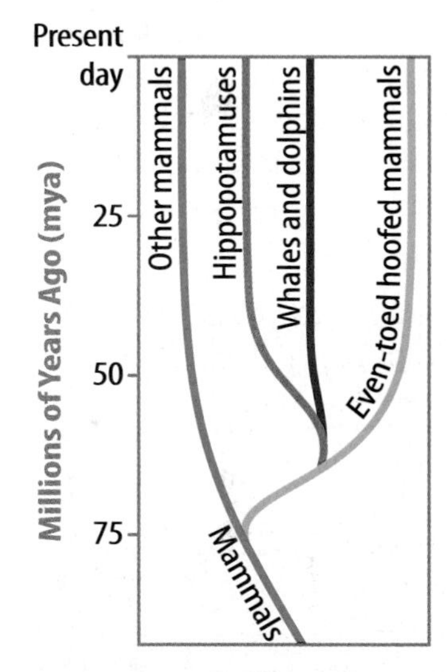

Figure 18 Whales and hippopotamuses share an ancestor that lived 50–60 mya.

Apply It! After you complete the lab, answer these questions.

1. **Infer** Which two organisms listed have the fewest differences between their cytochrome c genes?

2. **Predict** Using your knowledge of the physical appearance of the organisms listed in the graph, do you think cytochrome c is a good indicator of species relatedness? Why or why not?

Figure 19 Many scientists think that natural selection produces new species slowly and steadily. Other scientists think species exist stably for long periods, then change occurs in short bursts.

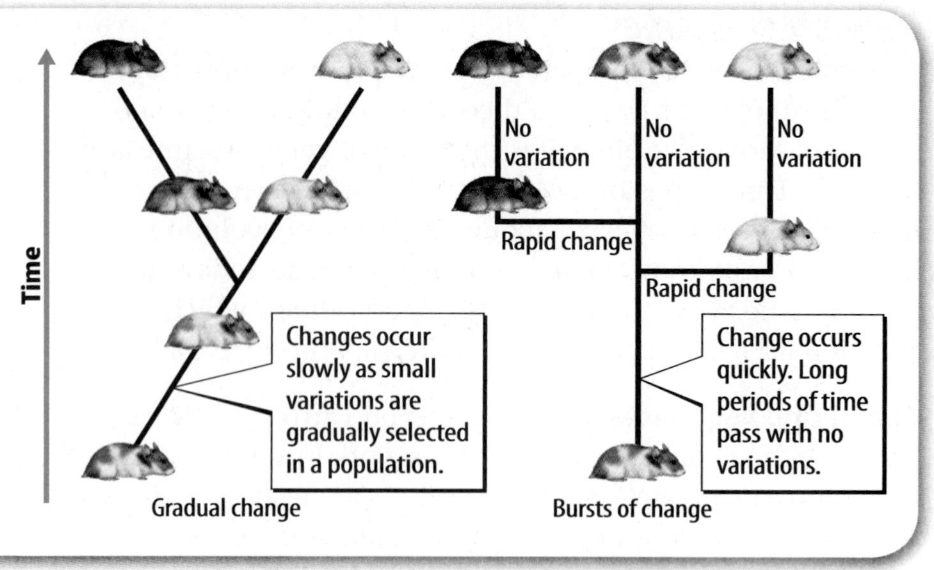

Figure 20 *Tiktaalik* lived 385–359 mya. Like amphibians, it had wrists and lungs. Like fish, it had fins, gills, and scales. Scientists think it is an intermediate species linking fish and amphibians.

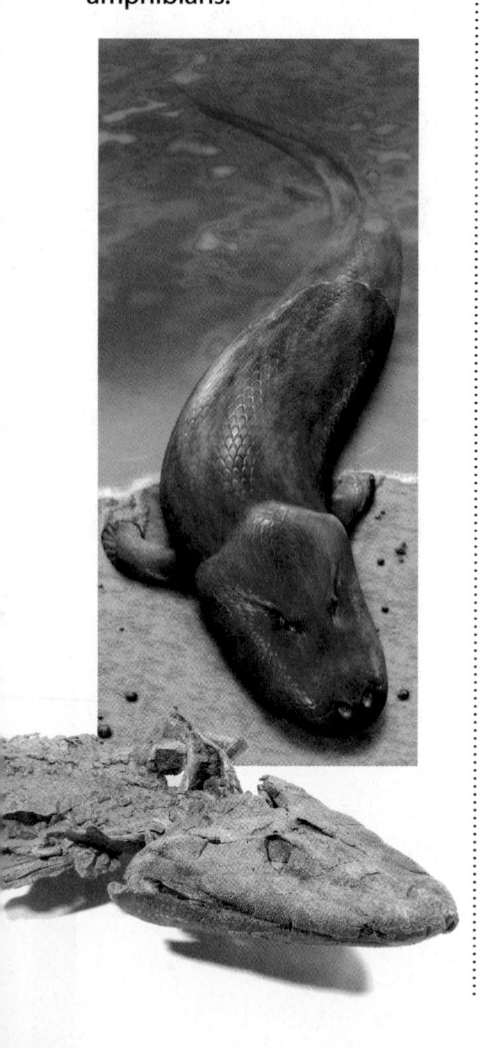

The Study of Evolution Today

The theory of evolution by natural selection is the cornerstone of modern biology. Since Darwin published his theory, scientists have confirmed, refined, and extended Darwin's work. They have observed natural selection in hundreds of living species. Their studies of fossils, anatomy, embryology, and molecular biology have all provided evidence of relatedness among living and extinct species.

How New Species Form

New evidence supporting the theory of evolution by natural selection is discovered nearly every day. But scientists debate some of the details. **Figure 19** shows that scientists have different ideas about the rate at which natural selection produces new species—slowly and gradually or quickly, in bursts. The origin of a species is difficult to study on human time scales. It is also difficult to study in the incomplete fossil record. Yet, new fossils that have features of species that lived both before them and after them are discovered all the time. For example, the *Tiktaalik* fossil shown in **Figure 20** has both fish and amphibian features. Further fossil discoveries will help scientists study more details about the origin of new species.

Diversity

How evolution has produced Earth's wide diversity of organisms using the same basic building blocks—genes—is an active area of study in evolutionary biology. Scientists are finding that genes can be reorganized in simple ways and give rise to dramatic changes in organisms. Though scientists now study evolution at the molecular level, the basic principles of Darwin's theory of evolution by natural selection have remained unchanged for over 150 years.

Lesson Review 3

Visual Summary

By comparing the anatomy of organisms and looking for homologous or analogous structures, scientists can determine if organisms had a common ancestor.

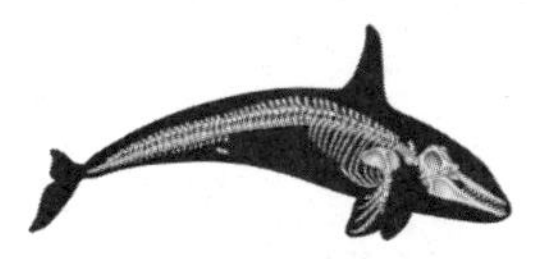

Some organisms have vestigial structures, suggesting that they descended from a species that used the structure for a purpose.

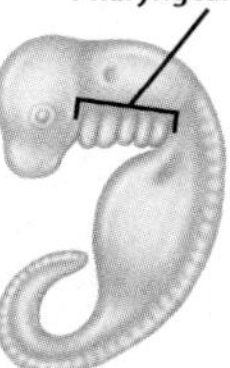

Scientists use evidence from developmental and molecular biology to help determine if organisms are related.

Use Vocabulary

1. Define *embryology* in your own words.

2. Distinguish between a homologous structure and an analogous structure.

3. Use the term *vestigial structure* in a complete sentence.

Understand Key Concepts

4. Scientists use molecular biology to determine how two species are related by comparing the genes in one species to genes
   - (A) in extinct species.
   - (B) in human species.
   - (C) in related species.
   - (D) in related fossils.

5. **Explain** why vestigial structures in whale fossils support the theory of evolution. SC.6.N.1.5

Interpret Graphics

6. **Assess** Fill in the graphic organizer below to identify four areas of study that provide evidence for evolution.

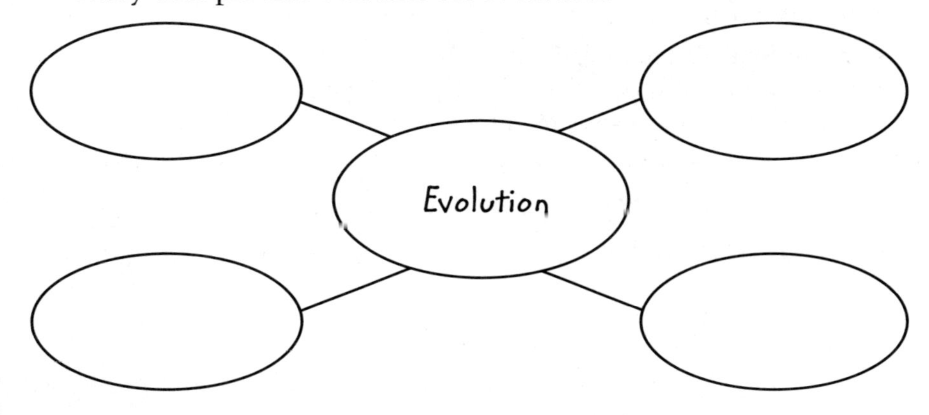

Critical Thinking

7. **Predict** what a fossil that illustrates the evolution of a bird from a reptile might look like. SC.8.N.1.6

Chapter 6 Study Guide

Think About It! Natural selection is a primary mechanism leading to change over time in organisms. Through natural selection, species adapt to changing environments.

Key Concepts Summary

LESSON 1 Fossil Evidence of Evolution

- Fossils form in many ways, including mineral replacement, carbonization, and impressions in sediment.
- Scientists can learn the ages of fossils by techniques of relative-age dating and absolute-age dating.
- Though incomplete, the **fossil record** contains patterns suggesting the **biological evolution** of related species.

Vocabulary

fossil record p. 207
mold p. 209
cast p. 209
trace fossil p. 209
geologic time scale p. 211
extinction p. 212
biological evolution p. 213

LESSON 2 Theory of Evolution by Natural Selection

- The 19th century **naturalist** Charles Darwin developed a theory of evolution that is still studied today.
- Darwin's theory of evolution by **natural selection** is the process by which populations with **variations** that help them survive in their environments live longer and reproduce more than those without beneficial variations. Over time, beneficial variations spread through populations, and new species that are adapted to their environments evolve.
- **Camouflage, mimicry,** and other **adaptations** are evidence of the close relationships between species and their changing environments.

naturalist p. 217
variation p. 219
natural selection p. 220
adaptation p. 221
camouflage p. 222
mimicry p. 222
selective breeding p. 223

LESSON 3 Biological Evidence of Evolution

- Fossils provide only one source of evidence of evolution. Additional evidence comes from living species, including studies in **comparative anatomy, embryology,** and molecular biology.
- Through evolution by natural selection, all of Earth's organisms are related. The more recently they share a common ancestor, the more closely they are related.

comparative anatomy p. 228
homologous structure p. 228
analogous structure p. 229
vestigial structure p. 229
embryology p. 230

Active Reading

FOLDABLES® Chapter Project

Assemble your lesson Foldables as shown to make a Chapter Project. Use the project to review what you have learned in this chapter.

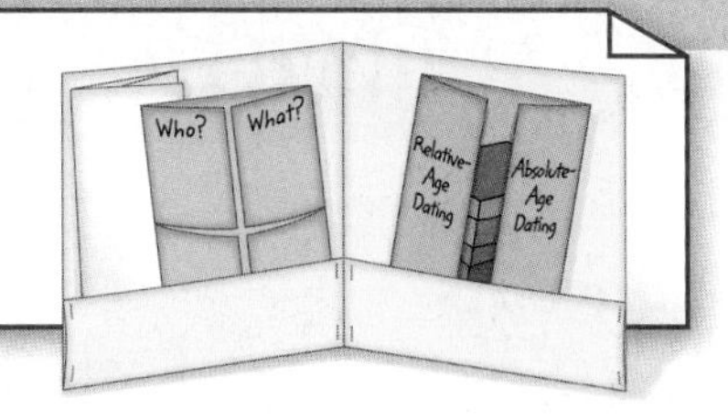

Use Vocabulary

Distinguish between the following terms.

1. absolute-age dating and relative-age dating

2. variations and adaptations

3. natural selection and selective breeding

4. homologous structure and analogous structure

5. vestigial structure and homologous structure

Link Vocabulary and Key Concepts

Use vocabulary terms from the previous page to complete the concept map.

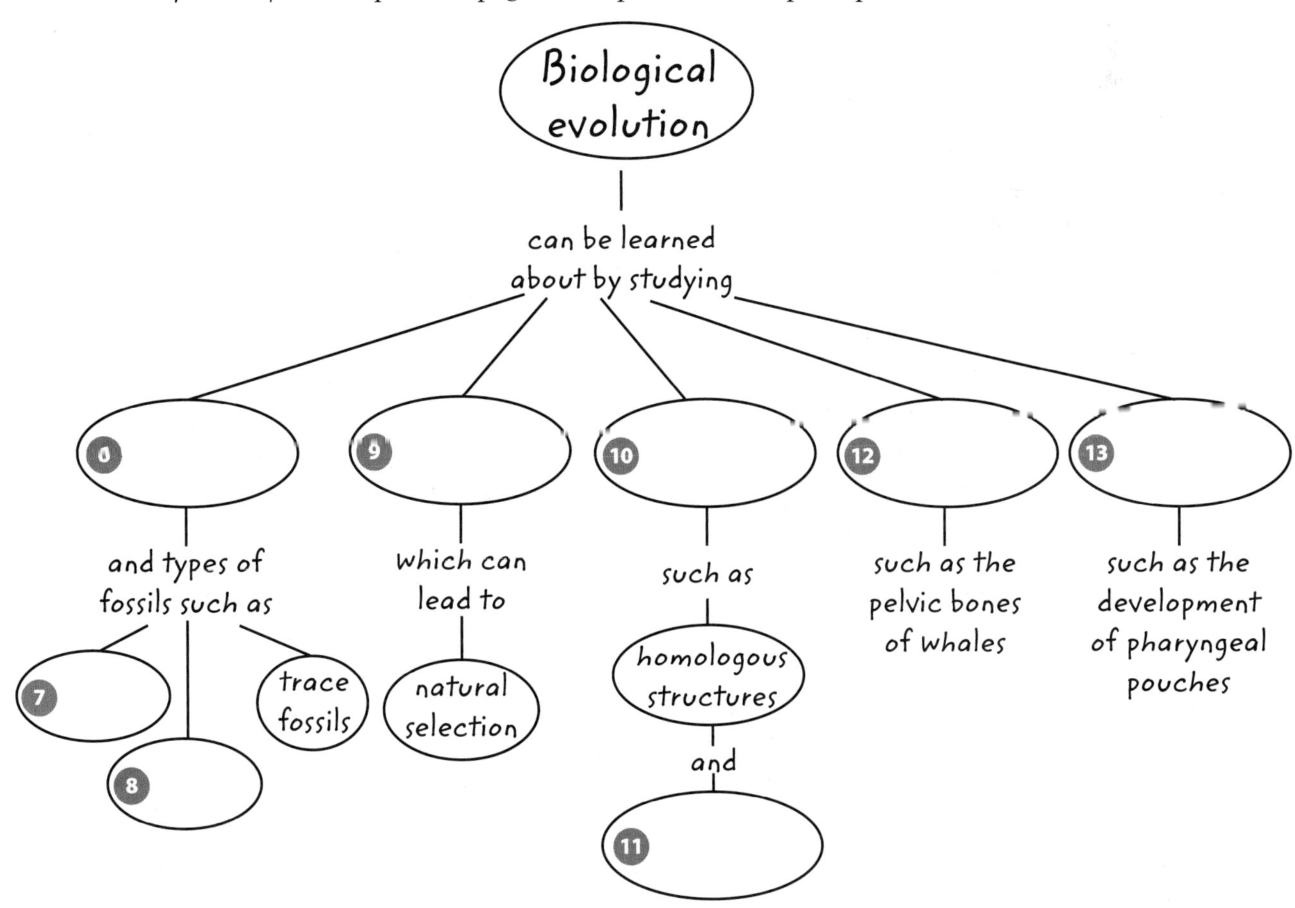

Chapter 6 Review

Fill in the correct answer choice.

Understand Key Concepts

1 Why do scientists think the fossil record is incomplete? **SC.6.N.1.5**

Ⓐ Fossils decompose over time.
Ⓑ The formation of fossils is rare.
Ⓒ Only organisms with hard parts become fossils.
Ⓓ There are no fossils before the Phanerozoic eon.

2 What do the arrows on the graph below represent? **SC.6.N.1.5**

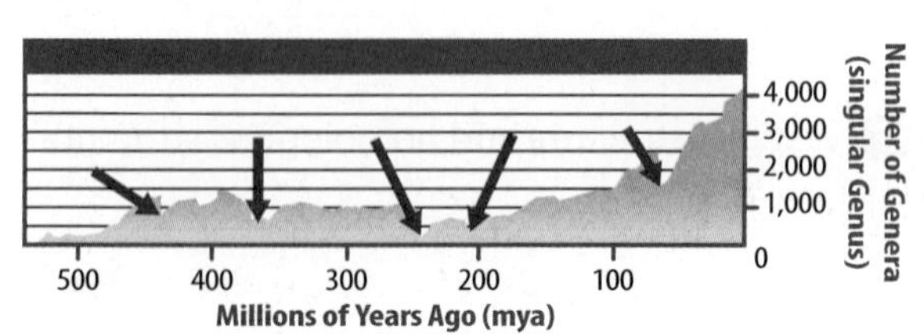

Ⓐ extinction events
Ⓑ meteorite impacts
Ⓒ changes in Earth's temperature
Ⓓ the evolution of a new species

3 What can scientists learn about fossils using techniques of absolute-age dating?

Ⓐ estimated ages of fossils in rock layers
Ⓑ precise ages of fossils in rock layers
Ⓒ causes of fossil disappearances in rock layers
Ⓓ structural similarities to other fossils in rock layers

4 Which is the sequence by which natural selection works? **SC.7.L.15.2**

Ⓐ selection → adaptation → variation
Ⓑ selection → variation → adaptation
Ⓒ variation → adaptation → selection
Ⓓ variation → selection → adaptation

5 Which type of fossil forms through carbonization?

Ⓐ cast
Ⓑ mold
Ⓒ fossil film
Ⓓ trace fossil

Critical Thinking

6 **Explain** the relationship between fossils and extinction events.

7 **Infer** In 2004, a fossil of an organism that had fins and gills, but also lungs and wrists, was discovered. What might this fossil suggest about evolution? **SC.7.L.15.1**

8 **Summarize** Darwin's theory of natural selection using the Galápagos tortoises or finches as an example. **SC.7.L.15.2**

9 **Assess** how the determination that Earth is 4.6 billion years old provided support for the idea that all species evolved from a common ancestor. **SC.7.L.15.1**

10 **Describe** how cytochrome c provides evidence of evolution.

11 **Explain** why the discovery of genes was powerful support for Darwin's theory of natural selection.

12 **Interpret Graphics** The diagram below shows two different methods by which evolution by natural selection might proceed. Discuss how these two methods differ. LA.6.2.2.3

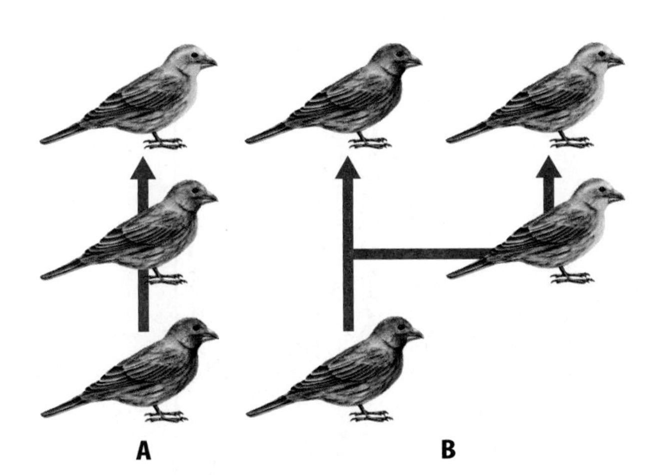

Writing in Science

13 **Write** a paragraph on a separate sheet of paper explaining how natural selection and selective breeding are related. Include a main idea, supporting details, and a concluding sentence. LA.6.2.2.3

Big Idea Review

14 How do species adapt to changing environments over time? Explain how evidence from the fossil record and from living species suggests that Earth's species are related. List each type of evidence and provide an example of each. SC.7.L.15.1

15 The photo on the chapter opener shows an orchid that looks like a bee. How might this adaptation be evidence of evolution by natural selection? SC.7.L.15.2

Math Skills MA.6.A.3.6

Use Scientific Notation

16 The earliest fossils appeared about 3,500,000,000 years ago. Express this number in scientific notation.

17 The oldest fossils visible to the unaided eye are about 565,000,000 years old. What is this time in scientific notation?

18 The oldest human fossils are about 1×10^4 years old. Express this as a whole number.

Fill in the correct answer choice.

Multiple Choice

1 According to the theory of natural selection, why are some individuals more likely than others to survive and reproduce? **SC.7.L.15.2**

Ⓐ They do not acquire any adaptations.

Ⓑ They are better adapted to exist in their environment than others.

Ⓒ They acquire only harmful characteristics.

Ⓓ The environment randomly decides which organisms will reproduce.

2 What do homologous structures, vestigial structures, and fossils provide evidence of? **SC.7.L.15.1**

Ⓕ analogous structures

Ⓖ food choice

Ⓗ populations

Ⓘ evolution

Use the figure below to answer question 3.

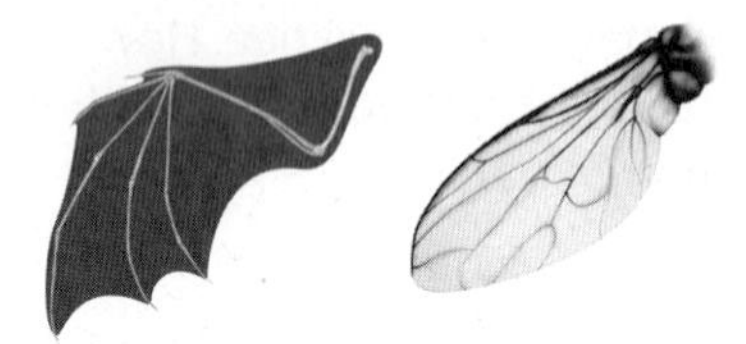

3 The analogous structures shown above are not related. However, they both evolved through natural selection. What are they both examples of? **SC.7.L.15.2**

Ⓐ vestigial organs

Ⓑ homologous structures

Ⓒ adaptations

Ⓓ variations

4 What is an adaptation? **SC.7.L.15.2**

Ⓕ a body part that has lost its original function through evolution

Ⓖ a characteristic that better equips an organism to survive in its environment

Ⓗ a feature that appears briefly during early development

Ⓘ a slight difference among the individuals in a species

5 What causes variations to arise in a population? **SC.7.L.15.2**

Ⓐ changes in the environment

Ⓑ competition for limited resources

Ⓒ random mutations in genes

Ⓓ rapid population increases

Use the image below to answer question 6.

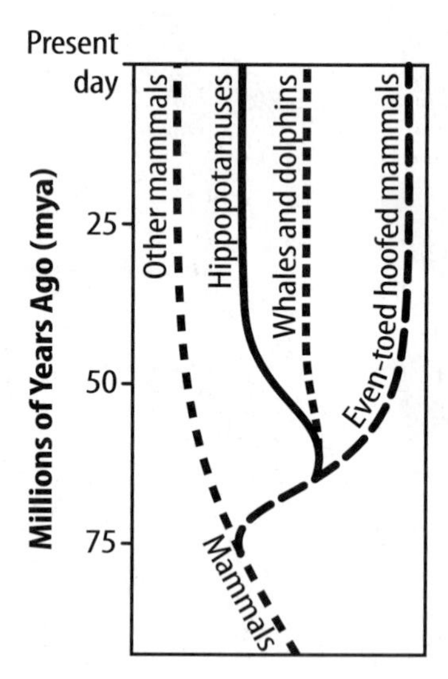

6 The image above shows that even-toed hoofed mammals and other mammals shared a common ancestor. When did this ancestor live? **SC.7.L.15.1**

Ⓕ 25–35 million years ago

Ⓖ 50–60 million years ago

Ⓗ 60–75 million years ago

Ⓘ 75 million years ago

7 What is an extinct species? **SC.7.L.15.3**

Ⓐ a species that blends in easily with its environment

Ⓑ a species that resembles another species

Ⓒ a species that was unable to adapt and died out

Ⓓ a species that is bred for specific characteristics

Use the figure below to answer question 8.

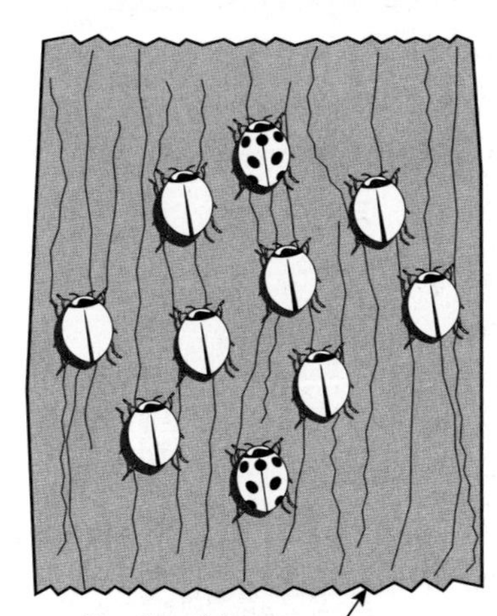

8 What unique characteristic would the beetles shown above develop through biological adaptation if, over a period of years, the bark on the trees shown became spotted? **SC.7.L.15.2**

Ⓕ The beetles would become spotted.

Ⓖ The beetles would become plain.

Ⓗ About half of the beetles would become spotted, and half would not.

Ⓘ There would be no change.

9 When compared with other fossils, what structure indicates the ancestors of whales used to walk on land? **SC.7.L.15.1**

Ⓐ pharyngeal pouches

Ⓑ camouflage

Ⓒ homologous wings

Ⓓ vestigial pelvic bones

Use the figure below to answer question 10.

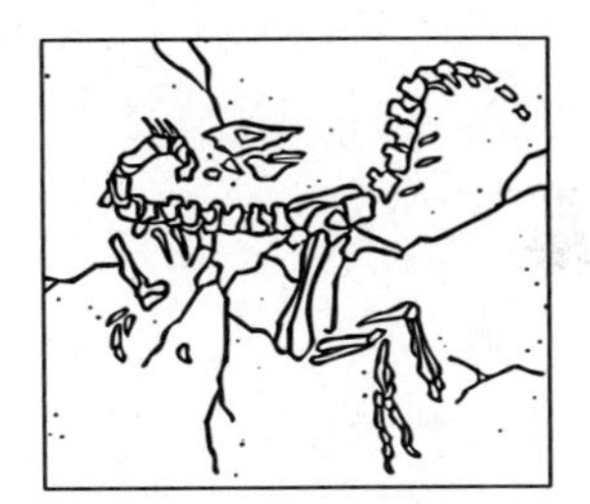

10 What method can scientists use to determine when the fossil above appeared on Earth? **SC.7.L.15.1**

Ⓕ mimicry

Ⓖ selective breeding

Ⓗ relative-age dating

Ⓘ biological evolution

11 Which is considered an important factor in natural selection? **SC.7.L.15.2**

Ⓐ limited reproduction

Ⓑ competition for resources

Ⓒ no variations within a population

Ⓓ plentiful food and other resources

Need Extra Help?

If You Missed Question...	1	2	3	4	5	6	7	8	9	10	11
Go to Lesson...	2	1,3	2	2	2	3	2	2	1,3	1	2

Name ______________________ Date ______________

Benchmark Mini-Assessment Chapter 6 • Lesson 1

FLORIDA NGSSS   mini BAT

Multiple Choice *Bubble the correct answer.*

1. Which is least likely to appear in the fossil record? **SC.7.L.15.1**

   (A) a bird's beak

   (B) a dinosaur's teeth

   (C) a mammal's skull

   (D) a worm's body

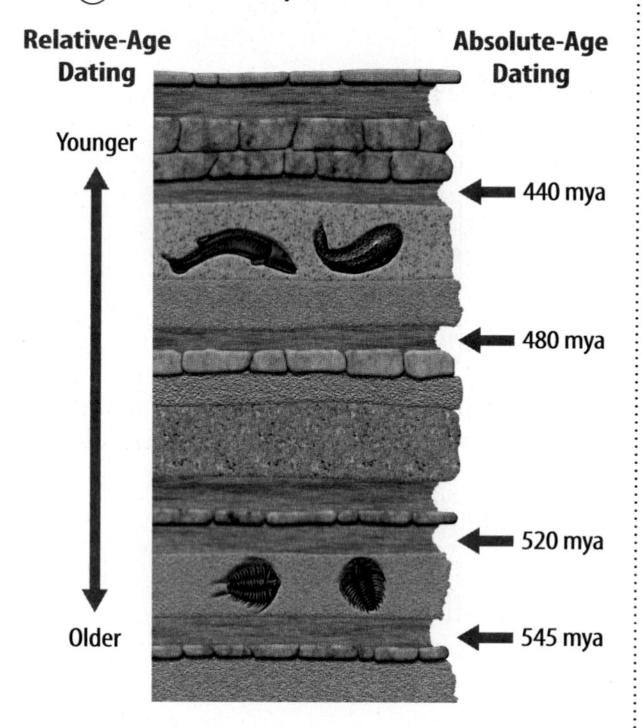

2. The dates shown in the diagram above refer to the ages of **SC.7.L.15.1**

   (F) original material.

   (G) trace fossils.

   (H) igneous rock layers.

   (I) sedimentary rock layers.

3. An organism dies, and its body leaves an impression in mud. Over time, the mud hardens into rock, and the impression becomes a fossil. Which kind of fossil was formed? **SC.7.L.15.1**

   (A) cast

   (B) mold

   (C) original material

   (D) trace fossil

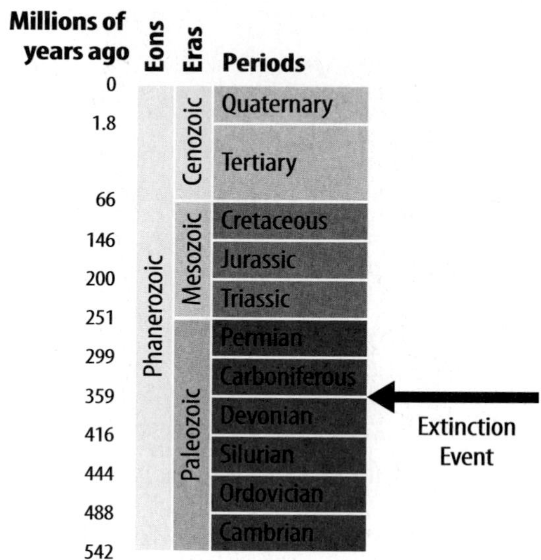

4. The extinction event shown in the chart above happened about how long ago? **SC.7.L.15.3**

   (F) 350 years ago

   (G) 350,000 years ago

   (H) 350 million years ago

   (I) 350 billion years ago

Name ______________________ Date ____________

Multiple Choice *Bubble the correct answer.*

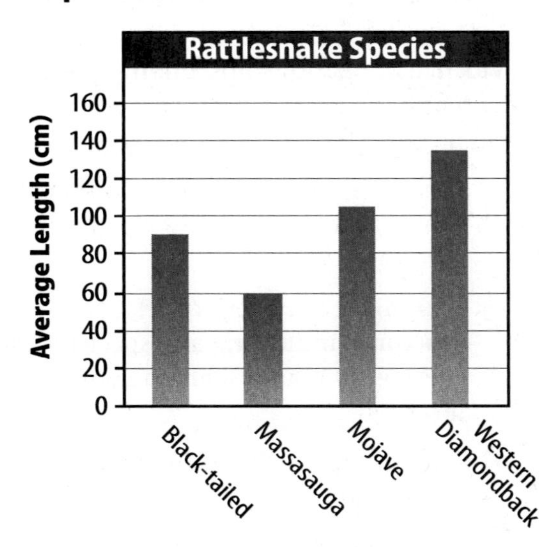

1. Which type of information about rattlesnakes does the bar graph above show? **SC.7.L.15.2**

Ⓐ adaptation
Ⓑ evolution
Ⓒ mimicry
Ⓓ variation

2. A theory is an explanation that has **SC.7.L.15.2**

Ⓕ been proven beyond a doubt.
Ⓖ little evidence to support it.
Ⓗ much evidence to support it.
Ⓘ no evidence to support it.

3. The alligator snapping turtle has a wormlike structure on its tongue. When this structure wiggles, fish are attracted to it and swim into the turtle's mouth. This wormlike appendage is an example of **SC.7.L.15.2**

Ⓐ camouflage.
Ⓑ mimicry.
Ⓒ behavioral adaptation.
Ⓓ functional adaptation.

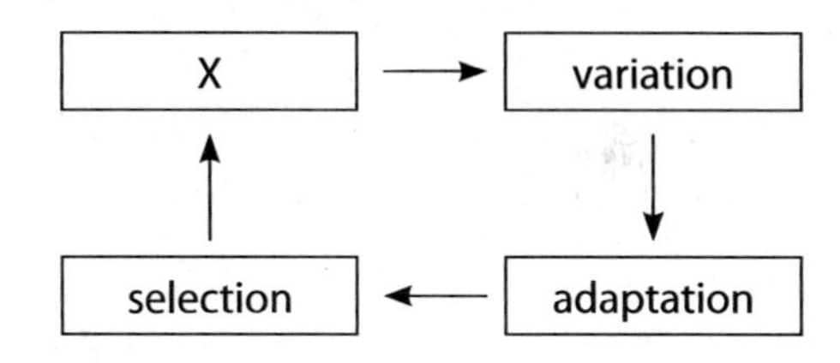

4. The diagram above shows the process of natural selection. What is X? **SC.7.L.15.2**

Ⓕ competition
Ⓖ evolution
Ⓗ mimicry
Ⓘ reproduction

Name ____________________ Date ____________

Multiple Choice *Bubble the correct answer.*

1. Frogs, robins, snakes, cows, and humans all have tongues. How do these homologous structures provide evidence for evolution? **SC.6.N.1.5**

Ⓐ Homologous structures do not provide evidence for evolution.

Ⓑ The more different a structure is in two living species, the more likely the species share a common ancestor.

Ⓒ The more similar a structure is in two living species, the less likely the species share a close common ancestor.

Ⓓ The more similar a structure is in two living species, the more likely the species share a close common ancestor.

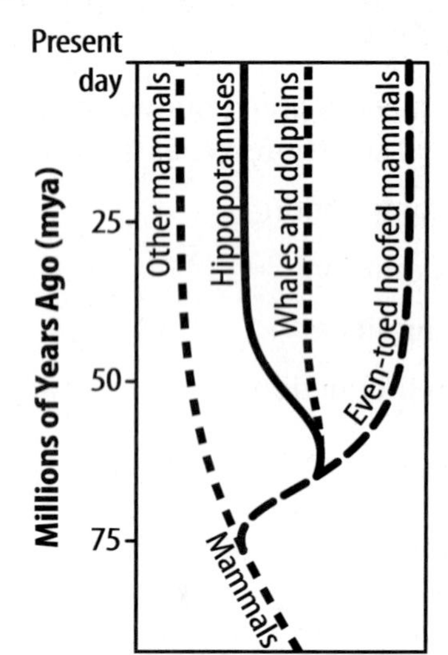

2. The diagram above shows how four species of mammals with a common ancestor diverged from one another through time. Which pair of animals is most closely related? **LA.6.2.2.3**

Ⓕ hippopotamuses and even-toed hoofed mammals

Ⓖ whales/dolphins and even-toed hoofed mammals

Ⓗ whale/dolphins and hippopotamuses

Ⓘ hippopotamuses and other mammals

3. How do pharyngeal pouches provide evidence of relationships among species? **SC.6.N.1.5**

Ⓐ Pharyngeal pouches are analogous structures, indicating any species with a pharyngeal pouch shares a common ancestor.

Ⓑ Pharyngeal pouches are vestigial structures, indicating any species with a pharyngeal pouch shares a common ancestor.

Ⓒ Pharyngeal pouches only develop in species directly related to each other.

Ⓓ Embryos of different species have pharyngeal pouches at different stages of development, indicating they share a common ancestor.

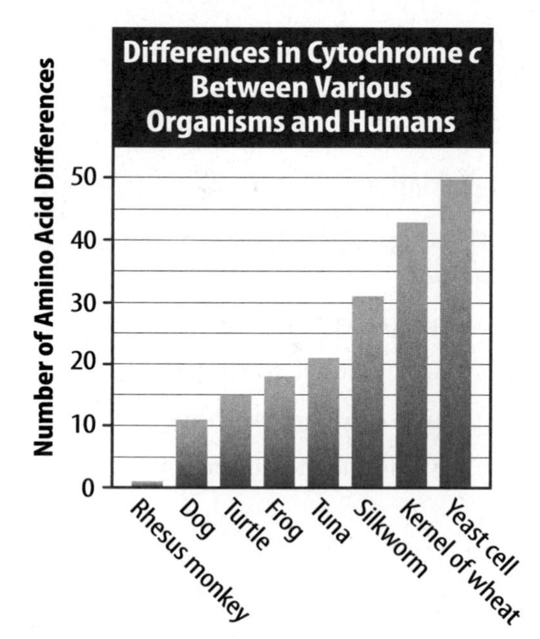

4. According to the information in the graph above, which organism is least like a human? **LA.6.2.2.3**

Ⓕ dog

Ⓖ frog

Ⓗ rhesus monkey

Ⓘ yeast cell

Name ______________________ Date ______________

Notes

Unit 2
From Bacteria to Animals

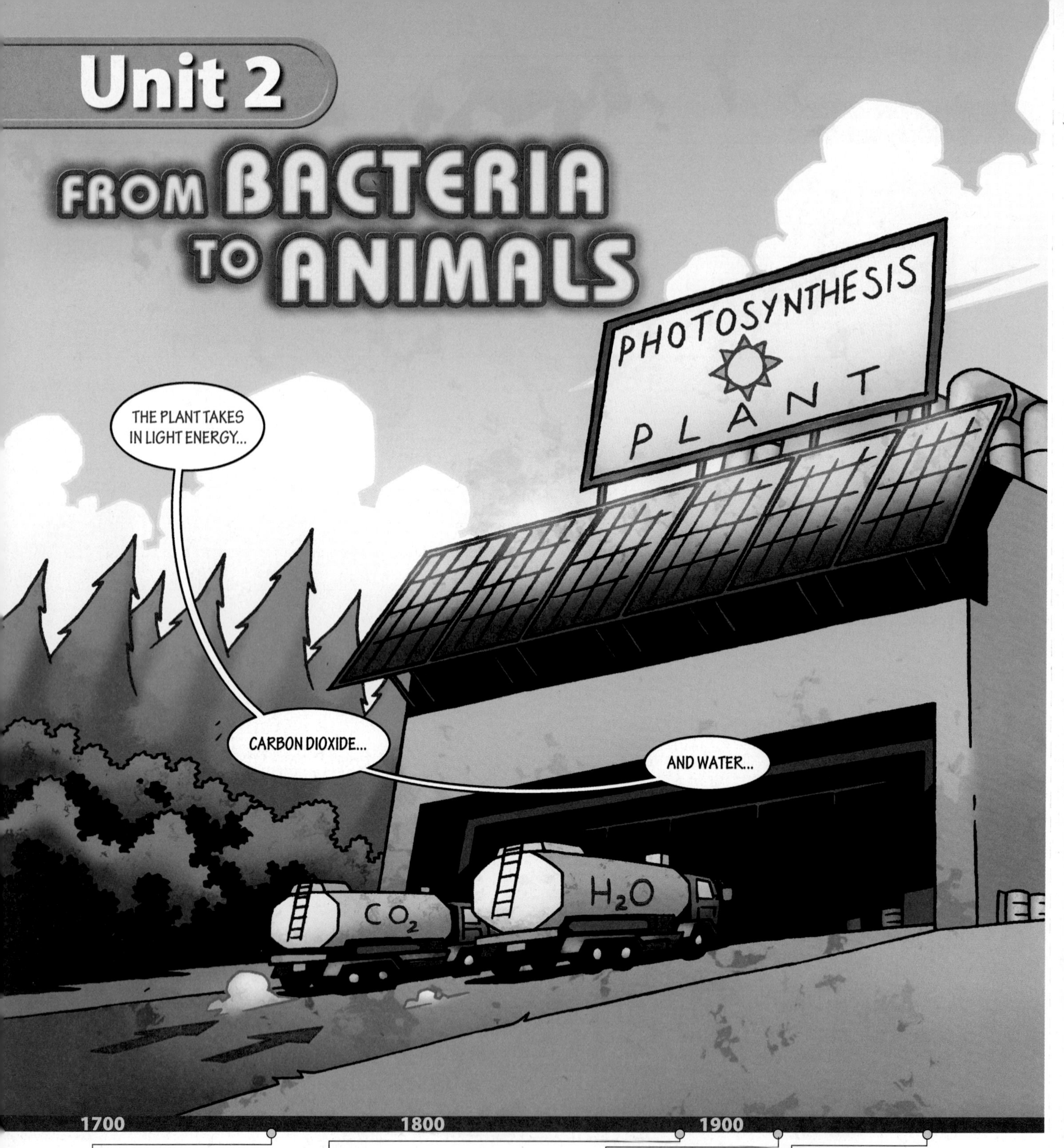

1753
Swedish botanist Carl Linnaeus publishes *Species Plantarum,* a list of all plants known to him and the starting point of modern plant nomenclature.

1892
Russian botanist Dmitri Iwanowski discovers the first virus while studying tobacco mosaic disease. Iwanowski finds that the cause of the disease is small enough to pass through a filter made to trap all bacteria.

1930s
The development of commercial hybrid crops begins in the United States.

1983
Luc Montagnier's team at the Pasteur Institute in France isolates the retrovirus now called HIV.

2000

2002
Some scientists estimate that there are around 420,000 different species of plants on Earth.

2007
Researchers discover fungi that uses radioactivity as an energy source.

2008
While in the Bahamas, biologist Mikhail Matz discovers ocean-dwelling protists the size of grapes that leave complex trails in the sand. These trails resemble those found in Precambrian fossils and shed new light on the mystery of early animal development.

? Inquiry
Visit ConnectED for this unit's **STEM** activity.

Unit 2

Nature of SCIENCE

Patterns

Have you ever seen an individual snowflake close-up? You might have heard someone say that no two snowflakes are alike. While this is true, it is also true that all snowflakes have similar patterns. A **pattern** is a consistent plan or model used as a guide for understanding or predicting things. Patterns can be created or occur naturally. The formation of snowflakes is an example of a pattern. They form as water drops in the air freeze into a six-sided crystal.

How Scientists Use Patterns

Studying and using patterns is useful to scientists because it can help explain the natural world or predict future events. A biologist might study patterns in DNA to predict what organisms will look like. A meteorologist might study cloud formation patterns to predict the weather. When doing research, scientists also try to match patterns found in their data with patterns that occur in nature. This helps to determine whether data are accurate and helps to predict outcomes.

Active Reading 1. **Review** How can patterns be useful to scientists?

____

____

Types of Patterns

Cyclic Patterns

A cycle, or repeated series of events, is a form of pattern. An organism's life cycle typically follows the pattern of birth, growth, and death. Scientists study an organism's life cycle to predict the life of its offspring.

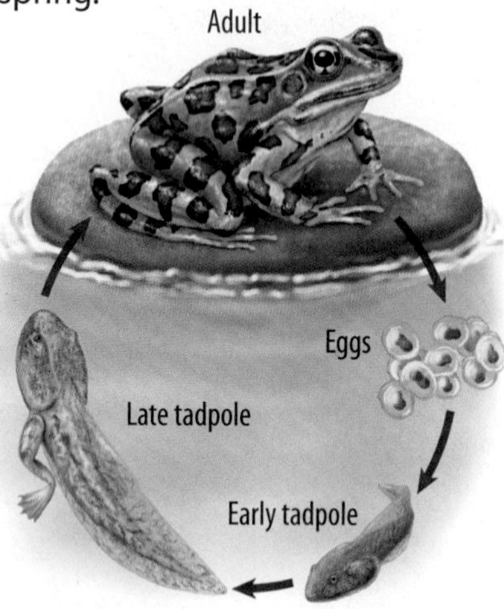

Physical Patterns

Physical patterns have an artistic or decorative design. Physical patterns can occur naturally, such as the patterns in the colors on butterfly wings or flower petals, or they can be created intentionally, such as a design in a brick wall.

Patterns in Life Science

Why do police detectives or forensic scientists take fingerprints at a crime scene? They know that every fingerprint is unique. Fingerprints contain patterns that can help detectives narrow a list of suspects. The patterns on the fingerprints can then be examined more closely to identify an individual. No two humans have the same set of fingerprints, just as no two zebras have the same stripe pattern.

Patterns are an important key to understanding science and the natural world. They are found across all classifications of science. Patterns help scientists understand the genetic makeup, lifestyle, and similarities of various species of plants and animals. Patterns help determine weather, the relative age of rocks, forces of nature, and the orbits of the planets. Look around you and observe patterns in your world.

2. Apply What patterns might a zoologist or botanist study?

____

____

Mathematical Patterns

Patterns are frequently applied in mathematics. Whenever you read a number, perform a mathematical operation, or describe a shape or graph, you are using patterns.

2, 5, 8, 11, ___, ___,

What numerals come next in this number pattern?

What will the next shape look like according to the pattern?

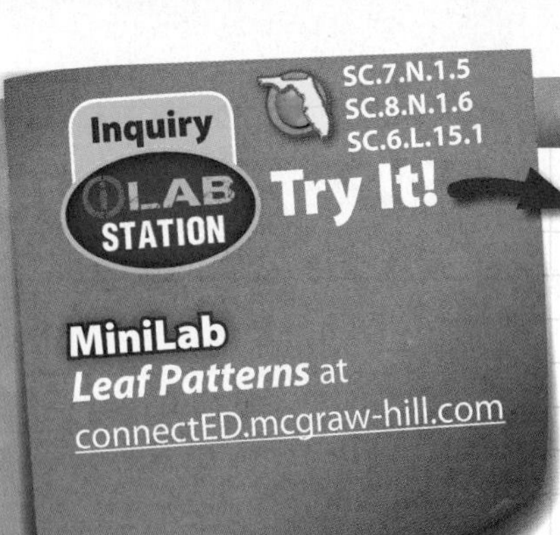

Apply It!

After you complete the lab, answer these questions.

1. Illustrate Design a pattern for each of the following types of patterns.

Cyclic Pattern

Physical Pattern

Mathematical Pattern

2. Extend What might a change in a young bird's feather pattern indicate? Support your reasoning.

Name ______ Date ______

Notes

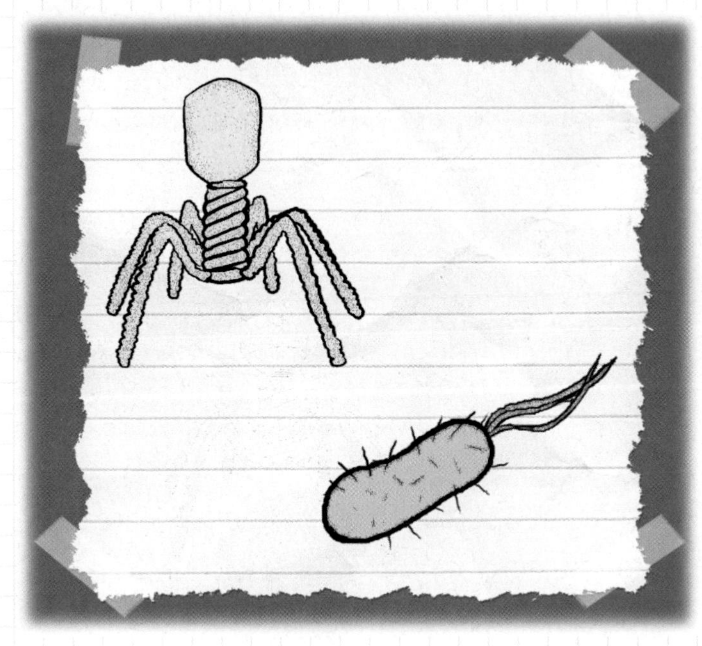

Is it an organism?

Janet wondered if bacteria and viruses are considered organisms. She asked her friends and this is what they said:

Tony: Bacteria are organisms, but viruses are not organisms.

Suze: Viruses are organisms, but bacteria are not organisms.

Lucas: Bacteria and viruses are both organisms.

Gina: Neither bacteria nor viruses are organisms.

Which friend do you agree with the most? ____________ Explain why you agree.

Bacteria and

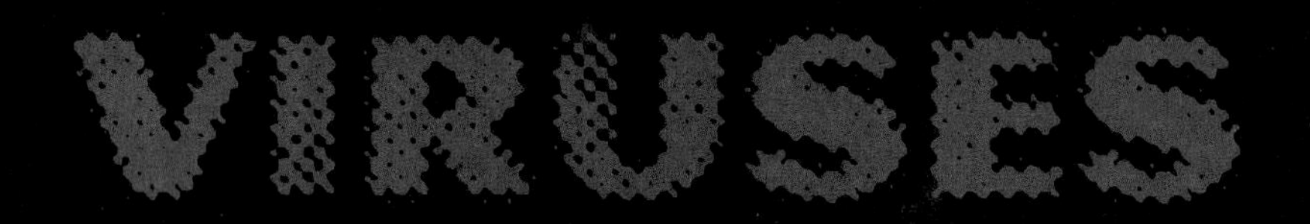

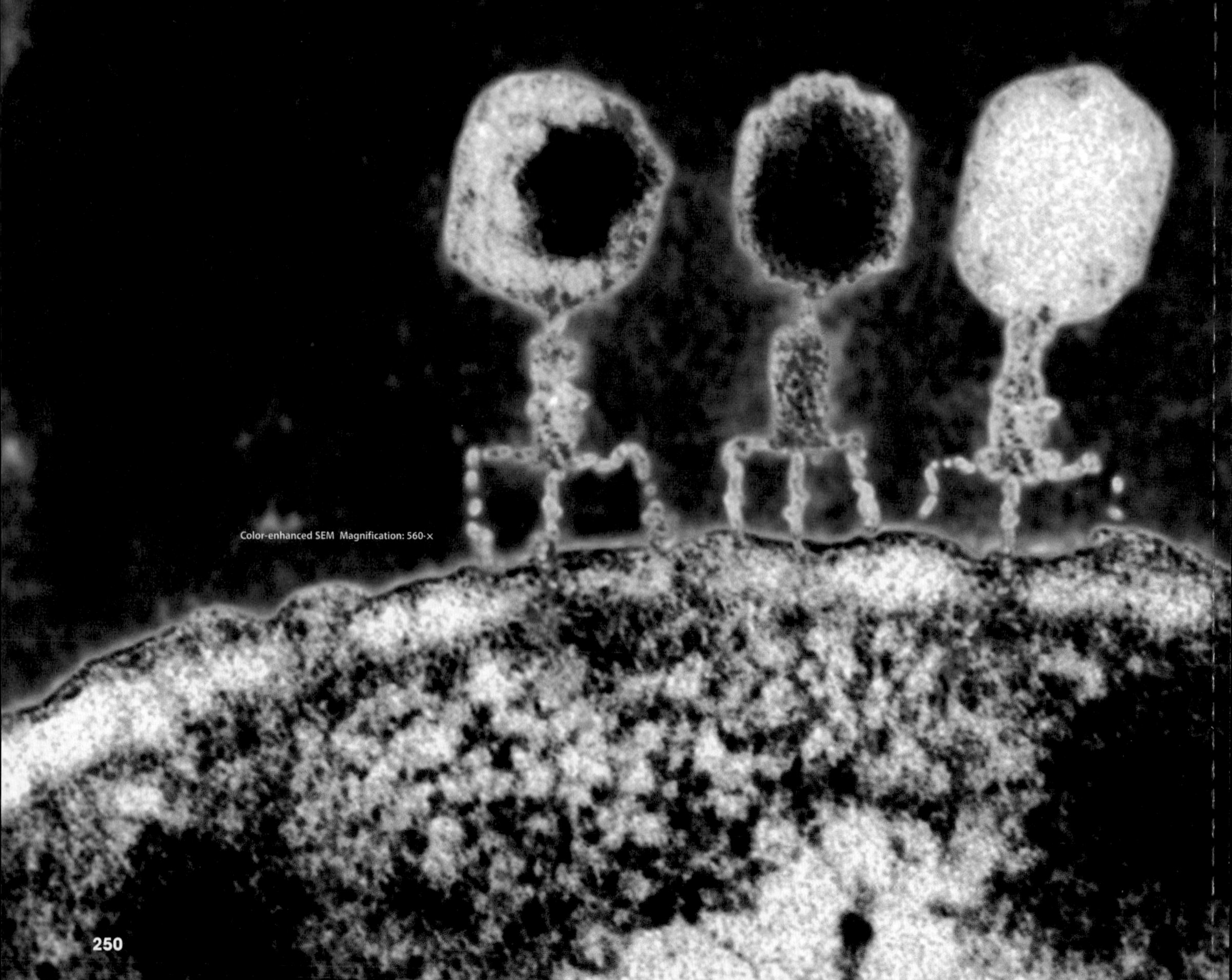

FLORIDA BIG IDEAS

1 The Practice of Science
2 The Characteristics of Scientific Knowledge
3 The Role of Theories, Laws, Hypotheses, and Models
14 Organization and Development of Living Organisms
16 Heredity and Reproduction

The Big Idea

Think About It!

What are bacteria and viruses and why are they important?

You might think this photo shows robots landing on another planet. Actually, this is a picture of viruses attacking a type of unicellular organism called a bacterium (plural, bacteria). Many viruses can attach to the surface of one bacterium.

1. Do you think the bacterium is harmful? Are the viruses?

2. What do you think happens after the viruses attach to the bacterium?

3. What are viruses and bacteria, and why do you think they are important?

Get Ready to Read

What do you think about bacteria and viruses?

Before you read, decide if you agree or disagree with each of these statements. As you read this chapter, see if you change your mind about any of the statements.

	AGREE	DISAGREE
1. A bacterium does not have a nucleus.	☐	☐
2. Bacteria cannot move.	☐	☐
3. All bacteria cause diseases.	☐	☐
4. Bacteria are important for making many types of food.	☐	☐
5. Viruses are the smallest living organisms.	☐	☐
6. Viruses can replicate only inside an organism.	☐	☐

There's More Online!
Video • Audio • Review • ⓘLab Station • WebQuest • Assessment • Concepts in Motion • Multilingual eGlossary

Lesson 1

What are BACTERIA?

ESSENTIAL QUESTIONS

What are bacteria?

Vocabulary

bacterium p. 253
flagellum p. 256
fission p. 256
conjugation p. 256
endospore p. 257

SC.6.N.1.5
SC.6.N.3.4

Inquiry Launch Lab

10 minutes

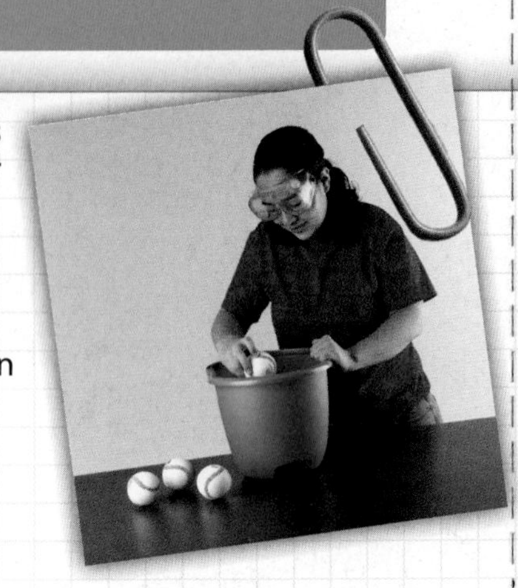

How small are bacteria?

Bacteria are tiny cells that can be difficult to see, even with a microscope. You might be surprised to learn that bacteria are found all around you, including in the air, on your skin, and in your body. One way of understanding how small bacteria are is to model their size.

Procedure 

1. Read and complete a lab safety form.
2. Examine the size of a **baseball** and a **2.5-gal. bucket.** Estimate how many baseballs you think would fit inside the bucket.
3. As a class, count how many baseballs it takes to fill the bucket.

Think About This

1. How much larger is the bucket than a baseball?

2. If your skin cells were the size of the bucket and bacteria were the size of the baseballs, how many bacterial cells would fit on a skin cell?

3. Key Concept Why do you think you cannot see bacteria on your skin or on your desk?

Florida NGSSS

LA.6.2.2.3 The student will organize information to show understanding (e.g., representing main ideas within text through charting, mapping, paraphrasing, summarizing, or comparing/contrasting);

LA.6.4.2.2 The student will record information (e.g., observations, notes, lists, charts, legends) related to a topic, including visual aids to organize and record information and include a list of sources used;

MA.6.A.3.6 Construct and analyze tables, graphs, and equations to describe linear functions and other simple relations using both common language and algebraic notation.

SC.6.N.2.1 Distinguish science from other activities involving thought.

SC.6.N.3.4 Identify the role of models in the context of the sixth grade science benchmarks.

SC.6.L.14.6 Compare and contrast types of infectious agents that may infect the human body, including viruses, bacteria, fungi, and parasites.

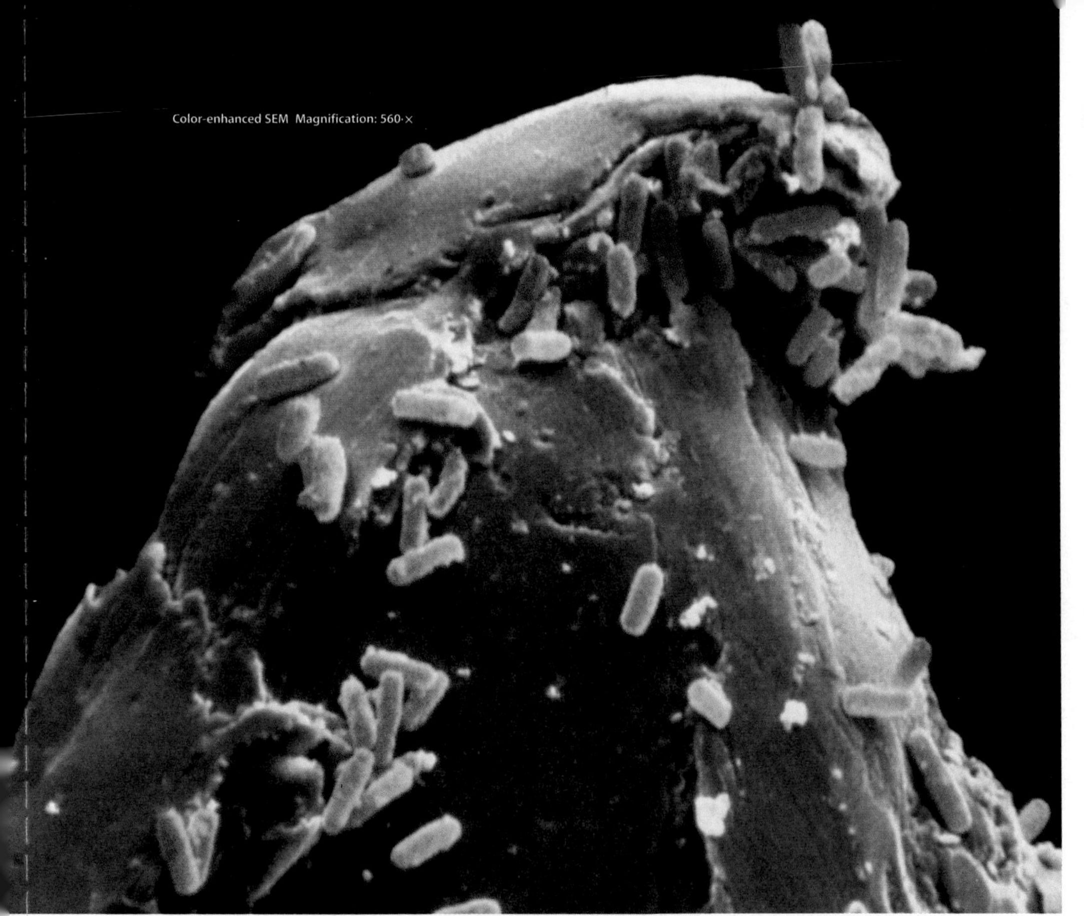

How clean is this surface?

1. This photo shows a microscopic view of the point of a needle. The small orange things are bacteria. Bacteria are everywhere, even on surfaces that appear clean. Do you think bacteria are living or nonliving? Why?

Characteristics of Bacteria

Did you know that billions of tiny organisms too small to be seen surround you? These organisms, called bacteria, even live inside your body. **Bacteria** (singular, bacterium) *are microscopic prokaryotes*. You might recall that a prokaryote is a unicellular organism that does not have a nucleus or other membrane-bound organelles.

Bacteria live in almost every habitat on Earth, including the air, glaciers, the ocean floor, and in soil. A teaspoon of soil can contain between 100 million and 1 billion bacteria. Bacteria also live in or on almost every organism, both living and dead. Hundreds of species of bacteria live on your skin. In fact, your body contains more bacterial cells than human cells! The bacteria in your body outnumber human cells by 10 to 1.

Active Reading **2. Explain** What are bacteria?

Other prokaryotes, called archaea (ar KEE uh; singular, archaean), are similar to bacteria and share many characteristics with them, including the lack of membrane-bound organelles. Archaea can live in places where few other organisms can survive, such as very warm areas or those with little oxygen. Both bacteria and archaea are important to life on Earth.

WORD ORIGIN

bacteria
from Greek *bakterion,* means "small staff"

Active Reading FOLDABLES® LA.6.2.2.3

Make a folded book from a sheet of notebook paper. Label it as shown. Use your book to organize your notes on the characteristics of bacteria.

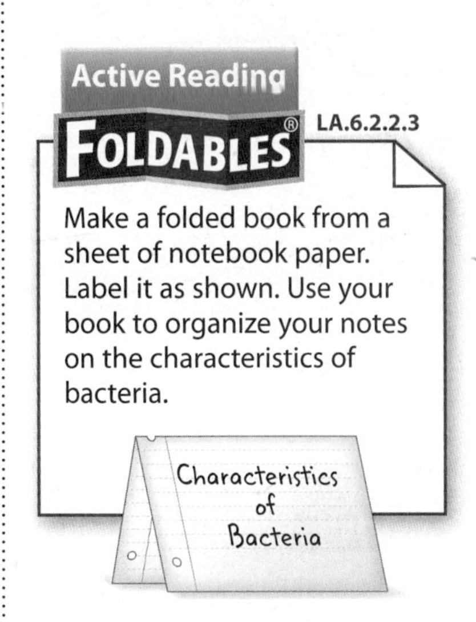

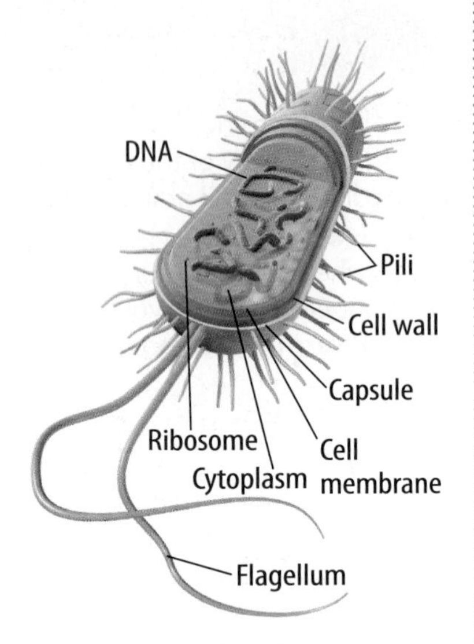

Figure 1 Bacteria have a cell membrane and contain cytoplasm.

Structure of Bacteria

A typical bacterium, such as the one shown in **Figure 1,** consists of cytoplasm and DNA surrounded by a cell membrane and a cell wall. The cytoplasm also contains ribosomes. Most bacteria have DNA that is one coiled, circular chromosome. Many bacteria also have one or more small circular pieces of DNA called plasmids that are separate from its other DNA.

Some bacteria have specialized structures that help them survive. For example, the bacterium that causes pneumonia (noo MOH nyuh), an inflammation of the lungs, has a thick covering, or capsule, around its cell wall. The capsule protects the bacterium from drying out. It also prevents white blood cells from surrounding and antibiotics from entering it. Many bacteria have capsules with hairlike structures called pili (PI li) that help the bacteria stick to surfaces.

Size and Shapes of Bacteria

Bacteria are much smaller than plant or animal cells. Bacteria are generally only 1–5 micrometers (μm) (1 m = 1 million μm) wide, while an average eukaryotic cell is 10–100 μm wide. Scientists estimate that as many as 100 bacteria could be lined up across the head of a pin. As shown in **Figure 2,** bacteria generally have one of three basic shapes.

Figure 2 Bacteria are generally shaped like a sphere, a rod, or a spiral.

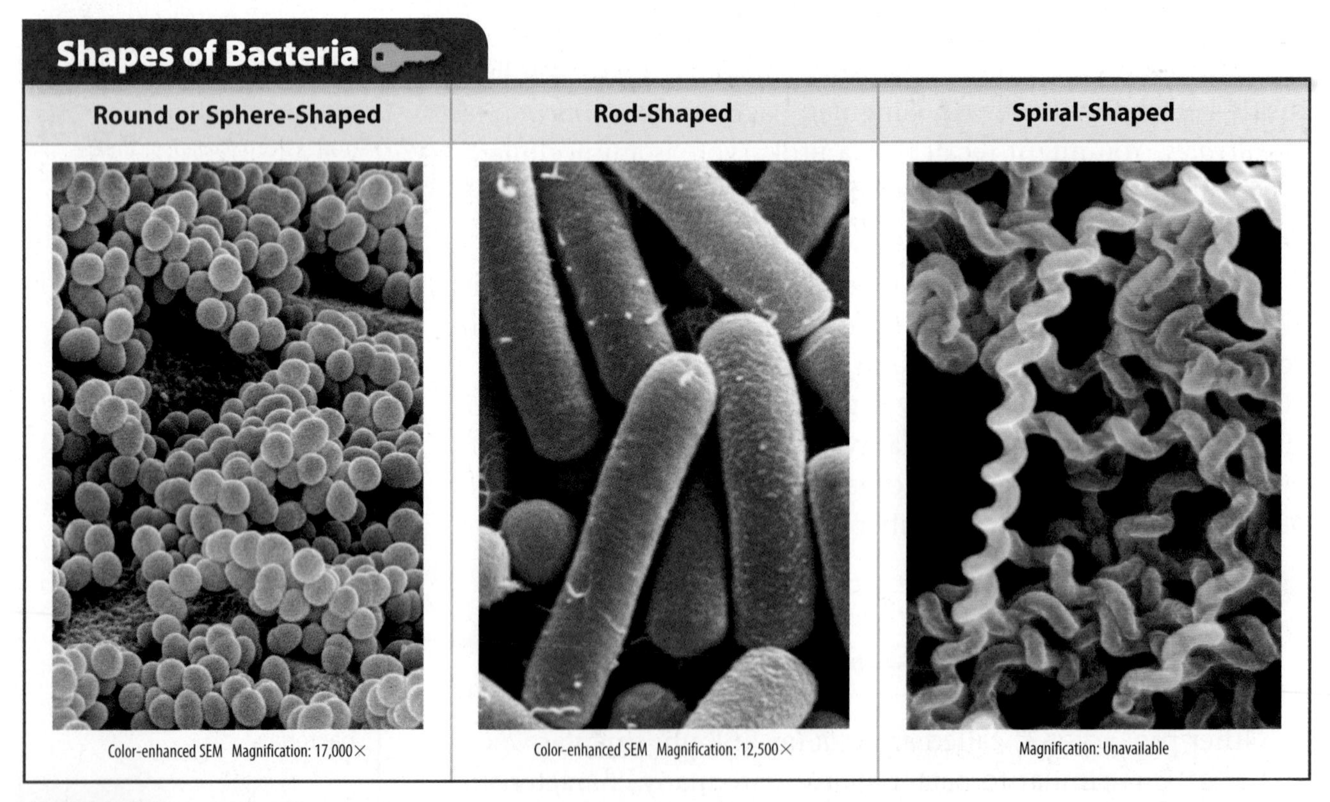

Active Reading 3. **Infer** Bacteria often form clusters. Which shape of bacteria do you think would likely form a cluster?

MiniLab ***How does a slime layer work?*** at connectED.mcgraw-hill.com

Apply It! After you complete the lab, answer this question.

1. How do you think Earth's nutrient cycles would be affected if bacteria cells did not have slime layers?

Obtaining Food and Energy

Bacteria live in many places. Because these environments are very different, bacteria obtain food in various ways. Some bacteria take in food and break it down and obtain energy. Many of these bacteria feed on dead organisms or organic waste, as shown in **Figure 3.** Others take in their nutrients from living hosts. For example, bacteria that cause tooth decay live in dental plaque on teeth and feed on sugars in the foods you eat and the beverages you drink.

Some bacteria make their own food. These bacteria use light energy and make food, like most plants do. These bacteria live where there is a lot of light, such as the surface of lakes and streams. Other bacteria use energy from chemical reactions and make their food. These bacteria live in places where there is no sunlight, such as the dark ocean floor.

Active Reading 4. **Restate** How do bacteria obtain food?

Figure 3 This banana is rotting because bacteria are breaking it down for food.

Most organisms, including humans, cannot survive without oxygen. However, certain bacteria do not need oxygen to survive. These bacteria are called anaerobic (a nuh ROH bihk) bacteria. Bacteria that need oxygen are called aerobic (er OH bihk) bacteria. Most bacteria in the environment are aerobic.

Active Reading 5. **Contrast** Complete the table by contrasting anaerobic bacteria with aerobic bacteria.

Aerobic Bacteria	Anaerobic Bacteria

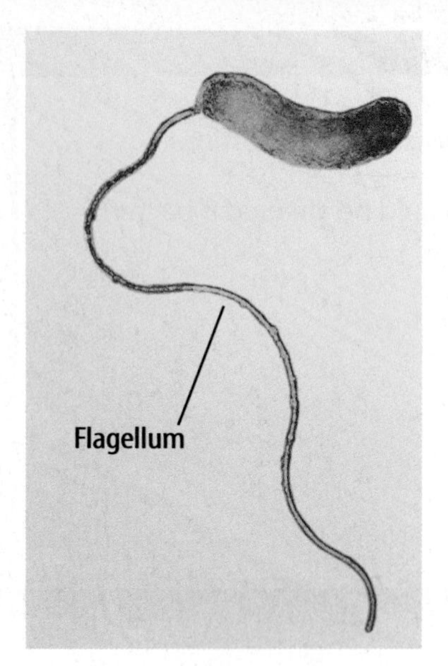

Figure 4 Some bacteria move using a flagellum.

Movement

Some bacteria are able to move around to find the resources that they need to survive. These bacteria have special structures for movement. *Many bacteria have long whiplike structures called* **flagella** (fluh JEH luh; singular, flagellum), as shown in **Figure 4.** Others twist or spiral as they move. Still other bacteria use their pili like grappling hooks or make threadlike structures that enable them to push away from a surface.

Reproduction

You might recall that organisms reproduce asexually or sexually. Bacteria reproduce asexually by fission. **Fission** *is cell division that forms two genetically identical cells.* Fission can occur quickly—as often as every 20 minutes under ideal conditions.

Bacteria produced by fission are identical to the parent cell. However, genetic variation can be increased by a process called conjugation, shown in **Figure 5.** *During* **conjugation** (kahn juh GAY shun), *two bacteria of the same species attach to each other and combine their genetic material.* DNA is transferred between the bacteria. This results in new combinations of genes, increasing genetic diversity. New organisms are not produced during conjugation, so the process is not considered reproduction.

Active Reading **6. Explain** How does conjugation increase the genetic diversity of bacteria?

____

____

____

Conjugation

Figure 5 Conjugation results in genetic diversity by transferring DNA between two bacterium cells.

Active Reading **7. Identify** (Circle) the structure that the donor cell uses to connect to the recipient cell.

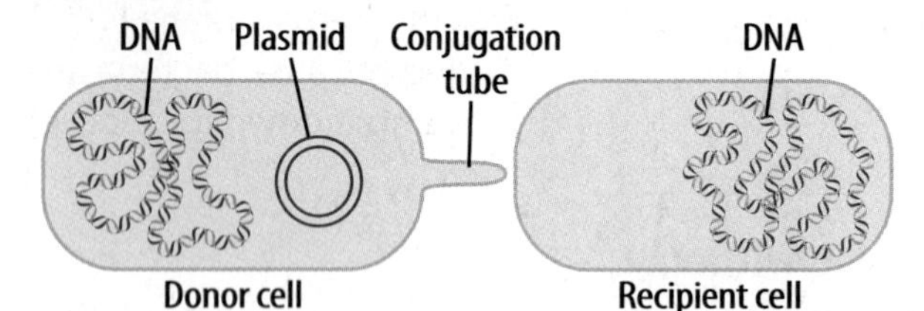

1 The donor cell and recipient cell both have circular chromosomal DNA. The donor cell also has DNA as a plasmid. The donor cell forms a conjugation tube and connects to the recipient cell.

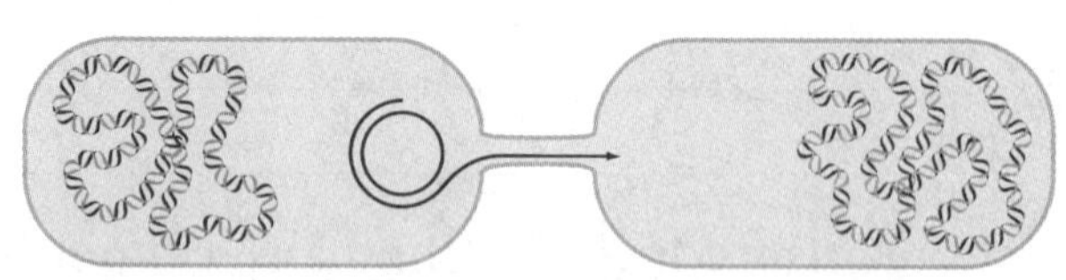

2 The conjugation tube connects both cells. The plasmid splits in two, and one plasmid strand moves through the conjugation tube into the recipient cell.

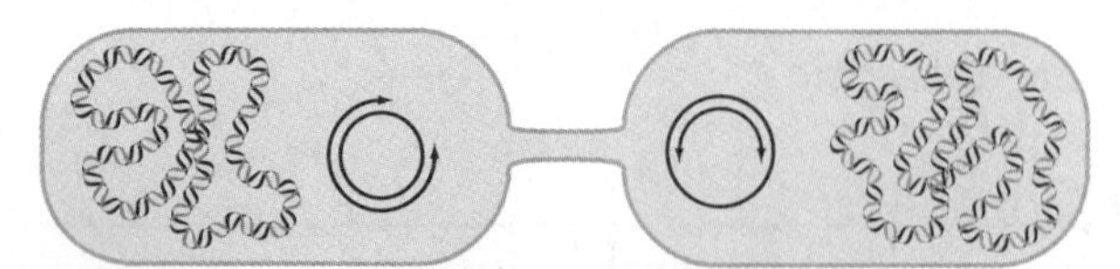

3 The complimentary strands of the plasmids are completed in both bacteria.

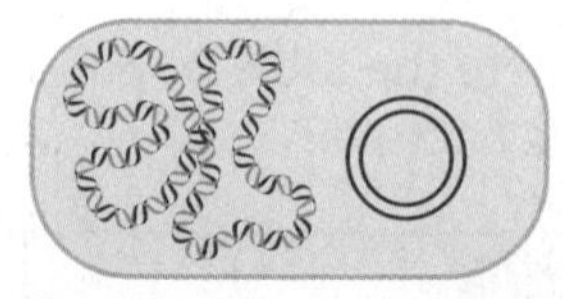
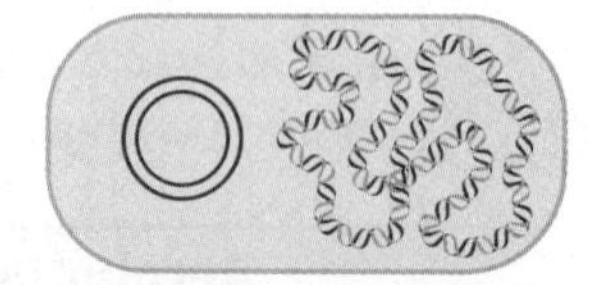

4 With the new plasmids complete, the bacteria separate from each other. The recipient cell now contains plasmid DNA from the donor cell as well as its own chromosomal DNA.

Endospore Formation **Figure 6**

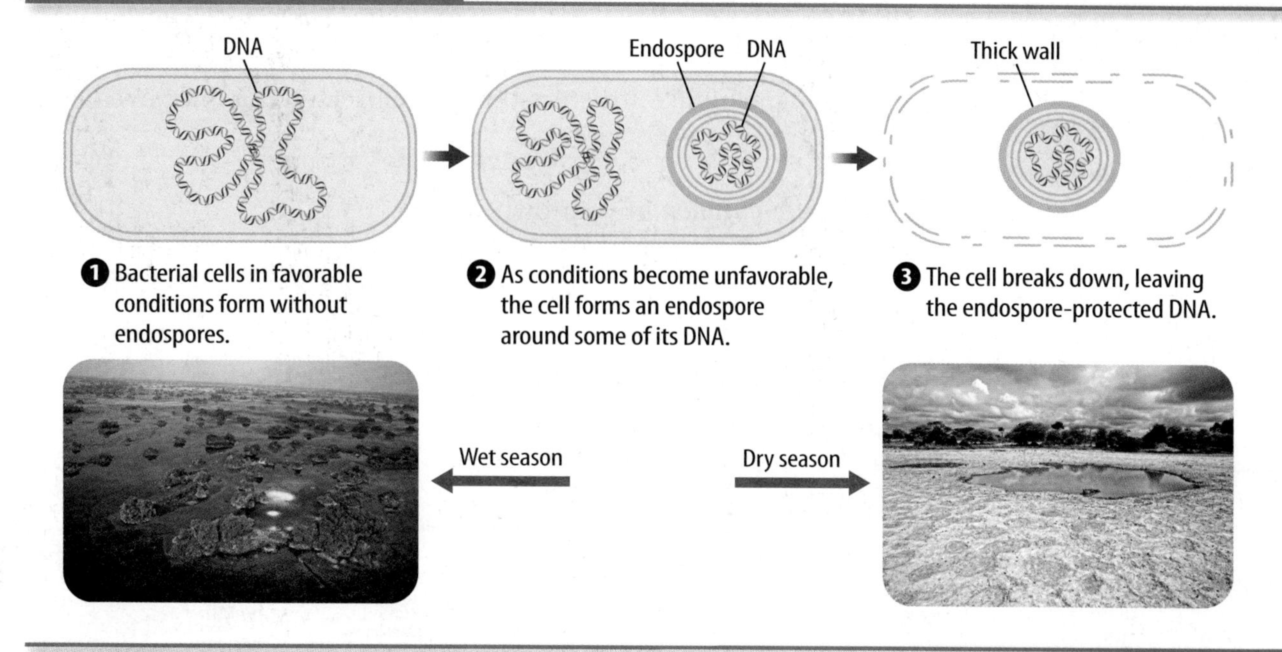

Active Reading **8. Label** An endospore can protect a ________________.

Endospores

Sometimes environmental conditions are unfavorable for the survival of bacteria. In these cases, some bacteria can form endospores. *An* **endospore** (EN doh spor) *forms when a bacterium builds a thick internal wall around its chromosome and part of the cytoplasm,* as shown in **Figure 6.** An endospore can protect a bacterium from intense heat, cold, or drought. It also enables a bacterium to remain dormant for months or even centuries. The ability to form endospores enables bacteria to survive extreme conditions that would normally kill them.

Archaea

Prokaryotes called archaea were once considered bacteria. Like a bacterium, an archaean has a cell wall and no nucleus or membrane-bound organelles. Its chromosome is also circular, like those in bacteria. However, there are some important differences between archaea and bacteria. The ribosomes of archaea more closely resemble the ribosomes of eukaryotes than those of bacteria. Archaea also contain molecules in their plasma membranes that are not found in any other known organisms. Archaea often live in extreme environments, such as hot springs and salt lakes. Some scientists refer to archaea as extremophiles (ik STREE muh filez)—a term that means "those that love extremes."

Math Skills MA.6.A.3.6

Use a Formula

Each time bacteria undergo fission, the population doubles. Use an equation to calculate how many bacteria there are:

$n = x \times 2^f$ where n is the final number of bacteria, x is the starting number of bacteria, and f is the number of times that fission occurs.

Example: 100 bacteria undergo fission 3 times.

$f = 3$, so 2^f is 2 multiplied by itself 3 times.

$(2 \times 2 \times 2 = 8)$

$n = 100 \times 8 = 800$ bacteria

Practice

9. How many bacteria would there be if 1 bacterium underwent fission 10 times?

Lesson Review 1

Visual Summary

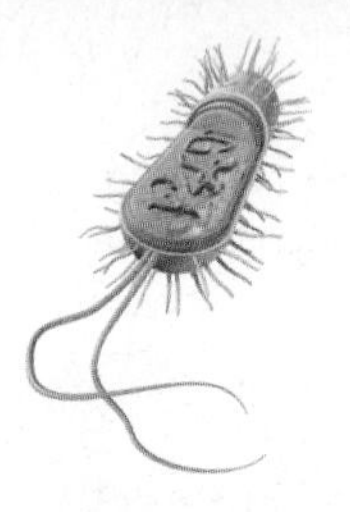

Bacteria are unicellular prokaryotes.

Many bacteria feed on dead organic matter.

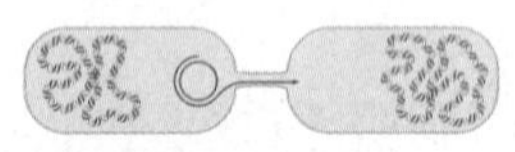

Bacteria can increase genetic diversity by sharing DNA through conjugation.

Use Vocabulary

1. The long, whiplike structure that some bacteria use for movement is a(n) ________________.

2. **Define** *conjugation* in your own words.

Understand Key Concepts

3. **Describe** a typical bacterium.

4. Which is NOT a common bacterium shape?
   - (A) rod
   - (B) sphere
   - (C) spiral
   - (D) square

Interpret Graphics

5. **Identify** Complete the table below to identify shapes of bacteria. LA.6.2.2.3

Bacterial Shapes	Illustration

Critical Thinking

6. **Analyze** how bacteria that can form endospores would have an advantage over bacteria that cannot form endospores.

Math Skills MA.6.A.3.6

7. How many bacteria would result if fission occurred 4 times with 1,000 bacteria?

Cooking Bacteria!

HOW NATURE WORKS

How Your Body Is Like Bleach

When it comes to killing germs, few things work as well as household bleach. How does bleach kill bacteria? Believe it or not, killing bacteria with bleach and boiling an egg involve similar processes.

After cooking, egg proteins become a tangled mass.

Eggs are made mostly of proteins. Proteins are complex molecules in all plant and animal tissues. Proteins have specific functions that are dependent on the protein's shape. A protein's function changes if its shape is changed. When you cook an egg, the thermal energy transferred to the egg causes changes to the shape of the egg's proteins. Think of the firm texture of a cooked egg. When the egg's proteins are heated, they become a tangled mass.

Before cooking, the proteins in eggs remain unfolded and change shape easily.

Bacteria also contain proteins that change shape when exposed to heat.

A common ingredient in bleach is also found in your body's immune cells.

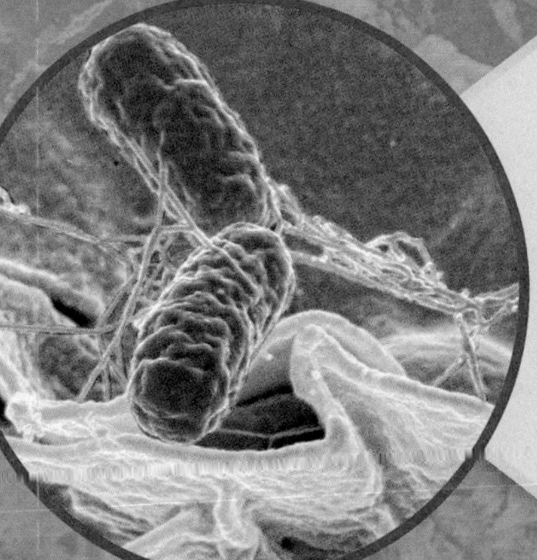

Like eggs, bacteria also contain proteins. When bacteria are exposed to high temperatures, their proteins change shape, similar to those in a boiled egg. But what is the connection with bleach? Scientists have discovered that an ingredient in bleach, hypochlorite (hi puh KLOR ite), also causes proteins to change shape. The bacterial proteins that are affected by bleach are needed for the bacterias' growth. When the shape of those proteins changes, they no longer function properly, and the bacteria die.

Scientists also know now that your body's immune cells produce hypochlorite. Your body protects itself with the same chemical you can use to clean your kitchen!

It's Your Turn

RESEARCH AND REPORT A bacterial infection often causes inflammation, or a response to tissue damage that can include swelling and pain. Research and report on what causes inflammation.

Lesson 2

Bacteria in NATURE

ESSENTIAL QUESTIONS

 How can bacteria affect the environment?

 How can bacteria affect health?

Vocabulary

decomposition p. 262
nitrogen fixation p. 262
bioremediation p. 263
pathogen p. 264
antibiotic p. 264
pasteurization p. 265

SC.6.N.1.5

Launch Lab

10 minutes

How do bacteria affect the environment?

Bacteria are everywhere in your environment. They are in the water, in the air, and even in some foods.

Procedure

1. Read and complete a lab safety form.
2. Carefully examine the contents of the two **bottles** provided by your teacher.
3. Record your observations below.

Data and Observations

Think About This

1. Compare your observations of bottle A to those of bottle B. Which one appears to have more bacteria in it? Support your answer.

2. Key Concept Based on your observations, how could bacteria affect the environment around you?

Florida NGSSS

LA.6.2.2.3 The student will organize information to show understanding (e.g., representing main ideas within text through charting, mapping, paraphrasing, summarizing, or comparing/contrasting);

SC.6.N.1.4 Discuss, compare, and negotiate methods used, results obtained, and explanations among groups of students conducting the same investigation.

SC.6.N.1.5 Recognize that science involves creativity, not just in designing experiments, but also in creating explanations that fit evidence.

SC.6.N.2.1 Distinguish science from other activities involving thought.

SC.6.L.14.6 Compare and contrast types of infectious agents that may infect the human body, including viruses, bacteria, fungi, and parasites.

SC.7.L.17.1 Explain and illustrate the roles of and relationships among producers, consumers, and decomposers in the process of energy transfer in a food web.

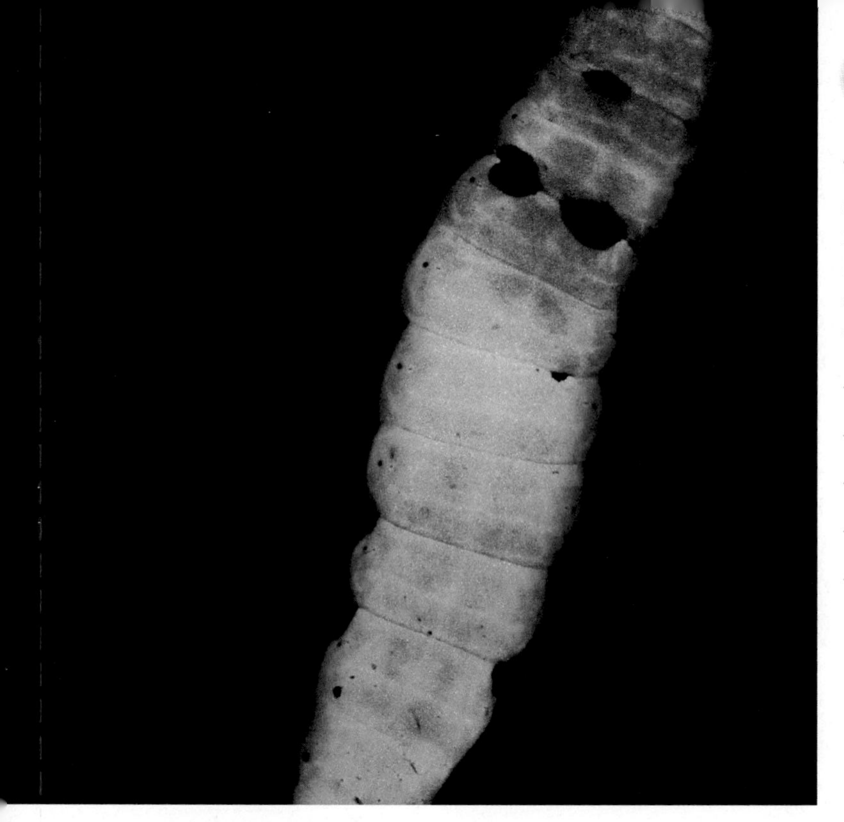

Inquiry Why does this larva glow?

1. Some bacteria have the ability to glow in the dark. The moth larva shown on this page is filled with many such bacteria. These bacteria produce toxins that can slowly kill the animal. A chemical reaction within each bacterium makes the larva's body appear to glow. Why can't you see the individual bacteria that cause the glowing of the moth larva?

Beneficial Bacteria

When you hear about bacteria, you probably think about getting sick. However, only a fraction of all bacteria cause diseases. Most bacteria are beneficial. In fact, many organisms, including humans, depend on bacteria to survive. Some types of bacterium help with digestion and other body processes. For example, one type of bacterium in your intestines makes vitamin K, which helps your blood clot properly. Several others help break down food into smaller particles. Another type of bacterium called *Lactobacillus* lives in your intestines and prevents harmful bacteria from growing.

Animals benefit from bacteria as well. Without bacteria, some organisms, such as the cow pictured in **Figure 7,** wouldn't be able to digest the plants they eat. Bacteria and other microscopic organisms live in a large section of the cow's stomach called the rumen. The bacteria help break down a substance in grass called cellulose into smaller molecules that the cow can use.

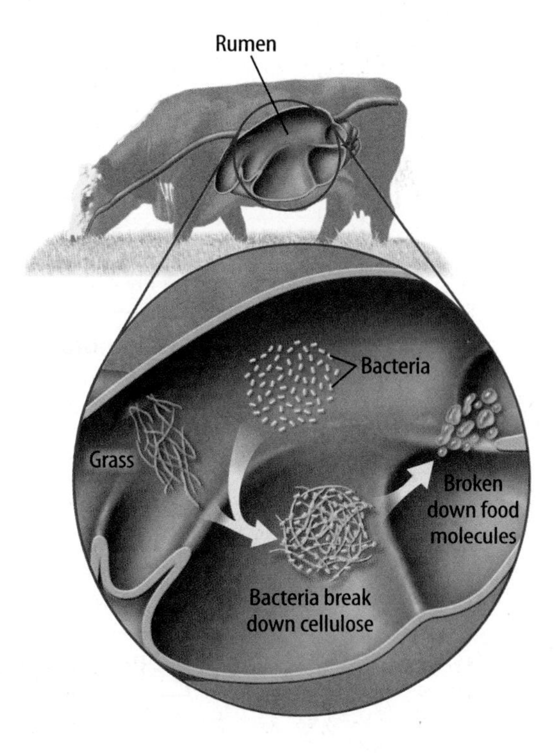

Figure 7 Cows get help digesting the cellulose in plants from the bacteria that live in their rumen—one of four stomach sections.

2. **Visual Check** Infer Examine **Figure 7.** What would happen if bacteria were not present?

Active Reading 3. **Explain** What role do bacteria play in a cow's digestion?

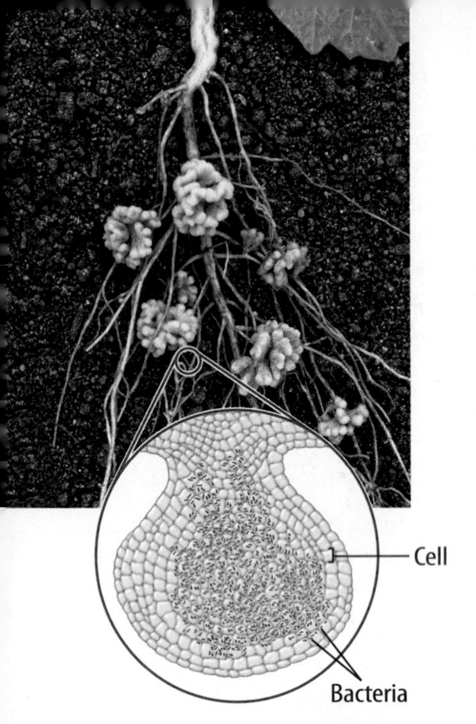

Figure 8 The roots of some plants have nodules that contain nitrogen-fixing bacteria.

Decomposition

What do you think would happen if organic waste such as food scraps and dead leaves never decayed? **Decomposition,** *the breaking down of dead organisms and organic waste,* is an important process in nature. When a tree dies, bacteria and other decomposing organisms feed on the dead organic matter. As decomposers break down the tree, they release molecules such as carbon and phosphorus into the soil that other organisms can then take in and use for life processes.

Nitrogen Fixation

Organisms use nitrogen to make proteins. Although about 78 percent of the atmosphere is nitrogen gas, it is in a form that plants and animals cannot use. Some plants can obtain nitrogen from bacteria. These plants have special structures called nodules, shown in **Figure 8,** on their roots. Bacteria in the nodules convert nitrogen from the atmosphere into a form usable to plants. **Nitrogen fixation** *is the conversion of atmospheric nitrogen into nitrogen compounds that are usable by living things.*

Active Reading **4. Explain** What is nitrogen fixation?

______

______

______

Active Reading **5. Explain** As you complete the lesson, provide examples of ways in which bacteria can be beneficial.

Benefit	Explanation and Example
Digestion	
Decomposition	
Nitrogen fixation	
Bioremediation	
Production of food	

MiniLab ***Can decomposition happen without oxygen?*** at connectED.mcgraw-hill.com

Apply It! After you complete the lab, answer this question.

1. Why might decomposers be described as the ultimate recyclers even if it takes many years for them to break down a tree?

Bioremediation

Can you imagine an organism that eats pollution? Some bacteria do just that. *The use of organisms, such as bacteria, to clean up environmental pollution is called* **bioremediation** (bi oh rih mee dee AY shun). These organisms often break down harmful substances, such as sewage, into less harmful material that can be used as landfill or fertilizers.

Other kinds of bacteria can help clean up radioactive waste, such as uranium in abandoned minefields. In many cases, without using bacteria, the substances would take centuries to break down and would contaminate soils and water. Bacteria are also commonly used to clean up areas that have been contaminated by oil, such as the Gulf of Mexico. Bacteria could aid in the long-term Gulf cleanup resulting from the explosion on the *Deepwater Horizons* oil platform in 2010. These natural oil-eating bacteria will help consume the oil and other pollutants in the water as shown in **Figure 9.**

Active Reading **6. Explain** Why might using bacteria to clean up environmental spills be a good option?

____

____

____

Active Reading **FOLDABLES®** LA.6.2.2.3

Make a four-door book and label it as shown. Use it to summarize the ways bacteria are beneficial to the environment.

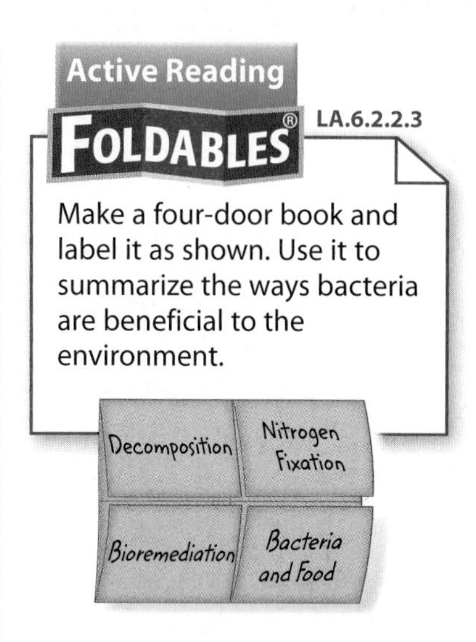

Figure 9 Some bacteria clean the environment by removing harmful pollutants from the water.

Active Reading **7. List** What events might cause oil contamination of water?

____

____

____

Bacteria and Food

Would you like a side of bacteria with that sandwich? If you have eaten a pickle lately, you might have had some. Some pickles are made when the sugar in cucumbers is converted into an acid by a specific type of bacteria. Pickles are just one of the many food products made with the help of bacteria. Bacteria are used to make foods such as yogurt, cheese, buttermilk, vinegar, and soy sauce. Bacteria are even used in the production of chocolate. They help break down the covering of the cocoa bean during the process of making chocolate. Bacteria are responsible for giving chocolate some of its flavor.

Harmful Bacteria

Of the 5,000 known species of bacteria, relatively few are considered **pathogens** (PA thuh junz)—*agents that cause disease.* Some pathogens normally live in your body but cause illness only when your immune system is weakened. For example, the bacterium *Streptococcus pneumoniae* lives in the throats of most healthy people. However, it can cause pneumonia if a person's immune system is weakened. Other bacterial pathogens can enter your body through a cut, the air you breathe, or the food you eat. Once inside your body, they can reproduce and cause disease.

Word Origin
pathogen
from Greek *pathos,* means "to suffer"; and *gen,* means "to produce"

8. **NGSSS Check** **Describe** Give an example of one way that bacteria can be harmful to your health. SC.6.L.14.6

____

____

____

Bacterial Diseases

Bacteria can harm your body and cause disease in one of two ways. Some bacteria make you sick by damaging tissue. For example, the disease tuberculosis, shown in **Figure 10,** is caused by a bacterium that invades lung tissue and breaks it down for food. Other bacteria cause illness by releasing toxins. For example, the bacterium *Clostridium botulinum* can grow in improperly canned foods and produce toxins. If the contaminated food is eaten, the toxins can cause food poisoning, resulting in paralyzed limbs or even death.

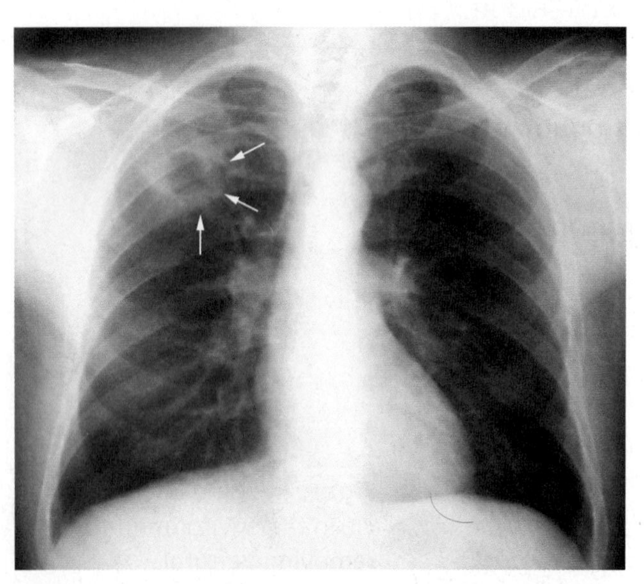

Figure 10 In an X-ray, the lungs of a person with tuberculosis may show pockets or scars where bacterial infection has begun.

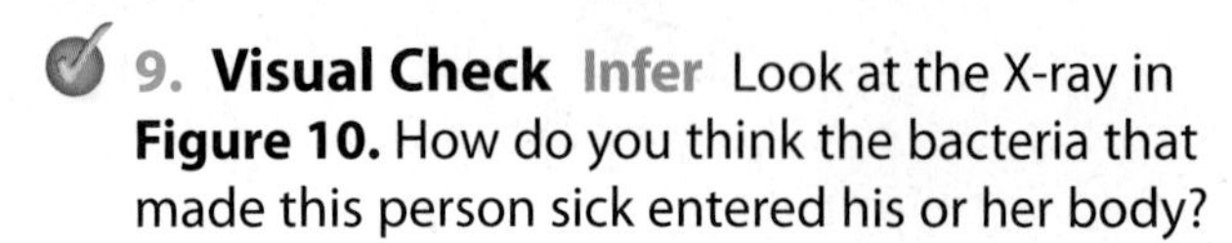

9. **Visual Check** **Infer** Look at the X-ray in **Figure 10.** How do you think the bacteria that made this person sick entered his or her body?

____

____

____

Treating Bacterial Diseases Most bacterial diseases in humans can be treated with antibiotics. **Antibiotics** (an ti bi AH tihks) *are medicines that stop the growth and reproduction of bacteria.* Many antibiotics work by preventing bacteria from building cell walls. Others affect ribosomes in bacteria, interrupting the production of proteins.

Many types of bacteria have become **resistant** to antibiotics over time. Some diseases, such as tuberculosis, pneumonia, and meningitis, are now more difficult to treat.

Science Use v. Common Use
resistance
Science Use the capacity of an organism to defend itself against a disease
Common Use the act of opposing something

Bacterial Resistance How do you think bacteria become resistant to antibiotics? This process, shown in **Figure 11,** can happen over a long or short period of time depending on how quickly the bacteria reproduce. Random mutations occur to a bacterium's DNA that enable it to survive or "resist" a specific antibiotic. If that antibiotic is used as a treatment, only the bacteria with the mutation will survive.

Over time, the resistant bacteria will reproduce and become more common. The antibiotic is no longer effective against that bacterium, and a different antibiotic must be used to fight the disease. Scientists are always working to develop more effective antibiotics to which bacteria have not developed resistance.

10. Describe Underline a section in this text discussing how bacteria develop resistance to antibiotics.

Food Poisoning

All food, unless it has been treated or processed, contains bacteria. Over time these bacteria reproduce and begin breaking down the food, causing it to spoil. As you read on the previous page, eating food contaminated by some bacteria can cause food poisoning. By properly treating or processing food and killing bacteria before the food is stored or eaten, it is easier to avoid food poisoning and other illnesses.

Pasteurization (pas chuh ruh ZAY shun) *is a process of heating food to a temperature that kills most harmful bacteria.* Products such as milk, ice cream, yogurt, and fruit juice are usually pasteurized in factories before they are transported to grocery stores and sold to you. After pasteurization, foods are much safer to eat. Foods do not spoil as quickly once they have been pasteurized. Because of pasteurization, food poisoning is much less common today than it was in the past.

11. Explain How does pasteurization affect human health?

___

___

How Resistance Develops

Figure 11 A population of bacteria can develop resistance to antibiotics after being exposed to them over time.

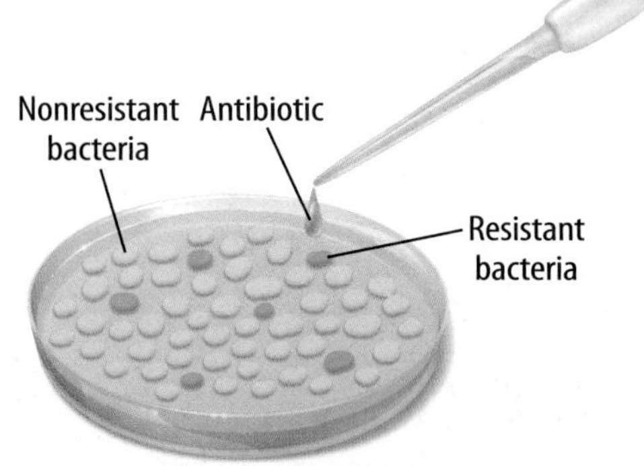

1. An antibiotic is added to a colony of bacteria. A few of the bacteria have mutations that enable them to resist the antibiotic.

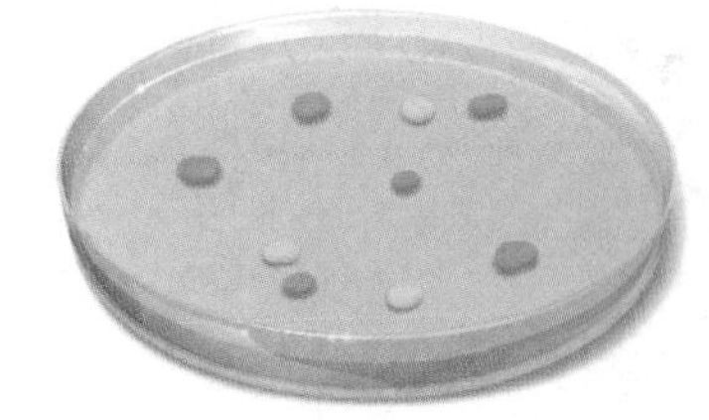

2. The antibiotic kills most of the nonresistant bacteria. The resistant bacteria survive and reproduce, creating a growing colony of bacteria.

3. Surviving bacteria are added to another plate containing more of the same antibiotic.

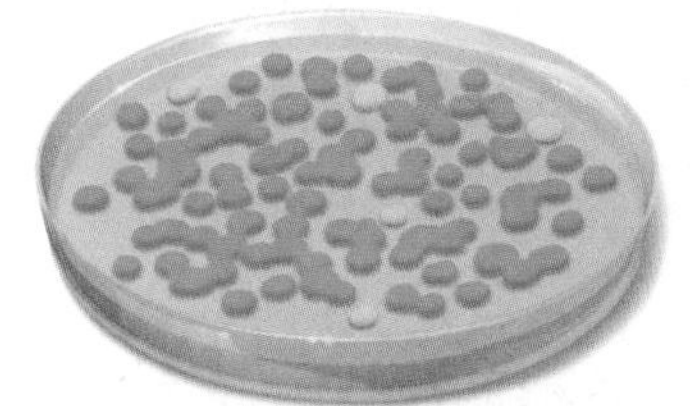

4. The antibiotic now affects only a small percentage of the bacteria. The surviving bacteria continue to reproduce. Most of the bacteria are resistant to the antibiotic.

Lesson Review 2

Visual Summary

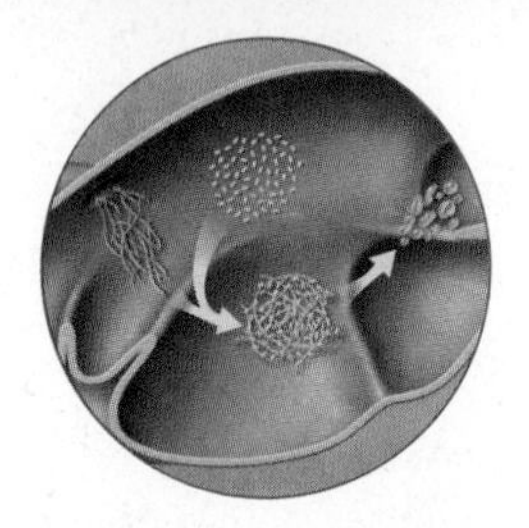

Bacteria can help some organisms, including humans and cows, digest food.

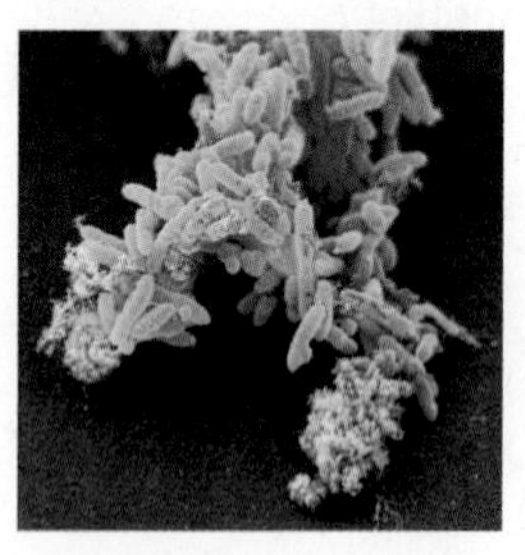

Bacteria can be used to remove harmful substances such as uranium.

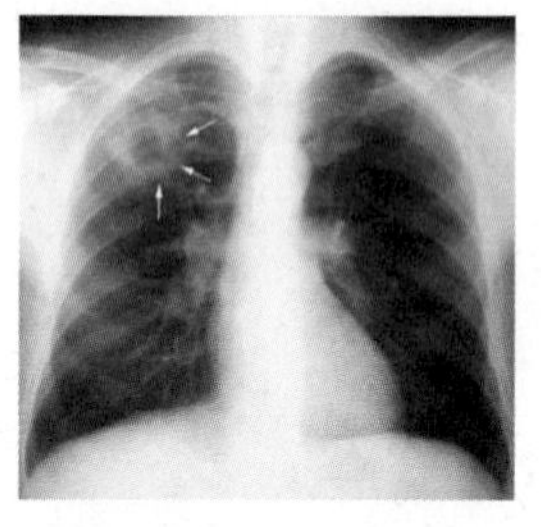

Some bacteria are pathogen, and cause diseases in humans and other organisms.

Use Vocabulary

1. **Distinguish** between an antibiotic and a pathogen.

2. **Define** *bioremediation* using your own words.

3. **Use the term** *pasteurization* in a sentence.

Understand Key Concepts

4. Which is NOT a beneficial use of bacteria? **SC.6.L.14.6**
   - (A) bioremediation
   - (B) decomposition
   - (C) food poisoning
   - (D) nitrogen fixation

5. **Compare** the benefits of nitrogen fixation and decomposition.

6. **Analyze** the importance of bacteria in food production.

Interpret Graphics

7. **Examine** the figure and describe what would happen if bacteria were not present. **SC.6.L.14.6**

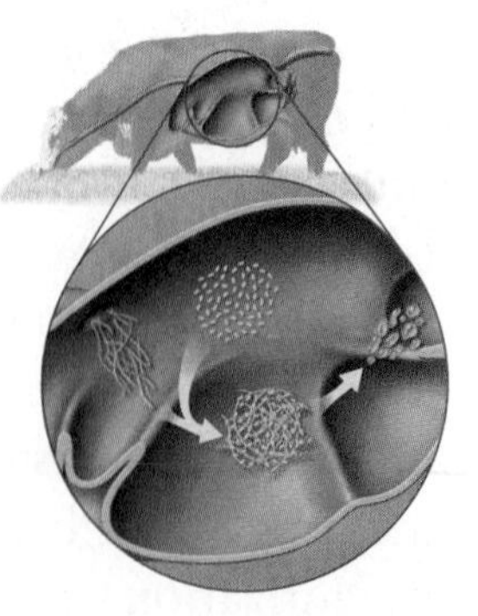

Critical Thinking

8. **Evaluate** the effect of all bacteria becoming resistant to antibiotics.

Model Complete the graphic organizer below to identify ways that bacteria can be beneficial. LA.6.2.2.3

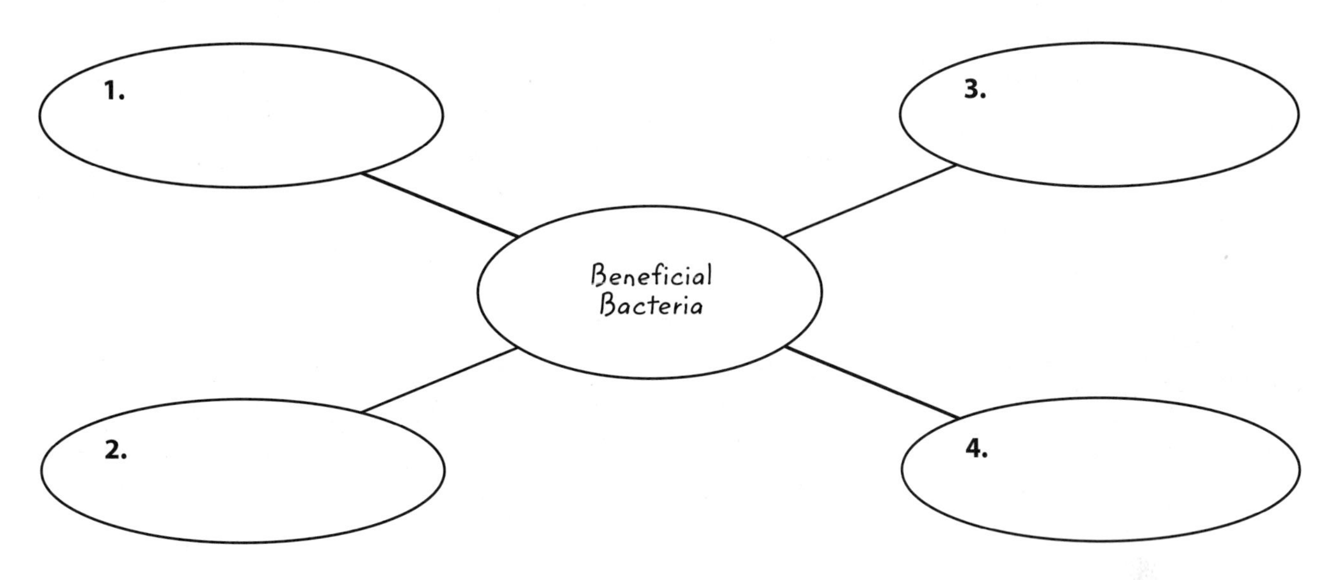

Sequence Complete the graphic organizer below to identify the development of antibiotic resistance.

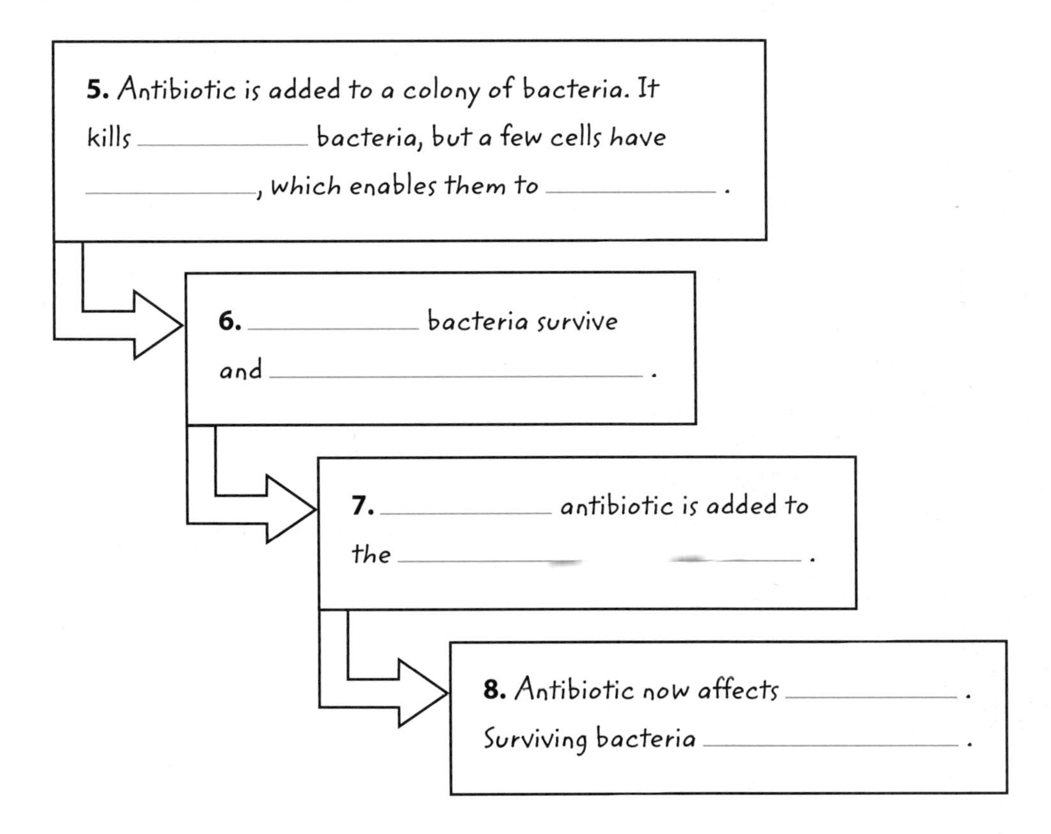

Lesson 3

What are VIRUSES?

ESSENTIAL QUESTIONS

What are viruses?

How do viruses affect human health?

Vocabulary

virus p. 269

antibody p. 273

vaccine p. 274

Florida NGSSS

LA.6.2.2.3 The student will organize information to show understanding (e.g., representing main ideas within text through charting, mapping, paraphrasing, summarizing, or comparing/contrasting);

SC.6.L.14.6 Compare and contrast types of infectious agents that may infect the human body, including viruses, bacteria, fungi, and parasites.

SC.6.N.1.1 Define a problem from the sixth grade curriculum, use appropriate reference materials to support scientific understanding, plan and carry out scientific investigation of various types, such as systematic observations or experiments, identify variables, collect and organize data, interpret data in charts, tables, and graphics, analyze information, make predictions, and defend conclusions.

SC.6.N.2.1 Distinguish science from other activities involving thought.

SC.6.N.3.4 Identify the role of models in the context of the sixth grade science benchmarks.

SC.7.L.16.4 Recognize and explore the impact of biotechnology (cloning, genetic engineering, artificial selection) on the individual, society and the environment.

SC.8.N.1.3 Use phrases such as "results support" or "fail to support" in science, understanding that science does not offer conclusive 'proof' of a knowledge claim.

SC.8.N.1.6 Understand that scientific investigations involve the collection of relevant empirical evidence, the use of logical reasoning, and the application of imagination in devising hypotheses, predictions, explanations and models to make sense of the collected evidence.

SC.6.N.1.5, SC.6.N.2.1

Inquiry Launch Lab

10 minutes

How quickly do viruses replicate?

One characteristic that viruses share is the ability to produce many new viruses from just one virus. In this lab you can use grains of rice to model virus replication. Each grain of rice represents one virus.

Procedure

1. Read and complete a lab safety form.
2. Estimate the number of **grains of rice** in the **fishbowl** and record this number for the first generation.
3. One student will add the contents of his or her **cup** to the fishbowl. Estimate how many viruses are now in the fishbowl and record your estimate for the second generation.
4. The rest of the class will add the contents of their cups to the fishbowl. Estimate the number of viruses and record that number of viruses for the third generation.

Data and Observations

Generation	First	Second	Third
Number of "viruses"			

Think About This

1. Recall that bacteria double every generation. How does the number of viruses produced in each generation compare with the number of bacteria produced in each generation?

2. Key Concept How could the rate at which viruses are produced affect human health?

Painted Flowers?

1. The streaking patterns on the petals of these tulips are not painted on but are caused by a virus. Tulips with these patterns are prized for their beautiful appearance. How do you think a virus could cause this flower's pattern? Do you think all viruses are harmful?

Characteristics of Viruses

Do chicken pox, mumps, measles, and polio sound familiar? You might have received shots to protect you from these diseases. You might have also received a shot to protect you from influenza, commonly known as the flu. What do these diseases have in common? They are caused by different viruses. *A* **virus** *is a strand of DNA or RNA surrounded by a layer of protein that can infect and replicate in a host cell.* If you have had a cold, you have been infected by a virus.

A virus does not have a cell wall, a nucleus, or any other organelles present in cells. The smallest viruses are between 20 and 100 times smaller than most bacteria. Recall that about 100 bacteria would fit across the head of a pin. Viruses can have different shapes, such as the crystal, cylinder, sphere, and bacteriophage (bak TIHR ee uh fayj) shapes shown in **Figure 12.**

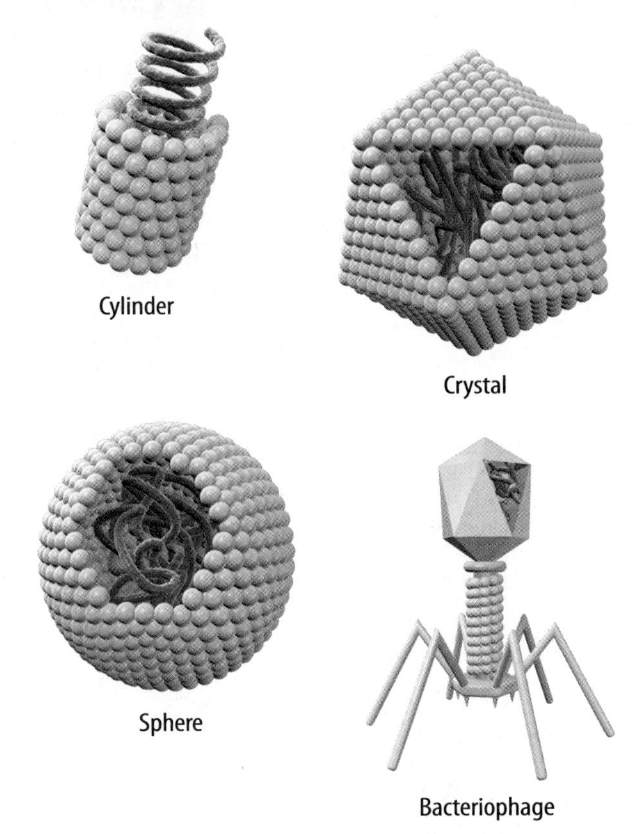

Figure 12 Viruses have a variety of shapes.

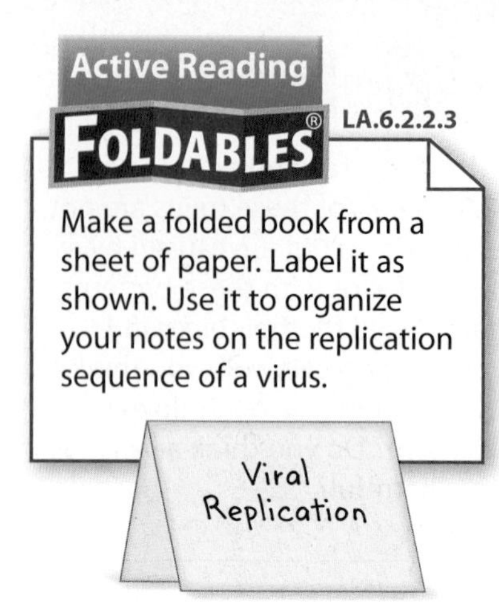

Active Reading **2. Infer** Are viruses alive? Explain why or why not.

Active Reading **3. Describe** What occurs when a virus becomes latent?

Dead or Alive?

Scientists do not consider viruses to be alive because they do not have all the characteristics of a living organism. Recall that living things are organized, respond to stimuli, use energy, grow, and reproduce. Viruses cannot do any of these things. A virus can make copies of itself in a process called replication, but it must rely on a living organism to do so.

Viruses and Organisms

Viruses must use organisms to carry on the processes that we usually associate with a living cell. Viruses have no organelles so they are not able to take in nutrients or use energy. They also cannot replicate without using the cellular parts of an organism. Viruses must be inside a cell to replicate. The living cell that a virus infects is called a host cell.

When a virus enters a cell, as shown in **Figure 13,** it can either be active or latent. Latent viruses go through an inactive stage. Their genetic material becomes part of the host cell's genetic material. For a period of time, the virus does not take over the cell to produce more viruses. In some cases, viruses have been known to be inactive for years and years. However, once it becomes active, a virus takes control of the host cell and replicates.

Figure 13 A virus infects a cell by inserting its DNA or RNA into the host cell. It then directs the host cell to make new viruses.

Active Reading **4. Model** Complete the chart below with the correct information.

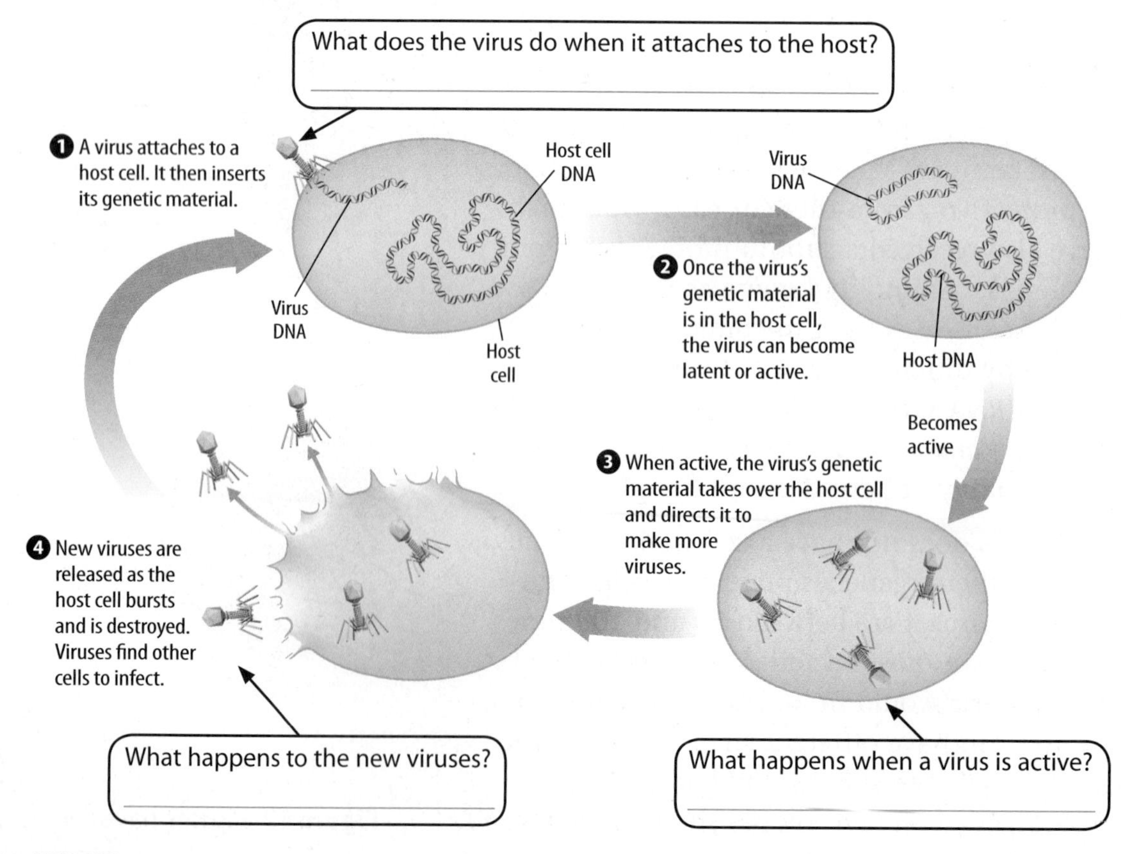

Replication

As you read earlier, a virus can make copies of itself in a process called replication, shown in **Figure 13.** A virus cannot infect every cell. A virus can only attach to a host cell with specific molecules on its cell wall or cell membrane. These molecules enable the virus to attach to the host cell. This is similar to the way that only certain electrical plugs can fit into an outlet on a wall. After a virus attaches to the host cell, its DNA or RNA enters the host cell. Once inside, the virus either starts to replicate or becomes latent, also shown in **Figure 13.** After a virus becomes active and replicates in a host cell, it destroys the host cell. Copies of the virus are then released into the host organism, where they can infect other cells.

Mutations

As viruses replicate, their DNA or RNA frequently mutates, or changes. These **mutations** enable viruses to adjust to changes in their host cells. For example, the molecules on the outside of host cells change over time to prevent viruses from attaching to the cell. As viruses mutate, they are able to produce new ways to attach to host cells. These changes happen so rapidly that it can be difficult to cure or prevent viral diseases before they mutate again.

REVIEW VOCABULARY

mutation
a change in genetic material

Active Reading **5. Describe** How does mutation enable viruses to continue causing disease?

Active Reading **6. Model** Complete the chart below with the correct information.

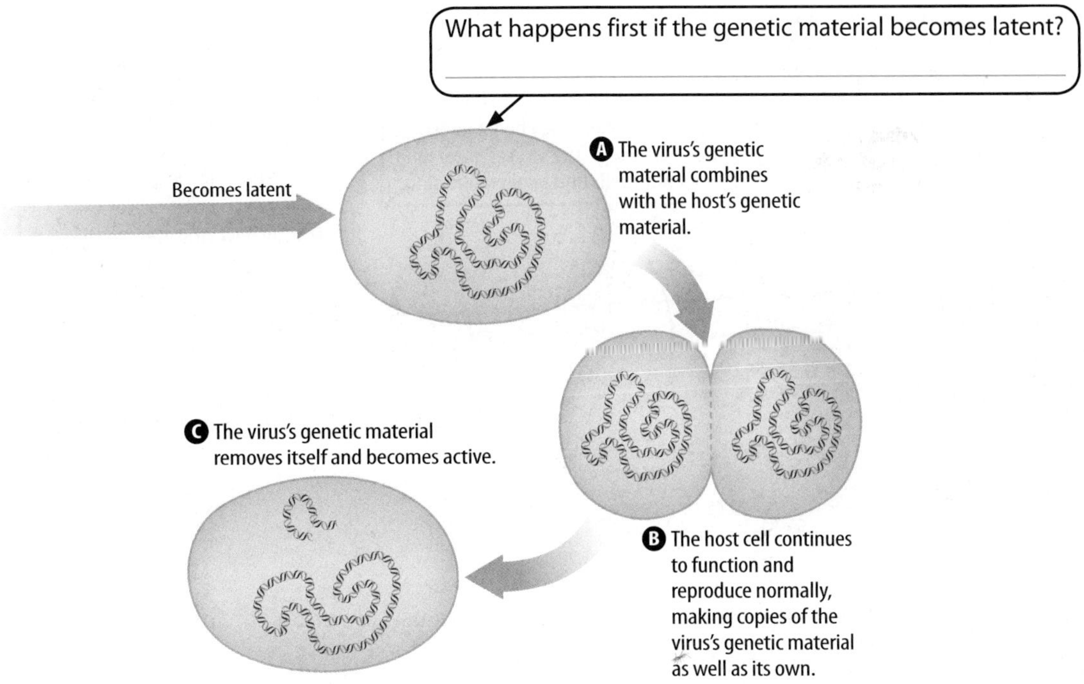

What happens first if the genetic material becomes latent?

Becomes latent

A The virus's genetic material combines with the host's genetic material.

B The host cell continues to function and reproduce normally, making copies of the virus's genetic material as well as its own.

C The virus's genetic material removes itself and becomes active.

Viral Diseases

You might know that viruses cause many human diseases, such as chicken pox, influenza, some forms of pneumonia, and the common cold. But viruses also infect animals, causing diseases such as rabies and parvo. They can infect plants as well—in some cases causing millions of dollars of damage to crops. The tulips shown at the beginning of this lesson were infected with a virus that caused a streaked appearance on the petals. Most viruses attack and destroy specific cells. This destruction of cells causes the symptoms of the disease.

Some viruses cause symptoms soon after infection. Influenza viruses that cause the flu infect the cells lining your respiratory system, as shown in **Figure 14.** The viruses begin to replicate immediately. Flu symptoms, such as a runny nose and a scratchy throat, usually appear within two to three days.

Other viruses might not cause symptoms right away. These viruses are sometimes called latent viruses. Latent viruses continue replicating without damaging the host cell. HIV (human immunodeficiency virus) is one example of a latent virus that might not cause immediate symptoms.

HIV infects white blood cells, which are part of the immune system. Initially, infected cells can function normally, so an HIV-infected person might not appear sick. However, the virus can become active and destroy cells in the body's immune system, making it hard to fight other infections. It can often take a long time for symptoms to appear after infection. People infected with latent viruses might not know for many years that they have been infected.

Active Reading **7. Infer** Why is HIV considered a latent virus?

____

____

____

The Flu **Figure 14**

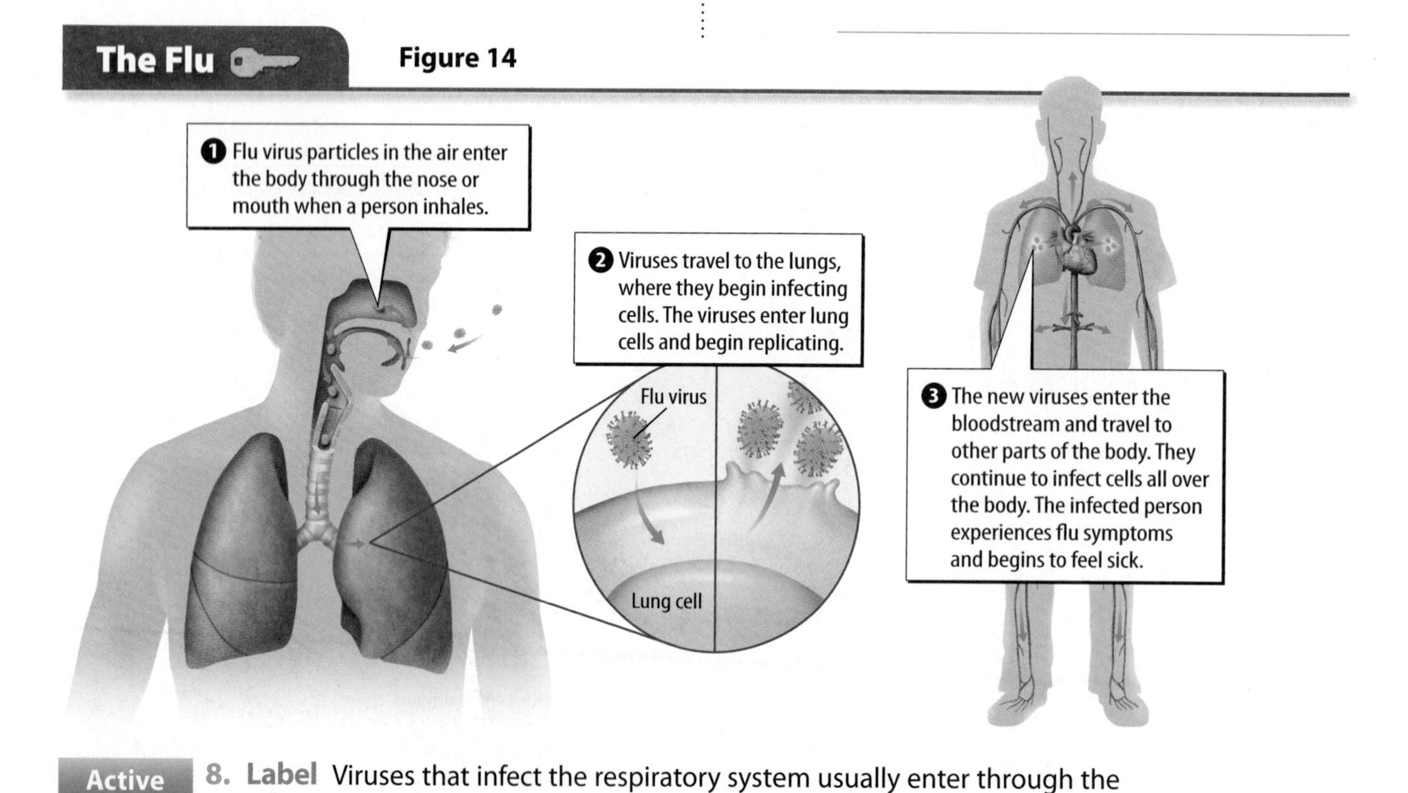

Active Reading **8. Label** Viruses that infect the respiratory system usually enter through the ______________________ or ______________________.

9. Visual Check Examine (Circle) the area where flu viruses replicate in the image above.

Treating and Preventing Viral Diseases

Since viruses are constantly changing, viral diseases can be difficult to treat. Antibiotics work only against bacteria, not viruses. Antiviral medicines can be used to treat certain viral diseases or prevent infection. These medicines prevent the virus from entering a cell or stop the virus from replicating. Antiviral medicines are specific to each virus. Like bacteria, viruses can rapidly change and become resistant to medicines.

Health officials use many methods to prevent the spread of viral diseases. One of the best ways to prevent a viral infection is to limit contact with an infected human or animal. The most important way to prevent infections is to practice good hygiene, such as washing your hands.

Immunity

Has anyone you know ever had chicken pox? Did the person get it more than once? Most people who became infected with chicken pox develop an immunity to the disease. This is an example of acquired **immunity**. When a virus infects a person, his or her body begins to make special proteins called antibodies. An **antibody** *is a protein that can attach to a pathogen and make it useless.* Antibodies bind to viruses and other pathogens and prevent them from attaching to a host cell, as shown in **Figure 15.** The antibodies also target viruses and signal the body to destroy them. These antibodies can multiply quickly if the same pathogen enters the body again, making it easier for the body to fight infection. Another type of immunity, called natural immunity, develops when a mother passes antibodies to her unborn baby.

Word Origin

immunity
from Latin *immunis,* means "exempt, free"

Antibodies

Figure 15 Antibodies bind to pathogens and prevent them from attaching to cells.

10. Visual Check **Interpret** How does the antibody prevent the virus from attaching to the host cell?

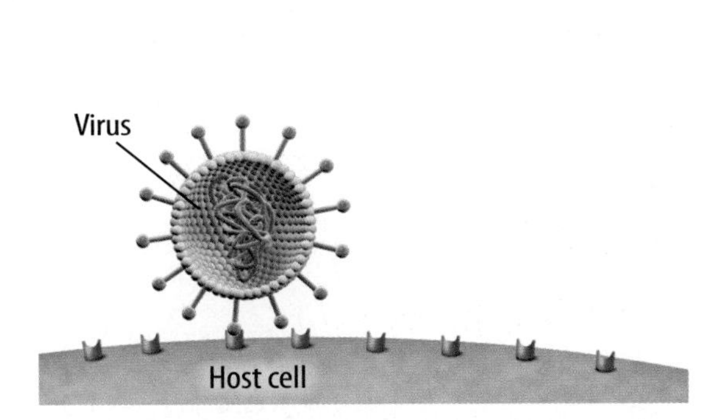

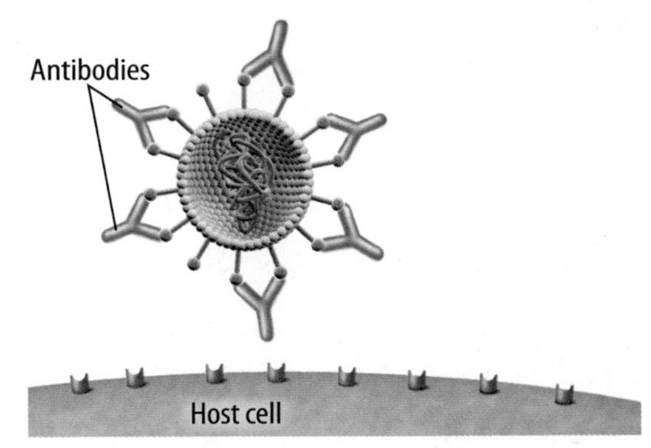

Vaccines

One way to prevent viral diseases is through vaccination. *A* **vaccine** *is a mixture containing material from one or more deactivated pathogens, such as viruses.* When an organism is given a vaccine for a viral disease, the vaccine triggers the production of antibodies. This is similar to what would happen if the organism became infected with the virus normally. However, because the vaccine contains deactivated pathogens, the organism suffers only mild symptoms or none at all. After being vaccinated against a particular pathogen, the organism will not get as sick if exposed to the pathogen again.

Vaccines can prevent diseases in animals as well as humans. For example, pet owners and farmers get annual rabies vaccinations for their animals. This protects the animals from the disease. Humans are then protected from rabies.

Research with Viruses

Scientists are researching new ways to treat and prevent viral diseases in humans, animals, and plants. Scientists are also studying the link between viruses and cancer. Viruses can cause changes in a host's DNA or RNA, resulting in the formation of tumors or abnormal growth. Because viruses can change very quickly, scientists must always be working on new ways to treat and prevent viral diseases.

You might think that all viruses are harmful. However, scientists have also found beneficial uses for viruses. Viruses may be used to treat genetic disorders and cancer using gene transfer. Scientists use viruses to insert normal genetic information into a specific cell. Scientists hope that gene transfer will eventually be able to treat genetic disorders that are caused by one gene, such as cystic fibrosis or hemophilia.

11. NGSSS Check Analyze How do viruses affect human health? SC.6.L.14.6

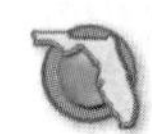

Apply It! After you complete the lab, answer these questions.

1. What is a vaccine?

2. What is a vaccine used for?

3. How is an organism's reaction to a vaccine different from the reaction that would occur if the organism became infected naturally?

Lesson Review 3

Visual Summary

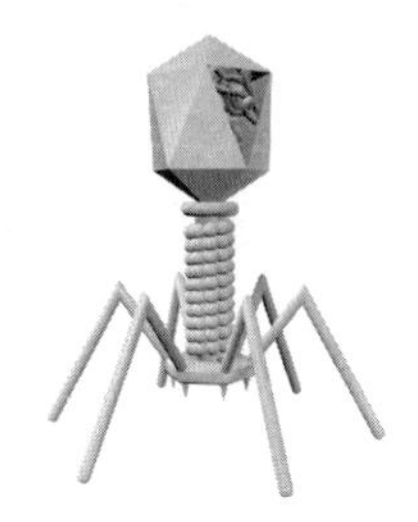

A virus is a strand of DNA or RNA surrounded by a layer of protein.

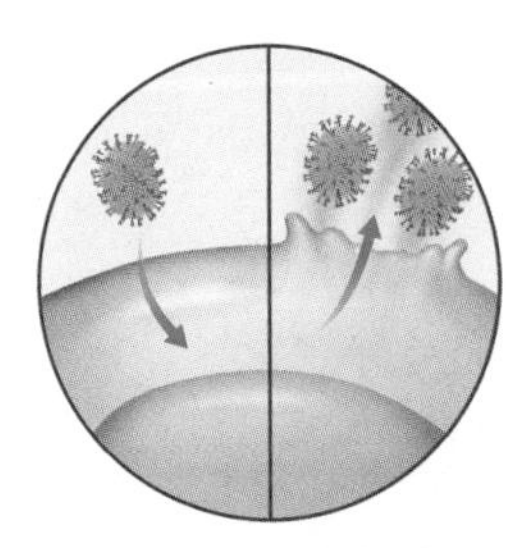

Viruses cause human diseases such as chicken pox and influenza.

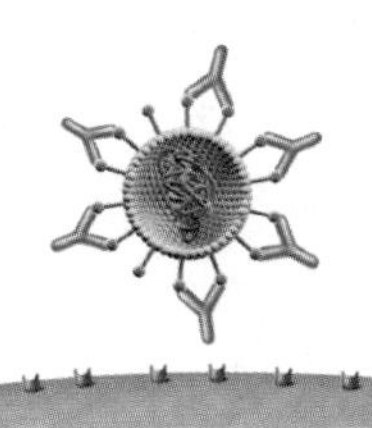

A person's body produces proteins called antibodies that prevent an infection by viruses.

Use Vocabulary

1. **List** the different virus shapes. SC.6.L.14.6

2. **Describe** in your own words how a vaccine works.

Understand Key Concepts 

3. **Describe** the structure of a virus. SC.6.L.14.6

4. Which is made by the body to fight viruses?
   - (A) antibody
   - (B) bacteria
   - (C) bacteriophage
   - (D) proteins

5. **Compare** a vaccine and an antibody.

Interpret Graphics

6. **Label** the the steps that occur when a virus infects a cell. LA.6.2.2.3

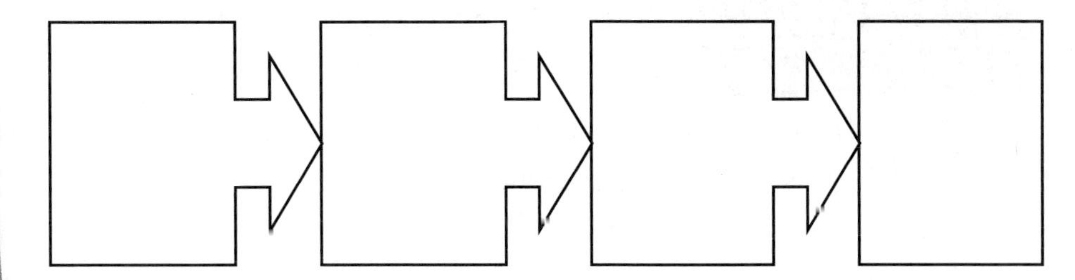

Critical Thinking

7. **Predict** the effect of preventing future mutations of the influenza virus. SC.6.L.14.6

Chapter 7 Study Guide

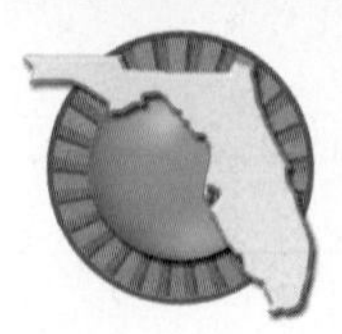

Think About It! Bacteria are unicellular prokaryotes, and viruses are small pieces of DNA or RNA surrounded by protein. Both bacteria and viruses may infect the human body; they can cause harmful diseases or can be useful.

Key Concepts Summary

Vocabulary

LESSON 1 What are bacteria?

- **Bacteria** and archeans are unicellular organisms without nuclei. They have structures for movement, obtaining food, and reproduction.
- Bacteria exchange genetic information in a process called conjugation. They reproduce asexually by fission.

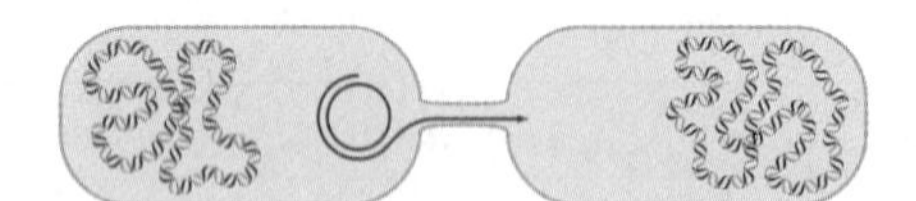

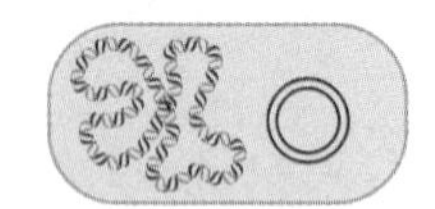

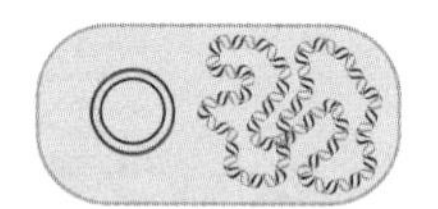

bacterium p. 253
flagellum p. 256
fission p. 256
conjugation p. 256
endospore p. 257

LESSON 2 Bacteria in Nature

- Bacteria decompose materials, play a role in the nitrogen cycle, clean the environment, and are used in food.
- Some bacteria cause disease, and others are used to treat it.

decomposition p. 262
nitrogen fixation p. 262
bioremediation p. 263
pathogen p. 264
antibiotic p. 264
pasteurization p. 265

LESSON 3 What are viruses?

- A **virus** is made up of DNA or RNA surrounded by a protein coat.
- Viruses can cause disease, can be made into vaccines, and are used in research.

virus p. 269
antibody p. 273
vaccine p. 274

Active Reading

FOLDABLES® Chapter Project

Assemble your lesson Foldables as shown to make a Chapter Project. Use the project to review what you have learned in this chapter.

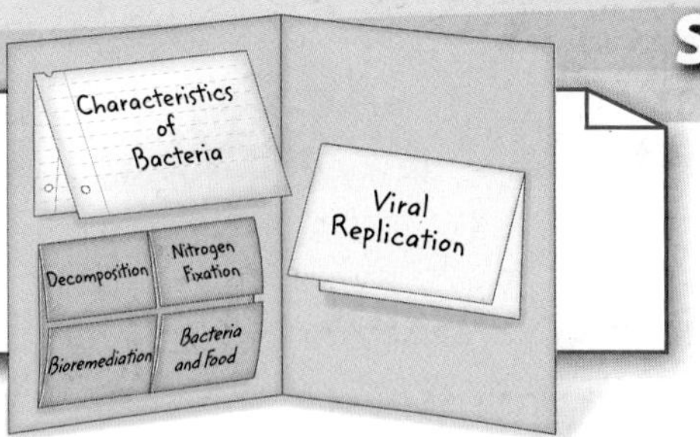

Use Vocabulary

1. Some bacteria have whiplike structures called ________________ that are used for movement.

2. Your body produces proteins called ________________ in response to infection by a virus.

3. Organisms that cause diseases are known as ________________.

4. The process of killing bacteria in a food product by heating it is called ________________.

5. Bacteria can form a(n) ________________ to survive when environmental conditions are severe.

6. A(n) ________________ is made by using pieces of deactivated viruses or dead pathogens.

Link Vocabulary and Key Concepts

Use vocabulary terms from the previous page and other terms from the chapter to complete the concept map.

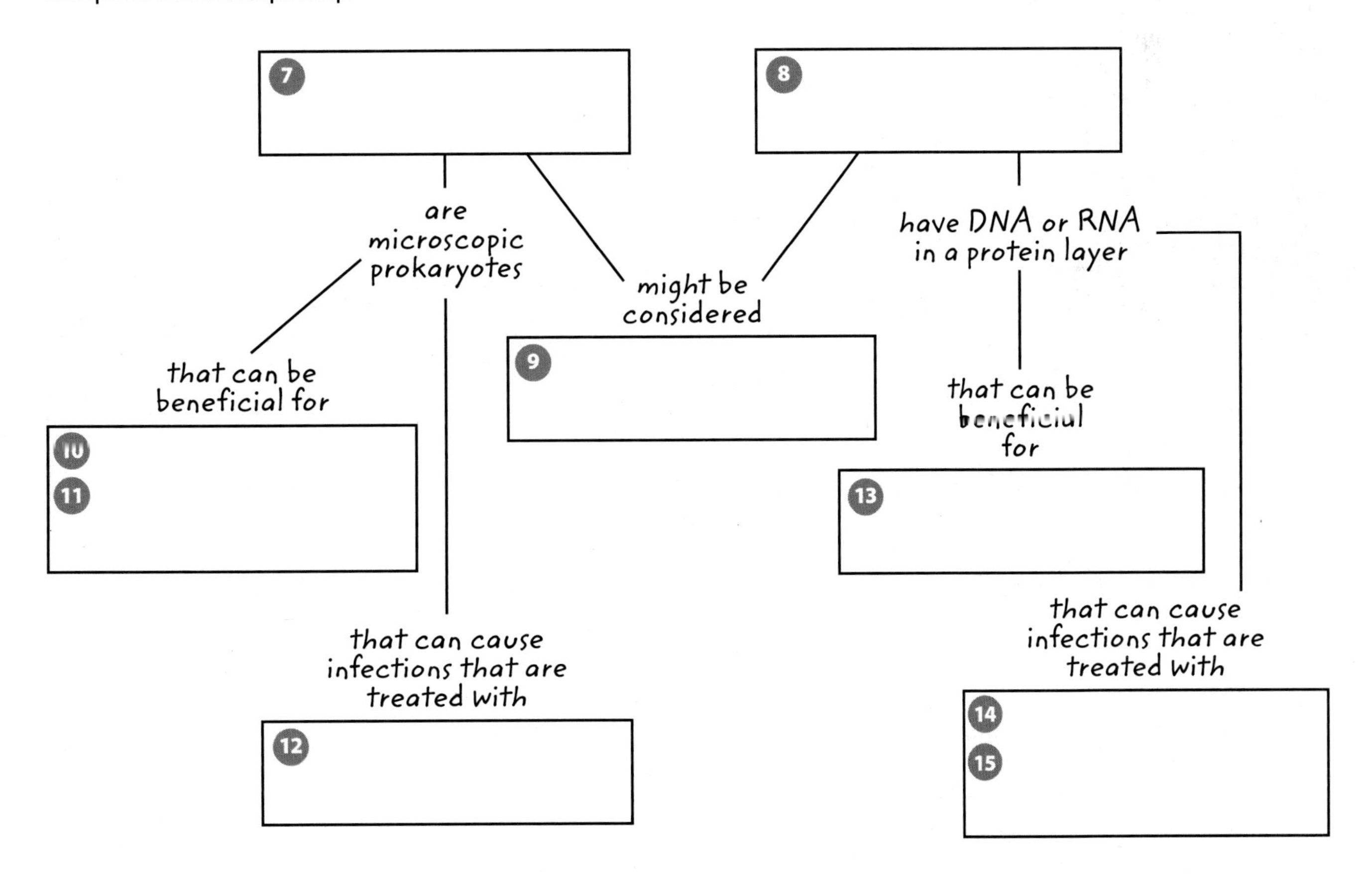

Chapter 7 Review

Understand Key Concepts

1. Which is NOT caused by a virus? **SC.6.L.14.6**
   Ⓐ chicken pox
   Ⓑ influenza
   Ⓒ rabies
   Ⓓ tuberculosis

2. What do vaccines stimulate the production of? **SC.6.L.14.6**
   Ⓐ antibodies
   Ⓑ DNA and RNA
   Ⓒ protein
   Ⓓ ribosomes

3. What process is occurring in the illustration below? **SC.6.L.14.6**

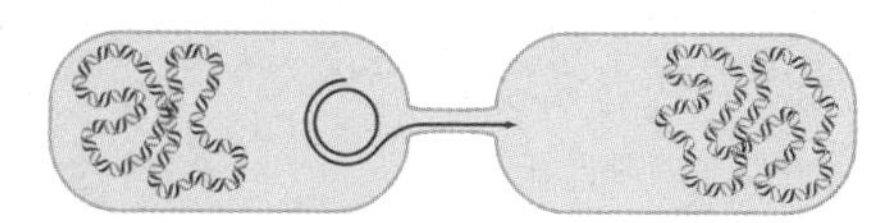

   Ⓐ budding
   Ⓑ conjugation
   Ⓒ fission
   Ⓓ replication

4. Which term describes how bacteria can be used to clean up environmental waste? **SC.6.L.14.6**
   Ⓐ bioremediation
   Ⓑ decomposition
   Ⓒ pasteurization
   Ⓓ nitrogen fixation

5. Which statement correctly describes pathogens? **SC.6.L.14.6**
   Ⓐ They are always bacteria.
   Ⓑ They are in your body only when you are sick.
   Ⓒ They break down dead organisms.
   Ⓓ They cause disease.

6. Which statement correctly describes antibiotics? **SC.6.L.14.6**
   Ⓐ They can kill any kind of bacterium.
   Ⓑ They help bacteria grow.
   Ⓒ They stop the growth and reproduction of bacteria.
   Ⓓ They treat all diseases.

Critical Thinking

7. **Compare and contrast** bacteria and archaea. **LA.6.2.2.3**

8. **Evaluate** the importance of bacterial conjugation. **SC.6.L.14.6**

9. **Model** the life of a bacterium that performs nitrogen fixation in the soil. **LA.6.2.2.3**

10. **Contrast** asexual reproduction in bacteria and replication in viruses. What are some advantages and disadvantages of each? **SC.6.L.14.6**

11. **Organize** the effects of bacteria on health in the table below. **SC.6.L.14.6**

Harmful Effects	Beneficial Effects

12. **Analyze** the importance of vaccines in preventing large outbreaks of influenza. **SC.6.L.14.6**

13 **Explain** what is happening in the petri dish shown below. How does this process eventually create new strains of bacteria that are resistant to antibiotics? SC.6.L.14.6

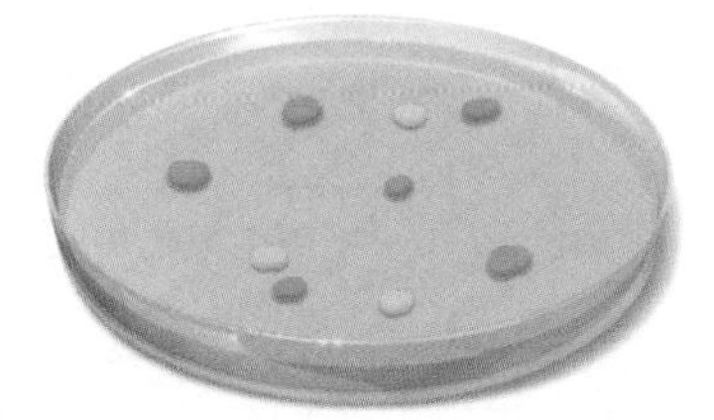

Writing in Science

14 **Summarize** On a separate piece of paper, write an argument that you could use to encourage all the families in your neighborhood to make sure their pets are vaccinated against rabies. SC.6.L.14.6

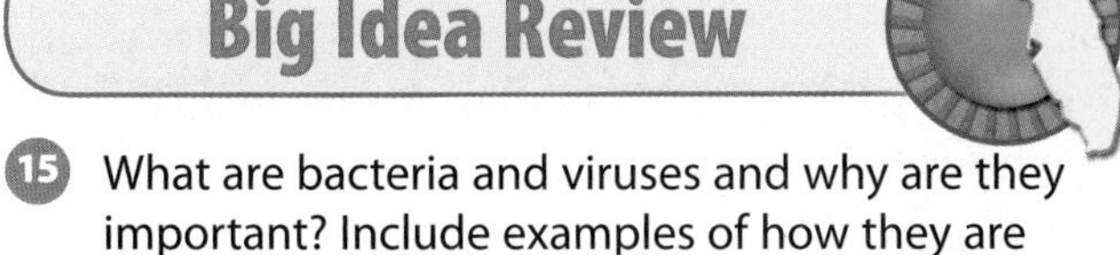

15 What are bacteria and viruses and why are they important? Include examples of how they are both beneficial and harmful to humans. SC.6.L.14.6

16 Describe what is happening in the photo below. Explain what is happening to both the bacterium and the virus. SC.6.L.14.6

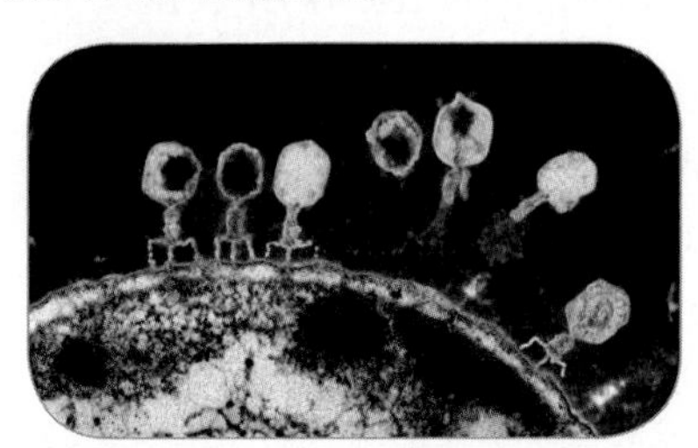

Math Skills MA.6.A.3.6

Use a Formula

17 How many bacteria would there be if 100 bacteria underwent fission 8 times?

18 If each fission cycle takes 20 minutes, how many cycles would it take for 100 bacteria to divide into 100,000?

19 A strain of bacteria takes 30 minutes to undergo fission. Starting with 500 bacteria, how many would there be after 4 hours?

Record your answers on the answer sheet provided by your teacher or on a sheet of paper.

Multiple Choice

1 Which is NOT a characteristic of bacteria? SC.6.L.14.6

Ⓐ They are microscopic.

Ⓑ They are unicellular.

Ⓒ They can live in many environments.

Ⓓ They have a membrane-bound nucleus.

2 Which is a characteristic of viruses? SC.6.L.14.6

Ⓕ unicellular prokaryotes

Ⓖ small pieces of DNA

Ⓗ no nucleus

Ⓘ spiral-shaped

3 Which disease is caused by bacteria? SC.6.L.14.6

Ⓐ chicken pox

Ⓑ influenza

Ⓒ tuberculosis

Ⓓ common cold

4 A chemical that harms only prokaryotic cells would affect which of the following? SC.6.L.14.6

Ⓕ viruses

Ⓖ plants

Ⓗ animals

Ⓘ bacteria

5 Bacteria and viruses have a variety of shapes. Which shape represents a virus? SC.6.L.14.6

Ⓐ bacteriophage

Ⓑ rod-shaped

Ⓒ spiral-shaped

Ⓓ round

6 How do bacteria reproduce? SC.6.L.14.6

Ⓕ conjugation

Ⓖ fission

Ⓗ mutation

Ⓘ pasteurization

Use the diagram below to answer question 7.

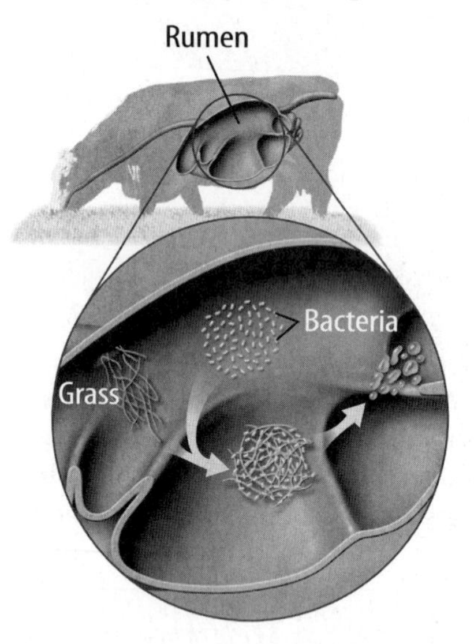

7 What role do bacteria play in the process shown above? SC.6.L.14.6

Ⓐ They break down cellulose.

Ⓑ They convert nitrogen in grass.

Ⓒ They prevent viruses from growing.

Ⓓ They remove harmful pollutants.

8 How does a virus infect the rest of a body once it replicates? SC.6.L.14.6

Ⓕ by attaching to the host cell

Ⓖ by entering the bloodstream

Ⓗ by becoming a latent virus

Ⓘ by disguising itself as a bacterium

Use the diagram below to answer question 9.

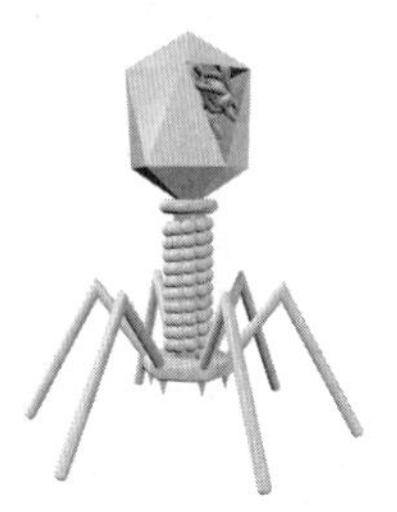

9 What is pictured in the diagram above? SC.6.L.14.6

Ⓐ an antibody

Ⓑ a bacteriophage

Ⓒ a bacterium

Ⓓ a plasmid

10 How do viruses reproduce? SC.6.L.14.6

Ⓕ conjugation

Ⓖ fission

Ⓗ mutation

Ⓘ replication

11 Which disease is NOT caused by a virus? SC.6.L.14.6

Ⓐ mumps

Ⓑ measles

Ⓒ pneumonia

Ⓓ polio

12 Which can be used to treat most bacterial diseases in humans? SC.6.L.14.6

Ⓕ antibiotics

Ⓖ bioremediation

Ⓗ pasteurization

Ⓘ pathogens

13 Which is NOT a reason viral diseases are difficult to cure or prevent? SC.6.L.14.6

Ⓐ Viruses often mutate every time they replicate.

Ⓑ Latent viruses cause immediate symptoms.

Ⓒ Antibiotics only work on bacteria.

Ⓓ Limiting contact between humans is very hard.

14 Which is a mixture that contains material from one or more deactivated pathogens, such as viruses? SC.6.L.14.6

Ⓕ antibiotic

Ⓖ endospore

Ⓗ flagellum

Ⓘ vaccine

15 What is the most important way to prevent infections? SC.6.L.14.6

Ⓐ Don't wash to build up immunity.

Ⓑ Practice good hygiene.

Ⓒ Take antibiotics even when healthy.

Ⓓ Skip all vaccination shots.

NEED EXTRA HELP?

If You Missed Question . . .	1	2	3	4	5	6	7	8	9	10	11	12	13	14	15
Go to Lesson . . .	1	3	2	1	3	1	2	3	3	3	3	2	3	3	3

Name ______________________ Date ____________

Multiple Choice *Bubble the correct answer.*

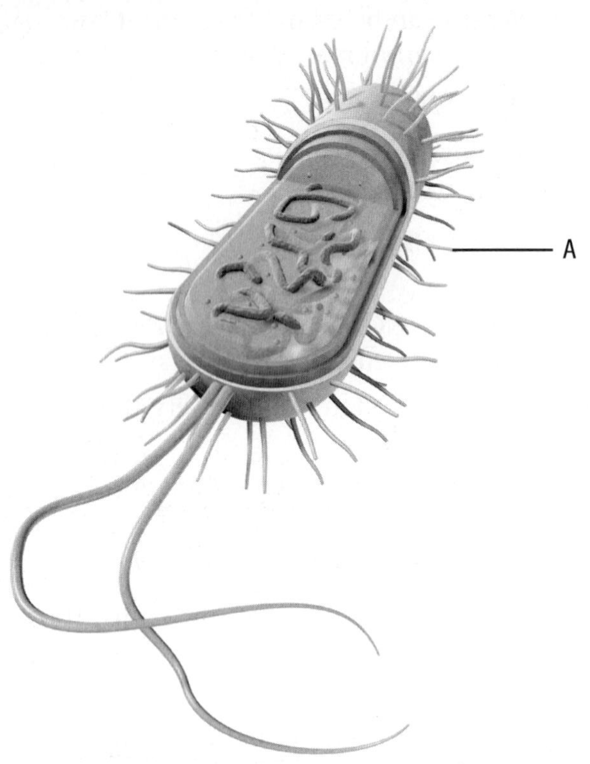

1. The diagram above shows a bacterial cell. What does the labeled structure do? **SC.6.L.14.6**

   (A) It controls all processes of the bacterial cell.

   (B) It helps the bacteria stick to surfaces.

   (C) It helps the bacterial cell move.

   (D) It keeps the bacterium from drying out.

2. Which type of bacteria needs oxygen to live? **SC.6.L.14.6**

   (F) aerobic

   (G) anaerobic

   (H) archaea

   (I) endospore

3. Which of the following is NOT a characteristic of an archaea cell? **SC.6.L.14.6**

   (A) It has a cell wall.

   (B) It has a nucleus.

   (C) It has a circular strand of DNA.

   (D) It has an extreme environment.

4. *Lactobacillus* is a beneficial bacterium that is rod-shaped. Which image below could be *Lactobacillus*? **SC.6.L.14.6**

   (F) 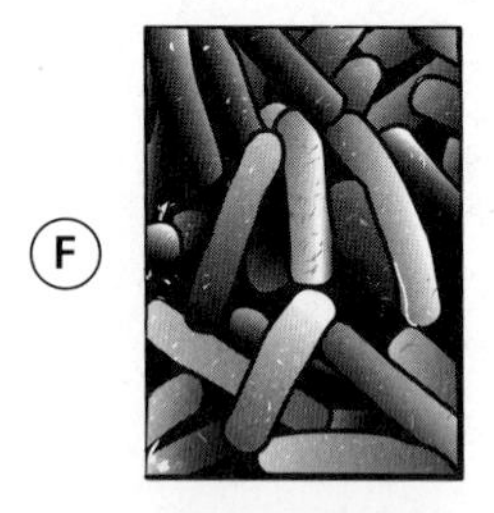

   (G) 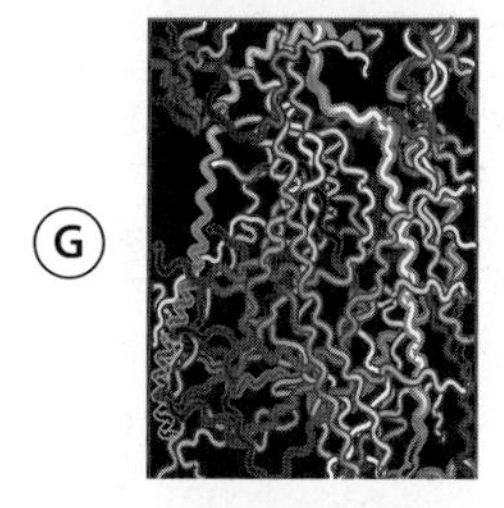

   (H) 

   (I) 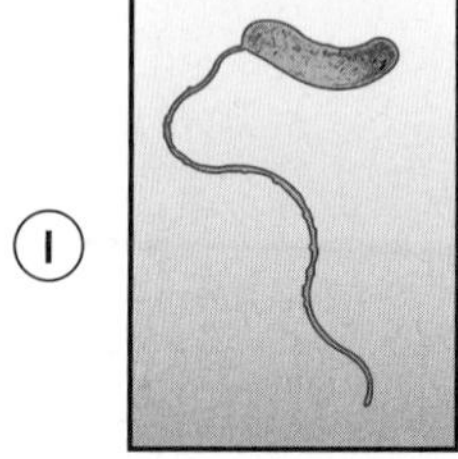

Name ______________________________ Date ______________

Multiple Choice *Bubble the correct answer.*

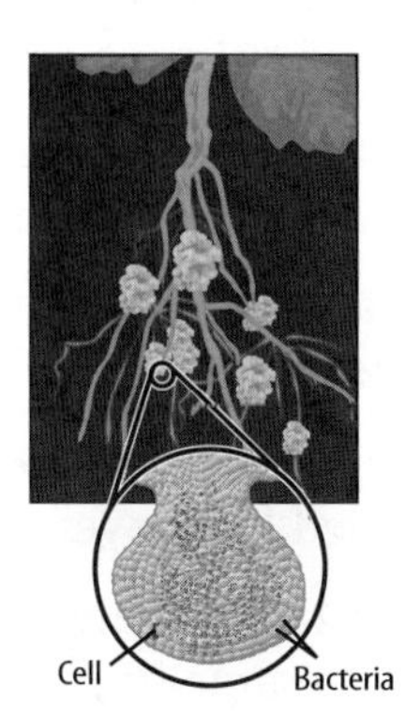

1. Examine the image above. What is the purpose of the structures in the image? **SC.6.L.14.6**

(A) to break down organic matter in the soil

(B) to convert nitrogen in air into a form that plants can use

(C) to fight pathogens that could invade the roots

(D) to steal nutrients away from the host plant

2. Leaves litter the ground in a maple forest. Bacteria break down these leaves through the process of **SC.6.L.14.6**

(F) bioremediation.

(G) decomposition.

(H) bacterial resistance.

(I) nitrogen fixation.

3. Which process is used to make dairy products safe to eat? **SC.6.L.14.6**

(A) bioremediation

(B) decomposition

(C) pasteurization

(D) pathogens

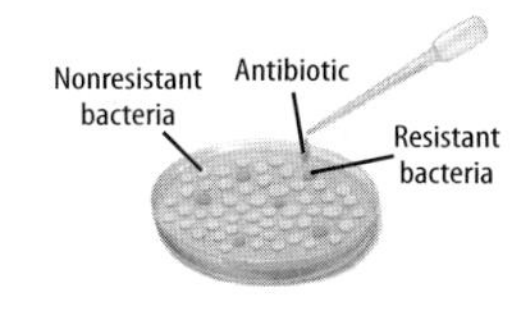

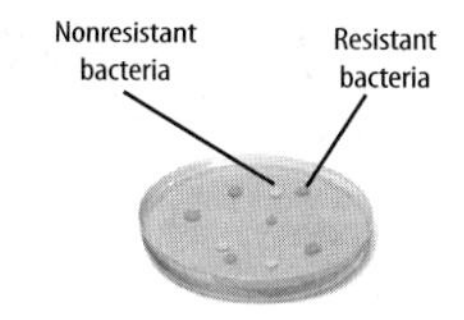

4. What is happening above in the image on the right? **SC.6.L.14.6**

(F) The antibiotic has no effect on the nonresistant bacteria.

(G) The antibiotic kills most of the nonresistant bacteria.

(H) The antibiotic kills most of the resistant bacteria.

(I) The antibiotic kills the nonresistant and resistant bacteria.

Multiple Choice *Bubble the correct answer.*

1. Which image shows the structure of a bacteriophage? **SC.6.L.14.6**

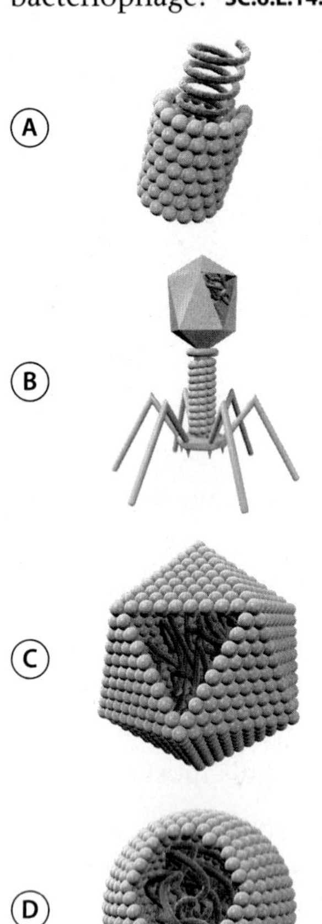

Ⓐ

Ⓑ

Ⓒ

Ⓓ

2. What happens within a dog's body when the dog is vaccinated for rabies? **SC.6.L.14.6**

Ⓕ Any rabies virus cells that are in the dog are killed.

Ⓖ Any rabies virus cells that are in the dog become latent.

Ⓗ Immune cells in the dog mutate and attack the virus in the vaccine.

Ⓘ Immune system of the dog produces antibodies to the rabies virus.

3. Which statement is true about viruses? **SC.6.L.14.6**

Ⓐ A virus can replicate on its own.

Ⓑ A virus is a living organism that responds to stimuli.

Ⓒ A virus is a strand of RNA or DNA surrounded by a layer of protein.

Ⓓ A virus is larger than bacteria and is always the same shape.

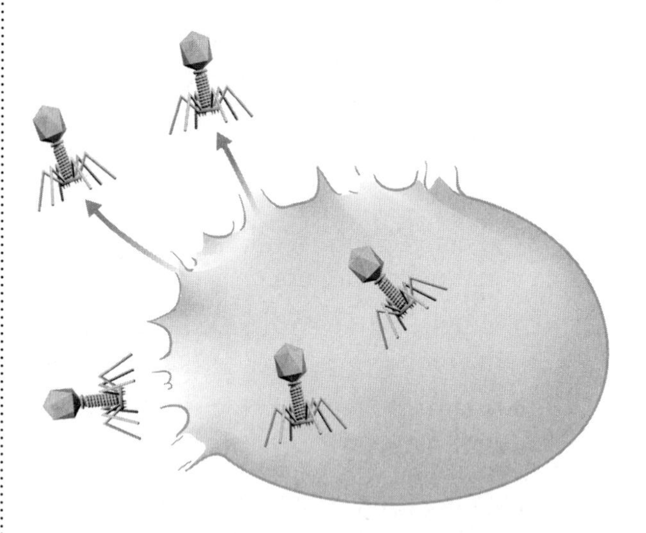

4. The image above shows one step in the process of viral replication. Which statement describes this step? **SC.6.L.14.6**

Ⓕ Latent virus DNA combines with the DNA of the cell.

Ⓖ New viruses are released into the bloodstream.

Ⓗ The virus's DNA directs the cell to make new viruses.

Ⓘ A virus can insert its DNA into a host cell.

Name ______________________ Date ______________

Notes

Name ______ Date ______

Notes

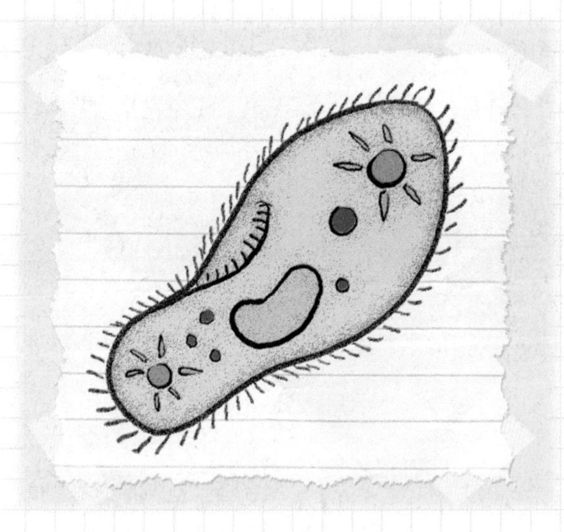

What are protists?

Protists are a diverse classification of organisms. When the students in Mrs. Applebee's science class were asked "What are protists?" they had different ideas. Here are some of their ideas:

Anna: I think they share most of their characteristics with plants.

Jordan: I think they share most of their characteristics with animals.

Ken: I think they share most of their characteristics with fungi.

Alysha: I think they share most of their characteristics with bacteria.

Joaquin: I think they share characteristics with animals, plants, and fungi.

LaVon: I don't think they share characteristics with any of the organisms you mentioned. They have their own unique characteristics.

Which student do you agree with the most? Explain why you agree.

Protists and FUNGI

FLORIDA BIG IDEAS

1 **The Practice of Science**
2 **The Characteristics of Scientific Knowledge**
3 **The Role of Theories, Laws, Hypotheses, and Models**
14 **Organization and Development of Living Organisms**

The Big Idea

Think About It!

What are protists and fungi, and how do they affect an environment?

These organisms are neither plant nor animal. Protists and fungi are two groups of living things that have characteristics similar to those of plants or animals.

1. How is the organism pictured similar to a plant? An animal?

____

____

2. How might this organism benefit its environment?

____

____

Get Ready to Read

What do you think about protists and fungi?

Before you read, decide if you agree or disagree with each of these statements. As you read this chapter, see if you change your mind about any of the statements.

	AGREE	DISAGREE
1 Protists are grouped together because they all look similar.	☐	☐
2 Some protists cause harm to other organisms.	☐	☐
3 Many protists make their own food.	☐	☐
4 Mushrooms and yeasts are two types of fungi.	☐	☐
5 Fungi are always helpful to plants.	☐	☐
6 Some fungi can be made into foods or medicines.	☐	☐

There's More Online!
Video • Audio • Review • ⓘLab Station • WebQuest • Assessment • Concepts in Motion • Multilingual eGlossary

Lesson 1

What are PROTISTS?

ESSENTIAL QUESTIONS

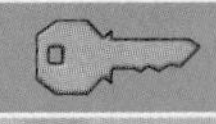
What are the different types of protists and how do they compare?

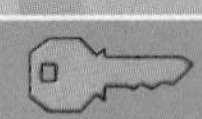
How are protists beneficial?

Vocabulary

protist p. 291
algae p. 292
diatom p. 293
cilia p. 296
protozoan p. 296
paramecium p. 296
amoeba p. 297
pseudopod p. 297

SC.6.N.1.5

Launch Lab

10 minutes

How does a protist react to its environment?

Like other organisms, protists can react to their environment in many ways. One type of protist called *Euglena* has specialized structures to move, perform photosynthesis, and react to light.

Procedure

1. Read and complete a lab safety form.
2. Place a **Petri dish** containing a ***Euglena* culture** on a piece of white **paper.** Using a **hand lens,** observe the *Euglena.*
3. Carefully cut a hole the size of a dime in a piece of **aluminum foil.** Place the foil on top of the dish so that the hole is centered over the top. Shine the light from a **desk lamp** at the hole.
4. At the end of class, remove the foil and observe the *Euglena* again.

Think About This

1. Where were the *Euglena* in the dish at the beginning of class? At the end?
2. Why do you think this behavior is beneficial to *Euglena?*
3. Key Concept What structures do you think help *Euglena* react to its environment?

Florida NGSSS

LA.6.2.2.3 The student will organize information to show understanding (e.g., representing main ideas within text through charting, mapping, paraphrasing, summarizing, or comparing/contrasting);

LA.6.4.2.2 The student will record information (e.g., observations, notes, lists, charts, legends) related to a topic, including visual aids to organize and record information and include a list of sources used;

SC.6.N.1.5 Recognize that science involves creativity, not just in designing experiments, but also in creating explanations that fit evidence.

SC.6.N.2.1 Distinguish science from other activities involving thought.

SC.6.L.14.6 Compare and contrast types of infectious agents that may infect the human body, including viruses, bacteria, fungi, and parasites.

SC.6.N.3.4 Identify the role of models in the context of the sixth grade science benchmarks.

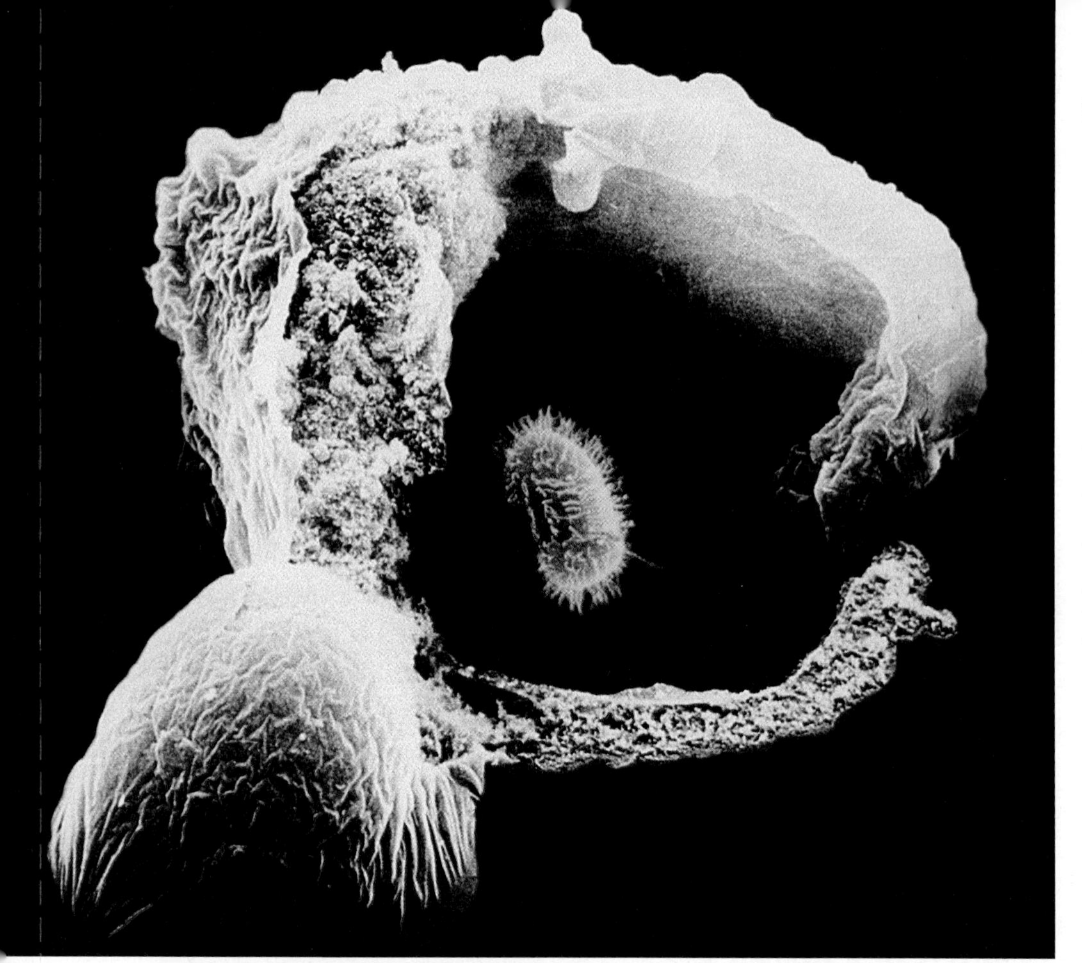

Inquiry Grabbing a Snack?

1. The protist group includes diverse organisms. What do you think the larger organism is doing in the photo? How is this organism similar to an animal?

______

______

______

______

______

______

What are protists?

When you see a living thing, one of the first questions you might have is whether it is a plant or an animal. You might recognize a dog as an animal because of its fur. You might know a flower is a plant because of its leaves. Besides appearance, organisms can also be classified by structures in their cells. A plant cell has a cell wall made of cellulose and a membrane made of flexible fats. A plant cell often contains chloroplasts, organelles that carry out photosynthesis. An animal cell also has a membrane made of flexible fats but does not contain chloroplasts or have a cell wall. These characteristics make it easy to identify both types of cells. However, some organisms, such as the protist shown in **Figure 1**, cannot be classified as easily.

A **protist** *is a member of a group of eukaryotic organisms, which have a membrane-bound nucleus.* Protists share some characteristics with plants, animals, or organisms known as fungi. However, they are not classified as any of these groups. Although protists are classified together, they are diverse and have different adaptations for movement and finding food.

Active Reading 2. **Explain** What is a protist?

______

______

Figure 1 Many photosynthetic algae look like plants.

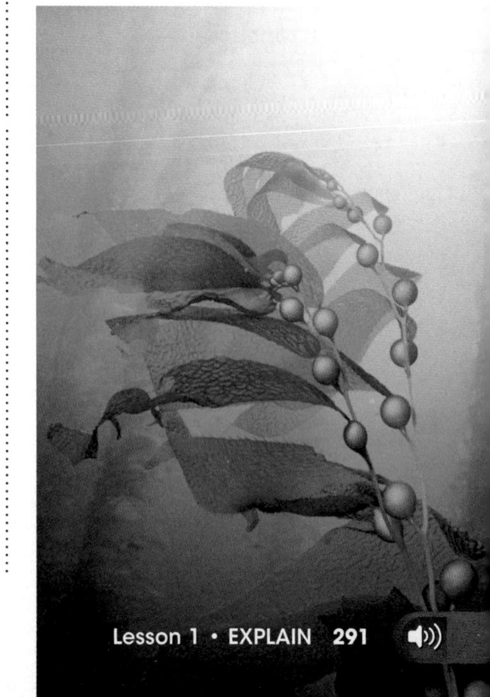

Reproduction of Protists

REVIEW VOCABULARY

asexual reproduction
a type of reproduction in which one parent reproduces without a sperm and an egg joining

Most protists reproduce asexually. What does the offspring of **asexual reproduction** look like? It is an exact copy of the parent. Asexual reproduction can create new organisms quickly. However, many protists can also reproduce sexually. Offspring of sexual reproduction are genetically different from the parents. Sexual reproduction takes more time, but it creates new organisms with a variety of characteristics.

Classification of Protists

Scientists usually classify organisms according to their similarities. However, protists are a unique and diverse classification of organisms. Typically, a protist is any eukaryote that cannot be classified as a plant, an animal, or a fungus. However, protists might look and act very much like these other types of organisms. Scientists classify protists as plantlike, animal-like, or funguslike based on which group they most resemble, as shown in **Table 1.**

Active Reading **3. Locate** Underline the three different types of protists.

Table 1 Protists Classified into One of Three Groups

Classification	Plantlike	Animal-like	Funguslike
Example	algae	paramecium	slime mold
Characteristics	• make their own food • unicellular or multicellular	• eat other organisms for food • mostly microscopic and unicellular	• break down organic matter for food • mostly multicellular

Active Reading **4. Interpret** List two types of protists that can be multicellular.

____

____

Plantlike Protists

You might have seen brown, green, or red seaweed at the beach or in an aquarium. These seaweeds are algae (AL jee; singular, alga), one type of plantlike protist. Why might they be classified as plantlike? **Algae** *are plantlike protists that produce food through photosynthesis using light energy and carbon dioxide.* Most plantlike protists, however, are much smaller than the multicellular algae shown in **Table 1.** You can't see most algae without a microscope.

Diatoms

A type of microscopic plantlike protist with a hard outer wall is a **diatom** (DI uh tahm). Diatoms are so common that if you filled a cup with water from the surface of any lake or pond, you would probably collect thousands of them. Look at the unicellular diatoms shown at the top of **Figure 2.** A diatom can resemble colored glass. In fact, the cell walls of diatoms contain a large amount of silica, the main mineral in glass.

Dinoflagellates

Can you guess how the protist in the middle of **Figure 2** moves? This organism is a dinoflagellate (di noh FLA juh lat), a unicellular plantlike protist that has flagella—whiplike parts that enable the protist to move. The flagella beat back and forth, enabling the dinoflagellate to spin and turn. Some of these protists glow in the dark because of a chemical reaction that occurs when they are disturbed.

Active Reading **5. Explain** What purpose do flagella serve?

____

____

____

Euglenoids

Another type of plantlike protist also uses flagella to move but has a unique structure covering its body. A euglenoid (yew GLEE noyd), shown at the bottom of **Figure 2,** is a unicellular plantlike protist with a flagellum at one end of its body. Instead of a cell wall, euglenoids have a rigid, rubbery cell coat called a pellicle (PEL ih kul). Euglenoids have eyespots that detect light and determine where to move. Euglenoids swim quickly and can creep along the surface of water when it is too shallow to swim. These protists have chloroplasts and make their own food. If there is not enough light for making food, they can absorb nutrients from decaying matter in the water. Animals such as tadpoles and small fish eat euglenoids.

Figure 2 All of these microscopic organisms are protists. The cell walls of diatoms contain silica. The dinoflagellate has two flagella that cause it to spin. The euglenoid has a flagellum and a rigid cell coat.

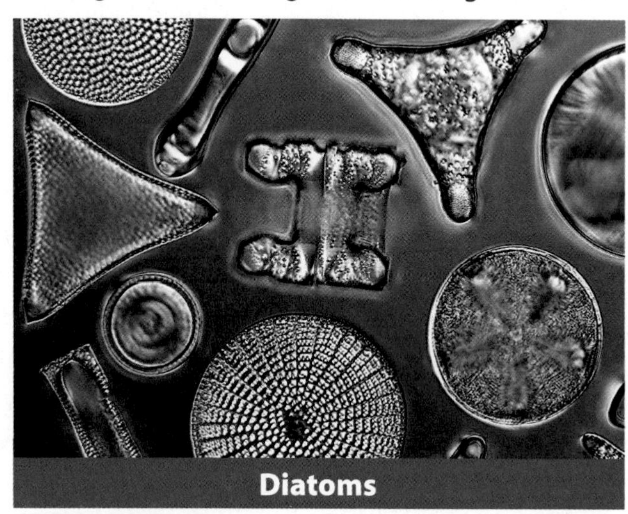

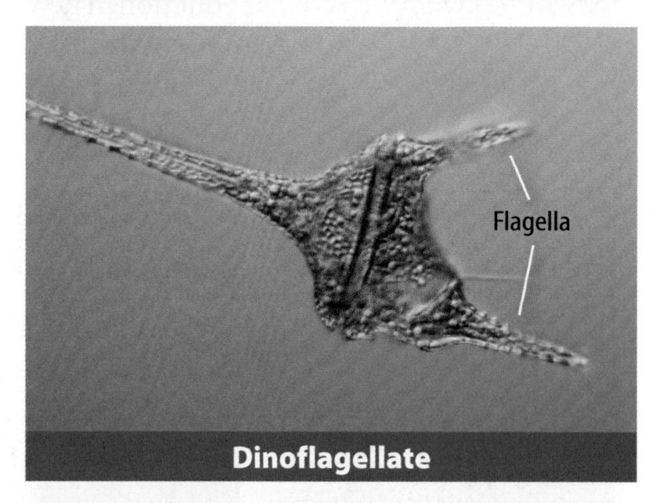

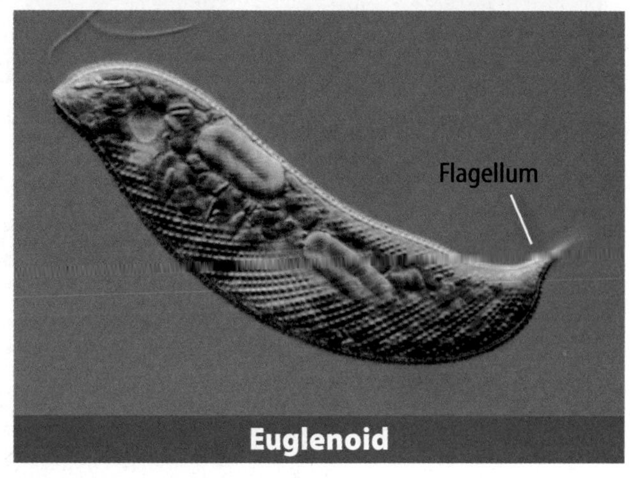

Active Reading **6. Compare** What characteristics do plantlike protists share with plants?

____

____

____

Active Reading **7. Describe** How do red and brown multicellular algae differ from plants?

Algae

Recall that algae are photosynthetic plantlike protists. Some algae are big and multicellular, like the seaweeds in **Figure 3.** Other algae are unicellular and can be seen only with a microscope. Algae are classified as red, green, or brown, depending on the pigments they contain.

Some types of red and brown algae appear similar to plants. Unlike plants, these algae do not have a complex organ system for transporting water and nutrients. Instead of roots, they have holdfasts, structures that secrete a chemical-like glue that fastens them to the rocks.

One unusual green alga is volvox. In **Figure 3** you can see that many volvox cells come together to form a larger sphere. These cells move together as one group and beat their flagella in unison. Some cells produce parts necessary for sexual reproduction. The volvox cells in the front of the group have larger eyespots that sense light for photosynthesis. Do you think volvox should be considered unicellular or multicellular?

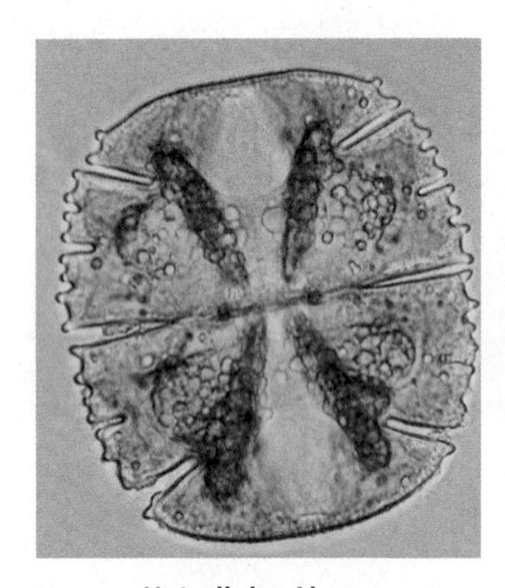

Unicellular Algae

Multicellular Algae

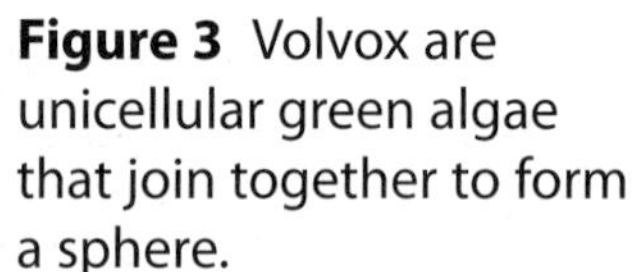

Figure 3 Volvox are unicellular green algae that join together to form a sphere.

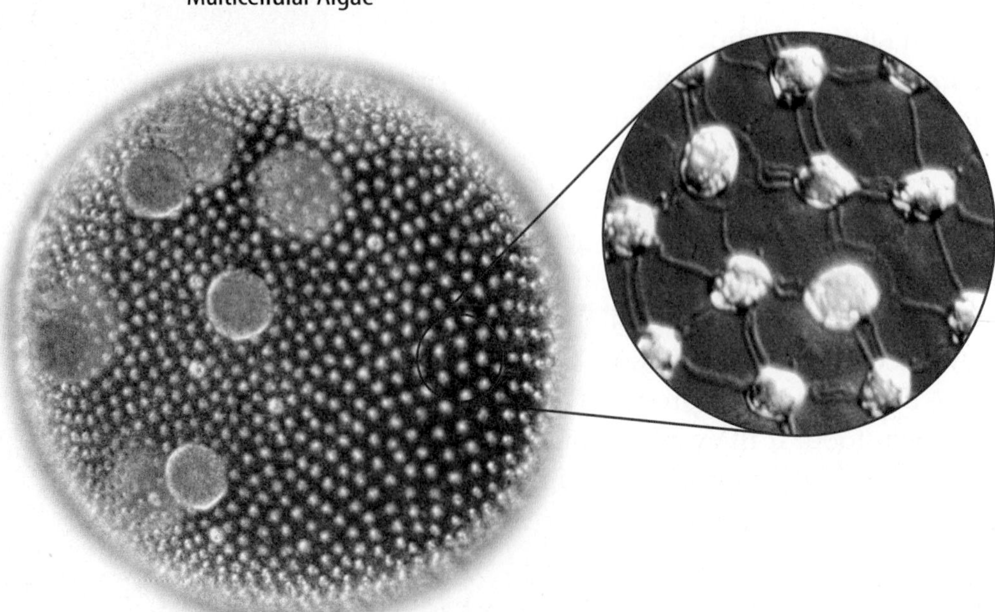

The Importance of Algae

Do you use algae in your everyday life? You might be surprised by all the materials you use that contain algae. You might be eating algae if you snack on ice cream, marshmallows, or pudding. Algae are a common ingredient in other everyday products, including toothpaste, lotions, fertilizers, and some swimming pool filters.

Algae and Ecosystems

Algae provide food for animals and animal-like protists. They also provide shelter for many aquatic organisms. In **Figure 4,** you can see that some brown algae grow tall. Thick groups of tall brown algae are called kelp forests. Sea otters and seals come to the kelp forest to eat smaller animals.

Active Reading **8. Explain** How are algae beneficial to an ecosystem?

____

____

Do you think algae ever cause problems in an ecosystem? Algae and other photosynthetic protists can help remove pollution from the water. However, this pollution can be a food source for the algae, allowing the population of algae to increase quickly. The algae produce wastes that can poison other organisms. As shown in **Figure 5,** when the number of these protists increases, the water can appear red or brown. This is called a red tide or a harmful algal bloom (HAB).

Active Reading **9. Explain** What causes a red tide?

____

____

Kelp Forest

Figure 4 Brown algae can form thick kelp forests that are home to many animals and other protists.

Figure 5 *Karenia brevis* is a dinoflagellate commonly known as the Florida red tide organism. The dinoflagellate species involved in Florida HABs is red or brown in color, giving a reddish tint to the water. Florida red tides occur in the Gulf of Mexico almost every year in the late summer or early fall. A bloom typically lasts three to five months and may have a negative effect on fish, birds, and marine mammals.

Active Reading **10. Evaluate** Complete the chart noting the importance of algae.

Human Uses of Algae	Algae and Ecosystems

Animal-like Protists

Some protists are similar to plants, but others are more like animals. **Protozoans** (proh tuh ZOH unz) *are protists that resemble tiny animals.* Animal-like protists all share several characteristics. They do not have chloroplasts or make their own food. Protozoans are usually microscopic and all are unicellular. Most protozoans live in wet environments.

Ciliates

Cilia (SIH lee uh) *are short, hairlike structures that grow on the surface of some protists.* Protists that have these organelles are called ciliates. Cilia cover the surface of the cell. They can beat together and move the animal-like protist through the water.

Active Reading **11. Explain** What function do cilia perform?

____

____

ACADEMIC VOCABULARY
process
(noun) an event marked by gradual changes that lead toward a particular result

A common protozoan with these cilia is the **paramecium** (pa ruh MEE see um; plural, paramecia)—*a protist with cilia and two types of nuclei.* One example of a paramecium is shown in **Figure 6.** A paramecium, like most ciliates, gets its food by forcing water into a groove in its side. The groove closes and a food vacuole, or storage area, forms within the cell. The food particles are digested, and the extra water is forced back out. Ciliates reproduce asexually, but they can exchange some genetic material through a **process** called conjugation (kahn juh GAY shun). This results in more genetic variation.

Figure 6 A paramecium, like the one shown below, has two nuclei and is covered with hairlike structures called cilia.

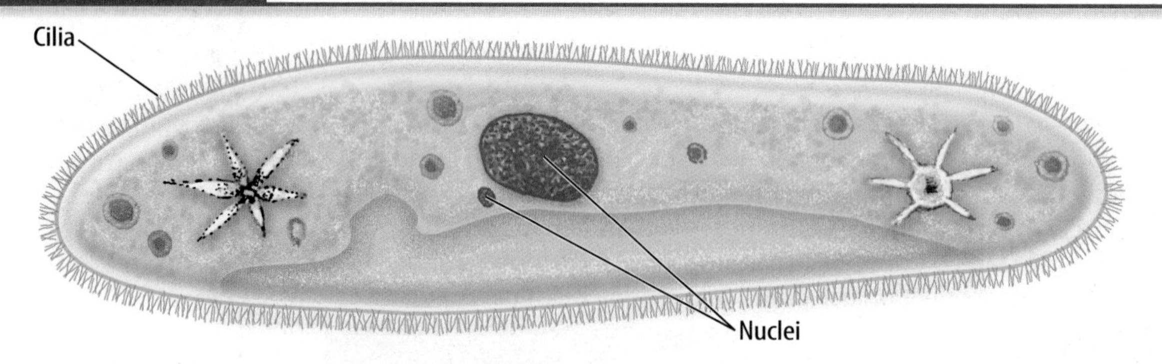

Flagellates

Recall that dinoflagellates, a type of plantlike protist, use one or more flagella to move. A type of protozoan also has one or more flagella—a flagellate. However, a flagellate does not always spin when it moves.

Flagellates eat decaying matter including plants, animals, and other protists. Many flagellates live in the digestive system of animals and absorb nutrients from food eaten by them.

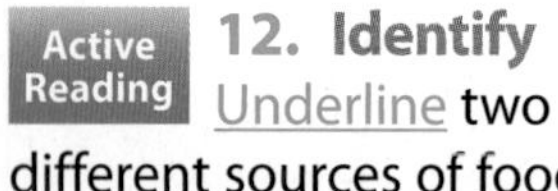

Active Reading **12. Identify** Underline two different sources of food for flagellates.

Sarcodines

Animal-like protists called sarcodines (SAR kuh dinez) have no specific shape. At rest, a sarcodine resembles a random cluster of cytoplasm, or cellular material. These animal-like protists can ooze into almost any shape as they slide over mud or rocks.

An **amoeba** (uh MEE buh) *is one common sarcodine with an unusual adaptation for movement and getting nutrients.* An amoeba moves by using a **pseudopod**, *a temporary "foot" that forms as the organism pushes part of its body outward.* It moves by first stretching out a pseudopod and then oozing the rest of its body up into the pseudopod. This movement is shown in **Figure 7.**

Amoebas also use pseudopods to get nutrients. An amoeba surrounds a smaller organism or food particle with its pseudopod and then oozes around it. A food vacuole forms inside the pseudopod where the food is quickly digested. You can see an amoeba capturing its prey in the photo at the beginning of this lesson.

Some sarcodines get nutrients and energy from ingesting other organisms, while others make their own food. Some sarcodines even live in the digestive systems of humans and get nutrients and energy from the human's body.

Figure 7 An amoeba moves by extending its body to create a temporary "foot."

Amoeba Movement

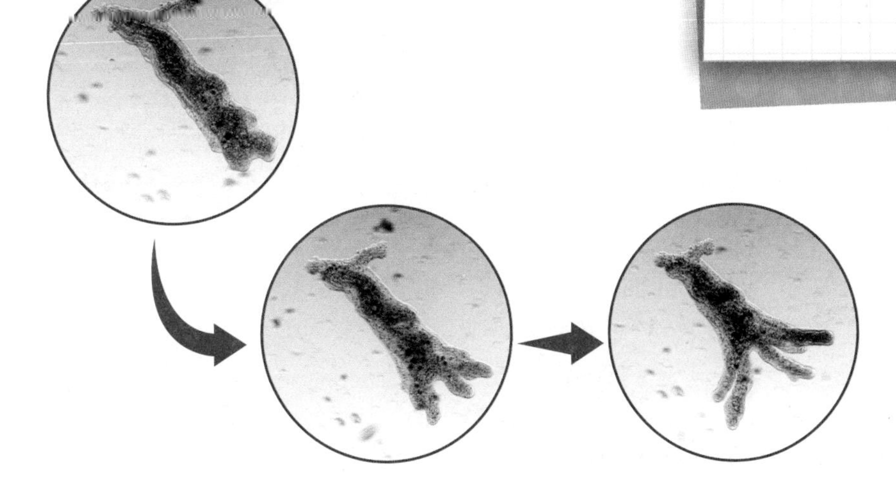

Active Reading **13. Create** Complete the spider map to identify the major characteristics of the three groups of protozoans. Record at least two characteristics of each group.

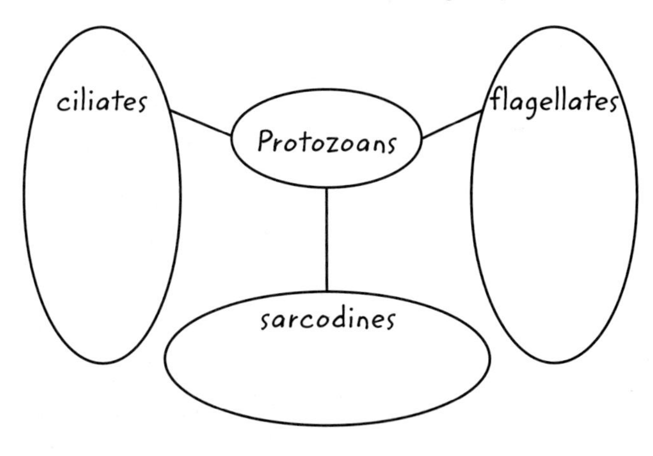

Apply It!

After you complete the lab, answer the questions below.

1. How does the amoeba use its pseudopod to move?

2. How do flagella differ from cilia?

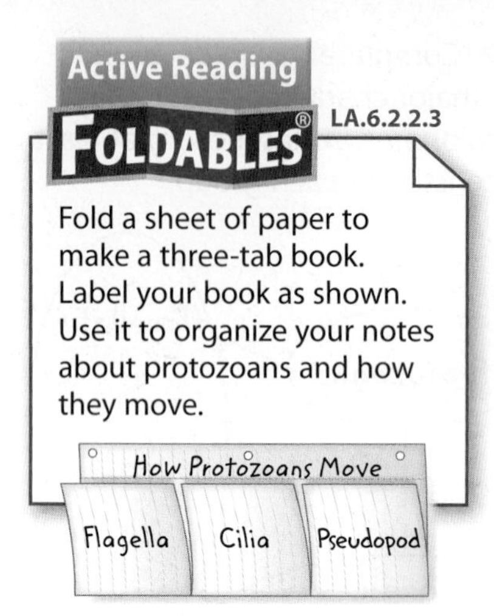

Active Reading **14. Identify** Circle the stage in **Figure 8** when the parasite transfers to a healthy human.

The Importance of Protozoans

Imagine living in a world without organisms that decompose other organisms. Plant material and dead animals would build up until the surface of Earth quickly became covered. Many protozoans are beneficial to an environment because they break down dead plant and animal matter. This decomposed matter is then recycled back into the environment and used by living organisms.

Some protozoans can cause disease by acting as parasites. These organisms can live inside a host organism and feed off it. Protozoan parasites are responsible for millions of human deaths every year.

One example of a disease caused by a protist is malaria. **Figure 8** illustrates how malaria develops and is spread to humans by mosquitoes. Protozoan parasites called plasmodia (singular, plasmodium) live and reproduce in red blood cells. Malaria kills more than one million people each year.

15. NGSSS Check Infer In what ways are protists helpful and harmful to humans? SC.6.L.14.6

____

____

____

Plasmodium Life Cycle

Figure 8 A small parasitic protozoan called plasmodium causes malaria. It is transferred among humans by mosquitoes.

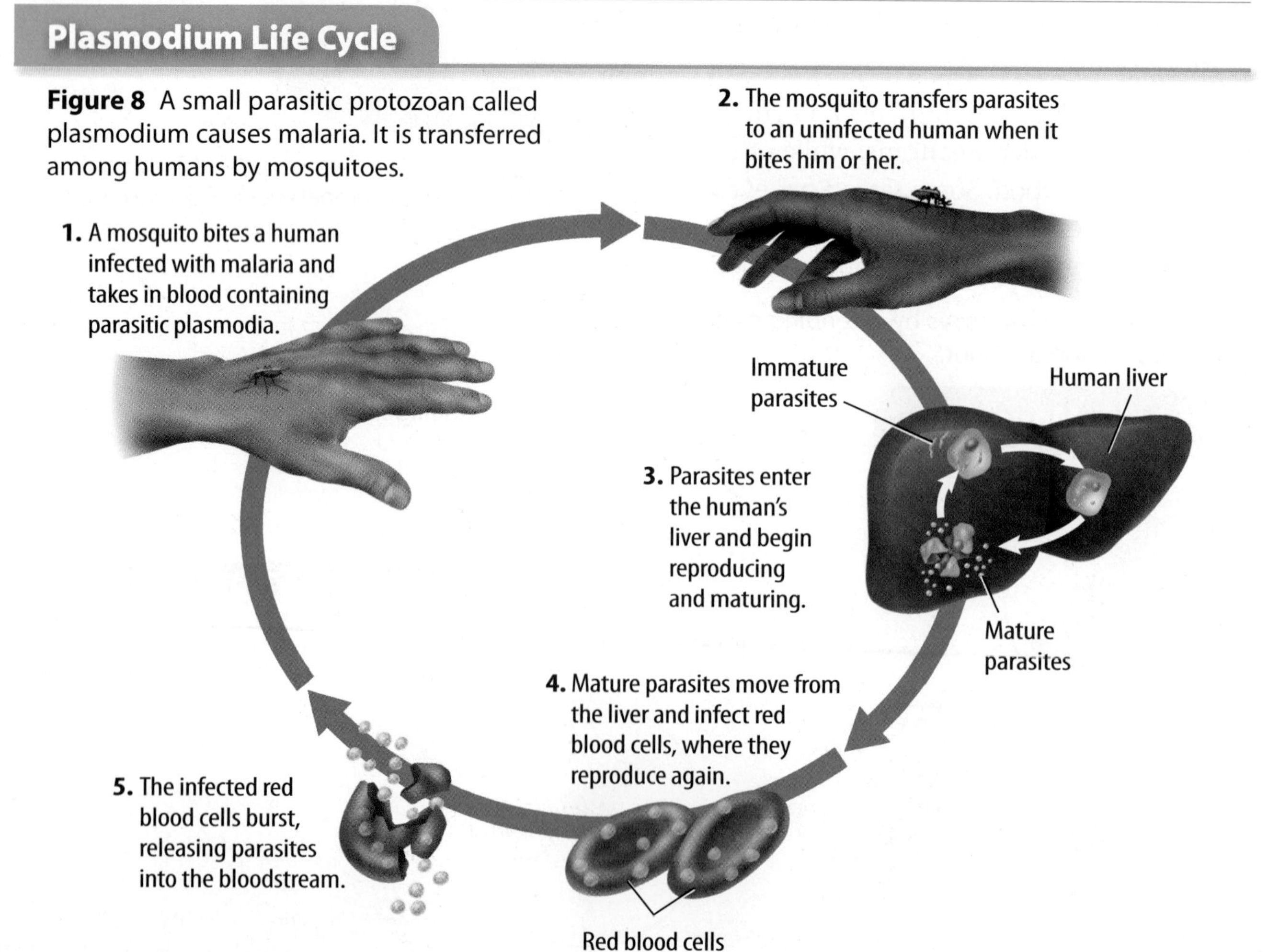

Funguslike Protists

In addition to plantlike and animal-like protists, there are funguslike protists. These protists share many characteristics with fungi. However, because they differ from fungi, they are classified as protists.

Slime and Water Molds

Have you ever seen a strange organism like the one shown in **Figure 9?** These funguslike protists, called slime molds, look like they could have come from another planet. The body of the slime mold is composed of cell material and nuclei floating in a slimy mass. Most slime molds absorb nutrients from other organic matter in their environment.

Active Reading **16. Explain** Where do slime molds get their nutrients?

A Funguslike Protist

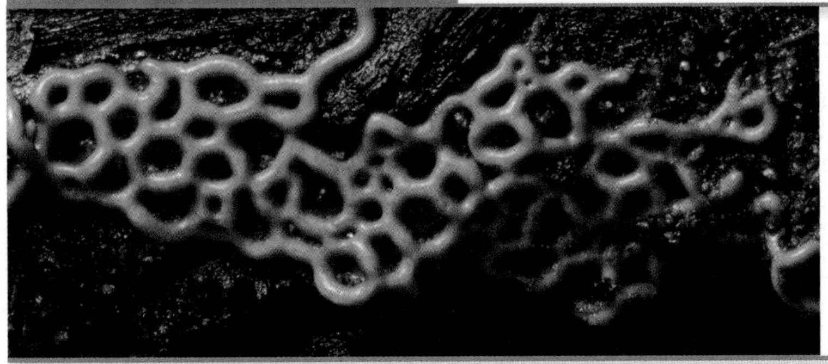

Figure 9 Slime molds come in a variety of colors and forms. These protists often live on the surfaces of plants.

A water mold is another kind of funguslike protist that lives as a parasite or feeds on dead organisms. Originally classified as fungi, water molds often cause diseases in plants.

Both slime molds and water molds reproduce sexually and asexually. The molds usually reproduce sexually when environmental conditions are harsh or unfavorable.

Active Reading **17. Analyze** Animals and animal-like protists share some characteristics. Tell how the organisms in these groups are the same. Then hypothesize why scientists have placed them in different groups.

Importance of Funguslike Protists

Funguslike protists play a valuable role in the ecosystem. They break down dead plant and animal matter, making the nutrients from these dead organisms available for living organisms. While some slime molds and water molds are beneficial, many others can be very harmful.

Many funguslike protists attack and consume living plants. The Great Irish Potato Famine resulted from damage by a funguslike protist. In 1845 this water mold destroyed more than half of Ireland's potato crop. More than one million people starved as a result.

Active Reading **18. Explain** How are funguslike protists beneficial to an environment?

Lesson Review 1

Visual Summary

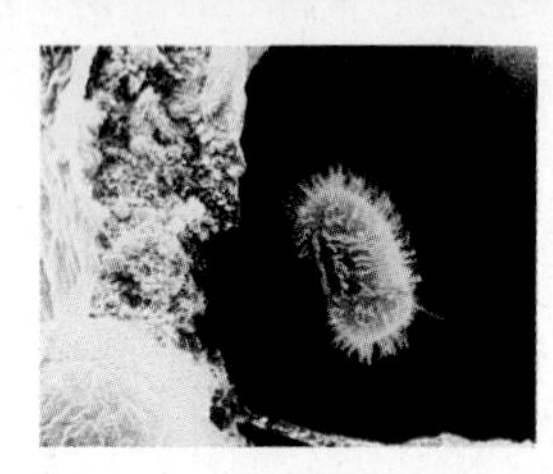

Protists are a diverse group of organisms that cannot be classified as plants, animals, or fungi.

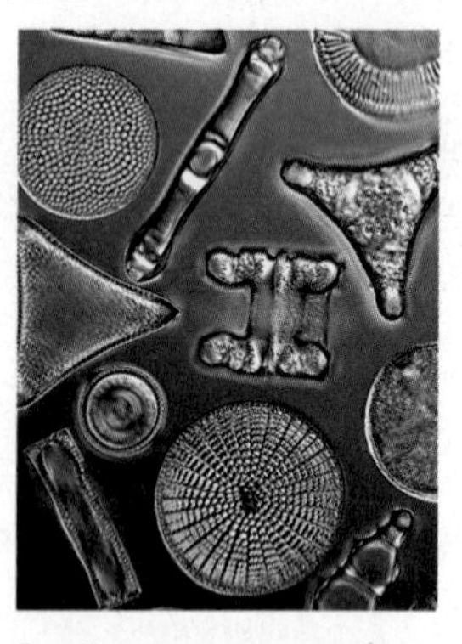

Protists are grouped according to the type of organisms they most resemble. Diatoms are one type of plantlike protist.

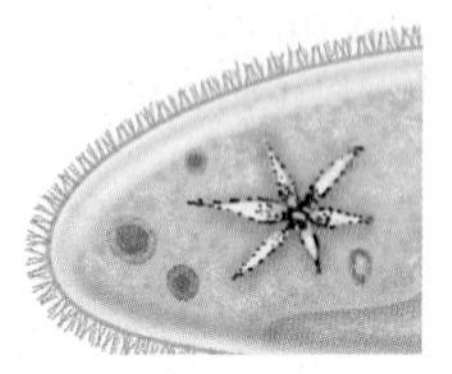

Some protists use hairlike structures called cilia to move.

Use Vocabulary

1. **Distinguish** between cilia and flagella.

2. **Define** *pseudopod* in your own words or with a drawing. LA.6.2.2.3

Understand Key Concepts

3. **List** three groups of animal-like protists and three groups of plantlike protists.

4. **Describe** one example of how protists benefit humans. SC.6.L.14.6

5. Identify which protist causes red tides. SC.6.L.14.6
   - (A) algae
   - (B) diatoms
   - (C) euglenoids
   - (D) paramecia

Interpret Graphics

6. **Identify** Fill in the graphic organizer with the three categories of protists. LA.6.2.2.3

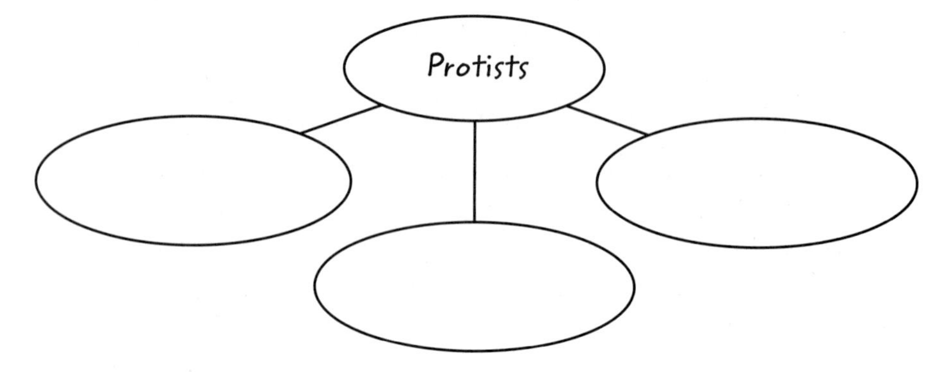

Critical Thinking

7. **Formulate** a plan for deciding how to classify a newly discovered protist.

FOCUS on FLORIDA

The Benefits of Algae

Big Benefits from Tiny Organisms

Algae are protists that can do more than just cover a pond as slimy scum. They release oxygen through photosynthesis. In fact, most of the oxygen in Earth's atmosphere comes from photosynthesis that occurs in algae, plants, and some bacteria. Algae also are food for many organisms, including humans.

Microalgae can grow outdoors in ponds or indoors under lights in photobioreactors. A photobioreactor is a tank filled with water and nutrients. Photosynthesis requires carbon dioxide. Instead of releasing carbon dioxide gas into the atmosphere, power plants can pump it into photobioreactors for microalgae to use.

A Florida-based energy company is using this technology to produce large amounts of protein-rich nutrients for food. This type of technology is beneficial because the algae are able to grow rapidly within the micro-crops without competing against other food crops for fertile land or irrigated water. The algae can be used as a protein enhancement in sports drinks, nutrition bars, nutritional supplements, and baked goods.

Processing plants using photobioreactors, such as this one, are a source of algae oil additives found in various food products.

It's Your Turn

RESEARCH Protists, including algae, are important sources of food. Research five types of organisms that depend on protists for food. Make a display of your results to share with your class. **LA.6.4.2.2**

Lesson 2

What are FUNGI?

ESSENTIAL QUESTIONS

 What are the different types of fungi and how do they compare?

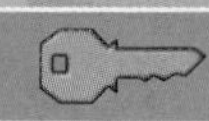 Why are fungi important?

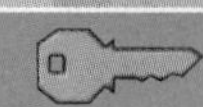 What are lichens?

Vocabulary

hyphae p. 303
mycelium p. 303
basidium p. 304
ascus p. 305
lichen p. 310
zygosporangia p. 305
mycorrhiza p. 308

Launch Lab

SC.6.N.1.5

10 minutes

Is there a fungus among us?

The mold you see on food is fungi that are consuming and decomposing it. Fungi are also found as molds or mushrooms on wood, mulch, and other organic materials.

Procedure

1. Read and complete a lab safety form.
2. Examine the different **samples of fungi** your teacher provides. Use a **magnifying lens** to observe similarities and differences among the samples.
3. Record your observations in the Data and Observation section below. Include drawings of the different structures or characteristics you notice.

Data and Observations

Think About This

1. What similarities did you see among the fungi samples?

2. Why do you think your teacher had the mold samples in closed containers?

3. Key Concept In what ways do you think the fungi you observed are helpful or not helpful to people?

Florida NGSSS

LA.6.2.2.3 The student will organize information to show understanding (e.g., representing main ideas within text through charting, mapping, paraphrasing, summarizing, or comparing/contrasting);

MA.6.A.3.6 Construct and analyze tables, graphs, and equations to describe linear functions and other simple relations using both common language and algebraic notation.

SC.6.N.2.1 Distinguish science from other activities involving thought.

SC.6.L.14.6 Compare and contrast types of infectious agents that may infect the human body, including viruses, bacteria, fungi, and parasites.

SC.6.N.1.5 Recognize that science involves creativity, not just in designing experiments, but also in creating explanations that fit evidence.

Inquiry Up in Smoke?

1. The organism pictured is a puffball mushroom, named for the puff of material that it releases. What do you think the material is? What is the purpose of the puff of material?

Active Reading **2. Describe** How are hyphae and mycelium related?

What are fungi?

What would you guess is the world's largest organism? A fungus in Oregon is the largest organism ever measured by scientists. It stretches almost 9 km^2. Fungi, like protists, are eukaryotes. Scientists estimate more than 1.5 million species of fungi exist.

Fungi form long, threadlike structures that grow into large tangles, usually underground. *These structures, which absorb minerals and water, are called* **hyphae** (HI fee). *The hyphae create a network called the* **mycelium** (mi SEE lee um), shown in **Figure 10.** The fruiting body of the mushroom, the part above ground, is also made of hyphae.

Fungi are heterotrophs, meaning they cannot make their own food. Some fungi are parasites, obtaining nutrients from living organisms. Fungi dissolve their food by releasing chemicals that decompose organic matter. Fungi then absorb the nutrients.

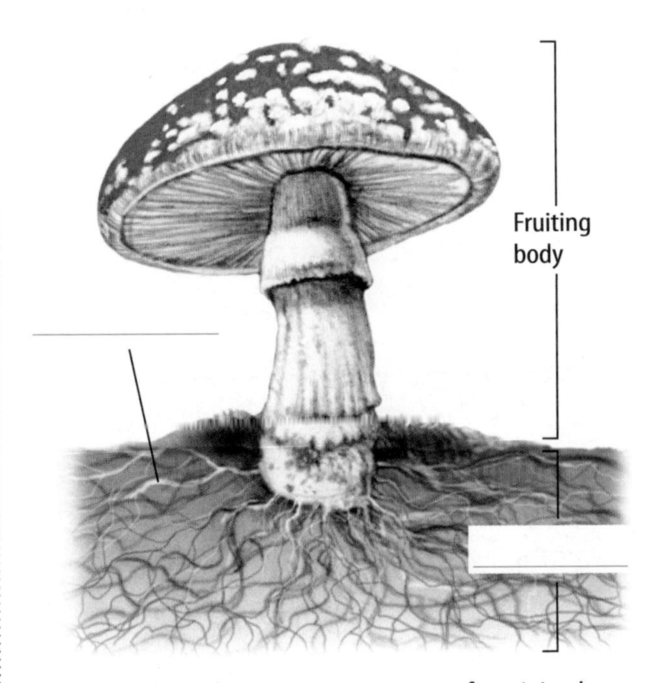

Figure 10 Mushrooms are common fungi. In the drawing, you can see mycelium, the network of hyphae. The hyphae release enzymes and absorb water and nutrients.

Active Reading **3. Identify** In **Figure 10** above, fill in the blanks with the correct labels.

Active Reading **4. Identify** What are the four groups of fungi?

____

____

____

____

Types of Fungi

Scientists group fungi based on how they look and how they reproduce. Although fungi can reproduce sexually or asexually, almost all reproduce asexually by producing spores. Spores are small reproductive cells with a strong, protective outer covering. The spores can grow into new individuals.

The classification of fungi often changes as scientists learn more about them. Today, scientists recognize four groups of fungi: club fungi, sac fungi, zygote fungi, and imperfect fungi. As technology helps scientists understand more about fungi, the categories might change.

Club Fungi

SCIENCE USE V. COMMON USE

mushroom

Science Use a type of club fungi

Common Use the part of a fungus above the ground

When you think of fungi, you might think of a **mushroom.** Mushrooms belong to the group called club fungi. They are named for the clublike shape of their reproductive structures. However, the mushroom is just one part of the fungus. The part of the mushroom that grows above ground is a structure called a basidiocarp (bus SIH dee oh karp). Inside the basidiocarp are the **basidia** (buh SIH dee uh; singular, basidium), *reproductive structures that produce sexual spores.* Most of a club fungus is a network of hyphae that grows underground and absorbs nutrients.

Active Reading **5. State** Where does most of a club fungus grow?

____

____

Many club fungi are named for their various shapes and characteristics. Club fungi include puffballs like those at the beginning of the lesson, stinkhorns, and the bird's nest fungi shown in **Figure 11.** There is even a club fungus that glows in the dark due to a chemical reaction in its basidiocarp.

Figure 11 Club fungi, such as this bird's nest fungus, use basidiospores to reproduce.

Club Fungi

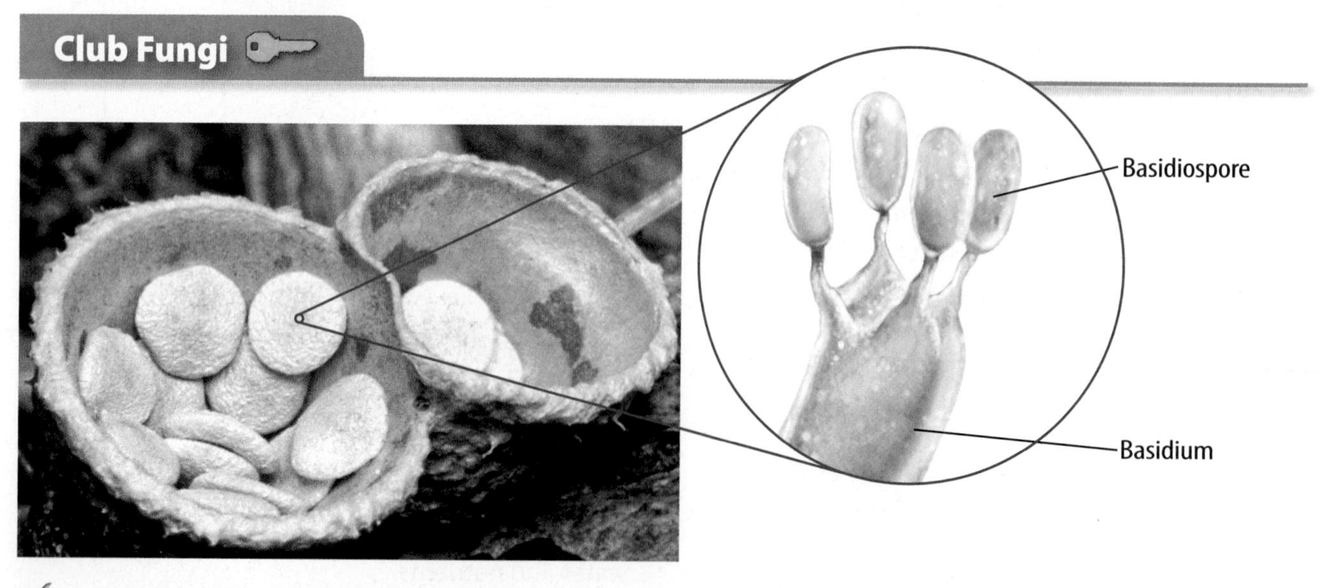

6. Visual Check Identify Which part of the fungus is club-shaped?

____

Sac Fungi

Do you know what bread and a diaper rash have in common? A type of sac fungus causes bread dough to rise. A different sac fungus is responsible for a rash that babies can develop on damp skin under their diapers. Many sac fungi cause diseases in plants and animals. Other common sac fungi, such as truffles and morels, are harvested by people for food.

Like club fungi, sac fungi are named for their reproductive structures. *The* **ascus** (AS kuhs; plural, asci) *is the reproductive structure where spores develop on sac fungi.* The ascus often looks like the bottom of a tiny bag or sack. The spores from sac fungi are called ascospores (AS kuh sporz). Sac fungi can undergo both sexual and asexual reproduction. Many yeasts are sac fungi, including the common yeast used to make bread, as shown in **Figure 12.** When the yeast is mixed with water and warmed, the yeast cells become active. They begin cellular respiration and release carbon dioxide gas. This causes the bread dough to rise.

Zygote Fungi

Another type of fungus can cause bread to develop mold. Bread mold, like the type shown in **Figure 12,** is caused by a type of fungus called a zygote fungus. You might also find zygote fungi growing in moist areas, such as a damp basement or on a bathroom shower curtain.

The hyphae of a zygote fungus grow over materials, such as bread, dissolving the material and absorbing nutrients. *Tiny stalks called* **zygosporangia** (zi guh spor AN jee uh) *form when the fungus undergoes sexual reproduction.* The zygosporangia release spores called zygospores. These zygospores then fall on other materials where new zygote fungi might grow.

Active Reading **7. Explain** How do sac and zygote fungi differ?

____

____

____

Figure 12 Some fungi can be used to make food, but other fungi can eat the food too.

Active Reading

FOLDABLES® LA.6.2.2.3

Fold a sheet of paper to make a four-door book. Label it as shown. Use your book to organize information about the characteristics of the different classifications of fungi.

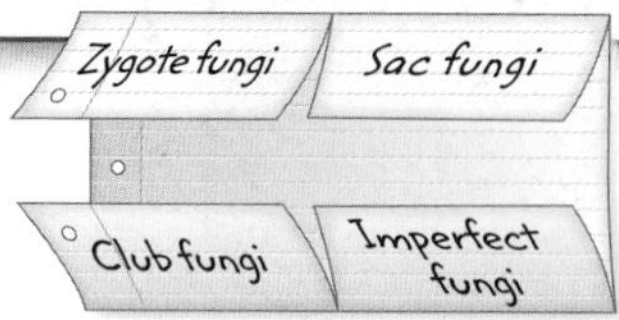

Active Reading **8. Interpret** Why are imperfect fungi classified that way?

Imperfect Fungi

How are itchy feet and blue cheese connected? They can both be caused by imperfect fungi. You might have had athlete's foot, an infection that causes flaking and itching in the skin of the feet. The imperfect fungus that causes athlete's foot grows and reproduces easily in the moist environment near a shower or in a sweaty shoe. The blue color you see in blue cheese comes from colonies of a different type of imperfect fungi. They are added to the milk or the curds during the cheesemaking process.

Imperfect fungi are named because scientists have not observed a sexual, or "perfect," reproductive stage in their life cycle. Since fungi are classified according to the shape of their reproductive structures, these fungi are left out, or labeled "imperfect." Often after a species of imperfect fungi is studied, the sexual stage is observed. The fungi is then classified as a club, sac, or zygote fungus based on these observations.

Active Reading **9. Compare** Complete the chart and compare the four groups of fungi.

Group	How They Reproduce Sexually	Examples
Club fungi		
Sac fungi		
Zygote fungi		
Imperfect fungi		

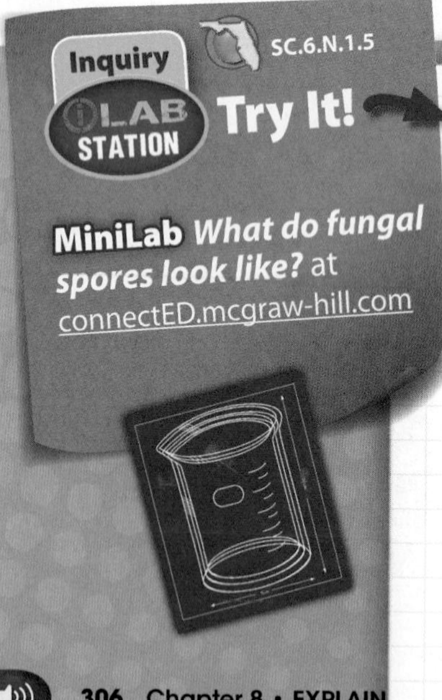

MiniLab ***What do fungal spores look like?*** at connectED.mcgraw-hill.com

Apply It! After you complete the lab, answer this question.

1. What characteristics of club fungi give the group its name?

Fungi Products

Figure 13 Products such as bread, cheese, and medicines are made using fungi.

The Importance of Fungi

Do you like chocolate, carbonated sodas, cheese, or bread? If so, you might agree that fungi are beneficial to humans. Fungi are involved in the production of many foods and other products, as shown in **Figure 13.** Some fungi are used as a meat substitute because they are high in protein and low in cholesterol. Other fungi are used to make antibiotics.

Decomposers

Fungi help create food for people to eat, but they are also important because of the things they eat. As you read earlier, fungi are an important part of the environment because they break down dead plant and animal matter, as shown in **Figure 14.** Without fungi and other decomposers, dead plants and animals would pile up year after year. Fungi also help break down pollution, including pesticides, in soil. Without fungi to destroy it, pollution would build up in the environment.

Living things need nutrients. The nutrients available in the soil would eventually be used up if they were not replaced by decomposing plant and animal matter. Fungi help put these nutrients back into the soil for plants to use.

Active Reading **10. Identify** Underline three benefits of fungi as decomposers.

Figure 14 Fungi help decompose dead organic matter, such as this rabbit.

May 8

October 6

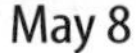

Math Skills MA.6.A.3.6

Using Fractions

Under certain conditions, 100 percent of the cells in fungus A reproduce in 24 hours. The number of cells of fungus A doubles once each day.

Day 1 = 10,000 cells

Day 2 = 20,000 cells

Day 3 = 40,000 cells

Day 4 = 80,000 cells

When an antibiotic is added to the fungus, the growth is reduced by 50 percent. Only half the cells reproduce each day.

Day 2 = 15,000 cells

Day 3 = 22,500 cells

Day 4 = 33,750 cells

Practice

11. Without an antibiotic, how many cells of fungus A would there be on day 6?

Fungi and Plant Roots

Plants benefit from fungi in other ways, too. Many fungi and plants grow together, helping each other. Recall that fungi take in minerals and water through the hyphae, or threadlike structures that grow on or under the surface. *The roots of the plants and the hyphae of the fungi weave together to form a structure called* **mycorrhiza** (mi kuh RI zuh; plural, micorrhizae).

Mycorrhizae can exchange molecules, as shown in **Figure 15.** As fungi break down decaying matter in the soil, they make nutrients available to the plant. They also increase water absorption by increasing the surface area of the plant's roots.

Fungi cannot photosynthesize, or make their own food using light energy. Instead, the fungi in mycorrhizae take in some of the sugars from the plant's photosynthesis. The plants benefit by receiving more nutrients and water. The fungi benefit and continue to grow by using plant sugars. Scientists suspect that most plants gain some benefit from mycorrhizae.

Active Reading 12. **Explain** How do mycorrhizae benefit both the plant and the fungus?

Figure 15 The roots of this buckthorn plant and the hyphae of fungi weave together, enabling the exchange of nutrients.

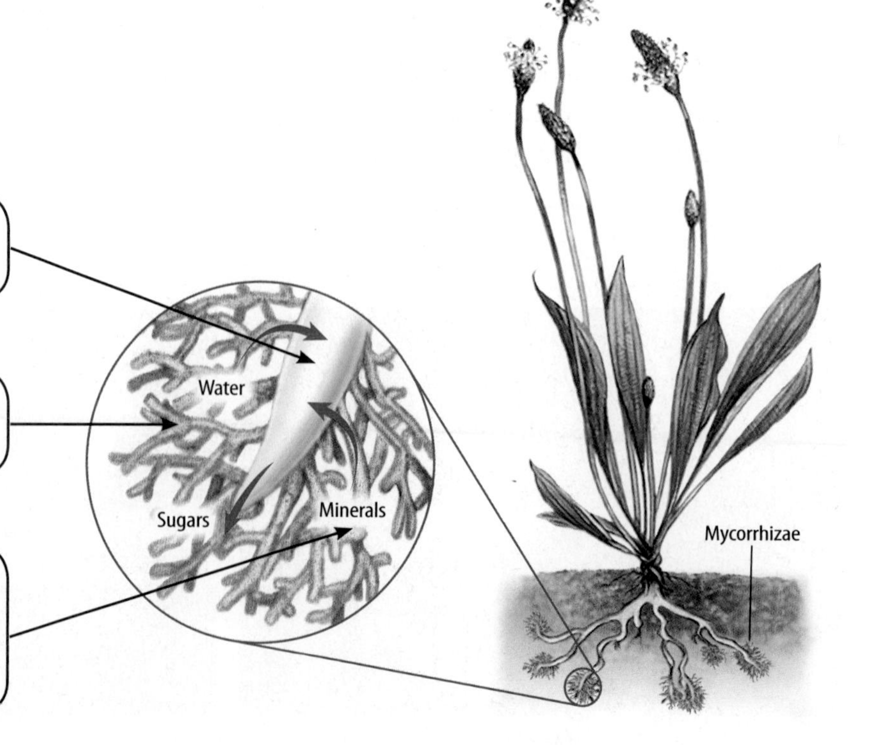

The whitish structure is part of the plant's root.

The threadlike structures are the hyphae of the fungus.

The arrows and labels show water and minerals moving into the root and sugars moving from the root to the hyphae.

Health and Medicine

You might recall that many protists can be harmful to humans and the environment. This is true of fungi as well. A small number of people die every year after eating poisonous mushrooms or spoiled food containing harmful fungi.

You do not have to eat fungi for them to make you sick or uncomfortable. You already read that fungi cause athlete's foot rashes and diaper rashes. Some fungi cause allergies, pneumonia, and thrush. Thrush is a yeast infection that grows in the mouths of infants and people with weak immune systems.

Although fungi can cause disease, scientists also use them to make important medicines. Antibiotics, such as penicillin, are among the valuable medications made from fungi. An accident resulted in the discovery of penicillin. Alexander Fleming was studying bacteria in 1928 when spores of *Penicillium* fungus contaminated his experiment and killed the bacteria. After years of research, this fungus was used to make an antibiotic similar to the penicillin used today. **Figure 16** illustrates how penicillin affects bacterial growth.

Active Reading **13. Distinguish** Tell the difference between helpful and harmful aspects of fungi in medicine.

Harmful
Helpful

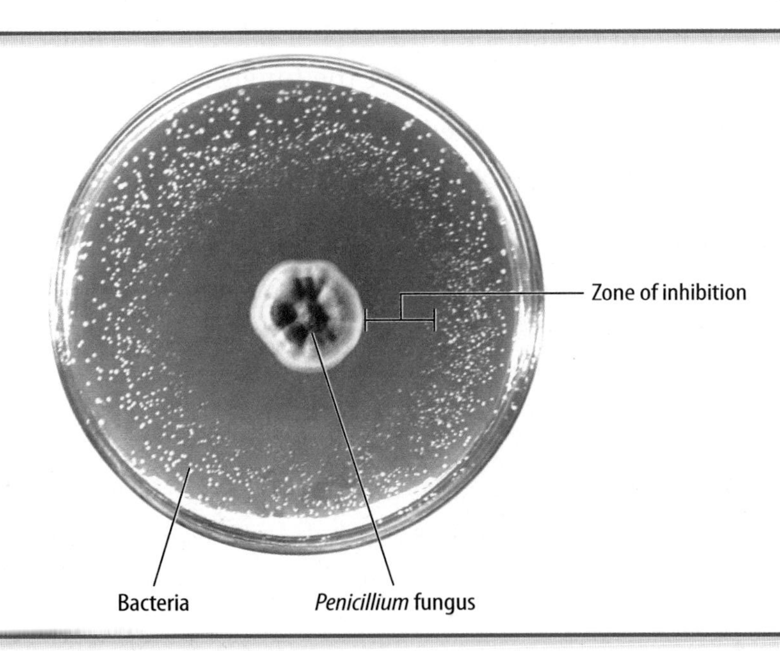

Figure 16 The *Penicillium* fungus that prevents bacteria from growing is used to make penicillin, an antibiotic medicine.

14. Visual Check
Infer How can you tell that the fungi are stopping the bacteria from growing?

Over time, some bacteria have become resistant to the many antibiotics used to fight illness. New antibiotics need to be developed to treat the same diseases. As new species of fungi are discovered and studied, scientists might find new sources of antibiotics and medicines.

Active Reading **15. Describe** Give two ways that fungi are important to humans.

What are lichens?

Word Origin

lichen
from Greek *leichen,* means "what eats around itself"

Active Reading **16. Identify** Underline which organisms usually grow together to form a lichen.

Do you recall the photo at the beginning of the chapter? The structure pictured is a lichen. *A* **lichen** (LI kun) *is a structure formed when fungi and certain other photosynthetic organisms grow together.* Usually, a lichen consists of a sac fungus or club fungus that lives in a partnership with either a green alga or a photosynthetic bacterium. The fungus's hyphae grow in a layer around the algae cells.

Green algae and photosynthetic bacteria are autotrophs, which means they can make their own food using photosynthesis. Lichens are similar to mycorrhizae because both organisms benefit from the partnership. The fungus provides water and minerals while the bacterium or alga provides the sugars and oxygen from photosynthesis.

The Importance of Lichens

Imagine living on a sunny, rocky cliff like the one in **Figure 17.** Not many organisms could live there because there is little to eat. A lichen, however, is well suited to this harsh environment. The fungus can absorb water, help break down rocks, and obtain minerals for the alga or bacterium. They can photosynthesize and make food for the fungus.

Once lichens are established in an area, it becomes a better environment for other organisms. Many animals that live in harsh conditions survive by eating lichens. Plants benefit from lichens because the fungi help break down rocks and create soil. Plants can then grow in the soil, creating a food source for other organisms in the environment.

Figure 17 Lichens are structures made of photosynthetic organisms and fungi that can live in harsh conditions.

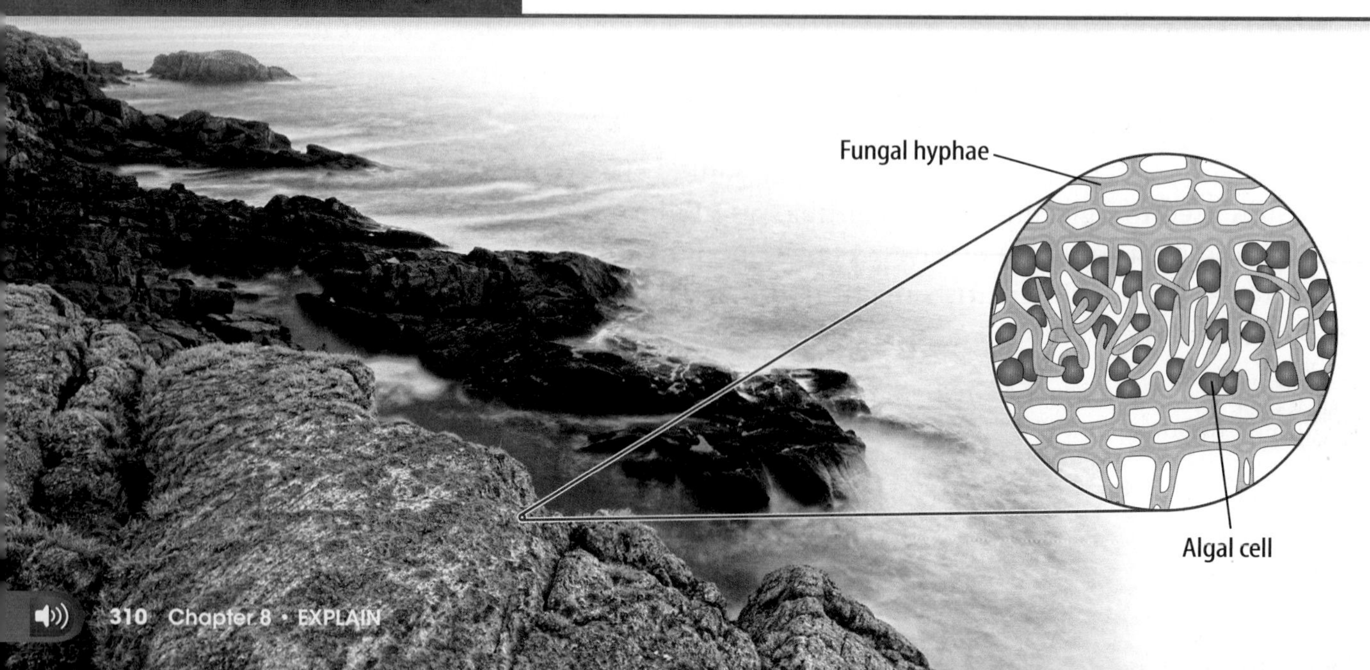

Lesson Review 2

Visual Summary

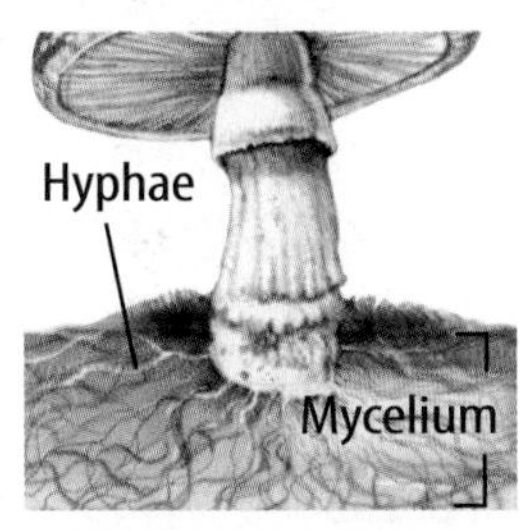

The body of a fungus is made up of threadlike hyphae that weave together to create a network of mycelium.

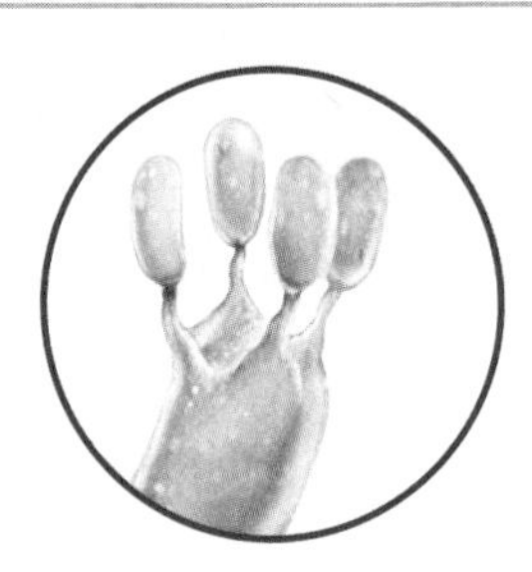

Club fungi produce sexual spores in the basidium.

A lichen is made of fungus and a photosynthetic bacterium or alga.

Use Vocabulary

1. **Distinguish** between a basidium and an ascus.

2. **Identify** the structure formed between fungal hyphae and plant roots.

Understand Key Concepts 

3. **List** the four groups of fungi.

4. Which disease is caused by a fungus? SC.6.L.14.6
   - (A) athlete's foot
   - (B) influenza
   - (C) malaria
   - (D) pneumonia

Interpret Graphics

5. **Compare and Contrast** Complete the table about sac fungi and zygote fungi. LA.6.2.2.3

Sac Fungi	Zygote Fungi

Critical Thinking

6. **Support** the claim that decomposition is important for the environment.

Math Skills MA.6.A.3.6

7. The number of cells in fungus X doubles every 2 hours. If you begin with 10 cells, how many would be present after 24 hours?

Chapter 8 Study Guide

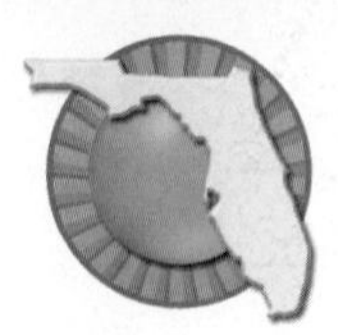

Think About It! Protists and fungi are diverse groups of organisms. Both may infect the human body. They are classified as neither plant nor animal and serve many functions in the ecosystem.

Key Concepts Summary

LESSON 1 What are protists?

- Scientists divide **protists** into three groups based on the type of organisms they most resemble. There are plantlike, animal-like, and funguslike protists.
- Protists are beneficial to humans in many ways. They are used to create many of the useful products you depend on. They also help decompose dead organisms and return nutrients to the environment.

Plantlike	Animal-like	Funguslike

Vocabulary

protist p. 291
algae p. 292
diatom p. 293
protozoan p. 296
cilia p. 296
paramecium p. 296
amoeba p. 297
pseudopod p. 297

LESSON 2 What are fungi?

- Scientists divide fungi into four groups, based on the type of structures they use for sexual reproduction. The four groups are club fungi, sac fungi, zygote fungi, and imperfect fungi.
- Fungi provide many foods and medicines that people use. In addition, fungi help break down dead organisms and recycle the nutrients into the environment.
- **Lichens** are structures made of a fungus and a photosynthetic organism. Both organisms work together to obtain food, water, and nutrients.

hyphae p. 303
mycelium p. 303
basidium p. 304
ascus p. 305
zygosporangia p. 305
mycorrhiza p. 308
lichen p. 310

Active Reading

FOLDABLES® Chapter Project

Assemble your lesson Foldables as shown to make a Chapter Project. Use the project to review what you have learned in this chapter.

Use Vocabulary

1. A protist that resembles a tiny animal is called a(n) ____________________.

2. A fungus and the roots of a plant form a structure called ____________________ that benefits both organisms.

3. The ____________________ is a saclike structure on a fungus that produces spores.

4. A(n) ____________________ is a microscopic, plantlike protist that can resemble glass or gems.

5. Short structures that cover the outside of some protists and help them move are called ____________________.

6. Fungi grow by extending threadlike body structures called ____________________.

Link Vocabulary and Key Concepts

Use vocabulary terms from the previous page and other terms from this chapter to complete the concept map.

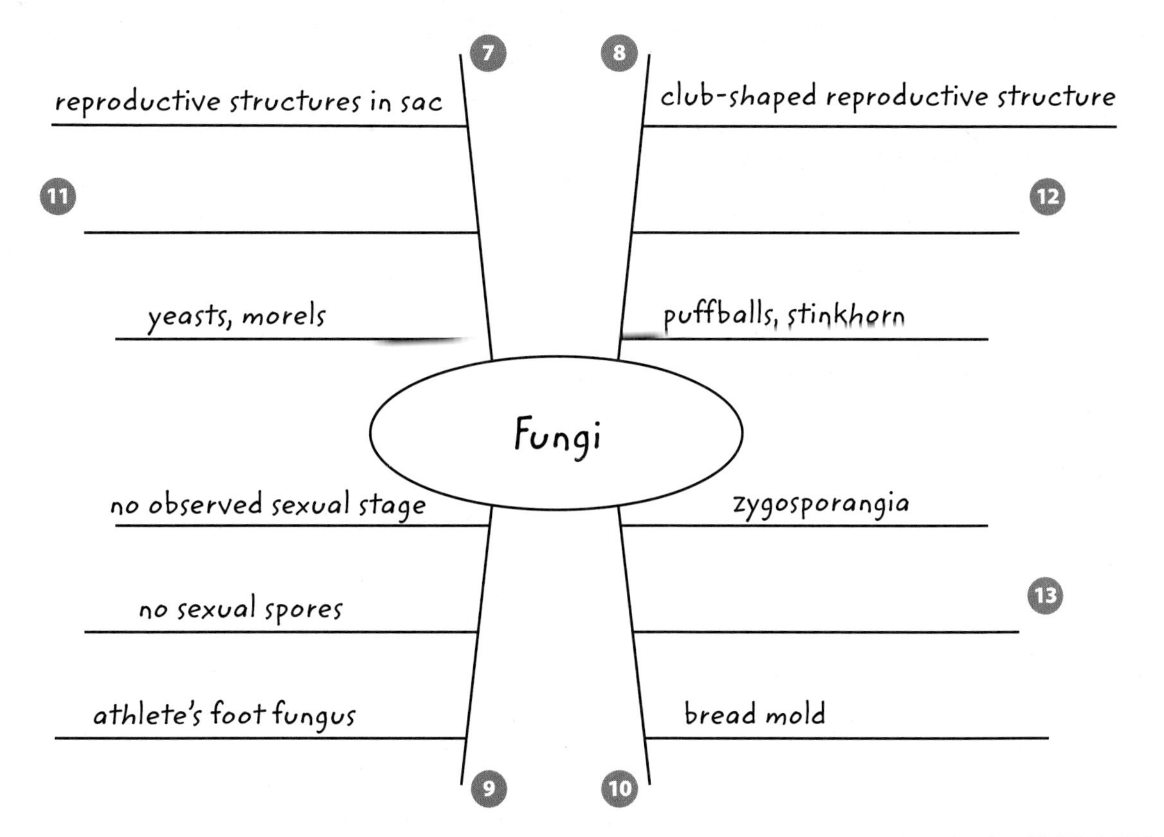

Chapter 8 Review

Fill in the correct answer choice.

Understand Key Concepts

1 Which organism causes red tides when found in large numbers? **SC.6.L.14.6**

Ⓐ algae
Ⓑ amoebas
Ⓒ ciliates
Ⓓ diatoms

2 Protists are a diverse group of organisms divided into what three categories? **SC.6.L.14.6**

Ⓐ animal-like, plantlike, protozoanlike
Ⓑ euglenoid, slime-mold, diatoms
Ⓒ plantlike, animal-like, and funguslike
Ⓓ green algae, red algae, and kelp

3 Which type of protist is commonly used in ice cream, toothpaste, soups, and body lotions? **SC.6.L.14.6**

Ⓐ algae
Ⓑ amoebas
Ⓒ ciliates
Ⓓ diatoms

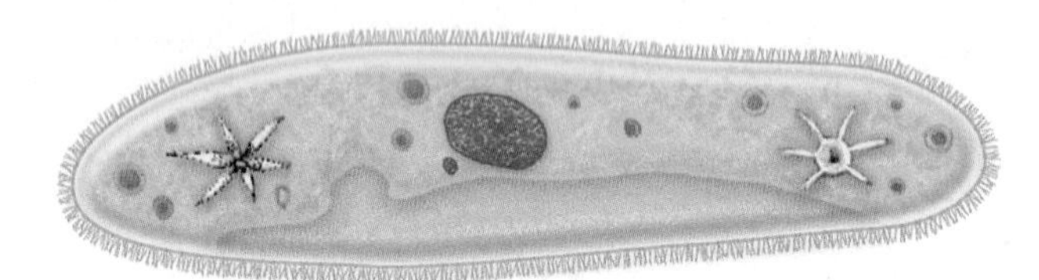

4 The organism in the figure above is a **SC.6.L.14.6**

Ⓐ ciliate.
Ⓑ diatom.
Ⓒ dinoflagellate.
Ⓓ kelp.

5 The main function of the hairlike structures surrounding the organism above is **SC.6.L.14.6**

Ⓐ decomposition.
Ⓑ movement.
Ⓒ photosynthesis.
Ⓓ reproduction.

6 What type of fungus is bread mold? **SC.6.L.14.6**

Ⓐ club
Ⓑ imperfect
Ⓒ sac
Ⓓ zygote

Critical Thinking

7 **Compare and contrast** different types of protists and fungi and how they can infect the human body. **SC.6.L.14.6**

8 **Evaluate** Imagine you are asked to justify removing kelp from an area of the ocean. Based on your knowledge of plantlike protists, what benefits or problems would you consider before you decide if the algae should be removed? **LA.6.2.2.3**

9 **Describe** Complete the table below with characteristics of the different types of animal-like protists. **LA.6.2.2.3**

	Number of nuclei	Method of eating	Method of movement
Ciliates			
Flagellates			
Sarcodines			

10 **Explain** how the movement of an amoeba differs from the movement of a dinoflagellate. **LA.6.2.2.3**

11 **List** several products you have used or seen that were made using fungi. SC.6.L.14.6

12 **Evaluate** how Alexander Fleming's experiments helped determine the importance of fungi to medicine. LA.6.2.2.3

Writing in Science

13 **Design** On a separate sheet of paper, design a brochure for a tour in which people could see several different types of lichens and fungi. What locations would be included, and which organisms would people be likely to observe? SC.6.L.14.6

Big Idea Review

14 **Explain** how decomposers such as protists and fungi play an important role in the environment. SC.6.L.14.6

15 What is the organism shown below, and how does it affect the environment? SC.6.L.14.6

Math Skills MA.6.A.3.6

Calculating Growth

16 The number of cells of Fungus Q doubles every three hours. If you begin with 1,000 cells, how many will there be after 12 hours?

17 Scientists want to know if an antibiotic is effective in treating a fungal infection. They start with two colonies of 100 cells each. The table shows what happens during the first two days of treatment.

	Day 2 Number of Cells	Day 3 Number of Cells
Untreated fungus	400	1600
Antibiotic A	200	300

a. How long does it take the untreated fungus to double in number?

b. What effect does the antibiotic have on the growth rate of the fungus?

Florida NGSSS Benchmark Practice

Fill in the correct answer choice.

Multiple Choice

1 How are malaria protozoan parasites spread from human to human? **SC.6.L.14.6**

Ⓐ through contaminated drinking water
Ⓑ through contaminated food
Ⓒ through animal bites
Ⓓ through mosquito bites

Use the diagram below to answer question 2.

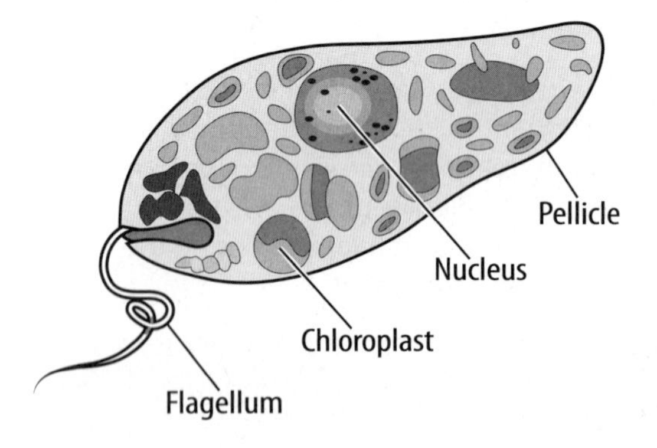

2 The euglenoid shown above does not infect humans. What structure indicates that it produces its own food? **SC.6.L.14.6**

Ⓕ chloroplast
Ⓖ flagellum
Ⓗ nucleus
Ⓘ pellicle

3 Which is an infection caused by a fungus that grows in the mouths of infants and people with weak immune systems? **SC.6.L.14.6**

Ⓐ athlete's foot
Ⓑ malaria
Ⓒ pneumonia
Ⓓ thrush

4 Which type of fungi can be poisonous to humans if eaten? **SC.6.L.14.6**

Ⓕ algae
Ⓖ paramecia
Ⓗ penicillin
Ⓘ mushrooms

5 Which animal-like protists can live in the digestive systems of humans and get nutrients and energy from the human's body? **SC.6.L.14.6**

Ⓐ sarcodines
Ⓑ dinoflagellates
Ⓒ ciliates
Ⓓ algae

6 Which small, parasitic protozoan causes malaria? **SC.6.L.14.6**

Ⓕ diatom
Ⓖ plasmodium
Ⓗ euglenoid
Ⓘ dinoflagellate

7 Where do protozoan parasites live? **SC.6.L.14.6**

Ⓐ in decomposing organisms
Ⓑ in host organisms
Ⓒ in soil
Ⓓ on leaves

8 In which human cells do plasmodia live and reproduce? **SC.6.L.14.6**

Ⓕ white blood cells
Ⓖ cancer cells
Ⓗ red blood cells
Ⓘ nerve cells

Use the diagram below to answer question 9.

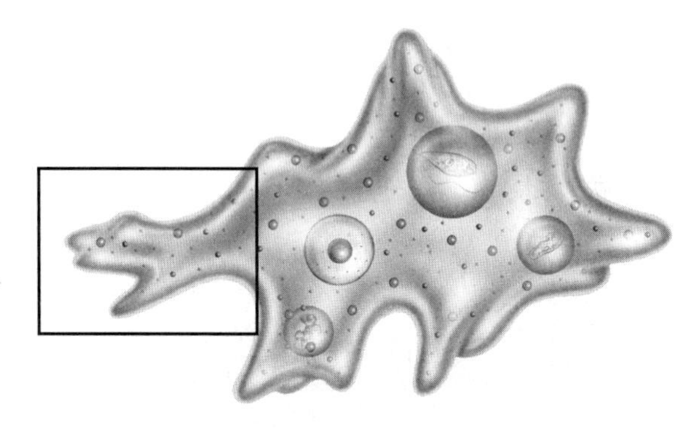

9 What is the function of the boxed area of this microscopic organism? **SC.6.L.14.6**

Ⓐ cellular respiration

Ⓑ defense

Ⓒ locomotion

Ⓓ photosynthesis

10 Which protist-caused disease is responsible for more than 1 million deaths each year? **SC.6.L.14.6**

Ⓕ thrush

Ⓖ athlete's foot

Ⓗ diaper rash

Ⓘ malaria

11 Which example of a sarcodine can live in human digestive tracts and move around using a pseudopod? **SC.6.L.14.6**

Ⓐ amoeba

Ⓑ dinoflagellate

Ⓒ diatom

Ⓓ paramecium

12 In which human organ do plasmodia spend part of their life cycle? **SC.6.L.14.6**

Ⓕ heart

Ⓖ stomach

Ⓗ liver

Ⓘ small intestine

13 Why are the fungi that cause athlete's foot labeled imperfect fungi? **SC.6.L.14.6**

Ⓐ The fungi do not show any normal characteristics of fungi.

Ⓑ Scientists classified imperfect fungi as protists.

Ⓒ The fungi reproduce in hot, dry environments.

Ⓓ Scientists have not observed a sexual reproductive stage in their life cycle.

14 Which type of fungi can cause diaper rash on babies? **SC.6.L.14.6**

Ⓕ club fungi

Ⓖ sac fungi

Ⓗ zygote fungi

Ⓘ imperfect fungi

NEED EXTRA HELP?

If You Missed Question...	1	2	3	4	5	6	7	8	9	10	11	12	13	14
Go to Lesson...	1	1	2	2	1	1	1	1	1	1	1	1	2	2

Name ______________________________ Date ______________

Benchmark Mini-Assessment Chapter 8 • Lesson 1

FLORIDA NGSSS mini BAT

Multiple Choice *Bubble the correct answer.*

Use the image below to answer questions 1 and 2.

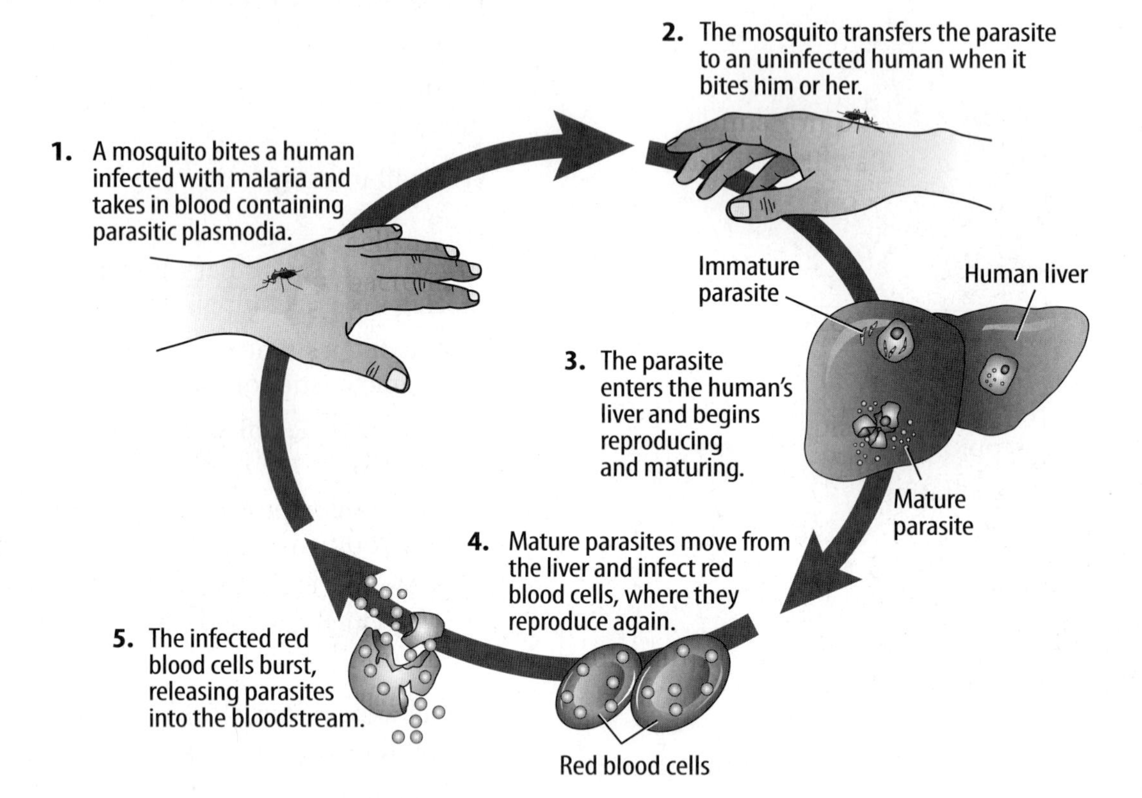

1. The parasite plasmodia causes malaria in humans. At which stage in the life cycle will the host be most affected by the symptoms of the disease? **SC.6.L.14.6**

Ⓐ Stage 2

Ⓑ Stage 3

Ⓒ Stage 4

Ⓓ Stage 5

2. At which point in the plasmodia life cycle can humans best interrupt the cycle and prevent the spread of malaria from person to person? **SC.6.L.14.6**

Ⓕ Stage 1

Ⓖ Stage 2

Ⓗ Stage 3

Ⓘ Stage 4

Name ______________________ Date ____________

Multiple Choice *Bubble the correct answer.*

Use the image below to answer questions 1 and 2.

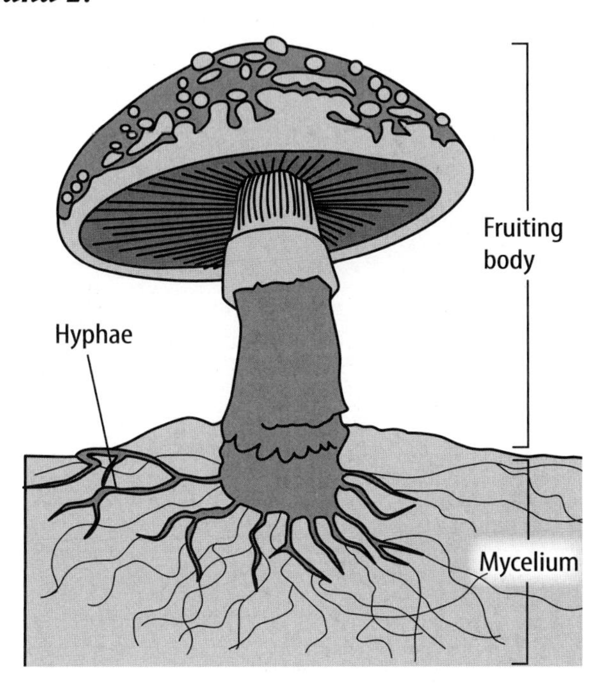

1. Which part of a typical fungus is responsible for taking in water and nutrients for the organism? **LA.6.2.2.3**

   Ⓐ cap
   Ⓑ gills
   Ⓒ hyphae
   Ⓓ mycelium

2. What type of fungus is shown above? **SC.6.L.14.6**

   Ⓕ club
   Ⓖ imperfect
   Ⓗ sac
   Ⓘ zygote

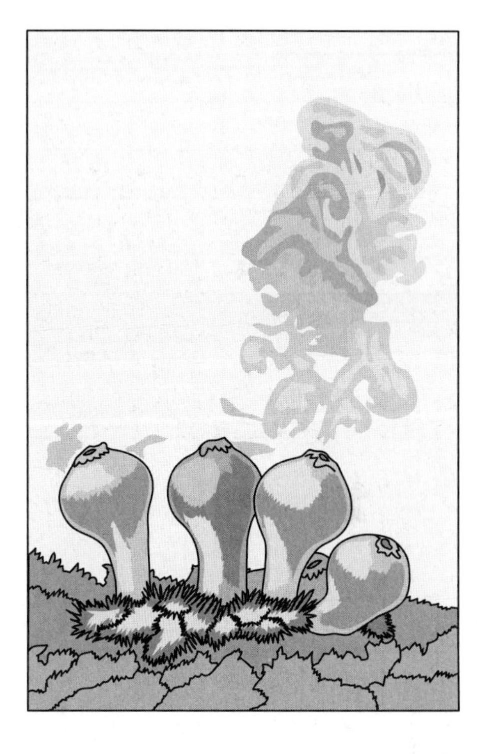

3. Look at the "smoke" rising from the puffballs above. This "smoke" is an example of **LA.6.2.2.3**

   Ⓐ asexual reproduction.
   Ⓑ fruiting body growth.
   Ⓒ hyphae growth.
   Ⓓ sexual reproduction.

4. How do fungi and other organisms sometimes interact? **SC.6.L.14.6**

   Ⓕ Fungi provide minerals to animals.
   Ⓖ Fungi provide sugars to plants.
   Ⓗ Fungi exchange minerals for sugars with animals.
   Ⓘ Fungi exchange minerals for sugars with plants.

Name ______________________ Date ____________

Notes

Plant Characteristics

There are many types of plants throughout the world. Although they may appear different, there are similarities among all plants. Put an *X* in front of the characteristics you think all plants share.

_____ A. They grow from seeds.

_____ B. They produce flowers.

_____ C. They make their own food.

_____ D. They can be one cell or many cells.

_____ E. They live only on land.

_____ F. They have roots, stems, and leaves.

_____ G. They need light, carbon dioxide, and water.

_____ H. They need oxygen.

_____ I. They have DNA in their cells.

_____ J. Their cells have cell walls.

Explain your thinking. Why did you choose some characteristics and not others?

FLORIDA

Chapter 9

Plant DIVERSITY

FLORIDA BIG IDEAS

1 The Practice of Science

2 The Characteristics of Scientific Knowledge

14 Organization and Development of Living Organisms

15 Diversity and Evolution of Living Organisms

The Big Idea

Think About It!

Why are plants in so many different environments on Earth?

There are many different kinds of plants growing here! What plant characteristics do you think enable such a wide variety of plants to grow in one place?

1 Where else do plants grow?

2 How are those plants different from the plants growing in this environment? How are they similar?

Get Ready to Read

What do you think about plant diversity?

Before you read, decide if you agree or disagree with each of these statements. As you read this chapter, see if you change your mind about any of the statements.

	AGREE	DISAGREE
1 All plants produce flowers and seeds.	☐	☐
2 Humans depend on plants for their survival.	☐	☐
3 Some plants move water only by diffusion.	☐	☐
4 Mosses can grow only in moist, shady places.	☐	☐
5 Some mosses and gymnosperms are used for commercial purposes.	☐	☐
6 All plants grow, flower, and produce seeds in one growing season.	☐	☐

There's More Online!
Video • Audio • Review • ⓘLab Station • WebQuest • Assessment • Concepts in Motion • Multilingual eGlossary

Lesson 1

What is a PLANT?

ESSENTIAL QUESTIONS

What characteristics are common to all plants?

What adaptations have enabled plant species to survive Earth's changing environments?

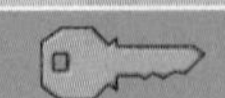
How are plants classified?

Vocabulary

producer p. 326

cuticle p. 327

cellulose p. 327

vascular tissue p. 328

Florida NGSSS

LA.6.2.2.3 The student will organize information to show understanding (e.g., representing main ideas within text through charting, mapping, paraphrasing, summarizing, or comparing/contrasting);

MA.6.A.3.6 Construct and analyze tables, graphs, and equations to describe linear functions and other simple relations using both common language and algebraic notation.

MA.6.S.6.2 Select and analyze the measures of central tendency or variability to represent, describe, analyze, and/or summarize a data set for the purposes of answering questions appropriately.

SC.6.N.1.5 Recognize that science involves creativity, not just in designing experiments, but also in creating explanations that fit evidence.

SC.6.L.14.4 Compare and contrast the structure and function of major organelles of plant and animal cells, including cell wall, cell membrane, nucleus, cytoplasm, chloroplasts, mitochondria, and vacuoles.

SC.6.L.15.1 Analyze and describe how and why organisms are classified according to shared characteristics with emphasis on the Linnaean system combined with the concept of Domains.

SC.7.N.1.6 Explain that empirical evidence is the cumulative body of observations of a natural phenomenon on which scientific explanations are based.

SC.6.N.1.5

Inquiry Launch Lab

15 minutes

What is a plant?

A plant often is described as a living thing that makes its own food, has leaves and stems, and is green in color. Even with the many different types of plants on Earth, this description often holds true. Or does it?

1. Examine the photos below.

1	2	3	4	5	6

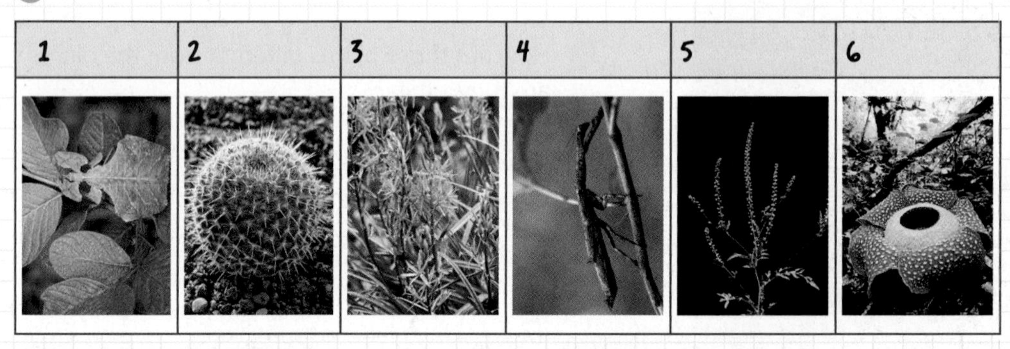

2. Next to each number above, write yes if you think the photo is of a plant or no if you think the photo is not of a plant.

Think About This

1. Which photos do you think are plants? Explain your choices.

2. Visualize each object without the background, and decide if you want to make changes in your list.

3. Key Concept What characteristics are common to each plant in these pictures?

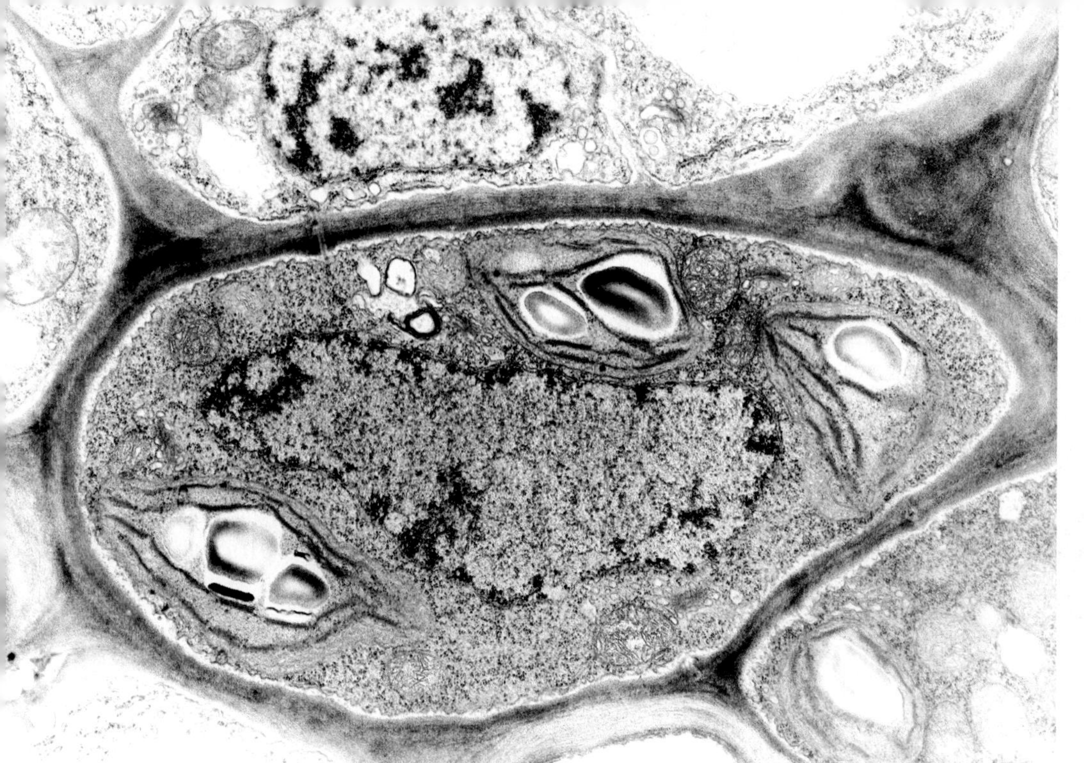

Why So Successful?

1. This plant cell has parts that animal cells don't have. How do you think those cellular parts help plants live? What other parts do plants have that enable them to be successful in so many diverse environments on Earth?

Characteristics of Plants

You might not think about plants often, but they are an important part of life on Earth. As you read this lesson, look for the characteristics that make plants so important to other organisms.

Cell Structure

Plants are made of eukaryotic cells. Recall that eukaryotic cells have membrane-bound organelles. Some of a plant cell's organelles are shown in **Figure 1.** A plant cell differs from an animal cell because it contains chloroplasts and a cell wall. Chloroplasts convert light energy to chemical energy. The cell wall provides support and protection. A mature plant cell also has one or two vacuoles that store a watery liquid called sap.

2. **NGSSS Check** Describe What structures in a plant cell are not in an animal cell? SC.6.L.14.4

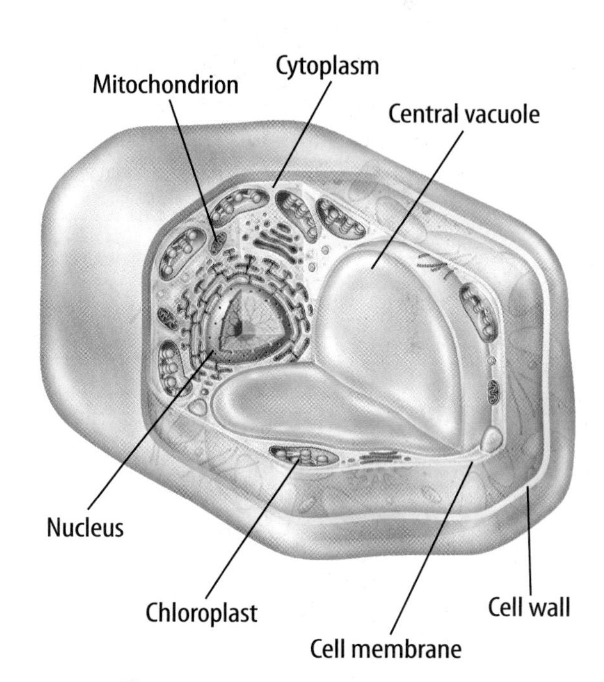

Figure 1 Plant cells and animals cells have many of the same organelles.

Active Reading 3. **Identify** Circle the organelle that converts light energy to chemical energy.

Figure 2  Some plants, such as the reproductive stage of some ferns, are microscopic. Other plants, such as redwood trees, are huge.

WORD ORIGIN

producer
from Latin *producer*, means "lead or bring forth; draw out"

Active Reading
FOLDABLES® LA.6.2.2.3

Make a vertical two-tab book and label it as shown. Use it to organize your notes on common plant characteristics and more-specific plant adaptations.

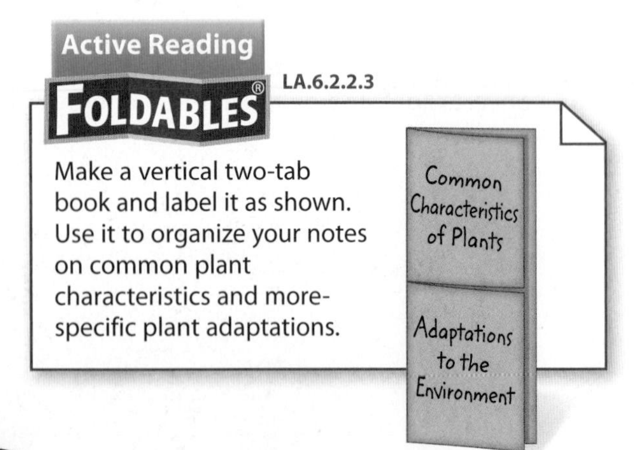

Multicellular

Plants are multicellular. This means they are made of many cells. The cells carry out specialized functions and work together to keep the plant alive. As shown in **Figure 2,** some plants are microscopic while others are some of the largest organisms on Earth.

Producers

Organisms that use an outside energy source, such as the Sun, to make their own food are **producers.** Plants are producers. They make their own food, a simple sugar called glucose, during a process called photosynthesis. All other organisms rely on producers, either directly or indirectly, for their sources of food.

Active Reading **4. Recall** List the characteristics that are common to all plants.

____

____

____

Plant Adaptations

Scientists hypothesize that present-day land plants and green algae evolved from a common ancestor. They base their hypothesis on chemical similarities between green algae and plants. Some of the pigments in green algae and land plants are the same kind. There also are DNA similarities between these two groups of organisms.

The first land plants probably lived in moist areas. Life on land would have provided some advantages. There would have been plenty of sunlight for photosynthesis to occur. The air that surrounded those plants would have been a mixture of gases, including carbon dioxide, which is also needed for photosynthesis. As land plants became more abundant, the amount of oxygen in the atmosphere increased because oxygen is a product of photosynthesis.

Plant species also had to adapt to survive without being surrounded by water. Many of the characteristics we now see in plants are adaptations to life on land.

Protection

One advantage to life on land is a constant supply of air that contains carbon dioxide. As you just read, carbon dioxide is needed for photosynthesis. Many plants have a *waxy, protective layer on their leaves, stems, and flowers called the* **cuticle.** It is made of a waxy substance that is secreted by the cells. Its waxy nature slows the evaporation of water from a plant's surface. This covering also provides some protection from insects that might harm a plant's tissues.

Active Reading **5. Determine** Highlight the ways a plant's cuticle protects it.

Support

The water that surrounds aquatic plants supports them. Land plants must provide their own support. Like all cells, a plant cell has a cell membrane. Recall that a rigid cell wall surrounds the cell membrane in a plant cell. The cell wall provides support and is made of cellulose. **Cellulose** *is an organic compound made of chains of glucose molecules.* Many land plants also produce a chemical compound called lignin (LIG nun). Lignin strengthens cellulose and makes it more rigid. The piece of wood shown in **Figure 3** is mostly made of cellulose and lignin.

Figure 3 The combined strength of all of a plant's cell walls provides support for the plant.

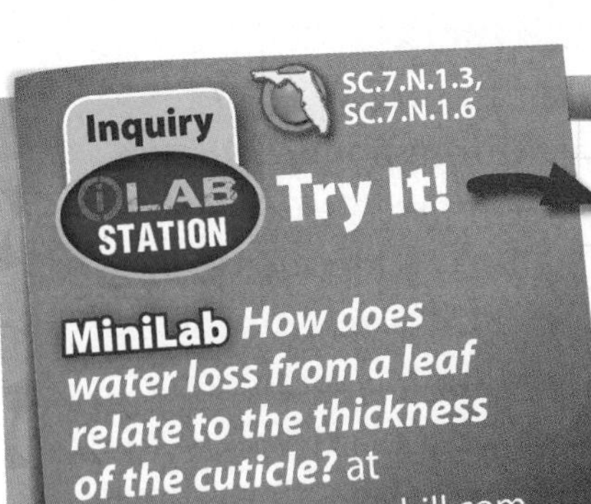

Apply It!

After you complete the lab, answer the questions below.

1. What type of environment do you think plants with thick cuticles live in? Explain.

2. A tropical rain forest can receive more than 25 cm of rain in one month. What type of cuticles would the plants in this environment have? Explain.

Active Reading **6. Identify** Fill in the table below with methods of support in land plants and aquatic plants. Use the word *cellulose* in your answer.

Land Plants	Aquatic Plants
supported by:	supported by

Transporting Materials

In order for a plant to survive, water and nutrients must move throughout its tissues. In some plants such as mosses, these materials can move from cell to cell by the processes of osmosis and diffusion. This means that water and other materials dissolved in water move from areas of a plant where they are more concentrated to areas where they are less concentrated. However, other plants such as grasses and trees have specialized tissues called vascular tissue. **Vascular tissue** *is composed of tubelike cells that* ***transport*** *water and nutrients in some plants.* Vascular tissue can carry materials throughout a plant—great distances if necessary, up to hundreds of meters. You will read more about vascular tissues in Lesson 3 of this chapter.

Reproduction

Water carries the reproductive cells of aquatic plants from plant to plant. How do you think land plants reproduce without water? Land plants evolved other strategies for reproduction. Some plants have water-resistant seeds or spores that are part of their reproductive process. Seeds and spores move throughout environments in different ways. These include animals and environmental factors such as wind and water. Several methods of seed dispersal are shown in **Figure 4.**

7. Summarize What adaptations of plants have enabled them to survive Earth's changing environments?

___

___

___

Adaptations for Seed Dispersal

Figure 4 Plants have developed different methods of dispersing their seeds.

8. Identify Label each photo as
A. carried by wind,
B. clinging to fur, or
C. floating in water.

Coconut seeds

Milkweed seeds

Burrs containing seeds

Figure 5 Liverworts and mosses reproduce by producing spores. Both pine and magnolia trees produce seeds.

Plant Classification

You might recall that kingdoms such as the animal kingdom consist of smaller groups called phyla. Members of the plant kingdom are organized into groups called divisions instead of phyla. Like all organisms, each plant has a two-word scientific name such as *Quercus rubra* for a red oak.

9. **NGSSS Check** **Explain** How are plants classified? SC.6.L.15.1

10. **Visual Check** **Compare and Contrast** How are the pine tree and the magnolia tree alike? How are they different?

Seedless Plants

Liverworts and mosses, such as the ones shown in **Figure 5,** reproduce by structures called spores. Plants that reproduce by spores often are called seedless plants. Seedless plants do not have flowers. Some seedless plants do not have vascular tissue and are called nonvascular plants. Others, such as ferns, have vascular tissue and are called vascular plants. Seedless plants are classified into several divisions.

Seed Plants

Most of the plants you see around you, such as pine trees, grasses, petunias, and oak trees, are seed plants. Almost all the plants we use for food are seed plants. Some seed plants have flowers that produce fruit with one or more seeds. Others, such as pine trees, produce their seeds in cones. Each seed has tissues that surround, nourish, and protect the tiny plant embryo inside it. It is thought that all present-day plants originated from a common ancestor, an ancient green algae, as shown in **Figure 6** on the next two pages.

Classification of Plants

Hornworts often grow in fields or along roads.

Ferns are vascular plants, but fern reproduction includes spores.

Mosses are the most abundant nonvascular plants. Mosses generally grow in shady, moist places.

Club mosses are spore-producing vascular plants.

Liverworts grow in almost every habitat on Earth.

Vascular Plants

Ancient green algae are thought to be the ancestors of land plants as well as present-day green algae.

Figure 6 This tree represents the evolutionary relationships among plant divisions.

11. **Visual Check** Identify What type of plant is thought to be the ancestor of land plants?

One stage of this plant's life cycle reminded people of a horse's tail. **Horsetails** also have been called scouring rushes due to their abrasive texture.

There is only one species of **ginkgo** alive today. It is often used as an ornamental tree.

Conifers are the largest and most diverse division of gymnosperms.

Seedless Plants

Cycads usually grow in tropical regions.

Gymnosperms

Seed Plants

Tulips grow best in climates that have long, cool springs.

Angiosperms

Grass flowers typically are small and easily overlooked.

Active Reading **12. Classify** Use the evolutionary tree in **Figure 6** to fill in the table below on nonvascular and vascular plants.

Nonvascular Plants	Vascular Plants		
	Seedless Plants	**Seed Plants**	
		Gymnosperms	*Angiosperms*

Lesson Review 1

Visual Summary

Plants are multicellular producers.

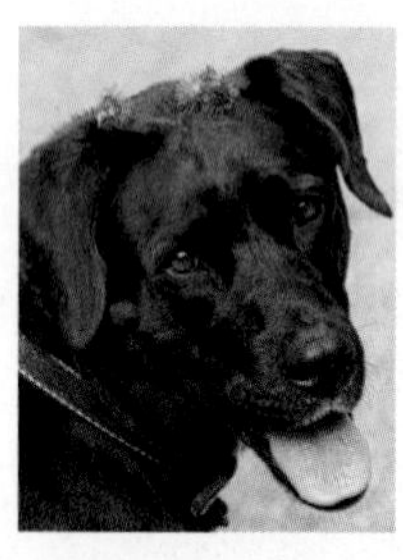

Water carries the reproductive cells of aquatic plants from plant to plant. Land plants evolved different reproductive strategies to ensure their survival.

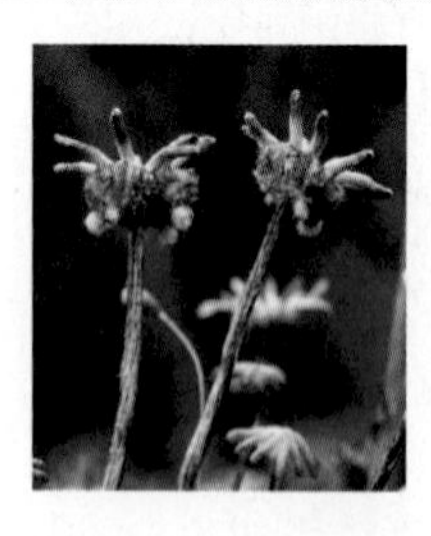

Members of the plant kingdom are classified into groups called divisions.

Use Vocabulary

1. Tubelike cells in some plants that transport water and nutrients are called ______.
2. An organic compound made of chains of glucose molecules is ______.

Understand Key Concepts

3. Which structure helps support plant cells? SC.6.L.14.4
   - (A) cell wall
   - (B) chloroplast
   - (C) mitochondria
   - (D) ribosome
4. **List** the common characteristics of plants.
5. **Describe** an example of a plant adaptation that helps plants survive on land.
6. **Distinguish** between seedless and seed plants. SC.6.L.15.1

Interpret Graphics

7. **Organize** Fill in the graphic organizer below with plant adaptations. LA.6.2.2.3

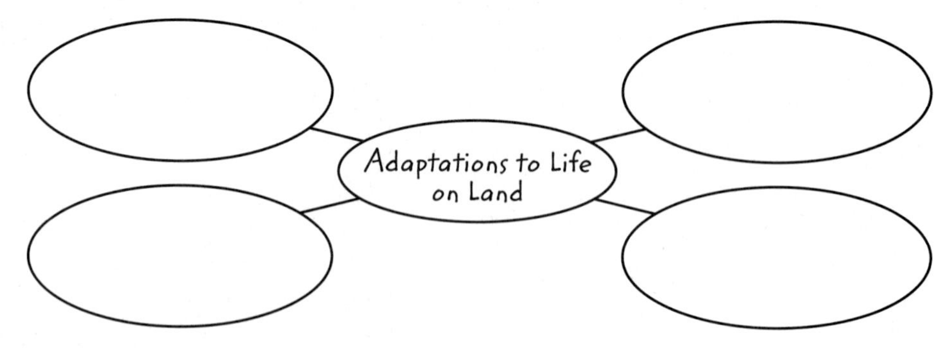

Critical Thinking

8. **Hypothesize** which of the adaptations to life on land might be missing in a plant that grows in shallow water.
9. **Assess** the importance of plants as producers.

Florida Trees In the Sea

How can a plant survive in salt water?

Mangrove forests thrive along Florida's warm coastlines. Birds perch in the treetops, while young fish and other animals hide among the widespread underwater roots. These amazing trees can survive where others cannot—in salt water. Salt water kills most plants. It causes freshwater to diffuse out of plant cells, and they collapse. In addition, most plants would not survive with their roots under water because they cannot absorb oxygen from water. But mangrove trees have special traits that help them survive along the ocean's edge.

1. Leaves

- **A mangrove's thick leaves store freshwater. The waxy coating, or cuticle, covering the leaves reduces the amount of water lost by evaporation.**
- **Some mangroves release salt through glands in their leaves. When the leaves drop off the plant, the salt goes with them.**

2. Seedlings

- **Unlike most other plant seeds that sprout in soil, mangrove seeds begin sprouting on the tree. They're ready to take root as soon as they drop.**
- **A mangrove seedling can survive floating in seawater for up to a year until it finds a suitable environment in which to grow.**

3. Bark

- **Some mangrove species store salt in bark that later peels away.**
- **Some mangrove bark also has tiny holes. Oxygen can enter a mangrove through these holes.**

4. Roots

- **Mangrove roots filter salt out of salt water.**
- **The roots arch over the water and absorb oxygen through tiny pores. These pores close when salt water covers the roots during a high tide.**

It's Your Turn

DIAGRAM Choose a species of tree in your area. Research the traits of that species, such as bark, roots, leaves, or seeds, that help it to be successful in its environment. Draw and label a diagram of the tree that illustrates your findings.

Lesson 2

Seedless PLANTS

ESSENTIAL QUESTION

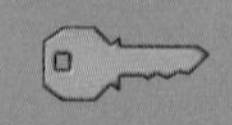

How are nonvascular and vascular seedless plants alike, and how are they different?

Vocabulary

rhizoid p. 335

frond p. 337

SC.7.N.1.3

Launch Lab

20 minutes

Which holds more water?

Peat moss is the common name of approximately 300 different types of mosses. The partially decomposed remains of these moss plants form peat moss. Some potting soils contain peat moss.

Procedure

1. Read and complete a lab safety form.
2. Fill a **250-mL beaker** with **potting soil** to about 3 cm from the top.
3. In a **tub,** mix equal parts **peat moss** and potting soil. Fill a second **250-mL beaker** to the same level with this mixture.
4. Pour 30 mL of water into each beaker. Examine the beakers after 5 min and record your observations below.
5. After another 5 min, place each beaker on its side in an **aluminum pie pan.** Record your observations.

Data and Observations

Think About This

1. How quickly did each soil mixture absorb the water?
2. What happened when you placed the beakers on their sides in the pie pans?
3. Why do you think peat moss is added to potting soil? How might this benefit plants?

Florida NGSSS

LA.6.2.2.3 The student will organize information to show understanding (e.g., representing main ideas within text through charting, mapping, paraphrasing, summarizing, or comparing/contrasting);

SC.6.L.15.1 Analyze and describe how and why organisms are classified according to shared characteristics with emphasis on the Linnaean system combined with the concept of Domains.

SC.7.N.1.3 Distinguish between an experiment (which must involve the identification and control of variables) and other forms of scientific investigation and explain that not all scientific knowledge is derived from experimentation.

Inquiry: Spores Instead of Seeds?

1. Ferns are vascular seedless plants. They reproduce using spores instead of seeds. This type of fern produces spores in clusters on one side of its fronds. What characteristics do you think all vascular seedless plants share?

REVIEW VOCABULARY

osmosis
the diffusion of water molecules

Nonvascular Seedless Plants

If someone asked you to make a list of plants, your list might include plants such as your favorite flowers or trees that grow near your home. You probably would not include any nonvascular seedless plants on your list.

Many scientists refer to all nonvascular seedless plants as bryophytes (BRI uh fites). These plants usually are small. Because they lack tubelike structures, called vascular tissue, that transport water and nutrients, the bryophytes usually live in moist environments. Materials move from cell to cell by diffusion and **osmosis.**

Because bryophytes do not have vascular tissue, they do not have roots, stems, or leaves. They have rootlike structures called rhizoids, shown in **Figure 7. Rhizoids** *are structures that anchor a nonvascular seedless plant to a surface.* Rhizoids can be unicellular—consisting of only one cell—or they can be multicellular. The photosynthetic tissue of bryophytes is often only one cell layer thick. This layer does not have a cuticle, which most other plants have. Reproduction is by spores and requires water. Mosses, liverworts, and hornworts are bryophytes.

2. NGSSS Check **Summarize** What characteristics are common in bryophytes? SC.6.L.15.1

Figure 7 Rhizoids anchor bryophytes to a surface, such as soil, rocks, and the bark of trees.

3. Visual Check **Explain** How is the structure of rhizoids well-suited to their function?

Active Reading

FOLDABLES® LA.6.2.2.3

Make a vertical, three-tab Venn diagram book. Label it as shown. Use it to compare and contrast vascular and nonvascular seedless plants.

Vascular Seedless Plants

Both

Nonvascular Seedless Plants

Mosses

You might be familiar with the most common bryophytes—the mosses. These small, green plants grow in forests, in parks, and sometimes even in the cracks of sidewalks. Mosses usually grow in shady, damp environments, but they are able to survive periods of dryness. As shown in **Figure 8,** mosses have leaflike structures that grow on a stemlike structure called a stalk. They have multicellular rhizoids.

Mosses play an important role in the ecosystem. They are often the first plants to grow in barren areas or after a natural disturbance such as a fire or a mud slide. The ability of mosses to retain large amounts of water makes peat moss a useful additive for potting soil, as you discovered in the Launch Lab. This moss has been used to enrich soil and as a heating source.

Active Reading **4. Describe** Underline why mosses are important to ecosystems.

Liverworts

Hundreds of years ago, people thought that this plant could be used to treat liver diseases. Liverwort also gets its name from its appearance—it resembles the flattened lobes of a liver. The rhizoids of liverworts are unicellular. The two common forms of liverworts are leafy and thallose (THA los) liverworts, as shown in **Figure 8.**

Hornworts

The long, hornlike reproductive structures shown in **Figure 8** give this group of plants its name. These reproductive structures produce spores. Hornworts are only about 2.5 cm in diameter. One unusual characteristic of hornworts is that each of its photosynthetic cells has only one chloroplast.

Figure 8 Mosses, liverworts, and hornworts all lack vascular tissue.

Nonvascular Seedless Plants

The green, leafy structures of mosses are not considered leaves because they do not contain vascular tissue.

A leafy liverwort, on the left, also lacks vascular tissue. The lobes of a thallose liverwort, on the right, can be as thin as one cell layer thick.

Until hornworts produce their reproductive structures, or "horns," they can easily be mistaken for liverworts.

Vascular Seedless Plants

Over 90 percent of plant species are vascular plants. Unlike nonvascular plants, they contain vascular tissue in their stems, roots, and leaves. Because vascular plants contain tubelike structures that transport water and nutrients, these plants generally are larger than nonvascular plants. However, present-day vascular seedless plants are smaller than their ancient ancestors. These ancient plants grew as tall as trees. Much of the fossil fuels that we use today came from the remains of these ancient plants.

Ferns

The **fronds**, *or leaves of ferns,* make up most of a fern. Ferns range in size from a few centimeters to several meters tall, such as the one in **Figure 9.** They grow in a variety of habitats, including damp, swampy areas and dry, rocky cliffs. Ferns often are houseplants.

Club Mosses

Unlike mosses, club mosses have roots, stems, and leaves. Club mosses, shown in **Figure 10,** are small plants that rarely grow taller than 50 cm. The stems often grow along the ground. The leaves are scalelike. The spores of club mosses make a fine powder that is so flammable that it has been used to make fireworks!

Horsetails

As shown in **Figure 11,** horsetails have small leaves growing in circles around the stems. Horsetail stems are hollow, and the tissues contain silica, a mineral in sand, that makes them abrasive. They once were used for scrubbing pots. Horsetails can be grown in water gardens but tend to spread rapidly.

5. **NGSSS Check** **Compare and Contrast** How are nonvascular and vascular seedless plants alike? How are they different? SC.6.L.15.1

Figure 9 Tree ferns, such as this one, once were the dominant plants on Earth.

Figure 10 Club mosses reproduce by producing spores in two or three cylindrical, yellow-green colored cones.

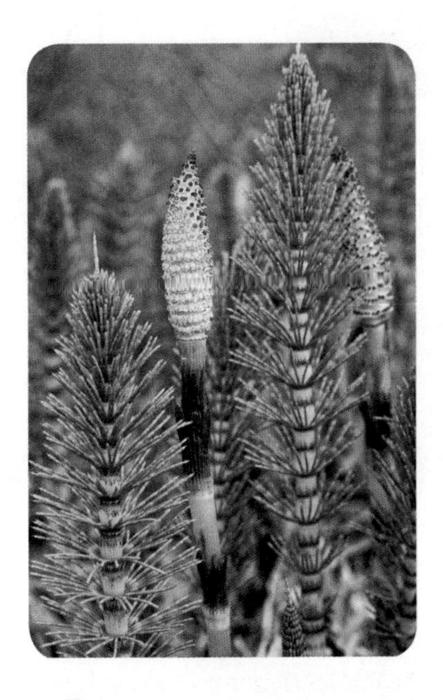

Figure 11 The hollow stem of a horsetail is its main photosynthetic structure.

Lesson Review 2

Visual Summary

Many scientists refer to all nonvascular seedless plants as bryophytes.

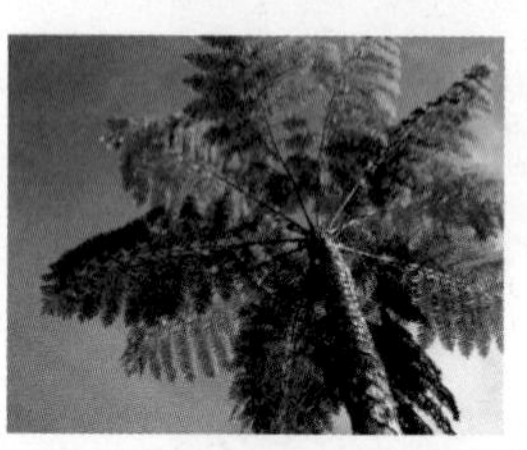

Vascular plants usually are larger than nonvascular plants.

Humans use both vascular and nonvascular plants for many purposes.

Use Vocabulary

1. **Write** a sentence using the term *rhizoid*.

Understand Key Concepts

2. Which does NOT belong with the others?
   - (A) club moss
   - (B) fern
   - (C) horsetail
   - (D) liverwort

3. **List** the different types of bryophytes. SC.6.L.15.1

Interpret Graphics

4. **Organize** Fill in the table below to summarize the different types of vascular and nonvascular seedless plants. SC.6.L.15.1

Types of Seedless Plants	
Vascular	Nonvascular

Critical Thinking

5. **Predict** how well moss plants would grow in the desert. Explain your reasoning.

6. **Compare and contrast** nonvascular and vascular seedless plants. SC.6.L.15.1

Organize information about bryophytes.

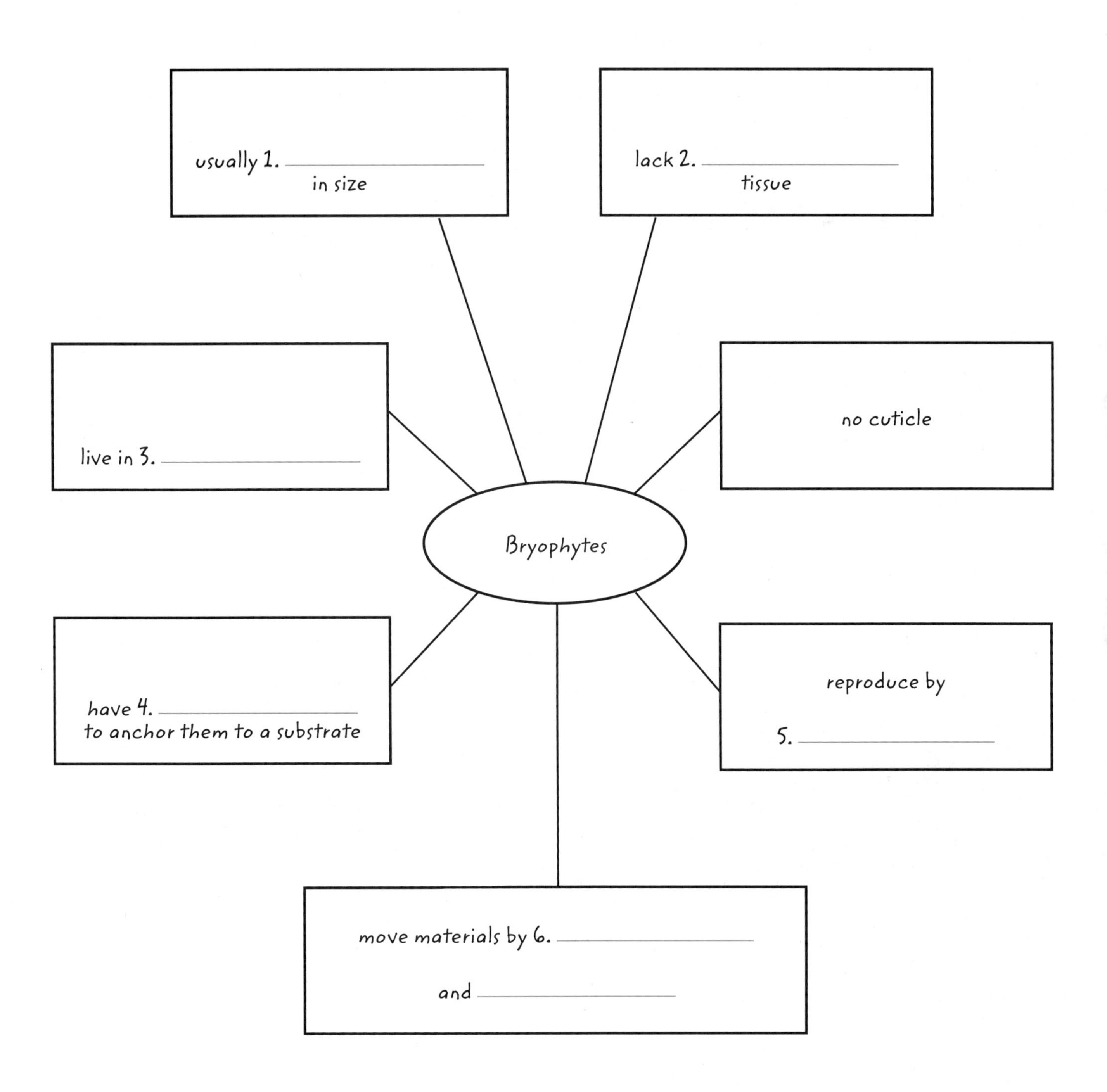

Lesson 3

SEED PLANTS

ESSENTIAL QUESTIONS

- What characteristics are common to seed plants?
- How do other organisms depend on seed plants?
- How are gymnosperms and angiosperms alike, and how are they different?
- What adaptations of flowering plants enable them to survive in diverse environments?

Vocabulary

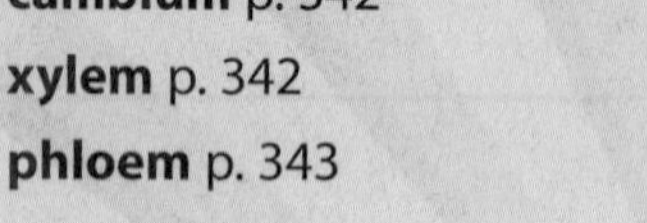

cambium p. 342
xylem p. 342
phloem p. 343
stoma p. 345

Florida NGSSS

LA.6.2.2.3 The student will organize information to show understanding (e.g., representing main ideas within text through charting, mapping, paraphrasing, summarizing, or comparing/contrasting);

MA.6.A.3.6 Construct and analyze tables, graphs, and equations to describe linear functions and other simple relations using both common language and algebraic notation.

SC.6.L.15.1 Analyze and describe how and why organisms are classified according to shared characteristics with emphasis on the Linnaean system combined with the concept of Domains.

SC.6.N.1.5 Recognize that science involves creativity, not just in designing experiments, but also in creating explanations that fit evidence.

SC.6.N.2.1 Distinguish science from other activities involving thought.

SC.7.N.1.3 Distinguish between an experiment (which must involve the identification and control of variables) and other forms of scientific investigation and explain that not all scientific knowledge is derived from experimentation.

SC.6.N.2.1

Inquiry Launch Lab

20 minutes

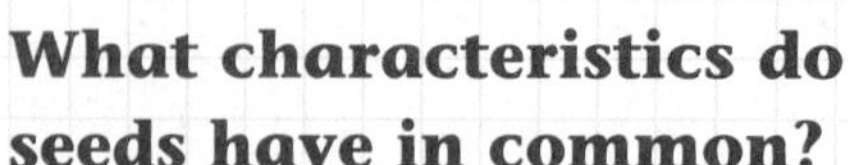

What characteristics do seeds have in common?

Seed plants have two characteristics in common: They have vascular tissue and seeds for reproduction. Do seeds also have common characteristics?

Procedure

1. Read and complete a lab safety form.
2. Examine the **several types of seeds** on your **tray.**
3. On a separate sheet of paper, make a grid of 2-cm by 2-cm squares to classify the seeds. Choose your own criteria to classify the seeds; for example, color, texture, size, or other special characteristics.
4. Label the columns and rows of your grid with the characteristics you chose.
5. Place the seeds on the section(s) of your grid where they belong based on the criteria you have chosen to classify them.

Think About This

1. Explain why you placed some of the seed samples in the same square.

2. Could some of the seeds have been placed in more than one square? Elaborate.

3. Do any of the seeds share common characteristics? Explain.

1. The berry the bird is eating has seeds in it. The bird depends on the tree for food—the berries. The tree helps the bird live. How do you think the bird is helping the tree live and be successful in its environment?

Characteristics of Seed Plants

Have you ever eaten corn, beans, peanuts, peas, or pine nuts, such as the ones shown in **Figure 12?** They are all examples of edible seeds. Recall that a seed contains a tiny plant embryo and nutrition for the embryo to begin growing. There are more than 300,000 species of seed plants on Earth.

Seed plants are organized into two groups—cone-bearing seed plants, or gymnosperms (JIHM nuh spurmz), and flowering seed plants, or angiosperms (AN gee uh spurmz). All seed plants have vascular tissue that transports water and nutrients throughout the plant. This means they also have roots, stems, and leaves. You will read more about the characteristics of seed plants in this lesson.

2. **NGSSS Check** **Recall** Highlight the characteristics that all seed plants have in common. SC.6.L.15.1

Corn kernels

Peas

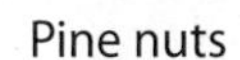

Beans

Figure 12 Of the major plant parts, seeds are the most important source of human food.

Science Use v. Common Use

tissue

Science Use a group of cells in an organism
Common Use a piece of absorbent paper

Word Origin

xylem

from Greek *xylon,* means "wood"

Vascular Tissue

All seed plants contain vascular **tissues** in their roots, stems, and leaves. This tissue transports water and nutrients throughout a plant. The two types of vascular tissue are xylem (ZI lum) and phloem (FLOH em). *The* **cambium** *is a layer of tissue that produces new vascular tissue and grows between xylem and phloem.* How do xylem and phloem differ? Keep reading to find out.

3. Identify What are the two types of vascular tissue?

____

In what structures of a vascular plant would such tissues be found?

____

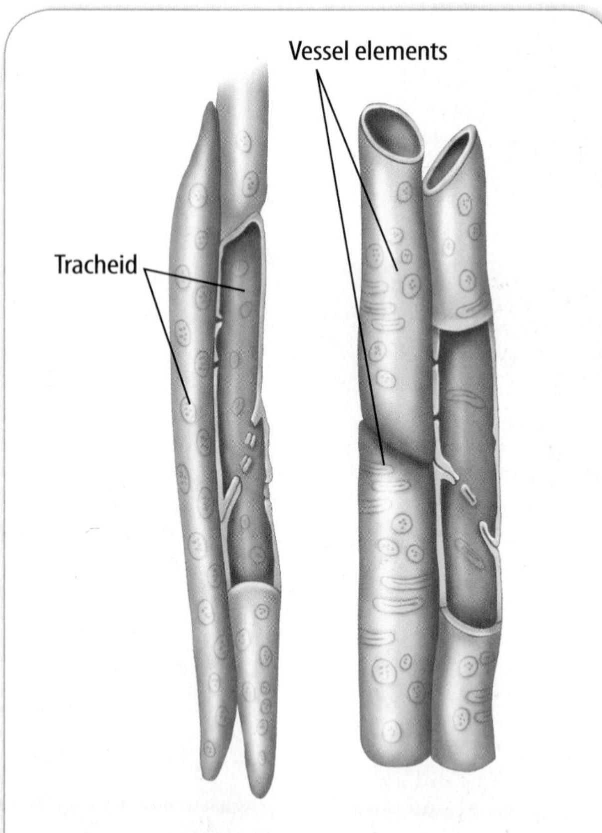

Figure 13 The cell walls between individual vessel elements have larger openings than the cell walls between tracheids.

4. Locate (Circle) the openings between tracheids and between vessel elements.

Xylem *One type of vascular tissue—***xylem***—carries water and dissolved nutrients from the roots to the stem and the leaves.* Due to the thickened cell walls of some xylem cells, this tissue also provides support for a plant.

Two kinds of xylem cells are tracheids (TRAY kee udz) and vessel elements. All vascular plants have xylem that is composed of tracheids. As shown in **Figure 13,** tracheid cells are long and narrow with tapered ends. The tracheid cells grow end-to-end and form a strawlike tube. Water can flow from one cell to another, passing through openings or pits in the end wall of each cell. Tracheid cells die at maturity, leaving a hollow tube. This enables water to flow freely through them.

In addition to tracheids, xylem in flowering plants includes a type of cell called a vessel element. The diameter of a vessel element is greater than that of a tracheid. The end walls of vessel elements have large openings where water can pass through, as shown in **Figure 13.** In some vessel elements, the end walls are completely open. Vessel elements are more efficient at transporting water than tracheids are.

Phloem *Another type of vascular tissue—***phloem***—carries dissolved sugars throughout a plant.* It is composed of two types of cells—sieve-tube elements and companion cells.

Sieve-tube elements are specialized phloem cells. These long, thin cells are stacked end-to-end and form long tubes. The end walls have holes in them, as shown in **Figure 14.** The cytoplasm of a sieve-tube element lacks many organelles, including a nucleus, mitochondria, and ribosomes.

Each sieve-tube element has a companion cell next to it that contains a nucleus. A companion cell helps control the functions of the sieve-tube element.

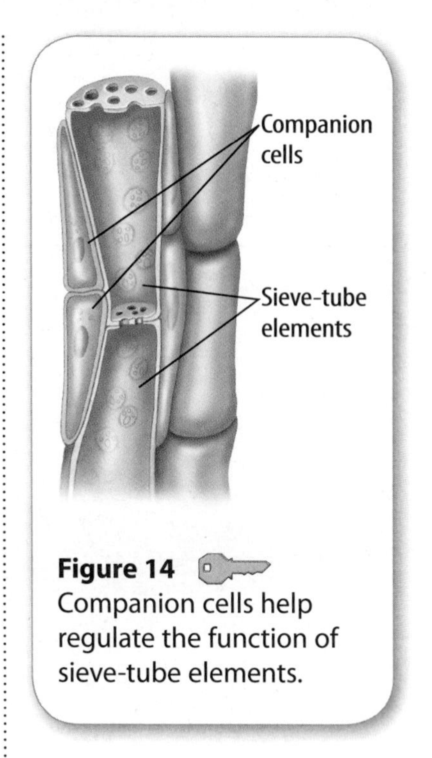

Figure 14 Companion cells help regulate the function of sieve-tube elements.

Active Reading **5. Describe** What is the function of each of the two types of vascular tissue in plants?

____

____

____

____

Roots

Even though the roots of most plants are never seen, they are vital to a plant's survival. Roots anchor a plant, either in soil or onto another plant or an object such as a rock. All roots help a plant stay upright. Some plants have roots that spread out in all directions several meters from a plant's stem. All root systems, such as the one shown in **Figure 15,** help a plant absorb water and other substances from the soil.

Plants such as radishes and carrots store food in their roots. This food can be used to grow new plant tissues after a dry period or a cold season. Sugar stored in the roots of sugar maple trees over the winter is converted to maple sap in the spring. Farmers drain some of the sap from these trees and boil it to make maple syrup.

Figure 15 Roots help anchor a plant and absorb water and minerals from the soil.

Figure 16 An herbaceous stem supports the black-eyed Susan flower. A woody stem supports this tree.

Stems

The part of a plant that connects its roots to its leaves is the stem. In plants such as the tree in **Figure 16,** the stem is obvious. Other plants, such as the potato and the iris, have underground stems that are often mistaken for roots.

Stems support branches and leaves. Their vascular tissues transport water, minerals, and food. Xylem carries water and minerals from the roots to the leaves. The sugar produced during photosynthesis flows through a stem's phloem to all parts of a plant. Another important function of stems is the production of new cells for growth, but only certain regions of a stem produce new cells.

Plant stems usually are classified as either herbaceous or woody. Woody stems, such as the one shown in **Figure 16,** are stiff and typically not green. Trees and shrubs have woody stems. Herbaceous stems usually are soft and green, as also shown in **Figure 16.**

Active Reading **6. Determine** (Circle) the paragraph above that describes the importance of a stem to a plant.

Inquiry SC.6.N.1.5

LAB STATION **Try It!**

MiniLab *How can you determine the stems, roots, and leaves of plants?* at connectED.mcgraw-hill.com

Apply It! After you complete the lab, answer these questions.

1. Were any of the plants you examined nonvascular plants? Explain.

2. **Classify** The coreopsis is Florida's state wildflower. How would you classify the roots, stems, and leaves of the coreopsis compared to the plants you examined?

Leaves

Leaves come in many shapes and sizes. Most leaves have an important function in common—they are the major site of photosynthesis for the plant. By capturing light energy and converting it to chemical energy, leaves provide the plant's food.

As shown in **Figure 17,** most leaves are made of layers of cells. The top and bottom layers of a leaf are made of epidermal (eh puh DUR mul) tissue. Epidermal cell walls are transparent, and light passes through them easily. These cells produce a waxy outer layer called the cuticle. The cuticle helps reduce the amount of water that evaporates from a leaf. *Most leaves have small openings in the epidermis called* **stomata** (STOH muh tuh; singular, stoma). When the stomata open, carbon dioxide, oxygen, and water vapor can pass through them. Two guard cells surround each stoma and control its size.

Below the upper epidermis are rows of tightly packed cells called palisade (pa luh SAYD) mesophyll (MEH zuh fil) cells. Photosynthesis mainly occurs in these cells. Under the palisade mesophyll cells is the spongy mesophyll layer. The arrangement of these cells enables gases to diffuse throughout a leaf. A leaf's xylem and phloem transport materials throughout the leaf.

Angiosperm and gymnosperm leaves each have some unique characteristics. An angiosperm leaf tends to be flat with a broad surface area. A gymnosperm leaf is usually needlelike or scale-like and often has a thick cuticle. Because gymnosperms often grow in drier areas, these characteristics help conserve water.

Figure 17 The structure of a leaf is well suited to its function of photosynthesis.

Active Reading **7. Locate** (Circle) the stomata in the figure below. What role do they play in photosynthesis?

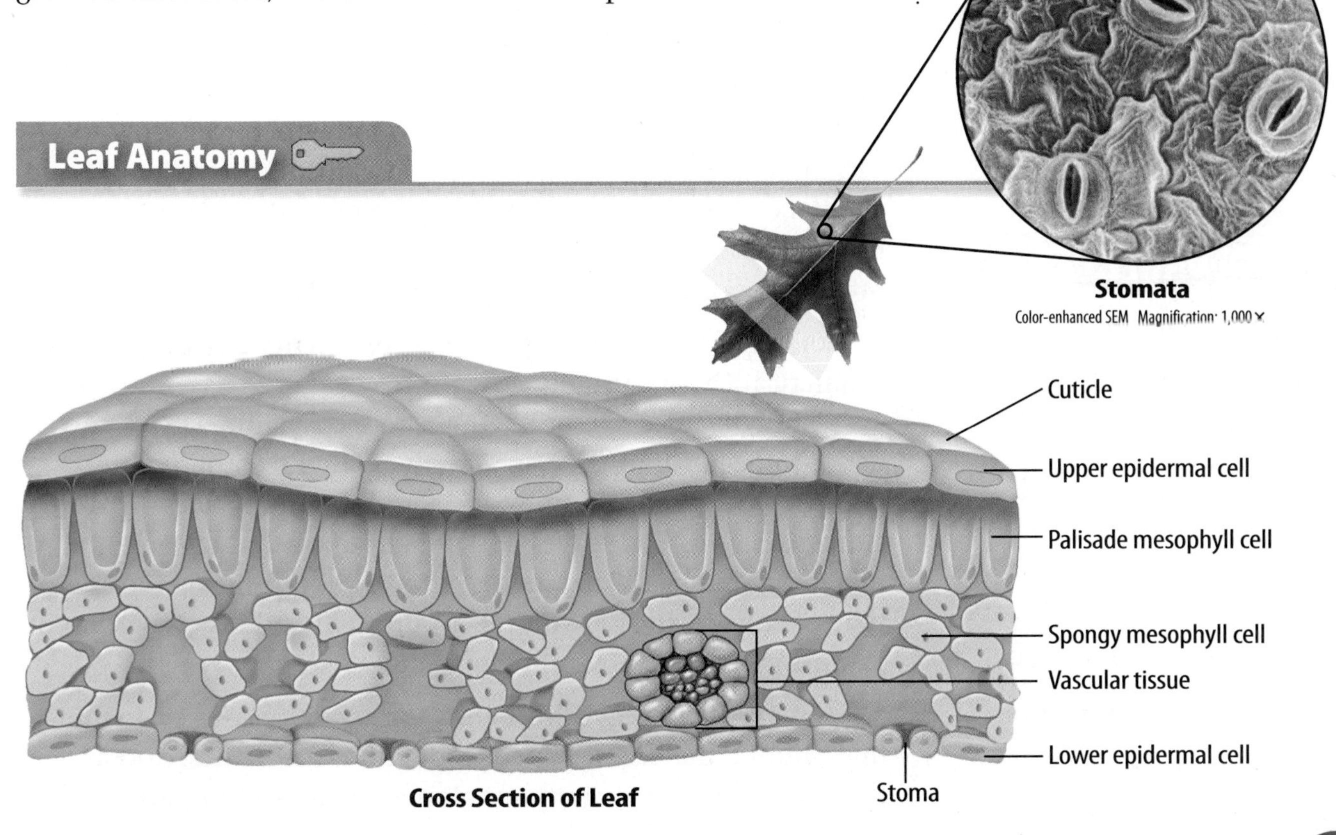

Types of Gymnosperms

Figure 18 Gymnosperms are a diverse group of plants, and all produce seeds without surrounding fruits.

Active Reading

FOLDABLES® LA.6.2.2.3

Make a three-tab Venn diagram book, and label it as shown. Use it to compare and contrast seed plants.

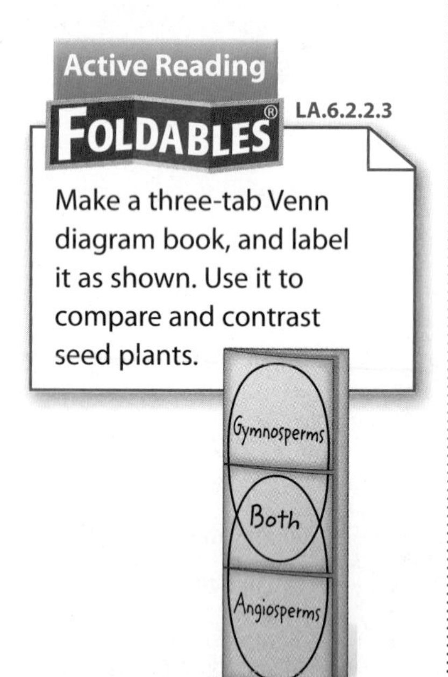

Gymnosperms

In gymnosperms, seeds are produced in a cone. This group includes the oldest plant (the bristlecone pine at 4,900 years old), the tallest plant (the coast redwood which can grow to 115 m), and perhaps Earth's largest organism (the sequoia). Conifers, such as spruces, pines, and redwoods, might be the most familiar gymnosperms to you. However, there are several other types of gymnosperms, as shown in **Figure 18.**

Conifers grow on all the world's continents except Antarctica. Cycads usually grow in tropical regions. Although cycads might resemble ferns, they are seed plants. DNA evidence indicates that they are closely related to other gymnosperms. One gymnosperm group has only one species—ginkgo. Ginkgos have broad leaves and are popular as an ornamental tree in urban areas. The gnetophytes (NEE tuh fites), another type of cone-bearing plant, are an unusual and diverse group of gymnosperms, as shown in **Figure 18.**

Humans rely on gymnosperms for a variety of uses, including building materials, paper production, medicines, and as ornamental plants in gardens, along streets, and in parks.

Angiosperms

There are more than 260,000 species of flowering plants, or angiosperms. Angiosperms began to flourish about 80 million years ago. They grow in a variety of habitats, from deserts to the tundra. Anytime you have stopped to smell a flower, you have enjoyed an angiosperm. Almost all of the food eaten by humans comes from angiosperms or from animals that eat angiosperms. Grains, vegetables, herbs, and spices are just a few examples of foods that come from angiosperms. Many other items, such as clothing, medicines, and building materials, also come from these plants.

Active Reading **8. Describe** Highlight the ways humans depend on flowering plants.

Flowers

Angiosperms produce seeds that are part of a fruit. This fruit grows from parts of a flower. All angiosperms produce flowers. Some flowers, such as tulips and roses, are beautiful and showy. You also might be familiar with other flowers, such as dandelions, because you have seen them growing in your neighborhood. However, some plants produce flowers that you might never have noticed. Grass flowers are tiny and not easily seen, as shown in **Figure 19.**

9. NGSSS Check **Compare and Contrast** How are angiosperms and gymnosperms alike, and how are they different? SC.6.L.15.1

____

____

____

Math Skills MA.6.A.3.6

Use Percentages

A percentage compares a part to a whole. For example, out of about 1,090 species of gymnosperms, 700 species are conifers. What percentage of gymnosperms are conifers?

Express the information as a fraction.

$$\frac{700 \text{ conifers}}{1{,}090 \text{ gymnosperms}}$$

Change the fraction to a decimal.

$$\frac{700}{1{,}090} = 0.64$$

Multiply by 100 and add a percent sign.

$$0.64 \times 100 = 64\%$$

Practice

10. Out of 1,090 species of gymnosperms, 300 0species are cycads. What percentage of gymnosperm species are cycads?

Types of Angiosperms

Figure 19 All angiosperms have flowers that contain reproductive organs. After pollination and fertilization, seeds are produced within a flower.

Table 1 Monocots v. Dicots

Monocots	Dicots
A. ______________________	
narrow with parallel veins	veins are branched
B. ______________________	
flower parts in multiples of three	flower parts in multiples of four or five
C. ______________________	
vascular tissue in bundles scattered throughout the stem	vascular tissue in bundles in rings
D. ______________________	
one cotyledon	two cotyledons

Table 1 Monocots and dicots differ in several ways.

11. Identify Fill in the table by identifying the structures that differ between monocots and dicots. Use the terms *Flowers, Leaves, Seeds,* and *Stems.*

Annuals, Biennials, and Perennials

Plants that grow, flower, and produce seeds in one growing season are called annuals. After one growing season, the plant dies. Examples include tomatoes, beans, pansies, and many common weeds.

Biennials complete their life cycles in two growing seasons. During the first year, the plant grows roots, stems, and leaves. The part of the plant that is above ground might become dormant during the winter months. In the second growing season, the plant produces new stems and leaves. It also flowers and produces seeds during this second growing season. After flowering and producing seeds, the plant dies. Carrots, beets, and foxglove are all biennials.

Perennial plants can live for more than two growing seasons. Trees and shrubs are perennials. The leaves and stems of some herbaceous perennial plants die in the winter. Stored food in the roots is used each spring for new growth.

12. Distinguish Underline the descriptions of the growing seasons of an annual, a biennial, and a perennial.

Monocots and Dicots

Flowering plants traditionally have been organized into two groups—monocots and dicots. These groups are based on the number of leaves in early development, or cotyledons (kah tuh LEE dunz), in a seed. Researchers have learned that dicots can be organized further into two groups based on the structure of their pollen. However, because these two groups of dicots share many characteristics, we will continue to refer to them just as dicots. Look carefully at **Table 1** to learn some of the differences between monocots and dicots.

13. Recall In the paragraph above, circle the name of the structure that is used to separate flowering plants into two groups.

Lesson Review 3

Visual Summary

Angiosperms are flowering plants.

Seed plants have many adaptations that enable them to survive in diverse environments.

Seed plants have many uses.

Use Vocabulary

1. The tissue that produces new xylem and phloem cells is the ______________________.

2. The vascular tissue that carries dissolved sugar throughout a plant is called ______________________.

Understand Key Concepts

3. Which cells carry on most of a plant's photosynthesis?
   - (A) guard cells
   - (B) xylem cells
   - (C) palisade mesophyll cells
   - (D) spongy mesophyll cells

4. **Contrast** gymnosperms with angiosperms. SC.6.L.15.1

____

____

Interpret Graphics

5. **Organize** Fill in the table below to describe the function of each plant structure. LA.6.2.2.3

Structure	Function
Roots	
Stem	
Leaves	

Critical Thinking

6. **Explain** why the placement of the companion cells in phloem is so critical to its function. SC.6.N.1.5

____

____

Math Skills MA.6.A.3.6

7. There are 300,000 species of seed plants. There are 9,000 species of grasses. What percentage of seed plants are grass species?

Chapter 9 Study Guide

Think About It! **Plant species are organized in a functional and structural hierarchy. Different plant species have different adaptations that enable them to survive in most of the environments on Earth.**

Key Concepts Summary

LESSON 1 What is a plant?

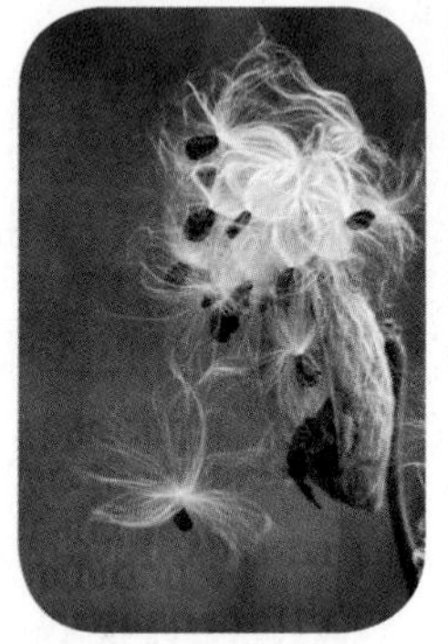

- Plants are multicellular **producers** composed of eukaryotic cells with cell walls composed of **cellulose.** Many plant cells have chloroplasts.
- Plants have developed adaptations, such as a cell wall for support, a **cuticle** to prevent water loss and to provide protection from insects, **vascular tissue** to transport materials, and numerous reproductive strategies, to survive in Earth's changing environments.
- Members of the plant kingdom are classified into groups called divisions, which are equivalent to phyla in other kingdoms. Plants have two-word scientific names.

Vocabulary

producer p. 326
cuticle p. 327
cellulose p. 327
vascular tissue p. 328

LESSON 2 Seedless Plants

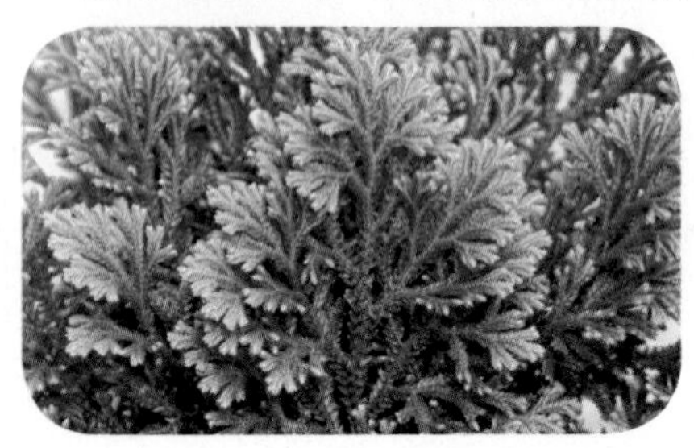

- Vascular and nonvascular seedless plants are multicellular producers composed of eukaryotic cells. Nonvascular seedless plants usually are smaller than vascular seedless plants, lack vascular tissue, and have **rhizoids** instead of roots to anchor them.

rhizoid p. 335
frond p. 337

LESSON 3 Seed Plants

- All seed plants make seeds and reproduce. Seed plants have leaves, stems, roots, and vascular tissue—**xylem** and **phloem.**
- Seed plants are important to other organisms for various reasons including for food, for the addition of oxygen to the environment, and for commercial uses.
- Gymnosperms and angiosperms are both seed plants. Angiosperms produce flowers; gymnosperms do not. The seeds of angiosperms are surrounded by fruit. The seeds of gymnosperms are not surrounded by fruit.
- Flowering plants have adaptations that enable them to survive in diverse environments. Such adaptations include leaves, stems, vascular tissue, roots, flowers, and seeds protected by a fruit.

cambium p. 342
xylem p. 342
phloem p. 343
stoma p. 345

Active Reading FOLDABLES® **Chapter Project**

Assemble your lesson Foldables as shown to make a Chapter Project. Use the project to review what you have learned in this chapter.

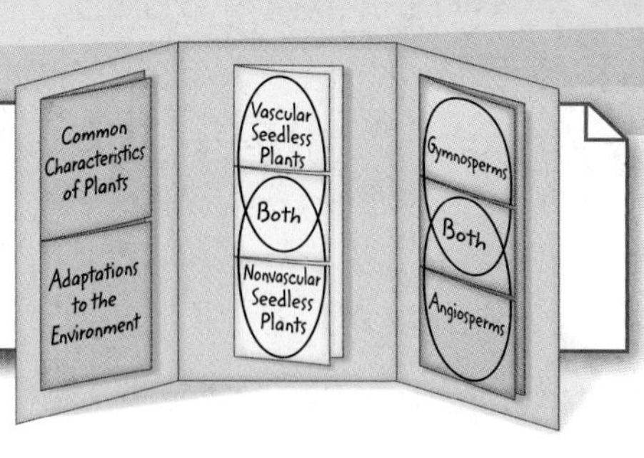

Use Vocabulary

1. Distinguish between xylem and phloem.

2. The openings in leaves that allow gases to pass into and out of the leaf are ____________.

3. Define the term *rhizoid* in your own words.

4. The tissue that produces new xylem and phloem cells is the ____________.

5. Write a sentence using the term *cuticle*.

6. Use the term *vascular tissue* in a sentence.

Link Vocabulary and Key Concepts

Use vocabulary terms from the previous page to complete the concept map.

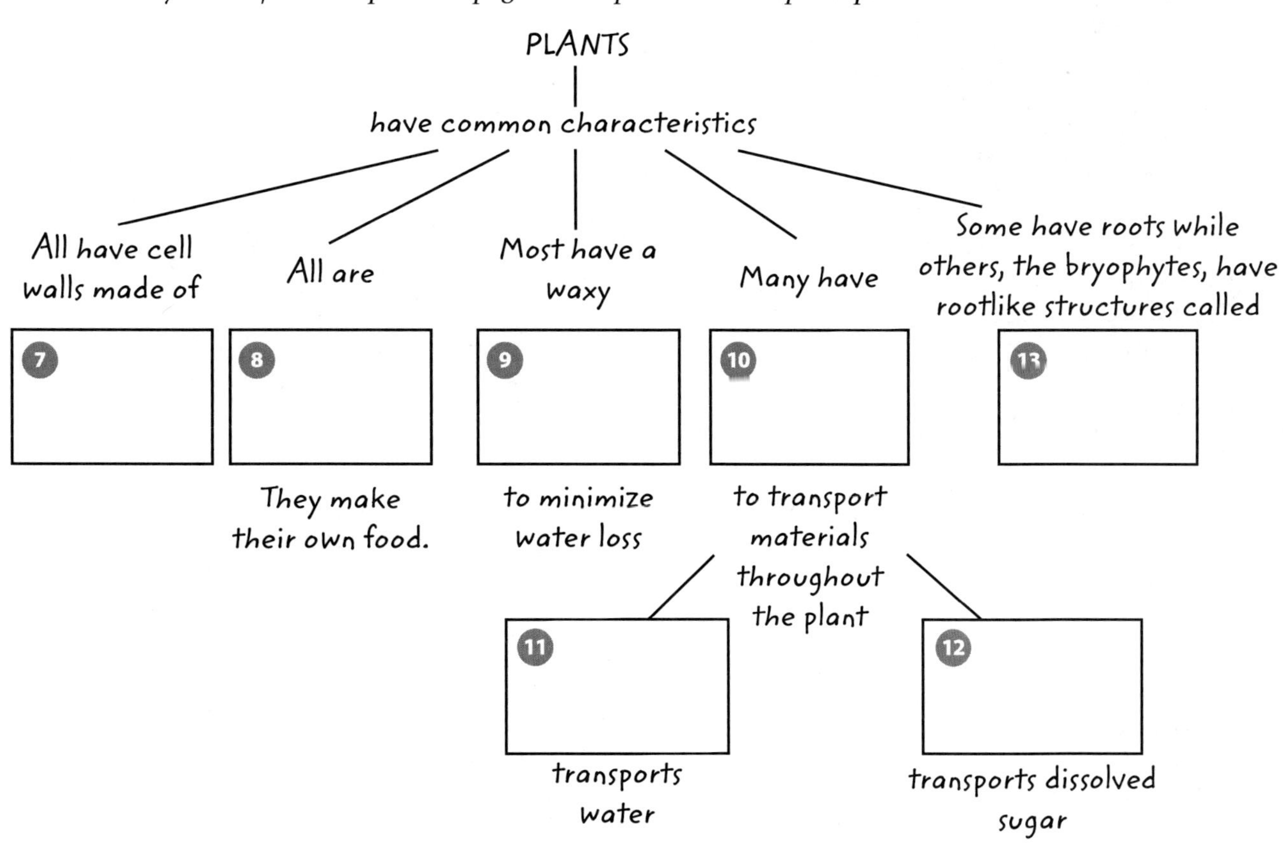

Chapter 9 Review

Fill in the correct answer choice.

Understand Key Concepts

1. Guard cells control the size of the opening of the **SC.7.N.1.1**
   Ⓐ cambium.
   Ⓑ cotyledon.
   Ⓒ stomata.
   Ⓓ xylem.

2. The major function of leaves is to **SC.7.N.1.1**
   Ⓐ anchor the plant.
   Ⓑ perform photosynthesis.
   Ⓒ shade the tree.
   Ⓓ support the stem.

3. Which is an angiosperm? **SC.6.L.15.1**
   Ⓐ fern
   Ⓑ moss
   Ⓒ pine
   Ⓓ tulip

4. Which does the plant below NOT have? **SC.6.L.15.1**

   Ⓐ chloroplasts
   Ⓑ rhizoids
   Ⓒ cell walls
   Ⓓ vascular tissue

5. Which is the plant part responsible for anchoring the plant in soil? **SC.7.N.1.1**
   Ⓐ flower
   Ⓑ leaf
   Ⓒ root
   Ⓓ stem

6. Which are vascular plants? **SC.6.L.15.1**
   Ⓐ ferns
   Ⓑ hornworts
   Ⓒ liverworts
   Ⓓ mosses

Critical Thinking

7. **Choose** one of the adaptations that plants have for living on land, and explain its significance. **SC.7.N.1.1**

8. **Suggest** additional uses for peat moss. **SC.6.N.1.5**

9. **Evaluate** the lack of cuticle in moss plants. **SC.6.L.15.1**

10. **Explain** the advantage of fruit production in angiosperms. **SC.6.L.15.1**

11. **Predict** the impact of a disease that killed all gymnosperms. Explain your reasoning. **SC.6.L.15.1**

12 **Explain** how the structure of a leaf as shown in the figure below is appropriate for its role in photosynthesis. SC.7.N.1.1

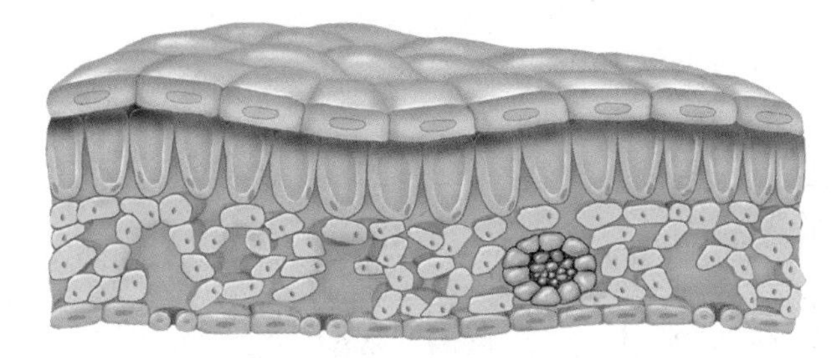

Writing in Science

13 **Write** On a separate sheet of paper, write a short story about a hiker's thoughts as he or she comes across different plants such as mosses, liverworts, ferns, horsetails, gymnosperms, and angiosperms. LA.6.2.2.3

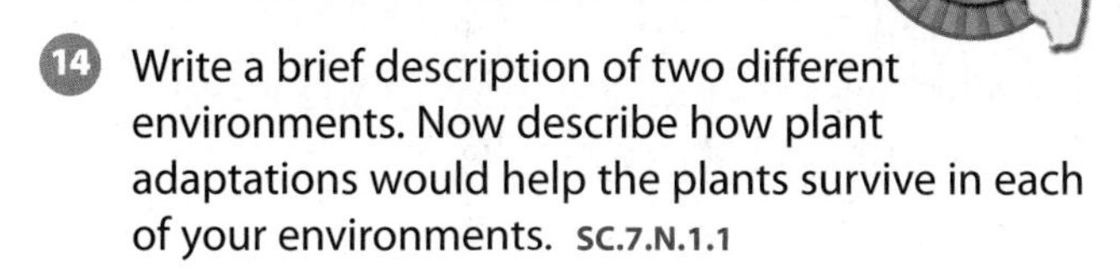

14 Write a brief description of two different environments. Now describe how plant adaptations would help the plants survive in each of your environments. SC.7.N.1.1

15 What structures enable the plants in the photo at the beginning of the chapter to live in this environment? SC.7.N.1.1

Math Skills MA.6.A.3.6

16 Some scientists estimate the total number of species of organisms on Earth to be about 14,000,000. If there are 300,000 species of seed plants, what percentage of Earth's species are seed plants?

17 Of the 1,090 species of gymnosperms, there is only one species of ginkgo. What percentage of gymnosperm species are ginkgoes?

18 Of the 1,090 species of gymnosperms, 90 species are gnetophytes. What percentage of gymnosperm species are gnetophytes?

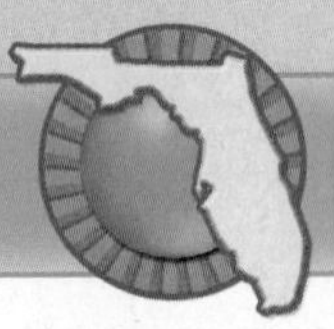

Florida NGSSS Benchmark Practice

Fill in the correct answer choice.

Multiple Choice

1 Which is NOT a characteristic of all plants? **SC.6.L.14.4**

Ⓐ They are multicellular.
Ⓑ They have vascular tissue.
Ⓒ They make their own food.
Ⓓ They undergo photosynthesis.

2 Which are seedless nonvascular plants? **SC.6.L.15.1**

Ⓕ conifers
Ⓖ grasses
Ⓗ mosses
Ⓘ tulips

Use the diagram below to answer question 3.

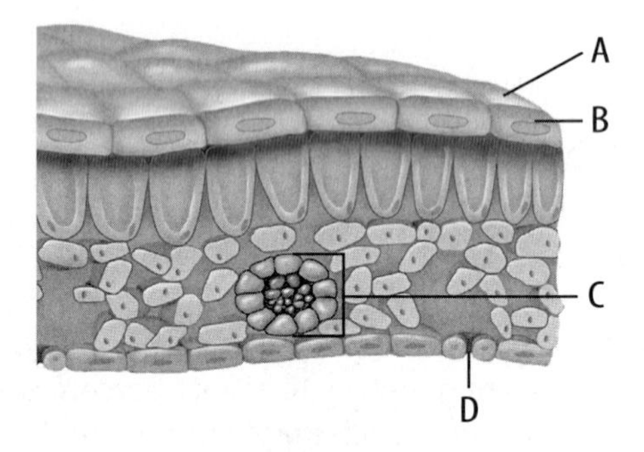

3 Which feature in the figure is a plant adaptation to life on land and reduces water loss through evaporation? **SC.7.N.1.1**

Ⓐ A
Ⓑ B
Ⓒ C
Ⓓ D

4 Which are NOT common to all seed plants? **SC.6.L.15.1**

Ⓕ cell walls
Ⓖ flowers
Ⓗ roots
Ⓘ vascular tissues

5 Which feature do scientists use to classify angiosperms as monocots or dicots? **SC.6.L.15.1**

Ⓐ embryonic leaves
Ⓑ flower shapes
Ⓒ seed numbers
Ⓓ xylem cells

Use the image below to answer question 6.

6 Which plants form seeds in structures such as the one shown in the figure? **SC.6.L.15.1**

Ⓕ biennials
Ⓖ conifers
Ⓗ maples
Ⓘ roses

7 Which plants provide most of the food humans eat? **SC.6.L.15.1**

Ⓐ angiosperms
Ⓑ gymnosperms
Ⓒ seedless nonvascular plants
Ⓓ seedless vascular plants

8 Which structures are in plant cells but are not in animal cells? **SC.6.L.14.4**

Ⓕ cell membranes and vacuoles
Ⓖ cell walls and chloroplasts
Ⓗ chloroplasts and mitochondria
Ⓘ vacuoles and mitochondria

Use the diagram below to answer question 9.

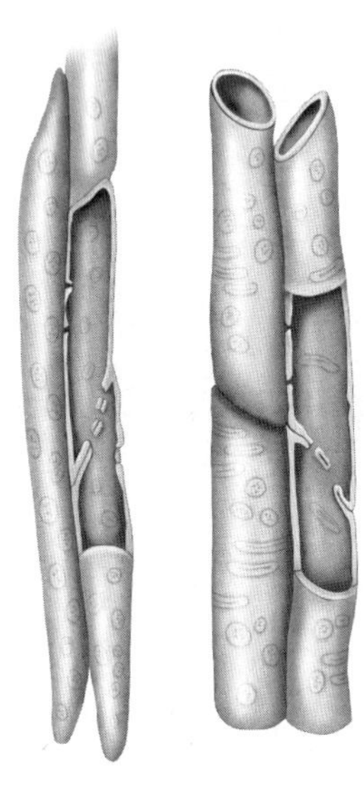

9 What type of plant tissue contains the cells shown in the figure? SC.7.N.1.1

Ⓐ cambium

Ⓑ mesophyll

Ⓒ phloem

Ⓓ xylem

10 An unknown plant is a few centimeters tall when fully grown. It lives only in moist areas and reproduces with spores. How would you classify this plant? SC.6.L.15.1

Ⓕ nonvascular seed plant

Ⓖ nonvascular seedless plant

Ⓗ vascular seedless plant

Ⓘ vascular seed plant

11 Only members of the plant kingdom are organized into SC.6.L.15.1

Ⓐ categories.

Ⓑ divisions.

Ⓒ groups.

Ⓓ phyla.

12 What is the function of the stomata? SC.7.N.1.1

Ⓕ to perform photosynthesis

Ⓖ to produce sugar

Ⓗ to allow water into the leaf

Ⓘ to enable gases to enter and leave

13 Which are produced by angiosperms but not by gymnosperms? SC.6.L.15.1

Ⓐ cones

Ⓑ flowers

Ⓒ leaves

Ⓓ seeds

14 What is the function of the structure shown below? SC.7.N.1.1

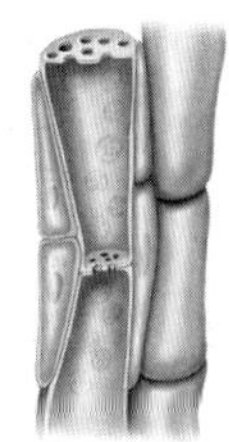

Ⓕ control gas exchange

Ⓖ perform photosynthesis

Ⓗ transport sugar

Ⓘ transport water

NEED EXTRA HELP?

If You Missed Question...	1	2	3	4	5	6	7	8	9	10	11	12	13	14
Go to Lesson...	1	2	1	3	1	3	3	1	3	1&2	1	3	3	3

Name ______________________ Date ______________

Multiple Choice *Bubble the correct answer.*

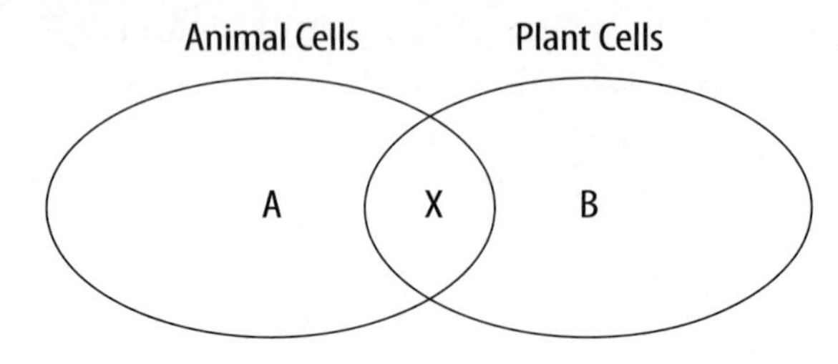

1. The Venn diagram above could be used to compare animal cells to plant cells. Which would be placed at X? **LA.6.2.2.3**

Ⓐ chloroplasts
Ⓑ nucleus
Ⓒ cell wall
Ⓓ central vacuole

2. When mountain pine beetles infest the trunk of a pine tree, they damage the plant's vascular tissues. Eventually the tree dies because it cannot **SC.6.L.15.1**

Ⓕ create new chloroplasts.
Ⓖ make its own food.
Ⓗ produce lignin.
Ⓘ transport water and nutrients.

3. Which is NOT true for all seed plants? **SC.6.L.15.1**

Ⓐ They are multicellular.
Ⓑ They are vascular.
Ⓒ They produce flowers.
Ⓓ They produce their own food.

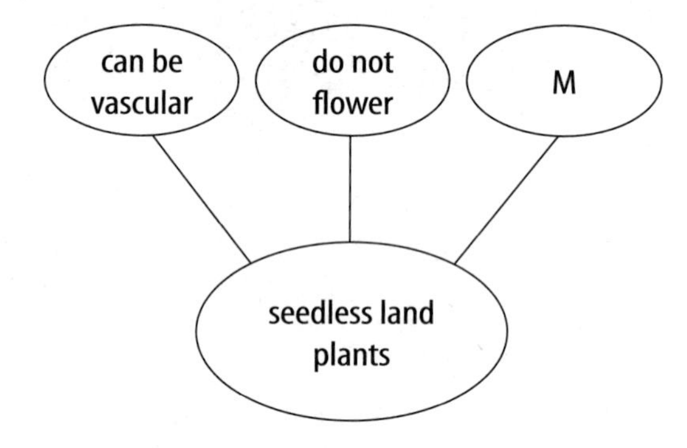

4. The graphic organizer above shows some characteristics of seedless plants. Which would be placed at M? **LA.6.2.2.3**

Ⓕ reproduce by burrs
Ⓖ reproduce by cones
Ⓗ reproduce by seeds
Ⓘ reproduce by spores

Name ______________________ Date ____________

mini BAT

Multiple Choice *Bubble the correct answer.*

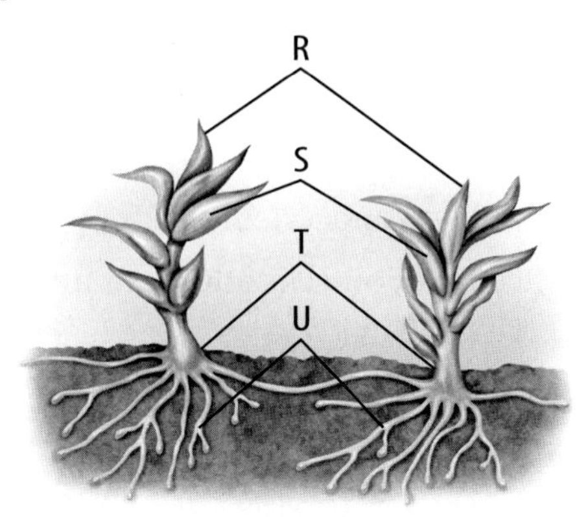

1. In the image above, which letter points to the rhizoids? **LA.6.2.2.3**

Ⓐ R

Ⓑ S

Ⓒ T

Ⓓ U

2. In which kind of environment do bryophytes do best? **SC.6.L.15.1**

Ⓕ damp

Ⓖ dry

Ⓗ sunny

Ⓘ underwater

3. Mosses' leaflike structures are not considered true leaves because they lack **SC.6.L.15.1**

Ⓐ cellulose.

Ⓑ chloroplasts.

Ⓒ green algae.

Ⓓ vascular tissue.

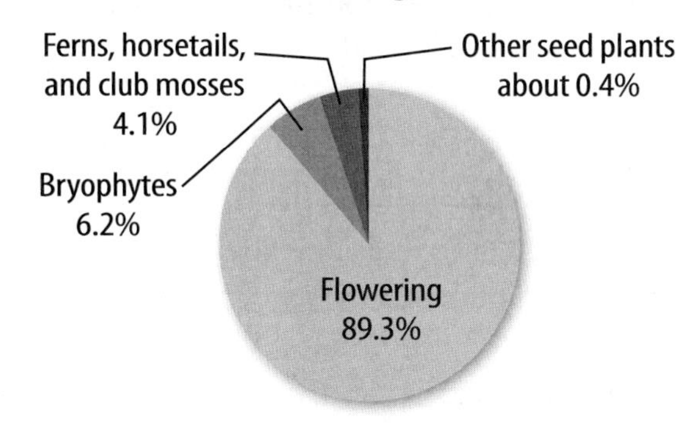

4. Which statement can be inferred about plants from the pie graph above? **SC.6.L.15.1**

Ⓕ 10.3% are nonvascular.

Ⓖ 93.8% are vascular.

Ⓗ 10.3% do not have true roots, stems, and leaves.

Ⓘ 95.5% have true roots, stems, and leaves.

Name ______________________ Date ______________

Multiple Choice *Bubble the correct answer.*

1. How would a disease that damages the palisade mesophyll cells of a plant affect the plant? **SC.6.L.15.1**

(A) The plant's roots would die.

(B) The plant would not flower.

(C) The plant would be unable to absorb water.

(D) The plant would have difficulty producing food.

Comparison of Gymnosperms and Angiosperms

Seed Plant	Flowers	Leaves	Cones	Spores	Vascular Tissue
Gymno-sperm	R	Yes	S	No	Yes
Angio-sperm	Yes	Yes	T	U	Yes

2. In the table above, four cells have been replaced by R, S, T, and U. Identify which cell should be *Yes*. **SC.6.L.15.1**

(F) R

(G) S

(H) T

(I) U

3. Which are the two types of vascular tissue found in seed plants? **SC.6.L.15.1**

(A) monocots and dicots

(B) phloem and xylem

(C) companion cells and sieve-tube elements

(D) tracheids and vessel elements

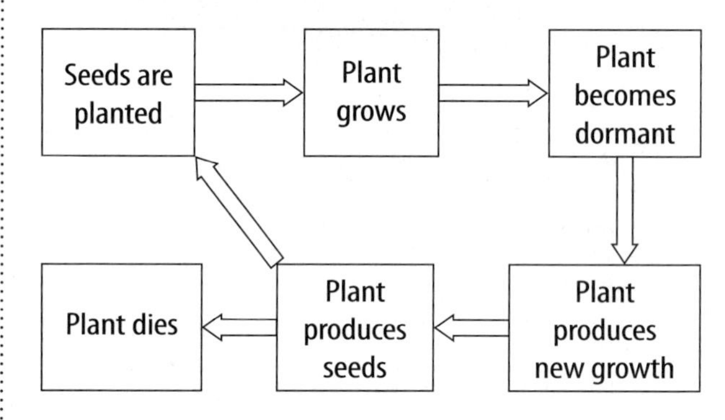

4. According to the information in the flow chart above, which kind of flowering plant is being described? **SC.6.L.15.1**

(F) annual

(G) biennial

(H) centennial

(I) perennial

Name ______________________________ Date ______________

Notes

Name ________________ Date ________________

Notes

What does a flower do?

Mrs. Gonzales has a flower garden. Her neighbors always remark how beautiful her flowers are. One day Mrs. Gonzales asked her neighbors, "Why do plants produce flowers? What is their main purpose?" They each had a different idea. This is what they said:

Mrs. Flores: I think plants produce flowers in order to attract bees.

Mr. Myer: I think plants produce flowers so they can make their own food.

Ms. Bricker: I think plants produce flowers so they will look pretty.

Mr. Pappas: I think plants produce flowers for reproduction.

Mr. Frist: I think plants produce flowers so they can respond to their environment.

Ms. Chai: I think plants produce flowers so people can use them for special occasions.

Mrs. Meehan: I think plants produce flowers so that other plants can detect them by smell.

Which person do you agree with the most? Explain your thinking about the function of a plant's flower.

Plant Processes and REPRODUCTION

FLORIDA BIG IDEAS

1 **The Practice of Science**

2 **The Characteristics of Scientific Knowledge**

3 **The Role of Theories, Laws, Hypotheses, and Models**

18 **Matter and Energy Transformations**

The Big Idea

Think About It!

What processes enable plants to survive and reproduce?

The common morning glory is native to the southeastern United States, including Florida. The tendril of this morning glory vine grows around the stem of another plant.

1. How do you think growing around another plant might help the morning glory survive?

____

____

____

2. Can you think of any other processes that enable plants to survive and reproduce?

____

____

____

Get Ready to Read

What do you think about plant processes?

Before you read, decide if you agree or disagree with each of these statements. As you read this chapter, see if you change your mind about any of the statements.

	AGREE	DISAGREE
1 Plants do not carry on cellular respiration.	☐	☐
2 Plants are the only organisms that carry on photosynthesis.	☐	☐
3 Plants do not produce hormones.	☐	☐
4 Plants can respond to their environments.	☐	☐
5 Seeds contain tiny plant embryos.	☐	☐
6 Flowers are needed for plant reproduction.	☐	☐

There's More Online!
Video • Audio • Review • ⓘLab Station • WebQuest • Assessment • Concepts in Motion • Multilingual eGlossary

Lesson 1

Energy Processing in PLANTS

ESSENTIAL QUESTIONS

- How do materials move through plants?
- How do plants perform photosynthesis?
- What is cellular respiration?
- What is the relationship between photosynthesis and cellular respiration?

Vocabulary

photosynthesis p. 366
cellular respiration p. 368

Florida NGSSS

SC.7.N.1.3 Distinguish between an experiment (which must involve the identification and control of variables) and other forms of scientific investigation and explain that not all scientific knowledge is derived from experimentation.

SC.8.N.1.1 Define a problem from the eighth grade curriculum using appropriate reference materials to support scientific understanding, plan and carry out scientific investigations of various types, such as systematic observations or experiments, identify variables, collect and organize data, interpret data in charts, tables, and graphics, analyze information, make predictions, and defend conclusions.

SC.8.N.1.6 Understand that scientific investigations involve the collection of relevant empirical evidence, the use of logical reasoning, and the application of imagination in devising hypotheses, predictions, explanations and models to make sense of the collected evidence.

SC.8.L.18.1 Describe and investigate the process of photosynthesis, such as the roles of light, carbon dioxide, water and chlorophyll; production of food; release of oxygen.

SC.8.L.18.2 Describe and investigate how cellular respiration breaks down food to provide energy and releases carbon dioxide.

Also covers: LA.6.2.2.3, LA.6.4.2.2

SC.7.N.1.3

Inquiry Launch Lab

20 minutes

How can you show the movement of materials inside a plant?

Most parts of plants need water. They also need a system to move water throughout the plant so cells can use it for plant processes.

Procedure

1. Read and complete a lab safety form.
2. Gently pull two stalks from the base of a bunch of **celery.** Leave one stalk complete. Use a **paring knife** to carefully cut directly across the bottom of the second stalk.
3. Put 100 mL of water into each of two **beakers.** Place 3–4 drops of **blue food coloring** into the water. Place one celery stalk in each beaker.
4. After 20 min, observe the celery near the bottom of each stalk. Observe again after 24 h. Record your observations in the space below.

Data and Observations

Think About This

1. What happened in each celery stalk?
2. Key Concept What did the colored water do? Why do you think this occurred?

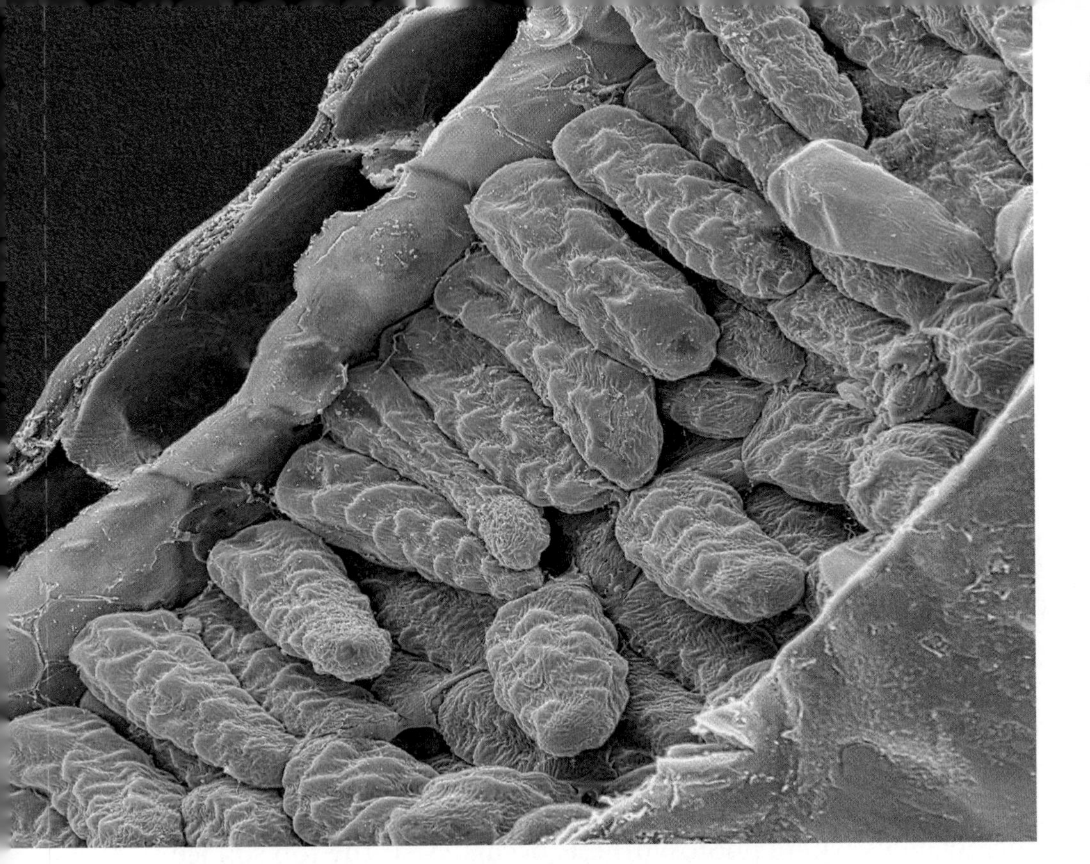

Inquiry All Leaf Cells?

1. You are looking at a magnified cross section of a leaf. As you can see, the cells in the middle of the leaf are different from the cells on the edges. What do you think this might have to do with the cellular processes a leaf carries out that enable a plant to survive?

Materials for Plant Processes

Food, water, and oxygen are three things you need to survive. Some of your organ systems process these materials, and others transport them throughout your body. Like you, plants need food, water, and oxygen to survive. Unlike you, plants do not take in food. Most of them make their own.

Moving Materials Inside Plants

You might recall reading about xylem (ZI lum) and phloem (FLOH em)—the vascular tissue in most plants. These tissues transport materials throughout a plant.

After water enters a plant's roots, it moves into xylem. Water then flows inside xylem to all parts of a plant. Without enough water, plant cells wilt, as shown in **Figure 1.**

Most plants make their own food a liquid sugar. The liquid sugar moves out of food-making cells, enters phloem, and flows to all plant cells. Cells break down the sugar and release energy. Some plant cells can store food.

Plants require oxygen and carbon dioxide to make food. Like you, plants produce water vapor as a waste product. Carbon dioxide, oxygen, and water vapor pass into and out of a plant through tiny openings in leaves.

Active Reading **2. Describe** How do materials move through plants?

Active Reading **3. Summarize** Make an outline of the information in the lesson. Use the main headings in the lesson as the main headings in your outline. Use your outline to review the lesson.

Figure 1 This plant wilted due to lack of water in the soil.

Active Reading

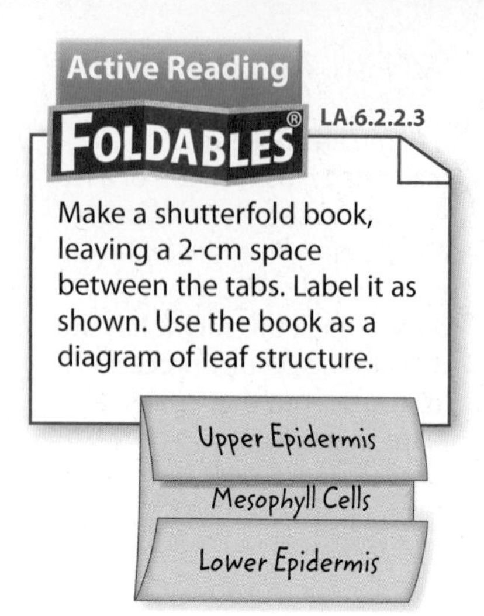

Make a shutterfold book, leaving a 2-cm space between the tabs. Label it as shown. Use the book as a diagram of leaf structure.

Photosynthesis

Plants need food, but they cannot eat as people do. They make their own food, and leaves are the major food-producing organs of plants. This means that leaves are the sites of photosynthesis (foh toh SIHN thuh sus). **Photosynthesis** *is a series of chemical reactions that convert light energy, water, and carbon dioxide into the food-energy molecule glucose and give off oxygen.* The structure of a leaf is well-suited to its role in photosynthesis.

Leaves and Photosynthesis

As shown in **Figure 2,** leaves have many types of cells. The cells that make up the top and bottom layers of a leaf are flat, irregularly shaped cells called epidermal (eh puh DUR mul) cells. On the bottom epidermal layer of most leaves are small openings called stomata (STOH muh tuh). Carbon dioxide, water vapor, and oxygen pass through stomata. Epidermal cells can produce a waxy covering called the cuticle.

Most photosynthesis occurs in two types of mesophyll (ME zuh fil) cells inside a leaf. These cells contain chloroplasts, the organelle where photosynthesis occurs. Near the top surface of the leaf are palisade mesophyll cells. They are packed together. This arrangement exposes the most cells to light. Spongy mesophyll cells have open spaces between them. Gases needed for photosynthesis flow through the spaces between the cells.

WORD ORIGIN

photosynthesis
from Greek *photo-,* means "light," and *synthesis,* means "composition"

Figure 2 Photosynthesis occurs inside the chloroplasts of mesophyll cells in most leaves.

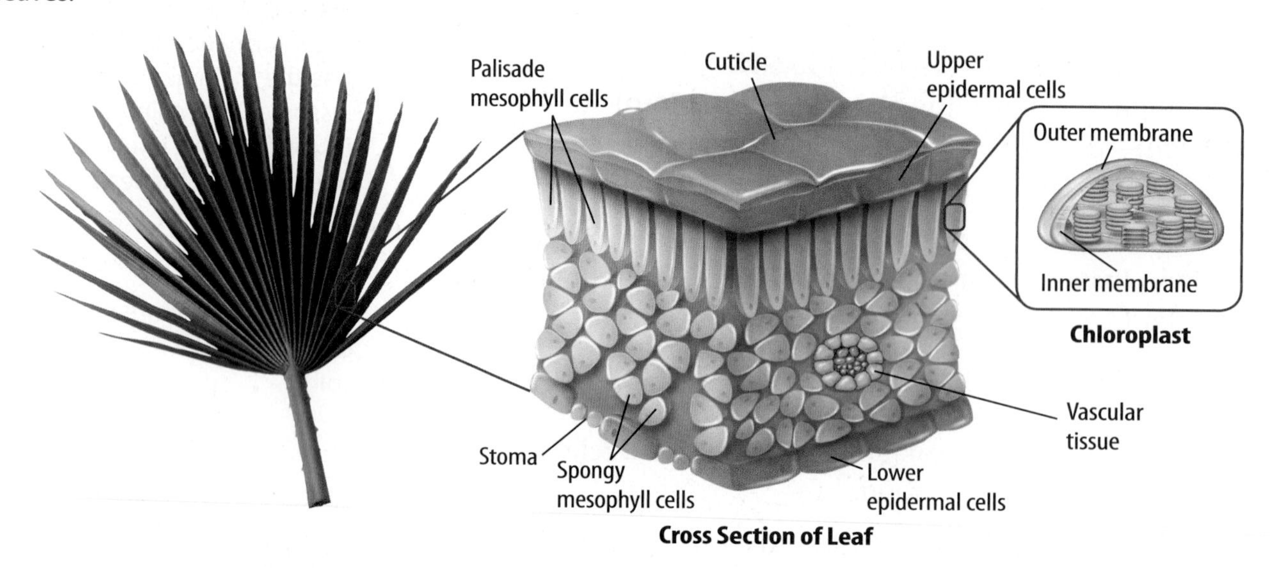

Active Reading **4. Identify** Which layer of cells contains vascular tissue? (Circle) this structure in **Figure 2.**

Capturing Light Energy

As you read about the steps of photosynthesis, refer to **Figure 3** to help you understand the process. In the first step of photosynthesis, plants capture the energy in light. This occurs in chloroplasts. Chloroplasts contain plant pigments. Pigments are chemicals that can absorb and reflect light. Chlorophyll, the most common plant pigment, is necessary for photosynthesis. Most plants appear green because chlorophyll reflects green light. Chlorophyll absorbs other colors of light. This light energy is used during photosynthesis.

Once chlorophyll traps and stores light energy, this energy can be transferred to other molecules. During photosynthesis, water molecules are split apart. This releases oxygen into the atmosphere, as shown in **Figure 3.**

5. Explain How do plants capture light energy?

Making Sugars

Sugars are made in the second step of photosynthesis. This step can occur without light. In chloroplasts, carbon dioxide from the air is converted into sugars by using the energy stored and trapped by chlorophyll. Carbon dioxide combines with hydrogen atoms from the splitting of water molecules and forms sugar molecules. Plants can use this sugar as an energy source, or they can store it. Potatoes and carrots are examples of plant structures where excess sugar is stored.

6. NGSSS Check Identify What are the two steps of photosynthesis? **SC.8.L.18.1**

Why is photosynthesis important?

Try to imagine a world without plants. How would humans and other animals get the oxygen they need? Plants help maintain the atmosphere you breathe. Photosynthesis produces most of the oxygen in the atmosphere.

Photosynthesis

Figure 3 Photosynthesis is a series of complex chemical processes. The first step is capturing light energy. In the second step, that energy is used for making glucose, a type of sugar.

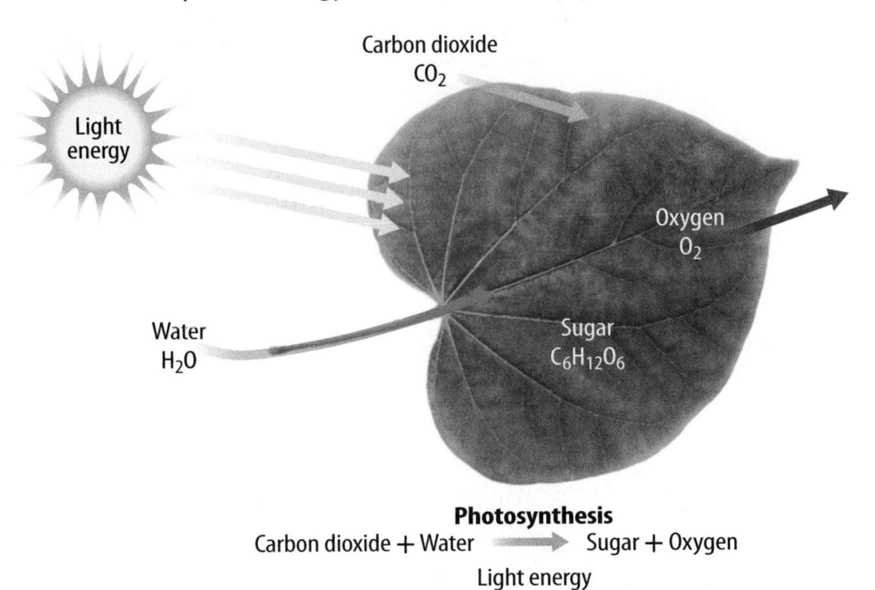

7. Visual Check Identify In **Figure 3,** what do the arrows illustrate?

Cellular Respiration

ACADEMIC VOCABULARY
energy
(noun) usable power

All organisms require **energy** to survive. Energy is in the chemical bonds in food molecules. A process called cellular respiration releases energy. **Cellular respiration** *is a series of chemical reactions that convert the energy in food molecules into a usable form of energy called ATP.*

Releasing Energy from Sugars

Glucose molecules break down during cellular respiration. Much of the energy released during this process is used to make ATP, an energy storage molecule. This process requires oxygen, produces water and carbon dioxide as waste products, and occurs in the cytoplasm and mitochondria of cells.

Why is cellular respiration important?

If your body did not break down the food you eat through cellular respiration, you would not have energy to do anything. Plants produce sugar, but without cellular respiration, plants could not grow, reproduce, or repair tissues.

8. **NGSSS Check** **Summarize** What is cellular respiration? SC.8.L.18.2

Inquiry LAB STATION **Try It!**
SC.7.N.1.1
SC.8.L.18.1
SC.8.L.18.2

MiniLab ***Can you observe plant processes?*** at connectED.mcgraw-hill.com

Apply It! After you complete the lab, answer these questions.

1. What gaseous element is required for cellular respiration?

2. What waste products does cellular respiration produce?

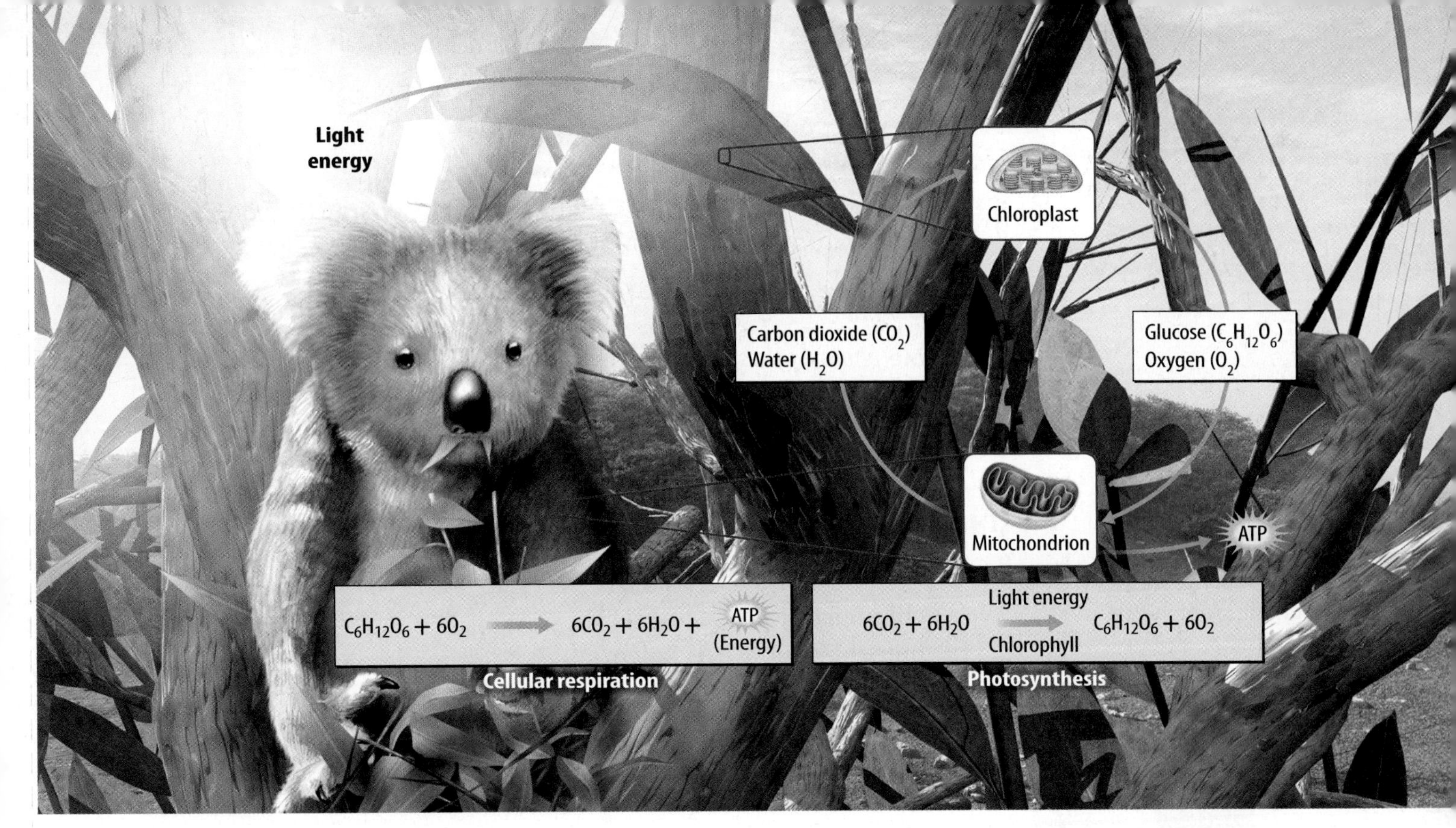

Figure 4 The relationship between cellular respiration and photosynthesis is important for life.

Comparing Photosynthesis and Cellular Respiration

Photosynthesis requires light energy and the reactants—substances that react with one another during the process—carbon dioxide and water. Oxygen and the energy-rich molecule glucose are the products, or end substances, of photosynthesis. Most plants, some protists, and some bacteria carry on photosynthesis.

Cellular respiration requires the reactants glucose and oxygen, produces carbon dioxide and water, and releases energy in the form of ATP. Most organisms carry on cellular respiration. Photosynthesis and cellular respiration are interrelated, as shown in **Figure 4.** Life on Earth depends on a balance of these two processes.

9. NGSSS Check
Summarize Fill in the table below to summarize the processes of photosynthesis and cellular respiration.
SC.8.L.18.1, SC.8.L.18.2

Photosynthesis v. Cellular Respiration

Process	Photosynthesis	Cellular Respiration
Reactants		
Products		
Organelle in which it occurs		
Type of organism	photosynthetic organisms including plants and algae	most organisms including plants and animals

Lesson Review 1

Visual Summary

Materials that a plant requires to survive move through the plant in the vascular tissue—xylem and phloem.

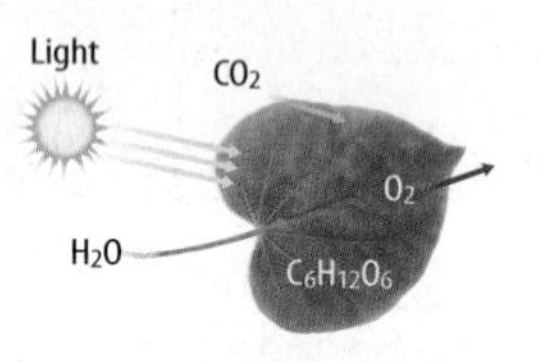

Plants can make their own food by using light energy, water, and carbon dioxide.

The products of photosynthesis are the reactants for cellular respiration.

Use Vocabulary

1. A series of chemical reactions that convert the energy in food molecules into a usable form of energy, called ATP, is called SC.8.L.18.2

____.

2. **Define** *photosynthesis* in your own words. SC.8.L.18.1

Understand Key Concepts

3. Which structure moves water through plants?
   - (A) chloroplast
   - (B) mitochondrion
   - (C) nucleus
   - (D) xylem

4. **Describe** how plants use chlorophyll for photosynthesis. SC.8.L.18.1

5. **Summarize** the process of cellular respiration. SC.8.L.18.2

Interpret Graphics

6. **Compare and Contrast** Fill in the table below to compare and contrast photosynthesis and cellular respiration.

Process	Similarities	Differences

Critical Thinking

7. **Predict** the effect of a plant disease that destroys all the chloroplasts in a plant.

8. **Evaluate** why plants perform cellular respiration. SC.8.L.18.2

GREEN SCIENCE

Deforestation and Carbon Dioxide in the Atmosphere

How does carbon dioxide affect climate?

What do you think when you hear the words *greenhouse gases?* Many people picture pollution from automobiles or factory smokestacks. It might be surprising to learn that cutting down forests affects the amount of one of the greenhouse gases in the atmosphere—carbon dioxide.

Deforestation is the term used to describe the destruction of forests. Deforestation happens because people cut down forests to use the land for other purposes, such as agriculture or building sites, or to use the trees for fuel or building materials.

▲ These cattle are grazing on land that was once part of a forest in Brazil.

Trees, like most plants, carry out photosynthesis and make their own food. Carbon dioxide from the atmosphere is one of the raw materials, or reactants, of photosynthesis. When deforestation occurs, trees are unable to remove carbon dioxide from the atmosphere. As a result, the level of carbon dioxide in the atmosphere increases.

Trees affect the amount of atmospheric carbon dioxide in other ways. Large amounts of carbon are stored in the molecules that make up trees. When trees are burned or left to rot, much of this stored carbon is released as carbon dioxide. This increases the amount of carbon dioxide in the atmosphere.

▲ In a process called slash-and-burn, forest trees are cut down and burned to clear land for agriculture.

Carbon dioxide in the atmosphere has an impact on climate. Greenhouse gases, such as carbon dioxide, increase the amount of the Sun's energy that is absorbed by the atmosphere. These gases also reduce the ability of heat to escape back into space. So, when levels of carbon dioxide in the atmosphere increase, more heat is trapped in Earth's atmosphere. This can lead to climate change.

It's Your Turn

RESEARCH AND REPORT How can we lower the rate of deforestation? What are some actions you can take that could help slow the rate of deforestation? Research to find out how you can make a difference. Make a poster to share what you learn. LA.6.4.2.2

Lesson 2

Plant RESPONSES

ESSENTIAL QUESTIONS

 How do plants respond to environmental stimuli?

 How do plants respond to chemical stimuli?

Vocabulary

stimulus p. 373
tropism p. 374
photoperiodism p. 376
plant hormone p. 377

Florida NGSSS

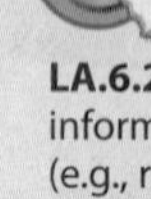

LA.6.2.2.3 The student will organize information to show understanding (e.g., representing main ideas within text through charting, mapping, paraphrasing, summarizing, or comparing/contrasting);

MA.6.A.3.6 Construct and analyze tables, graphs, and equations to describe linear functions and other simple relations using both common language and algebraic notation.

SC.6.N.2.1 Distinguish science from other activities involving thought.

SC.7.N.1.3 Distinguish between an experiment (which must involve the identification and control of variables) and other forms of scientific investigation and explain that not all scientific knowledge is derived from experimentation.

SC.7.N.1.4 Identify test variables (independent variables) and outcome variables (dependent variables) in an experiment.

SC.7.N.1.6 Explain that empirical evidence is the cumulative body of observations of a natural phenomenon on which scientific explanations are based.

SC.8.N.1.6 Understand that scientific investigations involve the collection of relevant empirical evidence, the use of logical reasoning, and the application of imagination in devising hypotheses, predictions, explanations and models to make sense of the collected evidence.

SC.7.N.1.3

Launch Lab

15 minutes

How do plants respond to stimuli?

Plants use light energy and make their own food during photosynthesis. How else do plants respond to light in their environment?

Procedure

1. Read and complete a lab safety form.
2. Choose a **pot of young radish seedlings.**
3. Place **toothpicks** parallel to a few of the seedlings in the pot in the direction of growth.
4. Place the pot near a **light source,** such as a gooseneck lamp or next to a window. The light source should be to one side of the pot, not directly above the plants.
5. Check the position of the seedlings in relation to the toothpicks after 30 min. Record your observations in the space below.
6. Observe the seedlings when you come to class the next day. Record your observations.

Data and Observations

Think About This

1. What happened to the position of the seedlings after the first 30 min? What is your evidence of change?

2. What happened to the position of the seedlings after a day?

3. Key Concept Why do you think the position of the seedlings changed?

Inquiry A Meat-Eating Plant?

1. This sundew plant is native to Florida. It has drops of a sticky, gel-like substance at the tip of each tentacle. The drops resemble morning dew. The sundew's main prey—insects—become trapped in the sticky gel. The plant's tentacles respond to being touched by wrapping around the trapped prey. To what other stimuli do you think plants might respond?

Stimuli and Plant Responses

Have you ever been in a dark room when someone suddenly turned on the light? You might have reacted by quickly shutting or covering your eyes. **Stimuli** (STIM yuh li; singular, stimulus) *are any changes in an organism's environment that cause a response.*

Often a plant's response to stimuli might be so slow that it is hard to see it happen. The response might occur gradually over a period of hours or days. Light is a stimulus. A plant responds to light by growing toward it, as shown in **Figure 5.** This response occurs over several hours.

In some cases, the response to a stimulus is quick, such as the Venus flytrap's response to touch. When stimulated by an insect's touch, the two sides of the trap snap shut immediately, trapping the insect inside.

Active Reading 2. **Explain** Why is it sometimes hard to see a plant's response to a stimulus?

Active Reading 3. **Identify** Write a phrase beside each paragraph that summarizes the main point of the paragraph. Use the phrases to review the lesson.

Figure 5 The light is the stimulus, and the seedlings have responded by growing toward the light.

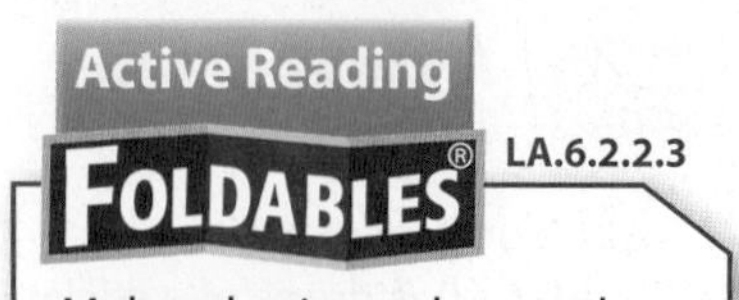

LA.6.2.2.3

Make a horizontal two-tab book and label it as shown. Record what you learn about the two types of stimuli that affect plant growth.

Environmental Stimuli

When it is cold outside, you probably wear a sweatshirt or a coat. Plants cannot put on warm clothes, but they do respond to their environments in a variety of ways. You might have seen trees flower in the spring or drop their leaves in the fall. Both are plant responses to environmental stimuli.

Growth Responses

Plants respond to a number of different environmental stimuli. These include light, touch, and gravity. *A* **tropism** (TROH pih zum) *is a response that results in plant growth toward or away from a stimulus.* When the growth is toward a stimulus, the tropism is called positive. A plant bending toward light is a positive tropism. Growth away from a stimulus is considered negative. A plant's stem growing upward against gravity is a negative tropism.

Word Origin

tropism
from Greek *tropos,* means "turn" or "turning"

Light The growth of a plant toward or away from light is a tropism called phototropism. A plant has a light-sensing chemical that helps it detect light. Leaves and stems tend to grow in the direction of light, as shown in **Figure 6.** This response maximizes the amount of light the plant's leaves receive. Roots generally grow away from light. This usually means that the roots grow down into the soil and help anchor the plant.

Active Reading **4. Explain** How is phototropism beneficial to a plant?

____

____

Response to Light

Figure 6 As a plant's leaves turn toward the light, the amount of light that the leaves can absorb increases.

Active Reading **5. Locate** Circle the plant that is responding to the light through phototropism.

Touch The response of a plant to touch is called a thigmotropism (thihg MAH truh pih zum). You might have seen vines growing up the side of a building or a fence. This happens because the plant has special structures that respond to touch. These structures, called tendrils, can wrap around or cling to objects, as shown in **Figure 7.** A tendril wrapping around an object is an example of positive thigmotropism. Roots display negative thigmotropism. They grow away from objects in soil, enabling them to follow the easiest path through the soil.

Gravity The response of a plant to gravity is called gravitropism. Stems grow away from gravity, while roots grow toward gravity. The seedlings in **Figure 8** are exhibiting both responses. No matter how a seed lands on soil, when it starts to grow, its roots grow down into the soil. The stem grows up. This happens even when a seed is grown in a dark chamber, indicating that these responses can occur independently of light.

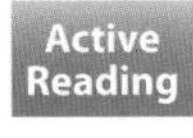

6. List What types of environmental stimuli do plants respond to? Give three examples.

____

____

____

Response to Touch

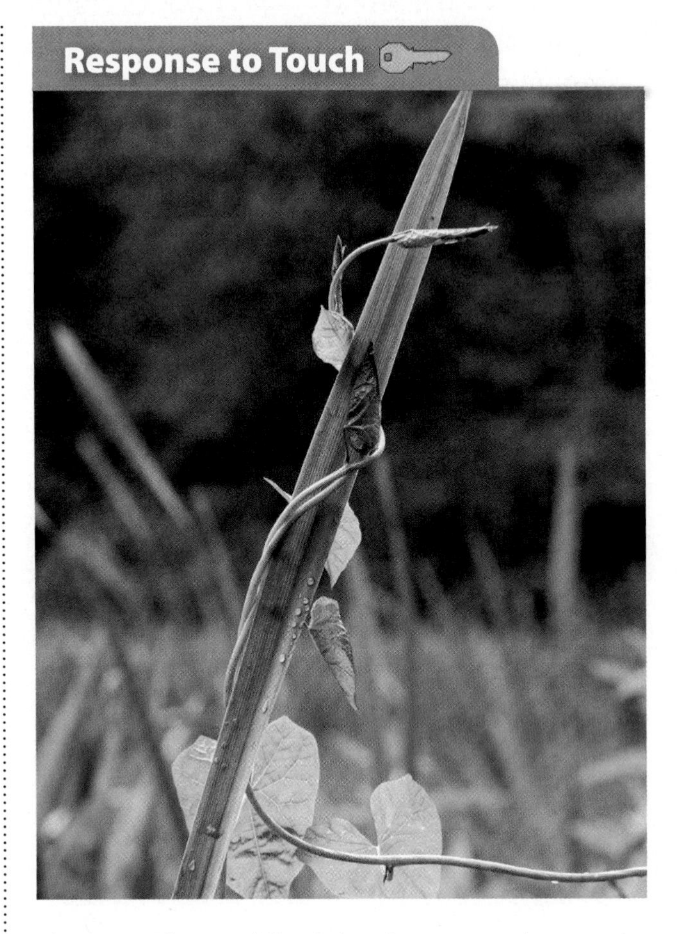

Figure 7 The tendrils of the vine respond to touch and coil around the blade of grass.

Response to Gravity

Figure 8 Both of these plant stems are growing away from gravity. The upward growth of a plant's stem is negative gravitropism, and the downward growth of its roots is positive gravitropism.

7. Visual Check Identify How is the plant on the left responding to the pot being placed on its side?

____

____

Flowering Responses

You might think all plants respond to light, but in some plants, flowering is actually a response to darkness! **Photoperiodism** *is a plant's response to the number of hours of darkness in its environment.* Scientists once hypothesized that photoperiodism was a response to light. Therefore, these flowering responses are called long-day, short-day, and day-neutral and relate to the number of hours of daylight in a plant's environment.

Long-Day Plants Plants that flower when exposed to less than 10–12 hours of darkness are called long-day plants. The carnations shown in **Figure 9** are examples of long-day plants. This plant usually produces flowers in summer, when the number of hours of daylight is greater than the number of hours of darkness.

Short-Day Plants Short-day plants require 12 or more hours of darkness for flowering to begin. An example of a short-day plant is the poinsettia, shown in **Figure 9.** Poinsettias tend to flower in late summer or early fall when the number of hours of daylight is decreasing and the number of hours of darkness is increasing.

Day-Neutral Plants The flowering of some plants doesn't seem to be affected by the number of hours of darkness. Day-neutral plants flower when they reach maturity and the environmental conditions are right. Plants such as the roses in **Figure 9** are day-neutral plants.

Active Reading **8. Infer** How is the flowering of day-neutral plants affected by exposure to hours of darkness?

____

____

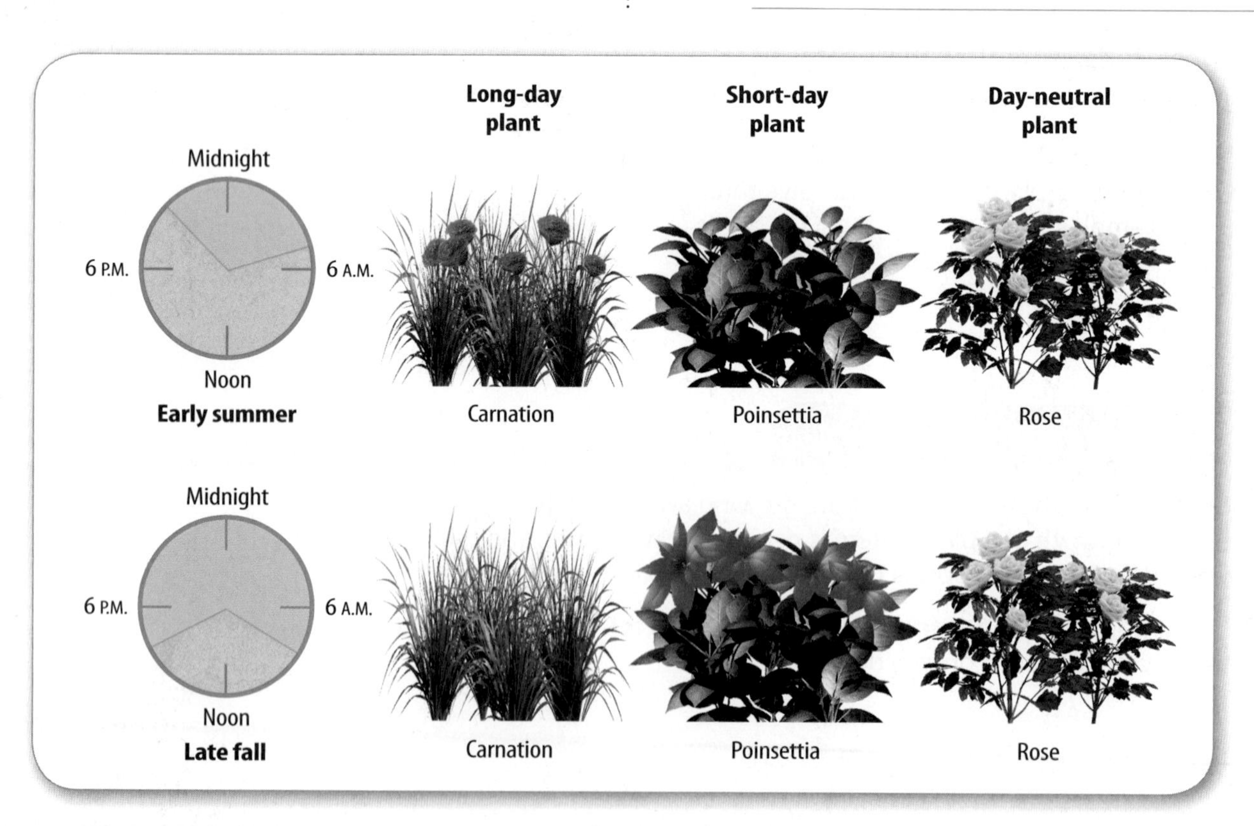

Figure 9 The number of hours of darkness controls flowering in many plants. Long-day plants flower when there are more hours of daylight than darkness, and short-day plants flower when there are more hours of darkness than daylight.

9. Visual Check Identify What time of year receives more darkness? ______________________________ (Circle) the plant(s) that produce(s) flowers during this time of year.

Chemical Stimuli

Plants respond to chemical stimuli as well as environmental stimuli. **Plant hormones** *are substances that act as chemical messengers within plants.* These chemicals are produced in tiny amounts. They are called messengers because they usually are produced in one part of a plant and affect another part of that plant.

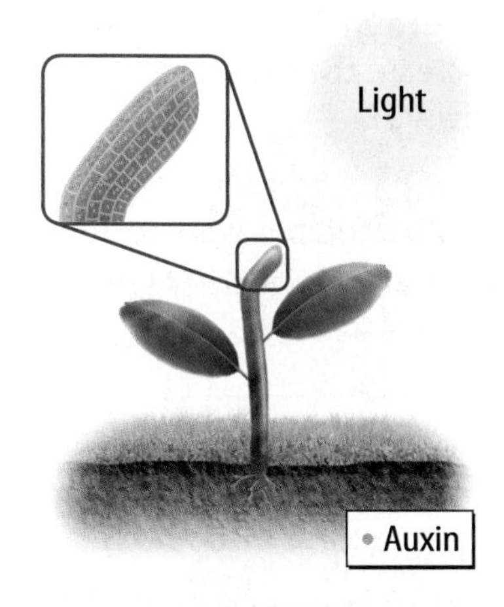

Figure 10 Auxin on the left side of the seedling causes more growth and makes the seedling bend to the right.

Auxins

One of the first plant hormones discovered was auxin (AWK sun). There are many different kinds of auxins. Auxins generally cause increased plant growth. They are responsible for phototropism, the growth of a plant toward light. Auxins concentrate on the dark side of a plant's stem, and these cells grow longer. This causes the stem of the plant to grow toward the light, as shown in **Figure 10.**

Ethylene

The plant hormone ethylene helps stimulate the ripening of fruit. Ethylene is a gas that can be produced by fruits, seeds, flowers, and leaves. You might have heard someone say that one rotten apple spoils the whole barrel. This is based on the fact that rotting fruits release ethylene. This can cause other fruits nearby to ripen and possibly rot. Ethylene also can cause plants to drop their leaves.

Active Reading 10. **Explain** How do plants respond to the chemical stimuli, or hormones, auxin and ethylene?

Inquiry SC.6.N.2.1

LAB STATION **Try It!**

MiniLab ***When will plants flower?*** at connectED.mcgraw-hill.com

Apply It! After you complete the lab, answer these questions.

1. In your own words, what is photoperiodism?

2. How do you think plants that flower well with any amount of darkness might have developed?

Figure 11 Gibberellins increase the rate of cell division.

Active Reading **11. Identify** Which grapes were treated with gibberellins: the ones on the left or right?

Math Skills MA.6.A.3.6

Use Percentages

A percentage is a ratio that compares a number to 100. For example, if a tree grows 2 cm per day with no chemical stimulus and 3 cm per day with a chemical stimulus, what is the percentage increase in growth?

Subtract the original value from the final value.

3 cm − 2 cm = 1 cm

Set up a ratio between the difference and the original value. Find the decimal equivalent.

$$\frac{1\text{ cm}}{2\text{ cm}} = 0.5\text{ cm}$$

Multiply by 100 and add a percent sign.

0.5 × 100 = 50%

Practice

12. Without gibberellins, pea seedlings grew to 2 cm in 3 days. With gibberellins, the seedlings grew to 4 cm in 3 days. What was the percentage increase in growth?

Response to Gibberellins

Gibberellins and Cytokinins

Rapidly growing areas of a plant, such as roots and stems, produce gibberellins (jih buh REL unz). These hormones increase the rate of cell division and cell elongation. This results in increased growth of stems and leaves. Gibberellins also can be applied to the outside of plants. As shown in **Figure 11,** applying gibberellins to plants can have a dramatic effect.

Root tips produce most of the cytokinins (si tuh KI nunz), another type of hormone. Xylem carries cytokinins to other parts of a plant. Cytokinins increase the rate of cell division, and in some plants, they slow the aging process of flowers and fruits.

Summary of Plant Hormones

Plants produce many different hormones. The hormones you have just read about are groups of similar compounds. Often, two or more hormones interact and produce a plant response. Scientists are still discovering new information about plant hormones.

Humans and Plant Responses

Humans depend on plants for food, fuel, shelter, and clothing. Humans make plants more productive using plant hormones. Some crops now are easier to grow because humans understand how they respond to hormones. As you study **Figure 12** on the next page, make a list of all the ways humans can benefit from understanding and using plant responses.

Active Reading **13. Locate** Underline the ways humans are dependent on plants.

Figure 12 Understanding how plants respond to hormones can benefit people in many ways.

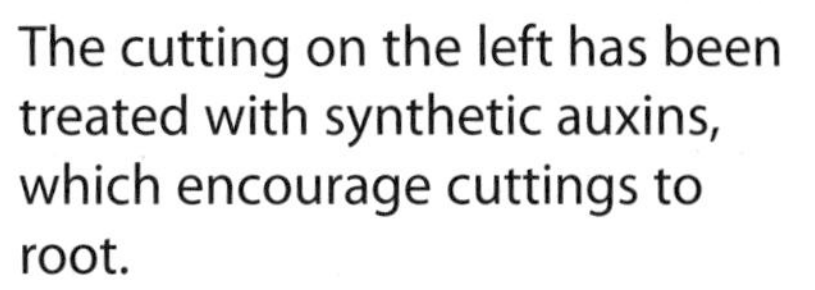

The cutting on the left has been treated with synthetic auxins, which encourage cuttings to root.

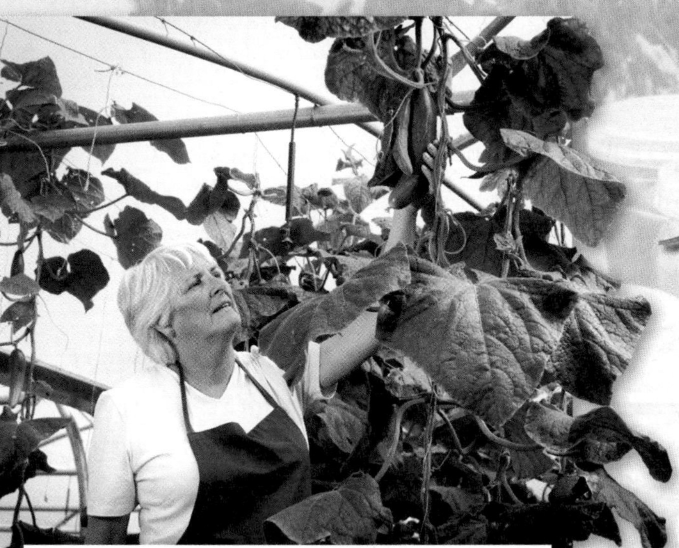

By choosing seeds that produce climbing cucumbers, farmers grow plants that are easier to pick. The cucumbers grow faster and bigger because they get more light.

Removing the apical bud of a plant suppresses auxin causing the plant to grow out instead of up, producing a fuller plant.

Bananas can be picked and shipped while still green and then be treated with ethylene to cause them to ripen.

The use of cytokinins helps scientists and horticulturists grow hundreds of identical plants.

Lesson Review 2

Visual Summary

Plants respond to stimuli in their environments in many ways.

Photoperiodism occurs in long-day plants and short-day plants. Day-neutral plants are not affected by the number of hours of darkness.

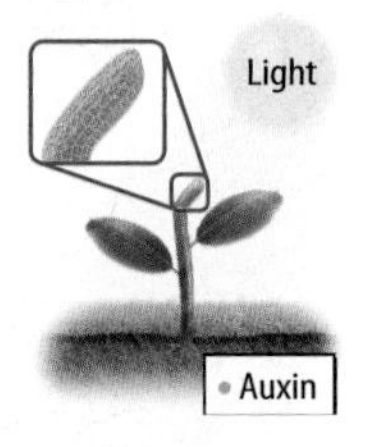

Plant hormones produce different responses in plants.

Use Vocabulary

1. **Define** *plant hormone* in your own words.

____

2. The response of an organism to the number of hours of darkness in its environment is called _______________.

Understand Key Concepts

3. **Compare** the effect of auxins and gibberellins on plant cells.

____

4. Which is NOT likely to cause a plant response?
   - (A) changing the amount of daylight
   - (B) moving plants away from each other
   - (C) treating with plant hormones
   - (D) turning a plant on its side

Interpret Graphics

5. **Identify** List the plant hormones mentioned in this lesson. Describe the effect of each on plants.

Hormone	Effect on Plants

Critical Thinking

6. **Infer** why the plant shown to the right is growing at an angle.

____

Math Skills

7. When sprayed with gibberellins, the diameter of mature grapes increased from 1.0 cm to 1.75 cm. What was the percent increase in size?

Summarize information about tropisms.

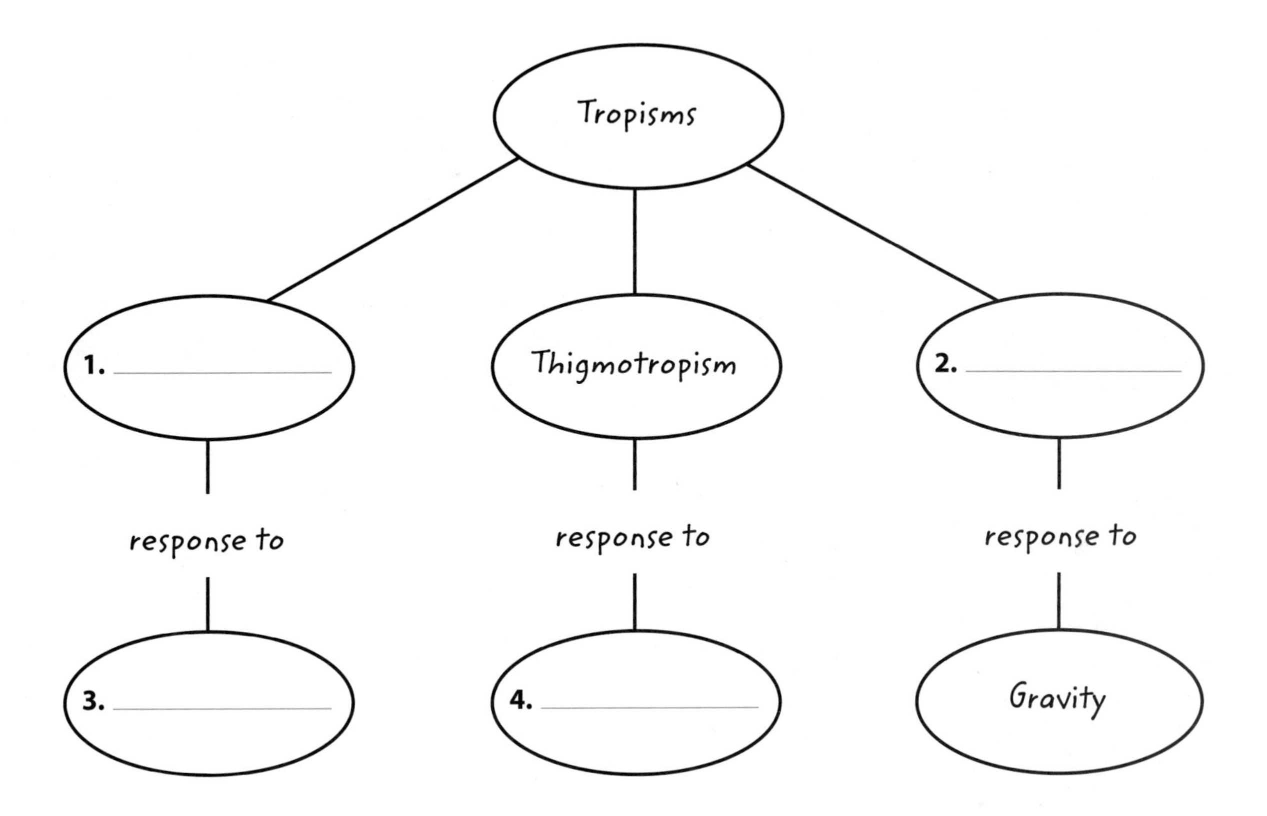

Review types of photoperiodism. Draw a line from each term on the left to all the descriptors that apply to it.

5. Day-neutral plants

6. Long-day plants

7. Short-day plants

- Carnation
- Poinsettia
- Rose
- Flower when exposed to less than 10–12 hours of darkness
- Flower when exposed to 12 or more hours of darkness
- Flowering is unaffected by the number of hours of darkness

Lesson 3

Plant REPRODUCTION

ESSENTIAL QUESTIONS

What is the alternation of generations in plants?

How do seedless plants reproduce?

How do seed plants reproduce?

Vocabulary

alternation of generations p. 384
spore p. 384
pollen grain p. 386
pollination p. 386
ovule p. 386
embryo p. 386
seed p. 386
stamen p. 388
pistil p. 388
ovary p. 388
fruit p. 389

Florida NGSSS

LA.6.2.2.3 The student will organize information to show understanding (e.g., representing main ideas within text through charting, mapping, paraphrasing, summarizing, or comparing/contrasting);

SC.6.N.2.1 Distinguish science from other activities involving thought.

SC.7.N.1.4 Identify test variables (independent variables) and outcome variables (dependent variables) in an experiment.

SC.8.N.1.6 Understand that scientific investigations involve the collection of relevant empirical evidence, the use of logical reasoning, and the application of imagination in devising hypotheses, predictions, explanations and models to make sense of the collected evidence.

SC.8.N.3.1 Select models useful in relating the results of their own investigations.

SC.6.N.2.1

Inquiry Launch Lab

15 minutes

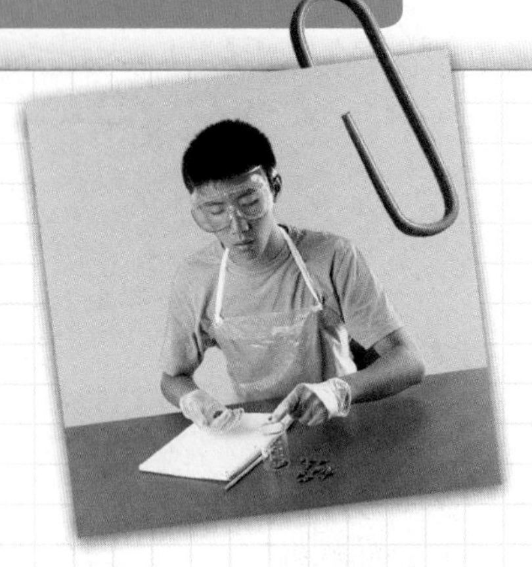

How can you identify fruits?

Flowering plants grow from seeds that they produce. Animals depend on flowering plants for food. The function of the fruit is to disperse the seeds for plant reproduction.

Procedure

1. Read and complete a lab safety form.
2. Make a two-column table on a separate sheet of paper. Label the columns *Fruits* and *Not Fruits.*
3. Examine a collection of **food items.** Determine whether each item is a fruit. Record your observations in your table.
4. Place each food item on a piece of **plastic wrap.** Use a **plastic or paring knife** to cut the items in half.
5. Examine the inside of each food item. Record your observations below.

Data and Observations

Think About This

1. What observations did you make about the insides of the food items? Would you reclassify any food item based on your observations? Explain.

2. How can the number of seeds or how they are placed in the fruit help with seed dispersal?

3. Key Concept What role do you think a fruit has in a flowering plant's reproduction?

Inquiry A Bee's-Eye View?

1. Bees can see ultraviolet (UV) light. We see a dandelion as yellow. Because of a bee's ability to see UV light, a bee sees a dandelion like the one at left. Why do you think bees see flowers differently than we do?

Asexual Reproduction Versus Sexual Reproduction

In early spring, you might see cars or sidewalks covered with a thick, yellow dust. Where did it come from? It probably came from plants that are reproducing. As in all living things, reproduction is part of the life cycle of plants.

Plants can reproduce asexually, sexually, or both ways. Asexual reproduction occurs when a portion of a plant develops into a separate new plant. This new plant is genetically identical to the original, or parent, plant. Some plants, such as irises and daylilies, can use their underground stems for asexual reproduction. Other plants, such as the houseleeks, or hens and chicks, in **Figure 13,** reproduce asexually using horizontal stems called stolons. One advantage of asexual reproduction is that just one parent organism can produce offspring. However, sexual reproduction in plants usually requires two parent organisms. Sexual reproduction occurs when a plant's sperm combines with a plant's egg. A resulting zygote can grow into a plant. This new plant is a genetic combination of its parents.

Active Reading **2. Identify** Underline the differences in asexual and sexual reproduction in plants.

Active Reading **3. Define** Make a vocabulary card for each bold term in this lesson. Write each term on one side of the card. On the other side, write the definition. Use these terms to review the vocabulary for the lesson.

Figure 13 Hens and chicks can reproduce asexually. New "chicks" can grow from the stolons on the main "hen" plant.

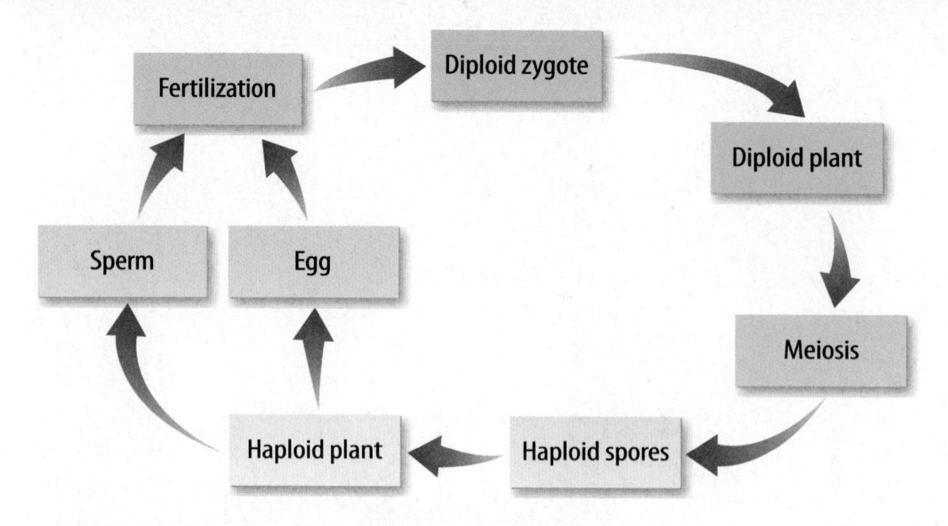

Figure 14 The life cycle of all plants includes an alternation of generations. The diploid generation begins with fertilization. The haploid generation begins with meiosis.

Active Reading **4. Identify** Circle the diploid generation in **Figure 14.**

Alternation of Generations

Your body is made of two types of cells—haploid cells and diploid cells. Most of your cells are diploid. The only human haploid cells are sperm and eggs. As a result, you will live your entire life as a diploid organism. To put it another way, your life cycle includes only a diploid stage. That isn't true for all organisms. Some organisms, including plants, have two life stages called **generations.** One generation is almost all diploid cells. The other generation has only haploid cells. **Alternation of generations** *occurs when the life cycle of an organism alternates between diploid and haploid generations,* as shown in **Figure 14.**

Science Use v. Common Use

generation

Science Use haploid and diploid stages in the life cycle of a plant

Common Use the average span of time between the birth of parents and their offspring

The Diploid Generation

When you look at a tree, you're seeing part of the plant's diploid generation. Meiosis occurs in certain cells in the reproductive structures of a diploid plant. *The daughter cells produced from haploid structures are called* **spores.** Spores grow by mitosis and cell division and form the haploid generation of a plant.

Word Origin

spore

from Greek *spora,* means "seed, a sowing"

5. **NGSSS Check Distinguish** Circle the correct word in the following statements: SC.7.L.16.3

- Daughter cells are produced in reproductive structures through *meiosis / mitosis.*
- The haploid generation forms as the spores grow by *meiosis / mitosis.*

The Haploid Generation

In most plants, the haploid generation is tiny and lives surrounded by special tissues of the diploid plant. In other plants, the haploid generation lives on its own. Certain reproductive cells of the haploid generation produce haploid sperm or eggs by mitosis and cell division. Fertilization takes place when a sperm and an egg fuse and form a diploid zygote. Through mitosis and cell division, the zygote grows into the diploid generation of a plant.

Reproduction in Seedless Plants

Not all plants grow from seeds. The first land plants to inhabit Earth probably were seedless plants—plants that grow from haploid spores, not from seeds. The mosses and ferns in **Figure 15** are examples of seedless plants found on Earth today.

Life Cycle of a Moss

The life cycle of a moss is typical for some seedless plants. The tiny, green moss plants that carpet rocks, bark, and soil in moist areas are haploid plants. These plants grow by **mitosis** and cell division from haploid spores produced by the diploid generation. They have male structures that produce sperm and female structures that produce eggs. Fertilization results in a diploid zygote that grows by mitosis and cell division into the diploid generation of moss, such as the one shown in **Figure 15.** A diploid moss is tiny and not easily seen.

Review Vocabulary

mitosis
the process during which a nucleus and its contents divide

Life Cycle of a Fern

An alternation of generations is also seen in the life cycle of a fern. The diploid generations are the green leafy plants often seen in forests. These plants produce haploid spores. The spores grow into tiny plants. The haploid plants produce eggs and sperms that can unite and form the diploid generations.

Active Reading 6. **Explain** How do seedless plants such as mosses and ferns reproduce?

____

____

____

Figure 15 Mosses and ferns usually grow in moist environments. Sperm must swim through a film of water to reach an egg.

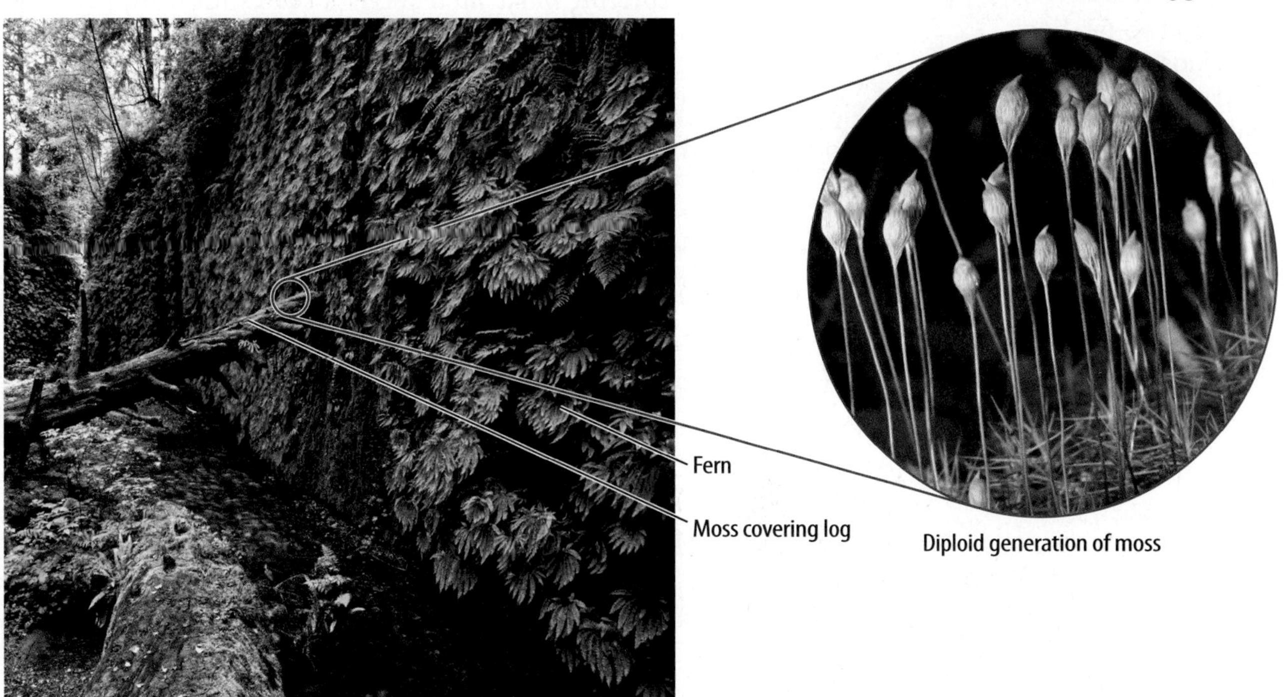

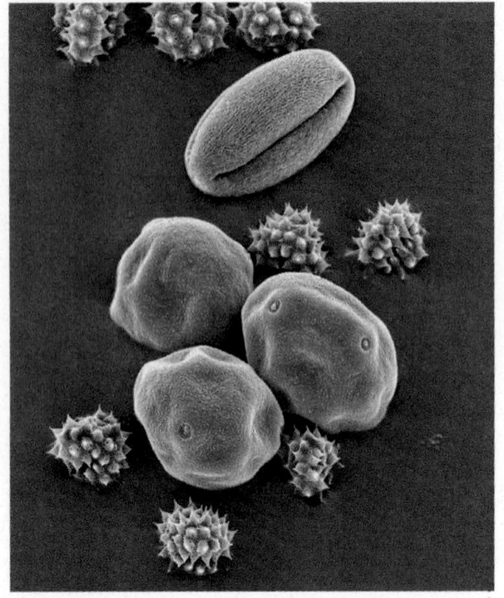

Color-enhanced SEM Magnification: 1,100×

Figure 16 Pollen grains of one type of plant are different from those of any other type of plant.

7. Visual Check
Calculate How many different types of pollen are visible in **Figure 16?**

How do seed plants reproduce?

Most land plants grow from seeds. There are two groups of seed plants—flowerless seed plants and flowering seed plants.

Unlike seedless plants, the haploid generation of a seed plant is within diploid tissue. Separate diploid male and diploid female reproductive structures produce haploid sperm and haploid eggs that join during fertilization.

The Role of Pollen Grains

A **pollen** (PAH lun) **grain** *forms from tissue in a male reproductive structure of a seed plant.* Each pollen grain contains nutrients and has a hard, protective outer covering, as shown in **Figure 16.** Pollen grains produce sperm cells. Wind, animals, gravity, or water currents can carry pollen grains to female reproductive structures.

Plants cannot move and find a mate as most animals can. Do you recall reading about the yellow dust at the beginning of this lesson? That dust is pollen grains. Male reproductive structures produce a vast number of pollen grains. **Pollination** (pah luh NAY shun) *occurs when pollen grains land on a female reproductive structure of a plant that is the same species as the pollen grains.*

The Role of Ovules and Seeds

The female reproductive structure of a seed plant where the haploid egg develops is called the **ovule.** Following pollination, sperm enter the ovule and fertilization occurs. A zygote forms and develops into an **embryo,** *an immature diploid plant that develops from the zygote.* As shown in **Figure 17,** *an embryo, its food supply, and a protective covering make up a* **seed.** A seed's food supply provides the embryo with nourishment for its early growth.

Figure 17 A seed contains a diploid plant embryo and a food supply protected by a hard outer covering.

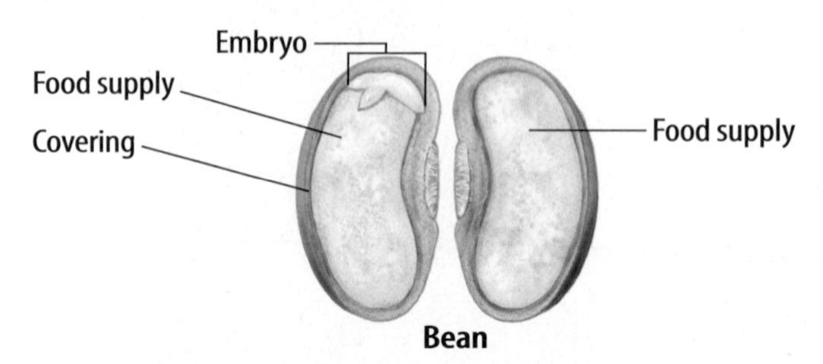

8. Visual Check **Infer** How does the size of the seed's embryo compare to the size of its food supply? Explain.

Reproduction in Flowerless Seed Plants

Flowerless seed plants are also known as gymnosperms (JIHM nuh spurmz). The word *gymnosperm* means "naked seed," and gymnosperm seeds are not surrounded by a fruit. The most common gymnosperms are conifers. Conifers, such as pines, firs, cypresses, redwoods, and yews, are trees and shrubs with needlelike or scalelike leaves. Most conifers are evergreens, which means they have leaves all year long. Conifers can live for many years. Bristlecone pines, such as the one shown in **Figure 18,** are among the oldest living trees on Earth.

Life Cycle of a Gymnosperm The life cycle of a gymnosperm, shown in **Figure 19,** includes an alternation of generations. Cones are the male and female reproductive structures of conifers. They contain the haploid generation. Male cones are small, papery structures that produce pollen grains. Female cones can be woody, berrylike, or soft, and they produce eggs. A zygote forms when a sperm from a male cone fertilizes an egg. The zygote is the beginning of the diploid generation. Seeds form as part of the female cone.

Active Reading **9. Recall** Highlight the name of the structures that contain the haploid generation of conifers.

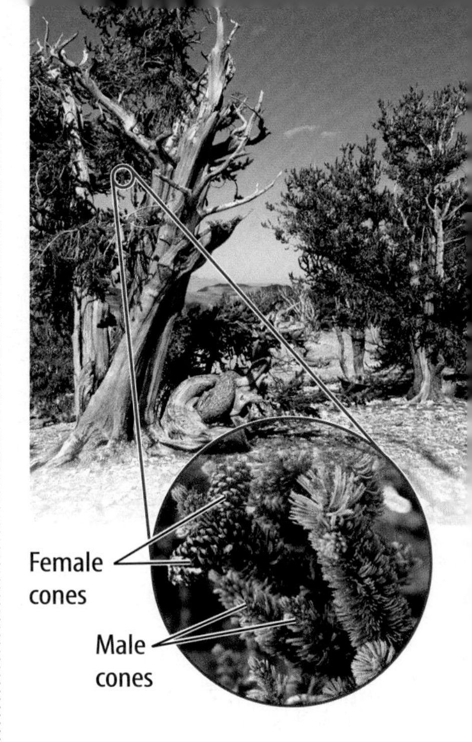

Figure 18 Seeds form at the base of each scale on a female cone.

Reproduction in Flowerless Seed Plants

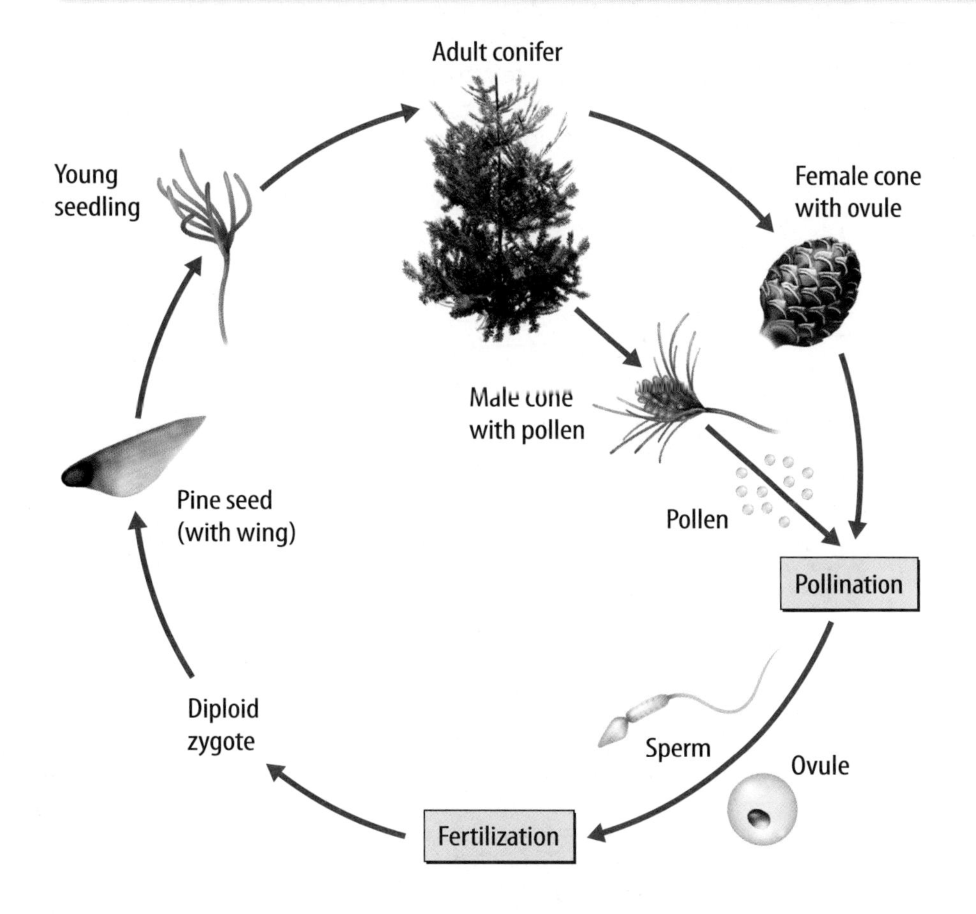

Figure 19 Male cones release clouds of pollen grains. Sperm from the pollen grains unite with eggs and form zygotes.

Active Reading **10. Identify** At what point in the cycle does the haploid generation become the diploid generation?

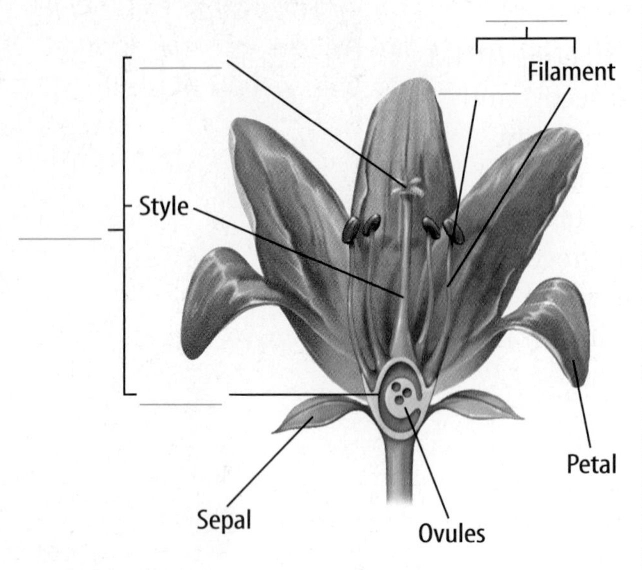

Figure 20 Typical flowers have both male and female structures.

Active Reading **11. Label** Complete the figure above using the following terms:

A. anther
B. ovary
C. pistil
D. stamen
E. stigma

Reproduction in Flowering Seed Plants

Most plants you see around you are angiosperms, or flowering plants. Fruits and vegetables come from angiosperms. Many animals depend on angiosperms for food.

The Flower Reproduction of an angiosperm begins in a flower. Most flowers have male and female reproductive structures, as shown in **Figure 20.**

The male reproductive organ of a flower is the **stamen.** Pollen grains form at the tip of the stamen in the anther. The filament supports the anther and connects it to the base of the flower. *The female reproductive organ of a flower is the* **pistil.** Pollen can land at the tip of the pistil, or stigma. The stigma is at the top of a long tube called the style. *At the base of the style is the* **ovary,** *which contains one or more ovules.* Recall that each ovule eventually will contain a haploid egg and might become a seed if fertilized.

Inquiry SC.6.N.2.1, SC.8.N.3.1

LAB STATION **Try It!**

MiniLab ***How can you model a flower?*** at connectED.mcgraw-hill.com

Apply It!

After you complete the lab, answer these questions.

 1. How does your model of a flower compare to flowers in the area of Florida where you live?

 2. How does your model differ from flowers in the area of Florida where you live?

Life Cycle of an Angiosperm A typical life cycle for an angiosperm is shown in **Figure 21.** Pollen grains travel by wind, gravity, water, or animal from the anther to the stigma, where pollination occurs. A pollen tube grows from the pollen grain into the stigma, down the style, to the ovary at the base of the pistil. Sperm develop from a haploid cell in the pollen tube. When the pollen tube enters an ovule, fertilization takes place.

As you read earlier, the zygote that results from fertilization develops into an embryo. Each ovule and its embryo will become a seed. *The ovary, and sometimes other parts of the flower, will develop into a* **fruit** *that contains one or more seeds.* The seeds can grow into new, genetically related plants that produce flowers, and the cycle repeats.

Active Reading **12. Select** Sperm develop *before / after* pollination.

Reproduction in Flowering Seed Plants

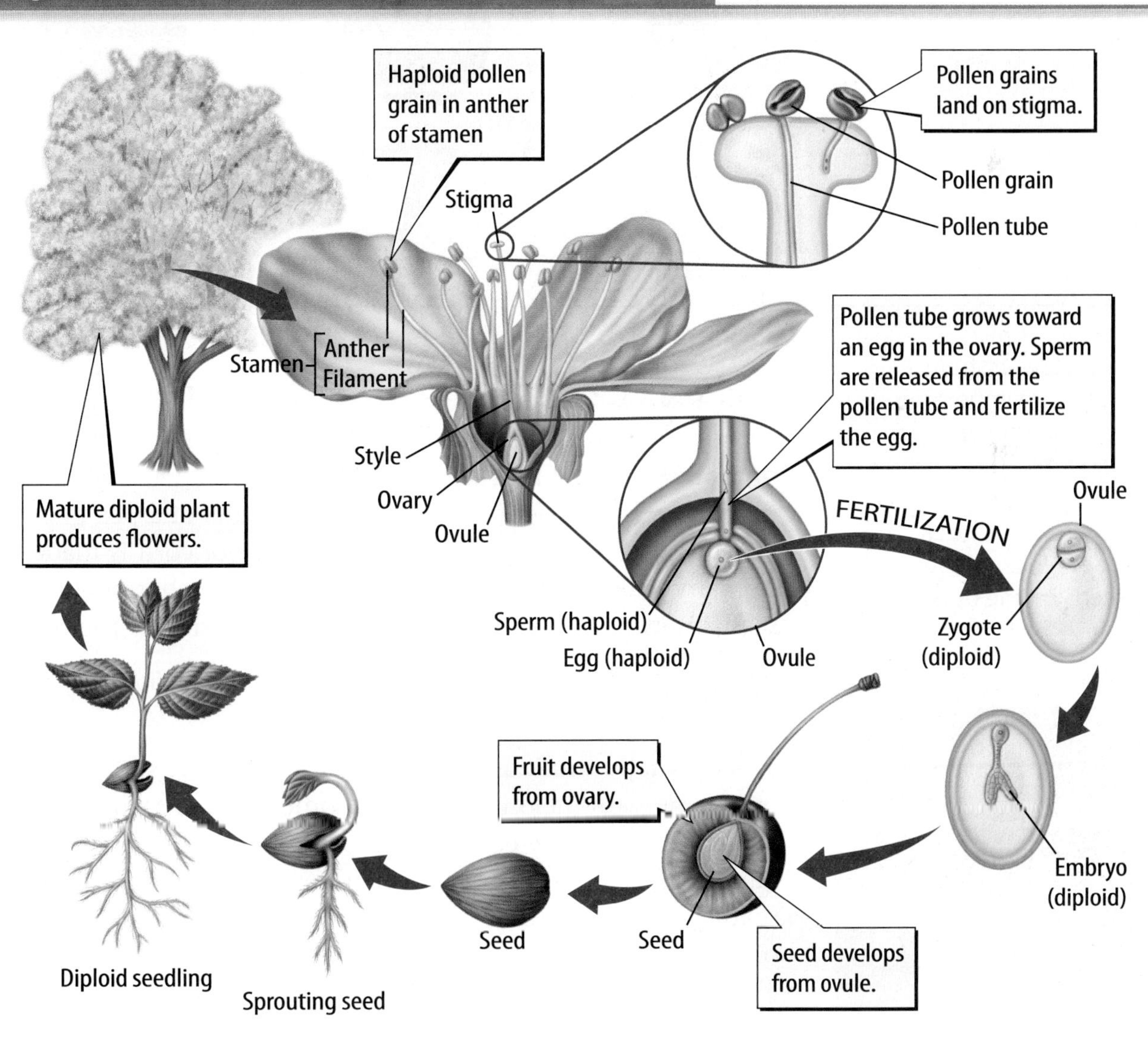

Figure 21 During the life cycle of a flowering plant, the haploid generation grows and develops inside the diploid plant.

13. Visual Check Describe How does sperm in a pollen grain reach an egg in the ovule?

Table 1 Flowers, Fruits, and Seeds of Common Plants

Plant	Flower	Fruit	Seed
Pea			
Corn			
Strawberry			
Dandelion			

Table 1 Flowers, fruits, and seeds are important for reproduction in angiosperms.

14. Visual Check Choose Which of these fruits has seeds on the outside?

Figure 22 The seeds will be excreted by the mouse and might grow into new blackberry bushes.

Fruit and Seed Dispersal Fruits and seeds, including the peas, corn, and strawberries shown in **Table 1,** are important sources of food for people and animals. In most cases, seeds of flowering plants are inside fruits. Pods are the fruits of a pea plant. The peas inside a pod are the seeds. An ear of corn is made up of many fruits, or kernels. The main part of each kernel is the seed. Strawberries have tiny seeds on the outside of the fruit.

Active Reading **15. Recall** Underline where the seeds of flowering plants are usually found.

We usually think of fruits as juicy and edible, such as an orange or watermelons. However, some fruits are hard and dry and not particularly edible. Each parachute-like structure of a dandelion is a dry fruit. Fruits help protect seeds and help scatter or disperse them. For example, some fruits such as that of a dandelion are light and float on air currents. When an animal eats a fruit, the fruit's seeds can pass through the animal's digestive system with little or no damage. Imagine what happens when an animal, such as the mouse shown in **Figure 22,** eats blackberries. The animal digests the juicy fruit but deposits the seeds on the soil with its wastes. By the time this happens, the animal might have traveled some distance away from the blackberry bush. This means the animal helped disperse the seeds away from the blackberry bush.

Lesson Review 3

Visual Summary

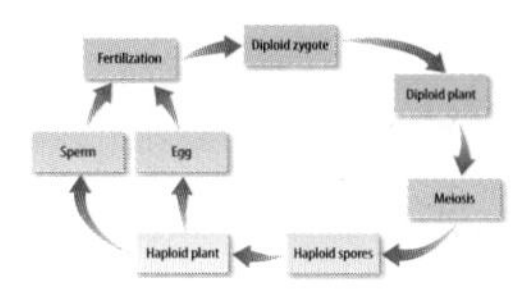

The life cycle of a plant includes an alternation of generations.

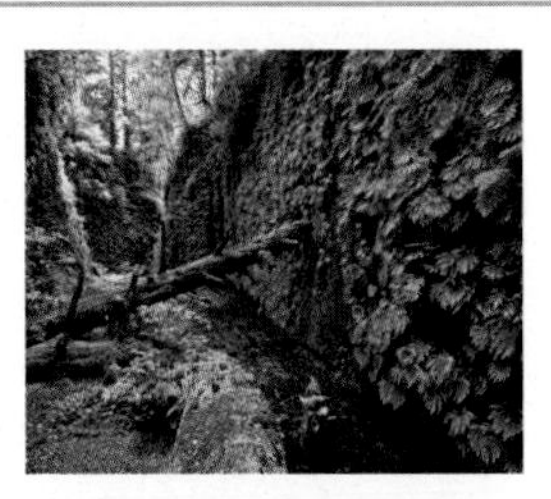

Seedless plants, such as ferns and mosses, grow from haploid spores.

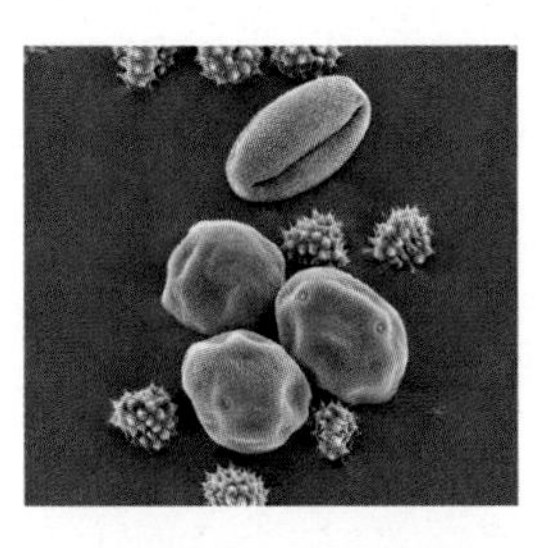

In seed plants, pollination occurs when pollen grains land on the female reproductive structure of a plant of the same species.

Use Vocabulary

1. The daughter cells produced from haploid structures are called _______________.

2. **Distinguish** between an ovule and an ovary.

3. **Define** *pollination* in your own words.

Understand Key Concepts 

4. Which is NOT part of the alternation of generations life cycle in plants?
   - (A) anther
   - (B) diploid
   - (C) haploid
   - (D) spore

5. **Contrast** the haploid generation of a moss with that of a fern.

6. **Describe** how a pollen tube carries sperm to the ovule in a flower.

Interpret Graphics

7. **Identify** Fill in the graphic organizer below to identify the female parts of a flower.

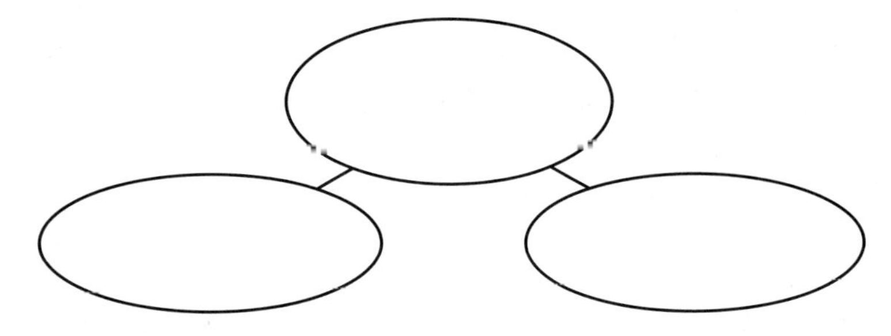

Critical Thinking

8. **Evaluate** the advantages of fruit production in plant reproduction.

Chapter 10 Study Guide

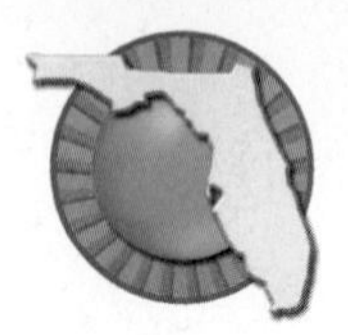

Think About It! Plants survive by maintaining homeostasis, responding to stimuli, and reproducing. In addition, they acquire the energy they need for life processes through photosynthesis and cellular respiration.

Key Concepts Summary

LESSON 1 Energy Processing in Plants

- The vascular tissue in most plants—xylem and phloem—moves materials throughout plants.
- In **photosynthesis,** plants convert light energy, water, and carbon dioxide into the food-energy molecule glucose through a series of chemical reactions. The process gives off oxygen.

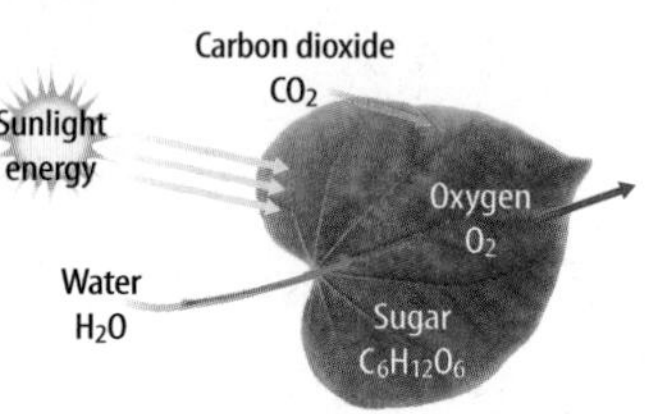

- **Cellular respiration** is a series of chemical reactions that convert the energy in food molecules into a usable form of energy called ATP.
- Photosynthesis and cellular respiration can be considered opposite processes of each other.

Vocabulary

photosynthesis p. 366
cellular respiration p. 368

LESSON 2 Plant Responses

- Although plants cannot move from one place to another, they do respond to **stimuli,** or changes in their environments. Plants respond to stimuli in different ways.
- **Tropisms** are growth responses toward or away from stimuli such as light, touch, and gravity. **Photoperiodism** is a plant's response to the number of hours of darkness in its environment.
- Plants respond to chemical stimuli, or **plant hormones,** such as auxins, ethylene, gibberellins, and cytokinins. Different hormones have different effects on plants.

stimulus p. 373
tropism p. 374
photoperiodism p. 376
plant hormone p. 377

LESSON 3 Plant Reproduction

- **Alternation of generations** is when the life cycle of an organism alternates between diploid and haploid generations.
- Seedless plants, such as ferns, reproduce when a haploid sperm fertilizes a haploid egg, forming a diploid zygote.
- Seed plants reproduce when **pollen grains,** which contain haploid sperm, land on the tip of the female reproductive organ. At the base of this organ is the **ovary,** which usually contains one or more **ovules.** Each ovule eventually will contain a haploid egg. If the sperm fertilizes the egg, an **embryo** will form within a **seed.**

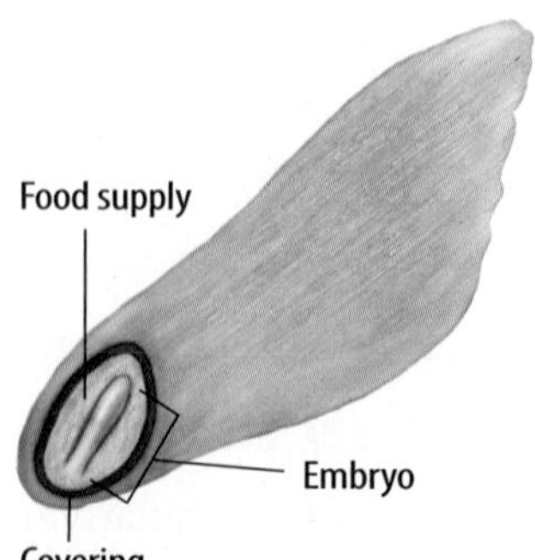

alternation of generations p. 384
spore p. 384
pollen grain p. 386
pollination p. 386
ovule p. 386
embryo p. 386
seed p. 386
stamen p. 388
pistil p. 388
ovary p. 388
fruit p. 389

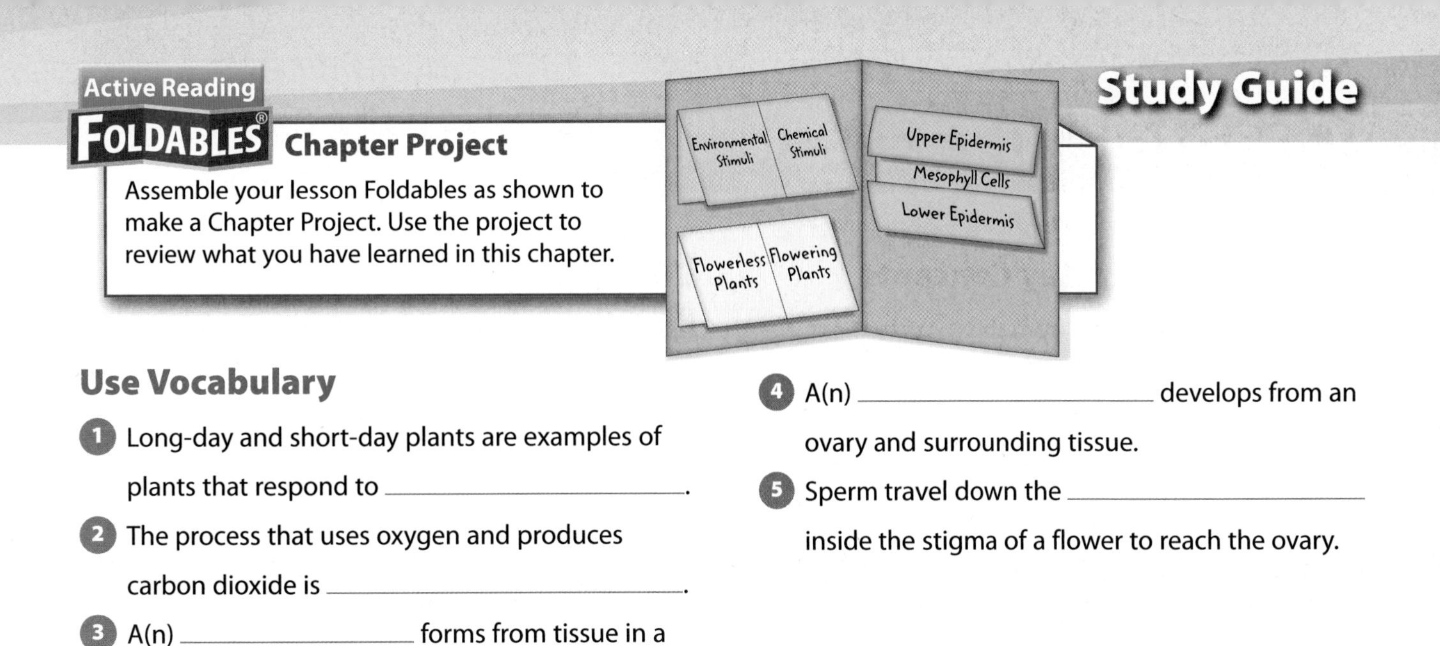

Active Reading **FOLDABLES®** **Chapter Project**

Assemble your lesson Foldables as shown to make a Chapter Project. Use the project to review what you have learned in this chapter.

Use Vocabulary

1. Long-day and short-day plants are examples of plants that respond to ____________.
2. The process that uses oxygen and produces carbon dioxide is ____________.
3. A(n) ____________ forms from tissue in a male reproductive structure of a seed plant.
4. A(n) ____________ develops from an ovary and surrounding tissue.
5. Sperm travel down the ____________ inside the stigma of a flower to reach the ovary.

Link Vocabulary and Key Concepts

Use vocabulary terms from the previous page to complete the concept map.

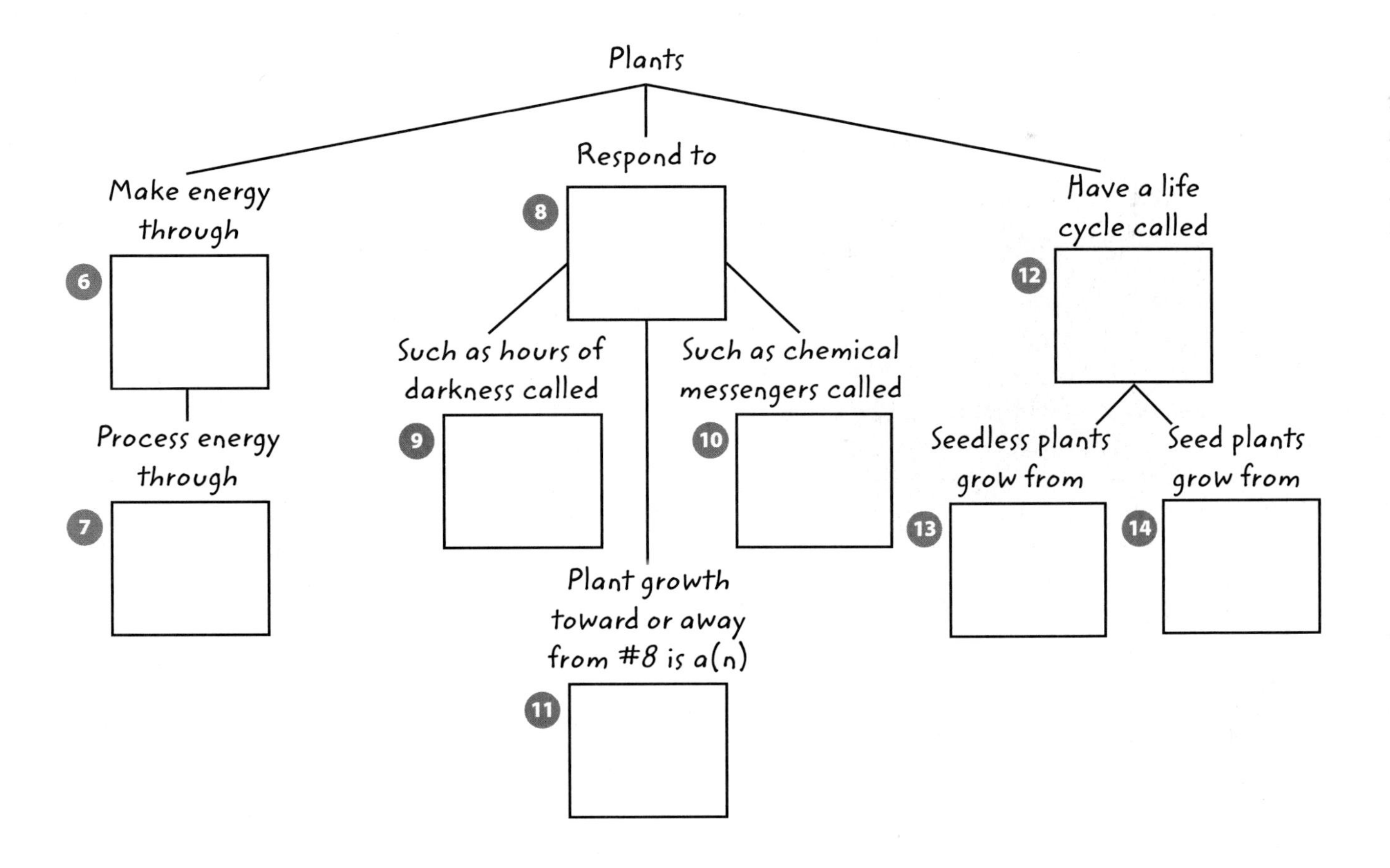

Chapter 10 Review

Fill in the correct answer choice.

Understand Key Concepts

1. Which material travels from the roots to the leaves through the xylem? **LA.6.2.2.3**
   Ⓐ oxygen
   Ⓑ sugar
   Ⓒ sunlight
   Ⓓ water

2. Which organelle is the site of photosynthesis? **SC.8.L.18.1**
   Ⓐ chloroplast
   Ⓑ mitochondria
   Ⓒ nucleus
   Ⓓ ribosome

3. Which is a product of cellular respiration? **SC.8.L.18.2**
   Ⓐ ATP
   Ⓑ light
   Ⓒ oxygen
   Ⓓ sugar

Use the image below to answer questions 4 and 5.

4. What type of plant-growth response is shown in the photo above? **LA.6.2.2.3**
   Ⓐ flowering
   Ⓑ gravitropism
   Ⓒ photoperiodism
   Ⓓ thigmotropism

5. Which stimulus is responsible for this type of growth? **LA.6.2.2.3**
   Ⓐ gravity
   Ⓑ light
   Ⓒ nutrients
   Ⓓ touch

Critical Thinking

6. **Infer** which came first—photosynthesis or cellular respiration. **SC.8.L.18.1, SC.8.L.18.2**

7. **Construct** a table to compare the reactants and products of photosynthesis and cellular respiration. **SC.8.L.18.1, SC.8.L.18.2**

8. **Evaluate** the internal structure of a leaf as a location for photosynthesis. **SC.8.L.18.1**

9. **Assess** the need for plants to respond to their environment. **LA.6.2.2.3**

10 **Infer** from the photo below where the light source is in relation to the plant. LA.6.2.2.3

11 **Evaluate** the importance of fruit production in flowering plants. LA.6.2.2.3

12 **Predict** the effect of cold temperature killing all the flowers on the fruit trees. LA.6.2.2.3

Writing in Science

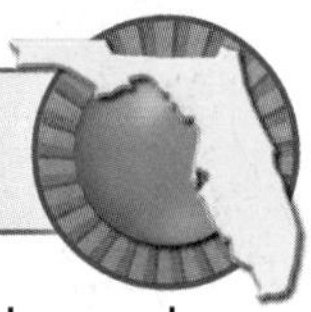

13 **Write** On a separate sheet of paper, write a five-sentence paragraph about the importance of plants in your life. Include a main idea, supporting details, and a concluding sentence. LA.6.2.2.3

Big Idea Review

14 Make a list of the plant processes you learned about in this chapter. How do these processes help a plant survive and reproduce? SC.8.L.18.1, SC.8.L.18.2

15 How does the process shown in the photo at the beginning of the chapter help a plant survive? LA.6.2.2.3

Math Skills MA.6.A.3.6

Use Percentages

16 Without treatment with gibberellins, 500 out of 1,000 grass seeds germinated. When sprayed with gibberellins, 875 of the seeds germinated. What was the percentage increase?

17 A bunch of bananas ripens (turns from green to yellow) in 42 hours. When the bananas are placed in a bag with an apple, which releases ethylene, the bananas ripen in 21 hours. What is the percentage change in ripening time?

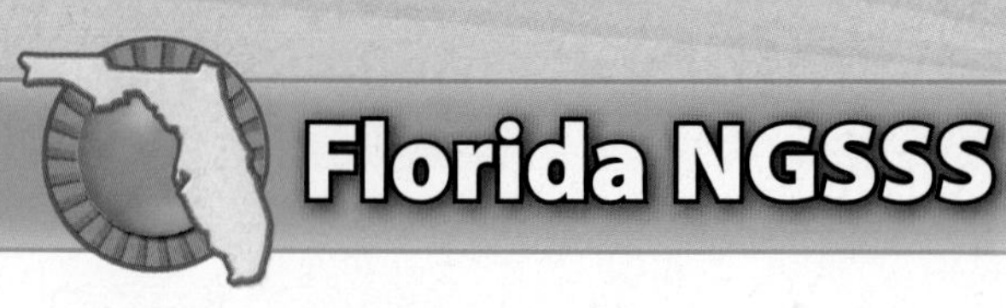

Fill in the correct answer choice.

Multiple Choice

1 Which structure transports sugars throughout a plant? **LA.6.2.2.3**

Ⓐ epidermis
Ⓑ phloem
Ⓒ stomata
Ⓓ xylem

2 What is one similarity between plants and animals? **SC.8.L.18.1, SC.8.L.18.2**

Ⓕ Both plants and animals carry on cellular respiration.
Ⓖ Both plants and animals carry on photosynthesis.
Ⓗ Both plants and animals have chloroplasts.
Ⓘ Both plants and animals use xylem and phloem to transport materials.

Use the diagram below to answer question 3.

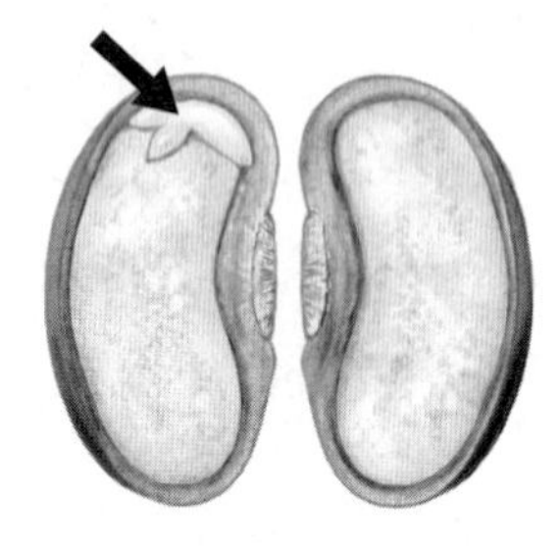

3 Look at the structure that is marked with an arrow in the image above. What will this structure become? **SC.7.L.16.3**

Ⓐ a diploid moss
Ⓑ a diploid seed plant
Ⓒ a haploid fern
Ⓓ a haploid flowerless seed plant

4 Which two plant hormones increase the rate of cell division? **LA.6.2.2.3**

Ⓕ auxins and cytokinins
Ⓖ cytokinins and gibberellins
Ⓗ ethylene and auxins
Ⓘ gibberellin and ethylene

5 Which is a product of photosynthesis? **SC.8.L.18.1**

Ⓐ carbon dioxide
Ⓑ glucose
Ⓒ light
Ⓓ water

Use the image below to answer question 6.

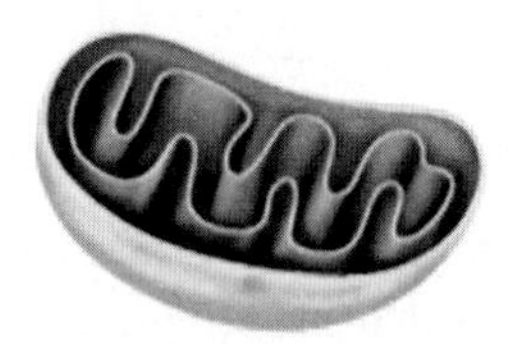

6 Which cellular process occurs within the organelle shown above? **SC.8.L.18.2**

Ⓕ photosynthesis
Ⓖ cellular respiration
Ⓗ transport of phloem
Ⓘ transport of xylem

7 Which plant has a diploid stage that is difficult to see? **SC.7.L.16.3**

Ⓐ conifer
Ⓑ cherry tree
Ⓒ dandelion
Ⓓ moss

8 How is cellular respiration related to photosynthesis? **SC.8.L.18.1, SC.8.L.18.2**

- Ⓕ Animals produce sugars through cellular respiration that are broken down by plants through photosynthesis.
- Ⓖ Animals use cellular respiration and plants use photosynthesis.
- Ⓗ Cellular respiration produces sugars, which are stored through photosynthesis.
- Ⓘ Photosynthesis produces sugars, which are broken down in cellular respiration.

Use the diagram below to answer question 9.

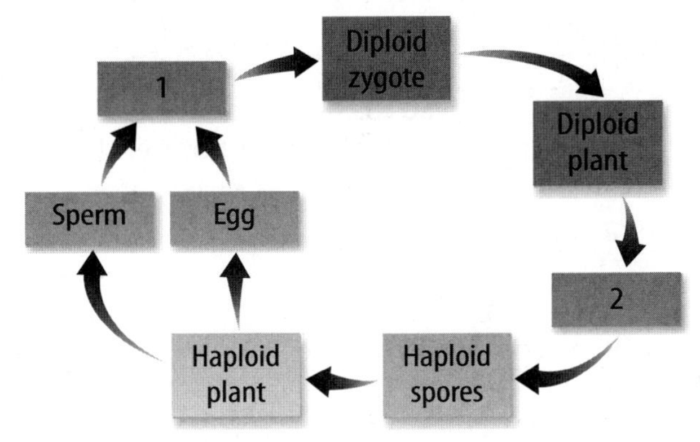

9 Which process occurs at the stage marked 2 on the plant life cycle diagram above? **SC.7.L.16.3**

- Ⓐ asexual reproduction
- Ⓑ fertilization
- Ⓒ meiosis
- Ⓓ mitosis

Use the image below to answer questions 10–12.

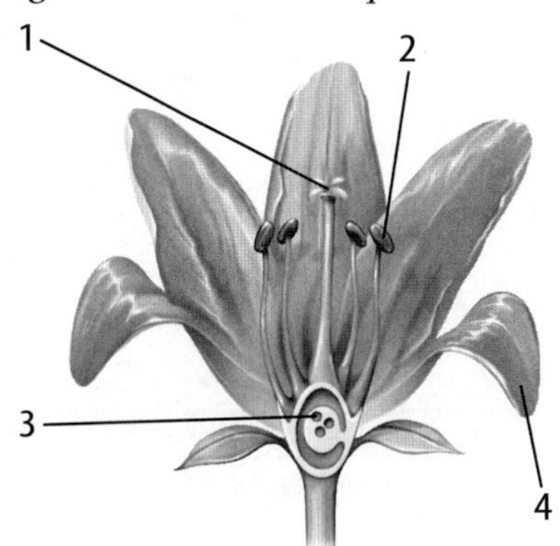

10 What is the name of structure number 3? **LA.6.2.2.3**

- Ⓕ anther
- Ⓖ ovule
- Ⓗ petal
- Ⓘ pistil

11 Where is pollen produced? **LA.6.2.2.3**

- Ⓐ 1
- Ⓑ 2
- Ⓒ 3
- Ⓓ 4

12 What part of a flower becomes a seed? **LA.6.2.2.3**

- Ⓕ 1
- Ⓖ 2
- Ⓗ 3
- Ⓘ 4

NEED EXTRA HELP?

If You Missed Question...	1	2	3	4	5	6	7	8	9	10	11	12
Go to Lesson...	1	1	3	2	1	1	3	1	3	3	3	3

Name ______________________________ Date ______________

Multiple Choice *Bubble the correct answer.*

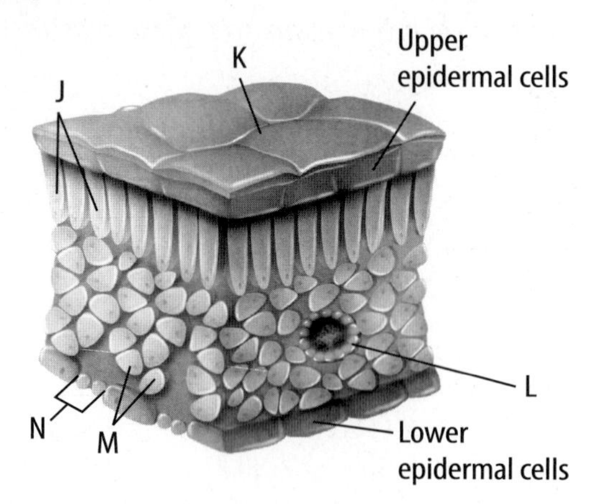

1. This cross section of a leaf shows the cells that make up a leaf's structure. Which two types of cells contain the organelles where photosynthesis occurs? **SC.8.L.18.1**

Ⓐ J and K

Ⓑ J and M

Ⓒ K and L

Ⓓ K and N

2. Chlorophyll is green because it absorbs **SC.8.L.18.1**

Ⓕ carbon dioxide.

Ⓖ water molecules.

Ⓗ all colors of light except green.

Ⓘ green light, and no other colors.

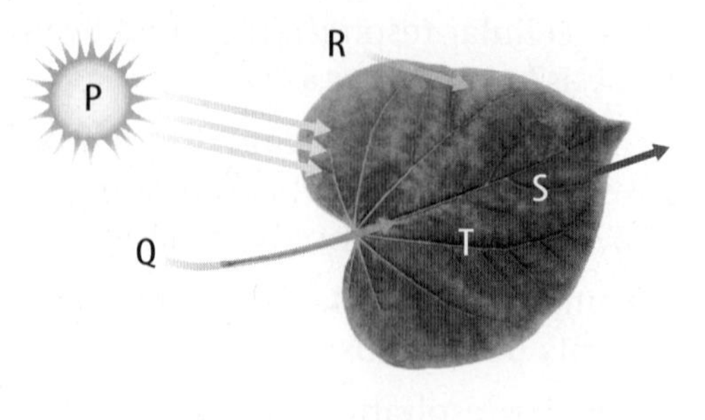

3. Based on the labels in the illustration above, which formula describes photosynthesis? **SC.8.L.18.1**

Ⓐ P + Q + R + chlorophyll → S + T

Ⓑ P + Q + R → S + chlorophyll

Ⓒ P + Q + R + T → S + chlorophyll

Ⓓ Q + R + T + chlorophyll → S

4. A plant is placed inside a dark closet for several days. Which part of photosynthesis can occur without light? **SC.8.L.18.1**

Ⓕ making oxygen

Ⓖ making sugar

Ⓗ making carbon dioxide

Ⓘ making sunlight energy

Name ______________________ Date ____________

Benchmark Mini-Assessment Chapter 10 • Lesson 2

Multiple Choice *Bubble the correct answer.*

1. Which type of plant response is shown in the illustration above? **SC.7.N.1.1**

Ⓐ gravitropism

Ⓑ photoperiodism

Ⓒ phototropism

Ⓓ thigmotropism

2. Which plant hormone causes leaves to grow toward light? **SC.7.N.1.1**

Ⓕ auxin

Ⓖ cytokinin

Ⓗ ethylene

Ⓘ gibberellin

	Plant Growth	Flower Size	Days Flowers Lasted
Solution R	12 cm	7 cm	4
Solution S	10 cm	4 cm	4
Solution T	10 cm	4 cm	6
Solution U	3 cm	4 cm	1

3. For two weeks, different solutions were applied to four identical plants. Each solution contained one kind of plant hormone. Based on the information in the chart above, which solution most likely contained ethylene? **MA.6.A.3.6**

Ⓐ Solution R

Ⓑ Solution S

Ⓒ Solution T

Ⓓ Solution U

4. Which plant hormones increase the rate of cell division and cell elongation? **SC.7.N.1.1**

Ⓕ auxins

Ⓖ cytokinins

Ⓗ ethylenes

Ⓘ gibberellins

Name ______________________ Date ______________

Multiple Choice *Bubble the correct answer.*

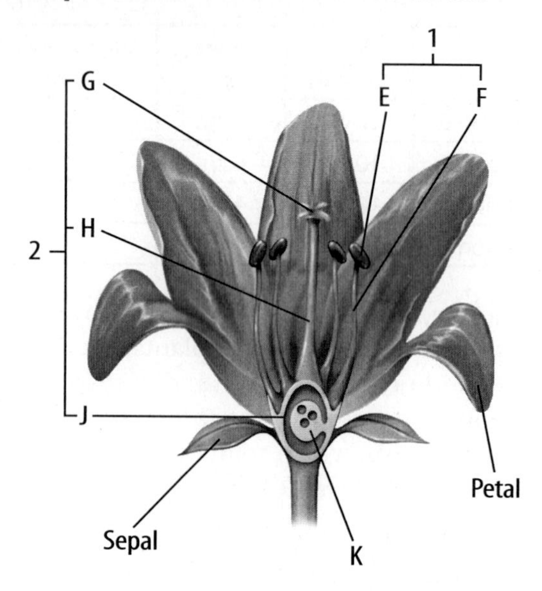

1. In order to reduce the amount of pollen a flower produces, which part would you remove from the flower in the image above? **SC.7.L.16.3**

Ⓐ E

Ⓑ H

Ⓒ J

Ⓓ K

2. To produce seedless fruit, which part of the flower is allowed to develop into fruit, and which part is NOT allowed to develop into seed? **SC.7.L.16.3**

Ⓕ The filament develops into fruit; the anther does not develop into seed.

Ⓖ The ovary develops into fruit; the ovule does not develop into seed.

Ⓗ The pistil develops into fruit; the stamen does not develop into seed.

Ⓘ The stamen develops into fruit; the pistil does not develop into seed.

3. A daylily's underground stems grow to produce several new, genetically identical daylily plants. This is an example of **LA.6.2.2.3**

Ⓐ alternation of fertilization.

Ⓑ alternation of generations.

Ⓒ asexual reproduction.

Ⓓ sexual reproduction.

	Flowers	Seeds	Leaves	Haploid Plants
Plant R			X	X
Plant S		X	X	
Plant T	X	X	X	

4. The table above shows characteristics of three different plants. Plant S is most likely **SC.7.L.16.3**

Ⓕ an angiosperm.

Ⓖ a gymnosperm.

Ⓗ a houseleek.

Ⓘ a housemoss.

Name ____________________ Date ____________

Notes

Name ______________________ Date ____________

Notes

Animal or Not?

Listed below are different types of organisms. Put an *X* next to each organism you think is classified as an animal.

____ elephant	____ whale	____ snail
____ sea anemone	____ earthworm	____ jellyfish
____ mouse	____ sea sponge	____ toad
____ hummingbird	____ lizard	____ lobster
____ slug	____ sea urchin	____ clam
____ spider	____ centipede	____ tapeworm
____ eel	____ honeybee	____ starfish

Explain your thinking about animals.

FLORIDA
Chapter 11
Animal
DIVERSITY

FLORIDA BIG IDEAS

1 **The Practice of Science**
2 **The Characteristics of Scientific Knowledge**
3 **The Role of Theories, Laws, Hypotheses, and Models**
14 **Organization and Development of Living Organisms**
15 **Diversity and Evolution of Living Organisms**

The Big Idea

Think About It!

What are the major groups of animals, and how do they differ?

What are the blue structures attached to these underwater rocks? Did someone spill paint on a clump of algae? This is a colony of animals called tunicates (TEW nuh kayts), also known as sea squirts. Believe it or not, they are classified in the same phylum as humans.

1. What characteristics do you think tunicates have in common with other animals?

2. How do you think tunicates differ from other animals?

Get Ready to Read

What do you think about animal diversity?

Before you read, decide if you agree or disagree with each of these statements. As you read this chapter, see if you change your mind about any of the statements.

	AGREE	DISAGREE
1 All animals digest food.	☐	☐
2 Corals and jellyfish belong to the same phylum.	☐	☐
3 Most animals have backbones.	☐	☐
4 All worms belong to the same phylum.	☐	☐
5 All chordates have backbones.	☐	☐
6 Reptiles have three-chambered hearts.	☐	☐

There's More Online!
Video • Audio • Review • ⓘLab Station • WebQuest • Assessment • Concepts in Motion • Multilingual eGlossary

Lesson 1

What defines an ANIMAL?

ESSENTIAL QUESTIONS

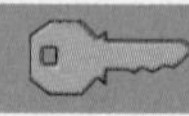 What characteristics do all animals have?

 How are animals classified?

Vocabulary

vertebrate p. 408
invertebrate p. 408
radial symmetry p. 409
bilateral symmetry p. 409
asymmetry p. 409

Florida NGSSS

LA.6.2.2.3 The student will organize information to show understanding (e.g., representing main ideas within text through charting, mapping, paraphrasing, summarizing, or comparing/contrasting);

LA.6.4.2.2 The student will record information (e.g., observations, notes, lists, charts, legends) related to a topic, including visual aids to organize and record information and include a list of sources used;

SC.6.L.15.1 Analyze and describe how and why organisms are classified according to shared characteristics with emphasis on the Linnaean system combined with the concept of Domains.

SC.6.N.1.5 Recognize that science involves creativity, not just in designing experiments, but also in creating explanations that fit evidence.

SC.6.N.2.3 Recognize that scientists who make contributions to scientific knowledge come from all kinds of backgrounds and possess varied talents, interests, and goals.

SC.7.N.1.3 Distinguish between an experiment (which must involve the identification and control of variables) and other forms of scientific investigation and explain that not all scientific knowledge is derived from experimentation.

Inquiry Launch Lab

SC.6.N.1.5

10 minutes

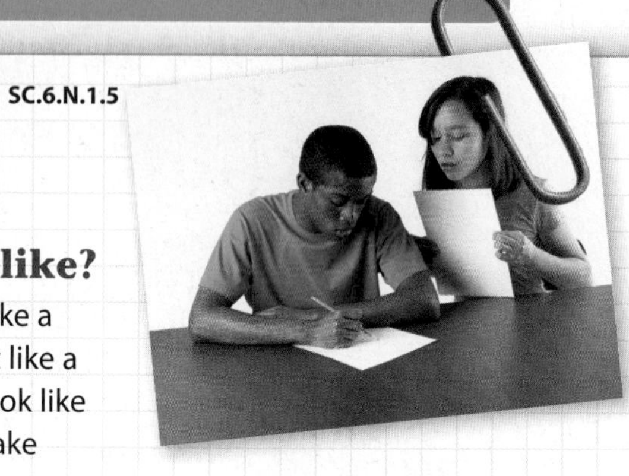

What does an animal look like?

Have you ever seen an animal that looks like a vase? How about an animal that looks just like a twig? There are even some animals that look like alien spaceships. The forms that animals take are almost as varied as your imagination.

Procedure

1. Look at a **photograph of an animal.** Without showing the picture to your partner, describe the animal in as much detail as possible.
2. Have your partner draw the animal using your description as a guide.
3. Compare the drawing to the photograph.

Think About This

1. Could someone looking at the drawing identify it as the same animal in the photograph? Why or why not?

2. Key Concept What characteristics do you think you and the animal you described have in common?

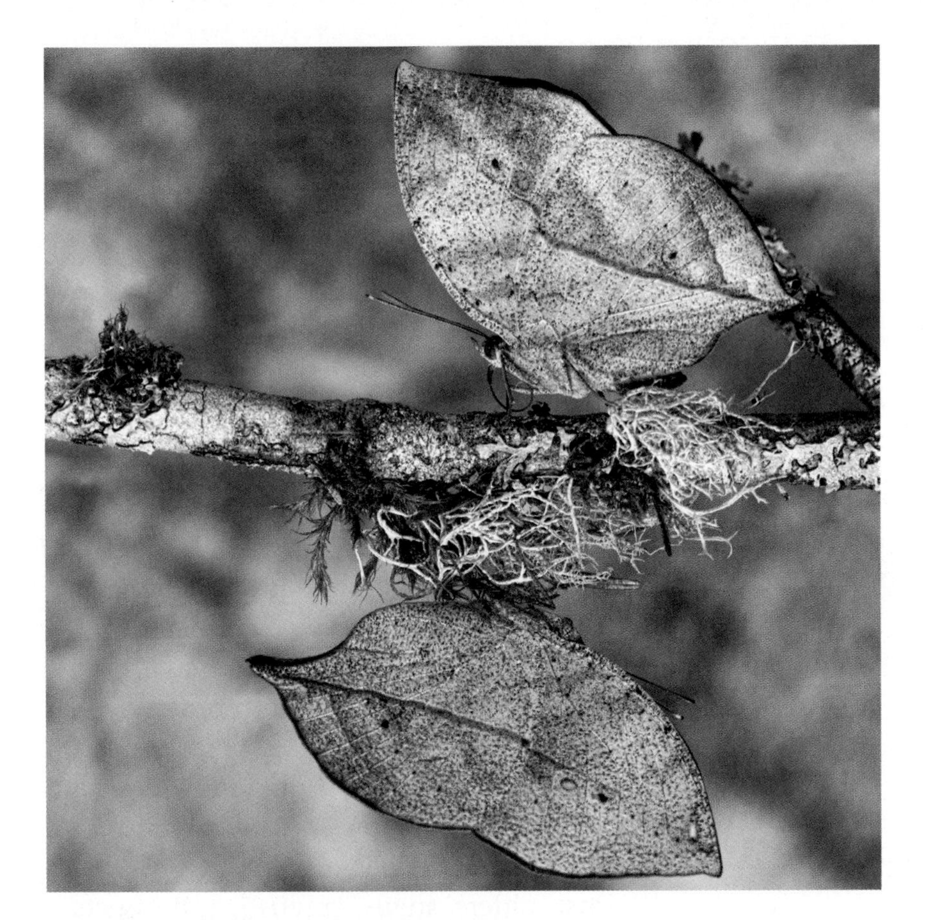

Inquiry A Pair of Leaves?

1. Although these organisms might look like leaves, they are actually butterflies, a type of animal. All animals share some characteristics that leaves do not. What makes leaves and butterflies different?

Animal Characteristics

When you look at an animal, what do you expect to see? Would you expect every animal to have legs and eyes? Ants and birds have legs, but the snake in **Figure 1** does not. Snails, spiders, and many other animals have eyes, but jellyfish do not. Although animals have many traits that make them unique, all animals have certain characteristics in common. Members of the Kingdom Animalia have the following characteristics:

- Animals are multicellular and eukaryotes.
- Animal cells are specialized for different functions, such as digestion, reproduction, vision, or taste.
- Animals have a protein, called collagen (KAHL uh juhn), that surrounds the cells and helps them keep their shape.
- Animals get energy for life processes by eating other organisms.
- Animals, such as the snake in **Figure 1,** digest their food.

In addition to the characteristics above, most animals reproduce sexually and are capable of movement at some point in their lives.

Active Reading 2. **List** What characteristics do all animals have?

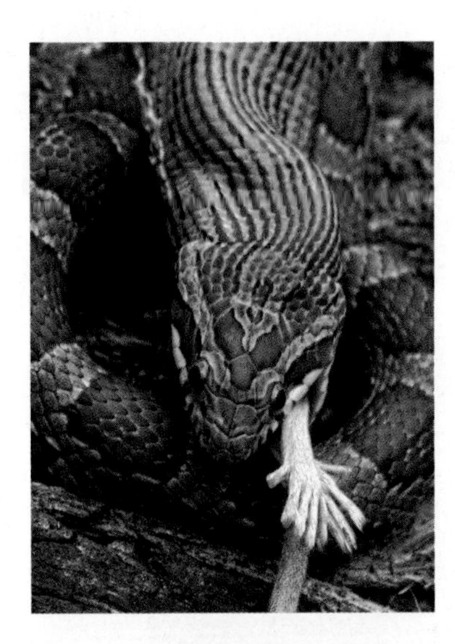

Figure 1 The snake began digesting its prey even before it finished swallowing.

Animal Classification

Scientists have described and named more than 1.5 million species of animals. Every year, thousands more are added to that number. Many scientists estimate that Earth is home to millions of animal species that no one has discovered—at least, not yet. What might happen if you discovered an animal no one else had ever seen? How would you begin to classify it?

Figure 2 A backbone, or spine, is part of a vertebrate's internal skeleton.

Vertebrates and Invertebrates

You could start classifying an animal by finding out if the animal has a backbone. Animals can be grouped into two large categories: vertebrates (VUR tuh brayts) and invertebrates (ihn VUR tuh brayts). *A* **vertebrate** *is an animal with a backbone.* Fish, humans, and the lizard shown in **Figure 2** are examples of vertebrates. *An* **invertebrate** *is an animal that does not have a backbone.* Worms, spiders, snails, crayfish, and insects are examples of invertebrates. Invertebrates make up most of the animal kingdom—about 95 percent.

Active Reading **3. Contrast** What is the difference between a vertebrate and an invertebrate?

Inquiry SC.7.N.1.3, SC.6.L.15.1

LAB STATION **Try It!**

MiniLab ***What is this animal?*** at connectED.mcgraw-hill.com

Apply It! After you complete the lab, answer these questions.

1. What other characteristics could you have used to classify the organisms?

2. How would you change the dichotomous key to include a classmate?

Animal Symmetry

Figure 3 Animals can be classified as having radial symmetry, bilateral symmetry, or asymmetry.

Active Reading **4. Identify** (Circle) the name of the type of symmetry that a bird has.

Radial symmetry

Asymmetry

Bilateral symmetry

Symmetry

Another step you could take to classify an animal is to determine what kind of symmetry it has. As shown in **Figure 3,** symmetry describes an organism's body plan. Symmetry can help identify the phylum to which an animal belongs.

An animal with **radial symmetry** *can be divided into two parts that are nearly mirror images of each other anywhere through its central axis.* A radial animal has a top and a bottom but no head or tail. It can be divided along more than one plane and still have two nearly identical halves. Examples include jellyfish, sea stars, and sea anemones.

An animal with **bilateral symmetry** *can be divided into two parts that are nearly mirror images of each other.* Examples include birds, mammals, reptiles, worms, and insects.

An animal with **asymmetry** *cannot be divided into any two parts that are nearly mirror images of each other.* An asymmetrical animal, such as the sponge in **Figure 3,** does not have a symmetrical body plan.

Word Origin

bilateral

from Latin *bi-*, means "two," and *latus*, means "side"

Active Reading **5. Explain** What is bilateral symmetry?

Grey-faced sengi

Vole

Elephant

Figure 4 The grey-faced sengi was first observed in Africa in 2006. Sengis look like voles, but molecular evidence shows that they are more closely related to elephants.

Molecular Classification

Molecules such as DNA, RNA, and proteins in an animal's cells also can be used for classification. For example, scientists can compare the DNA from two animals to determine if they are related. The more similar the DNA, the more closely the animals are related.

Molecular classification has led to new discoveries about relationships among species. Scientists used to classify the grey-faced sengi shown in **Figure 4** as a close relative of shrews and voles. Recently, molecular evidence has shown that sengis are more closely related to elephants and aardvarks.

Active Reading

FOLDABLES® LA.6.2.2.3

Make a small horizontal four-door book. Leave a 1-cm space between the tabs. Draw arrows and label the book as shown. Use it to record your notes about the classification of animals.

With or Without a Backbone

Body Symmetry

How Animals Are Classified

Cellular Characteristics

Body Structure

6. **NGSSS Check** Find How are animals classified? SC.6.L.15.1

Major Phyla

Scientists classify the members of the animal kingdom into as many as 35 phyla (singular, phylum). The nine major phyla, shown in **Figure 5,** contain 95–99 percent of all animal species. Animals belonging to the same phylum have similar body structures and other characteristics. For example, all sponges (the phylum Porifera [puh RIH fuh ruh]) have asymmetry, and their cells do not form tissues. Only one animal phylum, Chordata (kor DAH tuh), contains vertebrates, also shown in **Figure 5.** The other major phyla contain only invertebrates.

Kingdom Animalia

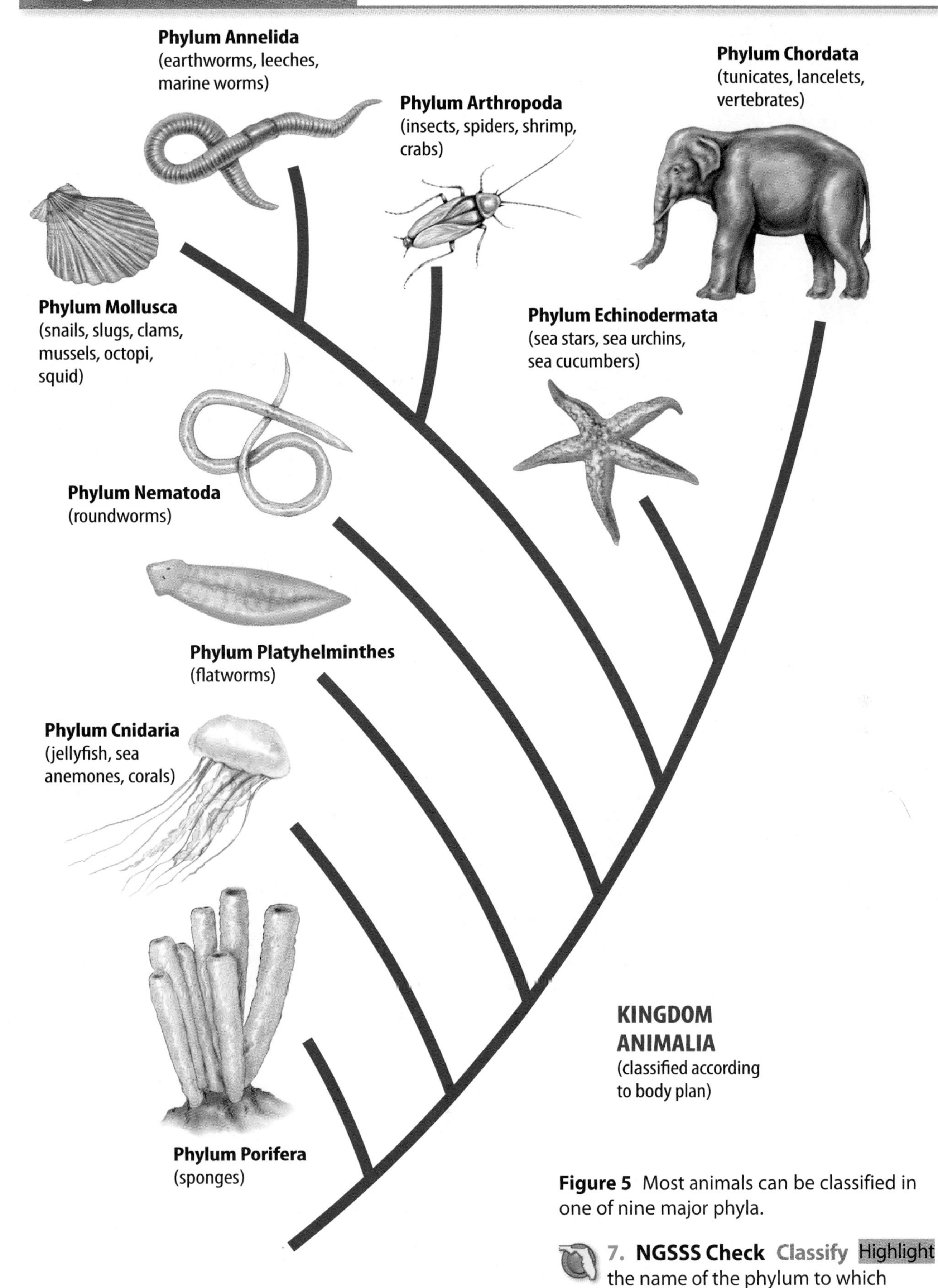

Figure 5 Most animals can be classified in one of nine major phyla.

7. NGSSS Check **Classify** Highlight the name of the phylum to which butterflies belong. SC.6.L.15.1

Lesson Review 1

Visual Summary

All animals share a series of characteristics.

Animals can be classified in several ways.

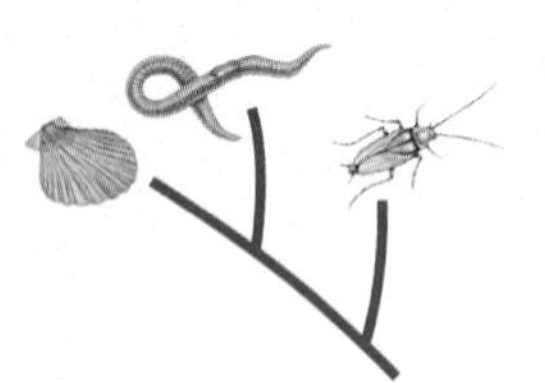

Animal classifications are always changing based on advanced technology.

Use Vocabulary

1. **Distinguish** between vertebrate and invertebrate animals.

2. **Compare and contrast** radial symmetry and bilateral symmetry.

Understand Key Concepts

3. **List** the characteristics that all animals have in common. **SC.6.L.15.1**

4. Which characteristic applies to a horse?
   - (A) asymmetry
   - (B) invertebrate
   - (C) spherical
   - (D) vertebrate

5. **Summarize Information** Use the graphic organizer below to summarize the ways animals can be separated into groups. **LA.6.2.2.3**

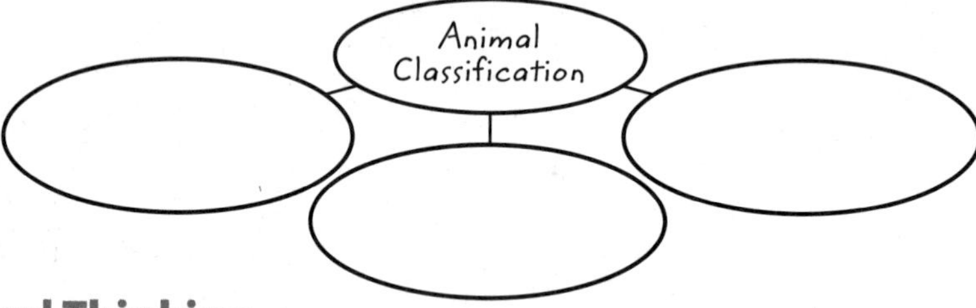

Critical Thinking

6. **Develop** a series of instructions that could be used to determine if an animal should be classified in the phylum Arthropoda, Echinodermata, or Chordata. **SC.6.L.15.1**

7. **Analyze** how the classification of the grey-faced sengi changed over time. How might technological advances change how other animals are classified?

CAREERS in SCIENCE

A Family Tree for Bats

Meet Nancy Simmons, a taxonomist who identifies bats.

When most people are going to bed, taxonomist Nancy Simmons is going to work. She's off to capture bats in a dense rain forest of South America. Because bats are most active at night, she and her team from the American Museum of Natural History work from dusk until dawn. They must capture, identify, and release the bats while it's dark.

Taxonomists study animals to see how they are related to each other and use that information to classify them. To classify a bat, Simmons carefully examines its body. She looks at characteristics such as wing size, fur color, and the shape of the bat's teeth. These characteristics help her classify each bat into a family or a group that shares physical features and behaviors.

In 1999 Dr. Simmons added a new member to the bat family tree. In the rain forest in Peru, her team discovered a species they named *Micronycteris matses,* the Matses' big-eared bat. Like other species in its genus, *M. matses* is small with large round ears, a long snout, and a fold of skin on its nose called a nose-leaf. *M. matses* is unique, however, because of its combination of dark brown fur, medium body size, small bottom front teeth, and short fur around its ears.

Dr. Simmons is looking for links between *M. matses* and other bat species. She compares their bodies, behavior, and even their DNA. Her goal is to create a family tree for all bats. With over a thousand species of bats worldwide, Dr. Simmons has plenty of work still to do.

All Kinds of Bats

Bats live on every continent except Antarctica, in areas ranging from tropical rain forests to chilly mountaintops. They also have an amazing variety of shapes and sizes. With over 1,100 species, bats make up one-fifth of the world's mammals.

Simmons holds a bat that she caught in her net.

▲ This species is the largest in the New World—it weighs about 150 g.

It's Your Turn

RESEARCH Investigate a species of bat in your area. Record where it lives in the environment and its characteristics, such as wingspan, fur color, and weight. With a partner, compare your bats. What do they have in common? What is different?

Lesson 2

Invertebrate PHYLA

ESSENTIAL QUESTIONS

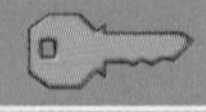
What are the characteristics of invertebrates?

How do the invertebrate phyla differ?

Vocabulary

exoskeleton p. 419

appendage p. 419

Florida NGSSS

LA.6.2.2.3 The student will organize information to show understanding (e.g., representing main ideas within text through charting, mapping, paraphrasing, summarizing, or comparing/contrasting);

SC.6.L.15.1 Analyze and describe how and why organisms are classified according to shared characteristics with emphasis on the Linnaean system combined with the concept of Domains.

SC.6.N.1.5 Recognize that science involves creativity, not just in designing experiments, but also in creating explanations that fit evidence.

SC.7.N.1.3 Distinguish between an experiment (which must involve the identification and control of variables) and other forms of scientific investigation and explain that not all scientific knowledge is derived from experimentation.

SC.7.N.1.5 Describe the methods used in the pursuit of a scientific explanation as seen in different fields of science such as biology, geology, and physics.

SC.6.N.1.5

Launch Lab

10 minutes

What does an invertebrate look like?

Some invertebrates have features that are similar to yours, such as eyes and legs. Others have little in common with you. What do you see when you look at invertebrates close-up?

Procedure 

1. Read and complete a lab safety form.
2. Examine a **collection of invertebrates,** and record your observations.
3. Use a **magnifying lens** to further examine the invertebrates. Record any additional observations.
4. Make a Venn diagram below to compare similarities and contrast differences among the invertebrates.

Data and Observations

Think About This

1. Which two invertebrates were the most dissimilar? Why?

2. What details did you see using a magnifying lens that you missed by looking just with your eyes?

3. Key Concept What characteristics do you think all the invertebrates you looked at have in common?

Inquiry How did it get there?

1. This octopus is alive and got inside the bottle by slowly pushing its soft, flexible body inside. Like many invertebrates, an octopus does not have a skeleton made of bone or other hard structures. How do you think an octopus moves without a skeleton?

Characteristics of Invertebrates

Can you imagine living without a backbone? Most animals do just that. As you have read, invertebrates are animals that lack a backbone. In most cases, invertebrates have no **internal** structures to help support their bodies. They also tend to be smaller and move more slowly than vertebrates. As shown in **Figure 6,** more than 95 percent of all animal species that have been recorded are invertebrates.

Academic Vocabulary
internal
(adjective) existing inside something

You probably could recognize a jellyfish or a clam if you saw one. What about an anemone or a sea cucumber? Invertebrates are a diverse group. Their physical characteristics range from the simple structures of sponges and jellyfish to the more complex bodies of worms, snails, and insects. Each invertebrate phylum contains animals with similar body plans and physical characteristics.

2. NGSSS Check Characterize What are the characteristics of invertebrates? SC.6.L.15.1

Invertebrate and Vertebrate Species

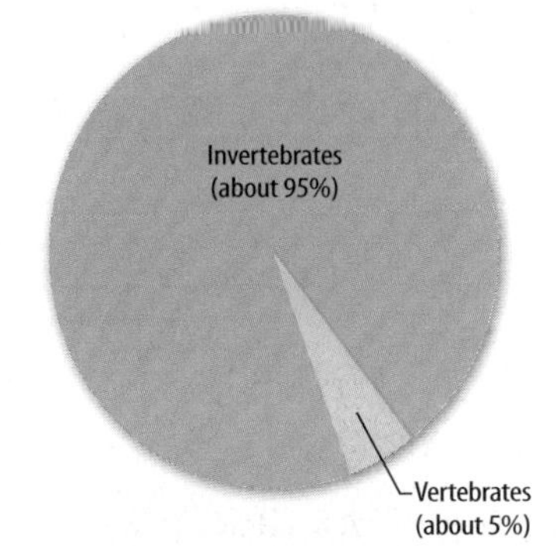

Figure 6 Invertebrates make up more than 95 percent of all living species on Earth.

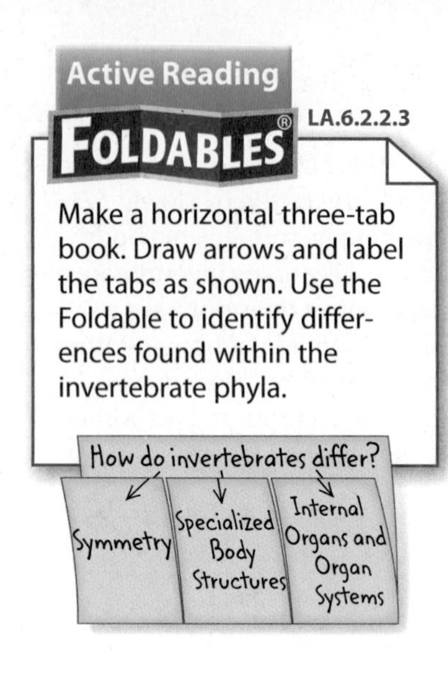

Sponges and Cnidarians

Have you ever seen a natural sponge from the Florida coast? The simplest of the invertebrates are the sponges, which belong to the phylum Porifera. All sponges share several characteristics.

All sponges are asymmetrical and have no tissues, organs, or organ systems. Their cells are specialized for capturing food, digestion, and reproduction. Other cells provide support inside the layers of the sponge. All sponges live in water, and most species live in ocean environments.

The phylum Cnidaria (ni DAR ee uh) includes jellyfish, sea anemones, hydras, and corals. Cnidarians, such as the sea anemone shown in **Figure 7,** differ from all other animals based on their unique characteristics.

Cnidarians have no organs or organ systems, but, unlike sponges, they have radial symmetry. They have a single body opening surrounded by tentacles. Simple tissues, including muscles, nerves, and digestive tissue, enable cnidarians to survive by moving, reacting to stimuli, and digesting food. They have specialized cells, called nematocysts (NE mah toh sihsts), that are used for defense and capturing food. Similar to sponges, most species of cnidarians live in ocean environments, and all live in water.

3. NGSSS Check Compare What characteristics do poriferans and cnidarians share? SC.6.L.15.1

____

____

____

Figure 7 The tentacles of all cnidarians contain stinging structures for capturing food and defending against predators.

4. NGSSS Check Recall Write in the appropriate descriptions of cnidarians below. SC.6.L.15.1

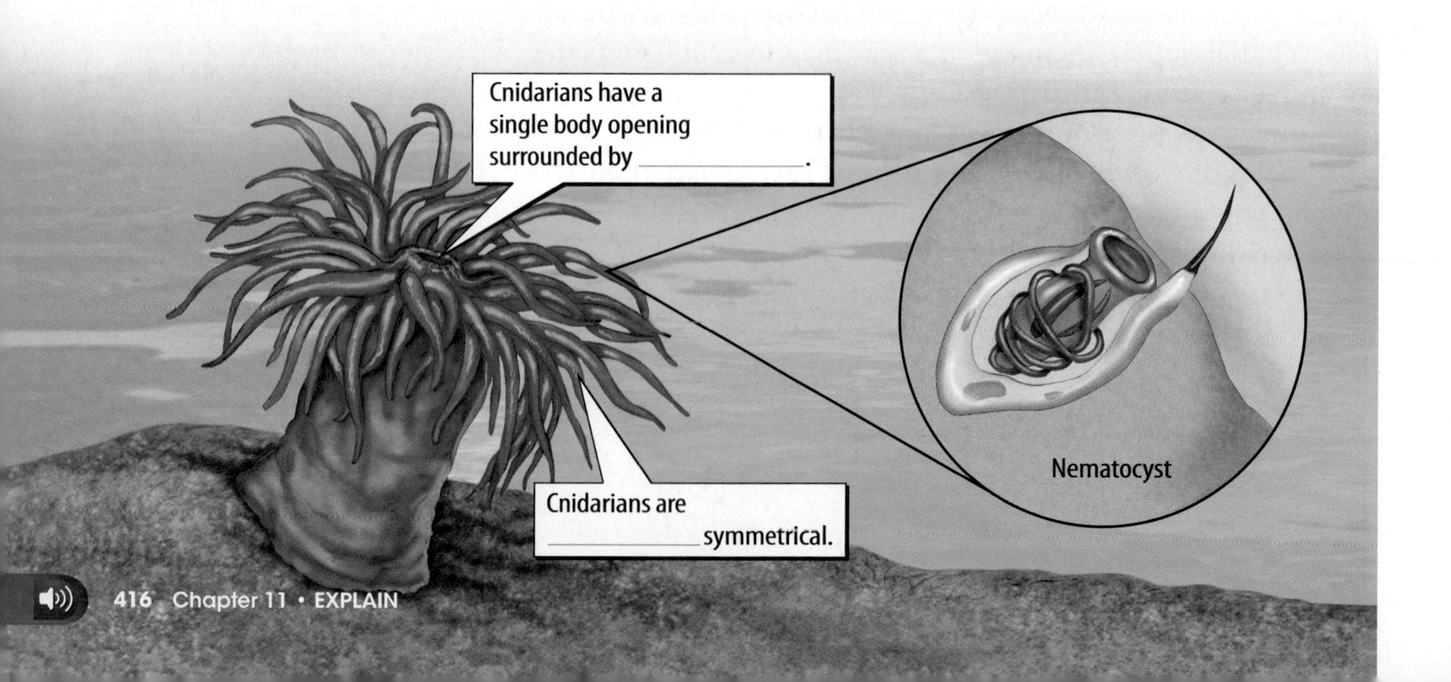

Flatworms and Roundworms

Flatworms are invertebrates that belong to the phylum Platyhelminthes (pla tih hel MIHN theez). All flatworms, including the tapeworm shown in **Figure 8**, have bilateral symmetry with nerve, muscle, and digestive tissues and a simple brain. They have soft and flattened bodies that are usually only a few cells thick. The digestive system of a flatworm has only one opening: a mouth.

Flatworms live in moist environments. Most, like tapeworms, are parasites that live in or on the bodies of other organisms and rely on them for food. Others are free-living, and many live in oceans or other marine environments.

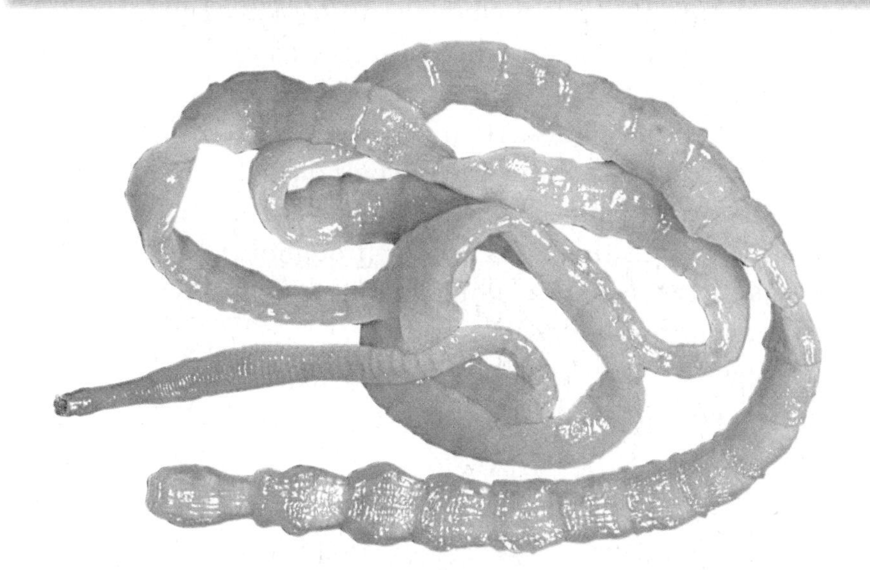

Figure 8 Most flatworm species, including this tapeworm, are parasites. They depend on other organisms for food and a place to live.

Active Reading **5. Describe** How would you describe a flatworm's body?

Roundworms, also called nematodes, belong to the phylum Nematoda (ne muh TOH duh). Roundworms, like flatworms, have bilateral symmetry with nerve, muscle, and digestive tissues and a simple brain. However, unlike flatworms, their bodies are round and covered with a stiff outer covering called a cuticle. A roundworm's digestive system has two openings: a mouth and an anus. Food enters the mouth and is digested as it travels to the anus, where wastes are excreted.

Roundworms live in moist environments. Some species are parasites that live in animals' digestive systems. Free-living roundworms like the one pictured in **Figure 9** eat material such as fecal matter and dead organisms.

6. NGSSS Check Contrast Highlight ways in which flatworms and roundworms differ. SC.6.L.15.1

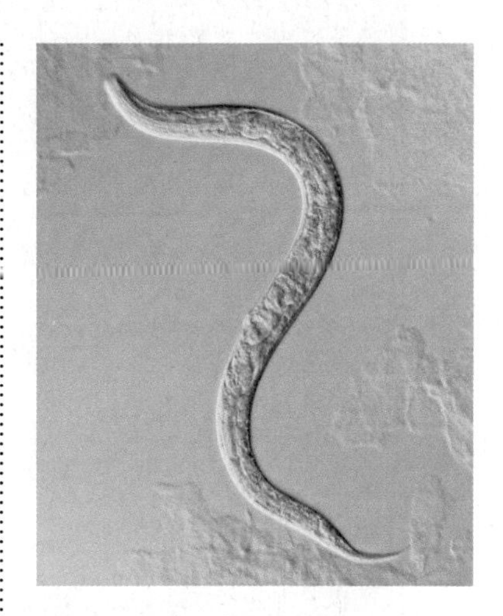

Figure 9 Roundworms are narrow and tapered at both ends. Most species are less than 1 mm long.

Figure 10 Snails have shells that protect their bodies.

Mollusks and Annelids

The phylum Mollusca (mah LUS kuh) includes snails, slugs, clams, mussels, octopi, and squid. All mollusks, including the snail shown in **Figure 10,** have bilateral symmetry. Their bodies are soft, and some species have hard shells that protect their bodies. You might have seen a slug slithering along the ground after a rainstorm. Slugs are one type of mollusk without a shell.

Mollusks have digestive systems with two openings. A body cavity contains the heart, the stomach, and other organs. The mollusk circulatory system contains blood but no blood vessels. Their nervous systems include eyes and other sensory organs as well as simple brains. Members of this phylum must remain wet and live in water or moist environments.

The phylum Annelida includes earthworms, leeches, and marine worms, like the ones living in Florida reefs. Annelid worms, including the one shown in **Figure 11,** have bilateral symmetry and soft bodies. Their bodies consist of repeating segments covered with a thin cuticle. Their digestive systems have two openings. Annelids have circulatory systems that are made up of blood vessels that carry blood throughout the body. Their nervous systems include a simple brain. Annelids live in water or moist environments such as soil.

7. NGSSS Check Compare What do mollusks and segmented worms have in common? SC.6.L.15.1

____

____

____

Figure 11 One characteristic that distinguishes annelids from other worms is their segments.

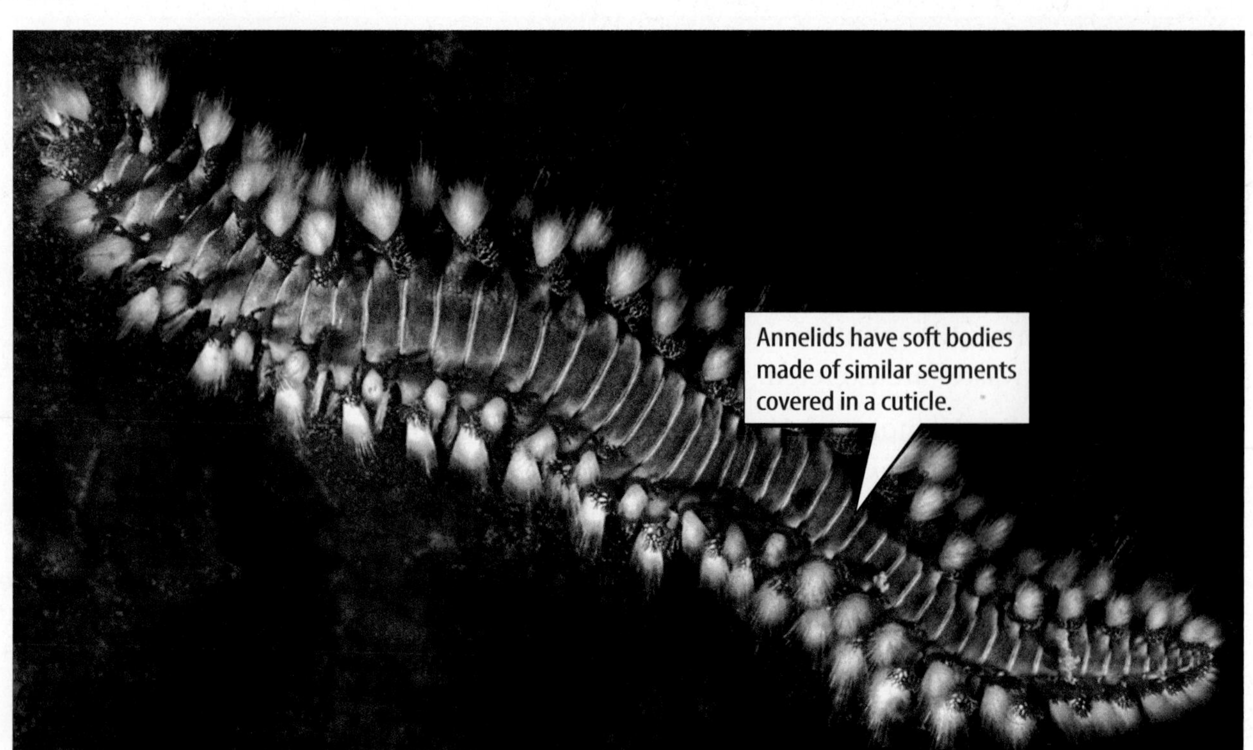

Arthropods

The phylum Arthropoda includes insects, spiders, shrimp, crabs, and their relatives. More species belong to this phylum than all the other animal phyla combined. There are more than 1 million identified species of arthropods.

All arthropods have bilateral symmetry. They also have **exoskeletons**—*thick, hard outer coverings that protect and support animals' bodies.* Arthropods have several pairs of jointed appendages. *An* **appendage** *is a structure, such as a leg or an arm, that extends from the central part of the body.* The body parts of arthropods are segmented and specialized for different functions such as flying and eating. Unlike many of the other animals you have read about so far, arthropods live in almost every environment on Earth.

Word Origin

appendage
from Latin *appendere,* means "to cause to hang from"

Active Reading **8. Explain** Underline the definition of an exoskeleton.

Insects

The largest order of arthropods is the insects, which includes the stag beetle shown in **Figure 12.** All insect species have three pairs of jointed legs, three body segments, a pair of antennae, and a pair of compound eyes. Many species also have one or two pairs of wings.

There are 16 major groups of insects. However, most insect species belong to one of five groups. Beetles form the largest group of insects. About 40 percent of all known species of insects are beetles.

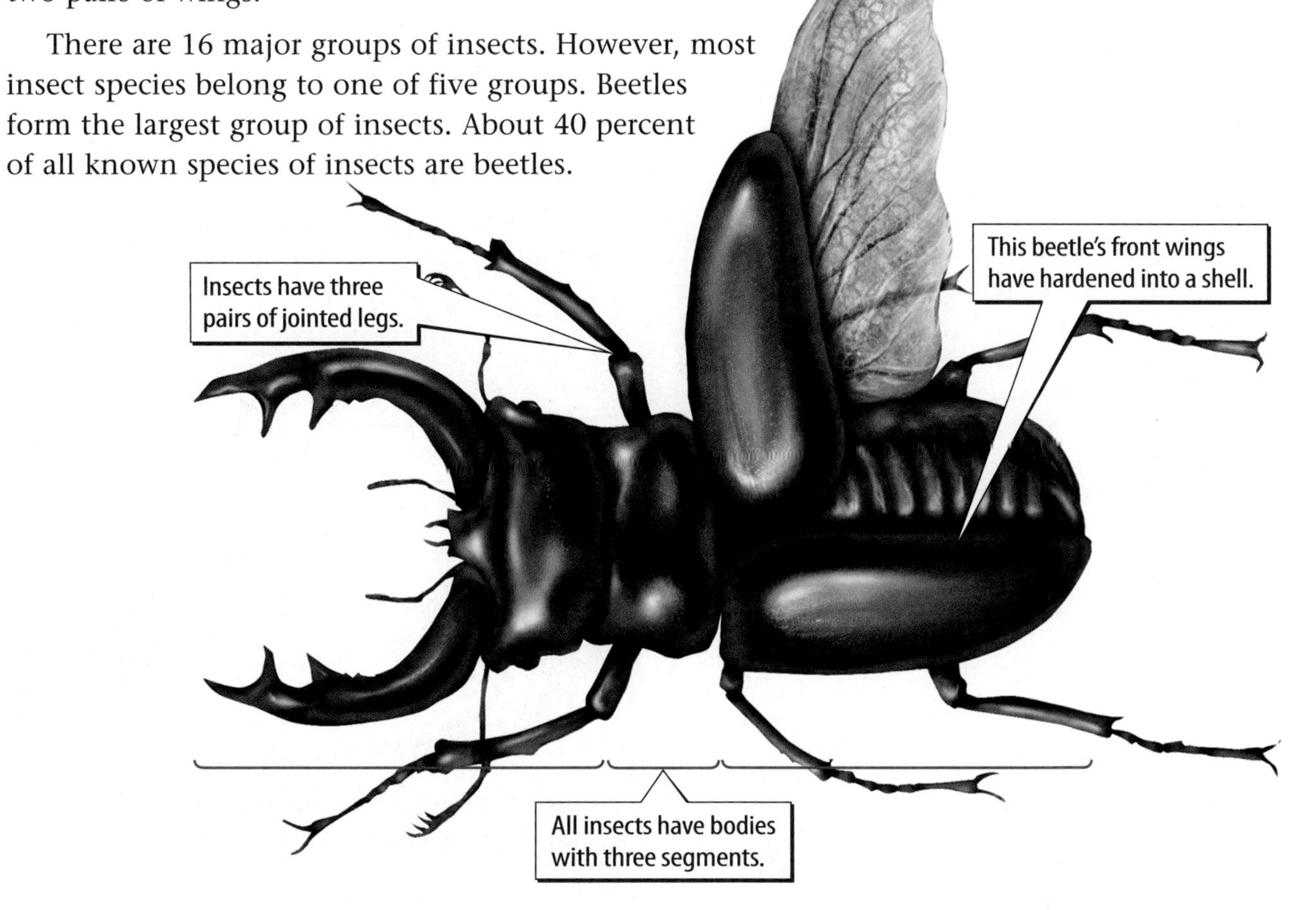

Figure 12 A stag beetle has characteristics common to all insect species.

MiniLab *How does your arm move?* at connectED.mcgraw-hill.com

Apply It!

After you complete the lab, answer these questions.

1. How are arthropod appendages different from mammal appendages, such as your arm?

2. How would your daily life change if you did not have jointed appendages?

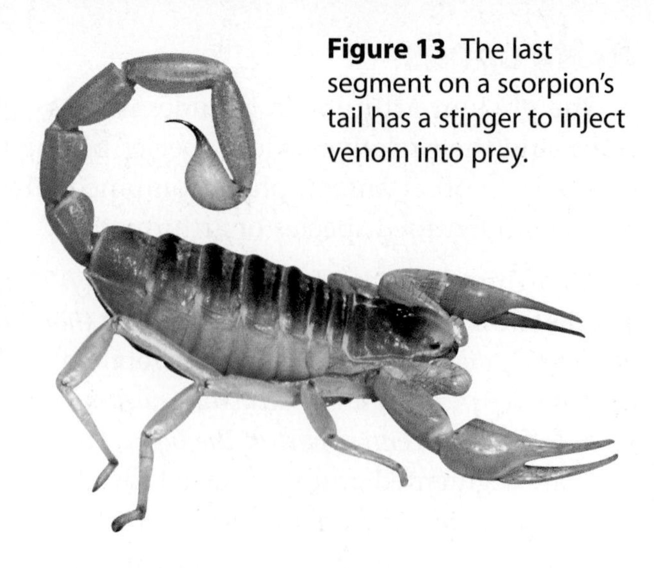

Figure 13 The last segment on a scorpion's tail has a stinger to inject venom into prey.

Arachnids

Spiders, ticks, and scorpions, such as the one shown in **Figure 13,** are arachnids (uh RAK nudz). All arachnids have four pairs of jointed legs and two body segments. They do not have antennae or wings.

Crustaceans

Crabs, shrimp, lobsters, and their close relatives are crustaceans (krus TAY shunz). All crustaceans have one or two pairs of antennae. They also have jointed appendages in the mouth area that are specialized for biting and crushing food. Many people like to eat crustaceans, including lobsters and crabs, such as the one shown in **Figure 14.**

9. **NGSSS Check** **Compare** How do arachnids and crustaceans differ? SC.6.L.15.1

Figure 14 Blue crabs live in many regions, including Florida and much of the East coast.

Echinoderms

The phylum Echinodermata (ih kin uh DUR muh tuh) includes sea stars, sea cucumbers, and sea urchins, such as the one shown in **Figure 15.** *Echinoderm* (ih KI nuh durm) means "spiny skin." Echinoderms have some unique features that are not in any of the other invertebrate phyla. They also are more closely related to vertebrates than to any other phyla.

All echinoderms have radial symmetry. Unlike any other phyla, echinoderms have hard plates embedded in the skin that support the body. Thousands of small, muscular, fluid-filled tubes, called tube feet, enable them to move and feed. They also have complete digestive systems including a mouth and an anus. Echinoderms live only in oceans. However, some can survive out of the water for short periods during low tides.

11. NGSSS Check **Describe** How do the invertebrate phyla differ? SC.6.L.15.1

____

____

____

____

Math Skills MA.6.A.3.6

Use Percentages

More than 95 percent of animals are invertebrates. This means that 95 out of every 100 animals is some type of invertebrate. Out of every 1,000 animal species, 48 are mollusks. What percentage of animal species are mollusks?

Express the information as a fraction.

$$\frac{48}{1{,}000}$$

Change the fraction to a decimal.

$$\frac{48}{1{,}000} = 0.048$$

Multiply by 100 and add a % sign.

$$0.048 \times 100 = 4.8\%$$

Practice

10. Out of 900,000 species of arthropods, 304,200 are beetles. What percentage of arthropods are beetles?

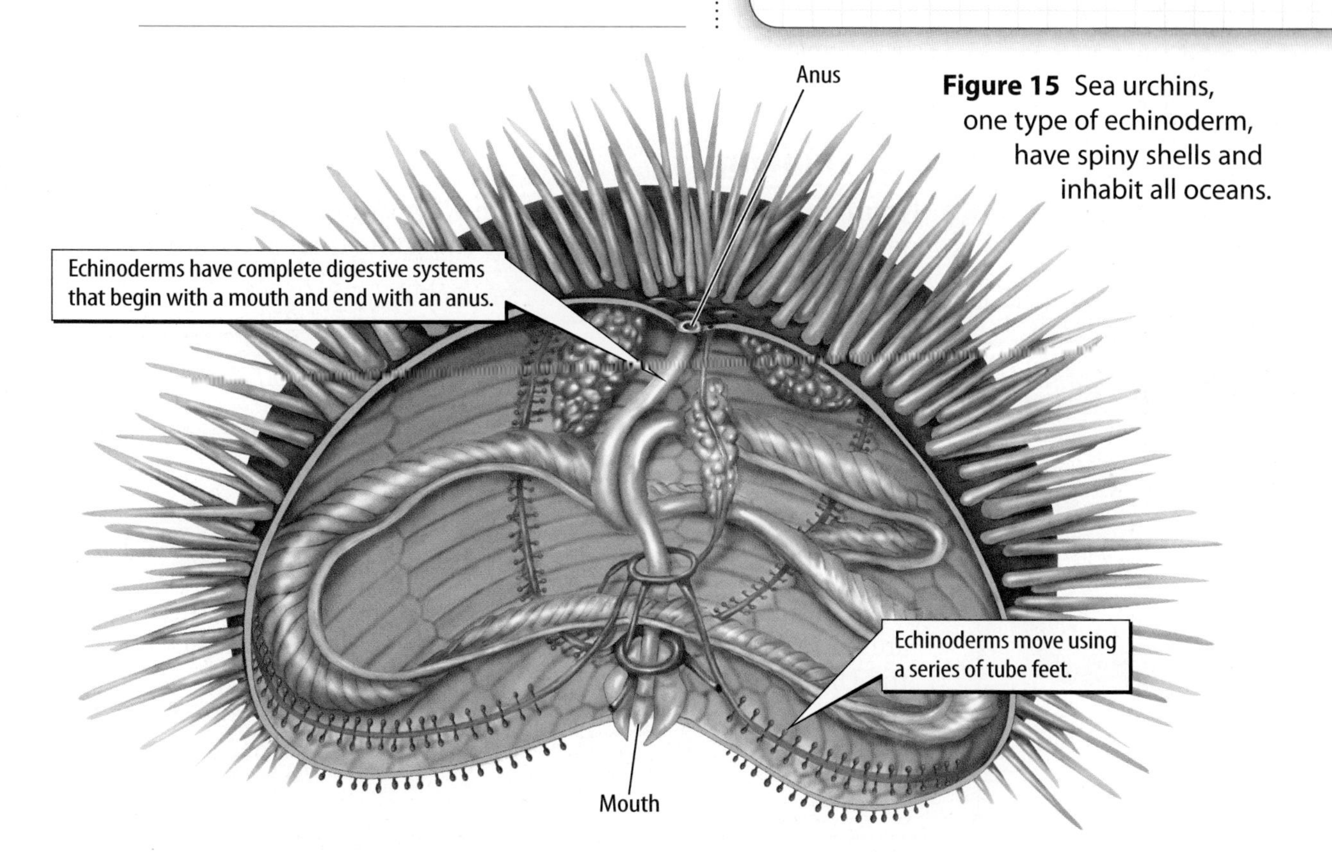

Figure 15 Sea urchins, one type of echinoderm, have spiny shells and inhabit all oceans.

Lesson Review 2

Visual Summary

Most invertebrates have no internal structures that support their bodies.

There are more arthropods than members of any other phyla.

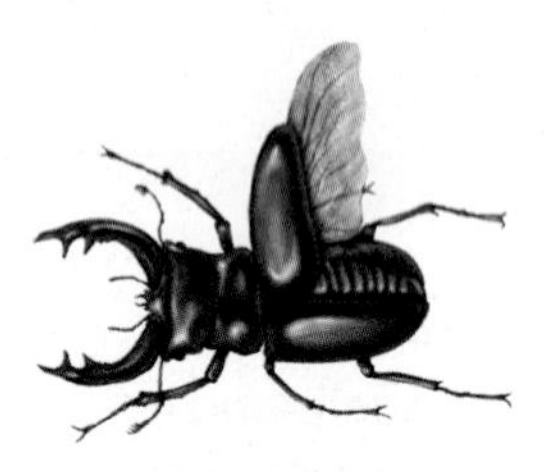

The largest order of arthropods is the insects.

Use Vocabulary

1. **Use the term** *appendage* in a sentence.

Understand Key Concepts

2. Which phylum contains asymmetrical invertebrates that have no tissues? SC.6.L.15.1
   - (A) Annelida
   - (B) Cnidaria
   - (C) Echinodermata
   - (D) Porifera

Interpret Graphics

3. **Summarize Information** Fill in the table below with the features common to the members of each invertebrate phylum. LA.6.2.2.3

Phylum	Characteristics	Example
Porifera		
Cnidaria		
Platyhelminthes		
Annelida		
Nematoda		
Arthropoda		
Echinodermata		

Critical Thinking

4. **Hypothesize** how a digestive system with two openings would enable an organism to absorb more nutrients than a digestive system with one opening.

Math Skills

5. About 11,000 species of Lepidoptera (butterflies and moths) have been identified in the United States. Only 679 of them are butterflies. What percentage of the Lepidoptera species in the United States are butterflies?

Bringing Back Beauty

FOCUS on FLORIDA

The Schaus Swallowtail Returns

Imagine walking through the hardwood hammocks of the upper Florida Keys. You notice a large brown and yellow butterfly flap its wings and fly into the trees. You just saw a Schaus swallowtail, a formerly endangered Florida species that has made a remarkable comeback.

When forests are cleared for homes or businesses, it can destroy plants that animals need for food and shelter. By 1984 so much of Florida's hardwood forests had been cleared that very few Schaus swallowtails were left. Scientists estimated that some areas contained only 50 individuals. In 1992, Hurricane Andrew destroyed even more land in southern Florida. To prevent the loss of more swallowtails, the state of Florida began a conservation program called the Multi-Species Recovery Plan. This program set out to protect existing land, as well as set aside new protected areas. Scientists also bred Schaus swallowtails in a safe environment and released them into Florida forests.

Today, the number of Schaus swallowtails in Florida is growing. Researchers continue to monitor the number of butterflies each year. Despite damage to forests due to weather and human activities, more Schaus swallowtails are surviving to adulthood than in previous years. With the help of protection plans, scientists hope that these beautiful butterflies are headed for a full recovery.

It's Your Turn

DEMONSTRATE How could you customize a butterfly house to protect a Schaus swallowtail from harmful chemicals?

Lesson 3

Phylum CHORDATA

ESSENTIAL QUESTIONS

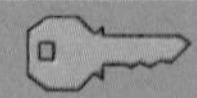 What are the characteristics of all chordates?

 What are the characteristics of all vertebrates?

 How do the classes of vertebrates differ?

Vocabulary

chordate p. 425

notochord p. 425

Florida NGSSS

LA.6.2.2.3 The student will organize information to show understanding (e.g., representing main ideas within text through charting, mapping, paraphrasing, summarizing, or comparing/contrasting);

SC.6.L.15.1 Analyze and describe how and why organisms are classified according to shared characteristics with emphasis on the Linnaean system combined with the concept of Domains.

SC.6.N.1.5 Recognize that science involves creativity, not just in designing experiments, but also in creating explanations that fit evidence.

SC.6.N.2.1 Distinguish science from other activities involving thought.

SC.6.N.2.3 Recognize that scientists who make contributions to scientific knowledge come from all kinds of backgrounds and possess varied talents, interests, and goals.

SC.7.N.1.3 Distinguish between an experiment (which must involve the identification and control of variables) and other forms of scientific investigation explain that not all scientific knowledge is derived from experimentation.

SC.8.N.3.1 Select models useful in relating the results of their own investigations.

SC.8.N.3.1

Launch Lab

10 minutes

How can you model a backbone?

All vertebrates have backbones. Most backbones are made out of a stack of short bones called vertebrae. Some vertebrae are shaped like discs with holes in the center. The largest structure passing through the center of the stack of vertebrae is the spinal cord. Between each of the vertebrae are padlike structures, called discs, that cushion the bones. Try building a model of a backbone.

Procedure

1. Read and complete a lab safety form.
2. Obtain **pasta wheels, circular gummy candies,** and a **chenille stem.** ⚠ Do not eat the lab materials.
3. Assemble the materials to make a model of a backbone.
4. Gently bend and move your model backbone. Observe how the parts move and interact with each other.

Think About This

1. When you bend your model backbone, how are the vertebrae, the discs, and the spinal cord affected?

2. When you compress your model backbone, how are the vertebrae, the discs, and the spinal cord affected?

3. Key Concept How do you think the structure of the backbone provides advantages to the body plan of vertebrates?

One of a Kind?

1. Several different types of animals come to this watering hole to get a drink. These elephants, antelopes, and birds look very different, but they are actually all related. All of these animals belong to the phylum Chordata and share several characteristics. What are some characteristics you think they share?

Characteristics of Chordates

Recall that one way to classify an animal is to check for a backbone and that animals with backbones are called vertebrates. Another way to classify animals is to look for the four characteristics of a chordate (KOR dat). *A* **chordate** *is an animal that has a notochord, a nerve cord, a tail, and structures called pharyngeal* (fer IN jee ul) *pouches at some point in its life.* In vertebrates, these characteristics are present only during embryonic development. *A* **notochord** *is a flexible, rod-shaped structure that supports the body of a developing chordate.* The nerve cord develops into the central nervous system. The pharyngeal pouches are between the mouth and the digestive system.

Most chordates are vertebrates, but the chordates also include two groups of invertebrates: tunicates and lancelets (LAN sluhts), shown in **Figure 16.** Invertebrate chordates are rarely more than a few centimeters long and live in salt water. In vertebrate chordates, such as humans, the notochord develops into a backbone during the growth of an embryo.

2. **NGSSS Check** Characterize What are the characteristics of chordates? SC.6.L.15.1

Figure 16 Lancelets can swim but spend most of their lives almost completely buried in sand.

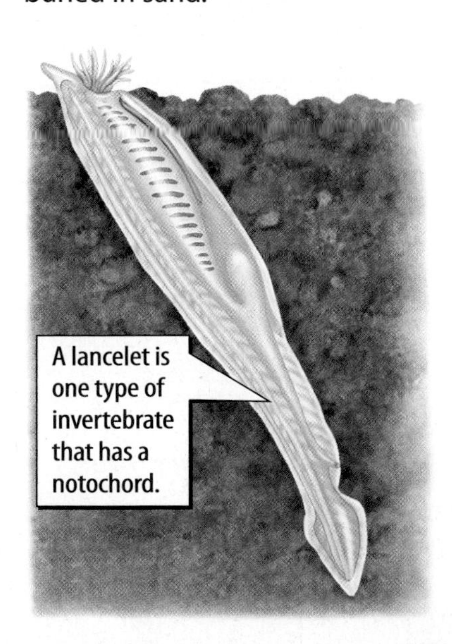

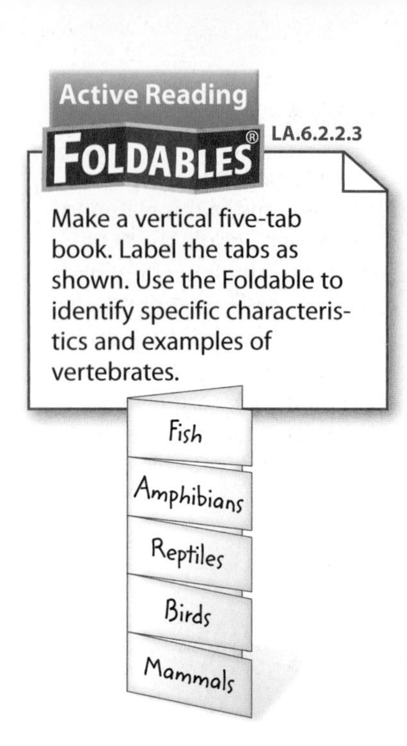

Characteristics of Vertebrates

Recall that all vertebrates have a backbone, also called a spinal column or spine. The backbone is a series of structures that surround and protect the nerve cord, or spinal cord. The spinal cord connects all the nerves in the body to the brain. Bones that form a backbone are called vertebrae (VUR tuh bray). If you gently touch the back of your neck, the bones you feel are some of your vertebrae.

Vertebrates have well-developed organ systems. All vertebrates have digestive systems with two openings, circulatory systems that move blood through the body, and nervous systems that include brains. The five major groups of vertebrates are fish, amphibians, reptiles, birds, and mammals.

3. **NGSSS Check** Characterize What are the characteristics of all vertebrates? SC.6.L.15.1

Fish

Most fish spend their entire lives in water. They have two important characteristics in common: gills for absorbing oxygen gas from water and paired fins for swimming. Fish are grouped into one of three classes.

Hagfish and lampreys are part of a group called jawless fish. Sharks, such as the one shown in **Figure 17,** skates, and rays are cartilaginous fish. They have skeletons made of a tough, fibrous tissue called cartilage (KAR tuh lihj). Both jawless and cartilaginous fish have internal structures made of cartilage.

Trout, guppies, perch, tuna, mackerel, and thousands of other species do not have cartilaginous skeletons. Instead, they have bones and are grouped together as bony fish.

Figure 17 Like all fish, sharks have gills and fins.

Amphibians

Frogs, toads, and salamanders belong to the class Amphibia, as shown in **Figure 18.** Most **amphibians** spend part of their lives in water and part on land. Their bodies change as they grow older. In many species, the young have different body forms than the adults do.

Amphibians have skeletons made of bone and have legs for movement. Their skin is smooth and moist, and their hearts have three chambers. Amphibians lay eggs that do not have hard protective coverings, or shells. Their eggs must be laid in moist environments, such as ponds. Young live in water and have gills; most adults develop lungs and live on land.

Word Origin

amphibian
from Greek *amphi-*, means "of both kinds" and *bios*, means "life"

4. NGSSS Check Contrast How do amphibians differ from fish? SC.6.L.15.1

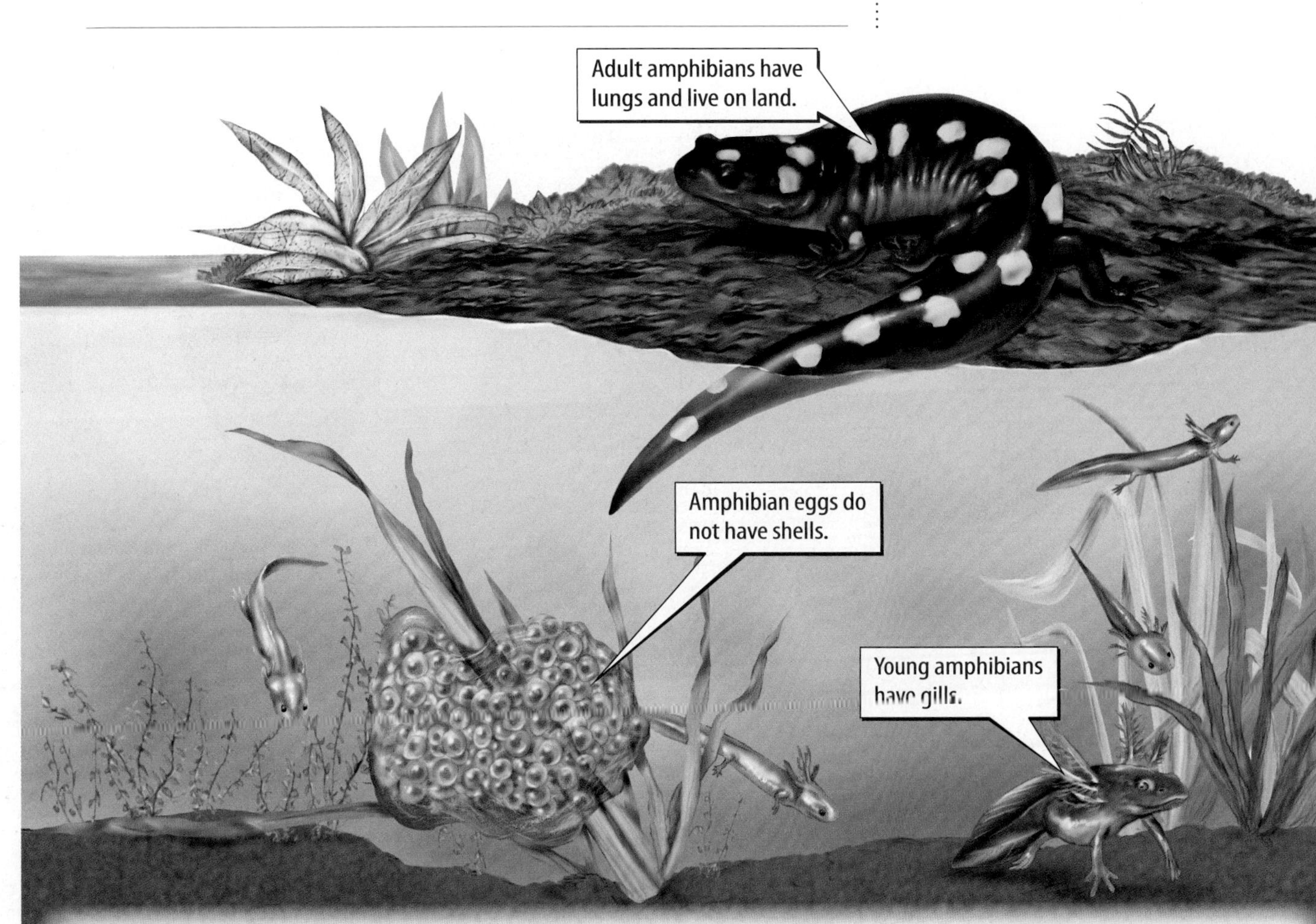

Figure 18 The body forms of many amphibians change as they grow.

5. NGSSS Check Describe How does the body form of this salamander change as it grows? SC.6.L.15.1

Reptiles

Lizards, snakes, turtles, crocodiles, and alligators belong to the class Reptilia. A leopard gecko, one example of this class, is shown in **Figure 19.**

SCIENCE USE V. COMMON USE

scale
Science Use small, flat plate that forms part of an animal's external covering
Common Use an instrument for measuring the mass of an object

All reptiles share several characteristics. Their skin is waterproof and covered in **scales.** Like amphibians, most reptiles have three-chambered hearts. Unlike amphibians, lizards and other reptiles have lungs throughout their lives.

Most reptiles lay fluid-filled eggs with leathery shells. Unlike amphibian eggs, reptile eggs are laid on land rather than in water. Young reptiles do not change form as they mature into adult reptiles.

6. NGSSS Check Contrast How do reptiles differ from amphibians? SC.6.L.15.1

____

____

____

Figure 19 This lizard has a three-chambered heart and lays fluid-filled eggs.

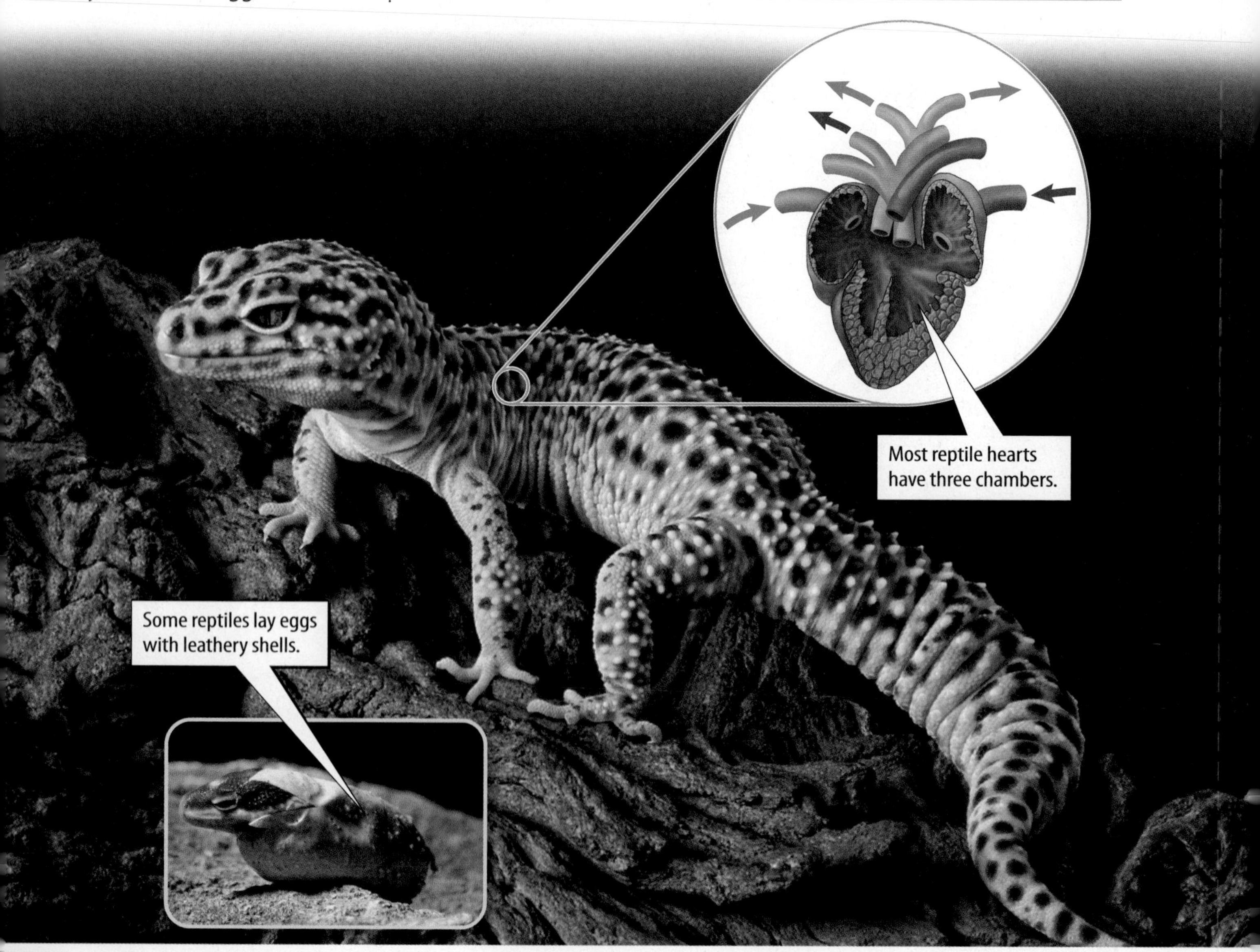

Birds

All birds, including the owl shown in **Figure 20,** are in the class Aves. Many birds make nests to hold their eggs, and many have unique calls or songs.

Birds have lightweight bones. Their skin is covered with feathers and scales. Birds also have two legs and two wings. Many birds can fly, and they have stiff feathers that enable them to move through the air. Birds that spend a lot of time in the water have oil glands that help water roll off their feathers.

Birds have beaks and do not chew their food. Instead, their digestive systems include gizzards, organs that help grind food into smaller pieces. Their circulatory systems include four-chambered hearts. Birds also lay fluid-filled eggs with hard shells and feed and care for their young

7. NGSSS Check Contrast How do birds differ from reptiles? SC.6.L.15.1

REVIEW VOCABULARY

egg
a female sex cell that forms in an ovary

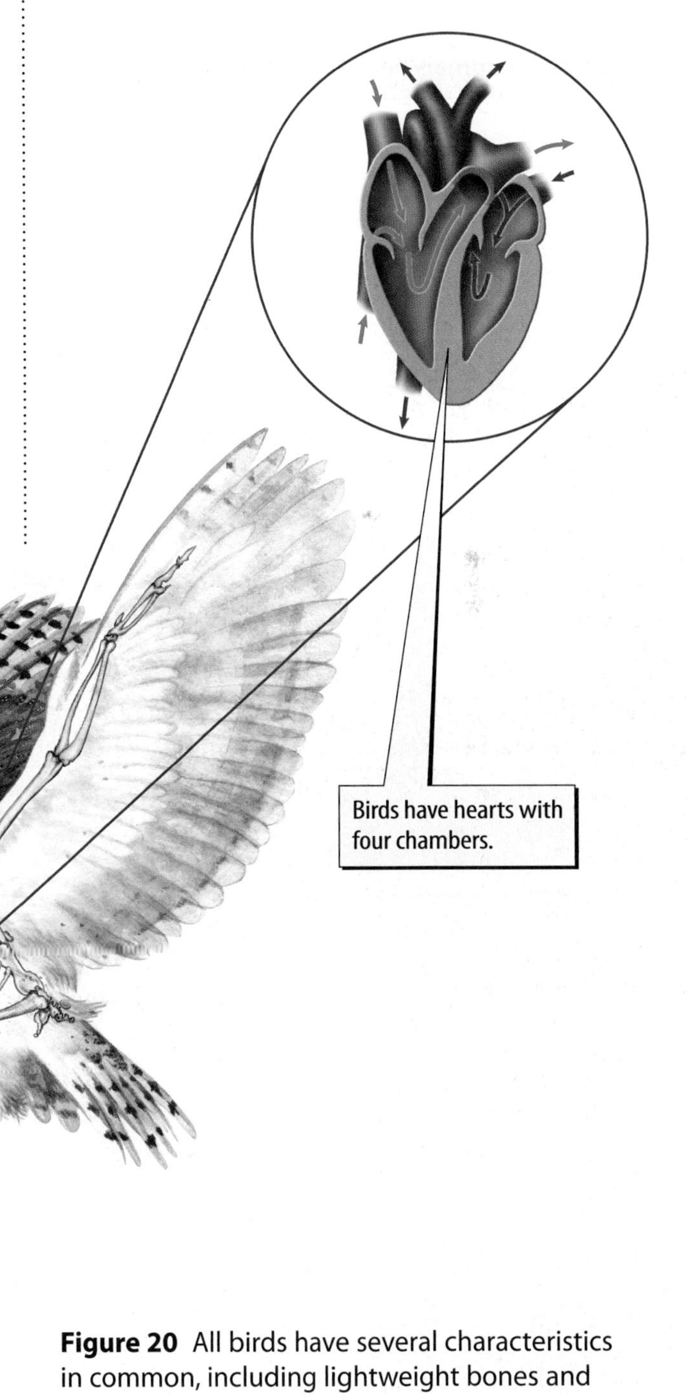

Figure 20 All birds have several characteristics in common, including lightweight bones and four-chambered hearts.

Figure 21 Mammals have hair or fur and mammary glands.

Mammals

Dogs, cats, goats, rats, seals, whales, and humans are among the many vertebrates belonging to the class Mammalia. All mammals have hair or fur covering their bodies. As shown in **Figure 21,** they tear and chew their food using teeth. Mammals have complete digestive systems, which include a mouth and an anus, and a complex nervous system including a brain.

The most notable characteristic of mammals, however, is the presence of mammary glands. These glands produce milk that feeds young mammals. Although many mammals have live young, a few species, including the duck-billed platypus, lay eggs.

8. **NGSSS Check** **Contrast** How do the classes of vertebrates differ? **SC.6.L.15.1**

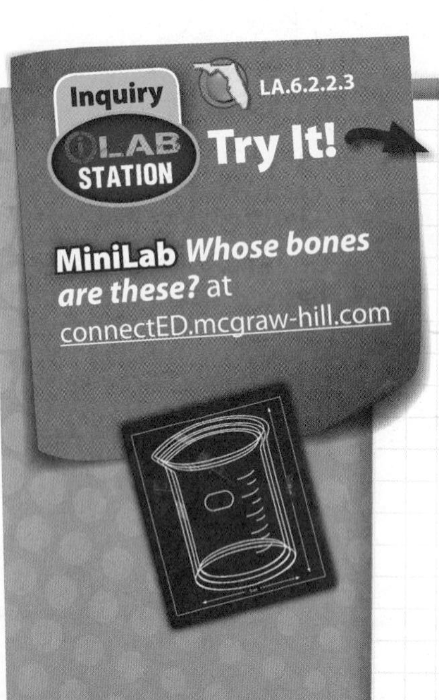

Inquiry LA.6.2.2.3

iLAB STATION **Try It!**

MiniLab ***Whose bones are these?*** at connectED.mcgraw-hill.com

Apply It! After you complete the lab, answer these questions.

1. What functions do you think the bones you examined have (for example, swimming, running, or flying)?

2. How might a flying animal's bones differ from those of a running animal?

Lesson Review 3

Visual Summary

Most chordates are vertebrates.

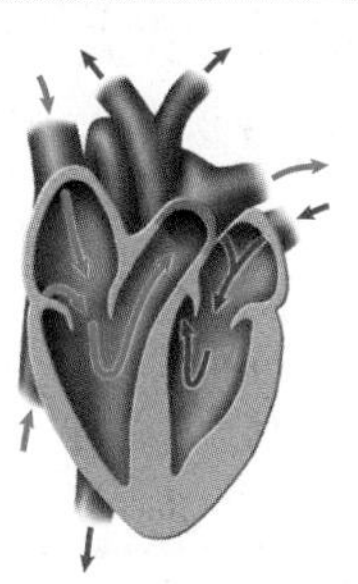

Vertebrates have well-developed organ systems including digestive systems with two openings, circulatory systems that move blood through the body, and nervous systems including brains.

Mammals produce milk to feed their young.

Use Vocabulary

1. **Distinguish** between reptiles and amphibians.

2. **Define** *notochord.*

Understand Key Concepts 

3. Which characteristic is common to all chordates?
   - (A) bones
   - (B) fur
   - (C) lungs
   - (D) notochord

4. **List** the characteristics common to all fish.

5. **Compare and contrast** birds and mammals.

Interpret Graphics

6. **Summarize Information** Fill in the table below with the features of each type of chordate.

Type of Animal	Characteristics	Example Animals
Invertebrate chordates		
Vertebrate chordates		

Critical Thinking

7. **Infer** What is the advantage of having bones that protect the central nerve cord?

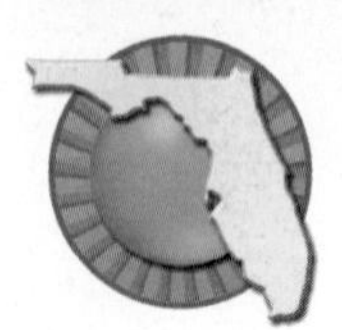

Chapter 11 Study Guide

Think About It! Animals are classified according to shared characteristics. The major groups include sponges, cnidarians, flatworms, roundworms, mollusks, segmented worms, arthropods, and chordates. They differ based on body structures and types of reproduction.

Key Concepts Summary

LESSON 1 What defines an animal?

- Animals are eukaryotic, multicellular organisms that eat other organisms, digest food, and have collagen to support cells. Most animals reproduce sexually and can move.
- Animals can be classified based on the presence of a backbone; body symmetry; the characteristics of proteins, DNA, and other molecules that make up their cells; and the kinds of body structures they possess.

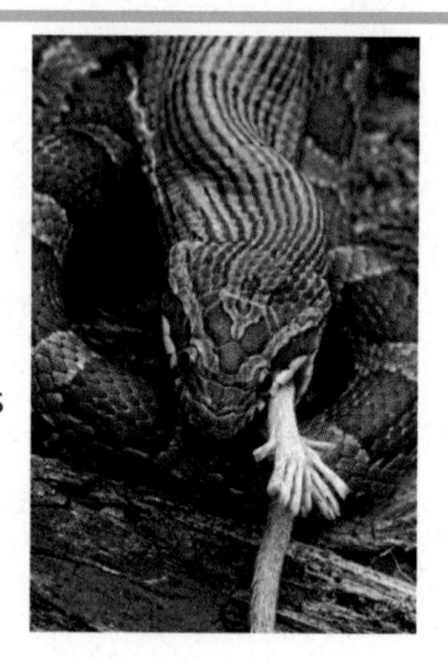

Vocabulary

vertebrate p. 408
invertebrate p. 408
radial symmetry p. 409
bilateral symmetry p. 409
asymmetry p. 409

LESSON 2 Invertebrate Phyla

- Invertebrates have no backbone or internal skeleton, and they tend to be smaller and slower-moving than vertebrates.
- Invertebrates differ based on symmetry, presence or absence of certain types of specialized body structures, and presence or absence of specific internal organs and organ systems.

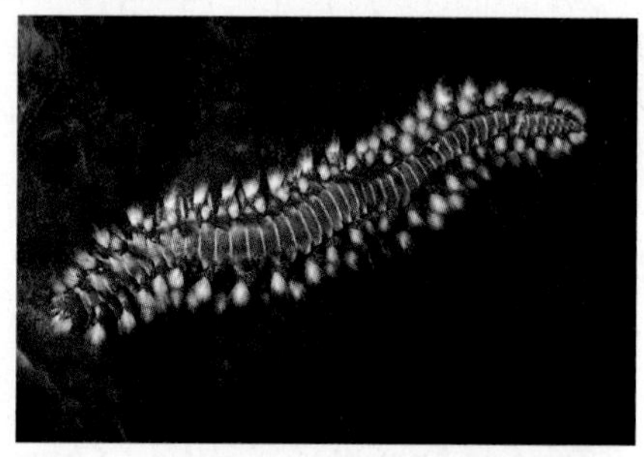

exoskeleton p. 419
appendage p. 419

LESSON 3 Phylum Chordata

- All **chordates** have a **notochord,** a nerve cord, pharyngeal pouches, and a tail at some time during their development.
- All vertebrates have a backbone and well-developed organs and organ systems.
- The classes of vertebrates differ based on presence or absence of characteristics such as gills, fins, scales, legs, wings, fur, and eggs.

notochord p. 425
chordate p. 425

Active Reading

FOLDABLES® Chapter Project

Assemble your lesson Foldables as shown to make a Chapter Project. Use the project to review what you have learned in this chapter.

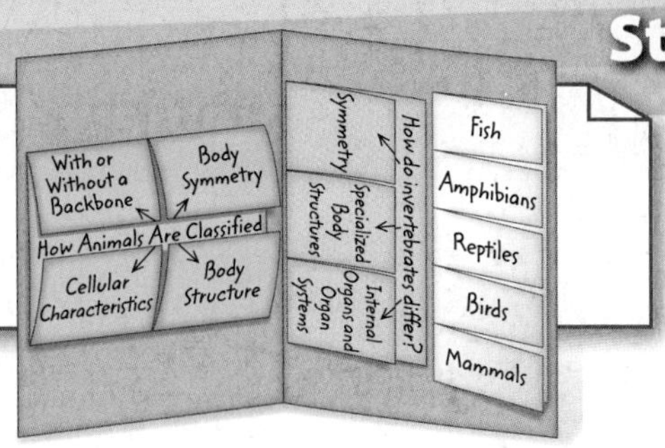

Use Vocabulary

Write the vocabulary term that best matches each phrase.

1. body plan that can be divided into two nearly equal parts anywhere through its central axis

2. body plan that cannot be divided into two nearly equal parts

3. structure that develops into a backbone in vertebrates

4. a structure such as a leg or an arm

5. two sides that are nearly mirror images of each other

6. a thick, hard covering on arthropods

Link Vocabulary and Key Concepts

Use vocabulary terms from the previous page to complete the concept map.

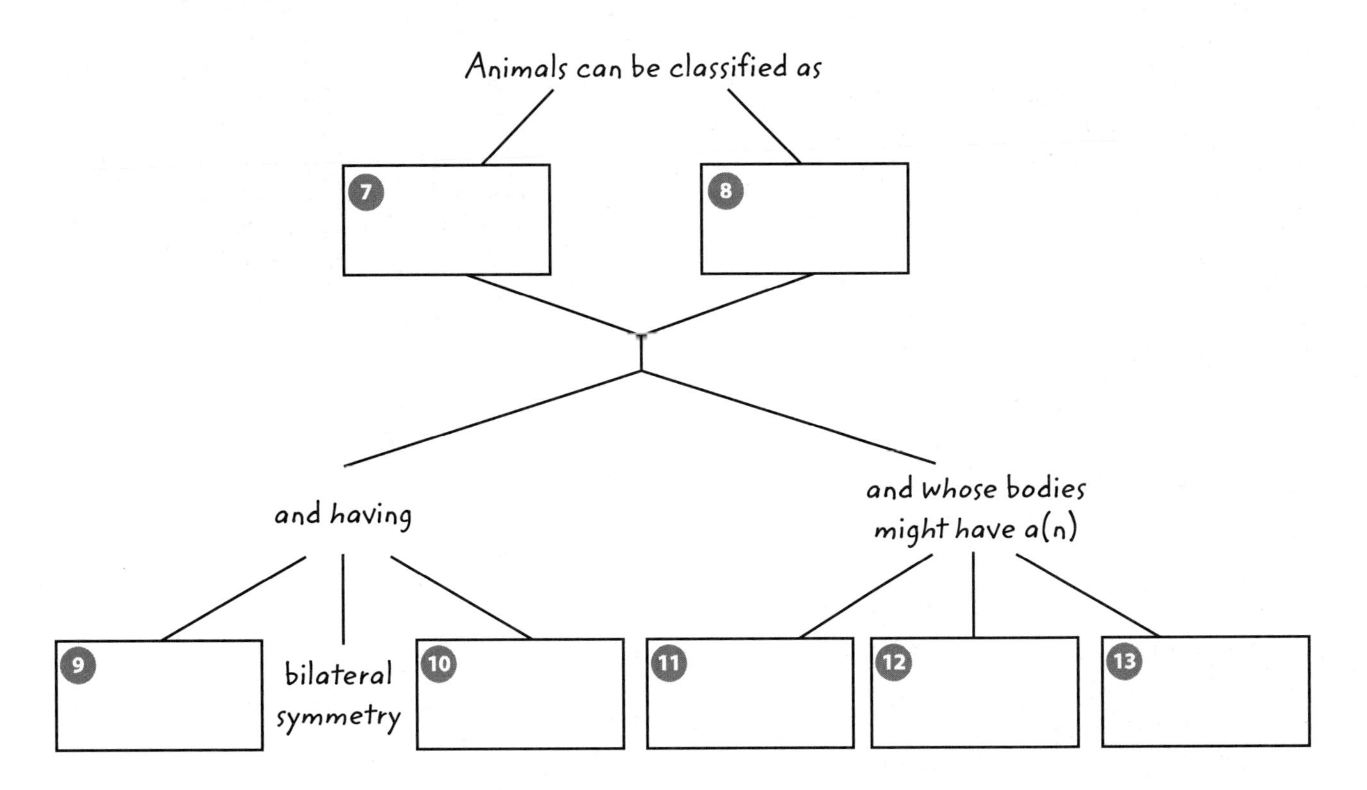

Chapter 11 Review

Understand Key Concepts

1. Which characteristic does NOT apply to animals? SC.6.L.15.1
   - Ⓐ collagen
   - Ⓑ photosynthesis
   - Ⓒ digestive system
   - Ⓓ eukaryotic cell

2. What characteristic applies to the animal shown below? SC.6.L.15.1

   - Ⓐ asymmetry
   - Ⓑ vertebrate
   - Ⓒ bilateral symmetry
   - Ⓓ radial symmetry

3. Which characteristic separates the animal kingdom into two categories? SC.6.L.15.1
   - Ⓐ backbone
   - Ⓑ DNA
   - Ⓒ notochord
   - Ⓓ symmetry

4. Which characteristics do all cnidarians have in common? SC.6.L.15.1
   - Ⓐ free swimming, radial symmetry
   - Ⓑ radial symmetry, attached to rocks
   - Ⓒ tentacles, bilateral symmetry
   - Ⓓ tentacles, stinging cells

5. Which characteristic is NOT shared by both flatworms and nematodes? SC.6.L.15.1
   - Ⓐ bilateral symmetry
   - Ⓑ a simple brain
   - Ⓒ live in moist environments
   - Ⓓ two openings in digestive system

Critical Thinking

6. **Create** a table below that compares the three types of symmetry in the animal kingdom. LA.6.2.2.3

7. **Analyze** Why is digestion important to animals but not to plants? SC.6.L.15.1

8. **Analyze** Explain how you could determine whether the animals shown below are from the same phylum or different phyla. SC.6.L.15.1

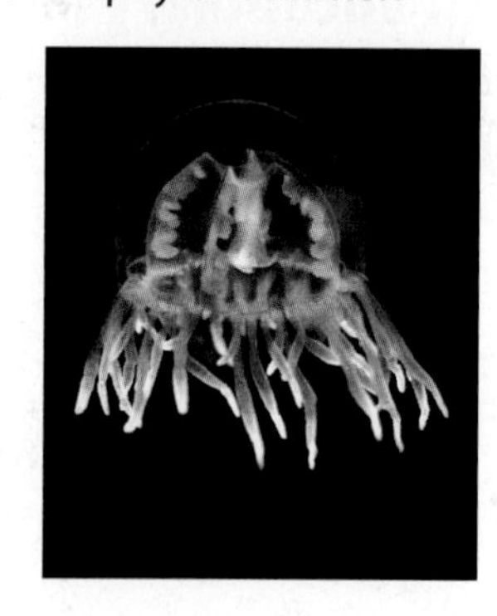

9. **Infer** Animals have cells specialized for different functions. Why is this feature an advantage for survival in a multicellular organism? SC.6.L.15.1

10 **Evaluate** Why are sponges, cnidarians, flatworms, and roundworms limited to life in water or moist environments? SC.6.L.15.1

11 **Compare** What characteristics do fish and lancelets have in common? How are they different? SC.6.L.15.1

12 **Infer** Why do most vertebrates have appendages such as fins, wings, and legs, whereas most invertebrates do not? SC.6.L.15.1

Writing in Science

13 **Write** a paragraph on a separate sheet of paper giving two ways a scuba diver could tell the difference between a sea anemone and a sea slug during an ocean dive. SC.6.L.15.1

Big Idea Review

14 In what ways does a body with an internal skeleton have an advantage over a body with no internal or external support? SC.6.L.15.1

15 What are the major groups of animals, and how do they differ? SC.6.L.15.1

Math Skills MA.6.A.3.6

Use Percentages

16 Worldwide, there are about 300,000 species of Lepidoptera, of which an estimated 14,500 are butterflies. What percentage of Lepidoptera species are butterflies?

17 Of the estimated 1.2 million species of invertebrates, about 40,000 are crustaceans. What percentage of invertebrates are crustaceans?

18 Of the estimated 1.2 million species of invertebrates, about 950,000 are insects. What percentage of invertebrates are insects?

Florida NGSSS Benchmark Practice

Fill in the correct answer choice.

Multiple Choice

1 Which is a characteristic of all vertebrates? **SC.6.L.15.1**

Ⓐ digestive system with one opening
Ⓑ offspring hatch from eggs
Ⓒ respiratory system with lungs
Ⓓ spinal cord enclosed in bones

Use the figure below to answer question 2.

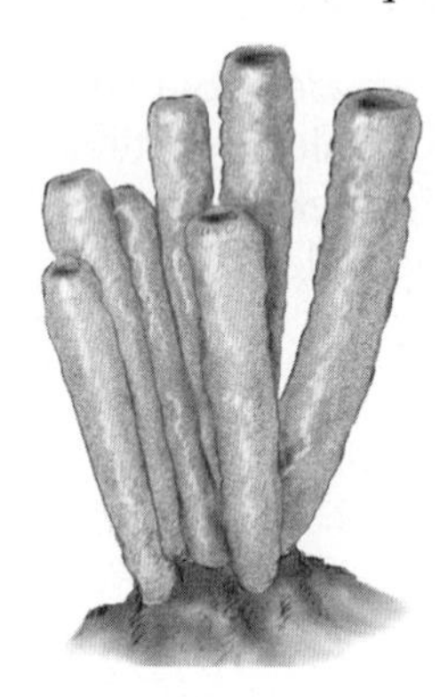

2 Which is true of the animal shown above? **SC.6.L.15.1**

Ⓕ It has bilateral symmetry.
Ⓖ It has radial symmetry.
Ⓗ It is an invertebrate.
Ⓘ It is a vertebrate.

3 Which animals make up Phylum Chordata? **SC.6.L.15.1**

Ⓐ insects, spiders, and crabs
Ⓑ mussels, octopuses, and squids
Ⓒ snails, slugs, and clams
Ⓓ vertebrates, lancelets, and tunicates

4 Which is a characteristic of all invertebrates? **SC.6.L.15.1**

Ⓕ backbone absent
Ⓖ exoskeleton present
Ⓗ symmetry absent
Ⓘ tube feet present

5 Which is NOT a characteristic of all animals? **SC.6.L.15.1**

Ⓐ digesting food
Ⓑ eating other organisms
Ⓒ having specialized cells
Ⓓ moving around

Use the figure below to answer question 6.

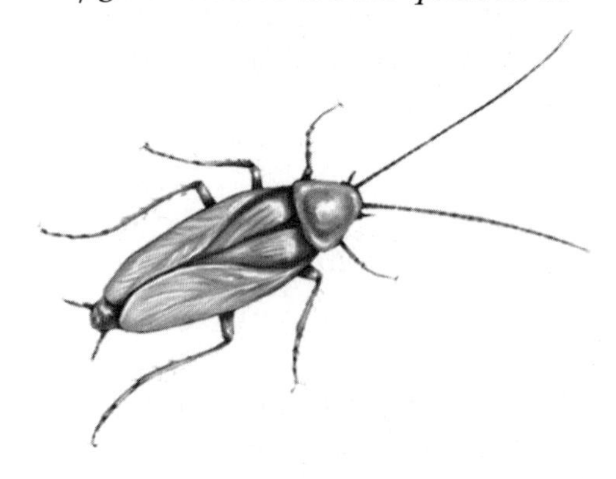

6 The animal in the figure above is classified in which phylum? **SC.6.L.15.1**

Ⓕ Arthropoda
Ⓖ Chordata
Ⓗ Echinodermata
Ⓘ Mollusca

7 Which animal has radial symmetry? **SC.6.L.15.1**

Ⓐ flatworm
Ⓑ jellyfish
Ⓒ octopus
Ⓓ sponge

Use the figure below to answer questions 8 and 9.

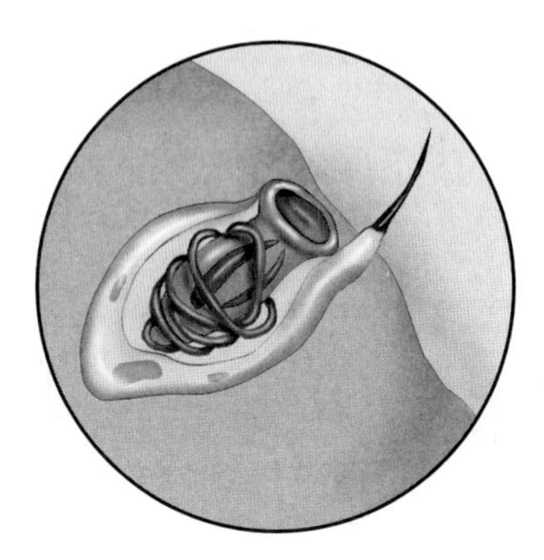

8 Which animals contain the structure shown in the figure? SC.6.L.15.1

Ⓕ bees and wasps

Ⓖ jellyfish and corals

Ⓗ octopuses and squids

Ⓘ sea stars and sea urchins

9 How do animals use the structure shown in the figure? SC.6.L.15.1

Ⓐ for breathing

Ⓑ for feeding

Ⓒ for mating

Ⓓ for seeing

10 Which class of vertebrates feeds milk to its young? SC.6.L.15.1

Ⓕ amphibians

Ⓖ fish

Ⓗ mammals

Ⓘ reptiles

11 To which phylum does the animal below belong? SC.6.L.15.1

Ⓐ Annelida

Ⓑ Mollusca

Ⓒ Nematoda

Ⓓ Platyhelminthes

Use the figure below to answer questions 12 and 13.

12 What characteristic distinguishes the animal shown above from a fish? SC.6.L.15.1

Ⓕ bones

Ⓖ notochord

Ⓗ nerve cord

Ⓘ no gills

13 To which phylum does this animal belong? SC.6.L.15.1

Ⓐ Annelida

Ⓑ Cnidaria

Ⓒ Chordata

Ⓓ Mollusca

Need Extra Help?

If You Missed Question...	1	2	3	4	5	6	7	8	9	10	11	12	13
Go to Lesson...	3	1	1	2	1	1	1	2	2	3	2	2,3	3

Name ________________________ Date ____________

Benchmark Mini-Assessment **Chapter 11 • Lesson 1** mini BAT

Multiple Choice *Bubble the correct answer.*

Use the image below to answer questions 1 through 4.

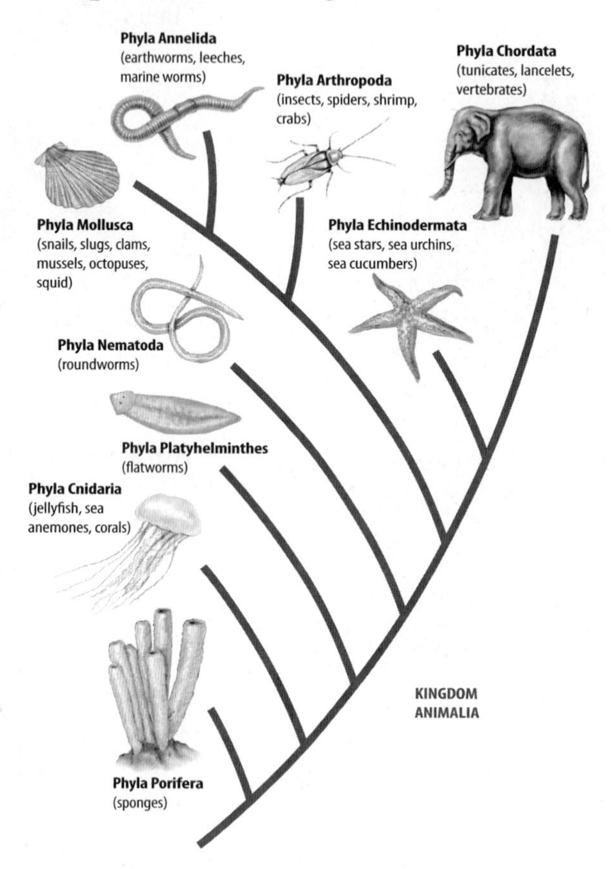

1. Which animal phyla are most closely related? **SC.6.L.15.1**

Ⓐ Arthropoda and Echinodermata
Ⓑ Echinodermata and Chordata
Ⓒ Mollusca and Annelida
Ⓓ Platyhelminthes and Nematoda

2. Which phyla include several kinds of organisms that have radial symmetry? **SC.6.L.15.1**

Ⓕ Annelida and Nematoda
Ⓖ Echinodermata and Cnidaria
Ⓗ Platyhelminthes and Arthropoda
Ⓘ Porifera and Mollusca

3. Which two phyla have the least similar DNA? **SC.6.L.15.1**

Ⓐ Arthropoda and Echinodermata
Ⓑ Cnidaria and Chordata
Ⓒ Nematoda and Mollusca
Ⓓ Porifera and Platyhelminthes

4. Which characteristic is true of members of phylum Porifera? **SC.6.L.15.1**

Ⓕ They are asymmetrical.
Ⓖ They are unicellular organisms.
Ⓗ They are vertebrates.
Ⓘ They are closely related to Echinodermata.

Name ______________________________ Date ______________

Multiple Choice *Bubble the correct answer.*

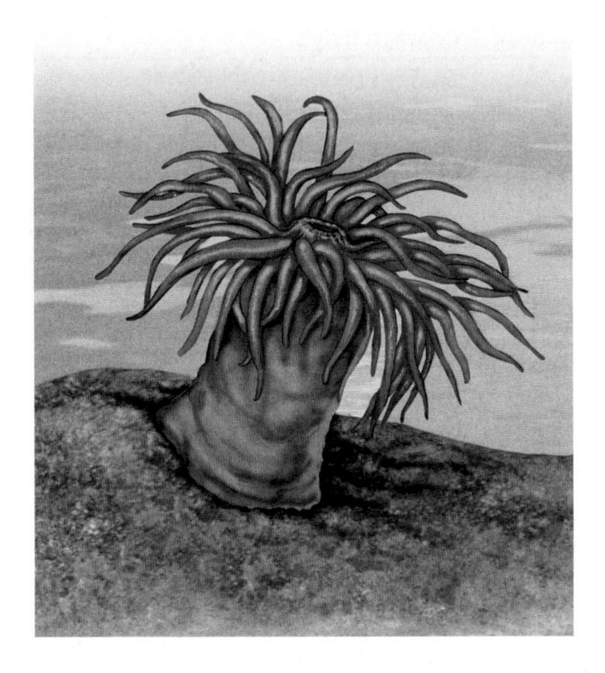

1. Which trait is true of the animal shown in the image above? **SC.6.L.15.1**
   - (A) It has an exoskeleton.
   - (B) It has radial symmetry.
   - (C) It is a vertebrate.
   - (D) It lacks symmetry.

2. Anish finds an organism in damp soil with bilateral symmetry and a long, soft, segmented body. It has two openings to its digestive system and a circulatory system made up of blood vessels. The organism most likely is a member of phylum **SC.6.L.15.1**
   - (F) Annelida.
   - (G) Mollusca.
   - (H) Platyhelminthes.
   - (I) Porifera.

3. To which phylum does the organism above belong? **SC.6.L.15.1**
   - (A) Annelida
   - (B) Arthropoda
   - (C) Cnidaria
   - (D) Echinodermata

4. Which phylum has members that are characterized by lack of symmetry? **SC.6.L.15.1**
   - (F) Echinodermata
   - (G) Mollusca
   - (H) Platyhelminthes
   - (I) Porifera

Multiple Choice *Bubble the correct answer.*

Use the images below to answer questions 1 and 2.

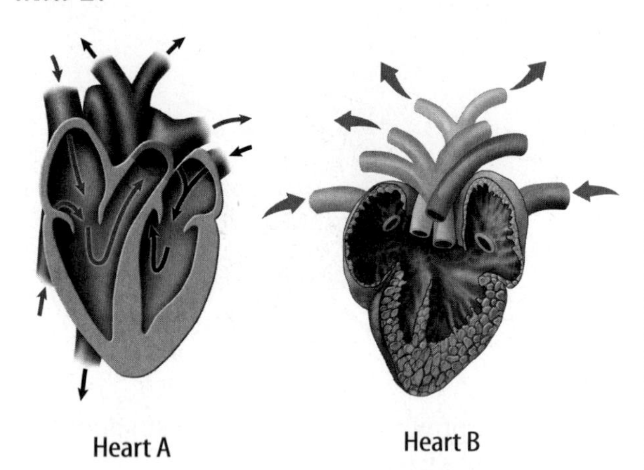

Heart A Heart B

1. How do the hearts in the images above differ? **SC.7.L.15.2**

(A) Heart A has four chambers while heart B has three chambers.

(B) Heart A has twice as many chambers as heart B has.

(C) Heart A is found in fish, but heart B is found in amphibians.

(D) Heart A is found in mammals while heart B is found in birds.

2. Animals in which group of organisms have a heart similar in structure to heart A? **SC.7.L.15.2**

(F) amphibians

(G) birds

(H) fish

(I) reptiles

Use the table below to answer questions 3 and 4.

Organism	Characteristics
Q	Smooth and moist skin, bony skeleton, adults have lungs
R	Feathers and scales, wings
S	Gills, fins, cartilaginous skeleton
T	Scales, lungs, fins
U	Hair or fur, glands that produce milk

3. Based on the table, which does NOT correctly identify the group to which the animal belongs? **LA.6.2.2.3**

(A) Q: amphibians

(B) R: birds

(C) T: reptiles

(D) U: mammals

4. An organism that has flippers and gives birth to live young belongs to which group? **LA.6.2.2.3**

(F) R

(G) S

(H) T

(I) U

Name ______________________ Date ______________

Notes

Unit 3

Human Body Systems

1800 | 1850 | 1900 | 1950

1818
The first well-documented case of a person-to-person blood transfusion is performed. British obstetrician James Blundell transfuses 4 oz. of blood from a man to his wife who had just given birth.

1823
German surgeon Christian Bünger performs the first autograft, replacing skin on a man's nose with some from his thigh.

1905
Dr. Eduard Zirm performs the first successful cornea transplant on patient Alois Glogar.

1954
The first successful organ transplant between living relatives (a kidney transplant between twins) is completed by Dr. Joseph Murray and Dr. David Hume in Boston.

1967
The first successful liver and heart transplants are completed. Dr. Thomas Starzl performs the liver transplant in Denver, Colorado. Dr. Christiaan Barnard performs the heart transplant in Cape Town, South Africa.

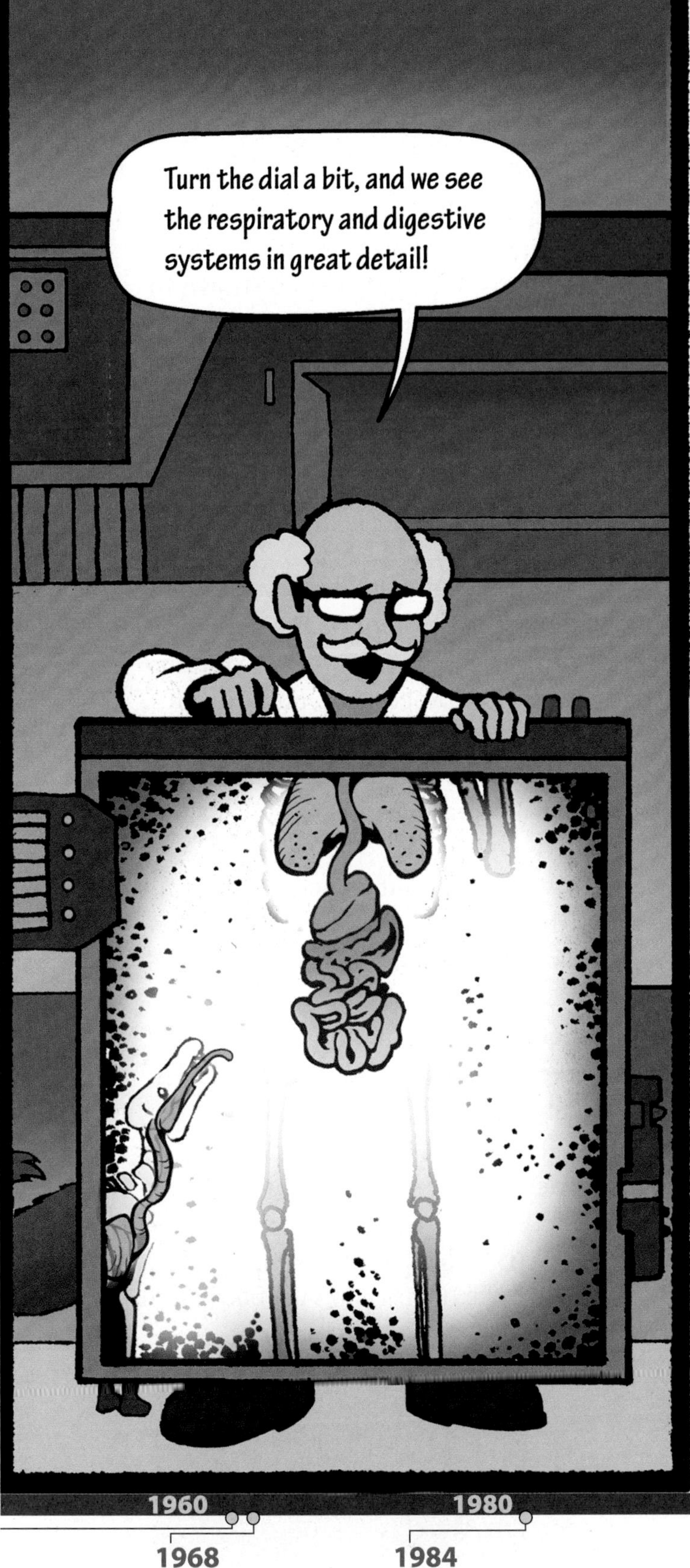

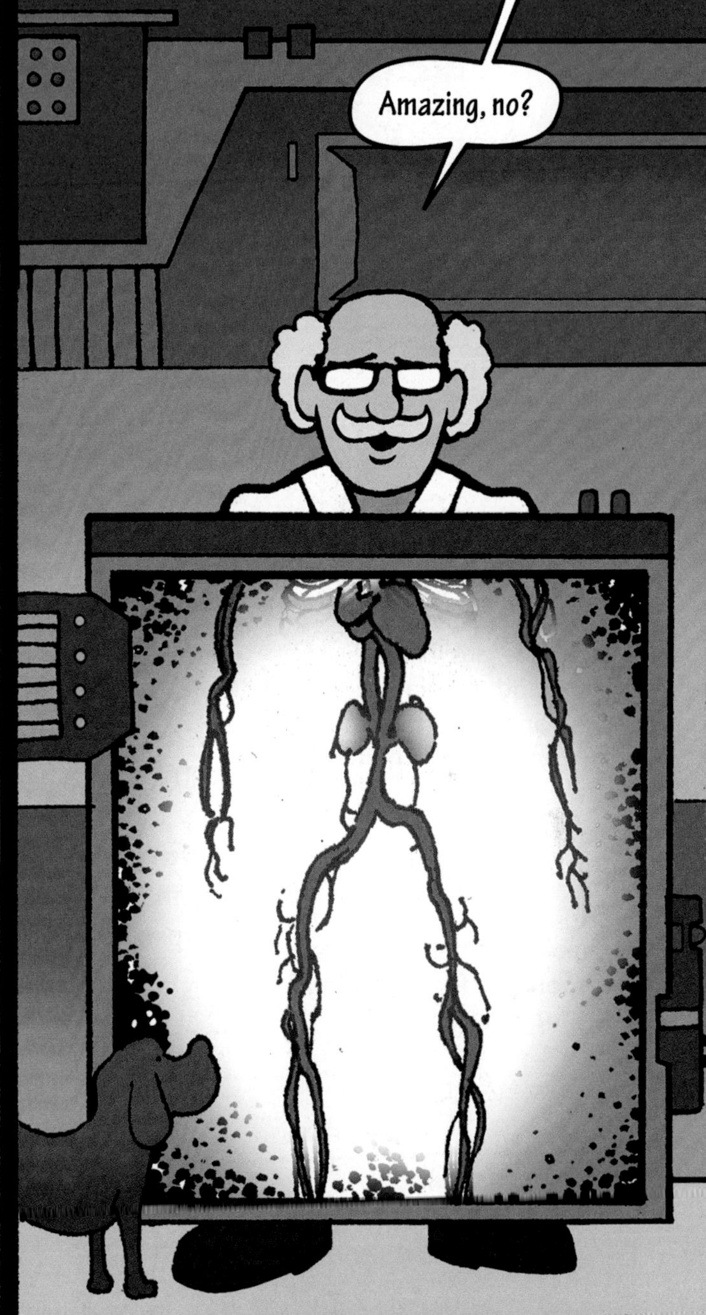

1960 1980 2000

1968
The Uniform Anatomical Gift Act establishes the Uniform Donor Card as a legal document that allows anyone 18 years of age or older to legally donate their organs upon death.

1984
The National Organ Transplant Act (NOTA) establishes a nationwide computer registry, authorizes financial support for organ procurement organizations, and prohibits the buying or selling of organs in the United States.

2001
Due to widespread use of advanced surgical techniques and higher success rates for surgeries, the number of living donors passes the number of deceased donors.

? Inquiry
Visit ConnectED for this unit's  STEM activity.

Systems

A **system** is a collection of parts that influence or interact with one another. For example, the human body is a large system made up of many smaller subsystems, such as the ones shown in **Figure 1.**

Like the human body system, complex systems often contain subsystems. The parts of each subsystem interact among themselves. Each subsystem has a different purpose, but interacts to keep the larger system working properly.

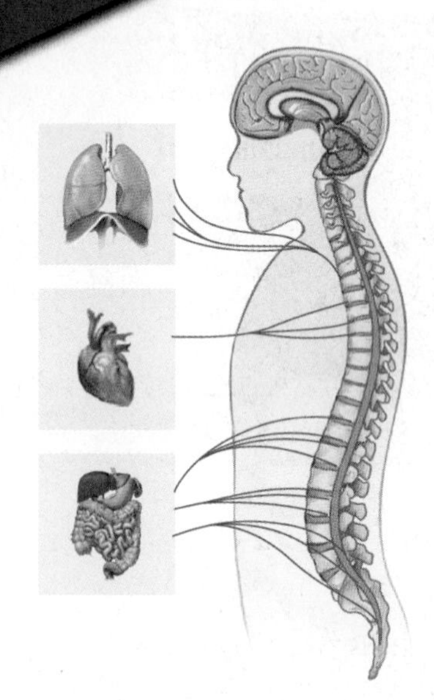

▲ **Figure 1** The nervous system, the respiratory system, the circulatory system, and the digestive system are subsystems of a larger system—the human body.

Parts of a System

Systems and subsystems work together to achieve a goal. For example, the nervous system, a subsystem of the body system, regulates body temperature, as shown in **Figure 2**. Systems often are described in terms of input, processing, and output.

Input is the matter, energy, or information that enters a system. When you exercise, one input to your nervous system is thermal energy. It is detected by brain and skin cells called receptors.

Processing is the changing of the input to achieve a goal. The hypothalamus processes the input from receptors. It sends electrical signals to other parts of the body. The signals tell the body it is too warm.

Output is the material, energy, or information that leaves a system. Outputs from the nervous system include sweat, goose bumps, and shivers, all of which can change body temperature.

Active Reading **1. Differentiate** Underline the terms *input, processing,* and *output* and their definitions.

Figure 2 The nervous system is responsible for regulating body temperature. ▼

Input: Thermal energy released by contracting muscles is detected by receptors.

Processing: Signals from receptors are sent to the brain. The brain then signals glands in the skin to produce sweat.

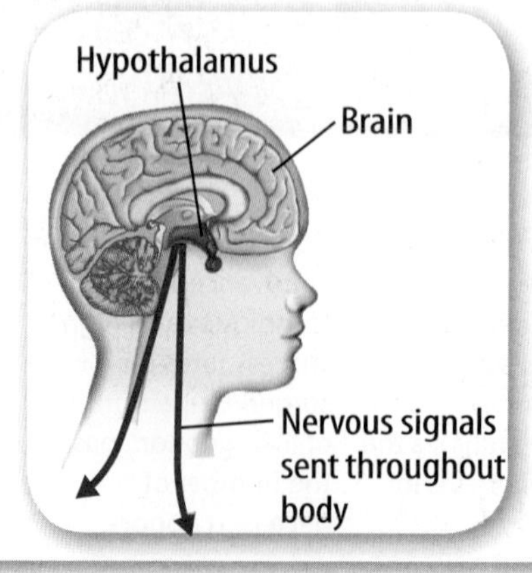

Output: Sweat forms on the skin. Then, it cools the body as it evaporates.

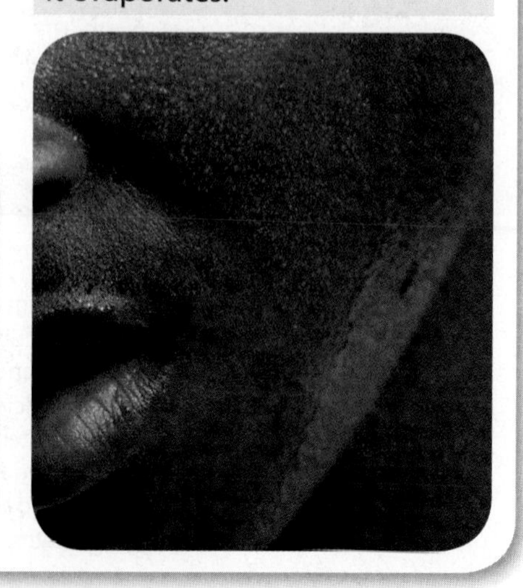

Feedback in Systems

Many systems use feedback to monitor and regulate input, process, and output. When you exercise, as shown in **Figure 3**, your muscles produce carbon dioxide as a waste product. Receptors detect higher levels of carbon dioxide in your blood. The brain processes this input and signals your nervous system to increase breathing. When you breathe harder and faster the levels of carbon dioxide in your blood decrease. Once this feedback is detected by receptors, your brain signals your nervous system to return to normal breathing.

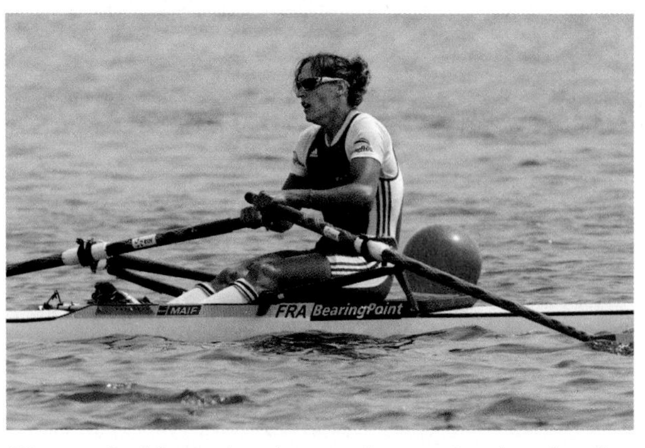

Figure 3 After a hard race, decreasing levels of carbon dioxide in a rower's blood act as feedback to the nervous system. This feedback signals the rower's brain to restore her breathing to a normal rate.

Cooperation, Order, and Change

Body subsystems work together to regulate temperature, remove waste from your blood, and respond to other changes in your body. A failure in one subsystem affects other subsystems. If a bad cold causes you to become congested, your respiratory system cannot efficiently exchange oxygen for carbon dioxide. Muscle cells no longer receive enough oxygen to function normally. As a result, you become tired and have trouble catching your breath.

Systems Thinking

Thinking in terms of systems might change the way you make choices. For example, some people think that if they reduce the amount they eat, they will lose weight. However, protein is necessary for your muscular system to function properly. Without an adequate amount of protein, muscle tissues begin to break down. Thinking about the interactions of the systems and subsystems in your body can lead to decisions that help you achieve long-term goals.

Active Reading **2. Explain** Why is "systems thinking" about interactions in your body essential to healthful life choices?

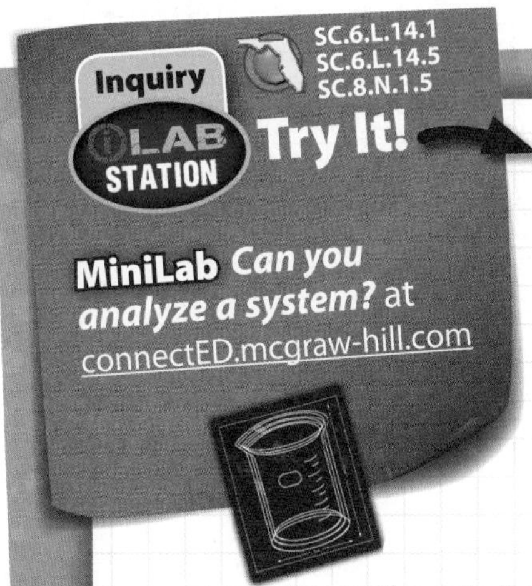

Apply It!

After you complete the lab, answer these questions.

1. **Design** Create a cycle flowchart to illustrate a system. Label *inputs, processes, outputs,* and *feedbacks* in your system.

Name ______________________________ Date ______________

Notes

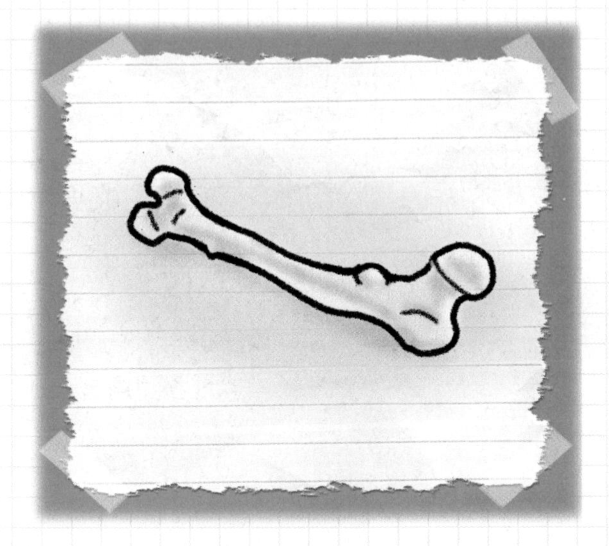

Is Bone Alive?

Five friends were arguing about bones. They each had different ideas about whether bones were living. This is what they said:

Mona: I think bones are living because they are inside our body.

Al: I think bones are living because they are made up of cells.

Bea: I think bones are nonliving because they are made up of cells that become hard bone after they die.

Mia: I think bones are nonliving because they are made up of minerals like calcium.

Tess: I think bones are nonliving because they exist even after a person dies.

Who do you agree with the most? Explain your thinking about bone as living or non-living material.

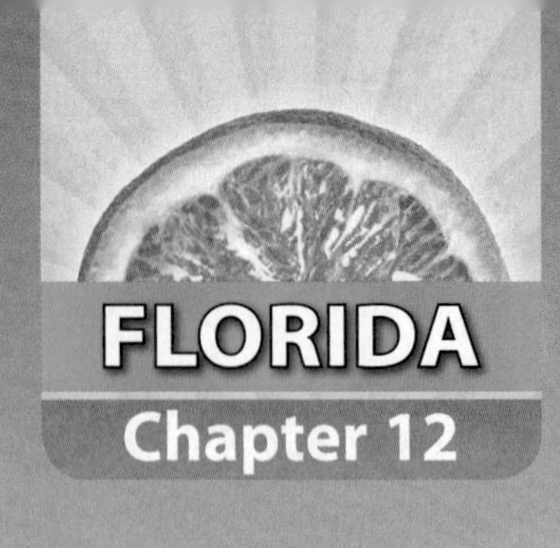

FLORIDA

Chapter 12

Structure and MOVEMENT

FLORIDA BIG IDEAS

1 The Practice of Science

14 Organization and Development of Living Organisms

The Big Idea

Think About It!

How do bones, muscles, and skin help maintain the body's homeostasis?

You could not play soccer if your body were made of only bones or only muscles. Bones, muscles, and skin work together and help you move and carry out processes that are necessary in order to live.

1. What would happen to your body if you had no skin? No muscles?

2. What functions do you think bones, muscles, and skin have?

3. How do bones, muscles, and skin work together to help your body function?

Get Ready to Read

What do you think about structure and movement?

Before you read, decide if you agree or disagree with each of these statements. As you read this chapter, see if you change your mind about any of the statements.

	AGREE	DISAGREE
1. Bones protect internal organs.	☐	☐
2. Bones do not change during a person's lifetime.	☐	☐
3. The same type of muscle that moves bones also pumps blood through the heart.	☐	☐
4. Muscles cannot push bones.	☐	☐
5. Skin helps regulate body temperature.	☐	☐
6. Skin is made of two layers of tissue.	☐	☐

There's More Online!
Video • Audio • Review • ⓘLab Station • WebQuest • Assessment • Concepts in Motion • Multilingual eGlossary

Lesson 1

The Skeletal SYSTEM

ESSENTIAL QUESTIONS

 What does the skeletal system do?

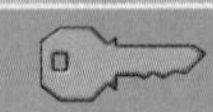 How do the parts of the skeletal system work together?

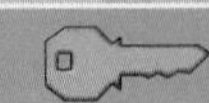 How does the skeletal system interact with other body systems?

Vocabulary

skeletal system p. 451
cartilage p. 454
periosteum p. 454
joint p. 455
ligament p. 455
arthritis p. 456
osteoporosis p. 456

Florida NGSSS

LA.6.2.2.3 The student will organize information to show understanding (e.g., representing main ideas within text through charting, mapping, paraphrasing, summarizing, or comparing/contrasting);

LA.6.4.2.2 The student will record information (e.g., observations, notes, lists, charts, legends) related to a topic, including visual aids to organize and record information and include a list of sources used;

SC.6.L.14.5 Identify and investigate the general functions of the major systems of the human body (digestive, respiratory, circulatory, reproductive, excretory, immune, nervous, and musculoskeletal) and describe ways these systems interact with each other to maintain homeostasis.

SC.6.N.1.1 Define a problem from the sixth grade curriculum, use appropriate reference materials to support scientific understanding, plan and carry out scientific investigation of various types, such as systematic observations or experiments, identify variables, collect and organize data, interpret data in charts, tables, and graphics, analyze information, make predictions, and defend conclusions.

SC.6.N.1.5 Recognize that science involves creativity, not just in designing experiments, but also in creating explanations that fit evidence.

Launch Lab

SC.6.N.1.5, SC.6.L.14.5

10 minutes

How are bones used for support?

If you have ever watched the construction of a building, you might have seen a wood or steel structure being used to provide support. In a similar way, bones support your body and the organs inside it.

Procedure

1. Read and complete a lab safety form.
2. Using pieces of **clay,** try to build a tower taller than your partner's. One person should use a **wooden dowel** to support his or her tower.
3. Measure and record the height of both towers. Find the class average for the height of towers with and without dowels.

Data and Observations

Think About This

1. Were towers with or without dowels generally higher?
2. Key Concept What do you think your body would be like if you had no bones?

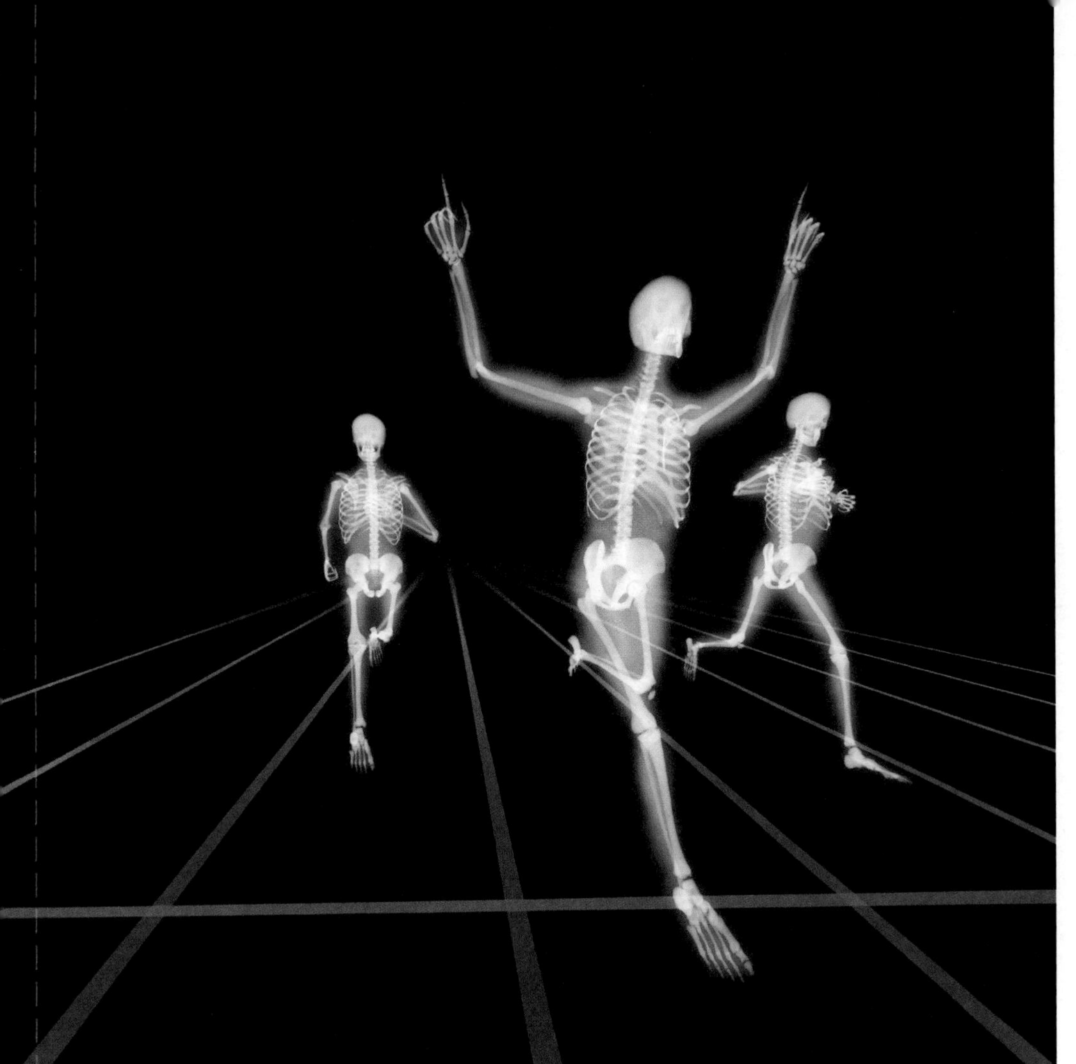

Inquiry **Why do you think bones move so easily?**

1. Bones work like mechanical joints, enabling you to turn, pivot, twist, and bend. So why do they move as easily as mechanical joints but do not need to be oiled?

Figure 1 Bones are part of the skeletal system.

Functions of the Skeletal System

Squeeze your hands and arms. The hard parts you feel are parts of your skeleton. When you think of your skeleton, you might think of bones, like those shown in **Figure 1.** These bones support your body and help you move. Your skeleton is part of your skeletal system and contains more than 200 bones. *The* **skeletal system** *contains bones as well as other structures that connect and protect the bones and that support other functions in the body.* In this lesson, you will learn how all parts of the skeletal system work together.

Support

Can you imagine trying to stack jiggling blocks of gelatin 1 m high? You would probably have a hard time because gelatin does not have any support structures inside it. Without bones, your body would be similar to gelatin. Bones provide support. They help you sit up, stand, and raise your arm over your head to ask a question.

Active Reading **2. Recall** Underline how bones act as a support system.

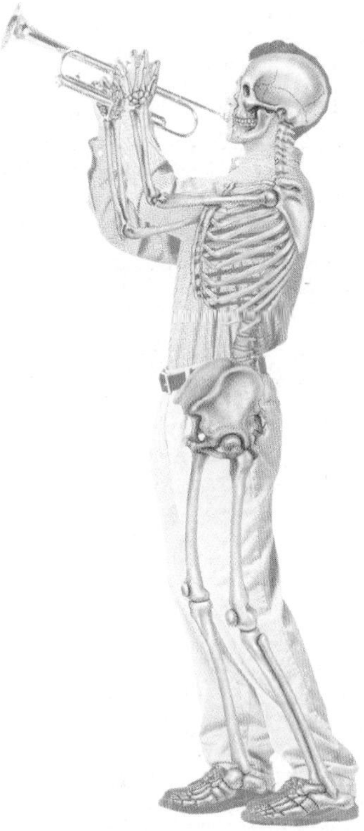

Science Use v. Common Use

tissue

Science Use a group of the same type of cells that perform a specific function within an organism

Common Use a soft, thin paper

Movement

The skeletal system enables different parts of the body to move in different ways, like when a person kicks a soccer ball, as shown in the beginning of this chapter. Bones can move because they are attached to muscles. The skeletal system and the muscular system work together and move your body.

Protection

Feel your head and then feel your stomach. Your stomach is softer than your head. The hard, rigid structure you feel in your head is your skull, shown in **Figure 2.** It protects the soft, fragile **tissue** of your brain from damage. Other bones protect the spinal cord, heart, lungs, and other internal organs.

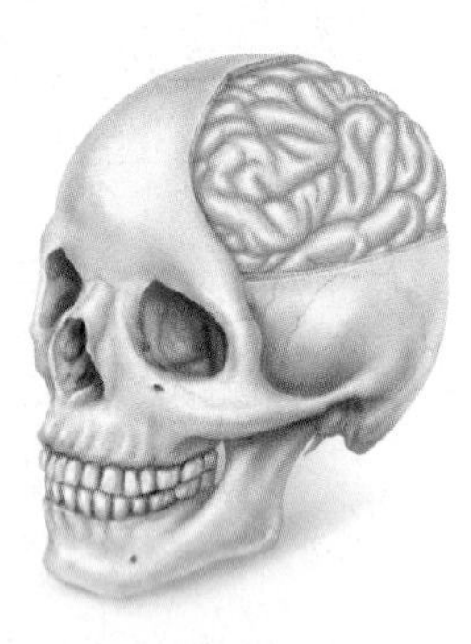

Figure 2 The skull protects the soft tissue of the brain.

Production and Storage

Another function of bones is to produce and store materials needed by your body. Red blood cells are produced inside your bones. Bones store fat and calcium. Calcium is needed for strong bones and for many cellular processes. When the body needs calcium, it is released from bones into the blood.

3. **NGSSS Check** **List** What are the major functions of the skeletal system? SC.6.L.14.5

Apply It! After you complete the lab, answer these questions.

1. Would you be able to pop any of the bubbles if they were placed in a tin can? Explain.

2. How might having a skeletal system as rigid as a tin can be a disadvantage in terms of movement?

Structure of Bones

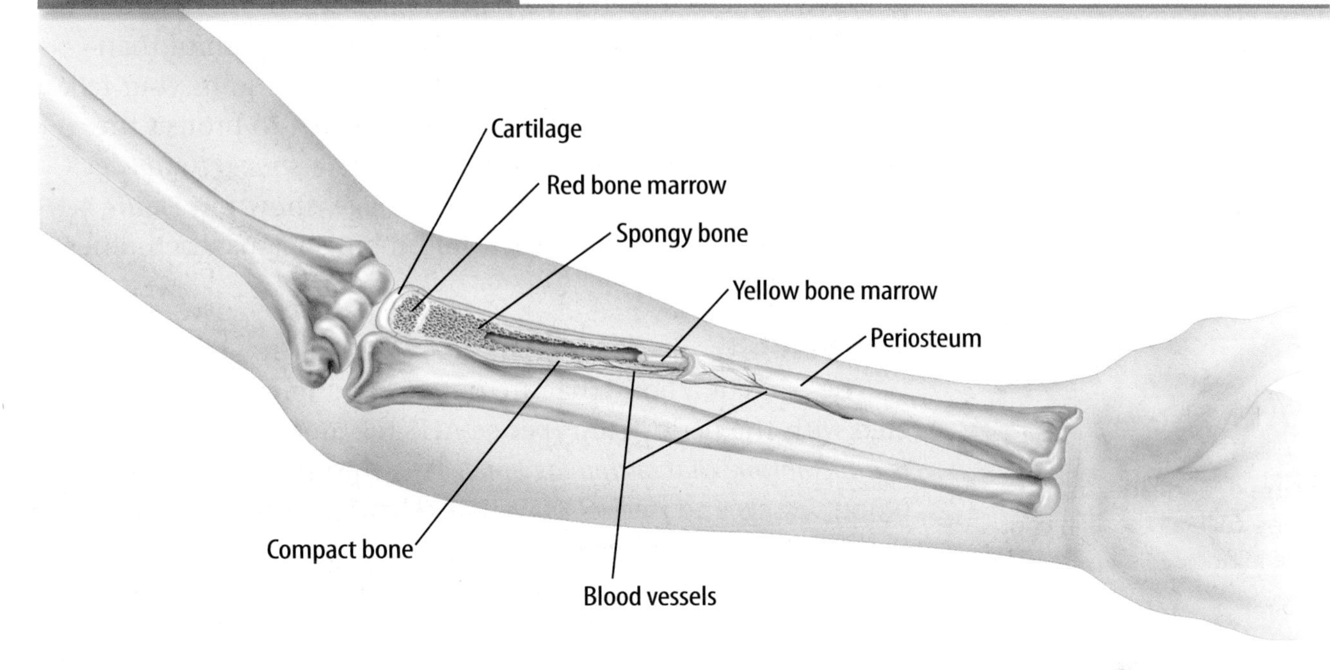

Figure 3 A bone contains many types of tissue.

Structure of Bones

A bone is an organ composed of living tissue. There are two main types of bone tissue: compact and spongy. A bone also contains other types of tissue, as you can see in the arm bone shown in **Figure 3.**

Compact Bone Tissue

The hard, outer portions of bones are made of compact bone tissue. Compact bone tissue is a dense web of fibers. A long bone, like the arm bone shown in **Figure 3,** is made mostly of compact bone tissue. However, in its ends, a long bone contains a different kind of tissue called spongy bone tissue.

Spongy Bone Tissue

The small holes in spongy bone tissue make it look like a sponge. Because of these holes, spongy bone is less dense than compact bone. A short bone, like one in your wrist, is mostly spongy bone tissue.

Bone Marrow

The insides of most bones contain a soft tissue called bone marrow (MER oh). There are two types of bone marrow. Red bone marrow is the tissue where red blood cells are made. It is found in the spongy ends of long bones and in some flat bones, such as the ribs. Yellow bone marrow stores fat and is found inside the longest part of long bones.

Active Reading **4. Break down** Highlight the difference between red bone marrow and yellow bone marrow.

Active Reading FOLDABLES® LA.6.2.2.3

Fold a sheet of paper in half to make a horizontal half book. Use your book to illustrate and describe the structure of bones.

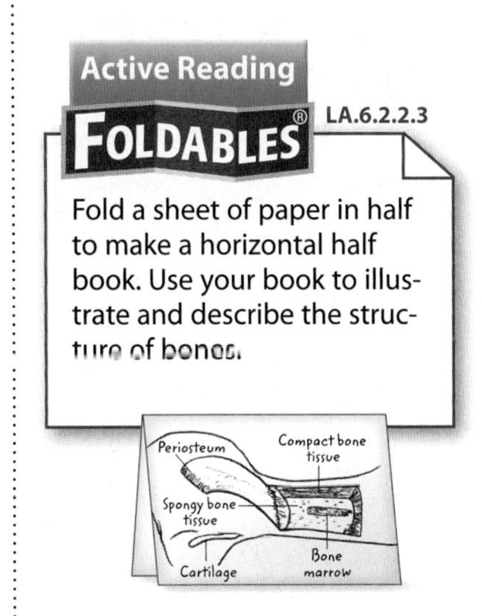

Cartilage

Have you ever fallen on a hard floor made of tiles or concrete? Falling on a hard surface is usually more painful than falling on a soft surface like a carpet, which cushions your fall. A special tissue in your body acts like a carpet to protect the skeletal system. **Cartilage** (KAR tuh lihj) *is a strong, flexible tissue that covers the ends of bones.* Cartilage, shown in **Figure 4,** prevents the surfaces of bones from rubbing against each other and reduces friction.

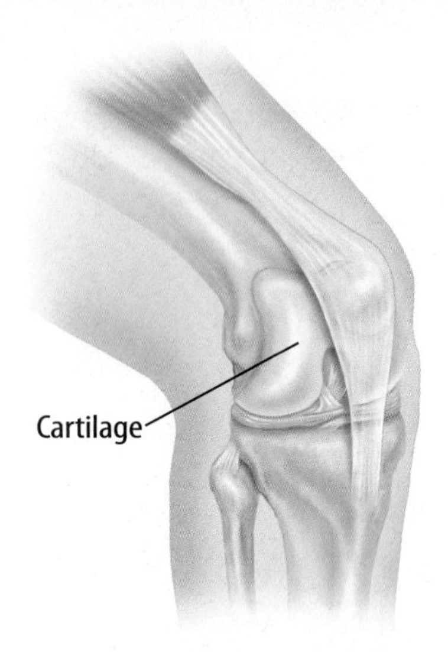

Figure 4 Cartilage protects bones, such as those in your knee joint.

Periosteum

The parts of a bone that are not covered in cartilage are covered with the periosteum (per ee AHS tee um). *The* **periosteum** *is a membrane that surrounds bone.* This thin tissue contains blood vessels and nerves as well as cells that produce new bone tissue. Periosteum nourishes bones and helps them function and grow properly, as well as heal after injury.

WORD ORIGIN

ligament
from Latin *ligare,* means "to bind, tie"

Formation of Bones

Before you were born, your skeleton was made mostly of cartilage. During your infancy and childhood, the cartilage was gradually replaced by bone, as shown in **Figure 5.** The long bones in children and young teens have regions of bone growth that produce new bone cells. These regions are called growth plates. A growth plate produces cartilage that is then replaced by bone tissue. A growth plate, shown in **Figure 5,** is the weakest part of an adolescent bone. Growth continues until adulthood, when most of the cartilage has turned to bone.

Figure 5 Bone gradually replaces cartilage as children grow.

Bone Development

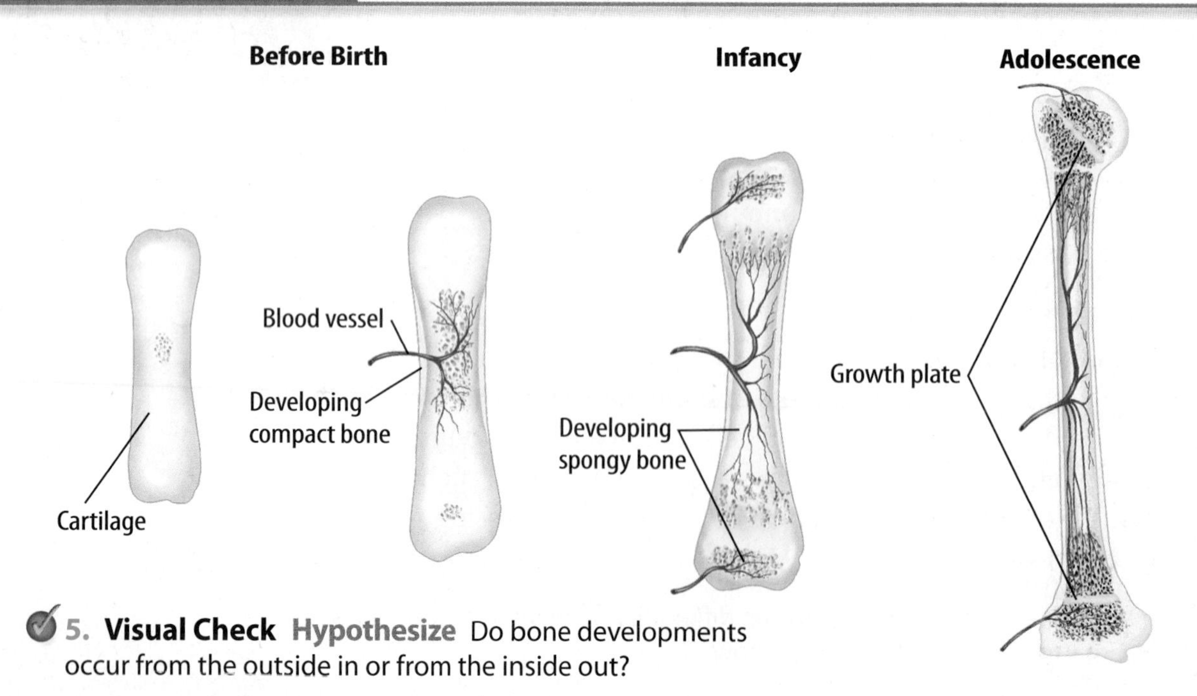

5. Visual Check Hypothesize Do bone developments occur from the outside in or from the inside out?

Joints

Your bones do not function independently. They work together at places called joints. *A* **joint** *is where two or more bones meet.* Joints provide flexibility and enable the skeleton to move. *Bones are connected to other bones by tissues called* **ligaments** (LIH guh munts). When the bones in joints move, ligaments stretch and keep the bones from shifting away from each other.

Ligaments connect bones at joints, but they do not protect bones. As you read earlier, cartilage protects the ends of bones, reducing friction between them so the bones move easily. Your skeletal system contains two types of joints—immovable joints and movable joints.

Immovable Joints

Some parts of your skeleton are made of bones that connect but do not move. These are immovable joints. Your skull contains several immovable joints.

Movable Joints

You are able to move your hand or bend your body because of movable joints. The body's movable joints allow for a wide range of motion. Three main types of movable joints and the ligaments that hold them together are shown in **Table 1.**

Active Reading **6. Differentiate** Underline how ligaments and cartilage help the skeletal system function.

Table 1 Types of Movable Joints

Joint	Description	Example
Ball and socket	allows bones to move and rotate in nearly all directions	hips and shoulders
Hinge	allows bones to move back and forth in a single direction	fingers, elbows, knees
Pivot	allows bones to rotate	neck, lower arm below the elbow

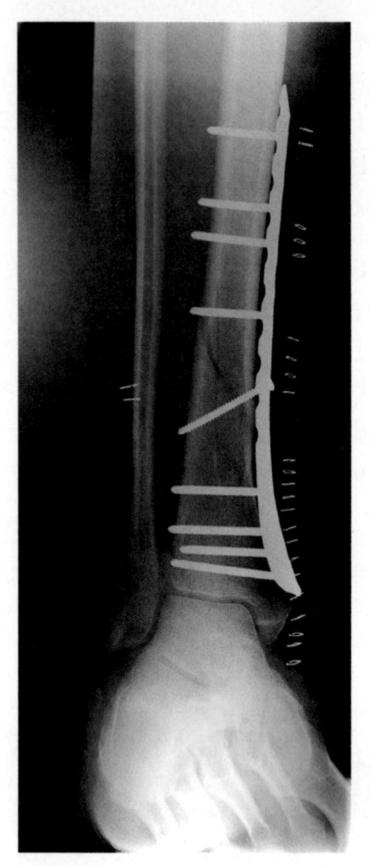

Figure 6 This X-ray was taken after screws and plates were placed in the bone to hold it together.

Bone Injuries and Diseases

Because bones are made of living tissues, they are at risk of injury and disease. A wooden board is hard and strong, but if enough pressure is applied to it, the board will break. The same is true for bones.

Broken Bones

A broken bone is called a fracture (FRAK chur). Broken bones are able to repair themselves, but it is a slow process. A broken bone must be held together while it heals, just as you hold two glued objects together while the glue dries. Often, a person wears a cast to keep broken bones in place. Sometimes metal plates and screws like those in **Figure 6** hold bones together while they heal.

Arthritis

You are able to move because your skeleton bends and rotates at joints. If the joints become irritated, it can be painful to move. **Arthritis** (ar THRI tus) *is a disease in which joints become irritated or inflamed, such as when cartilage in joints is damaged or wears away.* Arthritis is most common in adults, but it can also affect children.

Osteoporosis

Another common bone disease is **osteoporosis** (ahs tee oh puh ROH sus), *which causes bones to weaken and become brittle.* Anyone can develop osteoporosis, but it is most common in women over the age of 50. Osteoporosis can change a person's skeleton and cause fractures, as shown in **Figure 7.**

Osteoporosis

Figure 7 Osteoporosis can weaken the skeletal system over time.

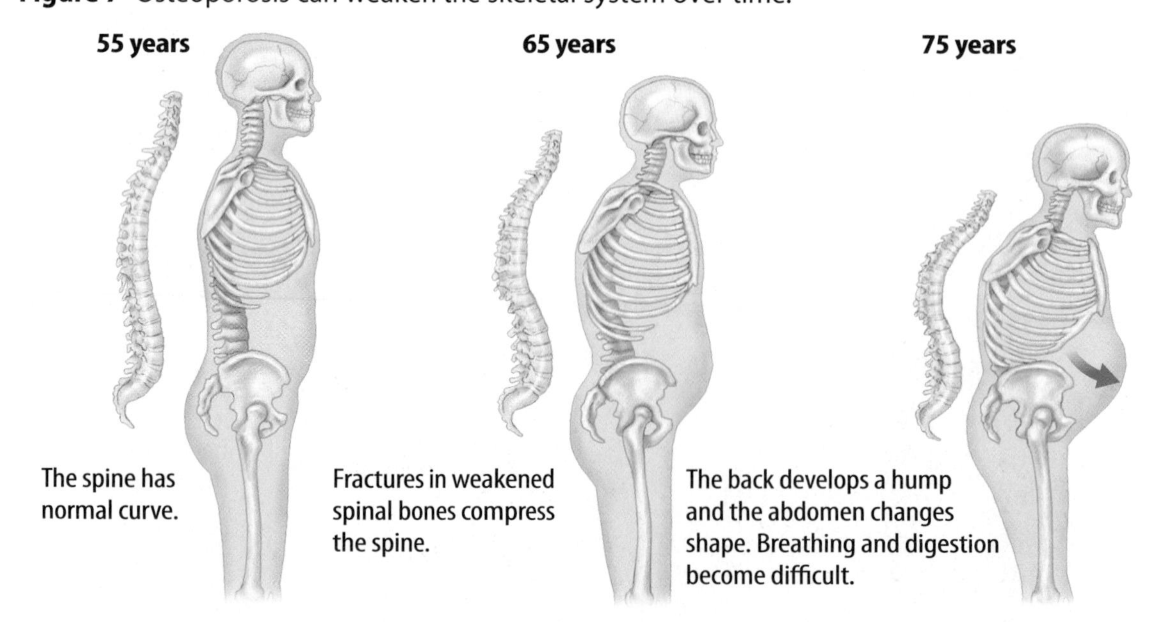

Running and walking build bone in your hips, legs, and feet.

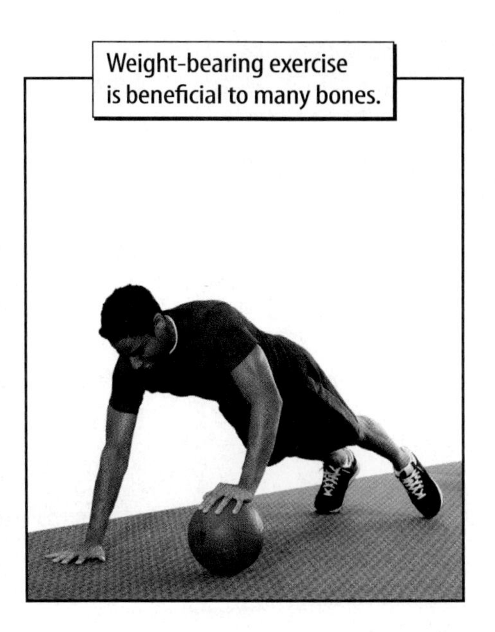

Figure 8 Exercises that place weight on bones or work bones against gravity help keep the skeletal system healthy.

Healthy Bones

One of the best ways to keep bones healthy is to exercise. Certain types of exercise, such as those shown in **Figure 8,** place weight on bones. This strengthens bones and builds new bone tissue. Without exercise, bones weaken and lose mass.

A balanced diet also keeps bones healthy. Bones especially need calcium and vitamin D. Calcium makes bones strong. It is also necessary for cellular processes in blood, nerve, and muscle cells. If you do not have enough calcium in your diet, your body will use the calcium stored in your bones. If your body uses too much of the stored calcium, your bones can become weak. Vitamin D is also important because it helps the body use calcium.

Active Reading **7. Consider** What are three things you did today to help your bones?

____

____

____

The Skeletal System and Homeostasis

You might recall that **homeostasis** is an organism's ability to maintain a stable internal environment. Homeostasis requires that all body systems function properly together. Because bones supply calcium to your nerves, muscles, and heart, a healthy skeletal system is important in maintaining your body's homeostasis.

REVIEW VOCABULARY

homeostasis
the ability of an organism to maintain a stable internal environment

Bones also help you respond to unpleasant stimuli, such as a mosquito bite. Working together with muscles, bones enable you to move away from unpleasant stimuli or danger, or even swat a mosquito.

8. NGSSS Check **Justify** Highlight how the skeletal system helps the body maintain homeostasis. SC.6.L.14.5

Lesson Review 1

Visual Summary

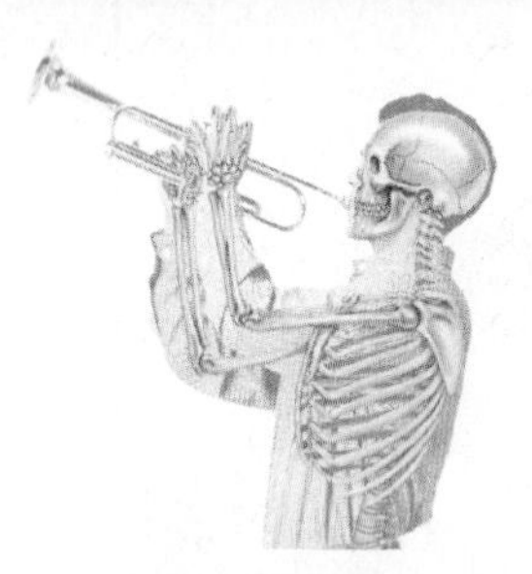

The skeletal system contains all of the bones, ligaments, and cartilage in the body.

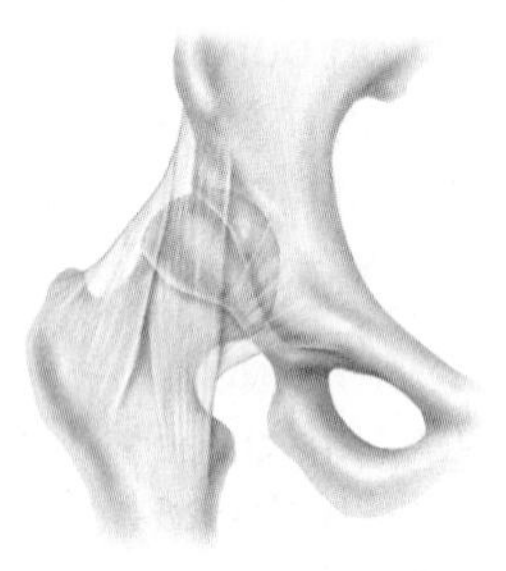

Bones connect to each other at joints.

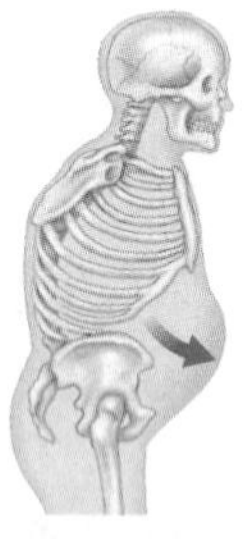

Osteoporosis is a disease that causes bones to weaken and become brittle.

Use Vocabulary

1. Strong, flexible tissue that covers the ends of bones is called ______________.
2. Bones connect to other bones at ______________.
3. A person with ______________ has irritated joints.

Understand Key Concepts

4. Which is NOT a part of the periosteum? SC.6.L.14.5
   - (A) blood vessels
   - (B) bone cells
   - (C) bone marrow
   - (D) nerves
5. **Distinguish** between cartilage and ligaments. SC.6.L.14.5
6. **Give** an example of how the skeletal system interacts with the nervous system. SC.6.L.14.5

Interpret Graphics

7. **Organize** Complete the graphic organizer below and, in the ovals, list the four major functions of the skeletal system. SC.6.L.14.5

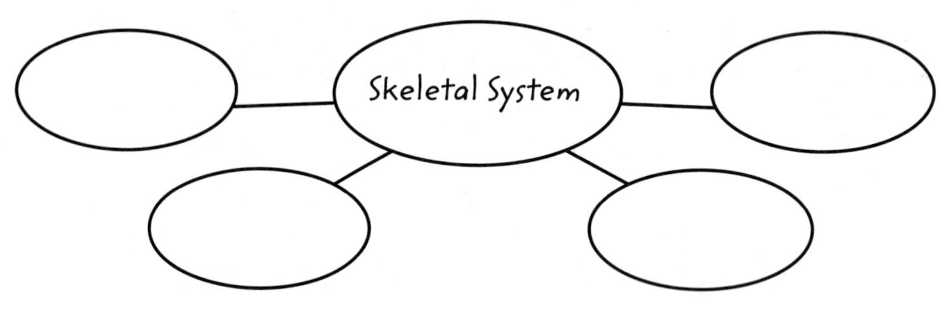

Critical Thinking

8. **Summarize** how bones help the body maintain homeostasis. SC.6.L.14.5
9. **Evaluate** why it is better for an adult skull to have immovable joints and not movable joints. SC.6.L.14.5

A Bionic Arm

HOW NATURE WORKS

How brains control mechanical arms

Imagine what your arms would be like without muscles. They would simply swing from your shoulders like pendulums. You would not be able to control them. For many years prosthetic, or artificial, arms looked real, but they didn't work like real arms. Recently scientists have developed a bionic, or mechanical, arm. Signals from the patient's brain control it.

1. **Doctors perform surgery and attach nerves that were once part of the damaged arm to chest muscles. These nerves sent signals to the patient's arm muscles.**
2. **When the patient's brain sends signals to move the arm or the hand, the signals travel from the brain to the chest muscles.**
3. **Electronic sensors in the bionic arm's harness detect the chest muscle moving. The sensors send corresponding signals down the bionic arm.** 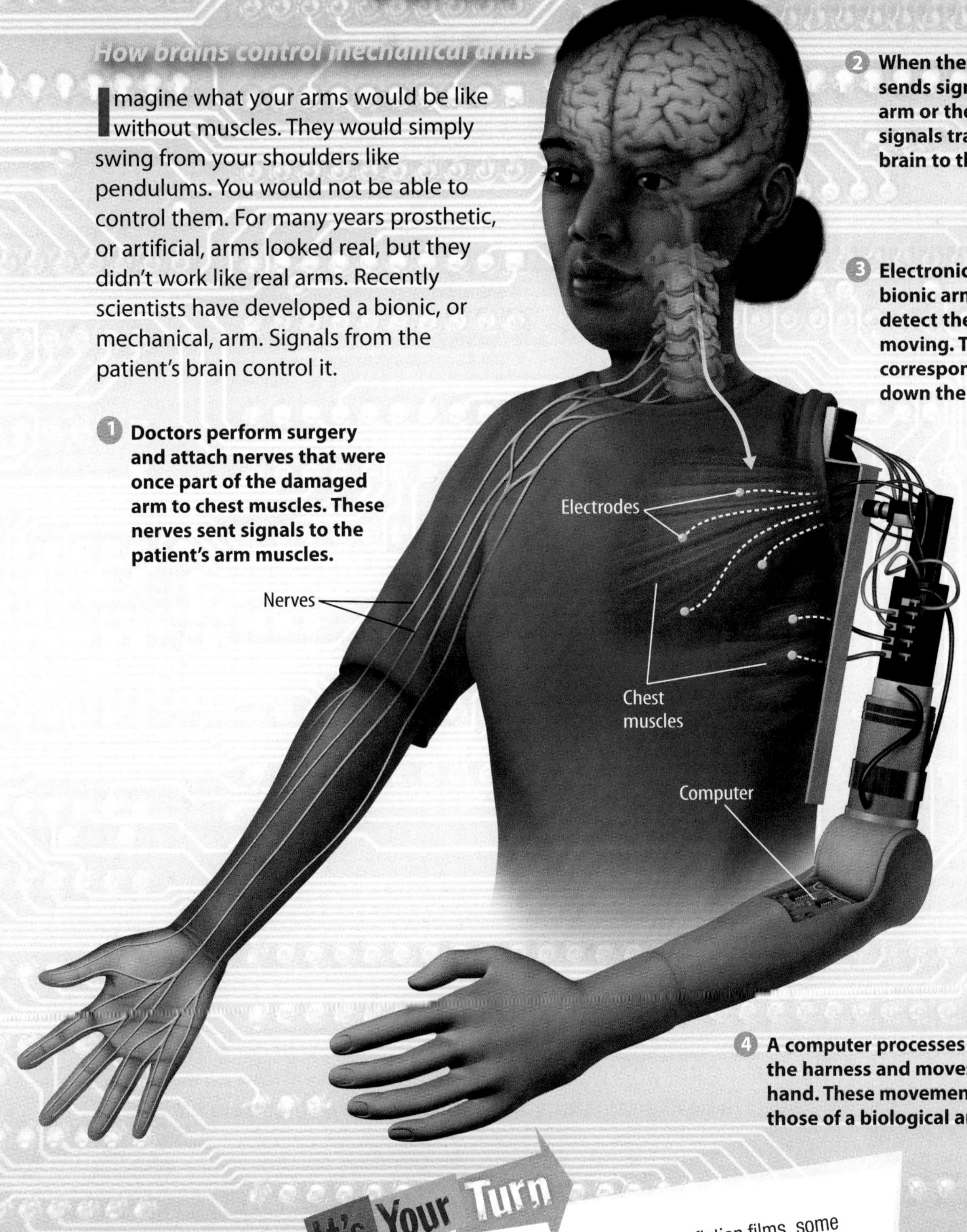
4. **A computer processes the signals from the harness and moves the arm and hand. These movements are similar to those of a biological arm and hand.**

It's Your Turn

RESEARCH AND REPORT In science-fiction films, some characters have bionic body parts such as ears and eyes. Are any other bionic body parts in development? Summarize your findings in a paragraph. **LA.6.4.2.2**

Lesson 2

The Muscular SYSTEM

ESSENTIAL QUESTIONS

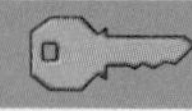
What does the muscular system do?

How do types of muscle differ?

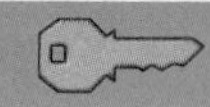
How does the muscular system interact with other body systems?

Vocabulary

muscle p. 461
skeletal muscle p. 463
voluntary muscle p. 463
cardiac muscle p. 464
involuntary muscle p. 464
smooth muscle p. 464

Florida NGSSS

LA.6.2.2.3 The student will organize information to show understanding (e.g., representing main ideas within text through charting, mapping, paraphrasing, summarizing, or comparing/contrasting);

SC.6.L.14.5 Identify and investigate the general functions of the major systems of the human body (digestive, respiratory, circulatory, reproductive, excretory, immune, nervous, and musculoskeletal) and describe ways these systems interact with each other to maintain homeostasis.

SC.6.N.1.5 Recognize that science involves creativity, not just in designing experiments, but also in creating explanations that fit evidence.

SC.7.N.1.3 Distinguish between an experiment (which must involve the identification and control of variables) and other forms of scientific investigation and explain that not all scientific knowledge is derived from experimentation.

SC.8.N.1.2 Design and conduct a study using repeated trials and replication.

Launch Lab

SC.6.N.1.1, SC.6.L.14.5

15 minutes

Can you control all your muscles?

Can you feel your heart beating? Your body contains many muscles that you can control by thinking about them. However, not all types of muscle are controllable.

Procedure

1. Shake hands with another student. Did you have to think about this action?
2. Rest your index and middle fingers on the thumb side of your wrist until you can feel your pulse. Can you change the speed of your pulse by thinking about it?

Think About This

1. Make a list of the muscles in your body that you can consciously control. What are their functions?

2. Key Concept Think of other muscles in your body, besides your heart, that work without you thinking about them. How do the functions of these muscles differ from the ones you consciously control?

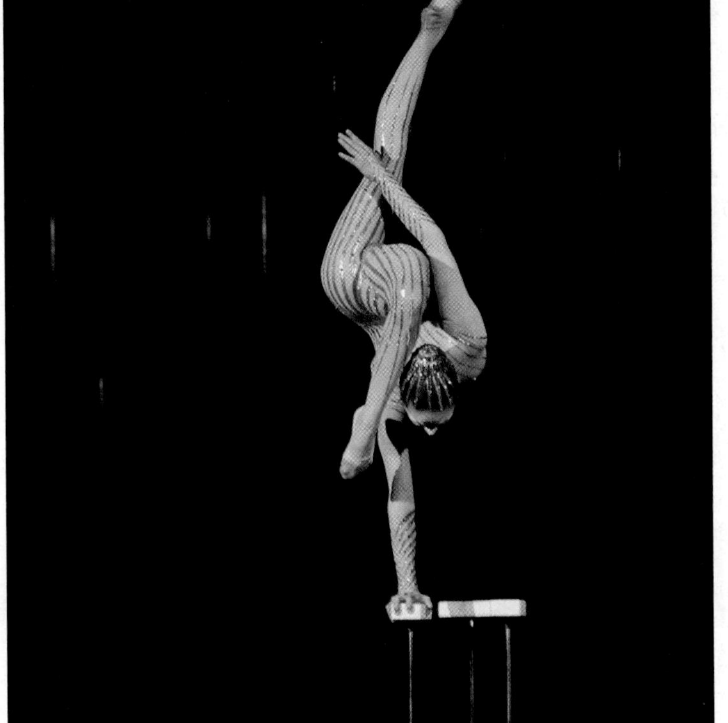

Inquiry How do you think she can do that?

1. This person is a contortionist (kun TOR shuh nist)—someone who can bend and flex his or her body in unusual ways. How is this person able to stay in position?

Functions of the Muscular System

What comes to mind when you think of muscles? Maybe you think about the muscles in your arms or legs that enable you to carry books or run fast. Movement is an important function of the muscular system. But muscles are also important for protection, stability, and maintaining body temperature.

Although muscles have different functions, all muscle tissues are made of cells that contract. *A* **muscle** *is made of strong tissue that can contract in an orderly way.* When a muscle contracts, the cells of the muscle become shorter, as shown in **Figure 9.** When the muscle relaxes, the cells return to their original length.

Movement

Many of your muscles are attached to bone and enable your skeleton to move. Bones move when these muscles contract. This movement can be fast, such as when you run, or slow, like when you stretch.

You also have many muscles in your body that are not attached to bones. The contractions in these muscles cause blood and food to move throughout your body. They also cause your heart to beat and the hair on your arms to stand on end when you get goose bumps.

2. **NGSSS Check Select** Circle one major function of the muscular system. SC.6.L.14.5

Figure 9 A muscle cell works by contracting and relaxing.

3. **Visual Check Choose** Which muscle is relaxed and which is contracted?

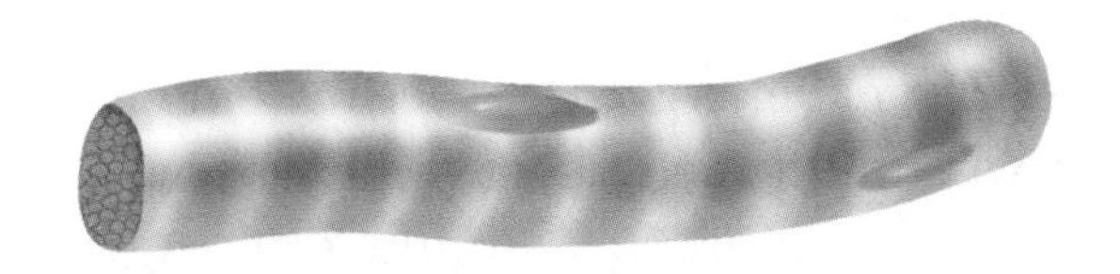

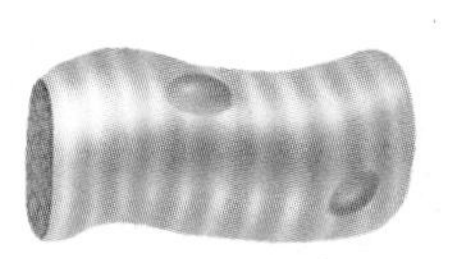

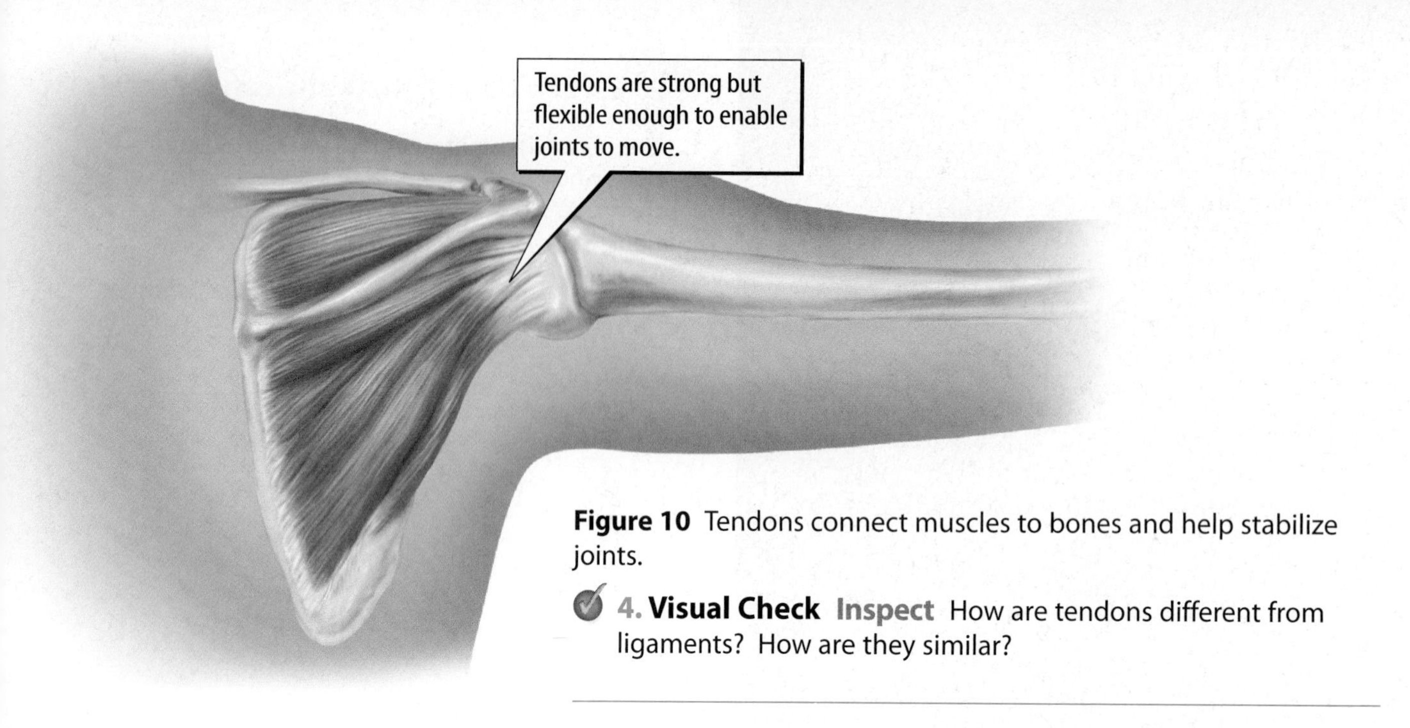

Figure 10 Tendons connect muscles to bones and help stabilize joints.

4. Visual Check Inspect How are tendons different from ligaments? How are they similar?

____

____

Stability

Muscles that are attached to bones support your body and help you balance. If your body becomes unstable, such as when you trip and lose your balance, your muscles pull in different directions and you regain your balance.

Tendons attach muscles to bones, as shown in **Figure 10.** Look at the back of your hand and move your fingers. Do you see cordlike structures moving under your skin? These are tendons. You can feel the Achilles tendon above either of your heels. Tendons work with muscles and keep joints in place when your body moves. Tendons also help hold your body in a proper posture, or shape.

Figure 11 Muscles cover the skeleton.

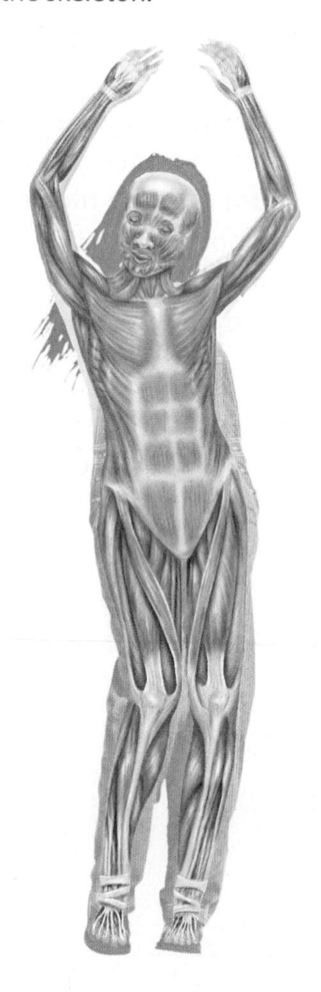

Protection

Muscles protect your body. As shown in **Figure 11,** muscles cover most of your skeleton. Muscles also cover most of the organs inside your body. Muscles are like a layer of padding. They surround your abdomen, chest, and back and protect your internal organs.

Temperature Regulation

Have you ever been in a cold environment and started shivering? Shivering is when muscles contract rapidly and change chemical energy to thermal energy. The thermal energy helps raise your body's temperature. This is important because a human's body temperature must stay around 37°C in order for the body to function properly. Muscles also change chemical energy to thermal energy during exercise. This is why you feel warm after physical activity.

Types of Muscles

Your body has three different types of muscles: skeletal, cardiac, and smooth. Each of these muscle tissues is specialized for a different function.

Skeletal Muscle

The type of muscle that attaches to bones is **skeletal muscle**. Skeletal muscles, as shown in **Figure 12,** are also called **voluntary muscles,** *which are muscles that you can consciously control.* Have you ever played with a puppet on a string? You controlled how the puppet moved. In a similar way, you control how skeletal muscles move. The contractions of skeletal muscles can be quick and powerful, such as when you run fast. However, contracting these muscles for long periods of time can exhaust or cramp them.

How Skeletal Muscles Work Skeletal muscles work by pulling on bones. Because muscles cannot push bones, they must work in pairs. **Figure 12** illustrates how an arm's biceps (BI seps) and triceps (TRI seps) muscles work as a pair.

Active Reading **5. Identify** Highlight why skeletal muscles must work in pairs.

Changes in Skeletal Muscles Your skeletal muscles can change throughout your lifetime. If you exercise, your muscle cells increase in size and the entire muscle becomes larger and stronger.

Active Reading

FOLDABLES® LA.6.2.2.3

Fold a sheet of paper into thirds to make a trifold book. Use it to organize your notes on the three types of muscle tissue.

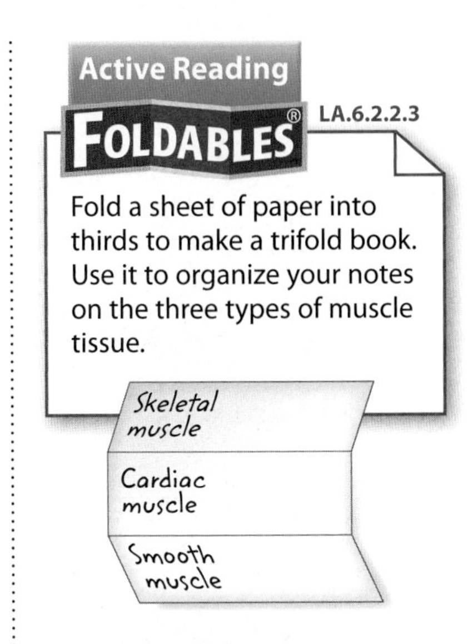

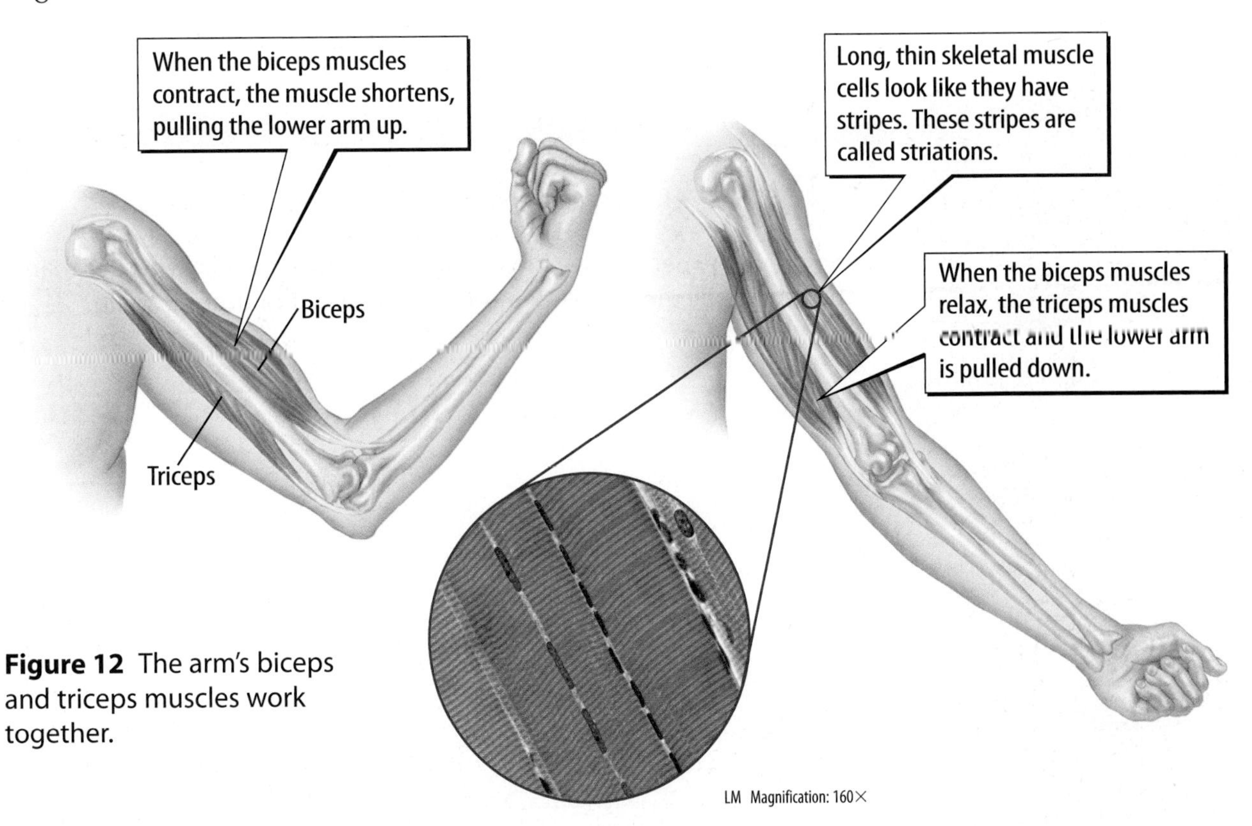

Figure 12 The arm's biceps and triceps muscles work together.

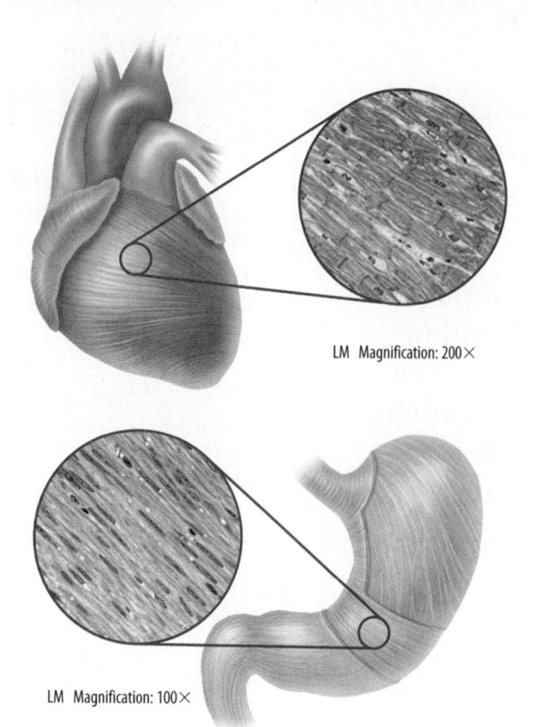

Figure 13 The heart is made of cardiac muscle. The stomach is lined with three layers of smooth muscle.

Cardiac Muscle

Your heart is made of **cardiac** (KAR dee ak) **muscles,** *which are found only in the heart.* A cardiac muscle is a type of **involuntary muscle,** *which is muscle you cannot consciously control.* When cardiac muscles contract and relax, they pump blood through your heart and through vessels throughout your body. Cardiac muscle cells, shown in **Figure 13,** have branches with discs at the ends. The discs send signals to other cardiac cells, and they all contract at nearly the same time.

Smooth Muscle

Blood vessels and many organs, such as the stomach shown in **Figure 13,** are lined with smooth muscles. **Smooth muscles** *are involuntary muscles named for their smooth appearance.* Contraction of smooth muscles helps move material through the body, such as food in the stomach, and controls the movement of blood through vessels.

6. List What are the three types of muscles?

____

____

____

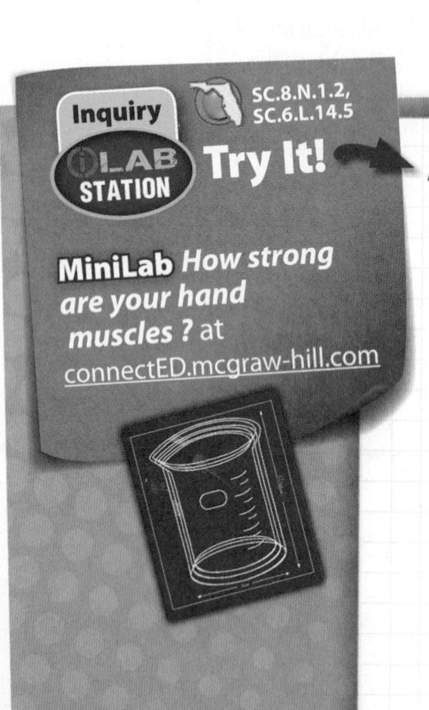

Apply It! After you complete the lab, answer this question.

1. Would you expect to be able to squeeze a material that is harder than the tennis ball more or less in the same amount of time? Explain.

Healthy Muscles

Recall that a good diet keeps your bones healthy. Your muscles benefit from a healthy diet, too. All muscles require energy to contract. This energy comes from the food you eat. Eating a diet full of nutrients such as protein, fiber, and potassium can help keep muscles strong.

Exercise also helps keep muscles healthy. Muscle cells decrease in size and strength without exercise, as shown in **Figure 14.** Decreased muscle strength can increase the risk of heart disease and bone injuries. It can also make joints less stable.

Word Origin
cardiac
from Greek *kardia,* means "heart"

Figure 14 Muscles lose size, strength, and mobility if they are not exercised.

Active Reading **7. Construct** Complete the missing information in the box below.

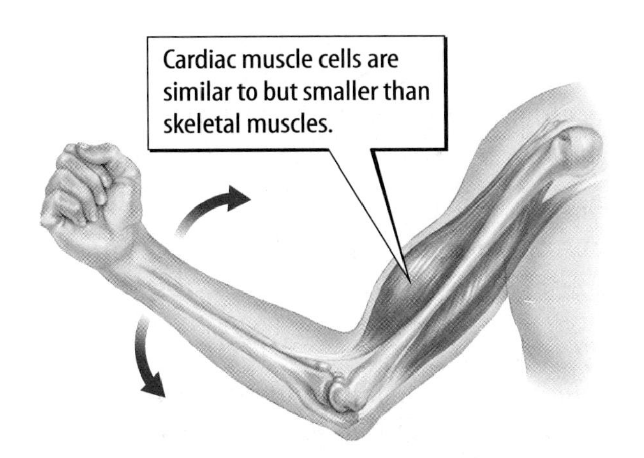

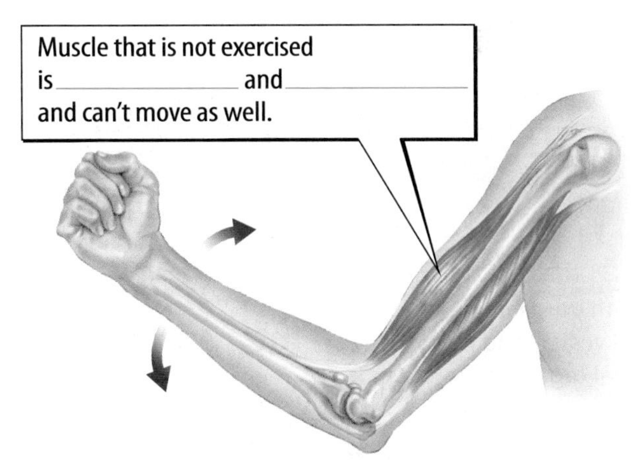

The Muscular System and Homeostasis

There are many ways the muscular system helps your body maintain homeostasis. You are probably sitting in a room where the temperature is somewhere between 21°C and 27°C. However, your body temperature is around 37°C. Your body must stay at this temperature to function well. As you have read, muscle contractions **convert** chemical energy to thermal energy and keep your body warm.

Academic Vocabulary
convert
(verb) to change something into a different form

When you exercise, your cells need more oxygen and release more waste, such as carbon dioxide. The cardiac muscles of your heart help maintain homeostasis by contracting more often. When it contracts faster, the heart pumps more blood, and more oxygen is carried to the cells.

8. NGSSS Check Compile Underline how muscle helps maintain homeostasis in the body. **SC.6.L.14.5**

Lesson Review 2

Visual Summary

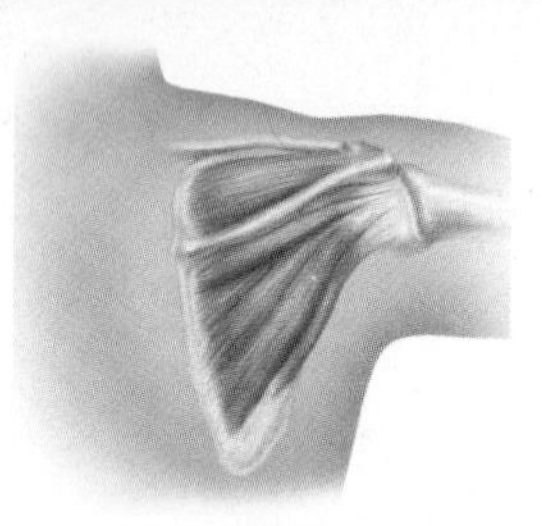

Your muscular system is made of different types of muscles. Skeletal muscles attach to bone and are muscles you can control.

Smooth muscles line blood vessels and many internal organs.

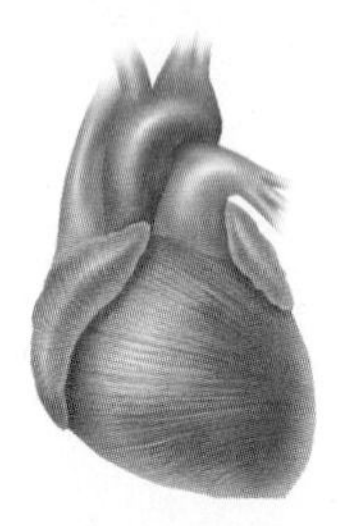

The heart is made of cardiac muscle.

Use Vocabulary

1. Muscle that pumps blood through the body is called ________.

2. A person is unable to control the contractions of ________.

3. Strong tissue that contracts is called ________.

Understand Key Concepts

4. Which is a voluntary muscle? SC.6.L.14.5
   - (A) biceps
   - (B) stomach
   - (C) blood vessel
   - (D) small intestine

5. **Explain** how the muscular system regulates body temperature. SC.6.L.14.5

6. **Distinguish** between a cardiac muscle and a smooth muscle. SC.6.L.14.5

Interpret Graphics

7. **Identify** Fill in the graphic organizer below to identify three functions of the muscular system. SC.6.L.14.5

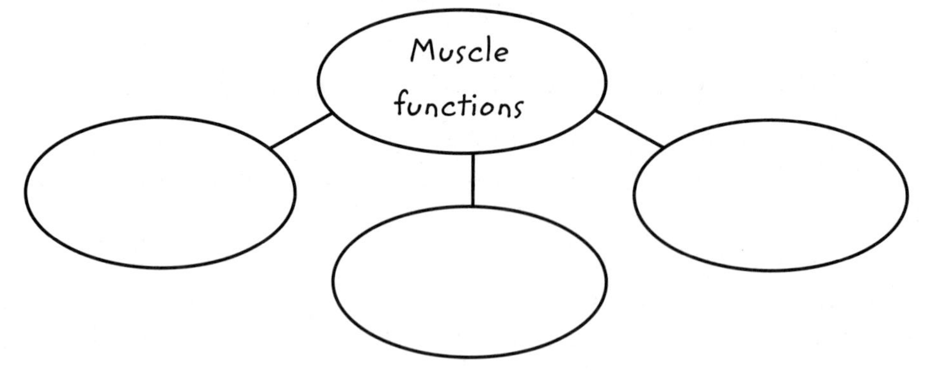

Critical Thinking

8. **Predict** what would happen if the smooth muscles in a person's body could not contract. SC.6.L.14.5

9. **Assess** the importance of exercise for muscle health.

Details

Identify how muscle contractions affect the body's temperature regulation.

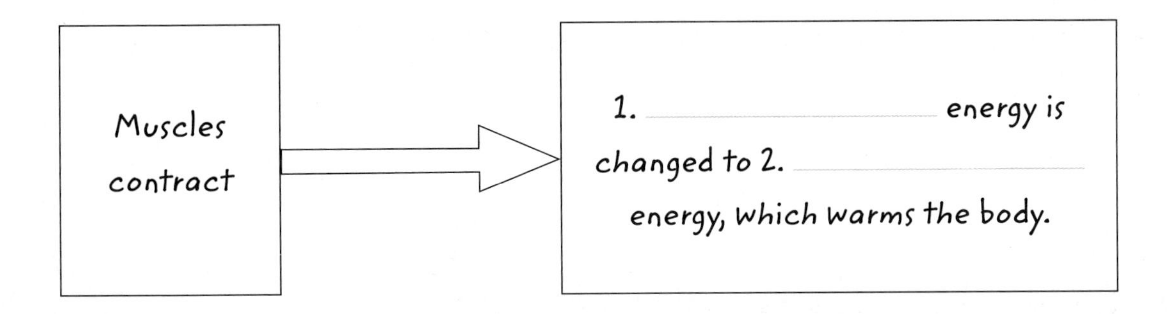

Sequence how the muscular system helps maintain homeostasis.

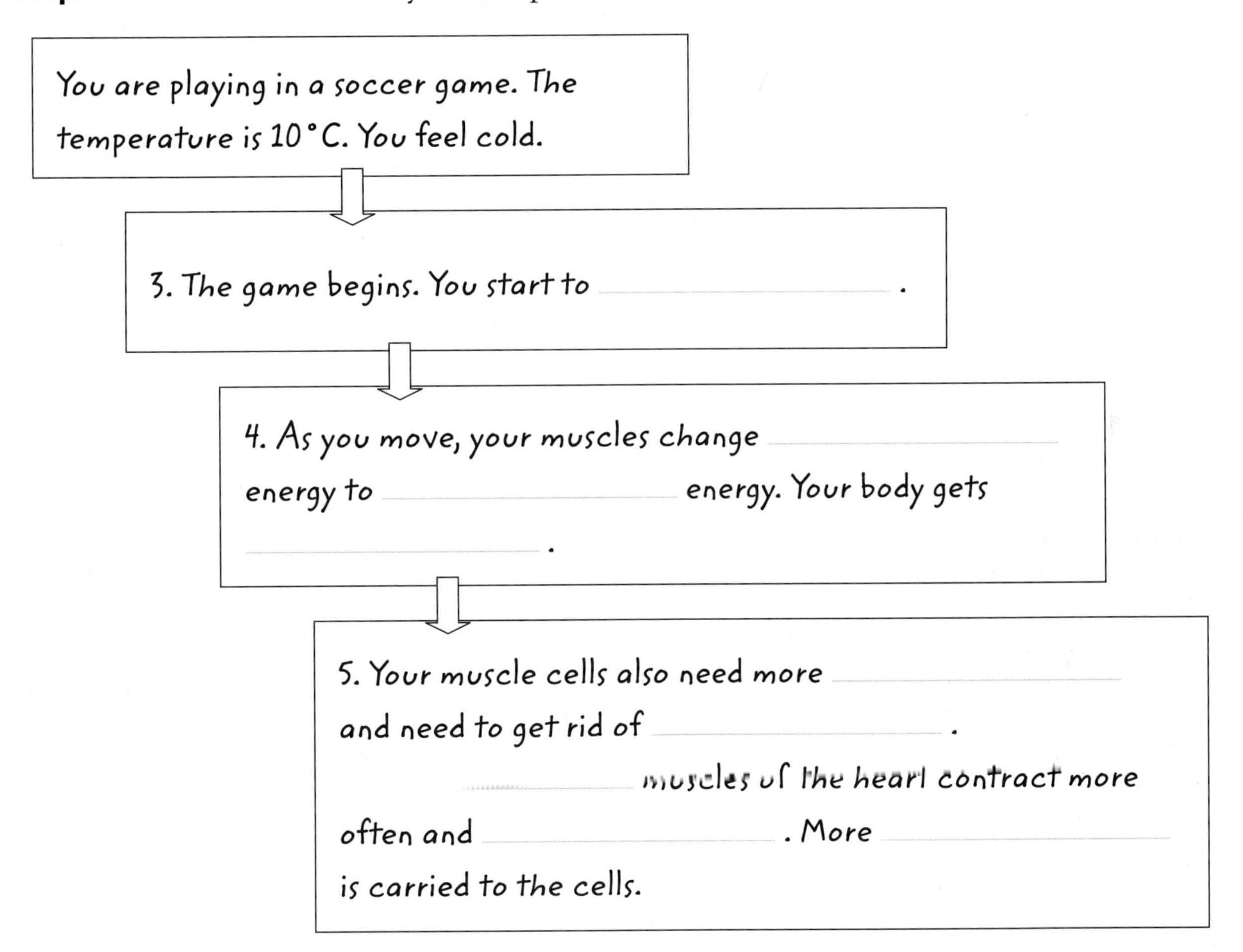

Lesson 3

The SKIN

ESSENTIAL QUESTIONS

What does the skin do?

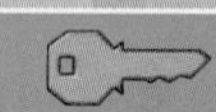
How do the three layers of skin differ?

How does the skin interact with other body systems?

Vocabulary

integumentary system p. 469
epidermis p. 471
melanin p. 471
dermis p. 471
bruise p. 472

Florida NGSSS

LA.6.2.2.3 The student will organize information to show understanding (e.g., representing main ideas within text through charting, mapping, paraphrasing, summarizing, or comparing/contrasting);

MA.6.A.3.6 Construct and analyze tables, graphs, and equations to describe linear functions and other simple relations using both common language and algebraic notation.

SC.6.L.14.5 Identify and investigate the general functions of the major systems of the human body (digestive, respiratory, circulatory, reproductive, excretory, immune, nervous, and musculoskeletal) and describe ways these systems interact with each other to maintain homeostasis.

SC.6.N.1.1 Define a problem from the sixth grade curriculum, use appropriate reference materials to support scientific understanding, plan and carry out scientific investigation of various types, such as systematic observations or experiments, identify variables, collect and organize data, interpret data in charts, tables, and graphics, analyze information, make predictions, and defend conclusions.

SC.7.N.1.3 Distinguish between an experiment (which must involve the identification and control of variables) and other forms of scientific investigation and explain that not all scientific knowledge is derived from experimentation.

Inquiry Launch Lab

SC.7.N.1.3, SC.6.L.14.5

5 minutes

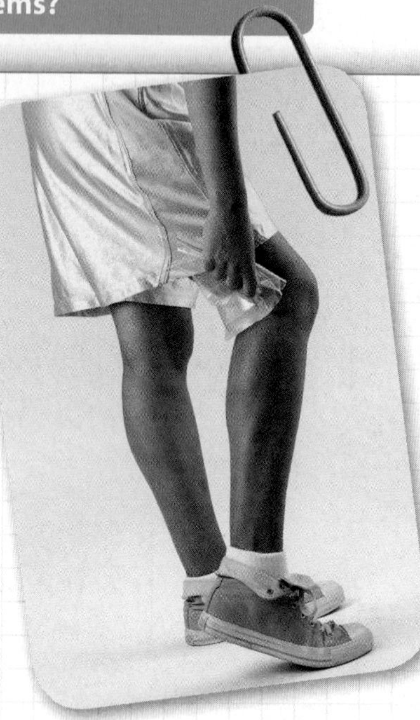

How does your skin protect your body?

Your skin is your body's first line of defense. When you touch something with your fingers, you can instantly tell if it is hot or cold. Are all parts of your body equally sensitive?

Procedure

1. Read and complete a lab safety form.
2. Touch the back of your hand with an **ice cube** in a **plastic bag.**
3. Now do the same to the back of your knee.

Analyze and Conclude

1. Which area was more sensitive to cold?

2. How do you think the skin senses temperature?

3. Key Concept How does sensitivity to temperature protect the body?

Inquiry **What do you think this is?**

1. You might think this is a picture of a landscape on another planet. However, this flaky image is what your skin looks like under a microscope. What do you think the spikes are?

Functions of the Skin

Touch your fingertips, your arm, and your face. The soft tissue you feel is the outermost layer of your skin. Skin is the largest organ of your body. It is part of the **integumentary** (ihn teh gyuh MEN tuh ree) **system,** *which includes all the external coverings of the body, including the skin, nails, and hair.* Most parts of the integumentary system are shown in **Figure 15.** Like bones and muscles, skin serves many different functions in your body.

WORD ORIGIN

integumentary
from Latin *integere,* means "to cover"

Protection

When you look at yourself in a mirror, you cannot see the bones, muscles, or other parts of your skeletal and muscular systems. Instead, you see your skin. Skin covers your bones and muscles and protects them from the external environment. It keeps your body from drying out in sunlight and wind. Skin also protects the cells and tissues under the skin from damage. Skin is the first line of defense against dirt, bacteria, viruses, and other substances that might enter your body.

Active Reading 2. **Infer** What would happen to your body if you had no skin?

Figure 15 The integumentary system includes skin, nails, and hair.

Sensory Response

Close your eyes, and feel the surface of your desk and the objects on top. Even with your eyes closed, you can tell the difference between the desk, a book, paper, and pencils. This is because your skin has special cells called sensory receptors that detect texture. Sensory receptors also detect temperature and sense pain. The more sensory receptors there are in an area of skin, the more sensitive it is.

Temperature Regulation

Skin helps control body temperature. When you exercise, sweat comes from tiny holes, or pores, on the skin's surface, as shown in **Figure 16**. Sweating is one way skin lowers your body temperature. As sweat evaporates, excess thermal energy leaves the body and the skin cools.

Another way that skin lowers body temperature is by releasing thermal energy from blood vessels. Has your face ever turned red while exercising? The girl in **Figure 16** has a red face because the blood vessels near the skin's surface dilated, or enlarged. This increases the surface area of the blood vessels and releases more thermal energy.

Active Reading 3. **Identify** Highlight how skin regulates body temperature.

Production of Vitamin D

If your skin is exposed to sunlight, it can make vitamin D. Your body needs vitamin D to help it absorb calcium and phosphorous and to promote the growth of bones. Your skin is not the only source of vitamin D. Vitamin D is usually added to milk and is found naturally in certain types of fish.

Elimination

Normal cellular processes produce waste products. The skin helps eliminate these wastes. Water, salts, and other waste products are removed through the pores. This occurs all the time, but you might only notice it when you sweat during exercise.

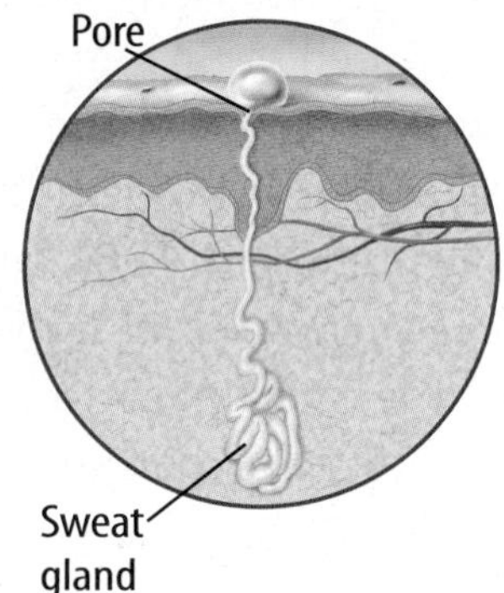

Figure 16 During exercise, sweat evaporates and blood vessels enlarge. This releases thermal energy.

Apply It! After you complete the lab, answer these questions.

1. Is this an example of elimination? Explain.

2. What is being eliminated from your body?

3. What would happen if you did not have pores?

Skin Structures

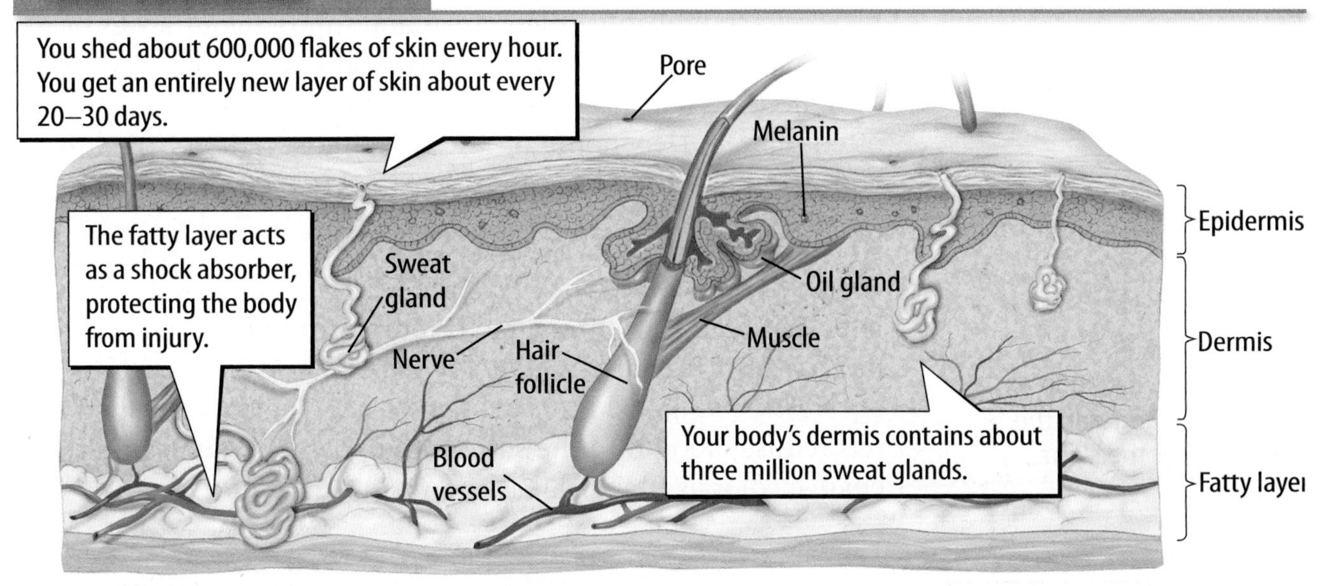

Figure 17 Skin has three layers, each with different functions.

Structures of the Skin

The skin that you see and feel on your body is the outermost layer of skin. Below it are two other layers of skin. These layers, shown in **Figure 17,** differ in structure and function.

Epidermis

The **epidermis** (eh puh DUR mus) *is the outermost layer of skin and the only layer in direct contact with the outside environment.* The epidermis is tough but thin. The epidermis on your eyelids is thinner than a sheet of paper. Cells of the epidermis are constantly shed and replaced by new cells.

One important function of the epidermis is the production of melanin (MEH luh nun). **Melanin** *is a pigment that protects the body by absorbing some of the Sun's damaging ultraviolet rays.*

Dermis

Below the epidermis is the dermis. The **dermis** *is a thick layer of skin that gives skin strength, nourishment, and flexibility.* The dermis contains sweat glands, blood vessels, nerves, hair follicles, and muscles, as shown in **Figure 17.** When the muscles in the dermis contract, you get goose bumps.

Fatty Layer

The innermost layer of skin insulates the body, acts as a protective padding, and stores energy. This layer is sometimes called the fatty layer. It can be very thin or very thick, depending on its location on the body.

Active Reading 5. **Distinguish** Underline how the three skin layers differ.

Math Skills MA.6.A.3.6

Using Proportions

The ratios $\frac{5}{1}$ and $\frac{25}{5}$ are equivalent, so they can be written as the proportion $\frac{5}{1} = \frac{25}{5}$. When ratios form a proportion, the cross products are equal. In the above proportion, $5 \times 5 = 25 \times 1$. You can use cross products to find a missing term. For example, if each 1 cm^2 of skin contains 300 pores, how many pores are there in 5 cm^2 of skin?

$$\frac{1\ cm^2}{300\ pores} = \frac{5\ cm^2}{n\ pores}$$

$$1 \times n = 300 \times 5;$$

$$n = 1{,}500\ pores$$

Practice

4. The palm of the hand has about 500 sweat glands per 1 cm^2. How many sweat glands would there be on a palm measuring 7 cm by 8 cm?

Skin Injuries and Repair

You have probably fallen down and injured your knees or other parts of your body. You might also have damaged your skin. Because skin is exposed to the outside environment, it is often injured. Depending on the type and severity of the injury, your body has different ways to repair skin.

Bruises

Have you ever bumped into the edge of a table and noticed later that your skin turned red or purple and blue? You probably had a bruise. *A* **bruise** *is an injury where blood vessels in the skin are broken, but the skin is not cut or opened.* The broken blood vessels release blood into the surrounding tissue. Bruises usually change color as they heal, as shown in **Figure 18.**

Figure 18 Bruises change color as they heal due to chemical changes in the blood under the skin's surface.

A new bruise is red because blood pools under the skin.

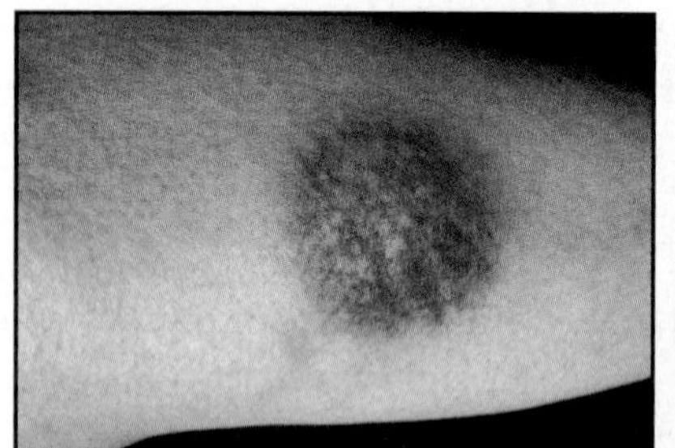

After 1–2 days, the red blood cells begin to break down, and the bruise darkens.

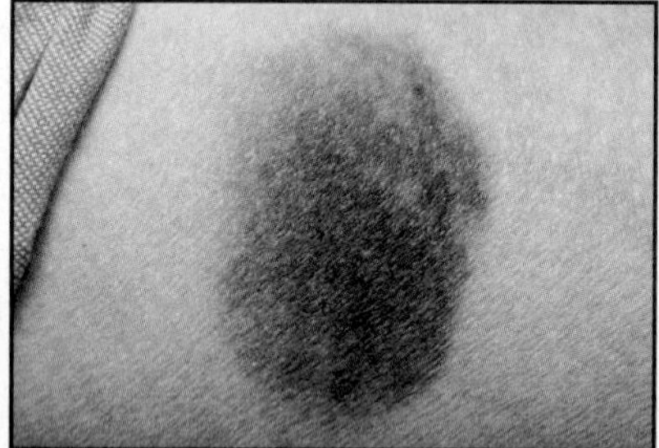

After 5–10 days, the bruise turns greenish-yellow. After about 2 weeks, it fades away.

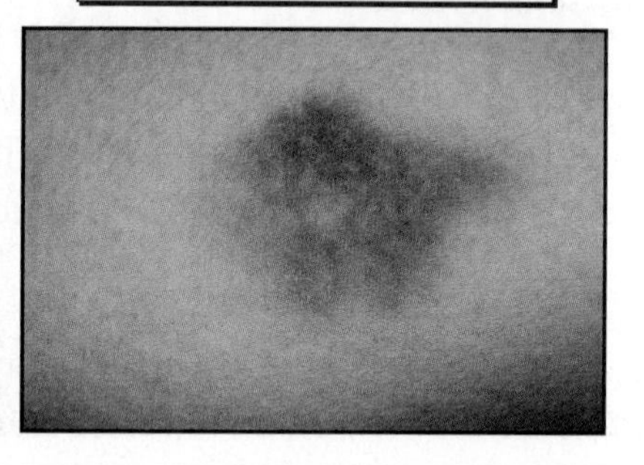

Cuts

When you break one or more layers of skin, it is called a cut. Cuts often cut blood vessels, too. The released blood will usually thicken and form a scab over the cut. The scab prevents dirt and other outside substances from entering the body. Skin heals by producing new skin cells that eventually repair the cut. Some cuts are too large to heal naturally. If that happens, stitches might be needed to close the cut while it heals.

Active Reading FOLDABLES® LA.6.2.2.3

Make a horizontal half book from a sheet of notebook paper. Use it to record information about the different types of skin injuries and how the body repairs them.

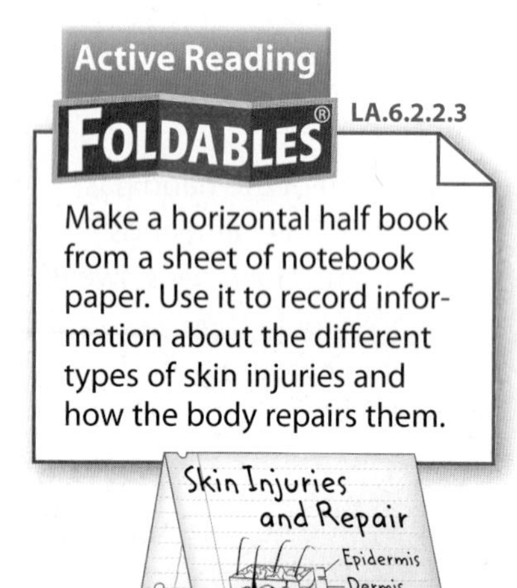

Burns

Have you ever injured your skin by contact with hot water or hot food? If so, then you had a burn. You might think that burns only occur from touching hot objects. However, burns can also be caused by touching extremely cold objects, chemicals, radiation (such as sunlight), electricity, or friction (rubbing). **Table 2** describes the three degrees, or levels, of burns.

Active Reading **6. Decide** (Circle) things that can cause your skin to burn.

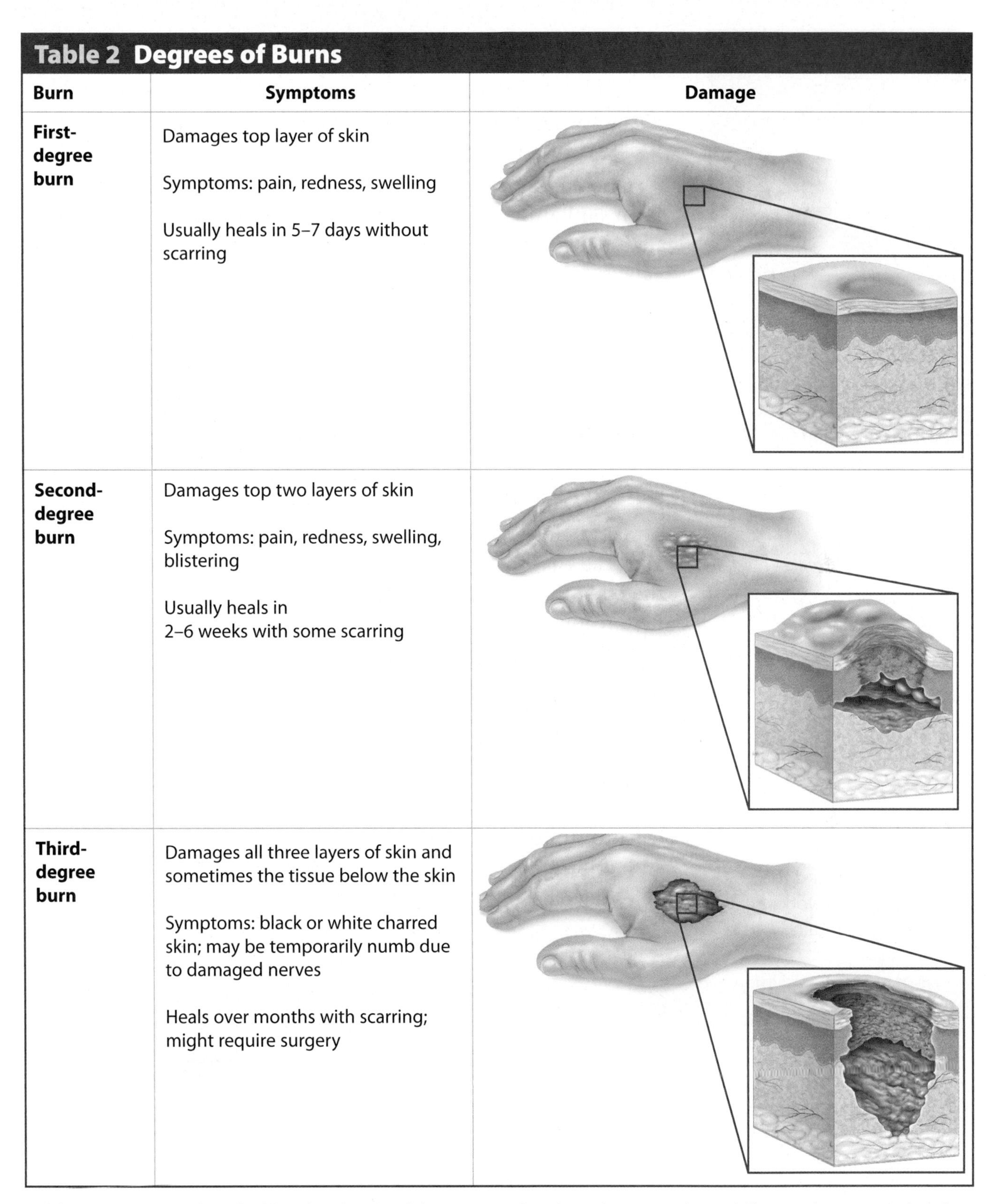

Table 2 Degrees of Burns

Burn	Symptoms	Damage
First-degree burn	Damages top layer of skin Symptoms: pain, redness, swelling Usually heals in 5–7 days without scarring	
Second-degree burn	Damages top two layers of skin Symptoms: pain, redness, swelling, blistering Usually heals in 2–6 weeks with some scarring	
Third-degree burn	Damages all three layers of skin and sometimes the tissue below the skin Symptoms: black or white charred skin; may be temporarily numb due to damaged nerves Heals over months with scarring; might require surgery	

Table 2 Burns are classified by the depth of damage to the skin. There are three different degrees, or levels, of burns.

7. **Visual Check** **Formulate** If swelling and blisters appear on the surface of a burned area of the skin, what degree burn would it be?

____

Figure 19 Protecting your skin from sunlight is important for maintaining skin health.

Healthy Skin

Just as diet and exercise are important to healthy skeletal and muscular systems, good life choices can keep your skin functioning well.

One important thing you can do for your skin is protect it from sunlight. The ultraviolet rays in sunlight can cause permanent damage to the skin, including wrinkles, dry skin, and skin cancer. Like the person in **Figure 19,** you can protect your skin by using sunscreen. You can also wear protective clothing and avoid outdoor activities during the middle of the day.

Another way to keep your skin healthy is to eat a balanced diet. You can also apply lotion to your skin to help keep it moist and use gentle soaps to clean it.

The Skin and Homeostasis

Skin helps your body maintain homeostasis in many ways. You read that the skin can make vitamin D and that it protects the body from outside substances. These functions of the skin help regulate the body's internal environment.

The skin also works with other body systems to maintain homeostasis. You read earlier how the skin and the circulatory system help cool the body when it becomes overheated. If the body becomes cold, blood vessels constrict, or narrow, reducing thermal energy loss. As you can see in **Figure 20,** the skin also works with the nervous and muscular systems to help the body react to stimuli.

8. **NGSSS Check** Combine Highlight how the skin interacts with other body systems to help maintain homeostasis. SC.6.L.14.5

Figure 20 The skin, the nervous system, and the muscular system work together to maintain homeostasis.

Reaction to Stimuli

1. Pain receptors in skin sense pain.

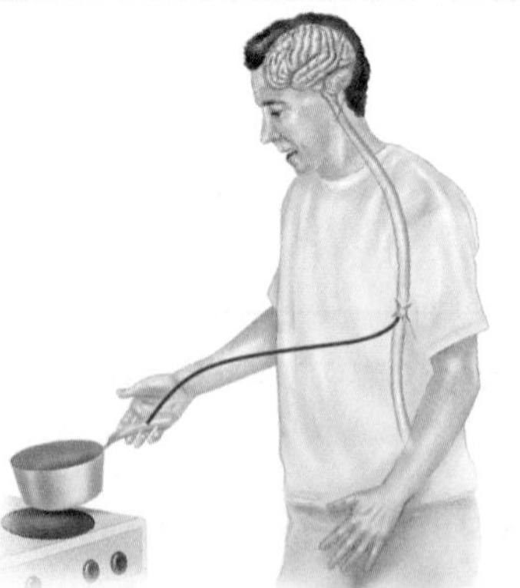

2. Nerve cells send signals to the nervous system.

3. Nerves send signals to the muscles to move your hand away.

Lesson Review 3

Visual Summary

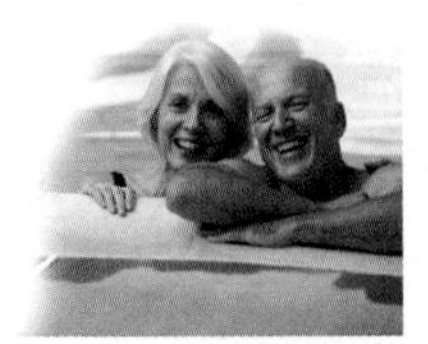

The integumentary system contains all the external coverings of the body.

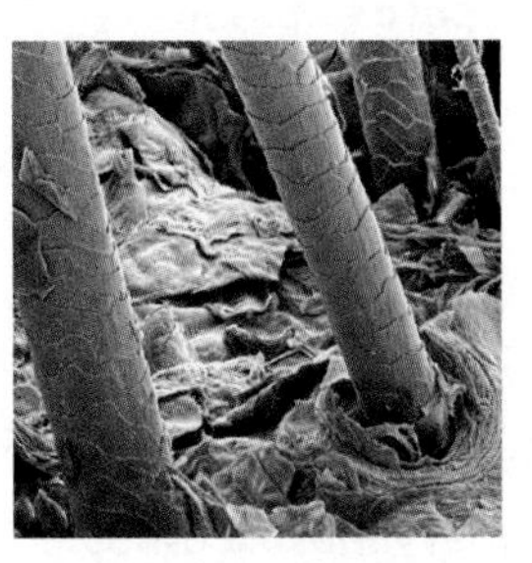

The epidermis is the outermost layer of skin.

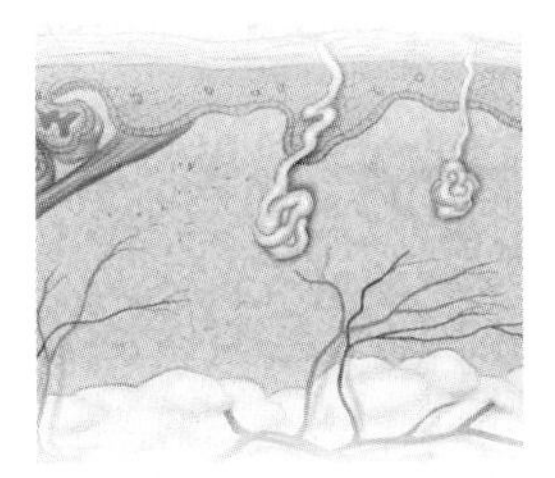

The dermis is the layer of skin that contains sweat glands, blood vessels, nerves, hair follicles, and muscles.

Use Vocabulary

1. A pigment that absorbs ultraviolet rays is called ____________.

2. The ____________ is the layer that nourishes the skin.

Understand Key Concepts

3. Which is the innermost layer of skin? SC.6.L.14.5
   - (A) dermis
   - (B) epidermis
   - (C) fatty layer
   - (D) melanin

4. **Give** an example of how the skin regulates body temperature. SC.6.L.14.5

5. **Compare** the dermis and epidermis. Explain the function of each. SC.6.L.14.5

Interpret Graphics

6. **Organize Information** Fill in the table below with details about the three degrees of burns. LA.6.2.2.3

First-degree burn	
Second-degree burn	
Third-degree burn	

Critical Thinking

7. **Evaluate** the importance of skin to homeostasis. SC.6.L.14.5

Math Skills

MA.6.A.3.6

8. The palms of the hands and soles of the feet have the highest concentration of sweat glands—about $500/cm^2$. How many sweat glands would there be on the sole of a foot measuring 10 cm × 27 cm?

Chapter 12 Study Guide

Think About It! The skeletal system, the muscular system, and the skin interact with each other to maintain homeostasis by moving, supporting, and protecting the body.

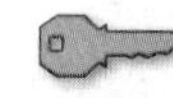

Key Concepts Summary

LESSON 1 The Skeletal System

- The **skeletal system** supports the body, helps it move, and protects internal organs. Bones store fat and calcium and make red blood cells.
- Bones are protected by **cartilage** and are connected to other bones by **ligaments** at **joints.**
- The skeletal system works with the muscular system and the skin to protect and support the body and enable it to move.

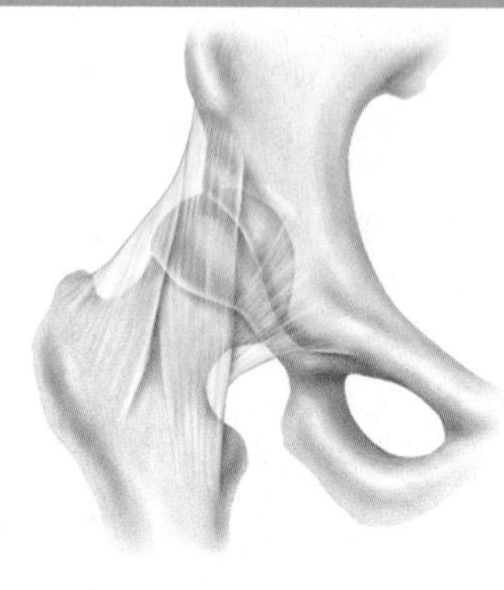

Vocabulary

skeletal system p. 451
cartilage p. 454
periosteum p. 454
joint p. 455
ligament p. 455
arthritis p. 456
osteoporosis p. 456

LESSON 2 The Muscular System

- **Muscles** support and stabilize the skeleton, enable bones and organs to move, protect the body, and regulate temperature.
- You can consciously control **skeletal,** or **voluntary, muscles.** You cannot consciously control **involuntary muscles,** which include the **cardiac muscle** that pumps blood and the **smooth muscles** that move food and blood.
- The muscular system interacts with other body systems to protect, support, and move the body.

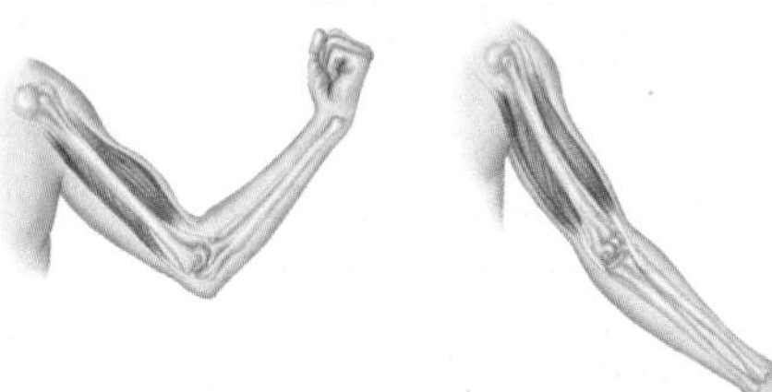

muscle p. 461
skeletal muscle p. 463
voluntary muscle p. 463
cardiac muscle p. 464
involuntary muscle p. 464
smooth muscle p. 464

LESSON 3 The Skin

- The skin protects the body, regulates temperature, contains receptors that respond to stimuli, makes vitamin D, and helps eliminate waste.
- The skin has three layers. The **epidermis** is the outermost layer. The **dermis** is below, and the fat layer is the inner layer. The epidermis produces **melanin,** which helps protect the body from ultraviolet rays.
- The skin works together with the skeletal and muscular systems to protect and support the body.

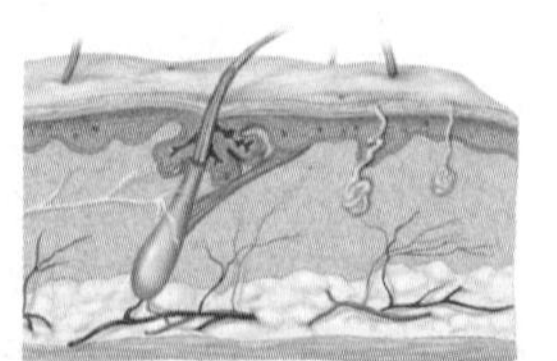

integumentary system p. 469
epidermis p. 471
melanin p. 471
dermis p. 471
bruise p. 472

Active Reading

FOLDABLES® Chapter Project

Assemble your lesson Foldables as shown to make a Chapter Project. Use the project to review what you have learned in this chapter.

Use Vocabulary

Explain the differences and similarities between the vocabulary words in each of the following sets.

1. arthritis, osteoporosis

2. ligament, joint

3. cartilage, periosteum

4. skeletal system, muscular system

5. epidermis, dermis

6. integumentary system, muscular system

Link Vocabulary and Key Concepts

Use vocabulary terms from the previous page and other terms from the chapter to complete the concept map.

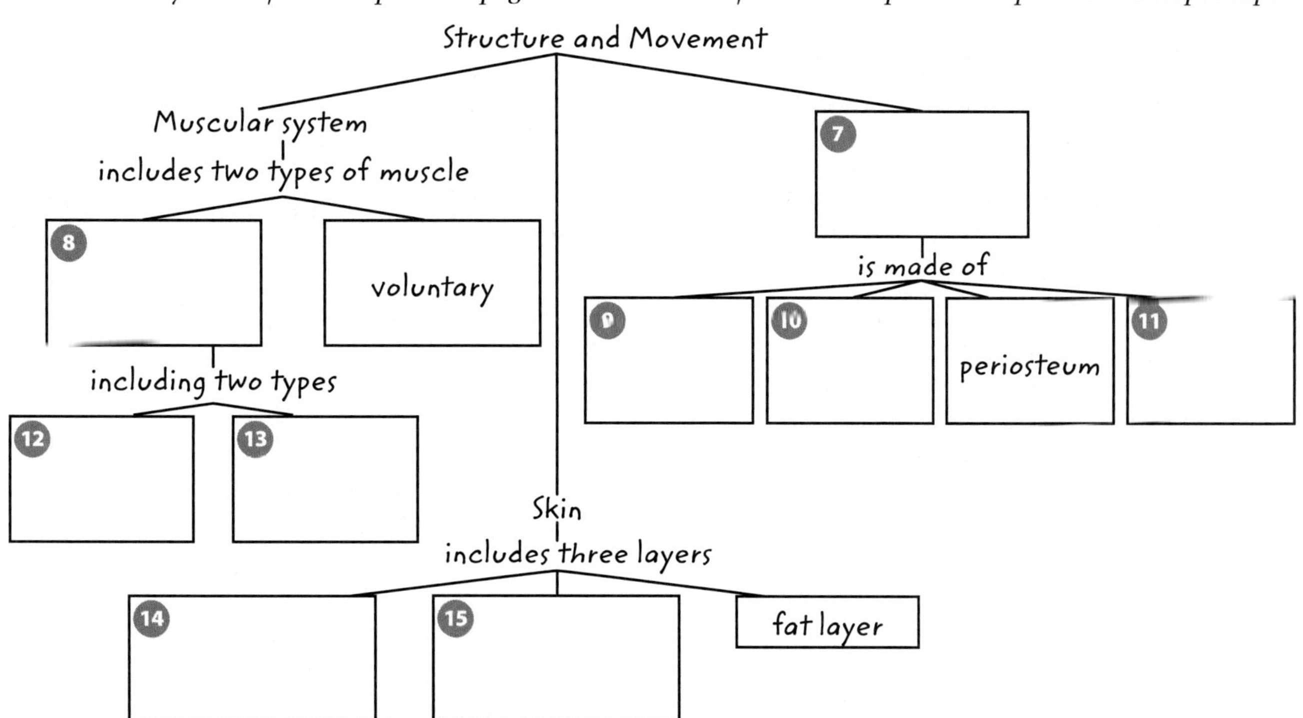

Chapter 12 Review

Fill in the correct answer choice.

Understand Key Concepts

1. What is the arrow pointing to in the figure below? **SC.6.L.14.5**

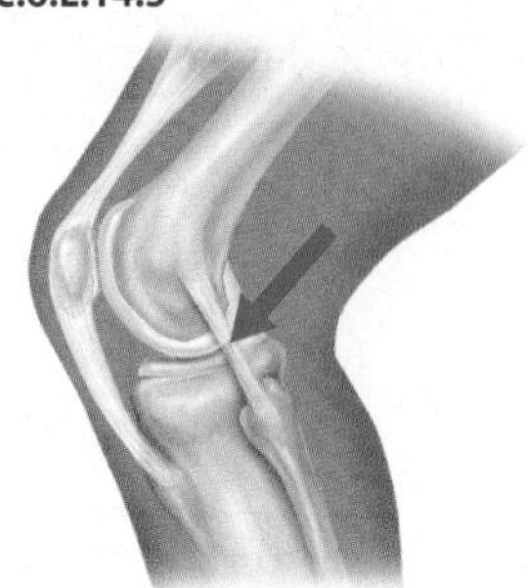

Ⓐ cartilage
Ⓑ ligament
Ⓒ periosteum
Ⓓ tendon

2. Which is NOT produced or stored inside bones? **SC.6.L.14.5**

Ⓐ calcium
Ⓑ fat
Ⓒ melanin
Ⓓ red blood cells

3. Which organ is NOT protected by the skeletal system? **SC.6.L.14.5**

Ⓐ brain
Ⓑ heart
Ⓒ lungs
Ⓓ skin

4. Which prevents bones from rubbing against each other? **SC.6.L.14.5**

Ⓐ cartilage
Ⓑ compact bone
Ⓒ ligament
Ⓓ spongy bone

5. What happens to a muscle when it contracts? **SC.6.L.14.5**

Ⓐ The muscle lengthens.
Ⓑ The muscle pushes on a bone.
Ⓒ The muscle pushes on another muscle.
Ⓓ The muscle shortens.

Critical Thinking

6. **Explain** how the muscular system and the skin regulate body temperature. **SC.6.L.14.5**

7. **Compare** spongy bone and compact bone. **SC.6.L.14.5**

8. **Assess** the importance of calcium in your diet, including the consequences of not getting enough. **SC.6.L.14.5**

9. **Give an example** of how the skeletal system maintains homeostasis. **SC.6.L.14.5**

10. **Predict** how your daily life would change if your skin contained no melanin. **SC.6.L.14.5**

11. **Discuss** why muscles must work in pairs. **SC.6.L.14.5**

12. **Evaluate** how the skin's ability to continually produce new skin cells helps protect your body. **SC.6.L.14.5**

13 **Identify** the kind of joint shown below. Give an example of where it is found in the body and what type of movement it enables the body to do. Record what structures you can identify and what system they belong to. SC.6.L.14.5

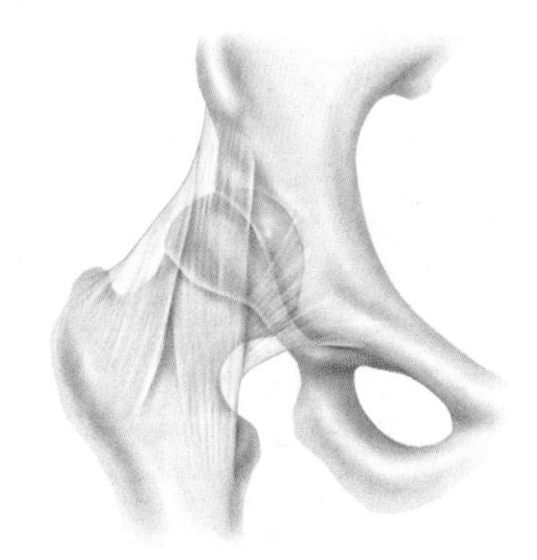

Writing in Science

14 **Write** a paragraph on a separate piece of paper about how bones grow, starting before birth. Hypothesize why children and teens are in greater danger of breaking a bone than adults. Include a main idea, supporting details and examples, and a concluding sentence. SC.6.L.14.5

Big Idea Review

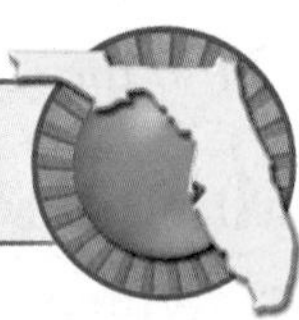

15 How do the skeletal system, the muscular system, and the skin help the body maintain homeostasis? Explain how each system works independently to maintain homeostasis, and give an example of how two of them work together. SC.6.L.14.5

Math Skills

MA.6.A.3.6

Use Proportions

16 A particular person produces about 95 mg of sweat per minute. How much sweat will the person produce in 1 hour?

17 The human body has approximately 1.8 m^2 of skin. (1 m^2 = 10,000 cm^2)

A What is the surface area of the human body in cm^2?

B. If the body averages 277 sweat glands per square centimeter, about how many sweat glands would the human body contain?

Florida NGSSS Benchmark Practice

Fill in the correct answer choice.

Multiple Choice

1 Which is part of the integumentary system? **SC.6.L.14.5**

Ⓐ fingernail
Ⓑ lung
Ⓒ muscle
Ⓓ nerve

Use the diagram below to answer question 2.

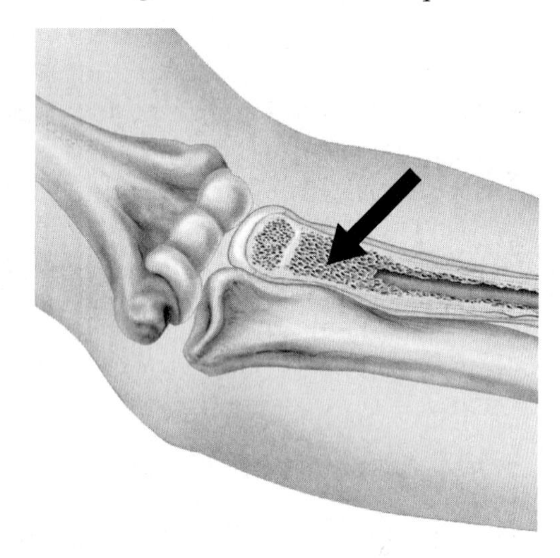

2 What does the arrow in the figure point to? **SC.6.L.14.5**

Ⓕ cartilage
Ⓖ compact bone
Ⓗ spongy bone
Ⓘ yellow marrow

3 Which does NOT help control a person's body temperature? **SC.6.L.14.5**

Ⓐ contracting muscles rapidly
Ⓑ enlarging blood vessels
Ⓒ producing melanin
Ⓓ sweating through pores

4 Which is a characteristic of skeletal muscles? **SC.6.L.14.5**

Ⓕ They are involuntary.
Ⓖ They are smooth muscle.
Ⓗ They contain discs.
Ⓘ They have striations.

5 Where is cardiac muscle? **SC.6.L.14.5**

Ⓐ heart
Ⓑ legs
Ⓒ lungs
Ⓓ stomach

Use the diagram below to answer question 6.

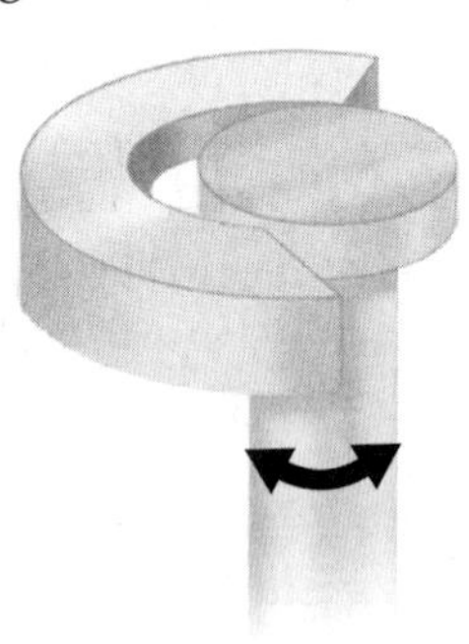

6 Where is this type of joint in the human body? **SC.6.L.14.5**

Ⓕ finger
Ⓖ knee
Ⓗ neck
Ⓘ shoulder

Use the figure below to answer questions 7 and 8.

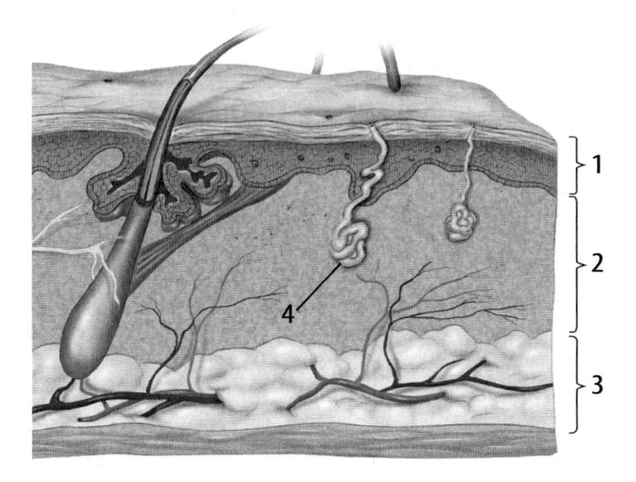

7 Which layer produces melanin? SC.6.L.14.5

Ⓐ 1
Ⓑ 2
Ⓒ 3
Ⓓ 4

8 Which layer gives the skin strength, flexibility, and nourishment? SC.6.L.14.5

Ⓕ 1
Ⓖ 2
Ⓗ 3
Ⓘ 4

9 Which bone material stores fat? SC.6.L.14.5

Ⓐ cartilage
Ⓑ compact bone
Ⓒ spongy bone
Ⓓ yellow bone marrow

10 Which type of muscle is responsible for movement of the arms and legs? SC.6.L.14.5

Ⓕ cardiac muscle
Ⓖ involuntary muscle
Ⓗ skeletal muscle
Ⓘ smooth muscle

11 A contraction of what type of muscle is not consciously controlled? SC.6.L.14.5

Ⓐ arm muscle
Ⓑ involuntary muscle
Ⓒ skeletal muscle
Ⓓ voluntary muscle

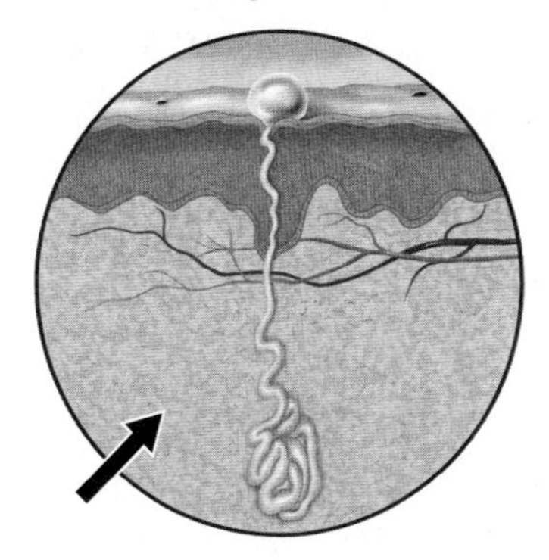

12 To which layer of skin is the arrow in the figure above pointing? SC.6.L.14.5

Ⓕ dermis
Ⓖ epidermis
Ⓗ fat layer
Ⓘ melanin

13 What is one function of sweating? SC.6.L.14.5

Ⓐ allows the body to move
Ⓑ enables calcium to be absorbed
Ⓒ increases body temperature
Ⓓ lowers body temperature

NEED EXTRA HELP?

If You Missed Question...	1	2	3	4	5	6	7	8	9	10	11	12	13
Go to Lesson...	3	1	2, 3	2	2	1	3	3	1	2	2	3	3

Name ______________________________ Date ______________

FLORIDA NGSSS

mini BAT

Multiple Choice *Bubble the correct answer.*

Use the image below to answer questions 1 through 3.

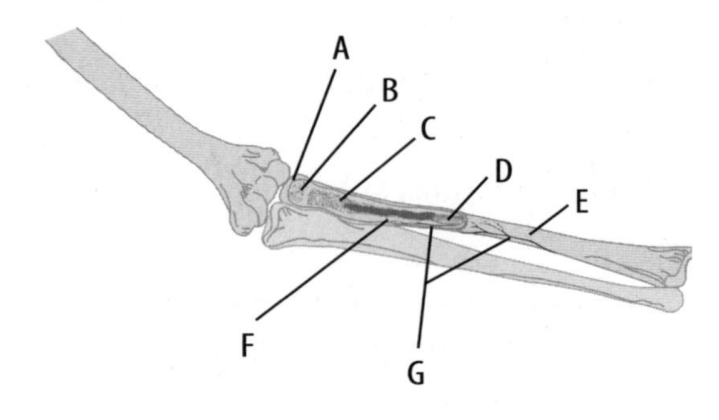

1. Which type of joint is shown in the image? **SC.6.L.14.5**

Ⓐ hinge

Ⓑ pivot

Ⓒ ball and socket

Ⓓ immovable

2. Which part of a bone is illustrated by the section marked E? **SC.6.L.14.5**

Ⓕ cartilage

Ⓖ periosteum

Ⓗ compact bone

Ⓘ spongy bone

3. Where on the image would a bone's growth plate be located? **SC.6.L.14.5**

Ⓐ between points A and B

Ⓑ between points B and C

Ⓒ between points C and F

Ⓓ between points D and E

4. Ligaments are tissues that connect bones to **SC.6.L.14.5**

Ⓕ bones.

Ⓖ cartilage.

Ⓗ muscles.

Ⓘ organs.

Multiple Choice *Bubble the correct answer.*

1. Which image shows smooth muscle? **SC.6.L.14.5**

Ⓐ 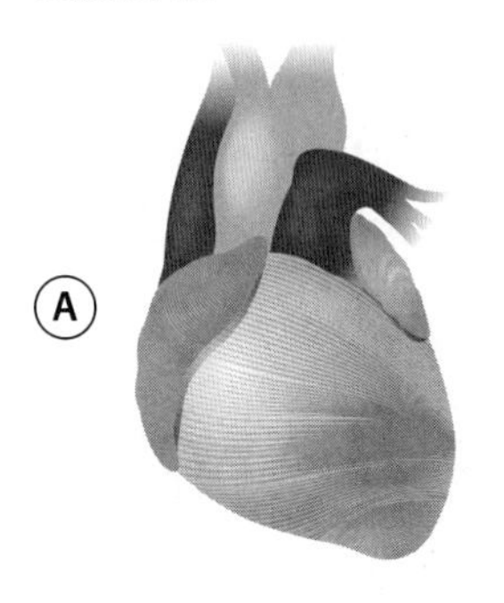

Ⓑ 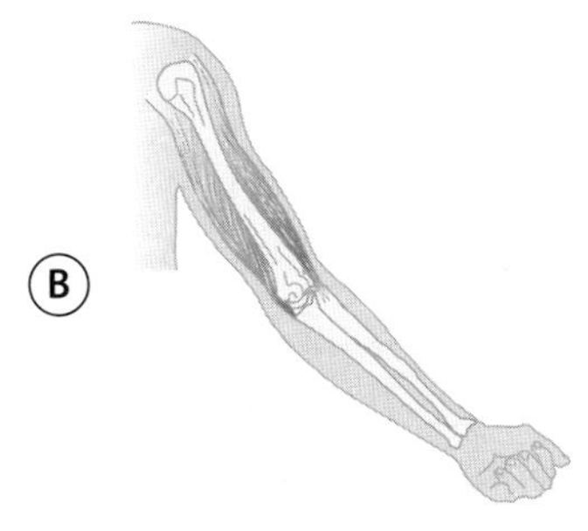

Ⓒ 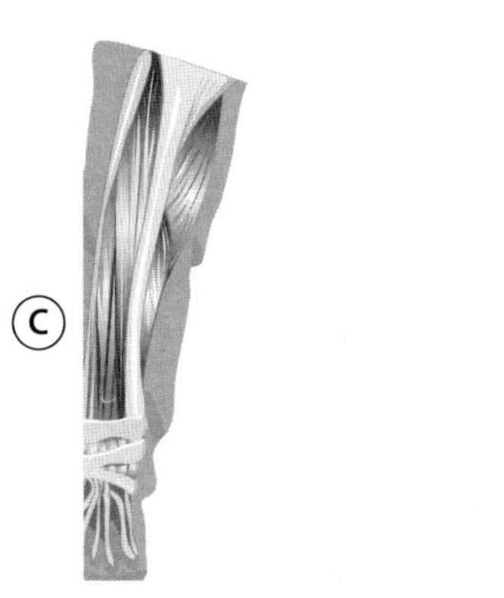

Ⓓ 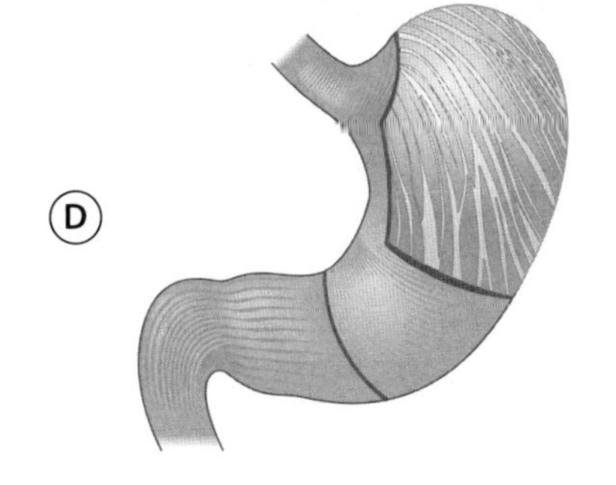

2. How is the contraction of smooth muscles different from the contraction of skeletal muscles? **SC.6.L.14.5**

Ⓕ We are not aware of the contraction of skeletal muscles.

Ⓖ We can always feel the contraction of smooth muscles.

Ⓗ We do not control the contraction of smooth muscles.

Ⓘ We do not control the involuntary contraction of skeletal muscles.

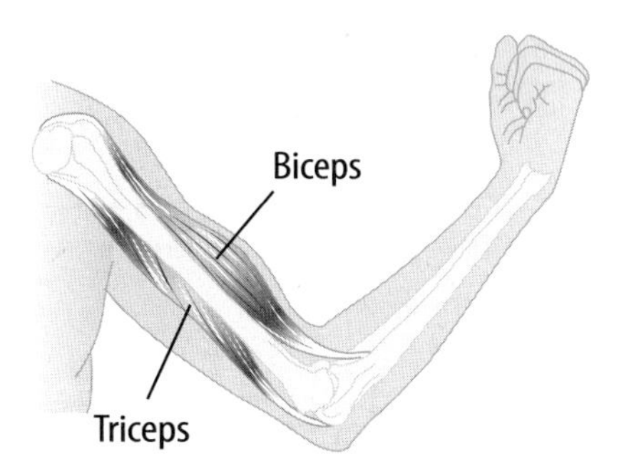

3. Based on the image above, which statement describes how the lower arm is raised? **SC.6.L.14.5**

Ⓐ The biceps muscle has shortened to raise the lower arm.

Ⓑ The triceps muscle has shortened to raise the lower arm.

Ⓒ The biceps muscle has pushed against the bone to raise the lower arm.

Ⓓ The triceps muscle has pushed against the bone to raise the lower arm.

Name ______________________________ Date ______________

Benchmark Mini-Assessment Chapter 12 • Lesson 3

mini BAT

Multiple Choice *Bubble the correct answer.*

Use the image below to answer questions 1 and 2.

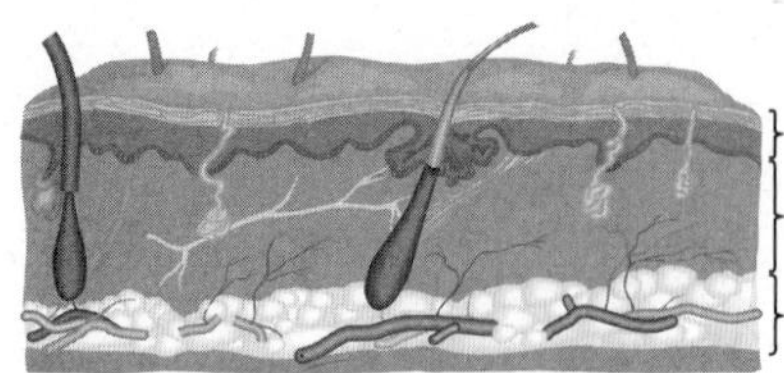

1. In which layer(s) of the skin are blood vessels found? **SC.6.L.14.5**

- (A) only in layer 1
- (B) only in layer 3
- (C) in layers 1 and 2
- (D) in layers 2 and 3

2. In which layer(s) of the skin is melanin produced? **SC.6.L.14.5**

- (F) only in layer 1
- (G) only in layer 3
- (H) in layers 1 and 2
- (I) in layers 2 and 3

3. What happens to hair follicles when the muscles in the dermis contract? **SC.6.L.14.5**

- (A) The follicles draw the hair inward.
- (B) The follicles draw the hair outward.
- (C) The follicles push the hair downward.
- (D) The follicles push the hair upward.

4. A purple area of the skin in which blood vessels are broken but the skin is not opened is a **SC.6.L.14.5**

- (F) bruise.
- (G) burn.
- (H) cut.
- (I) scab.

Name ______________________ Date ______________

Notes

Name ____________________ Date ____________________

Notes

Digestion and Food

The cells in our body need a source of energy to carry out their cell functions. They also need building blocks for growth and repair of tissues. The energy and building blocks come from food digested by the digestive system. Put an *X* next to all the things that our cells get from the digestive system to use for energy and building blocks.

_____ water

_____ molecules of sugar

_____ bread

_____ vitamins

_____ calcium

_____ banana

_____ carbon dioxide

_____ hamburger

_____ molecules of fat

_____ peanuts

_____ molecules of protein

_____ diet soda

_____ carrots

_____ rice

Explain your thinking. What rule or reasoning did you use to decide what cells use for energy and building blocks?

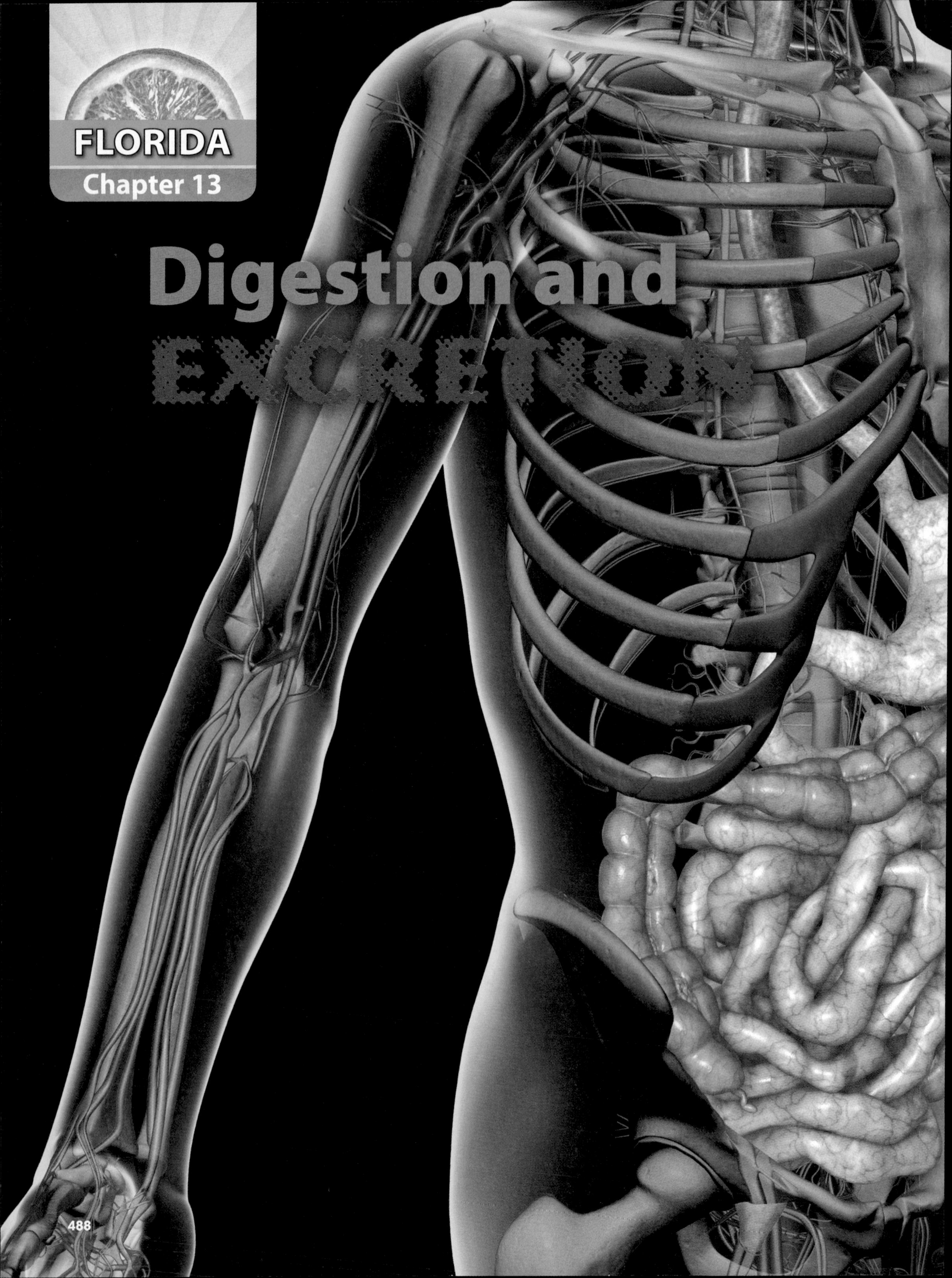

Digestion and EXCRETION

FLORIDA BIG IDEAS

1 The Practice of Science

3 The Role of Theories, Laws, Hypotheses, and Models

14 Organization and Development of Living Organisms

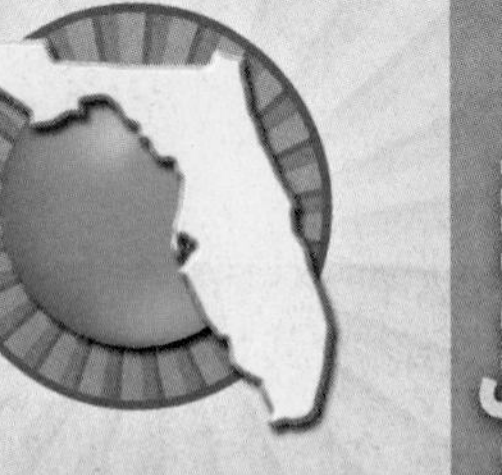

The Big Idea

Think About It!

How do the digestive and excretory systems help maintain the body's homeostasis?

This image shows parts of the digestive system. The small intestine is the structure that looks like a tangled-up rope. The small intestine can be up to 6 m long.

1. Why do you think the small intestine is so long?

____

2. What do you think the function of the small intestine is?

____

3. How might the digestive system help your body maintain homeostasis?

____

Get Ready to Read

What do you think about digestion?

Before you read, decide if you agree or disagree with each of these statements. As you read this chapter, see if you change your mind about any of the statements.

	AGREE	DISAGREE
1 An activity such as sleeping does not require energy.	☐	☐
2 All fats in food should be avoided.	☐	☐
3 Digestion begins in the mouth.	☐	☐
4 Energy from food stays in the digestive system.	☐	☐
5 Several human body systems work together to eliminate wastes.	☐	☐
6 Blood contains waste products that must be removed from the body.	☐	☐

There's More Online!

Video • Audio • Review • ⓘLab Station • WebQuest • Assessment • Concepts in Motion • Multilingual eGlossary

Lesson 1

NUTRITION

ESSENTIAL QUESTIONS

 Why do you eat?

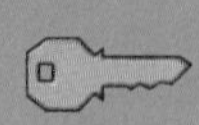 Why does your body need each of the six groups of nutrients?

 Why is eating a balanced diet important?

Vocabulary

Calorie p. 491
protein p. 492
carbohydrate p. 492
fat p. 493
vitamin p. 493
mineral p. 493

SC.6.N.1.5

Inquiry Launch Lab

20 minutes

How much energy is in an almond?

 *Food allergy*

Food contains energy. Is there enough energy in an almond to boil water?

Procedure

   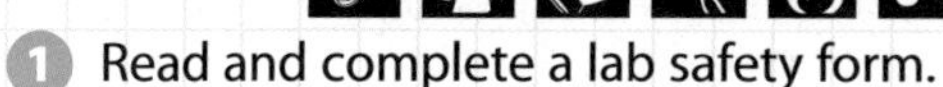

1. Read and complete a lab safety form.
2. Place a small amount of **clay** in a **shallow baking dish.** Straighten a **metal paper clip.** Insert one end into an unshelled **almond.** Anchor the other end in the clay.
3. Place a **25-mL test tube** in a **test-tube clamp.** Add 10 mL of **water** to the test tube.
4. Have your partner light the almond with a **long wooden match** until the almond starts burning on its own.
5. Gently swirl the test tube at an angle over the flame until the almond completely burns. Record your observations in the space below.

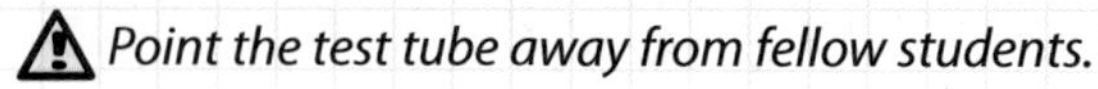 *Point the test tube away from fellow students.*

Data and Observations

Think About This

1. What happened to the water? Why did this happen?

2. Key Concept What do you think happens to your body when you eat an almond?

Florida NGSSS

LA.6.2.2.3 The student will organize information to show understanding (e.g., representing main ideas within text through charting, mapping, paraphrasing, summarizing, or comparing/contrasting);

SC.6.N.1.5 Recognize that science involves creativity, not just in designing experiments, but also in creating explanations that fit evidence.

Inquiry Time for Lunch?

1. This photo shows fried moth larvae on a banana leaf. It might not look appetizing, but it contains nutrients your body needs for energy and growth. Nutrients are in many different foods—from a cheeseburger to a fried insect. What kinds of nutrients do you think the food shown here contains?

____

____

____

Why do you eat?

How do you decide what to eat or when to eat? Although you can survive for weeks without food, you might become hungry within hours of your last meal. Hunger is your body's way of telling you that it needs food. Why does your body need food? Food provides your body with the energy and nutrients it needs to survive.

Energy

Every activity you do, such as riding a bike or even sleeping, requires energy. Your digestive system processes food and releases energy that is used for cellular processes and all activities that you do.

The amount of energy in food is measured in Calories. *A* **Calorie** *(Cal) is the amount of energy it takes to raise the temperature of 1 kg of water by 1°C.* How much energy do foods contain? Each food is different. One grape contains 2 Cal, but a slice of cheese pizza has 220 Cal. All foods give your body energy.

The amount of energy a person needs depends on several factors, such as weight, age, activity level, and gender. For example, a person with a mass of 68 kg usually burns more Calories than a person with a mass of 45 kg. Playing soccer requires more energy than playing a video game. How does the food you eat supply you with energy? The energy comes from nutrients.

Nutrients

Food is made of nutrients—substances that provide energy and materials for cell development, growth, and repair. The types and amounts of nutrients a person needs depend on age, gender, and activity level. Toddlers need more fat in their diets than older children do. Women need more calcium and iron than men do. Active people need more protein. Next, you'll read about the six groups of nutrients and their roles in maintaining your health.

Active Reading **2. Identify** Highlight why you eat.

Proteins

Carbohydrates

Figure 1 Good sources of protein include red meat, eggs, beans, and peanuts. Good sources of carbohydrates include red beans, fruits, and vegetables.

Active Reading **4. Describe** Give one example of a lunch that is high in proteins and carbohydrates.

____

____

____

____

WORD ORIGIN

protein
from Greek *proteios,* means "the first quality"

Groups of Nutrients

The six groups of nutrients are proteins, carbohydrates, fats, vitamins, minerals, and water. Each nutrient has a different function in the body. To be healthy, you need foods from each group every day.

Proteins

Most of the tissues in your body are made of proteins. *A* **protein** *is a large molecule that is made of amino acids and that contains carbon, hydrogen, oxygen, nitrogen, and sometimes sulfur.* Proteins have many functions, such as relaying signals between cells, protecting against disease, and providing support to cells. All of these functions are needed to maintain homeostasis, or the regulation of an organism's internal condition regardless of changes in its environment.

Combinations of 20 different amino acids make up the proteins in your body. Your cells can make more than half of these amino acids. The remaining amino acids must come from the foods you eat. Some foods that are good sources of protein are shown in **Figure 1.**

Active Reading **3. Explain** How does your body obtain amino acids that cannot be made in cells?

____

____

Carbohydrates

What do pasta, bread, and potatoes have in common? They have high levels of carbohydrates (kar boh HI draytс). **Carbohydrates** *are molecules made of carbon, hydrogen, and oxygen atoms and are usually the body's major source of energy.* They are commonly in one of three forms—starches, sugars, or fibers. All of them are made of sugar molecules that are linked together like a chain. It is best to eat foods that contain carbohydrates from whole grains because they are easier to digest. Also shown in **Figure 1** are foods that are high in carbohydrates.

Fats

You might think that fats in food are bad for you. But, you need a certain amount of fat in your diet and on your body to stay healthy. **Fats,** *also called lipids, provide energy and help your body absorb vitamins.* They are a major part of cell membranes. Body fat helps insulate against cold temperatures. Most people get plenty of fat in their diet, so deficiencies in fats are rare. But too much fat in your diet can lead to health problems. Only about 25–35 percent of the Calories you consume should be fats.

Fats are often classified as either saturated or unsaturated. A diet high in saturated fats can increase levels of cholesterol, which can increase the risk of heart disease. Most of the fat in your diet should come from unsaturated fats, such as those shown in **Figure 2.**

Figure 2 Fish, nuts, and liquid vegetable oils contain unsaturated fats.

Vitamins

Has anyone ever told you to eat certain foods because you need vitamins? **Vitamins** *are nutrients that are needed in small amounts for growth, regulating body functions, and preventing some diseases.* You can obtain most of the vitamins you need by eating a well-balanced diet. If you do not consume enough of one or more vitamins, then you might develop symptoms of vitamin deficiency. The symptoms depend on which vitamin you are lacking. **Table 1** lists some vitamins people need in their diet.

5. Identify Highlight reasons why you need vitamins in your diet.

Minerals

In addition to vitamins, you also need other nutrients called minerals. **Minerals** *are inorganic nutrients—nutrients that do not contain carbon—that help the body regulate many chemical reactions.* Similar to vitamins, if you do not consume enough of certain minerals, you might develop a mineral deficiency. **Table 1** also lists some minerals that you need in your diet.

Table 1 Vitamins and Minerals

Vitamin	Good Sources	Health Benefit
Vitamin B_2 (riboflavin)	milk, meats, vegetables	helps release energy from nutrients
Vitamin C	oranges, broccoli, tomatoes, cabbage	growth and repair of body tissues
Vitamin A	carrots, milk, sweet potatoes, broccoli	enhances night vision, helps maintain skin and bones
Mineral	**Good Sources**	**Health Benefit**
Calcium	milk, spinach, green beans	builds strong bones and teeth
Iron	meat, eggs, green beans	helps carry oxygen throughout the body
Zinc	meat, fish, wheat/grains	aids protein formation

Table 1 Vitamins and minerals are essential for maintaining a healthy body.

6. Interpret Circle what foods you would eat if you needed to improve your night vision and teeth.

FOLDABLES® LA.6.2.2.3

Fold a sheet of paper into a chart with three columns and two rows. Use your chart to organize information about the major food groups and to list examples of each.

Grains	Vegetables	Fruits
Oils	Milk	Meat and Beans

Water

You might recall that your body is mostly water. You need water for chemical reactions to occur in your body. Your body takes in water when you eat or drink. However, you lose water when you sweat, urinate, and breathe. To stay healthy, it is important to replace the water that your body loses. If you exercise, live in a warm area, or become sick, your body loses more water. When lost water is not replaced, you might become dehydrated. Symptoms of dehydration include thirst, headache, weakness, dizziness, and little or no urination.

Active Reading 7. **Expound** Why does your body need nutrients?

Healthful Eating

Imagine walking through a grocery store. Each aisle in the store contains hundreds of different foods. With so many choices, it's difficult to choose foods that are part of a healthful diet. Healthful eaters need to be smart shoppers. They make grocery lists beforehand and buy products that are high in nutrients. Nutritious foods come from the major food groups, which include grains, vegetables, fruits, oils, milk products, and meats and beans.

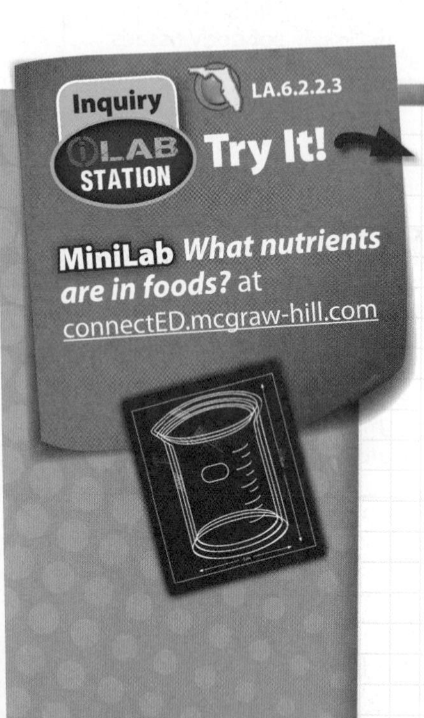

Apply It! After you complete the lab, answer these questions.

1. Were any of the foods you studied high in more than one nutrient? How might these foods be beneficial to your body?

2. How could your findings help you make changes to your own diet?

Table 2 Daily Recommended Amounts of Each Food Group for 9- to 13-Year-Olds

Food Group	Daily Amount Males, 9–13 years old	Daily Amount Females, 9–13 years old	Examples of Foods
Grains	6-ounce equivalents	5-ounce equivalents	whole-wheat flour, rye bread, brown rice
Vegetables	2 1/2 cups	2 cups	broccoli, spinach, carrots
Fruits	1 1/2 cups	1 1/2 cups	apples, strawberries, oranges
Fats	5 teaspoons or less	5 teaspoons or less	canola oil, olive oil, avocados
Milk	3 cups	3 cups	milk, cheese, yogurt
Meat and Beans	5 ounces or less	5 ounces or less	fish, beans, lean beef, lean chicken

A Balanced Diet

A healthful diet includes carbohydrates, proteins, fats, vitamins, minerals, and water. But how do you know how much of each food group you should eat? **Table 2** lists the daily recommended amounts of each food group for 9- to 13-year-olds.

The nutrient-rich foods that you choose might be different from the nutrient-rich foods eaten by people in China, Kenya, or Mexico. People usually eat foods that are grown and produced regionally. Regardless of where you live, eating a balanced diet ensures that your body has the nutrients it needs to function.

Active Reading **9. Explain** Why is eating a balanced diet important?

Food Labels

What foods would you buy to follow the recommended guidelines in **Table 2?** Most grocery stores sell many varieties of bread, milk, meat, and other types of food. How would you know what nutrients these foods contain? You can look at food labels, such as the one in **Figure 3.** Food labels help you determine the amount of protein, carbohydrates, fats, and other substances in food.

Figure 3 A food label lists a food's nutrients per serving, not per container.

Active Reading **8. List** What are the nutrients in this food?

Nutrition Facts
Serving Size 1/2 cup (130g)
Servings Per Container about 3

Amount Per Serving
Calories 30 Calories from Fat 0

	% Daily Value*
Total Fat 0g	0%
Cholesterol 0mg	0%
Sodium 290mg	12%
Total Carbohydrate 6g	2%
Dietary Fiber 1g	4%
Sugars 4g	
Protein 1g	

Vitamin A 15% • Vitamin C 35%
Calcium 2% • Iron 4%

Not a significant source of saturated fat and trans fat.
*Percent Daily Values are based on a 2,000 calorie diet.

Ingredients: Organic tomatoes and

Lesson Review 1

Visual Summary

People eat food to obtain the energy their bodies need to function.

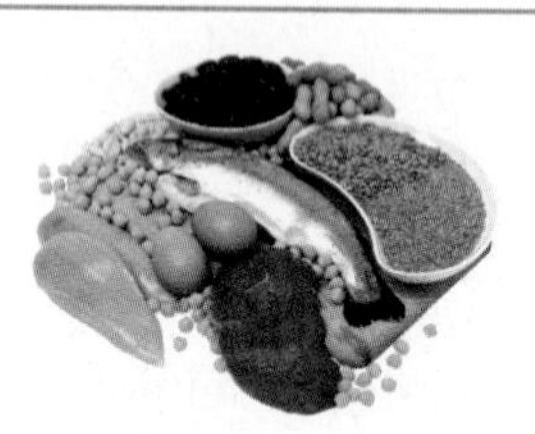

Proteins are one of the six groups of nutrients.

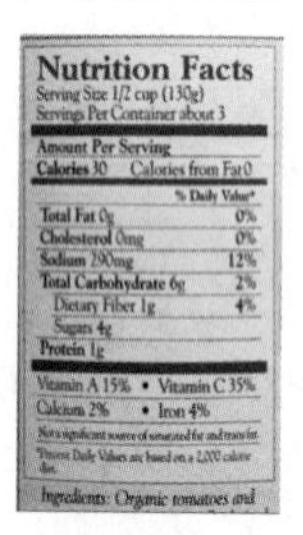

Evaluating food labels can help you eat a balanced diet.

Use Vocabulary

1. Nutrients made of long chains of amino acids are ________________.

2. The major source of energy in your diet comes from ________________.

3. The amount of energy in food is measured in ________________.

Understand Key Concepts

4. **Explain** why it is important to consume vitamins.

5. Which nutrient helps your body absorb vitamins?
   - (A) carbohydrate
   - (B) fat
   - (C) mineral
   - (D) protein

6. **Give an example** of when you might need to drink more water than usual.

Interpret Graphics

7. **Summarize** Copy and fill in the graphic organizer below to identify the six groups of nutrients. LA.6.2.2.3

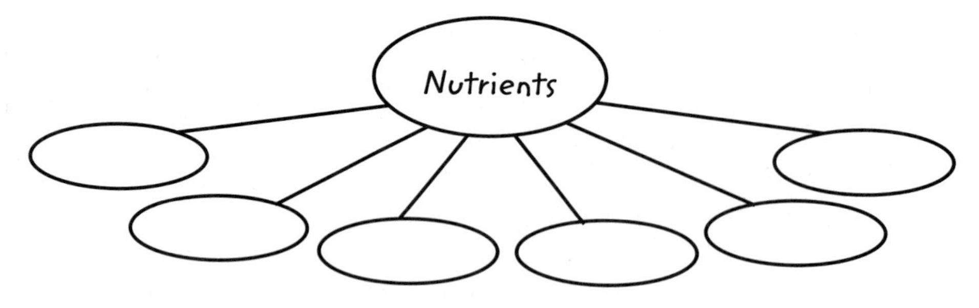

Critical Thinking

8. **Plan** a meal that contains a food from each of the six food groups.

9. **Analyze** One serving of a certain food contains 370 Cal, 170 Cal from fat, and 12 g of saturated fat (60% of the daily value). Is this food a good choice for a healthful lifestyle? Why or why not?

Nutrition

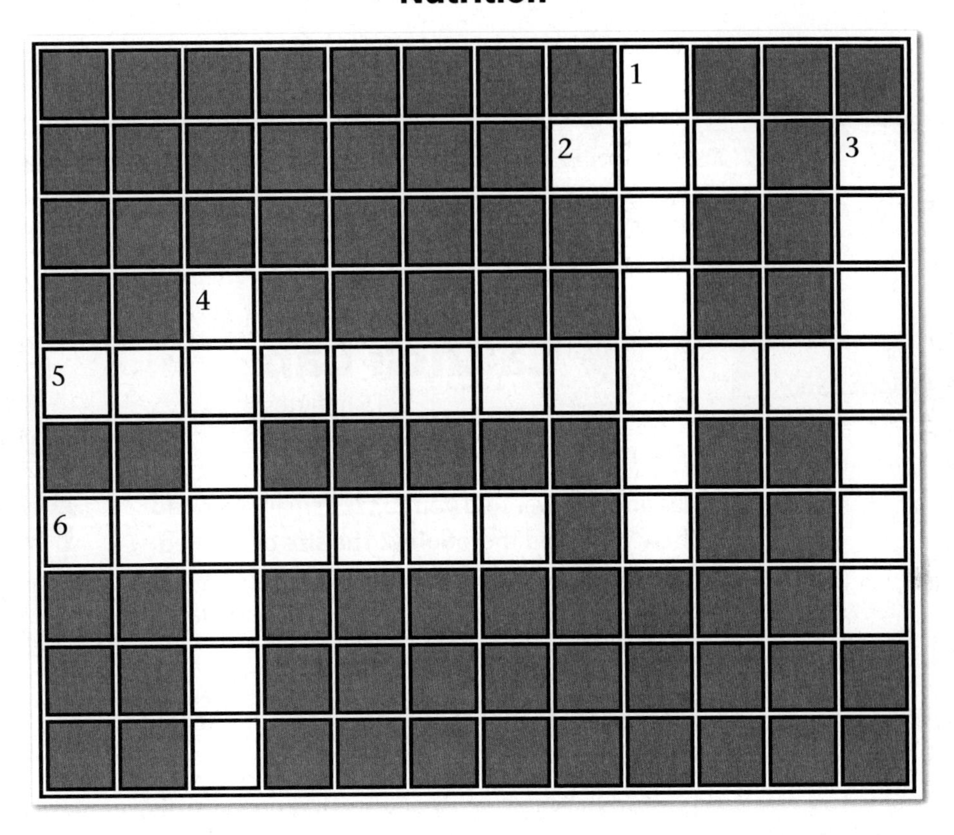

Across

2. provides energy and helps your body absorb vitamins

5. molecules that are usually the body's main source of energy

6. nutrients that are needed in small amounts for growth, regulation of body functions, and prevention of some diseases

Down

1. the amount of energy it takes to raise the temperature of 1 kg of water by 1°C

3. nutrients that do not contain carbon, which help the body regulate many chemical reactions

4. a large molecule that is made of amino acids

Lesson 2

The Digestive SYSTEM

ESSENTIAL QUESTIONS

What does the digestive system do?

How do the parts of the digestive system work together?

How does the digestive system interact with other systems?

Vocabulary

digestion p. 499
mechanical digestion p. 500
chemical digestion p. 500
enzyme p. 500
esophagus p. 502
peristalsis p. 502
chyme p. 503
villi p. 504

Florida NGSSS

LA.6.4.2.2 The student will record information (e.g., observations, notes, lists, charts, legends) related to a topic, including visual aids to organize and record information and include a list of sources used;

MA.6.A.3.6 Construct and analyze tables, graphs, and equations to describe linear functions and other simple relations using both common language and algebraic notation.

SC.6.L.14.5 Identify and investigate the general functions of the major systems of the human body (digestive, respiratory, circulatory, reproductive, excretory, immune, nervous, and musculoskeletal) and describe ways these systems interact with each other to maintain homeostasis.

SC.6.N.1.1 Define a problem from the sixth grade curriculum, use appropriate reference materials to support scientific understanding, plan and carry out scientific investigation of various types, such as systematic observations or experiments, identify variables, collect and organize data, interpret data in charts, tables, and graphics, analyze information, make predictions, and defend conclusions.

SC.6.N.1.5 Recognize that science involves creativity, not just in designing experiments, but also in creating explanations that fit evidence.

SC.6.N.3.4 Identify the role of models in the context of the sixth grade science benchmarks.

Also covers: LA.6.2.2.3

Inquiry Launch Lab

SC.6.N.1.5, SC.6.L.14.5

15 minutes

Which dissolves faster?

Has anyone ever told you to take small bites and chew your food thoroughly? The size of chewed food particles can affect how quickly food is digested. Similarly, the size of a sugar particle can affect how fast it dissolves in water.

Procedure

1. Read and complete a lab safety form.
2. Add the contents of one serving package of **granulated sugar** to a **500-mL beaker** containing 300 mL of **warm water.**
3. Gently stir the contents of the beaker with a **plastic spoon.** Have your partner use a **stopwatch** to time how long it takes the sugar to dissolve. Record the time below.
4. Add a **sugar cube** to another **500-mL beaker** containing 300 mL of warm water.
5. Repeat step 3.

Data and Observations

Think About This

1. Which dissolved faster—the granulated sugar or the sugar cube?
2. Why do you think particle size affects the rate at which sugar dissolves?
3. Key Concept How might food particle size affect how quickly food is digested?

Inquiry Under the Sea?

1. These colorful projections look like something you might see on the ocean floor, but they are found in your body. They line the walls of the small intestine, which is part of your digestive system. What do you think these projections do?

Functions of the Digestive System

Suppose you ate a cheeseburger and a pear for lunch. What happens to the food after it is eaten?

As soon as the food enters your mouth, it begins its journey through your digestive system. No matter what you eat, your food goes through four steps—ingestion, digestion, absorption, and elimination. All four steps happen in the organs and tissues of the digestive system in the following order.

- Food is ingested. Ingestion is the act of eating, or putting food in your mouth.
- Food is digested. **Digestion** *is the mechanical and chemical breakdown of food into small particles and molecules that your body can absorb and use.*
- Nutrients and water in the food are absorbed, or taken in, by cells. Absorption occurs when the cells of the digestive system take in small molecules of digested food.
- Undigested food is eliminated. Elimination is the removal of undigested food and other wastes from your body.

Word Origin
digestion
from Latin *digestus,* means "to separate, divide"

2. **NGSSS Check** Assess What does the digestive system do? SC.6.L.14.5

Types of Digestion

Before your body can absorb nutrients from food, the food must be broken down into small molecules by digestion. There are two types of digestion—mechanical and chemical. *In* **mechanical digestion**, *food is physically broken into smaller pieces.* Mechanical digestion happens when you chew, mash, and grind food with your teeth and tongue. Smaller pieces of food are easier to swallow and have more surface area, which helps with chemical digestion. *In* **chemical digestion**, *chemical reactions break down pieces of food into small molecules.*

Enzymes

Chemical digestion cannot occur without substances called enzymes (EN zimez). **Enzymes** *are proteins that help break down larger molecules into smaller molecules. Enzymes also speed up, or catalyze, the rate of* **chemical reactions.** Without enzymes, some chemical reactions would be too slow or would not occur at all.

REVIEW VOCABULARY

chemical reaction
process in which a compound is formed or broken down

There are many kinds of enzymes. Each one is specialized to help break down a specific molecule at a specific location.

Active Reading **3. Recall** What are enzymes?

______

______

______

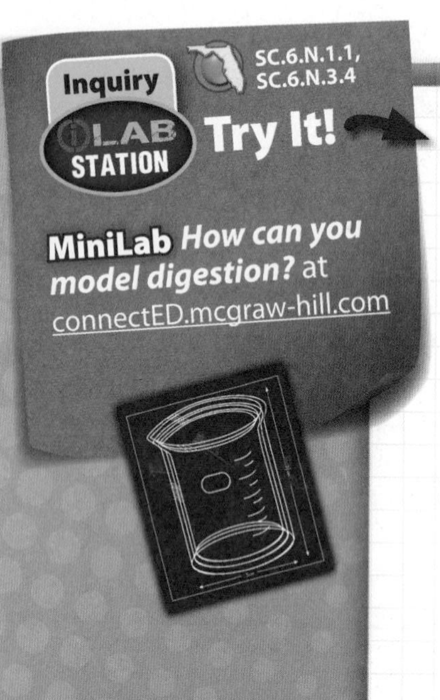

Apply It! After you complete the lab, answer these questions.

1. What type of digestion did you model in this lab?

2. How could you change the procedure to more closely model what happens when you eat?

The Role of Enzymes in Digestion

Nutrients in food are made of different molecules, such as carbohydrates, proteins, and fats. Many of these molecules are too large for your body to use. But, because these molecules are made of long chains of smaller molecules joined together, they can be broken down into smaller pieces.

The digestive system produces enzymes that are specialized to help break down each type of food molecule. For example, the enzyme amylase helps break down carbohydrates. The enzymes pepsin and papain help break down proteins. Fats are broken down with the help of the enzyme lipase. **Figure 4** illustrates how an enzyme helps break down food molecules into smaller pieces.

Notice in **Figure 4** that the food molecule breaks apart, but the enzyme does not change. The enzyme can immediately be used to break down another food molecule.

4. Explain What happens to an enzyme after it helps break down a food molecule?

____

____

____

Organs of the Digestive System

In order for your body to use the nutrients in the foods you eat, the nutrients must pass through your digestive system. Your digestive system has two parts: the digestive tract and the other organs that help the body break down and absorb food. These organs include the tongue, salivary glands, liver, gallbladder, and pancreas.

The digestive tract extends from the mouth to the anus. It has different organs connected by tubelike structures. Each organ is specialized for a certain function.

Recall the cheeseburger and pear mentioned at the beginning of this lesson. Where do you think digestion of this food begins?

Figure 4 An enzyme helps break down food molecules into smaller pieces.

Step 1
An enzyme attaches to a food particle.

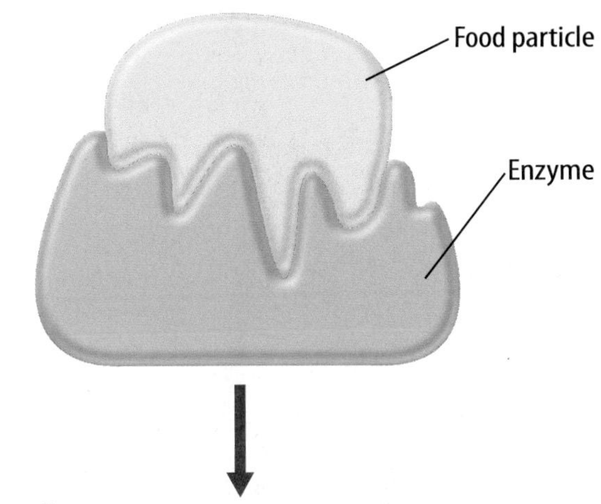

Step 2
The enzyme speeds up a chemical reaction that breaks down the food particle.

Step 3
The enzyme releases the broken-down food particle.

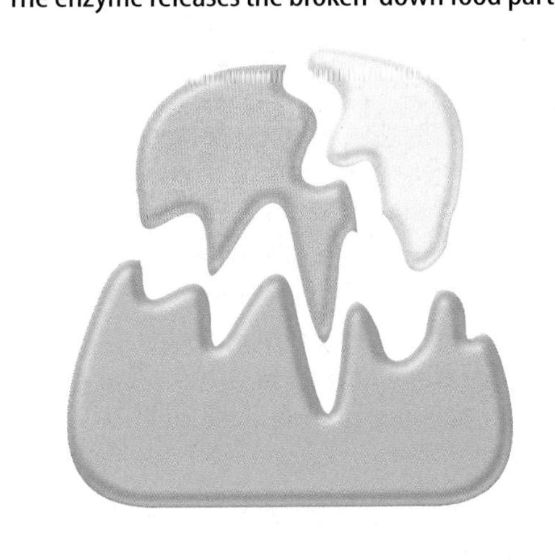

The Digestive System

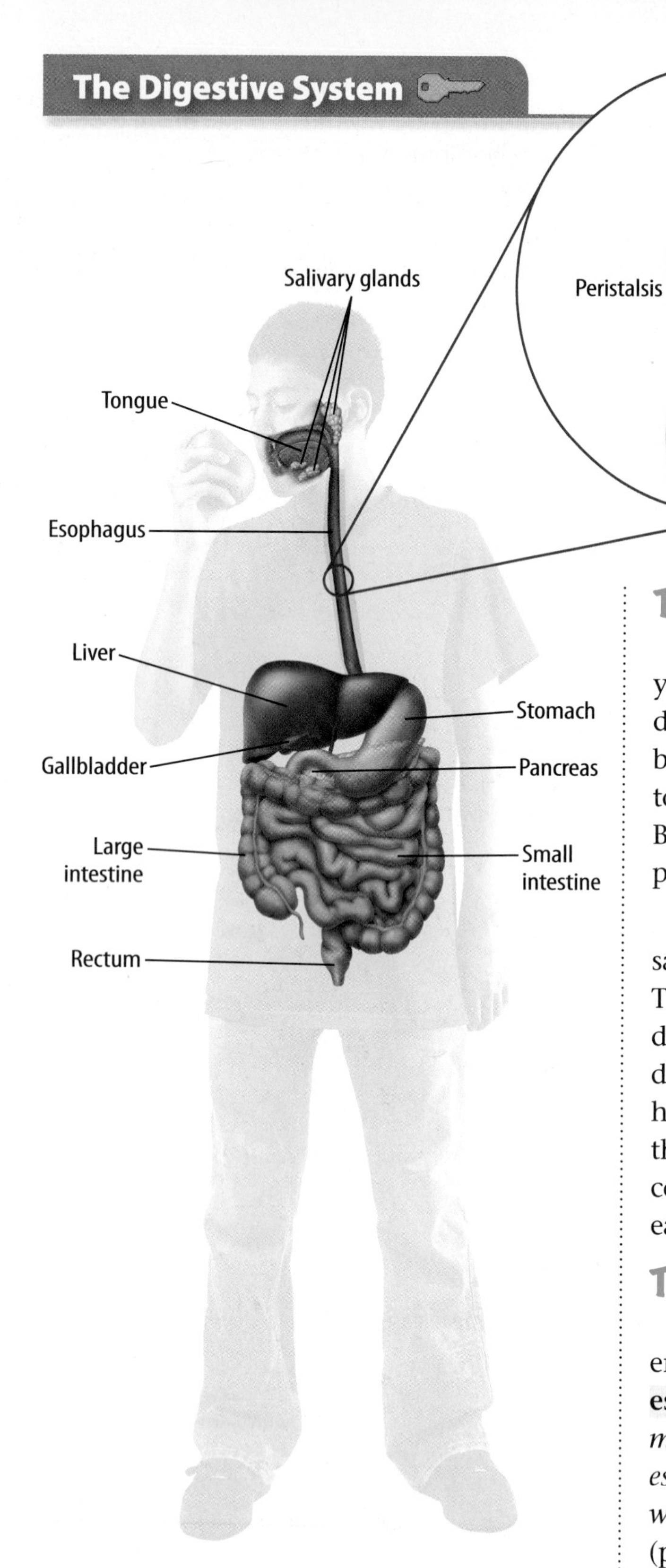

Figure 5 The digestive system includes the organs of the digestive tract, as well as other organs such as the tongue, salivary glands, liver, gallbladder, and pancreas.

Active Reading **5. Illustrate** Highlight the organ that connects the mouth to the stomach.

The Mouth

You can follow the path food takes through your digestive tract in **Figure 5.** Mechanical digestion of food, such as a pear or a cheeseburger, begins in your mouth. Your teeth and tongue mechanically digest food as you chew. But even before chewing begins, your mouth prepares for digestion.

Your salivary (SA luh ver ee) glands produce saliva (suh LI vuh) at the very thought of food. They produce more than 1 L of saliva every day. Saliva contains an enzyme that helps break down carbohydrates, such as those found in a hamburger bun. Saliva also contains substances that neutralize acidic foods. In addition, it contains a slippery substance that makes food easier to swallow.

The Esophagus

After you swallow a bite of your food, it enters your esophagus (ih SAH fuh gus). *The* **esophagus** *is a muscular tube that connects the mouth to the stomach. Food moves through the esophagus and the rest of the digestive tract by waves of muscle contractions, called* **peristalsis** (per uh STAHL sus).

Peristalsis is similar to squeezing a tube of toothpaste. When you squeeze the bottom of the tube, toothpaste is forced toward the top of the tube. As muscles in the esophagus contract and relax, partially digested food is pushed down the esophagus and into the stomach.

The Stomach

Once your partially digested food leaves the esophagus, it enters the stomach. The stomach is a large, hollow organ. One function of the stomach is to temporarily store food. This allows you to go many hours between meals. The stomach is like a balloon that can stretch when filled. An adult stomach can hold about 2 L of food and liquids.

6. **NGSSS Check** Describe Why is the stomach's ability to store food beneficial?

____

____

____

Another function of the stomach is to aid in chemical digestion. As shown in **Figure 6,** the walls of the stomach are folded. These folds enable the stomach to expand and hold large amounts of food. In addition, the cells in these folds produce chemicals that help break down proteins. For example, the stomach contains an acidic fluid called gastric juice. Gastric juice makes the stomach acidic. Acid helps break down some of the structures that hold plant and animal cells together, like the cells in hamburger meat, lettuce, tomatoes, and pears. Gastric juice also contains pepsin, an enzyme that helps break down proteins in foods into amino acids. Food and gastric juices mix as muscles in the stomach contract through peristalsis. As food mixes with gastric juice in the stomach, it forms *a thin, watery liquid called* **chyme** (KIME).

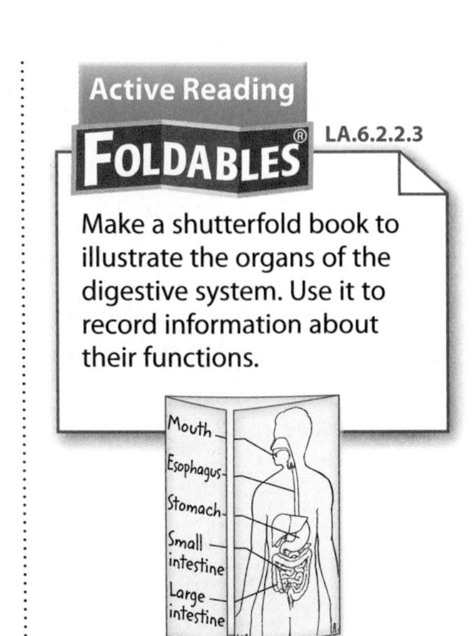

Figure 6 The stomach temporarily stores food and aids in chemical digestion.

Active Reading **7. Identify** Circle where food goes after it leaves the stomach.

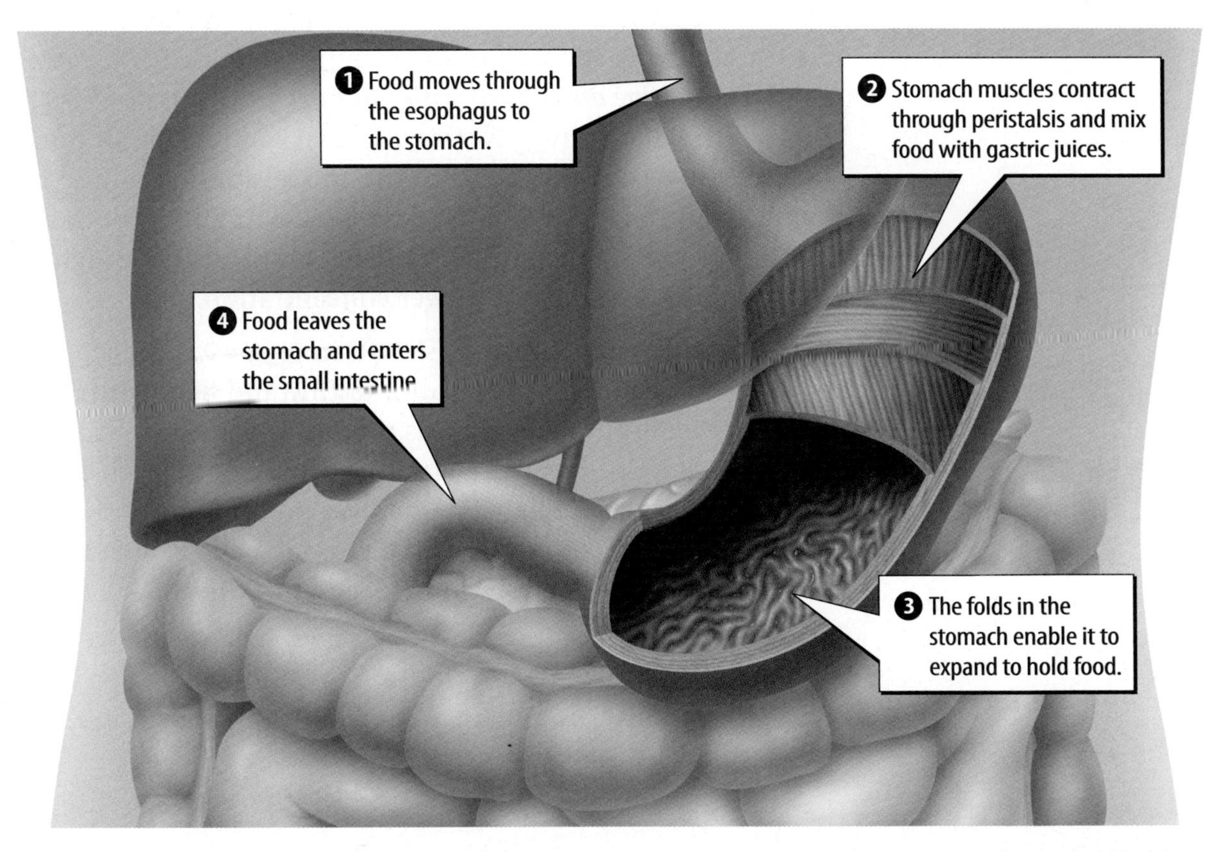

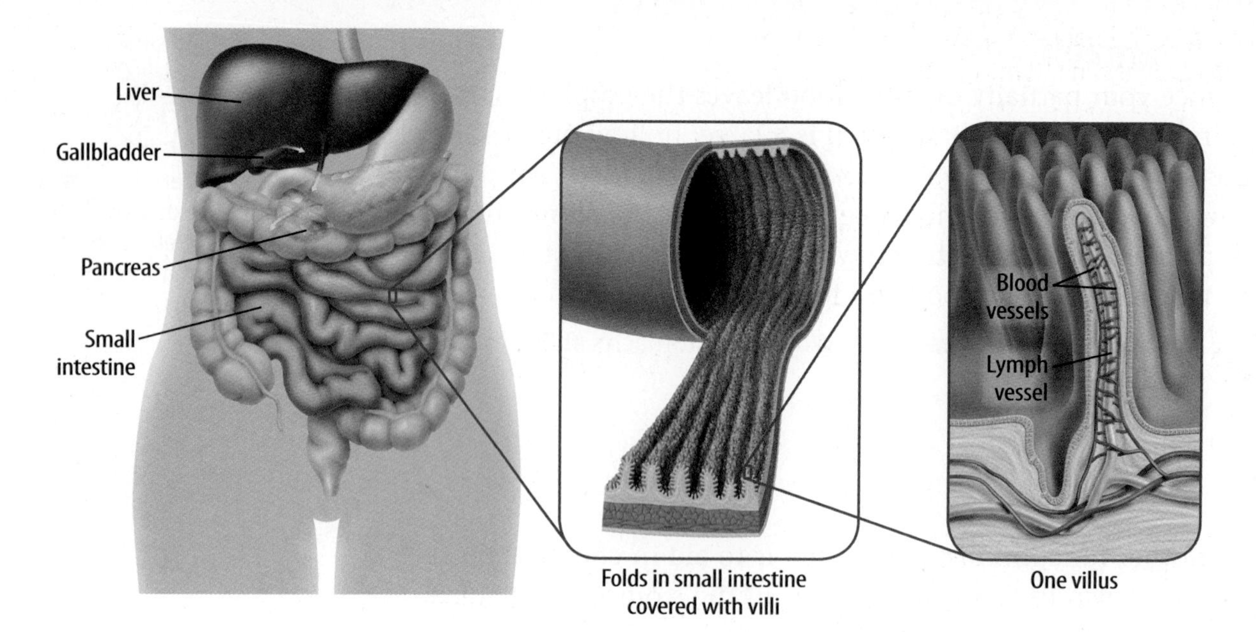

Folds in small intestine covered with villi

One villus

Figure 7 The walls of the small intestine are covered with villi that help move nutrients into the blood.

The Small Intestine

Chemical digestion of your cheeseburger and pear begins in the mouth and stomach. But most chemical digestion occurs in the small intestine. The small intestine is a long tube connected to the stomach. It is where chemical digestion and nutrient absorption occur. The small intestine is named for its small diameter—about 2.5 cm. It is about 7 m long.

Chemical digestion of proteins, carbohydrates, nucleic acids, and fats takes place in the first part of the small intestine, called the duodenum (doo uh DEE num). The remainder of the small intestine absorbs nutrients from food. Notice in **Figure 7** that, like the stomach, the wall of the intestine is folded. *The folds of the small intestine are covered with fingerlike projections called* **villi** (VIH li) (singular, villus). Notice also that each villus contains small blood vessels. Nutrients in the small intestine diffuse into the blood through these blood vessels. You might recall that diffusion is the movement of particles from an area of higher concentration to an area of lower concentration.

Science Use v. Common Use

substance

Science Use matter that has a particular chemical makeup

Common Use a fundamental quality

The pancreas and the liver, shown in **Figure 7,** produce **substances** that enter the small intestine and help with chemical digestion. The pancreas produces an enzyme called amylase that helps break down carbohydrates and a substance that neutralizes stomach acid. The liver produces a substance called bile. Bile makes it easier to digest fats. The gallbladder stores bile until it is needed in the small intestine.

8. NGSSS Check **Recall** Which organs work together to help with chemical digestion? **SC.6.L.14.5**

Figure 8 The bacteria shown here live in the large intestine. Without them, your food would not be digested well.

9. Choose Cocci bacteria are spherical, bacilli bacteria are rod-shaped, and spirilla bacteria are spiral-shaped. Circle which type of bacteria is shown in the photo.

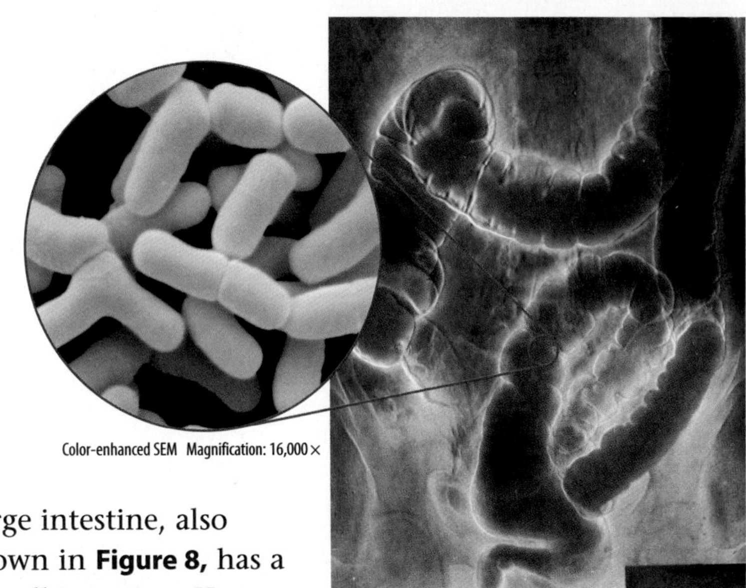

Color-enhanced SEM Magnification: 16,000×

The Large Intestine

The parts of your cheeseburger and pear that are not absorbed in the small intestine move by peristalsis into the large intestine, also called the colon. The large intestine, shown in **Figure 8,** has a larger diameter (about 5 cm) than the small intestine. However, at about 1.5 m long, it is much shorter than the small intestine.

Most of the water in ingested foods and liquids is absorbed in the small intestine. As food travels through the large intestine, even more water is absorbed. Materials that pass through the large intestine are the waste products of digestion. The waste products become more solid as excess water is absorbed. Peristalsis continues to force the remaining semisolid waste material into the last section of the large intestine, called the rectum. Muscles in the rectum and anus control the release of this semisolid waste, called feces (FEE seez).

Bacteria and Digestion

You might think that all bacteria are harmful. However, some bacteria have an important role in the digestive system. Bacteria, such as the ones shown in **Figure 8,** digest food and produce important vitamins and amino acids. Bacteria in the intestines are essential for proper digestion.

The Digestive System and Homeostasis

Recall that nutrients from food are absorbed in the small intestine. The digestive system must be functioning properly for this absorption to occur. These nutrients are necessary for other body systems to maintain homeostasis. For example, the blood in the circulatory system absorbs the products of digestion. The blood carries the nutrients to all other body systems, providing them with materials that contain energy.

11. Infer What might happen to other body systems if the digestive system did not function properly?

____

____

____

Math Skills MA.6.A.3.6

Use Percentages

A percentage is a ratio that compares a number to 100. For example, the total length of the intestines is about 8.5 m. That value represents 100%. If the rectum is 0.12 m long, what percentage of the intestines is made up of the rectum?

The ratio is $\frac{0.12 \text{ m}}{8.5 \text{ m}}$.

Find the equivalent decimal for the ratio.

$$\frac{0.12 \text{ m}}{8.5 \text{ m}} = 0.014$$

Multiply by 100.

$$0.014 \times 100 = 1.4\%$$

Practice

10. The total length of the intestines is about 8.5 m. If the small intestine is 7.0 m long, what percentage of the intestines is made up of the small intestine?

Lesson Review 2

Visual Summary

Enzymes in the digestive system break down food so nutrients can be absorbed by your body.

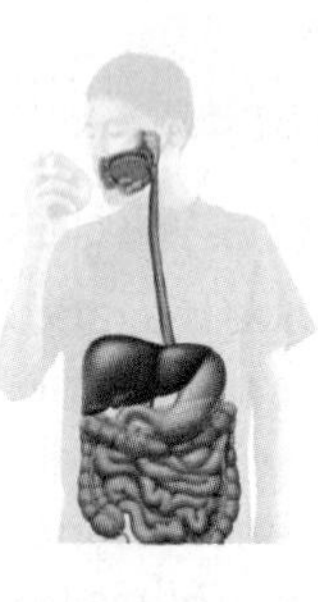

Food moves through the digestive tract by waves of peristalsis.

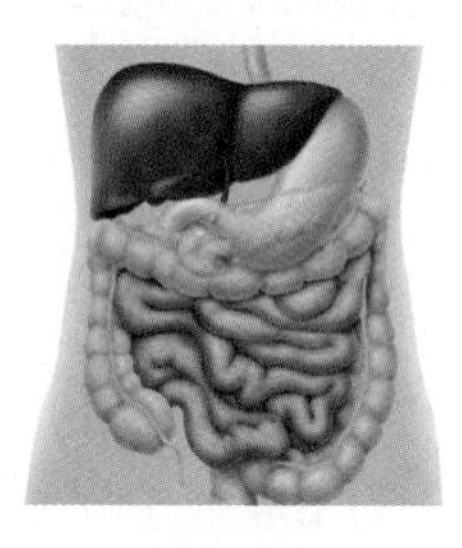

The liver and the pancreas produce substances that help with chemical digestion.

Use Vocabulary

1. **Define** *enzyme* in your own words. SC.6.L.14.5

2. **Distinguish** between absorption and digestion. LA.6.2.2.3

Understand Key Concepts

3. Where is the first place digestion occurs? SC.6.L.14.5
   - (A) mouth
   - (B) stomach
   - (C) large intestine
   - (D) small intestine

4. **Give an example** of how the digestive system affects other body systems. SC.6.L.14.5

Interpret Graphics

5. **Organize Information** Copy and fill in the graphic organizer below to show how food moves through the digestive tract. LA.6.2.2.3

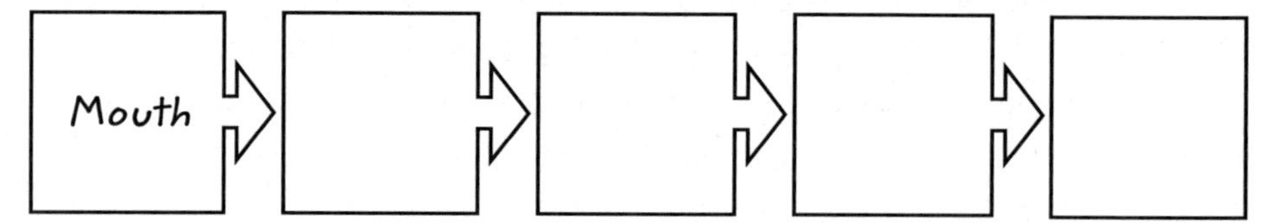

Critical Thinking

6. **Infer** what would happen if food passed more quickly than normal through the digestive system. SC.6.L.14.5

Math Skills

7. If the total length of the intestines is 8.5 m and the large intestine is 1.5 m long, what percentage of the intestines is made up of the small intestine? MA.6.A.3.6

SCIENCE & SOCIETY

Are digestive bacteria related to obesity?

Bacteria percentages might affect your health.

The worldwide rate of obesity greatly concerns medical and health professionals. New research reveals a possible link between bacteria in the human digestive tract and the risk of being overweight.

Bacteroides help digest food.

Your digestive system is home to between 10 and 100 trillion bacteria. That's ten times the number of cells in your body! Certain bacteria are necessary, however, for the digestion of food. Without "friendly" bacteria, you could eat all you wanted, but the food would pass through your intestines mostly undigested.

Recent studies suggest there might be a link between the bacteria in the human digestive tract and obesity. Some people have a type of bacteria that causes them to absorb more calories than normal from their food. They gain more weight than people with a different type of bacteria. In general, obese humans have a lower percentage of a group of bacteria called Bacteroidetes (BAK-tear-oid-dee-teez) and more of a group of bacteria called Firmicutes (fir-MIC-ku-teez). It is not clear whether Firmicutes bacteria make people obese, or whether obese people have more of this type of bacteria. But evidence supports the idea that changing the bacteria in someone's intestines and stomach—by means of diet or medications—might be an important weapon in the fight against obesity.

Additional research is needed to understand any link between digestive bacteria and obesity. But it is an exciting possibility that managing the bacteria in the digestive tract could be a new way to improve human health.

It's Your Turn

RESEARCH Find out more about the role of bacteria in human health. Research how the bacteria in your digestive tract help regulate your immune system. LA.6.4.2.2

Lesson 3

The Excretory SYSTEM

ESSENTIAL QUESTIONS

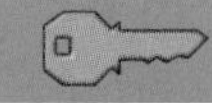
What does the excretory system do?

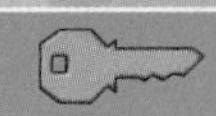
How do the parts of the excretory system work together?

How does the excretory system interact with other body systems?

Vocabulary

excretory system p. 509
kidney p. 511
nephron p. 511
urine p. 511
ureter p. 513
bladder p. 513
urethra p. 513

Florida NGSSS

SC.6.L.14.5 Identify and investigate the general functions of the major systems of the human body (digestive, respiratory, circulatory, reproductive, excretory, immune, nervous, and musculoskeletal) and describe ways these systems interact with each other to maintain homeostasis.

SC.6.N.1.1 Define a problem from the sixth grade curriculum, use appropriate reference materials to support scientific understanding, plan and carry out scientific investigation of various types, such as systematic observations or experiments, identify variables, collect and organize data, interpret data in charts, tables, and graphics, analyze information, make predictions, and defend conclusions.

SC.6.N.1.4 Discuss, compare, and negotiate methods used, results obtained, and explanations among groups of students conducting the same investigation.

SC.6.N.1.5 Recognize that science involves creativity, not just in designing experiments, but also in creating explanations that fit evidence.

SC.6.N.3.4 Identify the role of models in the context of the sixth grade science benchmarks.

SC.7.N.3.2 Identify the benefits and limitations of the use of scientific models.

SC.8.N.3.1 Select models useful in relating the results of their own investigations.

Also covers: LA.6.2.2.3

SC.6.N.1.5
SC.6.L.14.5

Launch Lab

10 minutes

What happens when you breathe out?

Look at the photo of the fingertip on the next page. The sweat glands in your skin are one way substances leave your body. Do substances also leave your body when you breathe out?

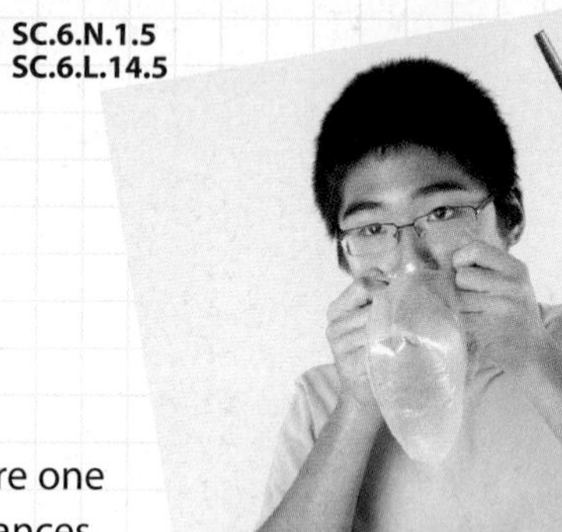

Procedure

1. Read and complete a lab safety form.
2. Take a deep breath and hold it.
3. Breathe out through your mouth into a **plastic bag.** Leave a small opening to allow some of the air to leave the bag as you blow into it.
4. Remove the bag from around your mouth. Let the air escape from the bag, but do not push the sides of the bag together.
5. Using the same plastic bag, repeat steps 2–4 three more times.
6. Observe the contents of the bag. Record your observations below.

Data and Observations

Think About This

1. Did the plastic bag look different after you breathed into it? Explain.

2. What do you think was in the plastic bag at the end of the activity?

3. Key Concept Based on your observations, do you think the respiratory system is part of the excretory system? Explain.

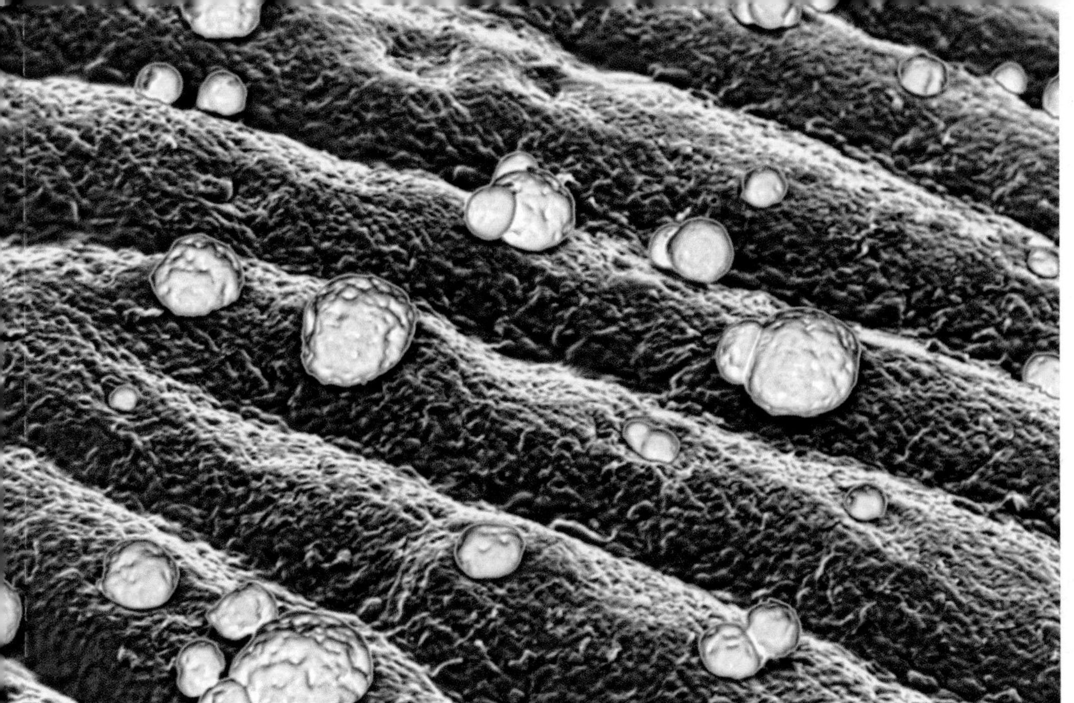

Inquiry A Sweaty Job?

1. Did you know that these are the ridges on a fingertip? The circular openings along the ridges are sweat glands. The sweat from these glands can leave a mark, or fingerprint, on objects that you touch. Why does sweat, or any material, leave your body?

Functions of the Excretory System

You have read about the nutrients in food that are necessary to maintain health. You have also read how the digestive system processes that food. Howzever, your body doesn't use all the food that you ingest. The unused food parts are waste products. What happens to the wastes? They are processed by the excretory system. *The* **excretory system** *collects and eliminates wastes from the body and regulates the level of fluid in the body.*

2. **NGSSS Check**
Explain What does the excretory system do? SC.6.L.14.5

Collection and Elimination

Your home probably has several places where waste is collected. You might have a trash can in the kitchen and another one in the bathroom. The furnace has an air filter that removes and collects dust from the air. Similarly, your body also collects wastes. The digestive system collects waste products in the intestines. The circulatory system collects waste products in the blood.

When the trash cans in your home fill up, you must take the trash outside. The same is true of the waste in your body. If waste is not removed, or eliminated, from your body, it could become toxic and damage your organs. You'll read about the different body systems that eliminate waste later in this lesson.

Regulation of Liquids

Another function of the excretory system is to regulate the level of fluids in the body. You might recall that water is an essential nutrient for your body. Some of the water in your body is lost when waste is eliminated. The excretory system controls how much water leaves the body through elimination. This ensures that neither too little nor too much water is lost.

Active Reading

FOLDABLES® LA.6.2.2.3

Make a four-door book to summarize information about the body systems that make up the excretory system. Label the front *The Excretory System.* Label the inside as shown.

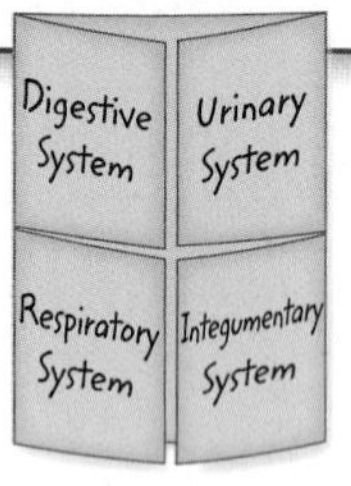

Types of Excretion

Your body excretes, or eliminates, different substances from different body systems. The excretory system is made of three body systems.

- The urinary system processes, transports, collects, and removes liquid waste from the body.
- The respiratory system removes carbon dioxide and water vapor from the body.
- The integumentary system, which includes the skin, secretes excess salt and water through sweat glands.

Figure 9 illustrates the body systems that make up the excretory system and identifies the substances they excrete. You read previously about how the organs of the respiratory system and the integumentary system eliminate waste products from the body. In this lesson, you will read about the organs of the urinary system and their roles in eliminating waste from the body.

3. NGSSS Check **Identify** Highlight the names of the systems that make up the excretory system. SC.6.L.14.5

The Excretory System

Figure 9 Several body systems make up the excretory system.

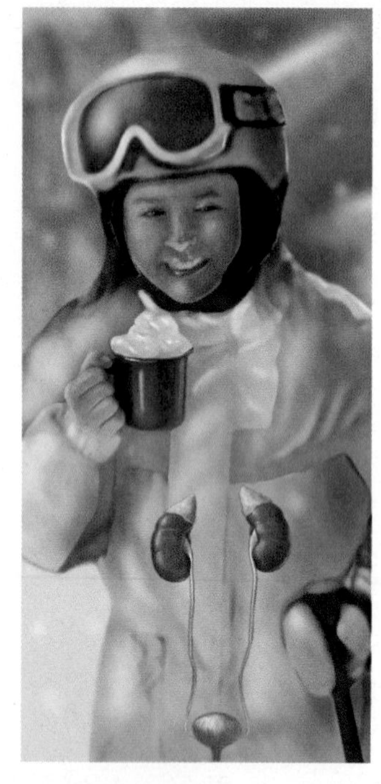

Urinary system
Removes liquid wastes

Integumentary system
Removes excess salt and water

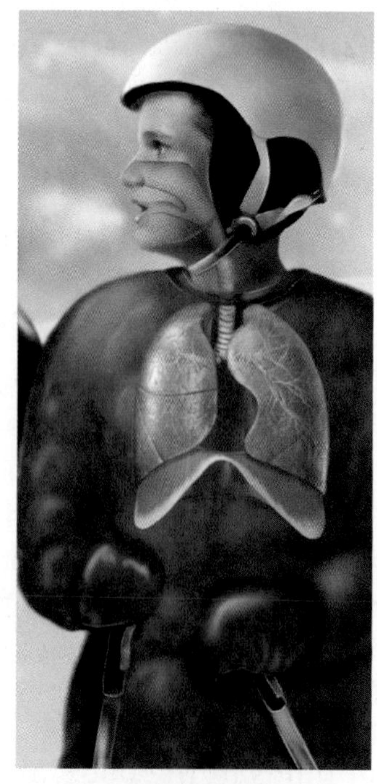

Respiratory system
Removes carbon dioxide and water

Organs of the Urinary System

The urinary system produces, stores, and removes liquid waste from the body and helps maintain homeostasis. The organs of the urinary system are shown in **Figure 10.** They include two kidneys, two ureters, the bladder, and the urethra. These organs work together to process, transport, collect, and excrete liquid waste.

The Kidneys

The bean-shaped organ that filters, or removes, wastes from blood is the **kidney.** You have two kidneys, one on each side of your body. They are near the back wall of your abdomen, above your waist, and below your rib cage. Each kidney is about the size of your fist. Kidneys are dark red in color because of the large volume of blood that passes through them.

The kidneys have several functions. This lesson will focus on the role of the kidneys in the urinary system. However, the kidneys also produce hormones that stimulate the production of red blood cells. In addition, they control blood pressure and help control calcium levels in the body.

The kidneys contain blood vessels and nephrons (NEH frahnz). **Nephrons** *are networks of capillaries and small tubes, or tubules, where filtration of blood occurs.* Each kidney contains about one million nephrons.

Blood contains waste products, salts, and sometimes toxins from cells that need to be removed from the body. These products are filtered from the blood as it passes through the kidneys. *When blood is filtered, a fluid called* **urine** *is produced.* The kidneys filter the blood and produce urine in two stages. You will read about this two-stage filtration process on the next page.

4. **NGSSS Check** Explain What is the function of the urinary system? SC.6.L.14.5

____

____

____

The Urinary System

Figure 10 Most functions of the urinary system occur in the kidneys. The kidneys connect to the ureters, then the bladder, and finally the urethra.

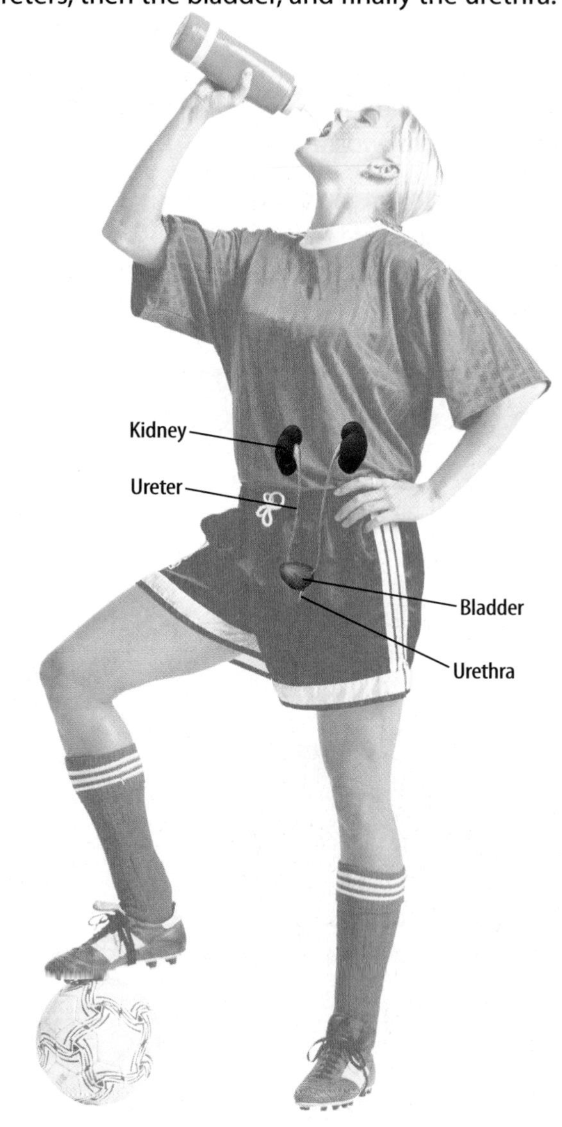

Word Origin

nephron
from Greek *nephros,* means "kidney"

First Filtration Blood is constantly circulating and filtering through the kidneys. In one day, the kidneys filter about 180 L of blood plasma, or the liquid part of blood. That's enough liquid to fill 90 2-L bottles. Your body contains about 3 L of blood plasma. This means your entire blood supply is filtered by your kidneys about 60 times each day. As shown in **Figure 11,** the first filtration occurs in clusters of capillaries in the nephrons. These clusters of capillaries filter water, sugar, salt, and wastes out of the blood.

Second Filtration If all of the liquid from the first filtration were excreted, your body would quickly dehydrate and important nutrients would be lost. To regain some of this water, the kidneys filter the liquid collected in the first filtration again. As shown in **Figure 11,** the second filtration occurs in small tubes in the nephrons. During the second filtration, up to 99 percent of the water and nutrients from the first filtration are separated out and reabsorbed into the blood. The remaining liquid and waste products form urine. On average, an adult excretes about 1.5 L of urine per day.

Figure 11 The kidneys produce urine in two stages.

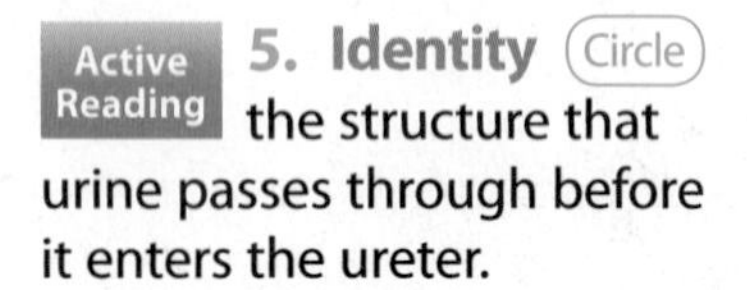

Active Reading **5. Identity** Circle the structure that urine passes through before it enters the ureter.

Filtration in the Kidneys

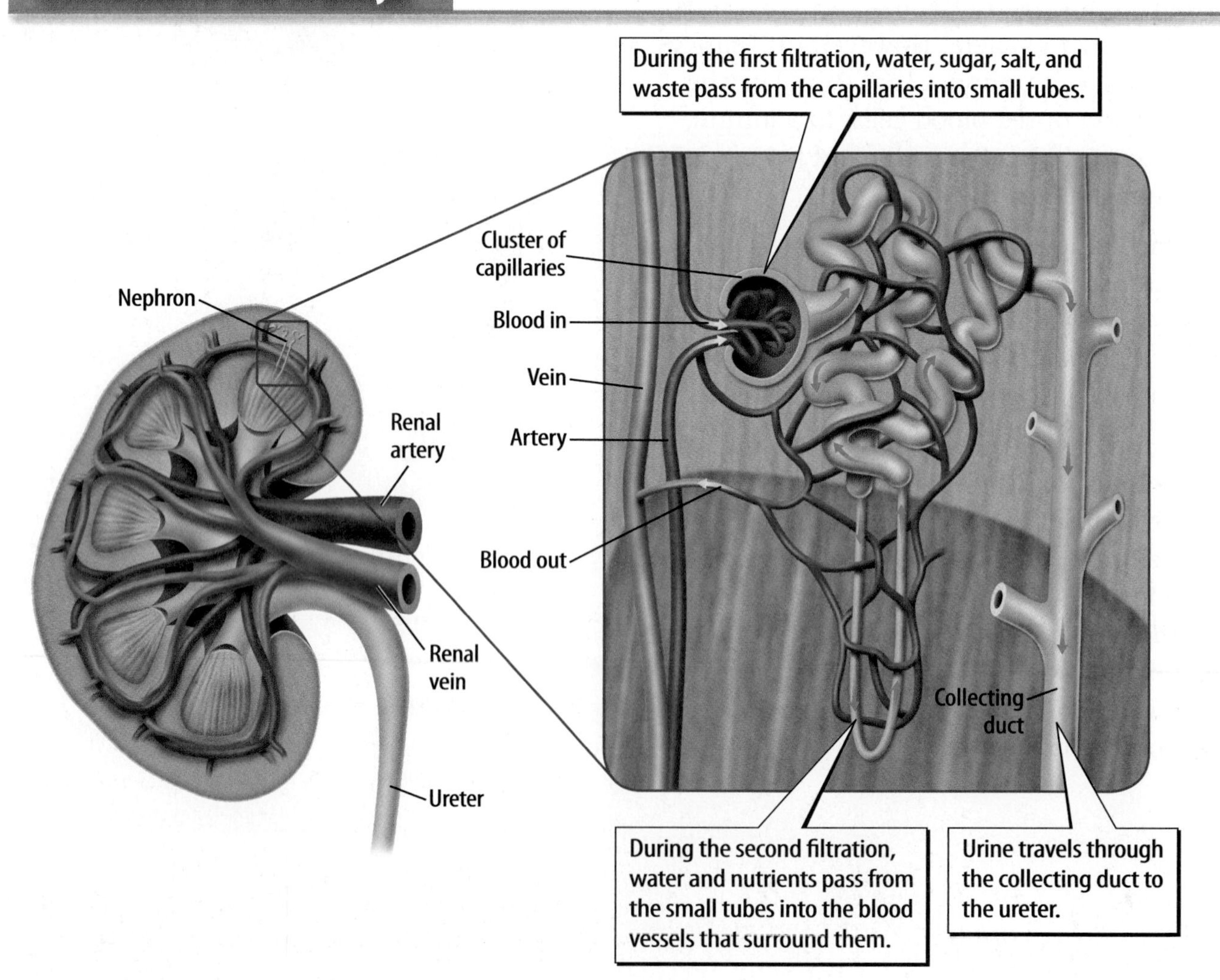

The Ureters, Bladder, and Urethra

Do you remember the trash can you read about earlier in this lesson? What would happen if you put garbage in the trash can but never emptied the trash can? The garbage would pile up. After a while, there would be too much garbage for the trash can to hold. To keep this from happening, you must empty the trash from the trash can. In a similar way, the urine produced by your body cannot stay in the kidney. *Urine leaves each kidney through a tube called the* **ureter** (YOO ruh tur). Refer back to **Figure 10** to see the locations of the ureter and other organs of the urinary system.

Word Origin

ureter
from Greek *ourethra,* means "passage for urine"

Both ureters drain into the bladder. *The* **bladder** *is a muscular sac that holds urine until the urine is excreted.* The bladder expands and contracts like a balloon when filled or emptied. An adult bladder can hold about 0.5 L of urine.

Urine leaves the bladder through a tube called the **urethra** (yoo REE thruh). The urethra contains circular muscles called sphincters (SFINGK turz) that control the release of urine.

Active Reading **6. Describe** How do the ureters, bladder, and urethra work together to excrete urine?

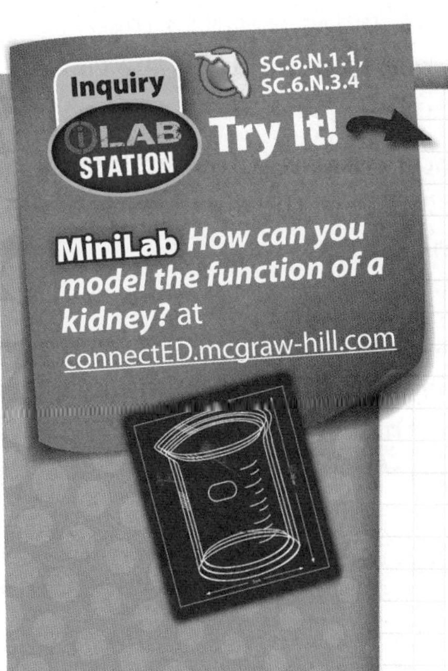

Apply It! After you complete the lab, answer these questions.

1. How would you change your model to also include the bladder?

2. How would the efficiency of your model change if you did not use filter paper?

Table 3 Urinary Disorders

Urinary Disorder	Description	Possible Causes
Kidney disease	The nephrons are damaged and the ability of the kidneys to filter blood is reduced. However, a person can have the beginning stages of kidney disease and experience no symptoms.	diabetes, high blood pressure, poisons, trauma
Urinary tract infection	Symptoms can include burning during urination, small and frequent urination, and blood in urine.	bacteria in the urinary system
Kidney stones	Kidney stones are solid substances that form in the kidney. Stones that pass through the urinary system can be very painful.	calcium buildup in the kidney

Urinary Disorders

A urinary disorder is an illness that affects one or more organs of the urinary system. Some urinary disorders are described in **Table 3.** Several of these disorders are relatively common. Urinary tract infections, for example, are a leading cause of doctor visits, second only to respiratory infections.

The Excretory System and Homeostasis

You have already read about some of the ways the excretory system helps maintain homeostasis. For example, the excretory system filters wastes from the blood. The blood is part of the circulatory system. If wastes were allowed to build up in the circulatory system, they would become toxic.

Another example of maintaining homeostasis is the removal of wastes from the digestive system. Similar to the circulatory system, wastes would damage your body if they were not removed from the digestive system by the excretory system.

The excretory system also interacts with the nervous system. The hypothalamus is an **area** of the brain that helps maintain homeostasis. One function of the hypothalamus is to control the secretion of some hormones. One such hormone causes the tubules in the kidney to absorb more water from the blood. This helps the body to regulate fluid levels. Water is retained in the blood instead of being excreted in the urine.

7. NGSSS Check
Explain How does the excretory system interact with the nervous system?

ACADEMIC VOCABULARY

area
(noun) a part of something that has a particular function

Lesson Review 3

Visual Summary

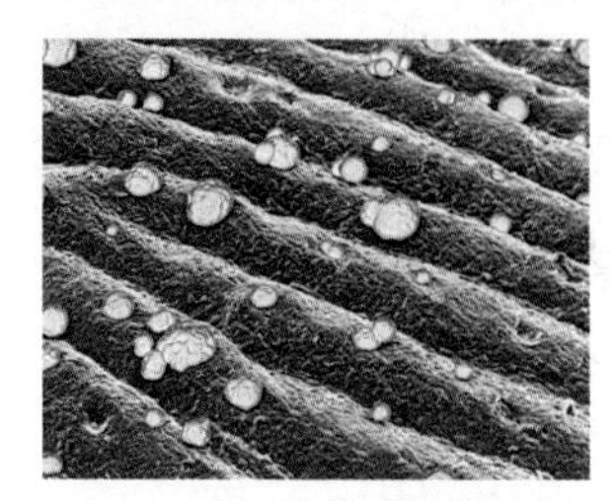

The excretory system collects and eliminates wastes from the body and regulates the level of fluid in the body.

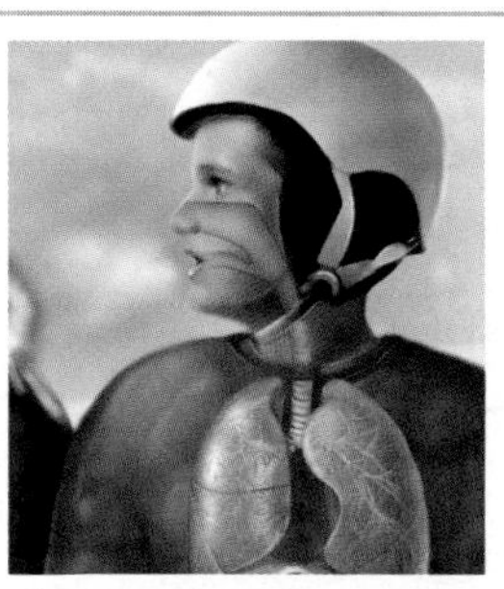

The respiratory system is one of the body systems that make up the excretory system.

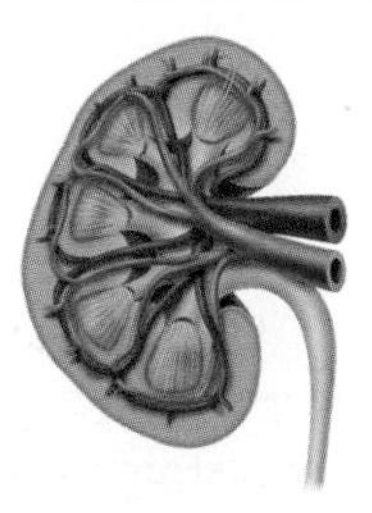

The organs of the urinary system process, transport, collect, and excrete waste.

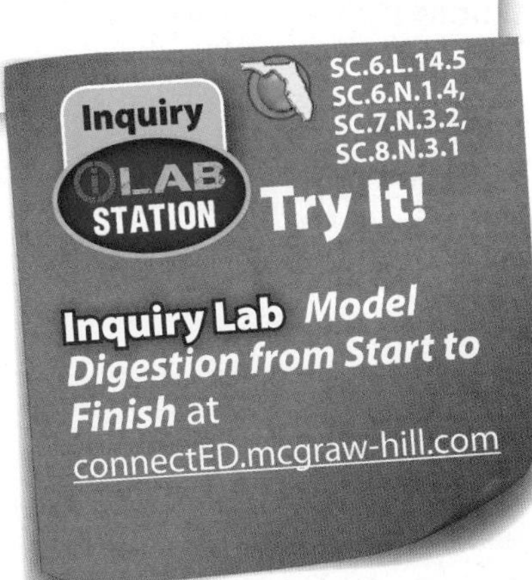

Use Vocabulary

1. **Define** the word *nephron* in your own words. SC.6.L.14.5

2. **Distinguish** between ureter and urethra. SC.6.L.14.5

3. **Use the term** *bladder* in a sentence. SC.6.L.14.5

Understand Key Concepts

4. The kidneys filter wastes from the SC.6.L.14.5
   - (A) blood.
   - (B) intestine.
   - (C) lungs.
   - (D) skin.

5. **Construct** a diagram of the urinary system showing the production and flow of urine on a separate piece of paper. SC.6.L.14.5

6. **Distinguish** between the excretory functions of the respiratory system and the integumentary system. SC.6.L.14.5

Interpret Graphics

7. **Organize Information** Fill in the table below with details about each organ of the urinary system. LA.6.2.2.3, SC.6.L.14.5

Organ	Structure and Function

Critical Thinking

8. **Hypothesize** What might happen if urine did not go through a second filtration? SC.6.L.14.5

Chapter 13 Study Guide

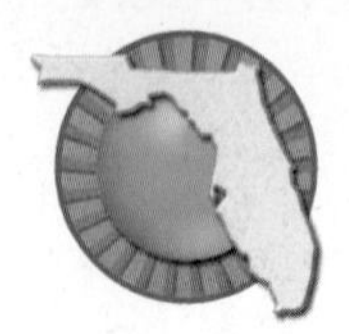

Think About It! **The digestive and excretory systems interact with each other to maintain homeostasis by moving materials through the body and removing wastes.**

Key Concepts Summary

LESSON 1 Nutrition

- People eat food to obtain the energy their bodies need to function. The amount of energy in food is measured in **Calories.**
- The types and amounts of nutrients a person needs depend on age, gender, and activity level.
- The six groups of nutrients are **proteins, carbohydrates, fats, vitamins, minerals,** and water.
- A balanced diet provides nutrients and energy for a healthful lifestyle.

Vocabulary

Calorie p. 491
protein p. 492
carbohydrate p. 492
fat p. 493
vitamin p. 493
mineral p. 493

LESSON 2 The Digestive System

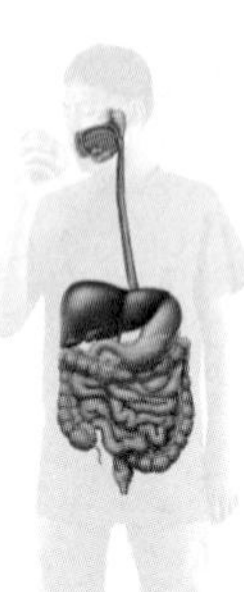

- The function of the digestive system is to break down food and absorb nutrients for the body.
- Organs of the digestive system include the mouth, **esophagus,** stomach, small intestine, and large intestine.
- The digestive system interacts with other body systems to maintain the body's internal balance.

digestion p. 499
mechanical digestion p. 500
chemical digestion p. 500
enzyme p. 500
esophagus p. 502
peristalsis p. 502
chyme p. 503
villi p. 504

LESSON 3 The Excretory System

- The function of the **excretory system** is to collect and eliminate wastes from the body and regulate the level of fluids in the body.
- The excretory system is made up of the digestive system, the respiratory system, the urinary system, and the integumentary system.
- The excretory system works with other body systems, including the nervous system, to maintain homeostasis.

excretory system p. 509
kidney p. 511
nephron p. 511
urine p. 511
ureter p. 513
bladder p. 513
urethra p. 513

Active Reading

FOLDABLES® Chapter Project

Assemble your lesson Foldables as shown to make a Chapter Project. Use the project to review what you have learned in this chapter.

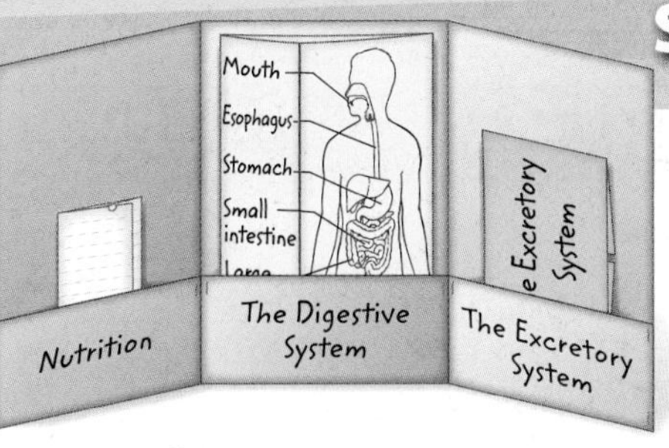

Use Vocabulary

1. About 25–35 percent of your total daily ______________ should be from fats.
2. One type of nutrient, ______________, is made of long chains of sugars.
3. Food moves down the esophagus by ______________.
4. The breakdown of food into small particles and molecules is called ______________.
5. A tube that connects a kidney to the bladder is called a(n) ______________.
6. Urine is stored in the ______________.

Link Vocabulary and Key Concepts

Use vocabulary terms from the previous page to complete the concept map.

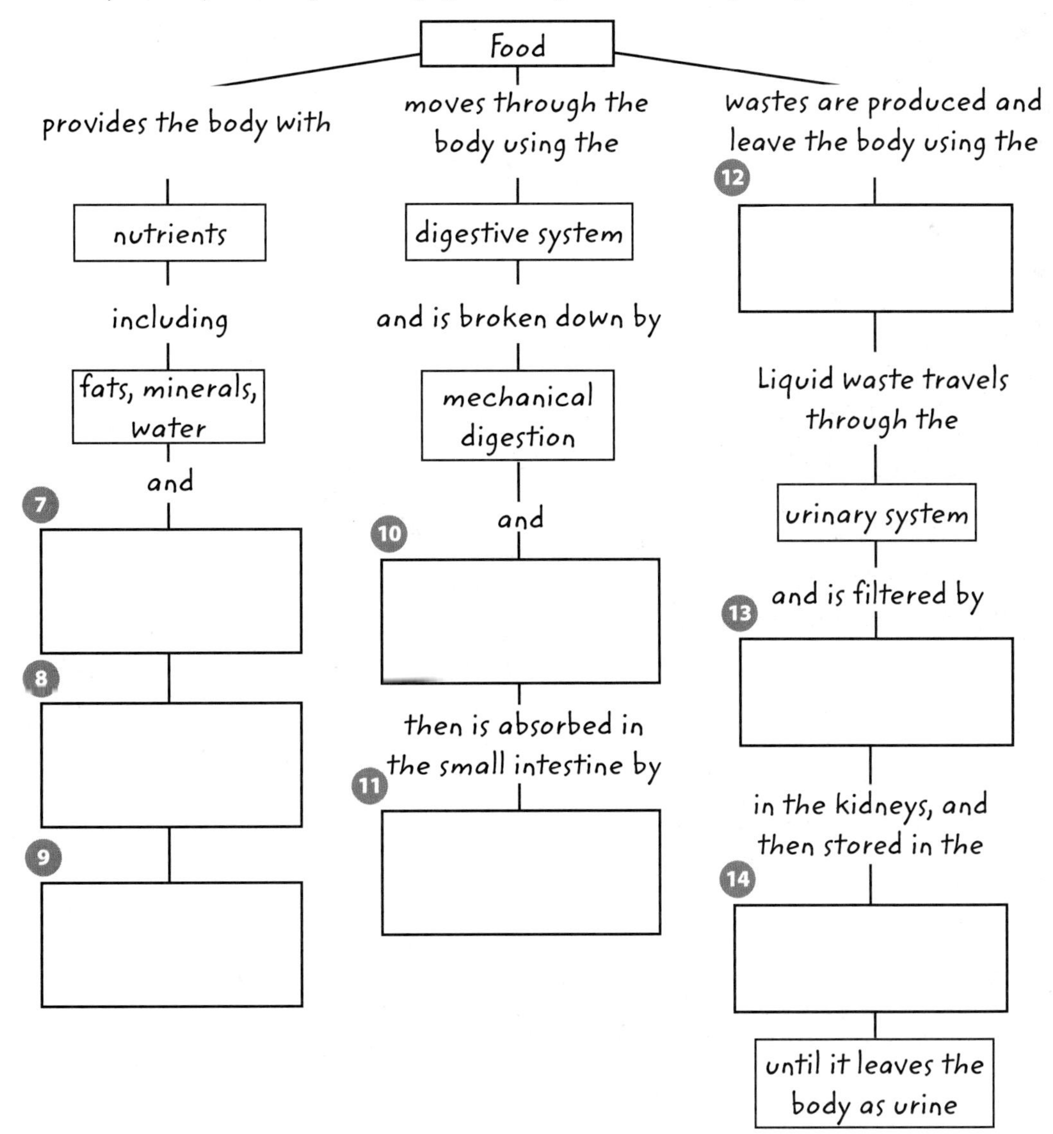

Chapter 13 Review

Fill in the correct answer choice.

Understand Key Concepts

1 What are proteins made of?
Ⓐ amino acids
Ⓑ minerals
Ⓒ sugars
Ⓓ vitamins

2 Look at the diagram below. Where does most absorption of nutrients occur? **SC.6.L.14.5**

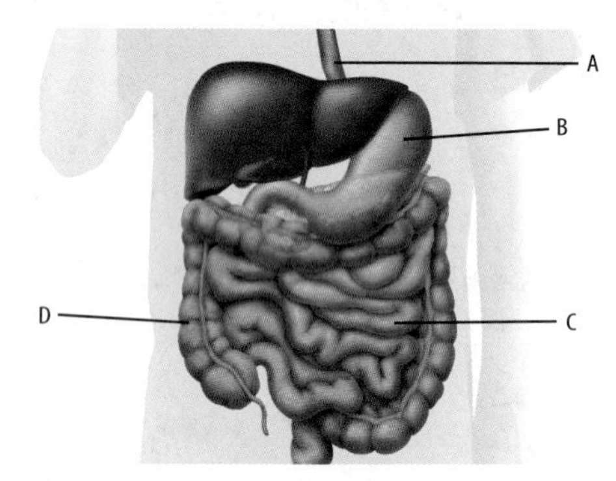

Ⓐ A
Ⓑ B
Ⓒ C
Ⓓ D

3 What is the correct order for how food is processed in the digestive system? **SC.6.L.14.5**
Ⓐ absorption, digestion, ingestion, elimination
Ⓑ elimination, ingestion, absorption, digestion
Ⓒ ingestion, absorption, digestion, elimination
Ⓓ ingestion, digestion, absorption, elimination

4 Which organ produces a substance that neutralizes acid from the stomach? **SC.6.L.14.5**
Ⓐ esophagus
Ⓑ gallbladder
Ⓒ liver
Ⓓ pancreas

5 Carbon dioxide is eliminated by which body system? **SC.6.L.14.5**
Ⓐ digestive system
Ⓑ integumentary system
Ⓒ respiratory system
Ⓓ urinary system

Critical Thinking

6 **Distinguish** between minerals and vitamins.

7 **Hypothesize** why a child might have different nutritional needs than an adult over the age of 60.

8 **Select** Study the nutrient information below. Circle the snack that would be a better choice as part of a healthful lifestyle. Explain your choice.

Nutrient Information	Tortilla Chips	
	Fried	**Baked**
Calories	150	110
Calories from fat	60	5
Total fat (g)	7	1
Saturated fat (g)	1	0
Sodium (mg)	135	200
Total carbohydrate (g)	22	24
Sugars	3	0
Protein	3	2

9 **Differentiate** Suppose your teacher showed you a diagram of a small intestine and a diagram of a large intestine. How might you distinguish between them? SC.6.L.14.5

10 **Hypothesize** How might digestion be affected if a person swallowed his or her food without first chewing it? SC.6.L.14.5

11 **Critique** the following statement: "Bacteria are harmful and should not be in the digestive system." SC.6.L.14.5

12 **Compare** the excretions of the urinary system and the digestive system. SC.6.L.14.5

Writing in Science

13 **Create** a commercial on a separate piece of paper to encourage people to eat a healthful amount from each food group. Include a setting and dialogue for your commercial.

Big Idea Review

14 Give examples of how the digestive system and excretory system help to maintain homeostasis. SC.6.L.14.5

15 What is the function of the small intestine? SC.6.L.14.5

Math Skills MA.6.A.3.6

Use Percentages

Use the table below to answer questions 16–18.

Location of Food	Time in Location (hrs)
Stomach	4
Small intestine	6
Large intestine	24

16 What percentage of the total digestive time does food spend in the stomach?

17 What percentage of the total digestive time does food spend in the large intestine?

18 What percentage of the total digestive time does food spend in the stomach and the small intestine combined?

Florida NGSSS Benchmark Practice

Fill in the correct answer choice.

Multiple Choice

1 Which process depends on enzymes? **SC.6.L.14.5**

Ⓐ chemical digestion
Ⓑ elimination
Ⓒ mechanical digestion
Ⓓ respiration

Use the diagram below to answer question 2.

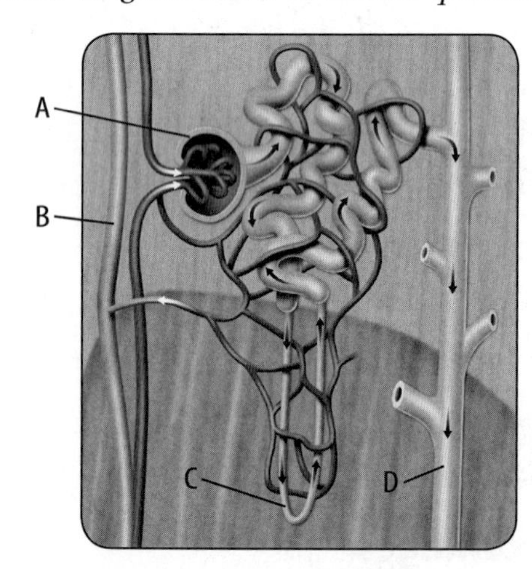

2 Where does the first filtration process occur in the nephron shown above? **SC.6.L.14.5**

Ⓕ A
Ⓖ B
Ⓗ C
Ⓘ D

3 Which factor does NOT influence how much energy a person needs?

Ⓐ age
Ⓑ gender
Ⓒ height
Ⓓ weight

Use the diagram below to answer questions 4 and 5.

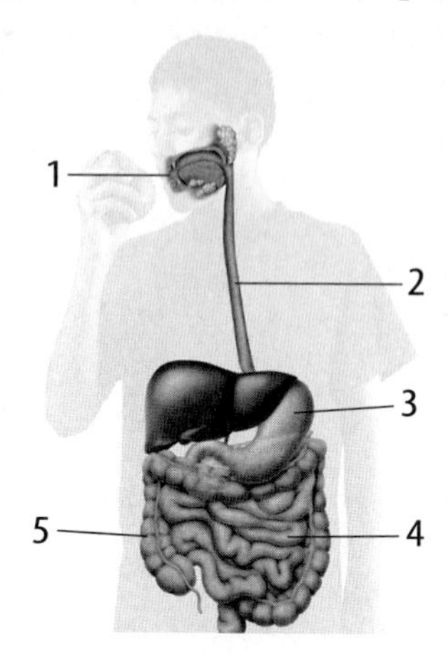

4 In which part of the system pictured above does chemical digestion begin? **SC.6.L.14.5**

Ⓕ 1
Ⓖ 2
Ⓗ 3
Ⓘ 4

5 In the diagram above, from which organ are nutrients absorbed into the bloodstream? **SC.6.L.14.5**

Ⓐ 2
Ⓑ 3
Ⓒ 4
Ⓓ 5

6 What is a main function of the excretory system? **SC.6.L.14.5**

Ⓕ fight diseases
Ⓖ move limbs
Ⓗ pump blood
Ⓘ remove wastes

7 Which part of the brain works with the urinary system to help maintain homeostasis? **SC.6.L.14.5**

Ⓐ cerebellum

Ⓑ cerebrum

Ⓒ hypothalamus

Ⓓ medulla

Use the diagram below to answer question 8.

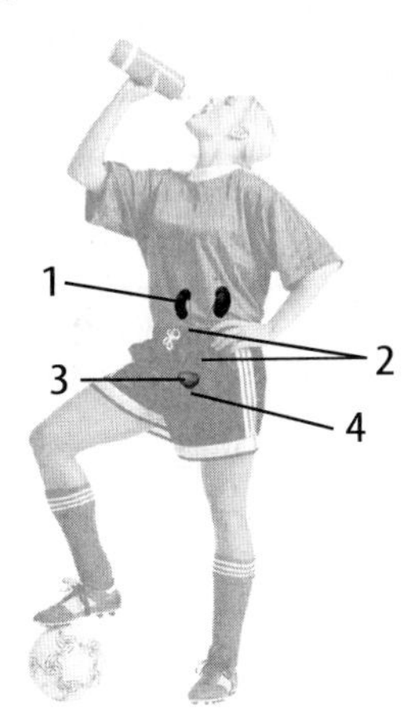

8 In the diagram above, where is urine produced? **SC.6.L.14.5**

Ⓕ 1

Ⓖ 2

Ⓗ 3

Ⓘ 4

9 Which system works with the digestive system to carry nutrients to the cells of the body? **SC.6.L.14.5**

Ⓐ circulatory

Ⓑ excretory

Ⓒ lymphatic

Ⓓ respiratory

10 Which would be considered a grain? **SC.6.L.14.5**

Ⓕ black beans

Ⓖ brown rice

Ⓗ canola oil

Ⓘ lean chicken

11 What is the main source of energy for your body? **SC.6.L.14.5**

Ⓐ carbohydrates

Ⓑ minerals

Ⓒ proteins

Ⓓ water

12 What organ is shown below? **SC.6.L.14.5**

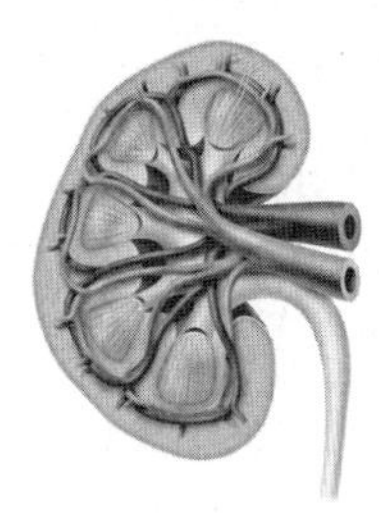

Ⓕ bladder

Ⓖ hypothalamus

Ⓗ kidney

Ⓘ ureter

13 What is produced by the urinary system? **SC.6.L.14.5**

Ⓐ blood

Ⓑ feces

Ⓒ perspiration

Ⓓ urine

NEED EXTRA HELP?

If You Missed Question...	1	2	3	4	5	6	7	8	9	10	11	12	13
Go to Lesson...	2	3	1	2	2	3	3	3	2	1	1	3	3

Name ______________________ Date ____________

Benchmark Mini-Assessment Chapter 13 • Lesson 1

mini BAT

Multiple Choice *Bubble the correct answer.*

Nutrition Facts	
Serving Size 1/2 cup (130g)	
Servings Per Container about 3	
Amount Per Serving	
Calories 30	Calories from Fat 0
	% **Daily Value***
Total Fat 0 g	**0%**
Cholesterol 0 mg	**0%**
Sodium 290 mg	**12%**
Total Carbohydrate 6 g	**2%**
Dietary Fiber 1 g	**4%**
Sugars 4 g	
Protein 1 g	
Vitamin A 15% • Vitamin C 35%	
Calcium 2% • Iron 4%	
Not a significant source of saturated fat and trans fat.	
*Percent Daily Values are based on a 2,000 calorie diet.	

1. Based on the nutrition facts label above, how much of the food in the can would a person have to eat to get his or her daily amount of vitamin C? **MA.6.A.3.6**

Ⓐ about ½ cup

Ⓑ about 1½ cups

Ⓒ almost 3 containers

Ⓓ almost 30 servings

2. From which food group should the smallest amount of food be consumed? **HE.6.C.1.3**

Ⓕ grains

Ⓖ milk

Ⓗ oils

Ⓘ vegetables

3. Which group of nutrients is the body's major source of energy? **HE.6.C.1.3**

Ⓐ carbohydrates

Ⓑ fats

Ⓒ proteins

Ⓓ vitamins

Vitamins and Minerals

Vitamin	**Good Sources**	**Health Benefit**
Vitamin B_2 (riboflavin)	milk, meats, vegetables	helps release energy from nutrients
Vitamin C	oranges, broccoli, tomatoes, cabbage	growth and repair of body tissues
Vitamin A	carrots, milk, sweet potatoes, broccoli	enhances night vision, helps maintain skin and bones
Mineral		
Calcium	milk, spinach, green beans	builds strong bones and teeth
Iron	meat, eggs, green beans	helps carry oxygen throughout the body
Zinc	meat, fish, wheat/grains	aids protein formation

4. Based on the table above, what might happen if you did not get enough vitamin C? **LA.6.2.2.3**

Ⓕ You might feel tired because of not getting enough energy.

Ⓖ You might have trouble seeing objects at night.

Ⓗ You might have weak bones and teeth.

Ⓘ You might take longer to heal when you are hurt.

Name ____________________ Date ____________

Multiple Choice *Bubble the correct answer.*

Use the flowchart below to answer questions 1 and 2.

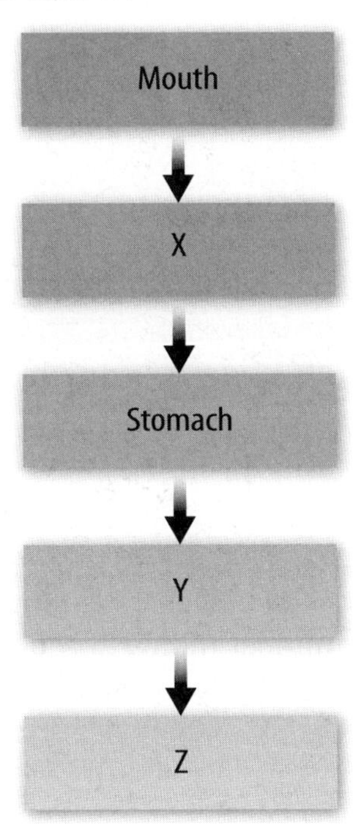

1. Which organ is indicated by the letter X? **LA.6.2.2.3**

Ⓐ esophagus

Ⓑ pancreas

Ⓒ large intestine

Ⓓ small intestine

2. During digestion, what takes place in the organ indicated by the letter Z? **LA.6.2.2.3**

Ⓕ Food is broken down mechanically during chewing.

Ⓖ The organ breaks down food by releasing gastric juices.

Ⓗ The organ releases amylase, which breaks down carbohydrates.

Ⓘ Water from the food is absorbed as it moves through the organ.

3. Which statement is an example of mechanical digestion? **SC.6.L.14.5**

Ⓐ Bile released by the liver breaks down fats.

Ⓑ Pepsin in the stomach breaks down proteins.

Ⓒ Enzymes released with saliva begin the breakdown of carbohydrates.

Ⓓ Peristalsis in the stomach breaks food into even smaller pieces.

4. Which organ of the digestive system stores bile, the enzyme that breaks down fats? **SC.6.L.14.5**

Ⓕ gallbladder

Ⓖ liver

Ⓗ pancreas

Ⓘ stomach

Multiple Choice *Bubble the correct answer.*

1. Which of the four body systems that make up the excretory system functions to secrete excess salt and water through sweat glands? **SC.6.L.14.5**

Ⓐ digestive system

Ⓑ integumentary system

Ⓒ respiratory system

Ⓓ urinary system

2. Which organ of the urinary system removes wastes from blood? **SC.6.L.14.5**

Ⓕ bladder

Ⓖ kidney

Ⓗ ureter

Ⓘ urethra

Use the image below to answer questions 3 and 4.

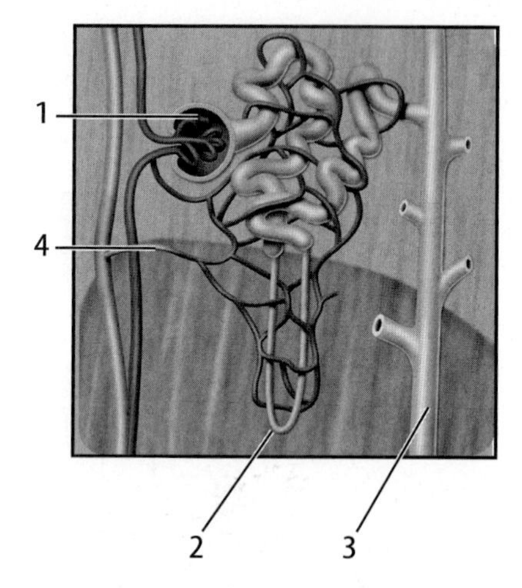

3. Which number indicates the area where first filtration takes place? **SC.6.L.14.5**

Ⓐ 1

Ⓑ 2

Ⓒ 3

Ⓓ 4

4. What is taking place at the area labeled 3? **SC.6.L.14.5**

Ⓕ Blood is entering the glomerulus from an artery.

Ⓖ Urine is traveling through the collecting duct to the ureter.

Ⓗ Water and nutrients are passing from the small tubes into the surrounding blood vessels.

Ⓘ Water, sugar, salt, and wastes are passing from the glomerulus into small tubes.

Name ______________________ Date ______________

Notes

Name ________________ Date ________________

Notes

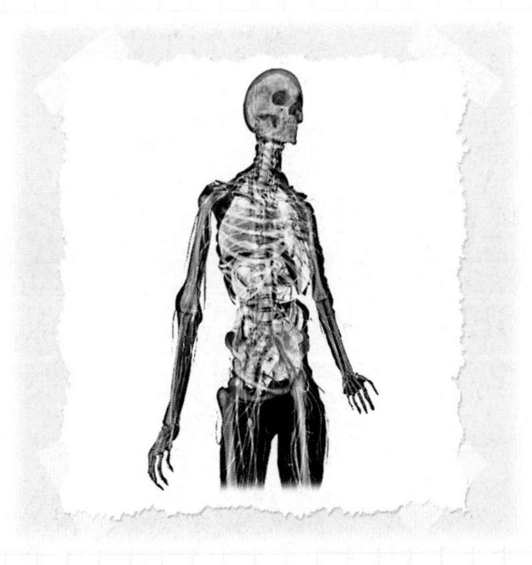

Oxygen in, Carbon Dioxide out

The human body has systems that perform specific functions. Place an *X* before the system or systems you think are responsible for moving the gases oxygen and carbon dioxide.

_____ muscular	_____ skeletal	_____ respiratory
_____ digestive	_____ endocrine	_____ immune
_____ excretory	_____ circulatory	_____ nervous

Explain your thinking. Why did you check off that system or systems?

Respiration and CIRCULATION

FLORIDA BIG IDEAS

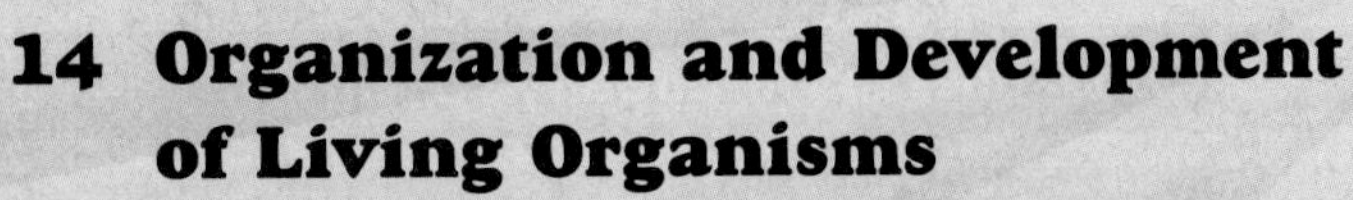

1 The Practice of Science
3 The Role of Theories, Laws, Hypotheses, and Models
14 Organization and Development of Living Organisms

Think About It! How do the respiratory and circulatory systems help maintain the body's homeostasis?

Scuba divers use special equipment to breathe under water. Notice the hose that runs from the air tank to the device in the diver's mouth. When she breathes in, air from the tank moves into her lungs.

1. Why do you think the diver needs air while she's under water?

____

____

2. Why do you think bubbles form when the diver breathes out?

____

____

3. How do you think your respiratory system helps your body maintain homeostasis?

____

____

Get Ready to Read What do you think about respiration and circulation?

Before you read, decide if you agree or disagree with each of these statements. As you read this chapter, see if you change your mind about any of the statements.

	AGREE	DISAGREE
1 Breathing and respiration are the same.	☐	☐
2 Lungs are the only parts of the body that use oxygen.	☐	☐
3 There are four chambers in a human heart.	☐	☐
4 Blood travels in both directions in veins.	☐	☐
5 All blood cells are red.	☐	☐
6 Blood plasma is just water.	☐	☐
7 Lymph nodes are located only in the neck.	☐	☐
8 The lymphatic system helps fight infections to maintain a healthy body.	☐	☐

Lesson 1

The Respiratory SYSTEM

ESSENTIAL QUESTIONS

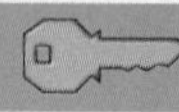 What does the respiratory system do?

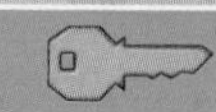 How do the parts of the respiratory system work together?

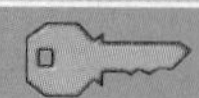 How does the respiratory system interact with other body systems?

Vocabulary

breathing p. 531
pharynx p. 532
larynx p. 532
trachea p. 532
bronchi p. 533
lungs p. 533
alveoli p. 533
diaphragm p. 534

Florida NGSSS

HE.6.C.1.8 Explain how body systems are impacted by hereditary factors and infectious agents.

LA.6.2.2.3 The student will organize information to show understanding (e.g., representing main ideas within text through charting, mapping, paraphrasing, summarizing, or comparing/contrasting);

MA.6.S.6.2 Select and analyze the measures of central tendency or variability to represent, describe, analyze, and/or summarize a data set for the purposes of answering questions appropriately.

SC.6.L.14.5 Identify and investigate the general functions of the major systems of the human body (digestive, respiratory, circulatory, reproductive, excretory, immune, nervous, and musculoskeletal) and describe ways these systems interact with each other to maintain homeostasis.

SC.8.N.1.2 Design and conduct a study using repeated trials and replication.

SC.8.N.3.1 Select models useful in relating the results of their own investigations.

SC.8.N.1.2

Inquiry Launch Lab

10 minutes

How much air is in a breath?

Do your lungs empty completely every time you breathe out? You can use a balloon to find out.

Procedure

1. Read and complete a lab safety form.
2. Place your hands on your ribs as you breathe in and out. Record your observations.
3. Breathe in normally. Breathe out normally into a **balloon.** Twist and hold the end of the balloon.
4. Have your partner use a **metric tape measure** to measure around the balloon at its widest point. Record the measurement. Let the air out of the balloon.
5. Breathe in normally again. Breathe out as much air as you can into the balloon. Twist and hold the end. Repeat step 4.
6. Switch roles with your partner, and repeat steps 2–5 using a different balloon.

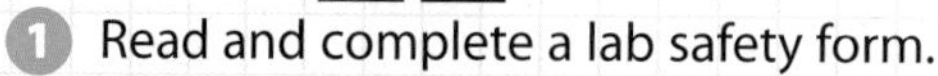
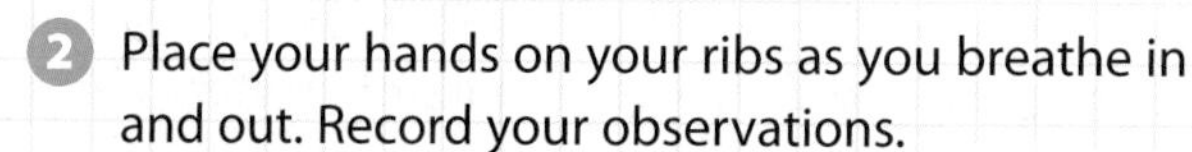

Think About This

1. Was there a difference in the two measurements? Why do you think this happened?

2. Key Concept How do your lungs interact with the bones and muscles of your chest?

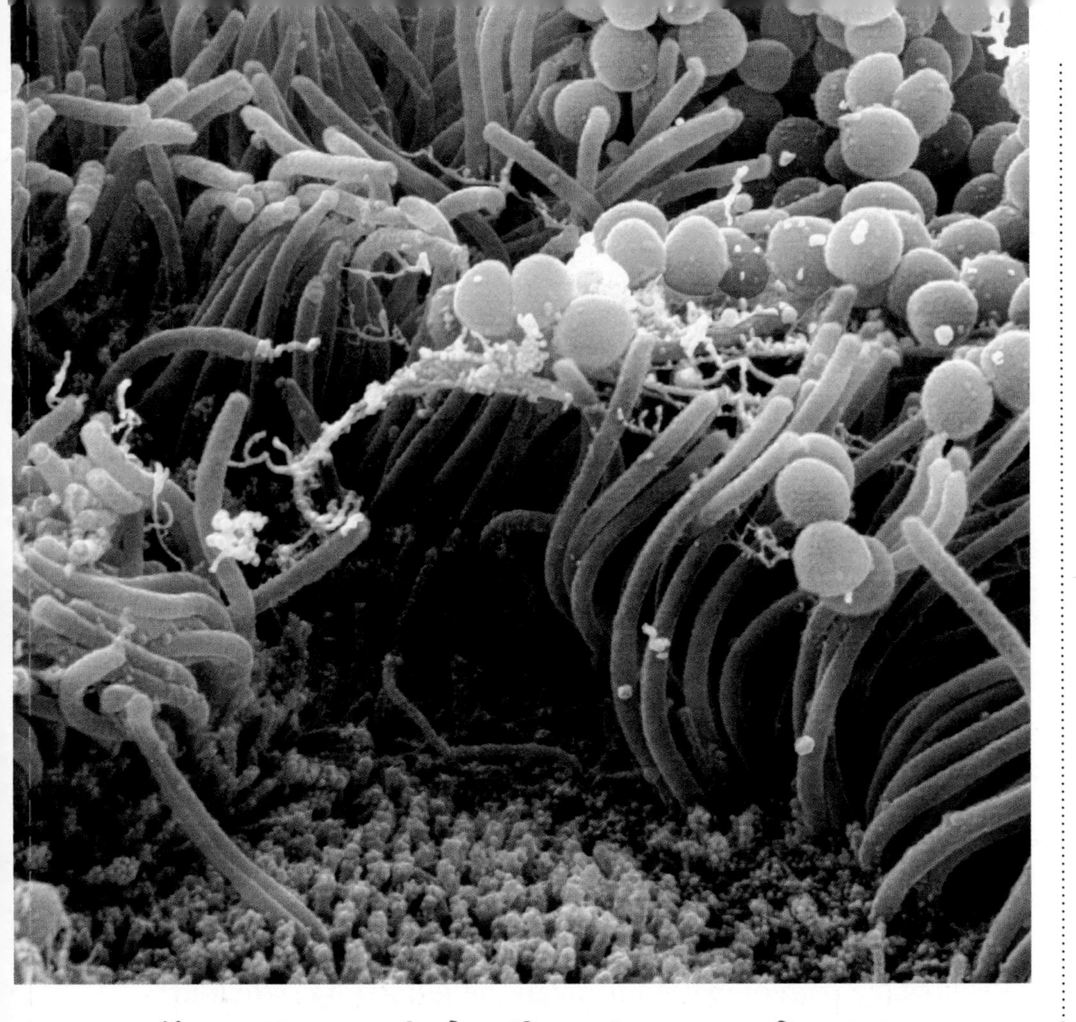

Inquiry Cleaning Up?

1. The hairlike structures here are called cilia (SIH lee uh). They move together in wave-like motions. The round particles on top of the cilia are bits of dust and other things that can block or irritate airways. What do you think these cilia are doing?

Functions of the Respiratory System

If you've ever held your breath, you probably took deep breaths afterward. That's your body's way of getting the oxygen it needs. **Breathing** *is the movement of air into and out of the lungs*. Breathing enables your respiratory system to take in oxygen and to eliminate carbon dioxide.

Taking in Oxygen

Think about the plumbing pipes that bring water into a house. Your respiratory system is similar. It is a system of organs that brings oxygen into your body. Oxygen is so important for life that your brain will tell your body to breathe even if you try not to. Why is oxygen so important? Every cell in your body needs oxygen for a series of chemical reactions called **cellular respiration**. During cellular respiration, oxygen and sugars react. This reaction releases energy a cell can use.

Eliminating Carbon Dioxide

The plumbing in a house also includes pipes that take away wastewater. In a similar way, your respiratory system removes carbon dioxide and other waste gases from your body. If waste gases are not removed, cells cannot function.

2. **NGSSS Check** Locate Highlight what the respiratory system does. SC.6.L.14.5

REVIEW VOCABULARY

cellular respiration
a series of chemical reactions that transform the energy in food molecules to usable energy

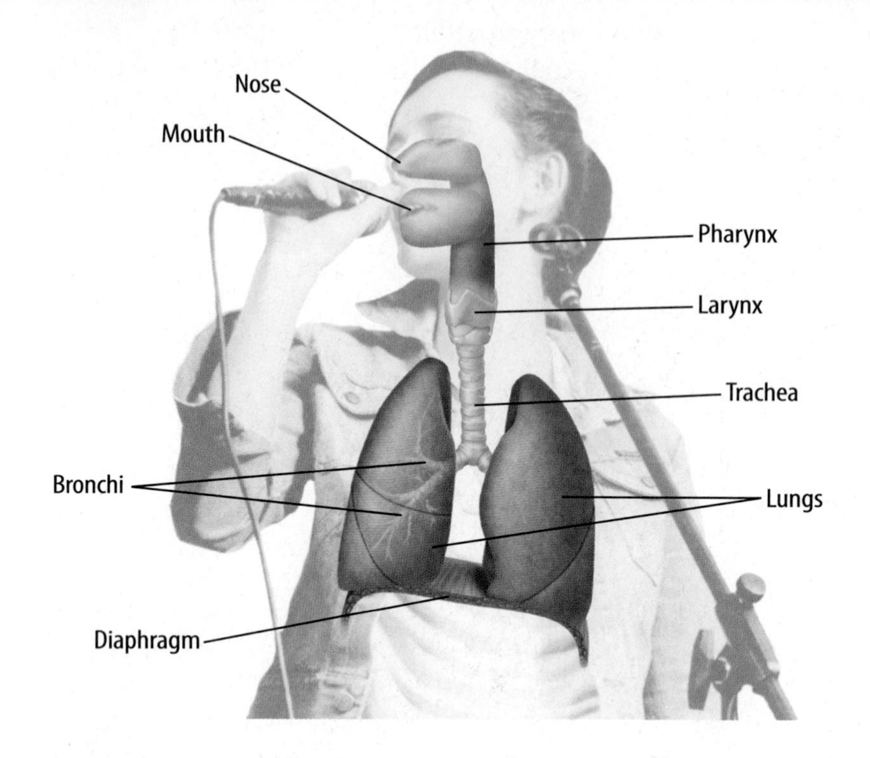

Figure 1 Air moves into and out of the lungs through the respiratory system.

3. Visual Check Survey Circle the part of respiratory system that contains bronchi.

Active Reading

FOLDABLES® LA.6.2.2.3

Make an eight-tab vocabulary book from a sheet of notebook paper. Use it to organize your notes on the organs of the respiratory system and their functions.

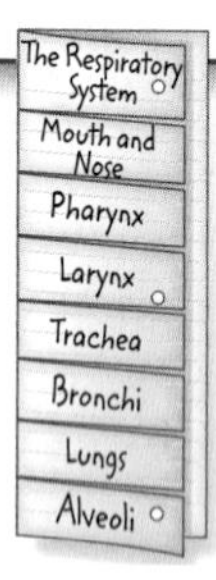

Organs of the Respiratory System

Follow the path of oxygen through the respiratory system in **Figure 1.** Air enters through the mouth and the nose. In the nose, the air is warmed and moistened. Hairs and sticky mucus in the nose help trap dust and dirt from the air. Cilia line the nose and most other airways in the respiratory system. Wavelike motions of the cilia carry trapped particles away from your lungs. The cilia help prevent harmful particles from getting very far into your respiratory system.

Active Reading **4. Evaluate** Underline the function cilia have in the respiratory system.

Pharynx

Air passes from the nose and mouth into the throat. *The* **pharynx** (FER ingks) *is a tubelike passageway at the top of the throat that receives air, food, and liquids from the mouth or nose.* The epiglottis (eh puh GLAH tus) is a flap of tissue at the lower end of the pharynx. It keeps food and liquids from entering the rest of the respiratory system.

Larynx and Trachea

Air passes from the pharynx into a triangle-shaped area called the voice box, or **larynx** (LER ingks). Two thick folds of tissue in the larynx—the vocal cords—vibrate and make sounds as air passes over them. Air then enters the **trachea** (TRAY kee uh), *a tube that is held open by C-shaped rings of cartilage.*

Bronchi and Lungs

The trachea branches into two narrower tubes called **bronchi** (BRAHN ki) (singular, bronchus) *that lead into the lungs.* **Lungs** *are the main organs of the respiratory system.* Inside the lungs, the bronchi continue to branch into smaller and narrower tubes called bronchioles.

Alveoli

In the lungs, the bronchioles end in *microscopic sacs, or pouches, called* **alveoli** (al VEE uh li; singular, alveolus), *where gas exchange occurs.* During gas exchange, oxygen from the air you breathe moves into the blood, and carbon dioxide from your blood moves into the alveoli.

Alveoli look like bunches of grapes at the ends of the bronchioles. Like tiny balloons, the alveoli fill with air when you breathe in. They contract and expel air when you breathe out. Notice in **Figure 2** how blood vessels surround an alveolus.

The walls of alveoli are only one cell thick. The thin walls and the large surface areas of the alveoli enable a high rate of gas exchange. If you could spread out all the alveoli in your lungs onto a flat surface, they would cover an area bigger than most classrooms. Every time you breathe, your alveoli enable your body to take in billions of molecules of oxygen and get rid of billions of molecules of carbon dioxide.

Active Reading 5. **Find** (Circle) the gases that are exchanged in the alveoli.

Word Origin

alveoli

from Latin *alveus,* means "cavity"

Figure 2 Red blood cells drop off carbon dioxide and pick up oxygen as they move through the small blood vessels that surround each alveolus.

6. **Visual Check**

Assess How many layers of cells form the walls of the alveolus shown in this figure?

____

Gas Exchange

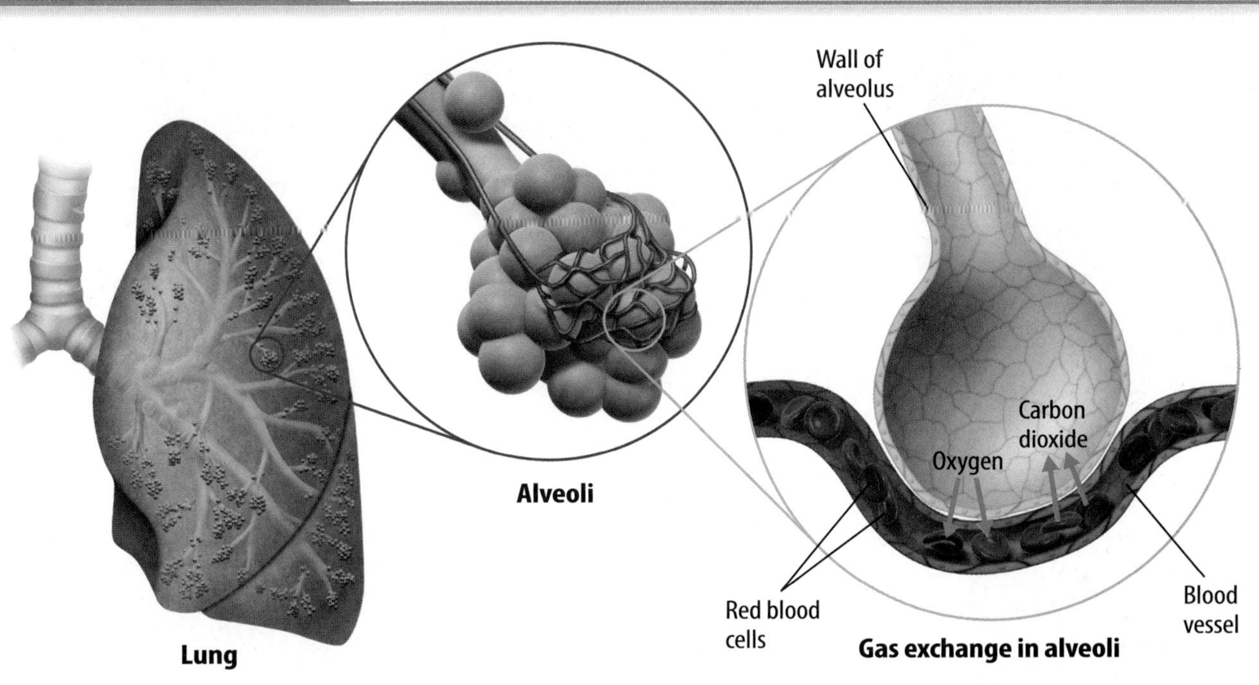

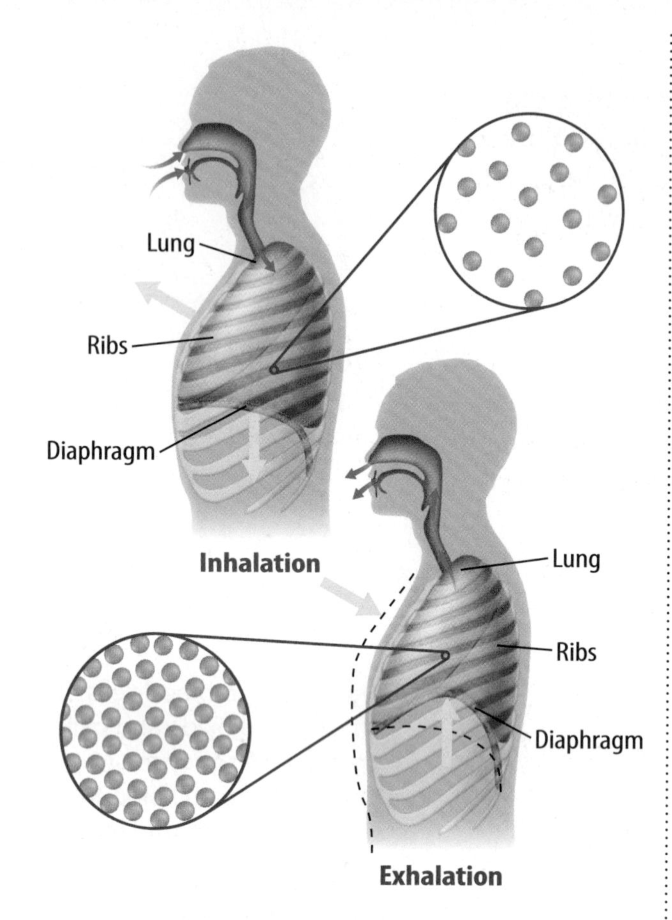

Figure 3 Your diaphragm contracts and moves down when you inhale. The chest cavity gets larger, and air rushes in to equalize the air pressure inside and outside the body. Your diaphragm relaxes and moves up as you exhale. Air rushes out to equalize air pressure.

Breathing and Air Pressure

How does your body know when to breathe? When high levels of carbon dioxide build up in your blood, the nervous system signals your body to breathe out, or exhale. After each exhalation, you breathe in, or inhale. How does this happen?

Below the lungs is a large muscle called the **diaphragm** (DI uh fram) *that contracts and relaxes and moves air in and out of your lungs.* The movement of your diaphragm, as shown in **Figure 3,** causes changes in the air pressure inside your chest. Breathing occurs because of these changes in air pressure.

During inhalation, the diaphragm contracts and moves down, enlarging the space around the lungs. The increased space reduces air pressure in the chest. Air rushes into your lungs until the air pressure inside your chest equals the air pressure outside it.

During exhalation, the diaphragm relaxes and moves up, reducing the space around the lungs. Air pressure in the chest increases. Waste gases rush out of your lungs.

Inquiry SC.8.N.1.2, SC.6.L.14.5

iLAB STATION **Try It!**

MiniLab ***How does exercise affect breathing rate?*** at connectED.mcgraw-hill.com

Apply It! After you complete the lab, answer these questions.

1. What are some other factors that can affect breathing rate?

2. Would a professional swimmer likely take more or fewer breaths per minute than you? Explain.

Respiratory Health

If you've ever had a cold, allergies, or asthma, you know what it's like to have a respiratory illness. A sore throat or a stuffed-up head makes breathing uncomfortable. Some respiratory illnesses make breathing difficult and can even become life-threatening. Common respiratory illnesses and their causes are listed in **Table 1**.

The best way to maintain good respiratory health is to stay away from irritants and air pollution. Don't smoke, and avoid secondhand smoke. On days when air quality is poor or pollen counts are high, it might be best to spend more time indoors.

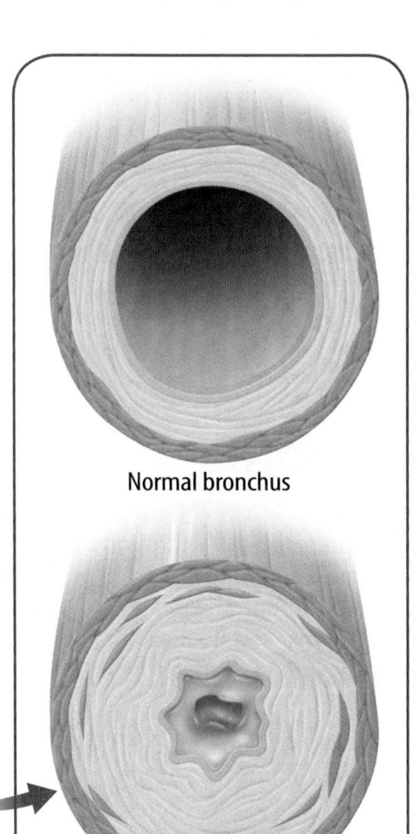
Normal bronchus

Bronchitis

Table 1 Respiratory Illnesses

Illness	Causes	Symptoms
Colds, flu	viruses	congestion, runny nose, watery eyes, coughing, sneezing
Bronchitis (brahn KI tus)	viruses, bacteria	coughing and fatigue due to mucus blocking the bronchi and bronchioles slows air movement
Pneumonia (noo MOH nyuh)	viruses, bacteria	difficulty breathing due to fluid in the alveoli, which slows gas exchange
Asthma (AZ muh)	dust, smoke, pollen, pollution	difficulty breathing due to swollen airways and increased mucus
Emphysema (em fuh SEE muh)	smoking	coughing, fatigue, loss of appetite, and weight loss due to destruction of alveoli
Lung cancer	smoking	coughing, difficulty breathing, and chest pain

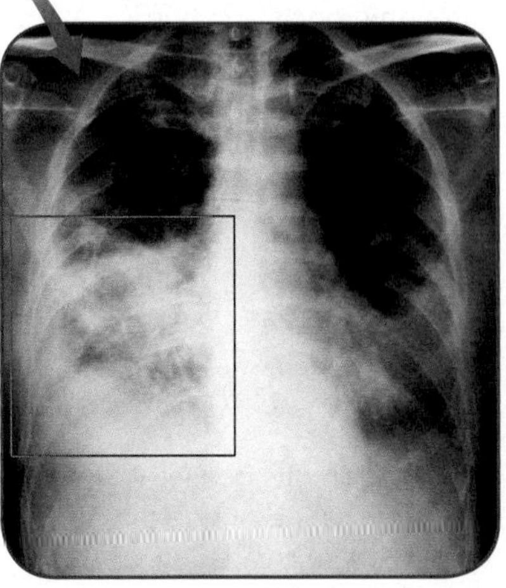
Pneumonia

The Respiratory System and Homeostasis

As you've read in this lesson, the muscular system interacts with the respiratory system so you can breathe. This interaction brings oxygen into your lungs and removes carbon dioxide from your lungs. In the next lesson, you'll read how the circulatory and respiratory systems work together to bring oxygen to body cells and remove carbon dioxide. All of these systems help maintain homeostasis.

7. **NGSSS Check** **Combine** Highlight how the respiratory and muscular systems work together to maintain your body's homeostasis. SC.6.L.14.5

Lesson Review 1

Visual Summary

Air enters the body through the nose and mouth. It passes through the pharynx, larynx, and trachea on its way into the lungs.

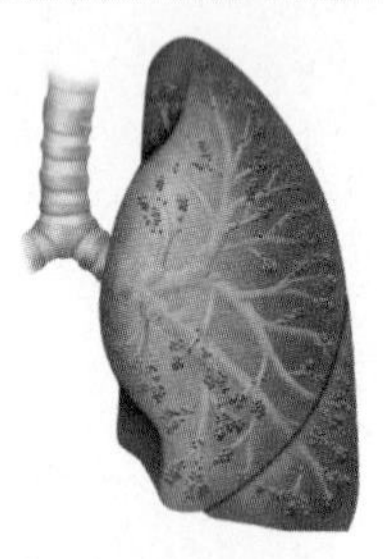

Inside the lungs, air moves through bronchi and bronchioles to the alveoli, where gas exchange takes place.

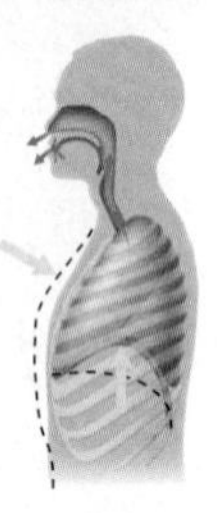

Breathing results from air pressure changes inside the chest that are created by the movement of the diaphragm muscle.

Use Vocabulary

1. The trachea branches into two narrower airways called ______________________.

2. Capillaries surround the ______________________, where gas exchange occurs.

3. **Distinguish** between breathing and respiration.

Understand Key Concepts

4. **Explain** how the nose helps clean air as the air enters the respiratory system. SC.6.L.14.5

5. **Describe** the functions of the respiratory system. SC.6.L.14.5

6. Which body system helps the respiratory system bring oxygen into the body? SC.6.L.14.5
   - (A) circulatory
   - (B) digestive
   - (C) excretory
   - (D) muscular

Interpret Graphics

7. **Compare** Fill in the Venn diagram below to explain the similarities and differences between the trachea and the bronchi. SC.6.L.14.5

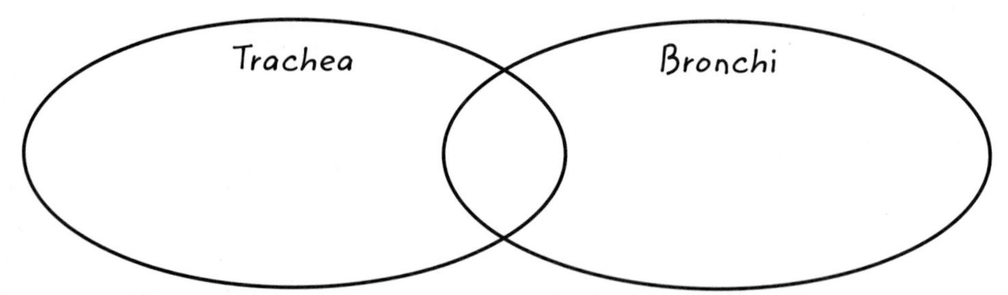

Critical Thinking

8. **Justify** Imagine that you answered a question in class by saying the contraction of the diaphragm causes a person to inhale. Another student disagrees. Justify your answer using words and a drawing. SC.6.L.14.5

Connect Key Concepts

Identify the action and functions of the respiratory system.

Function:	Action:	Function:
take in oxygen needed for 1. ______	2. ______	3. ______, a waste product of cellular respiration.

Sequence the path of air through the respiratory system, and write a short definition for each respiratory organ.

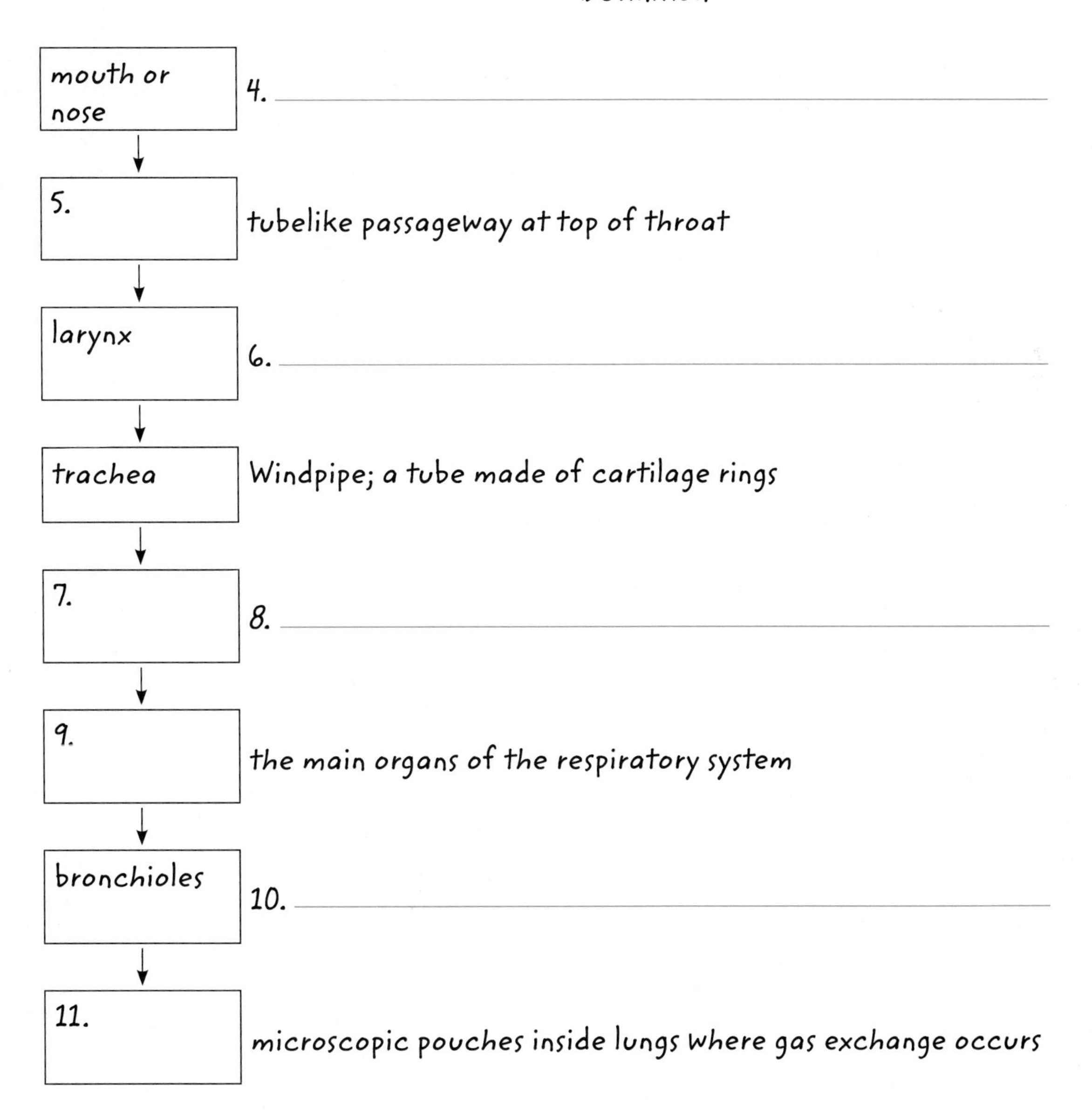

Lesson 2

The Circulatory SYSTEM

ESSENTIAL QUESTIONS

What does the circulatory system do?

How do parts of the circulatory system work together?

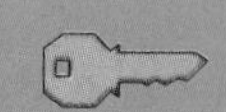
How does the circulatory system interact with other body systems?

Vocabulary

atrium p. 540
ventricle p. 540
artery p. 542
capillary p. 542
vein p. 542
systemic circulation p. 543
coronary circulation p. 543
pulmonary circulation p. 543
atherosclerosis p. 544

Florida NGSSS

HE.6.C.1.8 Explain how body systems are impacted by hereditary factors and infectious agents.

LA.6.2.2.3 The student will organize information to show understanding (e.g., representing main ideas within text through charting, mapping, paraphrasing, summarizing, or comparing/contrasting);

SC.6.L.14.5 Identify and investigate the general functions of the major systems of the human body (digestive, respiratory, circulatory, reproductive, excretory, immune, nervous, and musculoskeletal) and describe ways these systems interact with each other to maintain homeostasis.

SC.6.N.1.5 Recognize that science involves creativity, not just in designing experiments, but also in creating explanations that fit evidence.

SC.6.N.3.4 Identify the role of models in the context of the sixth grade science benchmarks.

Inquiry Launch Lab

SC.6.L.14.5, SC.6.N.1.5

10 minutes

How fast does your heart beat?

Have you ever felt your heartbeat speed up when you're exercising or when you're watching a scary movie? You can take your own pulse to find out how many times your heart beats every minute.

Procedure

1. Read and complete a lab safety form.
2. Sit quietly for 1 minute.
3. Feel your pulse by placing the middle and index fingers of one hand on an artery in your neck or an artery in your wrist.
4. While sitting quietly, count the number of heartbeats you feel in 30 seconds. Multiply this number by two to calculate your pulse. Record your data below.
5. Jog in place for 1 min.
6. Immediately repeat step 4.

Data and Observations

Think About This

1. How did your pulse after exercising compare to your resting pulse?

2. Key Concept Why do you think your pulse changed when you exercised?

Where To?

1. How does food get from where it's grown to your dinner table? Believe it or not, the vessels that carry blood through your body share similar principles.

Functions of the Circulatory System

Have you ever looked at a road map of the United States? A complex network of highways and roads crisscrosses the country. This road network is important for transporting people and materials from place to place. In a similar way, your circulatory system is important for transporting materials from one part of your body to another.

Transportation

Trucks haul food, fuel, and other products from factories and farms to markets and businesses around the country. Your circulatory system is like the network of roads, and your blood cells are like the vehicles that travel on those roads. Blood carries food, water, oxygen, and other materials through your circulatory system to your body's cells and tissues.

Elimination

Blood also carries away waste materials, just as garbage trucks haul away trash. As blood travels through the circulatory system, it picks up carbon dioxide produced during cellular respiration. It also picks up wastes produced by all the other chemical reactions that take place inside cells.

2. **NGSSS Check** Find Highlight what the circulatory system does. SC.6.L.14.5

Active Reading

FOLDABLES® LA.6.2.2.3

Make a horizontal two-tab book from a sheet of notebook paper. Label the front *The Circulatory System,* and label the inside as shown. Use it to organize your notes on the functions of the circulatory system and the organs associated with those functions.

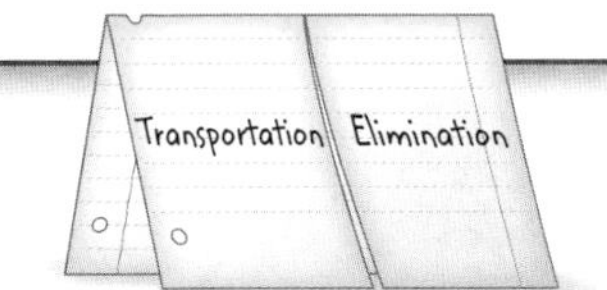

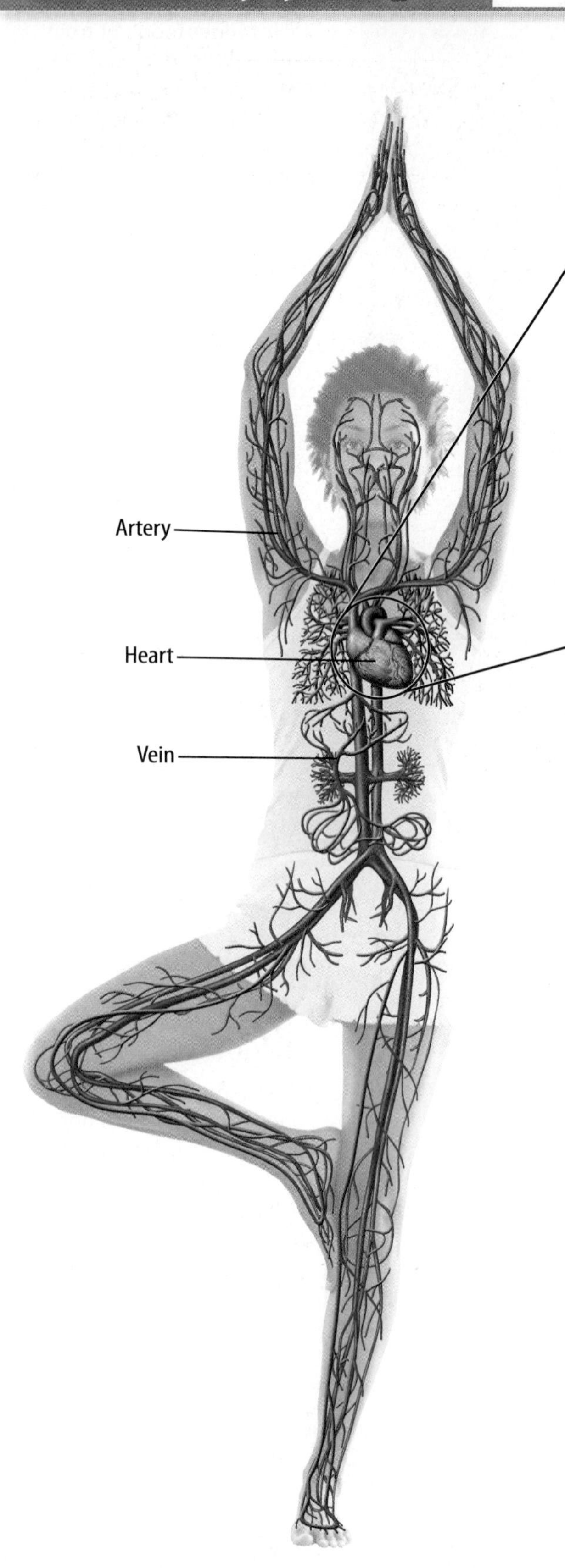

Figure 4 Your heart muscle is about the size of your fist. It acts as a pump that pushes blood through your circulatory system.

Circulatory System Organs

Highways connect and intersect and provide routes for traffic. **Figure 4** illustrates how your circulatory (SUR kyuh luh tor ee) system is similar. It provides routes for blood to flow through your body. Just as every vehicle on a highway is powered by its engine, your heart powers the flow of blood through your circulatory system.

The Heart

Your heart is always at work. The heart is a muscle that pushes blood through the circulatory system, as shown in **Figure 5**. On average, a human heart beats 70 to 75 times per minute, every minute of your life. It slows when you sleep. It speeds up when you exercise or are frightened.

Active Reading **3. Select** Underline what the heart does.

Look again at **Figure 4**. Notice that your heart has four chambers, two upper and two lower. *Blood enters the upper two chambers of the heart, called the* **atria** (AY tree uh; singular, atrium). *Blood leaves through the lower two chambers of the heart, called the* **ventricles** (VEN trih kulz).

The Path of Blood Through the Heart

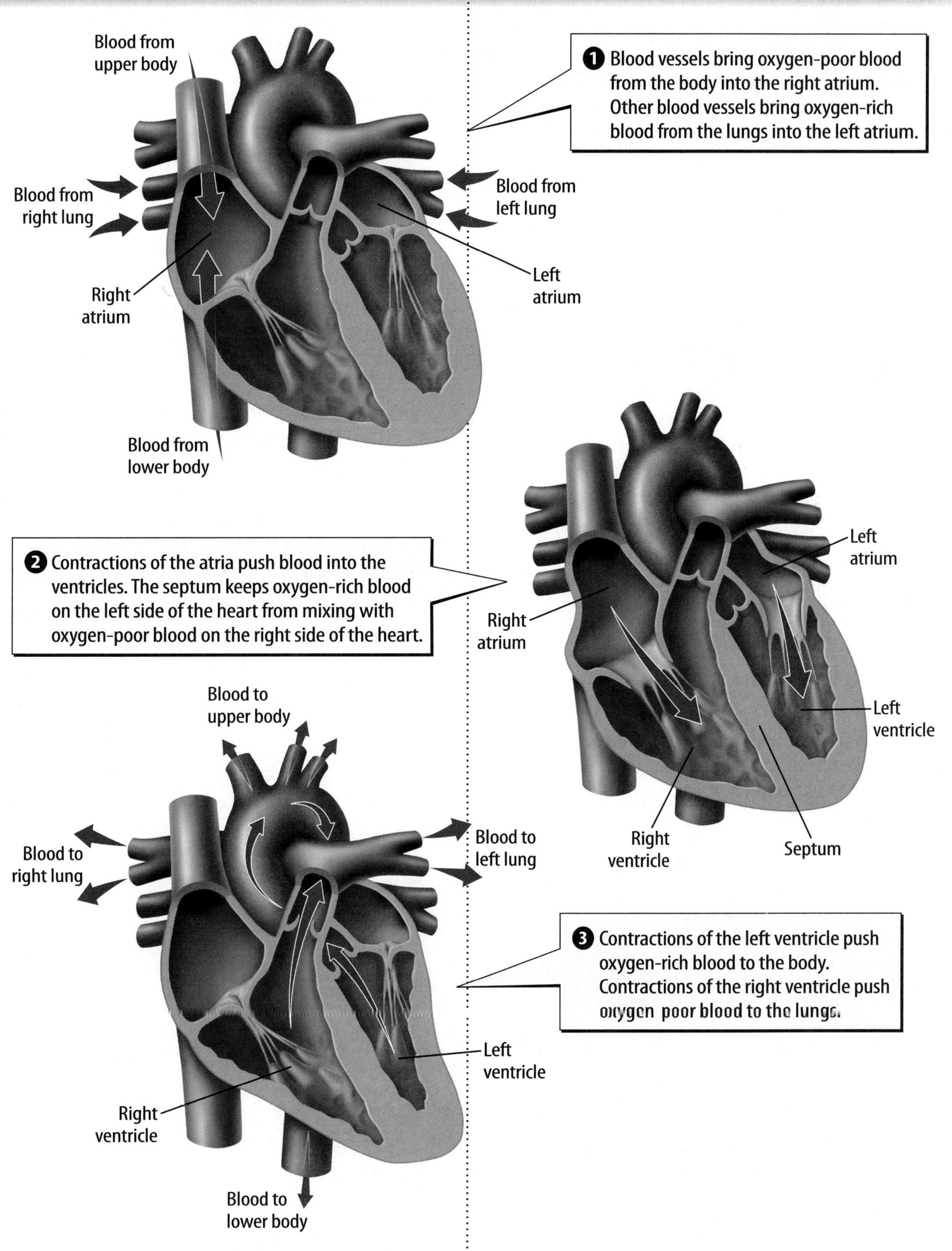

Figure 5 Veins deliver oxygen-poor blood from the body to the heart. The heart pumps this blood to the lungs, where it is resupplied with oxygen. The oxygen-rich blood then travels back to the heart and is pumped to the rest of the body.

4. Visual Check Determine What structure in the heart separates oxygen-poor blood from oxygen-rich blood?

Blood Vessels

If the circulatory system is like a network of roads for your body, then the different blood vessels are like different kinds of roads. Blood travels through blood vessels and reaches every cell in your body.

Arteries As shown in **Figure 6,** *a vessel that takes blood away from the heart is an* **artery.** Blood pressure in arteries is high because arteries are near the pumping action of the heart. Artery walls are thick and can withstand the high pressure of flowing blood.

The aorta is the largest artery. It carries a large volume of blood, just like freeways carry a high volume of traffic. Arteries branch into smaller vessels called arterioles.

Capillaries Notice in **Figure 6** that arterioles branch into **capillaries,** *tiny blood vessels that deliver supplies to individual cells and take away waste materials.* Capillaries are the smallest blood vessels in the circulatory system.

Many capillary walls are only one cell thick. This makes it possible for molecules of oxygen, food, water, wastes, and other materials to move between blood and body cells.

Veins *A vessel that brings blood toward the heart is a* **vein.** The pressure in veins is lower than in arteries. This is because capillaries separate veins from the pumping action of the heart. Because there is less pressure in veins, there is a greater chance that blood could flow backward. Veins have one-way valves that prevent blood from moving backward and keep it moving toward the heart.

Capillaries join and form larger vessels called venules. Venules join and form veins. The inferior vena cava is the largest vein. It carries blood from the lower half of your body to your heart.

Active Reading **5. Combine** How do the heart and blood vessels work together?

____

____

Figure 6 Arteries branch into arterioles and capillaries and bring oxygen-rich blood to cells. Capillaries combine to form venules and veins that carry oxygen-poor blood back to the heart.

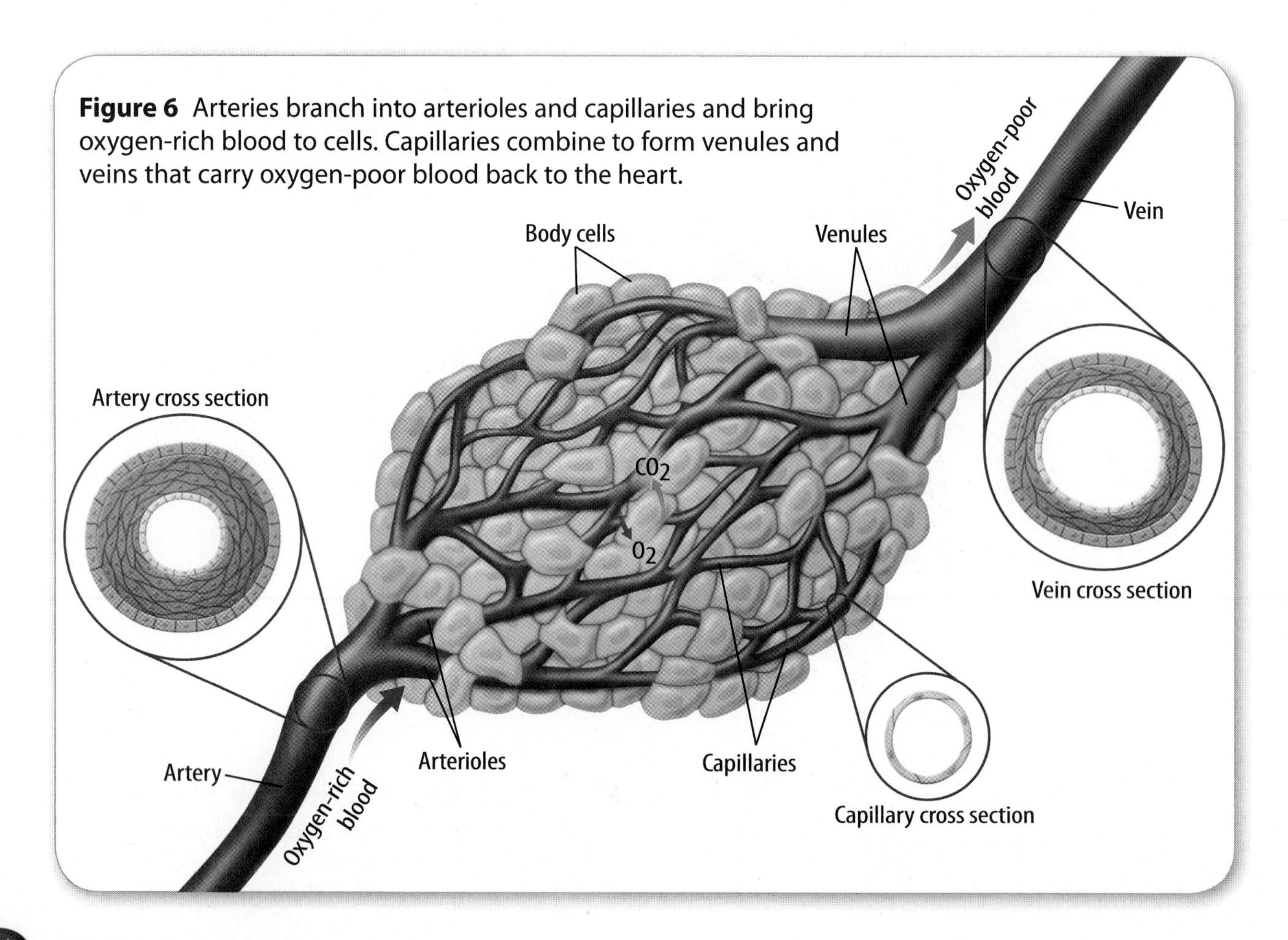

Types of Circulation

Your circulatory system is one large system that circulates blood throughout your entire body. However, when scientists and medical professionals discuss the circulatory system, they name three different types of circulation. One type supplies blood to the body. A second type supplies blood to the heart. A third type carries blood to and from the lungs.

Systemic Circulation

Blood leaves your heart through the aorta and travels to your arms, your toes, your head, and all other parts of your body, as shown by the orange vessels in **Figure 7.** **Systemic circulation** *is the network of vessels that carry blood from the heart to the body and from the body back to the heart.*

Coronary Circulation

You might think the cells of the heart get oxygen and nutrients from the blood that travels through the heart. However, the heart is a thick organ made up of many layers of cells. As a result, most heart cells don't come into contact with the blood inside the heart. *A network of arteries and veins called the* **coronary circulation** *supplies blood to all the cells of the heart.* Some of these vessels are on the outside of the heart, as shown in **Figure 7.**

Active Reading **6. Examine** What does coronary circulation do?

____

____

Pulmonary Circulation

The purple part of the circulation system shown in **Figure 7** illustrates how blood moves back and forth between the heart and the lungs. *The network of vessels that carries blood to and from the lungs is called* **pulmonary circulation**. Pulmonary circulation carries oxygen-poor blood from the heart to the lungs. It also carries oxygen-rich blood from the lungs back to the heart. Blood that enters the heart from the lungs is then pushed to the rest of the body.

Figure 7 Coronary circulation, shown in green, provides oxygen to heart cells. Pulmonary circulation, shown in purple, supplies blood with oxygen and removes carbon dioxide. Systemic circulation, shown in orange, supplies the rest of the body with oxygen and nutrients and removes wastes.

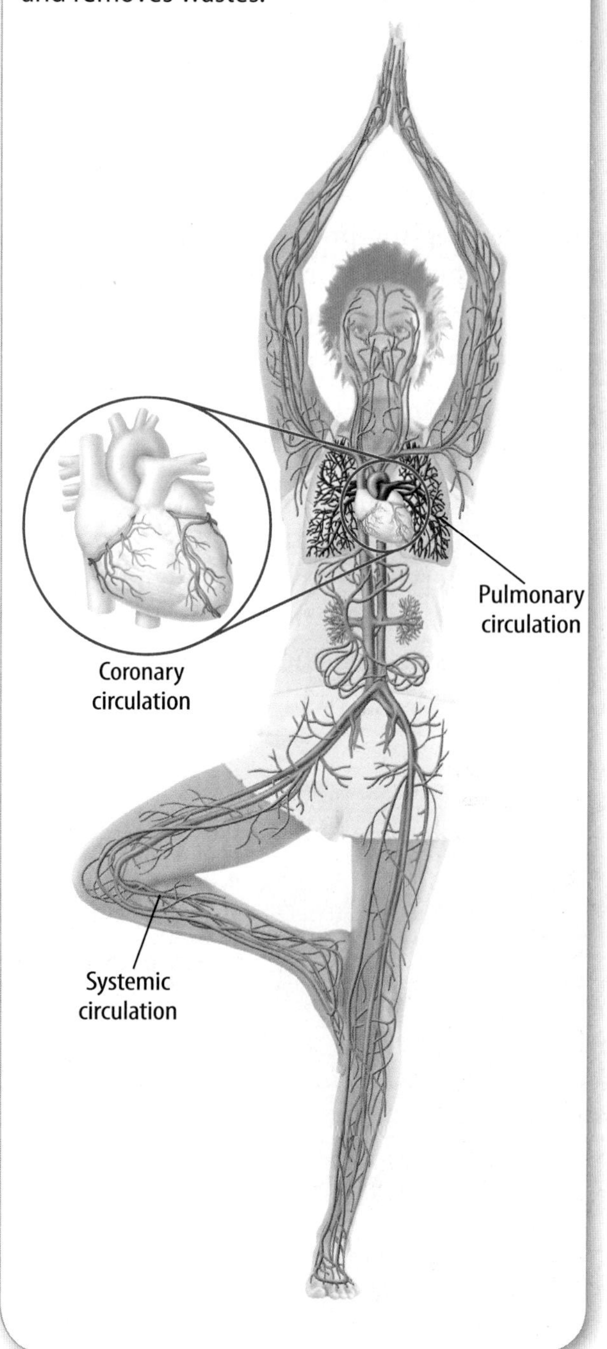

Word Origin

pulmonary
from Latin *pulmonarius*, means "of the lungs"

Circulatory System Health

Good health depends on a healthy circulatory system. Your heart muscle must be strong enough to push blood through all the blood vessels in your body. Your blood vessels must be flexible so that the volume of blood flowing through them can change. The valves in your heart and veins must work properly to keep blood from flowing in the wrong direction.

Circulatory diseases are illnesses that occur when some part of the circulatory system stops working properly. About one-third of all adults in the United States have some form of circulatory disease. Nearly 2,400 people die from it every day.

Hypertension

When the ventricles of the heart contract, they push blood into the arteries. When this happens, the arteries bulge a little. The bulging of an artery is what you feel when you check your pulse. The bulge happens because blood presses against the sides of the artery. That pressure is called blood pressure.

Have you ever had your blood pressure measured? Normal blood pressure is considered to be 120 mm Hg (millimeters of mercury) or less during the contraction of the ventricles. It is 80 mm Hg or less after the contraction. Normal blood pressure can be written as 120/80 mm Hg. Blood pressure higher than 140/90 mm Hg is known as hypertension, or high blood pressure. Hypertension can lead to weakened and less flexible artery walls.

Atherosclerosis

The buildup of fatty material within the walls of arteries is called **atherosclerosis** (a thuh roh skluh ROH sus). Fat deposits can interfere with the artery's blood flow. If a deposit breaks loose, it can flow to and block a narrower artery. A blockage in the heart can cause a heart attack. A blockage in a blood vessel in the brain can cause a stroke.

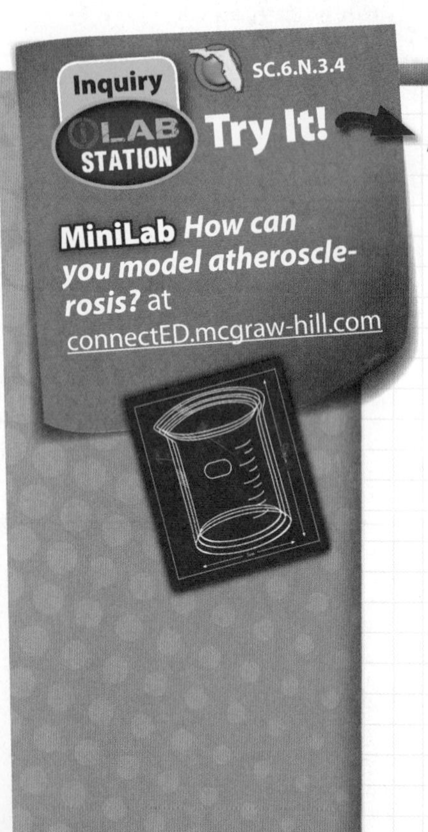

Apply It! After you complete the lab, answer this question.

1. What can a person do to decrease the risk of atherosclerosis?

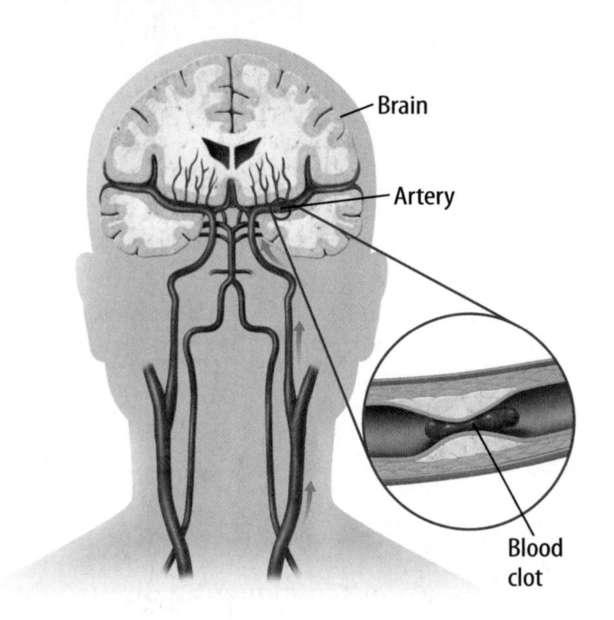

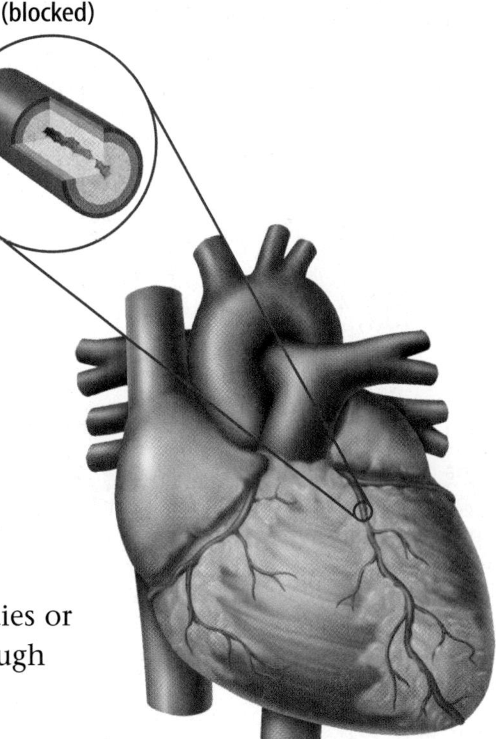

Heart Attacks, Strokes, and Heart Failure

A heart attack happens when part of the heart muscle dies or is damaged. A heart attack is usually caused when not enough oxygen reaches cells in the heart. Most heart attacks occur when a coronary vessel is blocked, as shown in **Figure 8.**

A stroke happens when part of the brain dies or is damaged. Most strokes are caused when not enough oxygen reaches cells in the brain. A stroke might occur if a blood clot blocks a blood vessel in the brain, also shown in **Figure 8.**

Heart failure occurs when the heart is not working efficiently. It can result from a previous heart attack, a problem with heart valves, or diseases that damage the heart.

Figure 8 Most heart attacks occur when a vessel of the coronary circulation system is blocked. Most strokes occur when a blood clot blocks a vessel in the brain.

Preventing Circulatory System Disorders

Some risk factors for circulatory system diseases cannot be avoided. For example, if one of your parents has a circulatory disease, you might have inherited a slightly higher risk of developing a similar disease. However, most risk **factors** can be controlled by making good life choices, like eating a healthful diet, controlling weight, exercising, and not smoking.

ACADEMIC VOCABULARY

factor
(noun) something that helps produce a result

The Circulatory System and Homeostasis

The circulatory system is closely connected with other body systems. Once oxygen enters your body, the respiratory system interacts with the circulatory system and transports oxygen to your body's cells. It also transports nutrients from the digestive system and hormones from the endocrine system. The nervous system regulates your heartbeat. Later in this chapter, you'll read how the circulatory and skeletal systems work together.

7. **NGSSS Check** **Support** Highlight how the circulatory and respiratory systems work together to maintain homeostasis. SC.6.L.14.5

Lesson Review 2

Visual Summary

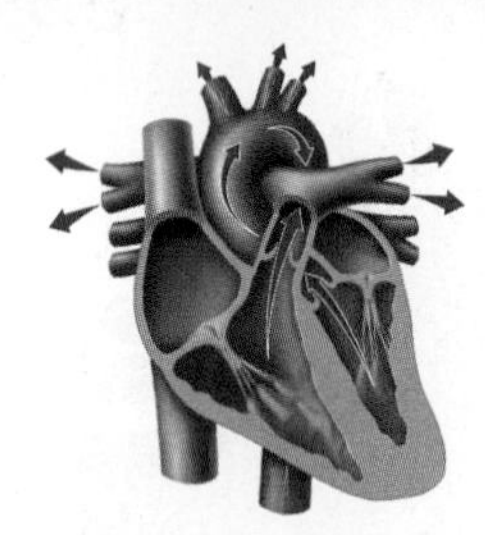

The contractions of the heart push blood through the circulatory system.

Arteries and veins carry blood throughout the body. Materials move between blood and cells through capillary walls.

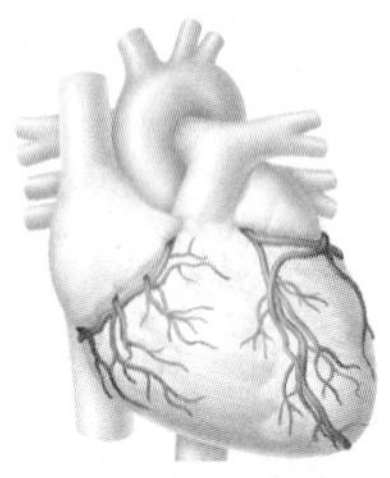

Coronary circulation supplies blood to heart cells.

Use Vocabulary

1. The narrow blood vessels where gas exchange occurs are ______________________.

2. The two lower chambers of the heart are ______________________.

3. **Distinguish** between veins and arteries.

Understand Key Concepts

4. **Explain** how blood keeps flowing continuously through the body. SC.6.L.14.5

5. A blockage of blood vessels in the brain can cause
   - (A) a heart attack.
   - (B) a stroke.
   - (C) heart failure.
   - (D) hypertension.

Interpret Graphics

6. **Summarize** Fill in the graphic organizer below to identify the three types of circulation. SC.6.L.14.5

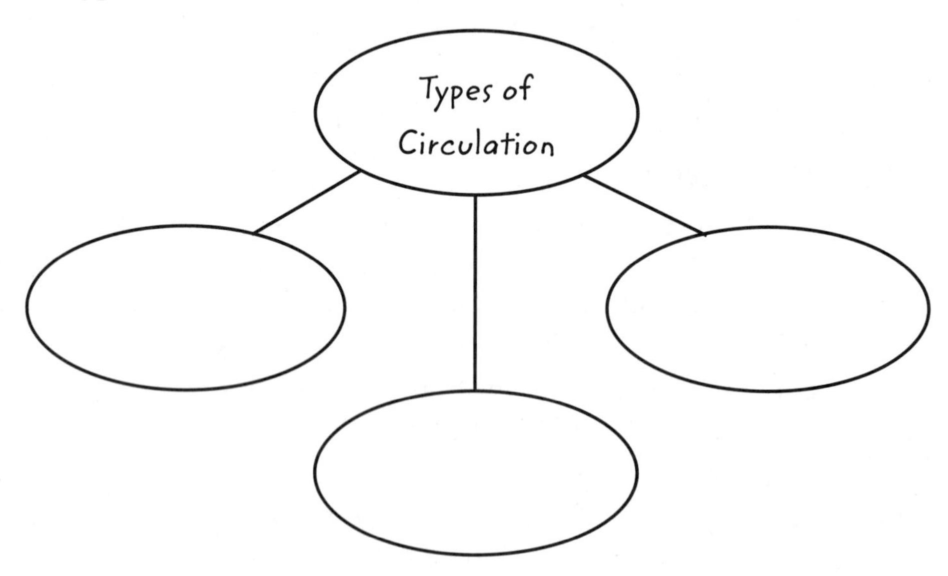

Critical Thinking

7. **Justify** A physician has a patient, age 42, whose blood pressure averages 141/89 mm Hg. Why might the physician recommend healthful life choices to the patient?

Summarize the functions of the circulatory system.

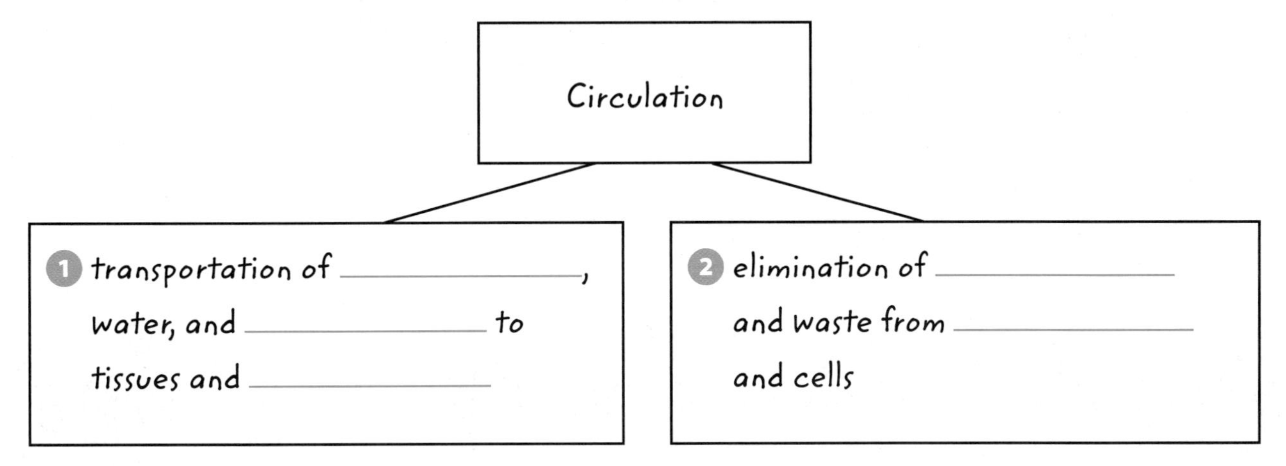

Sequence the journey of blood through the heart.

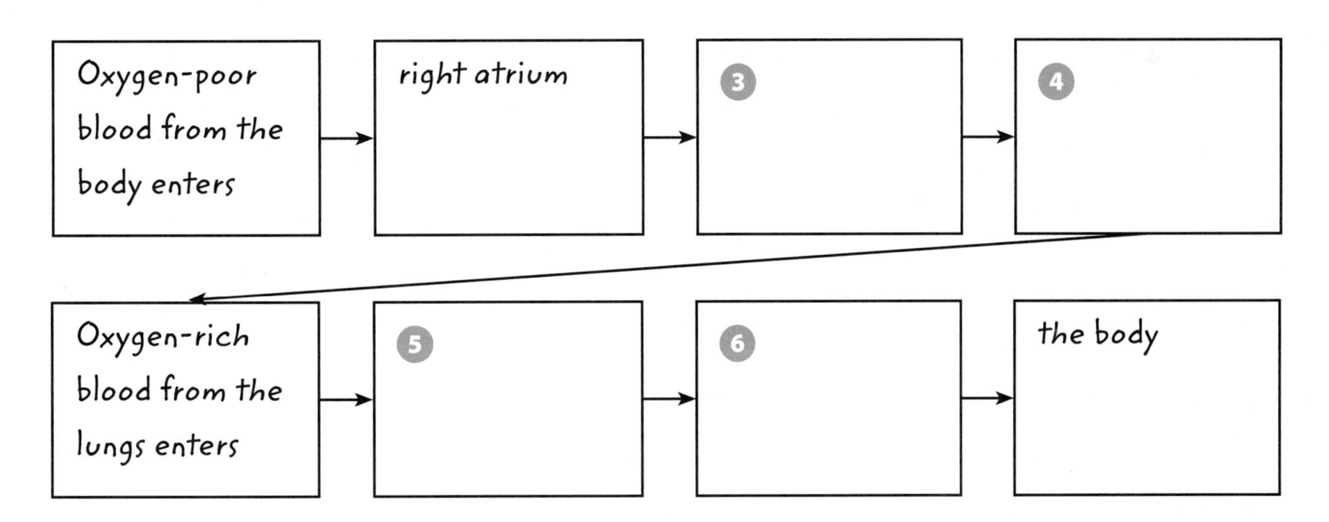

Distinguish among types of blood vessels.

Arteries	7	8
9	large vessels, carry blood toward heart	smallest vessels, deliver materials to and from individual cells

Classify types of circulation.

Type	Systemic	10	Pulmonary
Delivers blood to and from	11	the heart	12

Lesson 3

ESSENTIAL QUESTIONS

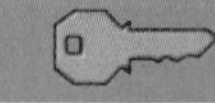
What does the blood do?

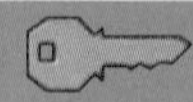
How do the parts of the blood differ?

Vocabulary

platelet p. 551

plasma p. 551

Rh factor p. 553

SC.6.N.1.5, SC.6.L.14.5

10 minutes

What do blood cells look like?

Like every tissue in your body, blood is a tissue made of different cells. Take a look in the microscope to see the different types of blood cells.

Procedure

1. Read and complete a lab safety form.
2. Observe a **prepared blood smear slide** under low power on a **microscope.**
3. Switch to high power, and observe the different cell types on the slide. In the space below, draw one example of each type of cell you see.
4. Return to low power, and remove the slide.

Data and Observations

Think About This

1. How many kinds of cells did you observe? How did their appearances differ?

2. Key Concept Why do you think there are different kinds of blood cells?

Florida NGSSS

HE.6.C.1.8 Explain how body systems are impacted by hereditary factors and infectious agents.

LA.6.2.2.3 The student will organize information to show understanding (e.g., representing main ideas within text through charting, mapping, paraphrasing, summarizing, or comparing/contrasting).

MA.6.A.3.6 Construct and analyze tables, graphs, and equations to describe linear functions and other simple relations using both common language and algebraic notation.

SC.6.L.14.5 Identify and investigate the general functions of the major systems of the human body (digestive, respiratory, circulatory, reproductive, excretory, immune, nervous, and musculoskeletal) and describe ways these systems interact with each other to maintain homeostasis.

SC.6.N.1.5 Recognize that science involves creativity, not just in designing experiments, but also in creating explanations that fit evidence.

Working Hard?

1. You might have noticed that the cheeks of some people turn bright red during vigorous exercise. Why does this happen? The red color comes from blood flowing near the surface of the skin. It helps release excess thermal energy from the body. What else do you think your blood does?

____

____

____

____

____

____

Functions of Blood

Have you ever had an injury that caused bleeding? Blood is a red liquid, slightly thicker than water. At the end of Lesson 2 you read that your circulatory system works closely with all your other body systems to maintain homeostasis. Blood is the link that connects the circulatory system with all other body systems. Blood transports substances throughout your body, helps protect your body against infection, and helps regulate your body's temperature.

Transportation

Blood transports many different substances throughout your body. You've read that blood carries oxygen to and carbon dioxide from the lungs. Blood also picks up nutrients in the small intestine and carries them to all body cells. It transports hormones produced by the endocrine system. Blood also carries waste products to the excretory system. Most of the substances carried in blood are dissolved in the liquid part of blood.

Protection

Some blood cells fight infection. They help protect you from harmful organisms, such as bacteria, viruses, fungi, and parasites. Blood also contains materials that help repair torn blood vessels and heal wounds. These materials help protect the body from losing too much blood.

Temperature Regulation

Blood helps your body maintain a steady temperature of about 37°C. When your body temperature rises, blood vessels near the surface of your skin widen. This increases blood flow to your skin's surface and releases more thermal energy to the air. Your body cools down. When your body temperature lowers, the vessels at your skin's surface get narrower. This decreases blood flow to the surface of your skin and reduces the amount of thermal energy that is lost to the air. Your body warms up.

2. **NGSSS Check** **List** What are the main functions of blood? SC.6.L.14.5

____

Parts of Blood

Blood is considered a tissue because it is made up of different kinds of cells that work together. As shown in **Figure 9,** blood consists of four main parts: red blood cells, white blood cells, platelets, and plasma. Most adults have about 70 mL of blood per kilogram of body weight. An average adult has about 5 to 6 liters of blood.

Red Blood Cells

Every cubic millimeter of your blood contains 4 to 6 million red blood cells, or erythrocytes (ih RIHTH ruh sites). Red blood cells contain hemoglobin (HEE muh gloh bun), iron-rich protein molecules. In the alveoli of the lungs, oxygen binds to the hemoglobin in red blood cells. The hemoglobin releases the oxygen when red blood cells enter the capillaries and come into close contact with body cells.

How would you describe the shape of the red blood cells in **Figure 9?** Some people describe them as a doughnut without a hole. This flattened disk shape gives them more surface area. This means red blood cells can carry more oxygen than if they were round like a ball. Your body produces new red blood cells all the time because they wear out in a few months.

White Blood Cells

Your blood also contains several kinds of white blood cells, or leukocytes (LEW kuh sites), shown in **Figure 9.** All white blood cells protect your body from illness and infection. Some attack viruses, bacteria, fungi, and parasites that might invade your body. Most white blood cells last only a few days and are constantly replaced. You have far fewer white blood cells—5,000 to 10,000 per cubic millimeter—than red blood cells.

Figure 9 Blood flows through blood vessels. It is made of liquid plasma, red blood cells, white blood cells, and platelets.

Parts of Blood

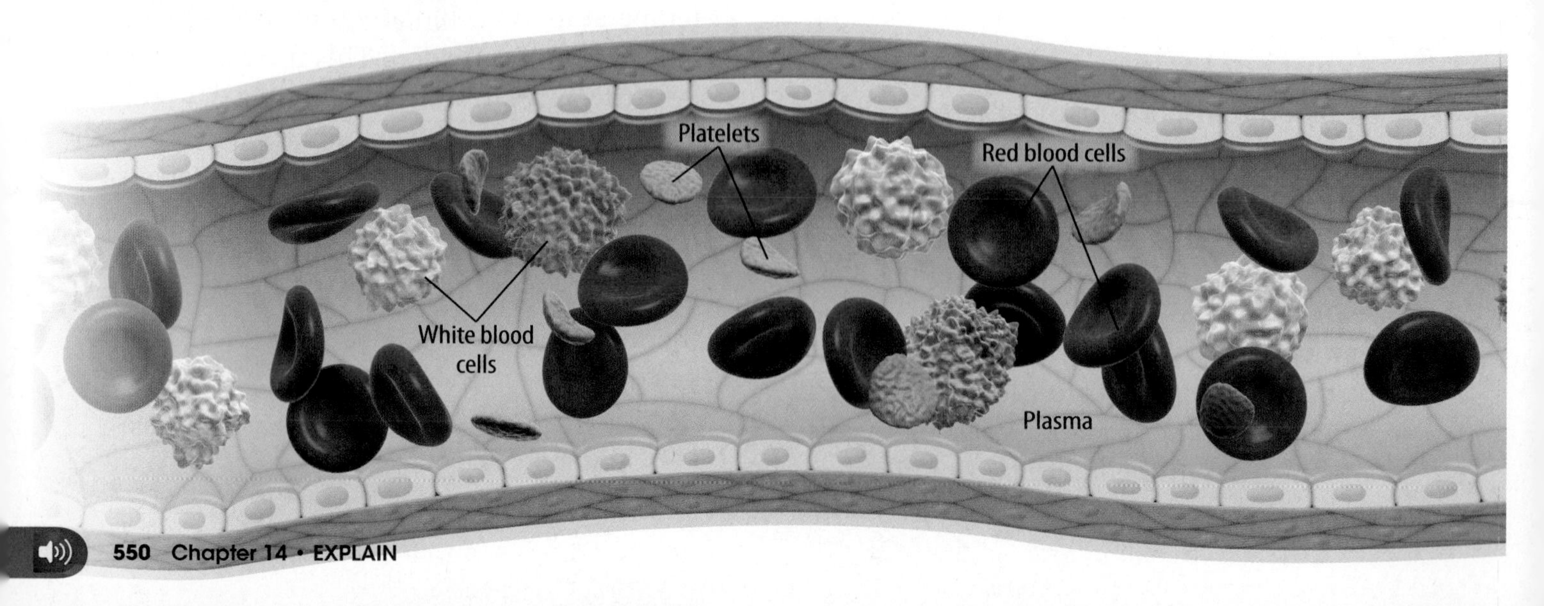

Platelets

What happens if you get a cut? The cut bleeds for a short time, and then the blood clots, as shown in **Figure 10.** **Platelets** *are small, irregularly shaped pieces of cells that plug wounds and stop bleeding.* Platelets produce proteins that help strengthen the plug. Without platelets, blood would not stop flowing. Your blood contains 150,000 to 440,000 platelets per cubic millimeter.

Plasma

The yellowish, liquid part of blood, called **plasma**, *transports blood cells.* Plasma is 90 percent water, which helps thin the blood. This enables it to travel through small blood vessels. Many molecules are dissolved in plasma. They include salts, vitamins, sugars, minerals, proteins, and cellular wastes.

Plasma also plays an important role in regulating the activities of cells in your body. For example, plasma carries chemical messengers that control the amounts of salts and glucose that enter cells.

Active Reading **3. Identify** Highlight how the parts of blood differ.

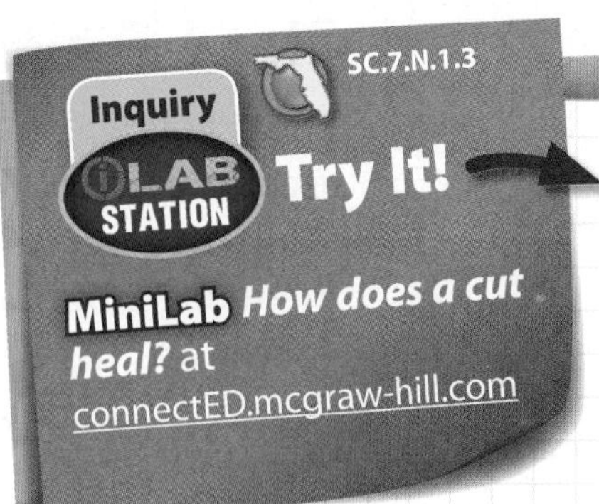

Apply It!

After you complete the lab, answer this question.

1. Hemophilia is a disease in which a person's blood has a low level of platelets. Why would a cut be particularly threatening to someone with this condition?

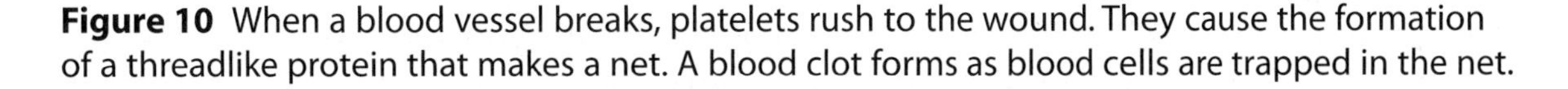

Figure 10 When a blood vessel breaks, platelets rush to the wound. They cause the formation of a threadlike protein that makes a net. A blood clot forms as blood cells are trapped in the net.

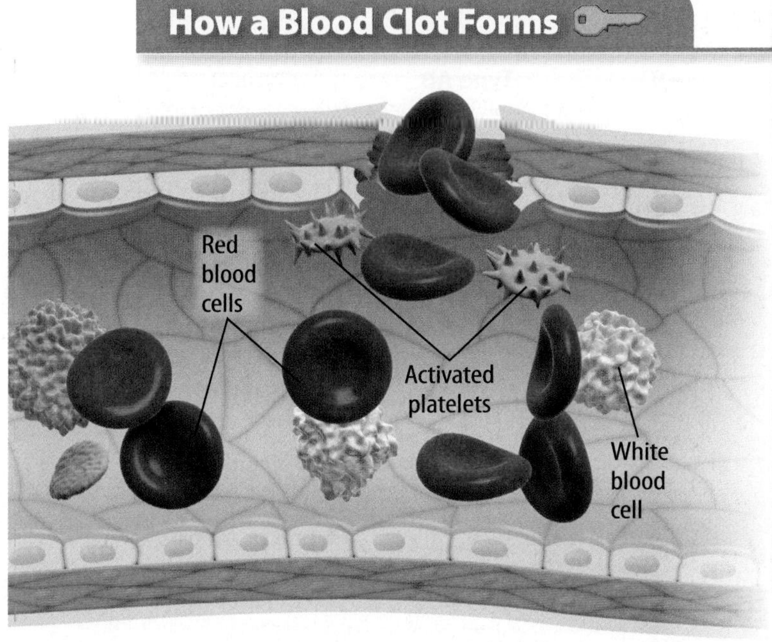

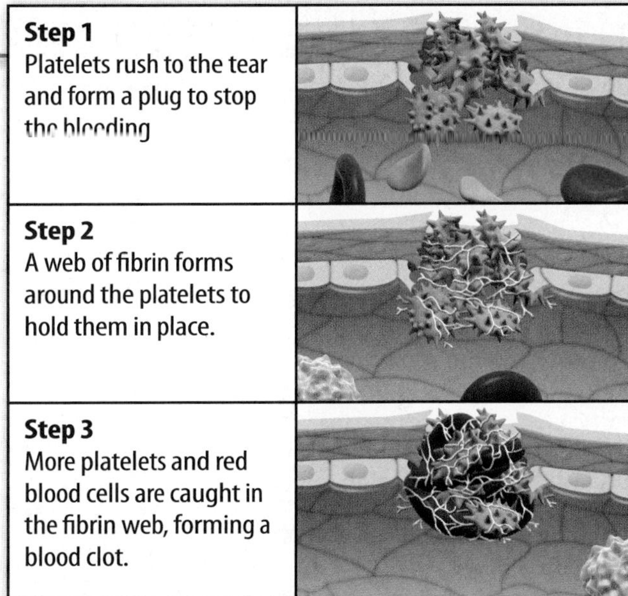

Step	
Step 1 Platelets rush to the tear and form a plug to stop the bleeding	
Step 2 A web of fibrin forms around the platelets to hold them in place.	
Step 3 More platelets and red blood cells are caught in the fibrin web, forming a blood clot.	

Blood Types

Do you know anyone who donates blood? Donated blood is used to help people who have lost too much blood from an injury or surgery and need a transfusion. A blood transfusion is the transfer of blood from one person to another. Even though your blood has the same four parts as everyone else's—red cells, white cells, platelets, and **plasma**—you can't receive blood from just anyone. Why? Because different people have different blood types.

WORD ORIGIN

plasma
from Greek *plassein,* means "to mold"

The ABO System

You inherited your blood type from your parents. Blood type refers to the type of proteins, or antigens, on red blood cells. The four human blood types are A, B, AB, and O. Type A blood cells have the A antigen. Type B cells have the B antigen. Type AB has both A and B antigens. Type O has no antigens.

Why is blood type important? Any time a blood transfusion brings foreign antigens to someone's blood, the red blood cells will clump together and lose their ability to function. Clumps form because of clumping proteins in blood plasma. As shown in **Table 2,** people with types A, B, and O blood have clumping proteins in their plasma. For example, a person with type A blood has anti-B clumping proteins. If given type B blood, his or her anti-B proteins would attack the type B antigens and cause the type B red blood cells to clump together.

Type AB blood has no clumping proteins. People with AB blood are known as "universal recipients" because they can receive transfusions of any blood type. Type O blood has clumping proteins that attack both A and B antigens. People with type O blood are known as "universal donors" because they can donate blood to anyone.

Table 2 People with blood types A, B, and O have clumping proteins in their plasma. These proteins determine what blood type a person can safely receive in a blood transfusion.

Table 2 Human Blood Types

Blood Type	**Type A**	**Type B**	**Type AB**	**Type O**
Antigens on red blood cells	A A A A A A	B B B B B B	B A A B B A	
Percentage of US population with this blood type	42	10	4	44
Clumping proteins in plasma	Anti-B	Anti-A	None	Anti-A and anti-B
Blood type(s) that can be RECEIVED in a transfusion	A or O	B or O	A or B or AB or O	O only
This blood type can DONATE TO these blood types	A or AB	B or AB	AB only	A or B or AB or O

The Rh Factor

Another protein found on red blood cells is a chemical marker called the **Rh factor.** People with blood cells that have this protein are Rh positive. People who do not have it are Rh negative. If Rh positive blood is mixed with Rh negative blood, clumping can result. Blood types usually have a plus (+) sign or a minus (–) sign to indicate whether a person is Rh positive or negative. For example, a person with an A+ blood type has red cells with A antigens and the Rh factor. Someone with O– blood has no antigens and no Rh factor.

Active Reading 4. **Name** What antigens are found in AB+ blood?

Blood Disorders

Some medical conditions disrupt the normal functions of blood. People with hemophilia lack a protein needed to clot blood. A person who has hemophilia bleeds at the same rate as other people, but the bleeding does not stop as quickly.

People suffering from anemia have low numbers of red blood cells or have red blood cells that do not contain enough hemoglobin. As a result, the blood might not carry as much oxygen as the body needs.

Bone marrow, the soft tissue in the center of bones, produces red blood cells. Cancer of the bone marrow is called leukemia. This kind of cancer can slow or prevent blood cell formation. Leukemia can lead to anemia and a damaged immune system.

People who inherit sickle-cell disease have red blood cells shaped like crescents, as shown in **Figure 11.** Sickle-shaped cells do not move through blood vessels as easily as normal disk-shaped cells. Sickle cells can prevent oxygen from reaching tissues and cause sickle-cell anemia.

Math Skills MA.6.A.3.6

Use Percentages

If percentages refer to the same factor, they can be added or subtracted. For example, you could add the percentages of people with each of the four blood types:

42% + 10% + 4% + 44% = 100%

You could also subtract to find what percentage of people do not have type O blood:

100% – 44% = 56%

Practice

5. Forty-four percent of people have type O blood. If 7 percent of people have type O blood and are Rh negative, what percent has type O Rh positive blood?

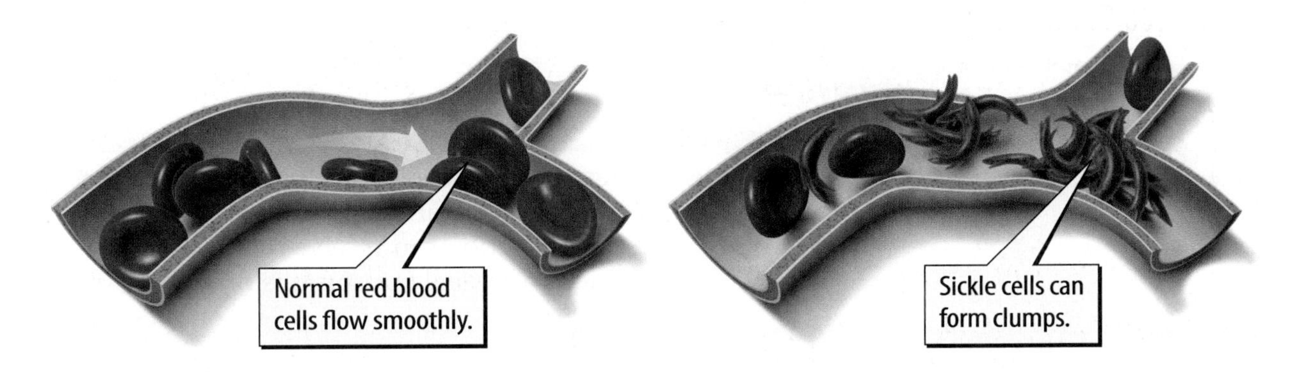

Figure 11 The crescent-shaped red blood cells in sickle-cell disorder form clumps that can block blood vessels.

Lesson Review 3

Visual Summary

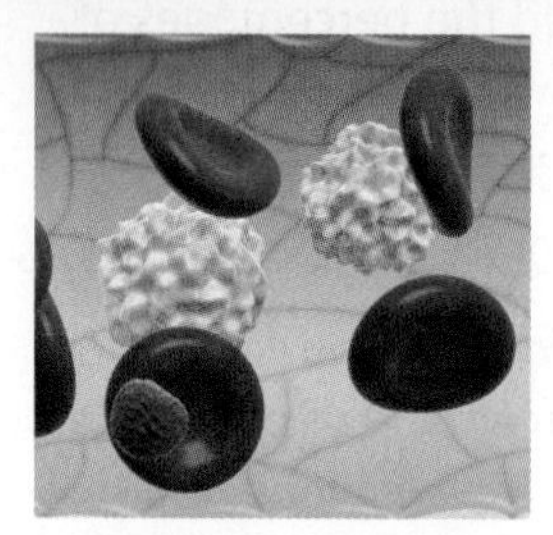

Red blood cells contain hemoglobin and carry oxygen. White blood cells help fight disease.

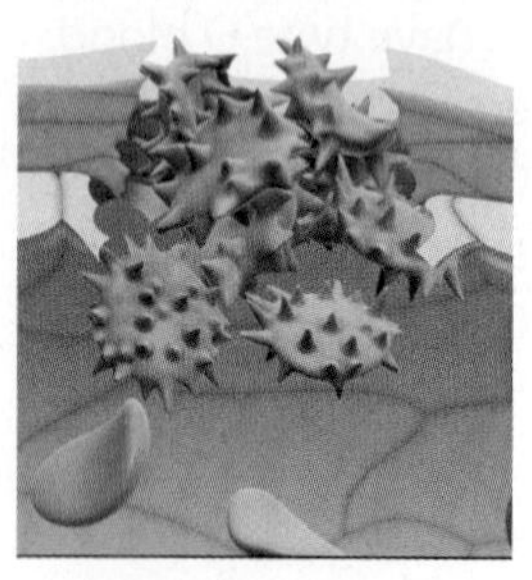

Platelets are pieces of cells that aid in blood clotting.

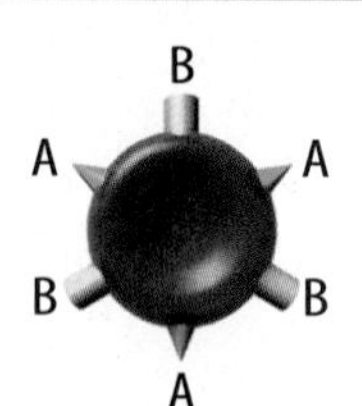

Blood type is determined by antigens on the surface of red blood cells and clumping proteins in blood plasma.

Use Vocabulary

1. **Identify** the yellowish liquid part of blood. ____________

2. **Distinguish between plasma and platelets.**

____

3. **Define** the term *Rh factor* using your own words.

____

Understand Key Concepts

4. **Give an example** of a blood disorder and explain how it can affect health. HE.6.C.1.8

____

5. Which part of the blood carries dissolved molecules such as glucose and salt? SC.6.L.14.5

   (A) plasma
   (B) platelets
   (C) red blood cells
   (D) white blood cells

Interpret Graphics

6. **Organize Information** List the parts of blood and their functions. SC.6.L.14.5

Part of Blood	Functions

Critical Thinking

7. **Design** and sketch a model to show how blood can help regulate body temperature. SC.6.L.14.5

Math Skills MA.6.A.3.6

8. In the United States, 42% of people have type A blood, 10% have type B, 4% have type AB, and 44% have type O. People with type B and type AB blood can receive type B blood. What percentage of people can receive transfusions of type B blood?

SCIENCE & SOCIETY

Very Special Blood Cells

Technicians remove only a small portion of the crabs' blood. After this procedure, the crabs are returned to the ocean. Their blood cell levels return to normal in a couple of weeks.

Horseshoe crabs, living relatives of extinct trilobites, have been gathering on beaches for 350 million years. They usually become food for fish and birds. Yet someday your life might depend on horseshoe crabs—or at least on their blood. Unlike human blood, horseshoe crab blood contains only one type of blood cell. If bacteria enter the crab's bloodstream from an open wound, its blood cells secrete a clotting factor. This secretion closes the wound, and the blood cells engulf the bacteria. When scientists saw that horseshoe crab blood turned to a gel in the presence of harmful bacteria, they realized its value. Today, medical professionals use an extract made from horseshoe crab blood to screen all intravenous medicines for bacteria. A quart of this special blood costs about $15,000!

The horseshoe crab blood can do even more. Another component of the blood can stop the human immunodeficiency virus (HIV) from replicating, or making copies of itself. Part of horseshoe crab blood can act as an antibiotic. Scientists also are using horseshoe crab blood in the development of a hand-held instrument that helps diagnose human illnesses. The instrument uses enzymes from the blood as illness detectors.

It's Your Turn

REPORT Medical professionals use certain types of snake venom to treat strokes. Conduct research to find other unusual animal products that have medical uses.

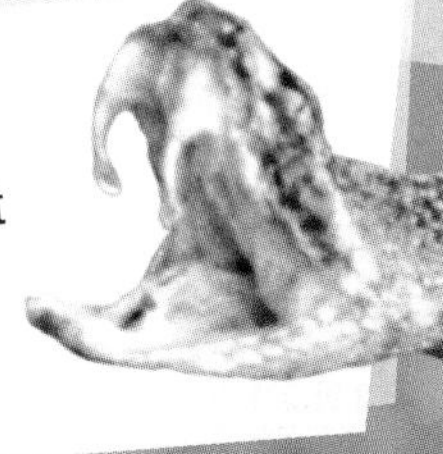

Lesson 4

The Lymphatic SYSTEM

ESSENTIAL QUESTIONS

What does the lymphatic system do?

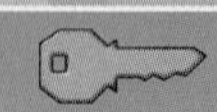
How do the parts of the lymphatic system work together?

How does the lymphatic system interact with other body systems?

Vocabulary

lymphatic system p. 557
lymph p. 558
lymph node p. 558
thymus p. 559
spleen p. 559

Florida NGSSS

HE.6.C.1.8 Explain how body systems are impacted by hereditary factors and infectious agents.

LA.6.2.2.3 The student will organize information to show understanding (e.g., representing main ideas within text through charting, mapping, paraphrasing, summarizing, or comparing/contrasting).

SC.6.L.14.5 Identify and investigate the general functions of the major systems of the human body (digestive, respiratory, circulatory, reproductive, excretory, immune, nervous, and musculoskeletal) and describe ways these systems interact with each other to maintain homeostasis.

SC.6.N.1.1 Define a problem from the sixth grade curriculum, use appropriate reference materials to support scientific understanding, plan and carry out scientific investigation of various types, such as systematic observations or experiments, identify variables, collect and organize data, interpret data in charts, tables, and graphics, analyze information, make predictions, and defend conclusions.

SC.6.N.3.4 Identify the role of models in the context of the sixth grade science benchmarks.

SC.8.N.1.6 Understand that scientific investigations involve the collection of relevant empirical evidence, the use of logical reasoning, and the application of imagination in devising hypotheses, predictions, explanations and models to make sense of the collected evidence.

SC.8.N.3.1 Select models useful in relating the results of their own investigations.

Launch Lab

SC.6.N.1.1, SC.6.N.3.4

10 minutes

How can you model a lymph node?

Fluid surrounds your body cells. Body cells absorb materials from and release materials into this fluid. Some of the fluid drains into vessels and then drains into spongy structures called lymph nodes. What happens in the lymph nodes?

Procedure

1. Read and complete a lab safety form.
2. Observe a **liquid** provided by your teacher. Record the observations below.
3. Use a **rubber band** to attach a square of **cheesecloth** to a **plastic drinking straw**. Hold the straw upright over a **paper plate**.
4. Use a **plastic dropper** to squeeze about 1 mL of the liquid into the open end of the straw.
5. Allow the liquid to drain from the cheesecloth and onto the plate. Observe the liquid. Record your observations below.

Data and Observations

Think About This

1. What differences did you observe in the liquid after it passed through the cheesecloth?

2. Key Concept What do you think the function of the lymph nodes might be?

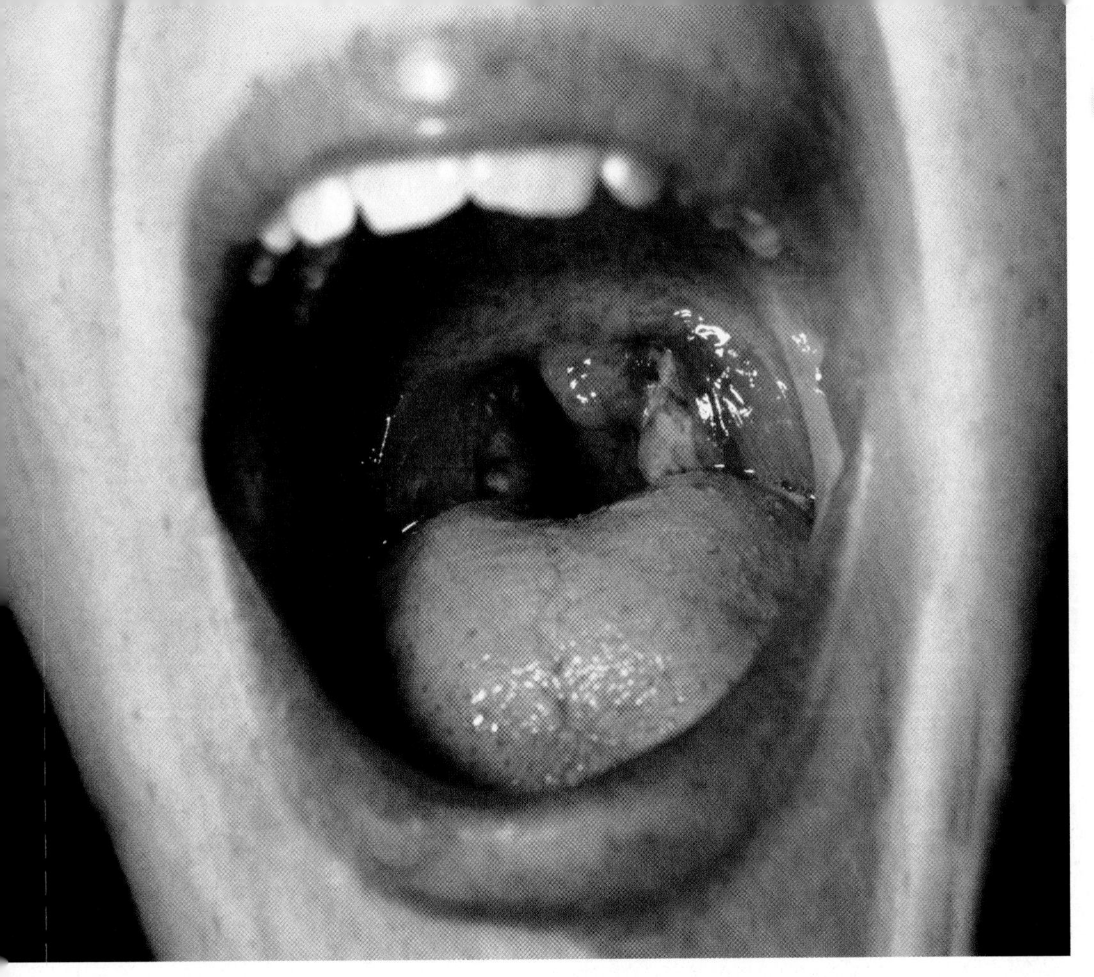

Healthy or Not?

1. Do you know anyone who has had his or her tonsils removed? Tonsils are clusters of lymph tissues that help the body fight off disease. Why do you think tonsils get swollen and inflamed, like the ones shown here?

Functions of the Lymphatic System

At times when you were sick, you might have noticed small, swollen structures under your jaw on each side of your neck. These structures can become swollen when they're working to fight off an infection in your body.

The **lymphatic system** *is part of the immune system and helps destroy microorganisms that enter the body.* The lymphatic system works closely with the circulatory system. Both systems move liquids through the body, and both contain white blood cells. However, their functions are different. There are four main functions of the lymphatic system.

- It absorbs some of the tissue fluid that collects around cells.
- It absorbs fats from the digestive system and transports them to the circulatory system.
- It filters dead cells, viruses, bacteria, and other unneeded particles from tissue fluid and then returns the tissue fluid to the circulatory system.
- It helps fight off illness and infections and includes structures in which white blood cells develop.

2. **NGSSS Check** Group Highlight the functions of the lymphatic system. SC.6.L.14.5

Active Reading FOLDABLES® LA.6.2.2.3

Fold a sheet of paper into an eight-page book. Use it to organize your notes about the parts of the lymphatic system and their functions.

The Lymphatic System

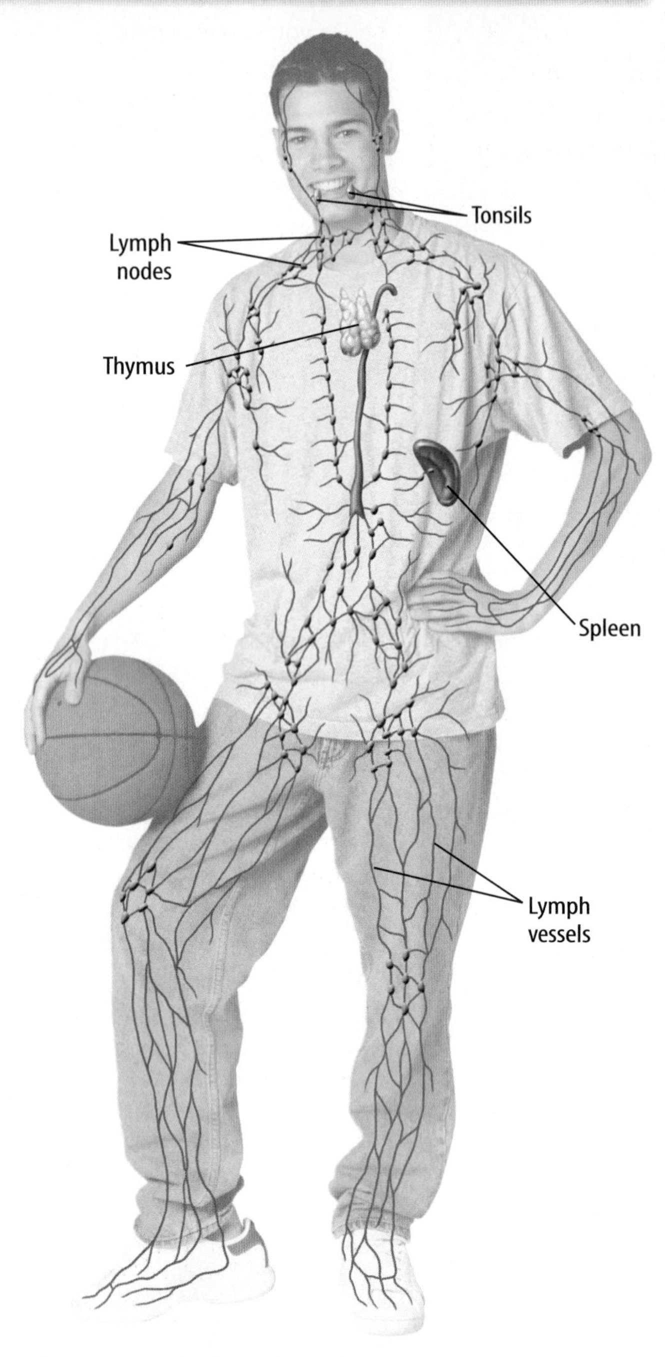

Figure 12 The lymphatic system is a network of vessels and organs. Vessels transport lymph. When it reaches the area beneath the collarbone, it re-enters the circulatory system.

4. **Visual Check** **Identify** What organs of the lymphatic system are in the throat?

Parts of the Lymphatic System

The lymphatic system, shown in **Figure 12**, includes lymph vessels and the fluid they carry. It also includes several other structures.

Lymph

Water, white blood cells, and dissolved materials such as salts and glucose leak out of capillary walls and into the spaces that surround tissue cells. This fluid is called tissue fluid. Cells absorb the materials they need from tissue fluid and release wastes into it. About 90 percent of the tissue fluid is reabsorbed by the capillaries. About 10 percent of the tissue fluid is absorbed by the lymph vessels and is called lymph.

Active Reading 3. **Review** Underline what lymph is.

Lymph Vessels

The lymphatic system forms a network of lymph vessels that look similar to the circulatory system's network of blood vessels. Lymph vessels absorb and transport lymph. The lymph is pushed through the lymph vessels by contractions of the muscles you use to move your body. Lymph is not pumped through the lymph vessels by the heart.

Lymph Nodes

Lymph vessels include *clusters of small, spongy structures called* **lymph nodes** *that filter particles from lymph.* Bacteria, viruses, fungi, and pieces of dead cells are trapped and removed from the lymph as it flows through a lymph node. Lymph nodes also store white blood cells that attack and destroy the trapped particles.

Large groups of lymph nodes are in the neck, the groin, and the armpits. When you have an infection, your body increases its production of white blood cells that fight the infection. Many of these white blood cells gather in your lymph nodes and cause the nodes to swell. The swelling disappears when the infection is gone.

Bone Marrow and Thymus

Lymphocytes (LIHM fuh sites) are white blood cells that destroy pathogens—infection-causing microorganisms such as viruses and bacteria. Bone marrow is the spongy center of bones where red and white blood cells, including lymphocytes, form. Lymphocytes include B cells and T cells. As shown in **Figure 13**, B cells mature in the bone marrow, and T cells mature in the thymus gland.

The **thymus** *is the organ of the lymphatic system in which T cells complete their development.* After immature T cells move from the bone marrow to the thymus, they develop the ability to recognize and destroy body cells that have been infected by microorganisms. Mature B cells and T cells move into the lymph and blood to help fight infection.

Active Reading **5. Summarize** Highlight how bone marrow and the thymus work together.

Spleen

You read earlier that the life of a red blood cell is only a few months. *The* **spleen** *is an organ of the lymphatic system that recycles worn-out red blood cells and produces and stores lymphocytes.* The spleen also stores blood and platelets. If a person is injured and loses a lot of blood, the spleen can release stored blood and platelets into the circulatory system.

Figure 13 Lymphocytes attack and destroy disease-causing microorganisms. B cells mature in the bone marrow. T cells mature in the thymus.

Where Lymphocytes Mature

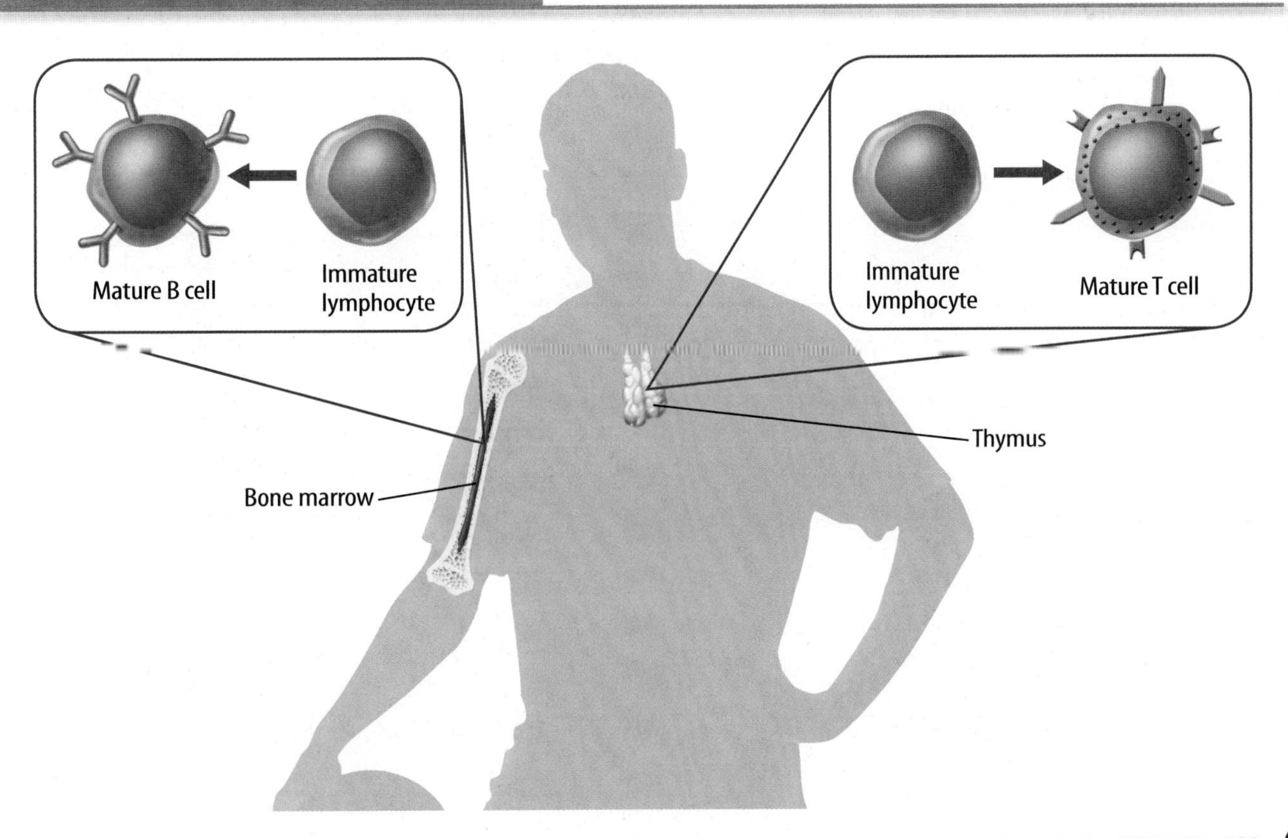

Tonsils

Your tonsils are clusters of lymph tissue on the sides of your throat. They help protect your body from infection by trapping and destroying bacteria and other pathogens that enter your nose and mouth. However, you can live without tonsils.

SCIENCE USE V. COMMON USE

vessel

Science Use a tube through which a body fluid travels

Common Use a container for holding something

Lymph Diseases and Disorders

Damage to the lymphatic system from injury or surgery can prevent tissue fluid from draining into lymph **vessels.** As a result, tissue fluid can build up around cells and cause swelling. Recall that the action of your body muscles pushes lymph through the lymph vessels. Inactivity can also cause lymph buildup and swelling.

Do you recall the swollen tonsils shown at the beginning of this lesson? If the cells of your tonsils become infected, you have tonsillitis—an inflammation of the tonsils.

The uncontrolled production of white blood cells is a type of cancer called lymphoma. Cancer of the lymph nodes is a related disease called Hodgkin's lymphoma.

The Lymphatic System and Homeostasis

The lymphatic system helps maintain your body's homeostasis by regulating fluid buildup around cells, as shown in **Figure 14.** It supports the circulatory system by cleaning fluids and replacing them in the bloodstream. It also supports overall health by helping to fight infection throughout the body.

6. NGSSS Check Combine Underline how the lymphatic system interacts with the circulatory system and immune system. SC.6.L.14.5

Figure 14 The lymphatic system helps maintain the body's homeostasis by preventing the buildup of excess tissue fluid, removing wastes, and fighting infection.

The Lymphatic System and Homeostasis

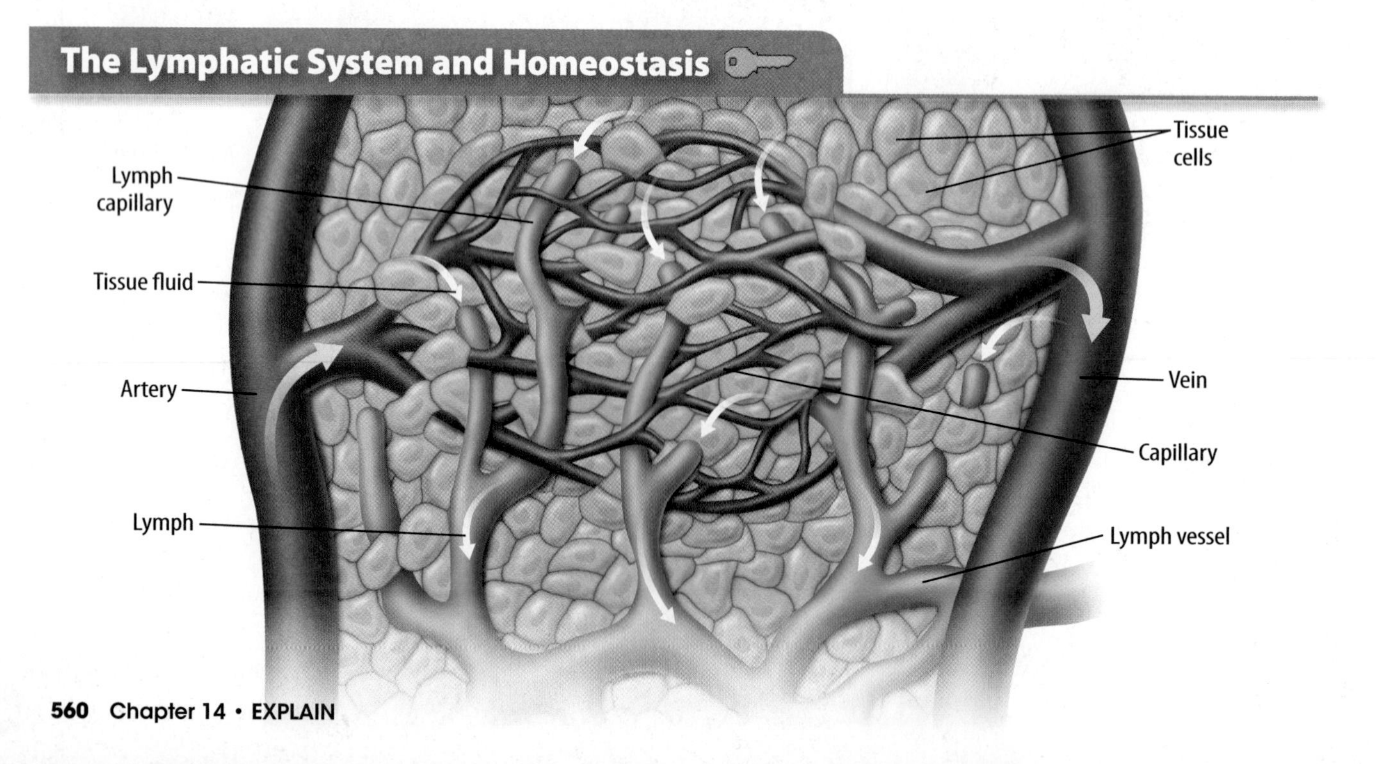

Lesson Review 4

Visual Summary

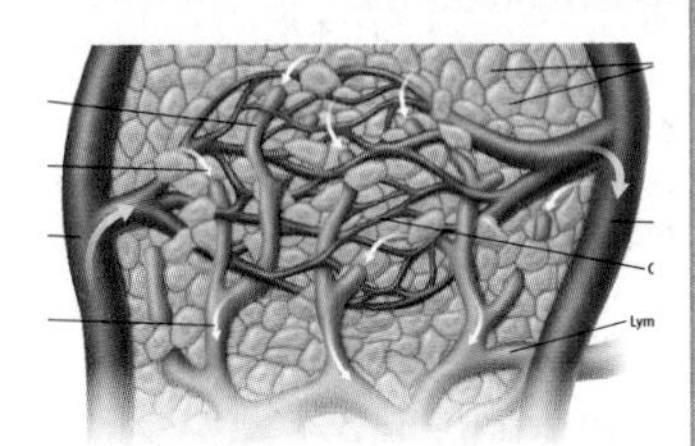

Tissue fluid that drains into the lymph vessels becomes lymph.

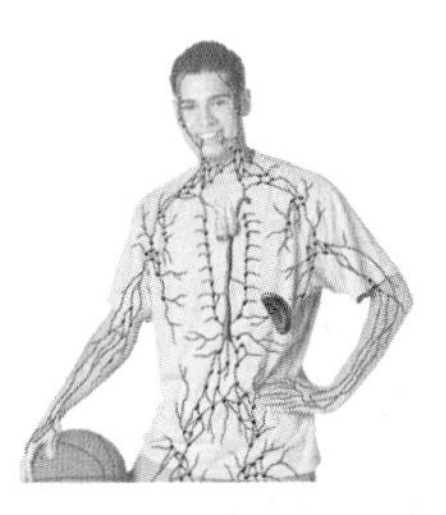

The lymphatic system consists of lymph nodes, lymph vessels, lymph, and several other organs.

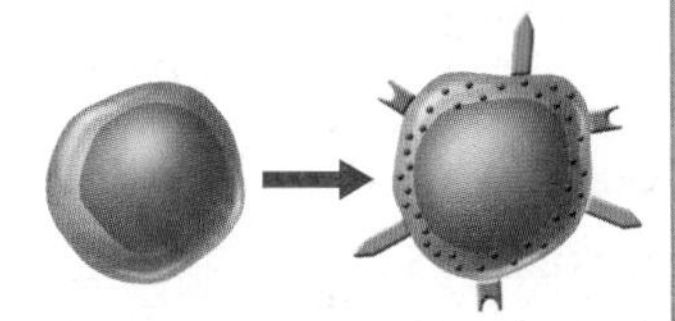

The lymphatic system cleans lymph, fights infection, and includes structures in which white blood cells develop.

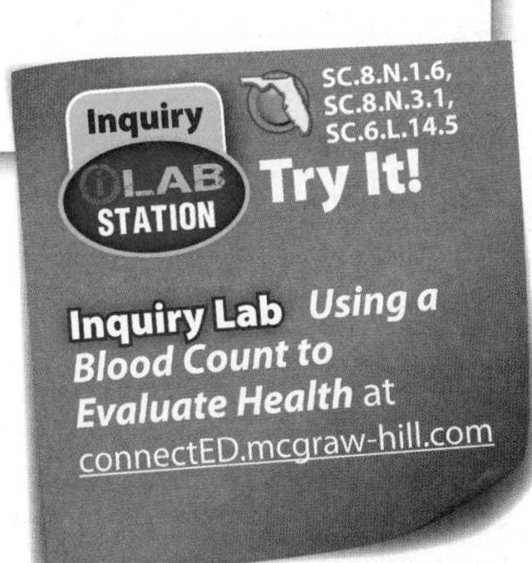

Use Vocabulary

1. **Define** the term *lymph*.

____

2. **Distinguish** between the spleen and the thymus.

____

____

3. Clusters of small, spongy structures that filter particles from lymph are called _______________.

Understand Key Concepts

4. **Describe** the function of the lymph nodes. SC.6.L.14.5

____

5. **Distinguish** between lymph and tissue fluid.

____

____

6. The lymphatic system cleans fluid for which system? SC.6.L.14.5
   - (A) circulatory
   - (B) digestive
   - (C) immune
   - (D) respiratory

Interpret Graphics

7. **Summarize** Fill in the graphic organizer below to identify the functions of the lymphatic system. SC.6.L.14.5

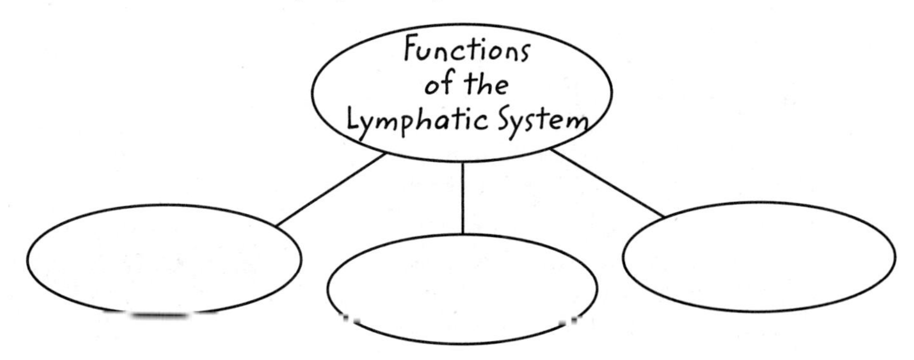

Critical Thinking

8. **Evaluate** In what ways are the circulatory and lymphatic systems similar? In what ways are they different?

____

____

____

Chapter 14 Study Guide

Think About It! **The respiratory and circulatory systems interact with each other to maintain homeostasis by moving materials through the body and removing wastes.**

Key Concepts Summary

LESSON 1 The Respiratory System

- The respiratory system provides the body with oxygen and removes carbon dioxide.
- In the **lungs,** oxygen is carried by the **bronchi** and the bronchioles to the **alveoli.**
- The respiratory system works with the circulatory and muscular systems to maintain homeostasis.

Vocabulary

breathing p. 531
pharynx p. 532
larynx p. 532
trachea p. 532
bronchi p. 533
lungs p. 533
alveoli p. 533
diaphragm p. 534

LESSON 2 The Circulatory System

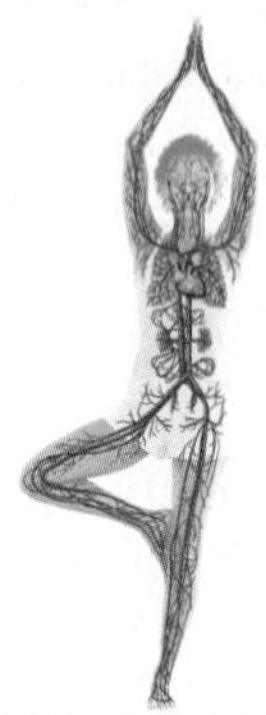

- The circulatory system moves materials throughout the body.
- **Arteries** carry blood away from the heart. **Capillaries** allow the exchange of materials between blood and body cells. **Veins** return blood to the heart.
- The circulatory system works with the respiratory, digestive, nervous, and endocrine systems to maintain homeostasis.

atrium p. 540
ventricle p. 540
artery p. 542
capillary p. 542
vein p. 542
systemic circulation p. 543
coronary circulation p. 543
pulmonary circulation p. 543
atherosclerosis p. 544

LESSON 3 Blood

- Blood transports oxygen, nutrients, and wastes; protects against illness and injury; and regulates body temperature.
- Red blood cells contain hemoglobin and carry oxygen. White blood cells fight infection. **Platelets** help stop bleeding. **Plasma** is the liquid portion of blood.

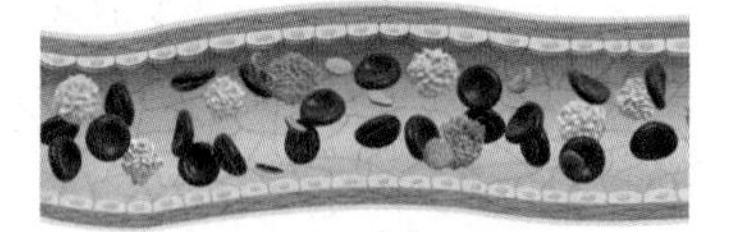

platelet p. 551
plasma p. 551
Rh factor p. 553

LESSON 4 The Lymphatic System

- The lymphatic system drains away excess tissue fluid and produces white blood cells that fight infection.
- **Lymph nodes** filter **lymph.** The **spleen** recycles worn-out red blood cells. B cells and T cells produced in the bone marrow fight disease-causing organisms. T cells mature in the **thymus.**
- The lymphatic system works together with the circulatory system to regulate the amount of fluid between cells.

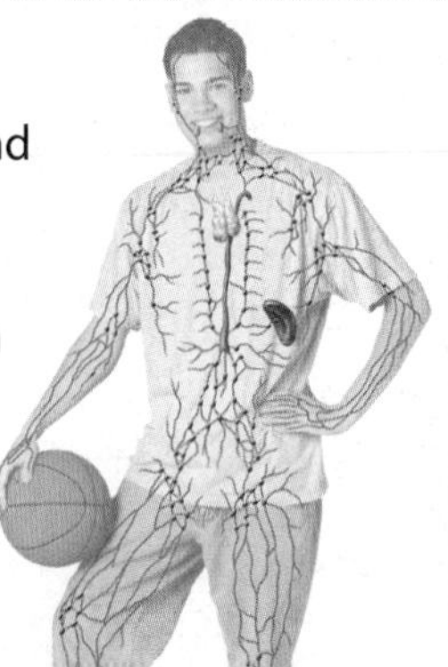

lymphatic system p. 557
lymph p. 558
lymph node p. 558
thymus p. 559
spleen p. 559

Active Reading

FOLDABLES® Chapter Project

Assemble your lesson Foldables as shown to make a Chapter Project. Use the project to review what you have learned in this chapter.

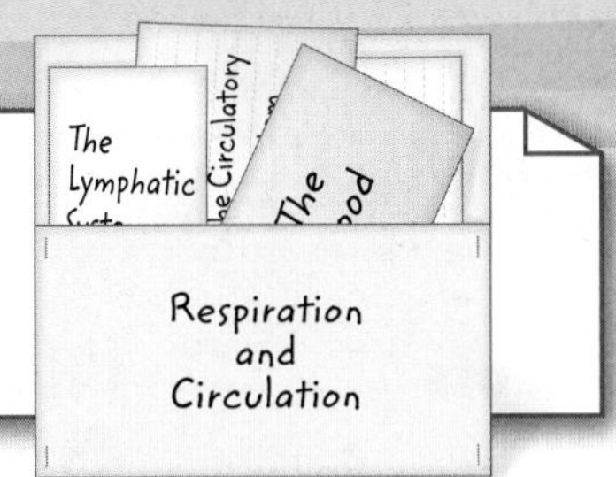

Use Vocabulary

1. The large muscle that contracts and relaxes to move gases into and out of the lungs is the ______________.

2. A respiratory infection in which the bronchi swell is ______________.

3. The smallest blood vessels are ______________.

4. The two lower chambers of the heart are called ______________.

5. Small, irregularly shaped pieces of cells in blood are ______________.

6. The organ that holds a reserve supply of blood and produces white blood cells is the ______________.

Link Vocabulary and Key Concepts

Use vocabulary terms from the previous page to complete the concept map.

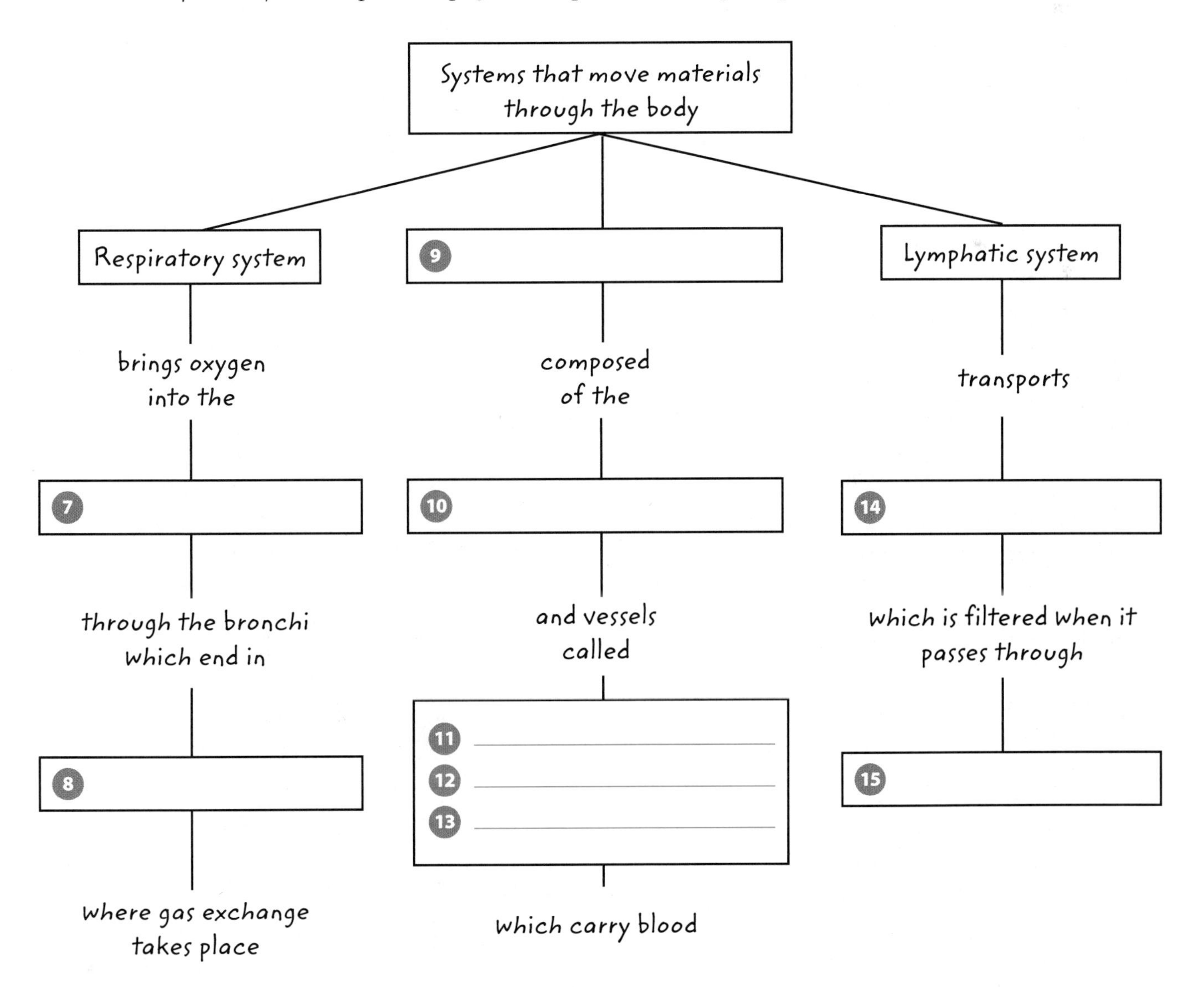

Chapter 14 Review

Fill in the correct answer choice.

Understand Key Concepts

1. Which process takes place in the structure shown below? SC.6.L.14.5

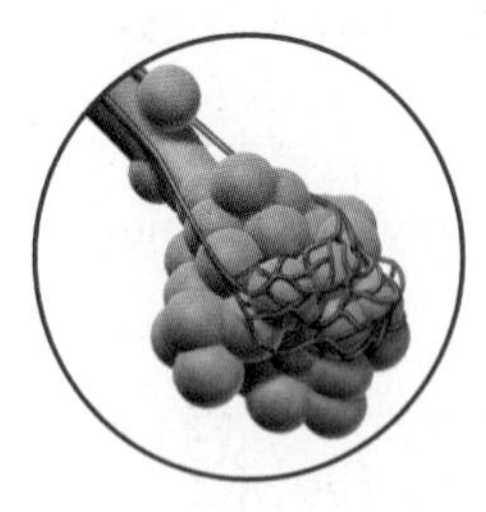

Ⓐ breathing
Ⓑ immunity
Ⓒ blood clotting
Ⓓ gas exchange

2. Which structure is held open by rings of cartilage? SC.6.L.14.5

Ⓐ alveolus
Ⓑ larynx
Ⓒ pharynx
Ⓓ trachea

3. What happens to the diaphragm during inhalation? SC.6.L.14.5

Ⓐ It contracts and moves down.
Ⓑ It contracts and moves up.
Ⓒ It relaxes and moves down.
Ⓓ It relaxes and moves up.

4. Which term describes the main function of pulmonary circulation? SC.6.L.14.5

Ⓐ fight infection
Ⓑ oxygenate blood
Ⓒ produce T cells
Ⓓ stop bleeding

5. Which type of circulation supplies oxygen to the cells of the heart? SC.6.L.14.5

Ⓐ coronary
Ⓑ lymphatic
Ⓒ pulmonary
Ⓓ systemic

Critical Thinking

6. **Illustrate** the path of air from the nose into the lungs. SC.6.L.14.5

7. **Compare** the structures and functions of the circulatory and lymphatic systems. SC.6.L.14.5

8. **Give an example** of a life choice that can harm the health of both the respiratory and circulatory systems. SC.6.L.14.5

9. **Interpret Graphics** The arrow below points to one of the chambers of the heart. Where does blood entering this chamber come from? Where does it go when it leaves this chamber? Is the blood oxygen-rich or oxygen-poor? SC.6.L.14.5

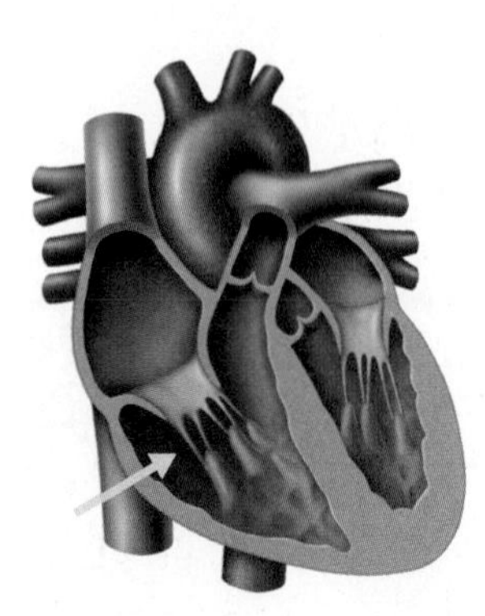

10 **Determine** A person with type AB blood regularly donates blood as a community service. Determine the blood type(s) that can receive this blood in a transfusion. SC.6.L.14.5

11 **Synthesis** On a separate sheet of paper, design a brochure for a tour through the lymphatic system. Include all the structures of the lymphatic system. SC.6.L.14.5

Writing in Science

12 **Write** a paragraph on a separate sheet of paper comparing the functions of the alveoli, the capillaries, and the lymph nodes. Your paragraph should have a topic sentence, supporting details, and a concluding sentence. LA.6.2.2.3

Big Idea Review

13 How does oxygen reach the cells of the body? Explain how the respiratory and circulatory systems work together to supply cells with the materials they need. SC.6.L.14.5

Math Skills MA.6.A.3.6

Use Percentages

The table below shows the percentages of the total population in the US with different blood types and with Rh– blood. Use the table to answer questions 14–16.

Blood type	A	B	AB	O
Percent with blood type	42	10	4	44
Percent who are Rh–	6	2	1	7

14 What percentage of the population has Rh+ blood?

15 What percentage of the total population has AB+ blood?

16 What percentage of people could donate blood to a person with O+ blood?

Florida NGSSS Benchmark Practice

Fill in the correct answer choice.

Multiple Choice

1 Where in the human body does gas exchange occur? **SC.6.L.14.5**

Ⓐ alveoli
Ⓑ bronchi
Ⓒ pharynx
Ⓓ trachea

Use the diagram below to answer question 2.

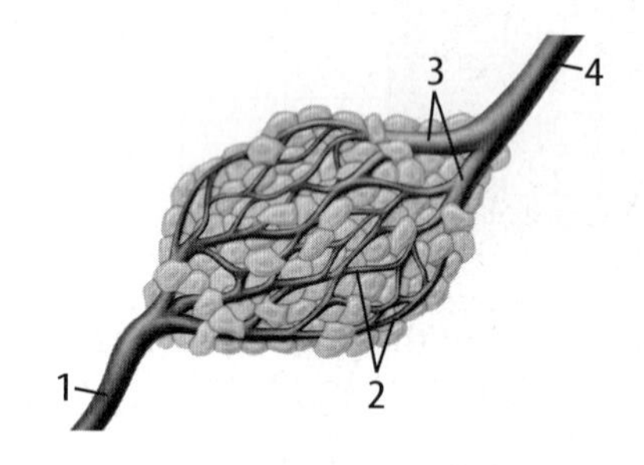

2 Which numbered blood vessel in the diagram above could be the aorta? **SC.6.L.14.5**

Ⓕ 1
Ⓖ 2
Ⓗ 3
Ⓘ 4

3 Which blood component stops the bleeding after a cut? **SC.6.L.14.5**

Ⓐ plasma
Ⓑ platelets
Ⓒ red blood cells
Ⓓ white blood cells

4 Which shows the general path of blood from the time it leaves the heart until it returns? **SC.6.L.14.5**

Ⓕ arteries → capillaries → veins
Ⓖ arteries → veins → capillaries
Ⓗ capillaries → arteries → veins
Ⓘ veins → capillaries → arteries

Use the diagram below to answer question 5.

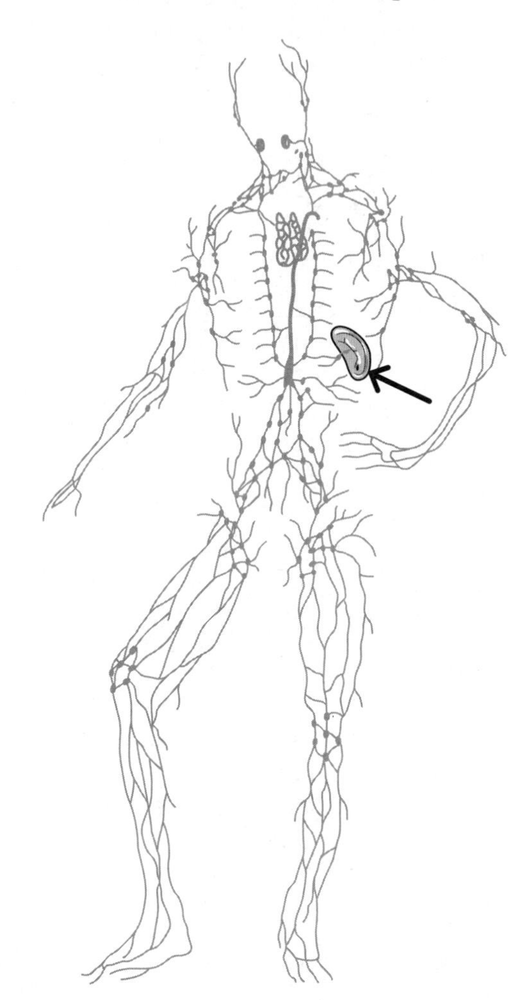

5 Which organ is highlighted in the diagram above? **SC.6.L.14.5**

Ⓐ heart
Ⓑ lung
Ⓒ spleen
Ⓓ stomach

6 Which blood type can be donated to all humans? **SC.6.L.14.5**

Ⓕ type A
Ⓖ type AB
Ⓗ type B
Ⓘ type O

7 Which is a function of the lymphatic system? **SC.6.L.14.5**

Ⓐ circulate blood

Ⓑ digest food

Ⓒ fight infection

Ⓓ transport gas

Use the diagram below to answer question 8.

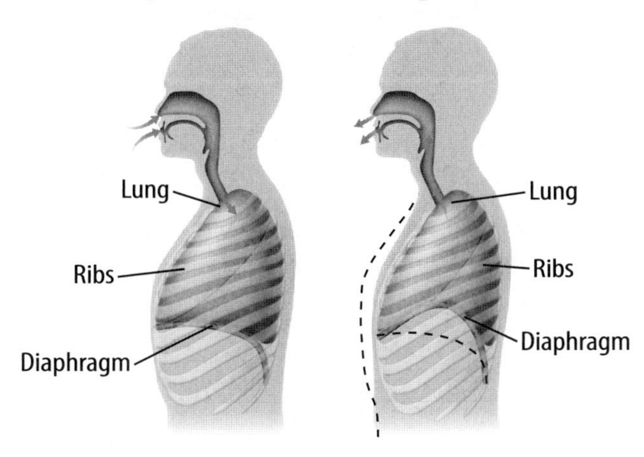

8 What process do the dotted lines in the diagram represent? **SC.6.L.14.5**

Ⓕ gas exchange

Ⓖ diaphragm contraction

Ⓗ muscle expansion

Ⓘ weight increase

9 Which contracts to move lymph through the lymphatic system? **SC.6.L.14.5**

Ⓐ heart

Ⓑ stomach

Ⓒ body muscle

Ⓓ heart muscle

10 Which organ keeps blood flowing through the body? **SC.6.L.14.5**

Ⓕ heart

Ⓖ lungs

Ⓗ spleen

Ⓘ thymus

11 The arrow in the diagram below points to which structure? **SC.6.L.14.5**

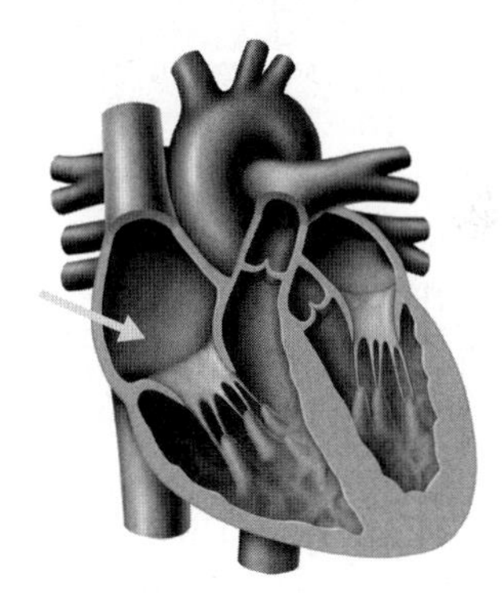

Ⓐ aorta

Ⓑ atrium

Ⓒ vein

Ⓓ ventricle

12 Which part of the blood helps defend the body from a virus infection? **SC.6.L.14.5**

Ⓕ plasma

Ⓖ platelets

Ⓗ red blood cells

Ⓘ white blood cells

13 What is the primary role of hemoglobin in blood? **SC.6.L.14.5**

Ⓐ attract platelets

Ⓑ blood typing

Ⓒ carry oxygen

Ⓓ fight parasites

Need Extra Help?

If You Missed Question...	1	2	3	4	5	6	7	8	9	10	11	12	13
Go to Lesson...	1	2	3	2	4	3	4	1	4	2	2	3	1

Name ______________________ Date ____________

Benchmark Mini-Assessment Chapter 14 • Lesson 1

Multiple Choice *Bubble the correct answer.*

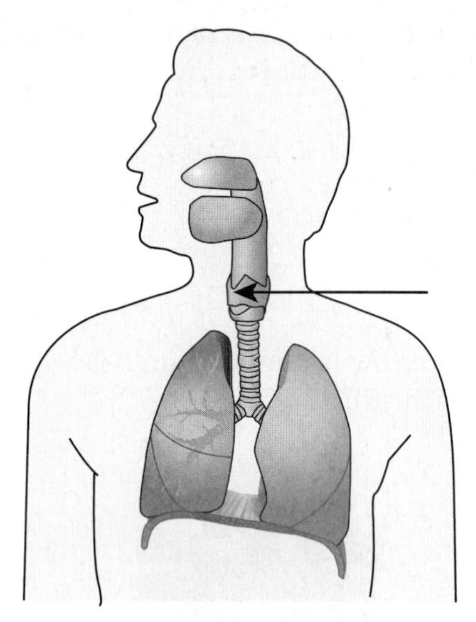

1. The diagram above shows the structures of the respiratory system. The arrow in the diagram is pointing to one of these structures. What does this structure do? **SC.6.L.14.5**

Ⓐ It allows gas exchange between the blood and the lungs.

Ⓑ It blocks food and liquids from entering the lungs.

Ⓒ It changes pressure inside your chest and allows you to breathe.

Ⓓ It vibrates and allows you to make sounds.

2. Which respiratory illness is caused by bacteria? **HE.6.C.1.8**

Ⓕ asthma

Ⓖ bronchitis

Ⓗ emphysema

Ⓘ flu

3. Which path below does air follow after it enters the nose? **SC.6.L.14.5**

Ⓐ larynx → pharynx → trachea → bronchi → bronchioles → alveoli

Ⓑ larynx → trachea → pharynx → bronchioles → bronchi – alveoli

Ⓒ pharynx → larynx → trachea → bronchi → bronchioles → alveoli

Ⓓ trachea → larynx → pharynx → bronchioles → bronchi → alveoli

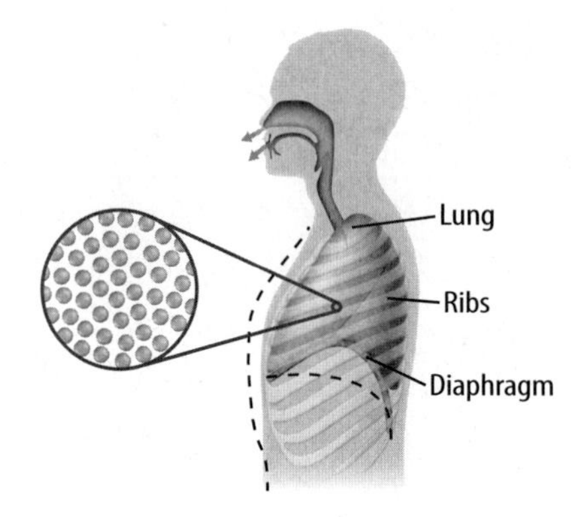

4. In the process shown by the image above, **SC.6.L.14.5**

Ⓕ the chest muscles are contracting.

Ⓖ the diaphragm is contracting.

Ⓗ the diaphragm is relaxing.

Ⓘ the ribs are expanding.

Name ______________________________ Date ______________

mini BAT

Multiple Choice *Bubble the correct answer.*

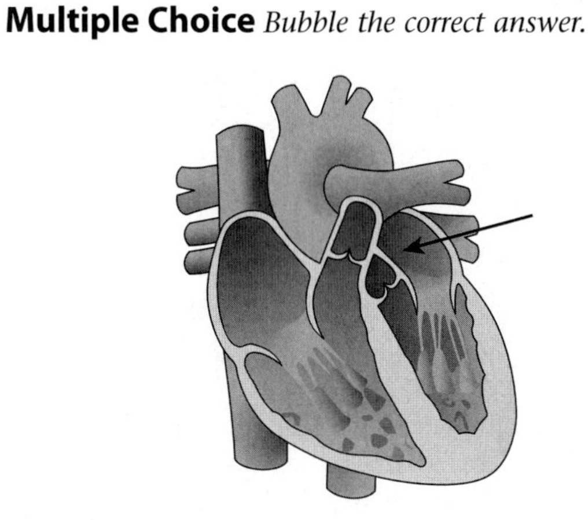

1. The diagram above shows the human heart. The chamber of the heart that the arrow points to in the diagram **SC.6.L.14.5**

   (A) pumps oxygen-poor blood to the lungs.
   (B) pumps oxygen-rich blood to the body.
   (C) receives oxygen-poor blood from the body.
   (D) receives oxygen-rich blood from the lungs.

2. Which type of circulation removes wastes from skeletal muscle cells? **SC.6.L.14.5**

   (F) coronary circulation
   (G) musculoskeletal circulation
   (H) pulmonary circulation
   (I) systemic circulation

3. Blood is most likely to slow down and pool in the **SC.6.L.14.5**

   (A) aorta.
   (B) arteries.
   (C) capillaries.
   (D) veins.

4. Which cross section below best represents a healthy artery? **SC.6.L.14.5**

(F) 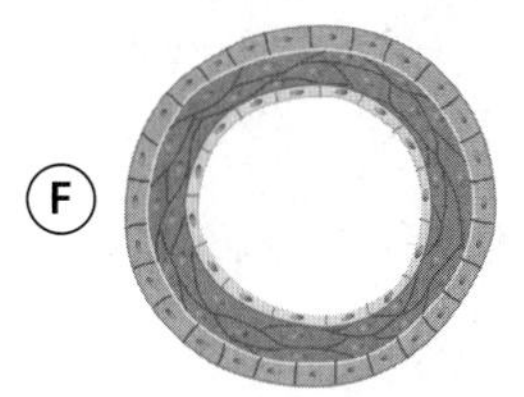

(G) 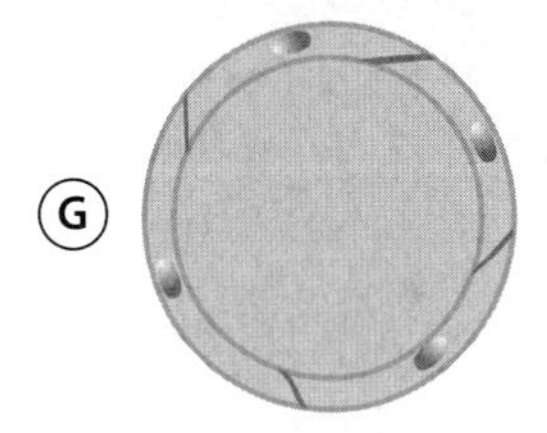

(H) 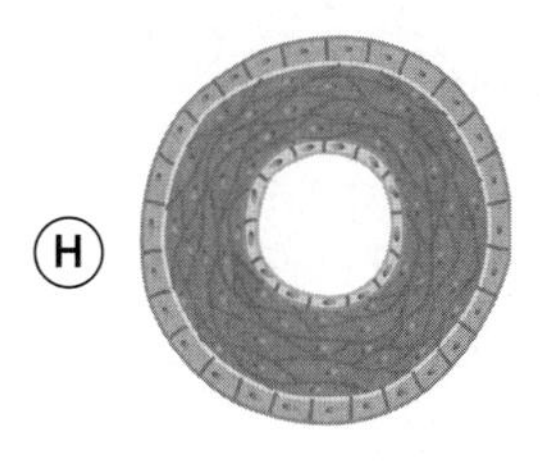

(I) 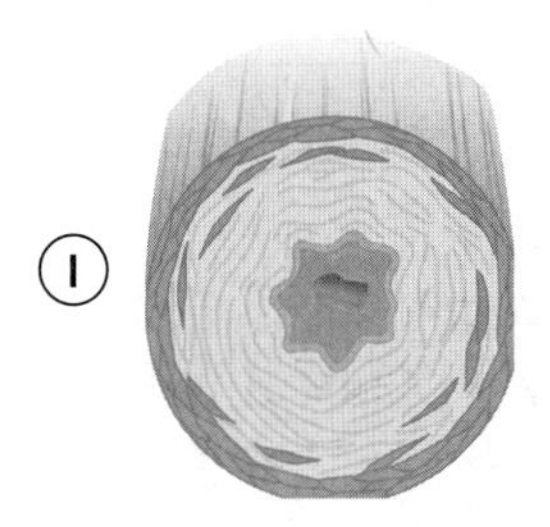

Name ______________________ Date ____________

Benchmark Mini-Assessment Chapter 14 • Lesson 3

mini BAT

Multiple Choice *Bubble the correct answer.*

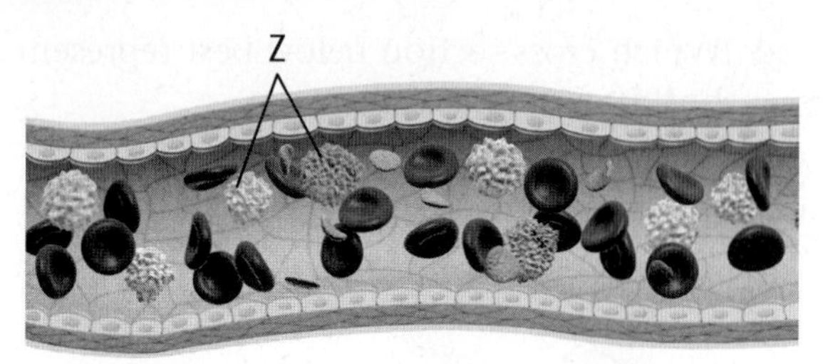

1. The diagram above shows parts of the blood. What is the function of the part that is labeled *Z* in the diagram? **SC.6.L.14.5**

- (A) to carry oxygen to cells throughout the body
- (B) to plug wounds and stop bleeding
- (C) to protect the body from illness and infection
- (D) to thin the blood and carry nutrients and waste products

2. Which part of the blood carries vitamin C? **SC.6.L.14.5**

- (F) plasma
- (G) platelets
- (H) red blood cells
- (I) white blood cells

3. Melissa has Type AB blood. Which clumping proteins are present in her plasma? **SC.6.L.14.5**

- (A) clumping protein A
- (B) clumping protein B
- (C) clumping proteins A and B
- (D) clumping proteins not present

4. Daniel was in an accident and lost a lot of blood. If Daniel has type A blood, which image represents the blood that he could safely receive in a transfusion? **SC.6.L.14.5**

(F) 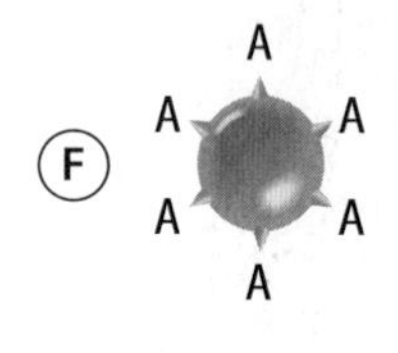

(G) 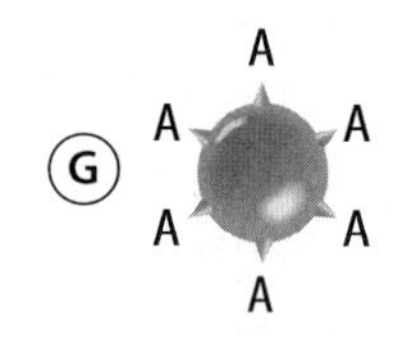

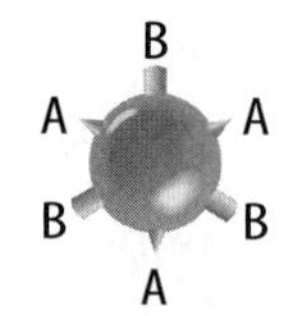

(H) 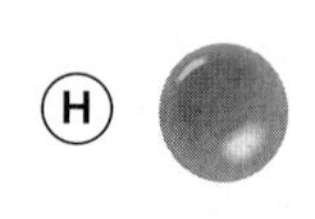
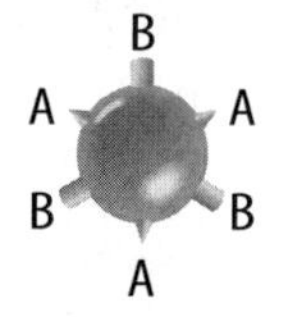

(I) 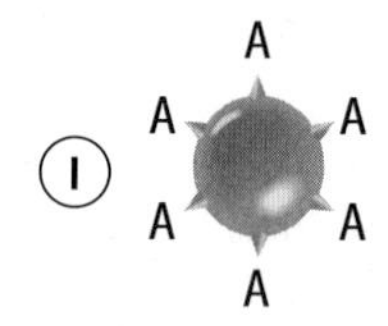

Name ______________________ Date ______________

Multiple Choice *Bubble the correct answer.*

Use the image below to answer questions 1 and 2.

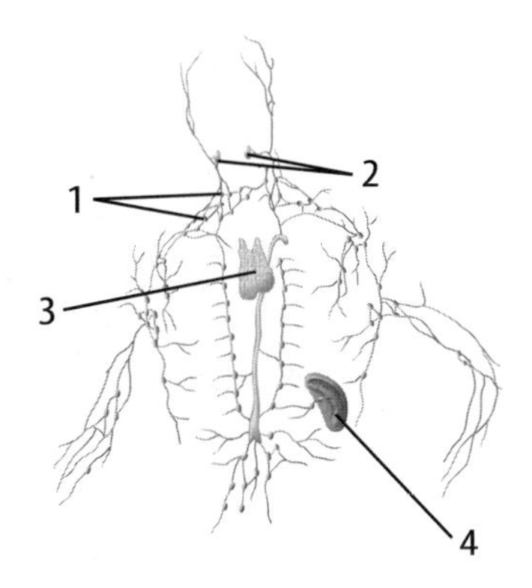

1. Which structure of the lymphatic system filters viruses from lymph? **HE.6.C.1.8**

Ⓐ 1

Ⓑ 2

Ⓒ 3

Ⓓ 4

2. When an injured person loses a lot of blood, which structure of the lymphatic system releases stored red blood cells into the circulatory system? **SC.6.L.14.5**

Ⓕ 1

Ⓖ 2

Ⓗ 3

Ⓘ 4

3. Where are red and white blood cells formed? **SC.6.L.14.5**

Ⓐ spleen

Ⓑ thymus

Ⓒ bone marrow

Ⓓ lymph vessels

4. The lymphatic system does all of the following EXCEPT **SC.6.L.14.5**

Ⓕ absorb fluid that collects outside of cells.

Ⓖ break down and recycle old red blood cells.

Ⓗ deliver nutrients to cells throughout the body.

Ⓘ trap and destroy the bacteria and viruses.

Name ____________________ Date ____________

Notes

Germ Fighters

Five friends talked about how their immune system attacks germs that get in their bodies that might make them sick. They each had different ideas. This is what they said:

Paula: I think the red blood cells are the ones that fight off germs.

Ryan: I think the white blood cells are the ones that fight off germs.

Simon: I think both the red blood cells and the white blood cells fight off germs.

Randy: I don't think germ-fighting cells come from the blood. The blood is part of the circulatory system.

Kara: I don't think cells attack germs. I think the immune system produces chemicals in the body that kill germs.

Circle the name of the friend you agree with the most. Explain why you agree. Describe your ideas about how the immune system fights germs that enter the body.

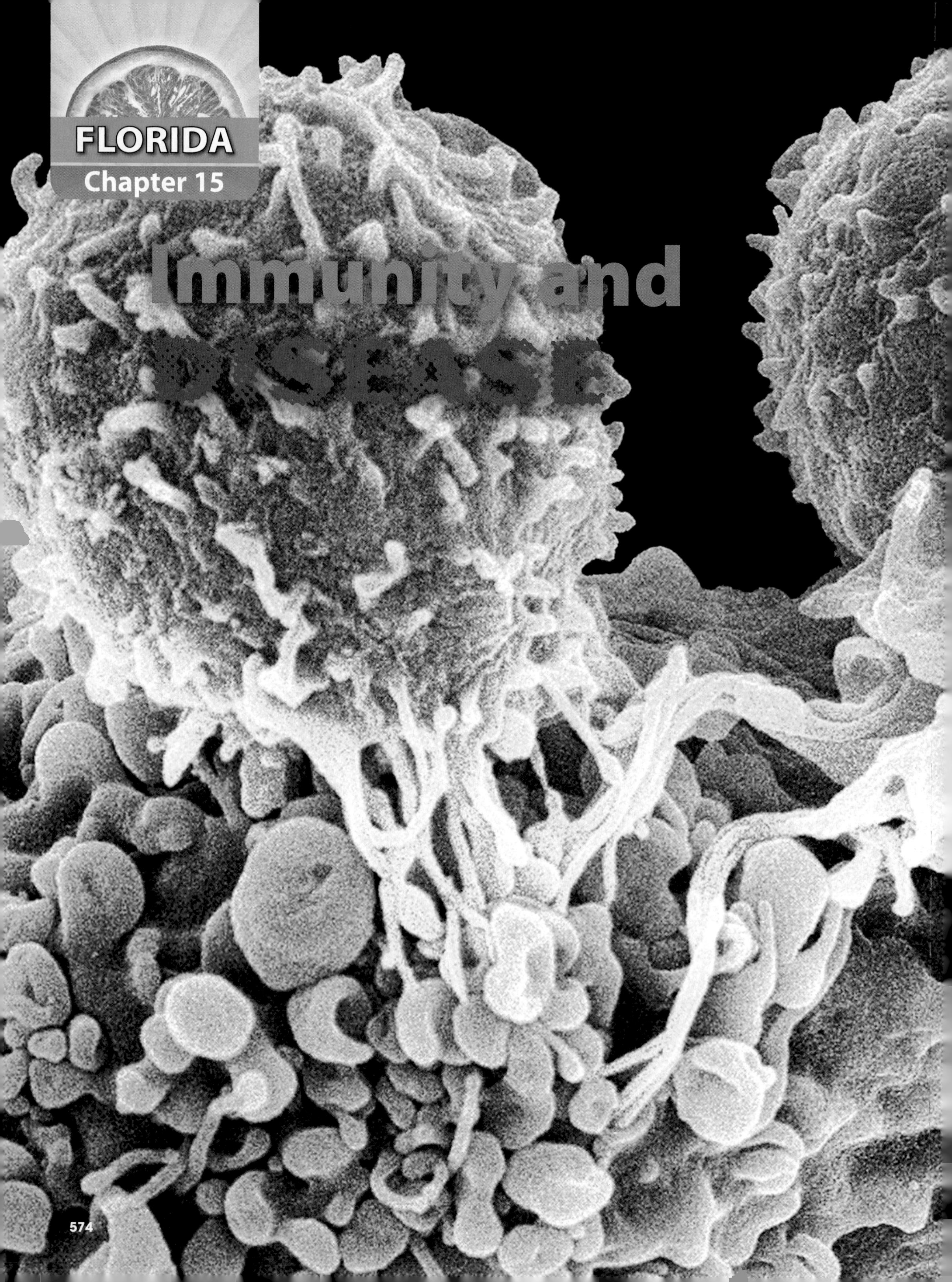

Immunity and Disease

FLORIDA BIG IDEAS

1 **The Practice of Science**

2 **The Characteristics of Scientific Knowledge**

3 **The Role of Theories, Laws, Hypotheses, and Models**

14 **Organization and Development of Living Organisms**

The Big Idea

Think About It!

How does the immune system help maintain the body's homeostasis?

Your immune system protects your body against invaders. Notice the two small brown-yellow cells on the large green cell.

1. Why do you think the small cells might be attacking the large cell?

______

2. How do you think your immune system helps your body maintain homeostasis?

______

Get Ready to Read

What do you think about diseases?

Before you read, decide if you agree or disagree with each of these statements. As you read this chapter, see if you change your mind about any of the statements.

	AGREE	DISAGREE
1 Some diseases are infectious, and others are noninfectious.	☐	☐
2 Cancer is an infectious disease.	☐	☐
3 The immune system helps keep the body healthy.	☐	☐
4 All immune responses are specific to the invading germs.	☐	☐
5 Exercise and sleep can help keep you healthy.	☐	☐
6 Chemicals make you sick and should not be used.	☐	☐

There's More Online!
Video • Audio • Review • ⓘLab Station • WebQuest • Assessment • Concepts in Motion • Multilingual eGlossary

Lesson 1

DISEASES

ESSENTIAL QUESTIONS

 Why do we get diseases?

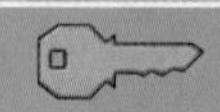 How do the two types of diseases differ?

Vocabulary

pathogen p. 577
pasteurization p. 579
infectious disease p. 581
vector p. 581
noninfectious disease p. 582
cancer p. 583

Florida NGSSS

LA.6.2.2.3 The student will organize information to show understanding (e.g., representing main ideas within text through charting, mapping, paraphrasing, summarizing, or comparing/contrasting);

SC.6.L.14.6 Compare and contrast types of infectious agents that may infect the human body, including viruses, bacteria, fungi, and parasites.

SC.6.N.1.5 Recognize that science involves creativity, not just in designing experiments, but also in creating explanations that fit evidence.

SC.8.N.3.1 Select models useful in relating the results of their own investigations.

SC.7.N.2.1 Identify an instance from the history of science in which scientific knowledge has changed when new evidence or new interpretations are encountered.

SC.6.N.2.2 Explain that scientific knowledge is durable because it is open to change as new evidence or interpretations are encountered.

SC.6.N.2.3 Recognize that scientists who make contributions to scientific knowledge come from all kinds of backgrounds and possess varied talents, interests, and goals.

SC.7.N.1.3 Distinguish between an experiment (which must involve the identification and control of variables) and other forms of scientific investigation and explain that not all scientific knowledge is derived from experimentation.

SC.7.N.1.7 Explain that scientific knowledge is the result of a great deal of debate and confirmation within the science community.

HE.6.C.1.8 Explain how body systems are impacted by hereditary factors and infectious agents.

Launch Lab

SC.6.N.1.5

15 minutes

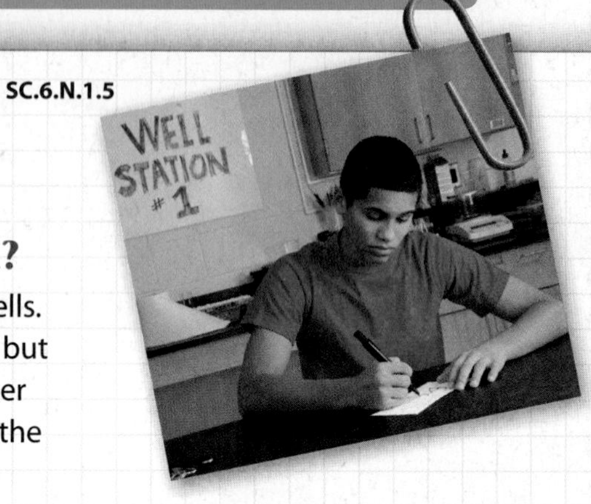

Which well is contaminated?

Imagine that you live in a town with four wells. You get your water regularly from one well, but sometimes you also drink water from another well. People are getting sick. Some suspect the water in one well is causing the sickness. Which well is contaminated?

Procedure

1. Take an **envelope** from your assigned well. Do not look inside the envelope.
2. Write your name on the envelope, and then pass it to another person from any well. You should also receive a different envelope from another person.
3. Repeat step 2.
4. Write your name on the third envelope and open it. If there is an "X" on the card inside, the three people who signed the envelope drank from the contaminated well and are sick. As a class, compile the results in a table.

Think About This

1. Which well was contaminated? How could you tell?

2. Key Concept What do you think you might do to stop the sickness from spreading?

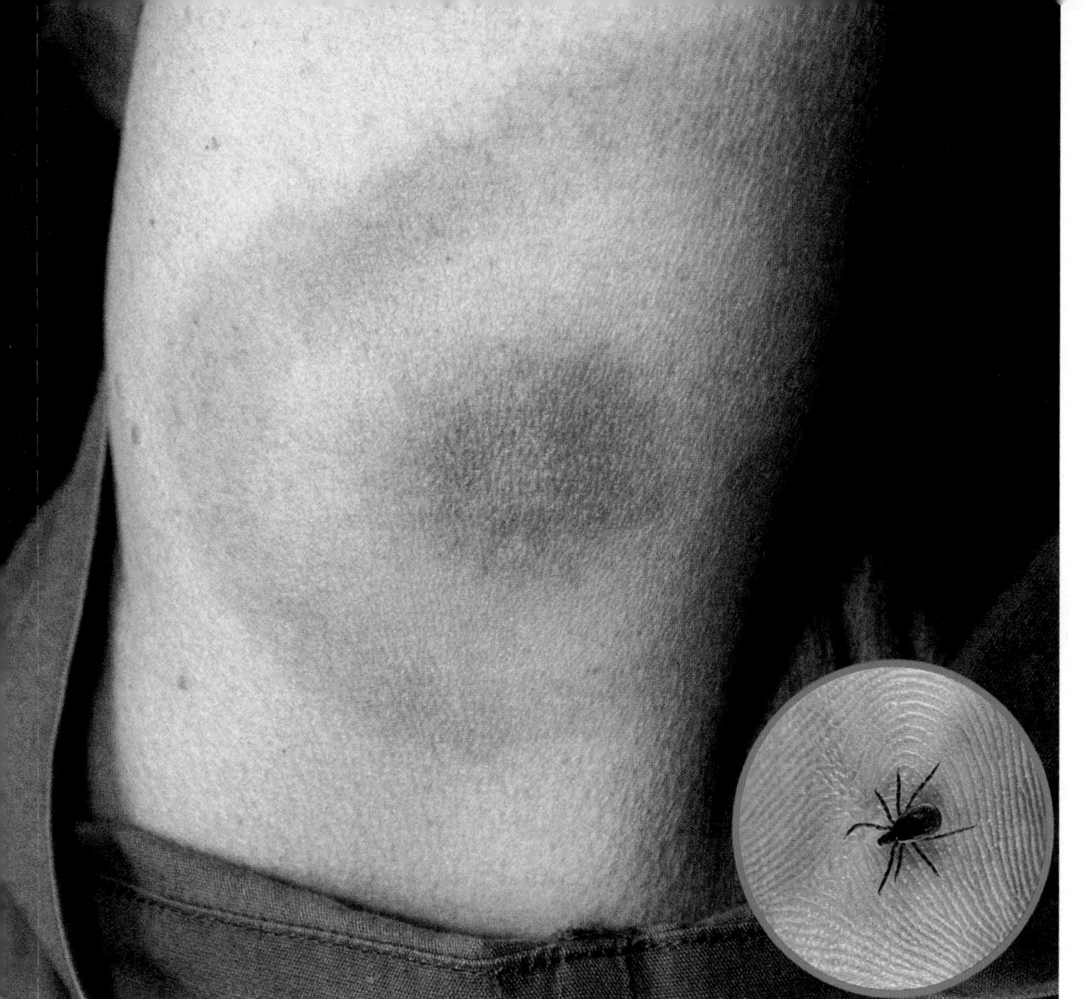

Inquiry A Bull's-Eye?

1. Have you ever seen a bull's-eye on someone's skin? It is a rash caused by Lyme disease. The disease is spread by a tick like the one shown here. Why does the skin turn red?

Disease Through History

Imagine you are sick, and your doctor suggests scraping your skull with a rock until a hole is created, as shown in **Figure 1.** The doctor tells you that the hole in your skull will allow the cause of your illness to escape. Today you might think this is strange. But thousands of years ago, this was an accepted treatment for disease.

Today we know that many diseases are caused by bacteria and viruses. *Disease causing agents, such as bacteria and viruses, are called* **pathogens.** Pathogens have always caused illnesses, but only in the last few hundred years has the relationship between pathogens and diseases been understood.

Before then little was known about disease and immunity, and superstitions were common. Today we know the hole-in-the-head treatment would be painful and would create an opportunity for even more pathogens to enter the body.

WORD ORIGIN

pathogen
from Greek *pathos,* means "disease"

Figure 1 Archaeologists have found skulls with smooth holes made while the patient was alive. Bone growth around the hole in this skull shows that the patient lived after the procedure.

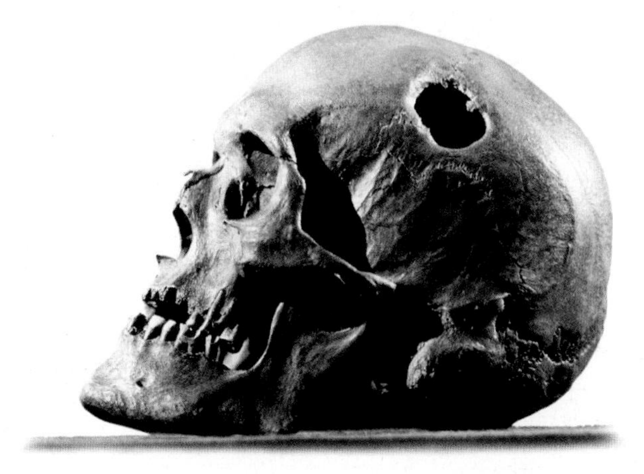

Early Research on Diseases

Despite limited technology and equipment, doctors in the eighteenth and nineteenth centuries learned a lot about the causes and treatments of diseases. The research and experiments performed by a few scientists saved many lives.

First Vaccination

In 1796 a doctor in England named Edward Jenner developed the first vaccination—a procedure that helps the body defend itself against disease. Jenner knew that women who milked cows often developed a mild disease called cowpox. However, these women were resistant to the deadly disease smallpox. He made a cut in the arm of a young boy and inserted pus from a cowpox sore. Two weeks later, he infected the boy with smallpox, but the boy did not develop smallpox. Although the smallpox vaccination saved many lives, scientists did not understand why or how it worked.

Connecting Disease with a Source

In the mid-1800s, people realized there was a connection between pathogens and disease. During this period, many people in London were dying from cholera, a bacterial disease of the intestinal tract. Dr. John Snow mapped outbreaks of the disease, as shown in **Figure 2**. He tracked the origin of one outbreak to a water pump. He had the pump closed, and new cases of cholera decreased immediately. John Snow thought a microscopic organism that he saw in the water—the cholera bacteria shown in **Figure 2**—caused the disease. Not everyone agreed, but people were beginning to think pathogens existed.

Figure 2 John Snow mapped the outbreak of cholera and realized the origin was the water from a specific pump.

Active Reading **2. Survey** Highlight how Snow used his map to identify the source of the cholera outbreak.

John Snow

- Cholera death
- Water pump
- ☆ Contaminated water pump

Cholera bacteria

Figure 3 Anton van Leeuwenhoek made many scientific discoveries using simple microscopes he designed. He named the moving organisms he saw in pond water *animalcules*. Today they are called bacteria and protozoa.

The Development of Microscopes

One of the reasons people were slow to accept the idea of pathogens was because they could not see them. The development of microscopes changed that. In the late 1600s, Dutch merchant Anton van Leeuwenhoek (LAY vun hook) made one of the first microscopes. He discovered bacteria in pond water, as illustrated in **Figure 3.** However, van Leeuwenhoek did not share how he made the lenses, so bacteria were not observed again until the nineteenth century.

Connecting Bacteria to Infections

When scientists first realized bacteria were present in wounds, they thought the wounds caused the bacteria to appear. When Louis Pasteur began doing experiments in the mid-1800s, he realized that this idea was backward. Instead, bacteria from outside the body caused the tissue in the wound to decay. Pasteur discovered that he could kill bacteria in boiling liquids. **Pasteurization** *is the process in which a food is heated to a temperature that kills most harmful bacteria.* It is based on the work of Pasteur.

Joseph Lister used Pasteur's discoveries to make surgery safer for patients. He found that carbolic acid killed bacteria. He developed a misting system to spray carbolic acid throughout an operating room during surgery. Infection and death from surgeries decreased greatly. In the late 1800s, doctors improved on Lister's idea. They used carbolic acid to sterilize tools before surgery and steam to sterilize the linens and clothes.

3. NGSSS Check Formulate Underline how Lister made surgery safer. SC.7.N.2.1

Koch's Rules

1. The bacterium must be found in all organisms suffering from the disease but not in healthy organisms.

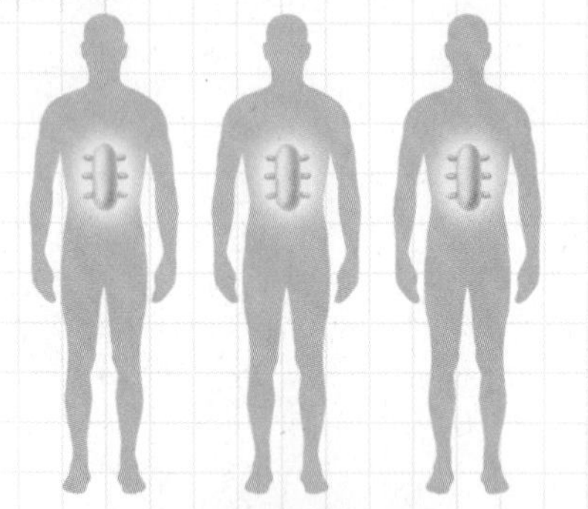

2. The bacterium must reproduce in the lab.

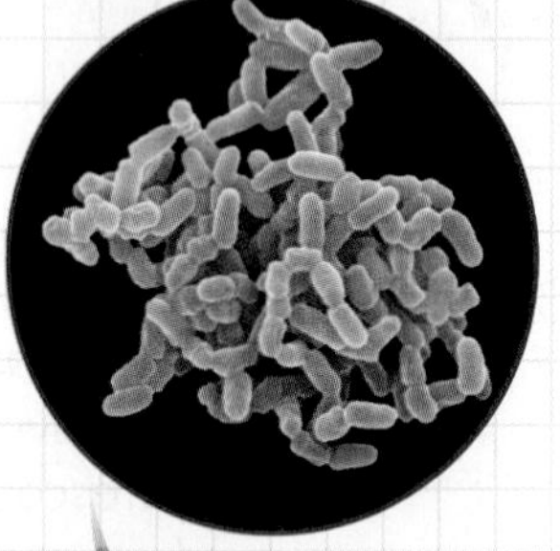

3. A sample of the newly grown pathogen must cause the illness when injected into a healthy animal.

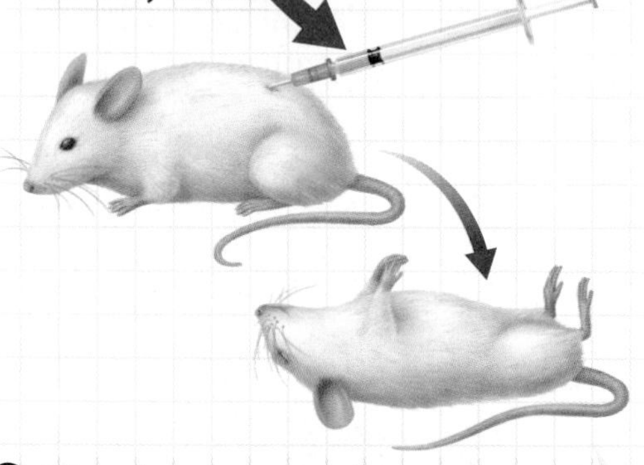

4. When the suspected pathogen is removed from the infected animal and grown in the lab, it must be identical to the original pathogen.

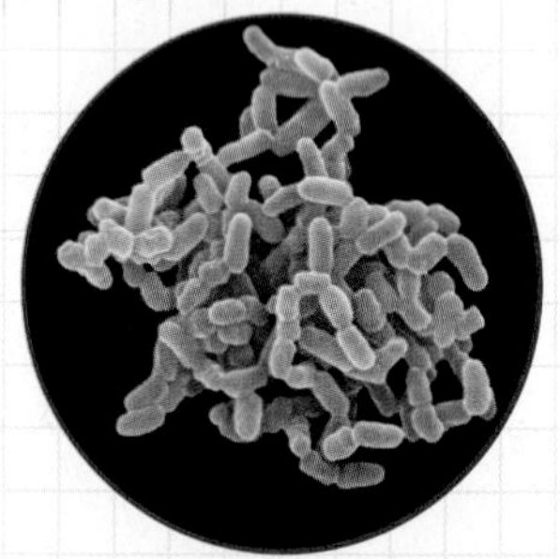

Figure 4 Koch developed a procedure to determine if a bacterium caused an illness.

Discovering Disease Organisms

Despite the research on bacteria in wounds, most people did not think bacteria could make a healthy person sick. In 1867, Robert Koch was one of the first scientists to argue that bacteria could cause illness in an animal as large as a cow. He developed a set of rules to determine if specific bacteria caused an illness. Koch's rules are illustrated in **Figure 4.** The research based on these rules convinced most scientists that some bacteria were disease-causing pathogens. Although the roles of pathogens in disease are not as simple as Koch thought, current understandings are based on his findings.

Active Reading **4. Determine** Highlight Koch's rules.

Bacteria are not the only pathogens that cause disease—viruses are others. However, they are so small that many years passed before scientists understood that viruses could be pathogens too. Some fungi and protists can also cause diseases. Some of the diseases in humans caused by different pathogens include the following:

- Viruses cause the flu, colds, chickenpox, and AIDS.
- Bacteria cause ear infections, strep throat, pneumonia, meningitis, whooping cough, and syphilis, a sexually transmitted disease.
- Fungi cause athlete's foot, ringworm, and yeast infections.
- Protists cause malaria, African sleeping sickness, and dysentery.

Pathogens can be transmitted through food and water and carried by insects. They also can be passed directly among people by physical contact, sneezing, coughing, or exchange of bodily fluids. Some pathogens, such as the bacterium that causes syphilis, require a host to reproduce.

Active Reading **5. Decide**  which form of pathogen chickenpox is.

Types of Diseases

Have you ever heard anyone say he or she "caught" a cold? The common cold is contagious. This means that the pathogens that cause the common cold can be passed from person to person. Not all diseases are caused by pathogens. Your inherited traits are responsible for some diseases. Others can be caused by external factors such as your environment and the choices you make about diet, exercise, and sleep.

6. NGSSS Check Hypothesize Why do we get diseases? HE.6.C.1.8

Infectious Diseases

Diseases caused by pathogens that can be transmitted from one person to another are **infectious diseases.** The way this happens can vary depending on the pathogen.

Flu and cold viruses can pass to others through direct contact, such as shaking hands. The human immunodeficiency virus (HIV) can pass through the exchange of blood or bodily fluids. HIV causes acquired immunodeficiency syndrome (AIDS), a disease that attacks the body system that fights pathogens.

The protist that causes malaria is transferred by a **vector,** *a disease-carrying organism that does not develop the disease.* The vector for malaria is a certain type of mosquito. The mosquito bites an animal that has the protist in its bloodstream. Then the pathogen enters the saliva of the mosquito, but the mosquito does not develop malaria. When the mosquito bites another animal, the pathogen moves into that animal's blood.

7. Support How does washing your hands help prevent the spread of infectious diseases?

Apply It! After you complete the lab, answer these questions.

1. Can you think of any ways you could prevent the disease from spreading so quickly?

2. What was the vector for the virus that spread?

3. If there were more vectors present, would you expect the disease to spread more or less quickly? Explain.

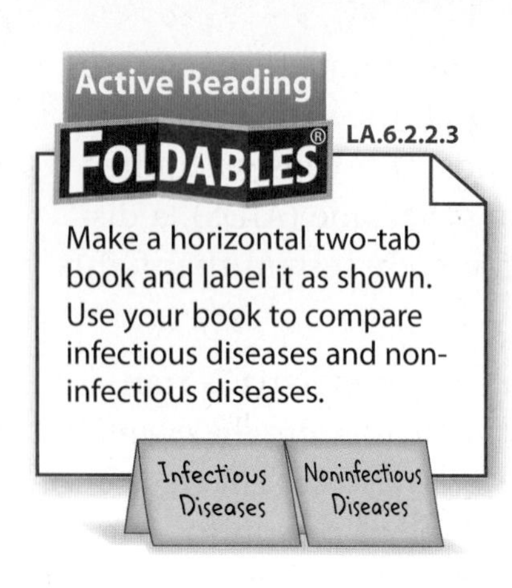

Noninfectious Diseases

A disease that cannot pass from person to person is a **noninfectious disease**. For example, you cannot catch lung cancer from another person. Pathogens do not directly cause noninfectious diseases. Two common causes of noninfectious diseases include:

- genetics, or traits inherited in your DNA from your biological parents, and
- environmental conditions, including lifestyle choices.

In many cases of noninfectious disease, a person has a genetic trait for a disease that environmental conditions make worse. It is the combination of genetics and environment that causes the disease to develop.

Childhood Diseases Noninfectious diseases that affect children are primarily due to genetics. One genetic disease is cystic fibrosis. It causes the body to produce mucus that is thicker than normal. This affects breathing and other body functions. Children with cystic fibrosis inherit a form of the gene that causes this disorder. It is a recessive trait, which means a person must inherit the gene from each parent, as shown in **Figure 5.** The parents might not have the disease, but they each must carry at least one gene form, or allele (uh LEEL), for cystic fibrosis. Like many genetic disorders in children, environmental conditions can make the disease worse. A poor diet, air pollution, and lack of exercise can make the symptoms of cystic fibrosis worse.

Figure 5 Diseases caused by genetic disorders are inherited.

8. Visual Check
Diagnose How many children inherited a gene for cystic fibrosis?

Inheritance of Cystic Fibrosis

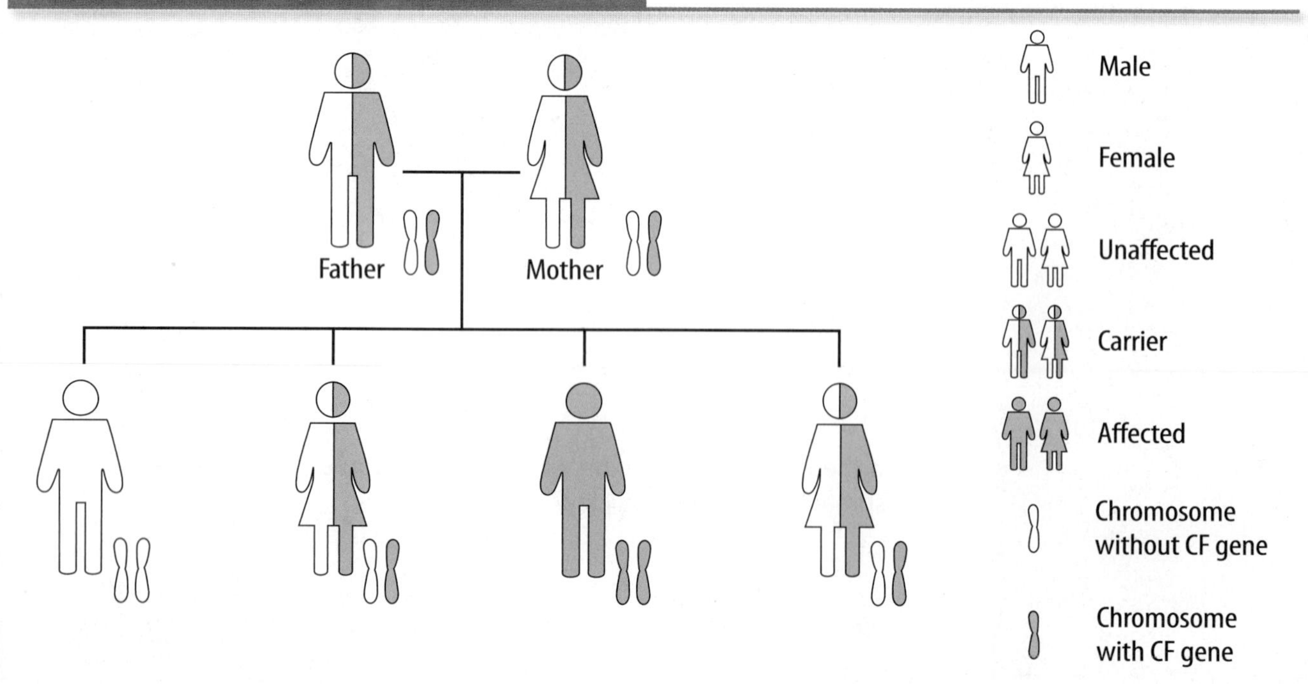

Other Diseases Many noninfectious diseases that affect adults are due primarily to environmental causes and lifestyle choices. For example, an unhealthful diet, obesity, a lack of regular exercise, and smoking cause most cases of heart disease. Osteoporosis is a disease in which bones become weak and less dense. People inherit a tendency to develop osteoporosis. However, years of poor lifestyle choices such as an unhealthful diet, lack of calcium and **vitamin** D, smoking, and a lack of exercise can all lead to weakened bones. There is also a type of diabetes that develops in adults that is strongly linked to environmental conditions, although there might also be a genetic link.

Review Vocabulary

vitamin
nutrient needed for growth, regulation of body functions, and prevention of some diseases

Active Reading **9. Develop** Highlight some causes of noninfectious diseases.

Cancer Tumors form when cells reproduce uncontrollably. **Cancer** *is a disease in which cells reproduce uncontrollably without the usual signals to stop.* For example, lung-cancer tumors form in the lungs and interfere with normal lung function. In **Figure 6,** notice the color difference in the lung that has not been functioning properly due to cancer. People can inherit forms of genes that make them more likely to develop lung cancer. However, if they are not exposed to such environmental conditions as poor air quality, or they do not smoke, they might not develop lung cancer.

11. NGSSS Check Differentiate Underline how infectious and noninfectious diseases differ. SC.6.L.14.6

Figure 6 Cancer cells in the lung form tumors and interfere with normal functioning.

10. Visual Check Judge What are the differences between the healthy lungs and the diseased lungs?

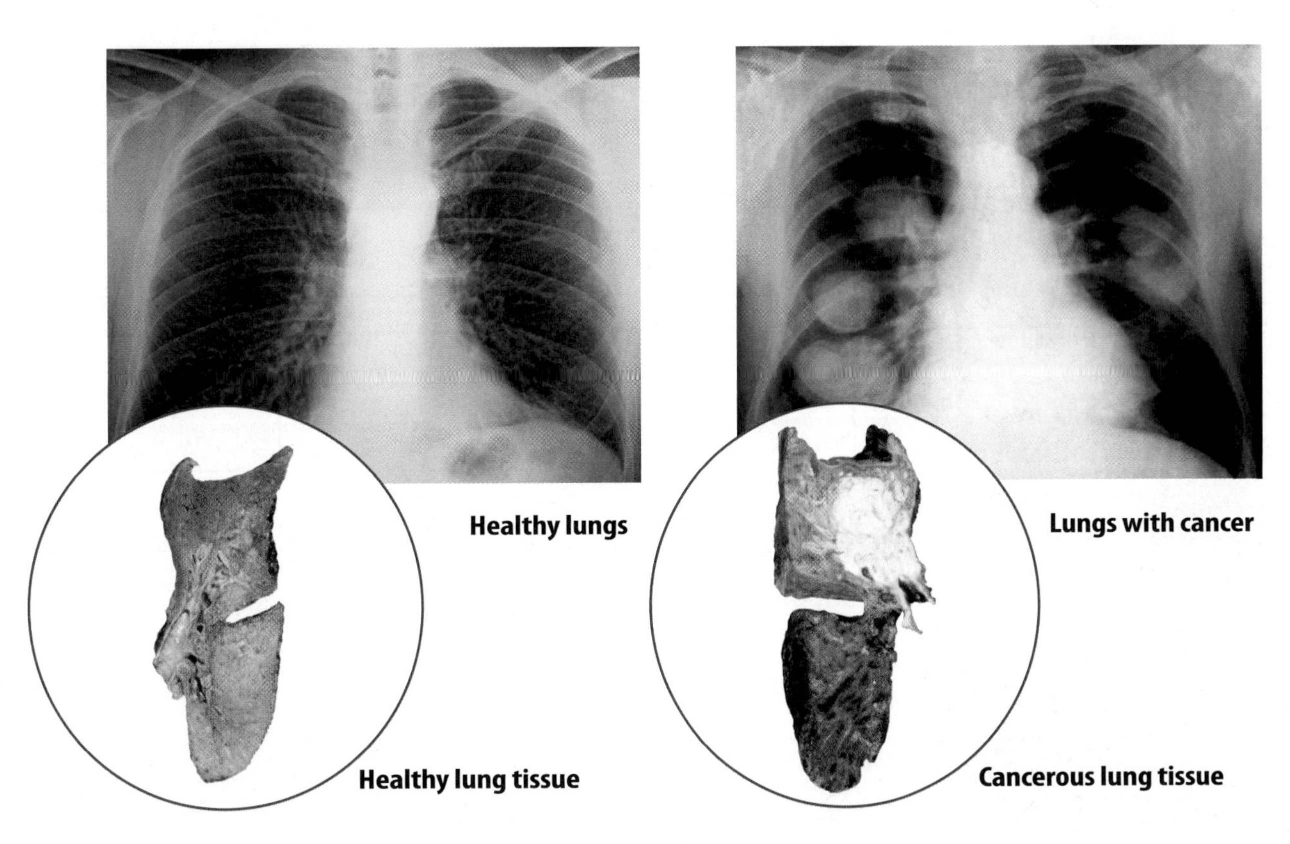

Lesson Review 1

Visual Summary

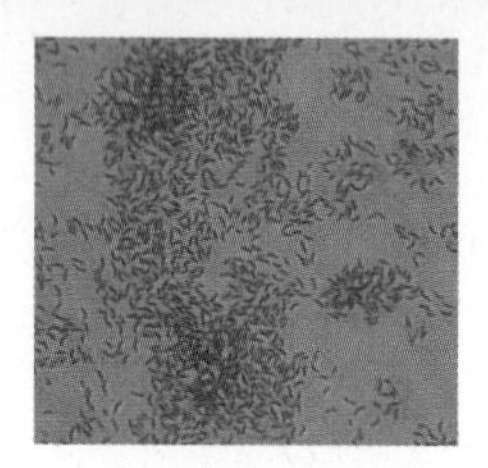

How a disease spreads depends on the pathogen. Some pathogens can be transmitted by a vector, such as a tick or a mosquito.

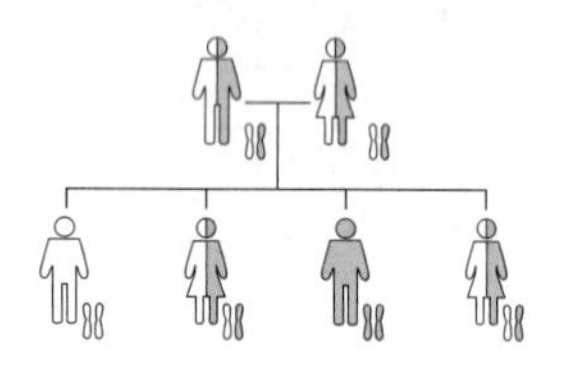

The two common causes of noninfectious diseases are environmental conditions and genetics.

People might inherit forms of genes that make them more likely to develop cancer.

Use Vocabulary

1. **Define** the term *pathogen* in your own words.

2. The process of boiling a liquid to kill bacteria and sealing it so bacteria cannot enter is called ______.

3. A host that transmits a pathogen but does not develop the disease is called a(n) ______.

Understand Key Concepts

4. **List** two main causes of disease. SC.6.L.14.6

5. Which is NOT a pathogen?
   - (A) bacterium
   - (B) fungus
   - (C) vector
   - (D) virus

Interpret Graphics

6. **Summarize** Fill in the table below using these terms: *viruses, bacteria, unhealthful diet, smoking, protists, gene forms, fungus.* LA.6.2.2.3, SC.6.L.14.6

Heredity	Environmental Conditions	Pathogens

Critical Thinking

7. **Support** the claim that genetics and environmental conditions can both contribute to a disease. HE.6.C.1.8

1. **Identify** ways in which pathogens can be transmitted.

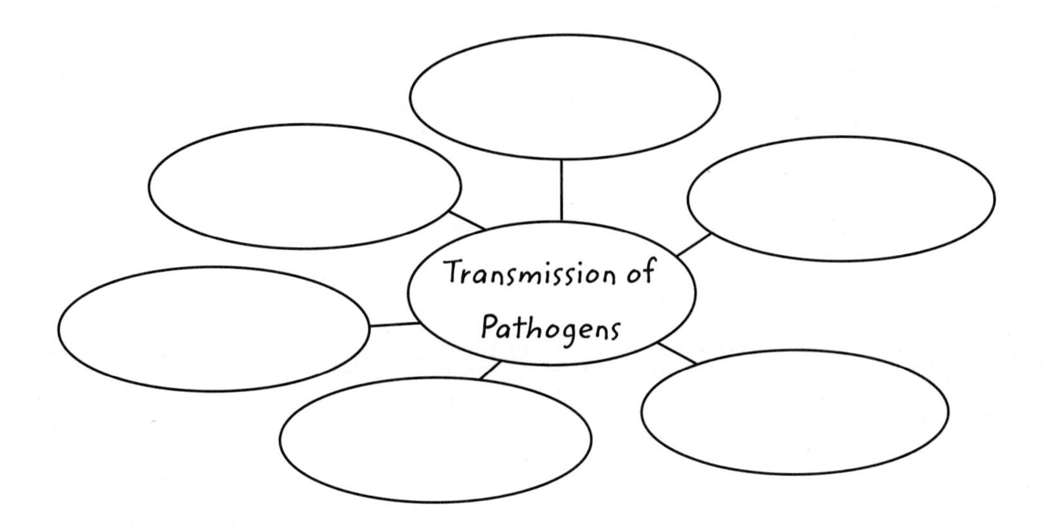

2. **Explain** the role of a vector in the transmission of infectious disease.

Compare types and causes of disease.

Type of Disease	Method of Transmission	Cause	Examples
Infectious	**3.**	direct contact	**4.**
		5. exchange of	**6.**
	vector	**7.**	**8.**
Noninfectious	**9.**	**10.** DNA from	**11.**
	environmental	lifestyle choices	**12.**

Lesson 2

The Immune SYSTEM

ESSENTIAL QUESTIONS

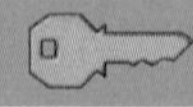
What does the immune system do?

How do the parts of the immune system work together?

How does the immune system interact with other body systems?

Vocabulary

inflammation p. 591
antigen p. 592
antibody p. 592
B cell p. 592
T cell p. 592
allergy p. 592
immunity p. 593
active immunity p. 593
vaccination p. 593
passive immunity p. 593

Florida NGSSS

MA.6.A.3.6 Construct and analyze tables, graphs, and equations to describe linear functions and other simple relations using both common language and algebraic notation.

SC.6.N.1.5 Recognize that science involves creativity, not just in designing experiments, but also in creating explanations that fit evidence.

SC.6.L.14.5 Identify and investigate the general functions of the major systems of the human body (digestive, respiratory, circulatory, reproductive, excretory, immune, nervous, and musculoskeletal) and describe ways these systems interact with each other to maintain homeostasis.

SC.8.N.3.1 Select models useful in relating the results of their own investigations.

HE.6.C.1.8 Explain how body systems are impacted by hereditary factors and infectious agents.

Also covers: LA.6.2.2.3, LA.6.4.2.2

SC.6.N.1.5, SC.6.L.14.5

Inquiry Launch Lab

10 minutes

Can you escape the pox?

The loffpox disease is an imaginary disease. How might it affect you?

Procedure

1. Your teacher will give you one of **three cards:** *healthy, in poor health,* or *pox.* Only one person will be given the *pox* card. Do not tell anyone which card you receive.
2. As you stand in a circle looking at each other, the person with the *pox* card will wink at the other students. If he or she winks at you and you have an *in poor health* card, you have caught the disease and you must sit down. If you have a *healthy* card, you do not catch the disease and you remain standing. However, if the person with the *pox* card winks at you a second time, you must sit down.

Think About This

1. Who is left standing?

2. Key Concept How does a person's state of health affect the pox disease? Why do you think it took the pox more than one wink to infect a healthy person?

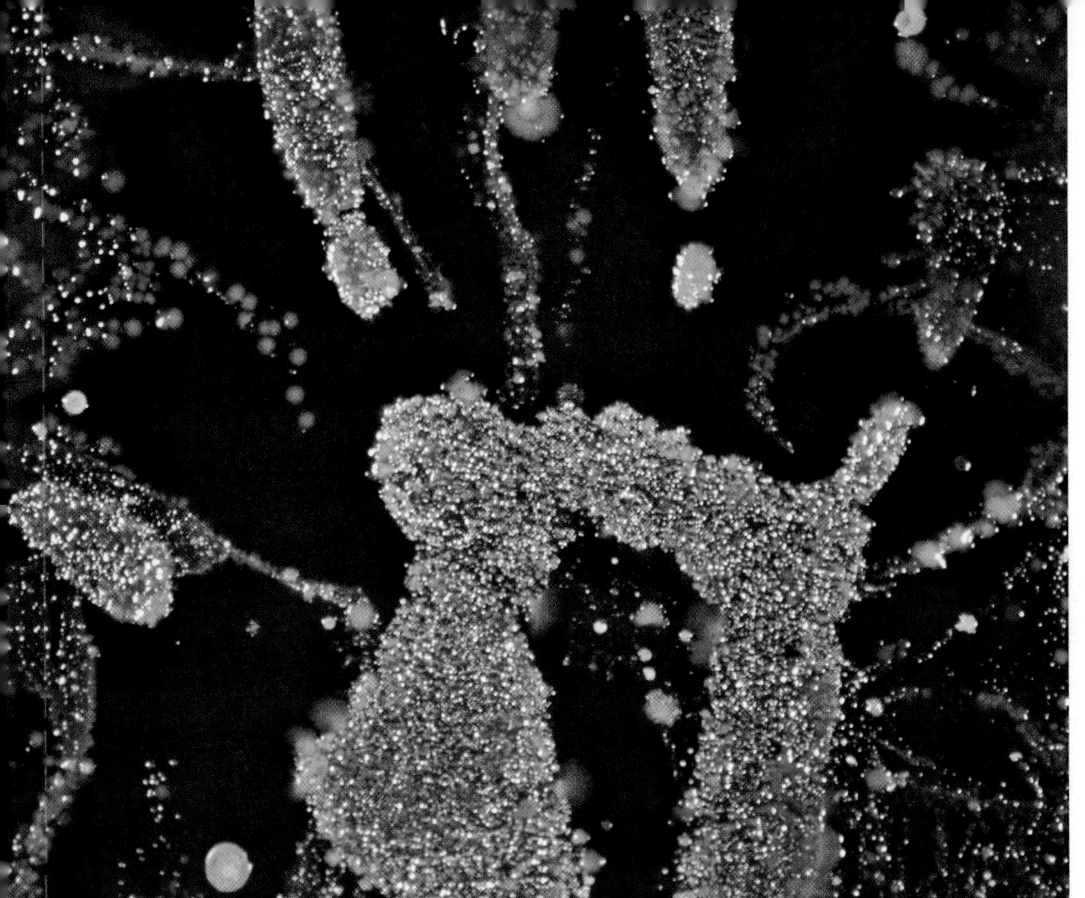

Inquiry Mysterious blobs?

1. The large yellow blobs you see are bacteria. The bacteria grew after a human hand touched the red agar plate. With all that bacteria on your hand, what keeps you from getting sick? How does the body protect itself?

Functions of the Immune System

Your body is constantly exposed to different pathogens. In Lesson 1 you read that disease-causing agents, such as bacteria and viruses, are pathogens. Pathogens also include fungi and protists. Pathogens are in the air, on objects, and in water. Like a spacesuit protects an astronaut, your immune system works to protect your body. There are many barriers to keep pathogens from entering your body.

Sometimes pathogens get past your body's initial barriers. When this happens, your immune system also has defenses to stop any pathogens that get past the barriers. For example, there are cells in your body that can destroy the pathogens. The immune system interacts with other body systems and helps keep you healthy, even as the environment outside your body changes.

2. **NGSSS Check** Generalize Highlight what the immune system does. SC.6.L.14.5

You can improve the effectiveness of these prevention methods by making healthful choices every day. Choices such as eating healthful food, getting enough sleep, exercising regularly, and using sunscreen support your immune system. As you read about the parts of the immune system, consider how the choices you make every day could affect how well your immune system functions.

Active Reading **FOLDABLES®** LA.6.2.2.3

Fold two sheets of paper into a layered book. Label it as shown. Use it to organize information about the immune system's lines of defense.

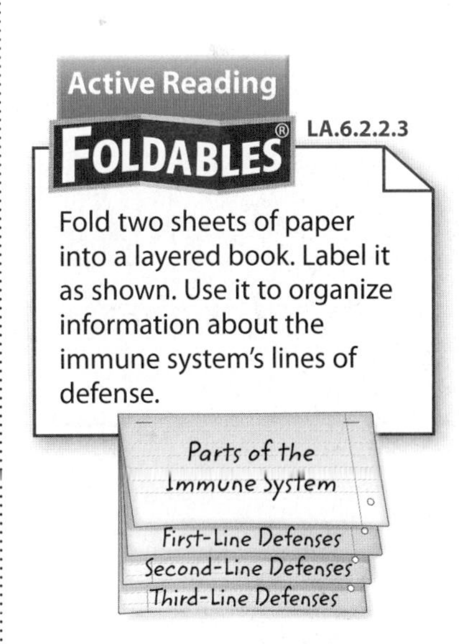

Parts of the Immune System

Different parts of your body work together to keep pathogens from making you sick. The integumentary system (skin), the respiratory system, the circulatory system, the digestive system, and the nervous system all work with the immune system to protect you against disease.

First-Line Defenses

Keeping germs from reaching the parts of your body where they can make you sick is the function of first-line defenses. Skin, hair, mucus, and acids are first-line defenses. They are effective against many types of pathogens. An immune defense that protects against more than one type of pathogen is a nonspecific defense.

Skin Often, the first nonspecific defense that protects you from pathogens is your skin, as shown in **Figure 7.** Your skin keeps dirt and germs from entering your body. Sweat and acids from skin cells kill some bacteria. Natural oils make skin waterproof so you can easily wash it.

You encounter pathogens every day, but your skin stops most of them from entering your body. Pathogens, such as cold and flu viruses, can survive for short periods on objects such as doorknobs or telephones. When you touch these objects, the pathogens can be transferred to your hand. If they reach your mouth, nose, eyes, or a cut, they can enter your body. Washing your hands often with soap and water easily removes most pathogens from your skin.

Your skin protects you from other dangers. It forms a chemical called melanin that protects you from the Sun's ultraviolet (UV) rays. Nerve endings in your skin can help you sense the warmth of a stove or the sharpness of a pin to protect you from injury.

3. Reason Why is your skin considered a first-line defense?

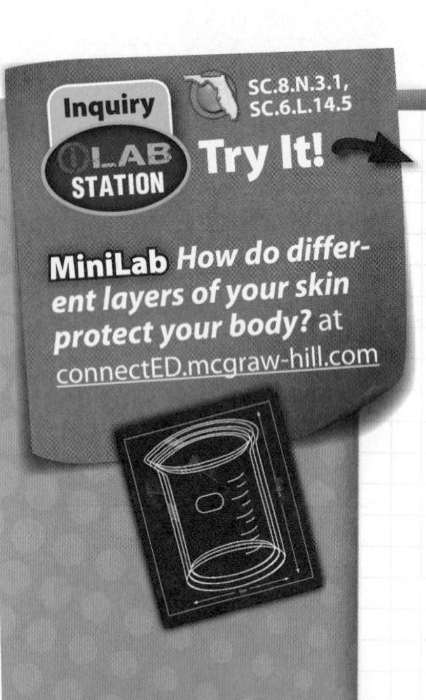

MiniLab ***How do different layers of your skin protect your body?*** at connectED.mcgraw-hill.com

Apply It! After you complete the lab, answer this question.

1. How does putting a bandage on a cut help reduce the risk of infection?

Respiratory System You can inhale pathogens from the air through your nose or mouth. The hairs in your nose help protect you by trapping dirt and pathogens. This keeps them from reaching the rest of your respiratory system. Small hairlike structures called cilia, shown in **Figure 7,** also trap pathogens and move them up and out of the upper respiratory system. If pathogens get past the cilia, they might encounter mucus. Mucus traps pathogens and enables your respiratory system to remove them by coughing, sneezing, or swallowing.

Digestive System Pathogens can enter your digestive system on or in the food you eat. The digestive system is effective at stopping bacteria from making you seriously ill. The stomach, also shown in **Figure 7,** contains strong acids. Stomach acids destroy many pathogens. Like the mucus in the respiratory system, mucus in the digestive system traps disease-causing bacteria and viruses, too.

Sometimes when you feel nauseated, it is actually your immune system clearing your body of pathogens. When disease-causing bacteria are not destroyed by stomach acids, your digestive system can reverse the usual direction of muscle contractions, and you vomit. Other times, muscle contractions speed up, and pathogens are removed through diarrhea.

Active Reading **4. Cite** Underline ways the digestive system helps defend against pathogens.

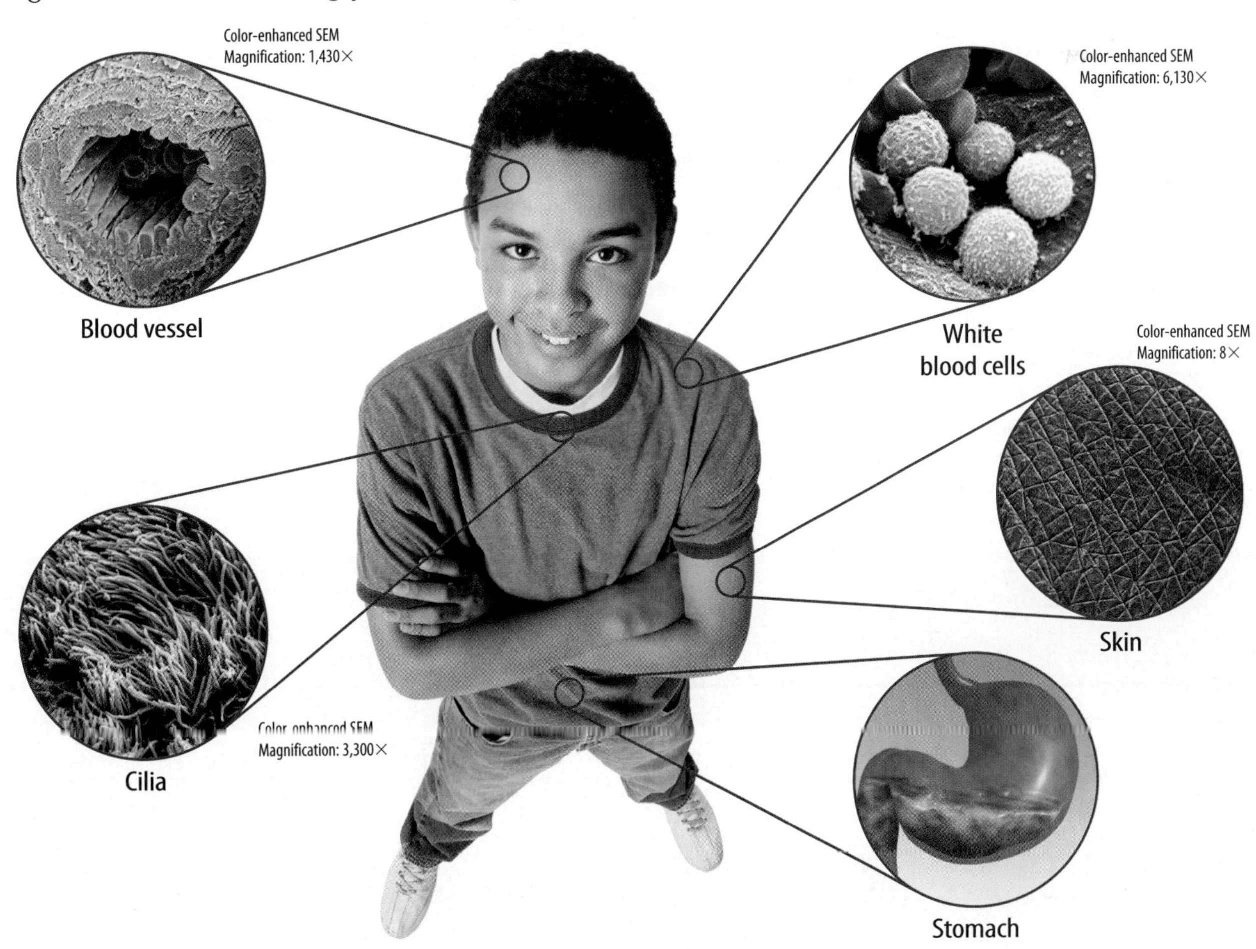

Figure 7 The skin, the respiratory system, the digestive system, and the circulatory system all support the immune system to provide the first line of protection against disease-causing pathogens. Blood vessels and white blood cells throughout the body help protect from pathogens.

5. Visual Check Determine How does the respiratory system trap pathogens?

Circulatory System and Nervous System Your circulatory system also protects you from pathogens. Pathogens can be moved through the circulatory system to organs that fight infection. The nervous system and the circulatory system also work together and increase the body's temperature to fight pathogens more effectively. Certain foreign substances trigger the brain to increase body temperature. When this **occurs,** blood vessels narrow and a fever develops. Many pathogens cannot survive at this higher temperature. For those that do survive, the fever brings another line of defense. The fever also stimulates white blood cells, which are part of the second-line defenses against pathogens.

ACADEMIC VOCABULARY

occur

(verb) to come into existence

Second-Line Defenses

Sometimes pathogens get past the defenses of the skin, the respiratory system, the digestive system, and the circulatory system. When they do, the next line of defense goes into action. Like the first-line defenses, second-line defenses are nonspecific, fighting against any type of pathogen.

White Blood Cells Recall that the spongy tissue in the center of your bones is called bone marrow. This is where white blood cells form. These cells attack pathogens. White blood cells flow through the circulatory system. However, they do most of their work attacking pathogens in the fluid outside blood vessels. They fight infection several different ways. Some white blood cells, such as the one shown in **Figure 8,** can surround and destroy bacteria directly. Others release chemicals that make it easier to kill the pathogens. Another type of white blood cell produces proteins that destroy viruses and other foreign substances that get past the first-line defenses.

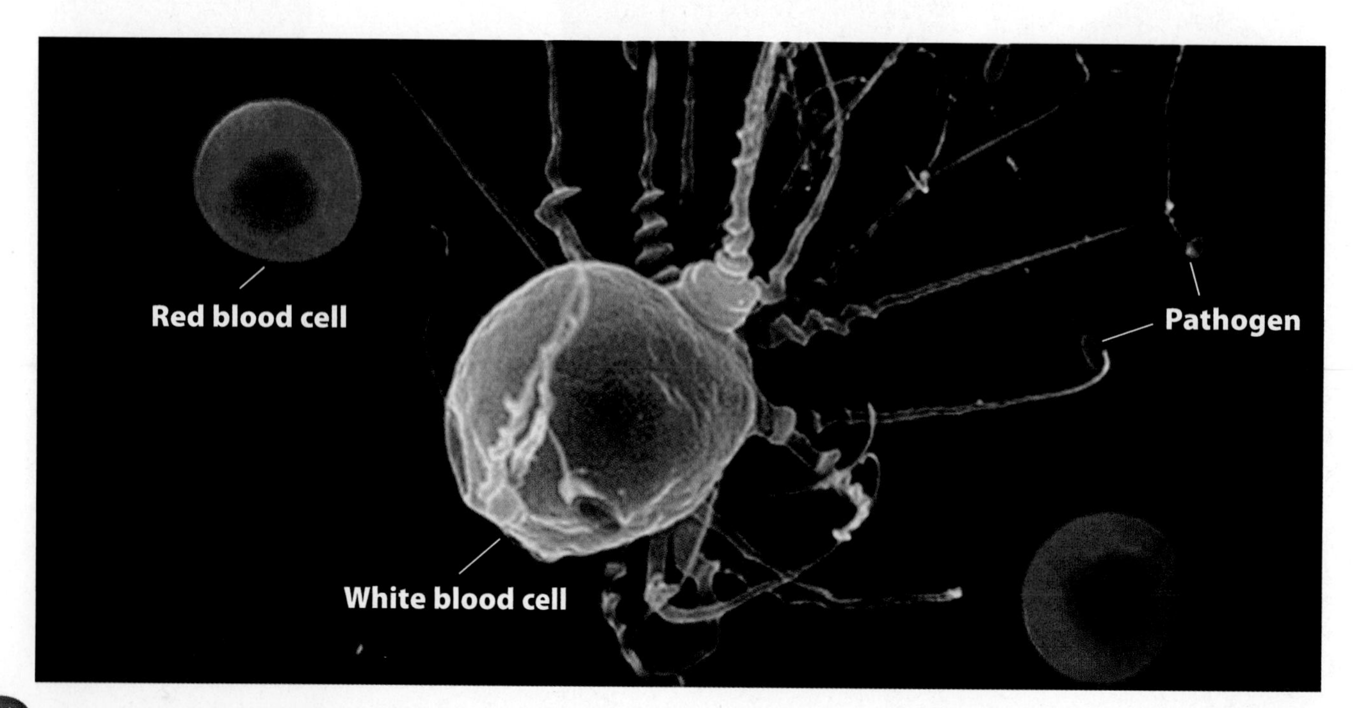

Figure 8 White blood cells fight pathogens that get past the first-line defenses. This white blood cell can digest pathogens and damaged cells.

The Inflammatory Response

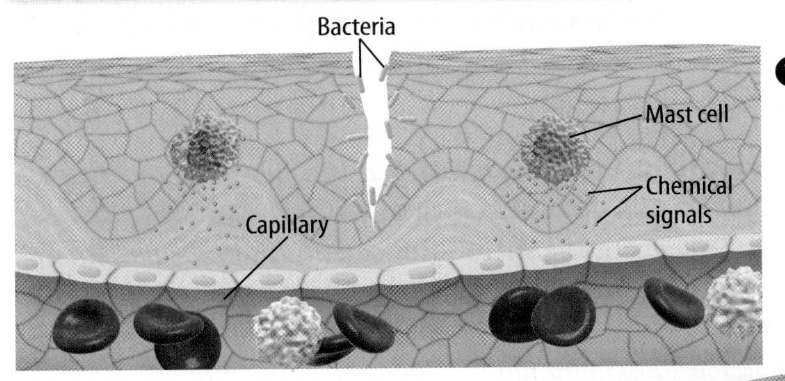

1 When skin is torn or cut, the damaged tissue triggers the inflammatory response. Special cells called mast cells release chemical signals into the surrounding tissue.

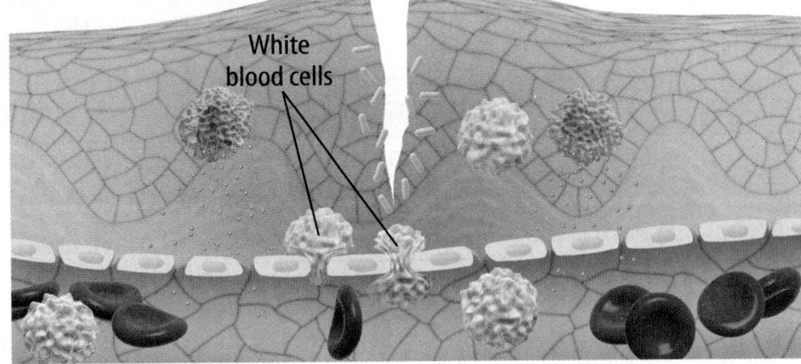

2 Some chemical signals attract more white blood cells to the area. Other chemical signals cause the capillary to dilate. White blood cells and plasma leak into the tissue, causing swelling.

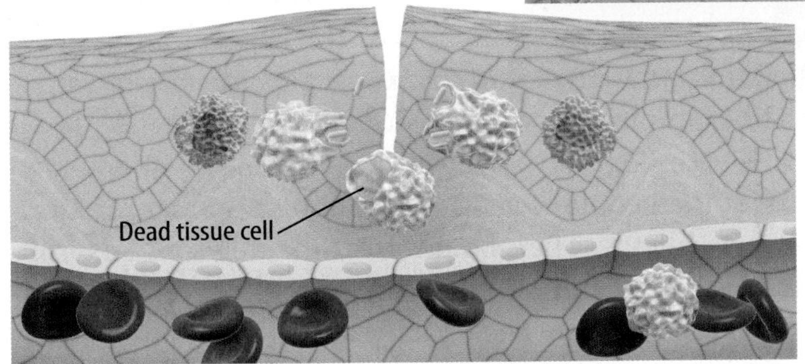

3 The white blood cells surround and take in the bacteria and any dead cells. As the tissue heals, the white blood cells and the plasma flow back into the capillary, and swelling decreases. The area returns to normal.

Inflammatory Response When you have an injury, your body produces an inflammatory response, causing inflammation. **Inflammation** *is a process that causes the area to become red and swollen.* If the injury is to the surface of the skin, you can observe the inflammatory response, as shown in **Figure 9.** First, damaged cells release a protein that signals the capillaries to dilate, or widen. Blood flow to the area increases, and the injury site becomes red and warmer than the surrounding area. Second, plasma and white blood cells leak into the area, causing swelling. Third, the white blood cells break down damaged cells and destroy any bacteria that might have entered the wound. The inflammatory response cleans the injured area and keeps the infection from spreading. The inflammation enables the damaged tissue to heal.

Active Reading **6. Summarize** Underline the text that describes inflammatory process.

Figure 9 Inflammation is another nonspecific response to pathogens.

WORD ORIGIN

inflammation
from Latin *inflammare,* means "to set on fire"

Third-Line Defenses

If first- and second-line defenses do not destroy all invading pathogens, another type of immune response occurs. Third-line defenses are specific to foreign substances. Often the three lines of defense work together.

Antigens and Antibodies *An* **antigen** *is a substance that causes an immune response.* An antigen can be on the surface of a pathogen. *Proteins called* **antibodies** *can attach to the antigen and make it useless.* Certain white blood cells, called B cells and T cells, form antibodies. **B cells** *form and mature in the bone marrow and secrete antibodies into the blood.* **T cells** *form in the bone marrow and mature in the thymus gland. They produce a protein antibody that becomes part of a cell membrane.* Antibodies match with specific antigens, as shown in **Figure 10.** Once your body has developed antibodies to an antigen, it can respond rapidly when the same pathogen invades your body again. This information is stored in antibodies on white blood cells called memory B cells.

Active Reading **7. Compile** Highlight how the parts of the immune system work together.

An **allergy** *is an overly sensitive immune response to common antigens.* For example, most people do not produce antibodies to the proteins in dog saliva. However, the antigens in dog saliva cause some people to have an immune response. These people have an allergy. Their bodies treat the dog saliva as if it were a pathogen. Inflammation and increased mucous production are common immune responses for people with allergies.

Figure 10 Antibodies are produced as a result of a specific immune response to particular antigens on pathogens.

Active Reading **8. Construct** Fill in the missing blanks below.

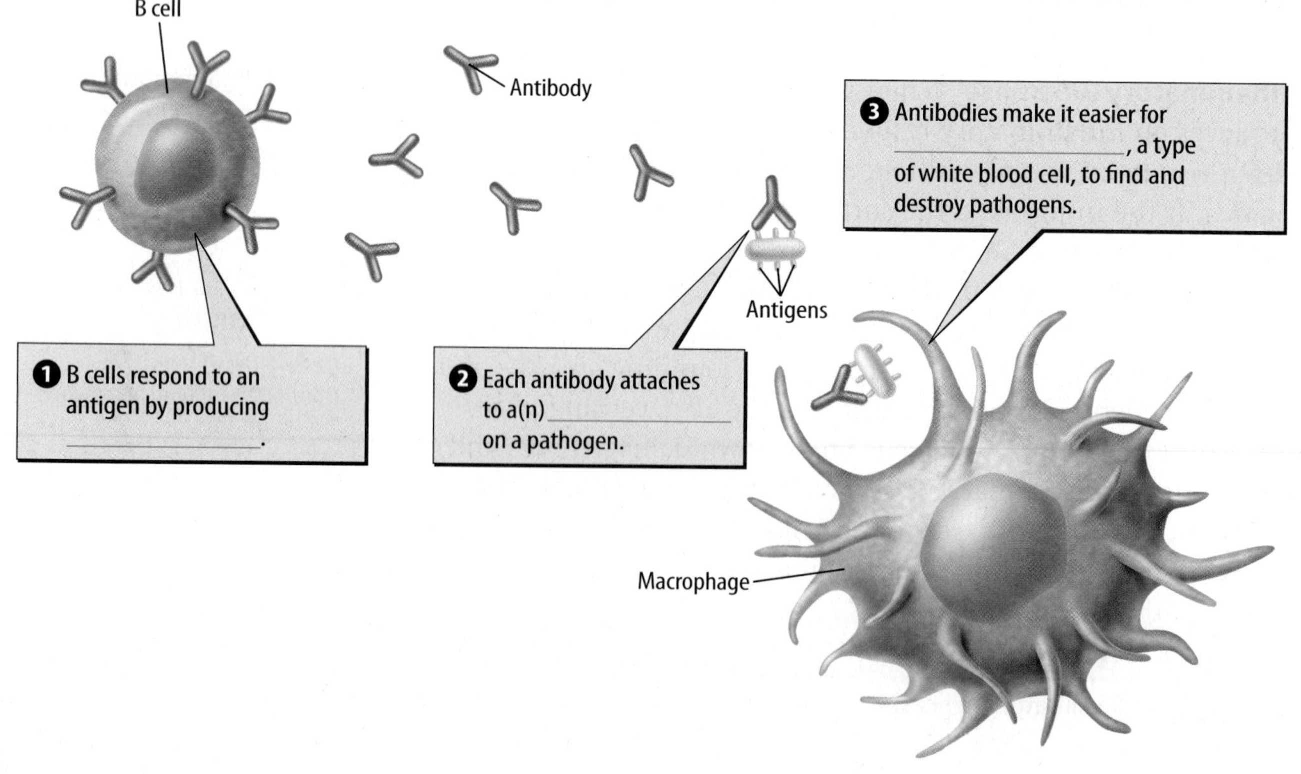

Immunity

The resistance to specific pathogens is **immunity.** There are two types of immunity—active immunity and passive immunity.

Active Immunity *Your body produces antibodies in response to an antigen in* **active immunity.** Your body recognizes the antigen, and the matching antibodies respond quickly. You can develop active immunity through illness or infection. After an illness or an infection is over, antibodies remain in your body. Because of this, you usually get certain diseases, such as chicken pox, only once. However, you can catch a cold many times because many different cold viruses cause similar symptoms.

You can also develop antibodies if you are exposed to an antigen through a vaccination. *A* **vaccination** *is weakened or dead pathogens placed in the body, usually by injection or by mouth.* A vaccination causes the body to develop specific antibodies that can rapidly fight a pathogen's antigens when exposed to them. **Table 1** lists the effects of vaccinations on the average annual number of cases of some diseases.

Passive Immunity You can also become resistant to specific antigens through passive immunity. **Passive immunity** *is the introduction of antibodies that were produced outside the body.* A fetus can get antibodies from its mother. Injections of some antibodies are available for adults. Passive immunity is temporary—the body does not continue to make these antibodies.

The Immune System and Homeostasis

You are exposed to many different pathogens every day. The immune system works to maintain your body's homeostasis. Body systems, including the circulatory system and respiratory system, work together and protect against invaders.

10. NGSSS Check **Construct** Underline why the immune system interacts with other body systems. SC.6.L.14.5

Math Skills MA.6.A.3.6

Use Percentages

Table 1 shows that the number of tetanus cases fell from 1,300 cases before the vaccine was developed to 34 cases after the vaccine was developed. What percent change does this represent?

Subtract the starting value from the final value.

$$34 - 1{,}300 = -1{,}266$$

Note that the value is negative because the number of cases decreased.

Divide the difference by the starting value.

$$-\frac{1{,}266}{1{,}300} = -0.974$$

Multiply the answer by 100 and add a percent sign.

$$-0.974 \times 100 = -97.4\%$$

Practice

9. Based on the data in **Table 1,** what was the percent change in the cases of polio after the vaccine was developed?

Table 1 Effects of Vaccinations

Disease	Cases Before Vaccination (annual average in the United States)	Year Vaccination Was Developed	Cases After Vaccination (annual average in the United States)
Tetanus	1,300	1927	34
Polio	18,000	1955/1962	8
Measles	425,000	1963	90
Mumps	200,000	1967	610
Rubella	48,000	1970	345

Lesson Review 2

Visual Summary

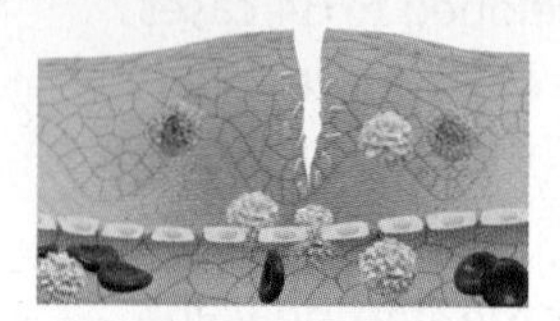

Inflammation may cause an injury to become warmer than the surrounding area due to increased blood flow to the area.

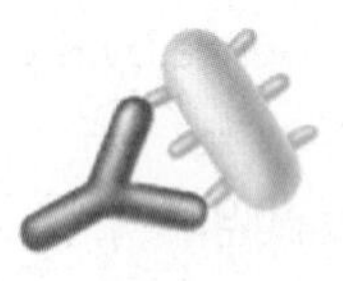

Antibodies produced by the white blood cells match with specific antigens, like a lock and key.

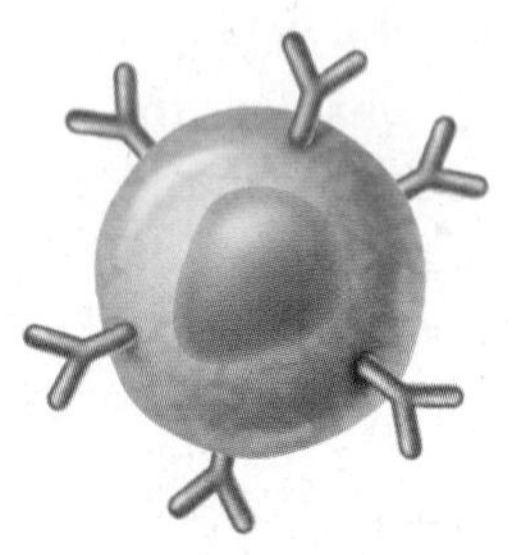

Immunity can be developed through different processes. If you have immunity to a particular pathogen, you will experience little or no effect from exposure to that pathogen.

Use Vocabulary

1. **Distinguish** between active immunity and passive immunity.

____

____

2. The _______________ response includes swelling and heat near injured tissue.

3. **Define** the term *vaccination* in your own words.

____

Understand Key Concepts

4. **List** three body systems that work with the immune system to form first-line defenses. **SC.6.L.14.5**

____

5. Which is a first-line defense? **SC.6.L.14.5**

Ⓐ antibody
Ⓑ hormone
Ⓒ inflammation
Ⓓ skin

6. **Explain** why antibodies are considered specific responses to pathogens. **SC.6.L.14.5**

____

____

Interpret Graphics

7. **Summarize** Fill in the graphic organizer below to summarize the steps in the inflammatory response. **LA.6.2.2.3**

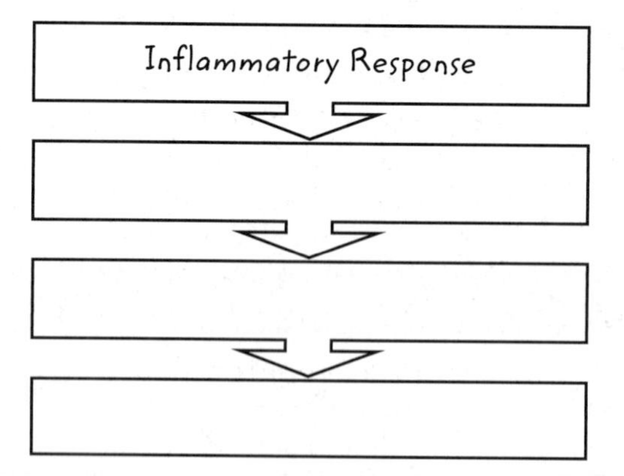

Critical Thinking

8. **Predict** what might happen to a person who had very few of the type of cell shown at right. **SC.6.L.14.5**

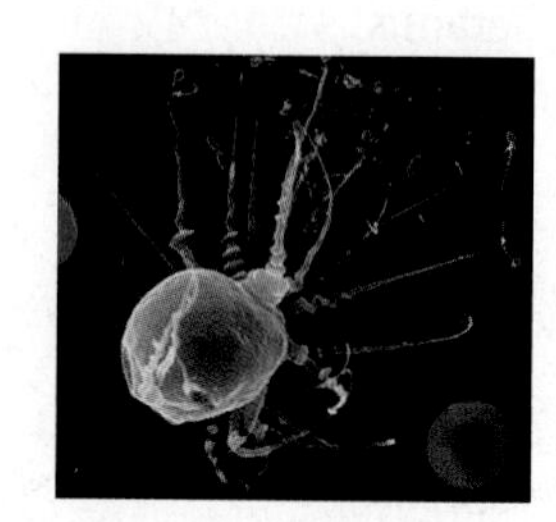

____

____

____

SCIENCE & SOCIETY

The Victory over Chicken Pox

Vaccines are helping win the war against viruses.

Until recently, having chicken pox was just a part of growing up. The disease commonly occurs in children, usually before the age of 15.

Chicken pox is highly contagious, passing easily from person to person. The chicken pox virus, varicella-zoster, produces a rash of red spots and small blisters that can appear all over the body. The fluid-filled blisters crust over and become very itchy. Within a few days, the spots and blisters disappear. It takes about one week for the disease to run its course.

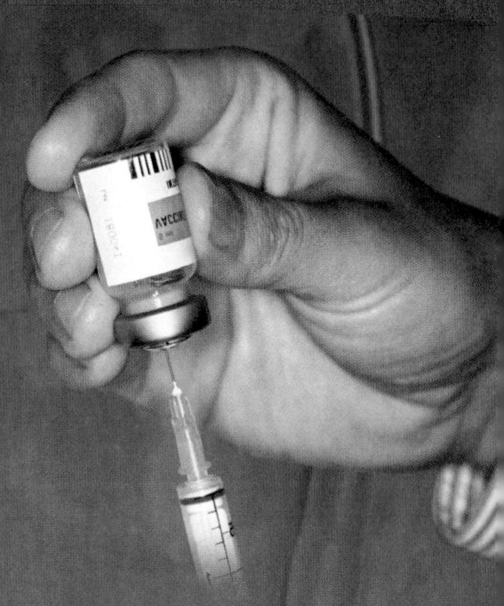

▲ **The chicken pox vaccine is 80 to 90 percent effective in preventing the disease.**

◄ **The chicken pox rash usually appears on the face, chest, and back.**

Chicken pox is contagious beginning 1–2 days before the rash appears until all the blisters form scabs. During this period, the virus can be spread through direct contact with the sores or through the air. Those who have not had chicken pox can become infected easily.

In 1995, a chicken pox vaccine was introduced in the United States. Since then, the number of cases of chicken pox reported each year has declined sharply. It is estimated that almost 80 percent of young children develop immunity after one dose of vaccine. More than 90 percent of older children and adults develop immunity after the second dose.

◄ **Varicella-zoster virus**

It's Your Turn

REPORT Identify three diseases that were common 50 years ago but are not common today due to vaccines. Write a report about the diseases, when vaccines were developed against them, and how the incidences of the diseases changed. LA.6.4.2.2

Lesson 3

Staying HEALTHY

ESSENTIAL QUESTIONS

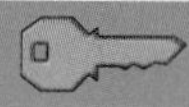 How can healthful habits and healthful choices affect diseases?

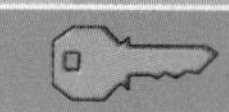 How do sanitation practices affect human health?

 How can chemicals affect the human body?

Vocabulary

antibiotic p. 600

chemotherapy p. 600

Florida NGSSS

LA.6.2.2.3 The student will organize information to show understanding (e.g., representing main ideas within text through charting, mapping, paraphrasing, summarizing, or comparing/contrasting);

SC.6.L.14.5 Identify and investigate the general functions of the major systems of the human body (digestive, respiratory, circulatory, reproductive, excretory, immune, nervous, and musculoskeletal) and describe ways these systems interact with each other to maintain homeostasis.

SC.6.N.1.1 Define a problem from the sixth grade curriculum, use appropriate reference materials to support scientific understanding, plan and carry out scientific investigation of various types, such as systematic observations or experiments, identify variables, collect and organize data, interpret data in charts, tables, and graphics, analyze information, make predictions, and defend conclusions.

SC.6.N.1.5 Recognize that science involves creativity, not just in designing experiments, but also in creating explanations that fit evidence.

SC.7.N.1.3 Distinguish between an experiment (which must involve the identification and control of variables) and other forms of scientific investigation and explain that not all scientific knowledge is derived from experimentation.

SC.7.N.1.4 Identify test variables (independent variables) and outcome variables.

SC.6.N.1.5

Launch Lab

10 minutes

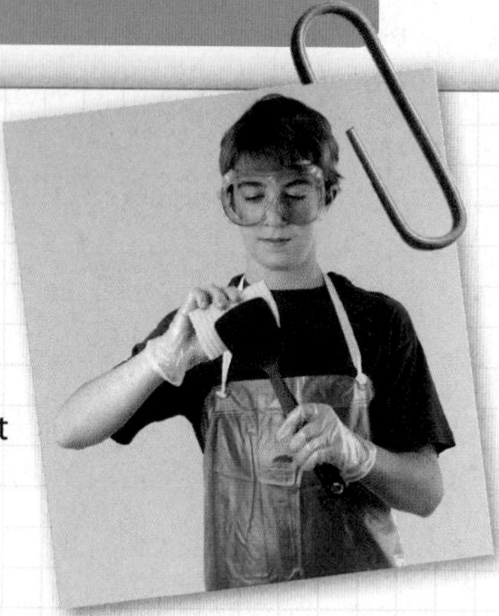

Where might bacteria be?

The people who work in the cafeteria in your school probably wear aprons, hairnets, and gloves. They might also wash dishes and wipe the counters with a bleach solution to kill bacteria. Why do they take such precautions?

Procedure

1. Read and complete a lab safety form.
2. Go to your assigned station where surfaces, **utensils,** and **other objects** representing cookware have been coated with a "bacterial" solution of invisible **fluorescent detergent.**
3. Using a **cloth** and a **cleaning solution containing bleach,** clean the objects at your station as well as you can in 5 min.

⚠ Be careful with the bleach solution; it will stain your clothes.

4. When you are finished, your teacher will examine your station with a **black light,** which will show any remaining "bacteria."

Think About This

1. How well did you clean your station? Did you miss any bacteria?

2. How could you change your cleaning methods to clean all surfaces and objects of bacteria?

3. Key Concept Why do you think it is important to keep a kitchen clean, especially in a school or a restaurant?

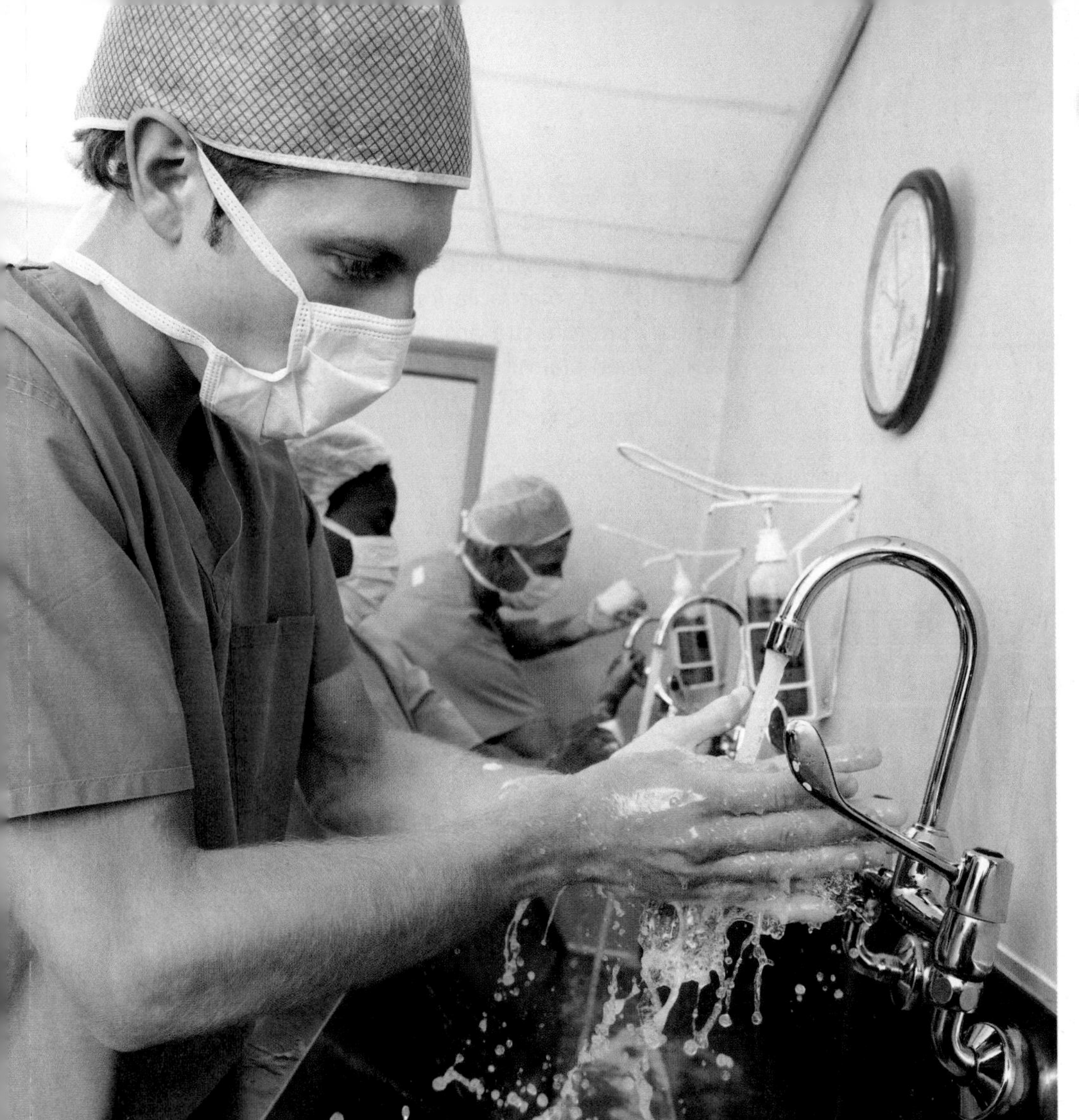

Why Wash?

1. Why do surgeons wash their hands before an operation even though they wear gloves? Can hand washing keep you healthy?

Healthful Habits

Imagine you are sitting in class, and the person next to you sneezes. Fortunately, she covers her nose and mouth with her hand. After sneezing, she picks up a pencil. Just then, you realize you need to borrow a pencil. She hands you her pencil. Will you get her cold? What could you do to make that less likely?

Pathogens passed from person to person make infectious diseases such as colds and flu very common. Personal hygiene can limit the spread of these pathogens. For example, good hygiene includes using a tissue or handkerchief when you sneeze and then washing your hands. This lessens the chance you will spread your germs to others. Good hygiene can protect you from getting an infectious disease, too.

Pathogens are less likely to get past your first-line defenses if you wash your hands before you eat and avoid putting objects, such as pencils, in your mouth. Why do you think surgeons scrub their hands, as shown above, even though they wear gloves during surgical procedures?

Active Reading

FOLDABLES® LA.6.2.2.3

Make a vertical half-book from a sheet of paper. Use it to record information about habits and choices that can help you stay healthy.

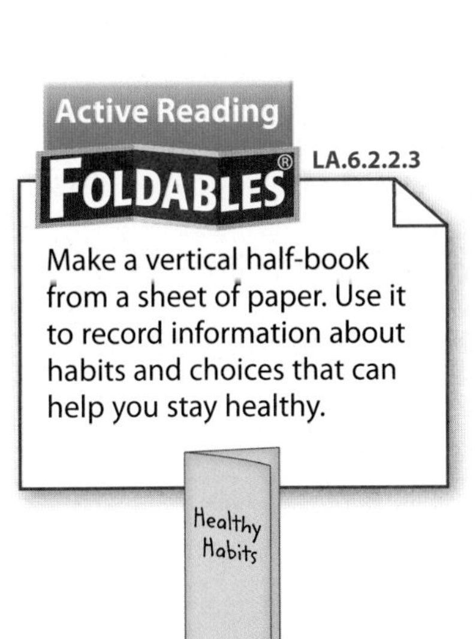

Figure 11 Because southern states such as Florida are located closer to the equator the amount of solar radiation received is more intense. Therefore sun protection should be a top priority when planning for a day at the beach.

2. Visual Check List Identify the healthful choices.

____

____

____

Healthful Choices

In addition to good personal hygiene, other everyday choices, like those shown in **Figure 11,** can help keep you healthy. Choices that affect environmental conditions can also protect you from many infectious and noninfectious diseases.

Diet Think about the foods you ate this week. If you eat a healthful diet, your immune system can react more efficiently against pathogens. A healthful diet can also protect you against noninfectious diseases such as osteoporosis and heart disease. A healthful diet, a healthful weight, and regular exercise have been linked with overall disease prevention.

Active Reading **3. Recommend** Underline ways that a healthful diet can protect against disease.

Sun Protection Skin cancer is a noninfectious disease. The ultraviolet (UV) rays from the Sun damage skin **cells** and can cause them to reproduce uncontrollably. Sunscreen blocks the UV rays and limits the damage from sunlight. Wearing a hat, long sleeves, pants, and sunglasses also helps protect you against UV damage.

Science Use v. Common Use
cell
Science Use the basic unit of life
Common Use a room in a monastery or prison

Alcohol and Tobacco Lung cancer is one of the most deadly cancers. Most cases of lung cancer are related to smoking or working in environments with poor air quality. Many other cancers are related to excessive alcohol use and to smoking. Healthful choices include not smoking or chewing tobacco and limiting or avoiding alcoholic beverages.

Active Reading **4. Explain** Highlight some healthful habits and choices that can help prevent disease.

Health and Sanitation

Improved cleanliness in schools, hospitals, and public areas has increased overall health in our communities. In the mid-1800s, hospitals were dirty, overcrowded places. Patients were rarely bathed, and linens were rarely washed. Pathogens caused infections in most patients. One nurse, Florence Nightingale, is credited with improving cleanliness in hospitals. She understood that there is a connection between cleanliness and health.

Figure 12 Sanitation has improved health by reducing exposure to pathogens.

Food Preparation

Improved sanitation in food preparation has also led to better health. Employees must wash their hands regularly, as indicated in **Figure 12,** and keep equipment clean. Inspections are performed regularly to catch problems early and protect consumers from most pathogens.

Waste Management

In the mid-1300s, there were no plumbing or sewer systems. People in European cities often dumped their personal waste and garbage in the streets. Today modern landfills and sewer systems keep our streets and households much cleaner. This cleanliness slows the spread of infectious diseases.

Active Reading **5. Justify** How do sanitation practices affect human health?

Apply It! After you complete the lab, answer this question.

1. How does washing your hands help improve health in places like your school and in other public areas?

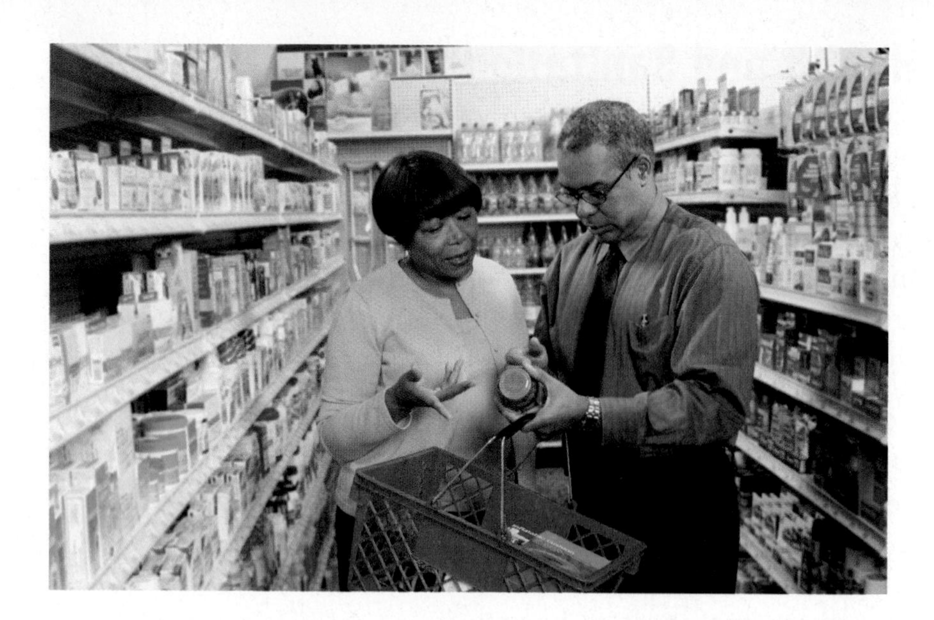

Figure 13 Many chemicals are beneficial to our health.

Health and Chemicals

Chemicals, like those shown in **Figure 13,** can be beneficial for people. Some chemicals, such as the ones in sunscreen that block UV rays, protect us from noninfectious diseases. Chemicals used to make vitamin supplements improve nutrition, which helps the immune system fight disease.

Word Origin

antibiotic

anti–, means "against"; and Greek *biotikos,* means "fit for life"

Other chemicals are used in medicines. **Antibiotics** *are medicines that stop the growth and reproduction of bacteria.* Chemicals are also used to destroy cancer cells. *These medicines, used in a type of treatment called* **chemotherapy**, *kill the cells that are reproducing uncontrollably.*

These and many other chemicals, such as paints and pesticides, might make our lives easier. But if they are not disposed of properly, they can harm our health. Some people choose to use chemicals that are harmful to their health. For instance, more than 50 of the chemicals in cigarettes have been linked to cancer.

Health and the Environment

Some chemicals that are harmful to our health, such as lead, are in our environment. Before 1978, lead was used in many paints. If the dried paint flaked, it released lead into the air. Inhaling lead-contaminated air can cause noninfectious kidney and nervous system diseases.

Some objects containing harmful chemicals are safe until the object is broken. For example, when ceiling and floor tiles containing asbestos are broken, asbestos fibers are released. People who are often exposed to such chemicals might develop cancer.

Active Reading **6. Determine** Highlight how chemicals can affect the human body.

Lesson Review 3

Visual Summary

Developing healthful habits and making healthful choices is one of the best ways you can stay healthy.

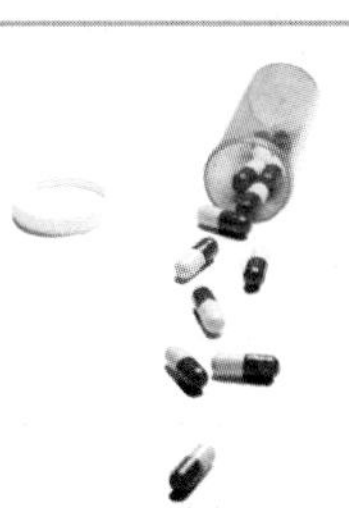

Chemicals are used in medicines such as antibiotics. These may be used for common bacterial infections, such as strep throat or ear infections.

Life choices, such as whether or not you eat a healthful diet, can influence the development and severity of diseases.

Use Vocabulary

1. A treatment that uses chemicals to kill cancer cells is called ____________________.

2. Medicines that kill bacteria are called ____________________.

Understand Key Concepts 

3. **List** two healthful life choices.

____

____

4. Which chemical can be harmful to your health?

   (A) antibiotic
   (B) asbestos
   (C) sunscreen
   (D) vitamin

Interpret Graphics

5. **Create** In the first column, list four chemicals or materials containing chemicals. In the second column, indicate whether the chemical is beneficial or harmful. In the third column, describe how the chemical benefits or harms people. LA.6.2.2.3

Chemical	Beneficial or Harmful?	How?

Critical Thinking

6. **Evaluate** your personal hygiene in your daily routine. What could you do to limit your exposure to pathogens?

____

____

____

Chapter 15 Study Guide

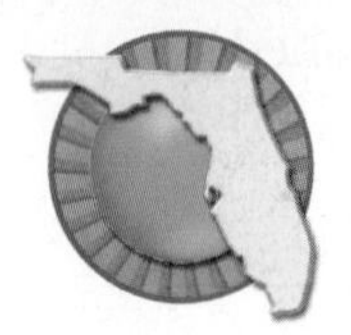

Think About It! The immune system maintains homeostasis by protecting the body against infections and diseases.

Key Concepts Summary

LESSON 1 Diseases

- Diseases can result from infection by **pathogens,** heredity, or the environment.
- **Infectious diseases** are caused by pathogens and are spread from an infected organism or the environment to another organism. **Noninfectious diseases** are not caused by pathogens and are not spread from one organism to another.

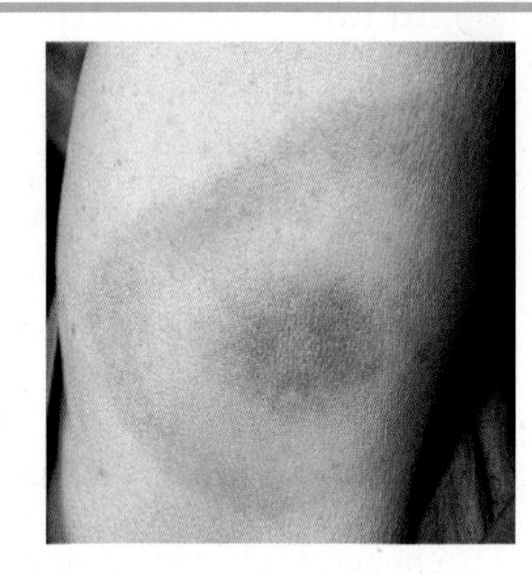

Vocabulary

pathogen p. 577
pasteurization p. 579
infectious disease p. 581
vector p. 581
noninfectious disease p. 582
cancer p. 583

LESSON 2 The Immune System

- The immune system protects against and defends the body from disease.
- Your body has first-line, second-line, and third-line defenses against pathogens.
- The immune system works with other body systems, including the circulatory system, the respiratory system, and the digestive system, to protect against invaders.

inflammation p. 591
antigen p. 592
antibody p. 592
B cell p. 592
T cell p. 592
allergy p. 592
immunity p. 593
active immunity p. 593
vaccination p. 593
passive immunity p. 593

LESSON 3 Staying Healthy

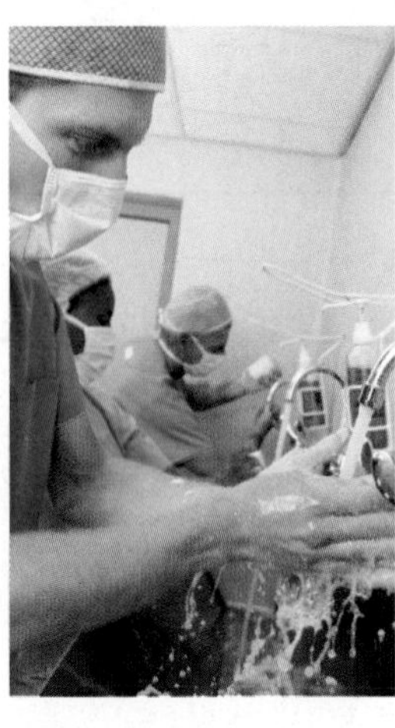

- Healthful habits, such as hand washing, can help prevent the spread of disease. Life choices, such as whether or not you eat a healthful diet or wear sunscreen, can influence the development and severity of diseases.
- Sanitation practices, such as safe food preparation and waste management, limit human exposure to pathogens and toxic substances.
- Chemicals can benefit human health when used as medicines, treatments for disease, and supplements. Some chemicals are harmful to human health and might cause diseases such as cancer.

antibiotic p. 600
chemotherapy p. 600

Active Reading

FOLDABLES® Chapter Project

Assemble your lesson Foldables as shown to make a Chapter Project. Use the project to review what you have learned in this chapter.

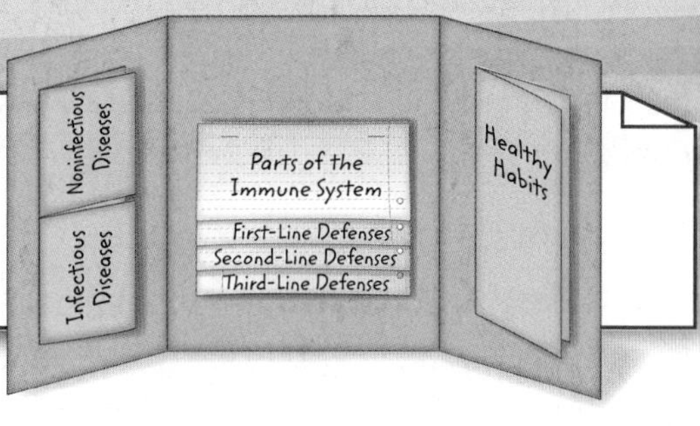

Use Vocabulary

1. A disease in which cells multiply uncontrollably is called ______.

2. Define *pasteurization* in your own words.

3. Distinguish between antibodies and antigens.

4. An overly sensitive immune response to common antigens is a(n) ______.

5. Use the term *antibiotics* in a sentence.

6. Differentiate between antibiotics and chemotherapy.

Link Vocabulary and Key Concepts

Use vocabulary terms from the previous page to complete the concept map.

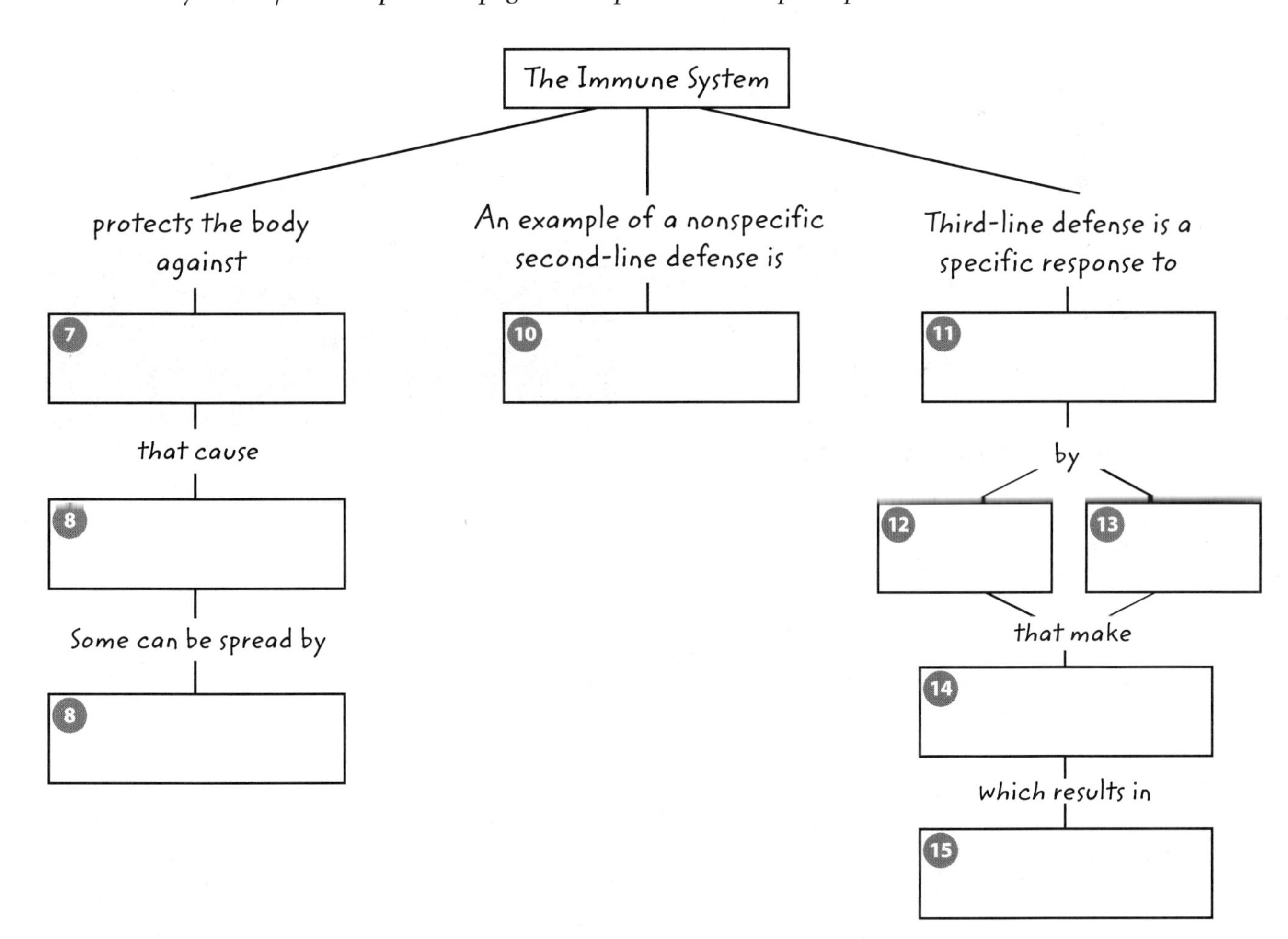

Chapter 15 Review

Fill in the correct answer choice.

Understand Key Concepts

1. Which part of the blood plays the most direct role in fighting pathogens? **SC.6.L.14.5**
   - Ⓐ plasma
   - Ⓑ platelets
   - Ⓒ red blood cells
   - Ⓓ white blood cells

2. What is illustrated below? **SC.6.L.14.6**

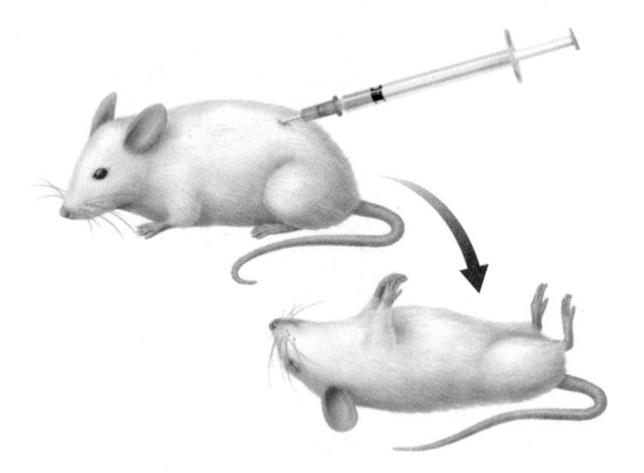

   - Ⓐ Bacteria can be treated with heat.
   - Ⓑ Bacteria do not infect other animals.
   - Ⓒ Bacteria cause disease in healthy animals.
   - Ⓓ One type of bacteria causes several different diseases.

3. Which has NOT led to improved health? **SC.7.N.1.4**
   - Ⓐ cleaner hospitals
   - Ⓑ lead paint
   - Ⓒ better waste management
   - Ⓓ cleaner food-preparation tools

4. Which chemical might be helpful to the human body? **SC.6.L.14.5**
   - Ⓐ asbestos
   - Ⓑ antibiotic
   - Ⓒ lead
   - Ⓓ pesticide

5. Which is NOT part of the nonspecific, first-line defenses against pathogens? **SC.6.L.14.5**
   - Ⓐ antibody
   - Ⓑ cilia
   - Ⓒ mucus
   - Ⓓ skin

Critical Thinking

6. **Compose** a letter to explain the causes and symptoms of cystic fibrosis to a friend of the family who is concerned about his or her child developing cystic fibrosis. **SC.6.L.14.5**

7. **Role-Play** Choose the most important researcher from among John Snow, Joseph Lister, Robert Koch, and Edward Jenner. Defend your choice. **SC.7.N.2.1**

8. **Evaluate** the following statement, using the data in the table below to support your conclusion: Every person has not been vaccinated for common diseases, such as tetanus, polio, measles, mumps, and rubella. **SC.7.N.2.1**

Effects of Vaccinations

Disease	Rate Before Vaccination	Rate After Vaccination Developed
Tetanus	1,300	34
Polio	18,000	8
Measles	425,000	90
Mumps	200,000	610
Rubella	48,000	345

9. **Categorize** these parts of the immune system as first-line, second-line, or third-line defenses: cilia, white blood cells, antibodies, skin, mucus, inflammation. SC.6.L.14.5

10. **Plan and implement** a survey to determine when and how often the students in your class wash their hands. SC.6.L.14.5

11. **Design and create** a poster to remind the students in your school to make healthful life choices. SC.6.L.14.5

Writing in Science

12. **Write** a paragraph on a separate piece of paper analyzing the differences in the causes of most childhood noninfectious diseases compared to other noninfectious diseases. LA.6.2.2.3

Big Idea Review

13. Explain how the immune system helps the body maintain homeostasis. SC.6.L.14.5

Math Skills MA.6.A.3.6

Use Percentages

14. The average annual number of rubella cases before the vaccine was developed was 48,000. There were 345 cases after the vaccine was developed. What percent change does this represent?

15. The average annual number of cases before vaccines were developed was 425,000 for measles and 200,000 for mumps. After the vaccines were developed, there were 90 cases of measles and 610 cases of mumps. Which vaccine was more effective in reducing cases of the disease?

Florida NGSSS Benchmark Practice

Fill in the correct answer choice.

Multiple Choice

1 Which would a doctor *exclude* as the cause of her patient's noninfectious disease? **SC.6.L.14.6**

Ⓐ environmental conditions

Ⓑ inherited traits

Ⓒ lifestyle choices

Ⓓ transmitted pathogens

Use the diagram below to answer question 2.

Immune Response

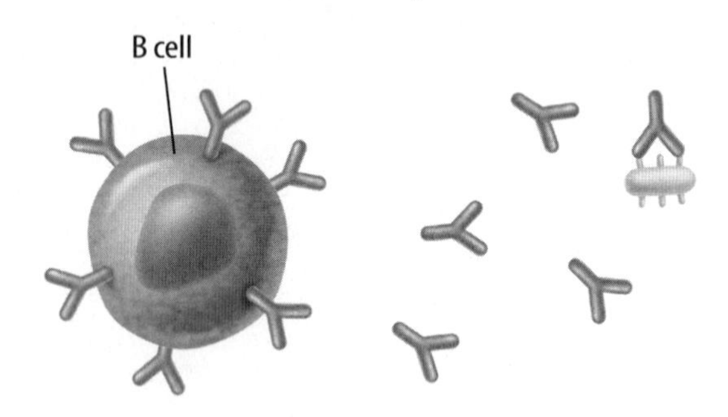

2 What are the Y-shaped objects in the diagram? **SC.6.L.14.6**

Ⓕ antibodies

Ⓖ antigens

Ⓗ bacteria

Ⓘ pathogens

3 Which first-line defense systems use acids to kill pathogens? **SC.6.L.14.5**

Ⓐ circulatory and respiratory

Ⓑ digestive and integumentary

Ⓒ nervous and circulatory

Ⓓ respiratory and digestive

4 Which is directly linked to increased risk of skin cancer in humans? **SC.6.L.14.5**

Ⓕ acid rain

Ⓖ asbestos insulation

Ⓗ sunlight exposure

Ⓘ water pollution

Use the diagram below to answer question 5.

Stage 1 — The Inflammatory Response

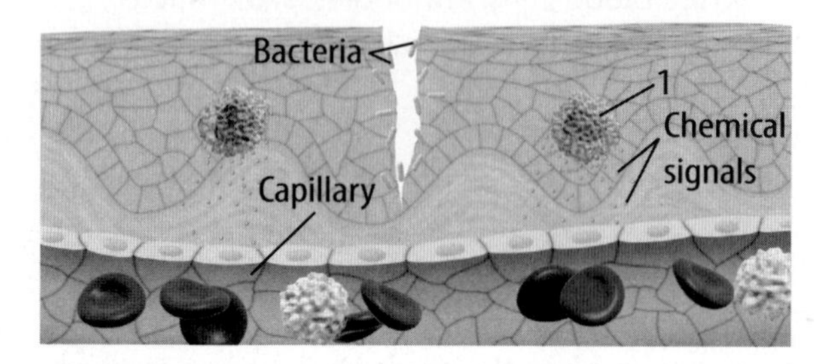

5 Which does *1* in the diagram represent? **SC.6.L.14.5**

Ⓐ host cell

Ⓑ mast cell

Ⓒ sheath cell

Ⓓ stem cell

6 In stage 2 of the inflammatory response, white blood cells and plasma leak into the area. What happens as a result? **SC.6.L.14.5**

Ⓕ bacteria destruction

Ⓖ capillary dilation

Ⓗ protein release

Ⓘ tissue swelling

7 How does sanitation improve health? **SC.6.L.14.5**

Ⓐ aids white blood cell production

Ⓑ increases vitamin absorption

Ⓒ reduces pathogen exposure

Ⓓ stimulates blood circulation

8 How do vitamin supplements contribute to health? **SC.6.L.14.5**

Ⓕ They block UV rays.

Ⓖ They improve nutrition.

Ⓗ They kill bacteria.

Ⓘ They kill cancer cells.

Use the table below to answer question 9.

Bacteria are present in ALL organisms with the disease but NOT in healthy organisms.
Bacteria must reproduce in the lab.
Sample of bacteria must cause disease in healthy animals.
Lab-grown pathogen is identical to original.

9 Which scientist developed the rules in the table above? **SC.7.N.2.1**

Ⓐ Koch

Ⓑ Lister

Ⓒ Pasteur

Ⓓ Snow

10 What kills bacteria in the pasteurization process? **SC.7.N.2.1**

Ⓕ antiseptic

Ⓖ heat

Ⓗ isolation

Ⓘ pressure

11 What is a disease-carrying organism that does NOT develop the disease? **SC.6.L.14.6**

Ⓐ antigen

Ⓑ B cell

Ⓒ T cell

Ⓓ vector

12 Inflammation is a common response to pathogens. Which would NOT be a common part of the inflammatory response at the site of an injury? **SC.6.L.14.5**

Ⓕ bruising

Ⓖ reddening

Ⓗ swelling

Ⓘ warmth

13 Why is the invention shown below considered one of the most important developments in early disease research? **SC.7.N.2.1**

Ⓐ It showed people how to pasteurize liquids.

Ⓑ It showed people that vaccines were effective.

Ⓒ It showed people that microorganisms existed outside of wounds.

Ⓓ It showed people that some diseases had a genetic component.

14 Which is NOT a healthful lifestyle choice? **SC.6.L.14.5**

Ⓕ exercising regularly

Ⓖ smoking cigarettes

Ⓗ using sunscreen

Ⓘ washing hands

NEED EXTRA HELP?

If You Missed Question...	1	2	3	4	5	6	7	8	9	10	11	12	13	14
Go to Lesson...	1	2	2	3	2	2	3	3	1	1	1	2	1	3

Name ______________________ Date ______________

Benchmark Mini-Assessment

Multiple Choice *Bubble the correct answer.*

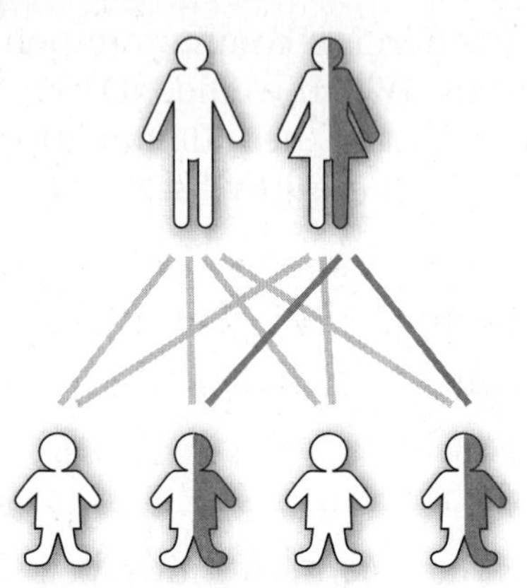

☐ Chromosome without sickle-cell gene
■ Chromosome with sickle-cell gene

1. In the image above, how many children have the sickle-cell trait? **HE.6.C.1.8**

(A) 1
(B) 2
(C) 3
(D) 4

2. Which is NOT a cause of infectious diseases? **SC.6.L.14.6**

(F) bacteria
(G) fungi
(H) genes
(I) viruses

1	The bacterium must be found in every individual with the disease.
2	The bacterium must reproduce in the lab.
3	A sample of the newly grown bacterium must cause the illness when injected into a healthy animal.
4	When the bacterium is removed from the infected animal and grown in the lab, it must be identical to the original bacterium.

3. Whose rules are shown in the table above? **SC.7.N.2.1**

(A) Koch
(B) Jenner
(C) Lister
(D) Pasteur

4. People contract malaria after they are bitten by a mosquito that is carrying an infectious protist. The mosquito is an example of a **HE.6.C.1.8**

(F) cancer.
(G) vector.
(H) bacterium.
(I) pathogen.

Name ______________________ Date ______________

Benchmark Mini-Assessment Chapter 15 • Lesson 2

FLORIDA NGSSS — mini BAT

Multiple Choice *Bubble the correct answer.*

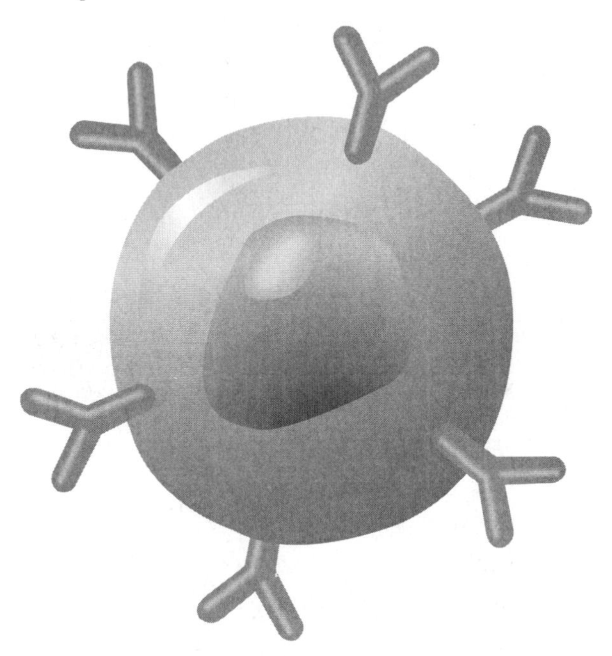

1. What is happening to the B cell in the image above? **SC.6.L.14.5**

(A) It is creating melanin.

(B) It is digesting pathogens.

(C) It is producing antibodies.

(D) It is releasing chemical signals.

2. Which is NOT part of the body's first line of defense against disease? **HE.6.C.1.8**

(F) mucus

(G) skin

(H) stomach

(I) antigens

3. Which is the first step of the inflammatory response? **SC.6.L.14.5**

(A) Bacteria will be destroyed in the infected area.

(B) White blood cells are attracted to the infected area.

(C) Damaged cells release a protein signaling capillaries to dilate.

(D) Surrounding tissue in the infected area begins to swell.

4. Which statement is an example of passive immunity? **SC.6.L.14.5**

(F) A child goes to the doctor for regular vaccinations.

(G) A person gets a shot that prevents illness after cutting his or her hand.

(H) A person is exposed to a pathogen and develops antibodies in response.

(I) A pregnant woman passes antibodies to her developing fetus.

Name ____ Date _______________

Benchmark Mini-Assessment Chapter 15 • Lesson 3

FLORIDA NGSSS mini BAT

Multiple Choice *Bubble the correct answer.*

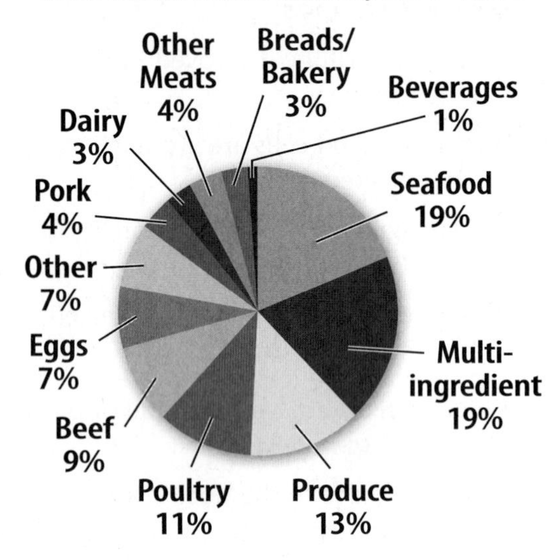

1. Based on the circle graph above, what percentage of food-borne illnesses was caused by animal sources between 1990 and 2004? **MA.6.A.3.6**

Ⓐ 19 percent
Ⓑ 24 percent
Ⓒ 43 percent
Ⓓ 57 percent

2. Which disease is usually associated with the use of tobacco? **HE.6.C.1.8**

Ⓕ cancer
Ⓖ cold
Ⓗ influenza
Ⓘ osteoporosis

3. Which of these is used to treat cancer? **HE.6.C.1.8**

Ⓐ antibiotics
Ⓑ chemotherapy
Ⓒ proteins
Ⓓ vitamins

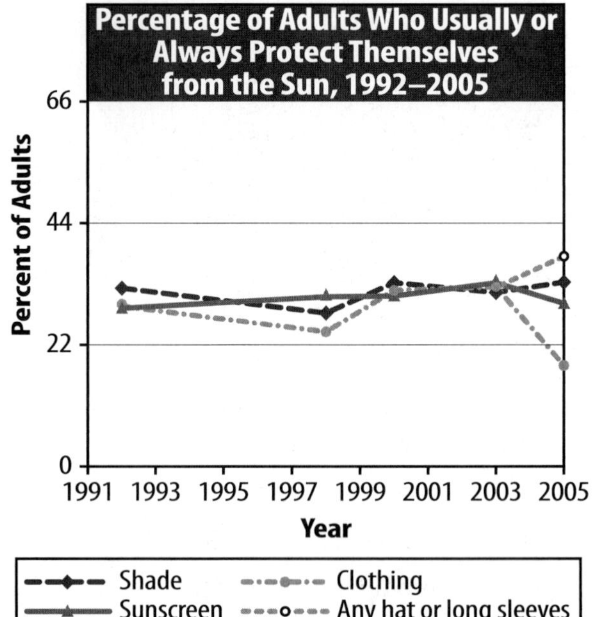

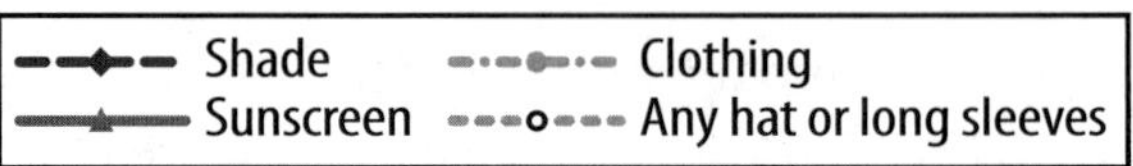

4. Based on the graph above, how did the percentage of adults who usually or always use sunscreen change over time? **MA.6.A.3.6**

Ⓕ The rate dropped, rose, and then dropped again after 2003.
Ⓖ The rate rose, dropped, and then rose again after 2003.
Ⓗ The rate rose steadily until 2003, when it dropped.
Ⓘ The rate stayed constant until 2003, when it dropped.

Name ______________________ Date ______________

Notes

Name ______ Date ______

Notes

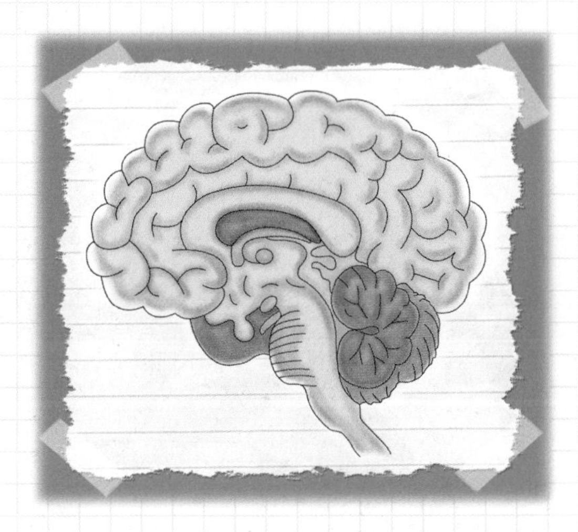

When do we use our brains?

Three friends argued about the brain. They had different ideas about when they use their brains. This is what they said:

Abby: I think we use our brains only when we are conscious. When we are unconscious or asleep, another part of our nervous system takes over.

Laura: I think we use our brains when we are conscious, but only when we are doing something like thinking, speaking, eating, moving, and other intentional things. When we are unconscious or asleep, another part of our nervous system takes over.

Tony: I think we use our brains all the time, even when we are unconscious or asleep. The brain is always working.

Circle the person you agree with the most. Explain your thinking about the brain.

Control and COORDINATION

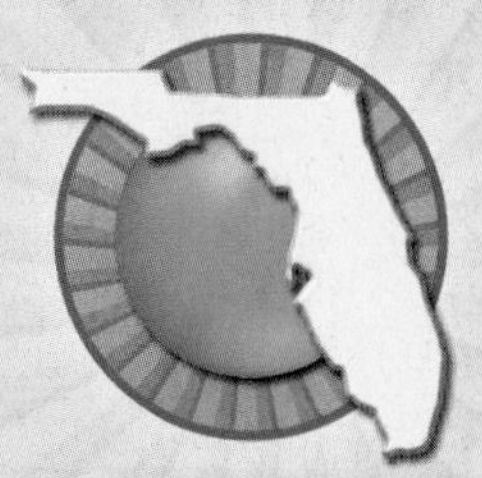

FLORIDA BIG IDEAS

1 **The Practice of Science**

3 **The Role of Theories, Laws, Hypotheses, and Models**

14 **Organization and Development of Living Organisms**

The Big Idea

Think About It!

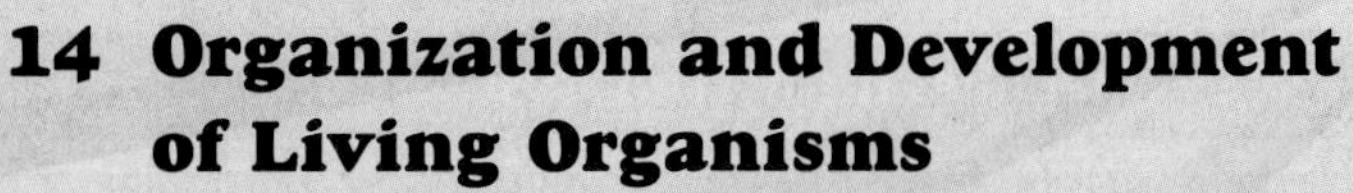

How do the nervous and endocrine systems help maintain the body's homeostasis?

This softball player must be able to sense activity and react quickly. The batter quickly swings the bat as the ball speeds toward her.

1. How do you think this softball player will react to try to hit the ball thrown toward her?

2. What do you think is happening in her brain and body to prepare for the action?

3. Why do you think these activities are essential for survival in daily life?

Get Ready to Read

What do you think about control and coordination?

Before you read, decide if you agree or disagree with each of these statements. As you read this chapter, see if you change your mind about any of the statements.

	AGREE	DISAGREE
1. The nervous system contains two parts—the central nervous system and the peripheral nervous system.	☐	☐
2. The autonomic nervous system controls voluntary functions.	☐	☐
3. A human has five senses that detect his or her environment.	☐	☐
4. The senses of smell and hearing work together.	☐	☐
5. Positive feedback systems in humans help maintain homeostasis.	☐	☐
6. Endocrine glands secrete hormones.	☐	☐

There's More Online!
Video • Audio • Review • ⓘLab Station • WebQuest • Assessment • Concepts in Motion • Multilingual eGlossary

Lesson 1

The Nervous SYSTEM

ESSENTIAL QUESTIONS

What does the nervous system do?

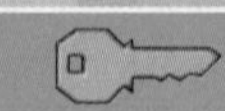
How do the parts of the nervous system work together?

How does the nervous system interact with other body systems?

Vocabulary

nervous system p. 617
stimulus p. 618
neuron p. 619
synapse p. 619
central nervous system p. 620
cerebrum p. 620
cerebellum p. 620
brain stem p. 621
spinal cord p. 621
peripheral nervous system p. 621
reflex p. 622

Florida NGSSS

LA.6.2.2.3 The student will organize information to show understanding (e.g., representing main ideas within text through charting, mapping, paraphrasing, summarizing, or comparing/contrasting);

MA.6.A.3.6 Construct and analyze tables, graphs, and equations to describe linear functions and other simple relations using both common language and algebraic notation.

SC.6.L.14.5 Identify and investigate the general functions of the major systems of the human body (digestive, respiratory, circulatory, reproductive, excretory, immune, nervous, and musculoskeletal) and describe ways these systems interact with each other to maintain homeostasis.

SC.8.N.1.2 Design and conduct a study using repeated trials and replication.

SC.8.N.3.1 Select models useful in relating the results of their own investigations.

MA.6.S.6.2 Select and analyze the measures of central tendency or variability to represent, describe, analyze, and/or summarize a data set for the purposes of answering questions appropriately.

Launch Lab

SC.6.L.14.5

5 minutes

Can you make your eyes blink or dilate?

Are you telling your body to breathe right now? Many functions the human body performs are not under your control. These functions, such as your heart beating, are usually important to your survival. Can you always control your muscles?

Procedure

1. Sit facing a partner. Take turns trying to blink your eyes. Record your observations in your Science Journal.
2. Face your partner again. Try to make your pupils dilate, or get bigger. Record your observations.
3. Cover your eyes with your hand for 30 seconds. Make sure to completely block the light from your eyes.
4. Have your partner look at your eyes as you take your hand away. Record the observation.
5. Repeat steps 3 and 4 using a **mirror** to observe your own eyes when you open them. Record what you observe.

Data and Observations

Think About This

1. Why do you think it might be important that you do not need to think about blinking or breathing, even though you can make yourself do those actions?

2. Key Concept What do you think the purpose of dilating the eye might be? Why might it be helpful to not have to think about dilating your eyes?

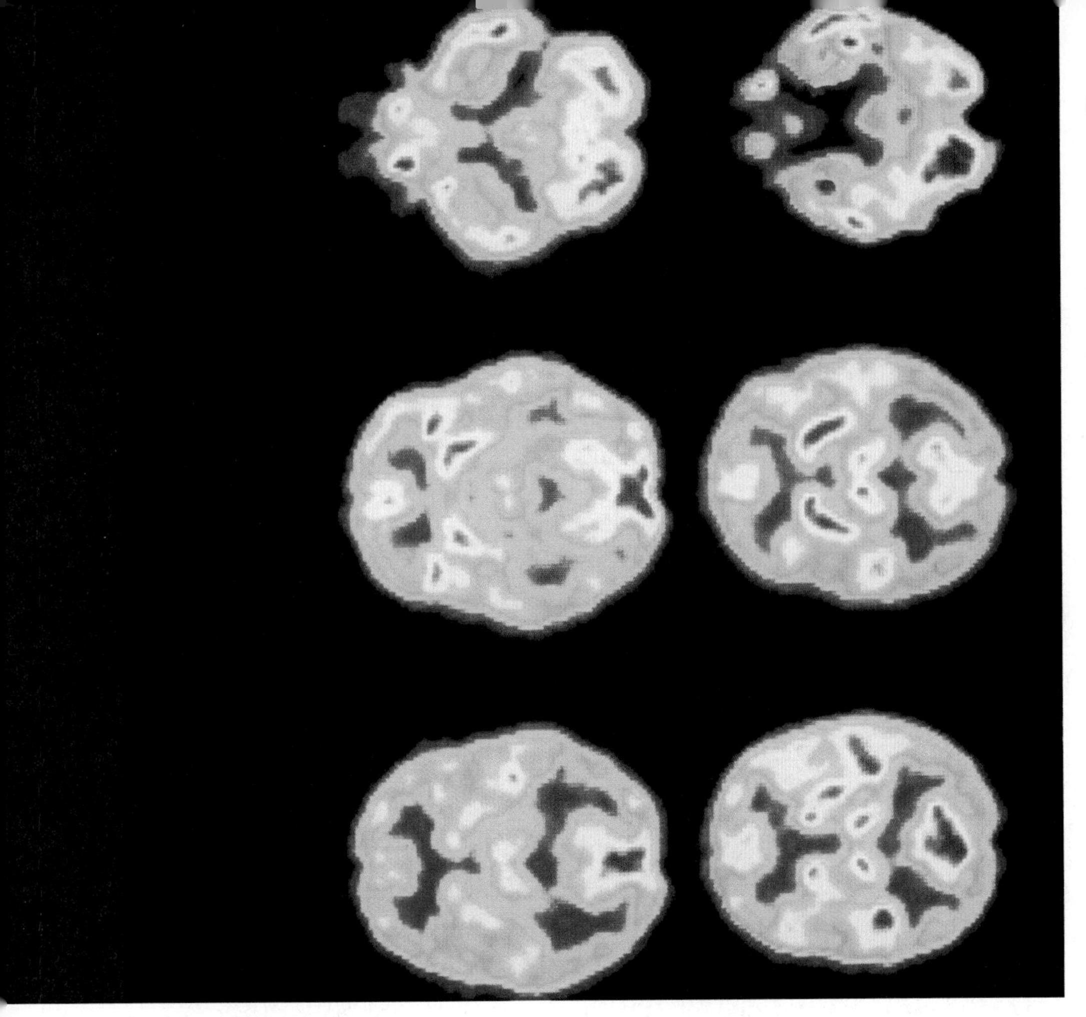

What is memory?

1. These brain scan images all belong to the same person. Activity occurring in the brain is colored red, yellow, and orange. The person being scanned was asked to listen to and memorize words as the scans were taking place. What might the scans tell you about how the brain is working as the person hears different words?

Functions of the Nervous System

Have you ever had goose bumps form on your arms when you were cold? These bumps form because muscle cells in your skin respond to the cold temperature. As the muscle cells contract, or shorten, bumps form, and the hairs on your arms rise up. The hairs trap air, which helps to insulate the skin. This helps you feel warmer. How did the muscle cells know to contract? When you first felt the cold, a message was sent to your brain. After the message was processed, the brain sent a message to your skin's muscle cells, and goose bumps formed.

The part of an organism that gathers, processes, and responds to information is called the **nervous system**. Your nervous system receives information from your five senses—vision, hearing, smell, taste, and touch. You will read more about the five senses in Lesson 2.

The nervous system functions very quickly. It can receive information, process it, and respond in less than one second. In fact, signals received by the nervous system can travel as fast as some airplanes. This is around 400 km/h.

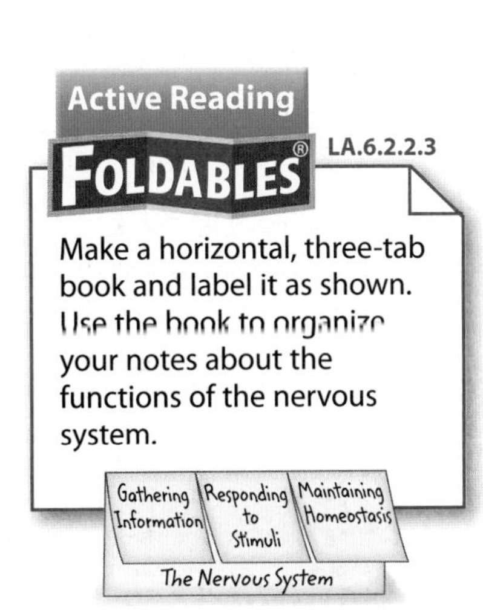

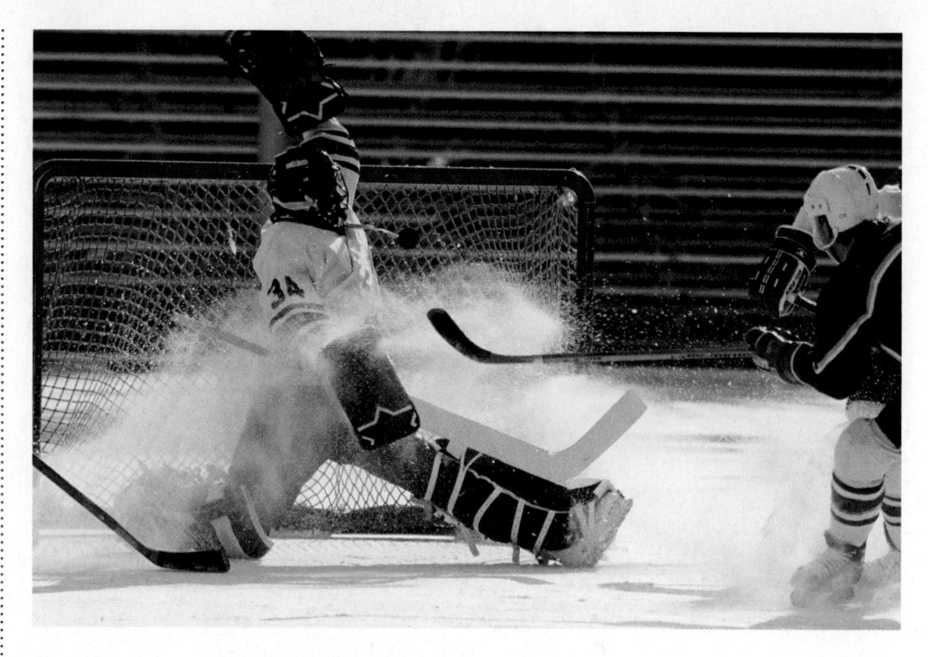

Figure 1 The goalie gathers information about the puck and responds by moving to block it from entering the goal. His body maintains homeostasis when his heart rate and breathing return to normal.

2. Interpret What is the stimulus that the goalie is reacting to?

Gathering Information

Have you ever seen a goalie react quickly to block a hockey puck from entering the goal? The sight of an object, such as the approaching puck in **Figure 1,** is a stimulus (STIHM yuh lus) (plural, stimuli). *A* ***stimulus*** *is a change in an organism's environment that causes a response.* The goalie's nervous system gathers and interprets the sight of the puck approaching and causes his body to react by raising his arm to block the shot.

WORD ORIGIN

stimulus
from Latin *stimulare*, means "goad, urge"

Responding to Stimuli

How would you react if you saw the puck approaching the goal in **Figure 1?** Some people might move quickly to block the shot, while others might turn away to avoid being hit. These reactions are ways that the nervous system enables people to respond to a stimulus from the environment. Since the nervous system receives many stimuli at the same time, the type of response depends on how the information is processed.

Maintaining Homeostasis

Think again about the event shown in **Figure 1.** While responding to the stimuli of the approaching puck, the goalie's nervous system causes his heart and breathing rates to increase. This helps make his reaction time faster. People continually react to changes in their environments. Their nervous systems help maintain homeostasis, or the regulation of their internal environments. For example, the goalie's nervous system must signal his heart and breathing to slow down to restore homeostasis once he has blocked the shot.

Active Reading **3. Organize** What are some of the tasks performed by the nervous system?

Neurons

The basic functioning units of the nervous system are called nerve cells, or **neurons** (NOO rahnz). Neurons help different parts of your body communicate with each other. Without looking down, how do you know whether you are walking on sand or pavement? Neurons in your feet connect to other neurons that send information to your brain about the surface. As shown in **Figure 2,** neurons have three parts—dendrites (DEN drites), a cell body, and an axon (AK sahn). A dendrite receives information from another neuron or from another cell in your body. A cell body processes information. An axon sends information out to another neuron or cell in your body.

Types of Neurons

There are three types of neurons that work together. They send and receive information throughout your body. Sensory neurons send information about your environment to your brain or spinal cord. Motor neurons send information from your brain or spinal cord to tissues and organs in your body. Interneurons connect sensory and motor neurons, much like a bridge connects two different areas of land.

Synapses

The gap between two neurons is called a **synapse** (SIH naps), as shown in **Figure 2.** Most neurons communicate across synapses by releasing chemicals. The chemicals carry information from the axon of one neuron to a dendrite of another neuron. This is similar to the way a baton is passed between runners in a relay race. Most synapses are between an axon of one neuron and a dendrite of another neuron. Information is usually transmitted in only one direction.

Figure 2 Information travels through the nervous system when chemical signals are released by the axon of one neuron and received by a dendrite on another neuron.

4. Name Label the parts of the nerve cell below.

Neurons and Synapses

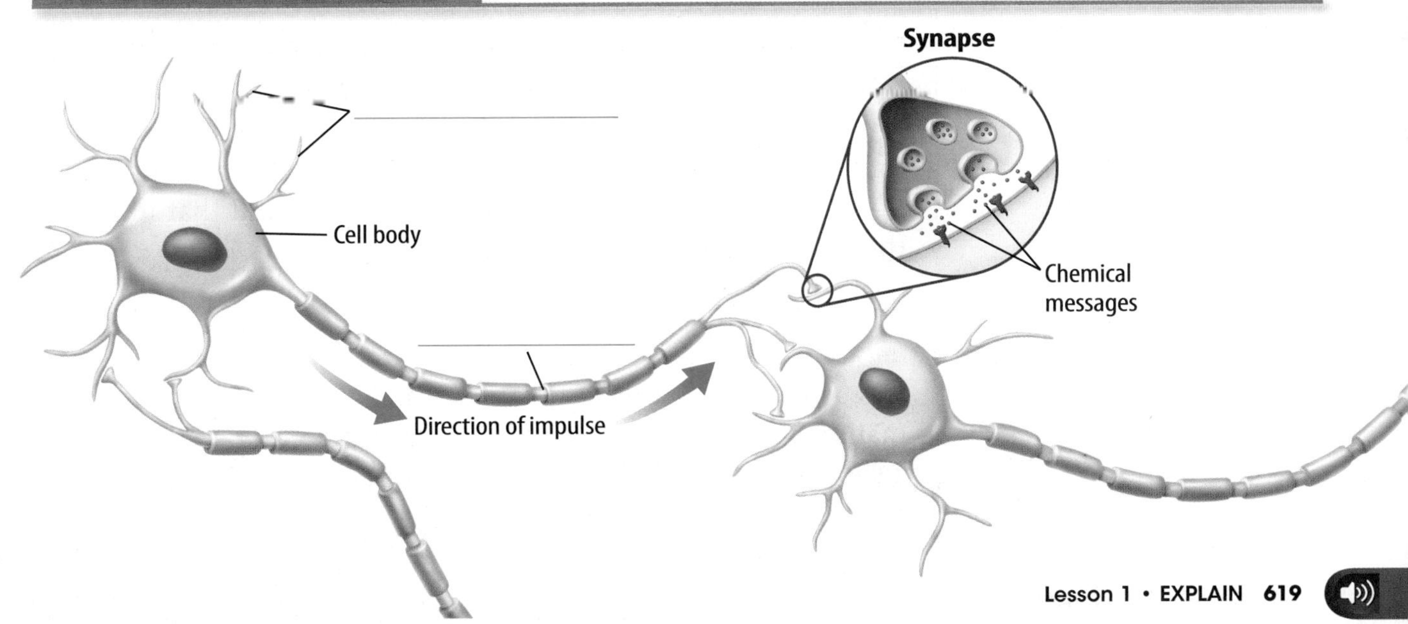

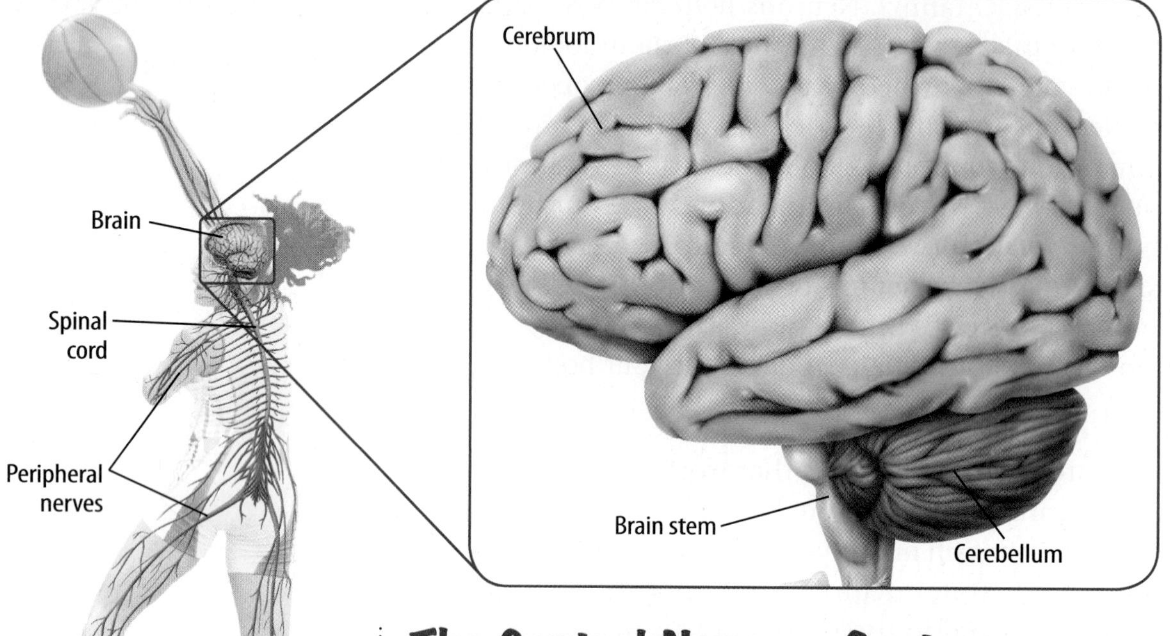

Figure 3 The nervous system consists of the central nervous system (CNS) and the peripheral nervous system (PNS). The brain is part of the CNS and has three main parts with specialized functions.

The Central Nervous System

As shown in **Figure 3,** your nervous system has two parts—the central nervous system and the peripheral (puh RIH frul) nervous system. You will read about the peripheral nervous system later in this lesson. *The* **central nervous system** (CNS) *is made up of the brain and the spinal cord.* The CNS receives, processes, stores, and transfers information.

The Brain

Much like a general is the commander of an army, the brain is the control center of your body. Your brain receives information, processes it, and sends out a response. The brain also stores some information as memories.

The Cerebrum *The part of the brain that controls memory, language, and thought is the* **cerebrum** (suh REE brum). The cerebrum also processes touch and visual information. It is the largest and most complex part of the brain. As shown in **Figure 3,** the surface of the cerebrum has many folds. These folds enable a large number of neurons to fit into a small space. If you could unfold the cerebrum, you would find that it has the surface area of a large pillowcase.

The Cerebellum *The part of the brain that coordinates voluntary muscle movement and regulates balance and posture is the* **cerebellum** (ser uh BEH lum). The cerebellum also stores information about movements that happen frequently, such as tying a shoe or pedaling a bicycle. This enables you to do repetitive things faster and with more accuracy.

The Brain Stem Some functions, such as digestion and the beating of your heart, are involuntary—they happen without your controlling them. *The area of the brain that controls involuntary functions is the* **brain stem,** shown in **Figure 4.** It also controls sneezing, coughing, and swallowing. It connects the brain to the spinal cord.

The Spinal Cord

The **spinal cord,** shown in **Figure 4,** *is a tubelike structure of neurons.* The neurons extend to other areas of the body. This enables information to be sent out and received by the brain. The spinal cord is like an information highway. Just like cars travel on a highway from one city to another, neurons in the spinal cord send information back and forth between the brain and other body parts. Bones called vertebrae protect the spinal cord.

Active Reading **5. Group** What are the two parts of the central nervous system?

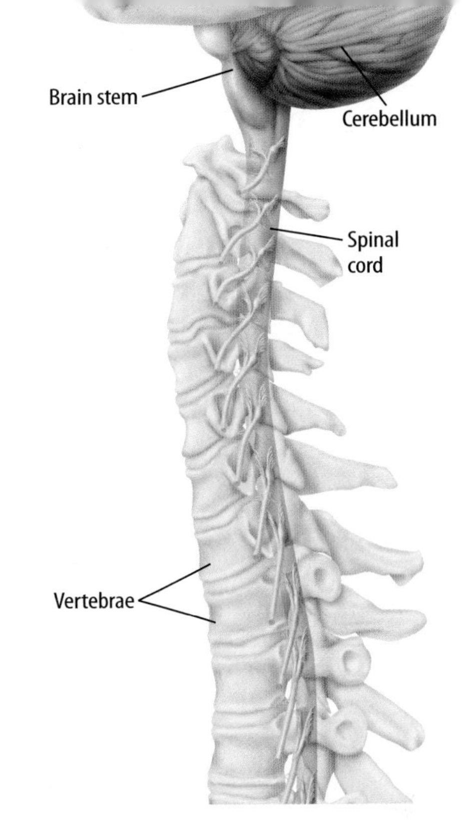

Figure 4 The spinal cord relays information between the brain and the PNS.

The Peripheral Nervous System

Recall that the nervous system is made of both the CNS and the peripheral nervous system. *The* **peripheral nervous system** (PNS), shown in **Figure 3,** *has sensory neurons and motor neurons that transmit information between the CNS and the rest of the body.*

Like the CNS, the PNS also has two parts—the somatic system and the autonomic system. The somatic system controls skeletal muscles. Neurons of the somatic system communicate between the CNS and skeletal muscles and cause voluntary movements, such as picking up a book. The autonomic system controls smooth muscles and cardiac muscles. It regulates involuntary actions, such as dilating blood vessels and the beating of your heart.

Active Reading **6. Combine** Underline how the PNS and the CNS interact.

Math Skills MA.6.A.3.6

Use Proportions

A nerve impulse from your hand travels at about 119 m/s. How long does it take the signal from your hand to reach your spinal cord if the distance is 0.40 m? You can use proportions to solve the problem.

Set up a proportion. $\frac{119 \text{ m}}{1 \text{ s}} = \frac{0.40 \text{ m}}{y \text{ s}}$

Cross multiply. $119 \text{ m} \times y \text{ s} = 0.40 \text{ m} \times 1 \text{ s}$

Solve for the y by dividing both sides by 119 m. $y = \frac{0.40 \text{ m·s}}{119 \text{ m}}$

$y = 0.003 \text{ s}$

Practice

7. One giraffe neuron has an axon 4.6 m long that extends from its toe to the base of its neck. How long will it take a nerve impulse to travel this distance at a speed of 75 m/s?

ACADEMIC VOCABULARY

physical
(adjective) relating to material things

Nervous System Health

A healthy nervous system is necessary for maintaining homeostasis. The nervous system can be damaged by infections and diseases. The most common way the nervous system is damaged is by **physical** injuries.

Physical Injuries

Falling, being in an automobile accident, and being hurt while participating in sports are some of the ways that you can injure and harm the nervous system. The injured nerves can no longer send and receive signals. This stops communication between the CNS and the PNS. When this happens, paralysis can occur. Paralysis is the loss of muscle function and sometimes a loss of feeling.

Preventing Injuries

Imagine that you are walking barefoot and you step on a sharp object. Without thinking, you quickly lift your foot. You do not think about moving your foot, it just happens. *An automatic movement in response to a stimulus is called a* **reflex.** Reflexes are fast because, in most cases, the information goes only to the spinal cord, not to the brain, as shown in **Figure 5.** This fast response protects us from injuries because it takes less time to move away from harm. A reflex often occurs before the brain knows that the body was in danger. However, once nerve signals reach the brain after the response, you feel pain.

Active Reading **8. Recall** Highlight why reflexes are fast.

Figure 5 A reflex enables you to respond to stimuli quickly.

The Path of a Reflex

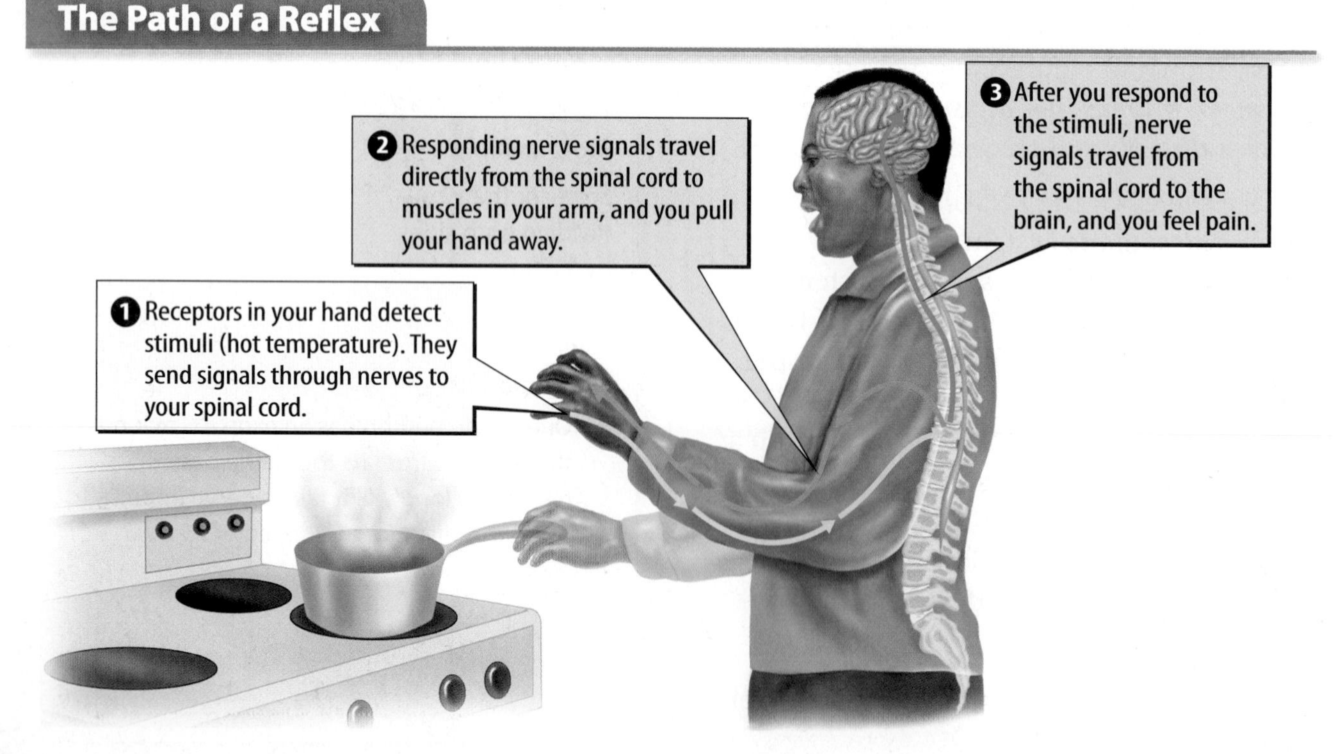

Drugs

In addition to physical injuries, the nervous system can also be affected by substances you take into your body. Drugs are chemicals that affect the body's functions. Many drugs affect the nervous system by either speeding up or slowing down the communication between neurons.

Some pain medicines slow down this communication so much that they stop pain stimuli from reaching the brain. A drug that slows down the communication between neurons is called a depressant. Some people avoid drinking beverages that contain caffeine in the evening because it keeps them awake. Caffeine speeds up the communication between neurons. A drug that speeds up neuron communication is called a stimulant.

The Nervous System and Homeostasis

Why do you shiver when you get cold? It's because your nervous system senses the cold temperature and signals your muscles to contract rapidly in order to warm your body. Your body maintains homeostasis by receiving information from your environment and responding to it. The nervous system is vital to sensing changes in your environment. The nervous system signals other systems, such as the digestive, endocrine, and circulatory systems, to make adjustments when needed.

9. NGSSS Check Develop Give an example of how the nervous system works with another body system to maintain homeostasis. SC.6.L.14.5

Apply It! After you complete the lab, answer these questions.

1. If you were to drink a beverage with caffeine before this lab, how would your reaction time change?

2. Were there any things you did to try to speed up your reaction time? What were they and did they work or not?

Lesson Review 1

Visual Summary

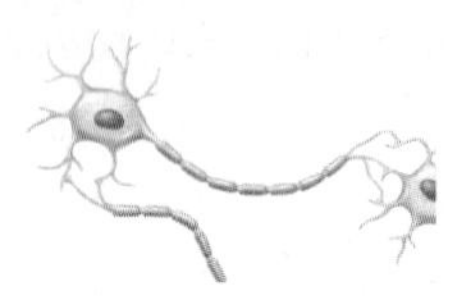

The nervous system gathers and interprets stimuli using a system of neurons that connect throughout the body.

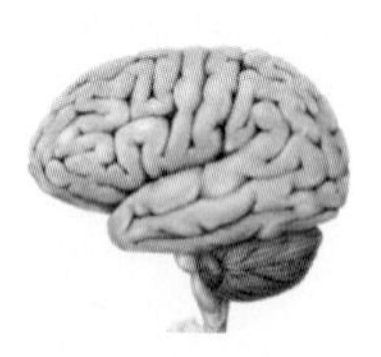

The central nervous system receives, processes, stores, and transfers information.

The peripheral nervous system is made up of the neurons that transmit information between the CNS and the rest of the body.

Use Vocabulary

1. **List** the parts of the central nervous system.

2. **Distinguish** between a neuron and a synapse.

3. **Use the terms** *autonomic* and *somatic* in a sentence.

Understand Key Concepts

4. Which causes an organism to react?
   - (A) CNS
   - (B) PNS
   - (C) stimulus
   - (D) synapse

5. **Compare** the functions of the three parts of the brain.

6. **Explain** how a spinal cord injury may prevent movement.

Interpret Graphics

7. **Summarize** Fill in the graphic organizer below to summarize how the nervous system receives, processes, and responds to a stimulus.

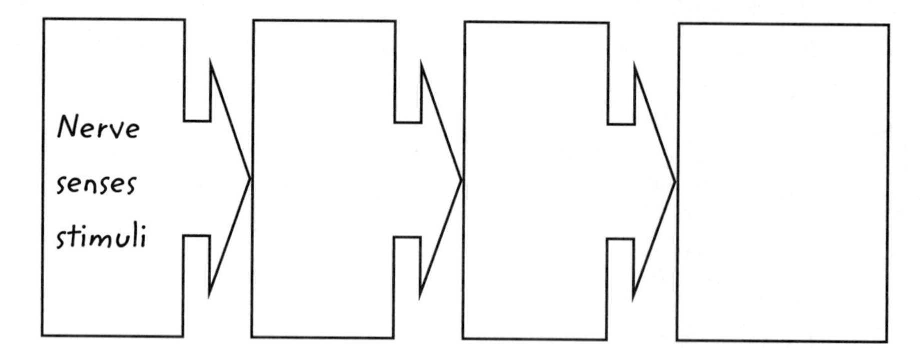

Critical Thinking

8. **Describe** Look at the synapse shown below. Describe the path of a signal from one neuron to another.

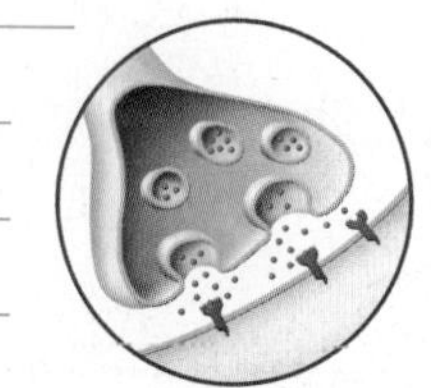

Connect Key Concepts

Examine the functions of the parts of the central nervous system (CNS).

The CNS is made up of:	Function
A. brain:	**1.**
a. cerebrum	**2.** ________________; processes touch and visual information
b. cerebellum	**3.** coordinates ________________ and regulates ________________; stores information about ________________ movements
c. brain stem	**4.** controls ________________ functions; connects the brain to the ________________
B. spinal cord	**5.** a tube-like structure of neurons; ________________

Analyze the parts of the peripheral nervous system (PNS).

PNS

6. made of ________________ that ________________ between the ________________

Somatic

7. ________________ ________________

8. ________________ controls involuntary actions; also controls ________________ ________________

Summarize the effects of physical injuries and drugs on the nervous system.

Cause	Effect
Physical injuries	**9.** harm the nervous system; stop ________________ between the ________________ and the PNS
Drugs	**10.** Stimulants ________________ communication between neurons; ________________ slow down communication between neurons.

Lesson 2

The SENSES

ESSENTIAL QUESTIONS

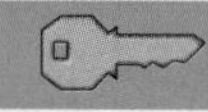 How do you learn about your environment?

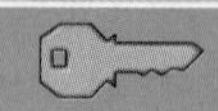 What is the role of the senses in maintaining homeostasis?

Vocabulary

sensory system p. 627
receptor p. 627
retina p. 628
eardrum p. 630

Florida NGSSS

LA.6.2.2.3 The student will organize information to show understanding (e.g., representing main ideas within text through charting, mapping, paraphrasing, summarizing, or comparing/contrasting);

SC.6.L.14.5 Identify and investigate the general functions of the major systems of the human body (digestive, respiratory, circulatory, reproductive, excretory, immune, nervous, and musculoskeletal) and describe ways these systems interact with each other to maintain homeostasis.

SC.6.N.1.1 Define a problem from the sixth grade curriculum, use appropriate reference materials to support scientific understanding, plan and carry out scientific investigation of various types, such as systematic observations or experiments, identify variables, collect and organize data, interpret data in charts, tables, and graphics, analyze information, make predictions, and defend conclusions.

SC.6.N.1.5 Recognize that science involves creativity, not just in designing experiments, but also in creating explanations that fit evidence.

Inquiry Launch Lab

SC.6.N.1.5, SC.6.L.14.5

10 minutes

Does your nose help you taste food?

Sometimes you might hold your nose when you don't want to taste something you're eating. Does this really work? Your senses all function because of nervous impulses that send information. Are your senses of taste and smell connected?

Procedure

1. Read and complete a lab safety form.
2. Hold your nose closed while your teacher walks around and places a **food cube** on your **plate.**
3. While still holding your nose, place the food cube in your mouth and chew it.
4. Write down what the food tasted like to you and what you think the food is.
5. Repeat steps 2–4 with a different food cube. Record your observations.
6. Let go of your nose and chew another sample of both food cubes. Record your observations.

Data and Observations

Think About This

1. What did the two food cubes taste like when you held your nose?
2. What did the two food cubes taste like when you were not holding your nose?
3. Key Concept How do both senses together affect your ability to learn about your environment?

Using All Five Senses?

1. Have you ever toasted marshmallows over a campfire? What did you see? What smells do you experience? What sounds did you hear? Which sensations did you feel? What tastes do you remember? How do these things together make up a campfire experience?

You and Your Environment

Recall from Lesson 1 that your nervous system enables your body to receive information about your environment, process the information, and react to it. Your nervous system is constantly responding to many different types of stimuli. However, your body has to receive a stimulus before it can respond to one. How does this happen?

The **sensory system** *is the part of your nervous system that detects or senses the environment.* A human uses five senses—vision, hearing, smell, taste, and touch—to detect his or her environment. What senses might you use around a campfire, like the one in the photo above. You might see the flames of the campfire, hear the crackle of the wood burning, smell the smoke, feel the warmth of the fire, and taste the cooked marshmallows. *All parts of the sensory system have special structures called* **receptors** *that detect stimuli.* Each of the five senses uses different receptors.

Word Origin

sensory
from Latin *sentire,* means "to perceive, feel"

Active Reading 2. **Describe** What is one way your senses help you to learn about the environment?

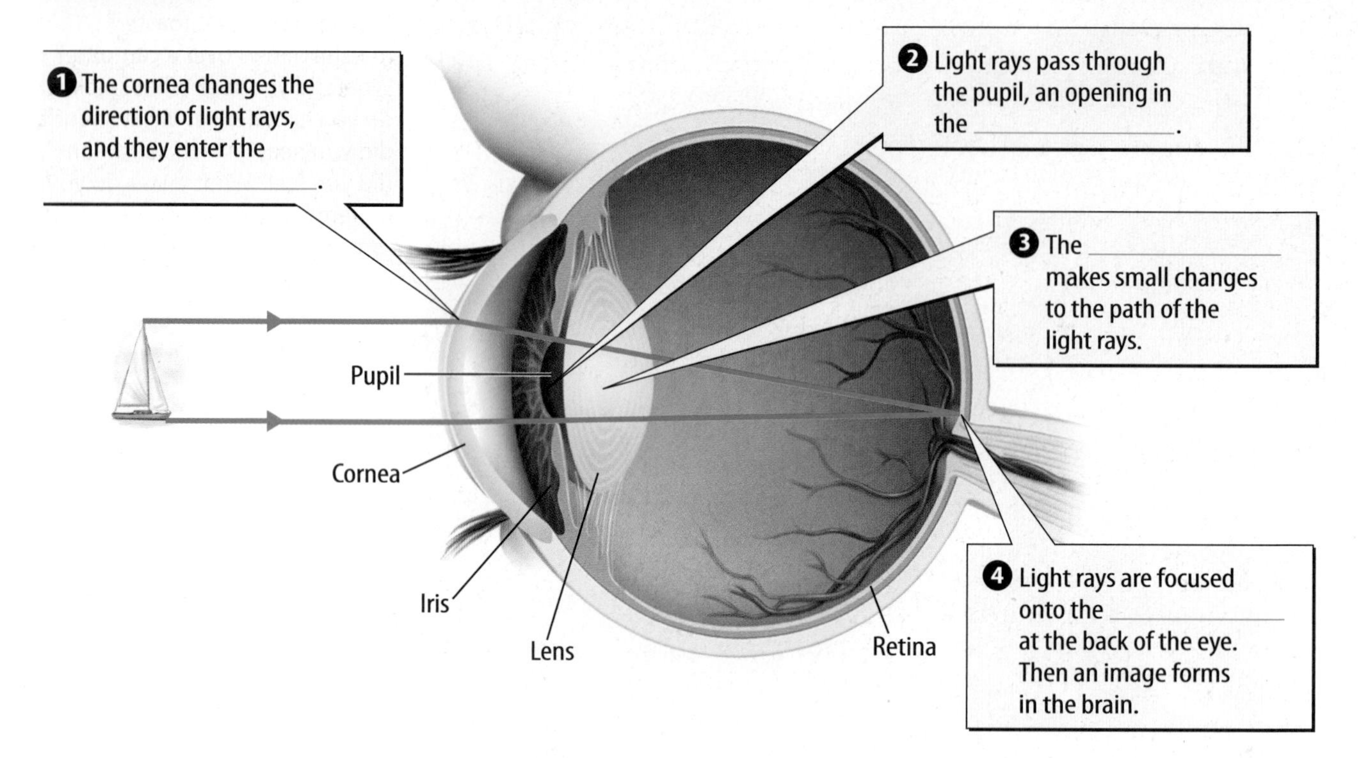

Figure 6 Your eyes contain photoreceptors that receive and interpret light signals from the environment.

Active Reading **3. Complete** Use the text to fill in the information in **Figure 6.**

Vision

Have you ever wondered how your eyes work? Your sense of vision lets you see things that are close, such as the words on this page, and objects that are far away, such as a star in the night sky. The visual system uses photoreceptors in the eye and detects light and creates vision.

The Parts of the Eye

As shown in **Figure 6,** light enters the eye through the cornea (KOR nee uh), a thin membrane that protects the eye and changes the direction of light rays. The colored part of your eye is the iris (I rus). After light passes through the cornea, it goes through an opening formed by the iris called the pupil. The iris controls the amount of light that enters the eye by changing the size of the pupil. In bright light, the iris constricts, making the pupil smaller and letting in less light. In dim light, the iris relaxes, making the pupil larger and letting in more light. Light then travels through a clear structure called the lens. As shown in **Figure 6,** the lens works with the cornea and focuses light. *The* **retina** (RET nuh) *is an area at the back of the eye that has two types of cells—rod cells and cone cells—with photoreceptors.* Rod cells detect shapes and low levels of light. They are important for night vision. Cone cells detect color and function best in bright light.

Active Reading

FOLDABLES® LA.6.2.2.3

Use three sheets of notebook paper to make a layered book. Label it as shown. Use it to record information about the five senses.

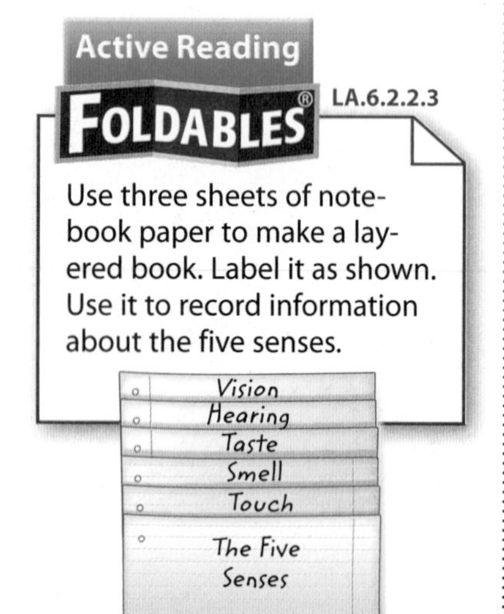

How You See

In order to see, light that enters your eyes has to be detected by the **rods** and cones in the retina. The rods and cones detect information about the colors and shapes of objects from the light that enters the eyes. The retina then sends that information as electric signals through the optic nerve to the brain. The brain uses the information and creates a picture of what you are seeing.

Science Use v. Common Use

rod

Science Use one of the photoreceptors in the eye that distinguish the shapes of objects

Common Use any long, cylinder-shaped object

Focusing Light

The lens and the cornea work together and change the direction of the light that enters the eye. As shown in **Figure 6,** both the lens and the cornea are curved. These curved shapes change the direction of light and focus it onto the retina. Why do some people need glasses to see well? If corneas or lenses are not curved exactly right, the eyes will have trouble focusing images, as shown in **Figure 7.** If a person's eyes are longer than normal, the person is nearsighted and has trouble seeing images that are far away. If a person's eyes are shorter than normal, the person is farsighted and has trouble seeing images that are close up. Glasses or contacts are used to correct vision problems by correctly focusing the light on the retina.

Active Reading 4. **Differentiate** Highlight which parts of vision rods and cones are responsible for.

Figure 7 The lens and the cornea focus light. Vision problems occur when they are not curved correctly.

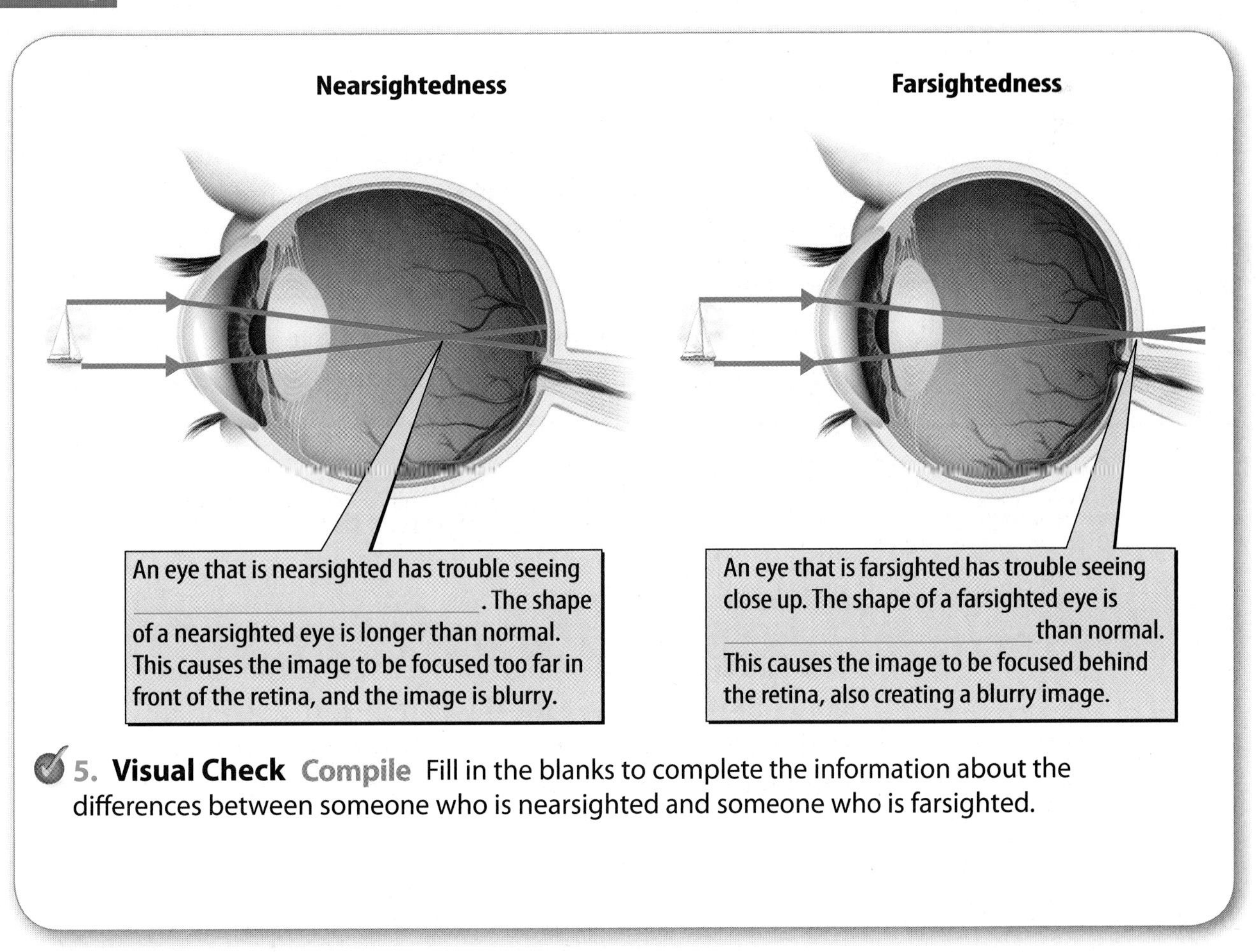

5. **Visual Check Compile** Fill in the blanks to complete the information about the differences between someone who is nearsighted and someone who is farsighted.

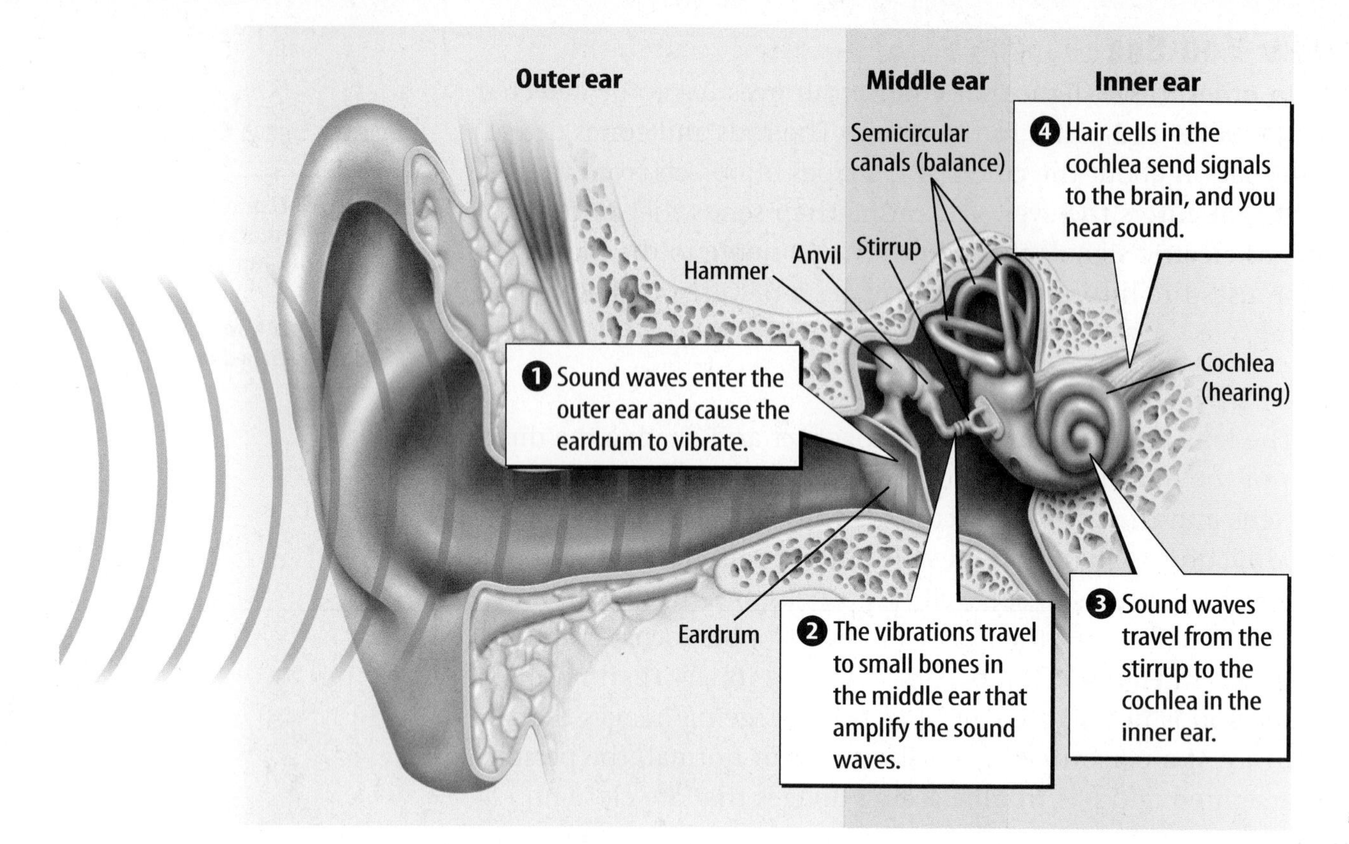

Figure 8 Scientists have identified three specialized regions in the human ear.

6. Visual Check
Identify Which structure contains hair cells which send signals to the brain?

REVIEW VOCABULARY

wave
a disturbance in a material that transfers energy without transferring matter

Hearing

How do you hear the sounds around you? The vibration of matter creates sound **waves** that travel through air and other substances. Sound waves that enter the ear are detected by auditory (AW duh tor ee) receptors. As waves travel within the ear, they are amplified, or increased, and move hair cells. The hair cells send information about the sound waves to the brain. The brain processes information about the loudness and tone of the sound, and you hear.

The Parts of the Ear

As shown in **Figure 8,** human ears have three areas—the outer ear, the middle ear, and the inner ear—each with a special function.

The Outer Ear The outer ear includes the parts of the ear that you can see. It collects sound waves that make the eardrum vibrate. *The* **eardrum** *is a thin membrane between the outer ear and the inner ear.*

The Middle Ear The vibrations pass from the eardrum to small bones in the middle ear—the hammer, the anvil, and the stirrup. The movement of these bones amplifies the sound waves.

The Inner Ear The part of the ear that detects sound is the inner ear. The inner ear converts sound waves into messages that are sent to the brain. The structure shown in **Figure 8** that looks like a snail shell is called the cochlea (KOHK lee ah). The cochlea is filled with fluid. When sound waves reach the cochlea, they make the fluid vibrate. The liquid in the cochlea moves much like the way that hot chocolate moves in a cup when you blow on it. The moving fluid bends hair cells, which send messages to the brain for processing.

Active Reading 7. **List** What are the three areas of the ear?

The Ear and Balance

In addition to detecting sound waves, the inner ear has another function. Parts of the inner ear, called the semicircular canals, help maintain balance. Like the cochlea, the semicircular canals contain fluid and hair cells. Whenever you move your head, the fluid moves, which moves the hair cells, as shown in **Figure 9.** Information about the movement of the hair cells is sent to the brain. The brain then signals muscles to move your head and body in order to maintain your balance.

Figure 9 Fluid in your inner ear helps maintain your balance by sensing changes in the position of your head.

The Ear and Balance

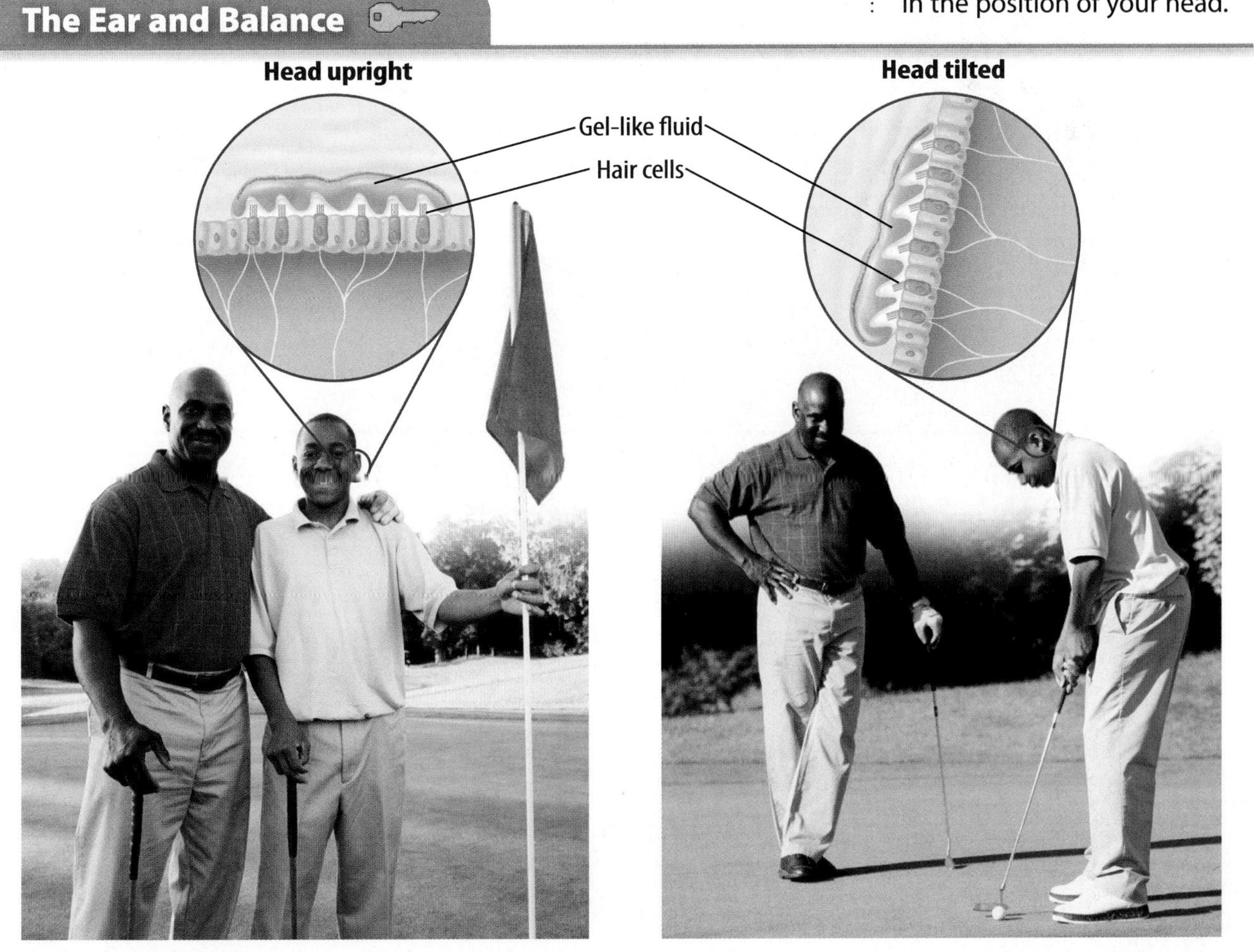

Smell

How are you able to distinguish between the citrus aroma of an orange and the sulfur odor of rotten eggs? Humans have hundreds of different receptors for detecting odors. Some dogs have over 1,000 different odor receptors and can be used to track things by smell. Odors are molecules that are detected by chemical receptors, called chemoreceptors (kee moh rih SEP turz), in your nose. These receptors send messages to the brain. The brain then processes the information about the odor. A smell might make you feel hungry, or it could trigger a strong memory or feeling.

Taste

The sense of taste also relies on chemoreceptors. Chemoreceptors on your tongue detect chemicals in foods and drinks. The receptors then send messages to the brain for processing. Chemoreceptors on the tongue are called taste buds. Taste buds, such as the ones shown in **Figure 10,** can detect five different tastes: bitter, salty, sour, sweet, and a taste called umami (oo MAH mee). Umami is the taste of MSG (monosodium glutamate), a substance often used in processed foods. **Figure 10** illustrates how the chemoreceptors in your nose and mouth work together to help you taste foods.

Taste and Smell

Figure 10 Taste and smell stimuli are both detected by chemoreceptors. Chemoreceptors detect chemicals in the substances you eat and drink and in the odors you breathe.

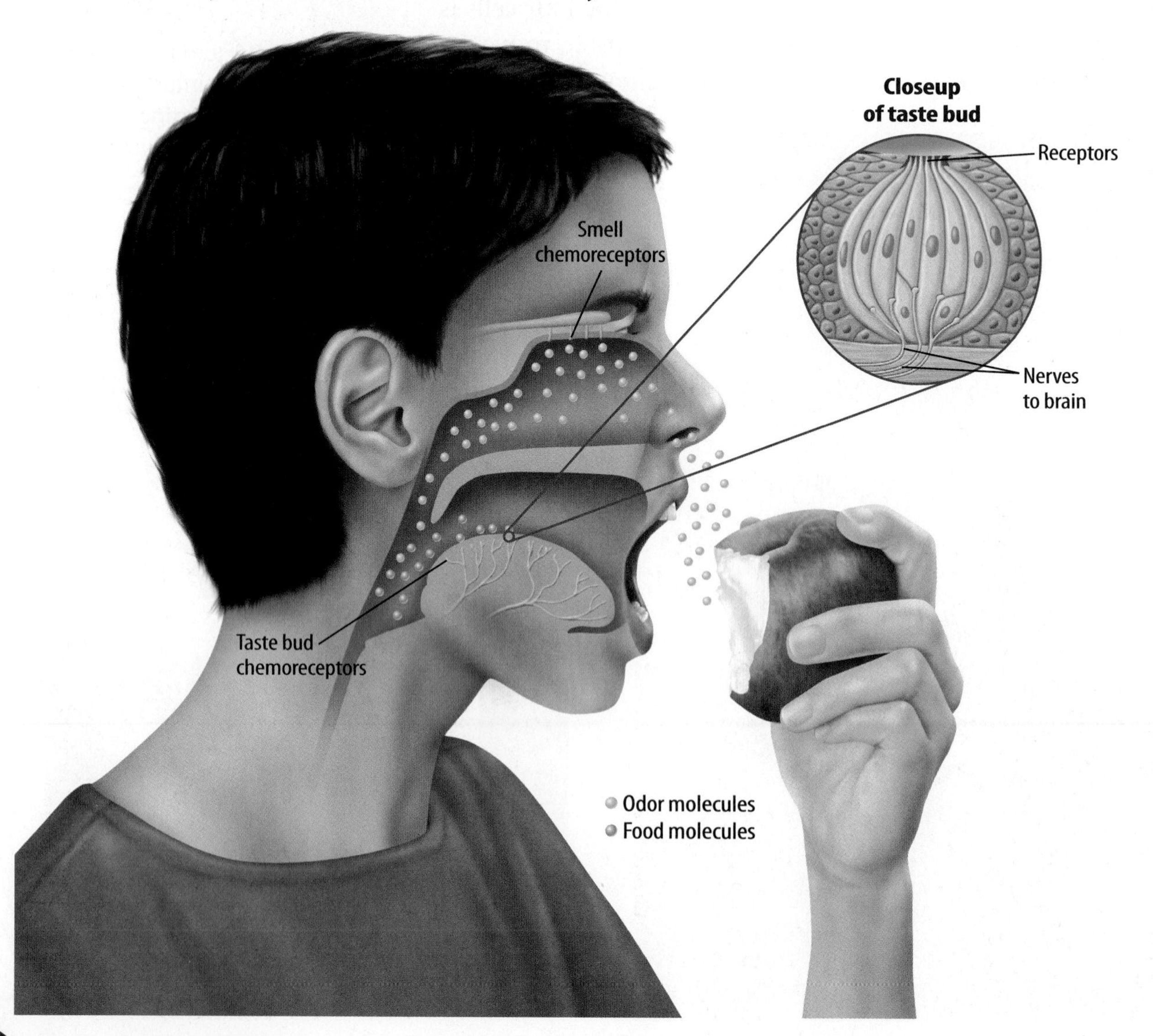

Touch

Like all the other senses, the sense of touch also uses special receptors that detect the environment. Touch receptors in your skin can detect temperature, pain, and pressure. For example, skin receptors can detect the difference between a light tap and a poke. These receptors are all over your body. Some areas, such as the palms of your hands and the soles of your feet, have lots of receptors. The middle of your back has fewer receptors. Just like the other four senses, touch receptors send messages to the brain for processing.

The Senses and Homeostasis

Gathering information about the environment is important for your survival. The five senses collect information about the environment and send it to your nervous system. Your brain is then able to respond and activate your body to maintain homeostasis. Whether it is sensing temperature changes, finding food and water, avoiding harmful environments, or detecting other important stimuli, your senses are the vital first step.

8. NGSSS Check Blend Highlight why senses are important to homeostasis?

Active Reading **9. Explain** how the senses help maintain the body's homeostasis.

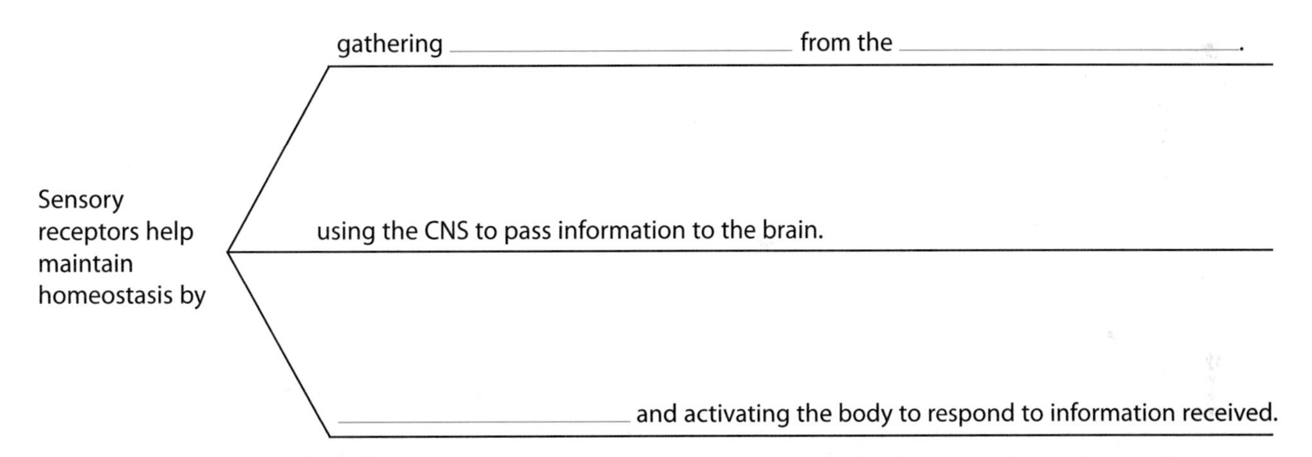

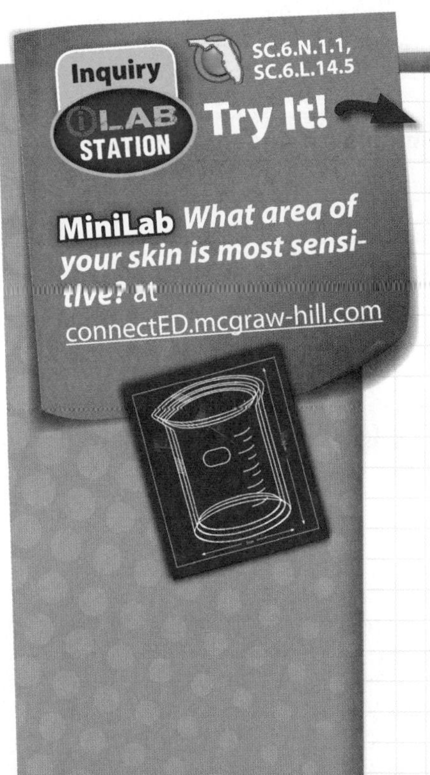

Apply It! After you complete the lab, answer this question.

1. Calluses, which are extra layers of epidermis, are produced when the skin is stressed and over stimulated from activities such as sports and playing guitar. How can developing calluses impact your ability to react to touch?

Lesson Review 2

Visual Summary

The sensory system is part of the nervous system. It uses receptors in five senses and collects information about the environment.

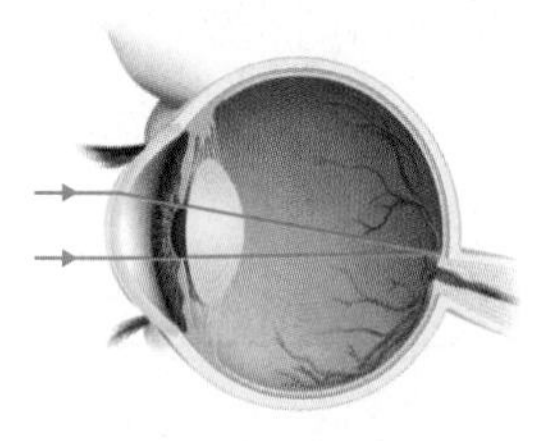

Your eyes detect light as it passes through the cornea, the pupil, and the lens, and then to the retina.

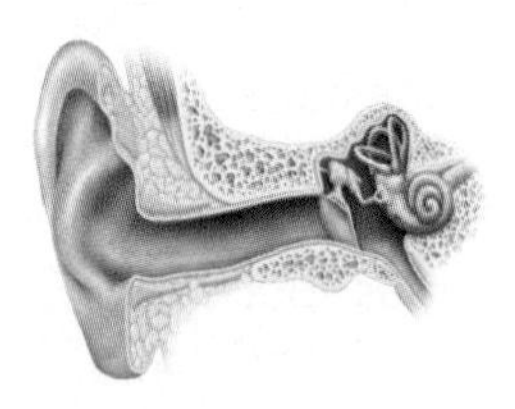

The eardrum is a structure between the outer and middle ear. It vibrates in the presence of sound waves.

Use Vocabulary

1. **Name** the part of the ear that collects sound waves from the environment.

2. **Define** the term *receptor* using your own words.

3. **Use the word** *retina* in a sentence.

Understand Key Concepts

4. Which is NOT a sensory system?
   - (A) digestion
   - (B) hearing
   - (C) smell
   - (D) vision

5. **Explain** the relationship between the five senses and homeostasis. SC.6.L.14.5

6. **Contrast** the different types of stimuli that can be detected by touch receptors.

Interpret Graphics

7. **Organize** Fill in the table below to organize information about the five senses.

Sense	Stimulus

Critical Thinking

8. **Examine** why it might be difficult to taste food when you have a cold.

9. **Infer** how an ear infection might lead to problems with balance.

Night Vision Goggles

HOW NATURE WORKS

Would you like to be able to see in the dark?

You might already know that vision begins with light entering your eyes. What happens when it is dark? If there is no light available to enter your eyes, how can any device make it possible to see?

Even when it is dark, there is light all around you. Although you can't see it, almost everything gives off, or emits, infrared light. Objects also reflect some infrared light, in the same way that they reflect visible light. Night vision goggles work by collecting that infrared and converting it to visible light.

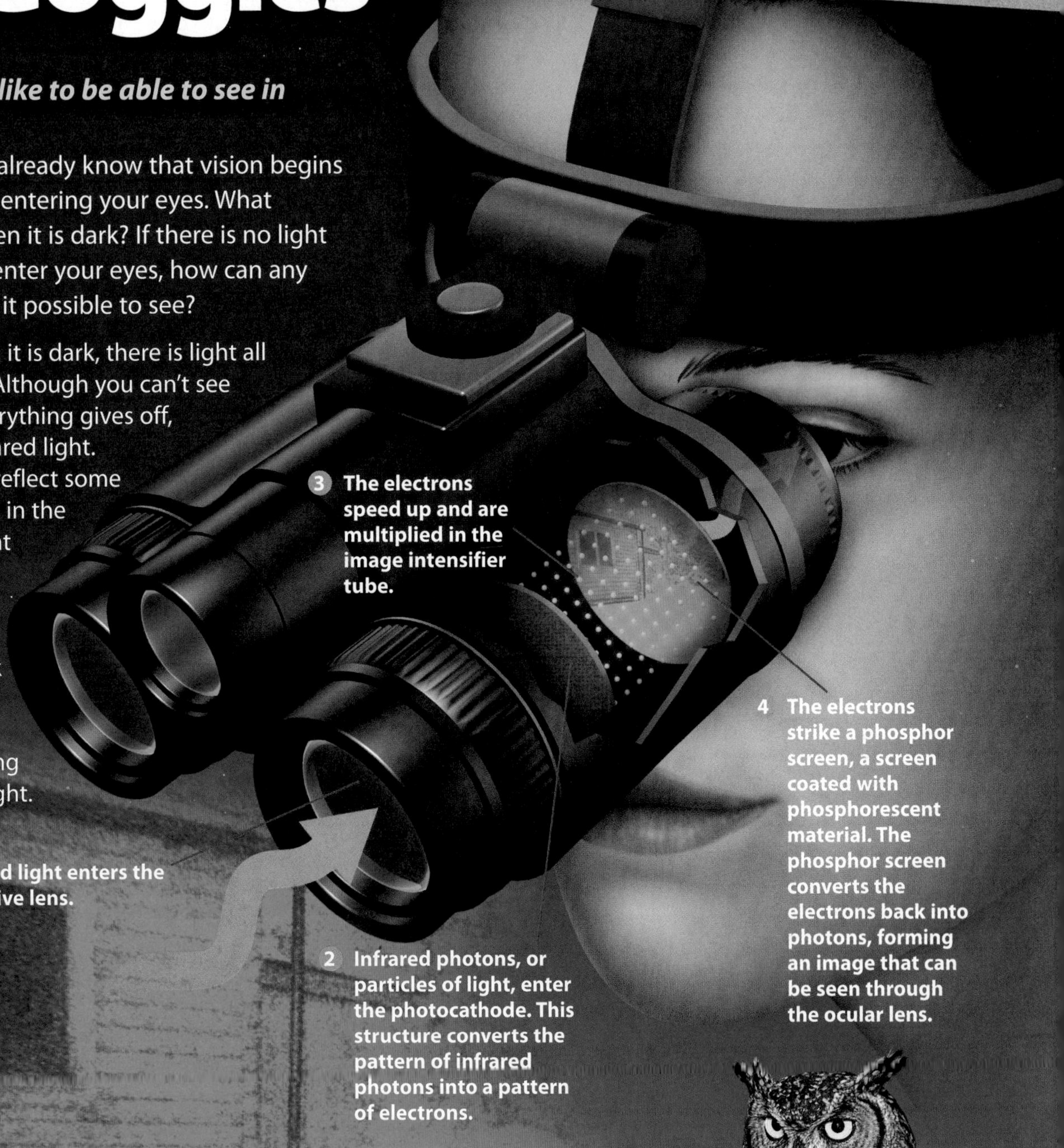

1 Infrared light enters the objective lens.

2 Infrared photons, or particles of light, enter the photocathode. This structure converts the pattern of infrared photons into a pattern of electrons.

3 The electrons speed up and are multiplied in the image intensifier tube.

4 The electrons strike a phosphor screen, a screen coated with phosphorescent material. The phosphor screen converts the electrons back into photons, forming an image that can be seen through the ocular lens.

It's Your Turn

REPORT How can owls see more clearly in dim light than humans can? Research rods and cones in the eye and what they have to do with vision. Use what you discover to make a "How It Works" diagram about rods and cones in the eye.

Lesson 3

The Endocrine SYSTEM

ESSENTIAL QUESTIONS

What does the endocrine system do?

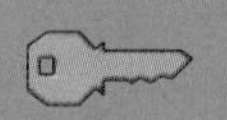
How does the endocrine system interact with other body systems?

Vocabulary

endocrine system p. 637

hormone p. 637

negative feedback p. 640

positive feedback p. 640

Inquiry Launch Lab

SC.7.N.1.1, SC.6.N.1.5, SC.6.L.14.5

10 minutes

What makes your heart race?

Have you ever felt your heart start to race? Maybe something startled you or you were nervous or frightened. Usually this sensation lasts only for a few minutes. What causes your heart to pound so fast at times?

Procedure

1. Read and complete a lab safety form.
2. Find your pulse by holding your first and second fingers of one hand on the inside of the wrist of your other hand. Count heartbeats for 15 seconds. Record this number. Multiply this number by four to find your average heartbeats per minute.
3. Your teacher will attempt to startle you. Immediately afterward, count your heartbeats again for 15 seconds and multiply by 4.
4. Wait five minutes. Calculate your heartbeats per minute again.

Think About This

1. What changes did you note in your heart rate after your teacher startled you?

2. What happened to your heartbeat when you checked five minutes later?

3. Key Concept Why do you think your heart reacted to the noise even though you knew it was coming? Why do you think this change occurred?

Florida NGSSS

LA.6.2.2.3 The student will organize information to show understanding (e.g., representing main ideas within text through charting, mapping, paraphrasing, summarizing, or comparing/contrasting);

SC.6.L.14.5 Identify and investigate the general functions of the major systems of the human body (digestive, respiratory, circulatory, reproductive, excretory, immune, nervous, and musculoskeletal) and describe ways these systems interact with each other to maintain homeostasis.

SC.6.N.1.5 Recognize that science involves creativity, not just in designing experiments, but also in creating explanations that fit evidence.

SC.8.N.3.1 Select models useful in relating the results of their own investigations.

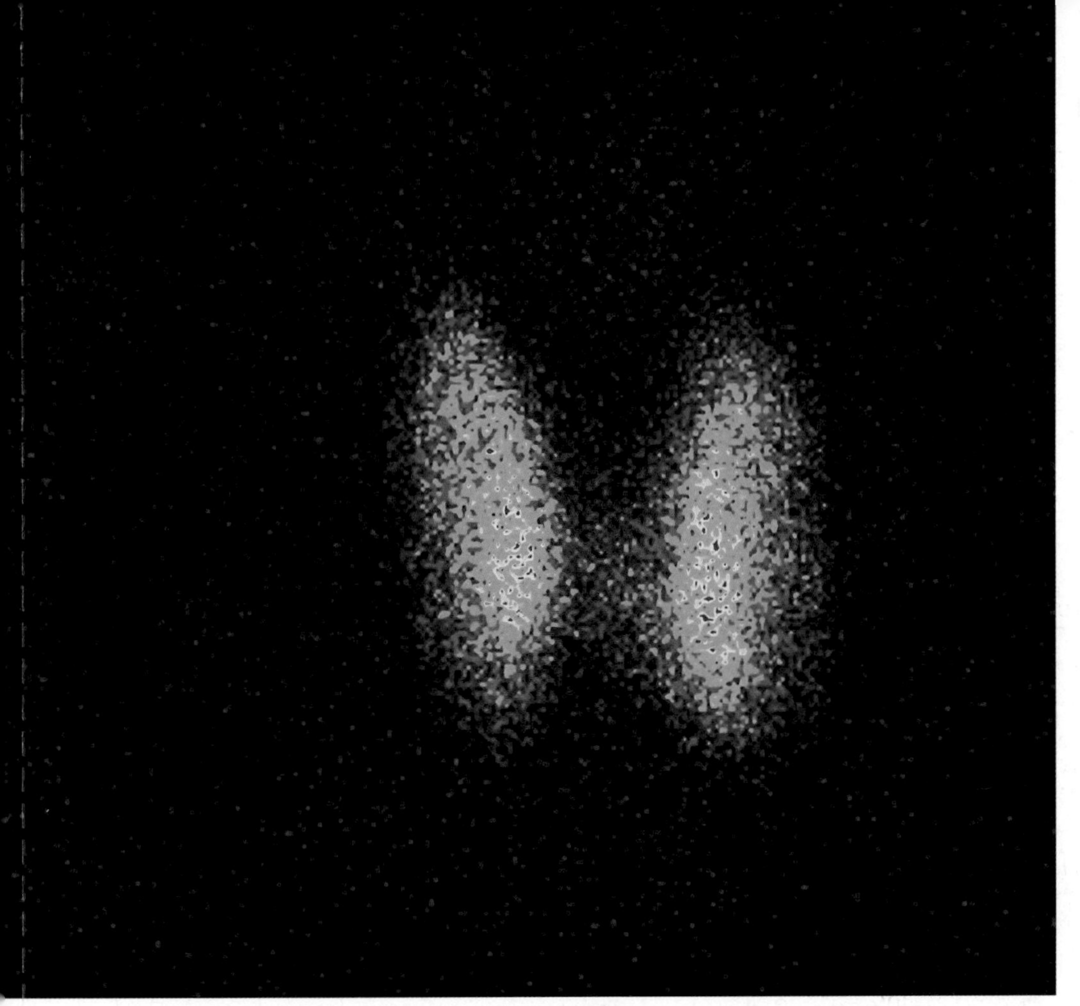

Inquiry Blue Butterfly?

1. This image might look like a blue and green butterfly, but it is actually an image of a person's butterfly-shaped thyroid gland. The image was taken after a dose of radioactive material was given to the patient. The radioactive material collected in the thyroid tissue. What do you think the thyroid gland does?

Functions of the Endocrine System

Like the nervous system, your body sends messages using another system called the endocrine (EN duh krun) system. *The* **endocrine system** *consists of groups of organs and tissues that release chemical messages into the bloodstream.*

Endocrine tissues that secrete chemical molecules are called endocrine glands. The thyroid gland on the previous page is one of the glands in the endocrine system. The chemical messages sent by the endocrine system are called hormones. *A* **hormone** *is a chemical that is produced by an endocrine gland in one part of an organism and is carried in the bloodstream to another part of the organism.*

WORD ORIGIN

hormone
from Greek *hormon,* means "that which sets in motion"

The messages sent by the endocrine system are transmitted less rapidly than messages sent by the nervous system. The chemical messages usually are sent to more cells and last longer than the messages sent by the nervous system. For example, a message from the nervous system might cause a body movement. A message from the endocrine system might cause your body to grow taller over a period of time.

2. **NGSSS Check** **Assess** Underline what the endocrine system does. SC.6.L.14.5

Endocrine Glands and Their Hormones

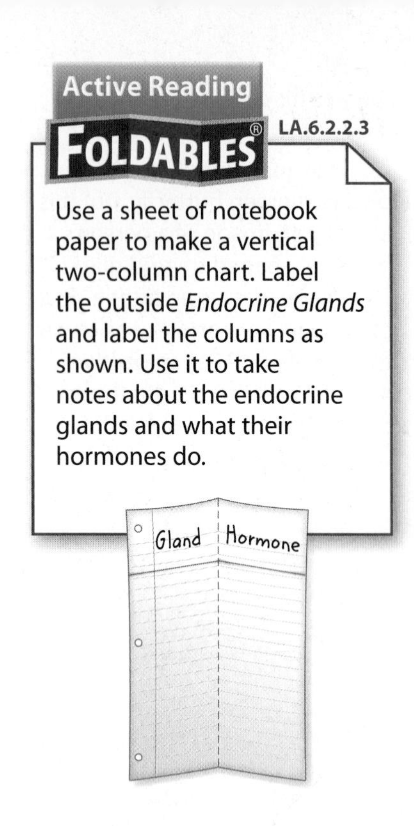

Hormones move from endocrine glands to other parts of the body in the bloodstream. Hormones affect the functions of organs and tissues by carrying messages to specific cells called target cells. Hormones recognize their target cells because the target cells have certain receptor proteins on or inside them. The hormone recognizes the receptor proteins and attaches to them. A hormone is like a key that wanders around and tries locks until it finds one that it fits and unlocks it. Once a hormone finds its target cell, it binds to a receptor protein and delivers its chemical message. The target cell responds by taking a specific action.

Figure 11 illustrates the path that a hormone takes from its production in an endocrine gland until it finds and delivers its message to the target cell. There are many different hormones with various purposes that are produced and distributed throughout your body. The endocrine glands and the hormones they produce are shown in **Figure 12.**

The Path of a Hormone

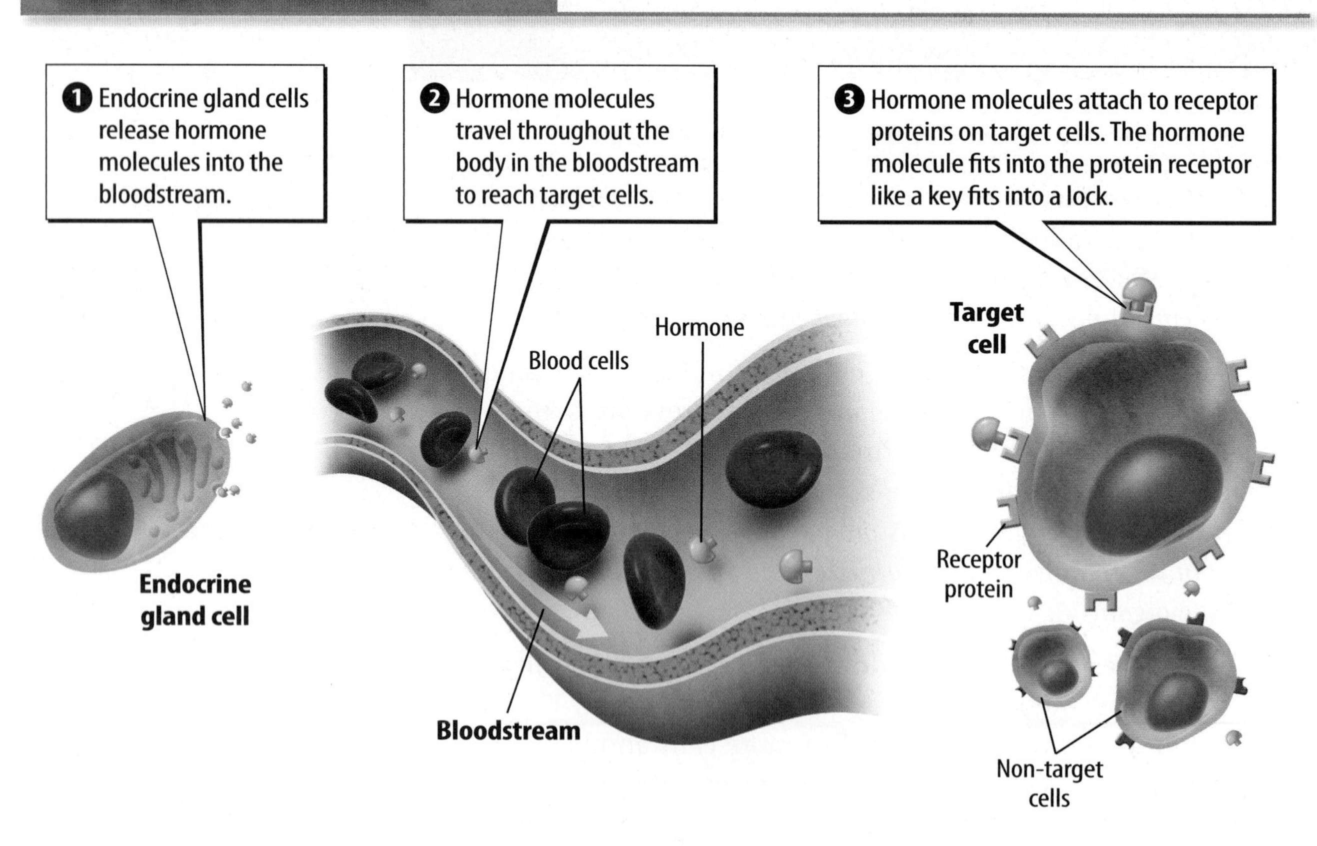

Figure 11 Hormones move through the bloodstream until they encounter the specific cells they are targeted to affect.

3. Visual Check Judge Why doesn't the green molecule attach to non-target cells?

Endocrine Glands and Hormones 

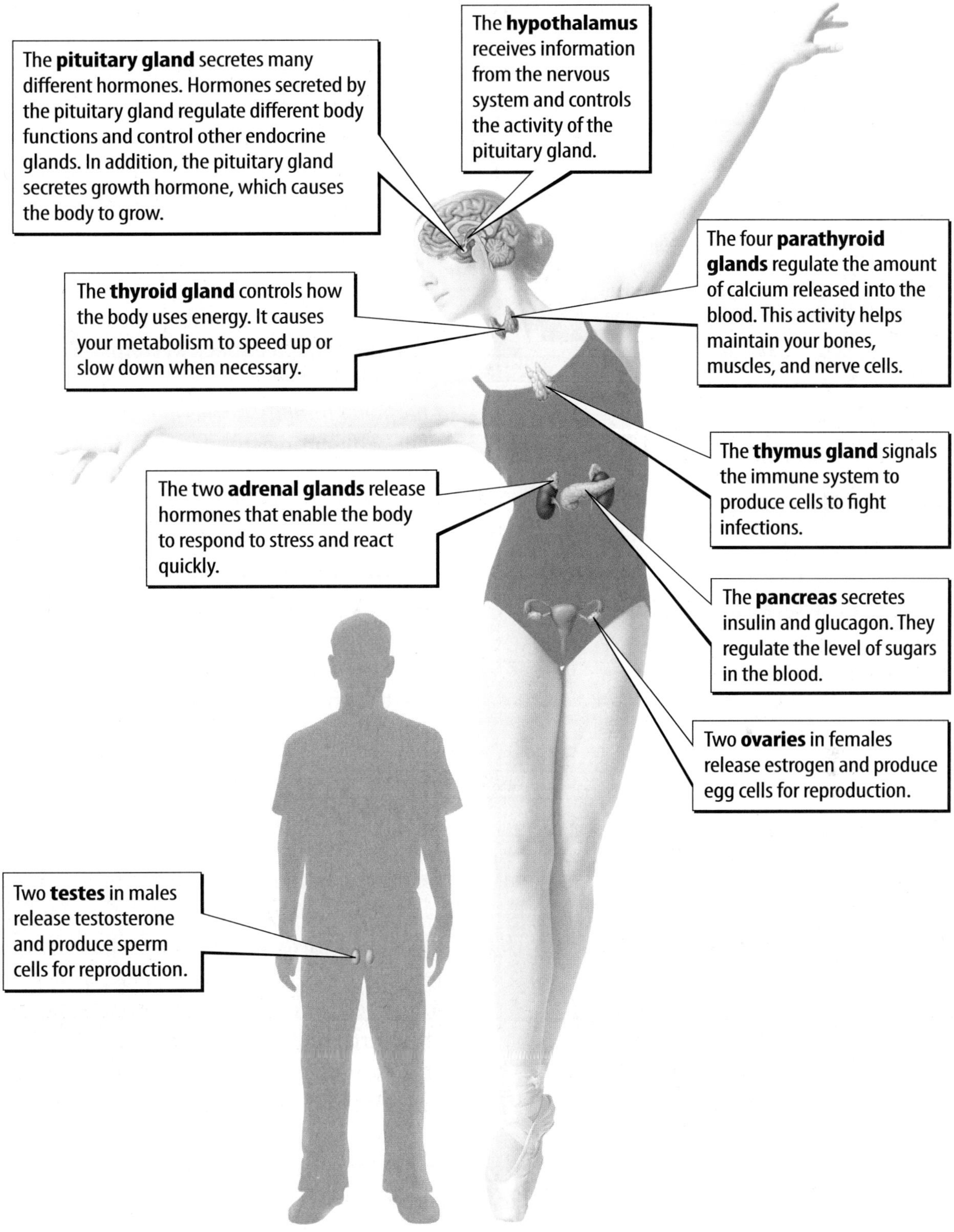

The **pituitary gland** secretes many different hormones. Hormones secreted by the pituitary gland regulate different body functions and control other endocrine glands. In addition, the pituitary gland secretes growth hormone, which causes the body to grow.

The **hypothalamus** receives information from the nervous system and controls the activity of the pituitary gland.

The **thyroid gland** controls how the body uses energy. It causes your metabolism to speed up or slow down when necessary.

The four **parathyroid glands** regulate the amount of calcium released into the blood. This activity helps maintain your bones, muscles, and nerve cells.

The **thymus gland** signals the immune system to produce cells to fight infections.

The two **adrenal glands** release hormones that enable the body to respond to stress and react quickly.

The **pancreas** secretes insulin and glucagon. They regulate the level of sugars in the blood.

Two **ovaries** in females release estrogen and produce egg cells for reproduction.

Two **testes** in males release testosterone and produce sperm cells for reproduction.

Figure 12 The endocrine system is made up of different glands that secrete hormones to maintain homeostasis in the body.

4. **Visual Check** Select Which endocrine gland affects metabolism?

Feedback Systems

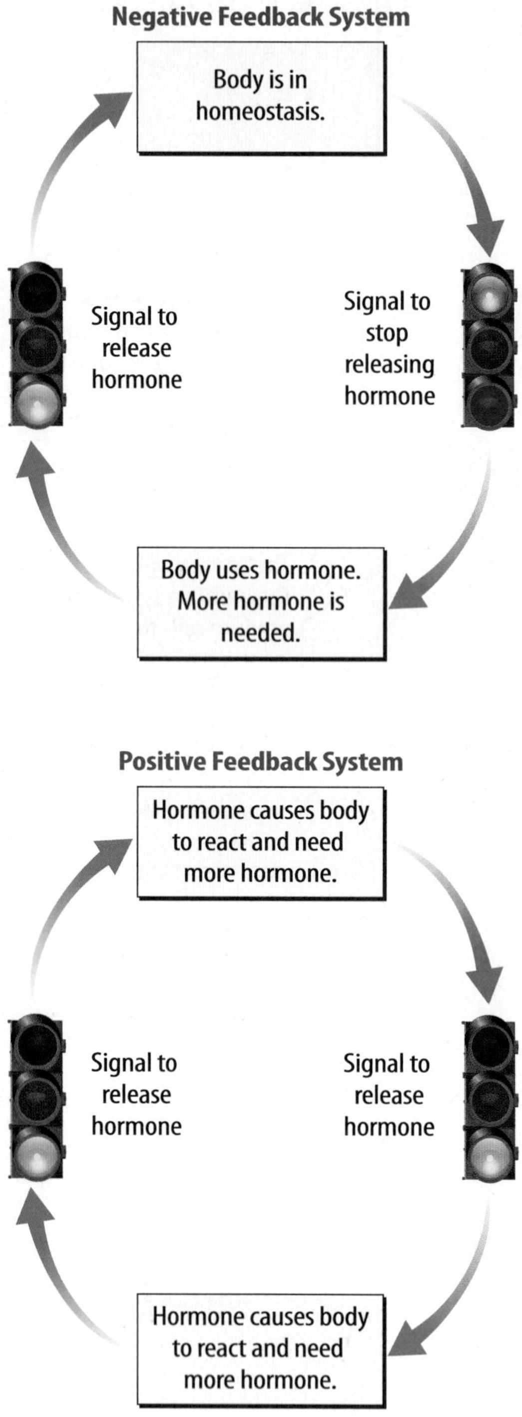

Negative Feedback System

Body is in homeostasis.

Signal to stop releasing hormone

Body uses hormone. More hormone is needed.

Signal to release hormone

Positive Feedback System

Hormone causes body to react and need more hormone.

Signal to release hormone

Hormone causes body to react and need more hormone.

Signal to release hormone

Figure 13 Negative feedback systems are used to maintain homeostasis while positive feedback systems increase the production of hormones.

5. Visual Check Inspect Underline how positive and negative feedback systems differ in the figure.

The Endocrine System and Homeostasis

You might recall from Lesson 1 that your nervous system responds to changes in your environment and maintains homeostasis. Glands also help maintain homeostasis by releasing hormones in response to stimuli. Hormones change the function of other tissues and organs in the body that regulate internal conditions.

Negative Feedback Systems

Organisms can maintain homeostasis using negative feedback systems. **Negative feedback** *is a control system where the effect of a hormone inhibits further release of the hormone.* The endocrine system uses negative feedback and controls the amount of hormone that a gland releases. This is similar to the way a thermostat in a house monitors temperature. The thermostat signals the furnace to turn on when the temperature in the house drops below a preset temperature. After the house warms to the preset temperature, the thermostat signals the furnace to shut off.

Positive Feedback Systems

Whereas negative feedback helps maintain homeostasis, positive feedback does not. **Positive feedback** *is a control system in which the effect of a hormone causes more of the hormone to be released.* Because positive feedback does not help maintain homeostasis, your body uses fewer positive feedback systems. Childbirth is one example of a positive feedback system. Labor begins when a hormone called oxytocin is released. Oxytocin causes contractions. The contractions cause more oxytocin to be released. **Figure 13** shows how both negative feedback systems and positive feedback systems work within the endocrine system.

6. NGSSS Check Highlight how glands work to maintain homeostasis. SC.6.L.14.5

Lesson Review 3

Visual Summary

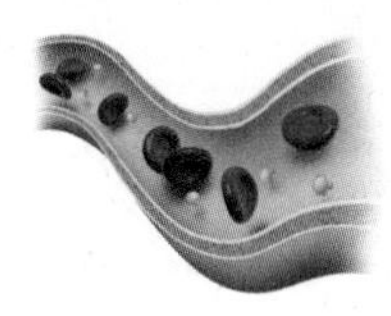

The endocrine system is made up of glands that secrete chemical hormones. They send messages throughout the body and maintain homeostasis.

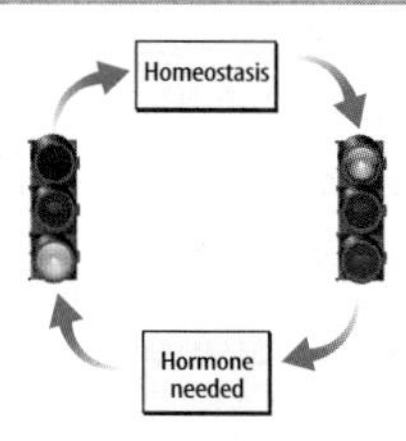

Negative feedback systems maintain homeostasis by inhibiting the release of a hormone as part of the effect of that hormone.

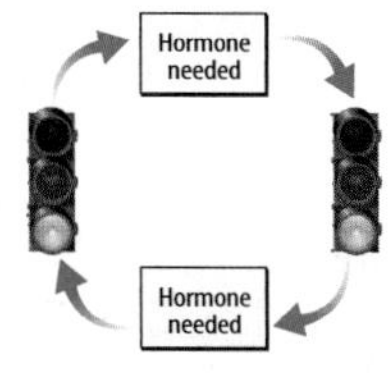

Positive feedback systems increase the effects caused by a hormone by signaling more of that hormone to be released.

Use Vocabulary

1. **Define** the phrase *endocrine system.*

2. **Distinguish** between negative feedback and positive feedback.

Understand Key Concepts

3. Hormones travel to other parts of the body using which system?
   - (A) circulatory
   - (B) digestive
   - (C) nervous
   - (D) skeletal

4. **Explain** the relationship between the nervous system and the endocrine system.

5. **Relate** negative feedback systems to maintaining homeostasis. **SC.6.L.14.5**

Interpret Graphics

6. **Explain** Fill in information about the role of oxytocin in the positive feedback system of childbirth.

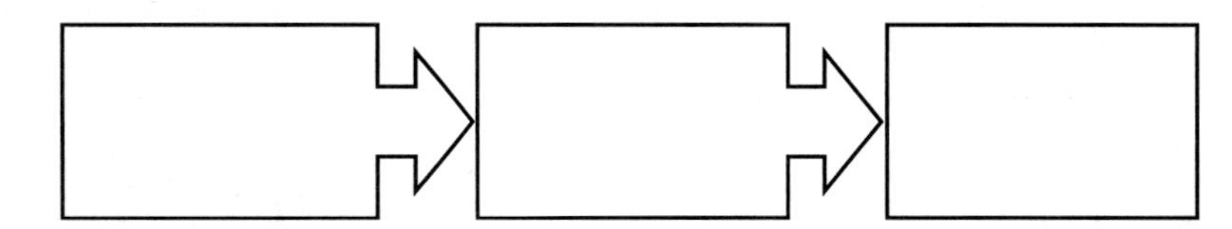

Critical Thinking

7. **Predict** what might happen if negative feedback did not function correctly in one of the endocrine glands.

8. **Explain** why doctors often check the activity of the thyroid glands of a person who is feeling tired and sluggish frequently.

Chapter 16 Study Guide

Think About It! The nervous and endocrine systems interact with other body systems to maintain homeostasis by sensing the environment and coordinating body functions.

Key Concepts Summary

LESSON 1 The Nervous System

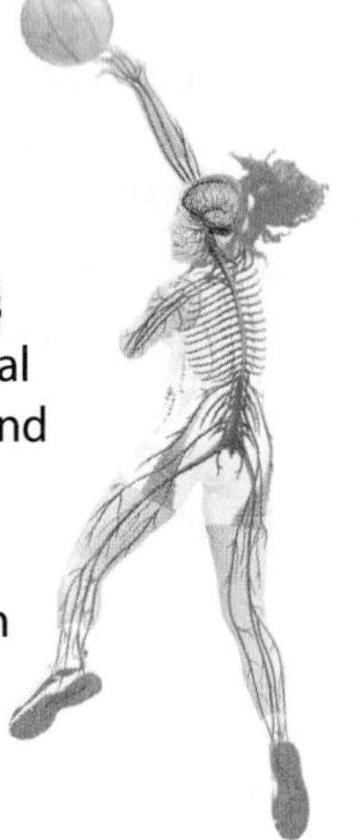

- The **nervous system** gathers information from the environment, processes it, and signals the body to respond.
- The nervous system is made up of the **central nervous system** and **peripheral nervous system.** The peripheral nervous system gathers and transmits information to and from the central nervous system, which processes the information.
- The nervous system helps other body systems maintain homeostasis by responding to the environment.

Vocabulary

nervous system p. 617
stimulus p. 618
neuron p. 619
synapse p. 619
central nervous system p. 620
cerebrum p. 620
cerebellum p. 620
brain stem p. 621
spinal cord p. 621
peripheral nervous system p. 621
reflex p. 622

LESSON 2 The Senses

- The **sensory system** provides information about the environment.
- The senses help maintain homeostasis by gathering information about changes in the environment so the body can respond.

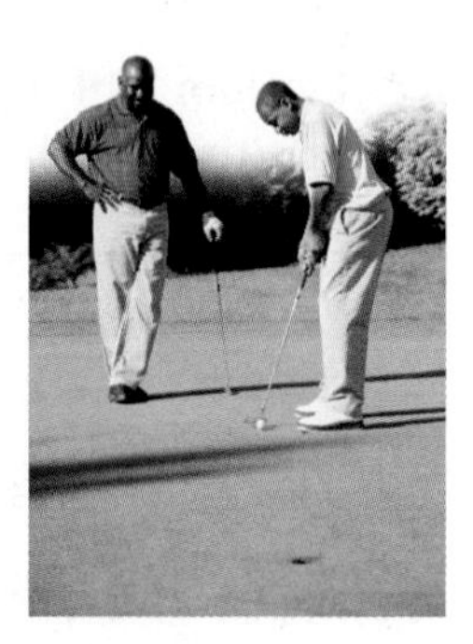

sensory system p. 627
receptor p. 627
retina p. 628
eardrum p. 630

LESSON 3 The Endocrine System

- The **endocrine system** releases chemical messages that control or affect body functions.
- Chemicals released by the endocrine system cause other body systems to react to changes in the environment to maintain homeostasis.

endocrine system p. 637
hormone p. 637
negative feedback p. 640
positive feedback p. 640

Active Reading

FOLDABLES® Chapter Project

Assemble your lesson Foldables as shown to make a Chapter Project. Use the project to review what you have learned in this chapter.

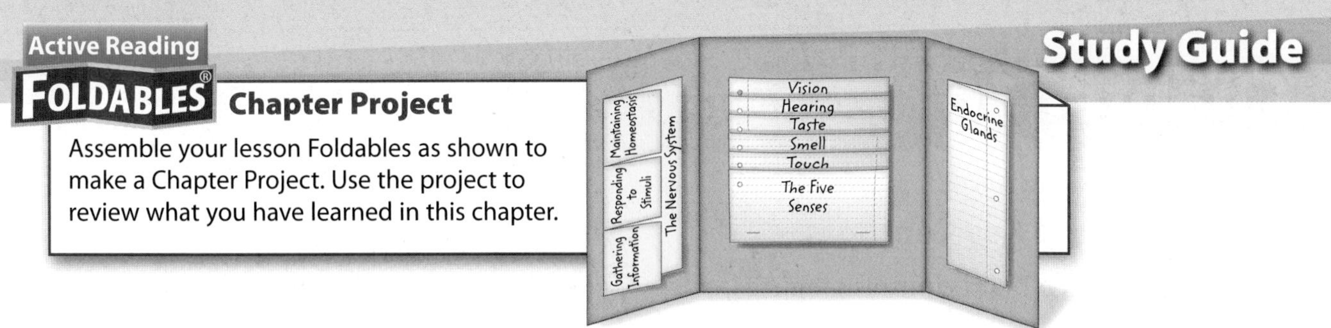

Use Vocabulary

1. The ______________ nervous system is made up of the somatic and autonomic systems.
2. Voluntary muscle movement and coordination is managed by the ______________ part of the brain.
3. The ______________ in the eye contains photoreceptors.
4. The fluid-filled structure in the inner ear that helps maintain balance is the ______________.
5. Chemicals produced by endocrine glands are called ______________.
6. The effect of a hormone on a ______________ feedback system is to cause it to release more of the hormone.

Link Vocabulary and Key Concepts

Use vocabulary terms from the previous page to complete the concept map.

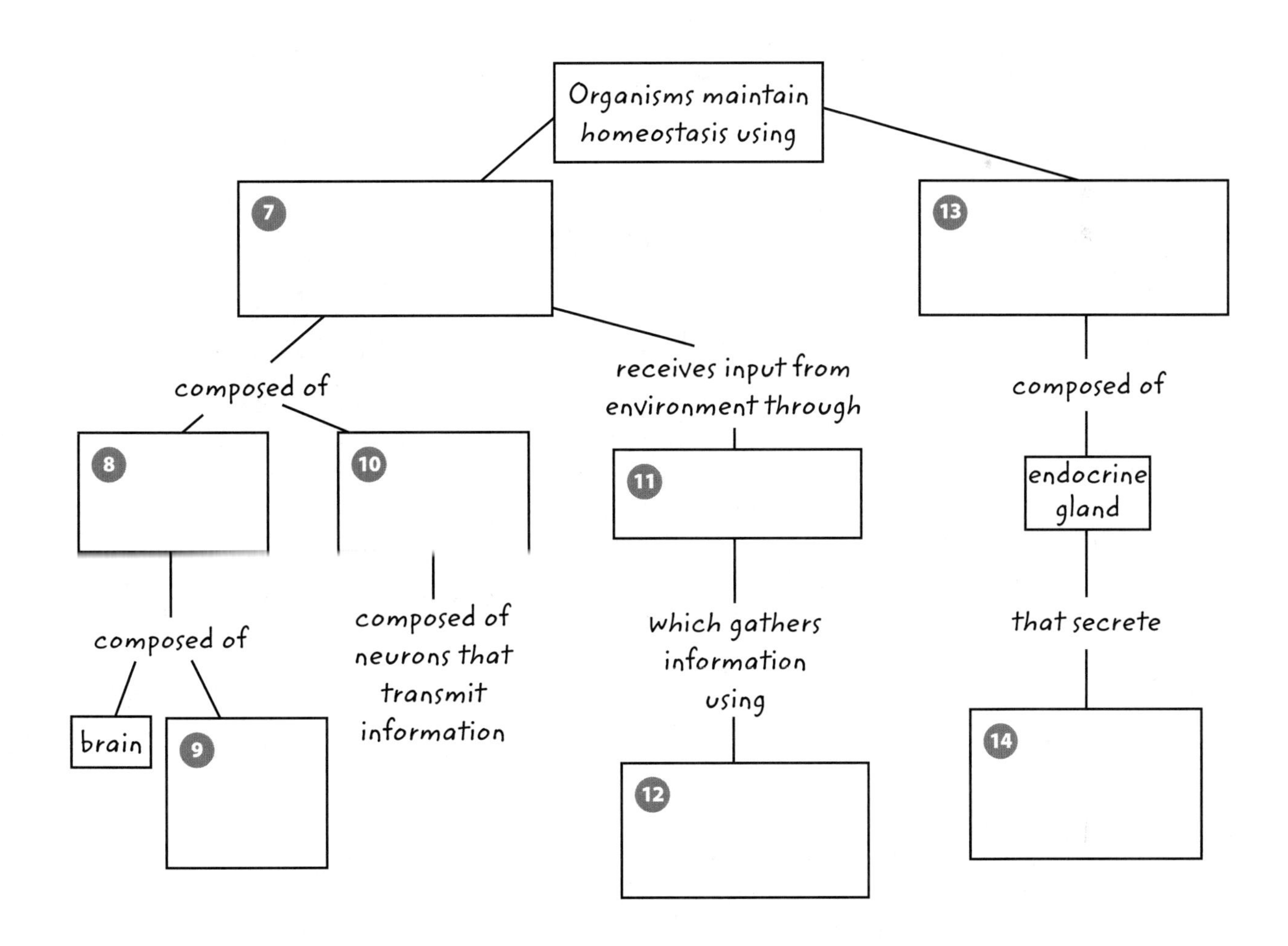

Chapter 16 Review

Fill in the correct answer choice.

Understand Key Concepts

1. Which part of the brain controls language, memory, and thought? **SC.6.L.14.5**
   - Ⓐ brain stem
   - Ⓑ cerebellum
   - Ⓒ cerebrum
   - Ⓓ spinal cord

2. Which is NOT a part of the CNS? **SC.6.L.14.5**
   - Ⓐ brain stem
   - Ⓑ cerebellum
   - Ⓒ somatic system
   - Ⓓ spinal cord

3. What is the gap between the two neurons shown below? **SC.6.L.14.5**

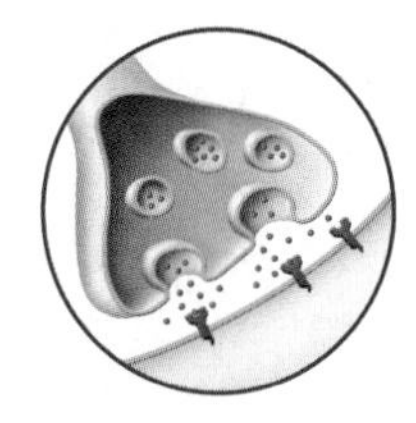

   - Ⓐ motor
   - Ⓑ reflex
   - Ⓒ stimulus
   - Ⓓ synapse

4. Which is NOT a part of the eye? **SC.6.L.14.5**
   - Ⓐ cochlea
   - Ⓑ cornea
   - Ⓒ pupil
   - Ⓓ retina

5. Which senses work together and determine how a person perceives flavor? **SC.6.L.14.5**
   - Ⓐ smell and taste
   - Ⓑ taste and hearing
   - Ⓒ taste and touch
   - Ⓓ vision and taste

6. Which part of the eye focuses light? **SC.6.L.14.5**
   - Ⓐ iris
   - Ⓑ lens
   - Ⓒ pupil
   - Ⓓ retina

Critical Thinking

7. **Describe** how reflexes prevent injuries. **SC.6.L.14.5**

8. **Compare and contrast** the somatic nervous system and the autonomic nervous system **SC.6.L.14.5**

9. **Investigate** the effect of a spinal cord injury on the function of the nervous system. **SC.6.L.14.5**

10. **Relate** how receptors in the nose take part in sensing food. **SC.6.L.14.5**

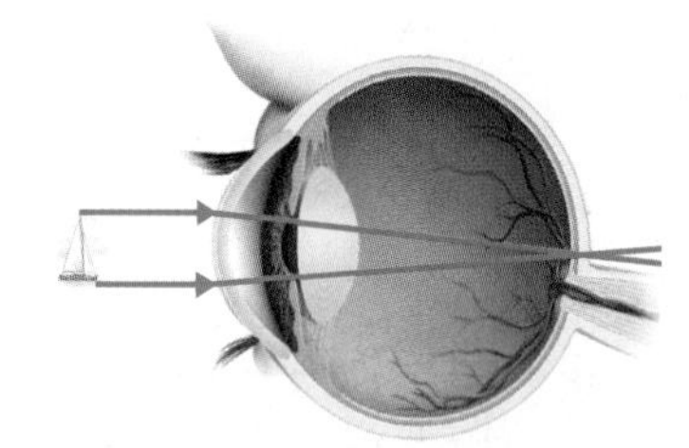

11. **Determine** Is the vision problem shown in the figure below nearsightedness or farsightedness? Explain your answer. **SC.6.L.14.5**

12. **Infer** why people sometimes feel sick when traveling on a boat. **SC.6.L.14.5**

13 **Evaluate** how negative feedback systems control the amount of hormone released by a gland. SC.6.L.14.5

14 **Infer** why positive feedback systems are not used to maintain homeostasis. SC.6.L.14.5

15 **Assess** why reflexes are controlled by the nervous system and not the endocrine system. SC.6.L.14.5

16 **Evaluate** how the endocrine system works with the circulatory system to maintain the body's homeostasis. SC.6.L.14.5

Writing in Science

17 **Write** a paragraph on a separate piece of paper that describes a time when your nervous system helped you to survive. Describe what likely occurred in your body, including the use of your senses and the endocrine system during the event. LA.6.2.2.3

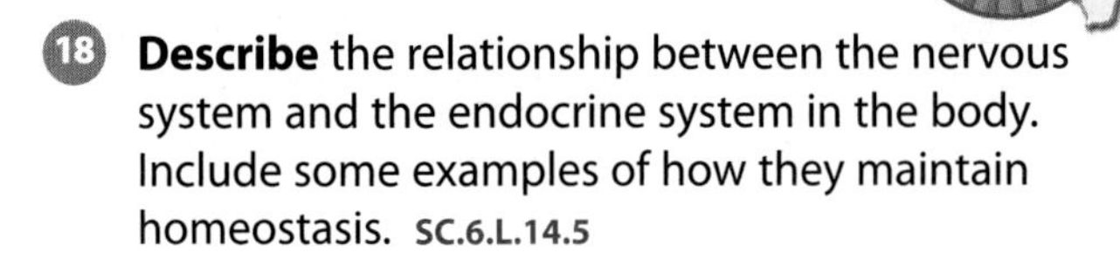

18 **Describe** the relationship between the nervous system and the endocrine system in the body. Include some examples of how they maintain homeostasis. SC.6.L.14.5

Math Skills MA.6.A.3.6

Use Proportions

Use the information below to answer questions 19 and 20.

A person's nerve axon from a toe to the spine is 1.1 m long. The nerve impulse that signals touch travels about 76.2 m/s.

19 How long does it take the nerve impulse for touch to travel from the toe to the spine?

20 Pain impulses travel more slowly—about 0.61 m/s. How long would it take the pain impulse to travel from the toe to the spine?

21 A reflex nerve impulse travels 100 m/s from a fingertip touching a hot object to the spine. The distance is 0.5 m. If a signal from the spine telling the finger to move travels back at the same speed, how long will it take the person to react?

Florida NGSSS Benchmark Practice

Fill in the correct answer choice.

Multiple Choice

Use the diagram below to answer question 1.

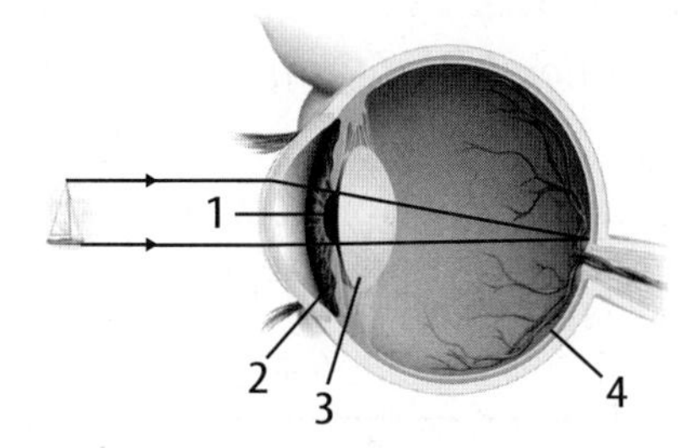

1 Which numbered structure focuses the light on the retina? **SC.6.L.14.5**

Ⓐ 1
Ⓑ 2
Ⓒ 3
Ⓓ 4

2 A doctor sees a patient who has a loss of balance from an illness. The doctor thinks the receptors for balance might be injured. Where are they located? **SC.6.L.14.5**

Ⓕ inner ear
Ⓖ middle ear
Ⓗ nasal cavity
Ⓘ spinal cord

3 Which statement about the nervous system is false? **SC.6.L.14.5**

Ⓐ It gathers information.
Ⓑ It maintains homeostasis.
Ⓒ It responds to stimuli.
Ⓓ It transports hormones.

4 Which relies on chemoreceptors? **SC.6.L.14.5**

Ⓕ hearing
Ⓖ sight
Ⓗ taste
Ⓘ touch

Use the graph below to answer question 5.

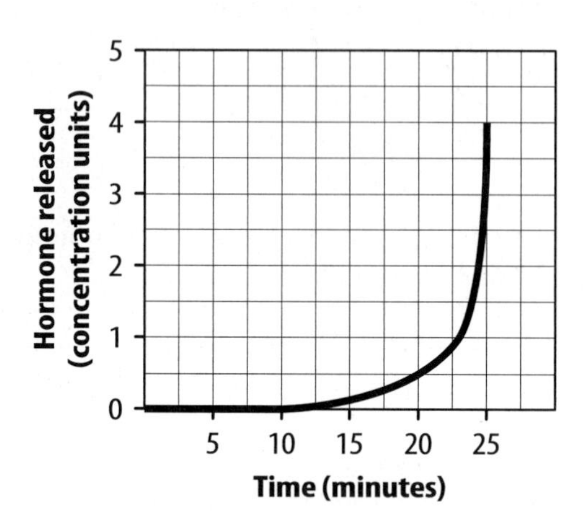

5 Which is likely to produce a curve similar to the one above? **SC.6.L.14.5**

Ⓐ a reflex
Ⓑ a synapse
Ⓒ negative feedback
Ⓓ positive feedback

6 Which system releases chemical messages to the bloodstream? **SC.6.L.14.5**

Ⓕ circulatory
Ⓖ digestive
Ⓗ endocrine
Ⓘ skeletal

7 Why does the nervous system tell the body to shiver? **SC.6.L.14.5**

Ⓐ to imprint a memory on the brain
Ⓑ to maintain homeostasis
Ⓒ to slow neuron communication
Ⓓ to stop PNS–CNS communication

Use the diagram below to answer question 8.

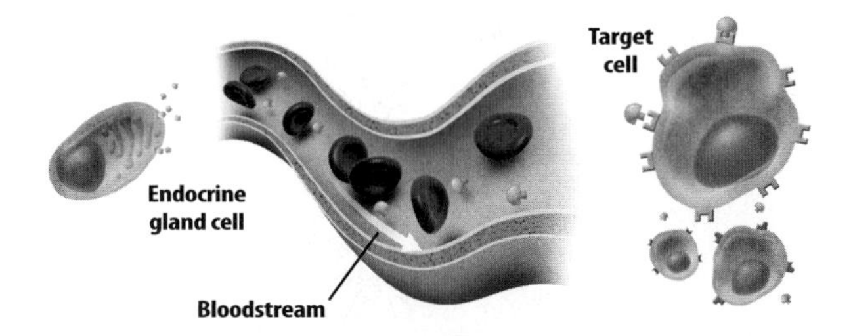

8 What is the path of a hormone as illustrated above? SC.6.L.14.5

Ⓕ bloodstream → endocrine cell → target cell

Ⓖ endocrine cell → bloodstream → target cell

Ⓗ endocrine cell → target cell → bloodstream

Ⓘ target cell → bloodstream → endocrine cell

9 What is the primary function of the senses? SC.6.L.14.5

Ⓐ choosing pleasant scents, sounds, tastes, and textures

Ⓑ gathering information about the environment

Ⓒ making each organism slightly different from others

Ⓓ providing ideas for organisms to communicate

10 Which system transports hormones to other parts of the body? SC.6.L.14.5

Ⓕ circulatory system

Ⓖ digestive system

Ⓗ nervous system

Ⓘ reproductive system

11 Which is NOT part of the endocrine system? SC.6.L.14.5

Ⓐ pancreas

Ⓑ ovary

Ⓒ skin

Ⓓ thymus

12 Which system is illustrated by the diagram below? SC.6.L.14.5

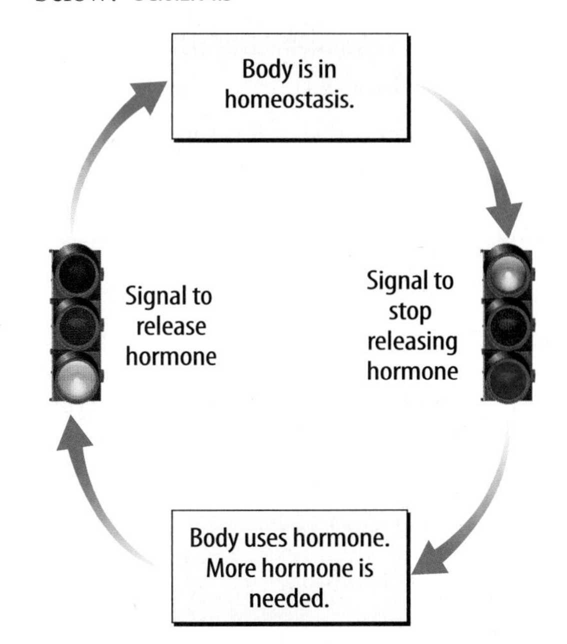

Ⓕ autonomic system

Ⓖ reflex system

Ⓗ negative feedback system

Ⓘ positive feedback system

NEED EXTRA HELP?

If You Missed Question...	1	2	3	4	5	6	7	8	9	10	11	12
Go to Lesson...	2	2	1	2	3	3	1	3	2	3	3	3

Name ______________________ Date ______________

Multiple Choice *Bubble the correct answer.*

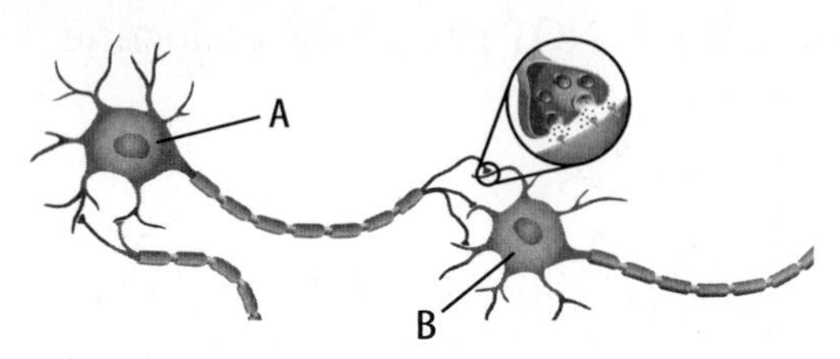

1. Using the image above, which statement explains how a signal moves between the neurons? **SC.6.L.14.5**

(A) A chemical signal is sent from the axon of neuron A to the dendrites of neuron B.

(B) A chemical signal is sent from the axon of neuron B to the dendrites of neuron A.

(C) An electric signal is sent from the dendrites of neuron A to the axon of neuron B.

(D) An electric signal is sent from the dendrites of neuron B to the axon of neuron A.

2. Which part of the nervous system controls sneezing? **SC.6.L.14.5**

(F) the cerebellum

(G) the cerebrum

(H) the brain stem

(I) the spinal cord

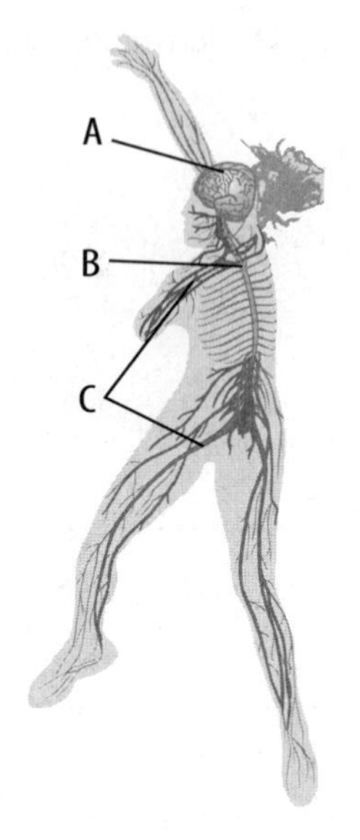

3. Which letters in the image above indicate structures that are part of the central nervous system? **SC.6.L.14.5**

(A) A and B

(B) A and C

(C) B and C

(D) A, B, and C

4. During an accident, a patient's cerebrum was injured. What might the patient experience as a result of the injury? **SC.6.L.14.5**

(F) The patient might have difficulty with his or her breathing.

(G) The patient might have difficulty moving his or her arms and legs.

(H) The patient might have trouble keeping his or her balance.

(I) The patient might have trouble with his or her memory.

Name ______________________________ Date ______________

Benchmark Mini-Assessment Chapter 16 • Lesson 2

FLORIDA NGSSS mini BAT

Multiple Choice *Bubble the correct answer.*

Use the following image to answer questions 1 and 2.

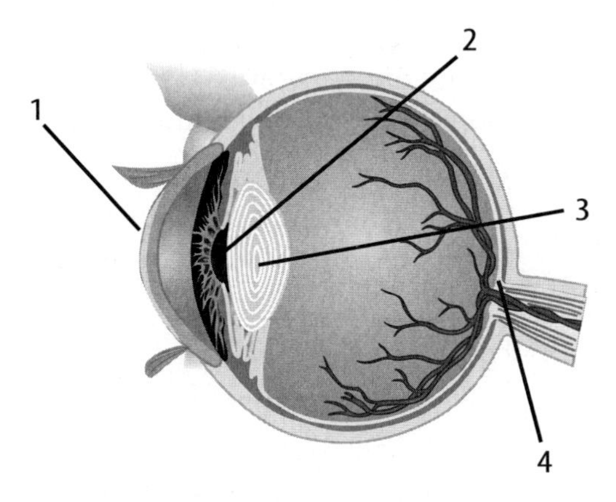

1. Identify which part of the eye bends light rays. **SC.6.L.14.5**

Ⓐ 1

Ⓑ 2

Ⓒ 3

Ⓓ 4

2. Identify which part of the eye is made of rod cells and cone cells. **SC.6.L.14.5**

Ⓕ 1

Ⓖ 2

Ⓗ 3

Ⓘ 4

3. Which senses rely on chemoreceptors? **SC.6.L.14.5**

Ⓐ sight and hearing

Ⓑ sight and smell

Ⓒ smell and taste

Ⓓ smell and touch

4. In addition to affecting the ability to hear, an ear injury also might affect **SC.6.L.14.5**

Ⓕ balance.

Ⓖ sight.

Ⓖ smell.

Ⓘ touch.

Name ______________________ Date ______________

Multiple Choice *Bubble the correct answer.*

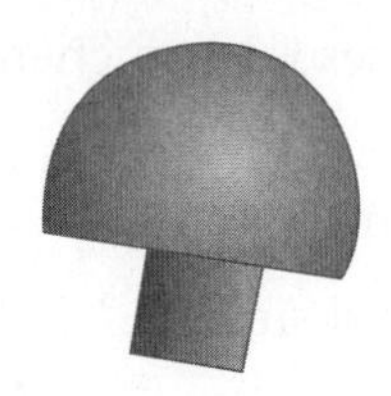

1. Which cell in the images below is the target cell for the hormone molecule shown in the image above? **SC.6.L.14.5**

(A)

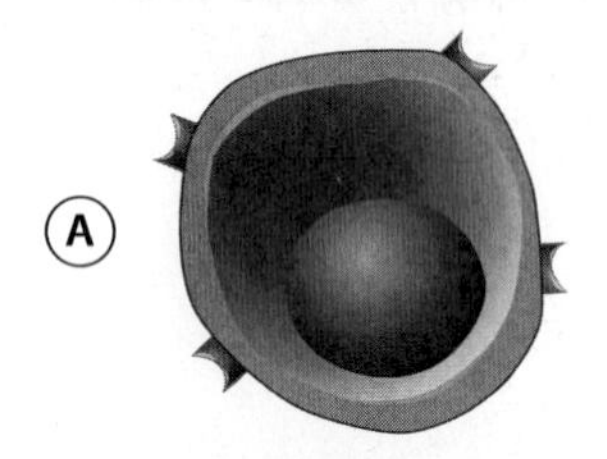

(B)

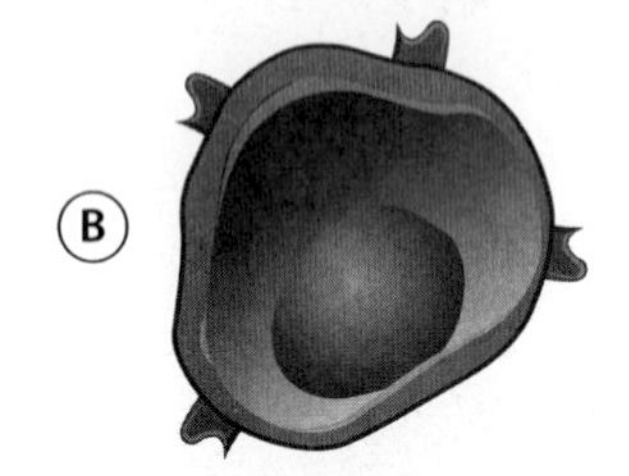

(C)

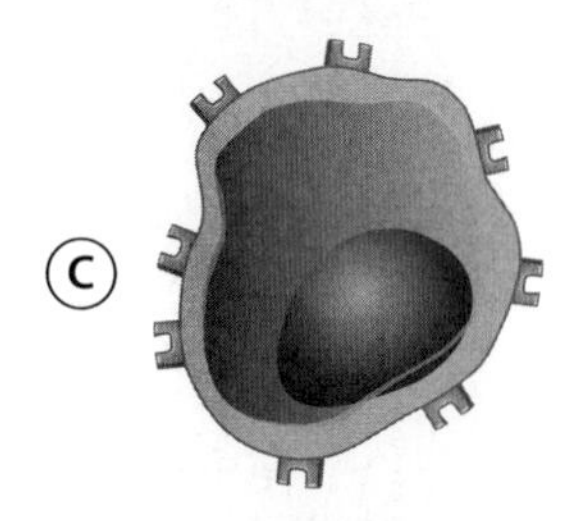

(D)

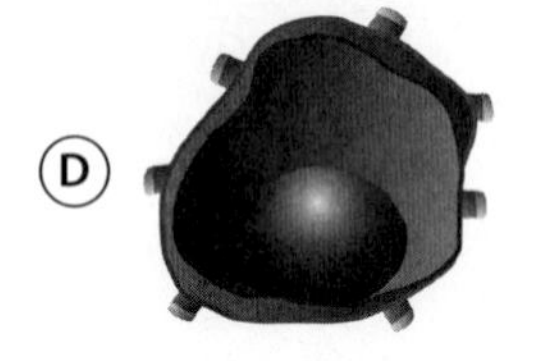

2. Which statement BEST compares the nervous and endocrine systems? **SC.6.L.14.5**

(F) The nervous system sends longer-lasting messages and sends messages to more cells than the endocrine system.

(G) The nervous system sends signals through the blood stream while the endocrine system sends messages through neurons.

(H) The nervous system transmits signals faster than the endocrine system does.

(I) The nervous system uses chemical messages while the endocrine system uses chemical and electric messages.

3. Which gland controls the activity of the pituitary gland? **SC.6.L.14.5**

(A) adrenal

(B) thyroid

(C) hypothalamus

(D) parathyroid

4. An illness that affects the thyroid gland might in turn affect **SC.6.L.14.5**

(F) metabolism.

(G) testosterone.

(H) the immune system.

(I) the response to stress.

Name ______________________ Date ______________

Notes

Name ______________________ Date ______________

Notes

Do they produce eggs?

Five friends looked at a chicken egg. They wondered if other organisms, besides chickens, produce eggs for reproduction. They each had a different idea.

Eddie: I think only birds produce eggs.

Vernon: I think only amphibians, reptiles, and birds produce eggs.

Forest: I think all animals, except mammals, produce eggs.

Angie: I think all animals, except humans, produce eggs.

Sophie: I think all animals, including humans, produce eggs.

Circle Who do you most agree with and explain why? Describe your ideas about eggs and reproduction.

Reproduction and DEVELOPMENT

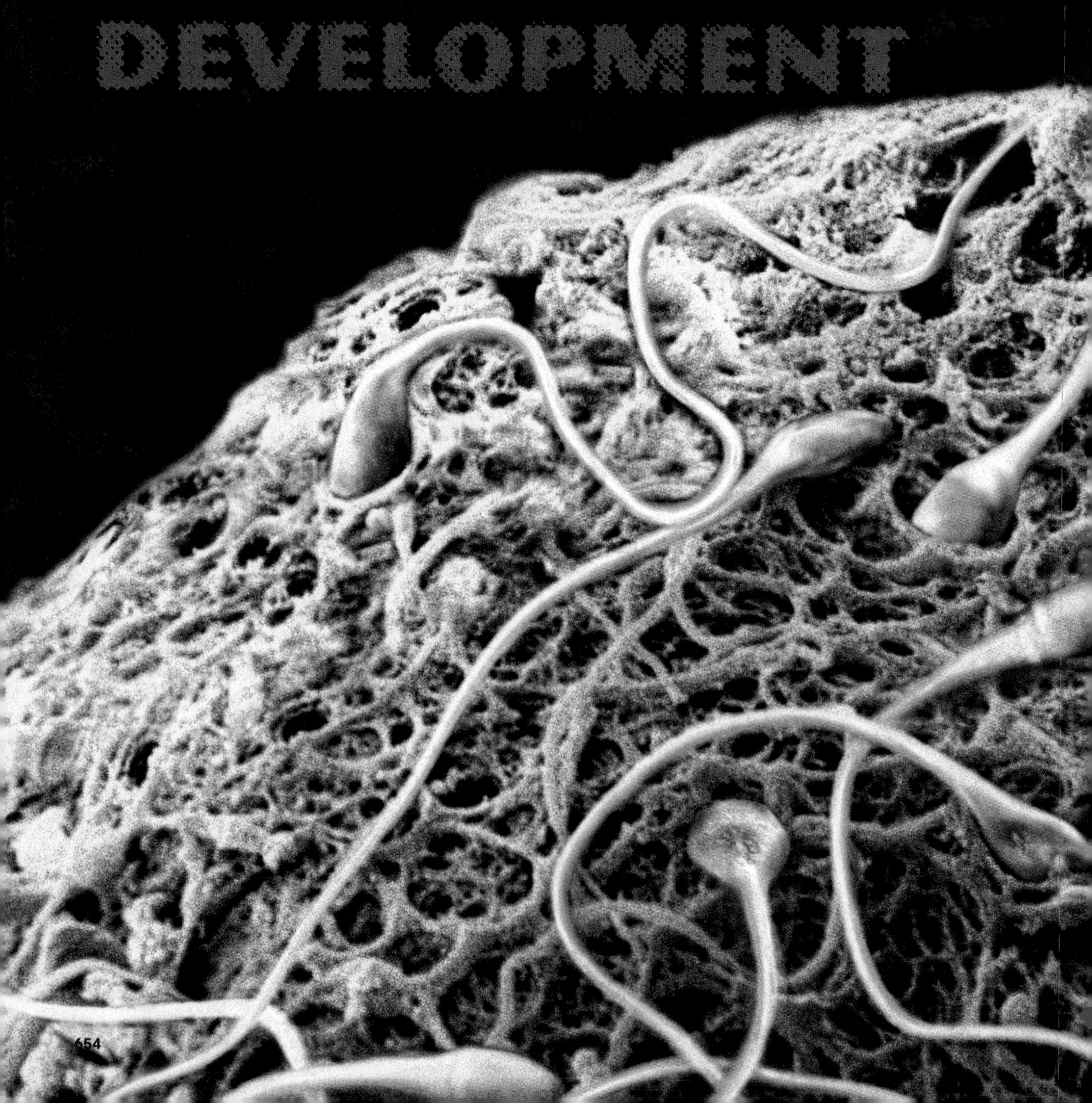

FLORIDA BIG IDEAS

1 **The Practice of Science**

2 **The Characteristics of Scientific Knowledge**

14 **Organization and Development of Living Organisms**

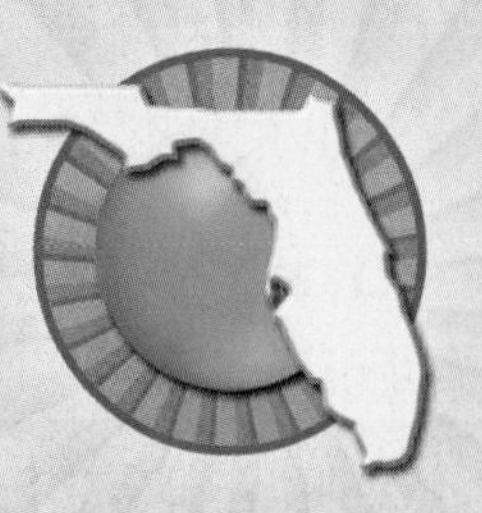

The Big Idea

Think About It!

What are the stages of human reproduction and development?

The things that look like strands of hair are male reproductive cells called sperm. They are covering a female reproductive cell called an egg.

1. What do you think the sperm are doing?

____

____

____

2. What stage of human reproduction do you think this is?

____

____

____

Get Ready to Read

What do you think about development?

Before you read, decide if you agree or disagree with each of these statements. As you read this chapter, see if you change your mind about any of the statements.

	AGREE	DISAGREE
1 Reproduction ensures that a species survives.	☐	☐
2 The male reproductive system has internal and external parts.	☐	☐
3 The menstrual cycle occurs in males and females.	☐	☐
4 Eggs are fertilized in the ovary.	☐	☐
5 Lead is a nutrient that helps a fetus develop.	☐	☐
6 Puberty occurs during adolescence.	☐	☐

There's More Online!
Video • Audio • Review • ⓘLab Station • WebQuest • Assessment • Concepts in Motion • Multilingual eGlossary

Lesson 1

The Reproductive SYSTEM

ESSENTIAL QUESTIONS

- What does the reproductive system do?
- 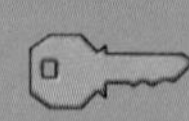 How do the parts of the male reproductive system work together?
- How do the parts of the female reproductive system work together?
-  How does the reproductive system interact with other body systems?

Vocabulary

sperm p. 657
egg p. 657
testis p. 658
semen p. 658
penis p. 658
ovary p. 660
vagina p. 660
menstrual cycle p. 662
ovulation p. 662
fertilization p. 663

Florida NGSSS

SC.6.L.14.5 Identify and investigate the general functions of the major systems of the human body (digestive, respiratory, circulatory, reproductive, excretory, immune, nervous, and musculoskeletal) and describe ways these systems interact with each other to maintain homeostasis.

SC.6.N.1.5 Recognize that science involves creativity, not just in designing experiments, but also in creating explanations that fit evidence.

SC.6.N.2.3 Recognize that scientists who make contributions to scientific knowledge come from all kinds of backgrounds and possess varied talents, interests, and goals.

Also covers: LA.6.2.2.3, LA.6.4.2.2

Inquiry Launch Lab

15 minutes

LA.6.2.2.3, SC.6.N.1.5, SC.6.L.14.5

How do male and female gametes compare?

Procedure 

1. Read and complete a lab safety form.
2. Obtain a **microscope** and **prepared slides of sperm cells and egg cells.** Carefully handle the microscope as instructed by your teacher.
3. Observe the slides under the magnification power specified.
4. Sketch a sperm cell and an egg cell below.

Data and Observations

Think About This

1. Compare and contrast the appearance of an egg cell and the appearance of a sperm cell.
2. Key Concept Why do you think there is a difference in the appearance of the male and female reproductive cells?

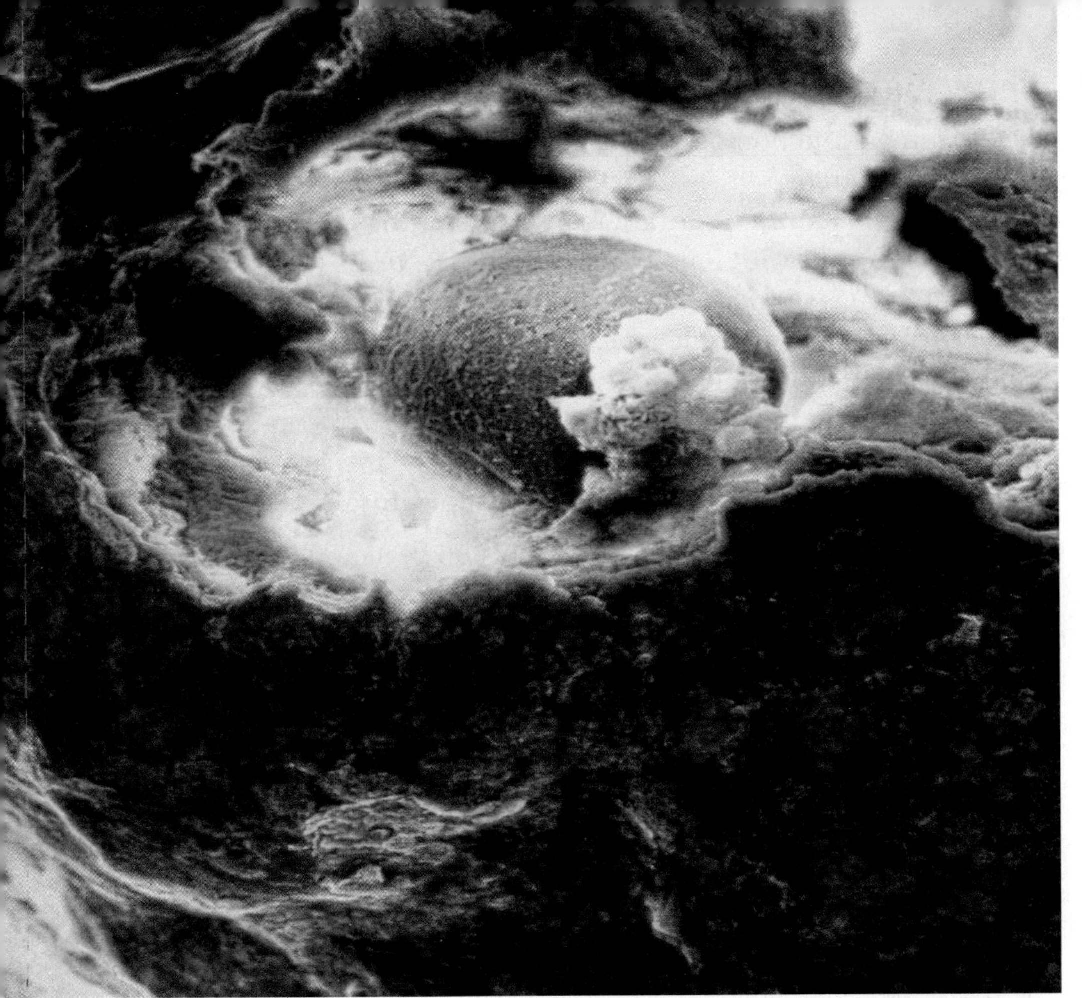

Inquiry A Red Ball?

1. The round object is a human egg being released from the ovary. How often are eggs released?

______

______

______

______

______

______

Active Reading **2. Build** As you read, make an outline to summarize the information in the lesson. Use the main headings in the lesson as the main headings in your outline. Use your outline to review the lesson.

Functions of the Reproductive System

You have read about many of the organ systems that enable you to grow and respond to changes in the environment. Organ systems such as the nervous system, the skeletal system, the circulatory system, and the digestive system are all important for an individual's survival. But what organ system ensures that the human species survives?

A reproductive system is a group of tissues and organs in humans. It enables the male and female reproductive cells to join and form new offspring. Like other animals, human males produce **sperm,** *the male reproductive cells,* and human females produce **eggs,** *the female reproductive cells.*

During reproduction, a sperm joins with an egg, as shown in **Figure 1**. This usually happens inside a female's reproductive system. Once joined, part of the female's reproductive system nourishes the developing human. In order to understand how humans develop, we must first learn the parts of the male and female reproductive systems.

Active Reading **3. Compile** Highlight what the reproductive system does.

Figure 1 When a sperm cell combines with an egg cell, development of a new human being begins.

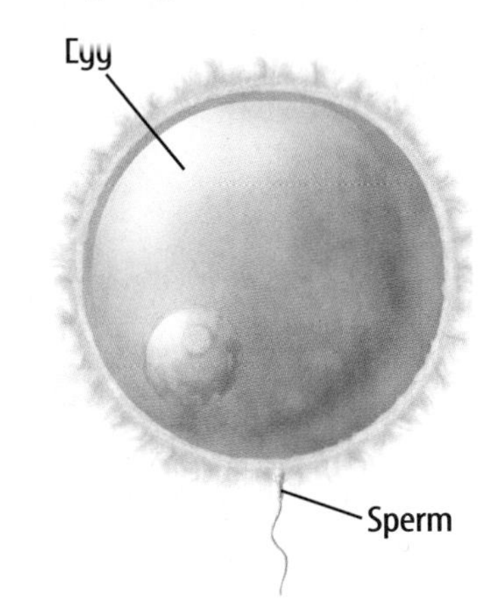

The Male Reproductive System

The main function of the male reproductive system is to produce and transport sperm to the female reproductive system. As shown in **Figure 2**, there are many parts of the male reproductive system, each with a unique function. *The* **testis** (TES tihs; plural, testes) *is the male reproductive organ that produces sperm.* A male's two testes are inside an external saclike **structure** called the scrotum (SKROH tum). Sperm development can occur only at a temperature that is lower than normal body temperature. Because the scrotum is outside the male's body, it is at a temperature slightly lower than normal body temperature.

Science Use v. Common Use
structure
Science Use cells and tissues arranged in a definite pattern
Common Use a building

Once sperm develop, they move to a tube called the sperm duct and are stored. During storage, sperm mature and develop the ability to swim. As mature sperm move through the remainder of the male reproductive system, they mix with fluids produced by several glands. This mixture of sperm and fluids is called **semen** (SEE mun). Semen also contains nutrients that provide sperm with energy. Semen leaves the body through the penis. *The* **penis** *is a tubelike structure that delivers sperm to the female reproductive system.* The path of sperm is shown in **Figure 3**.

Active Reading **4. Classify** Underline what semen contains.

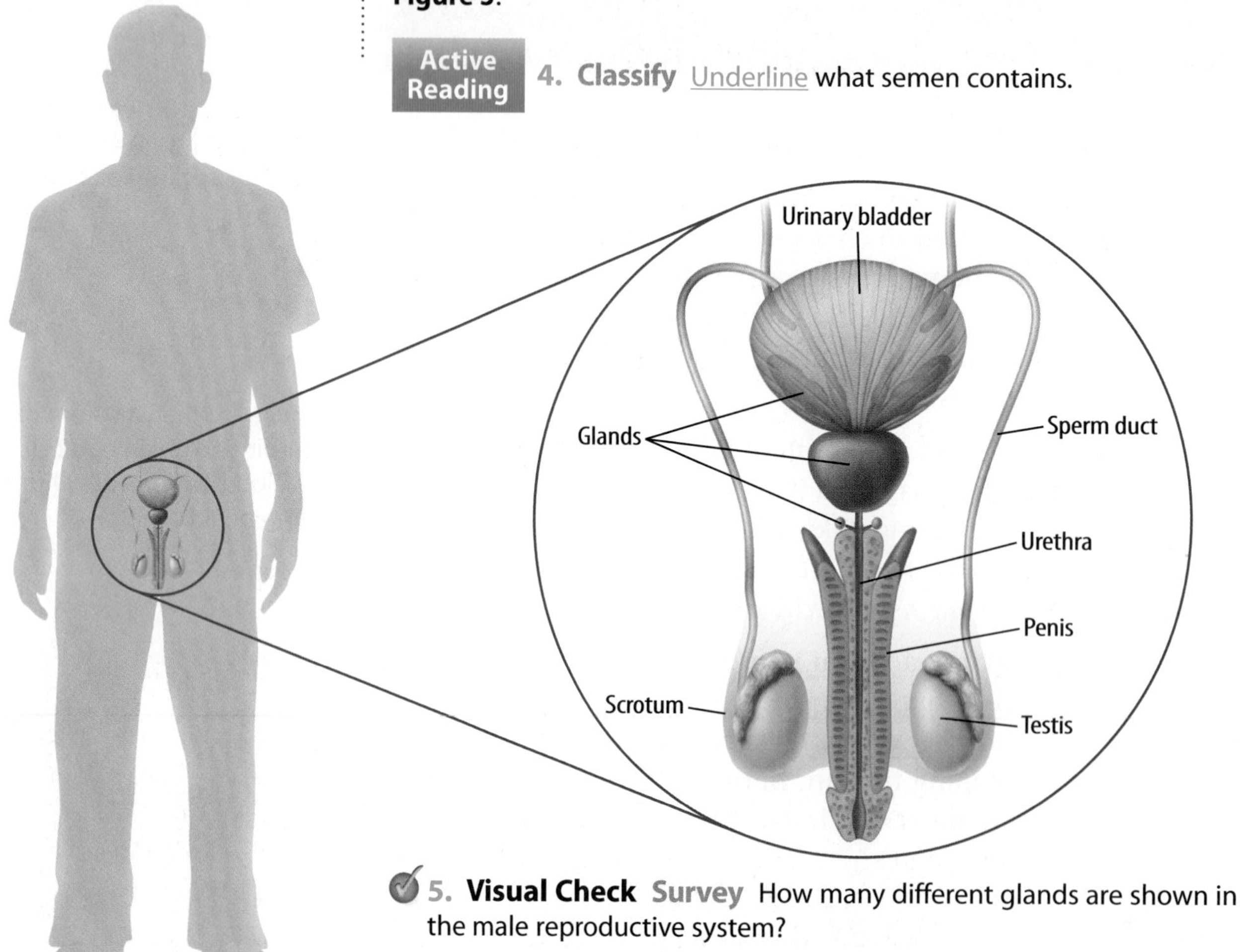

Figure 2 The male reproductive system has many parts, each with a unique function.

5. Visual Check Survey How many different glands are shown in the male reproductive system?

____

Sperm

Now that you have learned about the parts of the male reproductive system, let's take a closer look at sperm. When mature, sperm can join with an egg. As illustrated in **Figure 3,** a mature sperm cell has three main parts—a head, a midpiece, and a tail. The head contains DNA and substances that help the sperm join with an egg. The midpiece contains organelles called mitochondria. Recall that mitochondria are the cell organelles that process food molecules and release energy. This energy enables movement of a sperm's tail. The tail is a long, slender structure that whips back and forth and moves the sperm. Although semen contains millions of sperm, only one sperm joins with an egg.

Active Reading **6. Construct** How do the parts of the male reproductive system work together?

____

____

____

Figure 3 Mature sperm leave the body through the penis.

Male Reproductive System

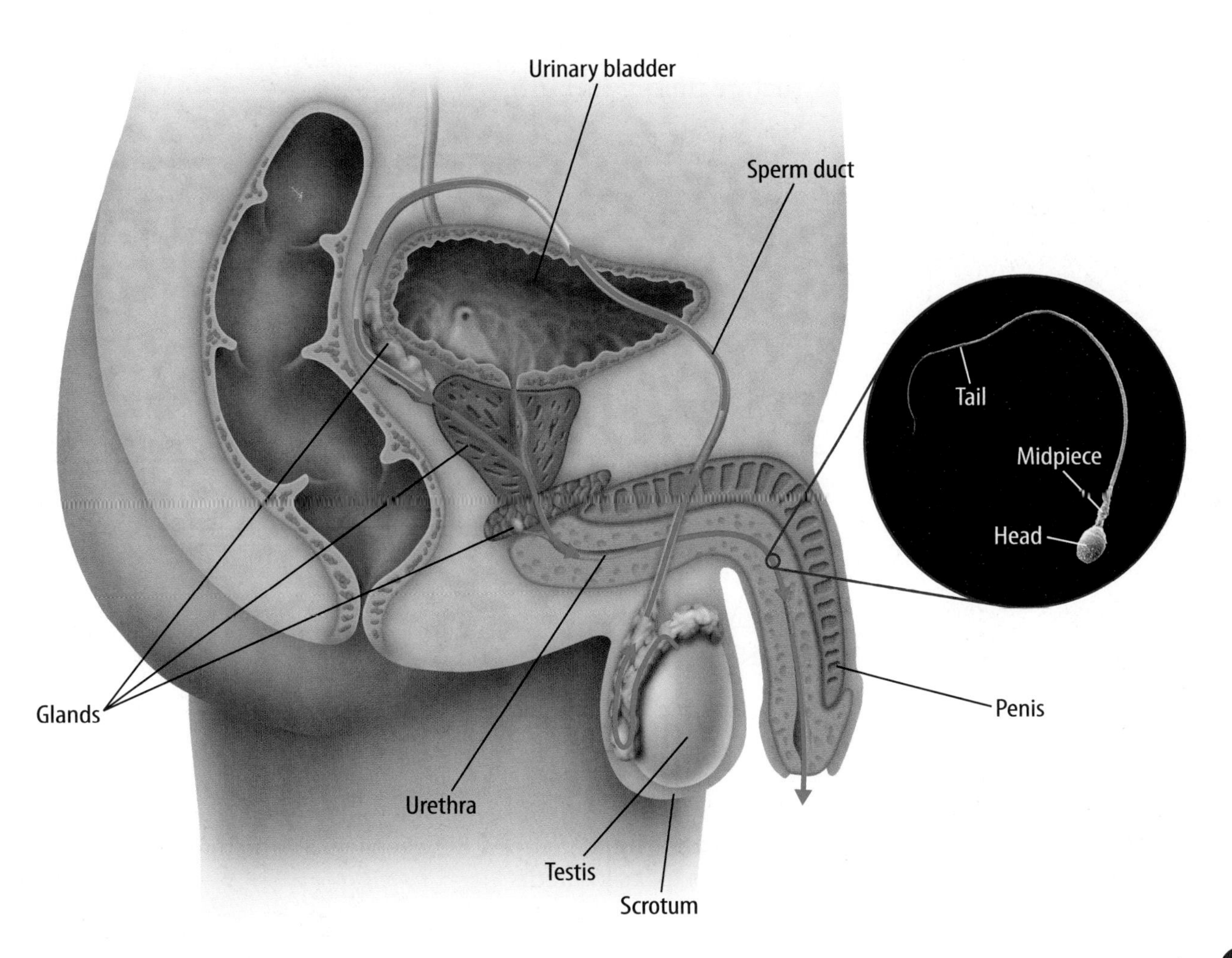

The Female Reproductive System

As shown in **Figure 4,** the female reproductive system also has many structures, each with a unique function. Recall from **Figure 2** that the male reproductive system has both internal and external parts. In contrast, all the parts of the female reproductive system are inside the body.

Word Origin
ovary
from Latin *ovum,* means "egg"

Recall that both the male and female reproductive systems produce reproductive cells. Males produce sperm, and females produce immature eggs called oocytes (OH uh sites). *An* **ovary** (OH vah ree) *is an organ where oocytes are stored and mature.* A mature oocyte is called an egg, or ovum (plural, ova). Just as males have two testes, females have two ovaries.

About once a month, an ovary releases an egg. After an egg is released, it enters the fallopian (fuh LOH pee un) tube. Short, hairlike structures called cilia move the egg through the fallopian tube toward the uterus. If the egg is fertilized by a sperm, the uterus provides a nourishing environment for a fertilized egg's development. You will read more about how humans develop in Lesson 2.

The part of the female reproductive system that connects the uterus to the outside of the body is the **vagina.** Sperm enter the female reproductive system through the vagina. The vagina is also called the birth canal because a baby moves through this structure during its birth.

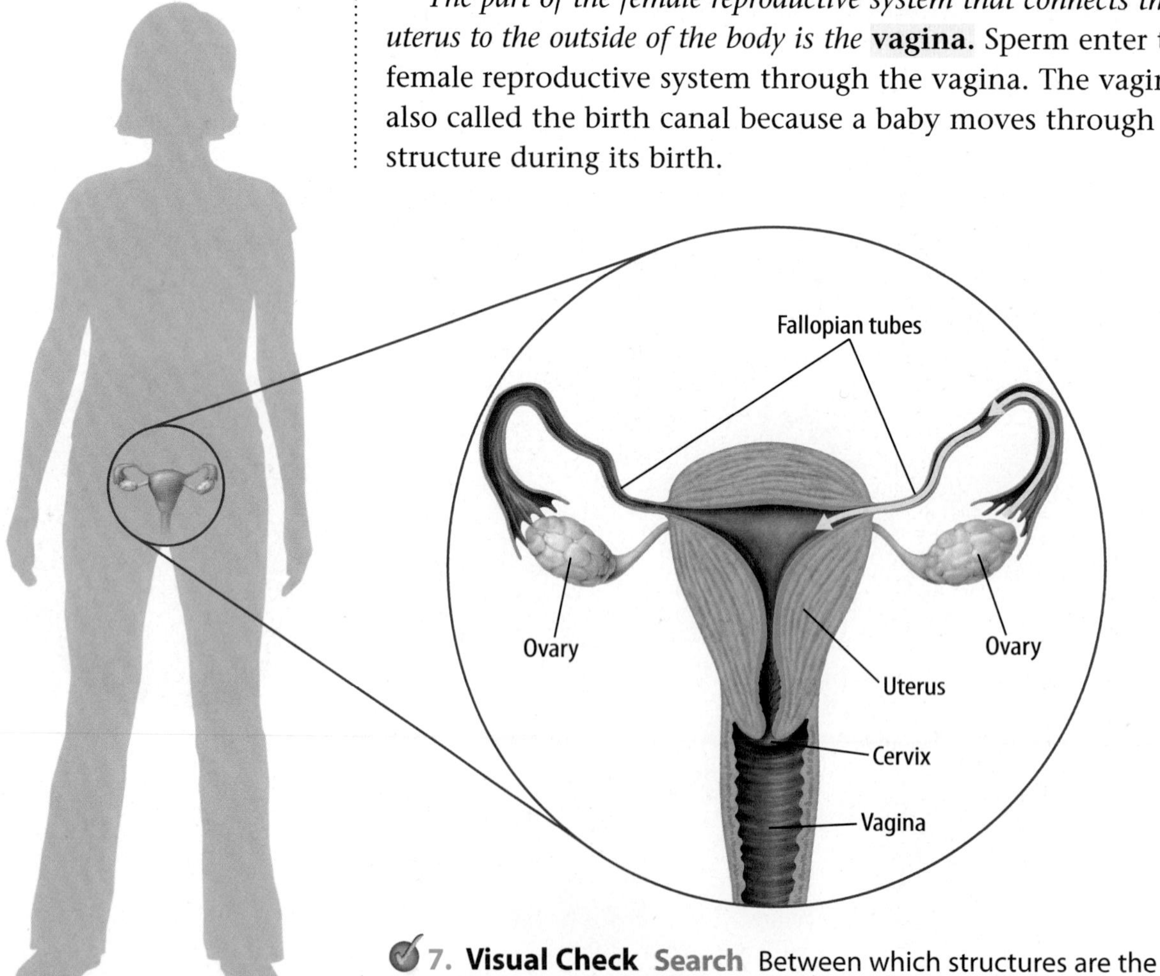

Figure 4 The female reproductive system has many parts, each with a unique function.

7. Visual Check Search Between which structures are the fallopian tubes located?

The Egg

Unlike sperm, which are long and slender when mature, eggs are large and round when mature. In fact, an egg is about 2,000 times larger than a sperm. Like sperm, eggs contain DNA. However, unlike sperm, an egg is filled with substances that provide it with nourishment. Another important difference between males and females is that a male releases millions of sperm in semen, but a female usually releases only one egg at a time. As shown in **Figure 5,** each oocyte in an ovary is surrounded by follicle cells. The follicle cells release hormones that help the oocytes develop into eggs. The changes that occur as oocytes develop and the release of an egg are also shown in **Figure 5.**

Active Reading 8. **Contrast** Highlight how mature sperm and eggs differ.

Figure 5 A follicle increases in size as it prepares to release an egg.

Female Reproductive System

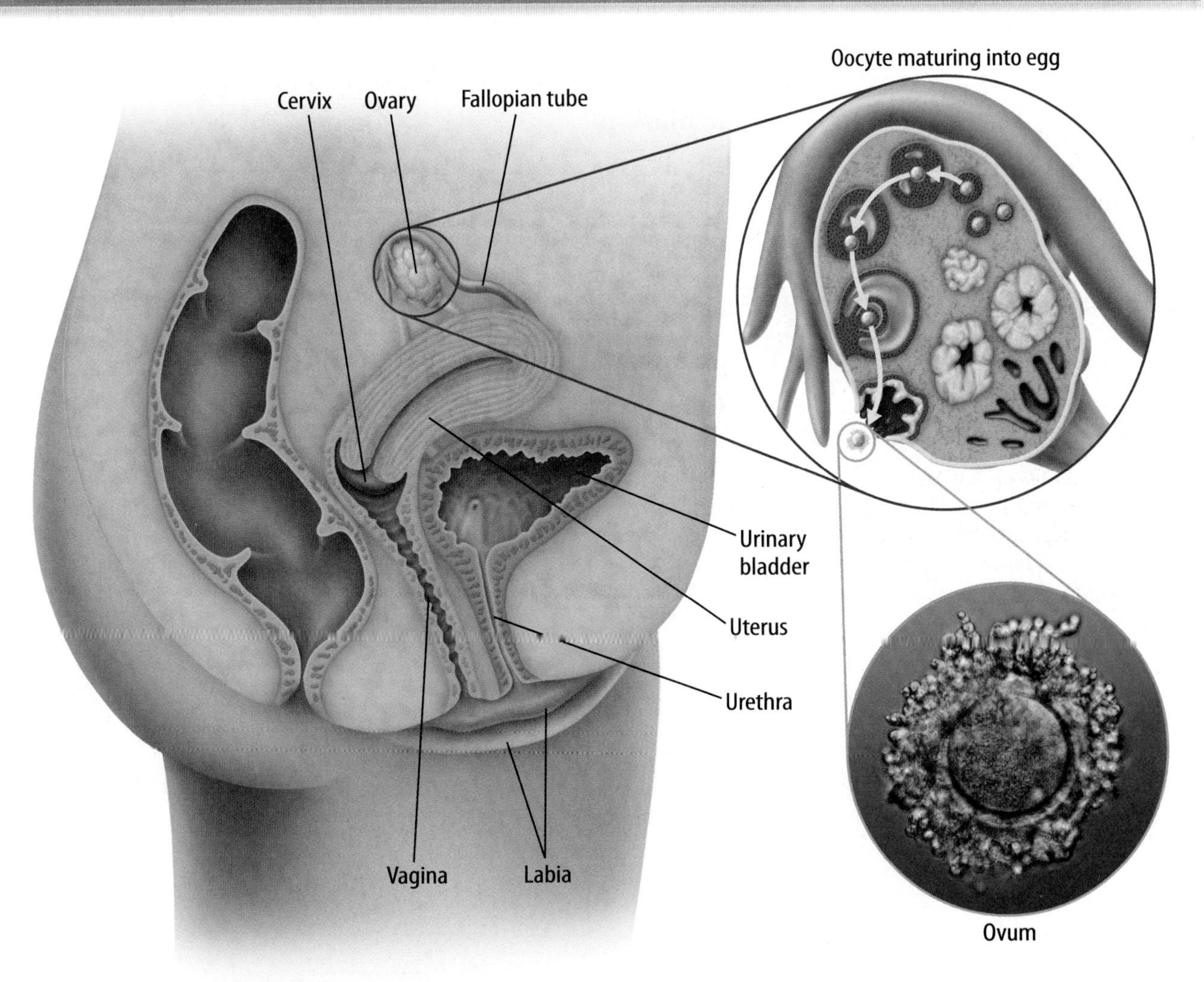

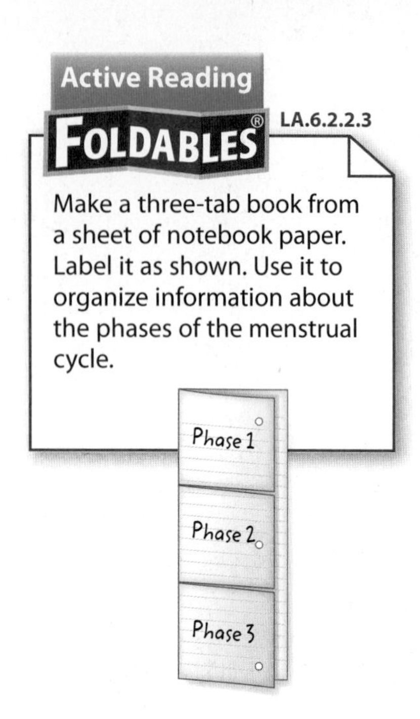

The Menstrual Cycle

A female usually releases one egg at a time. Because the function of the reproductive system is to produce a new human, an egg is released only when the uterus is prepared to nourish it. *The ovaries and the uterus go through reproductive-related changes called the* **menstrual** (MEN stroo ul) **cycle.** The menstrual cycle is caused by chemical signals called hormones. One menstrual cycle is about 28 days long and can be divided into three phases. It is called a cycle because the phases repeat in the same order and in about the same amount of time.

Phase 1 The menstrual cycle begins with a process called menstruation (men stroo WAY shun). During menstruation, tissue, fluid, and blood cells pass from the uterus through the vagina and are removed from the body. Menstruation usually lasts about five days.

Phase 2 In the next phase of the menstrual cycle, the tissue lining the uterus thickens, as shown in **Figure 6.** In the ovary, several oocytes begin maturing at the same time. After about a week, usually only one egg survives. *Near the end of this phase, hormones cause an egg to be released from the ovary in a process called* **ovulation.**

Figure 6 The tissue lining the uterus changes in thickness during the menstrual cycle.

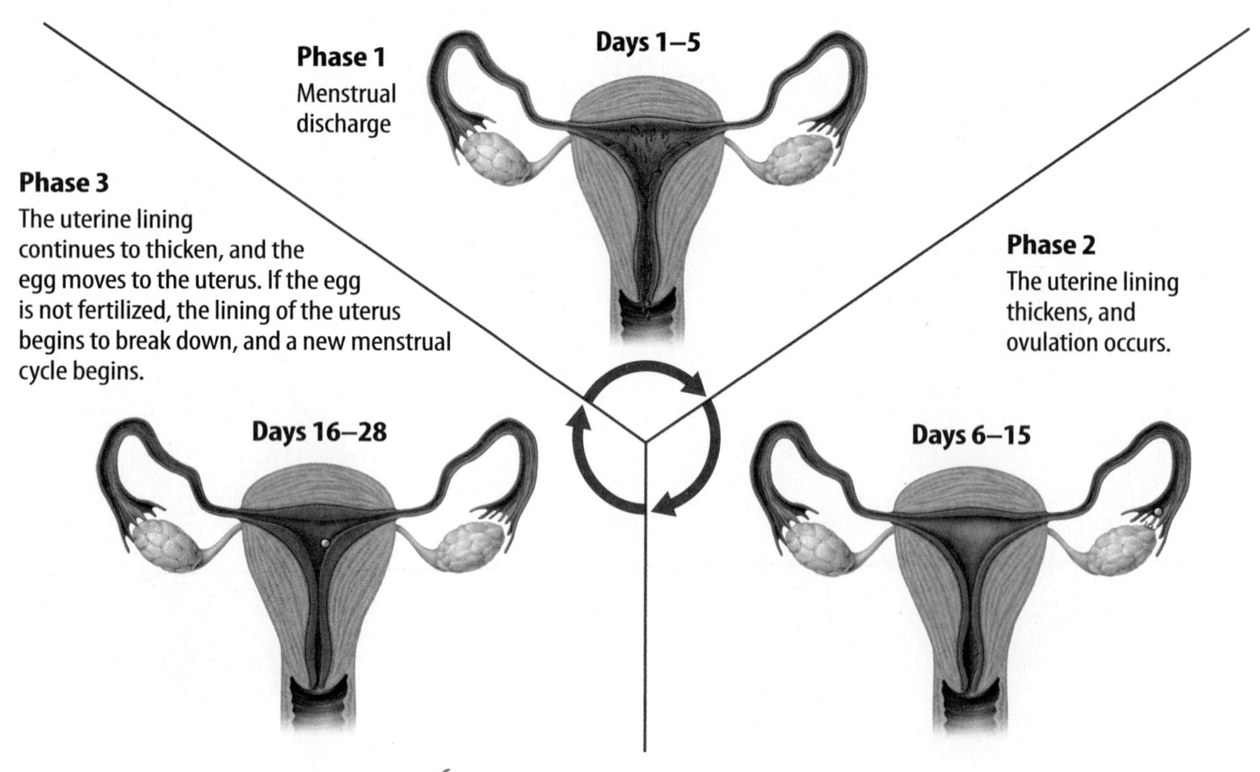

9. Visual Check **Interpret** Which are the days of menstrual flow?

Phase 3 During phase 3, the tissue that lines the uterus continues to thicken. *If sperm are present, the egg might join with a sperm in a process called* **fertilization.**

The tissue lining the uterus is called the endometrium (en doh MEE tree um). The endometrium provides a fertilized egg with nutrients and oxygen during its early development. If fertilization does not occur, the endometrium breaks down and the menstrual cycle repeats itself.

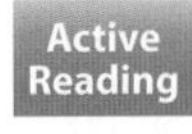

10. Combine Highlight how the parts of the female reproductive system work together.

Menopause When females get older, the reproductive system stops releasing eggs. When this happens, a woman reaches menopause—a time when the menstrual cycle stops. Menopause occurs because a woman's ovaries produce fewer hormones. There are not enough hormones to cause oocyte maturation and ovulation. Menopause usually happens between the ages of 45 and 55.

The Reproductive System and Homeostasis

Although the reproductive system does not help the body maintain homeostasis, reproduction is essential to ensure the survival of the human species. The reproductive systems of both males and females are controlled by the endocrine system.

Recall that hormones produced by the endocrine system cause the menstrual cycle. These hormones act by positive and negative feedback mechanisms and control when oocytes mature. Hormones also control sperm maturation in the male reproductive system. The reproductive system and the endocrine system work together and control when sperm and eggs mature.

11. NGSSS Check Blend How do body systems interact with the reproductive system? SC.6.L.14.5

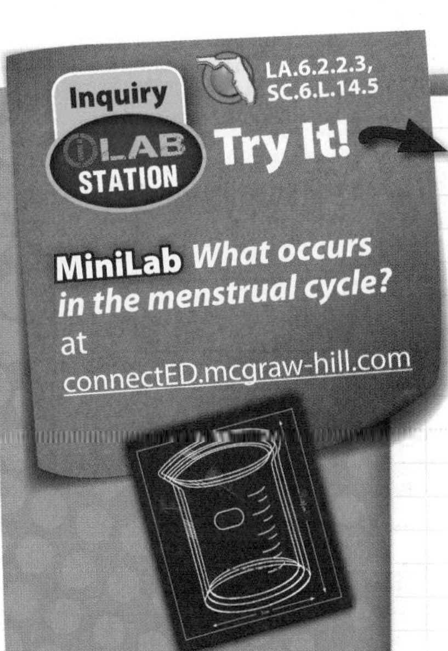

Apply It! After you complete the lab, answer these questions.

1. Why do you think menopause occurs later in life rather than earlier?

2. How do the hormones of the endocrine system influence the menstrual cycle?

Lesson Review 1

Visual Summary

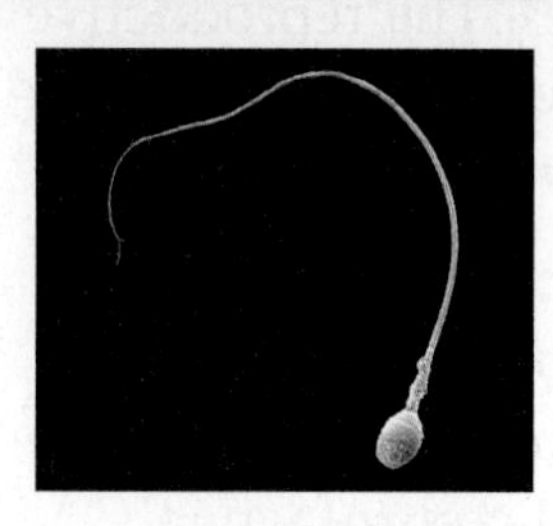

Sperm are male reproductive cells.

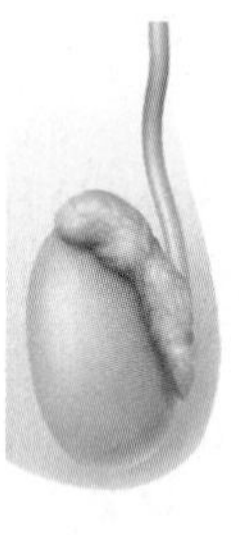

The testes produce sperm and are inside the scrotum.

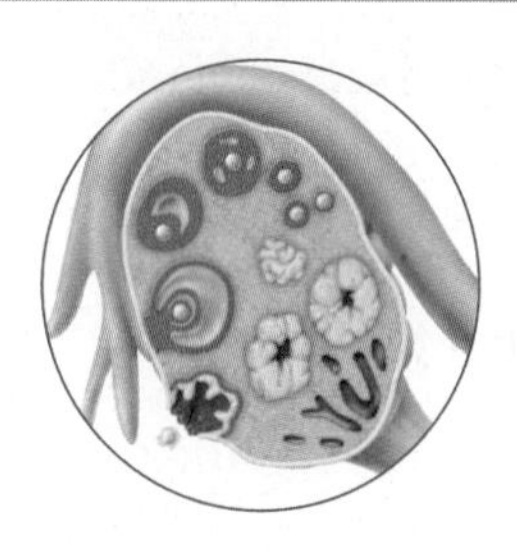

The ovary is the female reproductive organ where oocytes are stored and mature into eggs.

Use Vocabulary

1. **Use the terms** *testes* and *sperm* in a sentence.

2. **Distinguish** between ovulation and menstruation.

Understand Key Concepts

3. Which is a mixture of sperm, nutrients, and fluids? SC.6.L.14.5
   - (A) oocyte
   - (B) ovum
   - (C) penis
   - (D) semen

4. **Explain** why the scrotum is outside of the body. SC.6.L.14.5

5. **Compare** the structure of an egg to the structure of a sperm. SC.6.L.14.5

Interpret Graphics

6. **Summarize** Fill in the table below to list the parts and functions of the male reproductive system. SC.6.L.14.5

Part	Function

Critical Thinking

7. **Relate** the structures of the female reproductive system to their functions. SC.6.L.14.5

SCIENCE & SOCIETY

A Medical Breakthrough

Scientific research provides help for infertile couples.

Have you heard the term "test-tube baby"? Can a baby really come from a test tube? Three decades ago, two British scientists answered that question. Patrick Steptoe, an obstetrician and gynecologist, and Robert Edwards, a biologist and physiologist, developed a procedure that helped a British woman overcome infertility, or the inability to become pregnant. As a result of this procedure, the world's first test-tube baby, Louise Joy Brown, was born on July 25, 1978.

Prior to meeting, Steptoe and Edwards worked separately on human infertility. Steptoe developed a process called laparoscopy. Laparoscopy uses a narrow, tubelike instrument fitted with a fiber-optic light and a lens. This enables a doctor to examine a woman's ovaries, fallopian tubes, and uterus through a small incision in her abdomen. Edwards researched the fertilization of human eggs outside the body under laboratory conditions. As a team, Steptoe and Edwards developed a procedure now known as in vitro fertilization (IVF). In IVF, a doctor uses laparoscopy to remove mature eggs from a woman's ovaries. The eggs are fertilized in a Petri dish (the test tube) and grow into zygotes. A doctor transfers the zygotes into a woman's uterus, where they can continue to develop.

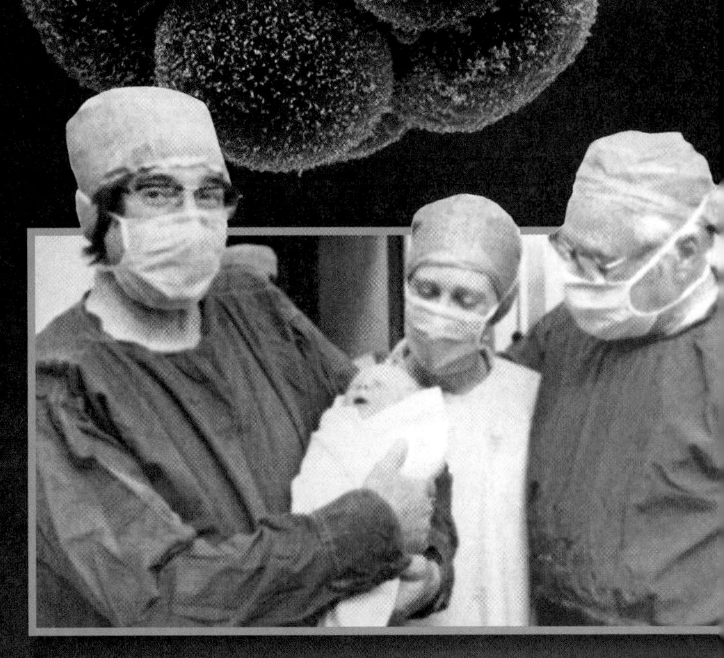

▲ **Dr. Steptoe (right) looks on as Dr. Edwards holds the world's first test-tube baby.**

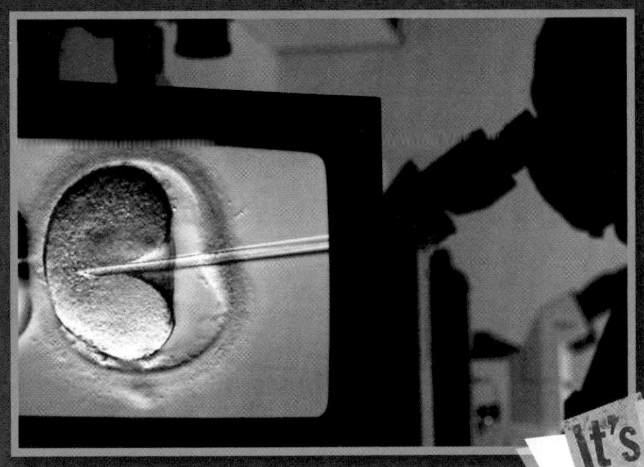

▲ **A doctor performs IVF by injecting a sperm cell into an egg cell.**

Steptoe and Edwards worked for ten years to perfect their procedure. The successful implantation and pregnancy achieved through IVF opened new doors for infertile couples. By 2006, it was estimated that as many as 3 million babies had been born using IVF since the birth of Louise Joy Brown in 1978.

It's Your Turn

REPORT How long has in vitro fertilization been practiced in the United States? What is its success rate? Research these questions and write a short report. LA.6.4.2.2

Lesson 2

Human Growth and DEVELOPMENT

ESSENTIAL QUESTIONS

 What happens during fertilization of a human egg?

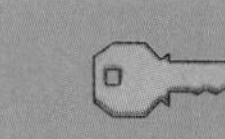 What are the major stages in the development of an embryo and a fetus?

 How do the life stages differ after birth?

Vocabulary

zygote p. 668
pregnancy p. 669
placenta p. 669
umbilical cord p. 669
embryo p. 670
fetus p. 670
cervix p. 672
puberty p. 674

Florida NGSSS

LA.6.2.2.3 The student will organize information to show understanding (e.g., representing main ideas within text through charting, mapping, paraphrasing, summarizing, or comparing/contrasting);

SC.6.L.14.5 Identify and investigate the general functions of the major systems of the human body (digestive, respiratory, circulatory, reproductive, excretory, immune, nervous, and musculoskeletal) and describe ways these systems interact with each other to maintain homeostasis.

SC.6.N.1.5 Recognize that science involves creativity, not just in designing experiments, but also in creating explanations that fit evidence.

SC.6.N.2.1 Distinguish science from other activities involving thought.

MA.6.A.3.6 Construct and analyze tables, graphs, and equations to describe linear functions and other simple relations using both common language and algebraic notation.

SC.6.N.1.5, SC.6.L.14.5

Inquiry Launch Lab

15 minutes

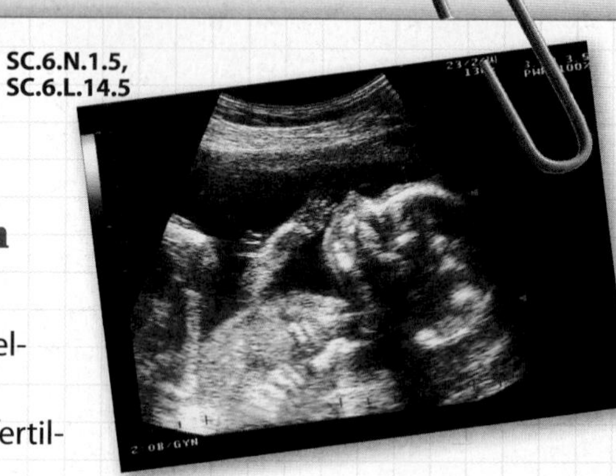

How does a fetus develop in the uterus?

Just as humans have several stages of development after birth, scientists have given names to the developmental stages from fertilization to birth.

Procedure

1. Observe the copy of a **sonogram** provided by your teacher.
2. Match the numbered structures on the sonogram to the **list of body parts** that your teacher has given you. Record your answers below.

Data and Observations

Think About This

1. What were the most difficult structures to identify? Why?

2. What questions do you have about fetal development?

3. Key Concept How do you think the stages of development after birth compare to the stages of development before birth?

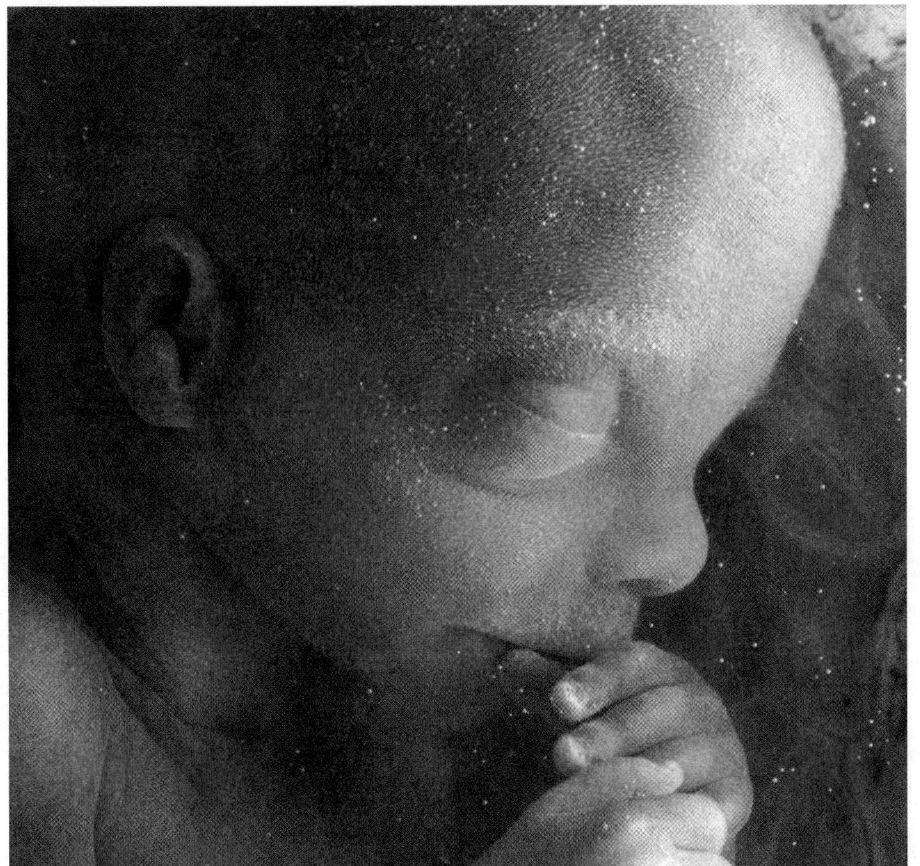

Inquiry When's my birthday?

1. This developing human is 12 weeks old. How did it develop from one cell?

Stages of Development

Have you ever been to a large family gathering? If you have, you might recall seeing people of every age. As shown in **Figure 7,** there probably were babies, children, teens, and adults.

As you might have noticed at the family gathering, people go through stages of development—infancy, childhood, adolescence (a duh LES unts), and adulthood. These stages are based on major developments that take place during each stage. In infancy there is rapid development of the nervous and muscular systems. During childhood, the abilities to speak, read, write, and reason develop. Adolescence is when a person becomes physically able to reproduce. Adulthood is the last stage—when growth of the muscular and skeletal systems stops.

Just as a human has several stages of development after he or she is born, the time before birth is divided into stages, beginning with fertilization.

Figure 7 People at this family reunion are in different stages of development. Humans grow and develop after birth.

Fertilization

If sperm enter the vagina, they can travel to the uterus and up into the fallopian tubes. The fallopian tubes are thin tubes that connect the uterus to the ovaries. Although millions of sperm are released into the vagina, most die before reaching the fallopian tubes.

A sperm contains substances that help its cell membrane join with the cell membrane of an egg. Once a sperm enters an egg, the egg's cell membrane changes rapidly. These changes ensure that only one male reproductive cell and one female reproductive cell combine to create a new human. When the nucleus of the sperm joins with the nucleus of the egg, fertilization is complete, as shown in **Figure 8.**

2. Describe Highlight what happens during fertilization.

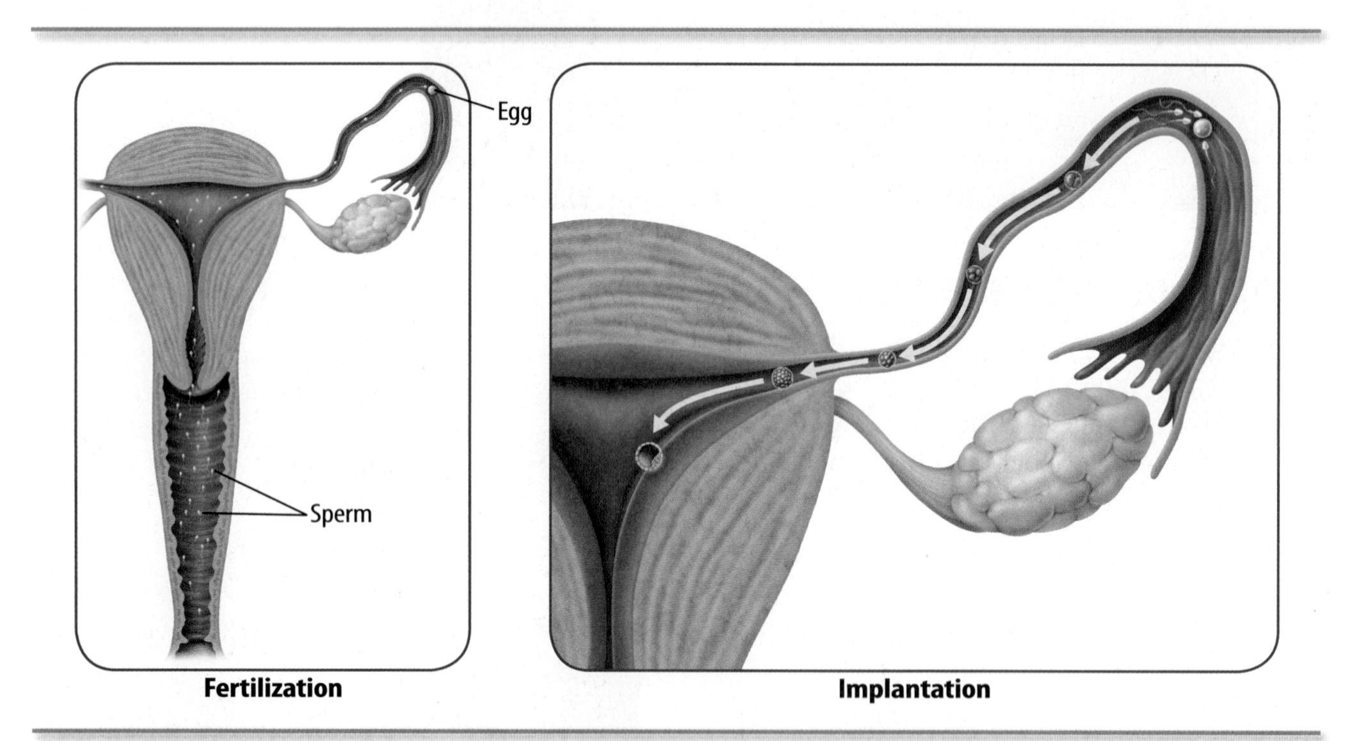

Figure 8 Sperm travel through the uterus and fallopian tubes to fertilize an egg. Eggs are fertilized in the fallopian tube.

Zygote Formation

A fertilized egg is called a **zygote** (ZI goht). Human zygotes contain 46 chromosomes of DNA—23 chromosomes from the sperm cell and 23 chromosomes from the egg cell. This means a zygote is a diploid cell. You might recall that reproductive cells form during **meiosis** and are haploid cells—they contain half the number of chromosomes of a diploid cell.

Review Vocabulary

meiosis
the process in which one diploid cell divides to form four haploid cells

The zygote moves through the fallopian tube to the uterus. As it moves toward the uterus, the zygote undergoes mitosis and cell division many times, developing into a ball of cells.

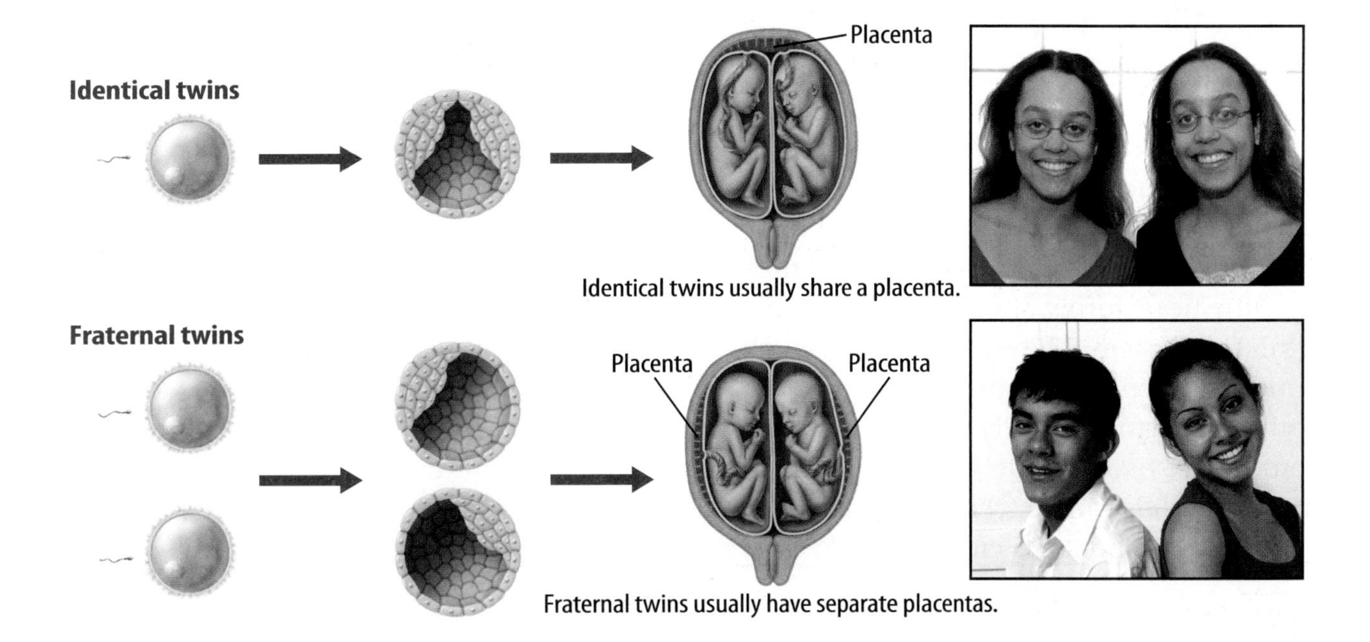

Figure 9 Each inner cell mass can develop into a baby.

Multiple Births

A zygote is now a ball of cells with a group of cells on the inside. This group of cells, called the inner cell mass, can develop into a baby. Sometimes the inner cell mass divides in two. When one zygote contains two inner cell masses, as shown in **Figure 9,** identical twins can develop. Identical twins are always of the same gender and usually look very similar. Multiple births can also occur when more than one egg is released and each is fertilized by a different sperm. This results in fraternal twins, also shown in **Figure 9.** Unlike identical twins, fraternal twins can be different genders.

Development Before Birth

Recall that the tissue lining the uterus, the endometrium, thickens during the menstrual cycle. About seven days after fertilization, the zygote enters the uterus and attaches to the thickened endometrium. As shown in **Figure 8,** this process is called implantation. After implantation, a zygote develops into a baby in about nine months. *The period of development from fertilized egg to birth is called* **pregnancy.**

After attaching to the uterus, *the outer cells of the zygote and cells from the uterus form an organ called the* **placenta** (pluh SEN tuh). *The outer zygote cells also form a rope-like structure, called the* **umbilical** (um BIH lih kul) **cord,** *which attaches the developing offspring to the placenta.* The developing offspring and the mother exchange materials through the umbilical cord. Nutrients and oxygen from the mother pass to the developing offspring, and waste and carbon dioxide are removed from it.

Math Skills MA.6.A.3.6

Use Percentages

In 2005, 32.2 sets of twins were born for every 1,000 births in the United States. You can express this rate as 32.2/1,000. To convert this to a percentage, first convert the fraction to a decimal:

$$\frac{32.2}{1,000} = 0.0322$$

Multiply the decimal by 100 and add a percent sign.

$$0.0322 \times 100 = 3.22\%$$

Practice

3. In 2005, there were 161.8 multiple births per 100,000 live births. What percent of live births were multiples?

From Zygote to Embryo

From the time the zygote attaches to the uterus until the end of the eighth week of pregnancy, it is called an **embryo.** During this time, cells divide, grow, and gain unique functions. As shown in **Figure 10,** the brain, heart, limbs, fingers, and toes start to form. Once the heart forms, the embryo develops a circulatory system. The embryo can now take in more nutrients and oxygen from its mother through the placenta. Bones and reproductive tissues begin to develop. The ears and eyelids can be seen. By eight weeks, the embryo is about 2.5 cm long.

4. Divide Highlight changes that take place in the embryo.

From Embryo to Fetus

During the time between nine weeks and birth, the developing offspring is called a **fetus.** Organ systems begin to function, and the fetus continues to grow in size, as shown in **Figure 10.** The fetus is now able to move its arms and legs. The heartbeat can be heard with a medical instrument called a stethoscope (STEH thuh skohp). During the remaining time until birth, the fetus grows rapidly. The bones fully develop but are still soft, and the lungs mature. The fetus can respond to sounds from outside the uterus, such as its mother's voice.

5. Section Underline the major stages in the development of a fetus.

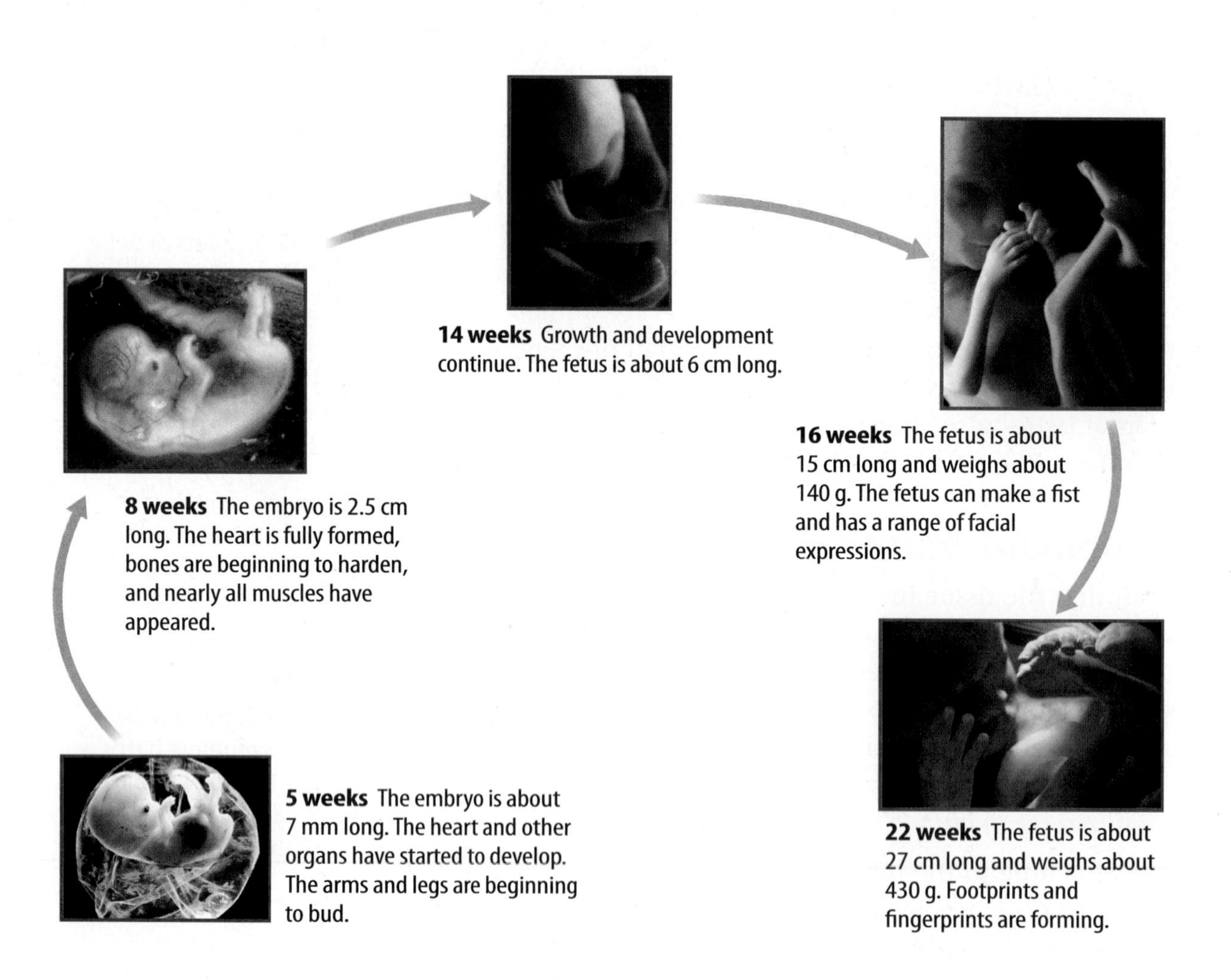

Figure 10 As an embryo develops into a fetus, it grows in size and its organ systems form.

6. Visual Check Distinguish What features do you recognize in the developing human?

Fetal Health

Because a fetus receives all nutrients from its mother, the fetus's growth and development depend on the food, water, and other nutrients that its mother eats and drinks. Other factors in a pregnant woman's environment, such as cigarette smoke or chemicals, can also **affect** the growth and development of her fetus.

ACADEMIC VOCABULARY

affect

(verb) to influence or alter an outcome

Nutrition It is important that a pregnant woman takes in enough protein and vitamins to provide her fetus with the nutrients it needs to grow and develop. Nutrients such as vitamin D, folic acid, and zinc are needed for fetal development of bones and the nervous system. Protein is needed for making all of the new cells as the fetus grows.

Environmental Factors A fetus is protected from many environmental factors because it develops in a woman's uterus. However, the mother's exposure to substances such as chemicals and smoke can harm her fetus. Heavy metals such as lead and mercury can also affect its growth and development.

Drugs and Alcohol When a woman drinks alcohol during pregnancy, the developing fetus can be harmed. This is because her fetus also takes in the alcohol through the placenta. When a fetus is exposed to alcohol, the baby that develops can be born with fetal alcohol syndrome, as shown in **Figure 11.** Fetal alcohol syndrome is a group of lifelong problems that include growth problems, vision and hearing problems, and delayed mental development. Drugs such as cocaine and the nicotine in tobacco can also have harmful effects on a developing fetus if a woman uses them during pregnancy.

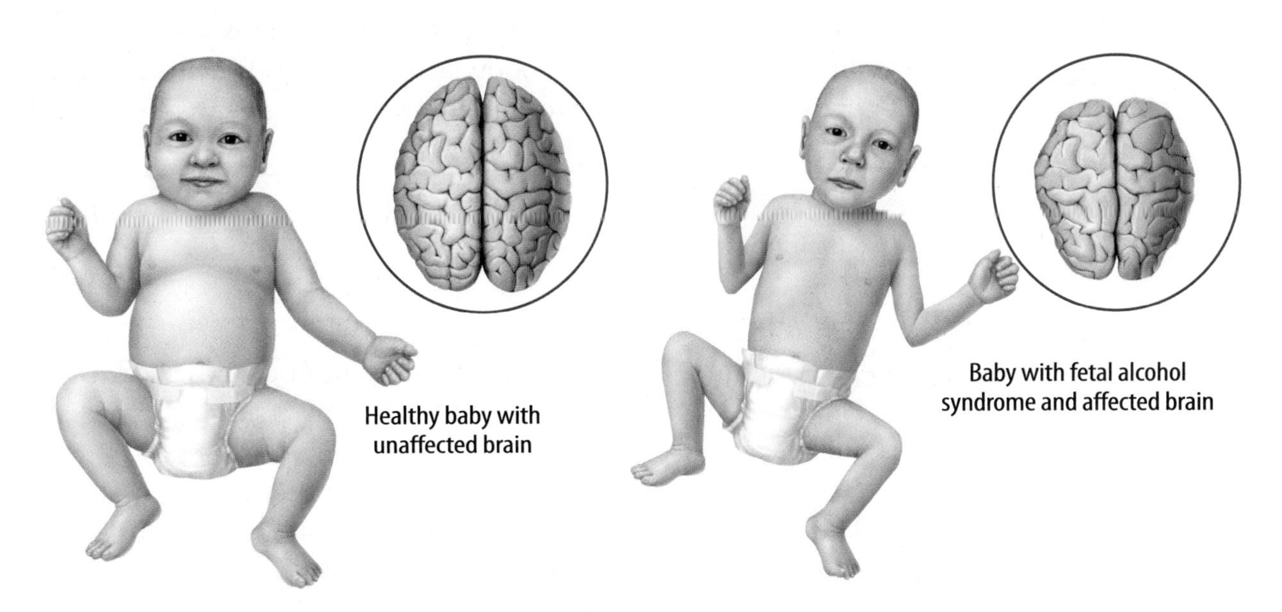

Figure 11 When a fetus is exposed to alcohol, many problems can occur. Some parts of the brain in the fetus with fetal alcohol syndrome are smaller than normal. This can cause learning and behavior problems.

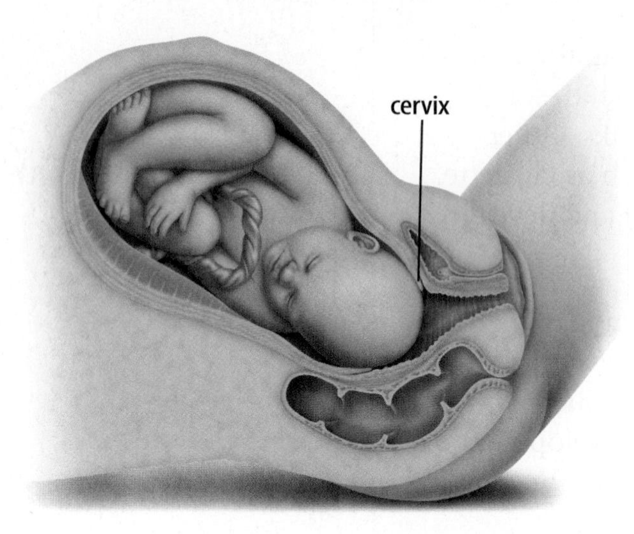

Stage 1 As the fetus moves into the birth canal, the opening to the uterus widens.

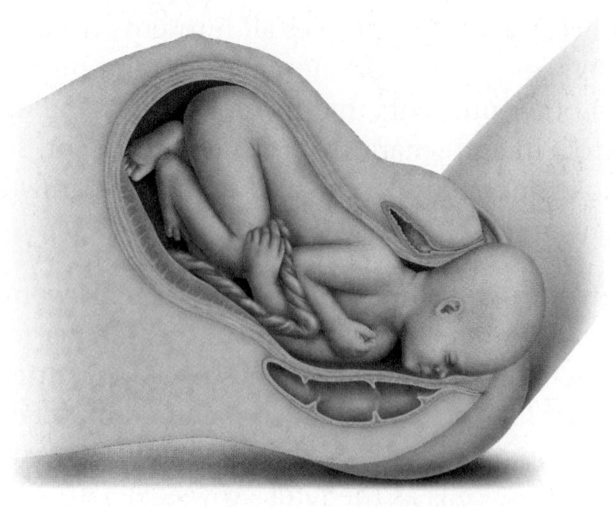

Stage 2 Muscle contractions help push the fetus out through the birth canal.

Figure 12 During birth, the head of the fetus moves toward the cervix. The fetus is delivered after the cervix opens.

Birth

A fetus leaves its mother's body and enters the world through a process called birth, as shown in **Figure 12.** Like the menstrual cycle, the birth of a fetus requires hormones. Hormones cause changes in the female reproductive system. These changes, called labor, help a fetus leave the uterus.

Labor and Delivery

Labor begins when hormones that are released by the endocrine system cause muscles in the uterus to contract. Also, *a small structure between the uterus and the vagina, called the* **cervix** (SUR vihks), begins to open. As more hormone is released, muscles in the uterus continue to contract faster and more strongly. As shown in **Figure 12,** the cervix opens wider to enable the fetus to leave the body. The contractions push a fetus into the vagina and out of the woman's body. After the fetus is delivered, the placenta breaks away from the uterus and also exits the woman's body through the vagina.

Word Origin
cervix
Latin, means "neck"

Active Reading **7. Review** Underline what happens during labor and delivery.

Cesarean Section

Sometimes delivery of a fetus does not occur as shown in **Figure 12.** Doctors can deliver the fetus by a surgical process called a cesarean (suh ZER ee un) section, or C-section. During a C-section, an incision is made in the mother's abdominal wall. Then another incision is made in the wall of the uterus. The baby is delivered through the openings in the uterine and abdominal walls. C-sections are often performed to prevent harm to a fetus and its mother.

Infancy

Once a baby is born, it stops depending directly on its mother for nutrients and oxygen and starts to function on its own. For the first time, the baby uses its own respiratory system and digestive system. The first two years of a newborn's life are called infancy. During infancy, the brain continues to develop, teeth form, and bones grow and get harder. As shown in **Figure 13,** an infant grows in size and learns to crawl, sit, walk, and speak. Organ systems continue to develop and mature, and the infant begins to eat solid food.

Childhood

The period between infancy and sexual maturity is called childhood. During this time, the brain continues to grow and develop and thinking improves. Muscle strength increases, and arms and legs grow longer.

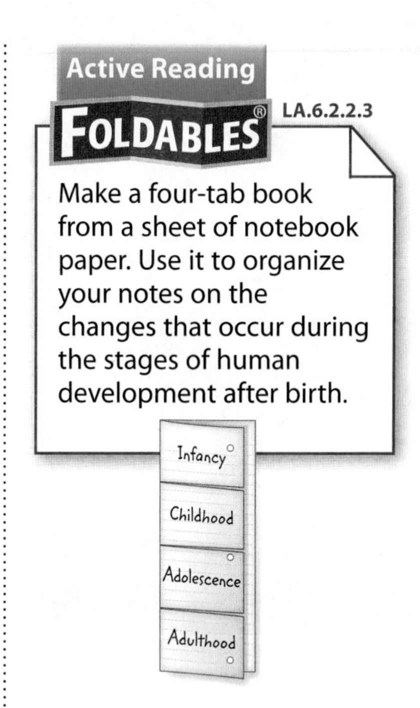

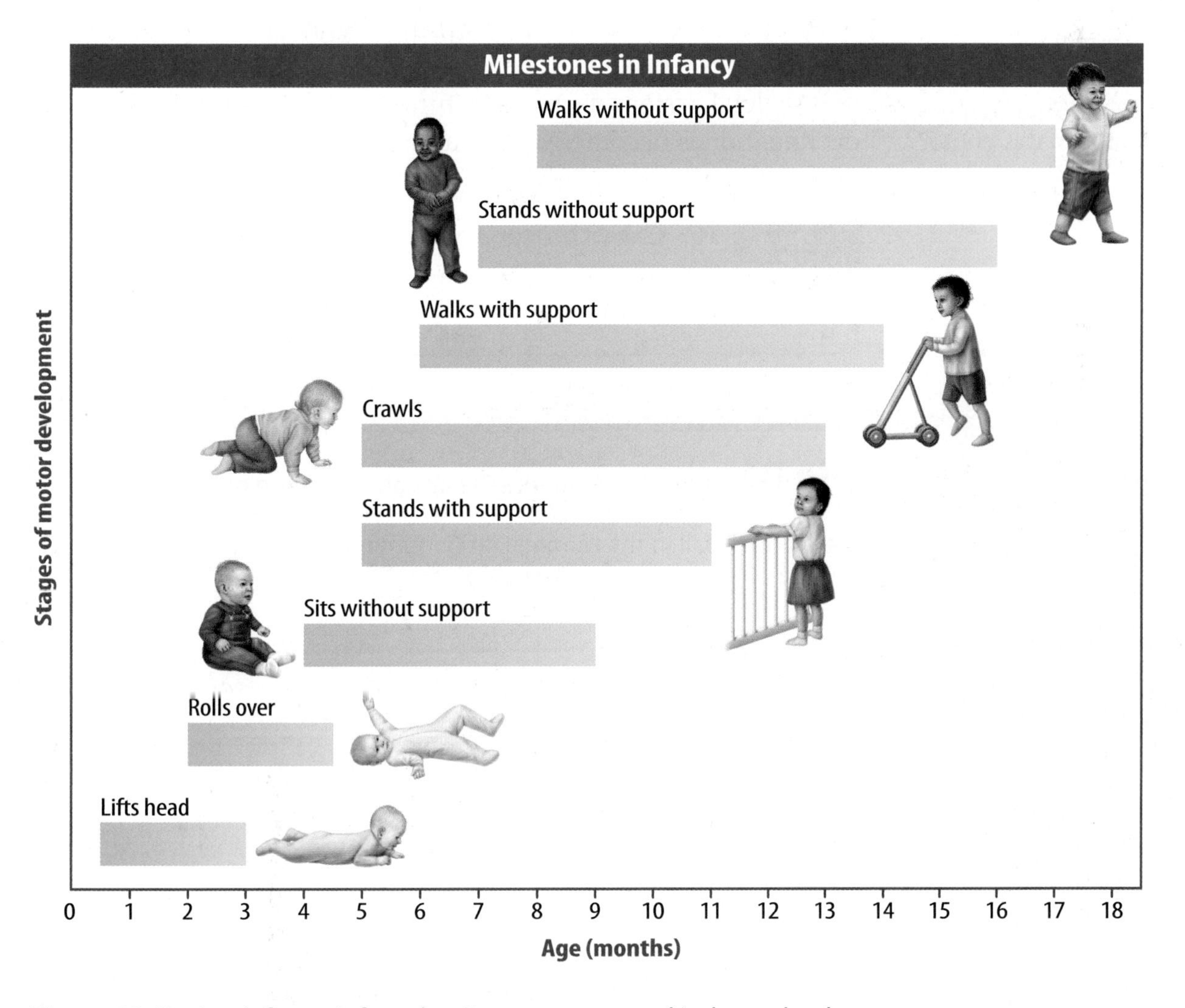

Figure 13 During infancy, infants begin to move around independently.

8. **Visual Check** Judge At what age do infants sit without support? ____________

Adolescence

Following childhood is a period of growth called adolescence. Both males and females grow taller as muscles and bones continue to grow. *During adolescence, the reproductive system matures in a process called* **puberty** (PYEW bur tee). Just as hormones are important for the menstrual cycle and for birth, hormones cause the changes that occur during puberty. In males, the voice deepens, muscles increase in size, and facial, pubic, and underarm hair grow. In females, breasts develop, pubic and underarm hair appear, and fatty tissue is added to the buttocks and thighs.

Figure 14 Many physical changes take place between adolescence and older adulthood.

9. Visual Check **Differentiate** What changes can you identify in this photo?

Adulthood and Aging

At the end of adolescence, a person enters adulthood, which continues through old age, as shown in **Figure 14.** Although adults will not grow taller, physical changes in body mass can still occur. Aging is the process of changes taking place in the body over time. Hair can turn white or gray and stop growing, and the skin wrinkles. As humans get older, the sensory system and skeletal system decrease in function. Vision and hearing decline, bones become weaker, and the digestive system slows down.

Active Reading **10. Characterize** What are the life stages after birth?

Inquiry SC.6.N.2.1

LAB STATION **Try It!**

MiniLab ***How do life stages after birth differ?*** at connectED.mcgraw-hill.com

Apply It! After you complete the lab, answer these questions.

1. Where do you fall in the Human Life Continuum?

2. Which stage of the Human Life Continuum has the most changes? Explain.

Lesson Review 2

Visual Summary

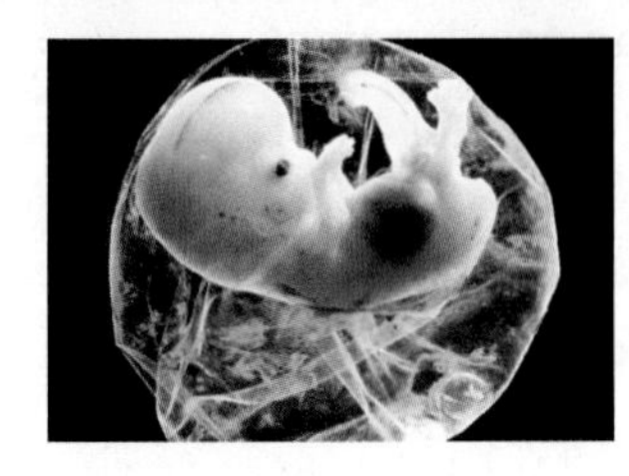

The developing zygote is called an embryo after it attaches to the uterus and until the end of the eighth week of pregnancy.

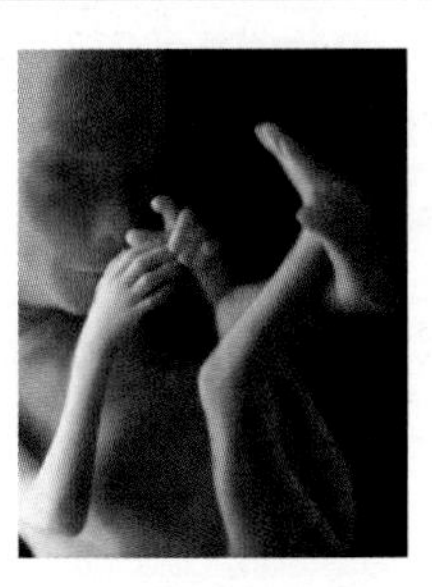

From the ninth week of pregnancy until birth, the developing embryo is called a fetus.

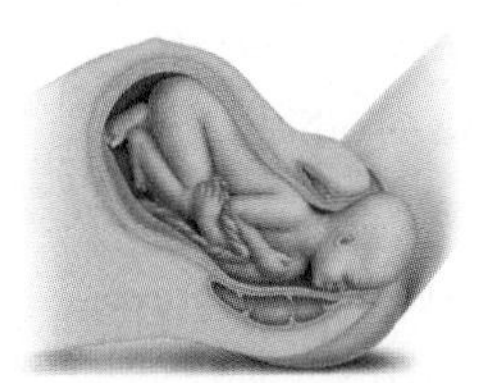

During labor, the fetus leaves the uterus through the cervix, a structure between the uterus and the vagina.

Use Vocabulary

1. **Distinguish** between a zygote and an embryo.

2. **Define** *umbilical cord* in your own words.

3. **Use the term** *puberty* in a sentence.

Understand Key Concepts

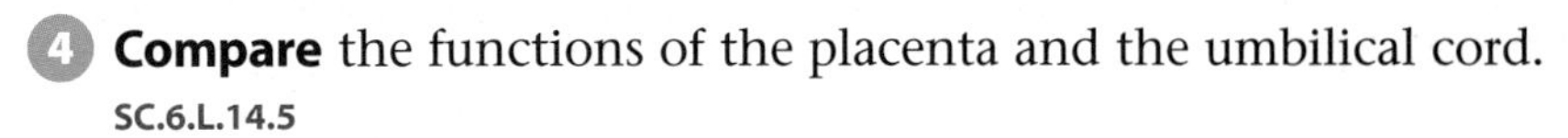

4. **Compare** the functions of the placenta and the umbilical cord. SC.6.L.14.5

5. Which does not happen during aging? SC.6.L.14.5
   - (A) Hair turns white.
   - (B) Skin wrinkles.
   - (C) An individual grows taller.
   - (D) Vision and hearing decline.

Interpret Graphics

6. **Summarize** Fill in the graphic organizer below to summarize the stages of development after an egg is fertilized and before birth. SC.6.L.14.5

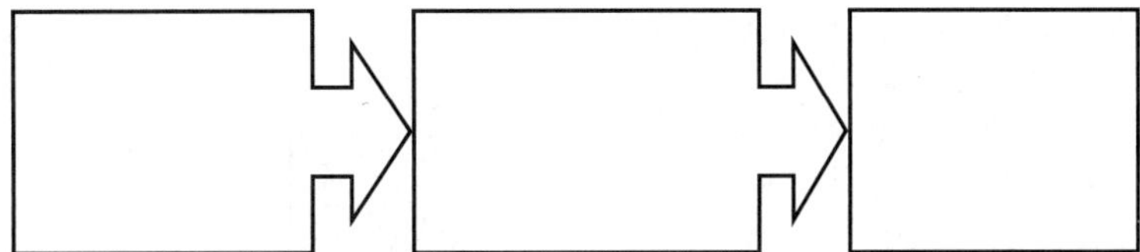

Critical Thinking

7. **Explain** why a developing human can be harmed during the stage of development shown in the figure below. SC.6.L.14.5

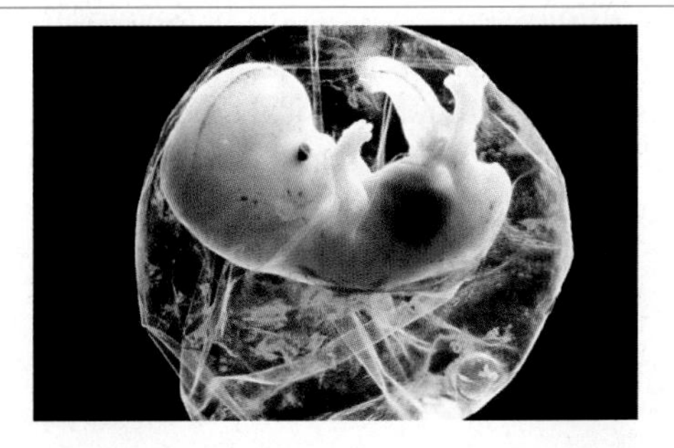

Chapter 17 Study Guide

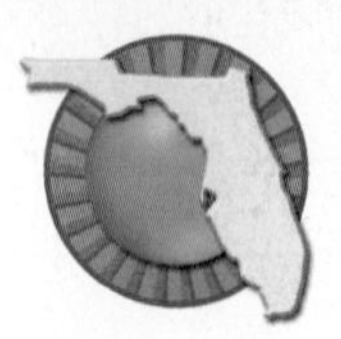

Think About It! **The general function of the human reproductive system is to ensure survival of species. Human reproduction and development begin when a sperm and an egg join. The zygote develops into an embryo and then a fetus. Human development continues after birth.**

Key Concepts Summary

LESSON 1 The Reproductive System

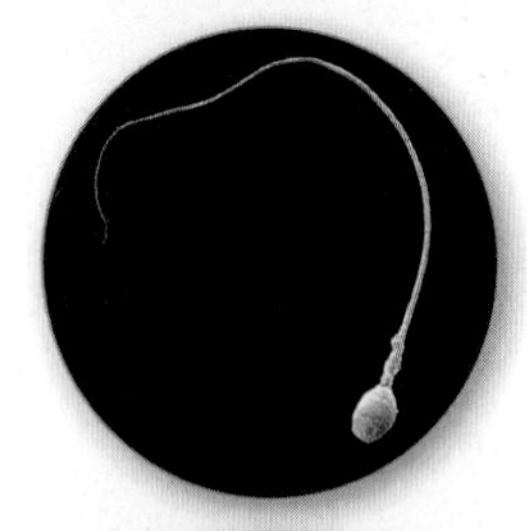

- The main function of the reproductive system is to ensure that the species survives. The male and female reproductive systems enable **sperm** and **egg** to join.
- The parts of the male reproductive system work together and produce and transport sperm to the female reproductive system.
- The parts of the female reproductive system work together and produce oocytes, provide a suitable environment for **fertilization** to occur, and nourish the developing offspring.
- The endocrine system interacts with the **testes** and **ovaries** and produces hormones that control and aid in sexual development.

Vocabulary

sperm p. 657
egg p. 657
testis p. 658
semen p. 658
penis p. 658
ovary p. 660
vagina p. 660
menstrual cycle p. 662
ovulation p. 662
fertilization p. 662

LESSON 2 Human Growth and Development

- After a sperm enters an egg, the nucleus of the sperm joins with the nucleus of the egg. This forms a fertilized egg, or **zygote.**
- After fertilization, the zygote travels to the uterus and attaches to the lining of the uterus. It is now known as an **embryo.** After the first two months of pregnancy, the developing embryo is called a **fetus,** and it can move its arms and legs.

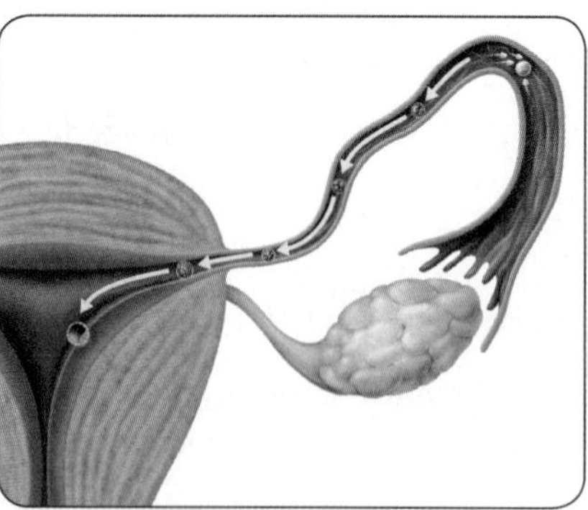

- Infancy is a period of rapid growth and development. During childhood, growth and development continue at a less rapid pace than in infancy. Adolescence is a period of development during which a person becomes physically able to reproduce. During adulthood, a person reaches his or her peak physical development and then continues to change as aging occurs.

zygote p. 668
pregnancy p. 669
placenta p. 669
umbilical cord p. 669
embryo p. 670
fetus p. 670
cervix p. 672
puberty p. 674

Active Reading

FOLDABLES® Chapter Project

Assemble your lesson Foldables as shown to make a Chapter Project. Use the project to review what you have learned in this chapter.

Phase 1
Phase 2
Phase 3
Infancy
Childhood
Adolescence
Adulthood

Use Vocabulary

Distinguish between the vocabulary words in each pair.

1. sperm and egg

2. ovulation and menstruation

3. uterus and ovary

Determine if each statement is true or false.

4. The umbilical cord attaches the developing offspring to the uterus.

5. Puberty occurs during childhood.

6. A human zygote contains 23 chromosomes.

Link Vocabulary and Key Concepts

Use vocabulary terms from the previous page to complete the concept map.

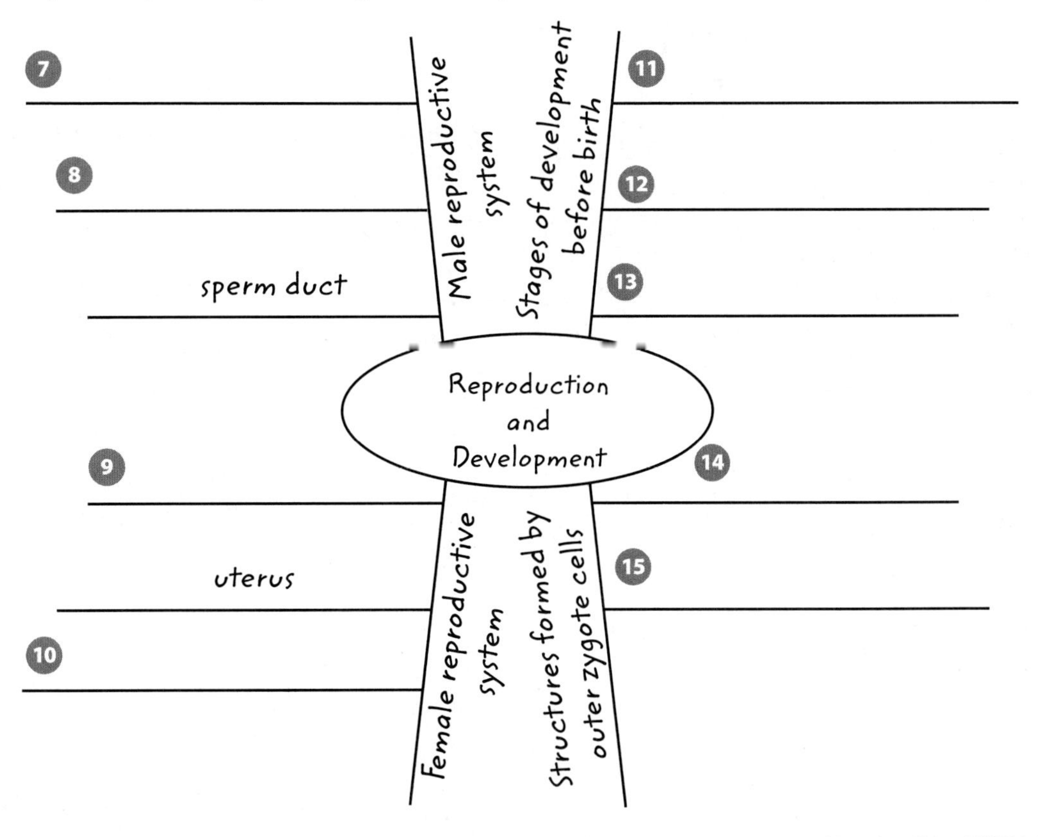

Chapter 17 Review

Fill in the correct answer choice.

Understand Key Concepts

1 Which part of the male reproductive system in the figure below produces sperm? SC.6.L.14.5

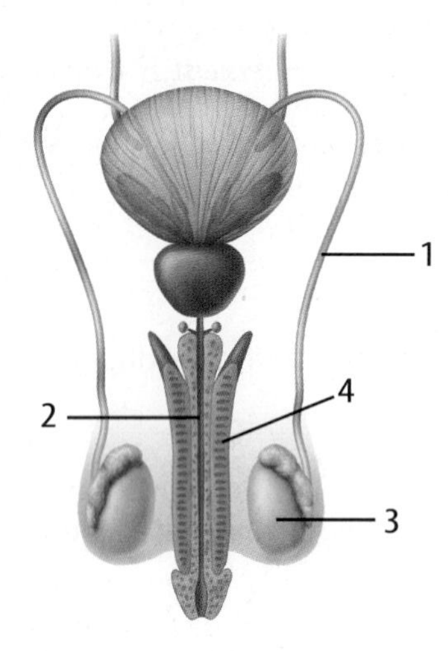

Ⓐ 1
Ⓑ 2
Ⓒ 3
Ⓓ 4

2 Which is NOT a part of sperm? SC.6.L.14.5

Ⓐ egg
Ⓑ head
Ⓒ midpiece
Ⓓ tail

3 Where do fertilized eggs develop? SC.6.L.14.5

Ⓐ ovary
Ⓑ scrotum
Ⓒ uterus
Ⓓ vagina

4 What happens to the menstrual cycle during menopause? SC.6.L.14.5

Ⓐ It begins.
Ⓑ It stops.
Ⓒ It gets faster.
Ⓓ It gets longer.

5 Which causes changes in male and female reproductive systems? SC.6.L.14.5

Ⓐ eggs
Ⓑ hormones
Ⓒ oocytes
Ⓓ sperm

Critical Thinking

6 **Infer** why semen contains millions of sperm. SC.6.L.14.5

7 **Compare** the parts of the male reproductive system to the parts of the female reproductive system. SC.6.L.14.5

8 **Infer** What would happen to sperm in a man whose testes were located inside his body instead of in the scrotum? SC.6.L.14.5

9 **Explain** why sperm and eggs contain only 23 chromosomes. SC.6.L.14.5

10 **Interpret** what is happening in the illustration below. SC.6.L.14.5

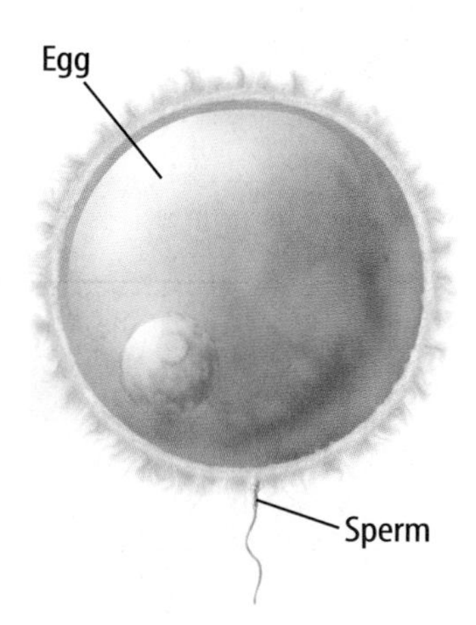

11. **Predict** how not having enough hormones affects puberty. **SC.6.L.14.5**

12. **Assess** the effect on the development of a fetus if a pregnant woman does not take in enough nutrients. **SC.6.L.14.5**

13. **Form** a hypothesis that explains why hormones are necessary for childbirth. **SC.6.L.14.5**

Writing in Science

14. **Write** a paragraph on a separate sheet of paper describing the changes in the uterus that lead to menstruation. Include a main idea, supporting details, and a concluding sentence. **SC.6.L.14.5**

Big Idea Review

15. What are the stages in human reproduction and development? Explain the stages before and after birth. **SC.6.L.14.5**

Math Skills MA.6.A.3.6

Use Percentages

16. Women who are themselves fraternal twins give birth to about 200 sets of twins in every 2,000 births. What is the probability of a fraternal twin having twins?

17. Women who are themselves identical twins give birth to about 12 sets of twins in every 2,000 births. What is the probability of an identical twin having twins?

18. The Yoruba people in West Africa have the highest rate of twin births in the world. About 23 sets of twins are born per 500 live births. What is the probability of a Yoruba woman having twins?

Florida NGSSS Benchmark Practice

Fill in the correct answer choice.

Multiple Choice

1 Which body system works with the reproductive system to determine when sperm and eggs mature? **SC.6.L.14.5**

Ⓐ circulatory

Ⓑ endocrine

Ⓒ lymphatic

Ⓓ nervous

Use the diagram below to answer question 2.

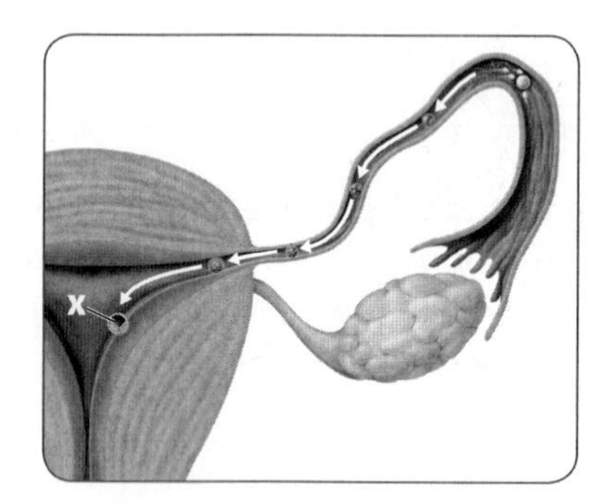

2 What takes place at the location marked *X* in the diagram above? **SC.6.L.14.5**

Ⓕ fertilization

Ⓖ implantation

Ⓗ menstruation

Ⓘ ovulation

3 How many chromosomes does each human zygote contain? **SC.6.L.14.5**

Ⓐ 1

Ⓑ 2

Ⓒ 23

Ⓓ 46

4 Which statement about fraternal twins is false? **SC.6.L.14.5**

Ⓕ They can resemble each other.

Ⓖ They must be different genders.

Ⓗ They result from multiple eggs.

Ⓘ They result from multiple sperm.

Use the table below to answer questions 5 and 6.

Milestone	Earliest Age	Latest Age
Rolls over	2 months	$4\frac{1}{2}$ months
Sits without support	4 months	9 months
Crawls	5 months	13 months
Stands alone	7 months	16 months
Walks alone	8 months	17 months

5 Which can be inferred from the above chart? **SC.6.L.14.5**

Ⓐ Age brings more independence.

Ⓑ All babies crawl before walking.

Ⓒ Infants develop at the same rate.

Ⓓ Muscles develop before the brain.

6 A parent wants to know when her baby might exhibit a certain skill. When is her baby most likely to begin crawling? **SC.6.L.14.5**

Ⓕ 4–9 months

Ⓖ 5–13 months

Ⓗ 7–16 months

Ⓘ 8–17 months

7 How are sperm and eggs similar? **SC.6.L.14.5**

Ⓐ Both are large and round when they mature.

Ⓑ Both are released one at a time.

Ⓒ Both contain DNA.

Ⓓ Both contain substances that nourish them.

8 Through which does nourishment pass directly to a fetus? **SC.6.L.14.5**

Ⓕ intestine

Ⓖ stomach

Ⓗ fallopian tube

Ⓘ umbilical cord

9 In which developmental stage do humans become physically capable of reproduction? **SC.6.L.14.5**

Ⓐ adolescence

Ⓑ adulthood

Ⓒ childhood

Ⓓ infancy

Use the diagram below to answer questions 10 and 11.

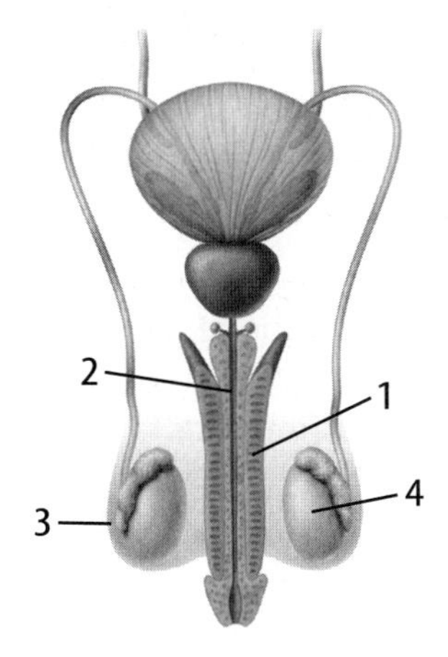

10 Which structure in the above diagram produces male reproductive cells? **SC.6.L.14.5**

Ⓕ 1

Ⓖ 2

Ⓗ 3

Ⓘ 4

11 Which structure in the diagram regulates temperature for sperm development? **SC.6.L.14.5**

Ⓐ 1

Ⓑ 2

Ⓒ 3

Ⓓ 4

12 How many diploid cells make up a zygote immediately following fertilization? **SC.6.L.14.5**

Ⓕ 0

Ⓖ 1

Ⓗ 2

Ⓘ 4

13 Fertilization of an egg takes place in which of the structures in the figure below? **SC.6.L.14.5**

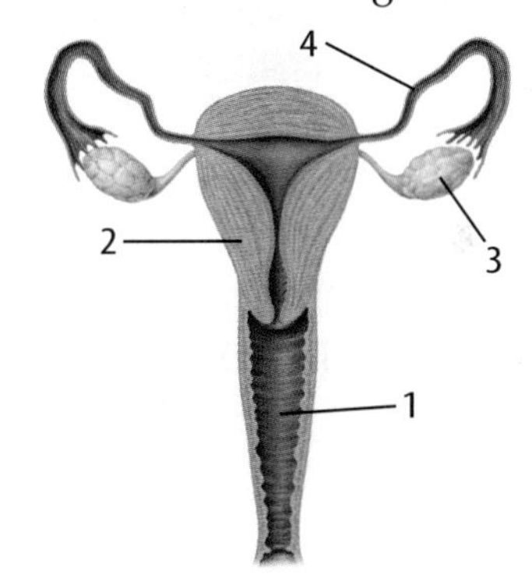

Ⓐ 1

Ⓑ 2

Ⓒ 3

Ⓓ 4

14 Which widens during delivery so the fetus can leave the uterus? **SC.6.L.14.5**

Ⓕ cervix

Ⓖ fallopian tube

Ⓗ ovary

Ⓘ uterus

15 Puberty occurs during which stage of development? **SC.6.L.14.5**

Ⓐ adolescence

Ⓑ adulthood

Ⓒ childhood

Ⓓ infancy

NEED EXTRA HELP?

If You Missed Question...	1	2	3	4	5	6	7	8	9	10	11	12	13	14	15
Go to Lesson...	1	2	2	2	2	2	1	2	2	1	1	2	2	2	2

Name ______________________ Date ______________

mini BAT

Multiple Choice *Bubble the correct answer.*

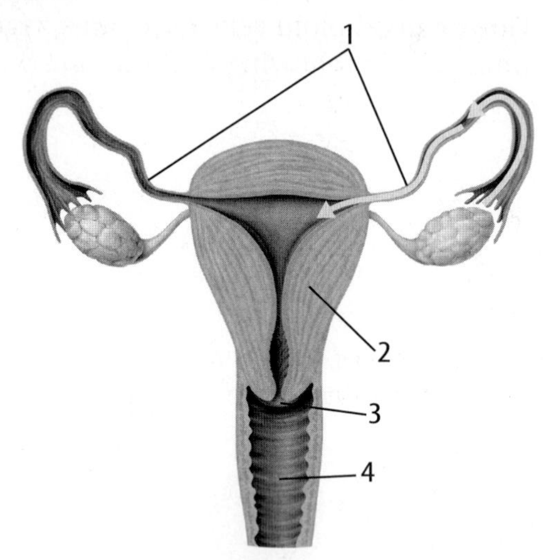

1. In the diagram above, which part of the female reproductive system is also known as the birth canal? **SC.6.L.14.5**

(A) 1

(B) 2

(C) 3

(D) 4

2. During which phase or phases of the menstrual cycle does fertilization occur? **SC.6.L.14.5**

(F) Phase 1

(G) Phase 3

(G) Phases 1 and 2

(I) Phases 3 and 4

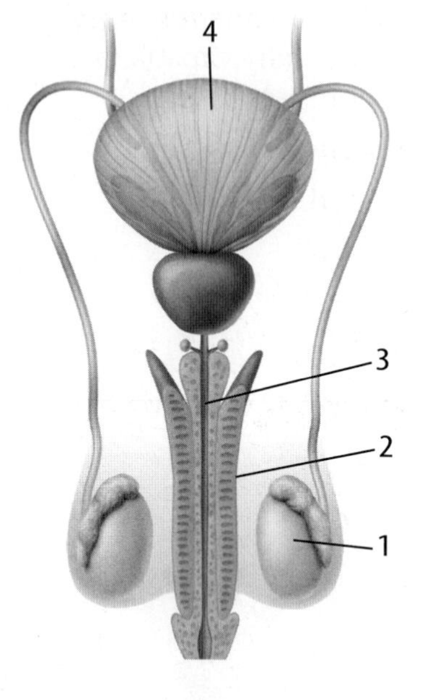

3. In the diagram above, which part of the male reproductive system stores sperm until they mature? **SC.6.L.14.5**

(A) 1

(B) 2

(C) 3

(D) 4

4. With what body system does the reproductive system work most closely to produce offspring? **SC.6.L.14.5**

(F) circulatory system

(G) digestive system

(H) endocrine system

(I) nervous system

Name ______________________________ Date ______________

Multiple Choice *Bubble the correct answer.*

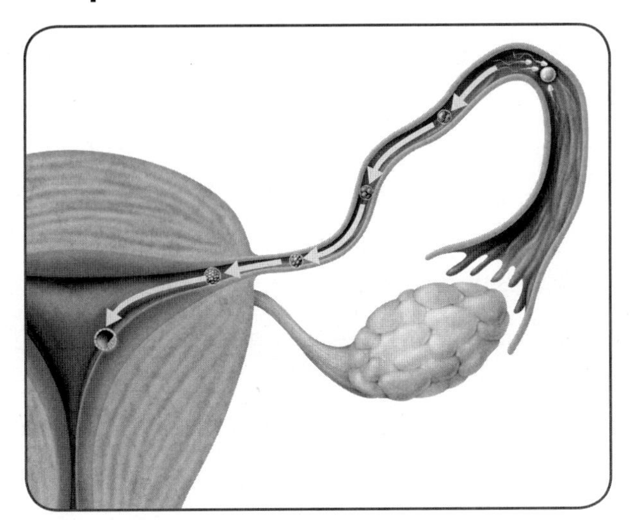

1. The diagram above shows a zygote becoming attached to the **SC.6.L.14.5**
   - (A) cervix.
   - (B) ovary.
   - (C) uterus.
   - (D) vagina.

2. What causes the beginning of labor in the mother? **SC.6.L.14.5**
   - (F) baby's hormones
   - (G) cesarean section
   - (H) mother's hormones
   - (I) organ development

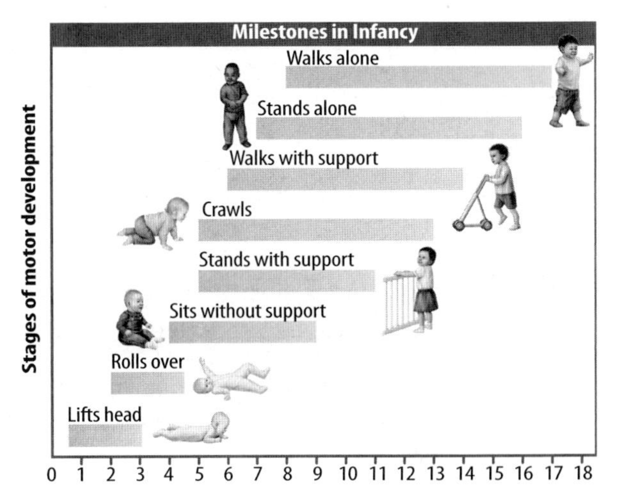

3. Look at the chart above. Based on the progress of a baby's physical development, which muscles tend to gain strength the most rapidly? **SC.6.L.14.5**
   - (A) arms
   - (B) back
   - (C) legs
   - (D) neck

4. Which of these should a developing fetus NOT be exposed to? **SC.6.L.14.5**
   - (F) mercury
   - (G) protein
   - (H) vitamins
   - (I) zinc

Name ______________________ Date ______________

Notes

Name ______________________ Date ______________

Notes

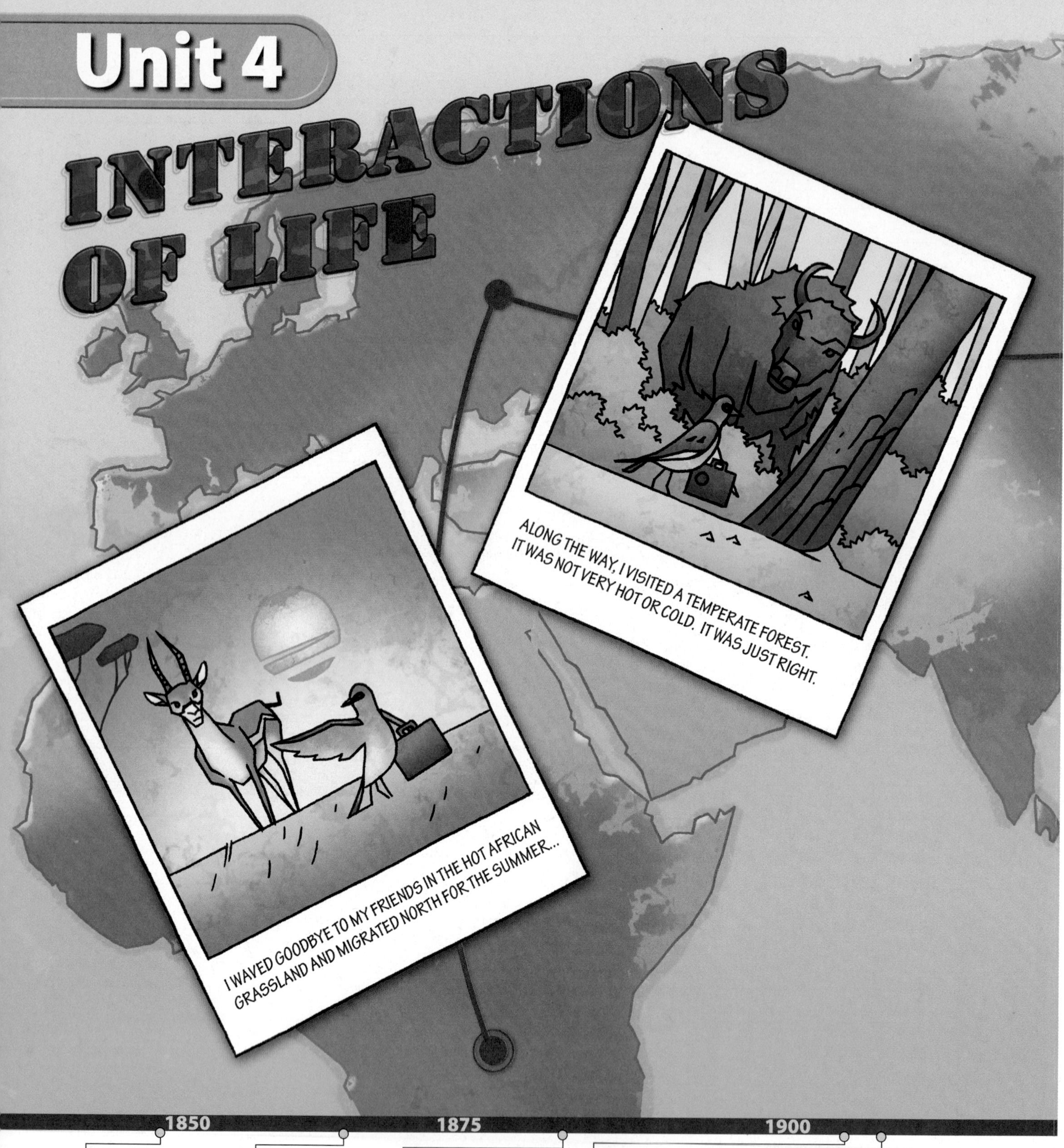

1849
The U.S. Department of Interior is established and is responsible for the management and conservation of most federal land.

1872
The world's first national park, Yellowstone, is created.

1892
The Sierra Club is founded in San Francisco by John Muir. It goes on to be the oldest and largest grassroots environmental organization in the United States.

1915
Congress passes a bill establishing Rocky Mountain National Park in Colorado.

1920
Congress passes the Federal Water Power Act. This act creates a Federal Power Commission with authority over waterways, and the construction and use of water-power projects.

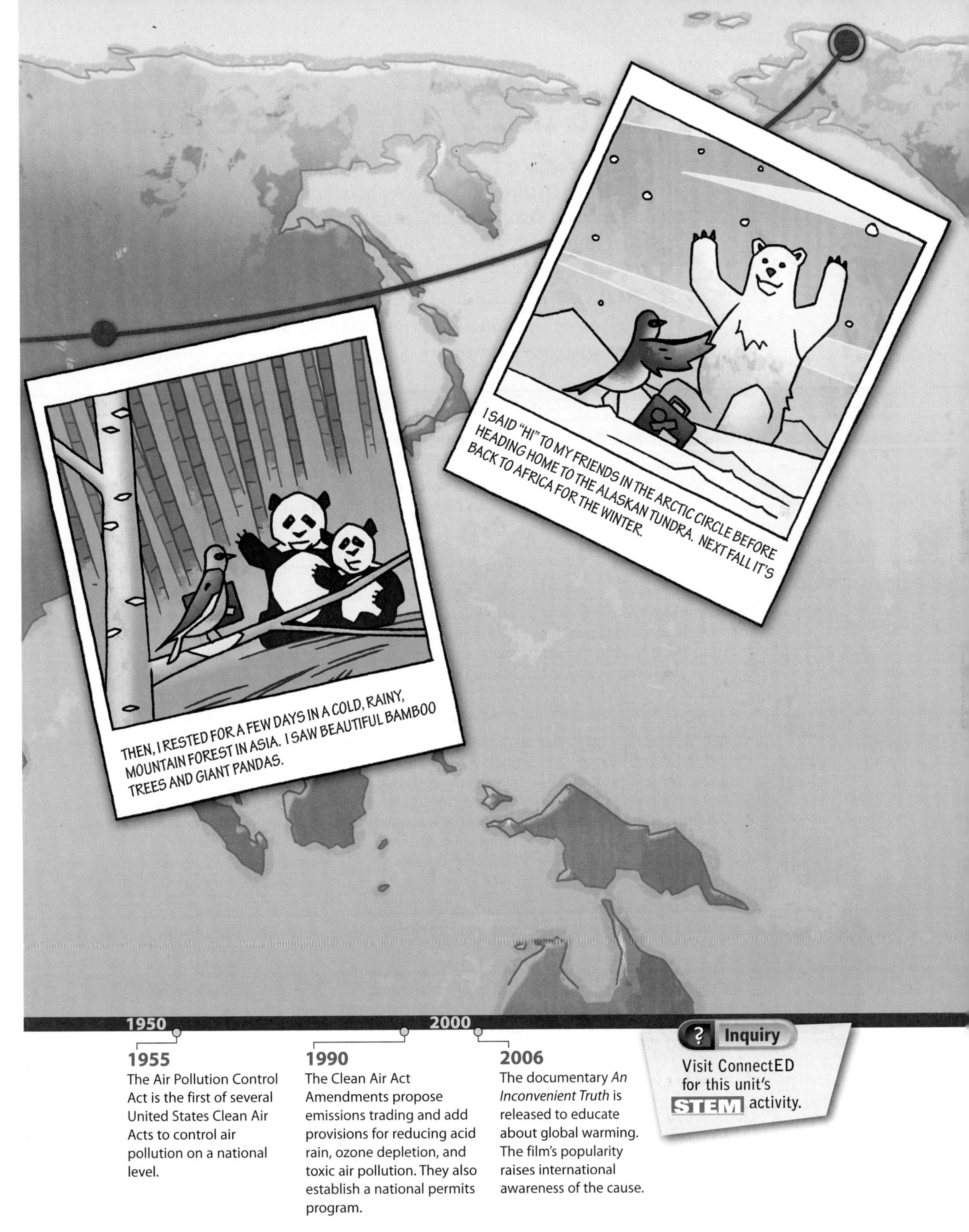

1955
The Air Pollution Control Act is the first of several United States Clean Air Acts to control air pollution on a national level.

1990
The Clean Air Act Amendments propose emissions trading and add provisions for reducing acid rain, ozone depletion, and toxic air pollution. They also establish a national permits program.

2006
The documentary *An Inconvenient Truth* is released to educate about global warming. The film's popularity raises international awareness of the cause.

Inquiry
Visit ConnectED for this unit's **STEM** activity.

History and Science

SC.7.N.2.1

Nearly 50,000 years ago, a group of hunter-gatherers might have roamed through the forest searching for food among the lush plants. The plants and animals that lived in that environment provided the nutritional needs of these people. Humans adapted to the nutrients that the wild foods contained.

Many of the foods you eat today are very different from those eaten by hunter-gatherers. **Table 1** shows how some of these changes occurred.

Table 1 How Science Has Changed Foods

	What?	Advantages	Disadvantages
	Gathering Wild Foods— Foods found in nature were the diet of humans until farming began around 12,000 years ago.	Wild foods provided all the nutrients needed by the human body.	Finding wild foods is not reliable. People moved from place to place in search of food. Sometimes they didn't find food, so they went hungry or starved.
	Farming—People grew seeds from the plants they ate. If soil conditions were not ideal, farmers learned to add water or animal manure to improve plant growth.	Farming allowed more food to be grown in less space. Over time, people learned to breed plants for larger size or greater disease resistance.	Farming practices began to affect the nutrient content of foods. Crops became less nutrient-rich. People began to suffer from nutrient deficiencies and became more prone to disease.
	Hybridizing Plants— Gregor Mendel crossed a plant with genetic material from another, producing a hybrid. A hybrid is the offspring of genetically different organisms.	Hybridization produced new plant foods that combined the best qualities of two plants. The variety of plants available for food increased.	Hybrid crops are prone to disease because of their genetic similarity. Seeds from hybrids do not always grow into plants that produce food of the same quality as the hybrid.
	Genetically Modified (GM) Foods—Scientists remove or replace genes in plants. Removing genes that control flowering in spinach results in more leaves.	GM plants can increase crop yields, nutrient content, insect resistance, and shelf-life of foods. The lettuce shown here has been modified to produce insulin.	Inserted genes might spread to other plants, producing "superweeds." Allergies to GM foods might increase. The long-term effect on humans is unknown.

Active Reading **1. Analyze** Identify two positive outcomes and two negative outcomes of how science has changed foods and food practices.

Positive Outcomes | **Negative Outcomes**

A Matter of Taste

In early history, food was eaten raw, just as it was found in nature. Cooking food probably occurred by accident. Someone might have accidentally dropped a root into a fire. When people ate the burnt root, it might have tasted better or been easier to chew. Over many generations, and with the influence of different cultures and their various ways to prepare food, the taste buds of people changed. People no longer enjoy as many raw foods.

Today, the taste buds of some people tempt them to eat high-calorie, low-nutrition, processed foods, as shown in **Figure 1.** These foods contain large amounts of calories, salt, and fat.

Active Reading **2. Assess** Underline the disadvantages of eating high-calorie, low-nutritional, processed foods.

In some parts of the world, people buy and prepare fresh fruits and vegetables every day, as shown in **Figure 2.** In general, these people have lower rates of obesity and fewer diseases that are common in people who eat more processed foods.

One scientist noted that people with a diet very different from their prehistoric ancestors are more susceptible to heart disease, cancer, diabetes, and other "diseases of civilization."

Active Reading **3. Distinguish** Circle the foods generally eaten by people with lower rates of obesity and fewer diseases than people who eat more processed foods.

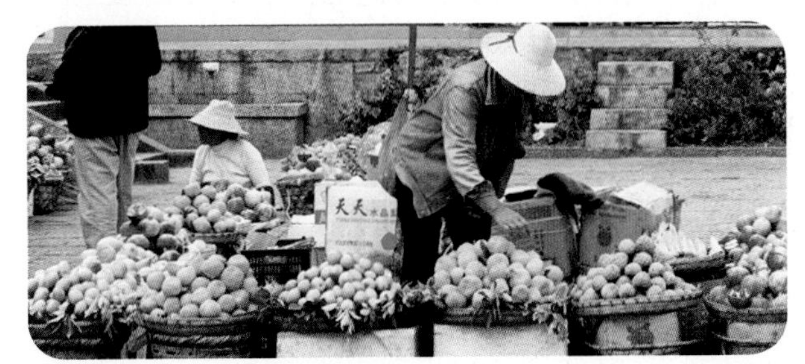

Figure 2 People in China shop in markets where farmers sell fresh produce that comes directly from the farms.

Figure 1 Processing foods increases convenience but removes nutrients and adds calories that could lead to obesity.

Apply It!

After you complete the lab, answer the questions below.

1. **Synthesize** Briefly summarize how the quality of food has changed over time.

2. **Evaluate** Discuss several things you can do to help insure you are eating healthy foods.

Name ______________________________ Date ______________

Notes

Cycling of Matter

Three friends were talking about carbon dioxide and oxygen in the ecosystem. They each had different ideas. This is what they said:

Flynn: I think animals take in oxygen and breathe out carbon dioxide. Plants then take in the carbon dioxide and release oxygen, and the cycle continues.

Jervis: I think both plants and animals take in oxygen and release carbon dioxide; but only the plants take in the carbon dioxide and release oxygen, and the cycle continues.

Melody: I think both plants and animals take in oxygen and release carbon dioxide. The oxygen is used up and carbon dioxide is not cycled again by living things.

Circle the name of the friend you agree with the most. Explain why you agree. Describe your ideas about the cycling of matter.

Matter and Energy in the ENVIRONMENT

FLORIDA BIG IDEAS

1 The Practice of Science

3 The Role of Theories, Laws, Hypotheses, and Models

17 Interdependence

18 Matter and Energy Transformations

The Big Idea

Think About It! How do living things and the nonliving parts of the environment interact?

The turtle needs food, air, water, and shelter to survive. The environment provides the turtle with all that it needs to survive.

1. How do you think the turtle depends on the nonliving things in the photo?

2. How might the turtle interact with living things in its environment?

Get Ready to Read What do you think about the environment?

Before you read, decide if you agree or disagree with each of these statements. As you read this chapter, see if you change your mind about any of the statements.

	AGREE	DISAGREE
1 The air you breathe is mostly oxygen.	☐	☐
2 Living things are made mostly of water.	☐	☐
3 Carbon, nitrogen, and other types of matter are used by living things over and over again.	☐	☐
4 Clouds are made of water vapor.	☐	☐
5 The Sun is the source for all energy used by living things on Earth.	☐	☐
6 All living things get their energy from eating other living things.	☐	☐

There's More Online!
Video • Audio • Review • ⓘLab Station • WebQuest • Assessment • Concepts in Motion • Multilingual eGlossary

Lesson 1

Abiotic FACTORS

ESSENTIAL QUESTIONS

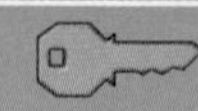

What are the nonliving parts of an environment?

Vocabulary

ecosystem p. 695
biotic factor p. 695
abiotic factor p. 695
climate p. 696
atmosphere p. 697

LA.6.4.4.2

Inquiry Launch Lab

10 minutes

Is it living or nonliving?

You are surrounded by living and nonliving things, but it is sometimes difficult to tell what is alive. Some nonliving things may appear to be alive at first glance. Others are alive or were once living, but seem nonliving. In this lab, you will explore which items are alive and which are not.

Procedure

1. Your teacher will provide you with a list of items. Decide if each item is living or nonliving, then write its name in the appropriate column in the table to the right.

Living	Nonliving

Think About This

1. What are some characteristics that the items in the *Living* column share?

2. Key Concept How might the nonliving items be a part of your environment?

Florida NGSSS

LA.6.2.2.3 The student will organize information to show understanding (e.g., representing main ideas within text through charting, mapping, paraphrasing, summarizing, or comparing/contrasting);

LA.6.4.2.2 The student will record information (e.g., observations, notes, lists, charts, legends) related to a topic, including visual aids to organize and record information and include a list of sources used;

Inquiry Why so Blue?

1. Have you ever seen a picture of a bright blue ocean like this one in Miami Beach, Florida? The water looks so colorful in part because of nonliving factors, such as matter in the water and the gases surrounding Earth. These nonliving things change the way you see light from the Sun, another nonliving part of the environment.

What is an ecosystem?

Have you ever watched a bee fly from flower to flower? Certain flowers and bees depend on each other. Bees help flowering plants reproduce. In return, flowers provide the nectar that bees use to make honey. Flowers also need nonliving things to survive, such as sunlight and water. For example, if plants don't get enough water, they can die. The bees might die, too, because they feed on the plants. All organisms need both living and nonliving things to survive.

An **ecosystem** *is all the living things and nonliving things in a given area.* Ecosystems vary in size. An entire forest can be an ecosystem, and so can a rotting log on the forest floor. Other examples of ecosystems include a pond, a desert, an ocean, and your neighborhood.

Biotic (bi AH tihk) **factors** *are the living things in an ecosystem.* **Abiotic** (ay bi AH tihk) **factors** *are the nonliving things in an ecosystem, such as sunlight and water.* Biotic factors and abiotic factors depend on each other. If just one factor—either abiotic or biotic—is disturbed, other parts of the ecosystem are affected. For example, severe droughts, or periods of water shortages, occurred in South Florida in 2001. Many fish in rivers and lakes died. Animals that fed on the fish had to find food elsewhere. A lack of water, an abiotic factor, affected biotic factors in this ecosystem, such as the fish and the animals that fed on the fish.

Active Reading **2. Write** What are three examples of biotic factors that you interact with daily?

___

___

___

WORD ORIGIN

biotic
from Greek *biotikos*, means "fit for life"

Figure 1 Abiotic factors include sunlight, water, atmosphere, soil, temperature, and climate.

What are the nonliving parts of ecosystems?

Some abiotic factors in an ecosystem are shown in **Figure 1.** Think about how these factors might affect you. You need sunlight for warmth and air to breathe. You would have no food without water and soil. These nonliving parts of the environment affect all living things.

The Sun

The source of almost all energy on Earth is the Sun. It provides warmth and light. In addition, many plants use sunlight and make food, as you'll read in Lesson 3. The Sun also affects two other abiotic factors—climate and temperature.

3. **NGSSS Check** Explain How do living things use the Sun's energy? LA.6.2.2.3

Climate

Alligator snapping turtles live in the Florida Panhandle. This area has a warm, moist climate. **Climate** *describes average weather conditions in an area over time.* These weather conditions include temperature, moisture, and wind.

Climate influences where organisms can live. A desert climate, for example, is dry and often hot. A plant that needs a lot of water could not survive in a desert. In contrast, a cactus is well adapted to a dry climate because it can survive with little water.

Temperature

Is it hot or cold where you live? Temperatures on Earth vary greatly. Temperature is another abiotic factor that influences where organisms can survive. Some organisms, such as alligators, thrive in hot conditions. Others, such as polar bears, are well adapted to the cold. Alligators don't live in cold ecosystems, and polar bears don't live in warm ecosystems.

Water

All life on Earth requires water. In fact, most organisms are made mostly of water. All organisms need water for important life processes, such as growing and reproducing. Every ecosystem must contain some water to support life.

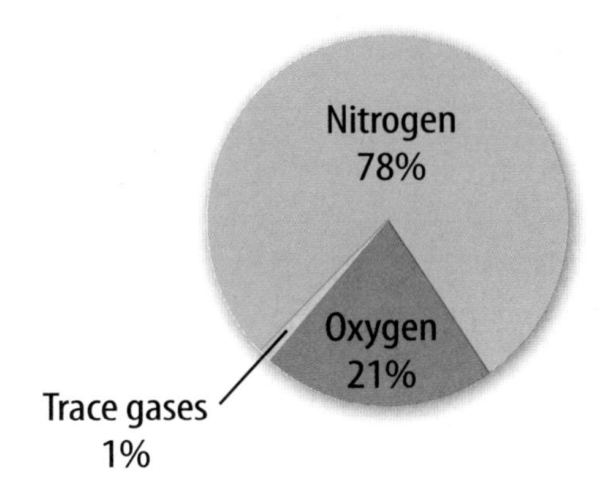

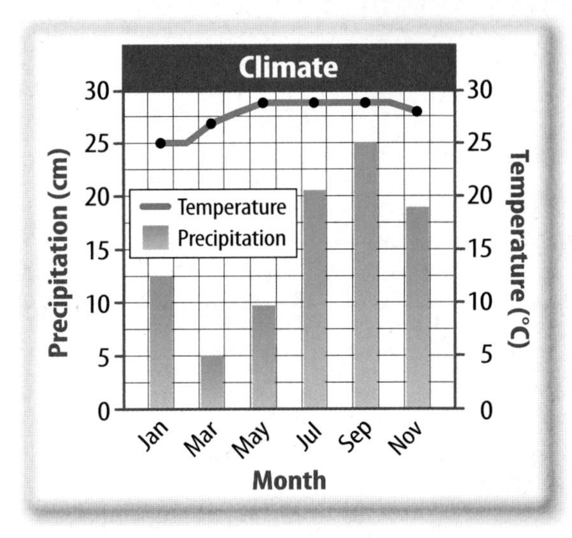

4. **Visual Check** How does the jaguar interact with the abiotic factors in its ecosystem?

Atmosphere

Every time you take a breath you are interacting with another abiotic factor that is necessary for life—the atmosphere. *The* **atmosphere** (AT muh sfir) *is the layer of gases that surrounds Earth.* The atmosphere is mostly nitrogen and oxygen with trace amounts of other gases, also shown in **Figure 1.** Besides providing living things with oxygen, the atmosphere also protects them from certain harmful rays from the Sun.

Soil

Bits of rocks, water, air, minerals, and the remains of once-living things make up soil. When you think about soil, you might picture a farmer growing crops. Soil provides water and nutrients for the plants we eat. However, it is also a home for many organisms, such as insects, bacteria, and fungi.

Factors such as water, soil texture, and the amount of available nutrients affect the types of organisms that can live in soil. Bacteria break down dead plants and animals, returning nutrients to the soil. Earthworms and insects make small tunnels in the soil, allowing air and water to move through it. Even very dry soil, like that in the desert, is home to living things.

Active Reading 5. **List** What are the nonliving things in ecosystems?

Active Reading
FOLDABLES® LA.6.2.2.3

Fold and cut a sheet of paper to make a six-door book. Label it as shown. Use it to organize information about the abiotic parts of an ecosystem.

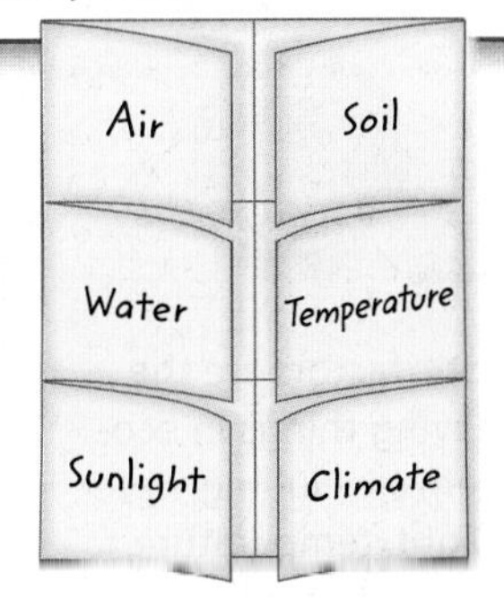

Lesson Review 1

Visual Summary

Ecosystems include all the biotic and abiotic factors in an area.

Biotic factors are the living things in ecosystems.

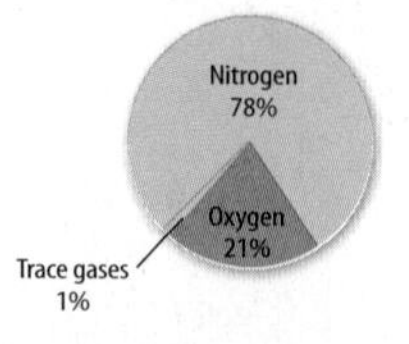

Abiotic factors are the nonliving things in ecosystems, including water, sunlight, temperature, climate, air, and soil.

Use Vocabulary

1. **Distinguish** between biotic and abiotic factors. LA.6.2.2.3

2. **Define** *ecosystem* in your own words.

3. **Use the term** *climate* in a complete sentence.

Understand Key Concepts

4. What role do bacteria play in soil ecosystems?
   (A) They add air to soil.
   (B) They break down rocks.
   (C) They return nutrients to soil.
   (D) They tunnel through soil.

5. **Explain** How would a forest ecosystem change if no sunlight were available to it? LA.6.2.2.3

Interpret Graphics

6. **Organize** Fill in each oval with an abiotic factor. LA.6.2.2.3

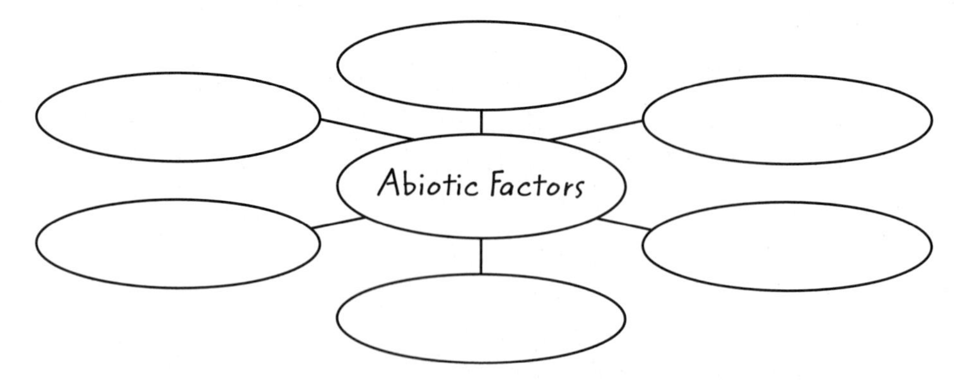

Critical Thinking

7. **Predict** Imagine that the soil in an area is carried away by wind and water, leaving only rocks behind. How would this affect the living things in that area?

Saving Florida's Citrus Crops

FOCUS on FLORIDA

Florida citrus growers are the leading citrus producers in the United States. They produce about 75 percent of the U.S. orange crop and about 80 percent of the U.S. grapefruit crop. Florida growers also produce lemons, limes, and tangerines.

Growing citrus fruit might seem simple, but the citrus growers deal with many issues including how to provide water and nutrients to their trees. The process of transporting water to fields for the purpose of helping crops grow is irrigation. Growers supply nutrients by applying fertilizers to the soil beneath the trees. Without fertilizers, the soil in which the citrus trees are grown would become depleted and the trees would not produce as much fruit.

Because Florida has a subtropical climate, it has high rainfall and high humidity. This environment is ideal for growing citrus. Unfortunately, these same conditions are a welcoming habitat for citrus diseases and pests—the two largest problems Florida citrus growers face. A habitat provides the food, shelter, moisture, temperature, and other factors necessary for an organism to survive.

Citrus trees are affected by pests such as fungi, bacteria, viruses, and insects. Infected trees can be treated with chemicals that kill the harmful organisms. Growers and researchers must take care to use treatments that will keep the trees healthy, but will not harm humans or the environment.

Citrus root weevils can damage leaves and harm plants.

It's Your Turn

RESEARCH Investigate one pest or disease that is harmful to Florida citrus crops and its method of treatment. Create a poster to share your findings.

Lesson 2

Cycles of MATTER

ESSENTIAL QUESTIONS

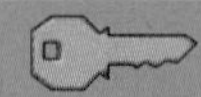

How does matter move in ecosystems?

Vocabulary

evaporation p. 702

condensation p. 702

precipitation p. 702

nitrogen fixation p. 704

Florida NGSSS

LA.6.2.2.3 The student will organize information to show understanding (e.g., representing main ideas within text through charting, mapping, paraphrasing, summarizing, or comparing/contrasting);

SC.6.N.1.4 Discuss, compare, and negotiate methods used, results obtained, and explanations among groups of students conducting the same investigation.

SC.6.N.3.3 Give several examples of scientific laws.

SC.7.N.1.4 Identify test variables (independent variables) and outcome variables (dependent variables) in an experiment.

SC.8.L.18.1 Describe and investigate the process of photosynthesis, such as the roles of light, carbon dioxide, water and chlorophyll; production of food; release of oxygen.

SC.8.L.18.3 Construct a scientific model of the carbon cycle to show how matter and energy are continuously transferred within and between organisms and their physical environment.

SC.8.L.18.4 Cite evidence that living systems follow the Laws of Conservation of Mass and Energy.

SC.8.N.1.2 Design and conduct a study using repeated trials and replication.

SC.8.N.1.6 Understand that scientific investigations involve the collection of relevant empirical evidence, the use of logical reasoning, and the application of imagination in devising hypotheses, predictions, explanations and models to make sense of the collected evidence.

SC.8.N.3.1 Select models useful in relating the results of their own investigations.

Inquiry Launch Lab

SC.8.N.1.6, SC.8.N.3.1, SC.8.L.18.4

15 minutes

How can you model raindrops?

Like all matter on Earth, water is recycled. It constantly moves between Earth and its atmosphere. You could be drinking the same water that a *Tyrannosaurus rex* drank 65 million years ago!

Procedure

1. Read and complete a lab safety form.
2. Half-fill a **plastic cup** with warm water.
3. Cover the cup with **plastic wrap.** Secure the plastic with a **rubber band.**
4. Place an **ice cube** on the plastic wrap. Observe the cup for several minutes.

Think About This

1. What did you observe on the underside of the plastic wrap? Why do you think this happened?

2. How does this activity model the formation of raindrops?

3. Key Concept Do you think other matter moves through the environment? Explain your answer.

Inquiry **Where does the water go?**

1. All water, including the water in this waterfall in Florida's Falling Waters State Park, can move throughout an ecosystem in a cycle. It also can change forms. What other forms do you think water takes as it move through an ecosystem?

How does matter move in ecosystems?

The water that you used to wash your hands this morning might have once traveled through the roots of a tree in Africa or even have been part of an Antarctic glacier. How can this be? Water moves continuously through ecosystems. It is used over and over again. Like water, other types of matter, such as carbon, oxygen, and nitrogen, go through physical and chemical changes. The total mass of the matter is the same before and after a change. This is known as the law of conservation of mass.

The Water Cycle

Look at a globe or a map. Notice that water surrounds the landmasses. Water covers about 70 percent of Earth's surface.

Most of Earth's water—about 97 percent—is in oceans. Water is also in rivers and streams, lakes, and underground reservoirs. In addition, water is in the atmosphere, icy glaciers, and living things.

Water continually cycles between Earth and its atmosphere. This movement of water is called the water cycle. It involves three processes: evaporation, condensation, and precipitation.

Active Reading **2. Identify** What are three Florida bodies of water near you?

The Water Cycle

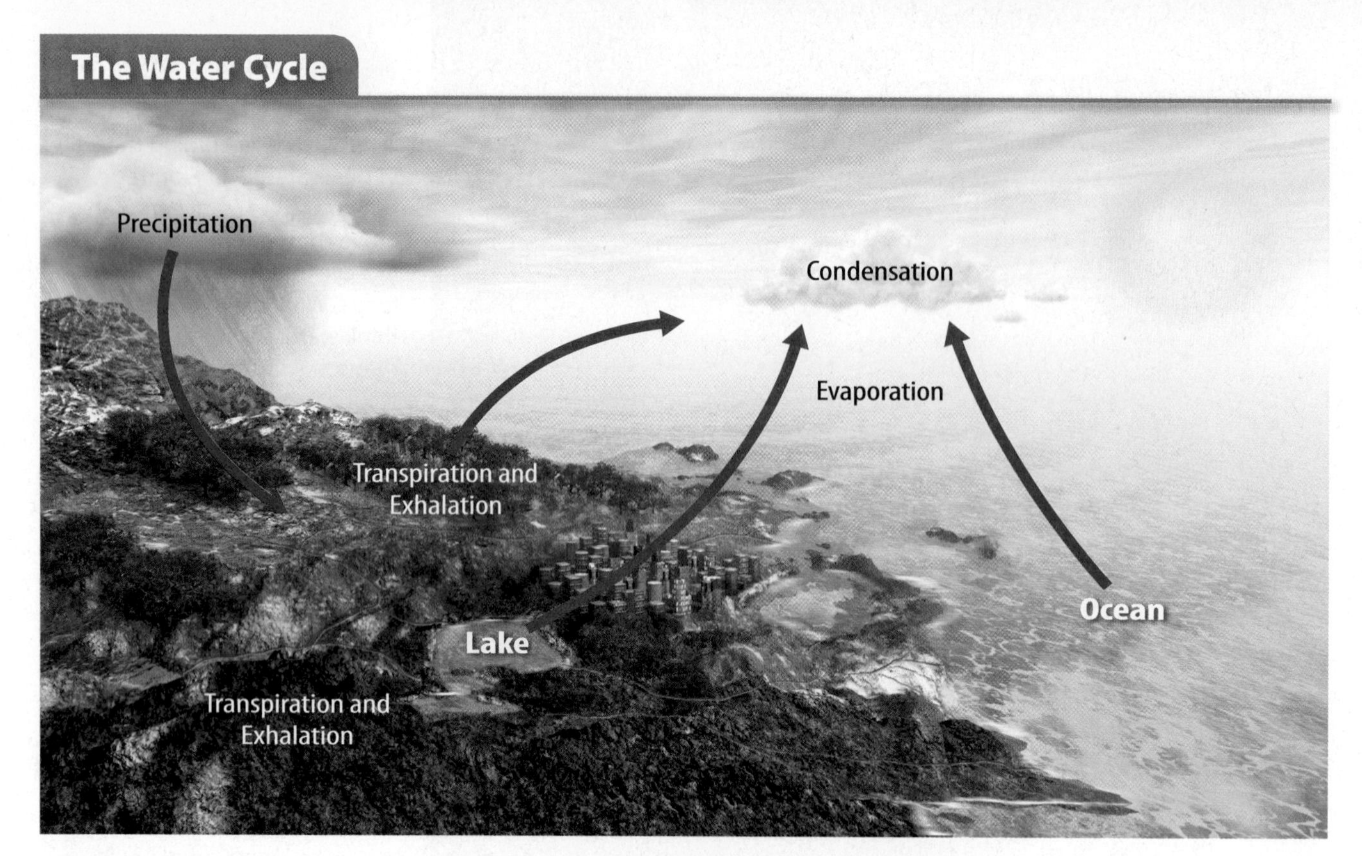

Figure 2 During the water cycle, the processes of evaporation, condensation, and precipitation move water from Earth's surface into the atmosphere and back again.

Evaporation

The Sun supplies the energy for the water cycle, as shown in **Figure 2.** As the Sun heats Earth's surface waters, evaporation occurs. **Evaporation** (ih va puh RAY shun), *is the process during which liquid water changes into a gas called water vapor.* This water vapor rises into the atmosphere. Temperature, humidity, and wind affect how quickly water evaporates.

Water is also released from living things. Transpiration is the release of water vapor from the leaves and stems of plants. Recall that cellular respiration is a process that occurs in many cells. A by-product of cellular respiration is water. This water leaves cells and enters the environment and atmosphere as water vapor.

Condensation

The higher in the atmosphere you are, the cooler the temperature is. As water vapor rises, it cools and condensation occurs.

Condensation (kahn den SAY shun), *is the process during which water vapor changes into liquid water.* Condensation causes clouds. Clouds are made of millions of tiny water droplets or crystals of ice. These form when water vapor condenses on particles of dust and other substances in the atmosphere.

Precipitation

Water that falls from clouds to Earth's surface is called **precipitation** (prih sih puh TAY shun). It enters bodies of water or soaks into soil. Precipitation can be rain, snow, sleet, or hail. It forms as water droplets or ice crystals join together in clouds. Eventually, these droplets or crystals become so large and heavy that they fall to Earth. Over time, living things use this precipitation, and the water cycle continues.

3. **NGSSS Check** **Determine** What forms does water take as it moves through ecosystems? SC.8.L.18.4

Apply It! After you complete the lab, answer these questions.

1. How could nitrogen help Florida farmers?

2. How might nitrogen levels differ throughout the state of Florida?

The Nitrogen Cycle

Just as water is necessary for life on Earth, so is the element nitrogen. It is an essential part of proteins, which all organisms need to stay alive. Nitrogen is also an important part of DNA, the molecule that contains genetic information. Nitrogen demonstrates the law of conservation of mass by cycling between Earth and its atmosphere and back again, as shown in **Figure 3.**

Active Reading **4. Interpret** Underline what living things use nitrogen for.

Figure 3 Different forms of nitrogen are in the atmosphere, soil, and organisms.

Bacteria in soil convert nitrogen compounds into nitrogen gas, which is released into the air.

Nitrogen gas in atmosphere

Lightning changes nitrogen gas in the atmosphere to nitrogen compounds. The nitrogen compounds fall to the ground when it rains.

Animals eat plants.

Nitrogen-fixing bacteria on plant roots convert unusable nitrogen in soil to usable nitrogen compounds.

Decaying organic matter and animal waste return nitrogen compounds to the soil.

Plants take in and use nitrogen compounds from the soil.

Nitrogen compounds in soil

From the Environment to Organisms

Recall that the atmosphere is mostly nitrogen. However, this nitrogen is in a form that plants and animals cannot use. How do organisms get nitrogen into their bodies? The nitrogen must first be changed into a different form with the help of certain bacteria that live in soil and water. These bacteria take in nitrogen from the atmosphere and change it into nitrogen compounds that other living things can use. *The process that changes atmospheric nitrogen into nitrogen compounds that are usable by living things is called* **nitrogen fixation** (NI truh jun • fihk SAY shun). Nitrogen fixation is shown in **Figure 4.**

Plants and some other organisms take in this changed nitrogen from the soil and water. Then, animals take in nitrogen when they eat the plants or other organisms.

Active Reading **5. Infer** Nitrogen fixation occurs in some types of sugar cane, a crop grown in Florida. How does nitrogen fixation benefit the sugar cane plant?

Figure 4 Certain bacteria convert nitrogen in soil and water into a form usable by plants.

From Organisms to the Environment

Some types of bacteria can break down the tissues of dead organisms. When organisms die, these bacteria help return the nitrogen in the tissues of dead organisms to the environment. This process is shown in **Figure 5.**

Nitrogen also returns to the environment in the waste products of organisms. Farmers often spread animal wastes, called manure, on their fields during the growing season. The manure provides nitrogen to plants for better growth.

Figure 5 Bacteria break down the remains of dead plants and animals.

Figure 6 Most oxygen in the air comes from plants and algae.

6. NGSSS Check Identify Fill in the blanks with the correct term and (circle) your part in the oxygen cycle. SC.8.L.18

The Oxygen Cycle

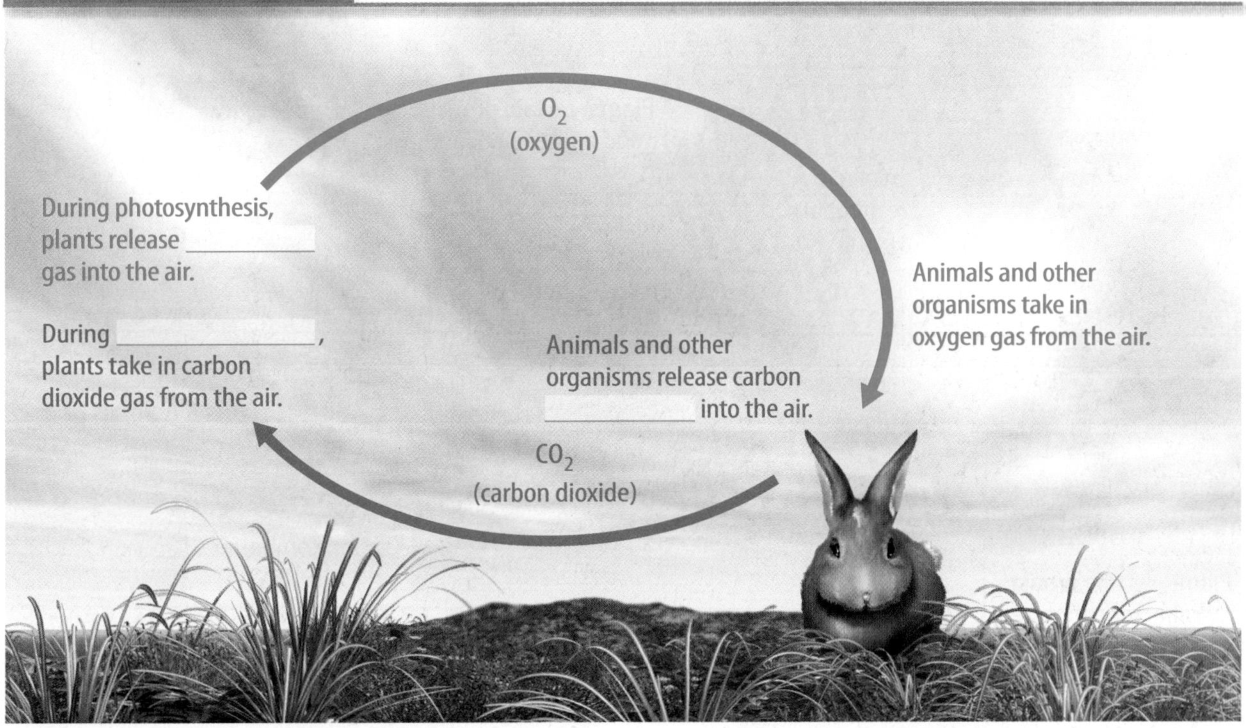

O_2
(oxygen)

During photosynthesis, plants release __________ gas into the air.

During __________, plants take in carbon dioxide gas from the air.

Animals and other organisms take in oxygen gas from the air.

Animals and other organisms release carbon __________ into the air.

CO_2
(carbon dioxide)

The Oxygen Cycle

Almost all living things need oxygen for cellular processes that release energy. Oxygen is also part of many substances that are important to life, such as carbon dioxide and water. Oxygen cycles through ecosystems, as shown in **Figure 6.**

Earth's early atmosphere probably did not contain oxygen gas. Oxygen might have entered the atmosphere when certain **bacteria** evolved that could carry out the process of photosynthesis and make their own food. A by-product of photosynthesis is oxygen gas. Over time, other photosynthetic organisms evolved and the amount of oxygen in Earth's atmosphere increased. Today, photosynthesis is the primary source of oxygen in Earth's atmosphere. Some scientists estimate that unicellular organisms in water, called phytoplankton, release more than 50 percent of the oxygen in Earth's atmosphere.

Many living things, including humans, take in the oxygen and release carbon dioxide. The interaction of the carbon and oxygen cycles is one example of a relationship between different types of matter in ecosystems. As the matter cycles through an ecosystem, both the carbon and the oxygen take different forms and play a role in the other element's cycle.

REVIEW VOCABULARY

bacteria
a group of microscopic unicellular organisms without a membrane-bound nucleus

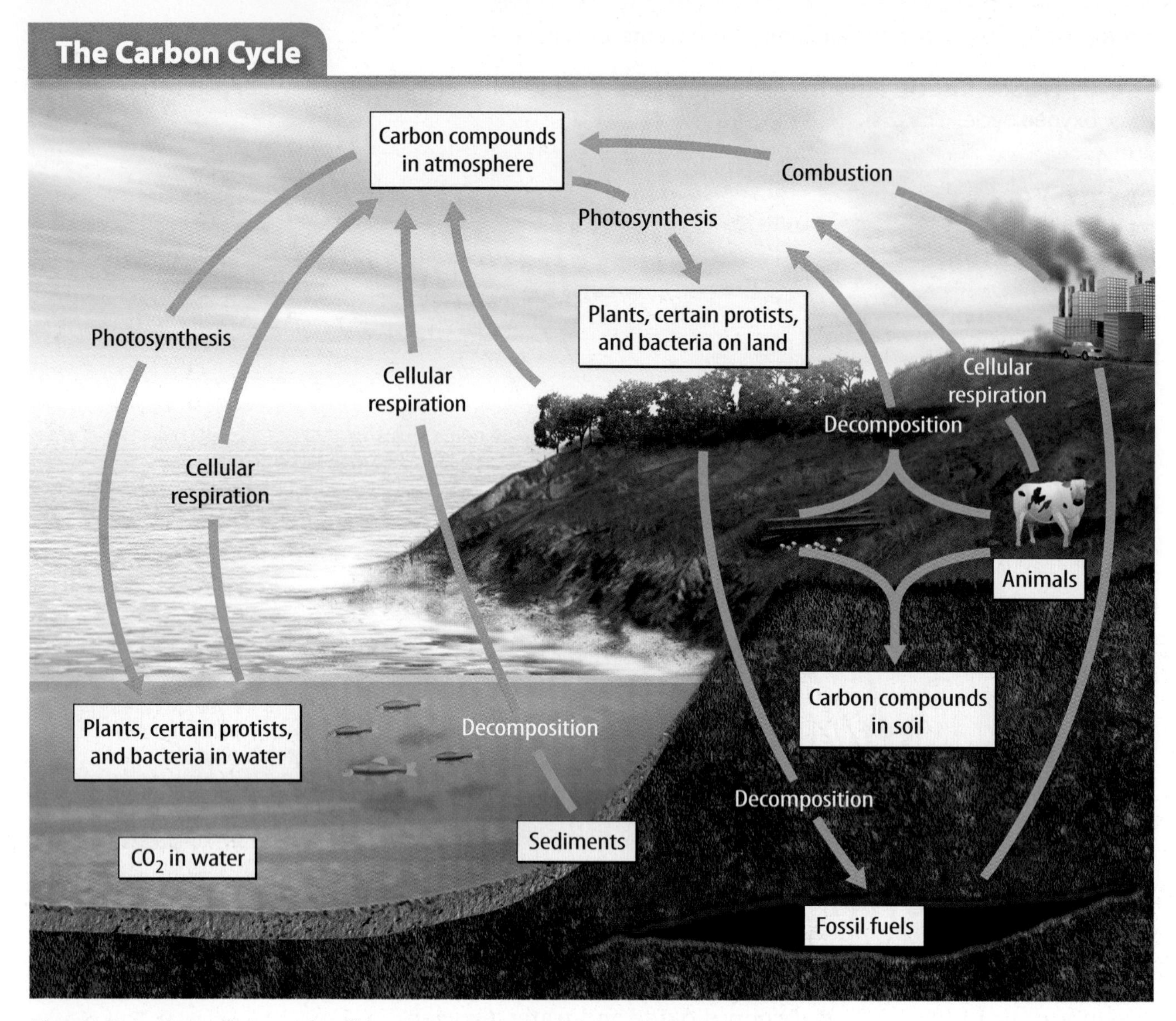

Figure 7 In the carbon cycle, all organisms return carbon to the environment.

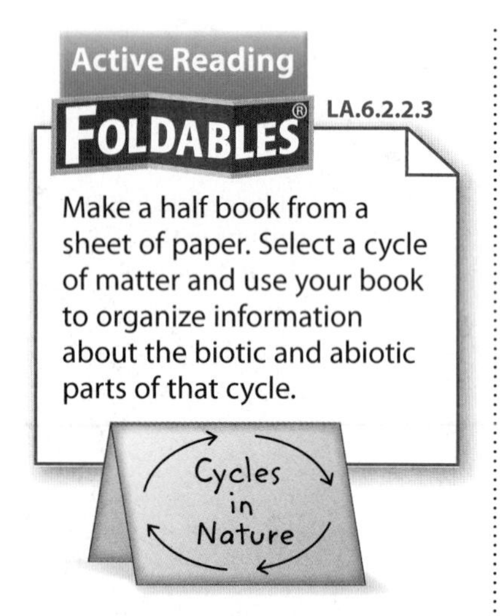

The Carbon Cycle

All organisms contain carbon. It is part of proteins, sugars, fats, and DNA. Some organisms, including humans, get carbon from food. Organisms, such as plants, get carbon from the atmosphere or bodies of water. Carbon is conserved as it cycles through ecosystems, as shown in **Figure 7.**

Carbon in Soil

Like nitrogen, carbon can enter the environment when organisms die and decompose. This returns carbon compounds to the soil and releases carbon dioxide (CO_2) into the atmosphere for use by other organisms. Carbon is also found in fossil fuels, which formed when decomposing organisms were exposed to pressure, high temperatures, and bacteria over hundreds of millions of years.

Carbon in Air

Recall that carbon is found in the atmosphere as carbon dioxide. Plants and other photosynthetic organisms take in carbon dioxide and water and produce energy-rich sugars. These sugars are a source of carbon and energy for organisms that eat photosynthetic organisms. When the sugar is broken down by cells and its energy is **released,** carbon dioxide is released as a by-product. This carbon dioxide gas enters the atmosphere, where it can be used again.

ACADEMIC VOCABULARY
release
(verb) to set free or let go

The Greenhouse Effect

Carbon dioxide is one of the gases in the atmosphere that absorbs thermal energy from the Sun and keeps Earth warm. This process is called the greenhouse effect. The Sun produces solar radiation, as shown in **Figure 8.** Some of this energy is reflected back into space, and some passes through Earth's atmosphere. Greenhouse gases in Earth's atmosphere absorb thermal energy that reflects off Earth's surface. The more greenhouse gases released, the greater the gas layer becomes and the more thermal energy is absorbed. These gases are one factor that keeps Earth from becoming too hot or too cold.

Active Reading **7. Identify** Underline What is the greenhouse effect?

While the greenhouse effect is essential for life, a steady increase in greenhouse gases can harm ecosystems. For example, carbon is stored in fossil fuels such as coal, oil, and natural gas. When people burn fossil fuels to heat homes, for transportation, or to provide electricity, carbon dioxide gas is released into the atmosphere. The amount of carbon dioxide in the air has increased due to both natural and human activities.

Figure 8 Some thermal energy remains close to the Earth due to greenhouse gases.

Active Reading **8. Infer** What might Florida be like if heat were not absorbed by greenhouse gases?

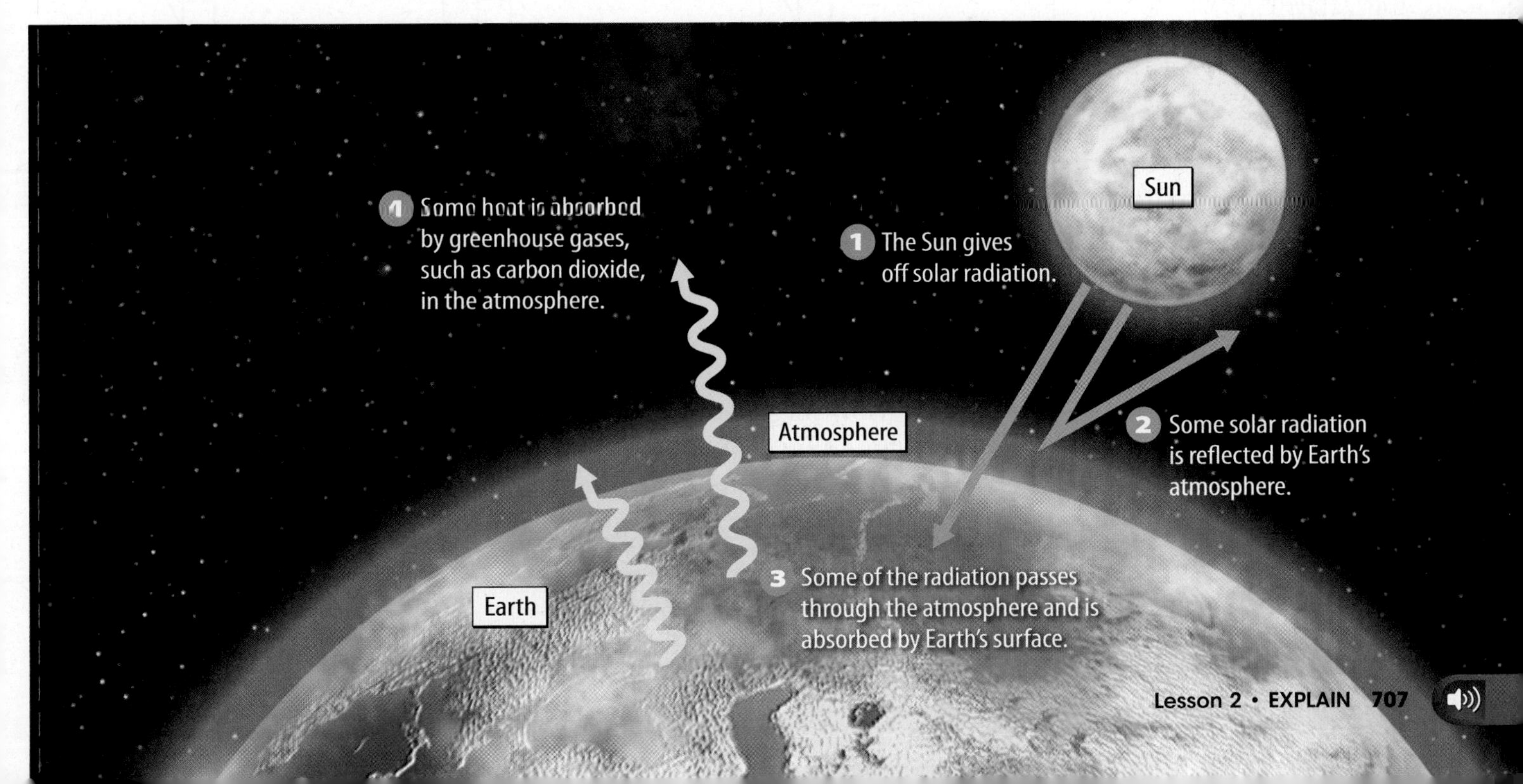

Lesson Review 2

Visual Summary

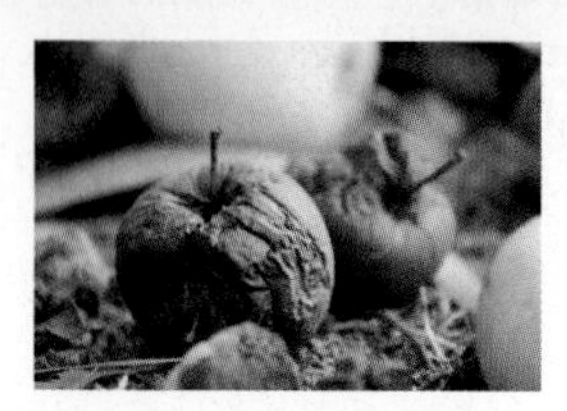

Matter such as water, oxygen, nitrogen, and carbon cycles through ecosystems.

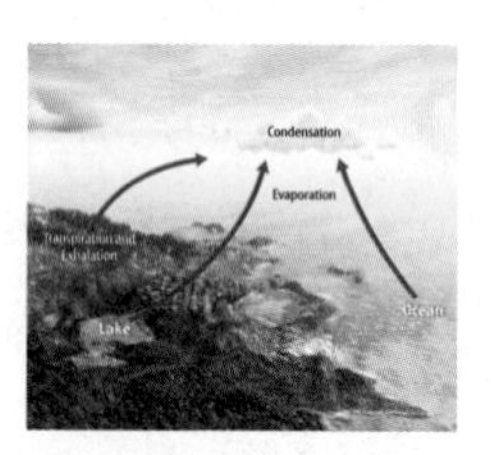

The three stages of the water cycle are evaporation, condensation, and precipitation.

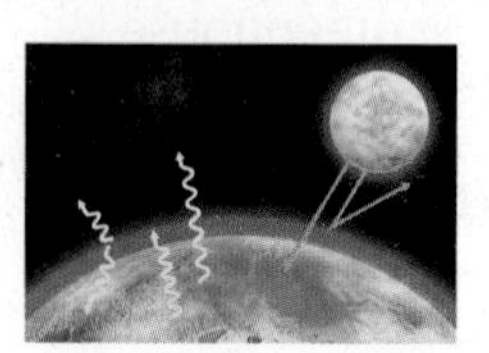

The greenhouse effect helps keep the Earth from getting too hot or too cold.

Use Vocabulary

1. **Distinguish** between evaporation and condensation.

2. **Define** *nitrogen fixation* in your own words.

3. Water that falls from clouds to Earth's surface is called ______.

Understand Key Concepts

4. What is the driving force behind the water cycle?
   - (A) gravity
   - (B) plants
   - (C) sunlight
   - (D) wind

5. **Infer** Farmers add nitrogen to their fields every year to help their crops grow. Why must farmers continually add nitrogen when this element recycles naturally? SC.8.L.18.4

Interpret Graphics

6. **Sequence** Draw a graphic organizer like the one below and sequence the steps in the water cycle. LA.6.2.2.3

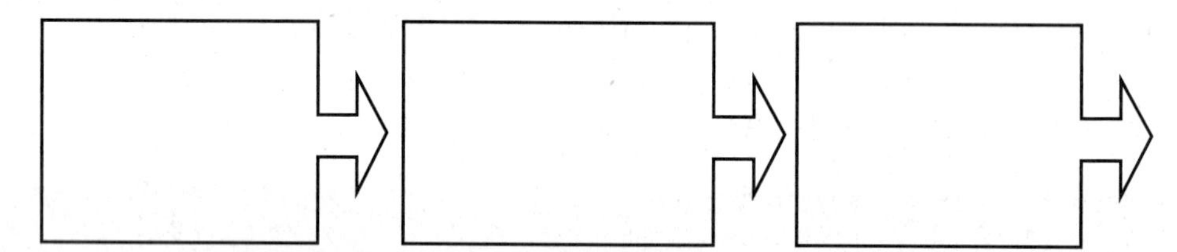

Critical Thinking

7. **Explain** how carbon cycles through the ecosystem in which you live. SC.8.L.18.3, SC.8.L.18.4

8. **Consider** How might ecosystems be affected if levels of atmospheric CO_2 continue to rise? SC.8.L.18.4

FOCUS on FLORIDA

Florida's Lake Wales Ridge

Cattle egret

Did you know that Florida has one of the most unique ecosystems in the world? Lake Wales Ridge is an ancient beach and sand dune system in central Florida. The approximately 100 miles of sand hills contain many organisms that are found only in the ridge, including 31 species of rare plants. Small shrubs and bushes, called "scrub," cover most of the area. These unique organisms are well-suited for living in a harsh environment like the Lake Wales Ridge.

The soil in the Ridge is very dry and sandy and does not contain many nutrients. Water drains out of the soil very quickly. Most plants would not be able to grow in this kind of environment. However, certain types of trees, such as shrubby oaks and Florida rosemary, grow well in it. These plants provide shelter and food for animals, such as Florida scrub jays and gopher tortoises.

Scientists estimate that about 80,000 acres of scrubland existed in Florida before European settlers arrived. Today, about 85 percent of the ridge has been cleared for use by agriculture, commercial, and residential developments. Government programs have protected the remaining scrubland in a wildlife refuge so generations to come can study this amazing ecosystem.

Sundew plant

It's Your Turn

RESEARCH Investigate how one type of matter moves throughout the Lake Wales Ridge ecosystem, and create a diagram showing the matter cycle.

Lesson 3

Energy in ECOSYSTEMS

ESSENTIAL QUESTIONS

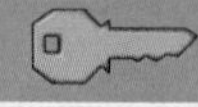 How does energy move in ecosystems?

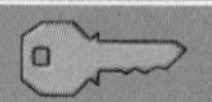 How is the movement of energy in an ecosystem modeled?

Vocabulary

photosynthesis p. 712
chemosynthesis p. 712
food chain p. 714
food web p. 715
energy pyramid p. 716

Florida NGSSS

SC.6.N.3.3 Give several examples of scientific laws.

SC.7.L.17.1 Explain and illustrate the roles of and relationships among producers, consumers, and decomposers in the process of energy transfer in a food web.

SC.7.N.1.3 Distinguish between an experiment (which must involve the identification and control of variables) and other forms of scientific investigation and explain that not all scientific knowledge is derived from experimentation.

SC.8.L.18.1 Describe and investigate the process of photosynthesis, such as the roles of light, carbon dioxide, water and chlorophyll; production of food; release of oxygen.

SC.8.L.18.3 Construct a scientific model of the carbon cycle to show how matter and energy are continuously transferred within and between organisms and their physical environment.

SC.8.L.18.4 Cite evidence that living systems follow the Laws of Conservation of Mass and Energy.

SC.8.N.1.1 Define a problem from the eighth grade curriculum using appropriate reference materials to support scientific understanding, plan and carry out scientific investigations of various types, such as systematic observations or experiments, identify variables, collect and organize data, interpret data in charts, tables, and graphics, analyze information, make predictions, and defend conclusions.

SC.8.N.3.1 Select models useful in relating the results of their own investigations.

Also covers: LA.6.2.2.3, MA.6.A.3.6

Launch Lab

10 minutes

SC.8.L.18.4
SC.7.N.1.3

How does energy change form?

Every day, sunlight travels hundreds of millions of kilometers and brings warmth and light to Earth. Energy from the Sun is necessary for nearly all life on Earth. Without it, most life could not exist.

Procedure

1. Read and complete a lab safety form.
2. Obtain **UV-sensitive beads** from your teacher. Write a description of them below.
3. Place half the beads in a sunny place. Place the other half in a dark place.
4. Wait a few minutes, and then observe both sets of beads. Record your observations.

Data and Observations

Think About This

1. Compare and contrast the two sets of beads after the few minutes. How are they different? How are they the same?
2. Hypothesize why the beads looked different.
3. Key Concept How do you think living things use energy?

Time for a snack?

1. All organisms need energy, and many get it from eating other organisms. How do you think each of the living things in this picture gets the energy it needs?

How does energy move in ecosystems?

When you see a picture of an ecosystem like Lake Okeechobee, Florida, it often looks quiet and peaceful. However, ecosystems are actually full of movement. Birds squawk and beat their wings, plants sway in the breeze, and insects buzz.

Each movement made by a living thing requires energy. All of life's functions, including growth and reproduction, require energy. The main source of energy for most life on Earth is the Sun. Unlike other resources, such as water and carbon, energy does not cycle through ecosystems. Instead, energy flows in one direction, as shown in **Figure 9**. In most cases, energy flow begins with the Sun and moves from one organism to another. Many organisms get energy by eating other organisms. Sometimes organisms change energy into different forms as it moves through an ecosystem. Not all the energy an organism gets is used for life processes. Some is released to the environment as thermal energy. You might have read that energy cannot be created or destroyed, but it can change form. This idea is called the law of conservation of energy.

2. **NGSSS Check** **Contrast** How do the movements of matter and energy differ? SC.8.L.18.4

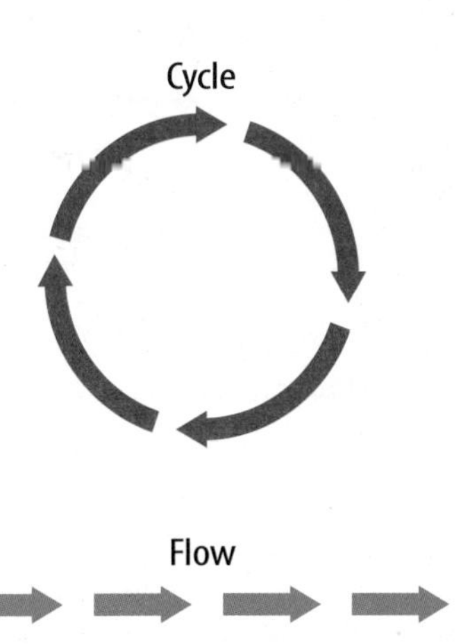

Figure 9 Matter moves in a cycle pattern, and energy moves in a flow pattern.

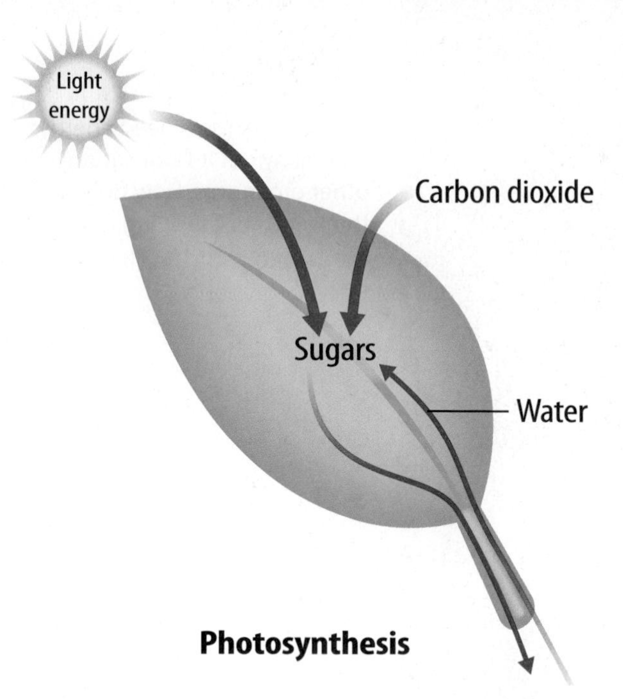

Figure 10 Most producers make their food through the process of photosynthesis.

WORD ORIGIN

photosynthesis
from Greek *photo*, meaning "light"; and *synthese*, meaning "synthesis"

Figure 11 The producers at a hydrothermal vent make their food using chemosynthesis.

Producers

People who make things or products are often called producers. In a similar way, living things that make their own food are called producers. Producers make their food from materials found in their environments. Most producers are photosynthetic (foh toh sihn THEH tihk). They use the process of photosynthesis (foh toh SIHN thuh sus), which is described below. Grasses, trees, and other plants are photosynthetic. Algae, some other protists, and certain bacteria are also photosynthetic. Other producers, including some bacteria, are chemosynthetic (kee moh sihn THEH tihk). They make their food using chemosynthesis (kee moh SIHN thuh sus).

Photosynthesis Recall that in the carbon cycle, carbon in the atmosphere cycles through producers, such as plants, into other organisms and then back into the atmosphere. This and other matter cycles involve photosynthesis, as shown in **Figure 10.** **Photosynthesis** *is a series of chemical reactions that convert light energy, water, and carbon dioxide into the food-energy molecule glucose and give off oxygen.*

Chemosynthesis As you read earlier, some producers make food using chemosynthesis. **Chemosynthesis** *is the process during which producers use chemical energy in matter rather than light energy and make food.* One place where chemosynthesis can occur is on the deep ocean floor. There, inorganic compounds that contain hydrogen and sulfur flow out from cracks in the ocean floor. These cracks are called hydrothermal vents. These vents, such as the one shown in **Figure 11,** contain chemosynthetic bacteria. These bacteria use the chemical energy contained in inorganic compounds and produce food.

Active Reading **3. Recall** What materials do producers use to make food during chemosynthesis?

____

____

Figure 12 Organisms can be classified by the type of food that they eat.

Consumers

Some consumers are shown in **Figure 12.** Consumers do not produce their own energy-rich food, as producers do. Instead, they get the energy they need to survive by consuming other organisms.

Consumers can be classified by the type of food they eat. Herbivores feed on only producers. For example, a deer is an herbivore because it eats only plants. Carnivores eat other animals. They are usually predators, such as lions and wolves. Omnivores eat both producers and other consumers. A bird that eats berries and insects is an omnivore.

Another group of consumers is detritivores (dih TRI tuh vorz). They get their energy by eating the remains of other organisms. Some detritivores, such as insects, eat dead organisms. Other detritivores, such as bacteria and mushrooms, feed on dead organisms and help decompose them. For this reason these organisms often are called decomposers. During decomposition, decomposers produce carbon dioxide that enters the atmosphere. Some of the decayed matter enters the soil. In this way detritivores help recycle nutrients through ecosystems. They also help keep ecosystems clean. Without decomposers dead organisms would pile up in an ecosystem.

Active Reading **4. Give Examples** List one herbivore, one carnivore, and one detritivore in a Florida ecosystem.

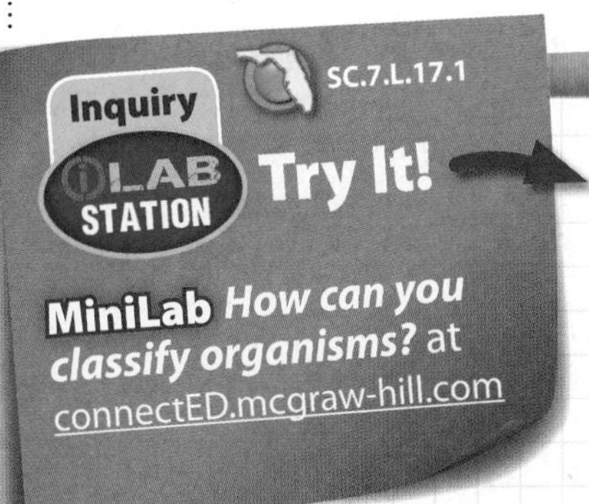

Apply It!

After you complete the lab, answer the question below.

1. Do you think any animals can be classified as more than one type of consumer? Why or why not?

Active Reading

FOLDABLES® LA.6.2.2.3

Make a pyramid book from a sheet of paper. Use each side to organize information about one of the ways energy flows in an ecosystem. You can add additional information on the inside of your pyramid book.

Modeling Energy in Ecosystems

Unlike matter, energy does not cycle through ecosystems because it does not return to the Sun. Instead, energy flows through ecosystems. Organisms store energy in their bodies as chemical energy and use some energy for life processes. When consumers eat these organisms, the energy is transferred to the consumer. However, some energy is changed to thermal energy in the process and enters the environment. Decomposers transfer energy back into the environment when organisms die. Scientists use models to study this flow of energy through an ecosystem. They use different models depending on how many organisms they are studying.

Food Chains

A **food chain** *is a model that shows how energy flows in an ecosystem through feeding relationships.* In a food chain, arrows show the transfer of energy. A typical food chain is shown in **Figure 13.** Notice that there are not many links in this food chain. That is because the amount of available energy decreases every time it is transferred from one organism to another.

5. **NGSSS Check** **Explain** How does a food chain model energy flow? SC.8.L.18.4

Figure 13 Energy moves from the Sun to a plant, a mouse, a snake, and a hawk in this food chain.

Food Chain

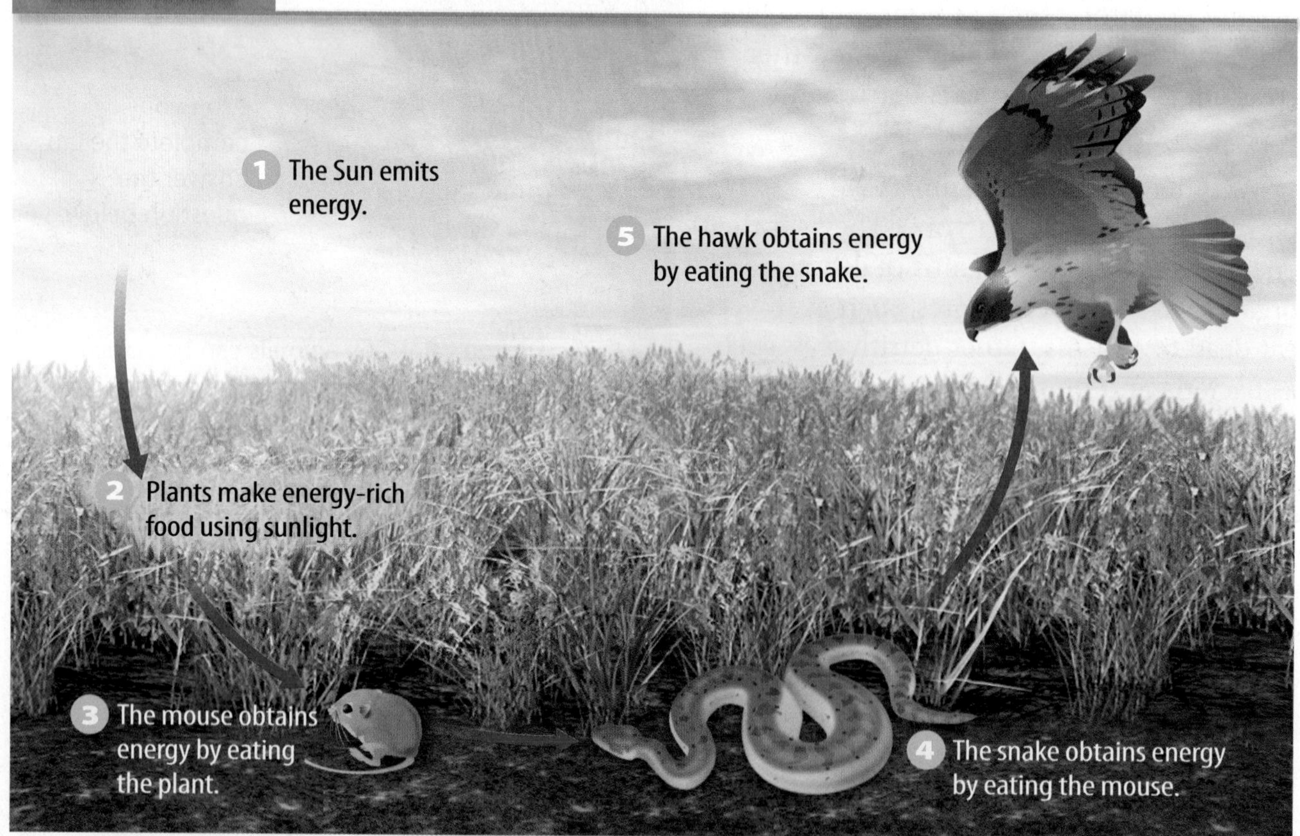

Food Webs

Imagine you have a jigsaw puzzle of part of the Everglades. Each piece of the puzzle shows only one small section of a wetland. A food chain is like one piece of an ecosystem jigsaw puzzle. It is helpful when studying certain parts of an ecosystem, but it does not show the whole picture.

In the food chain on the previous page, the mouse might also eat the seeds of several producers, such as corn, berries, or grass. The snake might eat other organisms such as frogs, crickets, lizards, or earthworms. The hawk hunts mice, squirrels, rabbits, and fish, as well as snakes.

Scientists use a model of energy transfer called a **food web** *to show how food chains in a community are interconnected,* as shown in **Figure 14.** You can think of a food web as many overlapping food chains. Like in a food chain, arrows show how energy flows in a food web. Some organisms in the food web might be part of more than one food chain in that web.

6. NGSSS Check **Assess** Underline What models show the transfer of energy in an ecosystem? **SC.8.L.18.4**

Figure 14 A food web shows the complex feeding relationships among organisms in an ecosystem.

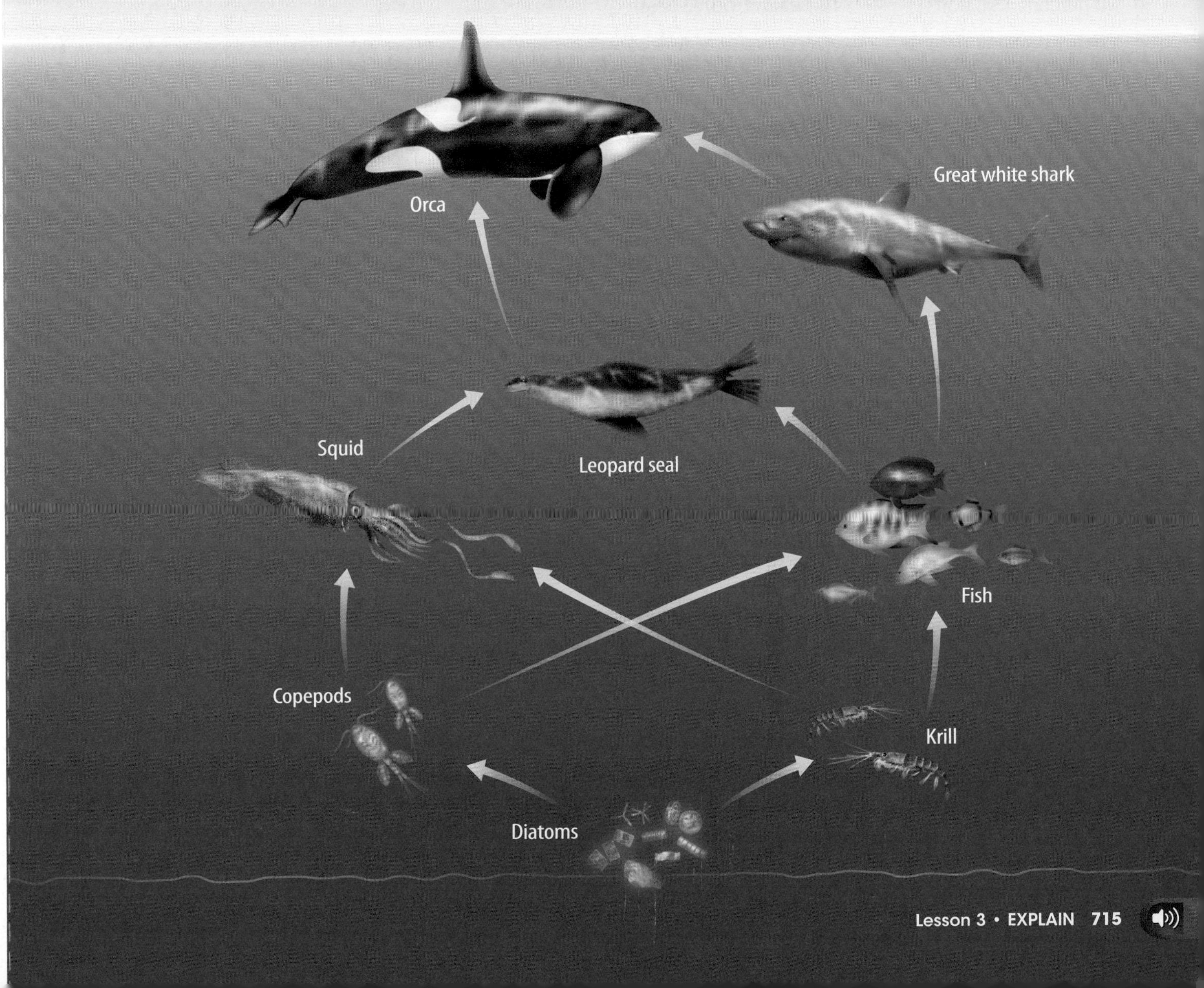

Orca

Great white shark

Squid

Leopard seal

Fish

Copepods

Krill

Diatoms

Math Skills MA.6.A.3.6

Use Percentages

The first trophic level—producers—obtains energy from the Sun. They use 90 percent of the energy for their own life processes. Only 10 percent of the energy remains for the second trophic level—herbivores. Assume that each trophic level uses 90 percent of the energy it receives. Use the following steps to calculate how much energy remains for the next trophic level.

First trophic level gets 100 units of energy.

First trophic level uses 90 percent = 90 units

Energy remaining for second trophic level = 10 units

Second trophic level uses 90 percent = 9 units

Energy remaining for third trophic level = 1 unit

Practice

7. If the first trophic level receives 10,000 units of energy from the Sun, how much energy is available for the second trophic level?

Energy Pyramids

Food chains and food webs show how energy moves in an ecosystem. However, they do not show how the amount of energy in an ecosystem changes. *Scientists use a model called an* **energy pyramid** *to show the amount of energy available in each step of a food chain,* as shown in **Figure 15.** The steps of an energy pyramid are also called trophic (TROH fihk) levels.

Producers, such as plants, make up the bottom trophic level. Consumers that eat producers, such as squirrels, make up the next trophic level. Consumers, such as hawks, that eat other consumers make up the highest trophic level. Notice that less energy is available for consumers at each higher trophic level. As you read earlier, organisms use some of the energy they get from food for life processes. During life processes, some energy is changed to thermal energy and is transferred to the environment. Only about 10 percent of the energy available at one trophic level transfers on to the next trophic level.

Figure 15 An energy pyramid shows the amount of energy available at each trophic level.

8. **Visual Check** **Analyze** How does the amount of available energy change at each trophic level?

Lesson Review 3

Visual Summary

Energy flows in ecosystems from producers to consumers.

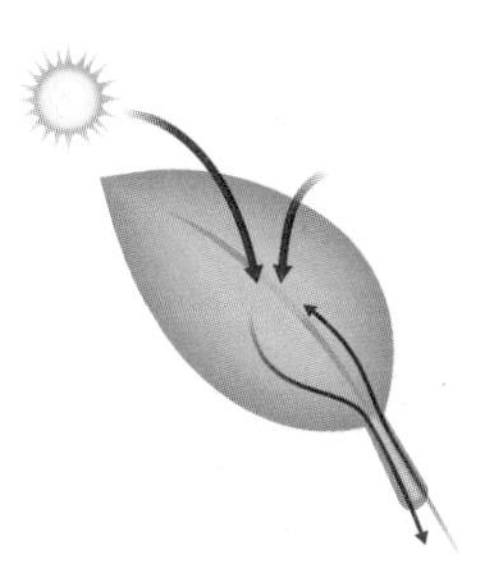

Producers make their own food through the processes of photosynthesis or chemosynthesis.

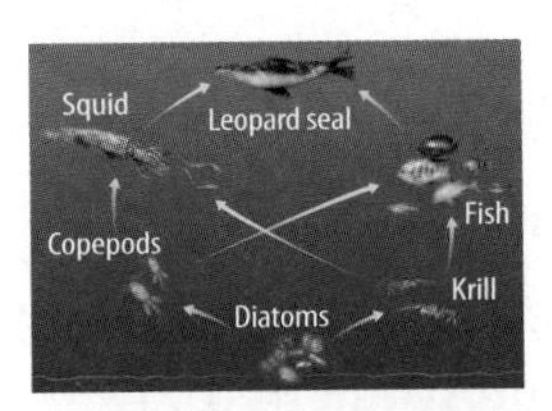

Food chains and food webs model how energy moves in ecosystems.

Inquiry Lab *Can you observe part of the carbon cycle?* at connectED.mcgraw-hill.com

Use Vocabulary

1. Scientists use a(n) ____________________ to show how energy moves in an ecosystem.

2. **Distinguish** between photosynthesis and chemosynthesis. SC.8.L.18.1

Understand Key Concepts 

3. Which organism is a producer?
   - (A) cow
   - (B) dog
   - (C) grass
   - (D) human

4. **Construct** a food chain with four links.

Interpret Graphics

5. **Assess** Which trophic level has the most energy available to living things?

Trophic level 3

Trophic level 2

Trophic level 1

Critical Thinking

6. **Recommend** A friend wants to show how energy moves in ecosystems. Which model would you recommend? Explain. SC.8.L.18.4

Math Skills MA.6.A.3.6

7. The plants in level 1 of a food pyramid obtain 30,000 units of energy from the Sun. How much energy is available for the organisms in level 2? Level 3?

Chapter 18 Study Guide

Think About It! Living things interact with and depend on each other and on the nonliving things in an ecosystem.

Key Concepts Summary

LESSON 1 Abiotic Factors

- The **abiotic factors** in an environment include sunlight, temperature, climate, air, water, and soil.

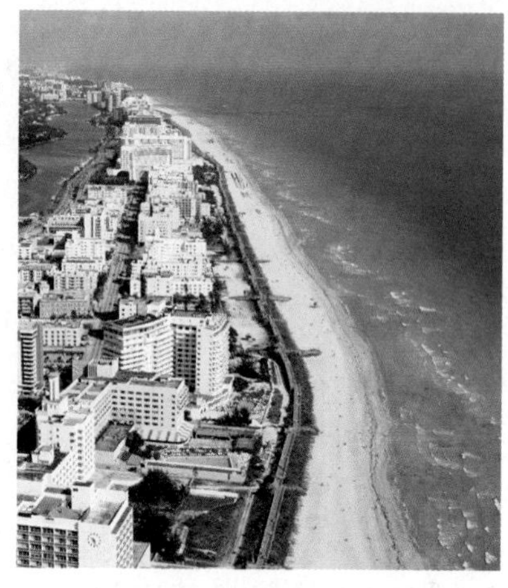

Vocabulary

ecosystem p. 695
biotic factor p. 695
abiotic factor p. 695
climate p. 696
atmosphere p. 697

LESSON 2 Cycles of Matter

- Matter such as oxygen, nitrogen, water, carbon, and minerals moves in cycles in the ecosystem.

evaporation p. 702
condensation p. 702
precipitation p. 702
nitrogen fixation p. 704

LESSON 3 Energy in Ecosystems

- Energy flows through ecosystems from producers to consumers.
- **Food chains, food webs,** and **energy pyramids** model the flow of energy in ecosystems.

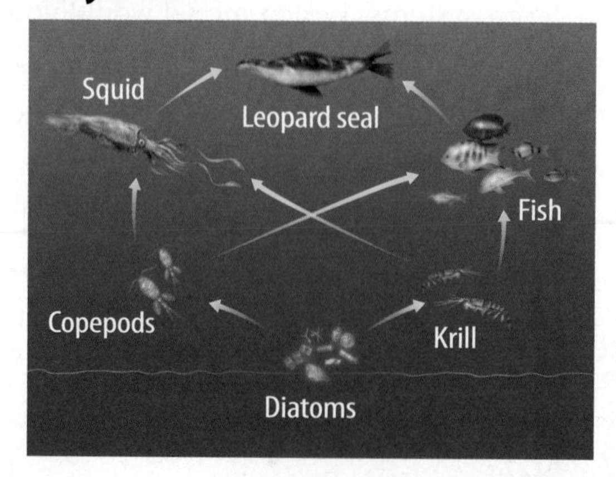

photosynthesis p. 712
chemosynthesis p. 712
food chain p. 714
food web p. 715
energy pyramid p. 716

Active Reading FOLDABLES® **Chapter Project**

Assemble your lesson Foldables as shown to make a Chapter Project. Use the project to review what you have learned in this chapter.

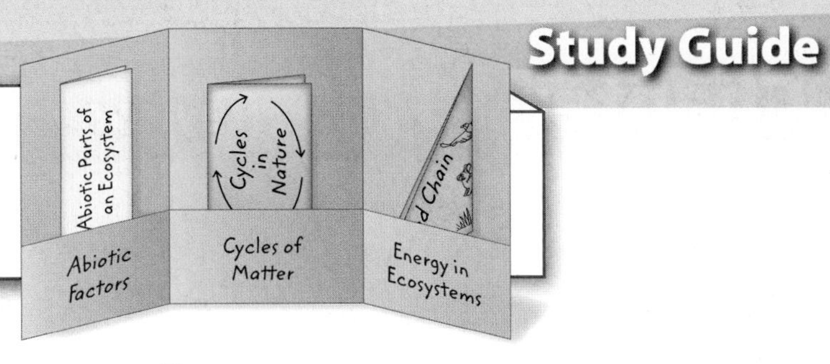

Use Vocabulary

1. Distinguish between climate and atmosphere.

____

____

2. The atmosphere is made mainly of the gases ______________ and ______________.

3. Living organisms in an ecosystem are called ______________, while the nonliving things are called ______________.

4. The process of converting nitrogen in the air into a form that can be used by living organisms is called ______________.

5. Use the word *precipitation* in a complete sentence.

____

____

6. Define *condensation* in your own words.

____

____

7. How does a food chain differ from a food web?

____

____

8. The process of ______________ uses energy from the Sun.

9. Define *chemosynthesis* in your own words.

____

____

Link Vocabulary and Key Concepts

Use vocabulary terms from the previous page and other terms from the chapter to complete the concept map.

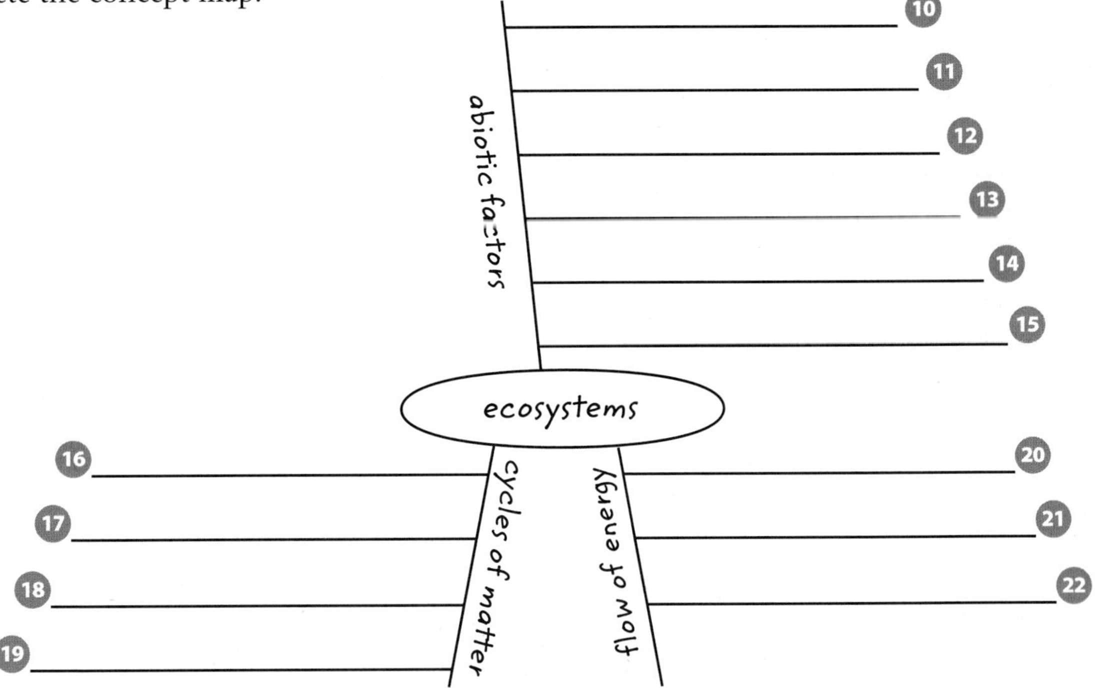

Chapter 18 Review

Fill in the correct answer choice.

Understand Key Concepts

1. What is the source of most energy on Earth? **SC.8.L.18.1**
   Ⓐ air
   Ⓑ soil
   Ⓒ the Sun
   Ⓓ water

2. Which is a biotic factor in an ecosystem?
   Ⓐ a plant living near a stream
   Ⓑ the amount of rainfall
   Ⓒ the angle of the Sun
   Ⓓ the types of minerals present in soil

3. Study the energy pyramid shown here.

   Which organism might you expect to find at trophic level I?
   Ⓐ fox
   Ⓑ frog
   Ⓒ grass
   Ⓓ grasshopper

4. Which includes both an abiotic and a biotic factor?
   Ⓐ a chicken laying an egg
   Ⓑ a deer drinking from a stream
   Ⓒ a rock rolling down a hill
   Ⓓ a squirrel eating an acorn

5. Which process helps keep temperatures on Earth from becoming too hot or too cold?
   Ⓐ condensation
   Ⓑ global warming
   Ⓒ greenhouse effect
   Ⓓ nitrogen fixation

Critical Thinking

6. **Compare and contrast** the oxygen cycle and the nitrogen cycle. **SC.8.L.18.4**

7. **Create** a plan for making an aquatic ecosystem in a jar. Include both abiotic and biotic factors.

8. **Recommend** a strategy for decreasing the amount of carbon dioxide in the atmosphere. **SC.8.L.18.4**

9. **Role-play** Working in a group, perform a skit about organisms living near a hydrothermal vent. Be sure to include information about how the organisms obtain energy. Write your notes below. **SC.7.L.17.1, SC.7.L.17.2**

10. **Assess** the usefulness of models as tools for studying ecosystems.

11 Study the food web below. Classify each organism according to what it eats and write its classification under its name. SC.7.L.17.1

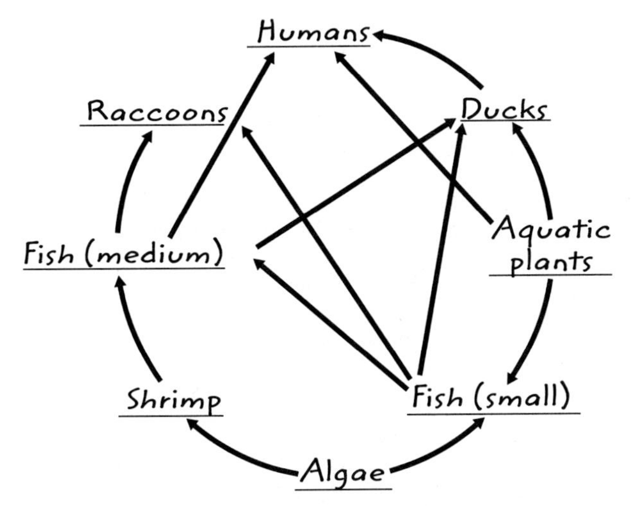

12 **Predict** what would happen if all the nitrogen-fixing bacteria in an ecosystem were removed.

Writing in Science

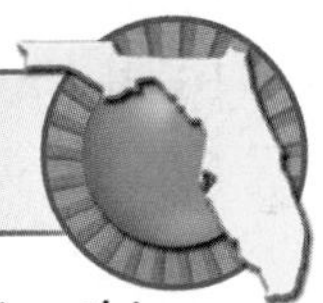

13 On a separate piece of paper, write an argument for or against the following statement. *The energy humans use in cars originally came from the sun.* SC.8.L.18.4

Big Idea Review

14 Describe an interaction between a living thing and a nonliving thing in an ecosystem.

15 How might the ram interact with nonliving things in its environment?

Math Skills MA.6.A.3.6

Use Percentages

16 A group of plankton, algae, and other ocean plants absorb 150,000 units of energy.

a. How much energy is available for the third trophic level?

b. How much energy would remain for a fourth trophic level?

17 Some organisms, such as humans, are omnivores. They eat both producers and consumers. How much more energy would an omnivore get from eating the same mass of food at the first trophic level than at the second trophic level?

Florida NGSSS Benchmark Practice

Fill in the correct answer choice.

Multiple Choice

1 Which process do producers complete to convert light energy, water, and carbon dioxide into glucose? **SC.8.L.18.1**

Ⓐ chemosynthesis
Ⓑ fermentation
Ⓒ carbon cycle
Ⓓ photosynthesis

Use the image below to answer question 2.

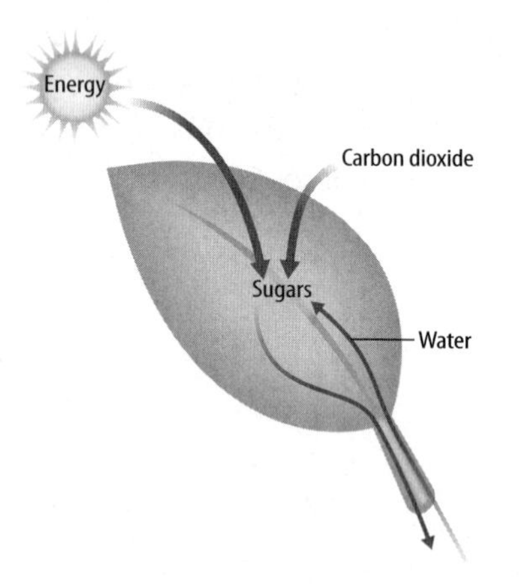

2 What process is shown above? **SC.8.L.18.1**

Ⓕ chemosynthesis
Ⓖ decomposition
Ⓗ nitrogen fixation
Ⓘ photosynthesis

3 Which process returns carbon compounds to the soil in the carbon cycle? **SC.8.L.18.3**

Ⓐ decomposition
Ⓑ transpiration
Ⓒ cellular respiration
Ⓓ nitrogen fixation

4 Which activity does NOT release carbon into the atmosphere during the carbon cycle? **SC.8.L.18.3**

Ⓕ cellular respiration
Ⓖ photosynthesis
Ⓗ humans breathing
Ⓘ fossil fuel combustion

Use the diagram below to answer question 5.

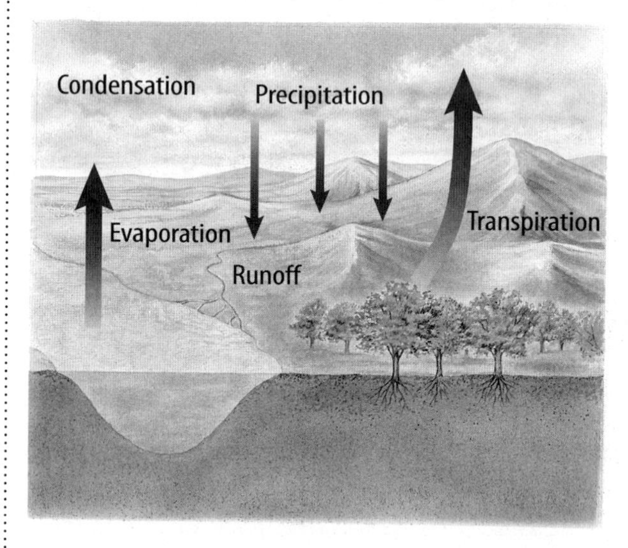

5 To complete the water cycle, which process causes water to fall from clouds to Earth's surface? **SC.8.L.18.4**

Ⓐ evaporation
Ⓑ condensation
Ⓒ exhalation
Ⓓ precipitation

6 Which is true of energy in ecosystems? **SC.8.L.18.4**

Ⓕ It never changes form.
Ⓖ It is both created and destroyed.
Ⓗ It flows in one direction.
Ⓘ It follows a cycle pattern.

7 Why is less energy available at each successive trophic level? **SC.8.L.18.4**

Ⓐ Some energy is given off as thermal energy.

Ⓑ Predators need less energy in higher trophic levels.

Ⓒ Some energy is destroyed in higher trophic levels.

Ⓓ Energy in higher trophic levels is unusable.

Use the diagram below to answer question 8.

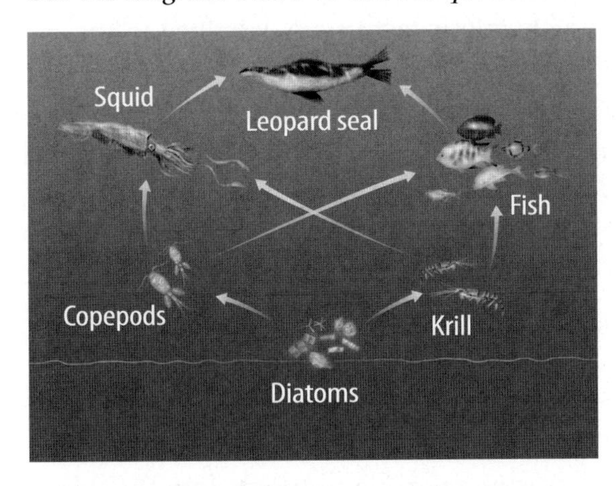

8 How does energy move in the food web pictured above? **SC.7.L.17.1**

Ⓕ from leopard seal to squid

Ⓖ from diatoms to krill

Ⓗ from fish to krill

Ⓘ from squid to diatoms

9 What three things do producers use during photosynthesis to make sugars? **SC.8.L.18.1**

Ⓐ oxygen, water, sunlight

Ⓑ oxygen, water, nitrogen

Ⓒ carbon dioxide, water, nitrogen

Ⓓ carbon dioxide, water, light energy

10 During which process is oxygen gas released into the atmosphere? **SC.8.L.18.1**

Ⓕ chemosynthesis

Ⓖ decomposition

Ⓗ photosynthesis

Ⓘ transpiration

11 Which is a by-product of photosynthesis? **SC.8.L.18.1**

Ⓐ carbon dioxide

Ⓑ nitrogen

Ⓒ water

Ⓓ oxygen

Use the diagram below to answer question 12.

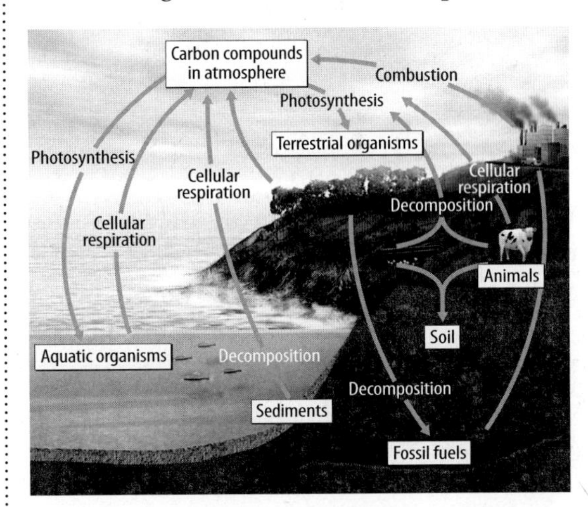

12 In the image of the carbon cycle shown above, which two processes return carbon from the atmosphere to the environment? **SC.8.L.18.3**

Ⓕ photosynthesis and decomposition

Ⓖ cellular respiration and decomposition

Ⓗ cellular respiration and photosynthesis

Ⓘ photosynthesis and combustion

Need Extra Help?

If You Missed Question...	1	2	3	4	5	6	7	8	9	10	11	12
Go to Lesson...	3	3	2	2	2	3	3	3	2	2	2	2

Name ______________________ Date ______________

Multiple Choice *Bubble the correct answer.*

1. Which abiotic factor probably has the greatest effect on the types of plants and animals that live in the environment shown above? **SC.7.L.17.3**
   - (A) air
   - (B) rocks
   - (C) soil
   - (D) water

2. Which of these is NOT an abiotic factor in the environment? **SC.7.L.17.3**
   - (F) air
   - (G) plants
   - (H) rocks
   - (I) sunlight

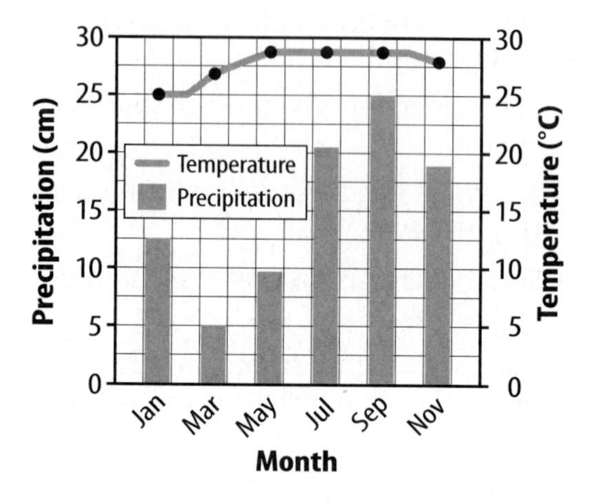

3. Which abiotic factors are described in the graph above? **SC.7.L.17.3**
   - (A) air and soil
   - (B) water and air
   - (C) precipitation and soil
   - (D) precipitation and temperature

4. In which way does the composition of soil, an abiotic factor, rely on living things? **SC.7.L.17.3**
   - (F) Bacteria break down dead matter and wastes, releasing nutrients into the soil.
   - (G) Bacteria erode rocks into smaller particles that are added to the soil.
   - (H) Insects make small tunnels in the soil. Bacteria use these tunnels to break down dead matter.
   - (I) Insects take in air and minerals and release these materials into the soil as wastes.

Name ______________________ Date ______________

Benchmark Mini-Assessment Chapter 18 • Lesson 2

FLORIDA NGSSS mini BAT

Multiple Choice *Bubble the correct answer.*

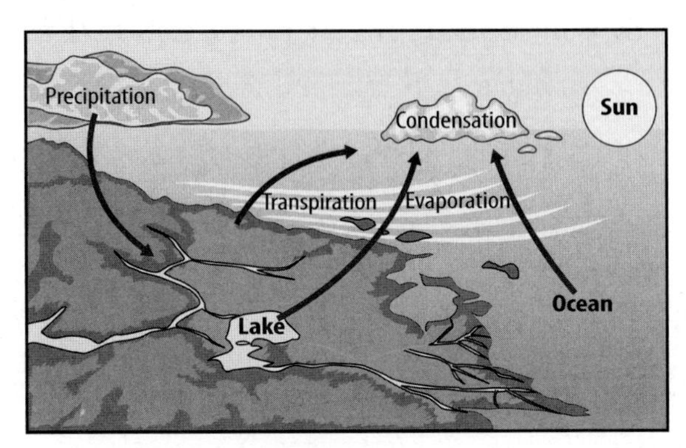

1. Based on the image above, which two processes add water to the atmosphere? **SC.7.L.17.3**

Ⓐ condensation and evaporation

Ⓑ condensation and precipitation

Ⓒ precipitation and evaporation

Ⓓ transpiration and evaporation

2. What might happen to the carbon cycle if dead organisms did not break down? **SC.8.L.18.3**

Ⓕ Bacteria would begin fixing carbon dioxide and adding it to the soil.

Ⓖ Carbon would not be recycled and added to soil, interrupting the cycle.

Ⓗ More carbon dioxide would be available for photosynthesis.

Ⓘ Other sources of carbon would need to be found to keep carbon in the cycle.

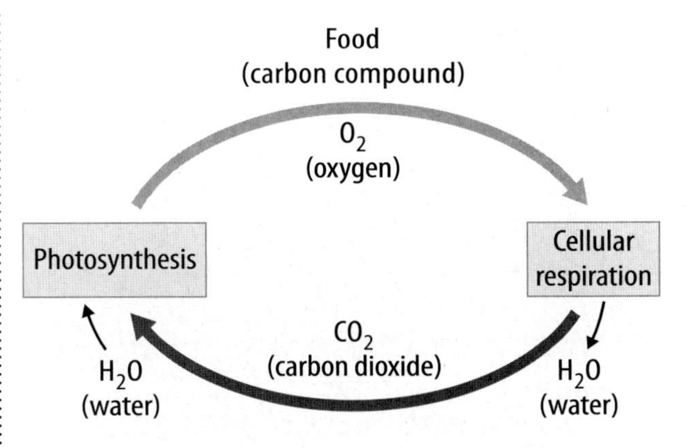

3. According to the image above, what substances are cycled during cellular respiration and photosynthesis? **SC.8.L.18.2**

Ⓐ oxygen and carbon dioxide

Ⓑ oxygen, carbon, and nitrogen

Ⓒ oxygen, carbon, and water

Ⓓ oxygen, nitrogen, and water

4. Without the greenhouse effect, temperatures on Earth would **SC.7.L.17.3**

Ⓕ be too cool or too hot.

Ⓖ become constant.

Ⓗ increase sharply.

Ⓘ vary too much.

Name ______________________ Date ______________

Multiple Choice *Bubble the correct answer.*

Use the image below to answer questions 1 and 2.

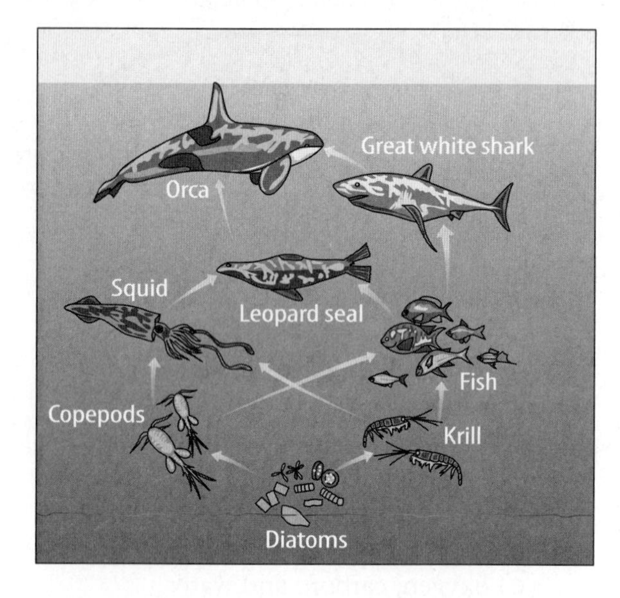

1. A food chain is one part of a food web. Which list is part of the food web shown above? **SC.7.L.17.1**

(A) diatoms→fish→leopard seal→orca

(B) diatoms→krill→fish→ great white shark→orca

(C) copepods→diatoms→krill→fish→ leopard seal→orca

(D) orca→leopard seal→squid→ copepods→diatoms

2. Which organism forms the base of the food web shown above? **SC.7.L.17.1**

(F) diatoms

(G) fish

(H) krill

(I) orcas

3. Which organisms are producers? **SC.7.L.17.1**

(A) algae

(B) amoebas

(C) mushrooms

(D) yeasts

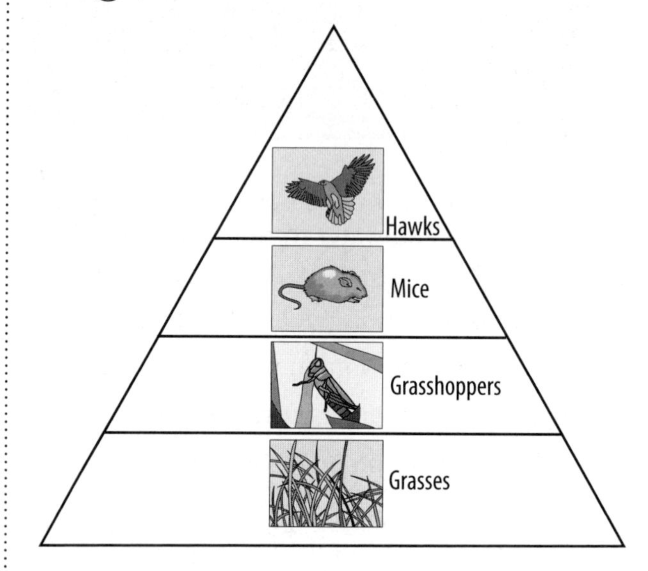

4. Which statement describes the amount of energy passed to the hawk in the energy pyramid above? **SC.8.L.18.4**

(F) The hawk has 50 percent of the energy available to it.

(G) The hawk has the least amount of energy available to it.

(H) The hawk has the most energy available to it.

(I) The hawk has no energy available to it.

Name ______________________ Date ______________

Notes

Name ______________________ Date ______________

Notes

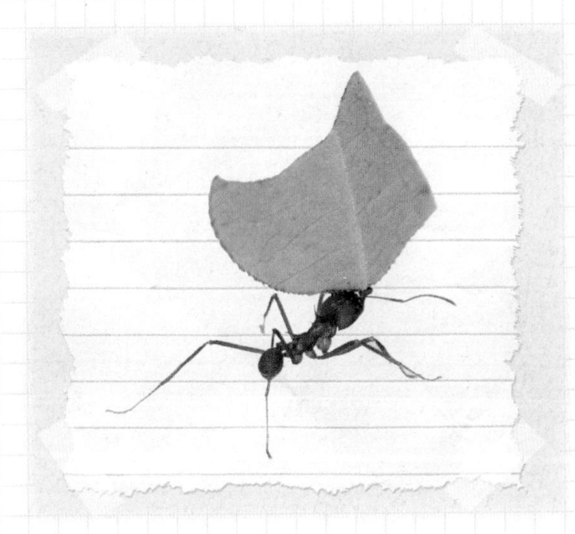

Populations and Communities

Scientists use the words *population* and *community* when they study ecosystems. Which best describes how scientists use these words?

A: *Population* is used to describe the number of organisms in an area; *community* describes the place where organisms live.

B: *Population* is used to describe all the different species living together in an area; *community* describes all the non-living features of the area species live in.

C: *Population* describes all the organisms of the same species living in the same area at the same time; *community* describes all the populations living together in the same area at the same time.

D: *Population* describes the number of different organisms living in the same area at the same time; *community* describes the area where these living things can be found.

E: *Population* describes the changing types and numbers of organisms in an area; *community* describes the types and numbers of organisms in an area that do not change.

F: *Population* describes all the organisms of the same species living in the same area at the same time; *community* describes how different species get along and interact with one another.

Explain your thinking. How do you use the words *population* and *community?*

Populations and COMMUNITIES

FLORIDA BIG IDEAS

1 **The Practice of Science**

2 **The Characteristics of Scientific Knowledge**

3 **The Role of Theories, Laws, Hypotheses, and Models**

17 **Interdependence**

The Big Idea

Think About It!

How do populations and communities interact and change?

This group of pigeons does not depend only on the environment for food. Tourists visiting the area also feed the pigeons. Because so much food is available, more pigeons than normal live in this part of the city.

1. Do you think this large number of pigeons affects other organisms in the area?

____

____

____

2. How do you think groups of pigeons and other organisms interact and change?

____

____

____

____

Get Ready to Read

What do you think about populations and communities?

Before you read, decide if you agree or disagree with each of these statements. As you read this chapter, see if you change your mind about any of the statements.

	AGREE	DISAGREE
1. Some life exists in the ice caps of the North Pole and the South Pole.	☐	☐
2. A community includes all organisms of one species that live in the same area.	☐	☐
3. Some populations decrease in numbers because of low birthrates.	☐	☐
4. An extinct species has only a few surviving individuals.	☐	☐
5. No more than two species can live in the same habitat.	☐	☐
6. A cow is a producer because it produces food for other organisms.	☐	☐

There's More Online!

Video • Audio • Review • ⓘLab Station • WebQuest • Assessment • Concepts in Motion • Multilingual eGlossary

Lesson 1

POPULATIONS

ESSENTIAL QUESTIONS

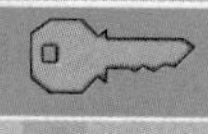
What defines a population?

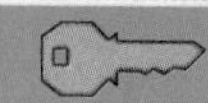
What factors affect the size of a population?

Vocabulary

biosphere p. 733
community p. 734
population p. 734
competition p. 735
limiting factor p. 735
population density p. 736
biotic potential p. 736
carrying capacity p. 737

Florida NGSSS

LA.6.2.2.3 The student will organize information to show understanding (e.g., representing main ideas within text through charting, mapping, paraphrasing, summarizing, or comparing/contrasting);

LA.6.4.2.2 The student will record information (e.g., observations, notes, lists, charts, legends) related to a topic, including visual aids to organize and record information and include a list of sources used;

SC.6.N.2.3 Recognize that scientists who make contributions to scientific knowledge come from all kinds of backgrounds and possess varied talents, interests, and goals.

SC.7.L.17.2 Compare and contrast the relationships among organisms such as mutualism, predation, parasitism, competition, and commensalism.

SC.7.L.17.3 Describe and investigate various limiting factors in the local ecosystem and their impact on native populations, including food, shelter, water, space, disease, parasitism, predation, and nesting sites.

SC.6.N.1.5 Recognize that science involves creativity, not just in designing experiments, but also in creating explanations that fit evidence.

SC.6.N.1.5

Launch Lab

15 minutes

How many times do you interact?

Every day, you interact with other people in different ways, including talking, writing, or shaking hands. Some interactions involve just one other person, and others happen among many people. Like humans, other organisms interact with each other in their environment.

Procedure

1. Make a list of all the ways you have interacted with other people today.
2. Use a **highlighter** to mark the interactions that occurred between you and one other person.
3. Use a **highlighter** of another color to mark interactions that occurred among three or more people.

Data and Observations

Think About This

1. Were your interactions mainly with one person or with three or more people?

2. Key Concept How might your interactions change if the group of people were bigger?

Inquiry **Looking for Something?**

1. Meerkats live in family groups. They help protect each other by watching for danger from eagles, lions, and other hunters of the Kalahari Desert. In what other ways might the meerkats interact?

The Biosphere and Ecological Systems

Imagine flying halfway around the world to Africa. When your plane flies over Africa, you might see mountains, rivers, grasslands, and forests. As you get closer to land, you might see a herd of elephants at a watering hole. You also might see a group of meerkats, like the ones shown above.

Now imagine hiking through an African forest. You might see monkeys, frogs, insects, spiders, and flowers. Maybe you catch sight of crocodiles sunning themselves by a river or birds perching on trees.

You are exploring Earth's **biosphere** (BI uh sfihr)—*the parts of Earth and the surrounding atmosphere where there is life.* The biosphere includes all the land of the continents and islands. It also includes all of Earth's oceans, lakes, and streams, as well as the ice caps at the North Pole and the South Pole.

Parts of the biosphere with large amounts of plants or algae often contain many other organisms as well. The biosphere's distribution of chlorophyll, a green pigment in plants and algae, is shown in **Figure 1**.

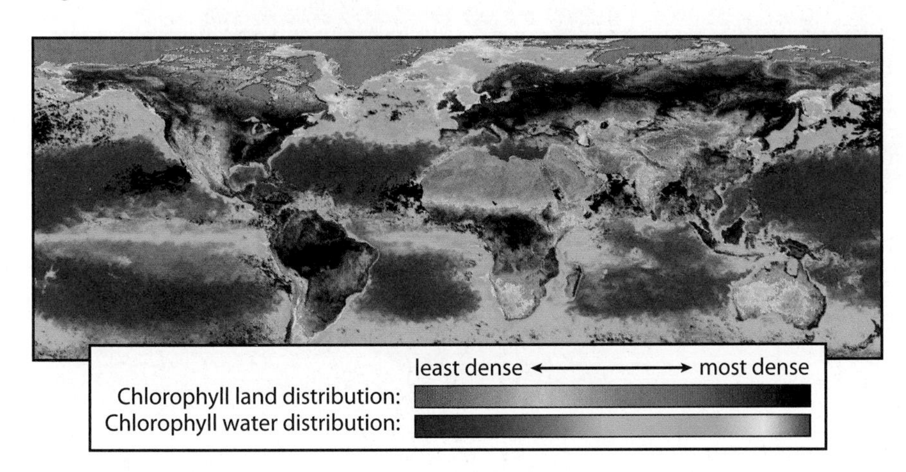

Figure 1 The colors in this satellite image represent the densities of chlorophyll, a green pigment found in plants and algae.

Active Reading **2. Identify** Circle the areas on the map that show the highest chlorophyll density.

What is a population?

The Kalahari Desert in Africa is a part of Earth's biosphere. A wildlife refuge in the Kalahari Desert is home to several groups of meerkats. Meerkats are small mammals that live in family groups and help each other care for their young.

Meerkats rely on interactions among themselves to survive. They sleep in underground burrows at night and hunt for food during the day. They take turns standing upright to watch for danger and call out warnings to others.

Meerkats are part of an ecosystem, as shown in **Figure 2.** An ecosystem is a group of organisms that lives in an area at one time, as well as the climate, soil, water, and other nonliving parts of the environment. The Kalahari Desert is an ecosystem. The study of all ecosystems on Earth is ecology.

Many species besides meerkats live in the Kalahari Desert. They include scorpions, spiders, insects, snakes, and birds such as eagles and owls. Also, large animals like zebras, giraffes, and lions live there. Plants that grow in the Kalahari Desert include shrubs, grasses, small trees, and melon vines. Together, all these plants, animals, and other organisms form a community. *A* **community** *is all the populations of different species that live together in the same area at the same time.*

All the meerkats in this refuge form a population. *A* **population** *is all the organisms of the same species that live in the same area at the same time.* A species is a group of organisms that have similar traits and are able to produce fertile offspring.

3. NGSSS Check Define Highlight the definition of *population.* SC.7.N.1.1

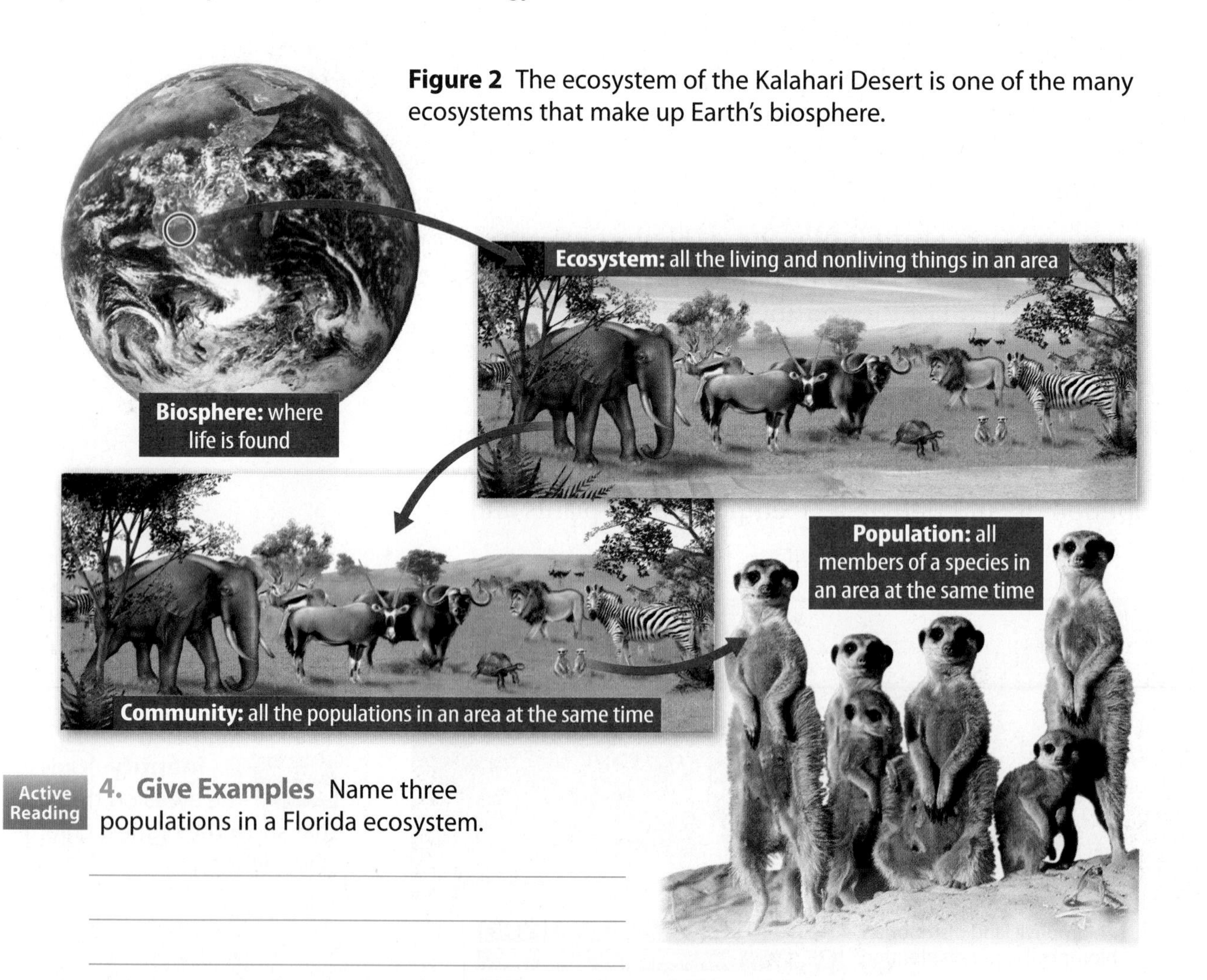

Figure 2 The ecosystem of the Kalahari Desert is one of the many ecosystems that make up Earth's biosphere.

Active Reading **4. Give Examples** Name three populations in a Florida ecosystem.

Competition

At times, not enough food is available for every organism in a community. Members of a population, including those in the Kalahari Desert, must compete with other populations and each other for enough food to survive. **Competition** *is the demand for resources, such as food, water, and shelter, in short supply in a community.* When there are not enough resources available to survive, there is more competition in a community.

Population Sizes

If the amount of available food decreases, what do you think happens to a population of meerkats? Some meerkats might move away to find food elsewhere. Female meerkats cannot raise as many young. The population becomes smaller. If there is plenty of food, however, the size of the population grows larger as more meerkats survive to adulthood and live longer. Changes in environmental factors can result in population size changes.

Limiting Factors

Environmental factors, such as available water, food, shelter, sunlight, and temperature, are possible limiting factors for a population. *A* **limiting factor** *is anything that restricts the size of a population.* Available sunlight is a limiting factor for most organisms. If there is not enough sunlight, green plants cannot make food by photosynthesis. Organisms that eat plants are affected if little food is available.

Temperature is a limiting factor for some organisms. When the temperature drops below freezing, many organisms die because it is too cold to carry out their life functions. Disease, predators—animals that eat other animals—and natural disasters such as fires or floods are limiting factors as well.

5. NGSSS Check Identify Underline examples of limiting factors. SC.7.N.1.1

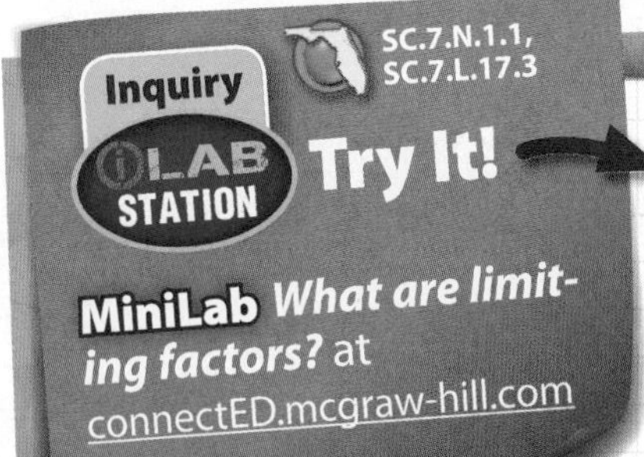

MiniLab ***What are limiting factors?*** at connectED.mcgraw-hill.com

Apply It!

After you complete the lab, answer the questions below.

1. How might space be a limiting factor for a population of rabbits?

2. What limiting factors affect the population of your school?

Figure 3 A sedated lynx is fitted with a radio collar and then returned to the wild.

Measuring Population Size

Sometimes it is difficult to determine the size of a population. How would you count scampering meerkats or wild lynx? One method used to count and monitor animal populations is the capture-mark-and-release method. The lynx in **Figure 3** is a member of a population in Poland that is monitored using this method. Biologists using this method sedate animals and fit them with radio collars before releasing them back into the wild. By counting how many observed lynx are wearing collars, scientists can estimate the size of the lynx population. Biologists also use the collars to track the lynx's movements and monitor their activities.

Suppose you want to know how closely together Cumberland azaleas (uh ZAYL yuhz), a type of flower, grow in the Great Smoky Mountains National Park. **Population density** *is the size of a population compared to the amount of space available.* One way of estimating population density is by sample count. Rather than counting every azalea shrub, you count only those in a representative area, such as 1 km^2. By multiplying the number of square kilometers in the park by the number of azaleas in 1 km^2, you find the estimated population density of azalea shrubs in the entire park.

WORD ORIGIN

population
from Latin *populus,* means "inhabitants"

density
from Latin *densus,* means "thick, crowded"

Active Reading **6. Identify** Highlight two examples of ways you can estimate population size.

Biotic Potential

Imagine that a population of raccoons has plenty of food, water, and den space. In addition, there is no disease or danger from other animals. The only limit to the size of this population is the number of offspring the raccoons can produce. **Biotic potential** *is the potential growth of a population if it could grow in perfect conditions with no limiting factors.* No population on Earth ever reaches its biotic potential because no ecosystem has an unlimited supply of natural resources.

Active Reading

FOLDABLES® LA.6.2.2.3

Make a horizontal half book and label it as shown. Use it to organize your notes on the relationship between population size and carrying capacity in an ecosystem.

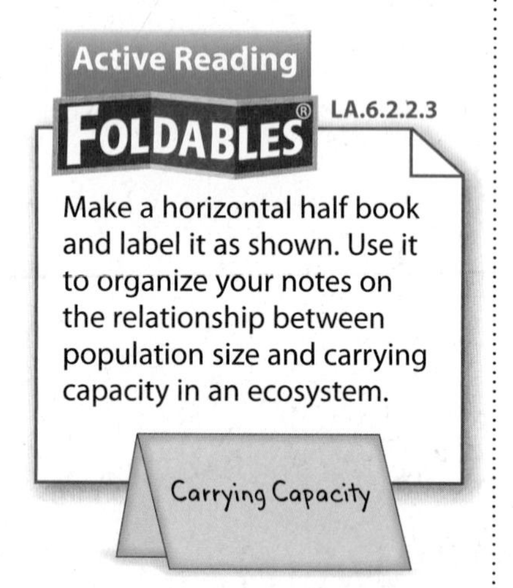

Carrying Capacity

What would happen if a population of meerkats reached its biotic potential? It would stop growing when it reached the limit of available resources such as food, water, or shelter. *The largest number of individuals of one species that an environment can support is the* **carrying capacity.** A population grows until it reaches the carrying capacity of an environment, as shown in **Figure 4.** Disease, space, predators, and food are some of the factors that limit the carrying capacity of an ecosystem. However, the carrying capacity of an environment is not constant. It increases and decreases as the amount of available resources increases and decreases. At times, a population can temporarily exceed the carrying capacity of an environment.

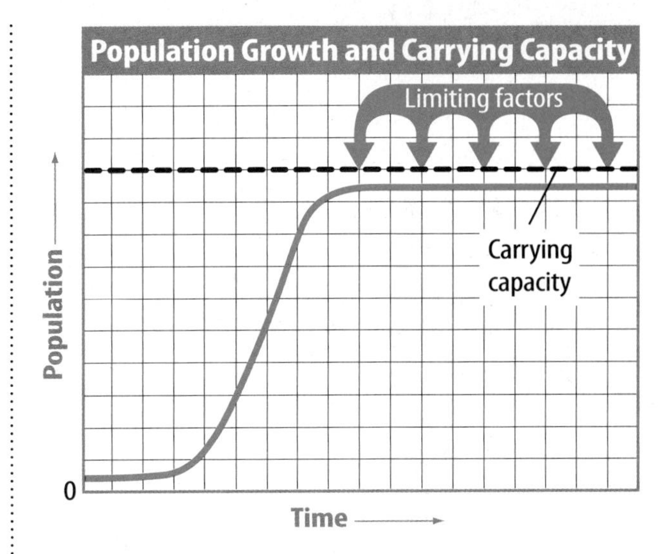

Figure 4 Carrying capacity is determined in part by limiting factors.

Active Reading **7. Modify** Draw what the population line would look like during overpopulation on the graph above.

Overpopulation

When the size of a population becomes larger than the carrying capacity of its ecosystem, overpopulation occurs. Overpopulation can cause problems for organisms. For example, meerkats eat spiders. An overpopulation of meerkats causes the size of the spider population in that community to decrease. Populations of birds and other animals that eat spiders also decrease when the number of spiders decreases.

Elephants in Africa's wild game parks provide another example of overpopulation. Elephants searching for food caused the tree damage shown in **Figure 5.** They push over trees to feed on the uppermost leaves. Other species of animals that use the same trees for food and shelter must compete with the elephants. The loss of trees and plants can also damage soil.

8. NGSSS Check **Describe** How can overpopulation affect a community?

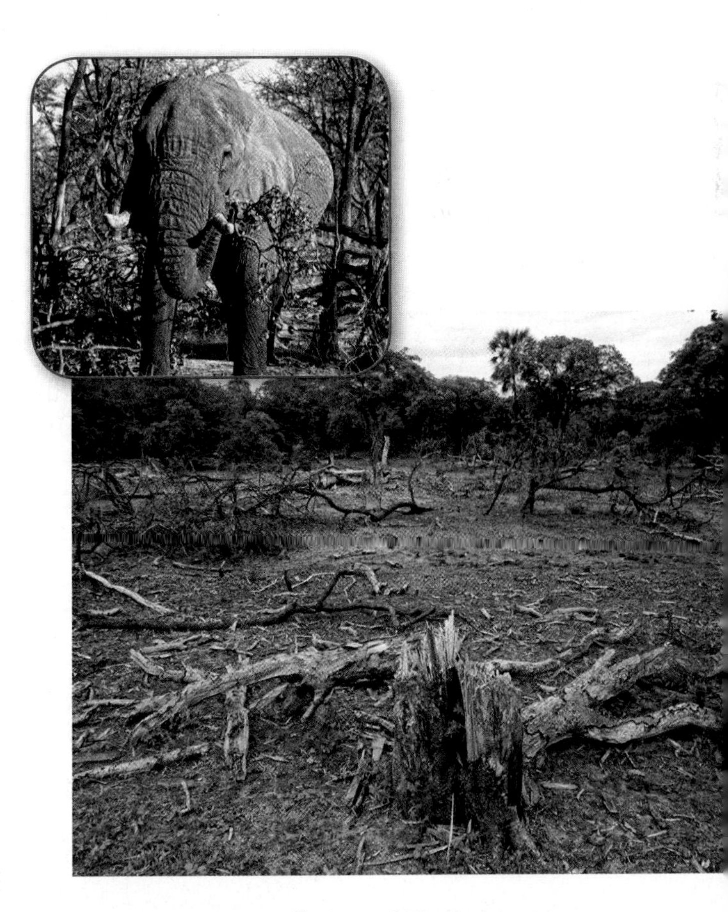

Figure 5 An overpopulation of elephants can cause damage as the herd searches for food.

Lesson Review 1

Visual Summary

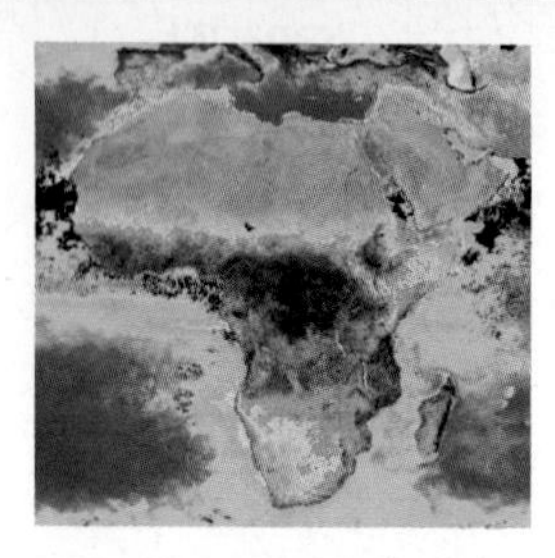

The population density of organisms, including green plants and algae, varies throughout the world.

A community is all the populations of different species that live together in the same area at the same time.

The number of individuals in a population varies as the amount of available resources varies.

Use Vocabulary

1. **Define** *population.*

2. **Distinguish** between carrying capacity and biotic potential.

3. Food, water, living space, and disease are examples of ______________. SC.7.L.17.3

Understand Key Concepts

4. **Explain** how competition could limit the size of a bird population. SC.7.L.17.3

5. One example of competition among members of a meerkat population is
   (A) fighting over mates.
   (B) warning others of danger.
   (C) huddling together to stay warm.
   (D) teaching young to search for food.

Interpret Graphics

6. **Sequence** Fill in the graphic organizer below to show the sequence of steps in one type of population study. LA.6.2.2.3

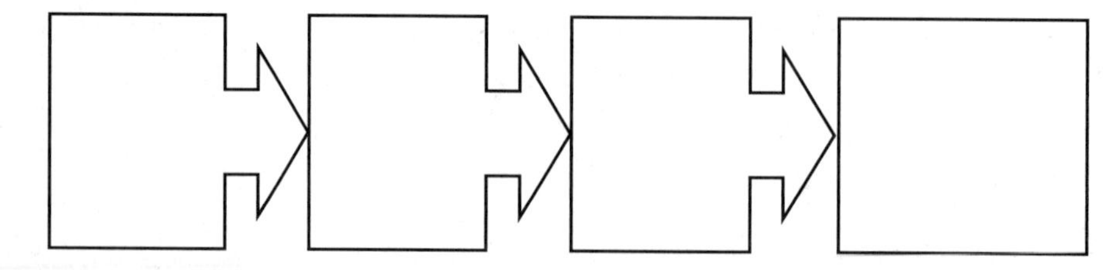

Critical Thinking

7. **Explain** Is the problem of elephants destroying trees in southern Africa overpopulation, competition, or both? SC.6.N.1.5

AMERICAN MUSEUM OF NATURAL HISTORY

CAREERS in SCIENCE

Familiar Birds in an Unlikely Place

Howling winds blow across the Altiplano—a cold plateau high in the Andes mountain range of South America. There, you might expect to see animals such as llamas, but Felicity Arengo travels to the Altiplano to observe flamingos.

Flamingos usually are associated with tropical regions. However, three species of flamingo—James, Andean, and Chilean—are adapted to the cold, barren Altiplano. Although this region differs from tropical areas where other flamingos live, both have food sources for the birds. The plateau is dotted with salty lakes containing brine shrimp, tiny organisms that flamingos eat.

Scientists have many questions about the flamingos that visit these salty lakes. How large are the flamingo populations? How do they survive when the lakes evaporate? To answer these questions, Arengo and her team visit the lakes and count flamingos there. They also tag flamingos with radio transmitters to track their movements. They have learned that one species, the Andean flamingo, is the rarest flamingo species in the world. Additionally, when the plateau lakes freeze, many flamingos fly to lakes in the lowlands of Argentina, Bolivia, Chile, and Peru.

As human activity changes the Altiplano, flamingos that live there might be in danger. On the plateau, mining operations use and pollute lake water. In the lowlands, ranchers often drain lakes for more land to grow crops or to feed animals.

Arengo and other scientists are working with organizations to protect flamingos' habitats. Scientists have trained park rangers to monitor flamingos' reproductive activities and protect nesting colonies. As scientists collect more data and find new ways to protect flamingos' habitats, a brighter future might be in store for the flamingos of the Altiplano.

▲ **Dr. Arengo tags a flamingo with a radio transmitter. Once she releases the flamingo, satellites will track the flamingo's movement.**

▲ **Flamingos' habitats cover four countries—Argentina, Bolivia, Chile, and Peru. As a well-known species, flamingos help motivate conservation efforts in these countries.**

It's Your Turn

BRAINSTORM With classmates, choose an ecosystem in your area that is in need of conservation. Brainstorm what animal would make a good species to represent the ecosystem, and create a poster designed to raise awareness.

LA.6.4.2.2

Lesson 2

Changing POPULATIONS

ESSENTIAL QUESTIONS

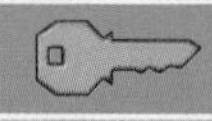 How do populations change?

 Why do human populations change?

Vocabulary

birthrate p. 741

death rate p. 741

extinct species p. 743

endangered species p. 743

threatened species p. 743

migration p. 744

Florida NGSSS

LA.6.2.2.3 The student will organize information to show understanding (e.g., representing main ideas within text through charting, mapping, paraphrasing, summarizing, or comparing/contrasting);

MA.6.A.3.6 Construct and analyze tables, graphs, and equations to describe linear functions and other simple relations using both common language and algebraic notation.

SC.7.N.1.1 Define a problem from the seventh grade curriculum, use appropriate reference materials to support scientific understanding, plan and carry out scientific investigation of various types, such as systematic observations or experiments, identify variables, collect and organize data, interpret data in charts, tables, and graphics, analyze information, make predictions, and defend conclusions.

SC.7.N.1.3 Distinguish between an experiment (which must involve the identification and control of variables) and other forms of scientific investigation and explain that not all scientific knowledge is derived from experimentation.

SC.7.L.17.3 Describe and investigate various limiting factors in the local ecosystem and their impact on native populations, including food, shelter, water, space, disease, parasitism, predation, and nesting sites.

SC.6.N.1.4 Discuss, compare, and negotiate methods used, results obtained, and explanations among groups of students conducting the same investigation.

SC.8.N.3.1 Select models useful in relating the results of their own investigations.

SC.7.N.1.1
SC.6.N.1.4

Inquiry Launch Lab

15 minutes

What events can change a population?

Populations can be affected by human-made and environmental changes, such as floods or a good growing season. A population's size can increase or decrease in response to these changes.

Procedure

1. Read and complete a lab safety form.
2. Record the number of **counting objects** you have been given. Each object represents an organism, and all the objects together represent a population.
3. Turn over one of the **event cards** you were given and follow the instructions on the card. Determine the event's impact on your population.
4. Repeat step 3 for four more "seasons," or turns.

Data and Observations

Think About This

1. Compare the size of your population with other groups. Do you all have the same number of organisms at the end of five seasons?

2. Key Concept What effect did the different events have on your population?

Same Mother?

1. Have you ever seen newly hatched baby spiders? Baby spiders can have hundreds or even thousands of brothers and sisters. What keeps the spider population from growing out of control?

How Populations Change

Have you ever seen a cluster of spider eggs? Some female spiders lay hundreds or even thousands of eggs in their lifetime. What happens to a population of spiders when a large group of eggs hatches all at once? The population suddenly becomes larger. It doesn't stay that way for long, though. Many spiders die or become food, like the one being eaten in **Figure 6,** before they grow enough to reproduce. The size of the spider population increases when the eggs hatch but decreases as the spiders die.

A population change can be measured by the population's birthrate and death rate. *A population's* **birthrate** *is the number of offspring produced over a given time period. The* **death rate** *is the number of individuals that die over the same time period.* If the birthrate is higher than the death rate, the population increases. If the death rate is higher than the birthrate, the population decreases.

2. **NGSSS Check** Describe What is one factor that might affect death rate? SC.7.N.1.1

Figure 6 Spiders have a high birthrate, but they usually have a high death rate too. Many spiders die or are eaten before they can reproduce.

SCIENCE USE V. COMMON USE

exponential

Science Use a mathematical expression that contains a constant raised to a power, such as 2^3 or x^2

Common Use in great amounts

Exponential Growth

When a population is in ideal conditions with unlimited resources, it grows in a pattern called **exponential** growth. During exponential growth, the larger a population gets, the faster it grows. *E. coli* bacteria are microscopic organisms that undergo exponential growth. This population doubles in size every half hour, as shown in **Figure 7.** It takes only 10 hours for the *E. coli* population to grow from one organism to more than 1 million. Exponential growth cannot continue for long. Eventually, limiting factors stop population growth.

Exponential Population Growth

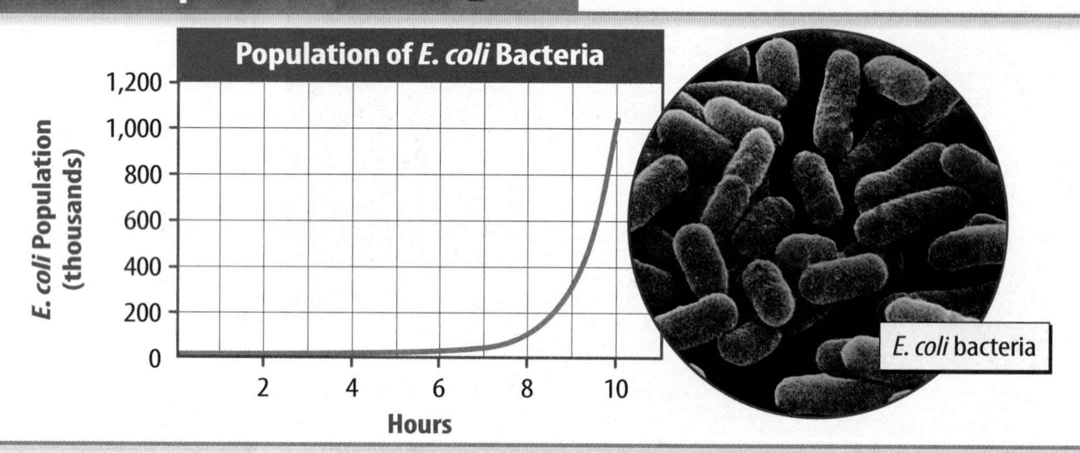

Figure 7 When grown in the laboratory, this population of *E. coli* bacteria is given everything it needs to briefly achieve exponential growth.

Population Size Decrease

Population size can increase, but it also can decrease. For example, a population of field mice might decrease in size in the winter because there is less food. Some mice might not be able to find enough food and will starve. More mice will die than will be born, so the population size decreases. When food is plentiful, the population size usually increases.

Natural disasters such as floods, fires, or volcanic eruptions also affect population size. For example, if a hurricane rips away part of a coral reef, the populations of coral and other organisms that live on the reef also decrease in size.

Disease is another cause of population decrease. In the mid-1900s, Dutch elm disease spread throughout the United States and destroyed many thousands of elm trees. Because of the disease, the size of the population of elm trees decreased.

Predation—the hunting of organisms for food—also reduces population size. For example, a farmer might bring cats into a barn to reduce the size of a mouse population.

3. NGSSS Check List Underline four reasons that a population might decrease in size. SC.7.N.1.1

Extinct The giant moa, a large bird that was nearly 4 m tall, was hunted to extinction.

Endangered Just over 700 mountain gorillas remain in the wild in Africa.

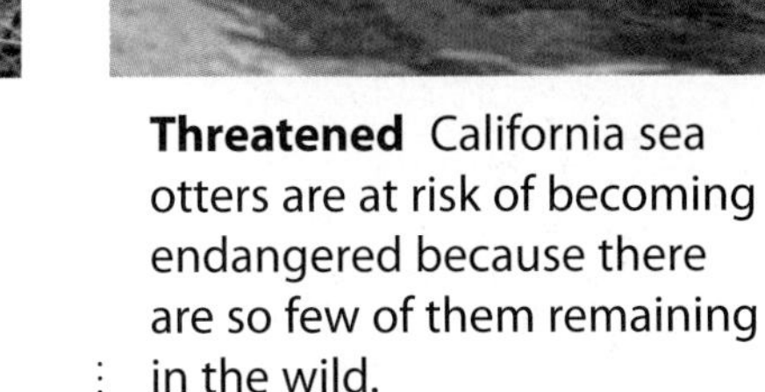

Threatened California sea otters are at risk of becoming endangered because there are so few of them remaining in the wild.

Figure 8 Organisms can be classified as extinct, endangered, or threatened.

Extinction If populations continue to decrease in numbers, they disappear. *An* **extinct species** *is a species that has died out and no individuals are left.* Extinctions can be caused by predation, natural disasters, or damage to the environment.

Some extinctions in Earth's history were large events that involved many species. Most scientists think the extinction of the dinosaurs about 65 million years ago was caused by a meteorite crashing into Earth. The impact would have sent tons of dust into the atmosphere, blocking sunlight. Without sunlight, plants could not grow. Animals, such as dinosaurs, that ate plants probably starved.

Most extinctions involve fewer species. For example, New Zealand was once home to a large, flightless bird called the giant moa, as shown in **Figure 8.** Humans first settled these islands about 700 years ago. They hunted the moa for food. As the size of the human population increased, the size of the moa population decreased. Within 200 years, all the giant moas had been killed and the species became extinct.

Word Origin

extinct

from Latin *extinctus,* means "extinguish"

Endangered Species The mountain gorillas shown in **Figure 8** are an example of a species that is endangered. *An* **endangered species** *is a species whose population is at risk of extinction.*

Threatened Species California sea otters almost became extinct in the early 1900s due to overhunting. In 1977, California sea otters were classified as a **threatened species**—*a species at risk, but not yet endangered.* Laws were passed to protect the otters, and by 2007 there were about 3,000 sea otters. Worldwide, there are more than 4,000 species that are classified as endangered or threatened.

Active Reading **4. Differentiate** What is the difference between an endangered species and a threatened species?

Apply It! After you complete the lab, answer these questions.

1. What might cause the birds to migrate?

2. Identify one migratory species in Florida. What is it, and where does it travel?

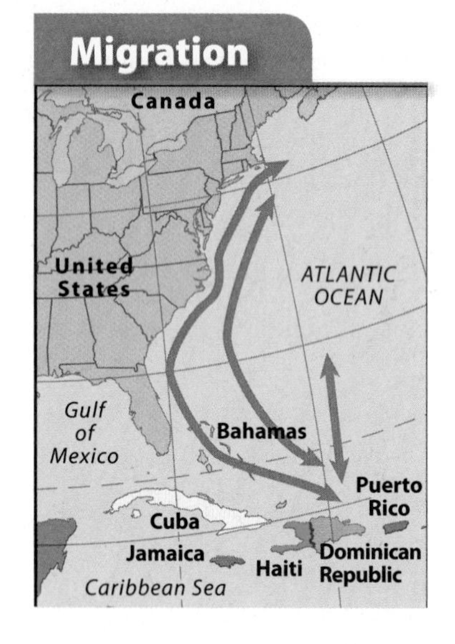

Figure 9 During the winter, humpback whales mate and give birth in warm ocean waters near the Bahamas. In the summer, they migrate north to food-rich waters along the coast of New England.

Movement

Populations also change when organisms move from place to place. When an animal population becomes overcrowded, some individuals might move to find more food or living space. For example, zebras might overgraze an area and move to areas that are not so heavily grazed.

Plant populations can also move from place to place. Have you ever blown on a dandelion puff full of seeds? Each tiny dandelion seed has a feathery part that enables it to be carried by the wind. Wind often carries seeds far from their parent plants. Animals also help spread plant seeds. For example, some squirrels and woodpeckers collect acorns. They carry the acorns away and store them for a future food source. The animal forgets some acorns, and they sprout and grow into new trees far from their parent trees.

Migration Sometimes an entire population moves from one place to another and later returns to its original location. **Migration** *is the instinctive seasonal movement of a population of organisms from one place to another.* Ducks, geese, and monarch butterflies are examples of organisms that migrate annually. Some fish, frogs, insects, and mammals—including the whales described in **Figure 9**—migrate to find food and shelter.

Active Reading 5. **List** What are three ways that populations change?

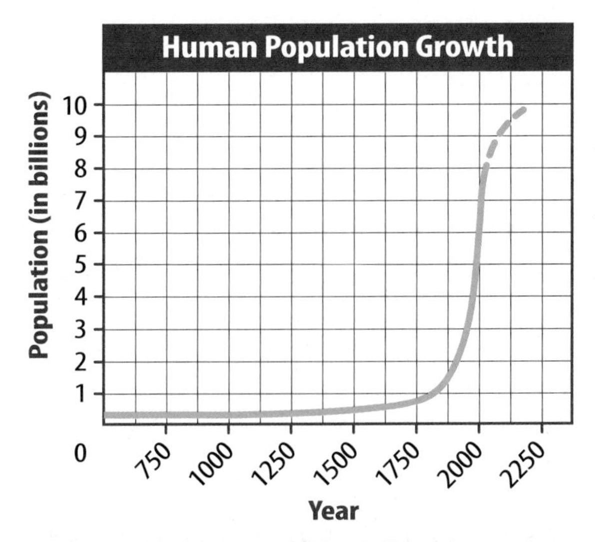

Figure 10 The human population has grown faster in the past 150 years than at any other time in Earth's history.

Human Population Changes

Human population size is affected by the same three factors that determine the sizes of all populations—birthrate, death rate, and movement. But, unlike other species, humans can increase the carrying capacity of their environment. Improved crop yields, domesticated farm animals, and methods of transporting resources enable people to survive in all types of environments.

Scientists **estimate** that there were about 300 million humans on Earth a thousand years ago. Today there are more than 6 billion humans on Earth, as shown in **Figure 10.** By 2050 there could be more than 9 billion. Some scientists estimate that Earth's carrying capacity is about 11 billion.

ACADEMIC VOCABULARY

estimate

(verb) to determine roughly the size, nature, or extent of something

As the human population grows, people need to build more houses and roads and clear more land for crops. This means less living space, food, and other resources for other species. In addition, people use more energy to heat and cool homes; to fuel cars, airplanes, and other forms of transportation; and to produce electricity. This energy use contributes to pollution.

One example of the consequences of human population growth is the destruction of tropical forests. Each year, humans clear thousands of acres of tropical forest to make room for crops and livestock, as shown in **Figure 11.** Clearing tropical forests is harmful because these forests contain a large variety of species that are not in other ecosystems.

6. **NGSSS Check** Explain How does human population growth affect other species? SC.7.N.1.1

Figure 11 Tropical forests are cleared for crops and livestock. The habitats of many organisms are destroyed, resulting in many species becoming endangered or extinct.

Math Skills MA.6.A.3.6

Use Graphs

Graphs are used to make large amounts of information easy to interpret. Line graphs show how data changes over a period of time. A circle graph, or pie graph, shows how portions of a set of data compare with the whole set. The circle represents 100 percent, and each segment represents one part making up the whole. For example, **Figure 14** shows all the moves made by people in the United States during 2004–2005. Fifty-seven percent of all moves were within the same county.

Practice

Based on **Figure 14,**

7. What percentage of the moves were from one state to another?

8. What percentage of the moves were within the same state?

Figure 12 Before vaccinations, many children died in infancy. The use of vaccines has significantly reduced death rates.

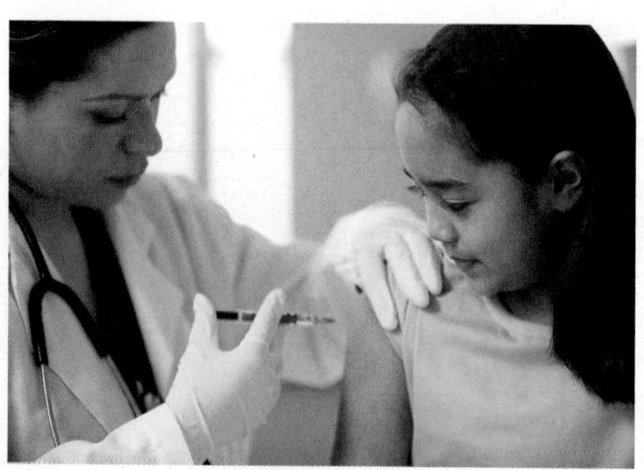

Population Size Increase

Do you know anyone who is more than 100 years old? In 2006, almost 80,000 people living in the United States were at least 100 years old. People are living longer today than in previous generations, and more children reach adulthood. Recall that when the birthrate of a population is higher than its death rate, the population grows. Several factors keep the human birthrate higher than its death rate. Some of these factors are discussed below.

Food For some, finding food might be as easy as making a trip to the grocery store, but not everyone can get food as easily. Advances in agriculture have made it possible to produce food for billions of people.

Resources Fossil fuels, cloth, metals, foods, and many other materials are easily transported around the world by planes, trains, trucks, or boats. Today, people have access to more resources because of better transportation methods.

Sanitation As recently as 100 years ago, diseases such as typhoid, cholera, and diphtheria were major causes of death. These diseases were spread through unclean water supplies and untreated sewage. Modern water-treatment technologies have reduced the occurrence of many diseases. Less expensive and more effective cleaning products are now available to help prevent the spread of disease-causing organisms. As a result, deaths from these illnesses are less common in many countries.

Medical Care Modern medical care is keeping people alive and healthy longer than ever before. As shown in **Figure 12,** scientists have developed vaccines, antibiotics, and other medicines that prevent and treat disease. As a result, fewer people get sick, and human death rates have decreased. Medical technologies and new medicines help people survive heart attacks, cancer, and other major illnesses.

Decreases in Human Population Size

Human populations in some parts of the world are decreasing in size. Diseases such as AIDS and malaria cause high death rates in some countries. Severe drought has resulted in major crop failures and lack of food. Floods, earthquakes, and other natural disasters can cause the deaths of hundreds or even thousands of people at a time. Damage from disasters, such as the damage shown in **Figure 13,** can keep people from living in the area for a long time. All of these factors cause decreases in human population sizes in some areas.

Figure 13 Natural disasters such as a tsunami can cause severe damage to people's homes, as well as drastically reduce the population size.

9. **NGSSS Check** **List** What are three events that can decrease human population size? SC.7.N.1.1

____

____

Population Movement

Have you ever moved to a different city, state, or country? The size of a human population changes as people move from place to place. The graph in **Figure 14** shows the percentages of each kind of move people make. Like other organisms, populations of humans might move when more resources become available in a different place.

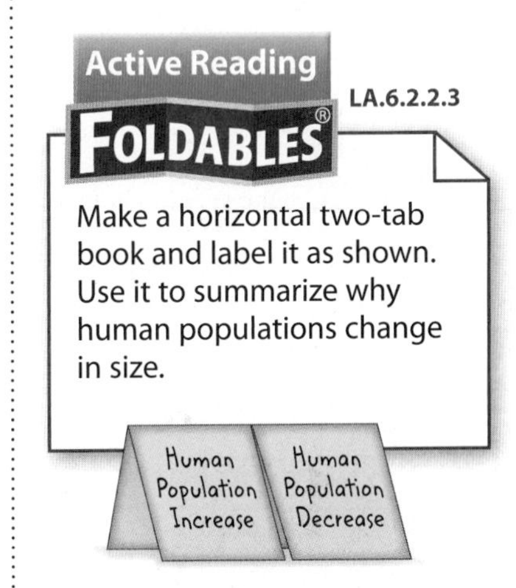

Did your parents, grandparents, or great-grandparents come to the United States from another country? Immigration takes place when organisms move into an area. Most of the U.S. population is descended from people who immigrated from Europe, Africa, Asia, and Central and South America.

Types of Moves, 2004–2005

From abroad 4.7%

Different state 18.7%

Same county 57.0%

Different county, same state 19.6%

Source: U.S. Census Bureau, Current Population Survey, 2005 Annual Social and Economic Supplement.

Figure 14 Populations can move between counties and states or even from another country.

Active Reading 10. **Identify** (Circle) the type of move made by the largest percentage of the population.

Lesson Review 2

Visual Summary

The birthrate and the death rate of any population affects its population size.

The giant moa is classified as an extinct species because there are no surviving members.

A population that is at risk but not yet endangered is a threatened species.

Skill Lab *How do populations change in size?* at connectED.mcgraw-hill.com

Use Vocabulary

1. **Define** *endangered species* in your own words.

2. **Distinguish** between birthrate and death rate.

3. The instinctive movement of a population from one place to another is ____________________.

Understand Key Concepts

4. Rabbits move into a new field where there is plenty of room to dig new burrows. This is an example of SC.7.N.17.3
   - (A) overpopulation.
   - (B) immigration.
   - (C) carrying capacity.
   - (D) competition.

Interpret Graphics

5. **Summarize** Fill in the graphic organizer below to identify the three major factors that affect population size. LA.6.2.2.3

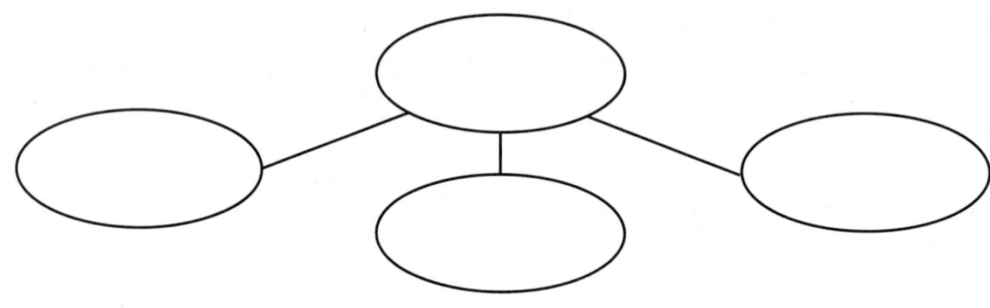

Critical Thinking

6. **Predict** what could happen to the size of the human population if a cure for all cancers were discovered. SC.7.N.1.3

7. **Recommend** an action humans could take to help prevent the extinction of tropical organisms.

FOCUS on FLORIDA

Florida's Panther Population

A Change for the Better

Have you ever heard of a Florida panther? These large cats weigh about 170 pounds and live in regions such as the Big Cypress National Preserve and Everglades National Park. For many years researchers have monitored their populations. In a relatively short period of time the Florida panther population size has drastically changed, a problem that scientists are trying to solve today.

For many years panthers were considered a danger to humans and animals. It was legal to hunt panthers in Florida until 1958. In 1967 they were listed as an endangered species, but the population continued to decline. Large amounts of land that could have been panther habitat were developed for human use. This led to the development of small populations which were separated from each other. Because adults had few choices for mates, many panther kittens were born with genetic health problems. A lack of habitat and increased health problems resulted in further population decline. In the 1990s some scientists estimated that only around 30 Florida panthers existed.

What can people do to help Florida's panther population recover? Scientists have developed the Florida Panther Recovery Plan to restore panther habitat, to introduce new animals to breeding populations, and to educate people about panthers. In 2008 scientists reported that at least 117 adult and subadult Florida panthers survived in the wild. They estimate that each group of Florida panthers will need to contain at least 240 individuals for a full recovery.

▲ Scientists examine a sedated Florida panther.

It's Your Turn

RELATE Write a letter to a community explaining why it is important to protect Florida panthers.

Lesson 3

COMMUNITIES

ESSENTIAL QUESTIONS

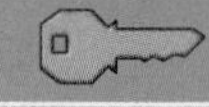
What defines a community?

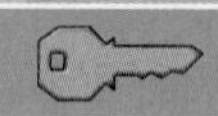
How do the populations in a community interact?

Vocabulary

habitat p. 751
niche p. 751
producer p. 752
consumer p. 752
symbiosis p. 755
mutualism p. 755
commensalism p. 756
parasitism p. 756

Florida NGSSS

LA.6.2.2.3 The student will organize information to show understanding (e.g., representing main ideas within text through charting, mapping, paraphrasing, summarizing, or comparing/contrasting);

SC.7.L.17.1 Explain and illustrate the roles of and relationships among producers, consumers, and decomposers in the process of energy transfer in a food web.

SC.7.L.17.2 Compare and contrast the relationships among organisms such as mutualism, predation, parasitism, competition, and commensalism.

SC.6.N.1.5 Recognize that science involves creativity, not just in designing experiments, but also in creating explanations that fit evidence.

SC.8.N.3.1 Select models useful in relating the results of their own investigations.

LA.6.2.2.3

15 minutes

Launch Lab

What are the roles in your school community?

Within a community, different organisms have different roles. Trees produce their own food from the environment. Then, they become food for other organisms. Mushrooms break down dead organisms and make the nutrients useful to other living things. Think about the members of your school community such as the students, teachers, and custodians. What roles do they have?

1. Fill in the table below with examples from your school.

Community Member	Role in the Community

Think About This

1. Are there any community members who have more than one role?

2. What is your role in the school community?

3. Key Concept Explain how it is beneficial for members of a community to have different roles.

Inquiry **Time for Lunch?**

1. This hoopoe (HOO poo) has captured its next meal. Some of the energy needed by this bird for its life processes will come from the energy stored in the body of the lizard. Where did the lizard get its energy?

Communities, Habitats, and Niches

High in a rain forest tree, a two-toed sloth munches leaves. Ants crawl on a branch, carrying away a dead beetle. Two birds build a nest. A flowering vine twists around the tree trunk. These organisms are part of a rain forest community. You read in Lesson 1 that a community is made up of all the species that live in the same ecosystem at the same time.

The place within an ecosystem where an organism lives is its **habitat.** A habitat, like the one in **Figure 15,** provides all the resources an organism needs, including food and shelter. A habitat also has the right temperature, water, and other conditions the organism needs to survive.

The rain forest tree described above is a habitat for sloths, insects, birds, vines, and many other species. Each species uses the habitat in a different way. *A* **niche** (NICH) *is what a species does in its habitat to survive.* For example, butterflies feed on flower nectar. Sloths eat leaves. Ants eat insects or plants. These species have different niches in the same environment. Each organism shown in **Figure 15** has its own niche on the tree. The plants anchor themselves to the tree and can capture more sunlight. Termites use the tree for food.

2. **NGSSS Check** **Define** What is a community?

WORD ORIGIN

habitat
from Latin *habitus,* means "to live, dwell"

Figure 15 This tree trunk is a habitat for ferns.

Energy in Communities

Sloths are the slowest mammals on Earth. They hardly make a sound, and they sleep 15 to 20 hours a day. Squirrel monkeys, however, chatter as they swing through treetops hunting for fruit, insects, and eggs. Sloths might appear to use no energy at all. However, sloths, squirrel monkeys, and all other organisms need energy to live. All living things use energy and carry out life processes such as growth and reproduction.

Energy Roles

How an organism obtains energy is an important part of its niche. Almost all the energy available to life on Earth originally came from the Sun. However, some organisms, such as those that live near deep-sea vents, are exceptions. They obtain energy from chemicals such as hydrogen sulfide.

Producers *are organisms that get energy from the environment, such as sunlight, and make their own food.* For example, most plants are producers that get their energy from sunlight. They use the process of photosynthesis and make sugar molecules that they use for food. Producers near deep-sea vents use hydrogen sulfide and carbon dioxide and make sugar molecules.

Consumers *are organisms that get energy by eating other organisms.* Consumers are also classified by the type of organisms they eat. Herbivores get their energy by eating plants. Cows and sheep are herbivores. Carnivores get their energy by eating other consumers. Harpy eagles, lions, and wolves are carnivores. Omnivores, such as most humans, get their energy by eating producers and consumers. Detritivores (dee TRI tuh vorz) get their energy by eating dead organisms or parts of dead organisms. Some bacteria and some fungi are detritivores.

3. **NGSSS Check** **Identify** Highlight a producer, an herbivore, a carnivore, and an omnivore. SC.7.L.17.1

Apply It! After you complete the lab, answer the questions below.

1. Which organisms in your web were producers?

2. Which organisms were consumers?

Energy Flow

A food chain is a way of showing how energy moves through a community. In a rain forest community, energy flows from the Sun to a rain forest tree, a producer. The tree uses the energy and grows, producing leaves and other plant structures. Energy moves to consumers, such as the sloth that eats the leaves of the tree, and then to the eagle that eats the sloth. When the eagle dies, detritivores, such as bacteria, feed on its body. That food chain can be written like this:

Sun ⟶ leaves ⟶ sloth ⟶ eagle ⟶ bacteria

A food chain shows only part of the energy flow in a community. A food web, like the one in **Figure 16,** shows many food chains within a community and how they overlap.

Figure 16 Organisms in a rain forest community get their energy in different ways.

Active Reading **4. List** Identify the members of two different food chains shown in the figure below.

____

____

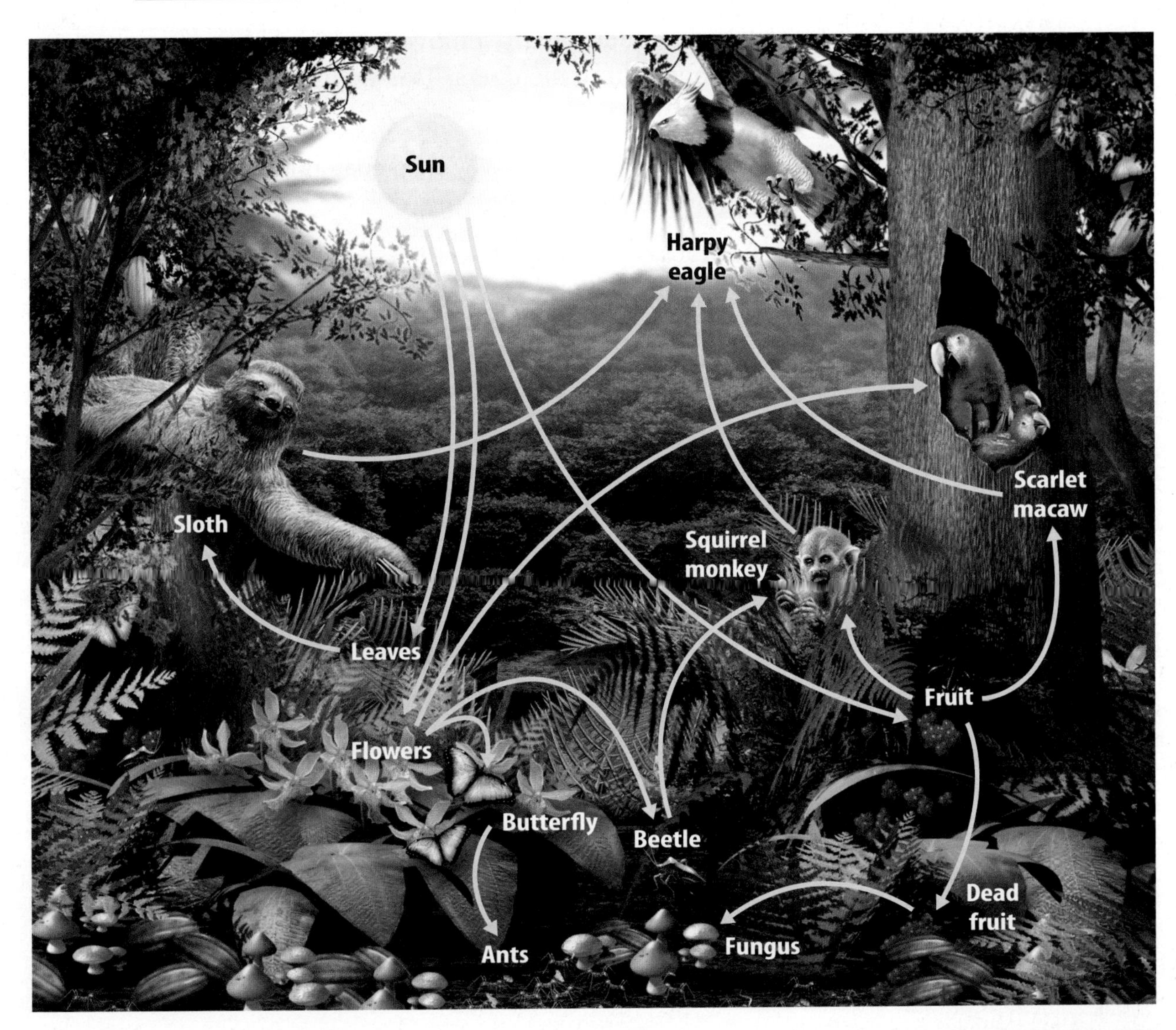

LA.6.2.2.3

Make a vertical three-tab book and label it as shown. Use it to organize information about the types of relationships that can exist among organisms within a community.

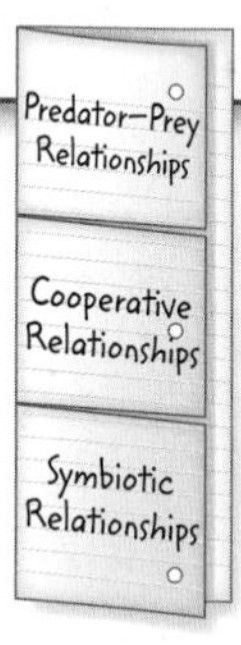

REVIEW VOCABULARY

predator
an organism that survives by hunting another

Relationships in Communities

The populations that make up a community interact with each other in a variety of ways. Some species have feeding relationships—they either eat or are eaten by another species. Some species interact with another species to get the food or shelter they need.

Predator-Prey Relationships

Hungry squirrel monkeys quarrel over a piece of fruit. They don't notice the harpy eagle above them. Suddenly, the eagle swoops down and grabs one of the monkeys in its talons. Harpy eagles and monkeys have a predator-prey relationship. The eagle, like other **predators,** hunts other animals for food. The hunted animals, such as the squirrel monkey or the lizard shown at the beginning of this lesson, are called prey.

As you read in Lesson 1, predators help prevent prey populations from growing too large for the carrying capacity of the ecosystem. The sand lizard, shown in **Figure 17,** is a predator in most of Europe. Like all predators, they often capture weak or injured individuals of a prey population. When the weak members of a population are removed, more resources are available for the remaining members. This helps keep the prey population healthy.

Active Reading **5. Describe** Why are predators important to a prey population?

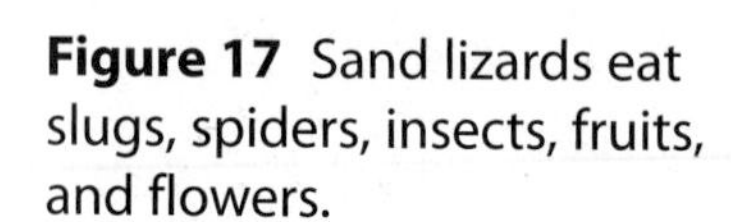

Figure 17 Sand lizards eat slugs, spiders, insects, fruits, and flowers.

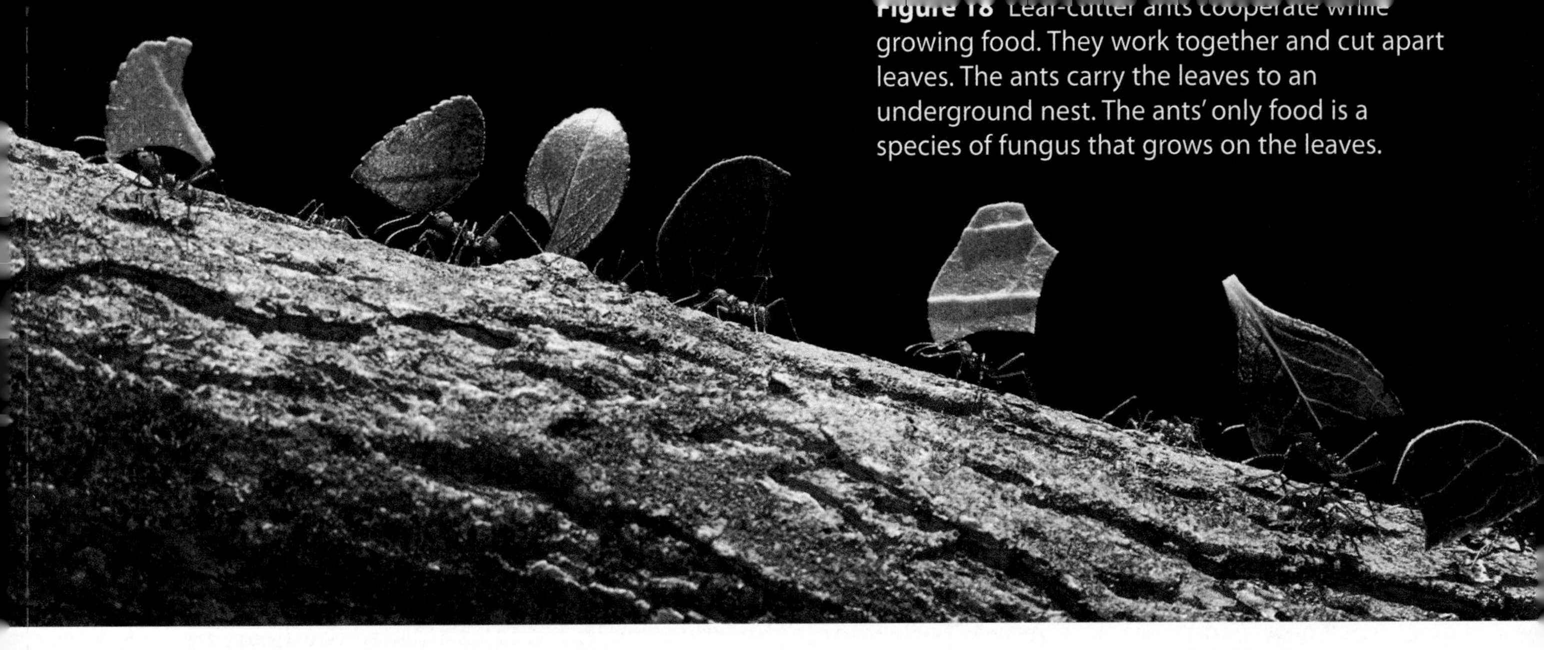

Figure 18 Leaf-cutter ants cooperate while growing food. They work together and cut apart leaves. The ants carry the leaves to an underground nest. The ants' only food is a species of fungus that grows on the leaves.

Cooperative Relationships

The members of some populations work together in cooperative relationships for their survival, like the leaf-cutter ants shown in **Figure 18.** As you read in Lesson 1, meerkats cooperate with each other as they raise young and watch for predators. Squirrel monkeys benefit in a similar way by living in groups. They cooperate as they hunt for food and watch for danger.

Symbiotic Relationships

Some species have such close relationships that they are almost always found living together. *A close, long-term relationship between two species that usually involves an exchange of food or energy is called* **symbiosis** (sihm bee OH sus). There are three types of symbiosis—mutualism, commensalism, and parasitism.

Mutualism Boxer crabs and sea anemones share a mutualistic partnership, as shown in **Figure 19.** *A symbiotic relationship in which both partners benefit is called* **mutualism.** Boxer crabs and sea anemones live in tropical coral reef communities. The crabs carry sea anemones in their claws. The sea anemones have stinging cells that help the crabs fight off predators. The sea anemones eat leftovers from the crabs' meals.

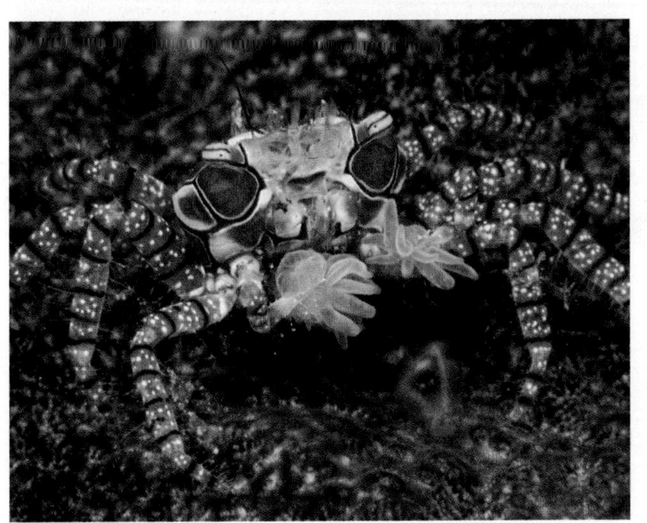

Figure 19 If a boxer crab loses its anemones, it can sometimes take a portion of another boxer crab's anemone.

6. NGSSS Check **Recall** What type of symbiotic relationship do the boxer crab and anemone have? SC.7.L.17.2

Figure 20 Epiphytes and trees share a commensal relationship.

7. NGSSS Check
Identify List ways species in a community interact. SC.7.L.17.1

Commensalism *A symbiotic relationship that benefits one species but does not harm or benefit the other is* **commensalism.** Plants called epiphytes (EH puh fites), shown in **Figure 20,** grow on the trunks of trees and other objects. The roots of an epiphyte anchor it to the object. The plant's nutrients are absorbed from the air. Epiphytes benefit from attaching to tree trunks by getting more living space and sunlight. The trees are neither helped nor harmed by the plants. Orchids are another example of epiphytes that have commensal relationships with trees.

Parasitism *A symbiotic relationship that benefits one species and harms the other is* **parasitism.** The species that benefits is the parasite. The species that is harmed is the host. Heartworms, tapeworms, fleas, and lice are parasites that feed on a host organism, such as a human or a dog. The parasites benefit by getting food. The host usually is not killed, but it can be weakened. For example, heartworms in a dog can cause the heart to work harder. Eventually, the heart can fail, killing the host. Other common parasites include the fungi that cause ringworm and toenail fungus. The fungi that cause these ailments feed on keratin (KER ah tihn), a protein in skin and nails.

The larvae of the hunting wasp is another example of a parasite. The female wasp, shown in **Figure 21,** stings a spider to paralyze it. Then she lays eggs in its body. When the eggs hatch into larvae, they eat the paralyzed spider's body. Another example of parasitism is the strangler fig. The seeds of the strangler fig sprout on the branches of a host tree. The young strangler fig sends roots into the tree and down into the ground below. The host tree provides the fig with nutrients and a trunk for support. Strangler figs grow fast, and they can kill a host tree.

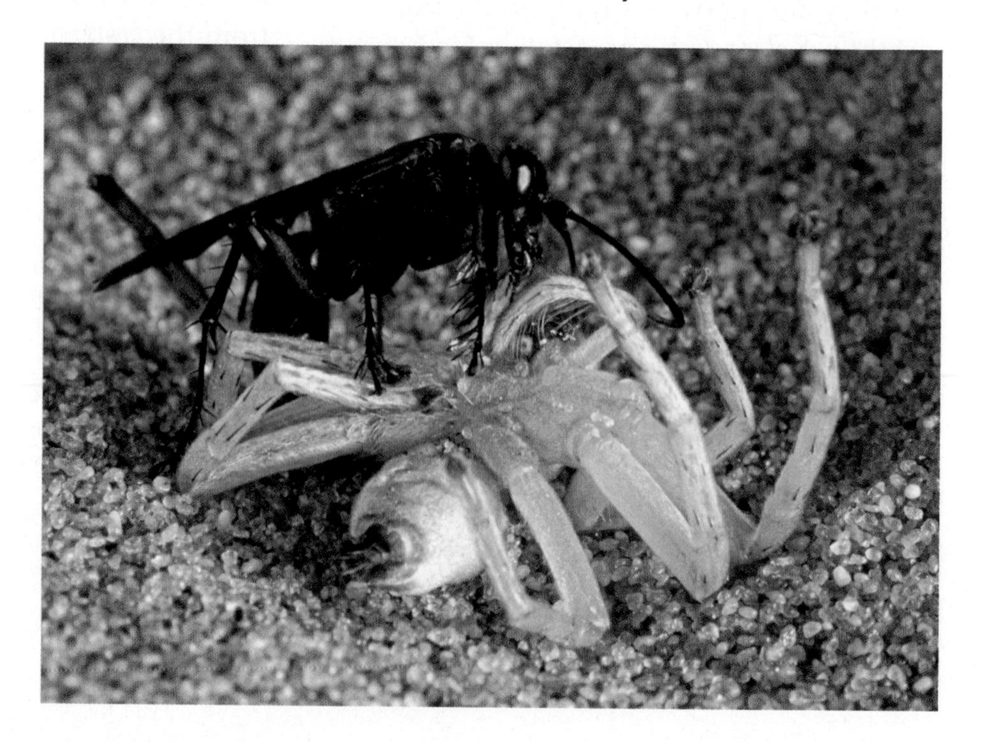

Figure 21 Hunting wasps are examples of parasites. The larvae use the paralyzed spider as food while they mature.

Lesson Review 3

Visual Summary

Each organism in a community has its own habitat and niche within the ecosystem.

Within a community, each organism must obtain energy for life processes. Some organisms are producers, and some are consumers.

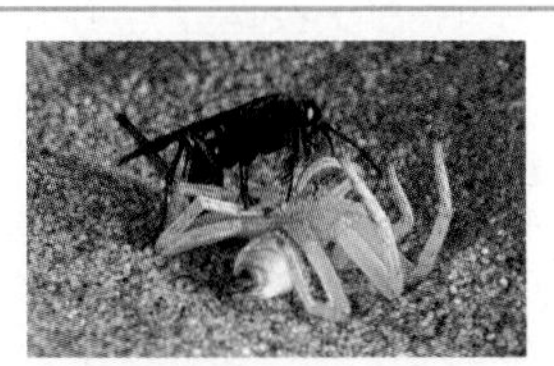

Some organisms have cooperative relationships, and some have symbiotic relationships. The hunting wasp and spider have a symbiotic relationship.

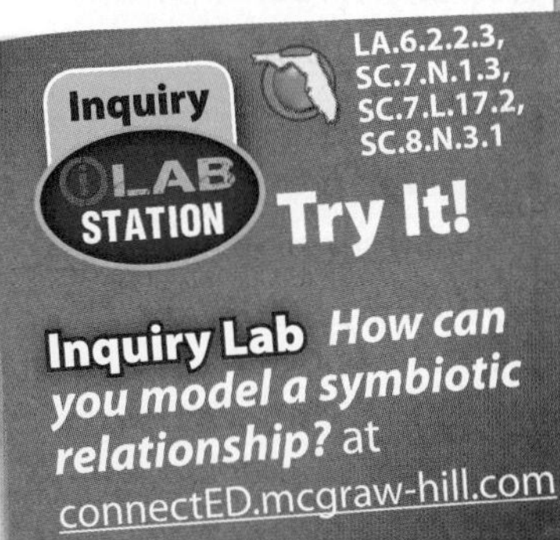

Use Vocabulary

1. **Define** *symbiosis.* SC.7.L.17.2

2. **Distinguish** between producers and consumers. SC.7.L.17.1

Understand Key Concepts

3. **Explain** how energy from the Sun flows through a rain forest community. SC.7.L.17.1

4. **Compare and contrast** predator-prey relationships and cooperative relationships. LA.6.2.2.3

5. A shrimp removes and eats the parasites from the gills of a fish. The fish stays healthier because the parasites are removed. This relationship is SC.7.L.17.2

   (A) commensalism. (C) mutualism.
   (B) competition. (D) parasitism.

Interpret Graphics

6. **Organize Information** Fill in the table below with details about the three different types of symbiosis. LA.6.2.2.3

Critical Thinking

7. **Predict** what could happen to a population of ants if anteaters, a predator of the ants, disappeared.

Chapter 19 Study Guide

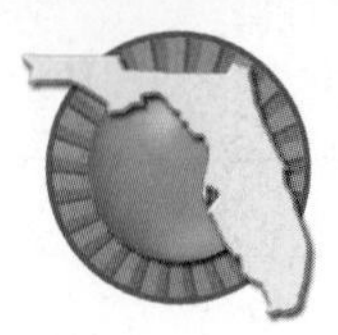

Think About It! A community contains many populations that interact through relationships such as mutualism, predation, parasitism, competition, and commensalism. Populations can increase, decrease, and move, affecting the community.

Key Concepts Summary

LESSON 1 Populations

- A **population** is all the organisms of the same species that live in the same area at the same time.
- Population sizes vary due to **limiting factors** such as environmental factors and available resources.
- Population size usually does not exceed the **carrying capacity** of the ecosystem.

Vocabulary

biosphere p. 733
community p. 734
population p. 734
competition p. 735
limiting factor p. 735
population density p. 736
biotic potential p. 736
carrying capacity p. 737

LESSON 2 Changing Populations

- Populations of living things can increase, decrease, or move.
- Populations can decrease until they are threatened, endangered, or extinct.
- Human population size is affected by the same three factors as other populations—**birthrate, death rate,** and movement.

birthrate p. 741
death rate p. 741
extinct species p. 743
endangered species p. 743
threatened species p. 743
migration p. 744

LESSON 3 Communities

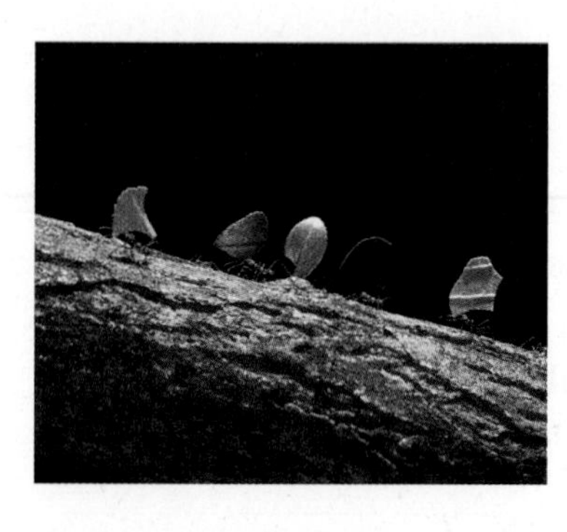

- A community is all the populations of different species that live together in the same area at the same time.
- The place within an ecosystem where an organism lives is its **habitat,** and what an organism does in its habitat to survive is its **niche.**
- Three types of relationships within a community are predator-prey, cooperative, and symbiotic.

habitat p. 751
niche p. 751
producer p. 752
consumer p. 752
symbiosis p. 755
mutualism p. 755
commensalism p. 756
parasitism p. 756

Active Reading

FOLDABLES® **Chapter Project**

Assemble your lesson Foldables as shown to make a Chapter Project. Use the project to review what you have learned in this chapter.

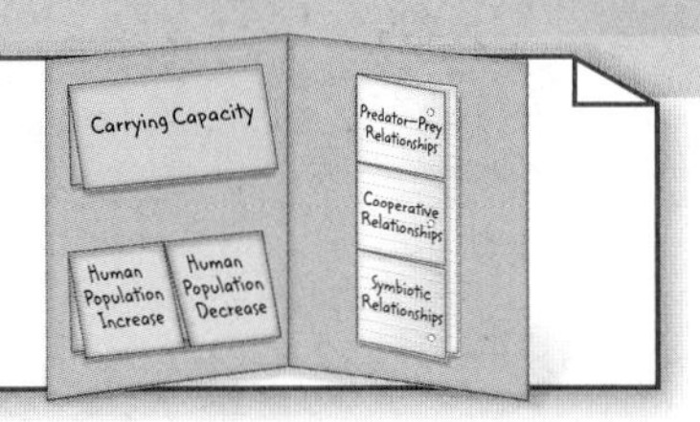

Use Vocabulary

1. The struggle in a community for the same resources is ______________.
2. The part of Earth that supports life is the ______________.
3. The instinctive movement of a population is ______________.
4. A(n) ______________ species is one at risk of becoming endangered.
5. A(n) ______________ is an organism that gets energy from the environment.
6. The largest number of offspring that can be produced when there are no limiting factors is the ______________.

Link Vocabulary and Key Concepts

Use vocabulary terms from the previous page to complete the concept map.

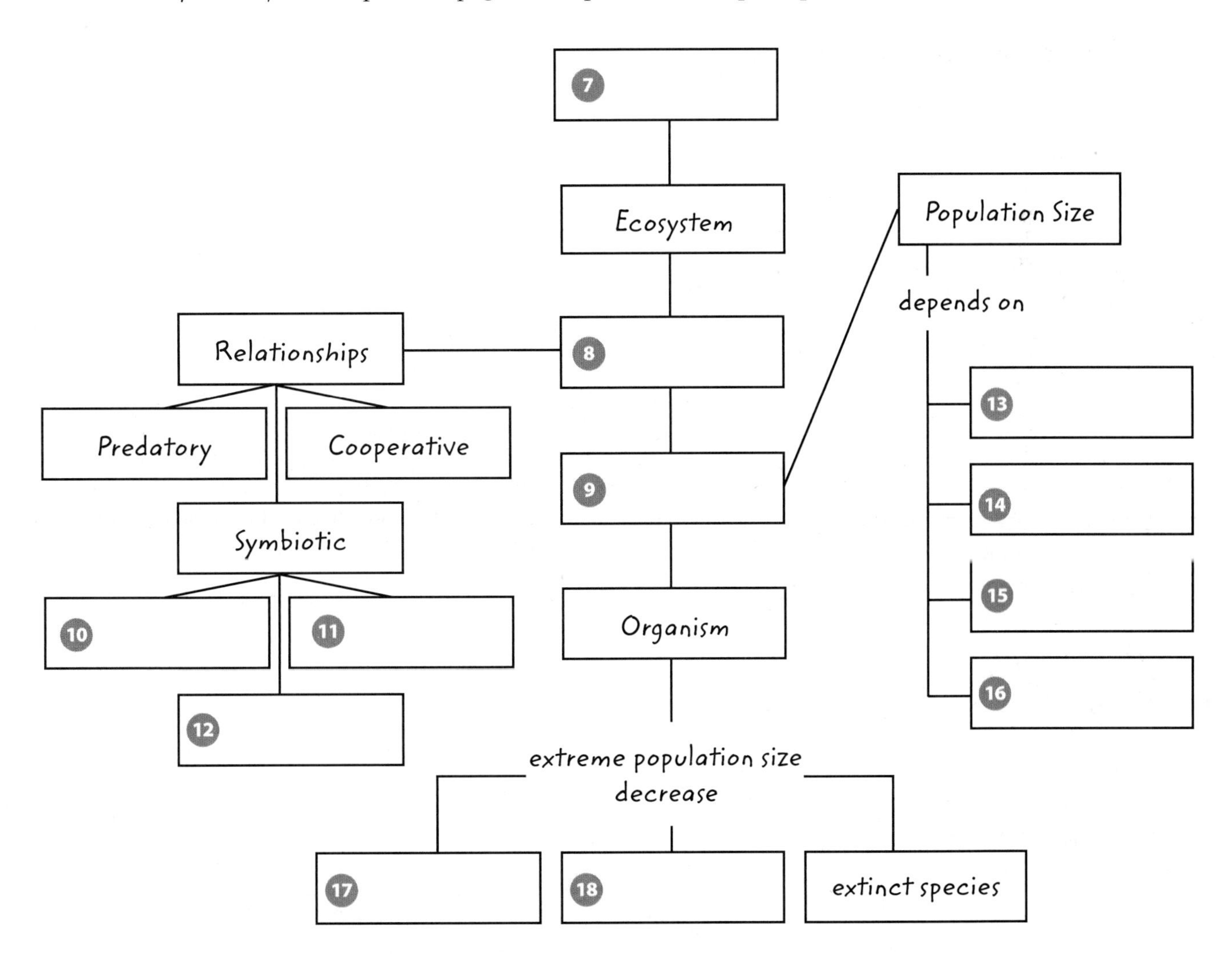

Chapter 19 Review

Fill in the correct answer choice.

Understand Key Concepts

1. What does the line indicated by the red arrow in the graph below represent? SC.7.L.17.3

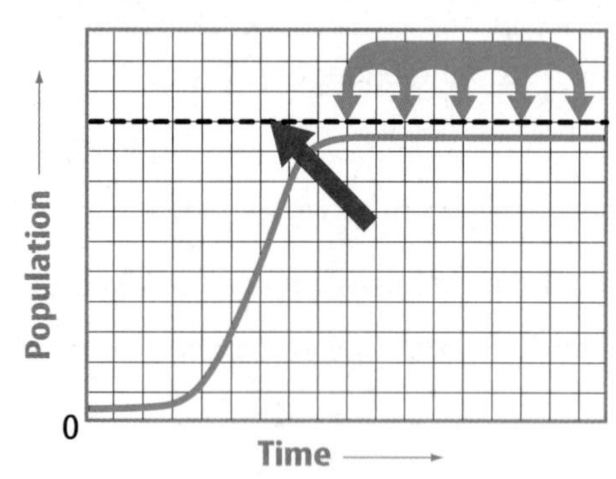

Ⓐ competition
Ⓑ biotic potential
Ⓒ carrying capacity
Ⓓ limiting factors

2. The need for organisms to rely on the same resources causes SC.7.L.17.3

Ⓐ competition.
Ⓑ biotic potential.
Ⓒ carrying capacity.
Ⓓ population growth.

3. An organism that uses sunlight to make food molecules is a(n) SC.7.L.17.1

Ⓐ carnivore.
Ⓑ consumer.
Ⓒ herbivore.
Ⓓ producer.

4. Which is a limiting factor for a cottontail rabbit population on the prairie in Oklahoma? SC.7.L.17.3

Ⓐ a large amount of food
Ⓑ a large amount of shelter space
Ⓒ an abundance of coyotes in the area
Ⓓ an unpolluted river in the ecosystem

5. Which factor does NOT normally affect human population size? SC.7.L.17.3

Ⓐ birthrate
Ⓑ death rate
Ⓒ population movement
Ⓓ lack of resources

Critical Thinking

6. **Select and draw** three food chains from the food web shown. SC.7.L.17.1

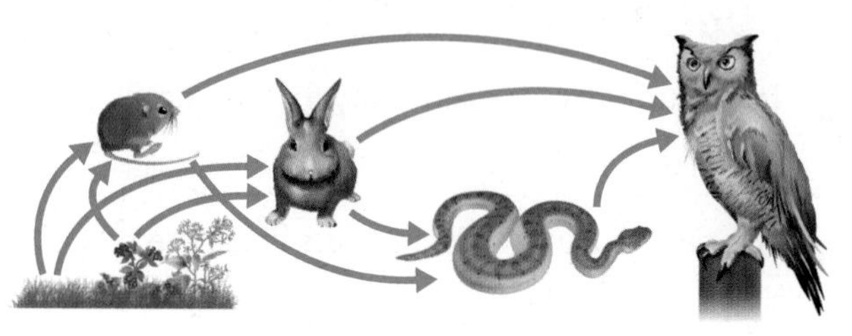

7. **Give an Example** What problems might result from overpopulation of pigeons in a city park?

8. **Describe** What are some possible solutions that a city might use to solve a pigeon overpopulation problem? SC.7.L.17.3

9. **Decide** Would sample counting or capture-mark-and-release at a specified time and place be the better method for measuring each of these populations: birds, whales, bluebonnet flowers, and oak trees? SC.6.N.1.5

10. **Compare and contrast** the feeding habits of carnivores, omnivores, and producers. SC.7.L.17.1

11 **Classify** Decide whether each of these relationships is mutualism, commensalism, or parasitism. SC.7.L.17.2

- Butterfly pollinates flower while drinking nectar.

- Tapeworm feeds on contents of dog's intestines.

- Fish finds shelter in coral reef.

Writing in Science

12 **Write** a story on a separate piece of paper that explains how an imaginary population becomes threatened with extinction. LA.6.2.2.3

Big Idea Review

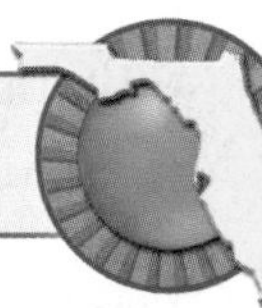

13 Describe three different types of relationships in a community between two different populations of organisms. SC.7.L.17.2

14 Do you think a large number of pigeons affects other organisms in the community? Explain your answer. SC.7.L.17.3

Math Skills MA.6.A.3.6

Use Graphs

Use the graph to answer the questions.

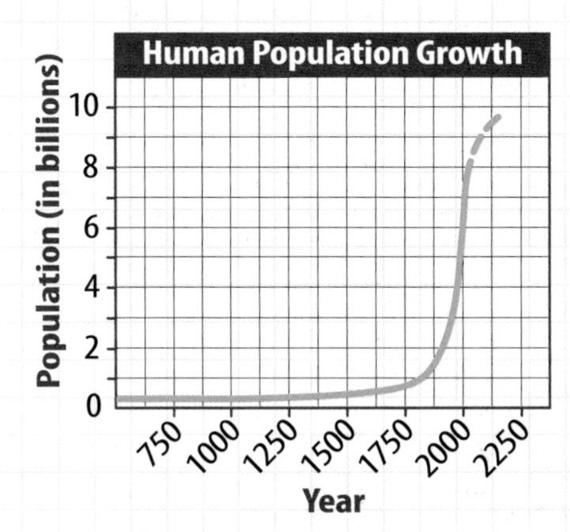

15 During what range of years did the population change the least? MA.6.A.3.6

16 The dotted line represents a prediction. What does it predict about population growth beyond the present time? MA.6.A.3.6

Florida NGSSS Benchmark Practice

Record your answers on the answer sheet provided by your teacher or on a sheet of paper.

Multiple Choice

1 Which is defined as the demand for resources in short supply in a community? **SC.7.L.17.3**

Ⓐ biotic potential

Ⓑ competition

Ⓒ density

Ⓓ limiting factor

Use the diagram below to answer questions 2 and 3.

2 In the diagram above, which number represents an ecosystem? **SC.7.N.1.1**

Ⓕ 1

Ⓖ 2

Ⓗ 3

Ⓘ 4

3 According to the arrows, how are the elements of the diagram organized? **SC.7.N.1.1**

Ⓐ endangered to overpopulated

Ⓑ farthest to nearest

Ⓒ largest to smallest

Ⓓ nonliving to living

4 Which is NOT a possible result of overpopulation? **SC.7.L.17.3**

Ⓕ damage to soil

Ⓖ increased carrying capacity

Ⓗ loss of trees and plants

Ⓘ more competition

Use the diagram below to answer question 5.

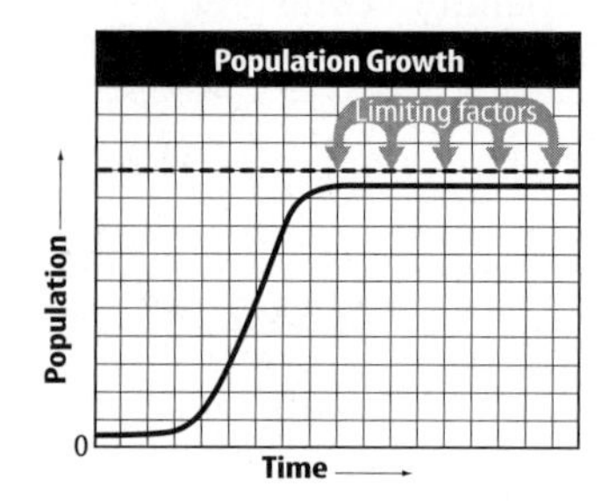

5 What does the dashed line in the diagram represent? **SC.7.N.1.1**

Ⓐ biotic potential

Ⓑ carrying capacity

Ⓒ overpopulation

Ⓓ population density

6 What does a population undergo when it has no limiting factors? **SC.7.L.17.3**

Ⓕ exponential growth

Ⓖ extinction

Ⓗ migration

Ⓘ population movement

7 What is the term for all species living in the same ecosystem at the same time? **SC.7.L.17.3**

Ⓐ biosphere

Ⓑ community

Ⓒ habitat

Ⓓ population

Use the diagram below to answer question 8.

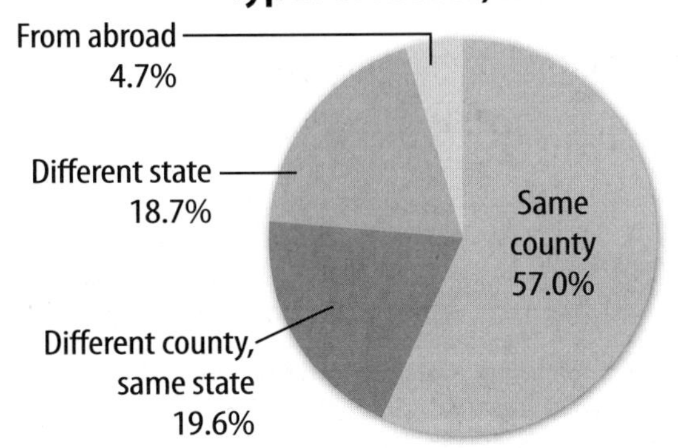

Source: U.S. Census Bureau, Current Population Survey, 2005 Annual Social and Economic Supplement.

8 According to the diagram, about how many of those who moved remained within the same state? **MA.6.A.3.6**

Ⓕ 20 percent
Ⓖ 39 percent
Ⓗ 57 percent
Ⓘ 77 percent

9 Which type of relationship includes mutualism and parasitism? **SC.7.L.17.2**

Ⓐ competition
Ⓑ cooperation
Ⓒ predation
Ⓓ symbiosis

10 Which is a population? **SC.7.L.17.3**

Ⓕ all meerkats in a refuge
Ⓖ all the types of birds in a foresta
Ⓗ all the types of cats in a zoo
Ⓘ all the types of insects in a swamp

11 The number of organisms in a specific area is **LA.6.2.2.3**

Ⓐ a community.
Ⓑ the carrying capacity.
Ⓒ the population density
Ⓓ the population growth.

12 The number of robins that hatch in a year is the population's **LA.6.2.2.3**

Ⓕ biotic potential.
Ⓖ birthrate.
Ⓗ carrying capacity.
Ⓘ exponential growth.

13 A robin population that reaches its biotic potential probably shows **LA.6.2.2.3**

Ⓐ exponential growth.
Ⓑ low growth.
Ⓒ negative growth.
Ⓓ no growth.

14 Which is NOT part of Earth's biosphere? **LA.6.2.2.3**

Ⓕ low atmosphere
Ⓖ surface of the Moon
Ⓗ bottom of the Pacific ocean
Ⓘ North American continent

15 What type of overall population change is shown at right? **LA.6.2.2.3**

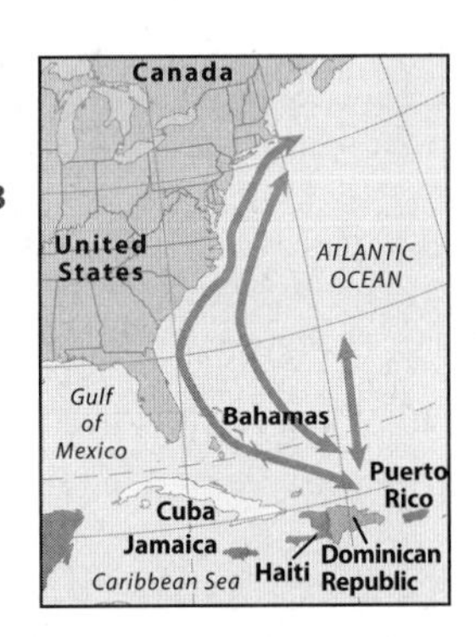

Ⓐ immigration
Ⓑ migration
Ⓒ population decrease
Ⓓ population increase

NEED EXTRA HELP?

If You Missed Question...	1	2	3	4	5	6	7	8	9	10	11	12	13	14	15
Go to Lesson...	1	1	1	1	1	2	1	2	3	1	1	2	2	1	2

Name ______________________ Date ____________

Multiple Choice *Bubble the correct answer.*

Use the graph below to answer questions 1 and 2.

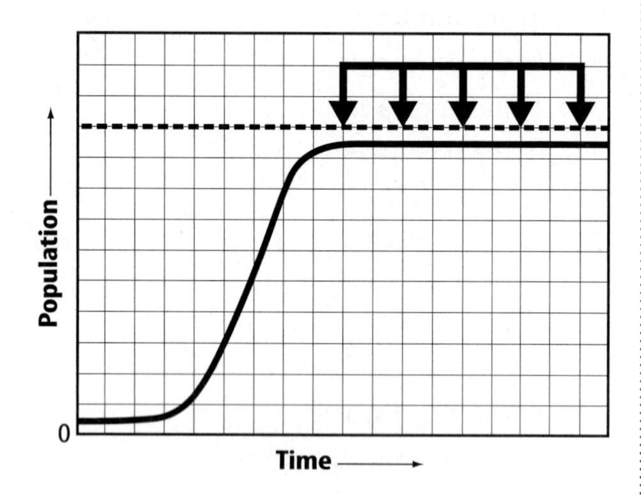

1. In the graph above, what does the dotted line represent? **MA.6.A.3.6**

Ⓐ biotic potential

Ⓑ carrying capacity

Ⓒ limiting factors

Ⓓ population density

2. In the graph above, what do the arrows represent? **SC.7.L.17.3**

Ⓕ biotic potential

Ⓖ carrying capacity

Ⓗ limiting factors

Ⓘ population density

3. Which list correctly describes the organization of the biosphere from most complex to least complex? **LA.6.2.2.3**

Ⓐ biosphere→community→ecosystem→population

Ⓑ biosphere→community→population→ecosystem

Ⓒ biosphere→ecosystem→community→population

Ⓓ biosphere→population→community→ecosystem

4. Which is an example of a limiting factor in an ecosystem? **SC.7.L.17.3**

Ⓕ After heavy rainstorms, a river floods a forest and part of a grassland.

Ⓖ Plants produce so much fruit that much of it falls to the ground uneaten.

Ⓗ Snowmelt refills an empty lake with enough water to support local populations.

Ⓘ Spring rains ensure that more than enough grass grows to support a gazelle population.

Name ______________________ Date ____________

Multiple Choice *Bubble the correct answer.*

1. Which graph shows exponential growth?
MA.6.A.3.6

Ⓐ 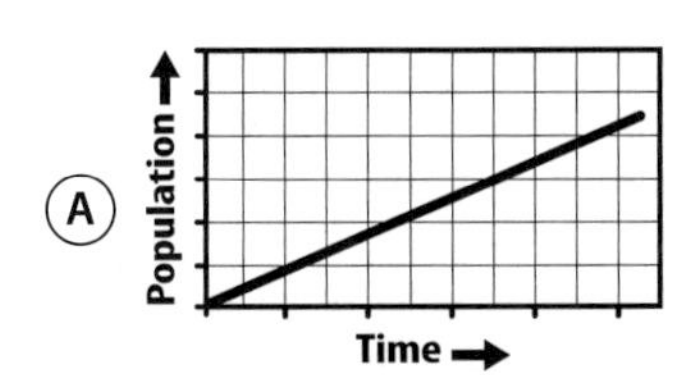

Ⓑ 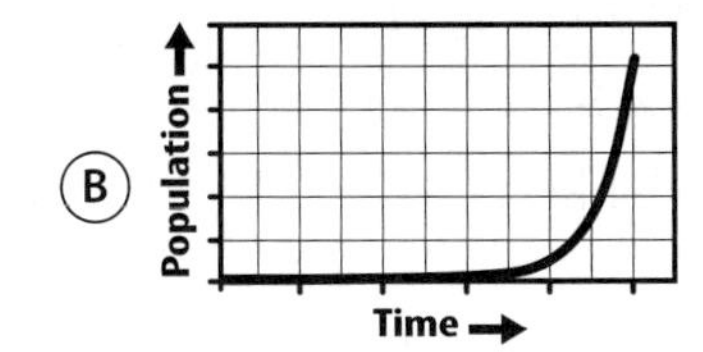

Ⓒ 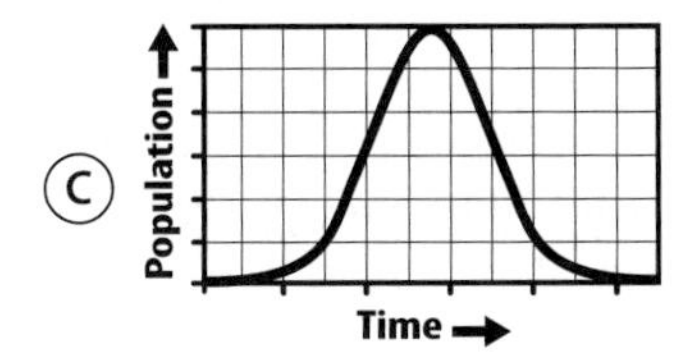

Ⓓ 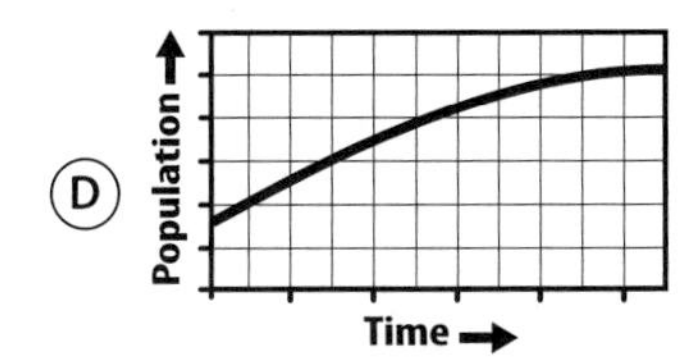

2. In which case would a population of mice increase? **SC.7.L.17.2**

Ⓕ A drought keeps grass from growing.

Ⓖ More mice die than are born.

Ⓗ The owl population increases.

Ⓘ The populations of insects increase.

3. In the 17th century, the last dodo bird died. The dodo is an example of a(n)
SC.7.L.15.3

Ⓐ endangered species.

Ⓑ extinct species.

Ⓒ migrating species.

Ⓓ threatened species.

4. Plant populations move from place to place through seed dispersal. Which of the following seeds is likely spread by wind?
SC.7.L.15.2

Ⓕ 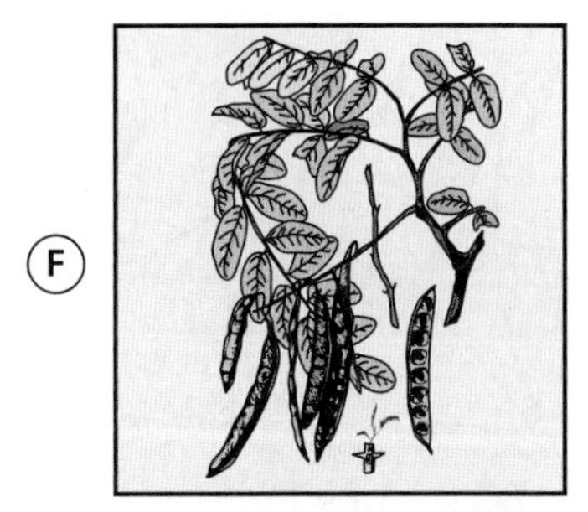

Ⓖ 

Ⓗ 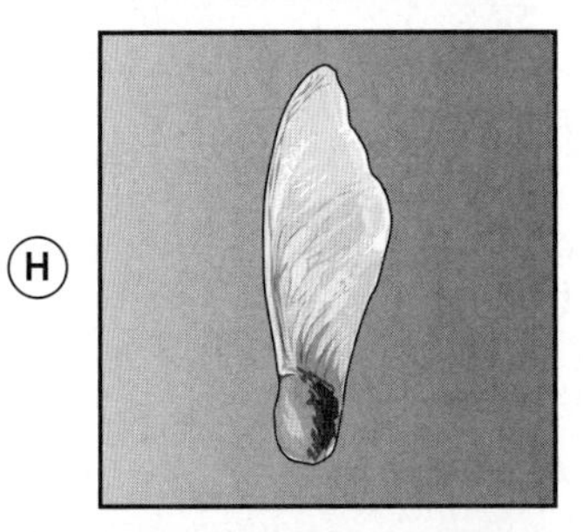

Ⓘ 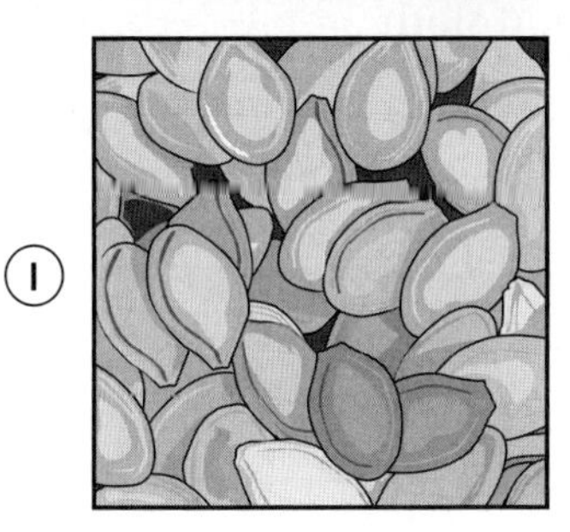

Name ______________________ Date ______________

Multiple Choice *Bubble the correct answer.*

Use the figure below to answer questions 1 and 2.

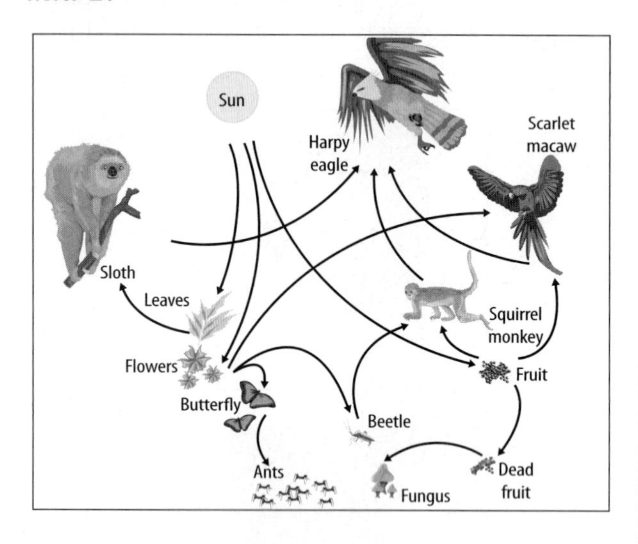

1. Which list describes a way in which energy flows in this ecosystem? **SC.7.L.17.1**

Ⓐ Sun→butterfly→ants→fungus

Ⓑ Sun→flowers→butterflies→ants

Ⓒ Sun→leaves→sloth→scarlet macaw

Ⓓ Sun→sloth→harpy eagle→fruit

2. In this ecosystem, what is an example of a predator-prey relationship? **SC.7.L.17.2**

Ⓕ butterfly and flowers

Ⓖ dead fruit and fungus

Ⓗ leaves and sloth

Ⓘ squirrel monkey and beetle

3. Which relationship is an example of mutualism? **SC.7.L.17.2**

Ⓐ Ants nest in the thorns of an acacia tree, get food from the tree, and protect the tree from herbivores.

Ⓑ Barnacles attach themselves to the hard shell of a mollusk, where the barnacles can gather food scraps.

Ⓒ A fig grows from the branches of a host plant to the ground, eventually blocking sunlight from the host plant.

Ⓓ Heartworms live in the heart of an animal, getting nutrients from the host and harming it in the process.

4. An organism that gets energy from both plants and animals is called a(n) **SC.7.L.17.1**

Ⓕ carnivore.

Ⓖ detritivore.

Ⓗ herbivore.

Ⓘ omnivore.

Name ______________________ Date ______________

Notes

Name ____________________ Date ____________

Notes

Desert Descriptions

Deserts are one of the seven major land biomes. Put an *X* next to any of the characteristics that can describe a desert.

_____ A. Earth's driest ecosystem

_____ B. Can be hot during the day and cold at night

_____ C. Can be very cold all the time

_____ D. Has soil that holds water

_____ E. Has plants that can store water

_____ F. Has plants with large leaves

_____ G. Can be near an ocean

_____ H. Only found in subtropical areas

_____ I. Are always sand-covered

_____ J. Lizards, bats, birds, and snakes live there.

Explain your thinking. Describe what makes a desert different from other biomes.

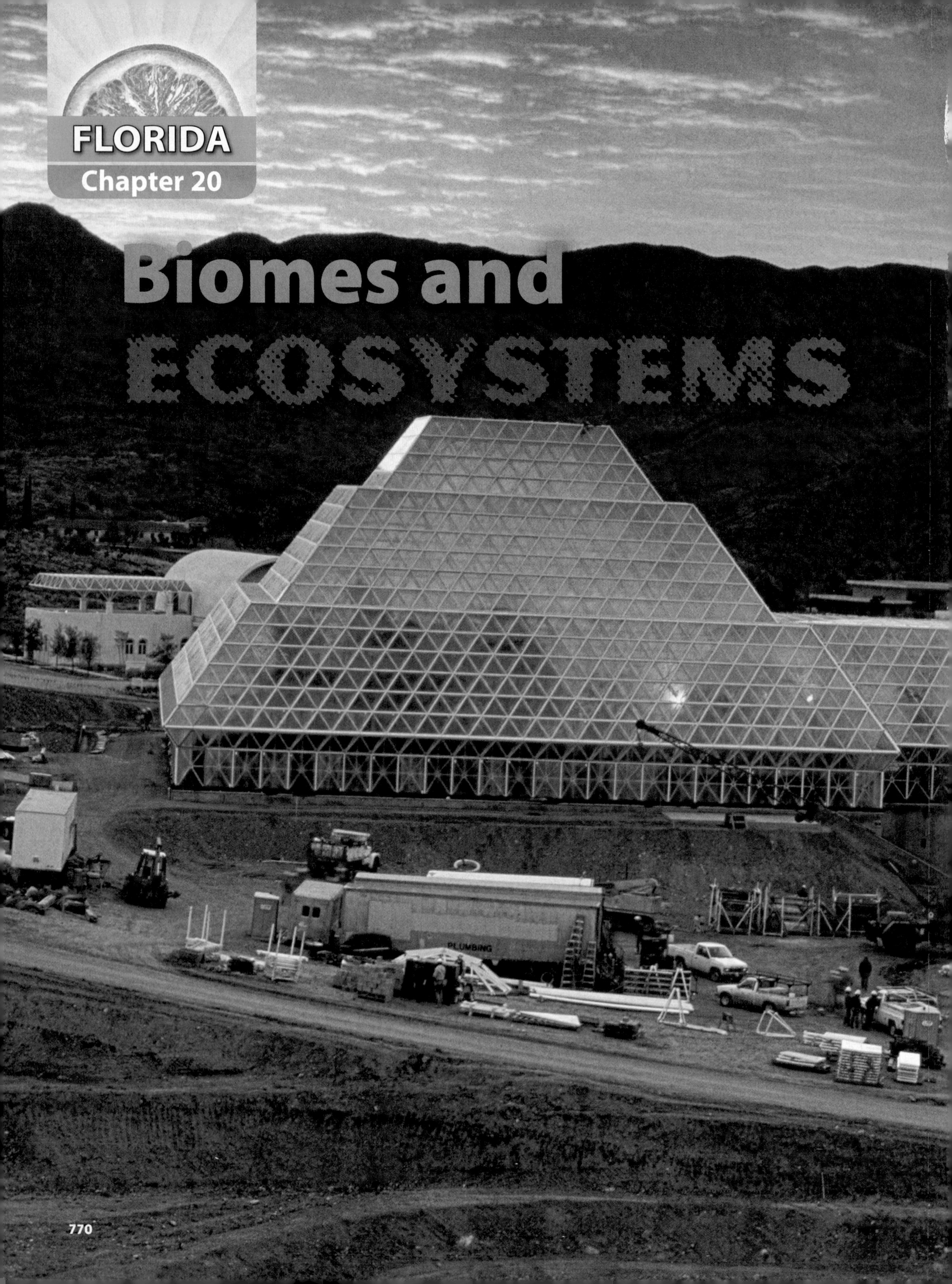

FLORIDA
Chapter 20

Biomes and ECOSYSTEMS

FLORIDA BIG IDEAS

1 **The Practice of Science**

3 **The Role of Theories, Laws, Hypotheses, and Models**

17 **Interdependence**

The Big Idea

Think About It!

How do Earth's biomes and ecosystems differ?

Although it might look like a piece of art, this structure was designed to replicate several ecosystems. When Biosphere 2 was built in the 1980s near Tucson, Arizona, it included a rain forest, a desert, a grassland, a coral reef, and a wetland. Today, it is used mostly for research and education.

1. How realistic do you think Biosphere 2 is?

2. Is it possible to make artificial environments as complex as those in nature?

3. How do Earth's biomes and ecosystems differ?

Get Ready to Read

What do you think about biomes and ecosystems?

Before you read, decide if you agree or disagree with each of these statements. As you read this chapter, see if you change your mind about any of the statements.

	AGREE	DISAGREE
1 Deserts can be cold.	☐	☐
2 There are no rain forests outside the tropics.	☐	☐
3 Estuaries do not protect coastal areas from erosion.	☐	☐
4 Animals form coral reefs.	☐	☐
5 An ecosystem never changes.	☐	☐
6 Nothing grows in the area where a volcano has erupted.	☐	☐

There's More Online!
Video • Audio • Review • ⓘLab Station • WebQuest • Assessment • Concepts in Motion • Multilingual eGlossary

Lesson 1

Land BIOMES

ESSENTIAL QUESTIONS

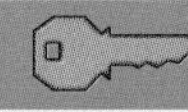 How do Earth's land biomes differ?

 How do humans impact land biomes?

Vocabulary

biome p. 773
desert p. 774
grassland p. 775
temperate p. 777
taiga p. 779
tundra p. 779

Florida NGSSS

LA.6.2.2.3 The student will organize information to show understanding (e.g., representing main ideas within text through charting, mapping, paraphrasing, summarizing, or comparing/contrasting);

SC.6.N.1.1 Define a problem from the sixth grade curriculum, use appropriate reference materials to support scientific understanding, plan and carry out scientific investigation of various types, such as systematic observations or experiments, identify variables, collect and organize data, interpret data in charts, tables, and graphics, analyze information, make predictions, and defend conclusions.

SC.7.N.1.1 Define a problem from the seventh grade curriculum, use appropriate reference materials to support scientific understanding, plan and carry out scientific investigation of various types, such as systematic observations or experiments, identify variables, collect and organize data, interpret data in charts, tables, and graphics, analyze information, make predictions, and defend conclusions.

SC.8.N.1.1 Define a problem from the eighth grade curriculum using appropriate reference materials to support scientific understanding, plan and carry out scientific investigations of various types, such as systematic observations or experiments, identify variables, collect and organize data, interpret data in charts, tables, and graphics, analyze information, make predictions, and defend conclusions.

SC.7.L.17.3 Describe and investigate various limiting factors in the local ecosystem and their impact on native populations, including food, shelter, water, space, disease, parasitism, predation, and nesting sites.

LA.6.2.2.3

Inquiry Launch Lab

10 minutes

What is the climate in China?

Beijing, China, and New York, New York, are about the same distance from the equator but on opposite sides of Earth. How do temperature and rainfall compare for these two cities?

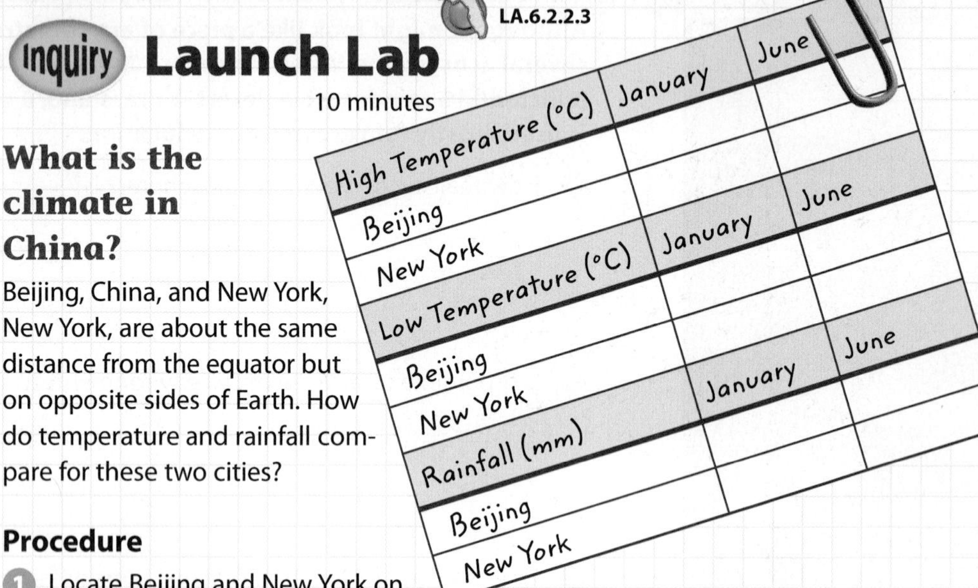

High Temperature (°C)	January	June
Beijing		
New York		

Low Temperature (°C)	January	June
Beijing		
New York		

Rainfall (mm)	January	June
Beijing		
New York		

Procedure

1. Locate Beijing and New York on a **world map.**
2. From the data and charts provided, find and record the average high and low temperatures in January and in June for each city.
3. Record the average rainfall in January and in June for each city.

Think About This

1. What are the temperature and rainfall ranges for each city?

2. Key Concept How do you think the climates of these cities differ year-round?

Plant or Animal?

1. Believe it or not, this is a flower. One of the largest flowers in the world, rafflesia (ruh FLEE zhuh), grows naturally in the tropical rain forests of southeast Asia. What do you think would happen if you planted a seed from this plant in a desert? Would it survive?

Land Ecosystems and Biomes

When you go outside, you might notice people, grass, flowers, birds, and insects. You also are probably aware of nonliving things, such as air, sunlight, and water. The living or once-living parts of an environment are the biotic parts. The nonliving parts that the living parts need to survive are the abiotic parts. The biotic and abiotic parts of an environment together make up an ecosystem.

Earth's continents have many different ecosystems, from deserts to rain forests. Scientists classify similar ecosystems in large geographic areas as biomes. *A* **biome** *is a geographic area on Earth that contains ecosystems with similar biotic and abiotic features.* As shown in **Figure 1,** Earth has seven major land biomes. Areas classified as the same biome have similar climates and organisms.

Figure 1 Earth contains seven major biomes.

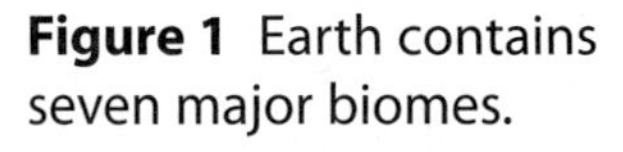

2. Identify (Circle) the name of the biome that contains most of Florida.

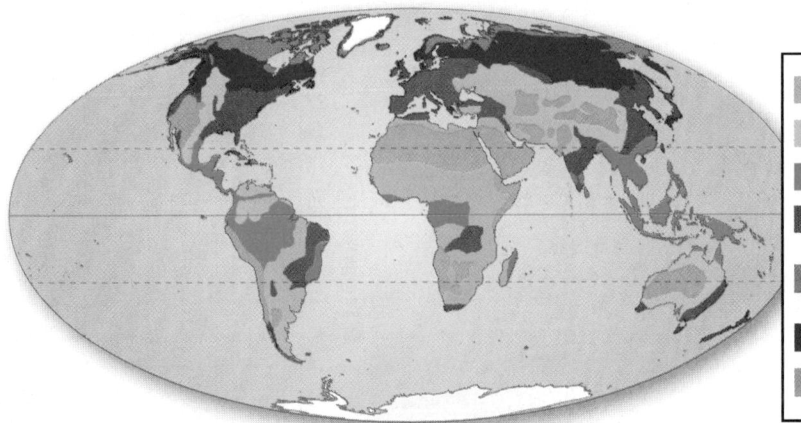

MiniLab ***How hot is sand?*** at connectED.mcgraw-hill.com

Apply It!

After you complete the lab, answer these questions.

1. Explain whether testing soil with little sand would give the same results.

2. Do you think organisms in your environment could survive in a desert? Why or why not?

Desert Biome

Deserts *are biomes that receive very little rain.* They are on nearly every continent and are Earth's driest ecosystems.

- Most deserts are hot during the day and cold at night. Others, like those in Antarctica, remain cold all of the time.
- Rainwater drains away quickly because of thin, porous soil. Large patches of ground are bare.

Biodiversity

- Animals include lizards, bats, woodpeckers, and snakes. Most animals avoid activity during the hottest parts of the day.
- Plants include spiny cactus and thorny shrubs. Shallow roots absorb water quickly. Some plants have accordion-like stems that expand and store water. Small leaves or spines reduce the loss of water.

Woodpeckers

Human Impact

- Cities, farms, and recreational areas in deserts use valuable water.
- Desert plants grow slowly. When they are damaged by people or livestock, recovery takes many years.

3. Infer Underline organism characteristics that might be beneficial in a desert biome.

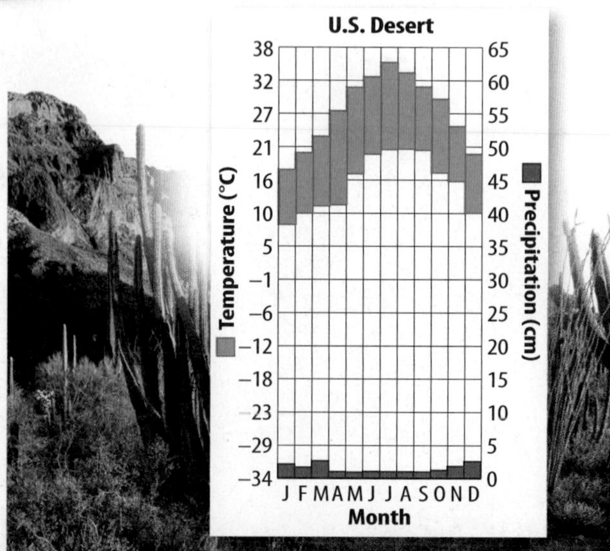

Grassland Biome

Grassland *biomes are areas where grasses are the dominant plants.* Also called prairies, savannas, and meadows, grasslands are the world's "breadbaskets." Wheat, corn, oats, rye, barley, and other important cereal crops are grasses. They grow well in these areas.

- Grasslands have a wet and a dry season.
- Deep, fertile soil supports plant growth.
- Grass roots form a thick mass, called sod, which helps soil absorb and hold water during periods of drought.

Active Reading **4. Find** Highlight reasons why plants grow well in grasslands.

Biodiversity

- Trees grow along moist banks of streams and rivers. Wildflowers bloom during the wet season.
- In North America, large herbivores, such as bison and elk, graze here. Insects, birds, rabbits, prairie dogs, and snakes find shelter in the grasses.
- Predators in North American grasslands include hawks, ferrets, coyotes, and wolves.
- African savannas are grasslands that contain giraffes, zebras, and lions. Australian grasslands are home to kangaroos, wallabies, and wild dogs.

Human Impact

- People plow large areas of grassland to raise cereal crops. This reduces habitat for wild species.
- Because of hunting and loss of habitat, large herbivores—such as bison—are now uncommon in many grasslands.

Black-footed ferret

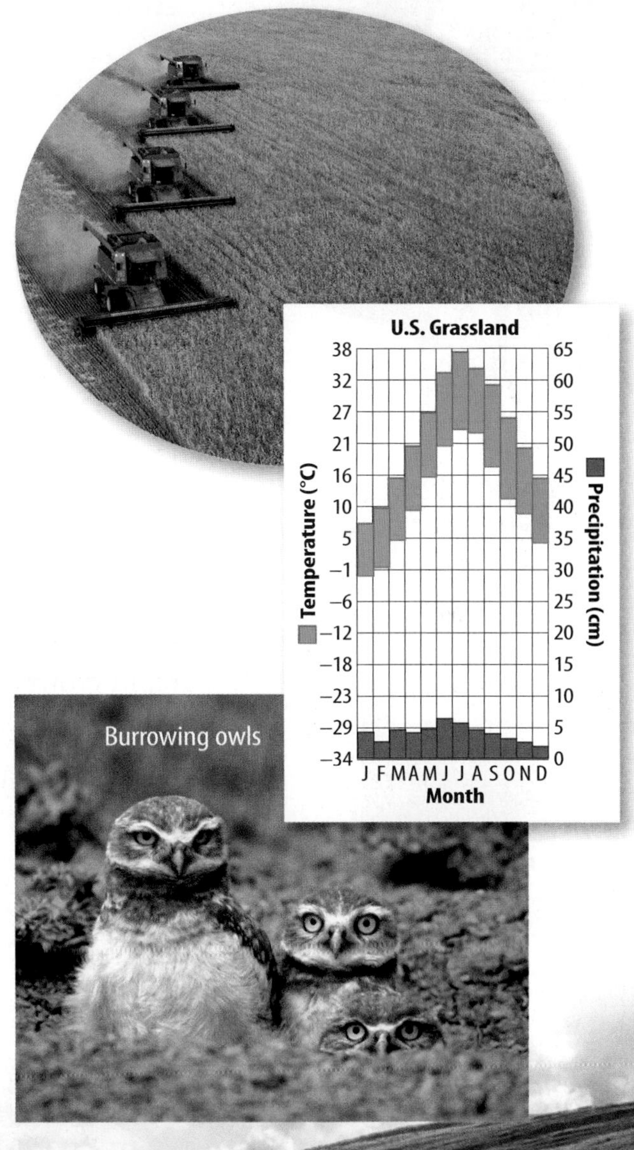

Burrowing owls

Tropical Rain Forest Biome

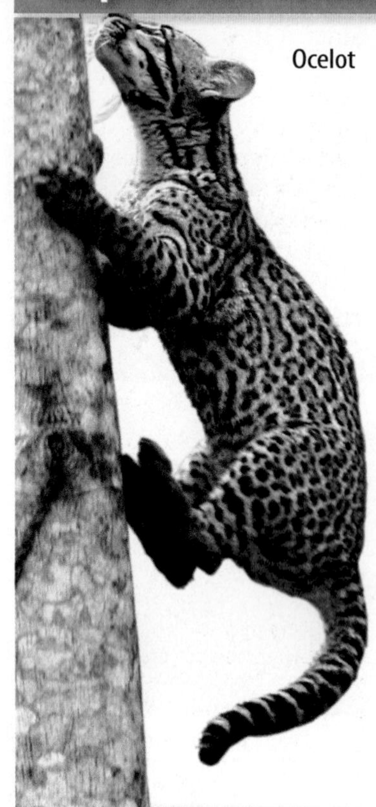
Ocelot

The forests that grow near the equator are called tropical rain forests. These forests receive large amounts of rain and have dense growths of tall, leafy trees.

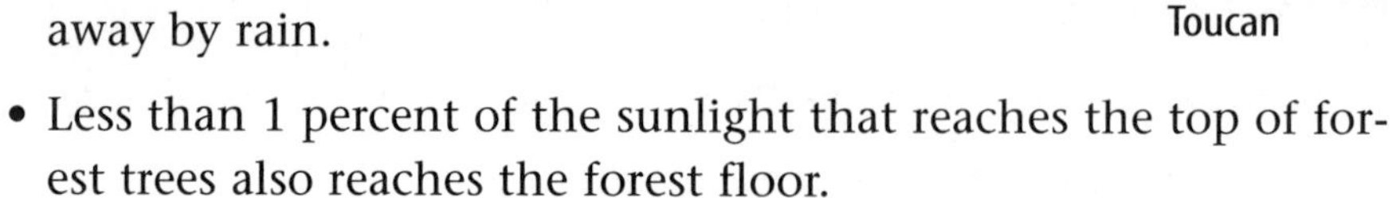
Toucan

- Weather is warm and wet year-round.
- The soil is shallow and easily washed away by rain.
- Less than 1 percent of the sunlight that reaches the top of forest trees also reaches the forest floor.
- Half of Earth's species live in tropical rain forests. Most live in the canopy—the uppermost part of the forest.

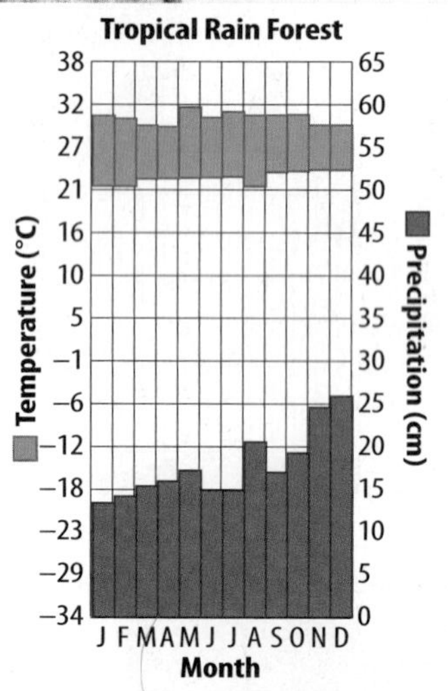

Biodiversity

- Few plants live on the dark forest floor.
- Vines climb the trunks of tall trees.
- Mosses, ferns, and orchids live on branches in the canopy.
- Insects make up the largest group of tropical animals. They include beetles, termites, ants, bees, and butterflies.
- Larger animals include parrots, toucans, snakes, frogs, flying squirrels, fruit bats, monkeys, jaguars, and ocelots.

Human Impact

- People have cleared more than half of Earth's tropical rain forests for lumber, farms, and ranches. Poor soil does not support rapid growth of new trees in cleared areas.
- Some organizations are working to encourage people to use less wood that has been harvested from rain forests.

5. NGSSS Check **Explain** What factors affect how many organisms can survive in a rain forest? SC.7.L.17.3

Temperate Rain Forest Biome

Regions of Earth between the tropics and the polar circles are **temperate** *regions.* Temperate regions have relatively mild climates with distinct seasons. Several biomes are in temperate regions, including rain forests. Temperate rain forests are moist ecosystems mostly in coastal areas. They are not as warm as tropical rain forests.

- Winters are mild and rainy.
- Summers are cool and foggy.
- Soil is rich and moist.

Biodiversity

- Forests are dominated by spruce, hemlock, cedar, fir, and redwood trees, which can grow very large and tall.
- Fungi, ferns, mosses, vines, and small flowering plants grow on the moist forest floor.
- Animals include mosquitoes, butterflies, frogs, salamanders, woodpeckers, owls, eagles, chipmunks, raccoons, deer, elk, bears, foxes, and cougars.

Human Impact

- Temperate rain forest trees are a source of lumber. Logging can destroy the habitat of forest species.
- Rich soil enables cut forests to grow back. Tree farms help provide lumber without destroying habitat.

Active Reading **6. Identify** On the graph to the right, circle the hottest and the driest months.

Active Reading

FOLDABLES® LA.6.2.2.3

Use a sheet of paper to make a horizontal two-tab book. Record what you learn about desert and temperate rain forest biomes under the tabs, and use the information to compare and contrast these biomes.

Desert Biome | Temperate Rain Forest Biome

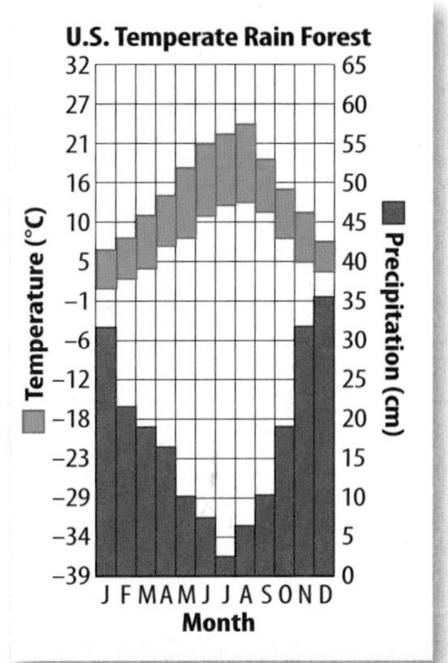

Elk

Temperate Deciduous Forest Biome

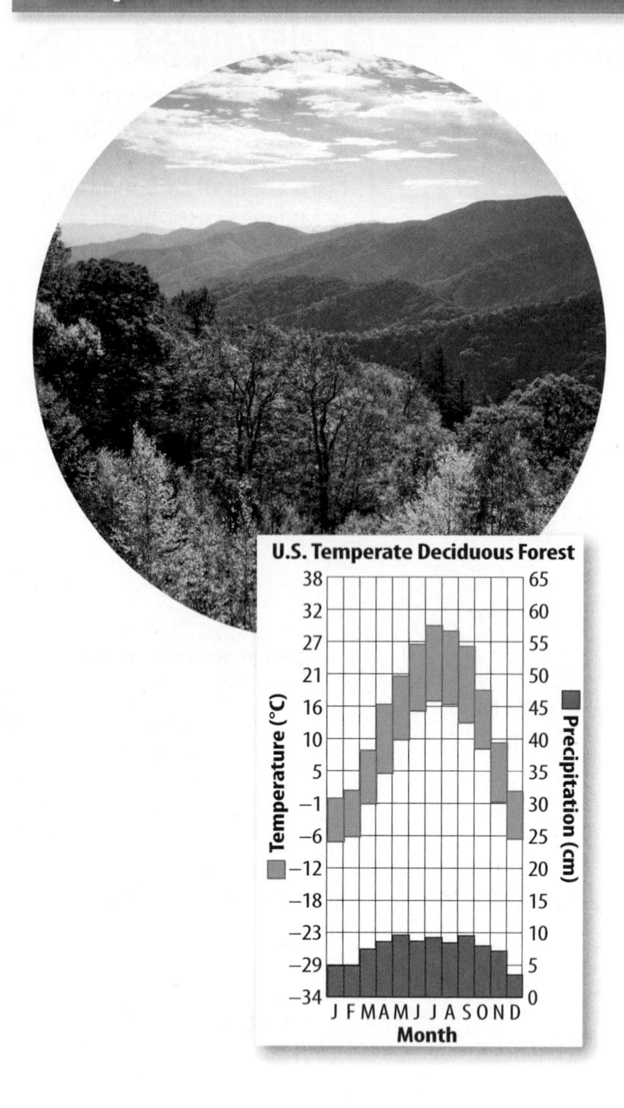

7. **NGSSS Check** **Explain** What factors affect how many organisms can survive in a temperate deciduous forest? SC.7.L.17.3

Temperate deciduous forests grow in temperate regions where winter and summer climates have more variation than those in temperate rain forests. These forests are the most common forest ecosystems in the United States. They contain mostly deciduous trees, which lose their leaves in the fall.

- Winter temperatures are often below freezing. Snow is common.
- Summers are hot and humid.
- Soil is rich in nutrients and supports a large amount of diverse plant growth.

Biodiversity

- Most plants, such as maples, oaks, birches, and other deciduous trees, stop growing during the winter and begin growing again in the spring.
- Animals include snakes, ants, butterflies, birds, raccoons, opossums, and foxes.
- Some animals, including chipmunks and bats, spend the winter in hibernation.
- Many birds and some butterflies, such as the monarch, migrate to warmer climates for the winter.

Human Impact

Over the past several hundred years, humans have cleared thousands of acres of Earth's deciduous forests for farms and cities. Today, much of the clearing has stopped and some forests have regrown.

Red fox

Taiga Biome

A **taiga** (TI guh) *is a forest biome consisting mostly of cone-bearing evergreen trees.* The taiga biome exists only in the northern hemisphere. It occupies more space on Earth's continents than any other biome.

- Winters are long, cold, and snowy. Summers are short, warm, and moist.
- Soil is thin and acidic.

Biodiversity

- Evergreen trees, such as spruce, pine, and fir, are thin and shed snow easily.
- Animals include owls, mice, moose, bears, and other cold-adapted species.
- Abundant insects in summer attract many birds, which migrate south in winter.

Human Impact

- Tree harvesting reduces taiga habitat.

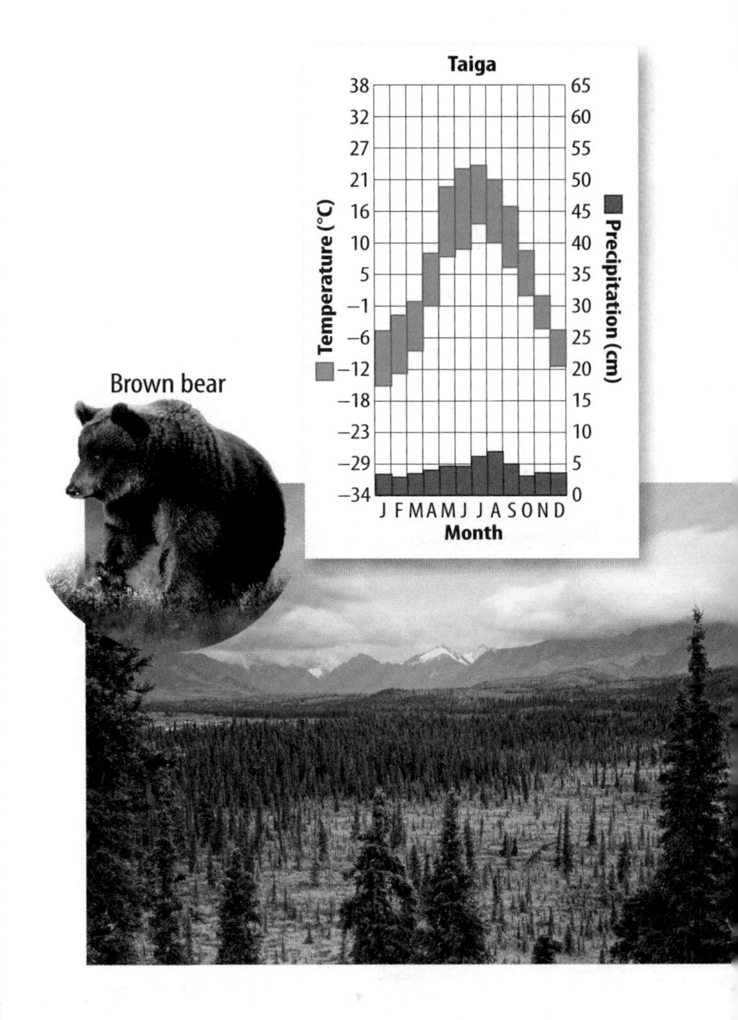

Tundra Biome

A **tundra** (TUN druh) *biome is cold, dry, and treeless.* Most tundra is south of the North Pole, but it also exists in mountainous areas at high altitudes.

- Winters are long, dark, and freezing; summers are short and cool; the growing season is only 50–60 days.
- Permafrost—a layer of permanently frozen soil—prevents deep root growth.

Biodiversity

- Plants include shallow-rooted mosses, lichens, and grasses.
- Many animals hibernate or migrate south during winter. Few animals, including lemmings, live in tundras year-round.

Human Impact

- Drilling for oil and gas can interrupt migration patterns.

Active Reading **8. Contrast** Underline factors that distinguish taigas from tundras.

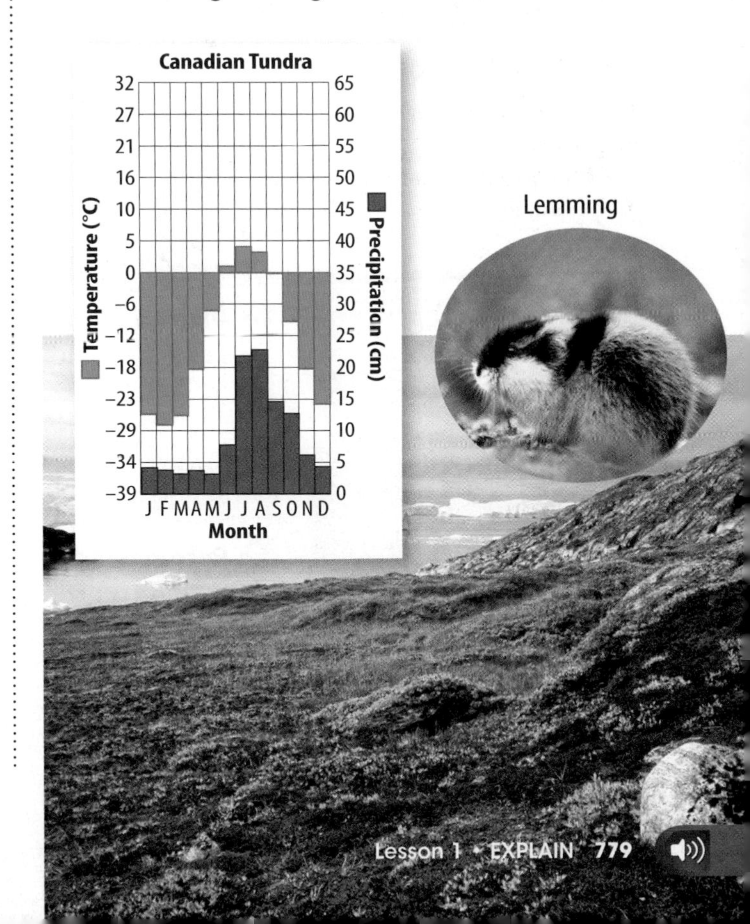

Lesson Review 1

Visual Summary

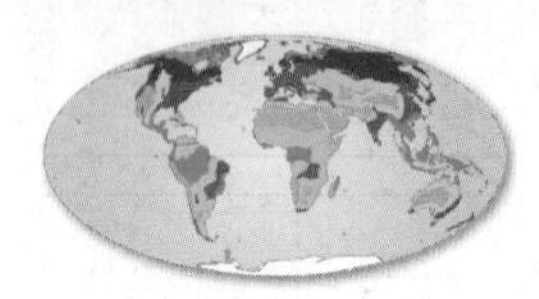

Earth has seven major land biomes, ranging from hot, dry deserts to cold, forested taigas.

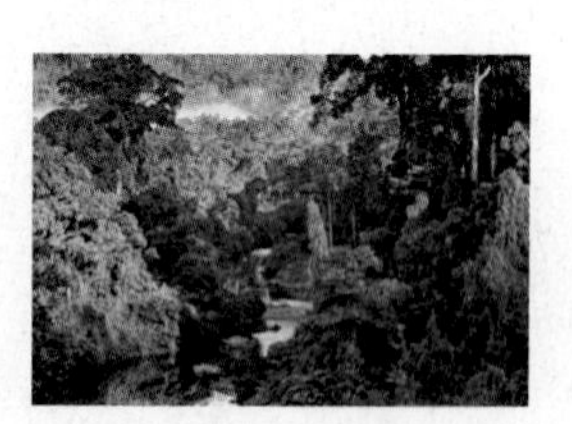

Half of Earth's species live in rain forest biomes.

Temperate deciduous forests are the most common forest biome in the United States.

Use Vocabulary

1. **Define** *biome* using your own words.

2. **Distinguish** between tropical rain forests and temperate rain forests.

3. A cold, treeless biome is a(n) ________.

Understand Key Concepts

4. **Explain** why tundra soil cannot support the growth of trees. SC.7.L.17.3

Interpret Graphics

5. **Summarize Information** Fill in the graphic organizer below with animals and plants of the biome you live in. LA.6.2.2.3

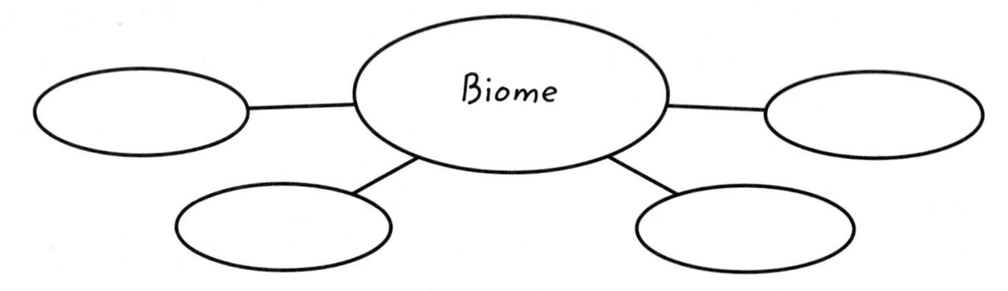

Critical Thinking

6. **Plan** an enclosed zoo exhibit for a desert ecosystem. What abiotic factors should you consider?

7. **Recommend** one or more actions people can take to reduce habitat loss in tropical and taiga forests. SC.7.L.17.3

Florida's Land Ecosystems

FOCUS on FLORIDA

A Diverse State

If you were to travel throughout Florida you would see a very diverse group of ecosystems from sandy beaches to exotic forests. Most of Florida is part of a temperate deciduous forest biome. However, each area contains regions with different ecosystems. Florida's ecosystems include prairies, dunes, scrub forests, and forests called hardwood hammocks. Just as the ecosystems differ, so do the limiting factors in each one.

Hardwood hammocks are small, dense groups of hardwood trees. Hardwood forest habitats are often destroyed so the land can be used for construction.

Prairies are a type of grassland that supports a large amount of grasses and shrubs with few or no trees. Prairies are susceptible to fires as well as organisms that eat the plants there.

Dunes form on beaches as the wind pushes sand into large mounds where certain organisms can live. Dune ecosystems are affected when people build homes or businesses near or on them.

Scrub forests contain small trees and large shrubs. Heat and lack of water as well as fires limit the growth of scrub forest ecosystems.

It's Your Turn

RESEARCH one Florida ecosystem near you and present your findings to the class.

Lesson 2

Aquatic ECOSYSTEMS

ESSENTIAL QUESTIONS

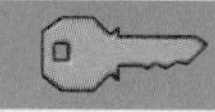
How do Earth's aquatic ecosystems differ?

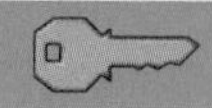
How do humans impact aquatic ecosystems?

Vocabulary

salinity p. 783
wetland p. 786
estuary p. 787
intertidal zone p. 789
coral reef p. 789

Florida NGSSS

LA.6.2.2.3 The student will organize information to show understanding (e.g., representing main ideas within text through charting, mapping, paraphrasing, summarizing, or comparing/contrasting);

MA.6.A.3.6 Construct and analyze tables, graphs, and equations to describe linear functions and other simple relations using both common language and algebraic notation.

SC.7.L.17.3 Describe and investigate various limiting factors in the local ecosystem and their impact on native populations, including food, shelter, water, space, disease, parasitism, predation, and nesting sites.

SC.6.N.1.5 Recognize that science involves creativity, not just in designing experiments, but also in creating explanations that fit evidence.

SC.6.N.1.5

Launch Lab

10 minutes

What happens when rivers and oceans mix?

Freshwater and saltwater ecosystems have different characteristics. What happens in areas where freshwater rivers and streams flow into oceans?

Procedure

1. Read and complete a lab safety form.
2. In a **plastic tub,** add 100 g of **salt** to 2 L of water. Stir with a **long-handled spoon** until the salt dissolves.
3. In another **container,** add 5 drops of **blue food coloring** to 1 L of water. Gently pour the colored water into one corner of the plastic tub. Observe how the color of the water changes in the tub.
4. Observe the tub again in 5 min.

Think About This

1. What bodies of water do the containers represent?

2. What happened to the water in the tub after 5 min? What do you think happened to the salt content of the water?

3. Key Concept How do you think the biodiversity of rivers and oceans differs? What organisms do you think might live at the place where the two meet?

Inquiry Floating Trees?

1. These plants, called mangroves, are one of the few types of plants that grow in salt water. They usually live along ocean coastlines in tropical ecosystems, such as Florida. What other organisms do you think live near mangroves?

Aquatic Ecosystems

If you've ever spent time near an ocean, a river, or another body of water, you might know that water is full of life. There are four major types of water, or aquatic, ecosystems: freshwater, wetland, estuary, and ocean. Each type of ecosystem contains a unique variety of organisms. Whales, dolphins, and corals live only in ocean ecosystems. Catfish and trout live only in freshwater ecosystems. Many other organisms that do not live under water, such as birds and seals, also depend on aquatic ecosystems for food and shelter.

Important abiotic factors in aquatic ecosystems include temperature, sunlight, and dissolved oxygen gas. Aquatic species have adaptations that enable them to use the oxygen in water. The gills of a fish separate oxygen from water and move it into the fish's bloodstream. Mangrove plants, pictured above, take in oxygen through small pores in their leaves and roots.

Salinity (say LIH nuh tee) is another important abiotic factor in aquatic ecosystems. **Salinity** *is the amount of salt dissolved in water.* Water in saltwater ecosystems has high salinity compared to water in freshwater ecosystems, which contains little salt.

Math Skills MA.6.A.3.6

Use Proportions

Salinity is measured in parts per thousand (PPT). One PPT water contains 1 g salt and 1,000 g water. Use proportions to calculate salinity. What is the salinity of 100 g of water with 3.5 g of salt?

$$\frac{3.5 \text{ g salt}}{100 \text{ g seawater}} = \frac{x \text{ g salt}}{1{,}000 \text{ g seawater}}$$

$$100x = 3500$$

$$x = \frac{3{,}500}{100} = 35 \text{ PPT}$$

Practice

2. A sample contains 0.1895 g of salt per 50 g of seawater. What is its salinity?

Great Blue Heron

Salmon

Freshwater ecosystems include streams, rivers, ponds, and lakes. Streams are usually narrow, shallow, and fast-flowing. Rivers are larger, deeper, and flow more slowly.

- Streams form from underground sources of water such as springs or from runoff from rain and melting snow.
- Stream water is often clear. Soil particles are quickly washed downstream.
- Oxygen levels in streams are high because air mixes into the water as it splashes over rocks.
- Rivers form when streams flow together.
- Soil that washes into a river from streams or nearby land can make river water muddy. Soil also introduces nutrients, such as nitrogen, into rivers.
- Slow-moving river water has higher levels of nutrients and lower levels of dissolved oxygen than fast-moving water.

Biodiversity

- Willows, cottonwoods, and other water-loving plants grow along streams and on riverbanks.
- Species adapted to fast-moving water include trout, salmon, crayfish, and many insects.
- Species adapted to slow-moving water include snails and catfish.

Stonefly larva

Human Impact

- People take water from streams and rivers for drinking, laundry, bathing, crop irrigation, and industrial purposes.
- Hydroelectric plants use the energy in flowing water to generate electricity. Dams stop the water's flow.
- Runoff from cities, industries, and farms is a source of pollution.

Freshwater: Ponds and Lakes

Ponds and lakes contain freshwater that is not flowing downhill. These bodies of water form in low areas on land.

- Ponds are shallow and warm.
- Sunlight reaches the bottom of most ponds.
- Pond water is often high in nutrients.
- Lakes are larger and deeper than ponds.
- Sunlight penetrates into the top few feet of lake water. Deeper water is dark and cold.

Biodiversity

- Plants surround ponds and lake shores.
- Surface water in ponds and lakes contains plants, algae, and microscopic organisms that use sunlight for photosynthesis.
- Organisms living in shallow water near shorelines include cattails, reeds, insects, crayfish, frogs, fish, and turtles.
- Fewer organisms live in the deeper, colder water of lakes where there is little sunlight.
- Lake fish include perch, trout, bass, and walleye.

Smallmouth bass

Active Reading **3. Find** Highlight reasons why few organisms live in the deep water of lakes.

Human Impact

- Humans fill in ponds and lakes with sediment to create land for houses and other structures.
- Runoff from farms, gardens, and roads washes pollutants into ponds and lakes, disrupting food webs.

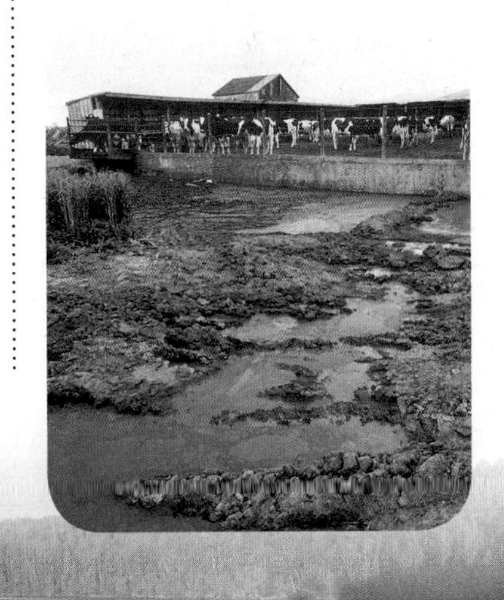

Active Reading **4. Contrast** Underline ways that ponds and lakes differ.

Common loon

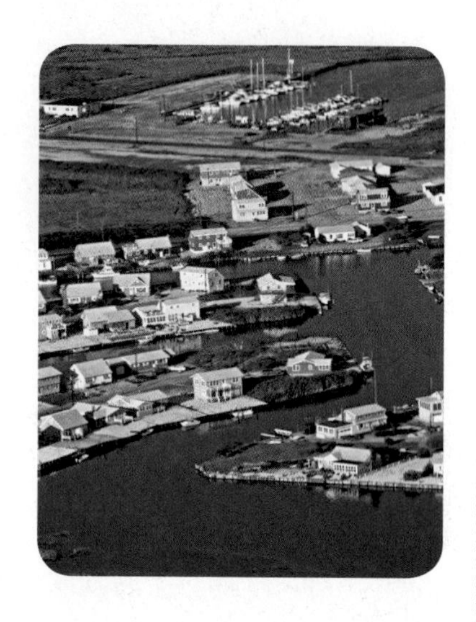

Some types of aquatic ecosystems have mostly shallow water. **Wetlands** *are aquatic ecosystems that have a thin layer of water covering soil that is wet most of the time.* Wetlands contain freshwater, salt water, or both. They are among Earth's most fertile ecosystems.

- Freshwater wetlands form at the edges of lakes and ponds and in low areas on land. Saltwater wetlands form along ocean coasts.
- Nutrient levels and biodiversity are high.
- Wetlands trap sediments and purify water. Plants and microscopic organisms filter out pollution and waste materials.

Biodiversity

- Water-tolerant plants include grasses and cattails. Few trees live in saltwater wetlands. Trees in freshwater wetlands include cottonwoods, willows, and swamp oaks.
- Insects are abundant and include flies, mosquitoes, dragonflies, and butterflies.
- More than one-third of North American bird species, including ducks, geese, herons, loons, warblers, and egrets, use wetlands for nesting and feeding.
- Other animals that depend on wetlands for food and breeding grounds include alligators, turtles, frogs, snakes, salamanders, muskrats, and beavers.

Human Impact

- In the past, many people considered wetlands as unimportant environments. Water was drained away to build homes and roads and to raise crops.
- Today, many wetlands are being preserved, and drained wetlands are being restored.

Active Reading **5. Determine** Highlight ways that humans impact wetlands.

Estuaries

Estuaries (ES chuh wer eez) *are regions along coastlines where streams or rivers flow into a body of salt water.* Most estuaries form along coastlines, where freshwater in rivers meets salt water in oceans. Estuary ecosystems have varying degrees of salinity.

- Salinity depends on rainfall, the amount of freshwater flowing from land, and the amount of salt water pushed in by tides.
- Estuaries help protect coastal land from flooding and erosion. Like wetlands, estuaries purify water and filter out pollution.
- Nutrient levels and biodiversity are high.

Biodiversity

- Plants that grow in salt water include mangroves, pickleweeds, and seagrasses.
- Animals include worms, snails, and many species that people use for food, including oysters, shrimp, crabs, and clams.
- Striped bass, salmon, flounder, and many other ocean fish lay their eggs in estuaries.
- Many species of birds depend on estuaries for breeding, nesting, and feeding.

Human Impact

- Large portions of estuaries have been filled with soil to make land for roads and buildings.
- Destruction of estuaries reduces habitat for estuary species and exposes the coastline to flooding and storm damage.

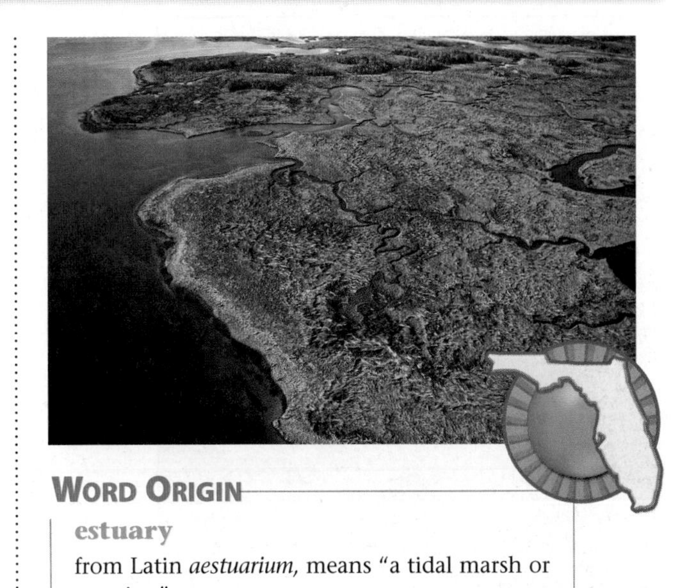

Word Origin

estuary
from Latin *aestuarium,* means "a tidal marsh or opening"

Active Reading

FOLDABLES® LA.6.2.2.3

Make a horizontal two-tab book and label it as shown. Use it to compare how biodiversity and human impact differ in wetlands and estuaries.

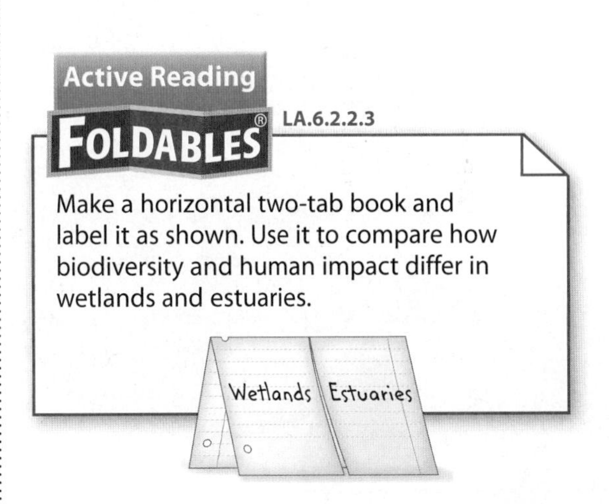

Harvest mouse

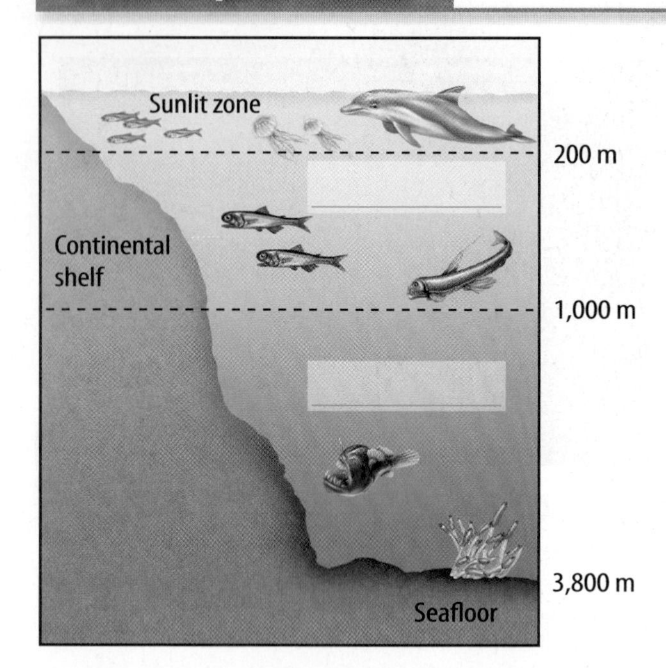

Active Reading **6. Compile** Fill in the missing ocean zone labels.

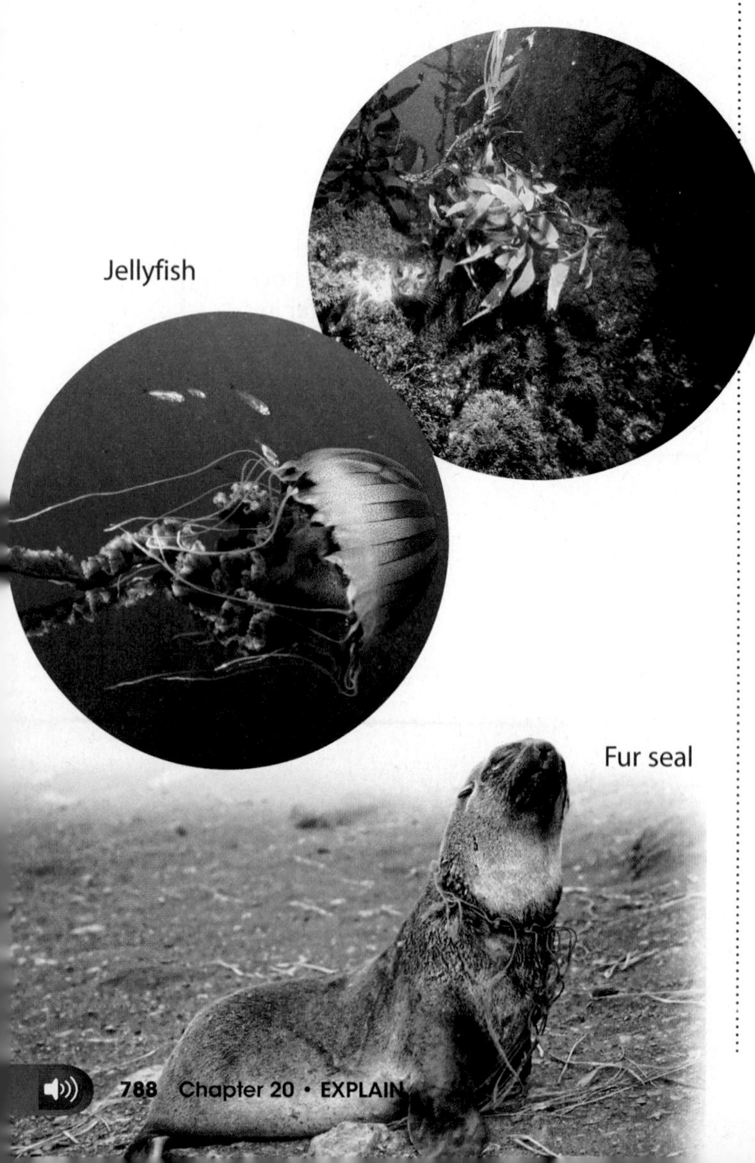

Most of Earth's surface is covered by ocean water with high salinity. The oceans contain different types of ecosystems. If you took a boat trip several kilometers out to sea, you would be in the open ocean—one type of ocean ecosystem. The open ocean extends from the steep edges of continental shelves to the deepest parts of the ocean. The amount of light in the water depends on depth.

- Photosynthesis can take place only in the uppermost, or sunlit, zone. Very little sunlight reaches the twilight zone. None reaches the deepest water, known as the dark zone.
- Decaying matter and nutrients float down from the sunlit zone, through the twilight and dark zones, to the seafloor.

Biodiversity

- Microscopic algae and other producers in the sunlit zone form the base of most ocean food chains. Other organisms living in the sunlit zone are jellyfish, tuna, mackerel, and dolphins.
- Many species of fish stay in the twilight zone during the day and swim to the sunlit zone at night to feed.
- Sea cucumbers, brittle stars, and other bottom-dwelling organisms feed on decaying matter that drifts down from above.
- Many organisms in the dark zone live near cracks in the seafloor where lava erupts and new seafloor forms.

Active Reading **7. Recall** Which organisms are at the base of most ocean food chains?

Human Impact

- Overfishing threatens many ocean fish.
- Trash discarded from ocean vessels or washed into oceans from land is a source of pollution. Animals such as seals become tangled in plastic or mistake it for food.

Ocean: Coastal Oceans

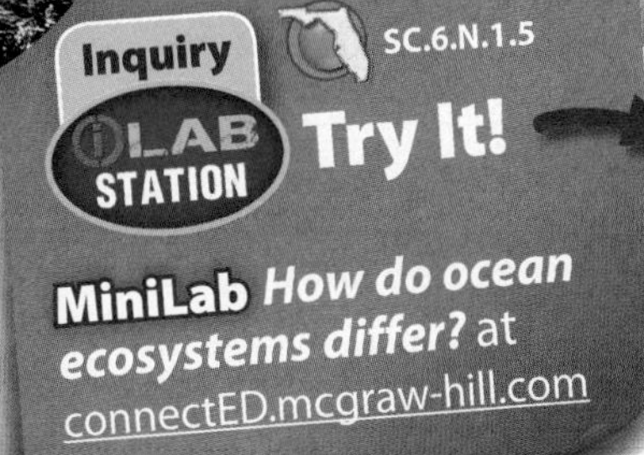
Sea stars

Coastal oceans include several types of ecosystems, including continental shelves and intertidal zones. *The* **intertidal zone** *is the ocean shore between the lowest low tide and the highest high tide.*

- Sunlight reaches the bottom of shallow coastal ecosystems.
- Nutrients washed in from rivers and streams contribute to high biodiversity.

Biodiversity

- The coastal ocean is home to mussels, fish, crabs, sea stars, dolphins, and whales.
- Intertidal species have adaptations for surviving exposure to air during low tides and to heavy waves during high tides.

Human Impact

- Oil spills, such as the spill off the coast of Florida, and other pollution harm coastal organisms.

Inquiry SC.6.N.1.5

LAB STATION **Try It!** → **Apply It!**

MiniLab ***How do ocean ecosystems differ?*** at connectED.mcgraw-hill.com

After you complete the lab, answer the question below.

1. Which ocean ecosystem would you expect to have more biodiversity? Explain your answer.

Ocean: Coral Reefs

Another ocean ecosystem with high biodiversity is the coral reef. *A* **coral reef** *is an underwater structure made from outside skeletons of tiny, soft-bodied animals called coral.*

- Most coral reefs form in shallow tropical oceans.
- Coral reefs protect coastlines from storm damage and erosion.

Biodiversity

- Coral reefs provide food and shelter for many animals, including parrotfish, groupers, angelfish, eels, shrimp, crabs, scallops, clams, worms, and snails.

Human Impact

- Pollution, overfishing, and harvesting of coral threaten coral reefs.

Lesson Review 2

Visual Summary

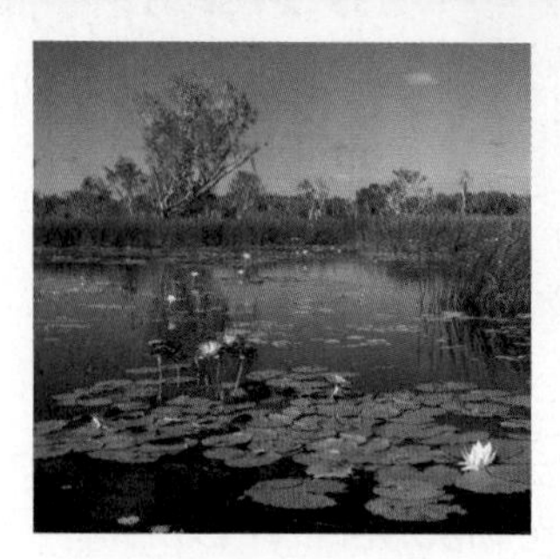

Freshwater ecosystems include ponds and lakes.

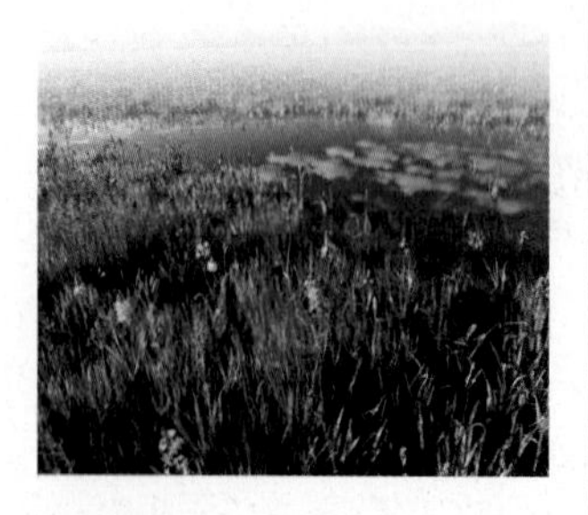

Wetlands can be saltwater ecosystems or freshwater ecosystems.

Coral reefs and coastal ecosystems have high levels of biodiversity.

Use Vocabulary

1. **Define** the term *salinity.*

2. **Distinguish** between a wetland and an estuary.

3. An ocean ecosystem formed from the skeletons of animals is a(n) ______________.

Understand Key Concepts

4. Which ecosystem contains both salt water and freshwater?
   - (A) estuary
   - (B) lake
   - (C) pond
   - (D) stream

5. **Describe** what might happen to a coastal area if its estuary were filled in to build houses. **SC.7.L.17.3**

Interpret Graphics

6. **Describe** Examine the drawing to the right and describe the characteristics of each zone. **LA.6.2.2.3**

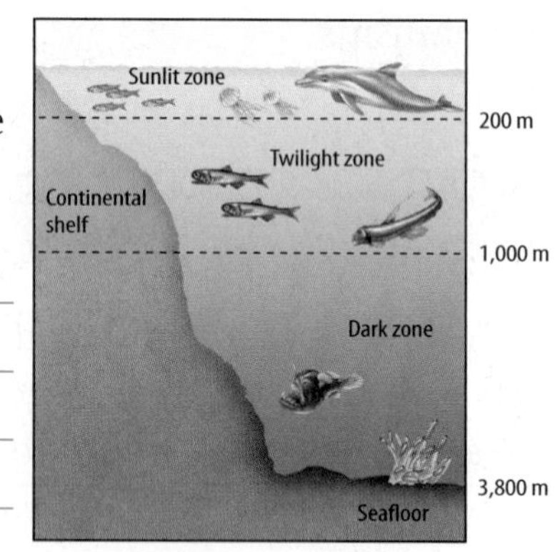

Critical Thinking

7. **Recommend** actions people might take to prevent pollutants from entering coastal ecosystems. **SC.7.L.17.3**

Math Skills MA.6.A.3.6

8. The salinity of the Baltic Sea is about 10 PPT. What weight of salt is present in 2,000 g of seawater?

SCIENCE & SOCIETY

Saving an Underwater Wilderness

A researcher takes a water sample from a marine reserve. ▼

How do scientists help protect coral reefs?

Pollution and human activities, such as mining and tourism, have damaged many ecosystems, including coral reefs. Scientists and conservation groups are working together to help protect and restore coral reefs and areas that surround them. One way is to create marine reserves where no fishing or collection of organisms is allowed.

A team of scientists, including marine ecologists Dr. Dan Brumbaugh and Kate Holmes from the American Museum of Natural History, are investigating how well reserves are working. These scientists compare how many fish of one species live both inside and outside reserves. Their results indicate that more species of fish and greater numbers of each species live inside reserves than outside—one sign that reefs in the area are improving.

Reef ecosystems do not have to be part of a reserve in order to improve, however. Scientists can work with local governments to find ways to limit damage to reef ecosystems. One way is to prevent overfishing by limiting the number of fish caught. Other ways include eliminating the use of destructive fishing practices that can harm reefs and reducing runoff from farms and factories.

By creating marine reserves, regulating fishing practices, and reducing runoff, humans can help reefs that were once in danger become healthy again.

Kate Holmes examines a coral reef. ►

AMERICAN MUSEUM OF NATURAL HISTORY

It's Your Turn

COMPOSE Write a letter to a town near a marine reserve describing why it is important to maintain a protected area.

SC.6.N.2.3

Lesson 3

How Ecosystems CHANGE

ESSENTIAL QUESTIONS

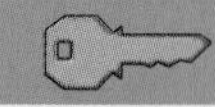 How do land ecosystems change over time?

 How do aquatic ecosystems change over time?

Vocabulary

ecological succession p. 793
climax community p. 793
pioneer species p. 794
eutrophication p. 796

Florida NGSSS

LA.6.2.2.3 The student will organize information to show understanding (e.g., representing main ideas within text through charting, mapping, paraphrasing, summarizing, or comparing/contrasting);

SC.7.L.17.3 Describe and investigate various limiting factors in the local ecosystem and their impact on native populations, including food, shelter, water, space, disease, parasitism, predation, and nesting sites.

SC.6.N.2.1 Distinguish science from other activities involving thought.

SC.7.N.1.1 Define a problem from the seventh grade curriculum, use appropriate reference materials to support scientific understanding, plan and carry out scientific investigation of various types, such as systematic observations or experiments, identify variables, collect and organize data, interpret data in charts, tables, and graphics, analyze information, make predictions, and defend conclusions.

SC.7.N.1.4 Identify test variables (independent variables) and outcome variables (dependent variables) in an experiment.

SC.8.N.1.6 Understand that scientific investigations involve the collection of relevant empirical evidence, the use of logical reasoning, and the application of imagination in devising hypotheses, predictions, explanations and models to make sense of the collected evidence.

SC.8.N.3.1 Select models useful in relating the results of their own investigations.

SC.6.N.2.1

Inquiry Launch Lab

15 minutes

How do communities change?

An ecosystem can change over time. Change usually happens so gradually that you might not notice differences from day to day.

Procedure

1. Your teacher has given you **two pictures of ecosystem communities.** One is labeled *A,* and the other is labeled *B.*
2. Imagine that community A changed and became like community B. On a blank piece of **paper,** draw what you think community A might look like midway in its change to becoming like community B.

Think About This

1. What changes did you imagine? How long do you think it would take for community A to become like community B?

2. Key Concept Summarize the changes you think would happen as the community changed from A to B.

Inquiry **How did this happen?**

1. This object was once part of a mining system used to move copper and iron ore. Today, so many forest plants have grown around it that it is barely recognizable. How do you think this happened? What do you think this object will look like after 500 more years?

How Land Ecosystems Change

Have you ever seen weeds growing up through cracks in a concrete sidewalk? If they were not removed, the weeds would keep growing. The crack would widen, making room for more weeds. Over time, the sidewalk would break apart. Shrubs and vines would move in. Their leaves and branches would grow large enough to cover the concrete. Eventually, trees could start growing there.

This process is an example of **ecological succession**—*the process of one ecological community gradually changing into another.* Ecological succession occurs in a series of steps. These steps can usually be predicted. For example, small plants usually grow first. Larger plants, such as trees, usually grow last.

The final stage of ecological succession in a land ecosystem is a **climax community**—*a stable community that no longer goes through major ecological changes.* Climax communities differ depending on the type of biome they are in. In a tropical forest biome, a climax community would be a mature tropical forest. In a grassland biome, a climax **community** would be a mature grassland. Climax communities are usually stable over hundreds of years. As plants in a climax community die, new plants of the same species grow and take their places. The community will continue to contain the same kinds of plants as long as the climate remains the same.

Active Reading **2. Define** Underline the definition of a climax community.

Active Reading

FOLDABLES® LA.6.2.2.3

Fold a sheet of paper into fourths. Use two sections on one side of the paper to describe and illustrate what land might look like before secondary succession and the other side to describe and illustrate the land after secondary succession is complete.

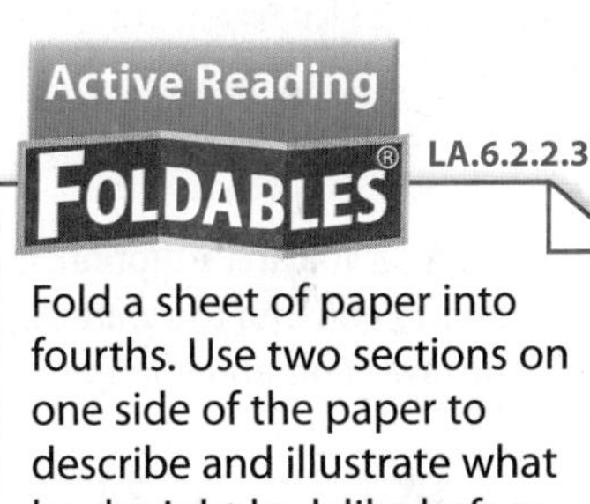

REVIEW VOCABULARY

community
all the organisms that live in one area at the same time

Science Use v. Common Use

pioneer

Science Use the first species that colonize new or undisturbed land

Common Use the first human settlers in an area

Primary Succession

What do you think happens to a lava-filled landscape when a volcanic eruption is over? As shown in **Figure 2,** volcanic lava eventually becomes new soil that supports plant growth. Ecological succession in new areas of land with little or no soil, such as on a lava flow, a sand dune, or exposed rock, is primary succession. *The first species that colonize new or undisturbed land are* **pioneer species.** The lichens and mosses in **Figure 2** are pioneer species.

Figure 2 Following a volcanic eruption, a landscape undergoes primary succession.

During a volcanic eruption, molten lava flows over the ground and into the water. After the eruption is over, the lava cools and hardens into bare rock.

Lichen spores carried on the wind settle on the rock. Lichens release acid that helps break down the rock and create soil. Lichens add nutrients to the soil as they die and decay.

Airborne spores from mosses and ferns settle onto the thin soil and add to the soil when they die. The soil gradually becomes thick enough to hold water. Insects and other small organisms move into the area.

After many years the soil is deep and has enough nutrients for grasses, wildflowers, shrubs, and trees. The new ecosystem provides habitats for many animals. Eventually, a climax community develops.

Secondary Succession

In areas where existing ecosystems have been disturbed or destroyed, secondary succession can occur. One example is forestland in New England that early colonists cleared hundreds of years ago. Some of the cleared land was not planted with crops. This land gradually grew back to a climax forest community of beech and maple trees, as illustrated in **Figure 3.**

Active Reading **3. Identify** (Circle) the description of where secondary succession occurs.

Figure 3 When disturbed land grows back, secondary succession occurs.

Aquatic Succession

Figure 4 The water in a pond is slowly replaced by soil. Eventually, land plants take over and the pond disappears.

How Freshwater Ecosystems Change

Like land ecosystems, freshwater ecosystems change over time in a natural, predictable process. This process is called aquatic succession.

Aquatic Succession

Aquatic succession is illustrated in **Figure 4.** Sediments carried by rainwater and streams accumulate on the bottoms of ponds, lakes, and wetlands. The decomposed remains of dead organisms add to the buildup of soil. As time passes, more and more soil accumulates. Eventually, so much soil has collected that the water disappears and the area becomes land.

Eutrophication

Word Origin
eutrophication
from Greek *eutrophos,* means "nourishing"

As decaying organisms fall to the bottom of a pond, a lake, or a wetland, they add nutrients to the water. **Eutrophication** (yoo troh fuh KAY shun) *is the process of a body of water becoming nutrient-rich.*

Eutrophication is a natural part of aquatic succession. However, humans also contribute to eutrophication. The fertilizers that farmers use on crops and the waste from farm animals can be high in nutrients. So can other forms of pollution. When fertilizers and pollution run off into a pond or lake, nutrient concentrations increase. High nutrient levels support large populations of algae and other microscopic organisms. These organisms use most of the dissolved oxygen in the water, so less oxygen is available for fish and other pond or lake organisms. As a result, many of these organisms die. Their bodies decay and add to the buildup of soil, speeding up succession.

4. **NGSSS Check** **Describe** What limiting factors might change during eutrophication? SC.7.L.17.3

Lesson Review 3

Visual Summary

Ecosystems change in predictable ways through ecological succession.

The final stage of ecological succession in a land ecosystem is a climax community.

The final stage of aquatic succession is a land ecosystem.

Use Vocabulary

1. **Define** *pioneer species* in your own words.

____

____

2. The process of one ecological community changing into another is ____.

3. **Compare and contrast** succession and eutrophication in freshwater ecosystems.

____

____

____

Understand Key Concepts

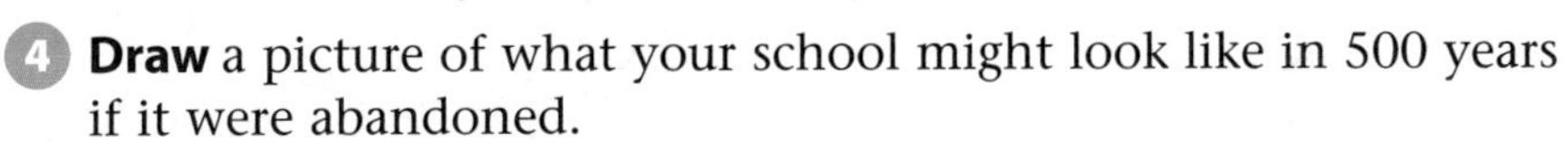

4. **Draw** a picture of what your school might look like in 500 years if it were abandoned.

5. Which process occurs after a forest fire?
   - (A) eutrophication
   - (B) photosynthesis
   - (C) primary succession
   - (D) secondary succession

6. **Summarize Information** Fill in the graphic organizer below with the types of succession an ecosystem can go through. LA.6.2.2.3

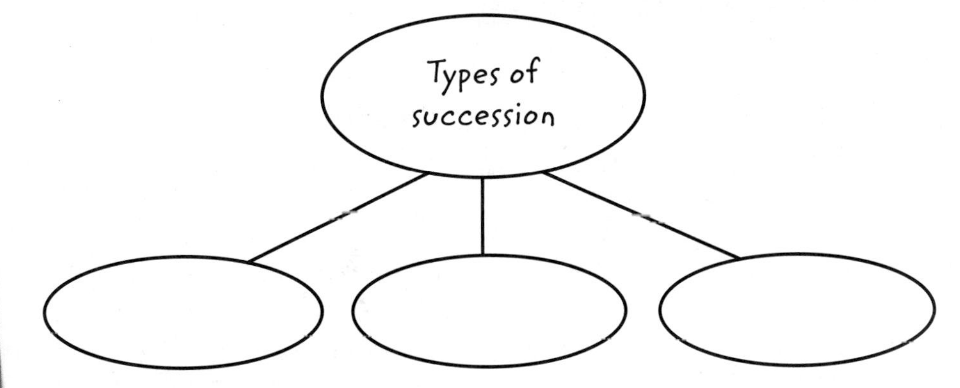

Critical Thinking

7. **Recommend** actions people can take to help prevent the loss of wetland and estuary habitats. SC.7.L.17.3

____

____

____

Chapter 20 Study Guide

Think About It! Each of Earth's land biomes and aquatic ecosystems is characterized by distinct environments and organisms. Both human activities and natural events can have major impacts on the environment.

Key Concepts Summary

LESSON 1 Land Biomes

- Each land **biome** has a distinct climate and contains animals and plants that are well adapted to the environment. Biomes include **deserts, grasslands,** tropical rain forests, **temperate** rain forests, deciduous forests, **taigas,** and **tundras.**
- Humans affect land biomes through agriculture, construction, and other activities.

Vocabulary

biome p. 773
desert p. 774
grassland p. 775
temperate p. 777
taiga p. 779
tundra p. 779

LESSON 2 Aquatic Ecosystems

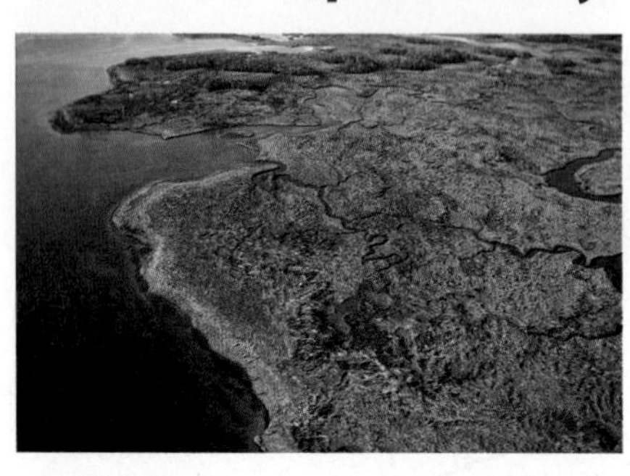

- Earth's aquatic ecosystems include freshwater and saltwater ecosystems. **Wetlands** can contain either salt water or freshwater. The **salinity** of **estuaries** varies.
- Human activities such as construction and fishing can affect aquatic ecosystems.

salinity p. 783
wetland p. 786
estuary p. 787
intertidal zone p. 789
coral reef p. 789

LESSON 3 How Ecosystems Change

- Land and aquatic ecosystems change over time in predictable processes of **ecological succession.**
- Land ecosystems eventually form **climax communities.**
- Freshwater ecosystems undergo **eutrophication** and eventually become land ecosystems.

ecological succession p. 793
climax community p. 793
pioneer species p. 794
eutrophication p. 796

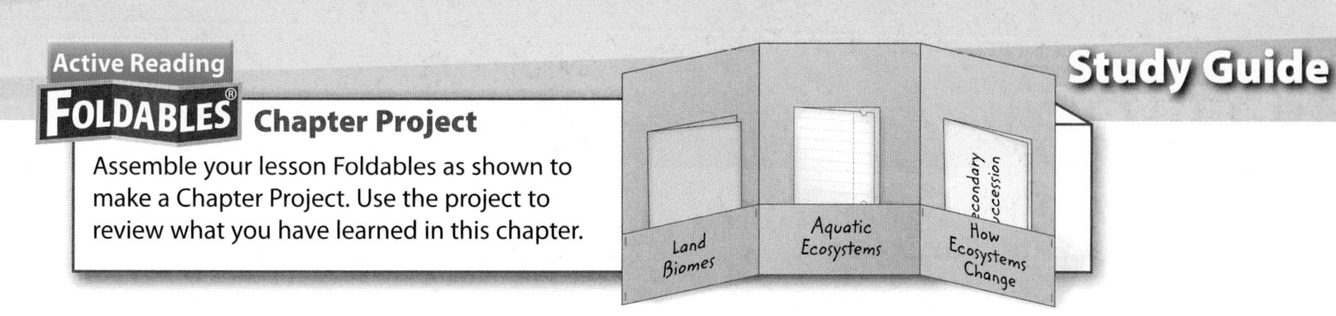

Chapter Project

Assemble your lesson Foldables as shown to make a Chapter Project. Use the project to review what you have learned in this chapter.

Use Vocabulary

Choose the vocabulary word that fits each description.

1. group of ecosystems with similar climate

2. area between the tropics and the polar circles

3. land biome with a layer of permafrost

4. the amount of salt dissolved in water

5. area where a river empties into an ocean

6. coastal zone between the highest high tide and the lowest low tide

7. process of one ecological community gradually changing into another

8. a stable community that no longer goes through major changes

9. the first species to grow on new or disturbed land

Link Vocabulary and Key Concepts

Use vocabulary terms from the previous page and other terms from this chapter to complete the concept map.

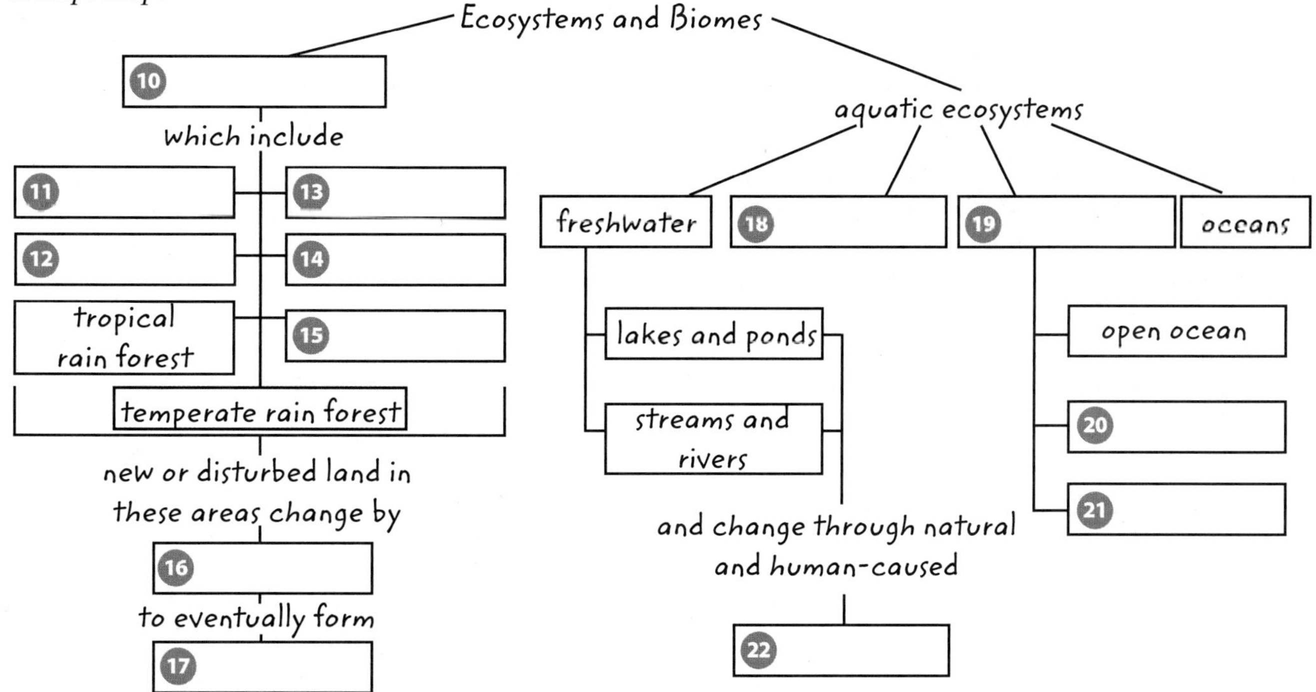

Chapter 20 Review

Fill in the correct answer choice.

Understand Key Concepts

1 Where would you find plants with stems that can store large amounts of water? SC.7.L.17.3

Ⓐ desert
Ⓑ grassland
Ⓒ taiga
Ⓓ tundra

2 What does the pink area on the map below represent? LA.6.2.2.3

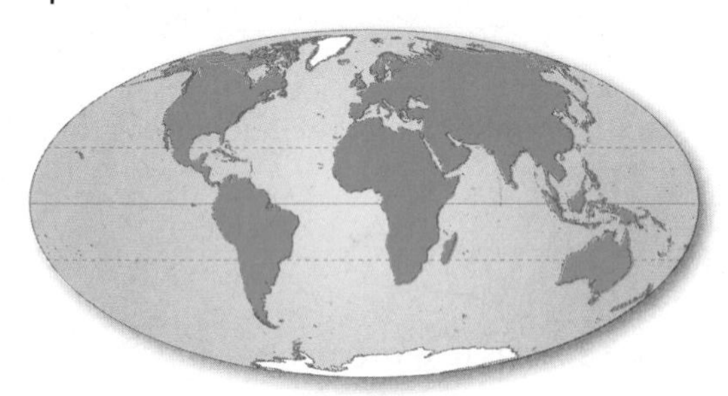

Ⓐ taiga
Ⓑ tundra
Ⓒ temperate deciduous forest
Ⓓ temperate rain forest

3 Which biomes have rich, fertile soil? LA.6.2.2.3

Ⓐ grassland and taiga
Ⓑ grassland and tundra
Ⓒ grassland and tropical rain forest
Ⓓ grassland and temperate deciduous forest

4 Which organism below would be the first to settle an area that has been buried in lava? LA.6.2.2.3

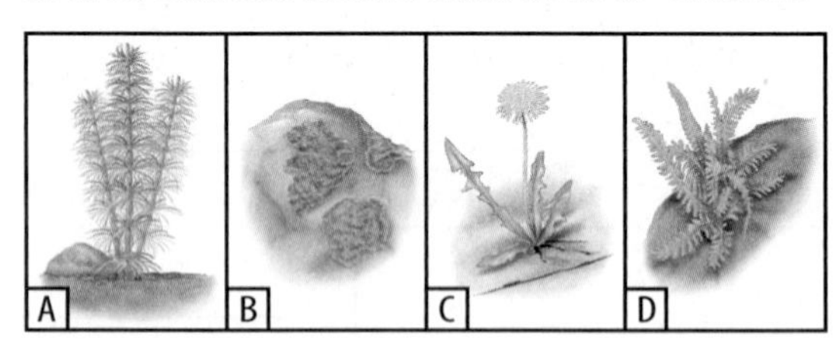

Ⓐ A
Ⓑ B
Ⓒ C
Ⓓ D

5 What is eutrophication? LA.6.2.2.3

Ⓐ decreasing nutrients
Ⓑ decreasing salinity
Ⓒ increasing nutrients
Ⓓ increasing salinity

Critical Thinking

6 **Compare** mammals that live in tundra biomes with those that live in desert biomes. What adaptations does each group have that help them survive? LA.6.2.2.3

7 **Analyze** You are invited to go on a trip to South America. Before you leave, you read a travel guide that says the country you will be visiting has hot summers, cold winters, and many wheat farms. What biome will you be visiting? Explain your reasoning. LA.6.2.2.3

8 **Contrast** How are ecosystems in the deep water of lakes and oceans different? LA.6.2.2.3

9 **Analyze** Which type of ocean ecosystem is likely to have the highest levels of dissolved oxygen? Why? SC.7.L.17.3

10 **Hypothesize** Why are the first plants that appear in primary succession small? SC.7.L.17.3

11 **Interpret Graphics** The following climate data were recorded for a forest ecosystem. To which biome does this ecosystem likely belong? LA.6.2.2.3

Climate Data	June	July	August
Average temperature (°C)	16.0	16.5	17.0
Average rainfall (cm)	3.0	2.0	2.0

Writing in Science

12 **Write** a paragraph on a separate sheet of paper explaining the succession process that might occur in a small pond in a cow pasture. Include a main idea, supporting details, and a concluding sentence. LA.6.2.2.3

Big Idea Review

13 Earth contains a wide variety of organisms that live in different conditions. How do Earth's biomes and ecosystems differ? SC.7.L.17.3

14 Biosphere 2 was built in Arizona as an artificial Earth. Imagine that you have been asked to build a biome of your choice for Biosphere 3. What biotic and abiotic features should you consider? SC.7.L.17.3

Math Skills MA.6.A.3.6

Use Proportions

15 At its highest salinity, the water in Utah's Great Salt Lake contained about 14.5 g of salt in 50 g of lake water. What was the salinity of the lake?

16 The seawater in Puget Sound off the coast of Oregon has a salinity of about 24 PPT. What weight of water is there in 1,000 g of seawater?

Florida NGSSS Benchmark Practice

Fill in the correct answer choice.

Multiple Choice

1 Which aquatic ecosystem contains a mixture of freshwater and salt water? **LA.6.2.2.3**

Ⓐ coral reef

Ⓑ estuary

Ⓒ pond

Ⓓ river

Use the diagram below to answer question 2.

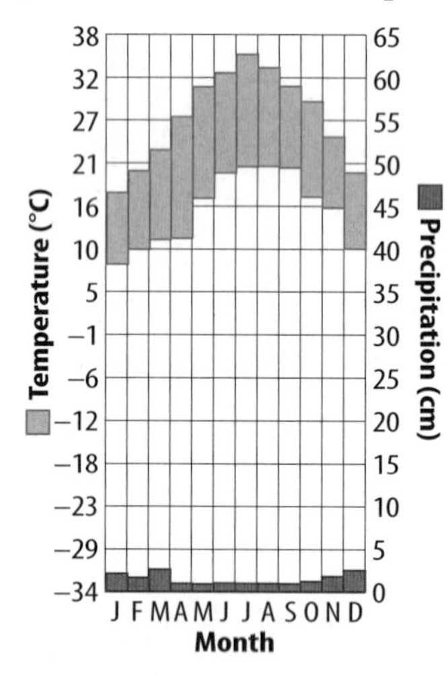

2 The diagram above most likely illustrates the climate of which biome? **LA.6.2.2.3**

Ⓕ desert

Ⓖ grassland

Ⓗ tropical rain forest

Ⓘ tundra

3 Which occurs during the first stage of ecological succession? **LA.6.2.2.3**

Ⓐ eutrophication

Ⓑ settlement

Ⓒ development of climax community

Ⓓ growth of pioneer species

4 In which biome is rain the most significant limiting factor? **SC.7.L.17.3**

Ⓕ grassland

Ⓖ taiga

Ⓗ temperate deciduous forest

Ⓘ tropical rain forest

Use the diagram below to answer question 5.

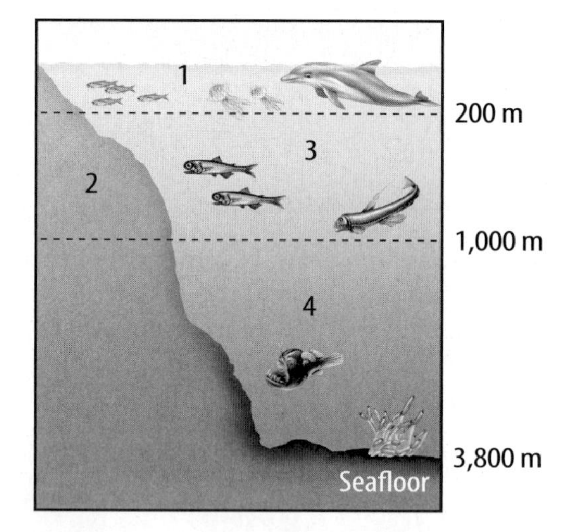

5 In the diagram above, where might you find microscopic photosynthetic organisms? **LA.6.2.2.3**

Ⓐ 1

Ⓑ 2

Ⓒ 3

Ⓓ 4

6 During aquatic succession, freshwater ponds **LA.6.2.2.3**

Ⓕ become salt water.

Ⓖ fill with soil.

Ⓗ gain organisms.

Ⓘ increase in depth.

Use the diagram below to answer question 7.

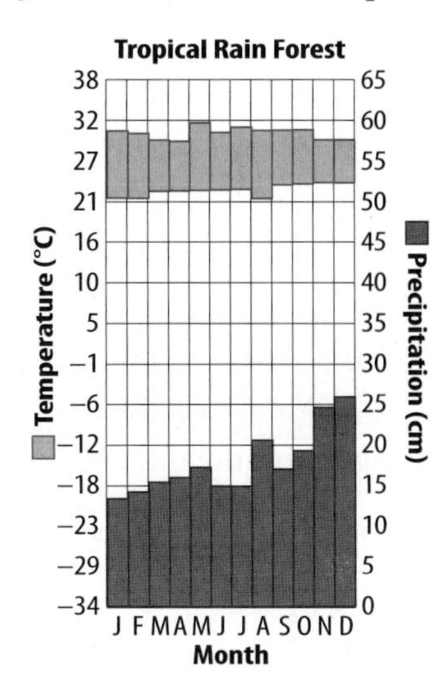

7 Based on the diagram above, which is true of the tropical rain forest biome? **LA.6.2.2.3**

Ⓐ Precipitation increases as temperatures rise.

Ⓑ Rainfall is greatest mid-year.

Ⓒ Temperatures rise at year-end.

Ⓓ Temperatures vary less than rainfall amounts.

8 In which ecosystem is it likely for runoff pollution to be a limiting factor? **SC.7.L.17.3**

Ⓕ coral reefs

Ⓖ intertidal zones

Ⓗ lakes

Ⓘ wetlands

9 Which is NOT a freshwater ecosystem? **SC.7.L.17.3**

Ⓐ oceans

Ⓑ ponds

Ⓒ rivers

Ⓓ streams

10 Where would you find species adapted to withstand strong wave action? **SC.7.L.17.3**

Ⓕ estuaries

Ⓖ wetlands

Ⓗ intertidal zone

Ⓘ twilight zone

11 Which ecosystem has flowing water? **SC.7.L.17.3**

Ⓐ estuary

Ⓑ lake

Ⓒ stream

Ⓓ wetland

12 Which ecosystems help protect coastal areas from flood damage? **SC.7.L.17.3**

Ⓕ estuaries

Ⓖ ponds

Ⓗ rivers

Ⓘ streams

13 What is a forest called that has had the same species of trees for 200 years? **SC.7.L.17.3**

Ⓐ climax community

Ⓑ pioneer species

Ⓒ primary succession

Ⓓ secondary succession

NEED EXTRA HELP?

If You Missed Question...	1	2	3	4	5	6	7	8	9	10	11	12	13
Go to Lesson...	2	1	2	3	2	2	2	2	1	1	2	2	3

Name ______________________________ Date ______________

Benchmark Mini-Assessment Chapter 20 • Lesson 1

Multiple Choice *Bubble the correct answer.*

1. Which climate graph shows the average temperatures and precipitation in a desert biome? **MA.6.A.3.6**

Ⓐ
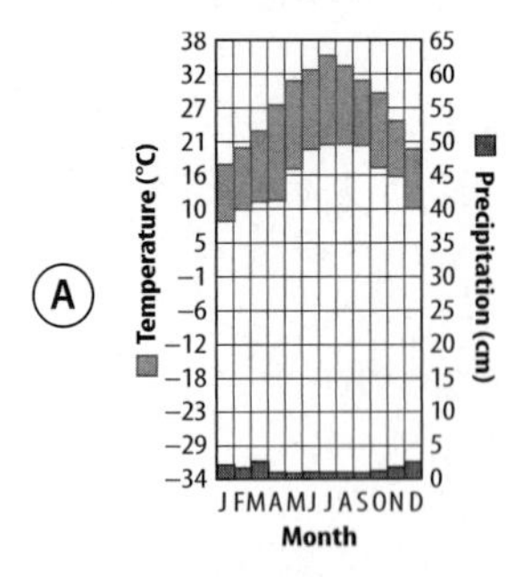

Ⓑ
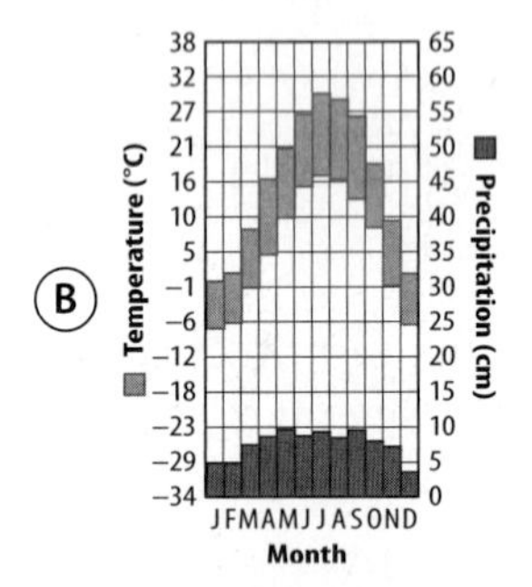

Ⓒ
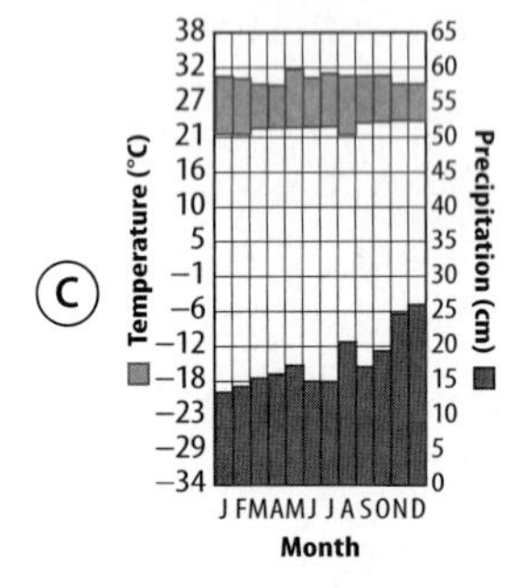

Ⓓ
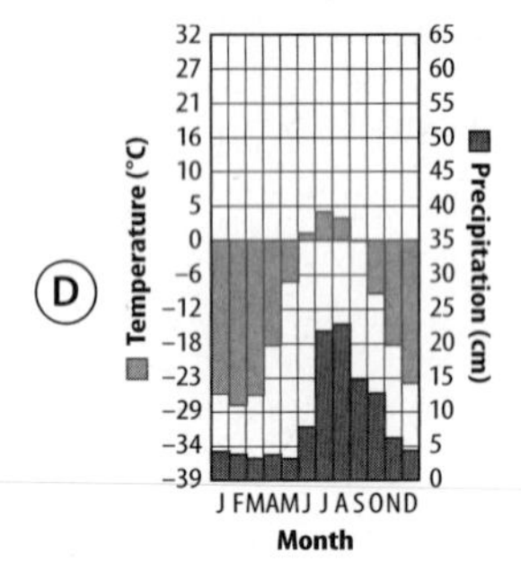

2. Which description correctly defines the term *biome*? **SC.7.L.17.3**

Ⓕ an area that is defined by the living things within it

Ⓖ a large geographic area that has similar biotic and abiotic factors

Ⓗ a small geographic area with unrelated biotic and abiotic factors

Ⓘ a small area of an environment, including its biotic and abiotic parts

3. In which biome would you expect to find organisms that hibernate through cold winters? **SC.7.L.17.3**

Ⓐ desert

Ⓑ taiga

Ⓒ temperate rain forest

Ⓓ tropical rain forest

4. Which biome has the greatest biodiversity? **SC.7.L.17.3**

Ⓕ desert

Ⓖ grassland

Ⓗ temperate deciduous forest

Ⓘ tropical rain forest

Multiple Choice *Bubble the correct answer.*

Use the image below to answer questions 1 and 2.

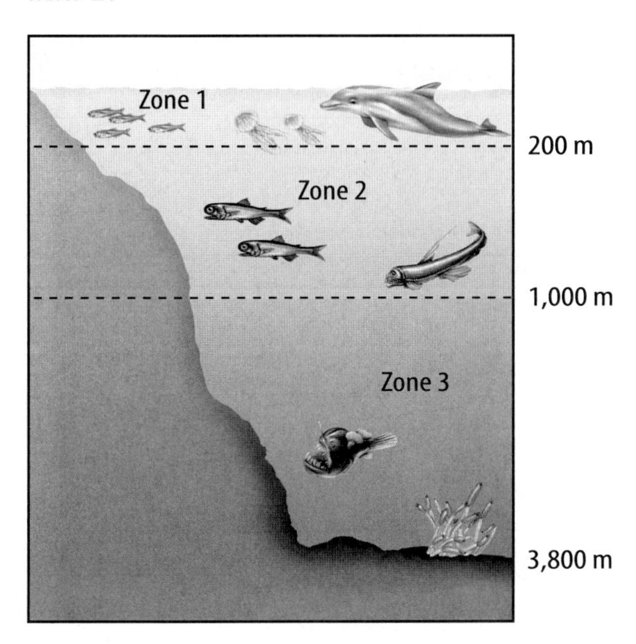

1. In which zone or zones would you expect to find photosynthetic organisms? **SC.8.L.18.1**

Ⓐ 1

Ⓑ 2

Ⓒ 2 and 3

Ⓓ 1 and 3

2. In which zone or zones would you expect to find a large number of organisms that feed on particles of decaying matter? **SC.7.L.17.1**

Ⓕ 1

Ⓖ 3

Ⓗ 1 and 2

Ⓘ 1 and 3

3. Which aquatic ecosystem is least likely to be affected by the abiotic factor of salinity? **SC.7.L.17.3**

Ⓐ estuary

Ⓑ lake

Ⓒ coral reef

Ⓓ open ocean

4. Which freshwater ecosystem plays a key role in flood control and filtering pollutants from water? **SC.7.L.17.3**

Ⓕ lake

Ⓖ pond

Ⓗ river

Ⓘ wetland

Name ______________________ Date ______________

Benchmark Mini-Assessment Chapter 20 • Lesson 3

Multiple Choice *Bubble the correct answer.*

1. In which area would you expect primary succession to take place? **SC.7.L.17.3**

Ⓐ 

Ⓑ 

Ⓒ 

Ⓓ 

2. Which of these are pioneer species? **SC.7.L.17.3**

Ⓕ flowers

Ⓖ grasses

Ⓗ lichens

Ⓘ trees

3. What is the last step of secondary succession? **SC.7.L.17.3**

Ⓐ A climax community of slow-growing trees forms.

Ⓑ A farmer abandons a field previously used for planting.

Ⓒ Fast-growing trees such as pines grow in a field.

Ⓓ Wildflowers and grasses sprout and grow in a field.

4. As a pond undergoes aquatic succession, it becomes smaller as it is filled with **SC.7.L.17.3**

Ⓕ algae and microscopic organisms.

Ⓖ dissolved oxygen and soil.

Ⓗ sediments and decayed organisms.

Ⓘ water-loving plants such as willows.

Name ______ Date ______

Notes

Glossary/Glosario

Multilingual eGlossary

A science multilingual glossary is available on the science website. The glossary includes the following languages.

Arabic	Hmong	Tagalog
Bengali	Korean	Urdu
Chinese	Portuguese	Vietnamese
English	Russian	
Haitian Creole	Spanish	

Cómo usar el glosario en español:

1. Busca el término en inglés que desees encontrar.
2. El término en español, junto con la definición, se encuentran en la columna de la derecha.

Pronunciation Key

Use the following key to help you sound out words in the glossary.

a . . . . back (BAK)
ay . . . . day (DAY)
ah . . . . father (FAH thur)
ow . . . . flower (FLOW ur)
ar . . . . car (CAR)
e . . . . less (LES)
ee . . . . leaf (LEEF)
ih . . . . trip (TRIHP)
i (i + com + e) idea (i DEE uh)
oh . . . . go (GOH)
aw . . . . soft (SAWFT)
or . . . . orbit (OR buht)
oy . . . . coin (COYN)
oo . . . . foot (FOOT)
ew . . . . food (FEWD)
yoo . . . . pure (PYOOR)
yew . . . . few (FYEW)
uh . . . . comma (CAH muh)
u (+ con) . . . . rub (RUB)
sh . . . . shelf (SHELF)
ch . . . . nature (NAY chur)
g . . . . gift (GIHFT)
j . . . . gem (JEM)
ing . . . . sing (SING)
zh . . . . vision (VIH zhun)
k . . . . cake (KAYK)
s . . . . seed, cent (SEED, SENT)
z . . . . zone, raise (ZOHN, RAYZ)

English — A — Español

abiotic factor (ay bi AH tihk • FAK tuhr): a nonliving thing in an ecosystem.

factor abiótico: componente no vivo de un ecosistema.

accuracy: a description of how close a measurement is to an accepted or true value.

exactitud: descripción de qué tan cerca está una medida a un valor aceptable.

active immunity: the process by which the human body produces antibodies in response to an antigen.

inmunidad activa: proceso por el cual el cuerpo humano produce anticuerpos en respuesta a un antígeno.

active transport: the movement of substances through a cell membrane using the cell's energy.

transporte activo: movimiento de sustancias a través de la membrana celular usando la energía de la célula.

adaptation (a dap TAY shun): an inherited trait that increases an organism's chance of surviving and reproducing in a particular environment.

adaptación: rasgo heredado que aumenta la oportunidad de un organismo de sobrevivir y reproducirse en su medioambiente.

alga (plural, algae): a plantlike protist that produces food through photosynthesis using light energy and carbon dioxide.

allele (uh LEEL): a different form of a gene.

allergy: an overly sensitive immune response to common antigens.

alternation of generations: process that occurs when the life cycle of organism alternates between diploid and haploid generations.

alveolus (al VEE uh lus; plural, alveoli): microscopic sacs or pouches at the end of the bronchioles where gas exchange occurs.

amoeba (uh MEE buh): one common sarcodine with an unusual adaptation for movement and getting nutrients.

analogous (uh NAH luh gus) structures: body parts that perform a similar function but differ in structure.

antibiotic (an ti bi AH tihk): a medicine that stops the growth and reproduction of bacteria.

antibody: a protein that can attach to a pathogen and makes it useless.

antigen: a substance that causes an immune response.

appendage: a structure, such as a leg or an arm, that extends from the central part of the body.

artery: a vessel that carries blood away from the heart.

arthritis (ar THRI tus): a disease in which joints become irritated or inflamed, such as when cartilage in joints is damaged or wears away.

ascus (AS kuhs): the reproductive structure where spores develop on sac fungi.

asexual reproduction: a type of reproduction in which one parent organism produces offspring without meiosis and fertilization.

alga (plural, algas): protista parecida a una planta que produce el alimento por medio de la fotosíntesis, usando la energía lumínica y el dióxido de carbono.

alelo: forma diferente de un gen.

alergia: respuesta inmune demasiado sensible a los antígenos comunes.

alternancia de generaciones: proceso que ocurre cuando el ciclo de vida de un organismo se alterna entre generaciones diploides y haploides.

alveolo (plural, alveolos): bolsas o sacos microscópicos en los extremos de los bronquiolos donde ocurre el intercambio de gas.

ameba: sarcodina común con una adaptación inusual para moverse y obtener nutrientes.

estructuras análogas: partes del cuerpo que ejecutan una función similar pero tienen una estructura distinta.

antibiótico: medicina que detiene el crecimiento y reproducción de las bacterias.

anticuerpo: proteína que se adhiere a un patógeno y lo hace inútil.

antígeno: sustancia que causa una respuesta inmune.

apéndice: estructura, como una pierna o un brazo, que se prolonga de la parte central del cuerpo.

arteria: vaso que lleva sangre fuera del corazón.

artritis: enfermedad en la que las articulaciones se irritan o inflaman, como cuando el cartílago en las articulaciones se lastima o desgasta.

ascus: estructura reproductiva donde se desarrollan las esporas en un hongo con saco.

reproducción asexual: tipo de reproducción en la cual un organismo parental produce crías sin mitosis ni fertilización.

asymmetry: a body plan in which an organism cannot be divided into any two parts that are nearly mirror images.

asimetría: plano corporal en el cual un organismo no se puede dividir en dos partes que sean casi imágenes al espejo una de otra.

atherosclerosis (a thuh roh skluh ROH sus): the buildup of fatty material within the walls of arteries.

arteriosclerosis: acumulación de material graso en el interior de las paredes de las arterias.

atmosphere (AT muh sfir): a thin layer of gases surrounding Earth.

atmósfera: capa delgada de gases que rodean la Tierra.

atria (AY tree uh; singular, atrium): the upper two chambers of the heart.

atrios (singular, atrio): las dos cámaras superiores del corazón.

B

B cell: a type of white blood cell that forms and matures in the bone marrow and secretes antibodies into the blood.

célula B: tipo de glóbulo blanco que se forma y madura en la médula ósea y secreta anticuerpos a la sangre.

bacterium: a microscopic prokaryote.

bacteria: procariota microscópica.

basidium (buh SIH dee uhm): reproductive structure that produces sexual spores inside the basidiocarp.

basidio: estructura reproductiva que produce esporas sexuales en el interior de un basidiocarpo.

bilateral symmetry: a body plan in which an organism can be divided into two parts that are nearly mirror images of each other.

simetría bilateral: plano corporal en el cual un organismo se puede dividir en dos partes que sean casi imágenes al espejo una de otra.

binomial nomenclature: a naming system that gives each organism a two-word scientific name.

nomenclatura binomial: sistema de nombrar que le da a cada organismo un nombre científico de dos palabras.

biological evolution: the change over time in populations of related organisms.

evolución biológica: cambio a través del tiempo en las poblaciones de organismos relacionados.

biome: a geographic area on Earth that contains ecosystems with similar biotic and abiotic features.

bioma: área geográfica en la Tierra que contiene ecosistemas con características bióticas y abióticas similares.

bioremediation (bi oh rih mee dee AY shun): the use of organisms, such as bacteria, to clean up environmental pollution.

biorremediación: uso de microorganismos, como bacterias, para limpiar la contaminación del medioambiente.

biosphere (BI uh sfihr): the parts of Earth and the surrounding atmosphere where there is life.

biosfera: partes de la Tierra y de la atmósfera que la rodea donde hay vida.

biotic factor (bi AH tihk • FAK tuhr): a living or once-living thing in an ecosystem.

factor biótico: vida cosa o anteriormente vida cosa en un ecosistema.

biotic potential: the potential growth of a population if it could grow in perfect conditions with no limiting factors.

potencial biótico: crecimiento potencial de una población si esta puede crecer en condiciones perfectas sin factores limitantes.

birthrate: the number of offspring produced by a population over a given time period.

tasa de nacimientos: número de crías que tiene una población durante un período de tiempo dado.

bladder: a muscular sac that holds urine until the urine is excreted.

vejiga: bolsa muscular que contiene la orina hasta que se excreta.

brain stem: the area of the brain that controls involuntary functions.

tallo cerebral: área del cerebro que controla funciones involuntarias.

breathing: the movement of air into and out of the lungs.

respiración: movimiento de aire hacia adentro y hacia afuera de los pulmones.

bronchus (BRAHN kus; plural, bronchi): one of two narrow tubes that carry air into the lungs from the trachea.

bronquio (plural, bronquios): uno de los dos tubos delgados que llevan aire de la tráquea a los pulmones.

bruise: an injury where blood vessels in the skin are broken, but the skin is not cut or opened.

moretón: herida en la cual los vasos sanguíneos de la piel se rompen, pero la piel no se corta ni abre.

budding: the process during which a new organism grows by mitosis and cell division on the body of its parent.

germinación: proceso durante el cual un organismo nuevo crece por medio de mitosis y división celular en el cuerpo de su progenitor.

C

Calorie: the amount of energy it takes to raise the temperature of 1 kg of water by 1°C.

caloría: cantidad de energía necesaria para aumentar la temperatura de 1 kg de agua a 1°C.

cambium: a layer of tissue that produces new vascular tissue and grows between xylem and phloem.

cámbium: capa de tejido que produce tejido vascular nuevo y crece en medio del xilema y el floema.

camouflage (KAM uh flahj): an adaptation that enables a species to blend in with its environment.

camuflaje: adaptación que permite a las especies mezclarse con su medioambiente.

cancer: a disease in which cells reproduce uncontrollably without the usual signals to stop.

cáncer: enfermedad en la cual las células se reproducen sin control sin las señales usuales para detenerse.

capillary: a tiny blood vessel that delivers supplies to an individual cell and takes away waste materials.

capilar: vaso sanguíneo diminuto que entrega suministros a una célula individual y extrae los materiales de desecho.

carbohydrate (kar boh HI drayt): a macromolecule made up of one or more sugar molecules, which are composed of carbon, hydrogen, and oxygen; usually the body's major source of energy.

carbohidrato: macromolécula constituida de una o más moléculas de azúcar, las cuales están compuestas de carbono, hidrógeno y oxígeno; usualmente es la mayor fuente de energía del cuerpo.

cardiac (KAR dee ak) muscle: muscle found only in the heart.

músculo cardíaco: músculo que sólo se encuentra en el corazón.

carrying capacity: the largest number of individuals of one species that an ecosystem can support over time.

capacidad de carga: número mayor de individuos de una especie que un medioambiente puede mantener.

cartilage (KAR tuh lihj): a strong, flexible tissue that covers the ends of bones.

cartílago: tejido fuerte y flexible que reviste los extremos de los huesos.

cast: a fossil copy of an organism made when a mold of the organism is filled with sediment or mineral deposits.

cell: the smallest unit of life.

cell cycle: a cycle of growth, development, and division that most cells in an organism go through.

cell differentiation (dihf uh ren shee AY shun): the process by which cells become different types of cells.

cell membrane: a flexible covering that protects the inside of a cell from the environment outside the cell.

cell theory: the theory that states that all living things are made of one or more cells, the cell is the smallest unit of life, and all new cells come from preexisting cells.

cellular respiration: a series of chemical reactions that convert the energy in food molecules into a usable form of energy called ATP.

cellulose: an organic compound made of chains of glucose molecules.

cell wall: a stiff structure outside the cell membrane that protects a cell from attack by viruses and other harmful organisms.

central nervous system (CNS): system made up of the brain and the spinal cord.

centromere: a structure that holds sister chromatids together.

cerebellum (ser uh BEH lum): the part of the brain that coordinates voluntary muscle movement and regulates balance and posture.

cerebrum (suh REE brum): the part of the brain that controls memory, language, and thought.

cervix (SUR vihks): a small structure between the uterus and the vagina.

chemical digestion: a process in which chemical reactions break down pieces of food into small molecules.

contramolde: copia fósil de un organismo compuesto en un molde de el organismo está lleno de sedimentos o los depósitos de minerales.

célula: unidad más pequeña de vida.

ciclo celular: ciclo de crecimiento, desarrollo y división por el que pasan la mayoría de células de un organismo.

diferenciación celular: proceso por el cual las células se convierten en diferentes tipos de células.

membrana celular: cubierta flexible que protege el interior de una célula del ambiente externo de la célula.

teoría celular: teoría que establece que todos los seres vivos están constituidos de una o más células (la célula es la unidad más pequeña de vida) y que las células nuevas provienen de células preexistentes.

respiración celular: serie de reacciones químicas que convierten la energía de las moléculas de alimento en una forma de energía utilizable llamada ATP.

celulosa: compuesto orgánico constituido de cadenas de moléculas de glucosa.

pared celular: estructura rígida en el exterior de la membrana celular que protege la célula del ataque de virus y otros organismos dañinos.

sistema nervioso central (SNC): sistema constituido por el cerebro y la médula espinal.

centrómero: estructura que mantiene unidas las cromátidas hermanas.

cerebelo: parte del cerebro que coordina el movimiento muscular voluntario y regula el equilibrio y la postura.

cerebrum: parte del cerebro que controla la memoria, el lenguaje y el pensamiento.

cérvix: estructura pequeña entre el útero y la vagina.

digestión química: proceso por el cual las reacciones químicas descomponen partes del alimento en moléculas pequeñas.

chemosynthesis (kee moh sihn THUH sus): the process during which producers use chemical energy in matter rather than light energy to make food.

chemotherapy: a type of cancer treatment in which chemicals are used to kill the cells that are reproducing uncontrollably.

chloroplast (KLOR uh plast): a membrane-bound organelle that uses light energy and makes food—a sugar called glucose—from water and carbon dioxide in a process known as photosynthesis.

chordate (KOR dat): an animal that has a notochord, a nerve cord, a tail, and structures called pharyngeal pouches at some point in its life.

chyme (KIME): a thin, watery liquid made of broken down food molecules and gastric juice.

cilia (SIH lee uh): short, hairlike structures that grow on the surface of some protists.

cladogram: a branched diagram that shows the relationships among organisms, including common ancestors.

climate: the long-term average weather conditions that occur in a particular region.

climax community: a stable community that no longer goes through major ecological changes.

cloning: a type of asexual reproduction performed in a laboratory that produces identical individuals from a cell or a cluster of cells taken from a multicellular organism.

codominance: an inheritance pattern in which both alleles can be observed in a phenotype.

commensalism: a symbiotic relationship that benefits one species but does not harm or benefit the other.

community: all the populations living in an ecosystem at the same time.

quimiosíntesis: proceso durante el cual los productores usan la energía química en la materia en vez de la energía lumínica, para elaborar alimento.

quimioterapia: tipo de tratamiento para el cáncer, en el cual se usan químicos para matar las células que se están reproduciendo sin control.

cloroplasto: organelo limitado por una membrana que usa la energía lumínica para producir alimento –un azúcar llamado glucosa– del agua y del dióxido de carbono en un proceso llamado fotosíntesis.

cordado: animal que en algún momento de su vida tiene notocordio, cordón nervioso, cola y estructuras llamadas bolsas faríngeas.

quimo: líquido diluido y acuoso constituido de moléculas de alimento descompuestas y jugos gástricos.

cilios: estructuras cortas parecidas a un cabello que crecen en la superficie de algunos protistas.

cladograma: diagrama de brazos que muestra las relaciones entre los organismos, incluidos los ancestros comunes.

clima: promedio a largo plazo de las condiciones del tiempo atmosférico de una región en particular.

comunidad clímax: comunidad estable que ya no sufrirá mayores cambios ecológicos.

clonación: tipo de reproducción asexual realizada en un laboratorio que produce individuos idénticos a partir de una célula o grupo de células tomadas de un organismo pluricelular.

condominante: patrón heredado en el cual los dos alelos se observan en un fenotipo.

comensalismo: relación simbiótica que beneficia a una especie pero no causa daño ni beneficia a la otra.

comunidad: todas las poblaciones que viven en un ecosistema, al mismo tiempo.

comparative anatomy: the study of similarities and differences among structures of living species.

competition: the demand for resources, such as food, water, and shelter, in short supply in a community.

compound microscope: a light microscope that uses more than one lens to magnify an object.

condensation (kahn den SAY shun): the process by which a gas changes to a liquid.

conjugation (kahn juh GAY shun): a process during which two bacteria of the same species attach to each other and combine their genetic material.

constants: the factors in an experiment that remain the same.

consumer: an organism that cannot make its own food and gets energy by eating other organisms.

coral reef: an underwater structure made from outside skeletons of tiny, soft-bodied animals called coral.

coronary circulation: the network of arteries and veins that supplies blood to all the cells of the heart.

critical thinking: comparing what you already know with information you are given in order to decide whether you agree with it.

cuticle: a waxy, protective layer on the leaves, stems, and flowers of plants.

cytokinesis (si toh kuh NEE sus): a process during which the cytoplasm and its contents divide.

cytoplasm: the liquid part of a cell inside the cell membrane; contains salts and other molecules.

cytoskeleton: a network of threadlike proteins joined together that gives a cell its shape and helps it move.

anatomía comparativa: estudio de las similitudes y diferencias entre las estructuras de las especies vivas.

competición: demanda de recursos, tales como alimento, agua y refugio, cuyo suministro es escaso en una comunidad.

microscopio compuesto: microscopio de luz que usa más de un lente para aumentar la imagen de un objeto.

condensación: proceso mediante el cual un gas cambia a líquido.

conjugación: proceso durante el cual dos bacterias de la misma especie se adhieren una a la otra y combinan sus material genético.

constantes: factores en un experimento que permanecen iguales.

consumidor: organismo que no puede hacer sus propios alimentos y obtiene energía comiendo otros organismos.

arrecife de coral: estructura bajo el agua formada por exoesqueletos de animales diminutos y de cuerpo blando.

circulación coronaria: red de arterias y venas que suministran sangre a todas las células del corazón.

pensamiento crítico: comparación que se hace cuando se sabe algo acerca de información nueva, y se decide si se está o no de acuerdo con ella.

cutícula: capa cerosa de protección que tienen las hojas, los tallos y las flores de las plantas.

citocinesis: proceso durante el cual el citoplasma y sus contenidos se dividen.

citoplasma: fluido en el interior de una célula que contiene sales y otras moléculas.

citoesqueleto: red de proteínas en forma de filamentos unidos que le da forma a la célula y le ayuda a moverse.

D

daughter cells: the two new cells that result from mitosis and cytokinesis.

death rate: the number of individuals in a population that die over a given time period.

decomposition: the breaking down of dead organisms and organic waste.

dependent variable: the factor a scientist observes or measures during an experiment.

dermis: a thick layer of skin that gives skin strength, nourishment, and flexibility.

description: a spoken or written summary of an observation.

desert: a biome that receives very little rain.

diaphragm (DI uh fram): a large muscle below the lungs that contracts and relaxes as air moves into and out of your lungs.

diatom (DI uh tahm): a type of microscopic plantlike protist with a hard outer wall.

dichotomous key: a series of descriptions arranged in pairs that lead the user to the identification of an unknown organism.

diffusion: the movement of substances from an area of higher concentration to an area of lower concentration.

digestion: the mechanical and chemical breakdown of food into small particles and molecules that your body can absorb and use.

diploid: a cell that has pairs of chromosomes.

DNA: the abbreviation for deoxyribonucleic (dee AHK sih ri boh noo klee ihk) acid, an organism's genetic material.

dominant (DAH muh nunt) trait: a genetic factor that blocks another genetic factor.

células hija: las dos células nuevas que resultan de la mitosis y la citocinesis.

tasa de mortalidad: número de individuos que mueren en una población en un período de tiempo dado.

descomposición: degradación de organismos muertos y desecho orgánico.

variable dependiente: factor que el científico observa o mide durante un experimento.

dermis: capa gruesa de piel que le proporciona a la piel fuerza, nutrimento y flexibilidad.

descripción: resumen oral o escrito de una observación de.

desierto: bioma que recibe muy poca lluvia.

diafragma: músculo grande debajo de los pulmones que se contrae y relaja a medida que el aire entra y sale a los pulmones.

diatomea: tipo de protista microscópico parecido a una planta que tiene una pared externa dura.

clave dicotómica: serie de descripciones organizadas en pares que dan al usuario la identificación de un organismo desconocido.

difusión: movimiento de sustancias de un área de mayor concentración a un área de menor concentración.

digestión: descomposición mecánica y química del alimento en partículas y moléculas pequeñas que el cuerpo absorbe y usa.

diploide: célula que tiene pares de cromosomas.

ADN: abreviatura para ácido desoxirribonucleico, material genético de un organismo.

rasgo dominante: factor genético que bloquea otro factor genético.

E

eardrum: a thin membrane between the outer ear and the inner ear.

ecological succession: the process of one ecological community gradually changing into another.

tímpano: membrana delgada en medio del oído externo y del oído interno.

sucesión ecológica: proceso en el que una comunidad ecológica cambia gradualmente en otra.

ecosystem: all the living things and nonliving things in a given area.

egg: the female reproductive, or sex, cell; forms in an ovary.

electron microscope: a microscope that uses a magnetic field to focus a beam of electrons through an object or onto an object's surface.

embryo: an immature diploid plant that develops from the zygote; a developing human from the time it attaches to the uterus until the eighth week of pregnancy.

embryology (em bree AH luh jee): the science of the development of embryos from fertilization to birth.

endangered species: a species whose population is at risk of extinction.

endocrine (EN duh krun) system: system consisting of groups of organs and tissues that release chemical messages into the bloodstream.

endocytosis (en duh si TOH sus): the process during which a cell takes in a substance by surrounding it with the cell membrane.

endospore (EN doh spor): a thick internal wall that a bacterium builds around its chromosome and part of its cytoplasm.

energy pyramid: a model that shows the amount of energy available in each link of a food chain.

enzyme (EN zime): a protein that helps break down larger molecules into smaller molecules and speeds up, or catalyzes, the rate of chemical reactions.

epidermis (eh puh DUR mus): the outermost layer of skin and the only layer in direct contact with the outside environment.

esophagus (ih SAH fuh gus): a muscular tube that connects the mouth to the stomach.

estuary (ES chuh wer ee): a coastal area where freshwater from rivers and streams mixes with salt water from seas or oceans.

ecosistema: todos los seres vivos y los componentes no vivos de un área dada.

óvulo: célula reproductiva femenina o sexual; forma en un ovario.

microscopio electrónico: microscopio que usa un campo magnético para enfocar un haz de electrones a través de un objeto o sobre la superficie de un objeto.

embrión: planta diploide inmadura que se desarrolla de un zigoto; ser humano en desarrollo desde el momento en que se adhiere al útero hasta la octava semana de embarazo.

embriología: ciencia que trata el desarrollo de embriones desde la fertilización hasta el nacimiento.

especie en peligro: especie cuya población se encuentra en riesgo de extinción.

sistema endocrino: sistema que consta de grupos de órganos y tejidos que liberan mensajes químicos en la corriente sanguínea.

endocitosis: proceso durante el cual una célula absorbe una sustancia rodeándola con la membrana celular.

endospora: pared interna gruesa que una bacteria produce alrededor del cromosoma y parte del citoplasma.

pirámide energética: modelo que explica la cantidad de energía disponible en cada vínculo de una cadena alimentaria.

enzima: proteína que descompone moléculas más grandes en moléculas más pequeñas y acelera, o cataliza, la velocidad de las reacciones químicas.

epidermis: capa más externa de la piel y la única capa que está en contacto directo con el medioambiente externo.

esófago: tubo muscular que conecta la boca al estómago.

estuario: zona costera donde el agua dulce de los ríos y arroyos se mezcla con el agua salada de los mares y los océanos.

eutrophication (yoo troh fuh KAY shun): the process of a body of water becoming nutrient-rich.

eutrofización: proceso por el cual un cuerpo de agua se vuelve rico en nutrientes.

evaporation (ih va puh RAY shun): the process of a liquid changing to a gas at the surface of the liquid.

evaporación: proceso de cambio de un líquido a un gas en la superficie del líquido.

excretory system: the system that collects and eliminates wastes from the body and regulates the level of fluid in the body.

sistema excretor: sistema que recolecta y elimina los desperdicios del cuerpo y regula el nivel de fluidos en el cuerpo.

exocytosis (ek soh si TOH sus): the process during which a cell's vesicles release their contents outside the cell.

exocitosis: proceso durante el cual las vesículas de una célula liberan sus contenidos fuera de la célula.

exoskeleton: a thick, hard outer covering; protects and supports an animal's body.

exoesqueleto: cubierta externa, gruesa y dura; protege y soporta el cuerpo de un animal.

explanation: an interpretation of observations.

explicación: interpretación que se hace de las observaciones.

extinction (ihk STINGK shun): event that occurs when the last individual organism of a species dies.

extinción: evento que ocurre cuando el último organismo individual de una especie muere.

extinct species: a species that has died out and no individuals are left.

especie extinta: especie que ha dejado de existir y no quedan individuos de ella.

facilitated diffusion: the process by which molecules pass through a cell membrane using special proteins called transport proteins.

difusión facilitada: proceso por el cual las moléculas pasan a través de la membrana celular usando proteínas especiales, llamadas proteínas de transporte.

fat: also called a lipid, a substance in the body that provides energy and helps your body absorb vitamins.

grasa: también llamada lípido, sustancia en el cuerpo que proporciona energía y ayuda al cuerpo a absorber vitaminas.

fermentation: a reaction that eukaryotic and prokaryotic cells can use to obtain energy from food when oxygen levels are low.

fermentación: reacción que las células eucarióticas y procarióticas usan para obtener energía del alimento cuando los niveles de oxígeno son bajos.

fertilization (fur tuh luh ZAY shun): a reproductive process in which a sperm joins with an egg.

fertilización: proceso reproductivo en el cual un espermatozoide se une con un óvulo.

fetus: term used to describe a developing human from the ninth week of the pregnancy until birth.

feto: término usado para describir al ser humano en desarrollo desde la novena semana de embarazo hasta el nacimiento.

fission: cell division that forms two genetically identical cells.

fisión: división celular que forma dos células genéticamente idénticas.

flagellum (fluh JEH lum): a long whiplike structure on many bacteria.

flagelo: estructura larga similar a un látigo que tienen muchas bacterias.

food chain: a model that shows how energy flows in an ecosystem through feeding relationships.

cadena alimentaria: modelo que explica cómo la energía fluye en un ecosistema a través de relaciones alimentarias.

food web: a model of energy transfer that can show how the food chains in a community are interconnected.

red alimentaria: modelo de transferencia de energía que explica cómo las cadenas alimentarias están interconectadas en una comunidad.

fossil record: record of all the fossils ever discovered on Earth.

registro fósil: registro de todos los fósiles descubiertos en la Tierra.

frond: a leaf of a fern.

fronda: hoja de un helecho.

fruit: plant structure that contains one or more seeds; develops from the ovary and sometimes other parts of the flower.

fruta: estructura de la planta que contiene una o más semillas; se desarrolla del ovario y algunas veces de otras partes de la flor.

G

gene (JEEN): a section of DNA on a chromosome that has genetic information for one trait.

gen: parte del ADN en un cromosoma que contiene información genética para un rasgo.

genetics: the study of how traits are passed from parents to offspring.

genética: estudio de cómo los rasgos pasan de los padres a los hijos.

genotype (JEE nuh tipe): the alleles of all the genes on an organism's chromosomes; controls an organism's phenotype.

genotipo: de los alelos de todos los genes en los cromosomas de un organismo, los controles de fenotipo de un organismo.

genus (JEE nus): a group of similar species.

género: grupo de especies similares.

geologic time scale: a chart that divides Earth's history into different time units based on changes in the rocks and fossils.

escala de tiempo geológico: tabla que divide la historia de la Tierra en diferentes unidades de tiempo, basado en los cambios en las rocas y fósiles.

glycolysis: a process by which glucose, a sugar, is broken down into smaller molecules.

glucólisis: proceso por el cual la glucosa, un azúcar, se divide en moléculas más pequeñas.

grassland: a biome where grasses are the dominant plants.

pradera: bioma donde los pastos son las plantas dominantes.

H

habitat: the place within an ecosystem where an organism lives; provides the biotic and abiotic factors an organism needs to survive and reproduce.

hábitat: lugar en un ecosistema donde vive un organismo; proporciona los factores bióticos y abióticos de un organismo necesita para sobrevivir y reproducirse.

haploid: a cell that has only one chromosome from each pair.

haploide: célula que tiene solamente un cromosoma de cada par.

heredity (huh REH duh tee): the passing of traits from parents to offspring.

herencia: paso de rasgos de los padres a los hijos.

heterozygous (he tuh roh ZI gus): a genotype in which the two alleles of a gene are different.

heterocigoto: genotipo en el cual los dos alelos de un gen son diferentes.

homeostasis (hoh mee oh STAY sus): an organism's ability to maintain steady internal conditions when outside conditions change.

homologous (huh MAH luh gus) chromosomes: pairs of chromosomes that have genes for the same traits arranged in the same order.

homologous (huh MAH luh gus) structures: body parts of organisms that are similar in structure and position but different in function.

homozygous (hoh muh ZI gus): a genotype in which the two alleles of a gene are the same.

hormone: a chemical signal that is produced by an endocrine gland in one part of an organism and carried in the bloodstream to another part of the organism.

hyphae (HI fee): long, threadlike structures that make up the body of fungi and also form an underground structure that absorbs minerals and water.

hypothesis: a possible explanation for an observation that can be tested by scientific investigations.

homeostasis: capacidad de un organismo de mantener las condiciones internas estables cuando las condiciones externas cambian.

cromosomas homólogos: pares de cromosomas que tienen genes de iguales rasgos dispuestos en el mismo orden.

estructuras homólogas: partes del cuerpo de los organismos que son similares en estructura y posición pero diferentes en función.

homocigoto: genotipo en el cual los dos alelos de un gen son iguales.

hormona: señal química producido por una glándula endocrina en una parte de un organismo y llevado en la corriente sanguínea a otra parte del organismo.

hifas: estructuras largas en forma de filamentos que constituyen el cuerpo de los hongos y que también forman una estructura subterránea que absorbe minerales y agua.

hipótesis: explicación posible de una observación que se puede probar por medio de investigaciones científicas.

I

immunity: the resistance to specific pathogens.

incomplete dominance: an inheritance pattern in which an offspring's phenotype is a combination of the parents' phenotypes.

independent variable: the factor that is changed by the investigator to observe how it affects a dependent variable.

infectious disease: a disease caused by a pathogen that can be transmitted from one person to another.

inference: a logical explanation of an observation that is drawn from prior knowledge or experience.

inflammation: a process that causes a bodily area to become red and swollen.

integumentary (ihn teg gyuh MEN tuh ree) system: the body system that includes all the external coverings of the body, including the skin, nails, and hair.

inmunidad: resistencia a patógeno específicos.

dominancia incompleta: patrón heredado en el cual el fenotipo de un hijo es una combinación de los fenotipos de los padres.

variable independiente: factor que el investigador cambia para observar cómo afecta la variable dependiente.

enfermedad infecciosa: enfermedad causada por un patógeno que se puede transmitir de una persona a otra.

inferencia: explicación lógica de una observación que se extrae de un conocimiento previo o experiencia.

inflamación: proceso que causa que un área del cuerpo se vuelva roja e hinchada.

sistema tegumentario: sistema corporal que comprende todas las coberturas externas del cuerpo, incluidos la piel, las uñas y el cabello.

International System of Units (SI): the internationally accepted system of measurement.

interphase: the period during the cell cycle of a cell's growth and development.

intertidal zone: the ocean shore between the lowest low tide and the highest high tide.

invertebrate (ihn VUR tuh brayt): an animal that does not have a backbone.

involuntary muscle: muscle you cannot consciously control.

Sistema Internacional de Unidades (SI): sistema de medidas aceptado internacionalmente.

interfase: período durante el ciclo celular del crecimiento y desarrollo de una célula.

zona intermareal: playa en medio de la marea baja más baja y la marea alta más alta.

invertebrado: animal que no tiene columna vertebral.

músculos involuntarios: músculo que se controla conscientemente.

J

joint: where two or more bones meet.

articulación: donde dos o más huesos se unen.

kidney: a bean-shaped organ that filters, or removes, wastes from blood.

riñón: órgano con forma de frijol que filtra, o extrae, los desechos de la sangre.

larynx (LER ingks): a triangle shaped area into which air passes from the pharynx; also called the voice box.

lichen (LI kun): a structure formed when fungi and certain other photosynthetic organisms grow together.

ligament (LIH guh munt): the tissue that connects bones to other bones.

light microscope: a microscope that uses light and lenses to enlarge an image of an object.

limiting factor: a factor that can limit the growth of a population.

lipid: a large macromolecule that does not dissolve in water.

lungs: the main organs of the respiratory system.

lymph: tissue fluid that has diffused into lymph vessels.

lymphatic system: part of the immune system that helps destroy microorganisms that enter the body.

lymph node: a small spongy structure that filters particles from lymph.

laringe: área en forma de triángulo dentro de la cual pasa el aire proveniente de la faringe; también se le llama caja sonora de voz.

líquen: estructura formada cuando crecen juntos los hongos y algunos organismos que realizan la fotosíntesis.

ligamento: tejido que conecta los huesos con otros huesos.

microscopio de luz: microscopio que usa luz y lentes para aumentar la imagen de un objeto.

factor limitante: factor que puede limitar el crecimiento de una población.

lípido: macromolécula extensa que no se disuelve en agua.

pulmones: órganos principales del sistema respiratorio.

linfa: fluido de los tejidos que se esparce en los vasos linfáticos.

sistema linfático: parte del sistema inmune que destruye los microorganismos que entran al cuerpo.

nódulo linfático: estructura pequeña y esponjosa que filtra partículas de la linfa.

macromolecule: substance that forms from joining many small molecules together.

mechanical digestion: a process in which food is physically broken into smaller pieces.

meiosis: a process in which one diploid cell divides to make four haploid sex cells.

melanin (MEH luh nun): a pigment that protects the body by absorbing some of the Sun's damaging ultraviolet rays.

menstrual (MEN stroo ul) cycle: a process of reproductive-related changes involving the ovaries and uterus.

migration: the instinctive, seasonal movement of a population of organisms from one place to another.

mimicry (MIH mih kree): an adaptation in which one species looks like another species.

mineral: any of several inorganic nutrients which help the body regulate many chemical reactions.

mitosis (mi TOH sus): a process during which the nucleus and its contents divide.

mold: the impression of an organism in a rock.

multicellular: a living thing that is made up of two or more cells.

muscle: strong body tissue that can contract in an orderly way.

mutation (myew TAY shun): a permanent change in the sequence of DNA, or the nucleotides, in a gene or a chromosome.

mutualism: a symbiotic relationship in which both organisms benefit.

mycelium (mi SEE lee um): an underground network of hyphae.

mychorrize (mi kuh RI zuh): a structure formed when the roots of a plant and the hyphae of a fungus weave together.

macromolécula: sustancia que se forma al unir muchas moléculas pequeñas.

digestión mecánica: proceso por el cual el alimento se descompone físicamente en pedazos más pequeños.

meiosis: proceso en el cual una célula diploide se divide para constituir cuatro células sexuales haploides.

melanina: pigmento que protege el cuerpo absorbiendo parte de los rayos ultravioleta dañinos del sol.

ciclo menstrual: proceso de cambios relacionados con la reproducción que involucra los ovarios y el útero.

migración: movimiento instintivo de temporada de una población de organismos de un lugar a otro.

mimetismo: una adaptación en el cual una especie se parece a otra especie.

mineral: cualquiera de los varios nutrientes inorgánicos que ayudan al cuerpo a regular muchas reacciones químicas.

mitosis: proceso durante el cual el núcleo y sus contenidos se divide.

molde: impresión de un organismo en una roca.

pluricelular: ser vivo formado por dos o más células.

músculo: tejido corporal fuerte que se contrae de manera sistemática.

mutación: cambio permanente en la secuencia de ADN, de los nucleótidos, en un gen o en un cromosoma.

mutualismo: relación simbiótica en la cual los dos organismos se benefician.

micelio: red subterránea de hifas.

micorriza: estructura formada cuando las raíces de una planta y las hifas de un hongo se entrelazan.

naturalist: a person who studies plants and animals by observing them.

natural selection: the process by which organisms with variations that help them survive in their environment live longer, compete better, and reproduce more than those that do not have the variations.

negative feedback: a control system in which the effect of a hormone inhibits further release of the hormone; sends a signal to stop a response.

nephron (NEH frahn): a network of capillaries and small tubes, or tubules, where filtration of blood occurs.

nervous system: the part of an organism that gathers, processes, and responds to information.

neuron (NOO rahn): the basic functioning unit of the nervous system; a nerve cell.

niche (NICH): the way a species interacts with abiotic and biotic factors to obtain food, find shelter, and fulfill other needs.

nitrogen fixation (NI truh jun • fihk SAY shun): the process that changes atmospheric nitrogen into nitrogen compounds that are usable by living things.

noninfectious disease: a disease that cannot pass from person to person.

notochord: a flexible, rod-shaped structure that supports the body of a developing chordate.

nucleic acid: a macromolecule that forms when long chains of molecules called nucleotides join together.

nucleotide (NEW klee uh tide): a molecule made of a nitrogen base, a sugar, and a phosphate group.

nucleus: part of a eukaryotic cell that directs cell activity and contains genetic information stored in DNA.

naturalista: persona que estudia las plantas y los animales por medio de la observación.

selección natural: proceso por el cual los organismos con variaciones que las ayudan a sobrevivir en sus medioambientes viven más, compiten mejor y se reproducen más que aquellas que no tienen esas variaciones.

retroalimentación negativa: sistema de control en el cual el efecto de una hormona inhibe más liberación de la hormona; envía una señal para detener una respuesta.

nefrona: red de capilares y tubos pequeños, o túbulos, donde ocurre la filtración de la sangre.

sistema nervioso: parte de un organismo que recoge, procesa y responde a la información.

neurona: unidad básica de funcionamiento del sistema nervioso; célula nerviosa.

nicho: forma de una especie interacciona con los factores abióticos y bióticos para obtener comida, encontrar refugio, y satisfacer otras necesidades.

fijación de nitrógeno: proceso por el cual el nitrógeno atmosférico se transforma en compuestos de nitrógeno que los seres vivos usan.

enfermedad no infecciosa: enfermedad que no se puede pasar de una persona a otra.

notocordio: estructura flexible con forma de varilla que soporta el cuerpo de un cordado en desarrollo.

ácido nucléico: macromolécula que se forma cuando cadenas largas de moléculas llamadas nucleótidos se unen.

nucelótido: molécula constituida de una base de nitrógeno, azúcar y un grupo de fosfato.

núcleo: parte de la célula eucariótica que gobierna la actividad celular y contiene la información genética almacenada en el ADN.

O

observation: the act of using one or more of your senses to gather information and take note of what occurs.

organ: a group of different tissues working together to perform a particular job.

organelle: membrane-surrounded component of a eukaryotic cell with a specialized function.

organism: something that has all the characteristics of life.

organ system: a group of organs that work together and perform a specific task.

osmosis: the diffusion of water molecules only through a membrane.

osteoporosis (ahs tee oh puh ROH sus): a bone disease that causes bones to weaken and become brittle.

ovary (OH vah ree): structure located at the base of the style of a flower that contains one or more ovules; the female reproductive organ that produces egg cells; stores oocytes which mature into ova.

ovulation: a process occurring near the end of phase 2 of the menstrual cycle in which hormones cause an egg to be released from the ovary.

ovule: female reproductive structure of a seed plant where the haploid egg develops.

observación: acción de usar uno o más sentidos para reunir información y tomar notar de lo que ocurre.

órgano: grupo de diferentes tejidos que trabajan juntos para realizar una función específica.

organelo: componente de una célula eucariótica rodeado de una membrana con una función especializada.

organismo: algo que tiene todas las características de la vida.

sistema de órganos: grupo de órganos que trabajan juntos y realizar una función específica.

ósmosis: difusión de las moléculas de agua únicamente a través de una membrana.

osteoporosis: enfermedad de los huesos que los debilita y los vuelve quebradizos.

ovario: estructura situado en la base del estilo de una flor que contiene uno o más óvulos; el órgano reproductivo femenino que produce óvulos; tiendas de ovocitos que maduran en los óvulos.

ovulación: proceso que ocurre cerca de la finalización de la segunda fase del ciclo menstrual en el cual las hormonas causan la liberación de un óvulo del ovario.

óvulo: estructura reproductiva femenina de la semilla de una planta donde el huevo haploide se desarrolla.

P

paramecium (pa ruh MEE see um): a protist with cilia and two types of nuclei.

parasitism: a symbiotic relationship in which one organism benefits and the other is harmed.

passive immunity: the introduction of antibodies that were produced outside the body.

passive transport: the movement of substances through a cell membrane without using the cell's energy.

Paramecio: protista con cilios y dos tipos de núcleos.

parasitismo: relación simbiótica en la cual se perjudica organismo se beneficia y el otro.

inmunidad pasiva: introducción de anticuerpos producidos fuera del cuerpo.

transporte pasivo: movimiento de sustancias a través de una membrana celular sin usar la energía de la célula.

pasteurization (pas chuh ruh ZAY shun): a process of heating food or liquid to a temperature that kills most harmful bacteria.

pathogen (PA thuh jun): an agent that causes disease.

penis: a tubelike structure that delivers sperm to the female reproductive system.

periosteum (per ee AHS tee um): a membrane that surrounds bone.

peripheral nervous system (PNS): system made of sensory and motor neurons that transmit information between the central nervous system (CNS) and the rest of the body.

peristalsis (per uh STAHL sus): waves of muscle contractions that move food through the digestive tract.

pharynx (FER ingks): a tubelike passageway at the top of the throat that receives air, food, and liquids from the mouth or nose.

phenotype (FEE nuh tipe): how a trait appears or is expressed.

phloem (FLOH em): a type of vascular tissue that carries dissolved sugars throughout a plant.

photoperiodism: a plant's response to the number of hours of darkness in its environment.

photosynthesis (foh toh SIHN thuh sus): a series of chemical reactions that convert light energy, water, and carbon dioxide into the food-energy molecule glucose and give off oxygen.

pioneer species: the first species that colonizes new or undisturbed land.

pistil: female reproductive organ of a flower.

placenta (pluh SEN tuh): an organ formed by the outer cells of the zygote and cells from the uterus.

plant hormone: a substance that acts as a chemical messenger within a plant.

plasma: the yellowish, liquid part of blood that transports blood cells.

platelet: a small, irregularly shaped piece of a cell that plugs wounds to stop bleeding.

pasteurización: proceso en el cual se calientan los alimentos o líquidos para matar la mayoría de bacterias dañinas.

patógeno: agente que causa enfermedad.

pene: estructura en forma de tubo que deposita esperma en el sistema reproductor femenino.

periostio: membrana que recubre los huesos.

sistema nervioso periférico (SNP): sistema formado por neuronas sensoriales y motoras que transmiten información entre el sistema nervioso central (SNC) y el resto del cuerpo.

peristalsis: ondas de contracciones musculares que mueven el alimento por el tracto digestivo.

faringe: pasadizo parecido a un tubo en la parte superior de la garganta que recibe el aire, el alimento y los líquidos provenientes de la boca o de la nariz.

fenotipo: forma como aparece o se expresa un rasgo.

floema: tipo de tejido vascular que transporta azúcares disueltos por toda la planta.

fotoperiodismo: respuesta de una planta al número de horas de oscuridad en su medioambiente.

fotosíntesis: serie de reacciones químicas que convierte la energía lumínica, el agua y el dióxido de carbono en glucosa, una molécula de energía alimentaria, y libera oxígeno.

especie pionera: primera especie que coloniza tierra nueva o tierra virgen.

pistilo: órgano reproductor femenino de una flor.

placenta: órgano formado por las células externas del cigoto y células del útero.

fitohormona: sustancia que actúa como mensajero químico dentro de una planta.

plasma: parte líquida y amarillenta de la sangre que transporta las células sanguíneas.

plaqueta: fragmento de una célula, pequeño y de forma irregular, que tapona las heridas para detener el sangrado.

pollen (PAH lun) grain: spore that forms from tissue in a male reproductive structure of a seed plant.

pollination (pah luh NAY shun): the process that occurs when pollen grains land on a female reproductive structure of a plant that is the same species as the pollen grains.

polygenic inheritance: an inheritance pattern in which multiple genes determine the phenotype of a trait.

population: all the organisms of the same species that live in the same area at the same time.

population density: the size of a population compared to the amount of space available.

positive feedback: a control system in which the effect of a hormone causes more of the hormone to be released; sends a signal to increase a response.

precipitation (prih sih puh TAY shun): water, in liquid or solid form, that falls from the atmosphere.

precision: a description of how similar or close measurements are to each other.

prediction: a statement of what will happen next in a sequence of events.

pregnancy: the period of human development from fertilized egg to birth.

producer: an organism that uses an outside energy source, such as the Sun, and produces its own food.

protein: a long chain of amino acid molecules; contains carbon, hydrogen, oxygen, nitrogen, and sometimes sulfur.

protist: a member of a group of eukaryotic organisms, which have a membrane-bound nucleus.

protozoan (proh tuh ZOH un): a protist that resembles a tiny animal.

pseudopod: a temporary "foot" that forms as the organism pushes part of its body outward.

grano de polen: espora que se forma de tejido en una estructura reproductiva masculina de una planta de semilla.

polinización: proceso que ocurre cuando los granos de polen posan sobre una estructura reproductiva femenina de una planta que es de la misma especie que los granos de polen.

herencia poligénica: patrón de herencia en el cual genes múltiples determinan el fenotipo de un rasgo.

población: todos los organismos de la misma especie que viven en la misma área al mismo tiempo.

densidad poblacional: tamaño de una población comparado con la cantidad de espacio disponible.

retroalimentación positiva: sistema de control en el cual el efecto de una hormona causa más liberación de la hormona; envía una señal para aumentar la respuesta.

precipitación: agua, en forma líquida o sólida, que cae de la atmósfera.

precisión: sescripción de qué tan similar o cercana están las mediciones una de otra.

predicción: afirmación de lo que ocurrirá a continuación en una secuencia de eventos.

embarazo: período del desarrollo del ser humano desde que se fertiliza el óvulo hasta el nacimiento.

productor: organismo que usa una fuente de energía externa, como el Sol, y fabricar su propio alimento.

proteína: larga cadena de aminoácidos; contiene carbono, hidrógeno, oxígeno, nitrógeno y, algunas veces, sulfuro.

protista: miembro de un grupo de organismos eucarióticos que tienen un núcleo limitado por una membrana.

protozoario: protista que parece un animal pequeño.

Seudópodo: "pata" temporal que se forma a medida que el organismo empuja parte del cuerpo hacia afuera.

puberty (PYEW bur tee): the process by which the reproductive system matures during adolescence.

pubertad: proceso por el cual el sistema reproductor madura durante la adolescencia.

pulmonary circulation: the network of vessels that carries blood to and from the lungs.

circulación pulmonar: red de vasos que lleva sangre hacia y desde los pulmones.

Punnett square: a model that is used to show the probability of all possible genotypes and phenotypes of offspring.

cuadro de Punnett: modelo que se utiliza para demostrar la probabilidad de que todos los genotipos y fenotipos posibles de cría.

R

radial symmetry: a body plan in which an organism can be divided into two parts that are nearly mirror images of each other anywhere through its central axis.

simetría radial: plano corporal en el cual un organismo se puede dividir en dos partes para que sean casi imágenes al espejo una de la otra, en cualquier parte del eje axial.

receptor: special structures in all parts of the sensory system that detect stimuli.

receptor: estructuras especiales en todas partes del sistema sensorial que detectan los estímulos.

recessive (rih SE sihv) trait: a genetic factor that is blocked by the presence of a dominant factor.

rasgo recesivo: factor genético boqueado por la presencia de un factor dominante.

reflex: an automatic movement in response to a stimulus.

reflejo: movimiento automático en respuesta a un estímulo.

regeneration: a type of asexual reproduction that occurs when an offspring grows from a piece of its parent.

regeneración: tipo de reproducción asexual que ocurre cuando un organismo se origina de una parte de su progenitor.

replication: the process of copying a DNA molecule to make another DNA molecule.

replicación: proceso por el cual se copia una molécula de ADN para hacer otra molécula de ADN.

retina (RET nuh): an area at the back of the eye that includes special light-sensitive cells—rod cells and cone cells.

retina: área en la parte posterior del ojo que incluye especiales sensibles a la luz –bastones y conos.

Rh factor: a protein found on red blood cells; a chemical marker.

factor Rh: proteína que se encuentra en los glóbulos rojos; es un marcador químico.

rhizoid: a structure that anchors a nonvascular plant to a surface.

rizoide: estructura que sujeta una planta no vascular a una superficie.

RNA: ribonucleic acid, a type of nucleic acid that carries the code for making proteins from the nucleus to the cytoplasm.

ARN: ácido ribonucleico, un tipo de ácido nucléico que contiene el código para hacer proteínas del núcleo para el citoplasma.

S

salinity (say LIH nuh tee): a measure of the mass of dissolved salts in a mass of water.

salinidad: medida de la masa de sales disueltas en una masa de agua.

science: the investigation and exploration of natural events and of the new information that results from those investigations.

ciencia: la investigación y exploración de los eventos naturales y de la información nueva que es el resultado de estas investigaciones.

scientific law: a rule that describes a pattern in nature.

scientific theory: an explanation of observations or events that is based on knowledge gained from many observations and investigations.

seed: a plant embryo, its food supply, and a protective covering.

selective breeding: the selection and breeding of organisms for desired traits.

semen (SEE mun): a mixture of sperm and fluids produced by several glands.

sensory system: the part of your nervous system that detects or senses the environment.

sexual reproduction: type of reproduction in which the genetic material from two different cells—a sperm and an egg—combine, producing an offspring.

significant digits: the number of digits in a measurement that are known with a certain degree of reliability.

sister chromatids: two identical chromosomes that make up a duplicated chromosome.

skeletal muscle: a type of muscle that attaches to bones.

skeletal system: body system that contains bones as well as other structures that connect and protect the bones and that support other functions in the body.

smooth muscle: involuntary muscle named for its smooth appearance.

species (SPEE sheez): a group of organisms that have similar traits and are able to produce fertile offspring.

sperm: a male reproductive, or sex, cell, forms in a testis.

spinal cord: a tubelike structure of neurons that sends signals to and from the brain.

spleen: an organ of the lymphatic system that recycles worn-out red blood cells and produces and stores lymphocytes.

spore: a daughter cell produced from a haploid structure.

stamen: the male reproductive organ of a flower.

ley científica: regla que describe un patrón dado en la naturaleza.

teoría científica: explicación de observaciones o eventos con base en conocimiento obtenido de muchas observaciones e investigaciones.

semilla: embrión de una planta, su suministro de alimento y cubierta protectora.

cría selectiva: selección y la cría de organismos para las características deseadas.

semen: mezcla de esperma y fluidos producidos por varias glándulas.

sistema sensorial: parte del sistema nervioso que detecta o siente el medioambiente.

reproducción sexual: tipo de reproducción en la cual el material genético de dos células diferentes de un espermatozoide y un óvulo se combinan, produciendo una cría.

cifras significativas: número de dígitos que se conoce con cierto grado de fiabilidad en una medida.

cromátidas hermanas: dos cromosomas idénticos que constituyen un cromosoma duplicado.

músculo esquelético: tipo de músculo que se adhiere a los huesos.

sistema esquelético: sistema corporal que comprende los huesos al igual que otras estructuras que conectan y protegen los huesos y que apoyan otras funciones en el cuerpo.

músculo liso: músculo involuntario llamado así por su apariencia lisa.

especie: grupo de organismos que tienen rasgos similares y que están en capacidad de producir crías fértiles.

esperma: célula reproductora masculina o sexual; forma en un testículo.

médula espinal: estructura de neuronas en forma de tubo que envía señales hacia y del cerebro.

bazo: órgano del sistema linfático que recicla los glóbulos rojos muertos y produce y almacena linfocitos.

espora: célula hija producida de una estructura haploide.

estambre: órgano reproductor masculino de una flor.

stem cell: an unspecialized cell that is able to develop into many different cell types.

stimulus (STIHM yuh lus): a change in an organism's environment that causes a response.

stoma (STOH muh): a small opening in the epidermis, or surface layer, of a leaf.

symbiosis (sihm bee OH sus): a close, long-term relationship between two species that usually involves an exchange of food or energy.

synapse (SIH naps): the gap between two neurons.

systemic circulation: the network of vessels that carry blood from the heart to the body and from the body back to the heart.

célula madre: célula no especializada que tiene la capacidad de desarrollarse en diferentes tipos de células.

estímulo: cualquier cambio en el medioambiente de un organismo que causa una respuesta.

estoma: abertura pequeña en la epidermis, o capa superficial, de una hoja.

simbiosis: relación intrínseca a largo plazo entre dos especies que generalmente involucra intercambio de alimento o energía.

sinapsis: espacio en medio de dos neuronas.

circulación sistémica: red de vasos que llevan sangre del corazón al cuerpo y de regreso del cuerpo al corazón.

T cell: a type of white blood cell that forms in the bone marrow and matures in the thymus gland; produces a protein antibody that becomes part of the cell membrane.

taiga (TI guh): a forest biome consisting mostly of cone-bearing evergreen trees.

technology: the practical use of scientific knowledge, especially for industrial or commercial use.

temperate: the term describing any region of Earth between the tropics and the polar circles.

testis (TES tihs): the male reproductive organ that produces sperm.

threatened species: a species at risk, but not yet endangered.

thymus: the organ of the lymphatic system in which T cells complete their development.

tissue: a group of similar types of cells that work together to carry out specific tasks.

trace fossil: the preserved evidence of the activity of an organism.

trachea (TRAY kee uh): a tube that is held open by C-shaped rings of cartilage; connects the larynx and the bronchi.

transcription: the process of making mRNA from DNA.

célula T: tipo de glóbulo blanco que se forma en la médula ósea y madura en la glándula del timo; produce un anticuerpo de proteína que se vuelve parte de la membrana celular.

taiga: bioma de bosque constituido en su mayoría por coníferas perennes.

tecnología: uso práctico del conocimiento científico, especialmente para uso industrial o comercial.

temperatura: término que describe cualquier región de la Tierra entre los trópicos y los círculos polares.

testículos: órgano reproductivo masculino que produce espermatozoides.

especie amenazada: especie en riesgo, pero que todavía no está en peligro.

timo: órgano del sistema linfático en el cual las células T completan su desarrollo.

tejido: grupo de tipos similares de células que trabajan juntas para llevar a cabo diferentes funciones.

traza fósil: evidencia conservada de la actividad de un organismo.

tráquea: tubo que los anillos en forma de C del cartílago mantienen abierto; este conecta la laringe y los bronquios.

transcripción: proceso por el cual se hace mARN de ADN.

translation: the process of making a protein from RNA.
tropism (TROH pih zum): plant growth toward or away from an external stimulus.
tundra (TUN druh): a biome that is cold, dry, and treeless.

traslación: proceso por el cual se hacen proteínas a partir de ARN.
tropismo: crecimiento de las plantas hacia o lejos de un estímulo externo.
tundra: bioma frío, seco y sin árboles.

U

umbilical (um BIH lih kul) cord: a rope-like structure formed by the outer zygote cells that attaches the developing offspring to the placenta.
unicellular: a living thing that is made up of only one cell.
ureter (YOO ruh tur): a tube through which urine leaves each kidney.
urethra (yoo REE thruh): a tube through which urine leaves the bladder.
urine: the fluid produced when blood is filtered by the kidneys.

cordón umbilical: estructura parecida a una cuerda formada por las células externas del cigoto que unen el hijo a la placenta.
unicelular: ser vivo formado por una sola célula.
uréter: tubo por el cual la orina sale de cada riñón.
uretra: tubo por el cual la orina sale de la vejiga.
orina: fluido que se produce cuando los riñones filtran la sangre.

V

vaccination: weakened or dead pathogens placed in the body, usually by injection or by mouth.
vaccine: a mixture containing material from one or more deactivated pathogens, such as viruses.
vagina: the part of the female reproductive system that connects the uterus to the outside of the body.
variable: any factor that can have more than one value.
variation (ver ee AY shun): a slight difference in an inherited trait among individual members of a species.
vascular tissue: specialized plant tissue composed of tubelike cells that transports water and nutrients in some plants.
vector: a disease-carrying organism that does not develop the disease.
vegetative reproduction: a form of asexual reproduction in which offspring grow from a part of a parent plant.

vacunación: patógenos debilitados o muertos introducidos en el cuerpo, generalmente por medio de una inyección o por la boca.
vacuna: mezcla que contiene material de uno o más patógenos desactivados, como los virus.
vagina: parte del sistema reproductor femenino que une el útero con el exterior del cuerpo.
variable: cualquier factor que tenga más de un valor.
variación: ligera diferencia en un rasgo hereditario entre los miembros individuales de una especie.
tejido vascular: tejido especializado de la planta compuesto de células tubulares que transportan agua y nutrientes en algunas plantas.
vector: organismo portador de una enfermedad pero que no la desarrolla.
reproducción vegetativa: forma de reproducción asexual en la cual el organismo se origina a partir de una planta parental.

vein: a vessel that carries blood toward the heart.

vena: vaso que lleva sangre hacia el corazón.

ventricles (VEN trih kulz): the lower two chambers of the heart.

ventrículos: las dos cámaras inferiores del corazón.

vertebrate (VUR tuh brayt): an animal with a backbone.

vertebrado: animal con columna vertebral.

vestigial (veh STIH jee ul) structure: body part that has lost its original function through evolution.

estructura vestigial: parte del cuerpo que a través de la evolución perdió la función original.

villus (VIH luhs): a fingerlike projection, many of which cover the folds of the small intestine.

vellosidad: proyección parecida a un dedo, muchas de las cuales cubren los pliegues del intestino delgado.

virus: a strand of DNA or RNA surrounded by a layer of protein that can infect and replicate in a host cell.

virus: filamento de ADN o de ARN rodeado por una capa de proteína que puede infectar una célula huésped y replicarse en ella.

vitamin: any of several nutrients that are needed in small amounts for growth, regulating body functions, and preventing some diseases.

vitamina: cualquiera de los varios nutrientes que se necesitan en cantidades pequeñas para el crecimiento, para regular las funciones del cuerpo y para prevenir algunas enfermedades.

voluntary muscle: muscle that you can consciously control.

musculares voluntarios: músculo que controlas conscientemente.

wetland: an aquatic ecosystem that has a thin layer of water covering soil that is wet most of the time.

Humedal: ecosistema acuático que tiene una capa delgada de suelo cubierto de agua que permanece húmedo la mayor parte del tiempo.

xylem (ZI lum): a type of vascular tissue that carries water and dissolved nutrients from the roots to the stem and the leaves.

xilema: tipo de tejido vascular que transporta agua y nutrientes disueltos desde las raíces hacia el tallo y las hojas.

zygosporangia (zi guh spor AN jee uh): tiny stalks formed when a zygote fungus undergoes sexual reproduction.

zigosporangia: tallos diminutos que se forman cuando un hongo zigoto se somete a reproducción sexual.

zygote (ZI goht): the new cell that forms when a sperm cell fertilizes an egg cell.

zigoto: célula nueva que se forma cuando un espermatozoide fecunda un óvulo.

Index

Italic numbers = illustration/photo **Bold numbers = vocabulary term**
lab = indicates entry is used in a lab on this page

A

B

C

D

E

I

J

K

L

Q

R

T

Z

Credits

COVER Klaus Honal/Photolibrary; **vii** The McGraw-Hill Companies; **NOS 2-NOS03** Images & Stories / Alamy; **NOS 4** (l)Kelly Jett / Alamy, (r)Natural Visions / Alamy; **NOS5** (t)MARK MOFFETT/MINDEN PICTURES/National Geographic Stock, (c)Frans Lanting/CORBIS, (b)Hank Morgan - Rainbow/ Science Faction/CORBIS; **NOS6** Lynn Keddie/Photolibrary; **NOS7** Jose Luis Pelaez, Inc./CORBIS; **NOS8** (t)David S. Holloway/Getty Images, (c)Klaus Guldbrandsen / Photo Researchers, Inc, (b)Giovanni Chaves-Portilla/ Fundación Ecodiversidad Colombia; **NOS9** Dennis Kunkel Microscopy, Inc. / PHOTOTAKE / Alamy; **NOS11** Plush Studios/Getty Images; **NOS12** Richard Peters / Alamy; **NOS13** PETE OXFORD/MINDEN PICTURES/National Geographic Stock; **NOS16** (l)Charles D. Winters / Photo Researchers, Inc., (r) Matt Meadows, (c)Louis Rosenstock/The McGraw-Hill Companies, (bl)David Chasey/Getty Images, (br)Biosphoto / NouN / Peter Arnold Inc.; **NOS17** (t) Lauren Burke/Getty Images, (b)Richard T. Nowitz/CORBIS; **NOS18** (tl) Stockbyte/Getty Images, (tr)Don Farrall/Getty Images, (bl)Medicimage/ Visulas Unlimited, Inc., (br)Hutchings Photography/Digital Light Source; **NOS19** (tl)Henry Groskinsky/Time & Life Pictures/Getty Images, (b) Georgette Douwma/Getty Image, (inset)Noel Hendrickson / Getty Images, (2)Hutchings Photography/Digital Light Source; **NOS20** NREL / US Department of Energy / Photo Researchers, Inc.; **NOS21** (l)akg-images, (r) Stefan Puchner/UPPA/Photoshot; **NOS22** (l)Tom & Therisa Stack/Tom Stack & Associates, (r)Jan Hinsch / Photo Researchers, Inc.; **NOS23** (t)Peter Ginter/ Getty Images; **NOS23** (c)Andrew Kaufman, (b)Ashley Cooper/Alamy; **NOS24** Colin Braley/Getty Images; **NOS26** Mark E. Gibson/CORBIS; **NOS27** Courtesy Seambiotic Ltd; **NOS28** (3)Hutchings Photography/Digital Light Source; **4** (tl)BSIP / Photo Researchers, Inc., (tr)Living Art Enterprises / Photo Researchers, Inc., (bl)Michael Matisse/Getty Images, (br) PHANIE / Photo Researchers, Inc.; **8-9** Vaughn Fleming/Garden Picture Library/ Photolibrary; **10** Hutchings Photography/Digital Light Source; **11** (t)Angela Wyant/Getty Images, (b)Andre Gallant/Getty Images, (b)Mark Gibson; **12** (l) BRUCE COLEMAN, INC./Alamy, (c)Mark Smith/Photo Researchers, (r)Gary Meszaros/Photo Researchers; **13** (l)Joe McDonald/CORBIS, (t)Gary Gaugler/ The Medical File/The Medical File/Peter Arnold, Inc., (r)BRUCE COLEMAN INC./Alamy; **14** John Kaprielian/Photo Researchers; **15** Michael Abbey/ Visuals Unlimited; **17** (t)Digital Vision/PunchStock, (tc)Robert Clay/Alamy, (c)John Kaprielian/Photo Researchers, (c)Splinter Images/Alamy, (b)Lee Rentz, (b)liquidlibrary/PictureQuest; **19** (inset)D. Finn/American Museum of Natural History, (bkgd)Reinaldo Minillo/Getty Images; **20** Hutchings Photography/Digital Light Source; **21** (t)Spencer Grant/Photolibrary, (b) PhotoLink/Getty Images; **23** Shin Yoshino/Minden Pictures/Getty Images; **24** Mark Steinmetz; **25** George Diebold/Getty Images; **27** Florida Fish and Wildlife Conservation Commission's Fish and Wildlife Research Institute, (bkgd)Photoshot Holdings Ltd/Alamy; **28** Hutchings Photography/Digital Light Source; **29** (t) Dr Jeremy Burgess/Photo Researchers, (b)Imagestate/ Photolibrary; **29** Eye of Science/Photo Researchers, Hutchings Photography/ Digital Light Source; **30** (l)Steve Gschmeissner/Photo Researchers, (r)JGI/ Getty Images; **31** (l)Stephen Schauer/Getty Images, (cl)ISM/Phototake, (cr) A. Syred/Photo Researchers, (b)age fotostock/SuperStock; **36** The McGraw-Hill Companies; **46-47** Dr. Dennis Kunkel/Visuals Unlimited/Getty Images; **49** (t)Tui De Roy/Minden Pictures, (c)Bon Appetit/Alamy, (b)Omikron/Photo Researchers; **50** (t)Hutchings Photography/Digital Light Source (c)Biophoto Associates/Photo Researchers, (b)Dr. Richard Kessel & Dr. Gene Shih/Visuals Unlimited/Getty Images; **51** FoodCollection/SuperStock; **52** Dr. Gopal Murti/ Photo Researchers; **55** (l)VICTOR SHAHIN, PROF. DR. H.OBERLEITHNER, UNIVERSITY HOSPITAL OF UENSTER/Photo Researchers, (r)NASA-JPL; **56** Hutchings Photography/Digital Light Source; **57** Eye of Science/Photo Researchers; **59** SPL/Photo Researchers; **61** Dr. Donald Fawcett/Visuals Unlimited/Getty Images; **62** (l)Dr. Donald Fawcett/Visuals Unlimited/Getty Images, (r)Dennis Kunkel / Phototake; **63** (l)Dr. R. Howard Berg/Visuals Unlimited/Getty Images, (r)Dennis Kunkel / Phototake; **66** The McGraw-Hill Companies; **67** LIU JIN/AFP/Getty Images; **74** Hutchings Photography/ Digital Light Source; **75** Colin Milkins/Photolibrary; **77** (t)Biology Media/ Photo Researchers, (b)Andrew Syred/Photo Researchers; **78** Michael & Patricia Fogden/Minden Pictures; **93-94** Robert Pickett/CORBIS; **94** Hutchings Photography/Digital Light Source; **95** Michael Abbey/Photo Researchers, (b)Bill Brooks/Alamy; **97** (t c)Dr. Richard Kessel & Dr. Gene Shih/Visuals Unlimited, (bl)Ed Reschke/Peter Arnold, Inc, (br)Biophoto Associates/Photo Researchers; **99** Don W. Fawcett/Photo Researchers; **100 101** (l)Ed Reschke/Peter Arnold, Inc.; **102** (l)P.M. Motta & D. Palermo/ Photo Researchers, (r)Manfred Kage/Peter Arnold, Inc.; **105** (l)Alex Wilson/ Marshall University/AP Images, (r)Bananastock/Alamy; **106** Hutchings Photography/Digital Light Source; **107** (t)Biosphoto/Bringard Denis/Peter Arnold, Inc, (c)Steve Gschmeissner/Photo Researchers, (b)Cyril Ruoso/ JH Editorial/Minden Pictures, (inset)Biosphoto/Lopez Georges/Peter Arnold, Inc.; **108** (t)Wim van Egmond/Visuals Unlimited/Getty Images, (c)Jeff Vanuga/CORBIS, (b)Alfred Pasieka/Photo Researchers; **111** (l)Biophoto Associates/Photo Researchers, (r)Andrew Syred/Photo Researchers; **112** (inset)Ed Reschke/Peter Arnold, Inc., (bkgd)Michael Drane/Alamy; **126-127** Brian J. Skerry/Getty Images; **128** Hutchings Photography/Digital Light Source; **129** Science Pictures Ltd./Photo Researchers; **132 133** (b)Ed Reschke/Peter Arnold, Inc.; **134** (t)Nature's Images/Photo Researchers, (b) Jeremy West/Getty Images; **136** (l)Rob Walls / Alamy, (r)Nigel Cattlin/ Alamy; **137** (tl)Stockbyte/Getty Images, (tr)Craig Lovell/CORBIS, (c)Piotr & Irena Kolasa/Alamy, (bl)image100/SuperStock, (br)Wally Eberhart/Visuals Unlimited/Getty Images; **139** (bl)Greg Broussard, (bkgd)David Thompson/ Photolibrary; **140** Horizons Companies; **141** Dr. Brad Mogen/Visuals Unlimited; **142** CNRI/Photo Researchers; **143** Biophoto Associates/Photo Researchers; **145** Wally Eberhart/Visuals Unlimited; **147** Roslin Institute; **148** Mark Steinmetz; **160-161** Tom and Pat Leesson; **162** (tl)Geoff du Feu/ Alamy, (tr)Ken Karp/McGraw-Hill Companies, (cl)Glow Images/Getty Images, (cr)Getty Images,(b)The Mcgraw-Hill Companies; **163** Nigel Cattlin/ Visuals Unlimited; **164** Wally Eberhart/Visuals Unlimited; **168** Kristina Yu/ Exploratorium Store; **170** (l)DEA PICTURE LIBRARY/Photolibrary, (r) WILDLIFE/Peter Arnold, Inc.; **171** (t)Pixtal/age Fotostock, (c)World History Archive/age Fotostock, (b)Tek Image/Photo Researchers, (bkgd)Jason Reed/ Getty Images; **172** Richard Hutchings/Digital Light Source; **173** (l to r, t to b) (1 2)Getty Images, (3)Punchstock, (4)CORBIS, (5-9)Getty Images, (10) CMCD/Getty Images, (11 12)CORBIS, (13)Biophoto Associates/Photo Researchers; **175** (t)Martin Shields/Photo Researchers, (b)Yann Arthus-Bertrand/CORBIS; **177** (l)Maya Barnes/The Image Works,(r)Bill Aron/ PhotoEdit; **178** (l)Peter Smithers/CORBIS, (tr)Geoff Bryant/Photo Researchers, (c)Bill Ross/CORBIS, (bl)J. Schwanke/Alamy; **179** Chris Clinton/ Getty Images; **180** (tl)Picture Net/CORBIS, (tr)June Green/Alamy Images, (c) Carolyn A. McKeone/Photo Researchers,(b)Hania Arensten-Berdys/www. gardensafari.net; **182** (inset)Charles Smith/CORBIS, Getty Images/ Photographer's Choice RF; **184** PHOTOTAKE Inc./Alamy; **203** Brand X Pictures/PunchStock; **204-205** William Osborn/Minden Pictures; **206** Hutchings Photography/Digital Light Source; **207** Florida Museum of Natural History photo by Eric Zamora ©2008; **208** (l)B.A.E. Inc./Alamy, (r) The Natural History Museum/Alamy; **209** (l)Mark Steinmetz, (c)Tom Bean/ CORBIS, (r)Dorling Kindersely/Getty Images; **215** (bkgd)Don Farrall/Getty Images, Florida Museum of Natural History, University of Florida; **216** Hutchings Photography/Digital Light Source; **217** DLILLC/CORBIS; **218** (t) Jeffrey Greenberg/Photo Researchers, (c)David Hosking/Alamy, (b)Mark Jones/Photolibrary; **219** Chip Clark; **221** (l)Carey Alan & Sandy/Photolibrary, (c)Robert Shantz/Alamy, (r)Stan Osolinski/Photolibrary; **222** (l)Paul Sutherland/National Geographic/Getty Images, (c)NHPA/Photoshot, (r)Kay Nietfeld/dpa/CORBIS, (b)Mitsuhiko Imamori/Minden Pictures; **223** (l) ARCO/D. Usher/age Fotostock, (t)Photodisc/Getty Images, (r)Joe Blossom/ NHPA/Photoshot; **225** (t)D. Parer & E. Parer-Cook, (b)B. Rosemary Grant/AP Images, (bkgd)Skan9/Getty Images; **226** Hutchings Photography/Digital Light Source; **227** Joseph Van Os/Getty Images; **232** T. Daeschler/The

Credits

Academy of Natural Sciences/VIREO; **246** (t) Ted Kinsman/Photo Researchers, (bl)Gary W. Carter/CORBIS, (br)Alamy Images; **250-251** Eye of Science/Photo Researchers; **252** Hutchings Photography/Digital Light Source; **253** Dr. Tony Brain/Photo Researchers; **254** (l)Dr. David M. Phillips/ Visuals Unlimited/Getty Images, (c)Medical-on-Line/Alamy, (r)Visuals Unlimited/CORBIS; **255** Stefan Sollfors/Getty Images; **256** Science VU/ Visuals Unlimited/Getty Images; **257** (l)Richard Du Toit/Minden Pictures, (r) Beverly Joubert/National Geographic/Getty Images; **259** (tl)Photodisc/Getty Images,(tr)Michael Rosenfeld/Getty Images, (c)Rich Legg/iStockphoto, (bl) Science Source/Photo Researchers, (br)Hercules Robinson/Alamy, (bkgd) Lawrence Lawry/Getty Images; **260** Hutchings Photography/Digital Light Source; **261** Dante Fenolio/Photo Researchers; **262** Wally Eberhart/Visuals Unlimited/Getty Images; **263** Spencer Platt/Getty Images; **264** Zephyr/ Photo Researchers; **266** Eye of Science/Photo Researchers; **269** Juliette Wade/Gap Photo/Visuals Unlimited; **288-289** Ed Reschke/Peter Arnold, Inc.; **290** Hutchings Photography/Digital Light Source; **291** (t)Visuals Unlimited/ CORBIS, (b)Flip Nicklin/Minden Pictures/Getty Images; **292** (l)Flip Nicklin/ Minden Pictures/Getty Images, (c)Michael Abbey/Photo Researchers, (r) Matt Meadows/Peter Arnold, Inc.; **293** (t)Manfred Kage/Peter Arnold, Inc., (b)Wim van Egmond/Visuals Unlimited/Getty Images; **294** (tl)Melba Photo Agency/PunchStock, (tr)WILDLIFE/Peter Arnold, Inc., (bl)M. I. Walker/Photo Researchers, (br)blickwinkel/Alamy; **295** (t)Images&Stories/Alamy, (bkgd) Associated Press; **297** Michael Abbey/Photo Researchers; **299** Mark Steinmetz; **301** (t)Global Warming Images / Alamy, (b)JUPITERIMAGES/ Brand X /Alamy, (bkgd)Kaz Chiba/Getty Images; **303** Felix Labhardt/Getty Images; **304** Mark Steinmetz; **305** (t)Hutchings Photography/Digital Light Source, (b)Gregory G. Dimijian/Photo Researchers; **307** (t)Hutchings Photography/Digital Light Source, (b)www.offwell.info; **309** C. James Webb / Phototake; **310** David Robertson/stockscotland; **322-323** Edmunds Dana/ Photolibrary; **324** (l to r)Art Wolfe/Getty Images, (1)Mark Steinmetz, (2) Photos Horticultural/Photoshot, (3)Beverly Joubert/National Geographic/ Getty Images, (4)SuperStock/SuperStock, (5)Renaud Visage/Getty Images; **325** Dr. Martha Powell/Visuals Unlimited/Getty Images; **326** (l)Biophoto Associates/Photo Researchers, (r)James Randklev/Photolibrary; **327** Siede Preis/Getty Images; **328** (l)H. Stanley Johnson/SuperStock, (c)Darryl Torckler/Getty Images, (r)Mark Steinmetz; **329** (tl)Biophoto Associates/ Photo Researchers, (tr)Darrell Gulin/CORBIS, (bl)David Maitland/Getty Images, (br)Chromorange RM/Photolibrary; **333** (t)TIM LAMAN/National Geographic Stock, (c)Gilbert S. Grant/Photo Researchers, (b)Tom Stack/Tom Stack & Associates; **334** Hutchings Photography/Digital Light Source; **336** (l)WILDLIFE/Peter Arnold, Inc.,(cl)David Whitaker/Alamy, (cr)Ed Reschke/ Peter Arnold, Inc., (r)Steven P. Lynch/The McGraw-Hill Companies, Inc.; **337** (t)Mark Bowler/Photoshot, (c)Mark Steinmetz, (b)Janis Burger/Bruce Coleman, Inc./Photoshot, Ern Mainka/Alamy; **340** studiomode/Alamy; **341** (t)Donald M. Jones/Minden Pictures, (tl)Comstock Images/Alamy, (tr)Burke Triolo Productions/Getty Images, (bc)Michael Rosenfeld/Getty Images, (bl) Bon Appetit/Alamy, (br)Walter Cimbal/Getty Images; **343** JIM RICHARDSON/ National Geographic Stock; **344** Mark Steinmetz; **345** Dr. Gerald Van Dyke/ Visuals Unlimited/Getty Images; **346** (t)Brian Gadsby/Photo Researchers, (tl)Robert Harding Picture Library/SuperStock, (tr)Steve Solum/Bruce Coleman, Inc./Photoshot, (b)JTB/Photoshot, (bl)WILDLIFE /Peter Arnold, Inc., (br)Dr. Carleton Ray/Photo Researchers; **347** (t)Mark Steinmetz, (c) Russell Illig/Photodisc/Getty Images, (b)Wally Eberhart/Visuals Unlimited/ Getty Images; **348** (tl)Ed Reschke/Peter Arnold, Inc., (tr)CORBIS, (cl)Clive Nichols/Getty Images, (cr)Travelscapes/Alamy, (bl)Steven P. Lynch/The McGraw-Hill Companies, Inc, (br)Al Telser/The McGraw-Hill Companies; **362-363** Brian North/Getty Images; **365** (t)Dr. Dennis Kunkel/Visuals Unlimited/Getty Images, (b)The McGraw-Hill Companies; **366** Brand X Pictures/PunchStock; **367** The Mcgraw-Hill Companies; **371** (t)Jacques Jangoux/Photolibrary, (b)MARK EDWARDS/Peter Arnold, Inc., (bkgd)Digital Vision/Getty Images; **372** Hutchings Photography/Digital Light Source; **373** (t)Jim Brandenburg/Getty Images, (b)Cathlyn Melloan/Getty Images; **375** (t)Stephen Dalton/Minden Pictures, (b)Martin Shields/Photo Researchers; **378** Sylvan Wittwer/Visuals Unlimited; **379** (tl)Monty Rakusen/Getty Images, (tr)Nigel Cattlin/Photo Researchers, (c)Nigel Cattlin/Alamy, (bl)B. ANTHONY STEWART/National Geographic Stock, (br)GardenPhotos.com/ Alamy; **382** Hutchings Photography/Digital Light Source; **383** (t)Bjorn Rorslett/Photo Researchers, (b)Wally Eberhart/Visuals Unlimited/Getty Images; **385** (l)CARR CLIFTON/ MINDEN PICTURES/National Geographic Stock, (r)Ed Reschke/Peter Arnold Inc.; **386** Eye of Science/Photo Researchers; **387** (t)Andrew Brown; Ecoscene/CORBIS, (b)Dr. Daniel Nickrent/Southern Illinois University; **390** (l to r, t to b)Don Klumpp/Getty Images, (1)Alan & Linda Detrick/Photo Researchers, (2)Comstock/ PunchStock, (3)Brad Mogen/Visuals Unlimited, (4)Walter H. Hodge/Peter Arnold, Inc., (5)Siede Preis/Getty Images, (6)Renee Morris/Alamy, (7) Elizabeth Whiting & Associates/Alamy, (8)USDA/Photo Researchers, (9) Adam Jones/Photo Researchers, (10)CORBIS, (11)Ingram Publishing/ SuperStock, (12)Gary Meszaros/Visuals Unlimited/Getty Images; **394** Scott Camazine/Photo Researchers; **404-405** Gary Bell/Getty Images; **406** Hutchings Photography/Digital Light Source; **407** (t)Darrell Gulin/Getty Images, (b)John Cancalosi/Alamy; **408** Geoff Brightling/Getty Images; **409** (tl)Tom Stack/Tom Stack & Associates, (tr)Brand X Pictures/PunchStock, (b) Paul Nicklen/National Geographic/Getty Images; **410** (l)Frank Greenaway/ Getty Images, (c)Francesco Rovero, (r)Ingram Publishing/Alamy **413** age fotostock/SuperStock, (inset)N. Simmons/American Museum of Natural History, (bkgd)age fotostock / SuperStock; **414** Hutchings Photography/ Digital Light Source; **415** Matthew Oldfield, Scubazoo/Photo Researchers; **417** (t)James H. Robinson/Photo Researchers, (b)Sinclair Stammers/Photo Researchers; **418** (t)Ingram Publishing/Alamy, (b)Doug Perrine/SeaPics. com; **420** (t)Ingram Publishing/Alamy, (b)Digital Zoo/Getty Images; **423** (bkgd)Ian Shive/Getty Images, James Balog/Getty Images; **424** Hutchings Photography/Digital Light Source; **425** FRANS LANTING/National Geographic Stock; **426** Pacific Stock/SuperStock; **427** JH Pete Carmichael/ Getty Images; **428** (inset)Chris Johns/Getty Images, The McGraw-Hill Companies, Inc./Barry Barker, photographe; **430** Steve Kaufman/CORBIS; **434** (l)Frederic Pacorel/Getty Images, (r)David Fleetham/Visuals Unlimited/ Getty Images; **444** Nick Laham/Getty Images; **445** AFP/Getty Images; **448-449** Charlie Schuck/Getty Images; **450** Hutchings Photography/Digital Light Source; **451** (t)Gustoimages/Photo Researchers, (b)CMCD/Getty Images; **456** Mark Steinmetz; **457** (l)Dennis MacDonald/PhotoEdit, (c)Daniel Dempster Photography/Alamy, (r)Bob Daemmrich/PhotoEdit; **459** Tomi/ PhotoLink/Getty Images; **460** Hutchings Photography/Digital Light Source; **461** Phil Loftus/Capital Pictures; **462** CMCD/Getty Images; **463** Innerspace Imaging/Photo Researchers; **464** (t)Dr, Gladden Willis/Visuals Unlimited/ Getty Images, (b)Dr, Gladden Willis/Visuals Unlimited/Getty Images; **468** Hutchings Photography/Digital Light Source; **469** (t)Steve Gschmeissner/ Photo Researchers, (b)Comstock/Jupiterimages; **470** Thomas Barwick/Getty Images; **472** (l)Stephen J. Krasemann/Photo Researchers, (c)Scientifica/ Visuals Unlimited, (r)Ian Leonard/Alamy; **474** Stockbyte/Getty Images; **488-489** MedicalRF.com/Getty Images; **490** Hutchings Photography/Digital Light Source; **491** Peter Menzel; **492** (l)Dorling Kindersley/Getty Images, (r) Hutchings Photography/Digital Light Source; **493** Michael Rosenfeld/Getty Images; **495** Mark Steinmetz; **498** Hutchings Photography/Digital Light Source; **499** Steve Gschmeissner/Photo Researchers; **502** Hutchings Photography/Digital Light Source; **505** (l)Scimat/Photo Researchers, (r) CNRI/Photo Researchers; **507** Dr. Dennis Kunkel/Visuals Unlimited; **508** Hutchings Photography/Digital Light Source; **509** Wolf Fahrenbach/Visuals Unlimited/Getty Images; **511** Photodisc/Getty Images; **528-529** Bo Tornvig/Getty Images; **530** Hutchings Photography/Digital Light Source; **531** Juergen Berger/Photo Researchers; **532** Holger Winkler/zefa/CORBIS; **535** James Cavallini/Photo Researchers; **538** Hutchings Photography/Digital Light Source; **539** Ron Chapple/Getty Images; **540** Image Source/

Credits

Jupiterimages; **547-575** Steve Gschmeissner/Photo Researchers; **548** Hutchings Photography/Digital Light Source; **549** Nati Harnik/AP Images; **555** (t)Andrew J. Martinez/Photo Researchers, (b)Steve Hamblin/Alamy, (bkgd)Jeffrey L. Rotman/CORBIS; **556** Hutchings Photography/Digital Light Source; **557** Dr P. Marazzi/Photo Researchers, Inc.; **558** C Squared Studios/Getty Images; **576** Hutchings Photography/Digital Light Source; **577** (t) Science Source/Photo Researchers, (b)Eurelios/Photo Researchers, (inset) Scott Camazine/Photo Researchers; **578** (l)Wellcome Library, London, (r) Michael Abbey/Photo Researchers; **579** The McGraw-Hill Companies; **580** Dr. Dennis Kunkel/Visuals Unlimited/Getty Images; **583** (tl)Tomi/PhotoLink/Getty Images, (tr)Zephyr/Photo Researchers, (bl)St Bartholomew's Hospital/Photo Researchers; **586** Hutchings Photography/Digital Light Source; **587** Science Pictures Ltd/Photo Researchers; **589** (tl)SPL Photo Researchers, (tr) Dr. Kessel & Dr. Kardon/Tissues & Organs/Visuals Unlimited/Getty Images, (c)RubberBall Productions, (cr)Eye of Science/Photo Researchers, (bl)Eye of Science/Photo Researchers, (br)Purestock/Getty Images; **590** PHOTOTAKE Inc./Alamy; **595** (t)Image Source/Getty Images, (b)George Musil/Visuals Unlimited/Getty Images, (bkgd)Stockbyte/SuperStock; **596** Hutchings Photography/Digital Light Source; **597** OJO Images/SuperStock; **598** (t)C Squared Studios/Getty Images, (bl)C Squared Studios/Getty Images; **600** JUPITERIMAGES/Thinkstock/Alamy; **614-615** Aflo Foto Agency/Photolibrary; **616** Hutchings Photography/Digital Light Source; **617** Dr. John Mazziotta et al./Photo Researchers; **618** David Stoecklein/CORBIS; **620** Doug Pensinger/Getty Images; **626** Hutchings Photography/Digital Light Source; **627** Steven Puetzer/Getty Images; **631** CORBIS Premium RF/Alamy; **634** Hutchings Photography/Digital Light Source; **635** (l)David J. Green - technology/Alamy, (r)Digital Vision/PunchStock; **636** Hutchings Photography/Digital Light Source; **637** ISM/Phototake; **639** Mark Andersen/Getty Images; **654-655** Yorgos Nikas/Getty Images; **656** Hutchings Photography/Digital Light Source; **657** P.M. Motta & J. Van Blerkom/Photo Researchers; **659** Eye of Science/Photo Researchers; **661** Yorgos Nikas/Getty Images; **665** (t)Y. Nikas/Biol Reprod/Getty Images, (c)Keystone/Getty Images, (b)Mauro Fermariello/Photo Researchers; **666** Vincenzo Lombardo/Getty Images; **667** (t)Steve Allen/Getty Images, (b)Yellow Dog Productions/Getty Images; **669** (t)Visual&Written SL/Alamy, (b)Tony Freeman/PhotoEdit; **670** (tl)Dr G. Moscoso/Photo Researchers, (tc tr)Anatomical Travelogue/Photo Researchers, (bl)Biophoto Associates/Photo Researchers, (br)Neil Bromhall/Photolibrary; **674** Digital Vision; **688** (t)Georg Gerster / Photo Researchers, Inc., (tc)Keren Su/Getty Images, (bc)Smneedham/Getty Images, (b)Courtesy of the University of Central Florida; **689** (t)SuperStock / SuperStock, (b) Stuart Fox/Getty Images; **692-693** Science Faction/Getty Images; **695** Scenics of America/PhotoLink/Getty Images; **696** Gerry Ellis/Minden Pictures/Getty Images, (inset)Riser/Getty Images, Weller/USDA; **700** Hutchings Photography/Digital Light Source; **701** Florida Department of Environmental Protection; **704** (t)Visuals Unlimited/CORBIS, (b) Melbourne Etc/Alamy; **709** (l)Van Hilliard Nature Photography/Getty Images, (r)Fuse/Getty Images, (bkgd)Florida Department of Environmental Protection; **710** Hutchings Photography/Digital Light Source; **711** Art Wolfe/Getty Images; **712** NOAA; **713** (l)Digital Vision/PunchStock, (cl) SA Team/ Foto Natura/Minden Pictures, (c)David Young-Wolff/PhotoEdit, (cr)Colin Young-Wolff/PhotoEdit, (r)Mark Steinmetz; **729** Redmond Durrell/Alamy; **730-731** Heidi & Hans-Jurgen Koch/Minden Pictures; **732** Hutchings Photography/Digital Light Source; **733** (t)Martin Harvey/CORBIS, (b)NASA Goddard Space Flight Center (NASA-GSFC); **734** (l)NASA, (r)Nigel J. Dennis/Gallo Images/CORBIS; **736** (t)Raymond Gehman/CORBIS, (b)Raymond Gehman/CORBIS; **737** (t) Wolfgang Kaehler, (b)Gerry Ellis/Minden Pictures/Getty Images; **739** (t) Andoni Canela/Photolibrary, (c)O. Rocha/American Museum of Natural History, (b)Pete Oxford/Minden Pictures; **740** Hutchings Photography/Digital Light Source; **741** (t)Dr Jeremy Burgess/Photo Researchers, (b) TomVezo.com; **742** Dr. David Phillips/Visuals Unlimited/Getty Images; **743** (t)Mary Evans Picture Library/Alamy, (c)Paul Souders/CORBIS, (b)Tom Brakefield/Getty Images; **745** Ricardo Beliel/BrazilPhotos; **746** Blend Images/Jupiterimages; **747** BEAWIHARTA/Reuters/CORBIS; **749** (l)Florida Museum of Natural History, University of Florida, (r)U.S. Fish & Wildlife Service/J & K Hollingsworth; **751** (t)Duncan Usher/Minden Pictures, (b) Jacques Jangoux/Photo Researchers; **754** Bach/zefa/CORBIS; **755** (t)CORBIS, (b)Mark Strickland/SeaPics.com; **756** (t)Carol & Don Spencer/Visual Unlimited/Getty Images, (b)Michael & Patricia Fogden/Minden Pictures/Getty Images; **770-771** Roger Ressmeyer/CORBIS; **773** Frans Lanting/CORBIS; **774** (t)Tom Vezo/Minden Pictures/Getty Images, (b)David Muench; **775** (t)Laura Romin & Larry Dalton/Alamy, (tc)Jim Brandenburg/Minden Pictures, (b)Chuck Haney / DanitaDelimont.com, (bc)Comstock/PunchStock; **776** (tl)age fotostock/Photolibrary, (tr)Gavriel Jecan/Getty Images, (bl) Jacques Jangoux/Mira.com, (br)Frans Lanting/CORBIS; **777** (t)M DeFreitas/Getty Images, (bl)Doug Sherman/Geofile, (br)Comstock Images/Alamy; **778** (t)Tom Till, (bl)Bruce Lichtenberger/Peter Arnold, Inc., (br)Russ Munn/CORBIS; **779** (t)age fotostock/SuperStock, (tc)Marvin Dembinsky Photo Associates/Alamy, (b)Andre Gallant/Getty Images, (bc)age fotostock/SuperStock; **782** Hutchings Photography/Digital Light Source; **783** Alexa Miller/Getty Images; **784** (tl)Comstock/PunchStock, (tr)Arthur Morris/Visuals Unlimited/Getty Images, (cl)Paul Nicklen/NGS/Getty Images, (cr) Stephen Dalton/Minden Pictures, (b)Medioimages/PunchStock, (bcl)Dr. Marli Mill/Visuals Unlimited/Getty Images; **785** (t)Jean-Paul Ferrero/Minden Picture, (tc)Lynn & Donna Rogers/Peter Arnold, Inc., (b)Image Ideas/PictureQuest, (bc)Photograph by Tim McCabe, courtesy USDA Natural Resources Conservation Service; **786** (t)Michael S. Quinton/National Geographic/Getty Images, (c)James L. Amos/Peter Arnold, Inc., (b)Steve Bly/Getty Images; **787** (t)Tom & Therisa Stack/Tom Stack & Associates, (bl) B. Moose Peterson, (br)David Noton Photography/Alamy; **788** (t)Camille Lusardi/Photolibrary, (c)Gregory Ochocki/Photo Researchers, (b)Doug Allan/Getty Images; **789** Gavriel Jecan/Photolibrary; **791** (t)K. Holmes/American Museum of Natural History, (c)K. Frey/American Museum of Natural History, (b)Stephen Frink/Getty Images, (bkgd)Wayne Levin/Getty Images; **793** Paul Bradforth/Alamy.

Name ______________________ Date ______________ Class ________

Science Benchmark Practice Test

FLORIDA NGSSS

Multiple Choice *Bubble the correct answer.*

1 A city resident posted a letter on the city's Web site. He said that he saw a spaceship land in a parking lot. The letter did not contain any photographs of the event. He said that although this happened once before, it can never happen again. What is one reason that this person's claim is pseudoscientific? **SC.8.N.2.1**

- (A) He posted it on a city Web site.
- (B) The city resident is not a scientist.
- (C) The alien landing is not reproducible.
- (D) There were no photographs of the event.

2 For hundreds of years, people believed that living things could appear out of nowhere. One piece of evidence for this idea was that insect larvae would eventually appear on rotting meat. Experiments showed that the larvae did not appear when there was a barrier between the rotting meat and insects. Today we understand that the larvae hatch from eggs that insects lay on the rotting meat. This story is an example of which concepts? **SC.7.N.2.1**

- (F) the benefits of scientific law
- (G) the benefits of scientific models
- (H) the difference between repetition and replication
- (I) scientific knowledge changing as a result of new evidence

3 Karim's theory is that his math teacher gives pop quizzes on Mondays. Amal's theory is that their teacher gives pop quizzes on Fridays. How are scientific theories different from Karim's and Amal's theories? **SC.6.N.3.1**

- (A) Only one individual can make a scientific theory about a topic.
- (B) Scientific theories must be voted on and win the majority vote.
- (C) Scientific theories are supported by evidence and are widely accepted.
- (D) There are usually more than two scientific theories about the same topic.

4 Scientific laws describe patterns in nature. Which areas of science are governed by scientific law? **SC.6.N.3.3**

- (F) gravity and conservation of energy
- (G) lab safety and experimental procedure
- (H) plate tectonics and natural selection
- (I) the use of hypotheses and inferences

NGSSS for Science Benchmark Practice continued

5 Amil hypothesized that plants exposed to rock music would grow faster than plants exposed to classical music. He used identical cuttings from a houseplant to grow 10 plants. He placed five of the plants in one room and five in another room. All conditions for the plants were identical, except he played rock music 24 hours per day in one room and classical 24 hours per day in the other room. He measured the starting height of each plant and then the final height of each plant after 30 days. The table below lists his data.

Rock Music	Plant Growth after 30 Days	Classical Music	Plant Growth after 30 Days
Plant 1R	1.8 cm	Plant 1C	2.0 cm
Plant 2R	2.5 m	Plant 2C	1.9 cm
Plant 3R	2.1 cm	Plant 3C	1.8 cm
Plant 4R	2.0 cm	Plant 4C	1.6 cm
Plant 5R	2.4 cm	Plant 5C	1.9 cm

Based on the data in this table, which conclusion can Emil make? **SC.8.N.1.3**

- (A) The hypothesis is true.
- (B) The results support the hypothesis.
- (C) The results prove that the hypothesis is correct.
- (D) The hypothesis must be revised to match the results.

6 Ivan studies the movement of plant cell components around a cell's central vacuole or storage compartment. He is devising a model to show this movement. He wants the model to reflect accurately his observations of real plant cells. What does Ivan need to do to make his model? **SC.8.N.1.6**

- (F) collect empirical evidence for at least one year
- (G) conduct experiments with controlled variables
- (H) form a conclusion based on similar results
- (I) use logical thinking and apply his imagination

7 While conducting research, Suzanne made observations, formed a hypothesis, designed an experiment, and collected data. She repeated her experiment several times. Each time, the results of her experiment did not support her hypothesis. What should she do next? **SC.8.N.2.2**

- (A) revise her data
- (B) revise her conclusion
- (C) revise her hypothesis
- (D) revise her observations

NGSSS for Science Benchmark Practice continued

8 A growing community needs more electrical power to meet increased demand. The power supplier has proposed three different routes for new electrical wires. The community hires a research group to survey wildlife abundance and diversity along each route. The community wants to choose a route that will cause the least harm to local wildlife. This story is an example of which concept?
SC.8.N.4.1

- (F) an idea of pseudoscientific science
- (G) new evidence changing scientific knowledge
- (H) the role of opinion in scientific investigation
- (I) the use of science to inform decision making

9 Researchers conduct an experiment to study the hatching rate of fruit flies. They have four trays of fruit fly eggs, each holding 100 eggs. Each tray is placed inside a different container. The first container is at constant temperature of 10°C; the second, 15°C; the third, 20°C; and the fourth, 25°C. The researchers count the number of eggs in each tray that hatch after 15 hours. In this experiment, which variable is the dependent variable? **SC.7.N.1.4**

- (A) number of trays
- (B) number of eggs
- (C) number of hours before the eggs
- (D) number of eggs that hatch after 15 hours

10 The cell theory says that all cells come from other cells and that all living things are made of cells. According to the cell theory, what is the most basic unit of life?
SC.6.L.14.2

- (F) atom
- (G) cell
- (H) molecule
- (I) organism

11 The illustration below shows a part of the human body.

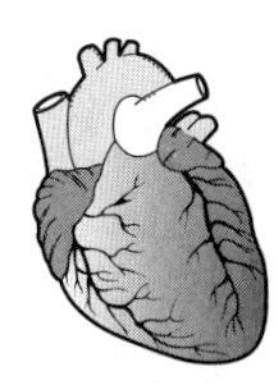

Which level of organization does this body part best represent? **SC.6.L.14.1**

- (A) organ
- (B) organism
- (C) organ system
- (D) tissue system

NGSSS for Science Benchmark Practice continued

12 Cellular respiration converts the energy in food molecules into a form of energy that a cell can use. This process begins with glycolysis, which occurs in the cytoplasm. Which organelle completes cellular respiration? **SC.6.L.14.4**

- (F) chloroplast
- (G) mitochondrion
- (H) nucleus
- (I) vacuole

13 Both single-celled organisms and the cells of multicellular organisms go through the process shown below.

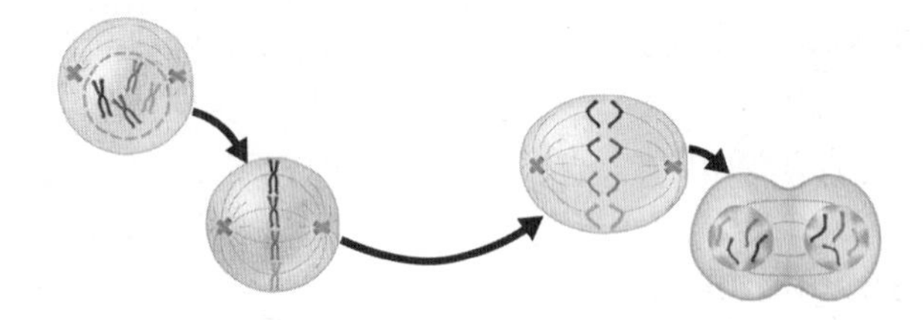

For a multicellular organism, this process can result in growth. For a single-celled organism, what is the result of this process? **SC.6.L.14.3**

- (A) growth
- (B) reproduction
- (C) wound healing
- (D) energy production

Name ______________________ Date ______________ Class ________

NGSSS for Science Benchmark Practice continued

14 All living things can be classified into one of three domains and then into one of six kingdoms. The table below shows which kingdoms are in each domain.

Domains	Kingdoms
Bacteria	Bacteria
Archaea	Archaea
Eukarya	Protista
	Fungi
	Plantae
	Animalia

If a new single-celled organism with a nucleus were discovered, into which domains might it be classified? **SC.6.L.15.1**

(F) Archaea or Bacteria

(G) Archaea or Eukarya

(H) Bacteria or Eukarya

(I) Bacteria, Archaea, or Eukarya

15 Sofia's 10-year study of a bird species has shown that its members make nests of twigs and dried grasses that measure 5 cm to 7 cm in diameter. This year, these birds built nests that were almost all over 7 cm in diameter. Over half of the nests observed contained bits of metal or silver-colored wrappers. What will Sofia need to do to develop an explanation that fits her observations? **SC.6.N.1.5**

(A) change results

(B) think creatively

(C) determine where she made an error

(D) make new observations more easily explained

16 An ulcer is a type of sore that occurs in a person's stomach. For many years, it was thought that several factors, including stress, caused ulcers. It is now thought that a bacterial infection, not stress, causes ulcers. Which could have led to this change in scientific understanding? **SC.6.N.2.2**

(F) new types of ulcers

(G) new viral outbreaks

(H) new opinions on ulcers

(I) new evidence about ulcers

NGSSS for Science Benchmark Practice continued

17 Although bacteria and viruses differ in many ways, they share some similarities. What is one similarity between a bacterium and a virus? **SC.6.L.14.6**

Ⓐ Both can be perform by mitosis.
Ⓑ Both can be destroyed by antibiotics.
Ⓒ Both can fix nitrogen in plant roots.
Ⓓ Both can cause illness in the human body.

18 This Punnett square shows a cross between two heterozygous parents.

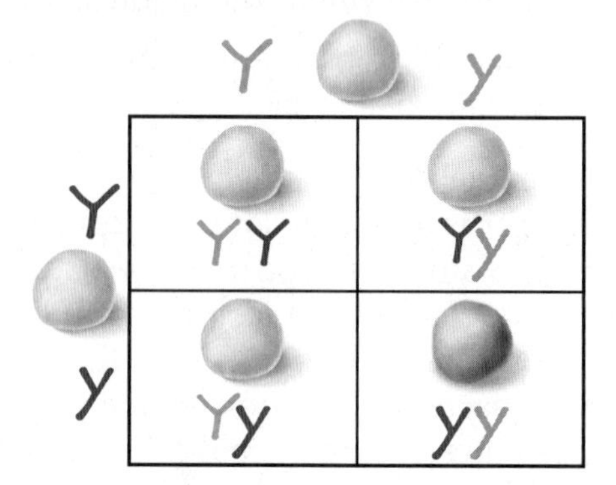

What is the probability that an offspring from this cross will have a heterozygous genotype? **SC.7.L.16.2**

Ⓕ 25 percent
Ⓖ 50 percent
Ⓗ 75 percent
Ⓘ 100 percent

19 Mitosis and meiosis are both processes that divide genetic material into new nuclei. These processes have many similarities. They also have important differences. Which can be a result of mitosis but not of meiosis? **SC.7.L.16.3**

Ⓐ the distribution of egg or sperm cells
Ⓑ a new organism that is identical to the parent cell
Ⓒ cells that contain half the genetic material that the parent cell contains
Ⓓ four cells that are all different from each other and from the parent cell

20 Mutualism, predation, parasitism, and commensalism are four types of symbiotic relationships. What is one way that mutualism is different from commensalism? **SC.7.L.17.2**

Ⓕ Both organisms benefit from a mutualistic relationship, while both organisms are harmed in a commensal relationship.
Ⓖ Both organisms are affected by a mutualistic relationship, while only one organism is affected by a commensal relationship.
Ⓗ Mutualism can occur only between plants, while commensalism can occur between both plants and animals.
Ⓘ Mutualistic relationships do not result in the death of either organism, while commensal relationships result in the death of at least one of the organisms.

NGSSS for Science Benchmark Practice continued

21 The illustration below shows a desert food web.

If a disease resulted in the death of most of the rattlesnakes in this food web, which population of organisms would decrease most? **SC.7.L.17.1, SC.7.L.17.3**

(A) birds

(B) coyotes

(C) lizards

(D) rodents

22 The diagram below shows the leg bone structures of horses and their ancestors. The drawings of bone structures are based on fossil evidence. Fossils provide evidence for biological evolution.

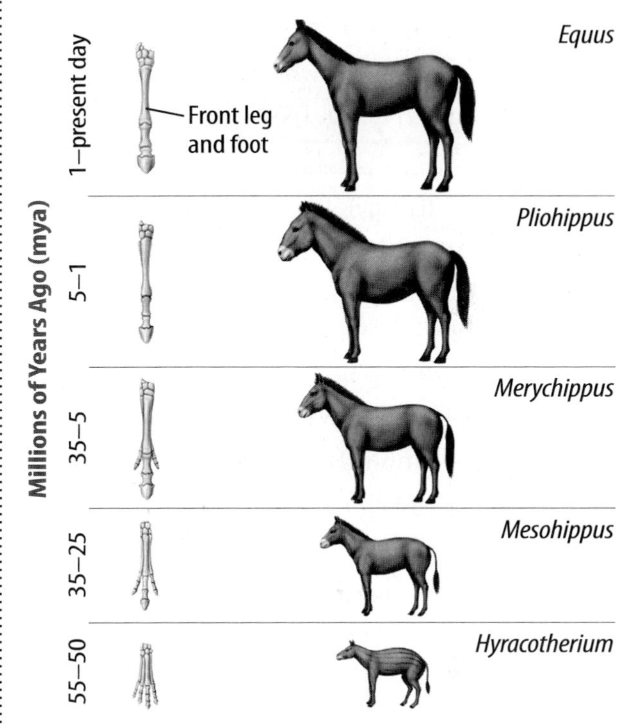

Which conclusion about the evolution of horses does fossil evidence support? **SC.7.L.15.1**

(F) All of the horse-like animals shown are now extinct.

(G) In the future, horses will be larger than they are today.

(H) Some ancestors of modern horses did not have hooves.

(I) Over millions of years, horses have evolved shorter hair.

NGSSS for Science Benchmark Practice continued

23 This organelle is involved in a process that cells of different organisms use to maintain homeostasis.

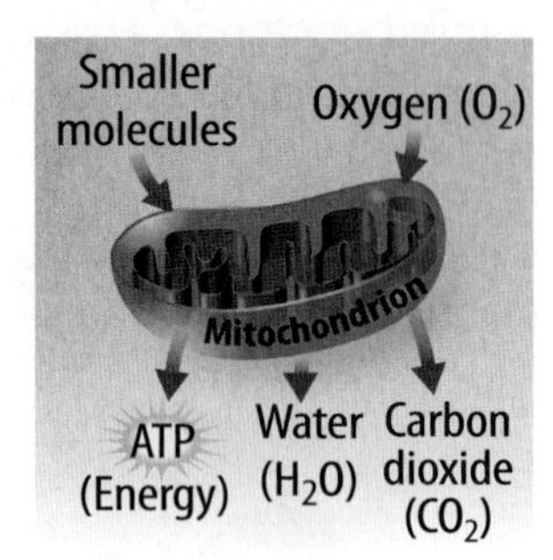

How does this organelle help maintain homeostasis? **SC.8.L.14.3**

(A) by forming sugars
(B) by obtaining water
(C) by getting rid of waste
(D) by releasing energy from food

24 Most of the structures in plant cells also are in animal cells.

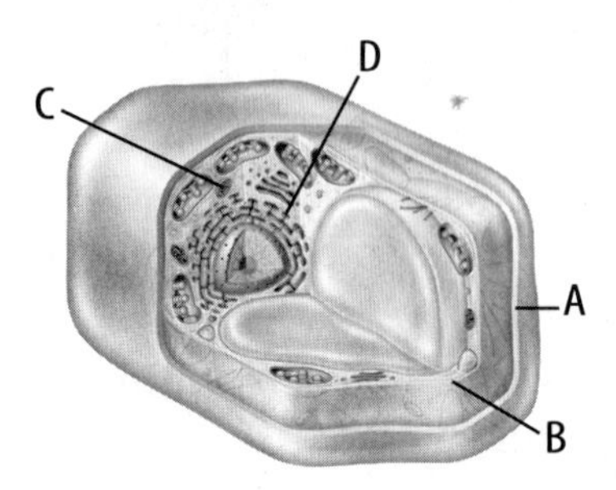

Which structure shown in this plant cell is not in animal cells? **SC.6.L.14.4**

(F) A
(G) B
(H) C
(I) D

25 This pedigree chart shows a family's phenotypes for earlobes.

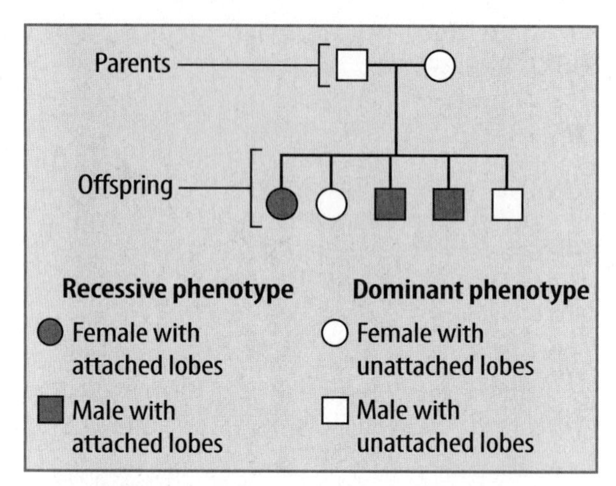

Based on this pedigree chart, what is the probability that a sixth child will have the recessive trait, attached earlobes? **SC.7.L.16.2**

(A) 10 percent
(B) 25 percent
(C) 50 percent
(D) 75 percent

PERIODIC TABLE OF THE ELEMENTS

Key	
Element	Hydrogen
Atomic number	1
Symbol	H
Atomic mass	1.01
State of matter	(Gas)

Gas
Liquid
Solid
Synthetic

A column in the periodic table is called a **group.**

A row in the periodic table is called a **period.**

	1	2	3	4	5	6	7	8	9
1	Hydrogen 1 H 1.01								
2	Lithium 3 Li 6.94	Beryllium 4 Be 9.01							
3	Sodium 11 Na 22.99	Magnesium 12 Mg 24.31							
4	Potassium 19 K 39.10	Calcium 20 Ca 40.08	Scandium 21 Sc 44.96	Titanium 22 Ti 47.87	Vanadium 23 V 50.94	Chromium 24 Cr 52.00	Manganese 25 Mn 54.94	Iron 26 Fe 55.85	Cobalt 27 Co 58.93
5	Rubidium 37 Rb 85.47	Strontium 38 Sr 87.62	Yttrium 39 Y 88.91	Zirconium 40 Zr 91.22	Niobium 41 Nb 92.91	Molybdenum 42 Mo 95.96	Technetium 43 Tc (98)	Ruthenium 44 Ru 101.07	Rhodium 45 Rh 102.91
6	Cesium 55 Cs 132.91	Barium 56 Ba 137.33	Lanthanum 57 La 138.91	Hafnium 72 Hf 178.49	Tantalum 73 Ta 180.95	Tungsten 74 W 183.84	Rhenium 75 Re 186.21	Osmium 76 Os 190.23	Iridium 77 Ir 192.22
7	Francium 87 Fr (223)	Radium 88 Ra (226)	Actinium 89 Ac (227)	Rutherfordium 104 Rf (267)	Dubnium 105 Db (268)	Seaborgium 106 Sg (271)	Bohrium 107 Bh (272)	Hassium 108 Hs (270)	Meitnerium 109 Mt (276)

The number in parentheses is the mass number of the longest lived isotope for that element.

Lanthanide series	Cerium 58 Ce 140.12	Praseodymium 59 Pr 140.91	Neodymium 60 Nd 144.24	Promethium 61 Pm (145)	Samarium 62 Sm 150.36	Europium 63 Eu 151.96
Actinide series	Thorium 90 Th 232.04	Protactinium 91 Pa 231.04	Uranium 92 U 238.03	Neptunium 93 Np (237)	Plutonium 94 Pu (244)	Americium 95 Am (243)

Metal

Metalloid

Nonmetal

Recently discovered

10	11	12	13	14	15	16	17	18
								Helium 2 **He** 4.00
			Boron 5 **B** 10.81	Carbon 6 **C** 12.01	Nitrogen 7 **N** 14.01	Oxygen 8 **O** 16.00	Fluorine 9 **F** 19.00	Neon 10 **Ne** 20.18
			Aluminum 13 **Al** 26.98	Silicon 14 **Si** 28.09	Phosphorus 15 **P** 30.97	Sulfur 16 **S** 32.07	Chlorine 17 **Cl** 35.45	Argon 18 **Ar** 39.95
Nickel 28 **Ni** 58.69	Copper 29 **Cu** 63.55	Zinc 30 **Zn** 65.38	Gallium 31 **Ga** 69.72	Germanium 32 **Ge** 72.64	Arsenic 33 **As** 74.92	Selenium 34 **Se** 78.96	Bromine 35 **Br** 79.90	Krypton 36 **Kr** 83.80
Palladium 46 **Pd** 106.42	Silver 47 **Ag** 107.87	Cadmium 48 **Cd** 112.41	Indium 49 **In** 114.82	Tin 50 **Sn** 118.71	Antimony 51 **Sb** 121.76	Tellurium 52 **Te** 127.60	Iodine 53 **I** 126.90	Xenon 54 **Xe** 131.29
Platinum 78 **Pt** 195.08	Gold 79 **Au** 196.97	Mercury 80 **Hg** 200.59	Thallium 81 **Tl** 204.38	Lead 82 **Pb** 207.20	Bismuth 83 **Bi** 208.98	Polonium 84 **Po** (209)	Astatine 85 **At** (210)	Radon 86 **Rn** (222)
Darmstadtium 110 **Ds** (281)	Roentgenium 111 **Rg** (280)	Copernicium 112 **Cn** (285)	Ununtrium * 113 **Uut** (284)	Ununquadium * 114 **Uuq** (289)	Ununpentium * 115 **Uup** (288)	Ununhexium * 116 **Uuh** (293)		Ununoctium * 118 **Uuo** (294)

* The names and symbols for elements 113-116 and 118 are temporary. Final names will be selected when the elements' discoveries are verified.

Gadolinium 64 **Gd** 157.25	Terbium 65 **Tb** 158.93	Dysprosium 66 **Dy** 162.50	Holmium 67 **Ho** 164.93	Erbium 68 **Er** 167.26	Thulium 69 **Tm** 168.93	Ytterbium 70 **Yb** 173.05	Lutetium 71 **Lu** 174.97
Curium 96 **Cm** (247)	Berkelium 97 **Bk** (247)	Californium 98 **Cf** (251)	Einsteinium 99 **Es** (252)	Fermium 100 **Fm** (257)	Mendelevium 101 **Md** (258)	Nobelium 102 **No** (259)	Lawrencium 103 **Lr** (262)